Yoshiki Oshida, Toshihiko Tominaga
NiTi Materials

Also of interest

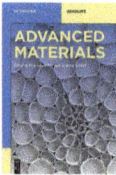

Advanced Materials
van de Ven, Soldera (Eds.), 2019
ISBN 978-3-11-053765-9, e-ISBN 978-3-11-053773-4

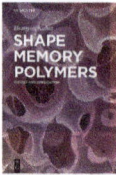

Shape Memory Polymers:
Theory and Application
Kalita, 2018
ISBN 978-3-11-056932-2, e-ISBN 978-3-11-057017-5

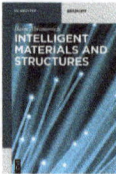

Intelligent Materials and Structures
Abramovich, 2016
ISBN 978-3-11-033801-0, e-ISBN 978-3-11-033802-7

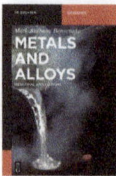

Metals and Alloys:
Industrial Applications
Benvenuto, 2016
ISBN 978-3-11-040784-6, e-ISBN 978-3-11-044185-7

Yoshiki Oshida, Toshihiko Tominaga

NiTi Materials

Biomedical Applications

DE GRUYTER

Author
Prof. Yoshiki Oshida
School of Dentistry
University of California San Francisco
513 Parnassus Ave
San Francisco
CA 94153-0340
USA

Dr. Toshihiko Tominaga
Department of Peridontology and Endodontology
Hokkaido University
Chome Kita 5
060-8648 Sapporo Kita Ward
Japan

ISBN 978-3-11-066603-8
e-ISBN (PDF) 978-3-11-066611-3
e-ISBN (EPUB) 978-3-11-066621-2

Library of Congress Control Number: 2020942826

Bibliographic information published by the Deutsche Nationalbibliothek
The Deutsche Nationalbibliothek lists this publication in the Deutsche Nationalbibliografie;
detailed bibliographic data are available on the Internet at http://dnb.dnb.de.

Prologue

This book is based on a review of about 3,500 carefully selected articles and presents itself as a typical example of evidence-based learning (EBL). Evidence-based literature reviews can provide foundation skills in research-oriented bibliographic inquiry, with an emphasis on such review and synthesis of applicable literature. Information is gathered by surveying a broad array of multidisciplinary research publications written by scholars and researchers.

In order for EBL to be used effectively, the content of every publication must be critically evaluated in terms of its degree of reliability. There is an established protocol for ranking the reliability of sources – this is an especially useful tool in the medical and dental fields. The ranking, from the most to least reliable evidence source, follows: (1) clinical reports using placebo and double blind studies; (2) clinical reports not using placebo, but conducted according to well-prepared statistical test plans; (3) study reports on time effect on one group of patients during predetermined period of time; (4) study/comparison reports, at one limited time, on many groups of patients; (5) case reports on a new technique and/or idea; and (6) retrospective reports on clinical evidence. Unfortunately, the number of published articles increases in this descending reliability order. The greatest advantage of the EBL review is that it helps identify common phenomena in a diversity of fields and literature. It is then possible to create synthetic, inclusive hypotheses that deepen and further our understanding.

As important as the diversity of source material is to the EBL review, it can pose unique challenges. The sources used in this review are mainly journal articles published in the medical/dental and engineering fields. These two groups' studies have very different analytical criteria and inherent issues.

In medical/dental journals, authors typically present statistically analyzed results; this lends these studies a high degree of reliability. Though we may be confident in the conclusions drawn from statistical analysis, it can be difficult to develop generalized ideas from this literature. Some controversy exists in the medical and dental fields on the relationship between in vitro and in vivo test results. This situation is further complicated and confusing because among the various in vivo tests exists a wide variety of animal model species. The in vitro studies are also not without weakness – it is almost impossible to broadly extrapolate in vitro test results, since it is very rare to find articles where identical test methods are employed.

In contrast to medical/dental studies, the data presented in the engineering journals are normally not subjected to statistical analysis. The characterization of data is considered more important to interpretation results because researchers try to explain phenomena, mechanisms, and kinetics. Discoveries in the engineering realm are vitally important to the advancement of medicine and dentistry: the materials and most of the technologies currently employed in medical and dental fields were originally developed in the engineering field. In this book, the authors draw upon

https://doi.org/10.1515/9783110666113-202

our unique experience in both fields to bridge the gap between medical/dental and engineering research.

For those who might be interested in literature survey in similar scope of this book, here is a lengthy list of journals: *Journal of Light Metals, Journal of Japan Institute of Light Metals, Journal of Advances Oral Research, Advances Dent Res, Journal of Less-Common Metals, Journal of Surgical Orthopaedic Advances, British Medical Journal, British Editorial Society of Bone and Joint Surgery, Cell, Clinical Implant Dentistry and Related Research, Clinical Oral Investigations, Critical Review in Oral Biology and Medicine, General Dentistry, Rare Metal Materials and Engineering, Werktoffe und Korrosion (Materials and Corrosion), International Dental Journal, International Journal of Oral & Maxillofacial Implants, International Journal of Oral Surgery, International Journal of Periodontics & Restorative Dentistry, International Journal of Prosthodontics, Implant Dentistry, International Journal of Molecular Sciences, Journal of Arthroplasty, Journal of Bacteriology, Journal of Bone and Joint Surgery, Journal of Bone and Mineral Research, Journal of Cardiovascular, Engineering & Technology, Journal of Chronic Diseases, Journal of Clinical Investigations, Journal of Clinical Pathology, Journal of Prosthetic Dentistry, Journal of Oral Implants, Journal of Oral Implantology, Journal of Oral Rehabilitation, Journal of Orthopaedic Trauma, Journal of Prosthodontics, Journal of Molecular Cell Biology, Journal of Orthopaedics, Oral Microbiology, Immunology, Proceedings of the Royal Society, Tissue Engineering, Toxicology, Dentistry and Practices, Journal of Dentistry and Oral Disorders, Journal of Dental and Oral Health, Orthodontic Waves, Progress in Organic Coatings, Quintessence International, Medical Devices & Technology, Journal of Medical Devices, Medical Devices: Evidence and Research, The Open Medical Devices Journal, Medical Science Monitor, The Journal of Adhesion, Journal of Japanese Society for Dental Materials and Devices, Journal of Alloys and Compounds, CrystEngComm, Journal of Electrochemical Society, The American Society for Metals, Wear, Journal of Vacuum Science and Technology, Journal of the Japan Institute of Metals, Dental Material Journal, Metallurgical and Materials Transaction A, Metallurgical and Materials Transaction B, Materials Letters, Material Science and Engineering: A, Material Science and Engineering: B, Material Performance, Material Science Forum, Transactional Society of Biomaterials, Thin Solid Films, Journal of Materials Chemistry, Journal of Material Science, Journal of Material Science Letters, Journal of Materials Processing Technology, Journal of Materials Science: Materials in Medicine, Journal of Nanoscience and Nanotechnology, Journal of Biochemical and Biophysical Methods, Journal of Metals, Journal of Biomechanics, Journal of Biomedical Materials Research A, Journal of Biomed Mater Res B, Journal of Bio-Medical Materials & Engineering, Journal of Applied Biomaterials, Journal of Association for Advancement of Medical Instrumentation, International Journal of Nanomedicine, Dental Materials Journal, Clinical Materials, Biomedical Materials, BioMedical Engineering OnLine, Biomedical Journal, CRC Critical Reviews in Biocompatibility, Annals of Biomedical Engineering, ACS Nano, ASM International, Applied Surface Science, Advanced Material Process, Journal of Materials and Manufacturing Processes, Journal of Colloid and*

Interface Science, Journal of the Less-Common Metals, Journal of Adhesion Science an Technology, Journal of the American Ceramic Society, Journal of Applied Physics, International Journal of Implant Dentistry, American Journal of Orthodontics and Dentofacial Orthopedic, The Angle Orthodontist, Journal of the Mechanical Behavior of Biomedical Materials, Materials and Manufacturing Processes, American Journal of Dentistry, Acta Biomaterialia, Acta Materialia, Biomaterials, Surface and Coatings Technology, Surface Technology, Tribology Industry, Biotribology, Digest Journal of Nanomaterials and Biostructures, Journal of Biomaterials Science-Polymer Edition, Journal of Material Design and Applications, Materials & Design, Journal of Engineering Tribology, Biomaterials Science: Processing, Properties and Applications, Scanning, Advanced Engineering Materials, Advanced Healthcare Materials, Advances in Materials Science and Engineering, Materials and Corrosion, Corrosion, Corrosion Science, Electrochimica Acta, Bioelectrochemistry, Biochemistry, Chemistry of Materials, International Archives of Allergy and Applied Immunology, Langmuir, Journal of Chemical Education, ACS Applied Materials and Interfaces, ACS Biomaterials Science and Engineering, Industrial and Engineering Chemistry Research, Faraday Discussions, Colloids Surface B: Biointerfaces, Journal of Electroceramics, Surface and Interface Analysis, Endodontic Topics, Contact Dermatitis, Journal of Endodontics, International Endodontic Journal, European Endodontic Journal, Journal of Investigative Dermatology, Tribology Letter, Transactions of the Electrochemical Society. We suspect that there should be more than the above-mentioned journals.

This book discusses the distinct and unique properties and applications of the shape-memory alloys. The first part (Chapters 1–5) discusses history of material's development, types of NiTi-based alloys, fundamental phenomena and mechanisms of shape-memory effect, superelasticity, phase transformation behavior, and heat treatment. The second part (Chapters 6 and 7) covers fabrication technologies to produce various forms of NiTi products and properties required when NiTi materials are used as biomaterials. The third part (Chapters 8 and 9) discusses basic properties in chemical and electrochemical reactions and mechanical properties (fatigue, torsion, and fracture). The fourth part (Chapter 10) touches high-temperature SMAs that are basically utilized in engineering and industrial fields; however, it should not be avoided from this book since the robots that had been developed in industry are employed in medical field. As discussed in the last chapter, the robots will be used mostly in AI-related medical field. The fifth part (Chapters 11–14) covers various properties and behaviors of NiTi materials when they are exposed to various biological environments and required surface modifications to improve surface characteristics. The sixth part (Chapters 15–17) is the main part of this book, covering medical and dental applications of NiTi biomaterials. This book ends with Chapter 18 on future perspective.

Table of Contents

Chapter 1
Introduction

There are various types of materials. About 100 pure elements, 783 out of possible 3,403 combinations of binary alloys, and 334 out of 91,881 possible tertiary alloys are considered as practically usable metallic materials. There are, furthermore, about 2,000 to 5,000 types of plastics, and about 10,000 kinds of ceramics (including oxides, nitrides, and carbides). Moreover, there are three major types of composites: metal-matrix compounds, plastic-matrix compounds, and ceramic-matric compounds. Among the above list of variety of materials, if you limit yourself to search for material(s) that is equally utilized in both industry/engineering field and medical/dental area, there are only two metallic materials selected: stainless steel and titanium materials [1, 2]. Of interest, with further narrow-down applications, these two metallic materials are also used in orthodontic appliances (such as archwires) and endodontic instruments (such as files and reamers).

In materials science, it is common to classify materials into two categories: structural materials and functional materials. Structural materials are nonactivatable materials that bear load. The key properties of such structural materials in relation to bearing load are elastic modulus, yield strength, ultimate tensile strength, hardness, ductility, fracture toughness, fatigue, and creep resistance. In addition, if materials corrode or wear, their ability to carry load will be degraded. Classical materials such as metals and alloys have played a significant role as structural materials for many centuries. Engineers have designed components and selected alloys by employing the classical engineering approach of understanding the macroscopic properties of the material and selecting the appropriate one to match the desired performances based on the application. With advancements in materials science and with increasing space and logistical limitations, scientists have been constantly developing high-performing materials for various applications [3]. On the other hand, functional materials possess properties that allow an energy conversion inside the material on the basis of physical effects. They can be used directly as material-based energy converter; in other words, it is possible to modify their material properties deliberately and reversibly. Functional materials are not assigned to one single material group. For listing just a few, examples of functional materials are as follows: piezoelectric materials (ZnO, $BaTiO_3$, etc.), optomechanical materials ($LiNbO_3$, KTN, ZnO, etc.), optical materials (SiO_2, GaAs, glasses, Al_2O_3, YAG, etc.), magnetic materials (Fe, Fe-Si, NiZn and MnZN ferrites, γ-Fe_2O_3, Co-Pt-Ta-Cr, etc.), shape-memory materials (NiTi, ZrO_2, AuCd, CuZnAl, polymer gels, magnetorheological fluids, etc.), electroactive polymers (carbon nanotubes), and energy technology and environment (UO_2, Ni–Cd, ZrO_2, $LiCoO_2$, etc.) [4]. This historical classification of materials appears to be unclear and some types of materials have been employed as a structural material while it exhibits

https://doi.org/10.1515/9783110666113-001

its unique functionality. As going on with this book, NiTi materials are the typical materials falling onto dual-characteristic materials.

By applying the surface technology (such as cladding, chemical/electrochemical treatment, and physiochemical/physical deposition), the surface zone can be modified to exhibit certain type(s) of function, the so-called functionalization. As we will discuss later, unique properties of NiTi [shape memory effect (SME) and/or superelasticity (SE)] can be added onto the surface layer of structural materials. This functionality can be manipulated on the surface zone to make a whole material as the functionally graded material (FGMs) which may be characterized by the variation in composition and structure gradually over volume, resulting in corresponding changes in the properties of the material. The materials can be designed for specific function and applications. Various approaches based on the bulk (particulate processing), preform processing, layer processing, and melt processing are used to fabricate the FGMs.

Biomaterials are those materials that are used in the human body. Biomaterials should have two important properties: biofunctionality and biocompatibility [1–5]. Good biofunctionality means that the biomaterial can perform the required function when it is used as a biomaterial. Biocompatibility means that the material should not be toxic within the body. Because of these two rigorous properties required for the material to be used as a biomaterial, not all materials are suitable for biomedical applications. The use of biomaterials in the medical field is an area of great interest as average life has increased due to advances in the use of surgical instruments and biomaterials [6, 7].

The NiTi alloys have been investigated extensively for 30 years after the establishment of basic understanding on the relationship among the microstructure, transformation behavior, and SME/SE phenomena. Many applications have been successfully developed in both engineering and medical fields. In particular, SE has been used for medical applications such as stents, guide wires and orthodontic arch wires, and endodontic files and reamers, as mentioned earlier.

The popularity of titanium biomaterials in both medical and dental fields can be recognized by counting numbers of peer-reviewed manuscripts published in variety of journals. In Figure 1.1, the total accumulated number of published articles for every 5 years is plotted. There are two straight lines in semilog scale. Top line is data for titanium biomaterials, which contain commercially pure titanium, Ti-based alloys such as Ti-6Al-4V alloy and NiTi-based alloys. The bottom line presents accumulated publication for each 5-year span on NiTi biomaterials, which contain NiTi, nitinol, and TiNi for medical/dental applications. Since the line for titanium biomaterials includes publications on NiTi biomaterials, it can be roughly said that about 20–30% of total publications on Ti materials are about NiTi materials. As indicated clearly in the figure, the exponentially increasing trend of published papers on medical and dental NiTi biomaterials may be attributed to different reasons that might include (1) needs from medical and dental sectors, (2) increased researchers and scientists involved in the NiTi biomaterials, and (3) expanded industrial scale being associated with the above two.

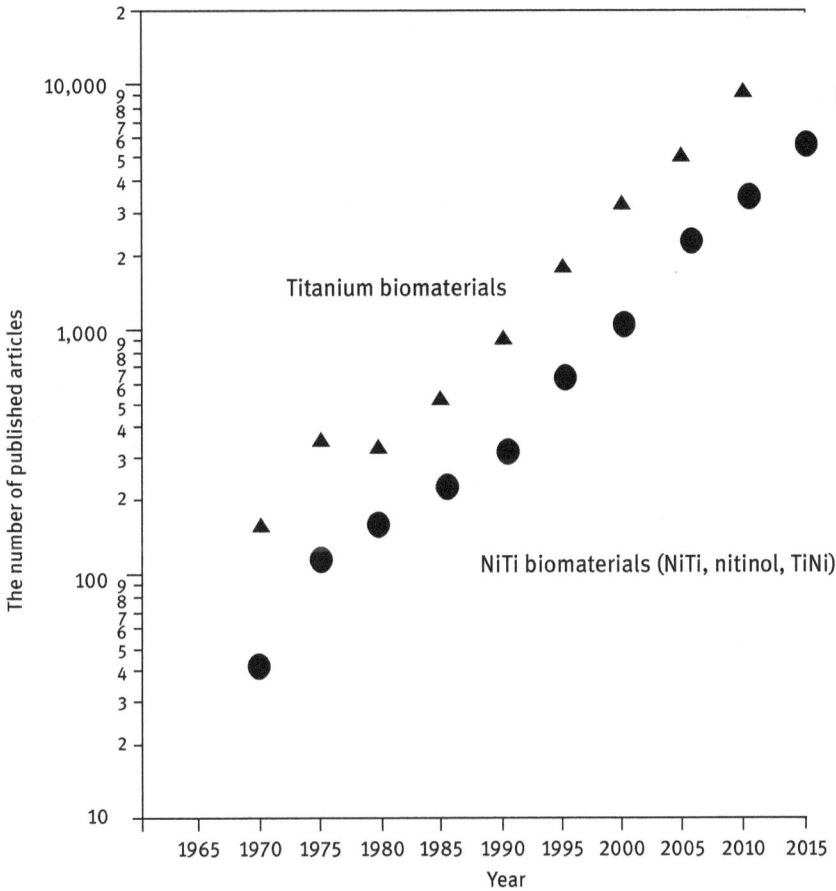

Figure 1.1: Accumulated number of published articles on titanium biomaterials and NiTi biomaterials. Titanium line includes publications on NiTi materials.

Although there are various manuscripts as book chapters on NiTi materials, we value the books [8, 9] covering basic mechanisms of SME* and SE** and applications using these interesting phenomena.

- * SME stands for shape-memory effect, and materials exhibiting this uniqueness are called as SMA (shape-memory alloys). Although SME does not, strictly speaking, describe the manner that such SMA exhibits an interesting phenomenon, rather it refers to the recovery of shape (or strain) after apparent "permanent" deformation which is previously induced at relatively cold temperatures by heating above a characteristic transformation temperature, so that it should be called as SRE (shape recovery effect). However, as majority of researchers and industry engineers utilize SME, in this book, we will follow the majority's judgment.

– **SE refers to superelasticity. Again, here is a controversial issue for naming. Although the term superelasticity implies that the extent of elasticity is great (super) from a viewpoint of phenomenology, it is not the fact that material's elasticity is super, rather it should refer to the isothermal recovery of relatively large (apparent elastic) strains during a mechanical load–unload cycle that occurs at temperatures above a characteristic transformation temperature. Hence, it should be called as pseudoelasticity. Due to a similar reason as SME, the term "super" can attract more people than the term "pseudo"; hence, we will follow the majority's naming.

Materials science and technology involved in titanium materials is typically interdisciplinary. When such titanium materials are treated as biomaterials, there should some limitations including biocompatibility and biofunctionality, so that certain types of titanium materials are accepted in medical and dental areas. Furthermore, when titanium biomaterials are employed as main materials for implant systems, additional requirements should be met before in vivo application. Such requirements should include osseointegration, biomechanical compatibility, and macro- and micromorphological compatibilities [1, 2, 10].

Figure 1.2 illustrates the complicated, yet nicely correlated interrelationship among different disciplines to establish a promising titanium materials science and engineering for assisting not only industry but also health providers, as well as the receiver of such services. In Figure 1.2, in order for engineered titanium materials to serve as titanium biomaterials, we will be discussing and reviewing numerous articles to prove that appropriate surface modifications and characterizations should be properly preformed and reflected to fabrication technologies and methods. Then, such titanium biomaterials are ready to be used in different dental and medical applications. We will review ever-growing Ti materials research and development for meeting specific aims, including V-free alloys, β-Ti alloys, Ti materials having better properties of fatigue, as well as wear, amorphatizable materials, materials exhibiting better superplastic formability, macro-, micro-, and nanoscale structure-controlled materials, SME and SE. These new Ti materials, along with conventional Ti materials, are characterized to evaluate whether they meet specific required characteristics. All these activities, as seen in the figure, are nicely correlated to establish titanium biomaterials, from which various dental and medical applications can be realized. Furthermore, currently and continuously in the future, with tremendous valuable and supportive technologies (including newly developed surface modification, near-net shape forming, better understating of bone healing mechanisms, and advanced tissue engineering materials and technologies) implant systems can be further developed to bring benefits to both patients by enhancing their quality of life level and clinicians' professional satisfaction.

Type/crystallography
- α type (HCP)
- Near α type
- (α+β) type
- β type (BCC)
- TiNi (β or martensite)
- TiAl

Surface modifications
- Sand-blasting (alumina, titania)
- Shot-peening
- Coating (hydroxyapatite, Ca-P, Ti beads, Ti)
- Gold-color TiN coating
- Silver-color Ti_2N coating
- Ti_2O (titania) film coating
- Porosity controlled surface
- Foamed titanium

Applications
- Denture
- Bridge/crown/coping
- Orthodontic archwires/brackets
- Endodontic files/reamers
- Post and cores
- Splint
- Orthopedic appliances (clamps/staples/stent)
- Total joint replacements (knee, hip)

Supportive technologies
- V-free new α/β alloy development
- Near-net shape (NNS) forming
  -- advanced SPF on composites
  -- laser forming
  -- injection forming
- Bone-healing mechanisms
- Tissue-engineering
  -- protein coating
  -- scaffold structure
  -- scaffold material

Implant application

Titanium biomaterials

Titanium materials

Shape/conditions
- Annealed
- Wrought
- Heat treated
- Sheet
- Rod
- Wire
- Plate
- Powder
- Beads

Characterization/compatibilities
- Corrosion resistance
- Passivation (auto healing effect)
- Oxidatioin
- Reaction with hydrogen
- Dissolution/metal ion release
- Wear resistance
- Biotribology/wear debris toxicity
- Toxicity
- Attachment
  -- cell attachment
  -- MIC (microbiologically-induced corrosion) product
- Compatibility
  -- biological
  -- mechanical
  -- morphological

Forming/machining
- Casting
- Machining
- EDM (electrical discharge machining)
- laser machining/forming
- Isothermal forming
- SPF (superplastic forming)
  -- ultrafine grain SPF
  -- transformation SPF
- DB (diffusion bond)
  -- SPF D/B
- P/M (powder metallurgy)
- MIM (metal injection molding) technique
- Laser welding
- Fusion welding
- Soldering
- Cementation

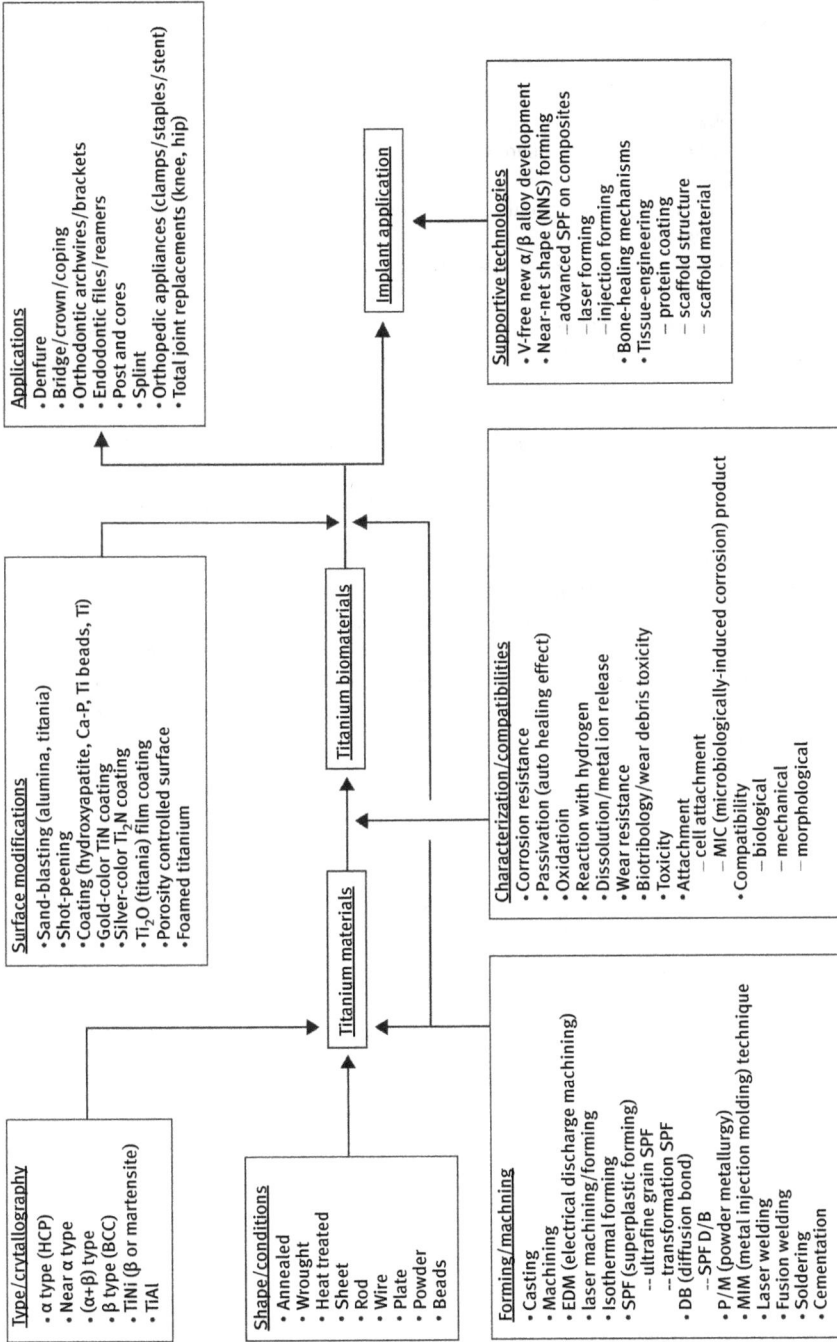

Figure 1.2: Titanium materials flow chart involving various materials science and engineering disciplines [1, 11].

References

[1]   Oshida Y. Bioscience and Bioengineering of Titanium Materials. Elsevier, Amsterdam,
       1st edition, 2007.
[2]   Oshida Y. Bioscience and Bioengineering of Titanium Materials. Elsevier, Amsterdam,
       2nd edition, 2013.
[3]   Rao A, Srinivasa AR, Reddy JN. Introduction to Shape Memory Alloys. In: Design of Shape
       Memory Alloy (SMA) Actuators. 2015, 1–31; doi: 10.1007/978-3-319-03188-0.
[4]   Askeland DR, Phulé PP. The Science and Engineering of Materials. Pacific Grove, CA, USA,
       Thomson Brooks/Cole, 2003.
[5]   Shabalovskaya SA. On the nature of the biocompatibility and on medical applications of NiTi
       shape memory and superelastic alloys. Biomed. Mater. Eng. 1996, 6, 267–89.
[6]   Tathe A, Ghodke M, Nikaljie AP. A brief review: biomaterials and their application. Int. J. Pharm.
       Pharm. Sci. 2010, 2, 19–23.
[7]   Rice C. Shape Memory Alloys, Applications. Encyclopedia of Smart Materials. 2002;
       https://doi.org/10.1002/0471216275.esm071.
[8]   Duerig TW, Melton KN, Stöckel D, Wayman CM. Engineering Aspects of Shape Memory Alloys.
       Butterworth-Heinemann, London, UK, 1990.
[9]   Yahia L. Shape Memory Implants. Springer-Verlag, Heidelberg, Germany, 2000.
[10]   Oshida Y, Miyazaki T, Tominaga T. Some biomechanistic concerns on newly developed
       implantable materials. J. Dent. Oral Health 2018, 4, 5 pages; https://scientonline.org/
       open-access/some-biomechanistic-concerns-on-newly-developed-implantable-materials.pdf.
[11]   Oshida Y, Tuna EB. Science and Technology Integrated Titanium Dental Systems. In: Basu et al.
       ed., Advanced Biomaterials – Fundamentals, Processing, and Applications. John Wiley & Sons,
       Hoboken, NJ, USA, 143–77.

Chapter 2
History of development and naming of NiTi materials

Since the discovery of shape-memory effect (SME) phenomena with NiTi alloy (in 1960s) [1], which was originally developed for the needs of high refractory metallic material having high damping capacity [2], there were no further remarkable development on NiTi materials till another unique property of superelasticity (in 1980s) was observed. Physical metallurgy investigation was carried out in detail on the dislocation mechanism, Ti_3Ni_4 precipitation, and R-phase transformation [3]. Once these two unique characteristics associated with NiTi were discovered and understood fully, further R&D has been advanced remarkably. First, in order to rise the effective temperature for SME phenomena high, 80°C, various NiTi-based alloy systems have been developed by adding Zr, Hf, or Nb. Then, the applications (including idea only as paper patents) were proposed in aerospace field, automotive engine, power generation station, nuclear power plant, and so on, exhibiting the typical seed-oriented development. This activity in R&D is shown in 1980s through early 1990s, as shown in Figure 2.1. Then, the second remarkable activity was found in advanced fabrication technology for, particularly, manufacturing various shapes of products such as foam metal with controlled porosity, thin film, hybrid, composites, as shown representatively with NiTi thin film in Figure 2.1. During the 1990s, new powerful actuators were developed to expand their applications in various sectors of engineering, robotics, and sensing engineering. Since 2000, after the biocompatibility and safety of NiTi were evaluated, a variety of applications in both medical and dental fields were explored [2, 3]. For future perspective, we discuss this at the last portion of this book.

The difference in denotation of NiTi and TiNi is more than individual's preference, rather there is a pure scientific reason. Scientists, engineers, and people involved in the research and development take more considerations on material properties on an atomic level when dealing with elemental materials, alloys, and composites. Therefore, they generally consider atomic interactions from one element to another, how a certain stoichiometric property alters as the composition is changed or an additive (as an alloying element) is introduced. So, in general, they deal materials on an atomic percentage (at%) or molecular percentage (mol%) level when considering such properties. Unless otherwise stated, liquids and solids are generally expressed in weight percentages (wt%), and gasses are expressed in volume percentage (vol%) or mol%. Atomic percentages are all well and good when the concern is how individual atoms from one species or another provides a solution to a problem according to how they interact on an atomic level, that is, when an optical, chemical, electrical, and physical evaluation is taken into consideration. However, for actual manufacturing, individual atoms cannot be measured very easily. Therefore, it is effective and useful to utilize the wt% when manufacturing an alloy or composite material.

https://doi.org/10.1515/9783110666113-002

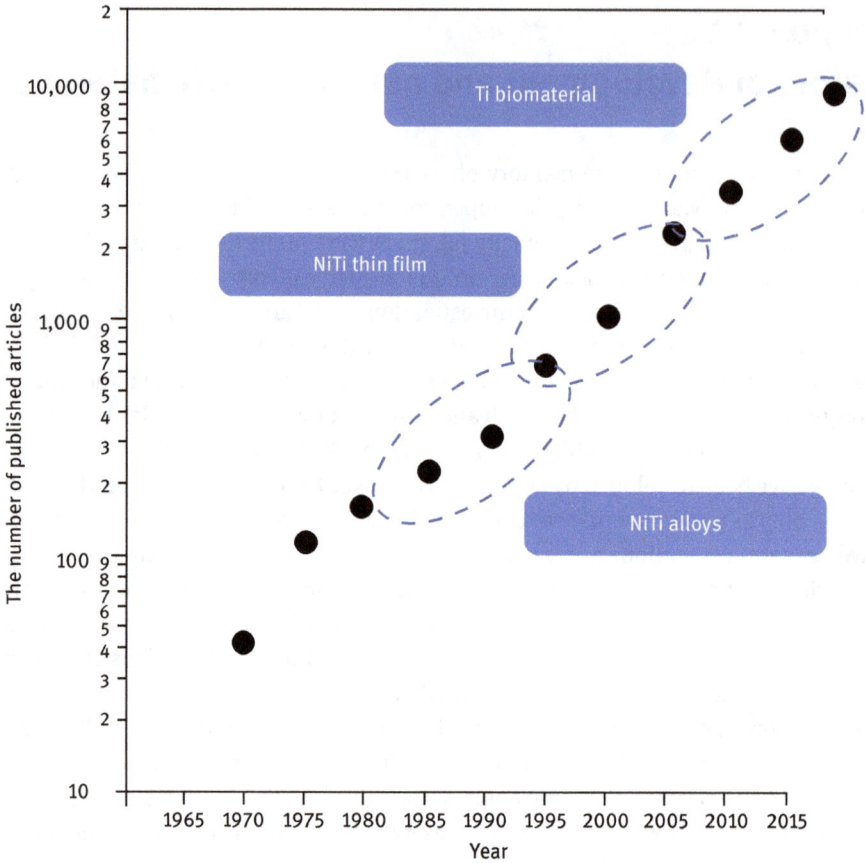

Figure 2.1: Three major noticeable activity eras in research and development history of NiTi materials.

Based on the above discussion, when we consider NiTi alloys, as we will see various physical, chemical, electrochemical, optical, and other properties to which if we know at% it would be much easier to understand the underlying reasons and mechanisms thereof. For example, there is a term "equiatomic" or "near-equiatomic" NiTi (as known as nitinol: an acronym of NiTi and Naval Ordnance Laboratory) [4–6], simply implying that it is a 50%Ni and 50%Ti in at% level. Referring to Figure 2.2, the equilibrium phase diagram is normally described with two compositional scales (wt% and at%), so that it is clearly understood that "equiatomic" alloy does not indicate an alloy with "equiweight" composition. The conversion between wt% and at% can be easily done by knowing atomic weight of each constituent elements. Let W_A and W_B be weight percentage of elements A and B, M_A and M_B be atomic percentage of elements A and B, and A_A and A_B be respective atomic weights, we have

(1) for converting from wt% to at%

at%A = $W_A/[W_A + (M_A/M_B)W_B] \times 100$

at%B = $W_B/[W_B + (M_B/M_A)W_A] \times 100 = 100 - $ at%A

(2) for converting from at% to wt%

wt%A = $(A_A/M_A)/(A_A M_A + A_B M_B) \times 100$

wt%B = $(A_B/M_B)/(A_A M_A + A_B M_B) \times 100 = 100 - $ wt%A

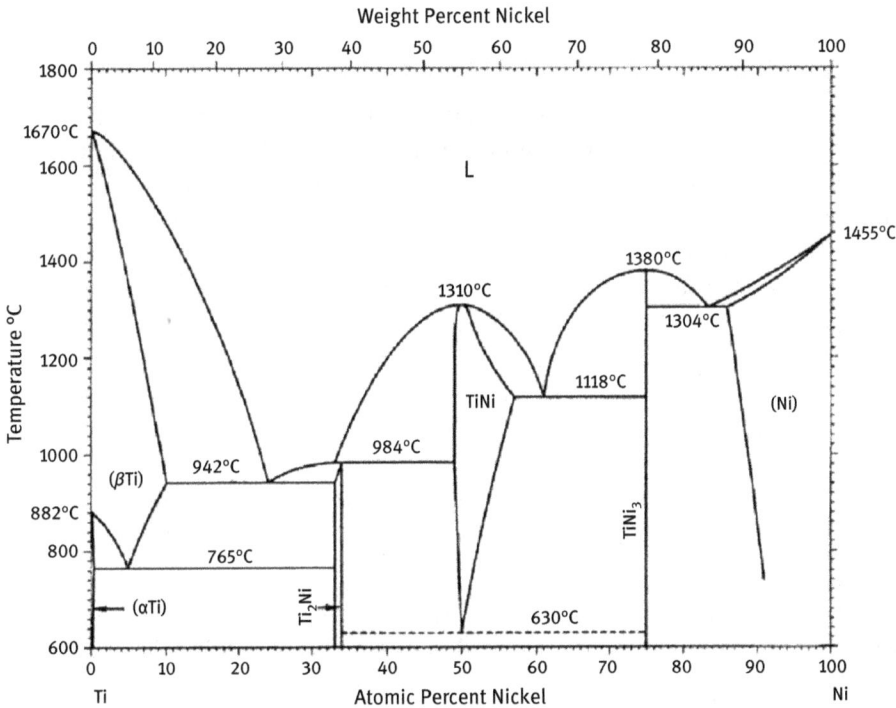

Figure 2.2: Equilibrium phase diagram of [Ti–Ni] alloy system with both wt% and at% scales.

Hence, 50–50 equiatomic NiTi alloy can be converted to 55.08wt%Ni and 44.92wt%Ti alloy, using atomic weight of 58.69 for Ti and 47.87 for Ni, respectively. To describe an alloy system, say A–B alloy, we normally put a base element (with majority of composition) in A position; therefore, the description of NiTi is normally practiced in wt% level; on the other hand, in at% level, since it is equiatomic, both NiTi or TiNi can be accepted. However, historically, there is another term "nitinol" for NiTi, because of this fact NiTi is the majority description of an alloy composed of Ni and Ti elements.

Now, move to the history of R&D for SME alloys (Figure 2.3) and NiTi-based alloys (including the binary alloy, ternary alloy, quaternary alloy, and quinary alloy, which are discussed in Chapter 3). Research and commercial applications include automobile, aerospace, robotic, and biomedical domains [5]. The demand for SMAs for

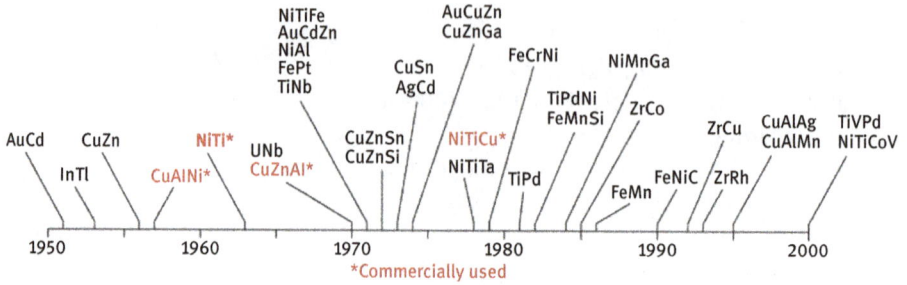

Figure 2.3: History of the discovery of shape-memory alloys [1].

engineering and technical applications has been increasing in numerous commercial fields; such as in consumer products and industrial applications [7–9], structures and composites [10], automotives [11], aerospace [12–15], mini actuators and microelectromechanical systems [14, 16], robotics [17], biomedical [18–23], and even in fashion [24]. Since the discovery of nitinol in 1963, the history of R&D of NiTi is that of development of devices and appliances employed in medical and dental fields, which will be discussed in later chapters.

References

[1]   Shaw J, Churchill C, Iadicola M. Tips and tricks for characterizing shape memory alloy wire: part 1 differential scanning calorimetry and basic phenomena. Exp. Techniques 2008, 32, 55–62.

[2]   Buehler WJ, Gilfrich JV, Wiley RC. Effect of low-temperature phase changes on the mechanical properties of alloys near composition TiNi. Appl. Phys. 1963, 34, 1475–7.

[3]   Miyazaki S, Otsuka K, Suzuki Y. Transformation pseudoelasticity and deformation behavior in a Ti-50.6 at% Ni alloy. Scripta. Met. 1981, 15, 287–92.

[4]   Liu Y, Galvin SP. Criteria for pseudoelasticity in near-equiatomic NiTi shape memory alloys. Acta Mater. 1997, 45, 4431–9.

[5]   Jani JM, Leary M, Subic A, Gibson MA. A review of shape memory alloy research, applications and opportunities. Mater. Design 2014, 56, 1078–113.

[6]   Kaufmann GB, Mayo I. The story of nitinol: the serendipitous discovery of the memory metal and its applications. Chem. Educat. 1997, 2, 1–21.

[7]   Wu, MH, Schetky IM. Industrial Applications for Shape Memory Alloys. In: International Conference on Shape Memory and Superelastic Technologies. Pacific Grove, CA, USA, 2000, 171–82.

[8]   Zider RB, Krumme JF. Eyeglass frame including shape-memory elements. US Patents 4772112: 1988.

[9]   Hautcoeur A, Eberthardt A. Eyeglass frame with very high recoverable deformability. US Patent 564027: 1997.

[10]   Furuya Y. Design and material evaluation of shape memory composites. Intell. Mater. Syst. Struct. 1996, 7, 321–30.

[11]   Stöckel D. Shape memory actuators for automotive applications. Mater. Des. 1990, 11, 302–7.

[12] Bill C, Massey K, Abdullah EJ. Wing morphing control with shape memory alloy actuators. J. Intell. Mater. System Struct. 2013, 24, 879–98.

[13] Hardt DJ, Lagoudas DC. Aerospace applications of shape memory alloys. Proc. Inst. Mech. Eng., Part G: J. Aerospace Eng. 2007, 221, 535–52.

[14] Humbeek JV. Non-medical applications of shape memory alloys. Mater. Sci. Eng. A, 1999, 134–48.

[15] McDonald SI. Shape memory alloy applications in space systems. Mater. Des. 1991, 12, 29–32.

[16] Sun I, Huang WM, Ding Z, Zhao Y, Wang CC, Purnawali H, Tang C, Huang WM, Wang CC. Stimulus-responsive shape memory materials: a review. Mater. Des. 2012, 33, 577–640.

[17] Furuya Y, Shimada H. Shape memory actuators for robotic applications. Mater. Des. 1991, 12–21–8.

[18] Petrini I, Magliavacca F. Biomedical applications of shape memory alloys. J. Metall. 2011, 2–11.

[19] Song C. History and current situation of shape memory alloys devices for minimally invasive surgery. Open. Med. Dev. J. 2010, 2, 24–31.

[20] Morgan NB. Medical shape memory alloy applications – the market and its products. Mater. Sci. Eng. A, 2004, 378, 16–23.

[21] Machado LG, Savi MA. Medical applications of shape memory alloys. Braz. J. Med. Biol. Res. 2003, 36, 693–91.

[22] Mantovani D. Shape memory alloys: properties and biomedical applications. JOM 2000, 52, 36–44.

[23] Duerig T, Pelton A, Stöckel D. An overview of nitinol medical applications. Mater. Sci. Eng. A, 1999, 273/275, 149–60.

[24] Langenhove LV, Hertleer C. Smart clothing: a new life. Int. J. Clothing Sci. Technol. 2004, 16, 63–72.

Chapter 3
NiTi-based alloys and alloying element effects

3.1 Ni or Ti elemental variation in binary NiTi alloy systems

NiTi is made of approximately equal amounts of nickel and titanium (at atomic %
level), and small variations in these proportions (in other words, Ni/Ti ratio) have
a radical effect on the properties of the alloy and, in particular, its transformation
temperature, metallurgical characteristics, physical and mechanical properties,
and chemical and electrochemical behaviors as well. For example, Sanjabi et al. [1],
studying mechanical and metallurgical properties in NiTi thin films, concluded that
the transformation from low-temperature martensitic phase (M-phase) to the high-
temperature parent phase took place below room temperature in Ni-rich NiTi while it
occurred above room temperature in Ti-rich and near-equiatomic NiTi. Nanoindentation tests demonstrated superelasticity (SE) in Ni-rich NiTi and martensitic deformation in Ti-rich and near-equiatomic NiTi compositions. Belyaev et al. [2] studied the
influence of the chemical composition of a NiTi alloy on the martensite stabilization
effect. The Ni_{50}-Ti, $Ni_{49.5}$-Ti, and $Ni_{49.0}$-Ti (at%) alloys were water-quenched from 900 °C
to exhibit the B2 $\leftrightarrow$ B19' transformation on cooling and heating without the intermediate R-phase formation. It was reported that the martensite stabilization effect
was observed in NiTi alloys, regardless of the chemical composition and value of the
preliminary strain. When the residual strain was less than 2.5%, the martensite stabilization effect values were close to each other, whereas if the residual strain exceeded
2.5%, the martensite stabilization effect values in the Ni_{50}-Ti and $Ni_{49.5}$-Ti were larger
than in the $Ni_{49.0}$-Ti (at%) alloys. TiNi films with different Ni/Ti ratios were prepared
by cosputtering technique [3] to investigate compositional effects on residual stress
evolution. It was found that for the film of Ti with 51.3 at%, a two-step transformation
was observed among martensite, R-phase, and austenite; and the residual stress was
quite low at room temperature. For the films with Ti contents of 47.3 at% and 53 at%,
residual stress was quite high due to the high intrinsic stress and partial relaxation
of stress caused by the R-phase transformation. When the films with Ti contents of
47.3 at% and 53 at% were annealed at 650 °C, residual stress in films decreased significantly, because postannealing could probably modify the film structure, reduce
the intrinsic stress, increase the transformation temperatures, and cause martensite
transformation above room temperature. Furthermore, Yoneyama et al. [4] investigated the ingots with 51.0 and 50.5 at% Ni by conducting tensile testing and differential scanning calorimetry (DSC). It was shown that Ni_{51}-Ti (at%) showed a brittle
property while $Ni_{50.5}$-Ti (at%) exhibited a low value of the apparent proof strength
with relatively large elongation, and residual strain increased with increasing titanium content. Yan et al. [5], investigating corrosion resistance laser spot-welded joint
NiTi wires in Hank's solution, found that the corrosion resistance improved due to

https://doi.org/10.1515/9783110666113-003

decrease of the surface defects and the increase of the Ti/Ni ratio, which may be attributed to more stable passive TiO_2 film formation.

The unique properties of shape-memory alloys (SMAs) are controlled by and are dependent on four external parameters: temperature (T), stress (σ), strain (ε), and time (t) [6]. These parameters cannot be changed independently. The complete mechanical behavior of SMAs has to be determined from a (T, σ, ε) diagram, in which the temperature axis covers the general temperature range from approximately 50 °C below and approximately 100 °C above M_S (upon cooling, the starting temperature for phase transformation from parent phase to M-phase). The value of a point along the (T, σ, ε) surface is not always constant and can move in any direction in that space. This time dependency can result from creep, stress relaxation, and changes due to variations in the chemical free energy of martensite and/or parent phases (which are stable at relatively high temperature and called as B2 crystal structure). During the phase transformation between B2 phase (A: austenite) and B19′ (M: martensite) phase, unique properties of shape-memory effect (SME) and SE take place. For enhancing these phenomena, normally postdeformation annealing, thermal/mechanical cycling, aging, and others are applied on NiTi materials [7–11]. By these additional thermal or thermomechanical treatment, the intermediate rhombohedral R-phase is frequently produced, from which an intermetallic compound, Ni_4Ti_3, is precipitated coherently and is normally not shown in the equilibrium phase diagram since it is in unstable phase, thereby resulting in generating an internal strain field. According to the appearance of R-phase, the transformation could be a multistep process; B2↔R↔M, and during the B2↔R transformation, SME and SE are also recognized [12]. The phase transformation temperature is very sensitive to alloy component, thermal/thermomechanical treatment conditions, and alloying elements [13].

With increasing temperature, the behavior changes from one-way effect (thermal memory effect) over SE (mechanical memory effect) to the stress–strain characteristic of conventional metal alloys. The position of the human body temperature on T-axis can be adjusted sensitively by the chemical composition and the thermomechanical treatment of the material [6].

In the following section, we divide the section into three major portions: Ni-rich NiTi, near- or equiatomic NiTi, and T-rich NiTi alloys.

3.1.1 Ni-rich NiTi alloy

The phase changing processes are executed by the forward and reverse transformation such as cubic austenitic phase (A-phase) B2 to monoclinic M-phase B19′ and vice versa. Phase transformation temperatures (A_S: on heating, transformation starting temperature from M-phase to parent A-phase, A_F: on heating, transformation finishing temperature from M-phase to parent A-phase, M_S: on cooling, transformation starting temperature from parent A-phase to M-phase, and M_F: on cooling, transformation finishing temperature from parent A-phase to M-phase) are decided by the range of phase limit. The state of phase presented in NiTi such as austenite or martensite decided the application. Therefore, controlling the microstructure of NiTi alloy through amendment of transformation temperatures by changing the chemical composition of NiTi alloy, and heat treatment process is a challenging mission in today's materials advancement.

Generally, aged Ni-rich NiTi alloys undergo martensitic transformations on cooling from high temperatures in two steps: B2 to R and then R to B19′ (normal behavior). However, under certain aging conditions, the transformation can also occur in three or more steps (unusual multiple step behavior) [14–18]. Aging of Ni-rich NiTi alloys was studied [19], by DSC and showed two transformation peaks on cooling after short aging times, three after intermediate aging times, and finally again two peaks after long aging times (2–3–2 transformation behavior). The three-step transformation was explained by two basic elements: (1) the composition inhomogeneity evolved during aging as Ni_4Ti_3 precipitates grow and (2) the difference between nucleation barriers for R-phase (small) and B19′ (large) [20]. The effect of 450 °C aging on the microstructure and on the martensitic transformations in a Ni-rich (50.8 at% Ni) NiTi SMA was studied [20] using transmission electron microscopy (TEM), X-ray diffraction (XRD), neutron diffraction, and DSC. It was found that on cooling from the high-temperature phase, two distinct peaks were observed after short-aging times, three peaks after intermediate-aging times, and two peaks again after long-aging times (2–3–2 transformation behavior). The first peak on cooling represents the formation of R-phase and the second peak is associated with the formation of M (B19′), suggesting that the burst-like transformation events during the growth of thermoelastic martensite and on the effect of oxidation are related to NiTi microstructures [20].

The effects of Ni concentration in Ni–Ti binary alloys on the multistep transformation were studied [21,22]. Using Ti–50.6, 50.8, and 51.0 at% Ni alloys, the effects of Ni concentration and aging conditions on the multistage martensitic transformation in aged Ni–Ti alloys were investigated by DSC and in situ scanning electron microscopy [22] by heat treating at 950 °C for 1 h and then aged at 500 °C for 1 h. It was found that although the triple-stage transformation appeared in the Ti–50.6 and 51 at% Ni alloys during cooling, the transformation sequence of the two alloys was completely different, and quadruple-stage transformation was observed in the Ti–50.8 at% Ni alloy [22].

The origin of the abnormal multistage martensitic transformations was investigated [23,24]. Followed by aging Ni-rich NiTi alloys, finely dispersed Ni_4Ti_3 particles embedded in B2 matrix are normally observed. With this situation, the B2 matrix (parent A-phase) normally undergoes two-stage martensitic transformation B2-R-B19′. However, as described earlier, there is also evidence of three-stage transformation. The origin of such abnormal three-stage transformation remains controversial. Fan et al. [23] conducted a comparative study between single crystals and corresponding polycrystals to find that all single crystals exhibit normal two-stage transformation, being independent of Ni content. It was further mentioned that, by comparison, polycrystals with low Ni content (50.6at%Ni) show three-stage transformation, but those with high Ni content (51.5at%Ni) again exhibit normal two-stage transformation. These new findings are consistent with a simple scenario that different transformation behaviors are a result of competition between preferential grain-boundary precipitation of Ni_4Ti_3 particles and a tendency for homogeneous precipitation when supersaturation of Ni is large [23]. Similar results were obtained by Zhou et al. [24]. According to the study, after aging at intermediate temperatures (400–500 °C), it was found that Ni-rich NiTi alloys undergo an abnormal three-stage martensitic transformation behavior (one-stage R and two-stage B19′), which stems from a preferential Ni_4Ti_3 precipitation around grain boundary. On the other hand, if aged at low temperatures (250–300 °C), they undergo two-stage R-phase transformation. Studying on low-Ni (50.6at%Ni, 51at%Ni) and high-Ni (52at%Ni) polycrystals, it was found that the former exhibited two-stage R-phase transformation, whereas the latter showed one-stage R-phase transformation. It was further concluded that the different transformation behavior of low-Ni and high-Ni polycrystals stems from a competition between two opposing tendencies: (1) for preferential precipitation in the grain boundary and (2) for homogeneous precipitation across the whole grain with high Ni content [24]. There are numerous articles supporting the fact that the precipitation of Ni_4Ti_3 phase is responsible for the multistage transformation in Ni-rich NiTi alloys, leading to SME and/or SE characteristics [25–31].

Chu et al. [32] investigated the effect of aging temperature on the reverse martensitic transformation in the Ni-rich NiTi alloy after the treatment of the solution at 1,050 °C for 4 h, using differential scanning calorimeter. It was reported that the type of reverse martensitic transformation changed from one step M-phase→A-phase (after solution treatment) through two steps M-phase→R→A-phase (after aged at 400, 450, and 475 °C) back to one step M-phase →A-phase (after aged at 500 °C). Both the austenite finish temperature (A_F) and the peak temperature corresponding to the transformation of R→A-phase decreased with the rise of the aging temperature. Zhang et al. [33] reported that during repeatedly imposed thermally induced martensitic transformations in NiTi SMAs, the M_S decreases. The temperature dependency of the phase transformation was investigated by Olbricht et al. [34]. Using binary ultra-fine-grained pseudoelastic NiTi wires, the phase transformations were studied in a wide temperature range by mechanical loading/unloading experiments, resistance measurements, DSC, thermal infrared imaging, and TEM. It was mentioned that the R-phase always forms prior to M-phase (B19′) when good pseudoelastic properties are observed; and the stress-induced A-phase (B2) to R-phase transformation occurred in a homogeneous manner, contrary to the localized character of the B2/R to B19′ transformations [34]. The electrical resistance variations of $Ni_{50.9}Ti_{49.1}$ shape-memory wires were studied during aging treatment at different temperatures via in situ electrical resistance measurement [35]. It was shown that during aging treatment, a cyclic behavior was observed in the electrical resistance variations, which could be related to the precipitation process. The evaluation of postaging transition temperatures was conducted using DSC analysis, and the precipitation process is found to occur in four different stages, suggesting that two-, three-, or four-step martensitic transformation could be observed, depending on the stress level around precipitates [35].

3.1.2 Equiatomic or near-equiatomic NiTi alloy

A small amount variation of Ni content from the equiatomic Ni/Ti ratio affects the changes in characteristic temperatures during the phase transformation; in particular, M_S and A_F (shape recovery temperature), as shown in Figure 3.1 [36–40]. It is clearly noticed that within the composition range at which the NiTi phase exists at ambient temperature, both M_S and A_F depend quite strongly on composition, particularly on the Ni-rich side, whereas Ti-rich alloys show less sensitivity. The composition dependency of M_S and A_F has important practical consequence, because the precise composition control is required when melting the alloys. It is normally believed that the NiTi alloy containing about 50.0 at%Ni shows typical SME, whereas NiTi alloy with more than 50.5 at%Ni exhibits SE phenomenon; hence, the SE appearance is more sensitive to the temperature control.

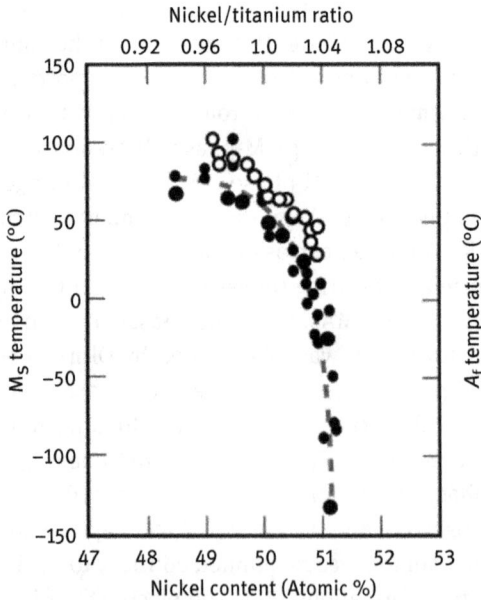

Figure 3.1: Effect of Ni contents on M_S and A_F temperatures in near-equiatomic NiTi alloy, where closed circles refer to M_S temperatures and open circles represent A_F temperatures.

Lin et al. [41] examined the effects of cold rolling on the martensitic transformation of an equiatomic TiNi alloy by internal friction and shear modulus measurements, hardness test, and TEM observation. It was found that the martensite stabilization can be induced by cold rolling at room temperature. Both deformed martensite structures and deform-induced dislocations/vacancies are considered to be related to the martensite stabilization. After the occurrence of the first reverse martensitic transformation of B19′ → B2, the martensite stabilization dies out and the transformation temperatures are depressed by retained dislocation on subsequent thermal cycles. The experimental results indicate that the martensite stabilization can depress the rate of martensitic transformation in the equiatomic TiNi alloy [41]. Martensite stabilization was also mentioned [42]. Structural changes with cooling and heating of Ni–Ti alloys involve three phases: austenite (B2), R-phase, and martensite (B19′). Jordan et al. [43], studying on an equiatomic $Ni_{50}Ti_{50}$, employed techniques such as DSC and internal friction to investigate structural transformation between any two phases, which is identified in these measurements by a peak in the studied property versus temperature plot. It was noted that, although, upon cooling, there is the evidence of two transformations, austenite → R-phase and R-phase → martensite, have been established in the literature during the reverse transformation from martensite on heating, the transformation can be associated with one or two peaks. It was observed that the reverse transformation of stress-induced martensite occurred at a temperature of approximately 20 K higher than that of thermal martensite. The increase in temperature for the reverse transformation was indicative of a stabilization effect, which is attributed to the change in the accommodation morphology of martensite variants from a self-accommodating state for the thermal martensite to an orientated state for the stress-induced martensite [44]. Lin et al. [45] studied the effect of cold rolling during the reverse transformation on the tensile behavior. It was reported that if the cold-rolled equiatomic TiNi alloy is subjected to a reversed martensitic transformation at temperature <300 °C, the strengthening effect induced by cold rolling can significantly improve the alloy's characteristic SME and pseudoelasticity (PE) by raising the critical shear stress for slip. If the cold-rolled specimens are subjected to an annealing at temperature ≥400 °C, the martensite accommodation or reorientation process can recover gradually due to the nullification of deformed martensite structures and defects. The recrystallization occurs at annealing temperature of ≥600 °C [46]. The deformation and transformation behavior associated with both the R and martensitic transformations in a near-equiatomic NiTi alloy were investigated using thermal cycling tests under constant applied load [46]. McCormick et al. [47] studied the effect of transformation cycling, with and without applied stress, on the martensitic transformation. Changes in transformation temperatures resulting from cycling are interpreted in terms of changes to the elastic strain energy and the irreversible energy associated with the transformation. Results indicated that the main effect of thermal cycling under zero stress is to increase the elastic strain energy, and with pseudoelastic cycling and extensive thermal cycling under stress, the elastic strain energy decreases with increasing cycling due to the development of a directional internal stress field associated with the formation of aligned variants [47].

Regarding the effect of heat treatment on the martensitic transformation, behavior of NiTi has been investigated [48], and it was shown that the occurrence of a two-stage parent-R–martensite transformation on cooling and a single-stage reverse transformation in specimens annealed at intermediate temperatures is a consequence of the irreversible energy loss associated with the martensitic transformation. The increase in the R transformation temperature in specimens that was heat treated at low temperature is shown to result from changes to the reversible, elastic, or chemical components of the transformation free energy change [48]. Sinha et al. [49], studying the influence of aging treatment on the work-hardening behavior of near-equiatomic NiTi alloy, reported that the maximum shape recovery is achieved at the peak-aged condition. The improvement in shape recovery has been attributed to the delayed onset of plasticity. As to a similar discussion to Ni-rich NiTi alloy, an importance of precipitate size and crystallographic orientation of Ni_4Ti_3 was found with equiatomic NiTi alloy [50, 51].

3.1.3 Ti-rich NiTi alloy

Using Ti(50.6at%)–Ni alloy, the effect of cold rolling followed by annealing on the martensitic transformation behavior was studied by means of the DSC, electron microscopy, and electrical resistivity measurements. It was found that after the applied treatment, both the B2 → R-phase and R → B19′ martensitic transformations are two-stage transformations [52]. Using Ti(50.4at%)-Ni alloy, the effect of thermal cycling on transformation temperature was studied by determining the electrical resistance, the internal friction, and the elastic modulus in terms of temperature. It was concluded that the pre M-phase becomes stable following the diminishment of the beginning of the martensite formation (M_S). The interrupted cooling has also shown that, contrary to the martensite, the R-phase exhibits no hysteretic behavior [53]. It is documented that the martensitic transformation start temperature (*Ms*) is higher than those of equiatomic or Ni-rich TiNi alloys. The Ti-rich TiNi alloys exhibit good shape recovery in spite of a great deal of second-phase Ti_2Ni or Ti_4Ni_2O existing around B2 grain boundaries. The nearly identical transformation temperatures indicate that the absorbed oxygen in Ti-rich TiNi alloys may react with Ti_2Ni particles, instead of the TiNi matrix, to form Ti_4Ni_2O [54]. Lin et al. [54], using $Ti_{51}Ni_{49}$ and $Ti_{50.5}Ni_{49.5}$ alloys, mentioned that martensite stabilization can be induced by cold rolling at room temperature. Thermal cycling can depress the transformation temperatures significantly and the R-phase transformation can be promoted by both cold rolling and thermal cycling in Ti-rich TiNi alloys due to introduced dislocations, depressing the *MS* temperature.

Phase transformations associated with SME in NiTi alloys can be one-stage, B19′ (martensite) ↔ B2 (austenite), two-stage including an intermediate R-phase stage, or multiple stage depending on the thermal and/or mechanical history of the alloy, including deformation and final annealing [55]. In general, it is believed that the near-equiatomic NiTi alloys are among the most important SMAs due to their outstanding mechanical properties, corrosion resistance, and biocompatibility. To modify the M-phase to A-phase transformation temperature, the thermal mechanical processing or additions of other elements are often performed [56]. Using Ti_{52}–Ni (at%) alloy that was cold worked under compression at various plastic strain levels of 0–20%, Lopez et al. [57] investigated aging effects (in the range 450–600 °C for 1 h) on transformation temperature. It was concluded that the aging promoted the precipitation of secondary phases in the cold-worked specimens. The extent of precipitation was apparently related to the aging temperature and the amount of cold work, resulting in affecting the martensitic M_S and M_F and austenitic A_S and A_F transformation temperatures. At temperatures between 550 and 600 °C, there was no consistent trend in the transformation temperatures, due to the inhomogeneous nature of the precipitation reaction [57]. A Ti_{52}–Ni alloy (at%), cold drawn to 30 at%, was annealed at 900 °C for 1 h, water quenched, and then subjected to DSC. It was reported that no evidence of the premartensitic R transformation was found during either the forward or the reverse transformation [58]. Besides nonequilibrium (unstable) Ni_4Ti_3 precipitated from NiTi alloy, there is equilibrium $NiTi_2$ or Ni_3Ti phase that is precipitated. Zhang et al. [59], heat-treating Ti-rich NiTi alloy, found spherical Ti_2Ni precipitates.

Karamn et al.[60] had deformed $Ti_{50.2}$-Ni (at%) alloy at three different temperatures; room temperature (below M_F), 50 °C (below A_S), and 150 °C (above A_F) because that the initial deforming phase (B2 austenite or B19′ martensite) and the initial governing deformation mechanism (martensite reorientation, stress-induced martensitic transformation, or dislocation slip in martensite) would be different. Based on results, it was indicated that although the M_S temperature did not change, the A_S temperature decreased significantly in all deformation conditions, probably because of the effect of the internal stress field caused by the deformed microstructure. All deformation conditions led to an increase in the strength levels and some deterioration of shape-memory characteristics. However, a subsequent low-temperature annealing treatment significantly improved pseudoelastic strain levels while preserving the ultrahigh strength levels, and concluded that the severe malforming could easily improve strength levels of NiTi alloys while preserving the shape-memory and PE characteristics and, thus, improve the thermomechanical fatigue behavior [60]. Tadayyon et al. [61] investigated the influence of the annealing temperatures on the thermomechanical behavior of Ti-rich NiTi alloy with regard to transformation temperatures, mechanical properties at room temperature and microstructure evolution under deformation. It was found that annealing above the recrystallization temperature (600 °C) modulated the mechanical behavior of the alloy significantly; and by increasing the annealing temperature, the shape-memory behavior of the alloys improved.

Besides, metallurgical manipulation for controlling the Ni/Ti ratio, Hassel et al. [62] treated NiTi in a complex inert gaseous atmosphere (a mixture of HCl and H_2O) at

600, 700, and 800 °C to investigate the selective dealloying phenomenon through the concurrent reaction of oxidation, chloridation, and evaporation of reaction products. It was reported that a stoichiometric titanium depletion took place to form the Ni_3Ti layer that was, in turn, covered by a thin oxide film of pure titanium oxide. It was mentioned, further, that selective dealloying can lead to effective surface modification exhibiting a high degree of biocompatibility.

As discussed later in this chapter for details, the Ni/Ti ratio can be manipulated by diffusion bonding, producing the so-called compositional gradation. Lim et al. [63] developed a unique actuator exhibiting a shape change over a wide temperature along the thickness direction by a spark plasma sintering after stacking ribbons of Ti-Ni_{51}, Ti-Ni_{50}, Ti-Ni_{49}, and Ti-Ni_{48} (at%) in sequence. It was found that the compositionally graded sample showed compositional variation of 1.5 at% Ti along the thickness direction (–120 μm) and a martensitic transformation temperature window as large as 91 K on cooling and 79 K on heating. A recoverable elongation of 0.9% was obtained under a stress of 80 MPa and the deformation rate, which is defined as the ratio of the recoverable elongation to the temperature range where the elongation occurred was 0.015%/K in the compositionally graded sample [63].

3.2 NiTi-X ternary alloys

The addition of the third or fourth alloying elements to NiTi-based alloys is done for the following purposes to control transformation temperatures, increase the stability of M_S with respect to thermal history, control the hysteresis width, increase the austenitic strength, reduce or increase martensitic strength, increase two-way effect ability, improve corrosion resistance, and suppress the R-phase [64]. Some of the additions giving particularly useful combinations of properties are copper, niobium, and precious metals.

For applications requiring M_S to be below room temperature, binary alloys show instability or an M_S dependence on prior thermal history. Furthermore, they tend to have poor ductility. Alloying elements such as Fe, Co, or Cu, which are known to depress M_S and substitute primarily Ni, can be used as additions to an approximately 50 at% Ti alloy. In this range (as shown in Figure 3.1), the sensitivity of M_S to Ni/Ti ratio variations is relatively low, but the M_S value corresponding to this plateau is lowered by the third element [40].

For materials scientists and engineers, it is very important to fully understand the alloying elements' effects to the base material. At the same time, the alloying effect of an element is not necessary to exhibit the same effect to the different base material. For example, manganese is recognized as an effective element for improving the hardenability to Fe-based alloy, while it enhances corrosion resistance, strength, and toughness for Al-based alloy. Ni element is well known as a powerful alloying element to Fe-based alloys for improvement of hardenability, while it enhances the high-temperature strength in Al-based alloys. NiTi alloys are rapidly being utilized as the material of choice in a variety of applications in the medical industry. It has been used for self-expanding stents, graft support systems, and various other devices for minimally invasive interventional and endoscopic procedures. However, the biocompatibility of this alloy remains a concern to many practitioners in the industry due to nickel sensitivity experienced by many patients. In recent times, several new NiTi alloys have been introduced with the addition of a ternary element.

Before discussing individual alloying elements to NiTi-based alloy, it is worth to review the overall alloying elements and their effects, and the site preference for, in particular, transition elements. Xi et al. [65] calculated lattice parameters, formation energy, elastic modulus, and deformation charge density of NiTi alloyed by 3d, 4d, 5d transition elements and discussed the site preference of transition elements in NiTi and their influence on the martensitic transformation temperature of NiTi. It was shown that, when the transition elements are added to NiTi alloy: the groups of V, Cr, Mn, Fe, Co, Pd, Cu prefer the Ni sites; Sc, Y, Zr, Hf prefer the Ti sites; Zn and Cd cannot form a stable structure. Furthermore, by analyzing the elastics modulus and the differential charge density map of NiTi, it was concluded that the replacement of Ni by the groups of V, Cr, Mn, Fe, Co or by Pd, Pt and the replacement of Ti by V, Cr, Mn, Fe will lower the transformation temperature M_S. Substitution of Hf, Zr, Ag, Au for Ni and substitution of Sc, Y, Hf, Zr for Ti will increase the transformation temperature Ms. The transformation temperature will be almost unchanged when Cu substitutes for Ni. Similar results are found elsewhere [65–72]. Furthermore, Novák et al. [73] added to NiTi alloy to lower the amount of $NiTi_2$ phase or at least to minimize undesirable effects on the alloy properties by various alloying elements (Al, Si, Mg, Fe, Nb, V).

In the following sections, we divide the section into three subsections. The first transition metal element group (3d elements: Sc, Ti, V, Cr, Mn, Fe, Co, Ni, Cu). The second transition metal element group (4d elements: Y, Zr, Nb, Mo, Tc, Ru, Rh, Pd, Ag). The third transition metal group (5d/4f elements: La, Ce, Pr, Nd, Pm, Sm, Eu, Gd, Tb, Dy, Ho, Er, Tm, Yb, Lu, Hf, Ta, W, Re, Os, Ir, Pt, Au). The underlined elements are actually alloyed to NiTi-based alloys and are cited in this book.

3.2.1 First transition metal elements

3.2.1.1 V

The effects of annealing temperature, annealing time, aging temperature, aging time, deformation temperature, and stress–strain cycle on SE of Ti-$Ni_{50.8}$-$V_{0.5}$ (at%) alloy were investigated by tensile tests and stress–strain cycle tests [74]. It was reported that with increasing the annealing temperature, the critical stress for inducing martensitic transformation of the alloy decreases first and then increases; the residual strain first increases, then decreases, and finally increases. In order to obtain an excellent SE at room temperature for the alloy, the annealing temperature should be 500–600 °C.

With increasing aging temperature, the stress-inducing martensitic transformation decreases, the residual strain increases, and the SE of the alloy becomes weak.

3.2.1.2 Cr

The transformation behavior of $Ni_{55.7}$-$Ti_{43.9}$-$Cr_{0.2}$ (wt%) alloy, in a temperature ranging from +90 to –170 °C has been studied as a function of heat treatment temperature by employing electrical resistivity probes [75]. It was concluded that the as-received 45% cold-worked, Ni–Ti–Cr alloy shows only the A↔R transformation over a wide temperature range between +90 and –170 °C, and the heat treatment temperature controls the range, the growth, and the decay of the R-phase in this alloy. It appeared that the tested Ni–Ti–Cr alloy exhibited distinct features of phase transformation in three regions of heat treatment temperatures: between room temperature to 400 °C, the A↔R transformation remains almost unaffected; between 400 and 600 °C, both A→R→M and M→R→A are present; and above 600 °C, only the A↔M transformation is present. Jeom et al. [76] investigated the compressive response of a Ni-Ti-$Cr_{0.2}$ (wt%) alloy at various initial temperatures and over a wide range of strain rates. It was found that the tested alloy exhibited SE over a range of initial temperatures and strain rates, for strains less than about 5%. The transition stress for the stress-induced M_S, the yield stress of the resulting martensites, and the yield stress of the parent austenite show strain-rate sensitivity, all increasing monotonically with the increasing strain rate. The strain rate significantly affects the superelastic and yielding behavior of this SMA within the superelastic temperature range [76].

Surface properties such as composition, roughness wettability, surface-free energy, and morphology will affect an implant material's physiochemistry [77, 78]. The thrombogenicity (which is directly related to the hemocompatibility of implants) is mainly dependent on its surface characteristics. The effect of magnetopolishing on $NiTiCr_{0.2}$ (wt%) was compared to mechanically polishing condition in terms of blood contact characteristics [79]. It was reported that the in vitro thrombogenicity tests revealed less platelet adherence on magnetopolished surface as compared to the mechanically polished surface. The formation of a dense and mixed hydrophobic oxide layer during the magnetopolishing was in response to an inhibition of the adhesion of negatively charged platelets. Hence, it is concluded that magnetopolished alloys can potentially be utilized for blood-contacting devices, where complications resulting from thrombogenicity can be minimized [79].

There are several researches on NiTi-based alloys with Cr and with other alloying element(s). Iijima et al. [80] studied corrosion behavior on Cu and Cr containing NiTi alloys in 0.9% NaCl and 1% lactic acid solutions. It was found that the addition of 0.19 at% Cr had little effect on the structure of the oxide films and the corrosion resistance of the Ni–Ti alloys. For Ni-Ti-$Cu_{5.0}$ (at%)-Cr(0.3at%) alloy, the metallic Cu was enriched at the alloy/oxide film interface, resulting in increased susceptibility to pitting corrosion above +1,000 mV. However, the passive current density and the amount of released Ni were not significantly increased by the addition of Cu. Zhou et al. [81], using Cr along with Co and Mn elements were added to NiTi-based as $Ti_{50}(Ni_{50-x}D_x)$ (D = Co, Cr, Mn), investigated the transition behavior of three different defect-doped systems $Ti_{50}(Ni_{50-x}Dx)$ (D = Co, Cr, Mn). They found a similarity in their transition behavior as a function of defect concentration x. Kök et al. [83] developed new NiTi-based alloys for biomedical applications, including $N_{45}Ti_{55}$, $Ni_{45}Ti_{50}Cr_{2.5}Cu_{2.5}$, $Ni_{48}Ti_{51}X$ (X = Mn, Sn, Co) alloys, in an arc melt furnace. It was reported that after the homogenization of these alloys, the M-phase transformation temperatures (which was determined with the differential scan calorimeter) were found to be below 37 °C (body temperature) in $Ni_{45}Ti_{50}Cr_{2.5}Cu_{2.5}$, $Ni_{48}Ti_{51}X$ (X = Mn, Co) alloys. The transformation temperature of the $N_{45}Ti_{55}$, $Ni_{48}Ti_{51}Sn$ alloys was found to be over 37 °C, and $Ni_{45}Ti_{50}Cr_{2.5}Cu_{2.5}$, $Ni_{48}Ti_{51}X$ (X = Mn, Co) alloys, which were in austenite phase at room temperature, included austenite B2 phase and Ti_2Ni precipitation phase, and the alloys that were in the M-phase at room temperature included martensite B19′ phase and Ti_2Ni phase.

3.2.1.3 Mn
It was found that, according to findings [81, 82], an alloying effect of Mn element is similar to those in Cr.

3.2.1.4 Fe
Normally, the addition of Fe to binary NiTi-based alloy would result in a dramatic decrease in transformation temperature, and a considerable decrease of hysteresis width; provoke suppression of the M-phase to the favor of R-phase transition and increasing ductility, resulting in good cycle stability. Furthermore, when the elastic characteristic after the work-hardening is desired, Fe element might be a powerful alloying element [83].

Kassab et al. [84] studied the corrosion behaviors of selected ternary nickel titanium (NiTi)-based alloys ($Ni_{45}Ti_{50}Cu_5$, $Ni_{47}Ti_{50}Fe_3$ and $Ni_{39}Ti_{50}Pd_{11}$) with a binary $Ni_{50.7}Ti_{49.3}$ alloy, using standard electrochemical techniques in a physiological solution (0.9% NaCl), simulating a body temperature of 37 ± 1 °C. It was found that the localized corrosion resistance of these ternary alloys is lower than the binary NiTi alloy, and the following relation has been proposed for their localized corrosion resistances: NiTiCu < NiTiFe < NiTiPd < NiTi.

In order to understand the influence of point defects (Fe element) on the martensitic transformation characteristics, Ramachandran et al. [85] prepared several $Ni_{50-x}Ti_{50}Fe_x$ (x = 2.0–10.0at%) alloys. It was reported that the $Ti_{50}Ni_{48}Fe_2$ and $Ti_{50}Ni_{47}Fe_3$ have a two-step martensitic transformation (B2 → R and R → B19′), while the $Ti_{50}Ni_{46}Fe_4$, $Ti_{50}Ni_{44.5}Fe_{5.5}$, and $Ti_{50}Ni_{44}Fe_6$ SMAs display a one-step martensitic transition (B2 → R). The compounds $Ti_{50}Ni_{42}Fe_8$ and $Ti_{50}Ni_{40}Fe_{10}$ show strain glass features (frozen strain-ordered state). The induced point defects significantly alter the martensitic transformation characteristics, namely transition temperature and width of thermal hysteresis during the transition. The evolution of phase transformation in the Fe-substituted TiNi alloys is presumably caused by the changes in local lattice structure via the induced local strain fields by Fe point defects. There are similar studies on effect of Fe addition on phase transformation characteristics; they should include $Ti_{50.75}Ni_{47.75}Fe_{1.5}$ [86], $Ti_{50}Ni_{44}Fe_6$ [87], $Ti_{50}Ni_{45}Fe_5$ [88], $Ti_{50}Ni_{47}Fe_3$ [89, 90], $TiNi_{50-x}Fe_x$ ($x = 2, 4, 6, 8$ in at%) [91, 92]. Summarizing all findings, it is indicated that Fe tends to stabilize the austenite against the martensite. The difference in cohesive energy of B2 and R-phases, and the martensitic transformation temperature as well, decreases with the increasing Fe concentration, until the Fe content exceeds a critical level at which point the R-phase turns into B2 phase automatically and the martensitic transformation is suppressed altogether. Since Fe atoms at different Ni sites in R-phase have divergent solution energy, the antiprecursor effect of Fe may vary at different spot in the material, and so will be the local transformation temperature. [85, 93, 94].

It is well known that plastic deformation at high temperatures is of great importance in manufacturing the products of NiTi-based SMAs, since it is characterized by the combination of dynamic recovery and dynamic recrystallization at high temperature [95, 96]. Static [97] or dynamic [98] crystallization on Fe-containing NiTi alloys was investigated. It was found that cold-canning compression and subsequent annealing can induce [111] direction fiber texture, which plays a dominant role in the compressive strength of NiTiFe. The intensity of [111] texture increases with increasing degree of deformation, which gives rise to the slip systems with larger Schmid factors when the alloy is compressed in [111] direction. Furthermore, [111] direction is the soft orientation of NiTiFe SMA, and it is the increase of [111] texture intensity that leads to the decrease in compressive strength of NiTiFe SMA [97]. Niu et al. [93] studied the behavior of substitutional Fe in both B2- and R-phases of $Ti_{50}Ni_{50-x}Fex$, at different concentrations. It was found that with no need of aggregation, an Fe atom can incur a drastic atomic scale local lattice distortion in the R-phase and makes its nearest neighbor environment like an intermediate structure between B2 and R. Its solution energy is lower in B2-phase than in R-phase, indicating that Fe tends to stabilize the austenite against the martensite, and the differences in the cohesive energy of B2- and R-phases, and the martensitic transformation temperature as well, decreases with the increasing Fe concentration, until the Fe content exceeds a critical level at which point the R-phase turns into B2-phase automatically, and the martensitic transformation is suppressed altogether.

3.2.1.5 Co

Alloying effect of Co element on phase transformation has been studied: $Ni_{47}Ti_{50}Co_3$ [43, 99,100] and $Ni_{51-x}Ti_{49}Co_x$ ($x = 0, 0.5, 1.5,$ and 4 at%) [101]. The common findings are although the matrix phase in the microstructure of $Ni_{51}Ti_{49}Co_0$ alloy is the austenite phase (B2) in addition to M-phase (B19′) and precipitates of NiTi intermetallic compounds, the parent phase in the Co-containing NiTi alloys is martensite. The hardness value of NiTi alloy is affected by Co additions. It was reported that the stiffness was improved by Co addition to NiTi alloy [102].

The electrochemical behavior of $Ti_{50}Ni_{47.2}Co_{2.8}$ alloy in deaerated artificial saliva solutions at 37 °C with binary NiTi alloy as reference was studied [103]. Potentiodynamic and potentiostatic tests results showed that the corrosion behavior of NiTiCo was similar to that of NiTi alloy. It was further mentioned that with the increase of pH value of the electrolytes, both corrosion potential (E_{corr}) and pitting corrosion potential (E_b) decreased. The outmost passive film consisted mainly of TiO_2, which were identical with that of NiTi alloy. The Ni ion release amount of NiTiCo was very close to that of NiTi alloy. Neither Ti nor Co ion was detected due to the detection limitation; thereby, concluding that the addition of Co had little effect on the corrosion behavior of NiTi as well as the formation of the passive film. The alloying effect of Co to NiTi alloys was investigated in terms of clinical properties in orthodontic mechanotherapy [104]. $Ti_{50}Ni_{47}Co_3$ (at%) alloys were prepared, followed by cold rolled at 30% reduction and heat-treated at 400 °C for 60 min. It was concluded that SE properties were confirmed in the manufactured commercial alloys at mouth temperature, and the difference of stress plateau in TiNi and TiNiCo, and commercial wires B at 25 °C changed significantly at various testing temperatures due to the combination of martensite and austenite phases. At certain temperatures, the alloys exhibited zero recovery stress at 2% strain and, consequently, produced zero activation force for moving teeth. The corrosion test showed that the addition of Cu and Co to TiNi alloys generates an increase in corrosion potential (E_{corr}) and corrosion current densities (I_{corr}), and addition of Co improved cell viability. These findings [104] and others [82] suggest that addition of an appropriate amount of a third alloying element such as Co element can help enhance the performances of TiNi in biomedical applications.

3.2.1.6 Cu

In this section, there are several things that are required to be mentioned prior to discussing alloying effects of Cu in detail. The Cu composition in Ni–Ti-based alloys is varied in a wide range from 0.2% to 25% or more. Cu element is substitutionally present at Ni site. It is expressed in either at% or wt%; however, since atomic weight of Cu (63.55) is very close to that of Ni (58.71), at% can be treated as wt% vice versa, as suggested in Chapter 2 for conversion between at% and wt%. In general, alloying effect (up to 15 at%) of Cu to NiTi-based alloys is recognized by the beneficial modification by Cu that there could be smaller differences in transformation temperatures during heating and cooling processes, resulting in the temperature hysteresis that gets narrower. Cu addition causes a lower martensitic yield strength. However, there is an adverse effect indicating that if Cu content increases, parent A-phase is embrittled, leading to worsen the hot workability, so that the upper limit of Cu addition is controlled up to 10 at%.

For relatively low Cu containing NiTi-based alloys, Alnomani et al. [105] studied the alloying effect of Cu for improving hardness. They found that during the sintering process, 1-h soaking at 900 °C of 2.5 wt% Cu containing NiTi alloy provided the highest hardness of 305 Hv. The increase of Cu content to 3 wt% led to decrease the hardness value of 295 Hv. Rondelli et al. [106] investigated the effect of Cu on the localized corrosion resistance of NiTi alloy ($Ni_{44}Ti_{51}Cu_5$) in 0.9% NaCl aqueous solution. They also reported that Ni-Ti-Cu and Ni–Ti showed low corrosion potentials (approximately 50–150 mV vs saturated calomel electrode (SCE)) inferior to that of Ti-Mo alloy (TMA), which proved to be immune to localized corrosion attacks up to 800 mV. Corrosion behavior was studied on Ni-Ti-5Cu-0.3Cr alloy in 0.9% NaCl and 1% lactic acid solutions [107]. It was reported that the addition of 0.19 at% Cr had little effect on the structure of the oxide films and the corrosion resistance of the Ni–Ti alloys. For Ni-Ti-5Cu-0.3Cr alloy, the metallic Cu was enriched at the alloy/oxide film interface, resulting in increased susceptibility to pitting corrosion above +1,000 mV; however, the passive current density and the amount of released Ni were not significantly increased by the addition of Cu, suggesting that small amounts of Cr and Cu added to change the superelastic characteristics do not change the corrosion resistance of the Ni–Ti alloy freely immersed in simulated physiological environments. The corrosion behavior of $Ni_{45}Ti_{50}Cu_5$ was compared with those of $Ni_{47}Ti_{50}Fe_3$ and $Ni_{39}Ti_{50}Pd_{11}$ and $Ni_{50.7}Ti_{49.3}$ alloy in a physiological solution (0.9% NaCl) simulating a body temperature of 37 ± 1 °C [108]. It was concluded that the localized corrosion resistance of ternary alloys was lower than the binary NiTi alloy. By comparing the different NiTi-based alloys, the following relation has been proposed for their localized corrosion resistances: NiTiCu < NiTiFe < NiTiPd < NiTi. The analysis of elemental depth revealed that the surface oxide film on all the investigated NiTi-based alloys was mainly of TiO_2; however, the NiTiPd and NiTiCu alloys showed metallic ternary element distributed within TiO_2 layer. These results confirm previous reports published by Oshida et al. [109, 110]. Phukaoluan et al. [104], evaluating the corrosion resistance of $Ti_{49}Ni_{46}Cu_5$, reported that the addition of Cu and Co to TiNi alloys generates an increase in corrosion potential (E_{corr}) and corrosion current densities (I_{corr}).

The effect of 5% Cu addition on SE of NiTi SMA was examined [111]. It was reported that the addition of copper was effective in narrowing the stress hysteresis and in stabilizing the SE characteristics against cyclic deformation, with the result that the slope of the load-deflection unloading curve of the alloy is lower than NiTi. Moreover, it produced greater stability of both the transformation temperature and the force applied to the teeth for a determined design and wire cross section. However, the presence of copper in NiTi orthodontic archwires reduced the aging effect. Grossmann et al. [112], testing functional fatigue behavior of NiTiCu with Cu of 5, 7.5, and 10at%, investigated the actuator stability during thermomechanical cycling. It was concluded that adding copper is more attractive than cold work, because it improves cyclic stability without sacrificing the exploitable actuator stroke. Cu reduces the width of the thermal hysteresis and improves geometrical and thermal actuator stability, because it results in a better crystallographic compatibility between the parent and the product phase. Es-Souni et al. [113], studying on $NiTi_{42}Cu_7$ and $NiTi_{42}$, mentioned that although adding Cu narrowed the hysteresis with superior mechanical properties including fatigue resistance, its cytotoxicity is higher than that of the binary alloy without Cu element; this might be due to the fact that it arises from the release of copper ions in the medium.

There are numerous studies conducted for TiNi materials containing Cu element ranging from 9 to 10 at%. Internal structure and shape-memory behavior of $Ti-Ni_{38.3}-Cu_{9.3}$ (at%) thin films was heat-treated at 600, 650, 700, and 800 °C and was subjected to thermal cycling tests [114]. It was reported that the Ti_2Ni precipitates increased in volume fraction with increasing heat-treatment temperature from 600 to 650 °C, then their volume fraction was almost kept constant above 650 °C. The recoverable strain decreased and the M_S increased with increasing heat-treatment temperature from 600 to 650 °C. Both the recoverable strain and the M_S became almost constant when the heat-treatment temperature was above 650 °C. Bending property of $Ti-Ni_{40.8}-Cu_{10.0}$ (mol%) alloy castings was investigated in a three-point bending test for orthodontic application in relation to the phase transformation [115]. The alloy was heat treated at 440, 480, and 520 °C for 30 min. The results show that the difference between the load values in the loading and the unloading processes was relatively small for Ti-Ni-Cu alloy. With respect to the residual deflection, there was no significant difference between Ti-Ni and Ti-Ni-Cu alloys with the same treatment condition. The load values in the loading and the unloading processes decreased by each heat treatment for Ti-Ni alloy; however, the decrease in the load values for Ti-Ni-Cu alloy was not distinct. Further, it is concluded that Ti-Ni-Cu alloy castings produce effective orthodontic force as well as stable low residual deflection, which is likely to be caused by the high and sharp thermal peaks during phase transformation. Single crystal specimens of $NiTiCu_{10}$ alloys were subjected to temperature cycling conditions under constant tensile and compressive stresses, and the transformation strains were monitored [116]. It was mentioned that the experimental transformation strains are higher in tension compared to compression for most single crystal orientations due to two factors: the additional strain associated with the detwinning of the B19′ (martensite) phase in the final microstructure, and the partial completion of the second step of the transformation limiting the compression strains. Samal et al. [117], utilizing $Ni_{50}Ti_{40}Cu_{10}$, studied the effect of Cu element on the phase evolution and transformation. It was reported that in the binary ($Ni_{60}Ti_{40}$) alloy, the ordered NiTi (B2) phase transforms to trigonal (R) phase followed by NiTi M-phase, i.e., B2 → R-phase → M-phase during solid-state cooling. However, Cu addition as an alloying element suppresses the martensitic transformation of the ordered NiTi (B2) dendrite; therefore, the ordered NiTi (B2) phase is transformed to only trigonal (R) phase, that is, B2 → R-phase. Cu-content dependence of the shape-memory characteristics in $Ti-Ni-Cu_{10}$ (at%) alloys were investigated by means of electrical resistivity measurements, DSC and constant load thermal cycling tests [118]. It was found that transformation start temperature for the B19′ martensite (M_S) decreased largely, whereas that for the B19′ martensite increased slightly with increasing Cu-content. The maximum recoverable elongation associated with the B2→B19 transformation decreased from 3.2% to 2.7%, and the hysteresis associated with the transformation decreased with increasing Cu-content from 10 to 20 at%. The temperature depen- dence of stress required to induce the B2→B19 transformation was nearly constant without regard to the Cu content for the alloys with above 10 at%Cu. The residual elongation decreased with increasing Cu content. The similar results were reported by Yang et al. [119] on $Ti_{50}Ni_{50-x}Cu_x$ (x = 10.4 at%).

There are several researches on surface characteristics. Using an atomic force microscopy roughness analysis, Persaud-Sharma et al. [120] observed the electropol- ished and magnetoelectropolished surfaces of $NiTiCu_{10}$ (wt%) and $NiTiTa_{10}$ (wt%). It was reported that $NiTiTa_{10}$ with an electropolished surface yielded the highest overall roughness, while the $NiTiCu_{10}$ had the lowest roughness. Aggregations of ternary ele- ments Cu at grain boundaries on both polished surfaces, causing an increase of cellular adhesion and accelerate surface endothelialization of endovascular stents, resulting in reducing the likelihood of in-stent restenosis and provide insight into hemodynamic flow regimes and the corrosion behavior of an implantable device influenced from such surface micropatterns [120]. Lin et al. [121] investigated the isothermal oxidation behav- ior of $Ti_{50}Ni_{40}Cu_{10}$ SMA in 700°–1,000 °C air. It was indicated that a multilayered oxide scale formed, consisting of an outermost $Cu_2O(Ni,Ti)$ layer, a layer of the mixture of TiO_2, $TiNiO_3$ and irregular small pores, a layer of the mixture of Ni(Ti,Cu), TiO_2 and irregu- lar large pores, a Ti(Ni,Cu)$_3$ layer and an innermost $Ti_{30}Ni_{43-47}Cu_{27-23}$ layer. The apparent activation energy for the oxidation reaction of this ternary alloy was 180 kJ/mol, and the oxidation rate follows a parabolic law.

$TiNi_{50}$ and $TiNi10_{Cu}$ (at%) alloys were subjected to a rotary bending fatigue tester to evaluate their fatigue lives [122]. The fatigue life decreased with increasing test temperature, in general. However, it became less sensitive to test temperature both in higher and lower temperature regions (in a range of 35, 50, 65, 85, and 125 °C). It was concluded that the fatigue life of the $TiNi_{50}$ is always longer than that of the $TiNi_{40}Cu_{10}$ (at%) if the fatigue life is plotted as a function of temperature difference between test temperature and M_S. The temperature dependence of the dynamic Young's modulus, the elastic energy dissipation coefficient and the heat flow were studied in an $Ni_{40}Ti_{50}Cu_{10}$ alloy [123]. It was reported that the modulus of elasticity exhibits softening when the start temperature M_S of the B2→B19 martensitic transition is approached on cooling and a much steeper modulus decreases between M_S and M_F. No internal friction peak occurs at the B2→B19 transition and the values of elastic energy dissipation coefficient are high in the B19 martensite. Internal friction was also investigated on $Ti_{50}Ni_{50-x}Cu_x$ (x = 16 and 20) [124]. It was reported that the broad peak appearing in B19′ martensite was confirmed to be a relaxation peak with an acti- vation energy of 0.76 and 0.67 eV for x = 16 and 20, respectively [124]. The influence of film thickness on the B2–B19 martensitic transformation properties of nanoscale $Ti_{51}Ni_{38}Cu_{11}$ thin films (from 750 to 50 nm thick) was investigated. Furthermore, it was reported that the substrate-attached $Ti_{51}Ni_{38}Cu_{11}$ thin films as thin as 50 nm showed reversible B2–B19 phase transformations. With decreasing film thickness, a change in the tetragonality of the B19 M-phase occurs [125].

By increasing Cu content to from 15% to 20%, structural dependent properties have been studied. Meng et al. [126] investigated the martensite structure in sputter-deposited thin films of $Ti_{48.6}Ni_{35.9}Cu_{15.5}$. It was found that the $Ti(Ni,Cu)_2$ phase precipitated during the annealing process. First, fine $Ti(Ni,Cu)_2$ precipitates can be deformed by the shear deformation of martensitic transformation, but they obstruct the movement of the twin boundaries to some extent. Second, coarse $Ti(Ni,Cu)_2$ precipitates seriously impede the growth of martensite plates and lead to a rectangular-cell-like structure of martensite in the film annealed at 600 °C. Third, the resistance of $Ti(Ni,Cu)_2$ precipitates to the growth of the martensite plates enhances with the coarsening of $Ti(Ni,Cu)_2$ precipitates, which is one of the reasons for the decrease in the maximum recoverable strain with increasing annealing temperature. Similar work on $Ti_{50.2}Ni_{30}Cu_{19.8}$ thin film deformed in the B19′ martensite state was investigated [127]. The studies on internal friction on $Ni_{30}Ti_{50}Cu_{20}$ [128] and those on $Ti_{50}Ni_{50-x}CuX$ ($x = 16$ and 20) [124] indicate that there are two essential ingredients to observe the high broad peak; twin boundaries and hydrogen.

For Cu_{25} containing NiTi-based alloys ($Ti_{50}Ni_{25}Cu_{25}$), during the annealing process, the sequential precipitation was reported: B11 TiCu → B11 TiCu + Ti_2(Ni, Cu) → Ti_2 (Ni, Cu) with increasing annealing temperature or duration [129–132]. Liu found that an excellent superelastic shape recovery of up to 10% strain while the hysteresis (in other words, the amount of energy dissipated during one cycle) decreases with $Ti_{50}Ni_{25}Cu_{25}$ which was annealed at 500 °C for 15 min [133].

In history of material development for orthodontic wires, there found Cu-NiTi alloys with Cu contents ranging from 3 to 10 wt%. Since wires utilized for performing the orthodontic mechanotherapy require specific properties including stiffness, formability, weldability, friction resistance, biocompatibility, relationship between human body temperature and material's A_S and A_F temperatures, there are a variety of researches reported. The force moment providing rotation of the tooth around the x-axis (buccal-lingual) is referred to as a torque expression. Archambault et al. [134] compared the torque expression of Cu–NiTi wire with stainless steels wire and TMA wire. It was found that the stainless steel has the largest torque expression, followed by TMA and then Cu–NiTi. Effect of Cu element on transformation was investigated [135, 136] and it was reported that the phase transformation temperatures clarified by DSC showed B2 ↔ B19(') transformation and the three-point bending test conducted at 37 °C showed the plastic deformation of the Cu–NiTi wire after annealing due to the effect of copper in the alloy composition.

Ni release from Cu–NiTi and NiTi should be minimized (or eliminated) as much as possible to avoid any unwanted adverse causes on human health. Gil et al. [137]

found that the surface treatment by nitrogen gas (i.e., nitrization) can inhibit the Ni release, and the reduction in this friction coefficient was achieved by nitrogen diffusion heat treatments. Besides Ni allergy or toxicity of Cu–NiTi or NTi, there is still a healthy issue on Cu–NiTi, namely Cu element, that possesses both advantage and disadvantage effects. NiTi biomaterials are widely used as implant materials in the human or animal body to repair organs and restore function, such as heart valves, meninges, peritoneum, and artificial organs. As has been described so far, it is well documented that alloying element affects the microstructure, mechanical property, corrosion resistance, and wear resistance. One more thing that we need to look at is something related to biocompatibility and biological activity of such alloying element(s). Recently, antibacterial metal alloys have shown great potential as a new kind of biomedical material, in which Cu has been widely used as antibacterial agent element [138]. In addition, biodegradable metal alloys, including magnesium alloy and zinc alloy, also have attracted much attention worldwide. Cu was also used as alloying element to adjust the degradation rate [139]. Thus, the role of Cu in the alloy design will be very important for the development of new alloy.

Although Cu element is recognized as an essential trace metal that is required for the catalysis of several important cellular enzymes, an excess of copper can also harm cells due to its potential to catalyze the generation of toxic reactive oxygen species. Therefore, transport of copper and the cellular copper content is tightly regulated [140]. Linder et al. [141] summarized alterations in copper metabolism associated with genetic and nongenetic diseases, including potential connections to inflammation, cancer, atherosclerosis, and anemia, and the effects of genetic copper deficiency (Menkes syndrome) and copper overload (Wilson's disease). Understanding these diseases suggests new ways of viewing the normal functions of copper and provides new insights into the details of copper transport and distribution in mammals [142]. Due to its redox activity, copper can also lead to the generation of toxic reactive oxygen species. Therefore, cellular uptake, storage as well as export of copper has to be tightly regulated in order to guarantee sufficient copper supply for the synthesis of copper-containing enzymes but also to prevent copper-induced oxidative stress. In brain, copper is of importance for normal development. In addition, both copper deficiency as well as excess of copper can seriously affect brain functions [142]. Klevay [143] mentioned that an increase in dietary copper intake with an increase in legume consumption also have contributed to their results because copper deficiency is the only nutritional insult that elevates cholesterol, blood pressure, and uric acid; has adverse effects on electrocardiograms; impairs glucose tolerance; and promotes thrombosis and oxidative damage. Hu [144] mentioned that Cu ions stimulate proliferation of human umbilical artery and vein endothelial cells but not human dermal fibroblasts or arterial smooth muscle cells and Cu-induced proliferation, along with Cu-induced migration of endothelial cells, may suggest a possible mechanism for the involvement of copper in the process of angiogenesis [136]. Abraham et al. [145] compared the adhesion of *Streptococcus mutans* to NiTi and Cu-NiTi archwires to correlate the adhesion to surface characteristics (surface free energy and surface roughness). Results indicate that *S. mutans* adhesion was more in Cu-NiTi archwires, which exhibited rougher surface and higher surface-free energy when compared to NiTi archwires, and a predominantly negative correlation was seen between the cycle threshold value of adherent bacteria and surface characteristics.

Among alloying elements with antibacterial properties, copper has shown superior in vitro antibacterial performance while maintaining an acceptable cytotoxicity profile. Cu could prevent early biofilm formation to limit periprosthetic infections [146]. Cu element is a well-known antibacterial alloying element, being similar to Ag effect. Furthermore, copper ions can alter the function and structure of proteins in the bacterial cell wall, and lead to its consequent rupture [147]. Moreover, they can bind and alter several bacterial enzymes, which are crucial for bacterial cellular respiration and metabolism [142]. The aforementioned multiple actions of copper ions explain the excellent antibacterial effects of copper alloying elements.

Summarizing alloying effect of Cu element to NiTi-based alloys, it should be noticed that it has been shown that adding Cu up to 10% into NiTi-based alloy would not damage the SME of TiNi alloys [138,148]. The metallography and microstructural studies reveal that Cu–NiTi alloys show microstructure with uniform equiaxial grains and clear grain boundaries with no clear influence on alteration of grain size. The microstructure is a mixture of NiTi austenite (B2 phase) and Ni_4Ti_3 precipitate. By increasing Cu content, the phases of NiTi martensite (B19′ phase) and Cu_4Ti_3 are present in $TiNi_{43.8}Cu_{7.0}$ and $TiNi_{40.8}Cu_{10}$ alloys, thereby indicating that Cu addition increased the M_S temperature and Cu probably substituted the position of Ni in NiTi alloys. As to the phase transformation of Cu-added NiTi-based alloy, it is known that for less Cu content NiTi alloys, a multiple stage transformation behavior is observed as follows: the three-stage transformation (B2 → R → B19 → B19′) during the cooling process takes place and two-stage transformation (B19′ → B19 → B2) during the heating in $TiNi_{49.8}Cu_{1.0}$ alloy. By increasing copper content to 4% ($TiNi_{46.8}Cu_{4.0}$) and 7% ($TiNi_{43.8}Cu_{7.0}$), two-stage transformation (B2 ↔ B19 ↔ B19′) has been observed. For high Cu content (10%) NiTi alloy, the typical phase transformation of a single-stage B2 ↔ B19′ takes place with M_S of 36 °C.

3.2.2 The second transition metal elements

3.2.2.1 Y

Several studies on alloying effect of Yttrium element on mechanical properties are reported. Alhumdany et al. [149, 150] prepared the Y and Ta co-added NiTi-based alloys at 1, 2, and 3 wt% of each element (substituted at the Ni site) in $Ni_{55}Ti_{45}$ alloy. Powders were compacted under 400, 500, 600, and 650 MPa pressure, followed by sintering at 500 °C for 2 h and then 850 °C for 6 h under vacuum conditions. The thus obtained compacts were subjected to the XRD testing, apparent density and porosity, hardness, and surface roughness tests. It was reported that the wear volume loss decreases with the addition of tantalum by 0.52% with 3% tantalum addition at 650 MPa compacting stress for a 15 N load; it also decreases by 0.48% with 2% Yttrium addition at 650 MPa for a 10 N load. The wear volume loss further decreases as the compacting stresses increase. Using similar test samples, fundamental mechanical test was conducted [150]. It was shown that the XRD diffraction tests of all samples have the same results of two phases: monoclinic NiTi and hexagonal Ni_3Ti, at room temperature. The specimen with 2wt% Y compacted at 650 MPa showed the hardness value of 275.7 Hv. The hardness value decreases at 3% and 1% additives at the same compacting stress to (272.26 and 200) Hv, respectively. The specimen of 2% Y compacted at 650 MPa showed 0.16-μm surface roughness, while the addition of tantalum element decreases the surface roughness from 0.36 to 0.15 μm at 3% tantalum compacted at 650 MPa. TiNi alloy has a high resistance to wear, and it could be an excellent candidate for various tribological applications. Ahmadi et al. [151] reported that by addition of yttrium, hardness properties and resistance to wear and corrosive wear of TiNi alloy were improved, and there was an optimum content for addition of yttrium between 2% and 5% (wt%), and above this content the improvement in properties of TiNi became minor.

3.2.2.2 Zr

Normally, it is believed that, by the addition of Zr, the phase-transformation temperature could be enhanced. Pu et al. [152], studying NiTiZr high-temperature alloys, it was reported that the martensite transformation temperature increases with Zr content when the Zr content is more than 10 at% in NiTiZr alloy. As the Zr content increases, the fully reversible strain of the alloys decreases; however, complete strain recovery behavior is exhibited, even those with a Zr content of 20 at%. Moreover, the stability of the NiTi–Zr alloys during thermal cycling test indicated that the NiTi–Zr alloys have poor stability against thermal cycling. Hsieh et al. [153] investigated the $Ti_{53-x}Ni_{47}Zr_x$ ($x = 5\sim20$ at %) SMAs and showed that the transformation temperatures increase linearly with increasing Zr content. There are three different phases observed in 900 °C homogenized alloys for Zr content $\geq$10 at %, including the grey (Ti,Zr)Ni matrix, the black $(Ti,Zr)_2Ni$ particles, and the white $\lambda 1$ phase. Differential thermal analysis (DTA) of $Ti_{38}Ni_{47}Zr_{15}$ alloy exhibits three endothermic peaks, in which the 930 °C endothermic peak is associated with the solid$\leftrightarrow$liquid transition of $\lambda 1$ phase. The peak at 990 °C is associated with the reverse peritectic transformation of $(Ti,Zr)_2Ni\rightarrow(Ti,Zr)Ni+$liquid and the one at 1,160 °C is the dissolution reaction of the (Ti,Zr)Ni$\rightarrow$liquid [156]. Similar research work on the martensitic transformation in $Ti_{50.5-x}Ni_{49.5}ZrX$ and $Ti_{51.5-x}Ni_{48.5}ZrX$ alloys ($x = 0$–25 at%) was conducted by using thermomechanical treatments [154]. It was reported that these alloys have a B2$\leftrightarrow$B199 transformation sequence, and their transformation peak temperature M^P can be raised to 50–450 °C by different additions of Zr. Although many second-phase particles exist around (Ti,Zr)Ni grain boundaries, these alloys still exhibit \$80% shape-memory recovery, and thermal cycling can depress the M^P temperature more significantly in the $Ti_{41.5}Ni_{48.5}Zr_{10.0}$ alloy than in the $Ti_{40.5}Ni_{49.5}Zr_{10.0}$ alloy in the first ten cycles, owing to the formers having greater hardness and more second-phase particles. Furthermore, it was mentioned that the martensite stabilization can be induced by cold rolling at room temperature for Ti-rich ternary TiNiZr alloys, and this stabilization may be due to the pinning effect on the interfaces of martensite plates by the point defects [154]. Martensitic transformation on $Ti_{40.5}Ni_{49.5}Zr_{10}$ alloy was investigated [155], and it was found that this alloy undergoes a B2$\leftrightarrow$B19' one-stage martensitic transformation at $M_S \approx 100$ °C. It exhibits about 90% shape-memory recovery even though many $(Ti,Zr)_2Ni$ particles exist around $(Ti,Zr)Ni$ grain boundaries. Thermal cycling can also decrease transformation temperatures while increasing the hardness of the alloy. Yang et al. [156], studying the martensitic transformation and phase stability of $Ni_{50}Ti_{50-x}Zrx$ SMAs, reported that the martensite monoclinic B19´ structure becomes more stable with increasing of the Zr content. The energy difference between austenite and martensite decreases slightly primarily when Zr < 10.4 at% and then increases sharply. This indicates that Zr addition increases martensitic transformation temperature dramatically. The thermal hysteresis increases with increasing the Zr element content.

As to precipitate formation during the heat treatments, McCluskey et al. [157] studied thin-film samples of Ni–Ti–Zr SMAs by combinatorial nanocalorimetry to determine the effects of high-temperature (900 °C) heat treatments and low-temperature (450 °C) thermal cycling on the characteristics of the martensite transformation. It was found that there are two precipitate types, a Ti_2Ni-base phase at low Zr concentration and a $Ni_{10}Zr_7$-base phase at high Zr concentration, affect the martensite transformation characteristics by altering the composition and the stress state of the shape-memory phase. During thermal cycling, the most stable sample demonstrates a transformation temperature reduction of 11 °C for 100 cycles, and this improved stability of the samples is attributed to the very small grain size of approximately 5–20 nm. It was reported that such precipitates are fully coherent with the austenite B2 matrix; however, upon martensitic transformation, they lose some coherency with the B19' matrix as a result of the transformation shear process in the surrounding matrix. The strain accommodation around the particles is much easier in the Ni–Ti–Zr-containing alloys than in the Ni–Ti–Hf system, which correlates well with the lower transformation strain and stiffness predicted for the Ni–Ti–Zr alloys [158]. Studying the microstructure, thermal cycling, and mechanical behavior of $Ni_{48.5}Ti_{31.5-X}Zr_{20}Al_x$ (x = 0, 1, 2, 3) alloys, Hsu et al. [159] found that the aluminum additions served to decrease transformation behaviors from 78 to 323 °C and reduce thermal stability. The addition of Al resulted in the refinement of the coarse, lenticular precipitates identified as $Ni_4(Ti,Zr)_3$. Kim et al. [160] studied the crystallization process of as-deposited NiTiZr (10.8–29.5) amorphous thin films. It was found that the sample with a low Zr content exhibited a single exothermic peak due to the crystallization of $(Ti,Zr)Ni$ with a B2 structure. Alternatively, a two-step crystallization process was observed in the Ti–Ni–Zr thin films with a high Zr content. The martensitic trans-formation start temperature increased with increasing Zr content until reaching the maximum value, then decreased with further increasing Zr content. The inverse dependence of transformation temperature on Zr content in the thin films with a high Zr content is due to the formation of a NiZr phase during the crystallization heat treatment. The formation of the NiZr phase increased the critical stress for slip but decreased the recovery strain [160]. Feng et al. [161] prepared ternary NiTi-based alloys with composition as $Ni_{50.6}Ti_{49.4-x}ZrX$ (x = 0, 1, 2, 3 at %) and investigated the influence of Zr additions on the properties of NiTi SMAs. It was shown that the trans-formation temperatures of alloys decrease first and then increase with the addition of Zr. Mechanical properties and shape-memory properties increased with increasing Zr content. The addition of Zr causes the appearance of the $(Ti,Zr)_2Ni$ precipitate, which should be response to the fracture of $NiTiZr_3$ alloy during the heat working process. The effect of isothermal aging on compression cycling, martensitic transformation temperatures and microstructure was investigated for a $Ni_{53}Ti_{41}Zr_6$ (mol%), and it was reported that the aging behavior of the microstructures was a two-stage process [162].

3.2.2.3 Nb

It is generally believed that, by increasing Nb content in NiTi-based alloy, although the transformation PE hysteresis gets broader, hysteresis due to temperature differ-ences (between M_F and A_F temperatures) will get wider, too. Although the hysteresis (between B2 and B19 transformation) in NiTi alloy is normally in a range of 30–50 °C, it can go to higher than 150 °C. This suggest that devices and materials for this purpose can be designed in such a way that the room temperature is within the hysteresis loop, so that the shape formed at cryogenic temperature can bring up to the room temperature. Occupation site of niobium (Nb) in NiTi-based alloy was investigated by aberration-corrected scanning TEM and precession electron diffraction [163]. It was mentioned that Nb atoms were found to prefer to occupy the Ti rather than Ni sites. Due to a higher melting point of Nb than those of Ni and Ti, it would be effec-tive to lower the niobium content in Ni–Ti–Nb alloys to reduce segregation [164]. He et al. [165] studying the martensitic transformation, found that $Ni_{50.1}Ti_{46.9}Nb_3$ (at%) alloy exhibited a sufficiently wide transformation temperature hysteresis which can be attained after deformation at (M_S +30 °C). It was furthermore reported that the mar-tensitic start temperature M_S, the martensitic finish temperature M_F, the austenitic start temperature A_S and the austenitic finish temperature A_F were −62, −117, −48, and 18 °C, respectively. As to microstructure of this alloy, it was mentioned that the dom-inant phase is the NiTi matrix phase, and β-Nb particles are difficult to find in scan-ning electron microscope observations. It is concluded that A_S temperature does not with increasing strain in this low niobium content alloy, however, the deformed NiTi matrix phase adjacent to the stress induced martensite will greatly increase the resis-tance of the thermal elastic martensite freely shrinkage upon heating, so the reverse transformation temperature interval was widened [165]. Wang et al. [166] investigated phase transformation and deformation behavior of the NiTiNb eutectic microstructure using TEM and cyclic loading–unloading tests. Obtained results showed that R-phase and B19′ martensite transformation are induced by plastic deformation, and R-phase transformation, which significantly contributes to SE, preferentially occurs at the interfaces between NiTi and eutectic region. An annealed commercial NiTiNb alloy (typically, $Ni_{46-50}Ti_{bal}Nb_{8-15}$) was investigated by scanning and TEM [167], and it was reported that the precipitates have a diameter of around 100 nm, which are faceted and have a cube-on-cube relation with the B2 matrix. In situ TEM cooling shows that the martensitic transformation is hampered by the presence of these precipitates. The latter could explain the increase in hysteresis when compared with the binary system. With this alloy with Nb containing about 9%, it was mentioned that the microstructure consists of the NiTi matrix and β-Nb phase [168]. It was further mentioned that the microstructure of the alloy consists of three phases: TiNi matrix, Nb-rich phase, and $Ti_3(Ni,Nb)_2$ compound (which is harder and embrittled phase). The Nb-rich phase is determined to be β-Nb with BCC structure containing a small amount of Ni and Ti, and the β-Nb is a soft phase, which forms a eutectic structure with TiNi phase during solidification. After hot working the soft β-Nb phase is dispersed in TiNi matrix and gives rise to a wide transformation hysteresis in the alloy [168]. Since Nb effect depends on whether it substitutes Ti or Ni sites, Piao et al. [169] designed three different alloy systems for NiTi alloys; $Ti_{50-x/2}Ni_{50-x/2}Nb_x$ (x = 0–25 at%), $Ti_{50}Ni_{50-x}Nb_x$ (x = 0–30 at%) and $Ti_{50-x}Ni_{50}Nb_x$ (x = 0–3 at%). The microstructure and the M_S temperature were investigated for these alloys. It was reported that the addition of Nb promotes phase separation; in $Ti_{50-x/2}Ni_{50-x/2}Nbx$ and $Ti_{50}Ni_{50-x}Nbx$ alloys, the microstructure changes with increasing Nb addition in the following way: B2 phase → primary B2 phase+eutectic → eutectic → primary BCC phase+eutectic, and a pseudobinary "TiNi-Nb" phase diagram was proposed to explain the above behavior. In $Ti_{50-x}Ni_{50}Nbx$ alloys, the microstructure was the primary B2 phase only within the observed composition range but slight microsegregation of Nb was observed. Furthermore, it was mentioned that the M_S temperatures of NiTiNb ternary alloys were measured for the three types of alloys as a function of Nb addition, and the result indicated that the M_S temperatures in the alloys are mainly dependent on the Ni/Ti ratio in the B2 phase [169] .

When NiTi SMA is utilized as coupling devices, it is important to notice that SME depends considerably on the phase transformation temperatures of the material. In particular, such devices should have an austenite start temperatures (A_S temperatures) above room temperature for storage purposes as well as low martensite start temperatures (M_S temperatures) in order to guarantee a high level of mechanical resistance service. Salvetr et al. [170], using $NiTi_{46}$ and $NiTi_X$ (x = Nb, V, or Si with 5 wt%, respectively) prepared by arc remelting technique, investigated microstructure, phase composition, hardness and transformation behavior. It was found that arc remelting produced a more homogeneous microstructure with a lower content of the Ti_2Ni phase than powder metallurgy methods. Considerably, a high amount of the secondary phase (a mixture of the Ni_2Ti, NiTi, and Ti_2Ni phases) was created in the microstructure of $NiTi_{46}$ wt% alloy and the addition of silicon caused an increase of transformation temperatures and hardness.

It is well known that the addition of Nb to NiTi alloy systems results in an increase in the width of the phase transformation temperature hysteresis, in particularly for the first phase transformation cycle [171–173], suggesting that NiTiNb alloys have great potential as mechanical components to join, fasten and seal. Cai et al. [174] mentioned that NiTi–X (NiTi–29Nb and NiTi–24V), which can be tuned to get highly ordered nanostructures. By directional solidification of these ternary alloys, nanowires/rods of Nb and V can be realized, although, addition of Nb, V in eutectic composition could hamper the shape-memory properties of the parental alloys, nevertheless, these systems would be beneficial to obtain nanowires of Nb and V, which is very tough otherwise. He et al. [164], studying low niobium content NiTi-based ternary alloy ($Ti_{46.9}Ni_{50.1}Nb_3$), reported that the alloy has excellent SME, and it can also exhibit enough wide transformation temperature hysteresis after deformation at M_S +30 °C.

There are additional researches on Nb-alloying effect. Zhang et al. [175] studied the abnormal wear behavior of NiTiNb alloy. It was mentioned that NiTiNb alloy exhibits a linear increase in wear with a number of impacts at the low impact energy density; however, a change in the volume wear occurs on the wear curve when impact energy density is increased. It is significant for practical applications because impact energy increases while the wear rate decreases It is also mentioned that the abnormal wear behavior of the TiNiNb alloy can be attributed to the excellent wear behavior of amorphous structure and the consumption of impact energy during amorphous structure production [175]. Mareci et al. [176] studied the electrochemical corrosion behavior of $Ni_{47.7}Ti_{37.8}Nb_{14.5}$ SMA in artificial saliva solution, designed for dental applications. It was reported that very low passive current densities were obtained from the anodic polarization curve, indicating a typical passive behavior for NiTiNb alloy. On the surface of the NiTiNb alloy, uniform corrosion appears, while in case of the NiTi alloy surface pitting corrosion is developed. The role that Nb plays as an alloying element is by increasing the resistance of NiTi alloy to localized corrosion, suggesting that NiTiNb surface is covered by a highly stable passive film.

3.2.2.4 Mo

Although there are several researches regarding effect of Mo, these are all related to surface modification and are not directly related to alloying effect. Accordingly, these important works are discussed in Chapter 12.

3.2.2.5 Pd

Normally, along with Pt element, Pd containing NiTi-based alloy can be considered as the high-temperature SMAs since replacing Ni either by Pd or Pt has similar effects on the shape-memory properties. Small additions of these elements cause a slight decrease of M_S temperature, and even shift it below 0 °C, but higher additions strongly elevate the transformation temperatures compared to other elements. The M_S temperature ranges from –26 °C (for 10 at% Pd) to 563 °C (for 50 at% Pd) [64].

Soga et al. [177] prepared NiTiPd (5, 7.5, 10, and 15 at%) alloyed to the equiatomic NiTi by the substitution for Ni and investigated the alloying effect of palladium (Pd) to improve the SE of the alloy castings at body temperature for dental application. It was reported that $TiNi_{42.5}Pd_{7.5}$ alloy castings showed good SE among the examined alloys from the viewpoint of residual strain and elongation, and $TiNi_{42.5}Pd_{7.5}$ alloy castings exhibited better superelastic flexibility than Ti–50.8Ni alloy, which is proven by lower apparent proof-stress and larger elongation, caused by its relatively high martensitic transformation starting temperature point. It is suggested that this flexibility with SE could widen the clinical application of the alloy casting in dentistry.

High-temperature SMAs have received increased attention as the demand for lightweight, compact, and efficient actuators has grown in recent years, in particular, alloying elements like Au, Pd, Pt, Hf, or Zr have been investigated its alloying effect and among these, NiTiPd alloy has been extensively studied [178,179]. Bigelow et al. [178] prepared NiTiPd (Pd contents ranging from 15 to 46 at%) to determine the transformation temperatures, transformation strain, and unrecovered strain per cycle (a measure of dimensional instability) as a function of stress for each alloy. It was reported that increasing the Pd contents resulted in a linear increase in transformation temperature. It was furthermore mentioned that, at a given stress level, work output decreased while the amount of unrecovered strain produced during each load-biased thermal cycle increased with increasing Pd content, during the initial thermal cycles. However, continued thermal cycling at constant stress resulted in a saturation of the work output and nearly eliminated further unrecovered strain under certain conditions, resulting in stable behavior amenable to many actuator applications [178]. Shimizu et al. [180] prepared nonequiatomic TiPdNi high-temperature SMAs (Ti:(Ni, Pd) ≠ 50:50) to improve the shape-memory characteristics by precipitation-hardening and conducted DSC, high-temperature tensile tests, and TEM. It was found that homogeneously distributed fine precipitates can be produced by aging treatment at the proper temperature of 500 °C for $Ti_{50.6}Pd_{30}Ni_{19.4}$ alloys, and these precipitates increase the critical stress for slip and improve the shape-memory characteristics substantially at high temperatures. Yang et al. [181] also found that Pd substitution can increase the transition temperature ranges to 500 °C for more than 15 at% Pd. Cai et al. [182] investigated the thermal cyclic characteristics under load were investigated in a $Ti_{50.6}Pd_{30}Ni_{19.4}$ alloy. It is shown that the M_S temperature, transformation strain and total plastic strain increase with increasing the number of the cycles and the increments increase with an increase of applied stress. These changes due to thermal cycling under load occur rapidly in the early cycles and then remain constant after a sufficiently large number of cycles, suggesting that the cyclic training is effective to stabilize the martensitic transformation and shape-memory characteristics of the alloy. Due to the temperature range of thermoelastic martensitic transformation, the practical usage of TiNi is limited by temperature to about 100 °C. Goldberg et al. [183] prepared $Ti_{50}Pd_{30}Ni_{20}$, which was cold rolled and heat treated, and reported that its martensitic transformation temperature was in a range from 217 to 247 °C, which is very close to the most hopeful interval for application of high-temperature SMA. A $Ti_{50}Pd_{40}Ni_{10}$ high-temperature SMA was studied and it was reported that the transformation temperature was in range from 300 to 500 °C [183].

Ma et al. [184] categorized SMAs exhibiting high transformation temperature into three groups in terms of their martensitic transformation temperatures: group I, transformation temperatures in the range of 100–400 °C; group II, in the range of 400–700 °C; and group III, above 700 °C. It is noticed that potential high-temperature SMAs must also exhibit acceptable recoverable transformation strain levels, long-term stability, resistance to plastic deformation and creep, and adequate environmental resistance and these criteria become increasingly more difficult to satisfy as their operating temperatures increase, due to greater involvement of thermally activated mechanisms in their thermomechanical responses.

Alloying effect on corrosion resistance of Pd element to NiTi-based alloy was tested in a physiological (0.9% NaCl) aqueous solution simulating a body temperature of 37 ± 1 °C [185], and it was reported that the corrosion resistance is ranked in the following order: NiTiCu < NiTiFe < NiTiPd < NiTi. It was also mentioned that the surface oxide film on all the investigated NiTi-based alloys is mainly of TiO_2; however, NiTiPd and NiTiCu alloys showed metallic ternary element distributed within TiO_2 layer. Tian et al. [186,187] studied the oxidation behavior of TiPdNi alloy. It was found that the outmost layer is mainly composed of rutile TiO_2 while $TiNiO_2$ and Ti_4Pd_2O phases exist in the inner scale. The alloying element addition (Ce: Cerium) in the TiPdNi alloy effectively impedes the outward diffusion of Ti ions and obviously improves the oxidation resistance. The addition of Ce in the TiNiPd alloy effectively impedes the Ti ions outward diffusion, causes a wide distribution of Ni and Pd over the scale depth, and obviously improves the oxidation resistance. Moreover, without Ce addition, it was reported that the outmost layer is mainly composed of rutile TiO_2 and $TiNiO_3$ and Ti_4Pd_2O oxides are formed on TiNiPd alloys, and these oxides act as important media during the ion exchanging process during the oxidation and cause different oxidation resistance.

3.2.2.6 Ag

Oh et al. [188] investigated the effect of silver addition to nickel–titanium alloys NiTiAg (15–20% Ag) for dental and medical application. It was mentioned that silver addition to nickel–titanium increased the transition temperature range to 100 °C and stabilized the M-phase (monoclinic structure) at room temperature, because the martensitic transformation starting temperature (M_S) was above room temperature. Ag addition was considered to improve the corrosion resistance and form a stable passive film. There was no toxicity in the NiTiAg alloys. Alloying effect of Ag to NiTi alloy was investigated [189]. It was reported that the typical microstructural feature of TiNiAg alloy at room temperature was tiny pure Ag particles (at submicrometer or micrometer scales with irregular shape) randomly distributed in the TiNi matrix phase, resulting in a slightly higher tensile strength and larger elongation of TiNiAg alloy in comparison with that of TiNi binary alloy. It was further mentioned that in electrochemical and immersion tests, TiNiAg alloy presented good corrosion resistance in simulated body fluid, comparable with that of CPT (commercially pure titanium) and TiNi alloy. The cytotoxicity evaluation revealed that TiNiAg alloy extract induced slight toxicity to cells, but the viability of experimental cells was similar to or higher than that of TiNi alloy extract. In vitro bacterial adhesion study indicated a significantly reduced number of bacteria (*S. aureus*, *S. epidermidis*, and *P. gingivalis*) on the TiNiAg alloy plate surface when compared with that on TiNi alloy plate surface, suggesting antibacterial function [189].

3.2.3 Third transition metal elements

3.2.3.1 Dy

Liu et al. [190] investigated the effect of rare earth element, dysprosium (Dy), addition on the microstructure and martensitic transformation behavior of $Ti_{49.3}Ni_{50.7}$ SMA by optical microscope, SEM, XRD, and DSC. It was reported that the microstructure of Dy-TiNi ternary alloy consists of Ti_2Ni phase, DyNi phase, and the matrix. One-step martensitic transformation is observed in quenched Dy–TiNi ternary alloys, which is the same as that in quenched TiNi binary alloys. The martensitic transformation temperatures increase evidently with Dy addition, and the maximum increase of M_S is about 110 °C.

3.2.3.2 Er

Tofail et al. [191] mentioned that a high X-ray visibility of surgical devices is critical during minimally invasive surgery for exact placement of these devices within the targeted anatomical site while keeping the X-ray exposure of the patient to a minimum. But, NiTi and common metallic biomaterials (e.g., stainless steel and Co–Cr alloys) used in minimally invasive surgery are inherently less visible due to their weak X-ray absorption ability (radiopacity). Accordingly, it is important to develop materials possessing a high X-ray visibility. It was reported that by adding rare earth lanthanide elements such as erbium (Er), NiTi alloy showed an X-ray higher visibility, so that patients' trauma during deployment can be reduced and radiation dose received by a patient in his lifetime will also be significantly reduced.

3.2.3.3 Hf

There are three kinds of Ni-Ti-Hf SMAs: (1) interstitial (Ti-rich $Ni_{50-x}Ti_{50}Hf_x$), (2) equiatomic ($Ni_{100-x/2}Ti_{100-x/2}Hf_x$ at $5 \leq x \leq 20$ at%), and (3) substitutional (Ni-rich $Ni_{50}Ti_{50-x}Hf_x$) [192]. Coppa et al. [193] studied the effect of Hf (0–1 at%) additions in a Ni–Ti–Pd alloy on P-phase precipitation and martensitic transformations. It was mentioned that the addition of hafnium (Hf) resulted in the refinement of precipitates with an increase in density. The overlapping strain fields, created due to the decrease in interprecipitate spacing, are suspected to reduce the matrix volume to be less than the critical free volume size needed for the martensitic transformation over the temperature range studied (90–300 °C), and Hf was also found to delay the aging time to achieve peak hardness, suggesting a reduction in growth and coarsening kinetics. Using a relatively new alloy ($Ni_{54}Ti_{45}Hf_1$), Casalena et al. [194] studied structure-property relationship and reported that the NiTiHf alloy exhibits strengths more than 40% greater than those of conventional NiTi-based SMAs (2.5 GPa in compression and 1.9 GPa in torsion) and retains those strengths during cycling. The superelastic hysteresis is very small and stable with cycling, and aging treatments are used to induce a very high density of Ni_4Ti_3 precipitates, which impede plasticity during cycling yet do not impart substantial dissipation to the reversibility of the phase transformation. Hornbuckle et al. [195], using a series of Ni-rich ternary alloys with dilute solute additions of Hf ($Ni_{54}Ti_{45}Hf_1$, $Ni_{55}Ti_{44}Hf_1$, $Ni_{54}Ti_{44}Hf_2$, and $Ni_{56}Ti_{40}Hf_4$ in at%), investigated alloying effect of Hf on precipitation and its related hardenability. It was concluded that the high hardness has been associated with the precipitation of a large volume fraction of Ni_4Ti_3 platelets, resulting in a matrix that consists of narrow B2 NiTi matrix channels. At aging times greater than ~100 h at 400 °C, all ternary alloys showed a slight secondary increase in hardness, which was attributed to H-phase (H-phase has a face-centered orthorhombic lattice and its composition is rich in Ni in NiTiHf alloy systems.) precipitation and growth within the B2 channels. In the particular case of the $Ni_{56}Ti_{40}Hf_4$ alloy, hardness increased with aging to a maximum value of 679 VHN, which was greater than all other binary or ternary alloys examined. The H-phase appeared to alter or delay the typical breakdown sequence of the metastable Ni_4Ti_3 strengthening phase by removing the excess Ni needed for its decomposition. Using the ternary $Ni_{49.8}Ti_{42.2}Hf_8$ SMA, Dalle et al. [196] studied the mechanical peculiarities compared with NiTi alloy and found that peculiarities are linked to the distortion of the habit plane of the ternary alloy and the observed (001) mechanical twinning, which helps to accommodate plastic deformation in $Ni_{49.8}Ti_{42.2}Hf_8$ and not in NiTi alloy.

Researches on transformation behaviors of NiTiHf with medium Hf contents are found. Thoma et al. [197] determined the influence of Ni (ranging from 490 to 50.1 at% in increment of 0.12 at%) and Ti content of $Ni_xTi_{90-x}Hf_{10}$ SMAs on the presence of a second phase $Ni(Ti+Hf)_2$ and transformation temperatures is determined. It was reported that the amount of second phase decreases with increasing Ni content and is least with a Ni content of 50.2 at%. The martensite to austenite and austenite to martensite transformation temperatures remain constant as the Ni content increases from 49.0 to 50.0 at%. A rapid decrease in the transformation temperatures is observed when the Ni content of the $Ni_xTi_{90-x}Hf_{10}$ alloy exceeds 50.0 at%. Using $Ti_{40.0}Ni_{47.5}Hf_{12.5}$ shape-memory thin-film system, structural, phase transformation, and functional fatigue properties were investigated [198]. It was found that temperature-dependent resistance measurements revealed a broad compositional region showing a reversible phase transformation. With increasing Ti content, the amount of the Laves phase increases, resulting in an increase in the thermal hysteresis and a simultaneous decrease in the transformation temperatures. Shape-memory properties were characterized by temperature-dependent stress change measurements using micromachined Si cantilever array wafers coated with Ti–Ni–Hf. The recovery stress was found to increase for small amounts of Laves phase precipitates. Strengthening of the matrix due to the Laves phase precipitates is concluded to be responsible for the observed increase in recovery stress and improved functional fatigue properties for (Ti,Hf)-rich alloy compositions ($Ti_{40.0}Ni_{47.5}Hf_{12.5}$). Using $Ni_{50.3}Ti_{34.7}Hf_{15}$ SMA, Evirgen et al. [199] investigated the effect of precipitation on the microstructure and shape-memory characteristics by TEM, DSC, and load-biased thermal cycling tests in tension. It was reported that the one-stage martensitic transformation from B2 austenite to B19′ martensite was observed in all the aged samples, but the transformation temperatures followed a more complicated trend depending on specific-aging conditions. The transformation temperatures decreased below room temperature; whereas, the transformation temperatures can be increased over a wide temperature range by increasing the precipitate size and volume fraction through aging for long durations or at higher temperatures. The alloy demonstrated excellent dimensional stability under stress levels as high as 300 MPa as a consequence of precipitation hardening, with a maximum fully recoverable strain of 3.3% after aging at 450 °C for 10 h. The transformation thermal hysteresis also decreased in the aged samples due to reduced defect generation in the precipitation-strengthened samples. The precipitate crystal structure was identified as the H-phase that was recently reported in Ni-rich NiTiHf and NiTiZr alloys, and not the Ni_4Ti_3-type structure as reported in other studies [199].

SMAs are materials that have the ability to retrieve their original shape under appropriate thermal conditions and the ability is possibly due to a solid state phase transformation (or thermoelastic martensitic transformation) with a capability of restoring of great deformations (between 8% and 10%) [192]. Some NiTi-based SMA systems are considered high-temperature SMAs, which are normally alloyed with several transition elements including Pd, Pt, Ta, Au, and Hf. Depending on the process and the amount added, NiTi–X alloy may result in martensitic transformation temperatures starting above 100 ºC. Examples of these alloys are: Ti–Ni–Pd, Ti–Ni–Pt, Ti–Ni–Ta, Ti–Ni–Au, and Ni–Ti–Hf. Microstructures and precipitates of $Ni_{49}Ti_{51-x}HfX$ (x = 10, 15, 20 at%) [200], NiTiHf (8–20 at% Hf) [201], $Ti_{36.5}Ni_{48.5}Hf_{15}$ [202], and NiTiHf alloy systems [158, 203] were investigated, and it was commonly found that there are various morphologies of martensite. The monoclinic B19′ martensite transforms in a reversible manner to the B2 parent phase at temperatures varying from 100 to 300 °C. The precipitates of $(Ti,Hf)_3Ni_4$ particles are fully coherent with the austenite B2 matrix; however, upon martensitic transformation, they lose some coherency with the B19′ matrix as a result of the transformation shear process in the surrounding matrix. Aging provided a significant rise in transformation temperatures until they reached their equilibrium states, corresponding to the equilibrium Ni content at each aging temperature, and the equilibrium transformation temperatures were higher when aging was performed at a lower temperature.

There are numerous studies investigating mechanical and functional behaviors of a Ni-rich $Ni_{50.3}Ti_{29.7}Hf_{20}$ high-temperature SMA. These properties were investigated under an isothermal tension and compression tests between room temperature and 260 °C, while isobaric thermomechanical cycling experiments were conducted at selected stresses up to 700 MPa [204]. Based on obtained results, it was indicated that the isothermal testing of the M-phase revealed no plastic strain up to the test limit of 1 GPa and near-perfect superelastic behavior up to 3% applied strain at temperatures above the austenite finish. Excellent dimensional stability with greater than 2.5% actuation strain without accumulation of noticeable residual strains (at stresses less than or equal to −400 compressive MPa) were observed during isobaric thermal cycling experiments. The absence of residual strain accumulation during thermomechanical cycling was confirmed by the lattice strains. Postdeformation cycling revealed the limited conditions under which a slight two-way SME was obtained, with a maximum of 0.34% two-way shape-memory strain after thermomechanical cycling under −700 compressive MPa. Similarly, $Ni_{50.3}Ti_{29.7}Hf_{20}$ was studied [205] in terms of thermal stability, with conclusions that under 200 MPa, 600 thermal cycles were sufficient to reach a two-way shape-memory strain as high as 2.95%, which was shown to be stable upon annealing up to 400 °C for 30 min. This two-way shape-memory strain was 85% of the maximum measured actuation strain under 200 MPa. Overall, it was found that nano-precipitation hardened $Ni_{50.3}Ti_{29.7}Hf_{20}$ shows relatively high two-way shape-memory strain and stable actuation response after much less number of training cycles as compared to binary NiTi and nickel lean NiTiHf compositions. Evirgen et al. [206] studied relationships between the crystallographic compatibility of austenite and martensite phases and the transformation thermal hysteresis of $Ni_{50.3}Ti_{29.7}Hf_{20}$. It was found that the alloy undergoes B2–B19′ martensitic transformation. Yang et al. [207] studied thermal aging of the $Ni_{50.3}Ti_{29.7}Hf_{20}$ (at%) high-temperature SMA to examine the role of a novel precipitate phase. The precipitate phase was investigated by conventional electron diffraction, high-resolution scanning TEM and three-dimensional atom probe tomography. Based on the proposed unrelaxed orthorhombic atomic structural model, it was reported that, as a result of the relaxation, atom shuffle displacements occur, and the relaxed structure, which is termed the H-phase, has also been verified to be thermodynamically stable at 0 K. Karaca et al. [208] characterized the shape-memory properties of a $Ni_{50.3}Ti_{29.7}Hf_{20}$ (at%) polycrystalline alloy after selected heat treatments. It was mentioned that the precipitation was found to alter the martensite morphology and significantly improve the shape-memory properties of the Ni-rich NiTiHf alloy. For the peak-aged condition, shape-memory strains of up to 3.6%, the lowest hysteresis, and a fully reversible superelastic response were observed at temperatures up to 240 °C. In general, the Ni-rich NiTiHf polycrystalline alloy exhibited a higher work output ($\approx$16.5 J cm^{-3}) than other NiTi-based high-temperature alloys. Stebner et al. [209] examined the transformation temperature, and the transformation strain behaviors of $Ni_{50.3}Ti_{29.7}Hf_{20}$ alloy were conducted. It was reported that the H-phase precipitates on the order of 10–30 nm were shown to increase transformation temperatures and also to narrow thermal hysteresis, compared to unaged material. The mechanical effects of increased residual stresses and numbers of transformation nucleation sites caused by the precipitates provide a plausible explanation for the observed transformation temperature trends. The work output and recoverable strain exhibited by the alloy were shown to approach maximal stresses of 500–800 MPa, suggesting these to be optimal working loads with respect to single cycle performance, the potential for transformation strain in single crystals of this material was calculated to be superior to binary NiTi in tension, compression, and torsion loading modes; however, the large volume fraction of precipitate phase, in part, prevents the material from realizing its full single crystal transformation strain potential in return for outstanding functional stability by inhibiting plastic strain accumulation during transformation [209]. Saghaian et al. [210] investigated the shape-memory properties of Ni-rich $Ni_{50.3}Ti_{29.7}Hf_{20}$ single crystals (which were solution treated and aged) along the different crystallographic orientations (i.e., [001], [011], and [111] directions in compression). It was found that aging at 550 °C for 3 h introduced coherent 10–20 nm precipitates in the matrix, which substantially improved the shape-memory and mechanical properties of the $Ni_{50.3}Ti_{29.7}Hf_{20}$ crystals. The [001]-oriented single crystals showed high-dimensional stability under stress levels as high as 1,500 MPa but with transformation strains of <2%. Thermal treatments can be used to tailor the transformation temperatures over a wide range with the M_S temperature varying from –25 °C in the solutionized case to 123 °C by aging at 650 °C for 3 h. When compared to the solutionized condition thermal hysteresis was reduced after aging at 550 °C/3 h, but increased with aging at 650 °C. Perfect SE with recoverable strain of >4% was observed for solutionized and 550 °C/3 h aged single crystals along the [011] and [111] orientations, and general superelastic behavior was observed over a wide temperature range [210]. Studying on Ni-rich $Ni_{51.2}Ti_{28.8}Hf_{20}$ alloy, Saghaina et al. [211] found that the martensitic transformation was severely suppressed in the solution treated condition (900 °C–3 h/water quench) and after aging at low temperatures, while the transformation temperatures were greater than 100 °C after 650 °C–3 h aging. The generation of nanosize precipitates (~20 nm in size) after 3 h aging at 450 and 550 °C improved the strength of the material, resulting in a near perfect dimensional stability during isobaric thermal cycling at stress levels of greater than 1,500 MPa, with work output of 20–30 J cm^{-3}. Superelastic behavior with 4% recoverable strain was demonstrated at low temperatures (–20 to 40 °C) after aging at 450 °C for 3 h and at elevated temperatures (120–160 °C) after aging at 550 °C for 3 h, with stresses reaching 2 GPa without the onset of plastic deformation. Manca et al. [212] studied the effect of aging in parent phase and martensite stabilization on melt spun $Ni_{50}Ti_{30}Hf_{20}$ ribbons by DSC and XRD. It was concluded that thermal treatments in parent phase generally shift the martensitic transformation range as far as an equilibrium state is reached. The equilibrium state, found here for a thermal treatment at 450 °C, was adopted as start state to investigate aging in martensite (commonly indicated as martensite stabilization). The martensite stabilization involves diffusion processes with an activation energy in the order of 1.2 eV. The effects of cold and warm rolling on the shape-memory response and thermomechanical cyclic stability of $Ni_{50}Ti_{30}Hf_{20}$ high-temperature SMA (cold rolled and annealed at 550 °C for 30 min) was studied [213]. It was shown that the rolling led to an increase in the resistance against defect generation accompanying martensitic transformation, resulting in a significant improvement in dimensional stability during thermal cycling. While transformation temperatures of all rolled samples were lower than those of the starting hot extruded sample; thermal hysteresis was notably higher in the rolled samples, which might be due to the increase in dislocation density and the change in the martensite microstructure with rolling. The functional behavior of the new $Ni_{51.2}Ti_{23.4}Hf_{25.4}$ high-temperature SMA was investigated along three crystal orientations in compression and compared with the polycrystal behavior [214]. Two aging treatments were then selected in order to fulfill the optimum superelastic behavior (500 °C/4 h), and the maximum transformation temperatures (550 °C/10 h) without compromising the alloy functionality. It was reported that the martensite structure was found to be the B19 orthorhombic, in contrast with monoclinic structure for lower Hf contents. The heterogeneity of the austenite → martensite transformation was observed.

A wide range of Hf contents and their alloying effects were investigated: $Ni_{50}Ti_{50-x}Hf_x$ (x = 8, 11, 14, 17, and 20 at%) alloys [192] and $Ni_{50}Ti_{50-x}Hf_x$, $Ti_{50}Ni_{50-x}Hf_x$, and $Ni_{(100-X)/2}Ti_{(100-X)/2}Hf_x$ ($5 \leq x \leq 20$) [215]. Common findings are the transformation temperatures increase with increasing the amount of Hf. A martensitic matrix is formed by two metastable phases: R and B19'. From the crystal structure refinement, Hf site preference and occupancy in Ni and Ti sites result in its preference for available sites in the structure. The B19' phase presented the highest percent fraction, and gradually adding promoted a slow increase of crystalline fraction of R-phase and a slow reduction of phase $(Ti,Hf)_2Ni$, located at grain boundaries.

Karakoc et al. [216,217] investigated the effects of upper cycle temperature on the actuation fatigue response of nanoprecipitation hardened $Ni_{50.3}Ti_{29.7}Hf_{20}$ high-temperature SMA, conducted under different constant tensile stresses while cycling temperature between two extreme martensitic transformation temperatures. It was reported that the samples subjected to 300 °C upper cycle temperature exhibit fatigue lives twice that of the samples with 350 °C, and those tested under 300 MPa with 300 °C upper cycle temperature withstand more than 10,000 cycles with actuation strains of 2–3%. Actuation strains remained constant or increased with thermal cycling in the 350 °C experiments, while those subjected to 300 °C exhibited decreasing actuation strains, at all stress levels. In the 300 °C tests, partial martensitic transformation becomes operative, resulting in a reduction of actuation strain in each cycle, and postpones damage accumulation; while, in the 350 °C cases, increase of actuation strain is attributed to the partial recovery of cyclically induced remnant deformation at higher temperatures and as a result, larger volume of transforming material [216]. It was furthermore mentioned that significantly high number of cycles to failure were observed: specimens tested under 200 MPa achieved ~21,000 cycles with the average actuation strain of ~2.15% while those tested under 500 MPa experienced ~2,100 cycles to failure with the average actuation strain of 3.22% [217].

3.2.3.4 Ta

Gong et al. [218] investigated the phase transformation and microstructure in solution-treated $Ni_{50}Ti_{47}Ta_3$, $Ni_{50}Ti_{45}Ta_5$, $Ni_{49}Ti_{46}Ta_5$ SMAs. It was found that two different second-phase particles, $(Ti,Ta)_2Ni$ and a tantalum-rich solid solution (β-Ta) mostly located at grain boundaries are observed. Tested alloys undergo a one stage B2 ↔ B19′ transformation, and the Ta atoms are substituted to Ti atoms, instead of Ni atoms, in these alloys. The transformation temperature depression of NiTiTa alloy is more than that of NiTi alloy under the same number of thermal cycles due to the presence of more second-phase particles. The above 4% deformation might induce plastic strain for the β-Ta, and the NiTiTa alloys exhibit a wide transformation hysteresis compared to NiTi binary alloys. Surface morphology of electropolished and magnetoelectropolished $NiTiTa_{10}$ (wt%) was studied and compared to polished $NiTiCu_{10}$ [120]. It was reported that the electropolished NiTiTa showed the highest overall roughness, while the NiTiCu alloy had the lowest roughness when analyzed over, thereby suggesting that such surface micropatterning on ternary NiTi alloys could increase cellular adhesion and accelerate surface endothelialization of endovascular stents. This results in reducing the likelihood of in-stent restenosis and provide insight into hemodynamic flow regimes and the corrosion behavior of an implantable device influenced from such surface micropatterns.

Surface properties such as composition, roughness wettability, surface free-energy, and morphology will affect the hemocompatibility of an implant material. Additionally, in the realm of metallic biomaterials, the specific composition of the alloy and its surface treatment are important factors that will affect the surface properties. Hemocompatibility of NiTiTa was investigated [219, 220]. Zhao et al. [219] formed a composite TiO_2/Ta_2O_5 nanofilm on NiTi SMA by Ta implantation, and the wettability, protein adsorption, platelets adhesion, and hemolysis tests are conducted to evaluate the hemocompatibility. It was reported that the contact angle measurements showed that the surface of the NiTi alloy kept hydrophilic before and after Ta implantation, although the water contact angle increased with the increasing of implantation current. The fibrinogen adsorption was enhanced by a high surface roughness or a large interfacial tension, whereas the albumin adsorption was insensitive to the surface modification. The platelet adhesion and activation were weakened, and the hemolysis rate was reduced at least 46% after Ta implantation due to the decreased surface energy and improved corrosion resistance ability, respectively. The thrombogenicity of a biomaterial is mainly dependent on its surface characteristics, which dictates its interactions with blood. Pulletikurthi et al. [220], using the magnetoelectropolished $NiTiTa_{10}$, investigated the blood contacting properties with comparison to those on mechanically polished surfaces. It was mentioned that the in vitro thrombogenicity tests revealed significantly less platelet adherence on ternary mechanically polished surfaces. The enhanced antiplatelet adhesive property of mechanically polished $NiTiTa_{10}$ was, in part, attributed to the Ta_2O_5 component of the alloy. Furthermore, the formation of a dense and mixed hydrophobic oxide layer during polishing is believed to have inhibited the adhesion of negatively charged platelets, concluding that mechanically polished NiTiTa alloy can potentially be utilized for blood-contacting devices, where complications resulting from thrombogenicity can be minimized.

Surface-enhanced NiTi by Ta element was investigated in terms of corrosion behavior and surface characterization [221]. Ta/NiTi was subjected to the polarization tests in Ringer's soliton at 37 °C, and it was reported that the Ta/TiNi sample with a moderate incident dose of 1.5×10^{17} ions/cm^2 exhibits the best corrosion resistance ability. The compact aggregates of nanograins uniformly disperse on the surface of the Ta/TiNi samples. The component of the surface layer is mainly composed of TiO$_2$ and Ta$_2$O$_5$, which is beneficial to the corrosion resistance ability and biocompatibility.

Alloying effects of Ta along with Y on mechanical properties were investigated [150]. It was reported that tested alloys have the same results of two phases of monoclinic NiTi and hexagonal Ni$_3$Ti. The result of adding Ta in the percentage of (1%, 2%, and 3%) in the expense of nickel, by increasing hardness value gradually to record the highest value (310) HV at 3% and also decrease the surface roughness from 0.36 to 0.15 μm at 3%Ta and 650 MPa. The wear behavior of NiTiTa was studied [149]. It was concluded that the wear volume loss decreases with the addition of 3% Ta at 650 MPa compacting stress for a 15 N load, the wear volume loss further decreases as the compacting stresses increase, and its rate increased as the load and time increased for all tested specimens.

3.2.3.5 W

Phase transformation of W containing NiTi alloy was investigated [222] and the obtained results were that annealed (Ti/Ni/W)$_n$ multilayer films formed a two-phase (B2-TiNi and β-W) system. The grain sizes revealed that B2-TiNi decreases with increasing W, due to the immiscible W layers obstructing its grain growth. With decreasing B2-TiNi grain size the R_S (B2–R) transformation temperature is not affected but the M_S (R–B19′) transformation temperature decreases significantly; thereby, suggesting that the addition of W to Ti–Ni is effective to induce the B2–R single-step transformation due to grain size effects. Ahadi et al. [223] studied the effect of minority W additions on the thermal stability of nanocrystalline NiTi thin films. It was reported that the films were produced in an amorphous state, and the addition of W was found to increase the activation energy for crystallization and led to finer grain sizes after crystallization. In the crystallized films, W was observed to both segregate to the NiTi grain boundaries and to phase out into fine precipitates. Together these effects contribute to the stability of the nanocrystalline state up to 1,200 °C. W segregation to grain boundaries increases with temperature, which contributes to an increased rate of coarsening and loss of stability against grain growth.

3.2.3.6 Re

El-Bagoury et al. [224] investigated the effect of aging on the structure and precipitation of second phases of Ni$_{52}$Ti$_{47.7}$Re$_{0.3}$ SMA, which was solution treated at 1,000 °C for 24 h before aging at various temperatures ranging from 300 to 600 °C for 3 h. It was found that the matrix phase in both solution treated and aged specimens was martensite. The Ti_2Ni phase was also present in the microstructure of both solution treated and aged specimens and its volume fraction decreased as the aging temperature increased. The Ni_4Ti_3 precipitate phase began in appearance by increasing aging temperature to 400 °C. Aging at 600 °C led to precipitation of Ni_3Ti phase in the microstructure.

3.2.3.7 Pt

Alloying effect of Pt element substituted for Ni site in NiTi alloys is known to increase the transformation temperature [185]. Using aged $Ti_{49.5}Ni_{29.5}Pt_{21}$ (at%) alloy, Yang et al. [225] found that at 500 °C aging, P-phase was formed stable up to 1,000 h, and a transformation from the P-phase to $Ti_2(Ni,Pt)_3$ precipitate was observed. As to the role the P-phase in the transformation behavior, Gau et al. [226] reported that the chemical nonuniformity and stress field associated with the fine coherent P-phase precipitates are in favor of the martensitic transformation. Their relative contributions to the increase in M_S temperature are quantified as a function of aging time.

Ternary NiTiPt high-temperature SMA was characterized [227], and it was reported that the transformation temperatures were depressed with initial Pt additions. However, at levels greater than 10 at%, the transformation temperature increased linearly with Pt content. The transformation temperatures were relatively insensitive to alloy stoichiometry within the range of alloys examined. The dependence of hardness on Pt content for a series of $Ni_{50-x}Pt_xTi_{50}$ alloys showed solution softening at low Pt levels, while hardening was observed in ternary alloys containing more than about 10 at% Pt. On either side of these "stoichiometric" compositions, hardness was also found to increase significantly [236]. Buchheit et al. [228], using a series of Ti-rich Ni–Ti–Pt ternary alloys with 13–18 at% Pt, studied the transformation behavior. It reported that the transformation took place between 175 and 225 °C and achieved recoverable strain exceeding 2%. From this broader set of compositions, three alloys containing 15.5–16.5 at% Pt exhibited transformation temperatures in the vicinity of 200 °C. Microstructural evaluation revealed a martensitic microstructure with small amounts of $Ti_2(Ni,Pt)$ particles. The room temperature mechanical testing gave a response characteristic of martensitic detwinning followed by a typical work-hardening behavior to failure. Alternatively, elevated mechanical testing, performed while the materials were in the austenitic state, revealed yield stresses of approximately 500 MPa and 3.5% elongation to failure. Noebe et al. [229] investigated the microstructure, transformation temperatures, basic tensile properties, and shape-memory behavior of $Ni_{30}Pt_{20}Ti_{50}$ alloy (transformation temperatures above 230 °C) and $Ni_{20}Pt_{30}Ti_{50}$ alloy (transformation temperatures above 530 °C). It was reported that both materials displayed shape-memory behavior and were capable of 100% (no-load) strain recovery for strain levels up to their fracture limit (3–4%) when deformed at room temperature. For the $Ni_{30}Pt_{20}Ti_{50}$ alloy, the tensile strength, modulus, and ductility dramatically increased when the material was tested just above the A_F temperature; while, for the $Ni_{20}Pt_{30}Ti_{50}$ alloy, a similar change in yield behavior at temperatures above the A_F was not observed. The $Ni_{30}Pt_{20}Ti_{50}$ alloy behaved similar to conventional binary NiTi alloys with work output due to the martensite-to-austenite transformation initially increasing with applied stress and the maximum work output measured in the $Ni_{30}Pt_{20}Ti_{50}$ alloy was nearly 9 J/cm^3 and was limited by the tensile ductility of the material; whereas, the martensite-to-austenite transformation in the $Ni_{20}Pt_{30}Ti_{50}$ alloy was not capable of performing work against any bias load. Kovarik et al. [230], studying the aging of the high-temperature SMA $Ti_{50}Ni_{30}Pt_{20}$ (at%), found that precipitate plays a key role in achieving desirable shape-memory properties. The precipitates have unique crystallography due to their nonperiodic character along one of the primary crystallographic directions, and the structure can be explained in terms of crystal intergrowth of monoclinic crystal, which is closely related to the high-temperature cubic B2 phase; the departure of the structure from the B2 phase can be attributed to ordering of Pt atoms on the Ni sublattice and relaxation of the atoms (shuffle displacements) from the B2 sites.

Using $NiTi_{49}$ and $NiPt_{30}Ti_{50}$ (at%) SMAs, Smialek et al. [231, 232] investigated the isothermal oxidation behaviors in air over the temperature range from 500 to 900 °C for 100 h. Based on obtained results, it was reported that a relatively pure TiO_2 rutile structure was identified as the predominant scale surface feature, typified by a distinct highly striated and faceted crystal morphology, with crystal size proportional to oxidation temperature. In general, graded mixtures of TiO_2, $NiTiO_3$, NiO, Ni(Ti), or Pt(Ni) metallic dispersoids, and continuous Ni_3Ti or Pt-rich metal depletion zones, were observed from the outer surface to the substrate interior. Hence, overall, substantial depletion of Ti occurred due to the formation of predominantly TiO_2 scales, and it was mentioned that the $NiPt_{30}Ti_{50}$ alloy oxidized more slowly than the binary $NiTi_{49}$ alloy by decreasing oxygen and titanium diffusion through the thin Pt-rich layer. It was furthermore mentioned that the activation energy was consistent with literature values for TiO_2-scale growth measured for elemental Ti and some NiTi alloys, at –210 to 260 kJ/mol.

3.2.3.8 Au

Martensitic transformations in $Ti_{50}Ni_{10}Au_{40}$ and $Ti_{50}Au_{50}$ alloys were investigated [233] and found that they are of the type B2 – B19 as found in a $AuCd_{47.5}$ alloy. The Ms temperatures of the $Ti_{50}Ni_{10}Au_{40}$ and $Ti_{50}Au_{50}$ alloys are 440 and 610 °C, respectively. The martensite in both alloys is orthorhombic and identified as a modified B19 structure and both $Ti_{50}Ni_{10}Au_{40}$ and $Ti_{50}Au_{50}$ alloys exhibit a one-way SME but not a two-way memory.

3.2.4 Others

3.2.4.1 Al

Kurita et al. [234] investigated the transformation behavior of ternary NiTiAl alloy with a differential scanning calorimeter in order to clarify the effect of aluminum substitution in NiTi, showing the shape memory due to the thermoelastic martensitic transformation between a high-temperature phase and a low-temperature phase. It was found that an intermediate phase appears due to aluminum substitution and its temperature region is broadening with increasing number of thermal cycles. By replacing titanium by aluminum, a unique behavior is observed in the transformation from the intermediate phase to the low-temperature phase. The transformation behavior of NiTiAl depends significantly on the concentration of the substituted aluminum, and the temperature region of the intermediate phase is controlled by the substituted aluminum. The SME also occurred in the transformation between the high temperature and the intermediate phases, concluding that the substitution of aluminum is useful for various applications of the transformation to the intermediate phase [234, 235]. Using NiTiAl (with 1–7 wt%Al), Salvetr et al. [236] studied the formation of intermetallics which might be in dependence on the heating rate (up to 110°C/min). It was indicated that thermal explosion self-propagating synthesis was initiated by exothermic reaction of nickel aluminides Ni_2Al_3 and $NiAl_3$ at the temperature of 535–610 °C. Comparing the reactive sintering mechanism of $NiTi_{46}$, it was reported that the changes in sintering duration were observed when the addition of aluminum into NiTi powder mixture was higher than 5 wt %. The diffusion-controlled reaction with the Ti_2Ni phase as a product occurs below the $\alpha{\to}\beta$ Ti transformation temperature (882 °C) and close to this temperature (860–890 °C). Thermal explosion self-propagating synthesis was initiated in $NiTi_{46}$, $NiTiAl_1$, and $NiTiAl_3$ wt% powder mixtures. The precipitation of L21–Ni_2TiAl phase from a supersaturated B2-TiNi matrix at 600 and 800 °C was reported [237].

The microstructure, thermal cycling, and mechanical behavior of $Ni_{48.5}Ti_{31.5-x}Zr_{20}Al_x$ (x = 0, 1, 2, 3) alloys were studied in the solution-treated and aged condition using microscopy techniques, DSC, and compression tests [159]. It was mentioned that there was a strong dependence of the transformation behavior on alloy chemistry and thermal cycling. The Al additions served to decrease transformation behaviors from 78 to 323 °C) and reduce thermal stability. Moreover, Al was shown to increase the plateau stress in the aged condition, whereas the formation of coarse-grained intermetallic phases caused the embrittlement of the microstructure, reducing its ductility. The addition of Al resulted in the refinement of the coarse, lenticular precipitates identified as $Ni_4(Ti,Zr)_3$. Hsieh et al. [238] studied the transformation sequence and hardening effects of $Ti_{47}Ni_{50.65}Al_{1.85}$ and $Ti_{49.5}Ni_{50.13}Al_{0.37}$ SMAs that were aged at 400 °C. It was reported that the hardening effects of $Ti_{47}Ni_{50.65}Al_{1.85}$ alloy are obvious and much higher than those of $Ti_{49.5}Ni_{50.13}Al_{0.37}$ alloy due to the former having the larger Ni/Ti ratio and a higher Al solute content in its matrix. The transformation sequence of 400 °C aged $Ti_{47}Ni_{50.65}Al_{1.85}$ alloy shows B2$\leftrightarrow$R-phase only for an aging time of more than 10 h and that of 400 °C aged $Ti_{49.5}Ni_{50.13}Al_{0.37}$ alloy shows the sequence B2↔R-phase↔B19′ or B2↔R-phase with different aging times. All of these characteristics are associated with $Ti_{11}Ni_{14}$ precipitates during the aging process. Meng et al. [239] investigated the mechanical properties of Ti-rich $Ni_{50-X}Ti_{50}Al_X$ (X = 6, 7, 8, 9) alloys by compression tests at room temperature and at high temperature from 400 to 800 °C. It was mentioned that the yield stress of 1,800 MPa and more than 10% of compression strain were achieved at room temperature. A yield stress of 400 MPa at 800 °C, controlling the shape, the volume percent, and the distribution of second phases in the matrix, is most important to obtain good mechanical properties in these alloys. The strengthening mechanism of aluminum addition on the mechanical properties was accompanied with precipitation hardening upon Al addition.

3.2.4.2 Si

Using $Ti_{51}Ni_{49-X}Si_X$ SMAs (x = 0, 1 and 2 at%), Ibrahim et al. [240] investigated the martensitic transformation behavior, second phases and hardness. It was found that transformation temperature of one stage martensitic reaction B2 ↔ B19′ is associated with the forward (M_S) and reverse (A_S) martensitic transformations, respectively. The martensitic transformation peaks (M^P) and reverse martensitic transformation peaks (A^P) are increased and became sharper with increasing Si content. The microstructure investigation revealed two types of precipitated second phase particles, including Ti_2Ni, which mainly located at grain boundaries and intermetallic compound of $Ti_2(Ni + Si)$ phase distributed inside the matrix, and the volume fraction of these two phases is increased with Si content. Furthermore, it was mentioned that a small amount of Si remained in solid solution of the matrix of $Ti_{51}Ni_{49-X}SiX$ alloys, and its hardness increased as the Si content increases [240]. Salvetr et al. [241] found that the addition of silicon (in an amount of 5 wt%) caused an increase of transformation temperatures and hardness.

3.2.4.3 Mg

The alloying effect of Mg to $NiTi_{46}$ (wt%) SMA was investigated [242] on improve the structural homogeneity. The NiTi alloy was produced by self-propagating high-temperature synthesis by lowering the amount of undesirable brittle Ti_2Ni phase. It was reported that magnesium was added to the $NiTi_{46}$ (wt.%) mixture in an amount of 5 wt.%, and the intermetallic phases were formed at temperatures of 505 °C, and the high-temperature synthesis reaction started at 981 °C. Li et al. [243] studied microstructures, electrochemical properties of Ti–Ni and ternary Ti–Ni–Mg alloy to examine the influence of milling time and Mg addition on the microstructures of the mechanically milled Ti–Ni alloys. It was found that the binary Ti–Ni alloy undergoes a refinement, dynamical recrystallization and amorphization process. With doping of Mg to the starting Ti–Ni powders, an FCC Ti–Mg structure was detected along with the main TiNi BCC phase; hence, concluding that the final product of milling Mg doped

Ti–Ni contains an FCC-structured $TiMg_3$ phase, which damages electrochemical performance in general as a result of coating effect on the TiNi phase.

3.2.4.4 Na, P

By improving the structure stabilization, it was noted that the performance efficiency of the sodium ion batteries application of NiTi was enhanced by adding Na and P [243].

3.3 NiTi–XY and NiTi–XYZ more complicated alloy systems

We have been looking at binary (NiTi alloy) and ternary (NiTi-based X alloy) systems. A majority of alloying element(s) may be substituted to either Ni site or Ti site to form NiTi-based alloy systems are transition elements. In the following, we will discuss quaternary (NiTi-based X–Y) alloy and quinary (NiTi-based X–Y–Z) alloy systems.

3.3.1 Quaternary alloy systems

Jung et al. [237] investigated an important role of precipitation (L21–Ni2TiAl phase from a supersaturated B2–TiNi matrix at 600 and 800 °C) to support the high strength TiNi-based SMAs. NiTiAlX (X = Hf, Pd, Pt, or Zr) alloys were prepared. It was reported that an analysis of the L21 precipitate size evolution suggests that in the case of alloys with Al, Zr, or Hf substitution for Ti, the precipitates follow coarsening kinetics at 600 °C and growth kinetics at 800 °C, while for alloys with Pd or Pt substitution for Ni, precipitates follow one kinetic behavior at both temperatures. The temperature-dependent partitioning behaviors of Hf, Pd, Pt, and Zr are analyzed and was mentioned that both Hf and Zr prefer to partition to the B2 phase at 800 °C while they exhibit reverse behavior at 600 °C, and Pt also partitions to B2 at 800 °C, while Pd partitions to the L21 phase at both 600 and 800 °C.

Karaca et al. [244] characterized shape-memory properties of a $Ni_{45.3}Ti_{29.7}Hf_{20}Pd_5$ (at%) alloy in compression after selected aging treatments. It was reported that precipitation hardening significantly improved the shape-memory properties of the alloy, under optimum aging conditions, shape-memory strains of up to 4% under 1 GPa were possible, and SE experiments resulted in full strain recovery without any plastic deformation, even at stress levels as high as 2 GPa, and very high damping capacity and work output indicate a capability to operate at high stress levels without significant plastic deformation and to a high mechanical hysteresis (>900 MPa) at temperatures ranging from 20 to 80 °C.

Microstructure, mechanical properties as well as phase transformation behavior of newly developed $Ni_{49}Ti_{36}Hf_{15-x}Ta_x$ (x = 0, 3, 6, 9,12) were studied [245]. It was reported that $Ni_{49}Ti_{36}Hf_{15}$ alloy consists of $(Ti,Hf)_2Ni$ phase, which is brittle and incoherent with the martensite matrix; however, addition of Ta in place of Hf suppresses the brittle $(Ti,Hf)_2Ni$ phase formation with white Ta rich phase. More than 70% reduction in area has been observed in Ta added $Ni_{49}Ti_{36}Hf_{15}$ high-temperature SMAs. The austenitic transformation temperature is well above 200 °C for lower Ta concentrations with up to 6 at%; however, at higher Ta concentrations (i.e., above 9 at%) decreases significantly, concluding that the addition of Ta greatly influences the hot deformation behavior and high-temperature shape-memory properties of $Ni_{49}Ti_{36}Hf_{15}$ alloy.

Li et al. [246] studied the effect of substitution of Nb by Mo in $Nb_{40}Ti_{30}Ni_{30}$ with respect to microstructural features and hydrogen dissolution, diffusion and permeation. It was mentioned that as-cast $Nb_{40-x}Mo_xTi_{30}Ni_{30}$ ($x = 0$, 5, 10) alloys consist of primary BCC-Nb phase and binary eutectic (BCC-Nb + B2-TiNi). The substitution of Nb by Mo reduces the hydrogen solubility in alloys, but may increase ($x = 5$) or decrease ($x = 10$) the apparent hydrogen diffusivity and permeability. The as-cast $Nb_{35}Mo_5Ti_{30}Ni_{30}$ exhibits a combined enhancement of hydrogen permeability and embrittlement resistance as compared to $Nb_{40}Ti_{30}Ni_{30}$. It is concluded that Mo is a desirable alloying element in Nb that can contribute to a reduction in hydrogen absorption and an increase in intrinsic hydrogen diffusion, thus improving embrittlement resistance with minimal permeability penalty.

Using a $Ti_{49.5}Ni_{25}Pd_{25}Sc_{0.5}$ high-temperature SMA, Atli et al. [247] studied the dimensional stability upon repeated thermal cycles under constant loads. It was found that the thermomechanical experiments reveal that the processed materials display enhanced shape-memory response, exhibiting higher recoverable transformation and reduced irrecoverable strain levels upon thermal cycling compared with the unprocessed material, and this improvement is attributed to the increased strength and resistance of the material against defect generation upon phase transformation as a result of the microstructural refinement.

Stebner et al. [248] discussed about the high-temperature $Ni_{19.5}Ti_{50.5}Pd_{25}Pt_5$ SMA with several excellent properties. This material is capable of being used in operating environments of up to 250 °C. It possesses very useful actuation capabilities, demonstrating repeatable strain recoveries up to 2.5% in the presence of an externally applied load, suggesting that it can be applied for a surge-control mechanism that could be used in the centrifugal compressor of a helicopter engine.

The electrochemical behavior of $Ti_{49.6}Ni_{45.1}Cu_5Cr_{0.3}$ alloy in artificial saliva solutions was investigated and the surface passive film after polarization tests was characterized [249]. It was reported that potentiodynamic test and potentiostatic test results showed that the corrosion behavior of this alloy was similar to that of NiTi alloy, and both corrosion potential (E_{CORR}) and pitting corrosion potential (E_B) showed a pH-dependent tendency that E_{CORR} and E_B decreased with the increase of the pH value. The passive film was identified as mainly of TiO_2 with a little amount of Ni oxides (NiO/Ni_2O_3) that was identical with NiTi alloy. It was concluded that the addition of Cu and Cr had little effect on the corrosion behavior of NiTi or on the composition and the structure of the passive film. Kök et al. [82] tested a slightly different composition of NiTiCrCu alloy with $Ni_{45}Ti_{50}Cr_{2.5}Cu_{2.5}$. It was found that after the homogenization of these alloys, the M-phase transformation temperatures were found to be below 37 °C (body temperature). $Ni_{45}Ti_{50}Cr_{2.5}Cu_{2.5}$ showed the austenite phase at room temperature, included B2 (NiTi) phase and Ti_2Ni precipitation phase, and the alloys that were in the M-phase at room temperature included B19′ (NiTi) phase and Ti_2Ni phase.

Samal et al. [117] prepared $Ni_{48}Cu_{10}Co_2Ti_{40}$ alloy, based on the material design concept that alloying elements such as Cu, Co, and Ta in the near-$Ni_{60}Ti_{40}$ eutectic alloy by replacing both Ni and Ti so that phase mixture in the microstructure remains the same from the binary to quinary alloy. The thus obtained alloy was subjected to evaluate the phase and microstructure evolution during solidification. It was shown that in the binary ($Ni_{60}Ti_{40}$) alloy, the ordered NiTi (B2) phase transforms to trigonal (R) phase followed by NiTi M-phase, that is, B2 → R-phase → M-phase during solid-state cooling; however, the addition of alloying elements such as Cu, Co to the binary ($Ni_{60}Ti_{40}$) alloy suppresses the martensitic transformation of the ordered NiTi (B2) dendrite. Thus, in the ternary and quaternary alloys, the ordered NiTi (B2) phase is transformed to only trigonal (R) phase, that is, B2 → R-phase. The secondary precipitate of Ti_2Ni has been observed.

Corrosion behavior of NiTiCuNb was investigated in 0.9% sodium chloride [250]. It was mentioned that copper and niobium addition did not have significant effects on the uniform corrosion characteristics, but significantly improved the pitting corrosion resistance. Both copper and niobium additions significantly increased the re-passivation potentials, while copper was observed to reduce the pitting hysteresis loop area. Alloys containing 15% copper and 2% niobium additions depicted the most improved pitting corrosion resistance, and increased the repassivation value from –315.60 mV to a high repassivation potential of 840.68 mV.

The microstructure, thermal cycling, and mechanical behavior of $Ni_{48.5}Ti_{31.5-x}Zr_{20}Al_x$ (x = 0, 1, 2, 3) alloys were studied in the solution-treated and aged condition [159]. It was reported that there was a strong dependence of the transformation behavior on alloy chemistry and thermal cycling. Al element additions served to decrease transformation behaviors from 78 to 323 °C) reduce thermal stability. Al was shown to increase the plateau stress in the aged condition, whereas the formation of coarse-grained intermetallic phases caused the embrittlement of the microstructure, reducing its ductility, and alloying with Al resulted in the refinement of the coarse, lenticular precipitates identified as $Ni_4(Ti,Zr)_3$.

Using the prepared $Ni_{50}Ti_{33}Zr_{16}Si_1$ alloy [251], it was found that a nanoscale amorphous phase in $Ni_{50}Ti_{33}Zr_{16}Si_1$ alloy was formed within the primary solidified $Ni_{50}(Ti,Zr)_{50}$ intermetallic compound with an isolated morphology. Coughlin et al. [252] tested NiTiSnAg alloy and reported that mechanical tensile testing of $NiTiSn_{3.5}Ag$ single fiber composites demonstrates superelastic behavior of the composite with 85% strain recovery. Fatigue experiments show an evolution in damage over cycles, and an S–N curve shows sharp transition between a nearly vertical low-cycle fatigue behavior and the high-cycle fatigue regime.

3.3.2 Quinary alloy systems

Kim et al. [251] investigated quinary $Ni_{40}Cu_{10}Ti_{33}Zr_{16}Si_1$ alloy and reported that local amorphization occurs at negatively curved interfacial areas between the primary $Ni_{50}(Ti,Zr)_{50}$ grains and the Ni_3Ti and NiTiZr matrix phases. Chemical analysis of the different phases in these alloys reveals that the amorphous phase is more Zr-rich compared to the primary $Ni_{50}(Ti,Zr)_{50}$ phase, indicating that the Zr content is very crucial for enhancing the glass-forming ability. However, Cu is decisive in controlling the volume fraction, the morphology, and the sites of the amorphous phase in these alloys.

Yim et al. [253], studying NiCuTiZrSi alloy, reported that the rapidly cooled cast strips show a primary Ni(Ti,Zr) B2 structure superimposed on the diffuse scattering maxima from the amorphous phase. Compression test results show that the composite starts to yield at 1,200MPa and fractures at 1,900MPa.

Samal et al. [254], studying the phase and microstructure evolution during solidification of $Ni_{48}Cu_{10}Co_2Ti_{38}Ta_2$ during suction casting. It was observed that the addition of alloying elements such as Cu, Co to the binary $(Ni_{60}Ti_{40})$ alloy suppresses the martensitic transformation of the ordered NiTi (B2) dendrite. $Ni_{48}Cu_{10}Co_2Ti_{38}Ta_2$ quinary alloy shows the disordered nature of NiTi dendrites. A novel Ni–Ti-based $Ni_{48}Cu_{10}Co_2Ti_{38}Ta_2$ alloy was investigated on its properties [269]. It was reported that a microstructure consists of nanostructured eutectic between cubic NiTi and hexagonal Ni_3Ti with micron-scale NiTi and cubic Ti_2Ni dendrites. The alloy possesses a high compressive strength, 2 GPa, with high plasticity of 13%.

With $Ti_{50}Ni_{46.7}Cu_{0.8}Fe_{2.3}Pd_{0.2}$ alloy, Xue et al. [255] found that this alloy shows very low thermal hysteresis (ΔT) of, at the smallest, ΔT (1.84 K).

Czeppe at el. [256] investigated two different Ni-based glasses $Ni_{58}Nb_{10}Zr_{13}Ti_{12}Al_7$ and $Ni_{58}Nb_{25}Zr_8Ti_6Al$ with form of amorphous melt-spun ribbons. It was found that the ribbon containing 10 at% Nb showed typical primary crystallization of the 50 nm grains of the NiTi(Nb) cubic phase; the ribbon containing 25 at% of Nb revealed high-thermal stability of the amorphous phase, which crystallized only in a small amount in the range of primary crystallization, preserving large fraction of the amorphous phase even high above the end of the crystallization. The ribbon containing 10 at% Nb showed stress relaxation and was maximally elongated up to 0.6%, whereas the ribbon with 25 at% Nb revealed a hardening effect and the slightly smaller maximal elongation following it.

3.4 Functionally graded NiTi materials

Materials is composed of multilayer, with each layer having unique characteristics, yet adjacent layers having some similarity is called functionally graded (or gradient) materials [257]. Although such functions can include various properties, it is limited to mechanical, physical, or thermal properties since other properties such as chemical or electrochemical are more likely important to the surface layer, and not related to bulky or semibulky behavior [272]. Functionally graded SMAs have the advantage of combining the functionalities of the SME and those of functionally graded structures. One obvious advantage of functionally graded SMAs is their widened transformation stress and temperature windows that provide improved controllability in actuating applications [258]. Shariat et al. [258–260] extensively investigated the application of concept of functionally gradient materials to NiTi. After examining thermomechanical deformation behavior of microstructurally graded, compositionally graded, and geometrically graded NiTi alloy components, it was reported that the martensitic transformation occurs partially over the nominal stress gradient. The property gradients created along the loading direction or perpendicular to the loading direction produce distinct thermomechanical behaviors. The property gradient along the loading direction provides stress gradient over stress-induced transformation, which can be adjusted by the property gradient profile. The property gradient through the thickness direction of plate specimens and perpendicular to the loading direction provides four-way shape-memory behavior during stress-free thermal cycling after tensile deformation [258–265].

Fabrication of net shape load bearing implants with complex anatomical shapes to meet desired mechanical and biological performance is still a challenge. The porosity is one of powerful factors to control mechanical property [259–265]. Bandyopadhyay et al. [266] employed the laser-engineered net shaping technology to control the porosity of NiTi. It was demonstrated that complex metallic implants with designed porosities up to 70 vol% to reduce stress-shielding. The effective modulus of NiTi was tailored to suit the modulus of human cortical bone by introducing 12–42 vol% porosity. Furthermore, laser-processed porous NiTi alloy samples show a 2–4% recoverable strain, a potentially significant result for load bearing implants.

Functionally gradient materials provide unique biological, chemical, or mechanical functionalities in medical devices and implants for bone replacement [267, 268]. It was introduced that an artificial biomaterial for knee joint replacement has been developed by building a graded structure consisting of ultra-high-molecular-weight polyethylene (UHMWPE) fiber reinforced high-density polyethylene combined with a surface of UHMWPE. The ingrowth behavior of titanium implants into hard tissue can be improved by depositing a graded biopolymer coating of fibronectin, collagen types I and III with a gradation, derived from the mechanisms occurring during healing in vivo. Functionally graded porous hydroxyapatite (HAP) ceramics can be produced using alternative routes, for example, sintering of laminated structures of

HAP tapes filled with polymer spheres or combining biodegradable polyesters such as polylactide, polylactide-*co*-glycolide, and polyglycolide, with carbonated nano-crystalline HAP, and HAP–collagen I scaffolds are an appropriate material for in vitro growth of bone [268].

The Ni/Ti ratio can be manipulated by diffusion bonding, producing the so-called compositional gradation. Lim et al. [63] aiming to develop a unique actuator exhibiting a shape change over a wide temperature along the thickness direction, prepared stacking ribbons of Ti–51Ni, Ti–50Ni, Ti–49Ni, and Ti–48Ni (at%) in sequence by spark plasma sintering. It was found that the compositionally graded sample showed compositional variation of 1.5 at% Ti along the thickness direction (–120 µm) and a martensitic transformation temperature window as large as 91 K on cooling and 79 K on heating. A recoverable elongation of 0.9% was obtained under a stress of 80 MPa and the deformation rate, which is defined as the ratio of the recoverable elongation to the temperature range where the elongation occurred was 0.015%/K in the compositionally graded sample [63]. An in situ TEM study of the B2 ↔ B19′ martensitic transformation in $Ti_{40.7}Hf_{9.5}Ni_{44.8}Cu_5$ SMA was carried out [269]. It was observed that the sequence of the martensite crystals shrinking on heating differed from the sequence of the martensite crystal appearance on previous cooling. This was shown that strain nanodomain formation on cooling prior to the forward martensitic transformation resulted in accumulation of the elastic energy. This led to the dependences of the elastic energy stored on cooling or released on heating on the volume fraction of the M-phase became different. In this case, at the same volume fraction of the M-phase, the configuration of the martensite crystals on cooling and heating was different and it was a reason for a violation of the sequence of the martensite crystal formation on cooling and its shrinking on heating. Shariat et al. [270] presented an analytical model describing the deformation behavior of functionally graded NiTi plates under tensile loading and their shape recovery during heating. The property gradient of the plate is achieved by either a compositional gradient or microstructural gradient through the thickness. Closed-form solutions are obtained for nominal stress–strain variations of such plates under uniaxial loading at different deformation stages. It is observed that the martensitic transformation occurs partially over the nominal stress gradient, unlike typical NiTi SMAs. The curvature–temperature relations are established for complex SME behavior of such plates during the recovery period and the analytical solutions are validated with relevant experimental results. Bogdanski et al. [271] investigated the biocompatibility of NiTi alloys by single-culture experiments on functionally graded samples with a stepwise change in composition from pure nickel to pure titanium, including a Ni–Ti SMA for a 50:50 mixture. This approach permitted a considerable decrease of experimental resources by simultaneously studying a full variation of composition. It was reported that a good biocompatibility for nickel content up to about 50% was indicated. The cells used in the biocompatibility studies comprised osteoblast-like osteosarcoma cells (SAOS-2, MG-63), primary human osteoblasts (HOB), and murine fibroblasts (3T3). Depending on the temperature and stress history, an SMA can be in a state of austenite (A), twinned martensite (TM), or detwinned martensite. Austenite and TM may coexist in the process of thermocycling.

References

[1]   Sanjabi S, Barber ZH. The effect of film composition on the structure and mechanical properties of NiTi shape memory thin films. Surf. Coat. Technol. 2010, 204, 1299–304.

[2]   Belyaev S, Resnina N, Iaparova E, Ivanova A, Rakhimov T, Andreev V. Influence of chemical composition of NiTi alloy on the martensite stabilization effect. J. Alloy Compd. 2019, 787, 1365–71.

[3]   Fu Y, Du H. Effects of film composition and annealing on residual stress evolution for shape memory TiNi film. Mater. Sci. Eng. A, 2003, 342, 236–44.

[4]   Yoneyama T, Doi H, Hamanaka H. Influence of composition and purity on tensile properties of Ni-Ti alloy castings. Dent. Mater. J. 1992, 11, 157–64.

[5]   Yan XJ, Yang DZ. Corrosion resistance of a laser spot-welded joint of NiTi wire in simulated human body fluids. J. Biomed. Mater. Res. A. 2006, 77, 97–102.

[6]   Mertmann M. Processing and Quality Control of Binary NiTi Shape Memory Alloys. In: Yahia L, ed., Shape Memory Implants. Springer, Berlin, Germany, 2000, 24–34.

[7]   Van Humbeeck J, Stalmans R, Chandrasekaran M, Delaey L. On the Stability of Shape Memory Alloy. In: Duerig TW, et al. ed., Engineering Aspects of Shape Memory Alloys. Butterworth-Heinemann, London, UK, 1990, 96–105.

[8]   Huang X, Liu Y. Effect of annealing on the transformation behavior and superelasticity of NiTi shape memory alloy. Scr. Mater. 2001, 45, 153–60.

[9]   Kim JI, Miyazaki S. Effect of nano-scaled precipitates on shape memory behavior of Ti-50.9 at.%Ni alloy. Acta Mater. 2005, 53, 4545–54.

[10]   Urbina C, de la Flor S, Ferrando F. Effect of thermal cycling on the thermomechanical behaviour of NiTi shape memory alloys. Mater. Sci. Eng. A 2009, A501, 197–206.

[11]   Urbina C, de la Flor S, Ferrando F. R-phase influence on different two-way shape memory training methods in NiTi shape memory alloys J. Alloy Compd. 2010, 490, 499–507.

[12]   Lygin K, Langbein S, Labenda P, Sadek T. A methodology for the development, production, and validation of R-Phase actuators. J. Mater. Eng. Perform. 2012, 21, 2657–62.

[13]   Wang XB, Verlinden B, Van Humbeeck J. R-phase transformation in NiTi alloys. Mater. Sci. Technol. 2014, 30, 1517–29.

[14]   Michutta J, Somsen C, Yawny A, Dlouhy A, Eggeler G. Elementary martensitic transformation processes in Ni-rich NiTi single crystals with Ni_4Ti_3 precipitates. Acta Mater. 2006, 54, 3525–42.

[15]   Carroll MC, Somsen C, Eggeler G. Multiple-step martensitic transformations in Ni-rich NiTi shape memory alloys. Scr. Mater. 2004, 50, 187–92.

[16]   Dlouhy A, Khalil-Allafi J, Eggeler G. Multiple-step martensitic transformations in Ni-rich NiTi alloys--an in-situ transmission electron microscopy investigation. J. Philos. Mag. A 2003, 83, 339–63.

[17]   Sitepu H, Schmahl WW, KAllafi J, Eggeler G, Dlouhy A, Toebbens DM, Tovar M. Neutron diffraction phase analysis during thermal cycling of a Ni-rich NiTi shape memory alloy using the Rietveld method. Scr. Mater. 2002, 46, 543–8.

[18]   Qin Q, Peng H, Fan Q, Zhang L, Wen Y. Effect of second phase precipitation on martensitic transformation and hardness in highly Ni-rich NiTi alloys. J. Alloys Compd. 2018, 739, 873–81.

[19]   Allafi JK, Ren X, Eggeler G. The mechanism of multistage martensitic transformations in aged Ni-rich NiTi shape memory alloys. Acta Mater. 2002, 50, 793–803.

[20]   Eggeler G, Khalil-Allafi J, Gollerthan S, Somsen C, Schmahl W, Sheptyakov D. On the effect of aging on martensitic transformations in Ni-rich NiTi shape memory alloys. Smart Mater. Struct. 2005, 14, 186–91.

[21]   Frenzel J, George EP, Dlouhy A, Somsen Ch, Wagner MF-X, Eggeler G. Influence of Ni on martensitic phase transformations in NiTi shape memory alloys. Acta Mater. 2010, 58, 3444–58.

[22]   Ravari BK, Farjami S, Nishida M. Effects of Ni concentration and aging conditions on multistage martensitic transformation in aged Ni-rich Ti–Ni alloys. Acta Mater. 2014, 69, 17–29.

[23]   Fan G, Chen W, Yang S, Zhu J, Ren X, Otsuka K. Origin of abnormal multi-stage martensitic transformation behavior in aged Ni-rich Ti–Ni shape memory alloys. Acta Mater. 2004, 52, 4351–62.

[24]   Zhou Y, Zhang J, Fan G, Ding X, Sun J, Ren X, Otsuka K. Origin of 2-stage R-phase transformation in low-temperature aged Ni-rich Ti–Ni alloys. Acta Mater. 2005, 53, 5365–77.

[25]   Saedi S, Turabi AS, Andani MT, Haberland C, Karaca H, Elahinia M. The influence of heat treatment on the thermomechanical response of Ni-rich NiTi alloys manufactured by selective laser melting. J. Alloys Compd. 2016, 677, 204–10.

[26]   Nishida M, Hara T, Ohba T, Yamaguchi K, Tanaka K, Yamauchi K. Experimental consideration of multistage martensitic transformation and precipitation behavior in aged Ni-Rich Ti-Ni shape memory alloys. Mater. Trans. 2003, 44, 2631–36.

[27]   Khalil-Allafi J, Amin-Ahmadi B. Multiple-step martensitic transformations in the Ni 51 Ti 49 single crystal. J. Mater. Sci. 2010, 45, 6440–5.

[28]   Priyadarshini BG, Aich S, Chakraborty M. An investigation on phase formations and microstructures of Ni-rich Ni-Ti shape memory alloy thin films. Metall. Mater. Trans. A 2011, 42, 3284–90.

[29]   Prokofiev EA, Burow JA, Payton EJ, Zarnetta R, Frenzel J, Gunderov DV, Valiev RZ. Suppression of Ni_4Ti_3 precipitation by grain size refinement in Ni-Rich NiTi shape memory alloys. Adv. Eng. Mater. 2010, 12, 747–53.

[30]   Bojda O, Eggeler G, Dlouhý A. Precipitation of Ni_4Ti_3-variants in a polycrystalline Ni-rich NiTi shape memory alloy. Scr. Mater. 2005, 53, 99–104.

[31]   Cao S, Nishida M, Schryvers D. Quantitative three-dimensional analysis of Ni_4Ti_3 precipitate morphology and distribution in polycrystalline Ni–Ti. Acta Mater. 2011, 59, 1780–9.

[32]   Chu CL, Chung CY, Lin PH. DSC study of the effect of aging temperature on the reverse martensitic transformation in porous Ni-rich NiTi shape memory alloy fabricated by combustion synthesis. Mater. Lett. 2005, 59, 404–7.

[33]   Zhang J, Somsen C, Simon T, Ding X, Hou S, Ren S, Ren X, Eggeler G, Otsuka K, Sun J. Leaf-like dislocation substructures and the decrease of martensitic start temperatures: a new explanation for functional fatigue during thermally induced martensitic transformations in coarse-grained Ni-rich Ti–Ni shape memory alloys. Acta Mater. 2012, 60, 1999–2006.

[34]   Olbricht J, Yawny A, Pelegrina JL, Dlouhy A, Eggeler G. On the stress-induced formation of R-phase in ultra-fine-grained Ni-Rich NiTi shape memory alloys. Metall. Mater. Trans. A 2011, 42, 2556–74.

[35]   Kazemi-Choobi K, Khalil-Allafi J, Elhami A, Asadi P. Influence of aging treatment on in-situ electrical resistance variation during aging of Nickel-Rich NiTi shape memory wires. Metall. Mater. Trans. A 2013, 44, 4429–33.

[36]   Farvini M. Challenges of suing elemental nickel and titanium powders for the fabrication of monolithic NiTi parts. Arch. Metall. Mater. 2017, 62, 1075–9.

[37]   Hanlon JE, Butler SR, Wasikewski RJ. Trans. AIME 1967, 239, 1323–7.

[38]   Otsuka K, Ren X. Physical metallurgy of Ti-Ni-based shape memory alloy. Progress. Mater.
        Sci. 2005, 50, 511–678.
[39]   Tang W. Thermodynamic study of the low-temperature phase B19' and the martensitic
        transformation in near-equiatomic Ti-Ni shape memory alloys. Metall. Mater. Trans. A 1997,
        28A, 537–54.
[40]   Melton KN. Ni-Ti based shape memory alloys. In: Duerig TW, et al. ed., Engineering Aspects
        of Shape Memory Alloys. Butterworth-Heinemann, London, UK, 1990, 21–35.
[41]   Lin HC, Wu SK, Chou TS, Kao HP. The effects of cold rolling on the martensitic transformation
        of an equiatomic TiNi alloy. Acta Metall. Mater. 1991, 39, 2069–80.
[42]   Liu Y, Liu Y, Humbeeck JV. Two-way shape memory effect developed by martensite
        deformation in NiTi. Acta Mater. 1998, 47, 199–209.
[43]   Jordan L, Chandrasekaran M, Masse M, Bouquet G. Study of the phase transformations in
        Ni-Ti based shape memory alloys. J. Phys. IV, 1995, 5, C489–94.
[44]   Liu Y, Calvin SP. Criteria for pseudoelasticity in near-equiatomic NiTi shape memory alloys.
        Acta Mater. 1997, 45, 4431–9.
[45]   Lin HC, Wu SK. The tensile behavior of a cold-rolled and reverse-transformed equiatomic
        TiNi alloy. Acta Metall. Mater. 1994, 42, 1623–30.
[46]   Stachowiak GB, McCornick PG. Shape memory behaviour associated with the R and
        martensitic transformations in a NiTi alloy. Acta Metall. 1988, 36, 291–7.
[47]   McCormick PG, Liu Y. Thermodynamic analysis of the martensitic transformation in NiTi: II.
        Effect of transformation cycling. Acta Metall. Mater. 1994, 42, 2407–13.
[48]   Liu Y, McCormick PG. Thermodynamic analysis of the martensitic transformation in NiTi: I.
        Effect of heat treatment on transformation behaviour. Acta Metall. Mater. 1994, 42,
        2401–6.
[49]   Sinha A, Datta S, Chakraborti PC, Chattopadhyay PP. Understanding the shape-memory
        behavior in Ti-(~49 At. Pct) Ni alloy by nanoindentation measurement. Metall. Mater. Trans.
        A 2013, 44, 1722–9.
[50]   Tillmann W, Momeni S. Tribological performance of near equiatomic and Ti-rich NiTi shape
        memory alloy thin films. Acta Mater. 2015, 92, 189–96.
[51]   Sehitoglu H, Karaman I, Anderson R, Zhang X, Gall K, Maier HJ, Chumlyakov Y. Compressive
        response of NiTi single crystals. Acta Mater. 2000, 48, 3311–26.
[52]   Chrobak D, Stroz D. Two-stage R phase transformation in a cold-rolled and annealed
        Ti–50.6 at.%Ni alloy. Scripta Mater. 2005, 52, 757–60.
[53]   Pelosin V, Riviere A. Effect of thermal cycling on the R-phase and martensitic
        transformations in a Ti-rich NiTi alloy. Metall. Mater. Trans. A 1998, 29, 1175–80.
[54]   Lin HC, Wu SK, Lin JC. The martensitic transformation in Ti-rich TiNi shape memory alloys.
        Mater. Chem. Phys. 1994, 37, 184–90.
[55]   Paula AS, Mahesh KK, dos Santos CML, Fernandes FMB, da Costa Viana CS.
        Thermomechanical behavior of Ti-rich NiTi shape memory alloys. Mater. Sci. Eng., A 2008,
        481/482, 146–50.
[56]   Lopez HF, Salinas-Rodriguez A, Rodriguez-Galicia JL. Microstructural aspects of precipitation
        and martensitic transformation in a Ti-rich Ni-Ti alloy. Scr. Mater. 1996, 34; doi:
        10.1016/1359-6462(95)00570-6.
[57]   Lopez HF, Rodriguez AS. Aging effects on transformation temperatures of cold worked Ni-52
        at.-%Ti shape memory alloy. Mater. Sci. Technol. 2002, 18, 268–72.
[58]   Lopez HF, Salinas A, Calderón H. Plastic straining effects on the microstructure of a Ti-rich
        NiTi shape memory alloy. Metall. Mater. Trans. A 2001, 32, 717–29.
[59]   Zhang JX, Sato M, Ishida A. On the Ti_2Ni precipitates and Guinier–Preston zones in Ti-rich
        Ti–Ni thin films. Acta Mater. 2003, 51, 3121–30.

[60] Karaman I, Karaca HE, Luo ZP, Maier HJ. The effect of severe marforming on shape memory characteristics of a Ti-rich NiTi alloy processed using equal channel angular extrusion. Metall. Mater. Trans. A 2003, 34, 2527–39.

[61] Tadayyon G, Mazinani M, Guo Y, Zebarjad SM, Tofail SAM, Biggs MJ. The effect of annealing on the mechanical properties and microstructural evolution of Ti-rich NiTi shape memory alloy. Mater. Sci. Eng., A 2016, 662, 564–77.

[62] Hassel AW, Neelakantan L, Zelenkevych A, Rush A, Spiegel M. Selective de-alloying of NiTi by oxochloridation. Corros. Sci. 2008, 50, 1368–75.

[63] Lim JH, Kim MS, Noh JP, Kim YW, Nam TH. Compositionally graded Ti-Ni alloys prepared by diffusion bonding. J. Nanosci. Nanotechnol. 2014, 14, 9042–6.

[64] Gallardo Fuentes JM, Gümpel P, Strittmatter J. Phase change behavior of nitinol shape memory alloys. Adv. Eng. Mater. 2002, 4, 437–52.

[65] Xi M, Shang J-X. First principle study on NiTi alloyed with transition elements. Rare Metal Mater. Eng. 2016, 45, 2041–5.

[66] Bozzolo G, Noebe RD, Mosca HO. Site preference of ternary alloying additions to NiTi: Fe, Pt, Pd, Au, Al, Cu, Zr and Hf. NASA Technical Reports Server (NTRS), 2004-01-01; https://archive.org/details/nasa_techdoc_20050198870/page/n13.

[67] Bozzolo G, Noebe RD, Mosca, HO. Site preference of ternary alloying additions to NiTi: Fe, Pt, Pd, Au, Al, Cu, Zr and Hf. J. Alloy Compd. 2005, 389, 80–94.

[68] Frenzel J Wieczorek A, Opahle I, Maaß B, Drautz R, Eggeler G. On the effect of alloy composition on martensite start temperatures and latent heats in Ni–Ti-based shape memory alloys. Acta Mater. 2015, 90, 213–31.

[69] Matsumoto H. Addition of an element to NiTi alloy by an electron-beam melting method. J. Mater. Sci. Lett. 1991, 10, 417–9.

[70] Singh N, Talapatra A, Junkaew A, Duong T, Gibbons S, Li S, Hassan Thawabi H, Olivos E, Arróyave R. Effect of ternary additions to structural properties of NiTi alloys. Comput. Mater. Sci. 2016, 112, 347–55.

[71] Salvetr P, Školáková A, Novák P. Effect of magnesium addition on the structural homogeneity of NiTi alloy produced by self-propagating high-temperature synthesis. Kovove Mater. 2017, 55, 379–83.

[72] Biesiekierski A, Wang J, Gepreel MAH, Wen C. A new look at biomedical Ti-based shape memory alloys. Acta Biomater. 2012, 8, 1661–9.

[73] Novák P, Salvetr P, Školáková A, Karlík M, Kopeček J. Effect of alloying elements on the reactive sintering behaviour of NiTi alloy. Mater. Sci. Forum 2017, 891, 447–51.

[74] Zhirong H. Superelasticity of Ti-Ni-V shape memory alloys. Rare Metal Mater. Eng. 2015, 44, 1639–42.

[75] Uchil J, Kumara KG, Mahesh KK. Effects of heat treatment temperature and thermal cycling on phase transformations in Ni–Ti–Cr alloy. J. Alloy Compd. 2001, 325, 210–4.

[76] Jeom SN-N, Choi Y. Strain rate dependence of deformation mechanisms in a Ni–Ti–Cr shape-memory alloy. Acta Mater. 2005, 53, 449–54.

[77] Lim YJ, Oshida Y. Initial contact angle measurements on variously treated dental/medical titanium materials. J. Bio-Med. Mat. Eng. 2001, 11, 325–41.

[78] Elias CN, Oshida Y, Lima JHC, Muller CA. Relationship between surface properties (roughness, wettability and morphology) of titanium and dental implant torque. J. Mech. Behav. Biomed. Mater. 2008, 1, 234–42.

[79] Pulletikurthi C, Munroe N, Stewart D, Haider W, Amruthaluri S, Rokicki R, Dugrot M, Ramaswamy S. Utility of magneto-electropolished ternary nitinol alloys for blood contacting applications. J. Biomed. Mater. Res. Part B: Appl. Biomater. 2015, 103B, 1366–74.

[80]   Iijima M, Endo K, Ohno H, Mizoguchi I. Effect of Cr and Cu addition on corrosion behavior of NiTi alloys. Dent. Mater. J. 1998, 17, 31–40.

[81]   Zhou Y, Xue D, Ding X, Wang Y, Zhang J, Zhang Z, Wang D, Otsuka K, Sun J, Ren X. Strain glass in doped $Ti_{50}(Ni_{50-x}Dx)$ (D = Co, Cr, Mn) alloys: implication for the generality of strain glass in defect-containing ferroelastic systems. Acta Mater. 2010, 58, 5433–42.

[82]   Kök M, Ateş G. The effect of addition of various elements on properties of NiTi-based shape memory alloys for biomedical application. Eur. Phys. J. Plus 2017, 132, 185–9.

[83]   Gallardo Fuentes JM, Gümpel P, Strittmatter J. Phase change behavior of nitinol shape memory alloys. Adv. Eng. Mater. 2002, 4, 437–52.

[84]   Kassab E, Neelakantan L, Frotscher M, Swaminathan S, Maaß B, Rohwerder M, Gomes J, Eggeler G. Effect of ternary element addition on the corrosion behaviour of NiTi shape memory alloys. Mater. Corros. 2014, 65, 18–22.

[85]   Ramachandran B, Chang P-C, Kuo Y-K, Chien C, Wu S-K. Characteristics of martensitic and strain-glass transitions of the Fe-substituted TiNi shape memory alloys probed by transport and thermal measurements.Sci. Rep. 2017, 7; doi: 10.1038/s41598-017-16574-0.

[86]   Hara T, Ohba T, Okunishi E, Otsuka K. Structural study of R-phase in Ti-50.23 at.% Ni and Ti-47.75 at.% Ni-1.50 at.% Fe alloys. Mater. Trans. Jpn. Inst. Metals 1997, 38, 11–7.

[87]   Yasuda Y, Terai T, Fukuda T, Kakeshita T. In situ transmission-electron-microscopy observation of solid-state amorphization behavior in $Ti_{50}Ni_{44}Fe_6$ alloy by high-voltage electron microscopy. Acta Mater. 2016, 104, 201–9.

[88]   Zhang J, Xue D, Cai X, Dig X, Ren X, Sun J. Dislocation induced strain glass in $Ti_{50}Ni_{45}Fe_5$ alloy. Acta Mater. 2016, 120, 130–7.

[89]   Salamon MB, Meichle ME, Wayman CM. Premartensitic phases of Ti50Ni47Fe3. Phys. Rev. B 1985, 31, 7306–15.

[90]   Hwang CM, Meichle ME, Salamon MB, Wayman CM. Transformation behaviour of a $Ti_{50}Ni_{47}Fe_3$ alloy II. Subsequent premartensitic behaviour and the commensurate phase. Phil. Mag. A 1983, 47, 31–62.

[91]   Choi MS, Fukuda T, Kakeshita T, Mori H. Incommensurate–commensurate transition and nanoscale domain-like structure in iron doped Ti–Ni shape memory alloys. Phil. Mag. 2006, 86, 67–78.

[92]   Wang D, Zhang Z, Zhang J, Zhou Y, Wang Y, Ding X, Wa Y, Ren X. Strain glass in Fe-doped Ti–Ni. Acta Mater. 2010, 58(18), 6206–15.

[93]   Niu JG, Geng WT. Anti-precursor effect of Fe on martensitic transformation in TiNi alloys. Acta Mater. 2016, 104, 18–24.

[94]   Frenzel J, Pfetzing J, Neuking K, Eggeler G. On the influence of thermomechanical treatments on the microstructure and phase transformation behavior of Ni–Ti–Fe shape memory alloys. Mater. Sci. Eng. A 2008, 481/482, 635–8.

[95]   Mirzadeh H, Parsa MH. Hot deformation and dynamic recrystallization of NiTi intermetallic compound. J. Alloys Compd. 2014, 614, 56–9.

[96]   Basu R, Jain L, Maji B, Krishnan M. Dynamic recrystallization in a Ni–Ti–Fe shape memory alloy: effects on austenite–M-phase transformation. J. Alloys Compd. 2015, 639, 94–101.

[97]   Zhang Y, Jiang S, Yan B, Wang M. Influence of degree of deformation on static recrystallization texture and compressive strength of NiTiFe shape memory alloy subjected to canning compression. Metall. Mater. Trans. A 2018, 49, 6277–89.

[98]   Yin X-Q, Park C-H, Li Y-F, Ye W-J, Zuo Y-T, Lee S-W, Yeom J-T, Mi X-J. Mechanism of continuous dynamic recrystallization in a 50Ti-47Ni-3Fe shape memory alloy during hot compressive deformation. J. Alloys Compd. 2017, 693, 426–31.

[99]   Jordan L, Masse M, Collier JY, Bouquet G. Effects of thermal and thermomechanical cycling on the phase transformations in NiTi and Ni-Ti-Co shape-memory alloys. J. Alloys Compd. 1994, 211, 204–7.

[100]   Lee JH, Park JB, Andreasen GF, Lakes RS. Thermomechanical study of Ni-Ti alloysJ. Biomed. Mater. Res. 1998, 22, 573–88.

[101]   El-Bagoury N. Microstructure and martensitic transformation and mechanical properties of cast Ni rich NiTiCo shape memory alloys. Mater. Sci. Tech. 2014, 30, 1795–800.

[102]   Fasching A, Norwich D, Geiser T, Paul GW. An evaluation of a NiTiCo alloy and its suitability for medical device applications. J. Mater. Eng. Perform. 2011, 20, 641–5.

[103]   Wang QY, Zheng YF. The electrochemical behavior and surface analysis of Ti50Ni47.2Co2.8 alloy for orthodontic use. Dent. Mater. 2008, 24, 1207–11.

[104]   Phukaoluan A, Khantachawana A, Kaewtatip P, Dechkunakorn S, Kajornchaiyakul J. Improvement of mechanical and biological properties of TiNi alloys by addition of Cu and Co to orthodontic archwires. Int. Orthod. 2016, 14, 295–310.

[105]   Alnomani SN, Fadhel EZ, Mehatlaf AA. Prepare nitinol alloys and improve their hardness using copper as an alloying element. IJAER 2017, 12, 4299–308.

[3106]   Rondelli G, Vicentini B. Effect of copper on the localized corrosion resistance of Ni-Ti shape memory alloy. Biomaterials 2002, 23, 639–44.

[107]   Iijima M, Endo K, Ohno H, Mizoguchi I. Effect of Cr and Cu addition on corrosion behavior of NiTi alloys. Dent. Mater. J. 1998, 17, 31–40.

[108]   Kassab E, Neelakantan L, Frotscher M, Swaminathan S, Maaß B, Rohwerder M, Gomes J, Eggeler G. Effect of ternary element addition on the corrosion behaviour of NiTi shape memory alloys. Mater. Corros. 2012, 65, 18–22.

[109]   Oshida Y, Sachdeva R, Miyazaki S.Changes in contact angles as a function of time on some pre-oxidized bio-materials. J. Mater. Sci.: Mater. Med. 1992, 3, 306–12.

[110]   Oshida Y, Sachdeva R, Miyazaki S. Microanalytical characterization and surface modification of TiNi orthodontic archwires. Bio-Med. Mater. Eng. 1992, 2, 51–69.

[111]   Gil FJ, Planell JA. Effect of copper addition on the superelastic behavior of Ni-Ti shape memory alloys for orthodontic applications. J. Biomed. Mater. Res. 1999, 48, 682–8.

[112]   Grossmann C, Frenzel F, Sampath V, Depka T, Eggeler G. Elementary transformation and deformation processes and the cyclic stability of NiTi and NiTiCu shape memory spring actuators. Metall. Mater. Trans. A 2009, 40, 2530–44.

[113]   Es-Souni M, Es-Souni M, Brandies HF. On the transformation behaviour, mechanical properties and biocompatibility of two niti-based shape memory alloys: NiTi42 and NiTi42Cu7. Biomaterials 2001, 22, 2153–61.

[114]   Tomozawa M, Kim HY, Miyazaki S. Shape memory behavior and internal structure of Ti–Ni–Cu shape memory alloy thin films and their application for microactuators. Acta Mater. 2009, 57, 441–52.

[115]   Yamamoto M, Kuroda T, Yoneyama T, Doi H. Bending property and phase transformation of Ti-Ni-Cu alloy dental castings for orthodontic application. J. Mater. Sci. Mater. Med. 2002, 13, 855–9.

[116]   Sehitoglu H, Karaman I, Zhang X, Viswanath A, Chumlyakov Y, Maier HJ. Strain–temperature behavior of NiTiCu shape memory single crystals. Acta Mater. 2001, 49, 3621–34.

[117]   Samal S, Biswas K, Phanikumar G. Solidification behavior in newly designed Ni-Rich Ni-Ti-Based alloys. Metall. Mater. Trans. A 2016, 47, 6214–23.

[118]   Nam TH, Saburi T, Shimizu K. Cu-Content dependence of shape memory characteristics in Ti–Ni–Cu alloys. Mater. Trans., JIM 1990, 31, 959–67.

[119]   Yang X, Ma L, Shang J. Martensitic transformation of $Ti_{50}(Ni_{50-x}Cu_x)$ and $Ni_{50}(Ti_{50-x}Zr_x)$ shape-memory alloys. Sci. Rep. 2019, 9, 3221; doi: 10.1038/s41598-019-40100-z.

[120] Persaud-Sharma D, Munroe N, McGoron A. Electro and magneto-electropolished surface micro-patterning on binary and ternary nitinol. Trends Biomater. Artif. Organs 2012, 26, 74–85.

[121] Lin K-N, Wu S-K. Oxidation behavior of $Ti_{50}Ni_{40}Cu10$ shapemMemory alloy in 700–1,000°C Air. Oxid. Met. 2009, 71, 187–200.

[122] Miyazaki S, Mizukoshi K, Ueki T, Sakuma T, Liu Y. Fatigue life of Ti–50 at.% Ni and Ti–40Ni–10Cu (at.%) shape memory alloy wires. Mater. Sci. Eng. A, 1999, A273/A275, 658–63.

[123] Mazzolai FM, Biscarini A, Campanella R, Coluzzi B, Mazzolai G, Rotini A, Tuissi A. Internal friction spectra of the $Ni_{40}Ti_{50}Cu_{10}$ shape memory alloy charged with hydrogen. Acta Mater. 2003, 51, 573–83.

[124] Fan G, Zhou Y, Otsuka K, Ren X, Nakamura K, Ohta T, Suzuki T, Yoshida I, Yin F. Effects of frequency, composition, hydrogen and twin boundary density on the internal friction of $Ti_{50}Ni_{50-x}Cux$ shape memory alloys. Acta Mater. 2006, 54, 5221–9.

[125] König D, Buenconsejo PJS, Grochla D, Hamann S, Pfetzing-Micklich J, Ludwig A. Thickness-dependence of the B2–B19 martensitic transformation in nanoscale shape memory alloy thin films: zero-hysteresis in 75 nm thick $Ti_{51}Ni_{38}Cu_{11}$ thin films. Acta Mater. 2012, 60, 306–13.

[126] Meng XL, Sato M, Ishida A. Structure of martensite in sputter-deposited (Ni,Cu)-rich Ti–Ni–Cu thin films containing $Ti(Ni,Cu)_2$ precipitates. Acta Mater. 2009, 57, 1525–35.

[127] Meng XL, Sato M, Ishida A. Structure of martensite in deformed Ti–Ni–Cu thin films. Acta Mater. 2011, 59, 2535–43.

[128] Biscarini A, Coluzzi B, Mazzolai G, Tuissi A, Mazzolai FM. Extraordinary high damping of hydrogen-doped NiTi and NiTiCu shape memory alloys. J. Alloys Compd. 2003, 355, 52–7.

[129] Miyamato H, Taniwaki T, Ohba T, Otsuka K, Nishigori S, Katc K. Two-stage B2–B19–B19' martensitic transformation in a $Ti_{50}Ni_{30}Cu_{20}$ alloy observed by synchrotron radiation. Scr. Mater. 2005, 53, 171–5.

[130] Tong Y, Liu Y, Xie Z, Zarinejad M. Effect of precipitation on the shape memory effect of $Ti_{50}Ni_{25}Cu_{25}$ melt-spun ribbon. Acta Mater. 2008, 56, 1721–32.

[131] Xie ZL, Cheng GP, Liu Y. Microstructure and texture development in $Ti_{50}Ni_{25}Cu_{25}$ melt-spun ribbon. Acta Mater. 2007, 55, 361–9.

[132] Xie ZL, Van Humbeeck J, Liu Y, Delaey L. TEM study of Ti50Ni25Cu25 melt spun ribbons. Scr. Mater. 1997, 37, 363–71.

[133] Liu Y. Mechanical and thermomechanical properties of a $Ti_{0.50}Ni_{0.25}Cu_{0.25}$ melt spun ribbon. Mater. Sci. Eng. A 2003, 354, 286–91.

[134] Archambault A, Major TW, Carey JP, Heo G, Badawi H, Major PW. A comparison of torque expression between stainless steel, titanium molybdenum alloy, and copper nickel titanium wires in metallic self-ligating brackets. Angle Orthod. 2010, 80, 884–9.

[135] Seyyed Aghamiri SM, Ahmadabadi MN, Raygan Sh. Combined effects of different heat treatments and Cu element on transformation behavior of NiTi orthodontic wires. J. Mech. Behav. Biomed. Mater. 2011, 4, 298–302.

[136] Seyyed Aghamiri SM, Nili Ahmadabadi M, Shahmir H, Naghdi F, Raygan Sh. Study of thermomechanical treatment on mechanical-induced phase transformation of NiTi and TiNiCu wires. J. Mech. Behav. Biomed. Mater. 2013, 21, 32–6.

[137] Gil FJ, Solano E, Mendoza A, Pena J. Inhibition of Ni release from NiTi and NiTiCu orthodontic archwires by nitrogen diffusion treatment. J. Appl. Biomater. Biomech. 2004, 2, 151–5.

[138] Li HF, Qiu KJ, Zhou FY, Li L, Zheng YF. Design and development of novel antibacterial Ti-Ni-Cu shape memory alloys for biomedical application. Sci. Rep. 2016, 6; doi: 10.1038/srep37475.

[139] Zhang E-L, Fu S, Wang R-X, Li H-X, Liu Y, Ma Z-Q, Liu G-K, Zhu C-S, Qin G-W, Chen D-F. Role of Cu element in biomedical metal alloy design. Rare Met. 2019, 38, 476–94.

[140] Scheiber I, Dringen R, Mercer JFB. Copper: effects of deficiency and overload. Metal Ions. Life Sci. 2013, 13, 359–87.

[141] Linder MC, Hazeghazam M. Copper biochemistry and molecular biology. Am. J. Clin. Nutr. 1996, 63, 797S-811S.

[142] Scheiber IF, Mercer JF, Dringen R. Metabolism and functions of copper in brain. Prog. Neurobiol. 2014, 116, 33–57.

[143] Klevay LM. Copper in legumes may lower heart disease risk. Arch. Intern. Med. 2002, 162, 1780–1.

[144] Hu GF. Copper stimulates proliferation of human endothelial cells under culture. J. Cell Biochem. 2015, 69, 32635.

[145] Abraham KS, Jagdish N, Kailasam V, Padmanabhan S. *Streptococcus mutans* adhesion on nickel titanium (NiTi) and copper-NiTi archwires: a comparative prospective clinical study. Angle Orthod. 2017, 87, 448–54.

[146] Norambuena, GA, Patel R, Karau M, Wyles CC, Jannetto PJ, Bennet JK, Hanssen AD, Sierra RJ. Antibacterial and biocompatible titanium-copper oxide coating may be a potential strategy to reduce periprosthetic infection: an *in vitro* study. Clin. Orthop. Relat. Res. 2016, 575, 722–32.

[147] Li, JY, Zhai D, Lv F, Yu Q, Ma H, Yin J, Yi Z, Liu M, Chang J, Wu C. Preparation of copper-containing bioactive glass/eggshell membrane nanocomposites for improving angiogenesis, antibacterial activity and wound healing. Acta Biomater. 2016, 36, 254–66.

[148] Moberly W, Proft J, Duerig T, Sinclair R. Twinless martensite in TiNiCu shape memory alloys. Mater. Sci. Forum 1991, 56, 605–10.

[149] Alhumdany AA, Abidali AK, Abdulredha HJ. Investigation of wear behaviour for NiTi alloys with yttrium and tantalum additions. IOP Conf. Ser.: Mater. Sci. Eng., 2018, ;https://iopscience.iop.org/volume/1757-899X/433 doi: 10.1088/1757-899X/433/1/012072.

[150] Alhumdany AA, Abidali AK, Abdulredha HJ. The effect of adding alloying element yttrium and tantalum on mechanical properties of NiTi shape memory alloy. J. Univ. Babylon, Eng. Sci. 2018, 26, 153.

[151] Ahmadi H, Nouri M. Effects of yttrium addition on microstructure, hardness and resistance to wear and corrosive wear of TiNi alloy. J. Mater. Sci. Technol. 2011, 27, 851–5.

[152] Pu ZJ, Tseng HK, Wu KH. Martensite transformation and shape-memory effect of NiTi-Zr high-temperature shape-memory alloys. In: SPIE Proceedings. 1995, 2441, 171; https://doi.org/10.1117/12.209815.

[153] Hsieh SF, Wu SK. Room-temperature phases observed in Ti Ni Zr high-temperature 53 2 x 47 x shape memory alloys. J. Alloys Compd. 1998, 266, 276–82.

[154] Hsieh SF, Wu SK. A study on ternary Ti-rich TiNiZr shape memory alloys. Mater. Charact. 1998, 41, 151–62.

[155] Wu SK, Hsieh SF. Martensitic transformation of a Ti-rich Ti40.5Ni49.5Zr10 shape memory alloy. J. Alloys Compd. 2000, 297, 294–302.

[156] Yang X, Ma L, Shang J. Martensitic transformation of $Ti_{50}(Ni_{50-x}Cu_x)$ and $Ni_{50}(Ti_{50-x}Zr_x)$ shape-memory alloys. Sci. Rep. 2019, 9, 3221; doi: 10.1038/s41598-019-40100-z.

[157] McCluskey PJ, Zhao C, Kfir O, Vlassak JJ. Precipitation and thermal fatigue in Ni–Ti–Zr shape memory alloy thin films by combinatorial nanocalorimetry. Acta Mater. 2011, 59, 5116–24.

[158] Santamarta R, Arróyave R, Pons J, Evirgen A, Karaman I, Karaca HE, Noebe RD. TEM study of structural and microstructural characteristics of a precipitate phase in Ni-rich Ni–Ti–Hf and Ni–Ti–Zr shape memory alloys. Acta Mater. 2013, 61, 6191–206.

[159]   Hsu DHD, Sasaki TT, Thompson GB, Manuel MV. The effect of aluminum additions on the microstructure and thermomechanical behavior of NiTiZr shape-memory alloys. Metall. Mater. Trans. A 2012, 43, 2921–31.

[160]   Kim HY, Mizutani M, Miyazaki S. Crystallization process and shape memory properties of Ti–Ni–Zr thin films. Acta Mater. 2009, 57, 1920–30.

[161]   Feng ZW, Gao BD, Wang JB, Qian DF, Liu YX. Influence of Zr additions on shape-memory effect and mechanical properties of Ni-rich NiTi alloys. Mater. Sci. Forum, 2002, 394/395, 365–8.

[162]   Sandu AM, Tsuchiya K, Yamamoto S, Todaka Y, Umemoto M. Influence of isothermal ageing on mechanical behaviour in Ni-rich Ti–Zr–Ni shape memory alloy. Scr. Mater. 2006, 55, 1079–82.

[163]   Shi H, Frenzel J, Martinez GT, Van Rompaey S, Bakulin A, Kullkova S, Van Aert S, Schryvers D. Site occupation of Nb atoms in ternary Ni–Ti–Nb shape memory alloys. Acta Mater. 2014, 74, 85–95.

[164]   He XM, Rong LJ, Yan DS, Li YY. TiNiNb wide hysteresis shape memory alloy with low niobium content. Mater. Sci. Eng.: A 2004, 371, 193–9.

[165]   He X, Rong L, Yan D, Jiang Z, Li Y. Effect of deformation on martensitic transformation behavior in $Ni_{50.1}Ti_{46.9}Nb_3$ shape memory alloy. J. Mater. Sci. 2005, 40, 5311–3.

[166]   Wang L, Wang C, Zhang LC, Chen L, Lu W, Zhang D. Phase transformation and deformation behavior of NiTi-Nb eutectic joined NiTi wires. Sci. Rep. 2016, 6. doi: 10.1038/srep23905.

[167]   Shi H, Pourbabak S, Humbeeck JV, Schryvers D. Electron microscopy study of Nb-rich nanoprecipitates in Ni–Ti–Nb and their influence on the martensitic transformation. Scr. Mater. 2012, 67, 939–42.

[168]   Zhang CS, Wang YQ, Cai W, Zhao LC. The study of constitutional phases in a $Ni_{47}Ti_{44}Nb_9$ shape memory alloy. Mater. Chem. Phys. 1991, 28, 43–50.

[169]   Piao M, Miyazaki S, Otsuka K, Nishida N. Effects of Nb addition on the microstructure of Ti–Ni alloys. Mater. Trans., JIM 1992, 33, 337–45.

[170]   Salvetr P, Školáková A, Kopeček J, Novák P. Properties of Ni-Ti-X shape memory alloys produced by arc re-melting. Acta Metall. 2017, 23, 141–6.

[171]   Siegert W, Neuking K, Mertmann M, Eggeler G. First cycle shape memory effect in the ternary NiTiNb system. Joiurnal de Physique IV 2003, 112, 739–42.

[172]   Zhao LC, Duerig TW, Justi S, Melton KN, Proft JL, Yu W, Wayman CM. The study of niobium-rich precipitates in a Ni-Ti-Nb shape memory alloy. Scr. Metall. Mater. 1990, 24, 221–6.

[173]   Melton KN, Proft JL, Duerig TW. Wide hysteresis shape memory alloys based on the Ni – Ti – Nb system. MRS 1989, 9, 165–70.

[174]   Cai W, Zhang CS, Zhao LC. Recovery stress of Ni-Ti-Nb wide-hysteresis shape memory alloy under constant strain and thermomechanical cycling. J. Mater. Sci. Lett. 1994, 13, 8–9.

[175]   Zhang J, Zhu J. Surface structure evolution and abnormal wear behavior of the TiNiNb alloy under impact load. Metall. Mater. Trans. A 2009, 40, 1126–30.

[176]   Mareci D, Cailean A, Sutiman D. Electrochemical characterization of $Ni_{47.7}Ti_{37.8}Nb_{14.5}$ shape memory alloy in artificial saliva. Werkst. Korros. (Mater. Corros.) 2012, 63, 807–12.

[177]   Soga Y, Doi H, Yoneyama T. Tensile properties and transformation temperatures of Pd added Ti-Ni alloy dental castings. J. Mater. Sci. Mater. Med. 2000, 11, 695–700.

[178]   Bigelow GS, Padula II SA, Garg A, Gaydosh D, Noebe RD. Characterization of ternary NiTiPd high-temperature shape-memory alloys under loadbiased thermal cycling. Metall. Mater. Trans. A 2010, 41, 3065–79.

[179]   Noebe RD, Biles T, Padula II SA. Advanced Structural Materials: Properties, Design Optimization, and Applications. Taylor & Francis CRC Press, Boca Raton, FL, 2007, 145–86.

[180]   Shimizu S, Xu Y, Okunishi E, Tanaka S, Otsuka K, Mitose K. Improvement of shape memory
        characteristics by precipitation-hardening of Ti-Pd-Ni alloys. Mater. Lett. 1998, 34, 23–9.

[181]   Yang WS, Mikkola DE. Ductilization of Ti-Ni-Pd shape memory alloys with boron additions.
        Scr. Metall. Mater. 1993, 28, 161–5.

[182]   Cai W, Tanaka S, Otsuka K. Thermal cyclic characteristics under load in a
        $Ti_{50.6}Pd_{30}Ni_{19.4}$ alloy. Mater. Sci. Forum, 2000, 327/328, 279–82.

[183]   Goldberg D, Xu Y, Murakami Y, Morito S, Otsuka K, Ueki T, Horikawa H. Improvement of a
        $Ti_{50}Pd_{30}Ni_{20}$ high temperature shape memory alloy by thermomechanical treatments. Scr.
        Metall. Mater. 1994, 30, 1349–54.

[184]   Kumar PK, Lagoudas DC, Zanca KJ, Lagoudas MZ. Thermomechanical characterization of
        high temperature SMA actuators. Proc. SPIE, 2006, 6170, 306–12.

[185]   Ma J, Karaman I, Noebe RD. High temperature shape memory alloys. Int. Mater. Rev. 2010,
        55, 257–315.

[186]   Tian Q, Wu J. Effect of the rare-earth element Ce on oxidation behavior of TiPdNi
        alloys. Mater. Sci. Forum 2002, 394/395, 455–8.

[187]   Tian Q, Chen J, Chen Y, Wu J. Oxidation behavior of TiNi–Pd shape memory alloys. Z.
        Metallkd. 2003, 94, 36–40.

[188]   Oh KT, Joo UH, Park GH, Hwang CJ, Kim KN. Effect of silver addition on the properties of
        nickel-titanium alloys for dental application. J. Biomed. Mater. Res. B Appl. Biomater. 2006,
        76, 306–14.

[189]   Zheng YF, Zhang BB, Wang BL, Wang YB, Li L, Yang QB, Cui LS. Introduction of antibacterial
        function into biomedical TiNi shape memory alloy by the addition of element Ag. Acta
        Biomater. 2011, 7(6), 2758–67.

[190]   Liu A, Gao Z, Gao L, Cai W, Wu Y. Effect of Dy addition on the microstructure and
        martensitic transformation of a Ni-rich TiNi shape memory alloy. J. Alloys Compd. 2007, 437,
        339–43.

[191]   Tofail SAM, Butler J, Gandhi AA, Carlson JM, Lavelle S, Carr S, Tiernan P, Warren G, Kennedy
        K, Biffi CA, Bassani P, Tuissi A. X-ray visibility and metallurgical features of NiTi shape
        memory alloy with erbium. Mater. Lett. 2014, 137, 450–4.

[192]   Soares RL, de Castro WB. Effects of composition on transformation temperatures and
        microstructure of Ni-Ti-Hf shape memory alloys. REM, Int. Eng. J. 2019, 72; http://dx.doi.
        org/10.1590/0370-44672018720072.

[193]   Coppa AC, Kapoor M, Hornbuckle BC, Weaver ML, Noebe RD, Thompson GB. Influence of
        dilute Hf additions on precipitation and martensitic transformation in Ni-Ti-Pd alloys. JOM
        2015, 67, 2244–50.

[194]   Casalena L, Bucsek AN, Pagan DC, Hommer GM, Bigelow GS, Obstalecki M, Noebe
        RD, Mills MJ, Stebner AP. Structure-property relationships of a high strength
        superelastic NiTi–1Hf alloy. Adv. Eng. Mater. 2018, 20; https://doi.org/10.1002/
        adem.201800046.

[195]   Hornbuckle BC, Noebe RD, Thompson GB. Influence of Hf solute additions on the
        precipitation and hardenability in Ni-rich NiTi alloys. J. Alloys Compd. 2015, 640, 449–54.

[196]   Dalle F, Perrin E, Vermaut P, Masse M, Portier R. Interface mobility in $Ni_{49.8}Ti_{42.2}Hf_8$ shape
        memory alloy. Acta Mater. 2002, 50, 3557–65.

[197]   Thoma PE, Boehm JJ. Effect of composition on the amount of second phase and
        transformation temperatures of NixTi90-xHf shape memory alloys. Mater. Sci. Eng. 1999,
        A273/A275, 385–9.

[198]   König D, Zarnetta R, Savan A, Brunken H, Ludwig A. Phase transformation, structural and
        functional fatigue properties of Ti–Ni–Hf shape memory thin films. Acta Mater. 2011, 59,
        3267–75.

[199] Evirgen A, Karaman I, Santamarta R, Pons J, Noebe RD. Microstructural characterization and shape memory characteristics of the $Ni_{50.3}Ti_{34.7}Hf_{15}$ shape memory alloy. Acta Mater. 2015, 83, 48–60.

[200] Liu M, Tu MJ, Zhang XM, Li YY, Shevlyakov AV. Microstructure of melt-spinning high temperature shape memory Ni-Ti-Hf alloys. J. Mater. Sci. Lett. 2001, 20, 827-30.

[201] Potapov P, Shelyakov A, Gulyaev A, Svistunova E, Matveeva N, Hodgson D. Effect of hf on the structure of Ni-Ti martensitic alloys. Mater. Lett. 1997, 32, 247–50.

[202] Han SD, Zou WH, Jin S, Zhang Z, Yang DH. The studies of the martensite transformations in a $Ti_{36.5}Ni_{48.5}Hf_{15}$ alloy. Scr. Metall. Mater. 1995, 32, 1441–6.

[203] Moshref-Javadi M, Hossein S, Mohammad S, Salehi T, Aboutalebi MR. Age-induced multi-stage transformation in a Ni-rich NiTiHf alloy. Acta Mater. 2013, 61, 2583–94.

[204] Benafan O, Garg A, Noebe RD, Bigelow GS, Padulall SA, Gaydosh DJ, Schell N, Mabe JH, Vaidyanathan R. Mechanical and functional behavior of a Ni-rich $Ni_{50.3}Ti_{29.7}Hf_{20}$ high temperature shape memory alloy. Intermetallics 2014, 50, 94–107.

[205] Hayrettin C, Karakoc O, Karaman I, Mabe JH, Santamarta R, Pons J. Two way shape memory effect in NiTiHf high temperature shape memory alloy tubes. Acta Mater. 2019, 163, 1–13.

[206] Evirgen A, Karaman I, Santamarta R, Pons J, Hayrettin C, Noebe RD. Relationship between crystallographic compatibility and thermal hysteresis in Ni-rich NiTiHf and NiTiZr high temperature shape memory alloys. Acta Mater. 2016, 121, 374–83.

[207] Yang F, Coughlin DR, Phillips PJ, Yang L, Devaraj A, Kovarik L, Noebe RD, Mills MJ. Structure analysis of a precipitate phase in an Ni-rich high-temperature NiTiHf shape memory alloy. Acta Mater. 2013, 61, 3335–46.

[208] Karaca HE, Saghaian SM, Ded G, Tobe H, Basaran B, Maier HJ, Noebe RD, Chumlyakov YI. Effects of nanoprecipitation on the shape memory and material properties of an Ni-rich NiTiHf high temperature shape memory alloy. Acta Mater. 2013, 61(19), 7422–31.

[209] Stebner AP, Biegelow GB, Yang J, Shukla DP, Saghaian SM, Rogers R, Garg A, Karaca HE, Chumlyakov Y, Bhattacharya KB, Noebe RD. Transformation strains and temperatures of a nickel–titanium–hafnium high temperature shape memory alloy. Acta Mater. 2014, 76, 40–53.

[210] Saghaian SM, Karaca HE, Tobe H, Souri M, Noebe RD, Chumlyakov YI. Effects of aging on the shape memory behavior of Ni-rich $Ni_{50.3}Ti_{29.7}Hf_{20}$ single crystals. Acta Mater. 2015, 87, 128–41.

[211] Saghaian SM, Karaca HE, Tobe H, Turabi AS, Saedi S, Saghaian SE, Chumlyakov YI, Noebe RD. High strength NiTiHf shape memory alloys with tailorable properties. Acta Mater. 2017, 134, 211–20.

[212] Manca A, Shelyakov AV, Airoldi G. Ageing in parent phase and martensite stabilization in a Ni50Ti30Hf20 alloy. Mater. Trans. 2003, 44, 1219–24.

[213] Babacan N, Bilal M, Hayrettin C, Liu J, Benafan O, Karamn I. Effects of cold and warm rolling on the shape memory response of $Ni_{50}Ti_{30}Hf_{20}$ high-temperature shape memory alloy. Acta Mater. 2018, 157, 228–44.

[214] Patriarca L, Sehitoglu H, Panchenko Y, Chumlyakov YI. High-temperature functional behavior of single crystal $Ni_{51.2}Ti_{23.4}Hf_{25.4}$ shape memory alloy. Acta Mater. 2016, 106, 333–43.

[215] Zarinejad M, Liu Y, White TJ. The crystal chemistry of martensite in NiTiHf shape memory alloys. Intermetallics 2008, 16, 876–83.

[216] Karakoc O, Hayrettin C, Bass M, Wang SJ, Canadinc D, Mabe JH, Lagoudas DC, Karaman I. Effects of upper cycle temperature on the actuation fatigue response of NiTiHf high temperature shape memory alloys. Acta Mater. 2017, 138, 185–97.

[217] Karakoc O, Hayrettin C, Canadinc D, Karaman I. Role of applied stress level on the actuation fatigue behavior of NiTiHf high temperature shape memory alloys. Acta Mater. 2018, 153, 156–68.

[218]   Gong C, Wang Y, Yang D. Phase transformation and second phases in ternary Ni-Ti-Ta shape memory alloys. Mater. Chem. Phys. 2006, 96, 183–7.

[219]   Zhao T, Li Y, Gao Y, Xiang Y, Chen H, Zhang T. Hemocompatibility investigation of the NiTi alloy implanted with tantalum. J. Mater. Sci. Mater. Med. 2011, 22, 2311–8.

[220]   Pulletikurthi C, Munroe N, Stewart D, Haider W, Amruthaluri S, Rokicki R, Dugrot M, Ramaswamy S. Utility of magneto-electropolished ternary nitinol alloys for blood contacting applications. J. Biomed. Mater. Res. Part B: Appl. Biomat. 2015, 103B, 1366–74.

[221]   Li Y, Wei SB, Cheng XQ, Zhang T, Cheng G. Corrosion behavior and surface characterization of tantalum implanted TiNi alloy. Surf. Coat. Technol. 2008, 202, 3017–22.

[222]   Buenconsejo PJS, Zarnetta R, Ludwig A. The effects of grain size on the phase transformation properties of annealed (Ti/Ni/W) shape memory alloy multilayers. Scr. Mater. 2011, 64, 1047–50.

[223]   Ahadi A, Kalidindi AR, Sakurai J, Matsushita Y, Tsuchiya K, Schuh CA. The role of W on the thermal stability of nanocrystalline NiTiW$_x$ thin films. Acta Mater. 2018, 142, 181–92.

[224]   El-Bagoury N. Precipitation of second phases in aged Ni rich NiTiRe shape memory alloy. Mater. High Temp. 2015, 32, 390–8.

[225]   Yang F, Noebe RD, Mills MJ. Precipitates in a near-equiatomic (Ni + Pt)-rich TiNiPt alloy. Scr. Mater., 2013, 69, 713–5.

[226]   Gau Y, Zhou N, Yang F, Cui Y, Kovarik L, Hatcher N, Noebe R, Mills MJ, Wang Y. P-phase precipitation and its effect on martensitic transformation in (Ni,Pt)Ti shape memory alloys. Acta Mater., 2012, 60, 1514–27.

[227]   Rios O, Noebe R, Biles T, Garg A, Palczer A, Scheiman D, Seifert HJ, Kaufman M. Characterization of ternary NiTiPt high-temperature shape memory alloys. NASA Technical Reports Server (NTRS), 2005-01-01.

[228]   Buchheit TE, Susan DF, Massad JE, McElhanon JR, Noebe RD. Mechanical and functional behavior of high-temperature Ni-Ti-Pt shape memory alloys. Metall. Mater. Trans. A 2016, 47, 1587–99.

[229]   Noebe R, Gaydosh D, Padula S, Garg A, Biles T, Nathal M. Properties and potential of two (Ni,Pt)Ti alloys for use as high-temperature actuator materials. *Proc. SPIE* 2005, 5761, 364–75.

[230]   Kovarik L, Yang F, Garg A, Diecrks D, Kaufman M, Noebe RD, Mills MJ. Structural analysis of a new precipitate phase in high-temperature TiNiPt shape memory alloys. Acta Mater. 2010, 58, 4660–73.

[231]   Smialek JL, Garg A, Rogers RB, Noebe RD. Oxide scales formed on NiTi and NiPtTi shape memory alloys. Metall. Mater. Trans. A 2012, 43, 2325–41.

[232]   Smialek JL, Humphrey DL, Noebe RD. Comparative oxidation kinetics of a NiPtTi high temperature shape memory alloy. Oxid. Met. 2010, 74, 125–44.

[233]   Wu SK, Wayman CM. Martensitic transformations and the shape-memory effect in Ti$_{50}$Ni$_{10}$Au$_{40}$ and Ti$_{50}$Au$_{50}$ alloys, Metallography 1987, 20, 359–76.

[234]   Kurita T, Matsumoto H, Sakamoto K, Tanji K, Abe H. Effect of aluminum addition on the transformation of NiTi alloy. J. Alloys Compd. 2005, 396, 193–6.

[235]   Ishii N, Matsumoto H. Intermediate phase on rapidly quenched Ni49Ti50Al1. J. Mater. Sci. Lett. 1999, 18, 1853–4.

[236]   Salvetr P, Školáková A, Hudrisier C, Novák P, Vojtěch D. Reactive sintering mechanism and phase formation in Ni-Ti-Al powder mixture during hating. Materials (Basel) 2018, 11, 689, 11 pages; doi: 10.3390/ma11050689.

[237]   Jung J, Ghosh G, Olson GB. A comparative study of precipitation behavior of Heusler phase (Ni2TiAl) from B2-TiNi in Ni-Ti-Al and Ni-Ti-Al-X (X = Hf, Pd, Pt, Zr) alloys. Acta Mater. 2003, 51, 6341–57.

[238]   Hsieh SF, Wu SK. A study on the nickel-rich ternary Ti–Ni–Al shape memory alloys. J. Mater.
         Sci. 1997, 32, 989–96.

[239]   Meng LJ, Li Y, Zhao XQ, Xu J, Xu HB. The mechanical properties of intermetallic Ni50–xTi50Alx
         alloys (x=6,7,8,9). Intermetallics, 2007, 15, 814–8.

[240]   Ibrahim KM, El Bagoury N, Fouad Y. Microstructure and martensitic transformation of cast
         TiNiSi shape memory alloys. J. Alloys Compd. 2011, 509, 3913–6.

[241]   Salvetr P. Školáková A, Kopeček J, Novák P. Properties of Ni-Ti-X shape memory alloys
         produced by arc re-melting. Acta Metall. 2017, 23, 141–6.

[242]   Li X, Elkedim D, Nowak O, Jurczyk M. Characterization and first principle study of ball milled
         Ti–Ni with Mg doping as hydrogen storage alloy. Int. J. Hydrogen Ener. 2014, 39, 9735–43.

[243]   Wang Y, Yang X, Zhao C, Li Y, Mi H, Zhang P. Improving the structure stabilization of red
         phosphorus anodes via the shape memory effect of a Ni-Ti alloy for high-performance
         sodium ion batteries. Chem. Commun. (Camb.) 2019, 55, 4659–62.

[244]   Karaca HE, Acar E, Ded GS, BasarBigelow G, Chumlyakov YI. Shape memory behavior of high
         strength NiTiHfPd polycrystalline alloys. Acta Mater. 2013, 61(13), 5036–49.

[245]   Prasad RVS, Park CH, Kim S-W, Hong JK, Yeom J-T. Microstructure and phase transformation
         behavior of a new high temperature NiTiHf-Ta shape memory alloy with excellent
         formability. J. Alloys Compd. 2017, 697, 55–61.

[246]   Li X, Liang X, Liu D, Chen R, Huang F, Wang R, Rettenmayr M, Su Y, Guo J, Fu H. Design of
         $(Nb, Mo)_{40}Ti_{30}Ni_{30}$ alloy membranes for combined enhancement of hydrogen permeability
         and embrittlement resistance. Sci. Rep. 2017, 7; doi: 10.1038/s41598-017-00335-0.

[247]   Atli KC, Karaman I, Noebe RD, Garg A, Chumlyakov YI, Kireeva IV. Shape memory
         characteristics of $Ti_{49.5}Ni_{25}Pd_{25}Sc_{0.5}$ high-temperature shape memory alloy after severe
         plastic deformation. Acta Mater. 2011, 59(12), 4747–60.

[248]   Stebner A, Padula II SA, Noebe RD, Quinn DD. Characterization of
         $Ni_{19.5}Ti_{50.5}Pd_{25}Pt_5$ high-temperature shape memory alloy springs and their potential
         applications in aeronautics. Proc. SPIE 6928, Active Passive Smart Struct. Int. Syst. 2008,
         69280X (18 April 2008); https://doi.org/10.1117/12.775805.

[249]   Zheng YF, Wang QY, Li L. The electrochemical behavior and surface analysis of
         Ti49.6Ni45.1Cu5Cr0.3 alloy for orthodontic usage. J. Biomed. Mater. Res. B Appl.
         Biomater. 2008, 86, 335–40.

[250]   Lethabane ML, Olubambi PA, Chikwanda HK. Corrosion behaviour of sintered Ti–Ni–Cu–Nb
         in 0.9% NaCl environment. J. Mater. Res. Technol. 2015, 4, 367–76.

[251]   Kim KB, Yi S, Choi-Yim H, Das J, Xu W, Johnson WL, Eckert J. Effect of Cu on local
         amorphization in bulk Ni–Ti–Zr–Si alloys during solidification. Acta Mater. 2006, 54,
         3141–50.

[252]   Coughlin JP, Williams JJ, Chawla N. Mechanical behavior of NiTi shape memory alloy fiber
         reinforced Sn matrix "smart" composites. J. Mater. Sci. 2009, 44, 700–7.

[253]   Yim HC, Conner RD, Johnson WL. In situ composite formation in the Ni–(Cu)–Ti–Zr–Si
         system. Scr. Mater. 2005, 53, 1467–70.

[254]   Samal S, Biswas K. Novel high-strength NiCuCoTiTa alloy with plasticity. J. Nanopart. Res.
         2013, 15, 1–11.

[255]   Xue D, Balachandran PV, Hogden J, Theiler J, Xue D, Lookman T. Accelerated search for
         materials with targeted properties by adaptive design. Nat. Commun. 2016, 7, 11241; doi:
         10.1038/ncomms11241.

[256]   Czeppe T, Ochin P, Sypień A, Major L. Microstructure and mechanical properties of the
         NiNbZrTiAl amorphous alloys with 10 and 25 at.% Nb content. J. Microsc. 2010, 237, 320–4.

[257]   Oshida Y. Surface Engineering and Technology for Biomedical Implants. Momentum Press,
         New York, NY, USA, 2014.

[258] Shariat BS, Meng Q, Mahmud AS, Wu Z, Bakhtiari R, Zhang J, Motazedian F, Yang H, Rio G, Nam T-H, Liu Y. Functionally graded shape memory alloys: design, fabrication and experimental evaluation. Mater. Des. 2017, 124, 225–37.

[259] Shariat BS, Liu Y, Meng Q, Rio G. Analytical modelling of functionally graded NiTi shape memory alloy plates under tensile loading and recovery of deformation upon heating. Acta Mater. 2013, 61, 3411–21.

[260] Shariat BS, Meng Q, Mahmud AS, Wu Z, Bakhtiari R, Zhang J, Motazedian F, Yang H, Rio G, Nam TH, Liu Y. Experiments on deformation behaviour of functionally graded NiTi structures. Data Brief 2017, 13, 562–8.

[261] Oh IH, Nomura N, Masahashi N, Hanada S. Mechanical properties of porous titanium compacts prepared by powder sintering. Scr. Mater. 2003, 49, 1197–1202.

[262] Assad M, Likibi F, Jarzem P, Leroux MA, Coillard C, Rivard Ch-H. Porous nitinol vs. titanium intervertebral fusion implants: Computer tomography, radiological and histological study of osseointegration capacity. Matialwissenschaft u. Werkstofftechnik (Mater. Sci. Eng. Technol.) 2004, 35; https://doi.org/10.1002/mawe.200400739.

[263] Korne L, Mentz J, Bram M, Buchkremer H, Stover D, Wagner M, Eggeler G, Christ D, Reese S, Bogdanski D, Koller M, Esenwein SA, Muhr G, Prymak O, Epple M. The potential of powder metallurgy for the fabrication of biomaterials on the basis of nickel-titanium: a case study with a staple showing shape memory behaviour. Adv. Eng. Mater. 2005, 7, 613–9.

[264] Krishna BV, Bose S, Bandyopadhyay A. Low stiffness porous Ti structures for load-bearing implants. Acta Biomater. 2007, 3, 997–1006.

[265] Zhang YP, Li DS, Zhang XP. Gradient porosity and large pore size NiTi shape memory alloys. Scr. Mater. 2007, 57, 1020–3.

[266] Bandyopadhyay A, Krishna BV, Xue W, Bose S. Application of Laser Engineered Net Shaping (LENS) to manufacture porous and functionally graded structures for load bearing implants. J. Mater. Sci.: Mater. Med. 2009, 20, S29-S34.

[267] Narayan RJ, Hobbs LW, in C, Rabiei A. The use of functionally gradient materials in medicine. JOM 2006, 58, 52–6.

[268] Pompe W, Worch H, Epple M, Friess W, Gelinsky M, Greil P, Hempel U, Scharnweber D, Schulte K. Functionally graded materials for biomedical applications. Mater. Sci. Eng. A 2003, 362, 40–60.

[269] Resnina N, Belyaev S, Shelyakov A, Ubyivovk E. Violation of the sequence of martensite crystals formation on cooling and their shrinking on heating during B2 $\leftrightarrow$ B19′ martensitic transformation in $Ti_{40.7}Hf_{9.5}Ni_{44.8}Cu_5$ shape-memory alloy. Phase Transitions 2017, 90, 1–20.

[270] Shariat BS, Liu Y, Meng Q, Rio G. Analytical modelling of functionally graded NiTi shape memory alloy plates under tensile loading and recovery of deformation upon heating. Acta Mater. 2013, 61, 3411–21.

[271] Bogdanski D, Köller M, Müller D, Muhr G, Bram M, Buchkremer HP, Stöver D, Choi J, Epple M. Easy assessment of the biocompatibility of Ni-Ti alloys by in vitro cell culture experiments on a functionally graded NiNiTi-Ti material. Biomaterials 2002, 23, 4549–55.

Chapter 4
Martensitic phase transformation and its related phenomena

4.1 Temperature-related transformation and stress-induced transformation

When we discuss basic features of both shape-memory effect (SME) or shape recovery effect and superelasticity (SE) or pseudoelasticity (PE) phenomena, we need to know the principal mechanism controlling these phenomena; that is, the martensitic transformation.

The term "martensitic transformation" can be defined as a phase transformation within a solid state, caused by shear deformation without the long-range diffusion (diffusionless) transition of constituent atoms. Such a diffusionless transformation does not change the composition of the parent phase, but rather only the crystal structure, so that it does not require long-range atomic movement. In such transformations, the new phase is formed through slight atomic shuffles of generally less than an atomic diameter, and atoms are cooperatively rearranged into a new, more stable crystal structure with the same chemical composition as the original parent phase. Because no atomic migration is necessary, the transformations usually progress in a time-independent fashion, with the speed of the interface between the two phases able to move at nearly the speed of sound. This type of transformation is referred to as an athermal transformation since it cannot progress at a constant temperature, but rather the amount of the new phase present depends only upon temperature, not time. On the other hand, diffusional transformations are often referred to as isothermal since they can progress with time at a constant temperature. The diffusional transformation takes place with no change in phase composition or number of phases present (e.g., melting, solidification of pure metal, allotropic transformations, and recrystallization.). Martensitic transformations are of the displacive variety, with the term martensite referring specifically to the lower temperature phase, and the term austenite (or parent phase) referring to the higher temperature phase from which martensite (resulting from any athermal, diffusionless phase transformation) is formed. The martensitic transformation in steel represents the most economically significant example of this category of phase transformations but an increasing number of alternatives, such as shape-memory alloys (SMAs), are becoming more important as well. And the martensitic transformation is reversible, so that when the martensite which was formed upon cooling is reheated, the parent phase is reformed. Both diffusional and diffusionless transformations take place in most commercially available Nitinol alloys, with the two processes competing to find the lowest energy state. In

https://doi.org/10.1515/9783110666113-004

such cases, quenching from the higher temperature phase usually suppresses diffusional transformations and decides the contest in favor of martensite.

As already noticed, the two terms "transition" and "transformation" have been used in literatures without any clear explanation for the term choice, although there is a profound discussion to differentiate the term "transition," which is based on the first-order and second-order phase transitions and "transformation." Simply here we decide to adopt the suggestion that the phrase "phase transformation" implies a structural change, while the phrase "phase transition" implies transformation without a structural change.

In the following, there are four characteristic temperatures involved in both SME and SE; they are A_S (austenitic transformation starting temperature), A_F (austenitic transformation finishing temperature), M_S (martensitic transformation starting temperature), and M_F (martensitic transformation finishing temperature). Referring to Figure 4.1, these four important temperatures are illustrated [1]. Due to unclear appearance of these temperatures, it is normally determined as illustrated. Among these four temperatures, particularly M_S and A_F are important and since these two temperatures do not coincident to each other, creating the temperature hysteresis, which might control the stability and performance efficacy of NiTi alloys in use.

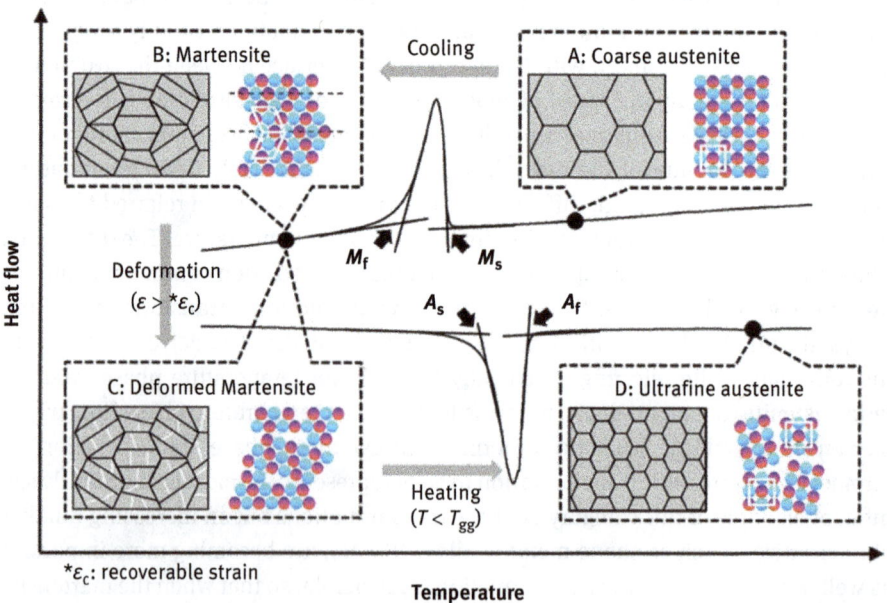

Figure 4.1: Determination of four important temperatures, characterizing austenitic and martensitic phase transformations [1].

A clear differentiation between SE and SME is illustrated in Figure 4.2 [2]. SE shows the shape recovery without heating procedure, while SME exhibits its shape recovery by heating process causing the reverse transformation of martensitic phase to the parent austenitic phase. Martensite's crystal structure (known as a monoclinic, or B19′ structure) has the unique ability to undergo limited deformation in some ways without breaking atomic bonds, known as the twinning, which consists of the rearrangement of atomic planes without causing slip or permanent deformation allowing to undergo about 6–8% strain in this manner. When martensite is reverted to austenite by heating, the original austenitic structure is restored, regardless of whether the martensite phase was deformed; hence, the name "shape memory" refers to the fact that the shape of the high-temperature austenite phase is "remembered," even though the alloy is severely deformed at a lower temperature.

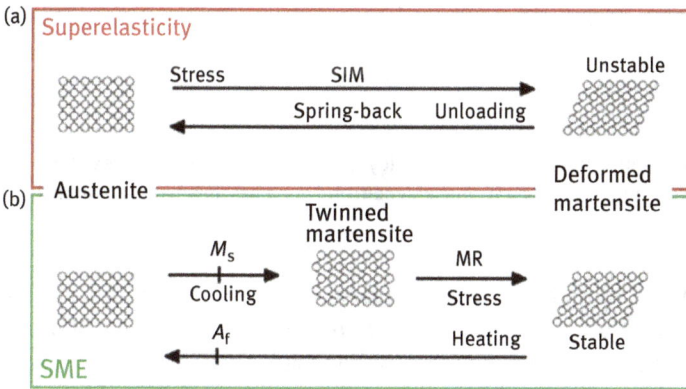

Figure 4.2: Schematic illustration of transformations generating strain-induced martensite or heat-assisted martensite [2].

The dual-phase microstructure of NiTi consists of two phases: the austenite phase (A), stable in high-energy levels with a body-centered cubic structure having low strain, and the martensite phase (M), stable in low-energy levels having transformation strain. Both SE and SME properties occur as a result of the austenite to martensitic transformation, which can be induced by stress or temperature. Hence, it can be more convenient to see stress–strain–temperature diagram (see Figure 4.3), where M_F is a martensite finish temperature upon cooling, A_S is a temperature at which the martensite (or R-phase) to austenite transformation begins upon heating; while A_F is the same completed transformation is , and M_D (martensite deformation temperature) refers to the highest temperature at which martensite will form from austenite phase in response to an applied stress and at the temperature above M_D (hence, M_D indicates the upper limit temperature at which austenite is subjected to stress-induced martensitic transformation); the NiTi SME alloy will not exhibit SE it will rather exhibit a typical elastic–plastic behavior when loaded, as indicated with black line in the figure. The temperature M_S (a temperature at which martensitic transformation starts, although it is not clearly determined) can be found in $M_F < M_S < A_S$. Moreover, A is an austenite phase, M is a martensite phase, and DM refers to deformed martensite and TM represents the twinned martensite (TM).

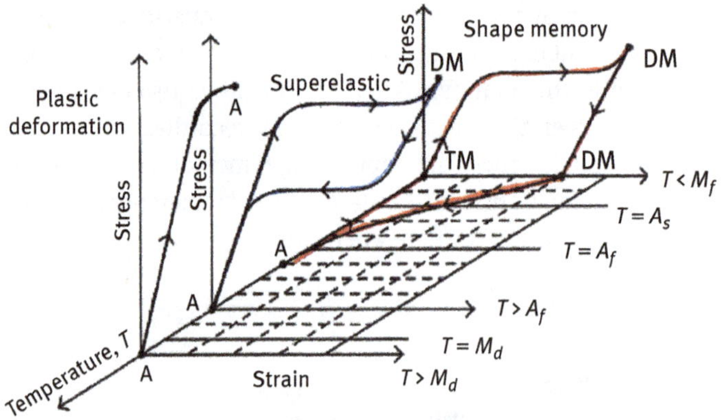

Figure 4.3: Stress–strain–temperature diagram of NiTi alloy [3]. Red-colored loop indicates SME phenomenon, blue-colored loop represents SE phenomenon, and the black line shows an ordinal stress–strain curve.

When the temperature is below, NiTi is in its TM phase. If a stress above a critical level is applied, it undergoes a phase transformation to detwinned martensite (DM) and stays in this phase even upon removal of the applied load. The material can regain its initial shape only when it is heated to a temperature above. Heating the NiTi alloy above not only results in shape recovery but also leads to the formation of the austenite phase. Through subsequent cooling, the SMA transforms to its initial TM phase with little or no residual deformation. This phenomenon is referred to as SME; that is, the ability of an SME-showing material (SMA) to recover its original shape through thermal cycling. On the other hand, when the temperature is above, the SMA is in its austenite phase. When a sufficiently high stress is applied, the SMA transforms into DM. Upon removal of the load, a reverse transformation to the austenite phase takes place, with the material undergoing shape recovery and exhibiting a noticeable hysteresis loop. This phenomenon is referred to as the SE effect, that is, the recovery of large strain as a result of the stress-induced phase transformations under a constant temperature. If the temperature is below Af but above Mf, there will only be a partial shape recovery after the load is removed, and the remaining strain could only be recovered via a heating and cooling cycle. Finally, when the temperature is above, the SMA is stable in its austenite phase and will behave like an ordinary metal that undergoes plastic deformation when stressed beyond a certain limiting value [3, 4].

Summarizing the above, in the stress–strain–temperature diagram, there are three distinct zones: (1) $T < M_S$: stable martensite being responsible to thermal memory effect; (2) $M_D < T < M_S$: metastable austenite being responsible to mechanical memory effect (or SE); and (3) $T > M_D$: stable austenite showing no related effects [5].

An important application of SMAs is a thermal actuator for which, in general, a small thermal hysteresis is required and although most martensitic transformation are associated with a temperature hysteresis in excess of 10 °C, the martensite-like phase possesses a temperature hysteresis as small as 1.5 °C, the so-called R-phase transition. The R-phase transition appears upon cooling, prior to the martensitic transformation in NiTi alloys [6]. When a Ni-rich NiTi (Ni: 51 mol%) which was solution-treated at 1,000 °C was cooled, the austenite (B2) → martensite (B19′) transformation takes place. If the alloy was annealed (after being subjected to cold working) at 400 °C (which is lower than recrystallization temperature), Ti_3Ni_4 precipitate and rhombohedral R-phase appears, while Ti_3Ni_4 was not precipitated with the equiatomic NiTi under the same heat treatment, and only R-phase was formed. The appearance of R-phase depends on the R-phase transformation temperature (T_R); by increasing TR, the transformation will be B2→R→M; while by decreasing T_R, the one-stage transformation will take place, B2→M. It is found that, for appearance of R-phase, the martensitic transformation temperature is needed to be lower, so that (i) increasing Ni content (increment of 100 K per 1 mol%Ni), (ii) Ti_3Ni_4 precipitate (by aging treatment), (iii) high dislocation density (by annealing after cold working), and (iv) addition of ternary alloying elements like Fe or Al [6–19]. Among these three phases, several transformation sequences are possible, stemming from various combinations of the A ↔ R, A ↔ M, and R ↔ M transformations, which all are martensitic in nature [10, 20, 21]. Figure 4.4 [22] compares experimental results obtained by the differential scanning calorimetry (DSC) and free-recovery heat flow in which an exothermic reaction indicates martensitic transformation while the endothermic one represents the austenitic transformation. Although the free recovery curve shows only the deflection during the heating, DSC indicates the appearance of R-phase upon cooling from the parent austenite phase. The stabilization of austenitic, rhombohedral, and martensitic phases is shown to critically depend on the temperatures of heat treatment by the analysis of temperature dependence of electrical resistivity in heating and cooling parts of the cycle. Using 40% cold-worked nitinol, Uchil et al. [23] investigated the phase transformations as a function of heat treatment using electrical resistivity variation with temperature. It was found that the R-phase has been found to form continuously with increasing heat treatment temperature starting from room temperature and to suddenly disappear beyond heat treatment at 410 °C.

In near-equiatomic NiTi alloys, the thermoelastic transformation between austenite and the R-phase shows unique properties, which make the R-phase transformation very promising for applications. Under certain conditions, NiTi SME alloy can exhibit two distinct transformations: (i) austenite → R-phase which has a transformation temperature difference ranging from 0 to 70 °C, a recovery strain of about 1%, and a temperature hysteresis between 2 and 3 °C, while (ii) austenite → martensite possesses a transformation temperature difference ranging from –10 to 100 C, a recovery strain of

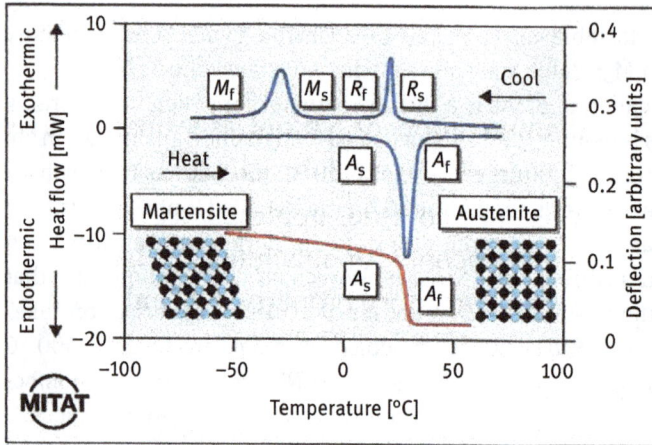

Figure 4.4: Comparison of results obtained from the differential scanning calorimetry and free recovery methods, depicting an appearance of R-phase on cooling from the parent austenite phase [22].

6–8%, and a temperature hysteresis from 20 to 40 °C. The R-phase is an intermediate phase between austenite and martensite; it has a rhombohedral structure that can be formed during forward transformation from martensite to austenite on heating and during reverse transformation from austenite to martensite on cooling [24].

In order to improve and optimize the performance of NiTi alloys, various thermomechanical treatments can be applied, for example, postdeformation annealing, thermal/mechanical cycling, and aging [25–29]. During and after these treatments, an intermediate rhombohedral phase (R-phase) between B2 and B19′ phase has been frequently observed, resulting in changing the transformation path from B2 → B19′ into B2 → R → B19′ [5, 25, 30–32]. As the B2 ↔ B19′ transformation, the B2 ↔ R also shows SME and PE [5, 30, 34]. Moreover, compared with the B19′ transformation, the R-phase transformation shows unique properties (e.g., narrow hysteresis and high fatigue life.), which attract widespread interest for practical applications. The R-phase transformation was characterized as a premartensitic transformation [35, 36]. Furthermore, it was considered as a second-order transformation [36]. Now it is clear that the B2 → R is an independent martensitic transformation created by the rhombohedral distortion of the B2 lattice that occurs before the B19′ transformation [33, 36–39], and it is mainly considered as a first-order transformation [40]. However, the R-phase also shows second-order transformation behavior, as after the B2 → R transformation, and the rhombohedral angle of the R-phase gradually decreases with decreasing temperature or increasing stress [34, 41, 42]. Detailed crystallography and basic mechanism of R-phase transformation are discussed in literatures [37, 43, 44–48].

NiTi SMAs are known to demonstrate three possible transformation paths between B2 and B19′ phases: B2→R→B19′, B2–B19′, and B2→B19→B19′, depending on their composition and thermo-mechanical treatment. Kustov et al. [48] studied the isothermal kinetics of accumulation of martensite/austenite for all types of martensitic transformations in NiTi and X-NiTi (X: Fe or Cu) has been studied by means of resistance measurements during interruption of cooling/heating scans. It was shown that (i) all transformations to the B19′ phase (B2→B19′, R→B19′, and B19→B19′) demonstrate a substantial isothermal accumulation of martensite during isothermal dwelling between the martensitic transformation start and finish temperatures, (ii) the reverse transformations B19′→R and B19′→B19 are also classified as isothermal, (iii) the isothermal accumulation of austenite detected during the reverse B19′–B2 transformation is much less intense, at least partially due to the low sensitivity of resistance to the martensite fraction variation during the reverse transformation and remains comparable with the resolution of the experimental set-up, and (iv) the transformations between the B2 and R as well as between the B2 and B19 phases are athermal concluding that the isothermal transformations possess a much broader hysteresis and transformation range compared with athermal ones. It was further mentioned that (i) the fact that the hysteresis of the transformation is influenced by the friction forces acting on involved interfaces and observation of the isothermal effects during reverse martensitic and intermartensitic transformations strongly support the interpretation of the observed isothermal effects in NiTi due to the diffusionless but thermally activated motion of interfaces during transformation, and (ii) the difference between the transformation to B19′ martensite (isothermal) and all others (athermal) is attributed to a distinction in strain accommodation [48].

Summarizing what we have been discussing about the martensitic transformation (for both one stage and two stage): (1) solution treatment at elevated temperatures induces single-stage A ↔ M transformation, and critical temperatures decrease with increasing Ni content at above 50.0 at%, (2) aging of Ni-rich NiTi alloys leads to the progressive precipitation of Ti_3Ni_4, Ti_2Ni_3, and $TiNi_3$ and precipitates in order with increasing aging time, temperature, and Ni content [49]. Ti_3Ni_4 is coherent with the matrix [50–54] and is most influential in affecting the transformation behavior and mechanical properties of the alloys [56], (3) aging at 400–450 °C of Ti–$Ni_{50.8}$ (at%) for 1.8–3.6 ks appears to be the optimum conditions for obtaining PE [49], (4) aging at low temperatures (typically below 460 °C) results in multiple stage transformation sequences, involving double-stage A → R transformations or double-stage R → M transformations [55], and (5) aging under the influence of bias stresses is able to create aligned Ti_3Ni_4 precipitates [54, 56], which in turn creates anisotropic internal stress fields and leads to the occurrence of an "all-round SME" [57, 58].

During and after heat treatments, an intermediate rhombohedral phase (R-phase) between A-phase (B2) and M-phase (B19′) has been frequently observed, resulting in changing the transformation path from A→M to A→ R→ M [25, 32, 50, 59]. As the A ↔ M transformation, the A ↔ R also shows SME and PE [5, 59, 60–62]. Moreover, compared with the M transformation, the R-phase transformation shows unique properties (e.g., narrow hysteresis and high fatigue life.), which attract widespread interest for practical applications. The R-phase transformation was characterized as a premartensitic transformation [35, 63]. It was furthermore considered as a second-order transformation. [36, 64]. Now it is clear that the A→ R is an independent martensitic transformation created by the rhombohedral distortion of the A lattice that occurs before the M transformation [37, 39, 60, 65, 66], and it is mainly considered as a first-order transformation [40, 65]. However, the R-phase also shows second-order transformation behavior, as after the A→ R transformation, the rhombohedral angle of the R-phase gradually decreases with decreasing temperature or increasing stress [41, 42, 61, 63]. The recovery stress associated with the R-phase transformation in TiNi SMA was investigated experimentally concerning constant residual strain [67]. The results are summarized as follows: (i) the recovery stress in the heating process is larger than the R-phase transformation stress in the loading process at low temperature, (ii) in the heating process, the recovery stress occurs at about 47 °C and takes a maximum value at about 62 °C, and the recovery stress increases along the reverse transformation line on the stress–temperature plane, (iii) the higher the shape-memory processing temperature, the larger the recovery stress, and (iv) the larger the maximum strain at low temperature, the larger the recovery stress.

4.2 Stress-induced transformation

There is another type of phase transformation, involved in NiTi-based alloy, that is, stress-induced martensitic transformation. For example, when unstable austenitic stainless steel (such as 18Cr–8Ni stainless steel) is subjected to severe plastic deformation, such deformed portion exhibits the martensitic transformation which can be easily recognized by examining with a magnet because the paramagnetic unstable austenite transformed to stress-assisted or strain-induced martensite is now turned to be ferromagnetism.

Shaw et al. [68] demonstrated an experimental methodology for simultaneous full-field monitoring of the deformation and thermal changes in NiTi during mechanically unstable regimes associated with the pseudoelastic material response. It is mentioned that (i) during loading, nucleation of martensite in an austenitic region is a distinct event requiring a higher stress than the stress required subsequently to continue the transformation, while (ii) the nucleation stress of austenite in a martensitic region during unloading is lower than the stress required to continue the transformation. It is further described that (iii) this distinction between the nucleation and propagation stresses, coupled with the local temperature change caused by the latent heats of the two transformations govern the number of nuclei of a new phase, and (iv) coexisting transition fronts tend to propagate at the same speed which is inversely proportional to the number of fronts in the specimen [68]. The comparison of stress–strain curves between stress-free and stressed conditions is clearly illustrated [69]. Figure 4.5(a) shows phase transformation as a function of amount of martensitic phase under the

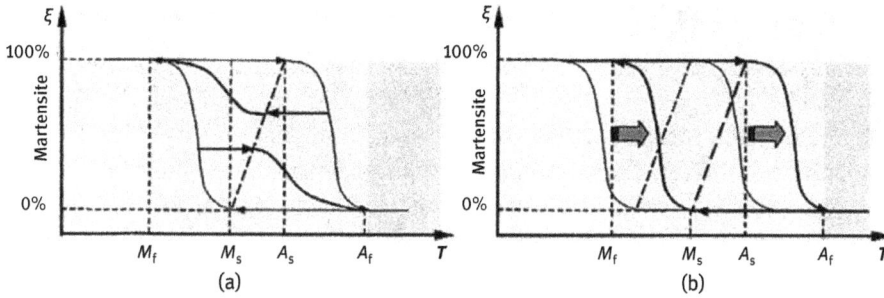

Figure 4.5: Transformation temperatures versus amount of martensite: (a) under the stress-free condition, and (b) under influence of externally applied stress [69].

stress-free condition (in other words, unconstrained condition). Under the stress-free situation, the strain is a function of change in phase composition during forward and reverse transformations. The determination of each temperature depends on the relevant phenomena. Because the energy required for initiating transformations is directly related to the relative involved mass fractions of austenite and martensite, there are possibilities that forward transformation could start at much higher temperatures than M_S and reverse transformation could start at temperatures much lower than A_S, respectively [69]. If externally applied stress is involved, the situation would differ from this. As shown in Figure 4.5(b), it is indicated that, under the influence of externally applied stress, the temperature hysteresis loop moves to the right; in other words, the higher the applied stress the higher the transition temperatures are.

As a result of a martensite to austenite transformation process, the intermetallic compounds are precipitated from NiTi matrix, which are believed to responsible to SME and SE phenomena. It is also known that the retention of these intermetallic compounds is a function of stresses, and Ni-rich precipitates can generate internal stress which can be relieved by an appropriate annealing heat treatment. A lengthy history of research on effects of applied stress on transformation behaviors in NiTi alloys can be recognized [70–78]. Mukherjee et al. [71] studied the stability of the β-phase TiNi alloy with respect to an applied constant stress in the vicinity of diffusionless transformation temperature. It was reported that (i) the temperature (at which the hardness minimum occurs with the M_S temperature) decreases with increasing number of transformation cycles, (ii) the hardness of the β-phase at a fixed temperature above the A_F increases with increasing number of transformation cycles, and (iii) a remarkable linear relationship between the hardness of the reverted β-phase and the diffusionless transformation temperature was recognized. Hedayat et al. [75], using the ultra-low-temperature isotropic carbon which was vapor deposited on a near-equiatomic $Ti_{49.9}Ni_{50.1}$ (at%) alloy, investigated the effects of surface constraint on the phase transformation. It was found that (i) the adhering carbon film, and carbide layer formed after annealing, introduced a surface constraint, and (ii) a marked decrease in the A_S and M_F temperatures of such a surface-constrained alloy during phase transformation was noted. Orgeas et al. [76] investigated the thermo-mechanical behavior of stress-induced martensitic transformation in an equiatomic NiTi alloy with respect to different deformation modes including uniaxial tension, compression, and shear of plate specimens at different temperatures above the M_S temperature. It was found that loading conditions have significant influences on the deformation behavior of the alloy; in particular, deformation behavior was observed to be asymmetric in tension and in compression. Lopez et al. [77] conducted anneal heat treatment Ti-rich NiTi52.0 (at%) alloy which was cold drawn to 30% at 900 °C for 1 h, followed by the water-quenching. It was found that (i) no evidence of the premartensitic R transformation was found during either the forward or the reverse transformation, and (ii) the x-ray diffraction indicated the presence of B2 phase when the alloy was plastically deformed in compression suggesting that the plastic deformation induces the reverse transformation to the B2 phase in highly stressed local regions. Sehitoglu et al. [78] studied the stress-induced transformation in single crystal NiTi alloys with respect to tension–compression asymmetry, crystal orientation effects on stress–strain response, the role of precipitates on transformation and slip resistance, and elastic moduli measurements on austenite, martensite, and the two-phase microstructure. It was reported that (i) the coherent precipitates lower the transformation stress at small strains while increasing the slip resistance of austenite and martensite domains at higher strains, and (ii) the martensite modulus is significantly higher than the austenite modulus for the single crystal orientations studied.

Research on stress influences on transformation phenomena is continued [79–83]. He et al. [79] investigated deformation dependence of the reverse martensitic transformation behavior in $Ni_{50.1}Ti_{46.9}Nb_3$ (at%) alloy, which was solution treated at 860 °C for about 30 min, followed by water-quenching. It was found that (i) the characteristic transformation temperatures of M_S, M_F, A_S, and A_F were −62, −117, −48, and 18 °C, respectively, (ii) with increasing deformation, the transformation peaks of stress-induced martensite become clear, and the transformation proceeds over a wider temperature range while those of thermal-induced martensite become diffused, and (iii) A_S does not obviously increase with increasing strain in this low niobium content alloy, while the deformed NiTi matrix phase adjacent to the stress-induced martensite will greatly increase the resistance of the thermal elastic martensite freely shrinkage upon heating freely, so the reverse transformation temperature interval of the Nb–NiTi alloy was widened. It is known that NiTi SMAs undergo relatively large recoverable inelastic deformations through a stress-induced martensitic phase transformation. Frick et al. [81] adopted the nanoindentation technique to study the fundamentals of stress-induced martensitic phase transformations in NiTi SMAs. It was reported that (i) an evidence of discrete forward and reverse stress-induced thermoelastic martensitic transformation in nanometer-scaled volumes of material was for the first time presented, (ii) stress-induced martensitic phase transformations nucleate at relatively low stresses at nanometer scales suggesting a fundamental departure from traditional size scale effects observed in metals deforming by dislocation plasticity, and (iii) the local material structure can be utilized to modify transformation behavior at nanometer scales yielding an insight into the nature of stress-induced martensitic phase transformations at small scales and providing an opportunity for the design of nanometer-sized NiTi actuators. Gollerthan et al. [82] investigated the crack loading and crack extension in pseudoelastic binary $TiNi_{50.7}$ (at%) SMA, using infrared thermography during in situ loading and unloading, which allowed for the observation of heat effects associated with the stress-induced transformation of martensite from B2 to B19′ during loading and the reverse transformation during unloading. It was mentioned that the crack growth occurs into a stress-induced martensitic microstructure, which immediately retransforms to austenite in the wake of the crack. Olbricht et al. [83] studied the phase transformations in binary ultra-fine-grained pseudoelastic NiTi wires using mechanical loading/unloading experiments, resistance measurements, DSC, thermal infrared imaging, and transmission electron microscopy. It was mentioned that (i) the stress–strain response of the R-phase can be isolated from the overall stress-strain data, (ii) the R-phase always forms prior to B19′ when good pseudoelastic properties are observed, and (iii) the stress-induced B2→R-phase transition occurs in a homogeneous manner, contrary to the localized character of the B2→R→B19′ transformations.

Khalil [84] illustrates stress–strain relationship under influences of concurrent transformation behavior, as shown in Figure 4.6. In Stage I, as ordinal other materials, NiTi behaves in a linear elastic manner until the initiation of martensite formation at a strain of about 1% followed by a plateau region (Stage II) characterizes the transformation process and usually continues to a strain of about 6% (at which the material is nearly fully martensite). The lattice structure responds elastically with continued deformation (Stage III) until a critical stress is reached where yielding occurs and plastic deformation continues until fracture (Stage IV) [84, 85]. There is a controversial issue that the phase transformation behavior under tension and compression differs from each other; under uniaxial tensile loading, the pseudoelastic SMAs exhibit a localized phase transformation, while the transformation proceeds homogeneously under compressive loading. Elibol et al. [86] investigated the martensitic transformation under tension, compression, and compression–shear conditions in pseudoelastic NiTi alloy and mentioned that (i) the deformation behavior of NiTi differs between tension and compression, (ii) the material exhibits a distinct mode of the formation and growth of localized martensite bands when under uniaxial tension, while under simple compression, the deformation proceeds homogeneously; however, (iii) a deformation inhomogeneity can also be observed during compression-shear loading.

Although many models for SMA available in the literature consider only fully reversible phase transformations (i.e., no permanent inelastic strains), which are proved by experiments to be sometimes not a fully realistic approximation, Auricchio et al. [87] demonstrated permanent inelasticity and degradation effects of repeated cycles, as shown in Figure 4.7. This figure depicts experimental results on a SMA NiTi

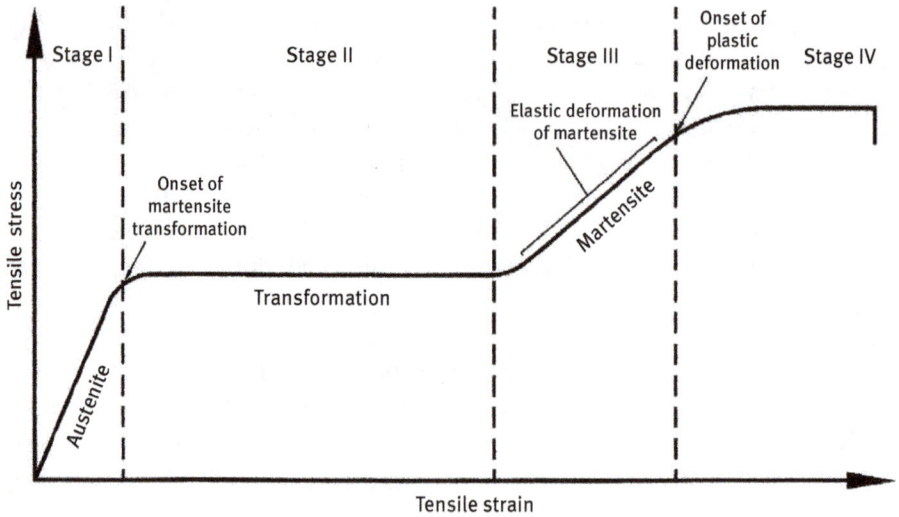

Figure 4.6: Simplified schematic of stress-induced martensite transformation [84].

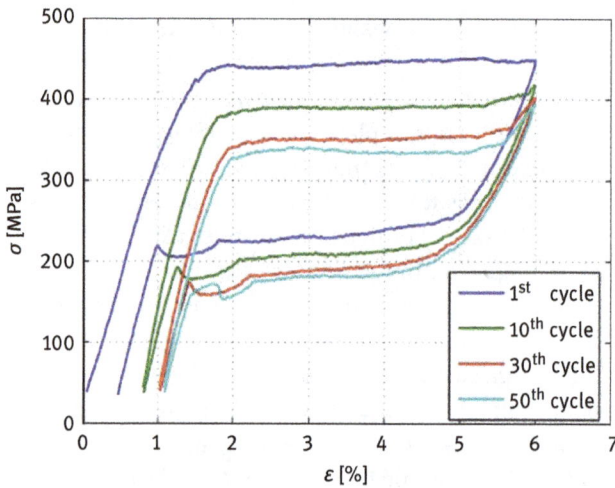

Figure 4.7: Cyclic stress–strain curves as a function of repeated number [87].

wire which was subjected to cyclic tension test up to 6% in strain. It is clearly shown that an increasing level of permanent inelasticity saturates on a stable value after a certain number of cycles, suggesting that degradation effects should be taken into account as well.

Addition to stress influences, effects such as thickness and size on martensitic transformation were reported. König et al. [88] studied the influence of film thickness on the B2→B19 martensitic transformation properties of nanoscale $Ti_{51}Ni_{38}Cu_{11}$ thin films with thicknesses ranging from 750 to 50 nm. It was mentioned that, with these thin films, (i) an unexpected behavior of the phase transformation temperatures was observed: A_F and O_S initially decrease with decreasing film thickness but increase sharply again for thicknesses <100 nm, (ii) the phase transformation temperatures range from 58 to 35 °C (A_F), (iii) substrate-attached Cu–NiTi thin films as thin as 50 nm show reversible B2–B19 phase transformations, and (iv) with decreasing film thickness a change in the tetragonality of the B19 martensite phase occurs. Mao et al. [89] prepared ultrathin NiTi miniature strips of 40–83 nm in thickness by means of focused ion beam milling from a polycrystalline NiTi SMA. The NiTi strips were subjected to tensile deformation inside a transmission electron microscope to observe the effect of thickness on the stress induced martensitic transformation behavior in the strips. It was found that (i) the transformation was completely suppressed in a strip of 40 nm in thickness whereas it was possible in thicker strips; in these strips, the stress induced martensitic transformation was found to commence sequentially in thicker strips first and then in thinner strips at higher strain (stress) levels, demonstrating the size effect, and (ii) this size effect is attributed to the effect of damaged surfaces suggesting that the observed "size effect" is not an intrinsic behavior of the martensitic transformation in NiTi but of extrinsic influences [89].

4.3 Multiple step transformation

The relationships of these three phases and their six possible transformations are presented in Figure 4.8 [90]. The six transformations are as follows: (1) austenite transforms into DM upon loading (A → DM), (2) DM undergoes reverse transformation into austenite after unloading at a high temperature or after heating without load or with a very low load (DM →A), (3) DM (variant k) transforms into another DM (variant l) after application of load (DM_k → DM_l), (4) TM transforms into DM upon loading (DM → TM), (5) austenite transforms into TM upon cooling (A → TM), and (6) TM transforms into austenite upon heating (TM → A).

So far we have been discussing two-stage R-phase transformation, particularly Ni-rich NiTi alloy [18, 91–93]. Kim et al. [18] investigated the effects of low temperature aging on the transformation behavior of a $Ni_{50.9}Ti$ (at%) alloy. It was reported that (i) aging in the temperature range of 200–300 °C induced a two-stage R-phase transformation, which was followed by a single-stage martensitic transformation at a lower temperature, (ii) it is identified that the first R-phase transformation on cooling was associated with the formation of the precipitates whereas the second R-phase transformation at a lower temperature was from the matrix away from precipitates, and (iii) the martensitic transformation was associated with the second R-phase transformation at the lower temperature, that is, it was an incomplete transformation from the regions away from precipitates. It is generally believed that there are three ways for NiTi-based alloy to exhibit the phase transform between

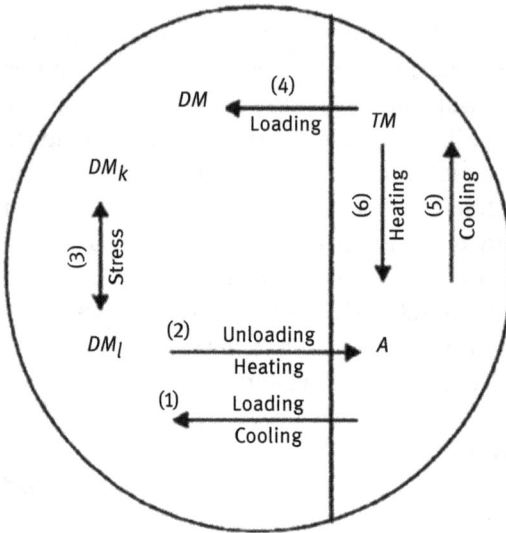

Figure 4.8: Three phases and six transformations: austenite, twinned martensite, and detwinned martensite; the subscripts "*l*" and "*k*" are used for the orientation variants of martensite [90].

the austenite and martensite phases: (1) direct transformation, with no evidence of R-phase during the forward or reverse transformation (cooling or heating), occurs in Ti-rich NiTi alloys and fully annealed conditions, (2) the symmetric R-phase transformation occurs when the R-phase intervenes between austenite and martensite on both heating and cooling, and (3) the asymmetric R-phase transformation is by far the more common transformational route, where the R-phase occurs during cooling, but not upon heating, due to the large hysteresis of the austenite–martensite transformation – by the time one reaches a sufficiently high temperature to revert martensite, the R-phase is no longer more stable than austenite, and thus the martensite reverts directly to austenite.

Multiple step martensitic transformations in Ni-rich NiTi SMAs have so far been rationalized, based on dislocation stress fields, coherency stress fields around Ni_4Ti_3 precipitates, and evolving Ni concentrations between precipitates during aging. Dlouhy et al. [19] studied the NiTi alloy to examine the occurrence of the multiple step transformation owing to heterogeneous microstructures that form during aging of solution annealed defect-free NiTi alloys. It was reported that on cooling from the B2 regime, the one-step transformation that was observed on heating process was lost (after solution annealing at 850 °C for 15 min) and evolved from two-step (after aging at 500 °C for 1 h) to three-step (after aging at 500 °C for 10 h) transformations. The change from one-step to a two-step transformation in NiTi has been well understood because both R-phase and B19′-phase are potential martensite candidates [94–98]. Figure 4.9 illustrates three-step martensitic transformation accompanied with Ni_4Ti_3 precipitate [99]. The coherent Ni_4Ti_3 precipitates

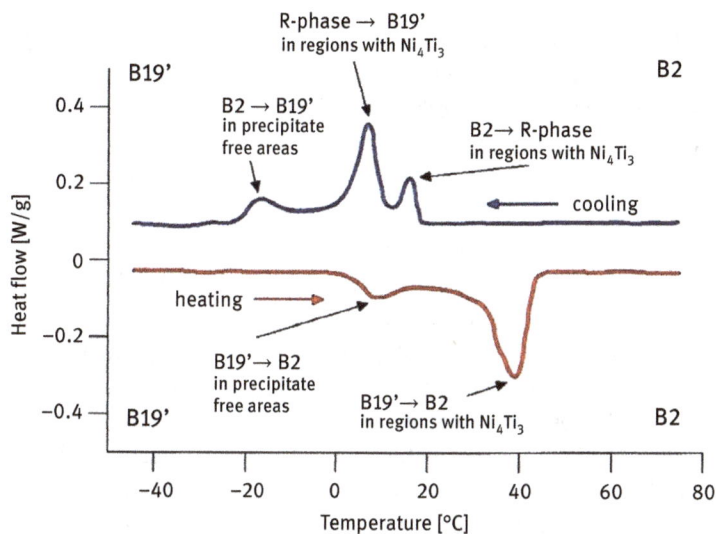

Figure 4.9: DSC chart for multiple step martensitic transformation in NiTi alloy with Ni_4Ti_3 precipitates [99].

resist large deformations associated with the formation of B19′ habit plane variants. The growing R-phase produces a significantly smaller deformation and is much less affected by particles. Therefore, the presence of precipitates favors the formation of R-phase, which results in the first transformation step B2→R and only later at lower temperature the second transformation step R→B19′ is observed. However, recent careful neutron diffraction experiments revealed that already in the temperature range between the first and second transformations the alloy microstructure consists of a mixture of R-phase and B19′-phase. In this case the $TiNi_{50.7}$ (at%) alloy was subjected to an annealing treatment (850 °C/900 s/water quench) and subsequently aged at 400 °C for 20 h [99].

The role of Ni_4Ti_3 precipitates responsible to multiple steps (more than two steps) can be found in numerous researches [11, 31, 54, 100, 102–107]. Not only three-step martensitic transformation, Kazemi-Choobi et al. [108], using $Ni_{50.9}Ti_{49.1}$ (at%) wires, studied multiple step transformations during aging treatment at different temperatures via in situ electrical resistance measurement. It was found that (i) during aging treatment, a cyclic behavior was observed in the electrical resistance variations, which could be related to the precipitation process, and (ii) depending on the stress level around precipitates, two-, three-, or four-step martensitic transformation could be observed in DSC curves. Although many research works were conducted, the explanation for the existence of multiple step martensitic transformations appears to be a controversy issue. Here are main different explanations for the presence of multiple step martensitic transformations in Ni-rich NiTi alloys.

In a course of detailed investigations on multiple step transformations through electrical resistivity measurements, microstructural observations, DSC measurements, structural diffraction studies, and others, the explanation for the existence of the multiple step transformations appears not to be well established yet. By the Morawiec dislocation theory [12,109–112], R-phase formation during both cooling and heating depends strongly on two martensitic transformations associated with a heterogeneous dislocation obstacles like substructure and subgrain boundaries. By the Bataillard growing behavior theory [113–115], the presence of R-phase and B19′-phase was recognized with different growing behavior; R-phase grows smoothly without apparent interruption (responding to the first step), while the B19′-phase martensite nucleates in sudden burst and grows rapidly to a significant size, requiring undercooling before consuming the rest of the matrix. As a result, the transformation temperatures in regions near particles (governed by high coherency stresses) are different than in regions far from particles (where coherency stresses are not important); hence these regions transform in two steps at a higher (near particles) and lower (far from particles) temperature. By the Khalil-Allafi inhomogeneous growth theory [19, 54, 116, 117], the 2-3-2 transformation behavior consists of (i) the homogeneous composition that is diffusion-controlled during aging during the precipitation of Ni4Ti3 and (ii) the difference between nucleation barriers for R-phase (small) and B19′-phase (large).

References

[1]   Park CH, Han SH, Kim SW, Hong JK, Nam T, Yeom JT. An effective approach to produce a nanocrystalline Ni–Ti shape memory alloy without severe plastic deformation. J. Alloy Compd. 2016, 654, 379–83.

[2]   Zupanc J, Vahdat-Pajouh N, Schäfer E. New thermomechanically treated NiTi alloys – a review. Int. Endod. J. 2018, 51(10), 1088–103.

[3]   Tang W, Lui EM, Hybrid recentering energy dissipative device for seismic protection. J. Struct. 2014; http://dx.doi.org/10.1155/2014/262409.

[4]   Shaw JA, Kyriakides S (1995) Thermomechanical aspects of NiTi. J. Mechan. Phys. Solids 1995, 43, 1243–81.

[5]   Mertmann M. Processing and Quality Control of Binary NiTi Shape Memory Alloys. In: Yahia L., ed. Shape Memory Implants. Springer-Verlag, Berlin, Germany, 2000, 24–34.

[6]   Otsuka K. Introduction of the R-Phase Transition. In: Duerig TW, et al. ed. Engineering Aspects of Shape Memory Alloys. Butterworth-Heinemann, London, UK, 1990, 36–45.

[7]   Zheng Y, Jiang F, Li L, Yang H, Liu Y. Effect of ageing treatment on the transformation behaviour of Ti–50.9 at.% Ni alloy. Acta Mater. 2008, 56, 736–45.

[8]   Otsuka K, Ren X. Recent developments in the research of shape memory alloys. Intermetallics 1999, 7, 511–28.

[9]   Liu Y, Favier D. Stabilisation of martensite due to shear deformation via variant reorientation in polycrystalline NiTi. Acta Mater. 2000, 48, 3489–99.

[10]   Liu Y, Kim JI, Miyazaki S. Thermodynamic analysis of ageing-induced multiple-stage transformation behaviour of NiTi. Philos. Mag. A 2004, 20, 2083; https://doi.org/10.1080/1 4786430410001678262.

[11]   Fan G, Chen W, Yang S, Zhu J, Ren X, Otsuka K. Origin of abnormal multi-stage martensitic transformation behavior in aged Ni-rich Ti–Ni shape memory alloys. Acta Mater. 2004, 52, 4351–62.

[12]   Il'czuk J, Morawiec H. Effect of dislocation inhomogeneity on the martensitic transformation in NiTi alloys. Met. Sci. Heat Treat. 1998, 40, 152–4.

[13]   Liu Y, Blanc M, Tan G, Kim JI, Miyazaki S. Effect of ageing on the transformation behaviour of Ti–49.5 at.% Ni. Mater. Sci. Eng. A 2006, 438/440, 617–21.

[14]   Khalil-Allafi J, Eggeler G, Dlouhy A, Schmahl WW, Somsen C. On the influence of heterogeneous precipitation on martensitic transformations in a Ni-rich NiTi shape memory alloy. Mater. Sci. Eng. A 2004, 378, 148–51.

[15]   Duerig TW, Bhattacharya K. The influence of the R-phase on the superelastic behavior of NiTi. Shape Mem. Superelastic. 2015, 1, 153–61.

[16]   Carroll MC, Somsen Ch, Eggeler G. Multiple-step martensitic transformations in Ni-rich NiTi shape memory alloys. Scr. Mater. 2004, 50, 187–92.

[17]   Su PC, Wu SK. The four-step multiple stage transformation in deformed and annealed Ti$_{49}$Ni$_{51}$ shape memory alloy. Acta Mater. 2004, 52, 1117–22.

[18]   Kim JI, Liu Y, Miyazaki S. Ageing-induced two-stage R-phase transformation in Ti-50.9at.%Ni. Acta Mater. 2004, 52, 487–99.

[19]   Dlouhy A, Khalil-Allafi J, Eggeler G. Multiple-step martensitic transformations in Ni-rich NiTi alloys – an in-situ transmission electron microscopy investigation. Philos. Mag. 2003, 83, 339–63.

[20]   Liu Y, McCormick PG. Criteria of transformation sequences in NiTi shape memory alloys. Mater. Trans. JIM 1996, 37, 691–6.

[21]   Liu Y, Chen X, McCormick. Effect of low temperature ageing on the transformation behaviour of near-equiatomic NiTi. J. Mater. Sci. 1997, 32, 5979–84.

[22]   Pelton AR, DiCello J, Miyazaki S. Optimisation of processing and properties of medical grade Nitinol wire. MITAT 2000, 9, 107–18.

[23]   Uchil J, Mohanchandra KP, Kumara KG, Mahesh KK. Study of critical dependence of stable phases in Nitinol on heat treatment using electrical resistivity probe. Mater. Sci. Eng. A, 1998, 251, 58–63.

[24]   Wang XB, Verkinden B, Van Humbeeck J. R-Phase transformation in NiTi alloys. Mater. Sci. Technol. 2014, 30, 1517–29.

[25]   Huang X, Liu Y. Effect of annealing on the transformation behavior and superelasticity of NiTi shape memory alloy. Scr. Mater. 2001, 45, 153–60.

[26]   Kim JI, Miyazaki S. Effect of nano-scaled precipitates on shape memory behavior of Ti-50.9 at.%Ni alloy. Acta Mater. 2005, 53, 4545–54.

[27]   Urbina C, de la Flor S, Ferrando F. Effect of thermal cycling on the thermomechanical bahaviour 9NiTi shape memory alloys. Mater. Sci. Eng. A 2009, A501, 197–206.

[28]   Urbina C, de la Flor S, Ferrando F. R-phase influence on different two-way shape memory training methods in NiTi shape memory alloys. J. Alloys Compd. 2010, 490, 499–507.

[29]   Lygin K, Langbein S, Labenda P, Sadek T. A methodology for the development, production, and validation of R-phase actuators. J. Mater. Eng. Perform. 2012, 21, 2657–62.

[30]   Miyazaki S, Otsuka K. Deformation and transition behavior associated with the *R*-phase in Ti-Ni alloys. Metall. Trans. A 1986, 17A, 53–63.

[31]   Bataillard J, Bidaux J-E, Gotthardt R. Interaction between microstructure and multiple-step transformation in binary NiTi alloys using *in-situ* transmission electron microscopy observations. Philos. Mag. A 1998, 78A, 327–44.

[32] Uchil J, Kumara KG, Mahesh KK. Effect of thermal cycling on R-phase in a NiTi shape memory alloy. Mater. Sci. Eng. A, 2002, A332, 25–28.

[33] Miyazaki S, Kimura S, Otsuka K. Shape-memory effect and pseudoelasticity associated with the R-phase transition in Ti-50·5 at.% Ni single crystals. Philos. Mag. A 1988, 57A, 467–78.

[34] Ling HC, Kaplow R. Phase transitions and shape memory in NiTi. Metall. Mater. Trans. A 1980, 11A, 77–83.

[35] Duerig TW. Some unsolved aspects of Nitinol. Mater. Sci. Eng. A, 2006, A438–A440, 69–74.

[36] Miyazaki S, Wayman CM. The R-phase transition and associated shape memory mechanism in Ti-Ni single crystals. Acta Metall. 1988, 36, 181–92.

[37] Otsuka K, Ren X. Physical metallurgy of Ti–Ni-based shape memory alloys. Prog. Mater. Sci. 2005, 50, 511–678.

[38] Šittner P, Novák V, Lukáš P, Landa M. Stress-strain-temperature behavior due to B2-R-B19′ transformation in NiTi polycrystals. J. Eng. Mater. Technol. 2006, 128, 268–78.

[39] Lukáš P, Šittner P, Lugovoy D, Neov D, Ceretti M. In situ neutron diffraction studies of the R-phase transformation in the NiTi shape memory alloy. Appl. Phys. A 2002, 74A, s1121–3.

[40] Fukuda T, Saburi T, Doi K, Nenno S. Nucleation and self-accommodation of the R-phase in Ti–Ni alloys. Mater. Trans. JIM, 1992, 33, 271–7.

[41] Šittner P, Landa M, Lukáš P, Novák V. R-phase transformation phenomena in thermomechanically loaded NiTi polycrystals. Mech. Mater. 2006, 38, 475–92.

[42] Šittner P, Sedlák P, Landa M, Novák V, Lukáš P. *In situ* experimental evidence on R-phase related deformation processes in activated NiTi wires. Mater. Sci. Eng. A 2006, A438–40, 579–84.

[43] Zhang X, Sehitoglu H. Crystallography of the B2 → R → B19′ phase transformations in NiTi. Mater. Sci. Eng. A 2004, A374, 292–302.

[44] Hara T, Ohba T, Okunishi E, Otsuka K. Structural study of R-phase in Ti-50.23 at.%Ni and Ti-47.75 at.%Ni-1.50 at.%Fe Alloys. Mater. Trans. JIM 1997, 38, 11–7.

[45] Jin YM, Weng GJ. A direct method for the crystallography of martensitic transformation and its application to TiNi and AuCd. Acta Mater. 2002, 50(11), 2967–87.

[46] Khalil-Allafi J, Schmahl WW, Toebbens DM. Space group and crystal structure of the R-phase in binary NiTi shape memory alloys. Acta Mater. 2006, 54(12), 3171–5.

[47] Gao Y, Zhou N, Wang D, Wang Y. Pattern formation during cubic to orthorhombic martensitic transformations in shape memory alloys. Acta Mater. 2014, 68, 93–105.

[48] Kustov S, Salas D, Santamarta R, Humbeeck JV. Isothermal and athermal martensitic transformations in Ni–Ti shape memory alloys. Acta Mater. 2012, 60(6–7), 2578–92.

[49] Nishida M, Wayman CM, Honma T. Precipitation processes in near-equiatomic TiNi shape memory alloys. Metall. Trans. A 1986, 17, 1505–15.

[50] Bataillard L, Bidaux J-E, Gotthardt R. Interaction between microstructure and multiple-step transformation in binary NiTi alloys using *in-situ* transmission electron microscopy observations. Philos. Mag. 1998, 78, 327–44.

[51] Tirry W, Schryvers D. In situ transmission electron microscopy of stress-induced martensite with focus on martensite twinning. Mater. Sci. Eng. A 2008, 481/482, 420–5.

[52] Michutta J, Somsen Ch, Yawny A, Dlouhy A, Eggeler G. Elementary martensitic transformation processes in Ni-rich NiTi single crystals with Ni_4Ti_3 precipitates. Acta Mater. 2006, 54, 3525–42.

[53] Tirry W, Schryvers D. Quantitative determination of strain fields around Ni_4Ti_3 precipitates in NiTi. Acta Mater. 2005, 53, 1041–9.

[54] Khalil-Allafi J, Dlouhy A, Eggeler G. Ni_4Ti_3-precipitation during aging of NiTi shape memory alloys and its influence on martensitic phase transformations. Acta Mater. 2002, 50, 4255–74.

[55]   Zhou Y, Zhang J, Fan G, Ding X, Sun J, Ren X, Otsuka K. Origin of 2-stage R-phase transformation in low-temperature aged Ni-rich Ti–Ni alloys. Acta Mater. 2005, 53, 5365–77.

[56]   Bojda O, Eggeler G, Dlouhy A. Precipitation of Ni_4Ti_3-variants in a polycrystalline Ni-rich NiTi shape memory alloy. Scr. Mater. 2005, 53, 99–104.

[57]   Nishida M, Honma T. All-round shape memory effect in Ni-rich TiNi alloys generated by constrained aging. Scr. Metall. 1984, 18, 1293–8.

[58]   Nishida M, Wayman CM, Honma T. Electron microscopy studies of the all-around shape memory effect in a Ti-51.0 at .%Ni alloy. Scr. Metall. 1984, 18, 1389–94.

[59]   Miyazaki S, Otsuka K. Deformation and transition behavior associated with the R-phase in Ti-Ni alloys. Metall. Trans. A 1986, 17A, 53–63.

[60]   Miyazaki S, Kimura S, Otsuka K. Shape-memory effect and pseudoelasticity associated with the R-phase transition in Ti-50·5 at.% Ni single crystals. Philos. Mag. A 1988, 57A, 467–78.

[61]   Ling HC, Kaplow R. Phase transitions and shape memory in NiTi. Metall. Mater. Trans. A, 1980, 11A, 77–83.

[62]   Stachowiak GB, McCormick PG. Shape memory behaviour associated with the R and martensitic transformations in a NiTi alloy. Acta Metall. 1988, 36, 291–7.

[63]   Salamon MB, Meichle ME, Wayman CM. Premartensitic phases of $Ti_{50}Ni_{47}Fe_3$. Phys. Rev. B, 1985, 31B, 7306–15.

[64]   Dautovich DP, Purdy GR. Phase transformations in NiTi. Can. Metall. Q. 1965, 4, 129–43.

[65]   Goo E, Sinclair R. The B2 to R transformation in $Ti_5ONi_{47}Fe_3$ and $Ti_{49.5}Ni_{50.5}$ alloys. Acta Metall. 1985, 33, 1717–23.

[66]   Šittner P, Lukáš P, Landa M, Novák V. Stress-strain-temperature behavior due to B2-R-B19′ transformation in NiTi polycrystals. J. Eng. Mater. Technol. 2006, 128, 268–78.

[67]   Tobushi H, Tanaka K, Sawada T, Hattori T, Lexcellent C. Recovery stress associated with R phase transformation in NiTi shape memory alloy. Trans. Jpn. Soc. Mech. Eng. A 1993, 59, 171–5.

[68]   Shaw JA, Kyriakides S. On the nucleation and propagation of phase transformation fronts in a NiTi alloy. Acta Mater. 1997, 45, 683–700.

[69]   Saadat S, Salichs J, Noori M, Hou Z, Davoodi H, Bar-on I, Suzuki Y, Masuda A. An overview of vibration and seismic applications of NiTi shape memory alloy. Smart Mater. Struct. 2002, 11, 218–29.

[70]   Wasilewski RJ. The effects of applied stress on the martensitic transformation in TiNi. Met. Trans. 1971, 2, 2973–81.

[71]   Mukherjee J, Milillo F, Chandrasekaran M. Effects of stress and transformation cycling on the transition behavior of a nearly stoichiometric TiNi alloy. Mater. Sci. Eng. 1974, 14, 143–7.

[72]   Ling HC, Kaplow R. Stress-induced shape changes and shape memory in the R and martensite transformations in equiatomic NiTi. Metall. Trans. A, 1981, 12, 2101–11.

[73]   Goldstein D, Kabacoff L, Tydings J. Stress effects on Nitinol phase transformations. J. Metals 1987, 39, 19–26.

[74]   Goldstein DM. Nitinol strain effects. J. Metals 1987, 39, 23–7.

[75]   Hedayat A, Rechtien J, Mukherjee K. The effect of surface constraint on the phase transformation of Nitinol.J. Mater. Sci. 1992, 27, 5306–14.

[76]   Orgeas L, Favier D. Stress-induced martensitic transformation of a NiTi alloy in isothermal shear, tension and compression. Acta Mater. 1998, 46, 5579–91.

[77]   Lopez HF, Salinas A, Calderón H. Plastic straining effects on the microstructure of a Ti-rich NiTi shape memory alloy. Metall. Mater. Trans. A 2001, 32, 717–29.

[78]   Sehitoglu H, Zhang XY, Chumlyakov YI, Karaman I, Gall K, Maier HJ. Observations on stress-induced transformations in NiTi alloys. IUTAM Symp. Mech. Martensitic Phase Transform. Sol. 2002, 103–9.

[79] He X, Rong L, Yan D, Jiang Z, Li Y. Effect of deformation on martensitic transformation behavior in $Ni_{50.1}Ti_{46.9}Nb_3$ shape memory alloy. J. Mater. Sci. 2005, 40, 5311–3.

[80] Paiva A, Savi MA, Braga AMB, Pacheco PMCL. A constitutive model for shape memory alloys considering tensile – compressive asymmetry and plasticity. Int. J. Sol. Struct. 2005, 42, 34357.

[81] Frick CP, Lang TW, Spark K, Gall K. Stress-induced martensitic transformations and shape memory at nanometer scales. Acta Mater. 2006, 54, 2223–34.

[82] Gollerthan S, Young ML, Neuking K, Ramamurty U, Eggeler G. Direct physical evidence for the back-transformation of stress-induced martensite in the vicinity of cracks in pseudoelastic NiTi shape memory alloys. Acta Mater. 2009, 57, 5892–7.

[83] Olbricht J, Yawny A, Pelegrina JL, Dlouhy A, Eggeler G. On the stress-induced formation of R-phase in ultra-fine-grained Ni-rich NiTi shape memory alloys. Metall. Mater. Trans. A 2011, 42, 2556–74.

[84] Khalil H. Changes in the mechanical behavior of nitinol following variations of heat treatment duration and temperature. Thesis, Georgia Institute of Technology, 2009; https://smartech.gatech.edu/bitstream/handle/1853/31852/khalil_heidi_f_200912_mast.pdf.

[85] Gall K, Tyber J, Brice V, Frick CP, Maier HJ, Morgan N. Tensile deformation of NiTi wires. J. Biomed. Mater. Res. 2005, 75A, 810–23.

[86] Elibol C, Wagner MFX. Investigation of the stress-induced martensitic transformation in pseudoelastic NiTi under uniaxial tension, compression and compression-shear. Mater. Sci. Eng. A 2015, 621, 76–81.

[87] Auricchio F, Realia A, Stefanelli U. A phenomenological 3D model describing stress-induced solid phase transformations with permanent inelasticity. Topics. Math. Smart Syst. 2007, 1–14; https://doi.org/10.1142/9789812706874_0001.

[88] König D, Buenconsejo PJS, Grochla D, Hamann S, Pfetzing-Micklich J, Ludwig A. Thickness-dependence of the B2–B19 martensitic transformation in nanoscale shape memory alloy thin films: zero-hysteresis in 75 nm thick $Ti_{51}Ni_{38}Cu_{11}$ thin films. Acta Mater. 2012, 60, 306–13.

[89] Mao S, Li H, Liu Y, Deng Q, Wang L, Zhang Y, Zhang Z, Han X (2013) Stress-induced martensitic transformation in nanometric NiTi shape memory alloy strips: an in situ TEM study of the thickness/size effect. J Alloy Compd. 2013, 579, 100–11.

[90] Sun L, Huang WM. Nature of the multiplestage transformation in shape memory alloys upon heating. Metal. Sci. Heat Treat. 2009, 51, UDC 669.14.018.6:620.181.4.

[91] Kim J, Liu Y, Miyazaki S. Ageing-induced two-stage R-phase transformation in Ti–50.9at.%Ni. Acta Mater. 2004, 52, 449–87.

[92] Zhou Y, Zhang J, Fan G, Ding X, Sun J, Ren X, Otsuka K. Origin of 2-stage R-phase transformation in low-temperature aged Ni-rich Ti–Ni alloys. Acta Mater. 2005, 53, 5365–77.

[93] Chrobak D, Stróż D. Two-stage R phase transformation in a cold-rolled and annealed Ti–50.6 at.%Ni alloy. Scr. Mater. 2005, 52, 757–60.

[94] Moine P, Allain J, Renker B. Observation of a soft-phonon mode and a pre-martensitic phase in the intermetallic compound $Ti_{50}Ni_{47}Fe_3$ studied by inelastic neutron scattering. J. Phys. F – Metal Phys. 1984, 14, 2517–23.

[95] Tietze H, Müller Y, Renker B. Dynamical properties of premartensitic NiTi. J Phys. C – Sol. State Phys. 1984, 17, L529–32.

[96] Ren X, Muirs N, Zhang J, Otsuka K, Tanaka K, Koiwa M, Suzuki S, Chumlyakov YI, Asai M. A comparative study of elastic constants of TiNi based alloys prior to martensitic transformation. Mat. Sci. Eng. 2001, A312, 196–206.

[97] Liu Y, Yang H, Voigt A. Thermal analysis of the effect of aging on the transformation behaviour of Ti-50.9 at%Ni. Mat. Sci. Eng. 2003, A360, 350–5.

[98]   Carrol MC, Somsen Ch, Eggeler GF. Multiple-step martensitic transformations in Ni-rich NiTi
       alloys – an in-situ transmission electron microscopy investigation. Phil. Mag. 2003, 83, 339–63.
[99]   Holec D. On the precipitation in NiTi based shape memory alloys. Thesis, Masaryk
       University, 2005; http://www.physics.muni.cz/ufkl/diplomas/DavidHolec-diplomka.pdf.
[100] Khalil-Allafi J, Ren X, Eggeler G. The mechanism of multistage martensitic transformations in
       aged Ni-rich NiTi shape memory alloys. Acta Mater. 2002, 50, 793–803.
[101] Nishida M, Hara T, Ohba T, Yamaguchi K, Tanaka K, Yamauchi K. Experimental consideration
       of multistage martensitic transformation and precipitation behavior in aged Ni-Rich Ti-Ni
       shape memory alloys. Mater. Trans. 2003, 44, 2631–6.
[102] Liu Y, Kim JI, Miyazaki S. Thermodynamic analysis of ageing-induced multiple-stage
       transformation behaviour of NiTi. Philos. Mag. 2004, 84, 2083–102.
[103] Chang S, Wu S, Chang G. Grain size effect on multiple-stage transformations of a cold-rolled
       and annealed equiatomic TiNi alloy. Scr. Mater. 2005, 52, 1341–6.
[104] Wagoner Johnson AJ, Sehitoglu H, Hamilton RF, Biallas G, Maier HJ, Chumlyakov YI, Woo HS.
       Analysis of multistep transformations in single-crystal NiTi. Metall. Mater. Trans. A 2005,
       36, 919–28.
[105] Fan G, Zhou Y, Chen W, Yang S, Ren X, Otsuka K. Precipitation kinetics of Ti_3Ni_4 in polycrys-
       talline Ni-rich TiNi alloys and its relation to abnormal multi-stage transformation behavior.
       Mater. Sci. Eng. A 2006, 438/440, 622–6.
[106] Wu SL, Liu XM, Chu PK, Chung CY, Chu CL, Yeung KWK. Phase transformation behavior of
       porous NiTi alloys fabricated by capsule-free hot isostatic pressing. J. Alloys Compd. 2008,
       449, 139–43.
[107] Moshref-Javadi M, Hossein S, Mohammad S, Salehi T, Aboutalebi MR. Age-induced
       multi-stage transformation in a Ni-rich NiTiHf alloy. Acta Mater. 2013, 61(7), 2583–94.
[108] Kazemi-Choobi K, Khalil-Allafi J, Elhami A, Asadi P. Influence of aging treatment
       on *in-situ* electrical resistance variation during aging of Nickel-Rich NiTi shape memory
       wires. Metall. Mater. Trans. A 2013, 44, 4429–33.
[109] Morawiec H, Stroz D, Gorczka T, Chrobak D. Two-stage martensitic transformation in a
       deformed and annealed NiTi alloys. Scr. Mater. 1996, 35, 485–90.
[110] Morawiec H, Stroz D, Chrobak D. Effect of deformation and thermal treatment of NiTi alloy on
       transition sequence. J. Phys. IV (Proceed.) 1995, 5; doi: 10.1051/jp4:1995232.
[111] Chrobak D, Morawiec H. Thermodynamic analysis of the martensitic transformation in
       plastically deformed NiTi alloy. Scr. Mater. 2001, 44, 725–30.
[112] Goryczka T, Morawiec H. Structure studies of the R-phase using the X-ray and electron
       diffraction method. J. Alloys Compd. 2004, 367, 137–41.
[113] Bataillard L, Gotthardt R. Influence of thermal treatment on the appearance of a three step
       martensitic transformation in NiTi. J. Phys. IV (Proceed.) 2014, 5, C8/647–C8/652.
[114] Bataillard L. Interaction between microstructure and multiple-step transformation in binary
       NiTi alloys using in-situ transmission electron microscopy observations. Philos. Mag.
       A 1998, 78, 327–44.
[115] Bataillard L, Bidaux J-E, Mari D, Bataillard L, Dunand D, Gotthard R. Martensitic
       transformation of NiTi and NiTi-TiC composites. J. Phys. IV (Proceed.) 1995, 5, 659; doi:
       10.1051/jp4/199558659.
[116] Shakeri MS, Khalil-Allafi J, Abbasi-Chianeh V, Ghabchi A. The influence of Ni4Ti3
       precipitates orientation on two-way shape memory effect in a Ni-rich NiTi alloy. J. Alloys
       Compd. 2009, 485, 320–3.
[117] Khalil-Allafi J, Dlouhý A, Neuking K, Eggeler GF. Influence of precipitation and dislocation
       substructure on phase transformation temperatures in a Ni-rich NiTi-shape memory alloy. J.
       Phys. IV France 2001, 11, 529–34.

Chapter 5
Heat treatments and their related phenomena

It is, in general, established that there are three ways to memorize (train) the shape of (potentially shape recoverable) NiTi alloy: (1) hardening by a cold working, followed by medium-temperature treatment at a range from 400 to 500 °C for 1 h, (2) annealing at temperature higher than 800 °C, formed and treatment at relatively low-temperature ranging from 200 to 300 °C, and (3) solution treatment followed by aging at 400 °C for several hours. Hence, heat treatments are diverse and appropriate selection and proper practice, thereof, should become a very crucial issue to make either/both shape-memory effect (SME) and superelasticity (SE) phenomena into more efficient and efficacious manner. In addition to these, a special concern regarding its stability should be indispensable. Since the martensitic transformation exhibit the temperature dependency, time dependency, as well as stress dependency, all these martensitic transformation-related phenomena and other effects will be discussed in this chapter.

5.1 Stability

During the phase transformation between B2 phase (A: austenite) and B19′ (M: martensite) phase, unique properties of SME and SE are realized. For enhancing these phenomena, normally post-deformation annealing, thermal/mechanical cycling, aging and others are applied to NiTi materials. By these additional thermal- or thermomechanical treatment, the intermediate rhombohedral R-phase is frequently produced, from which an intermetallic compound Ni_4Ti_3 is precipitated coherently (and is normally not shown in the equilibrium phase diagram since it is unstable phase) resulting in generating an internal strain field. According to the appearance of R-phase, the transformation could be a multistep process; B2↔R↔M, and during the B2↔R transformation, SME and SE are also recognized [1]. The unique properties of shape-memory alloys (SMAs) are controlled by and are dependent on four external parameters: temperature (T), stress (σ), strain (ε), and time (t). These parameters cannot be changed independently. The complete mechanical behavior of SMAs should be determined from [T, σ, ε]-diagram in which the temperature axis covers the general temperature range from about M_S−50 °C to M_S+100 °C [1, 2]. The value of a point along the [T, σ, ε]-surface is not always constant and can move in any direction in that shape. This time dependency can result from creep, stress relaxation, and changes due to variations in the chemical free energy of transformed martensite and/or parent austenite phases [1]. The hysteresis of the martensitic transformation and stability of the martensitic phase are factors influencing instabilities of SMAs [3]. These are directly and indirectly controlled by transformation temperatures, transformation stresses,

https://doi.org/10.1515/9783110666113-005

and transformation strains [4]. As Burow et al. pointed out [5], the martensitic transformations in NiTi alloys are strongly depend on the microstructure. Effect of various types of ultrafine-grained microstructures resulting from various processing routes on the martensitic transformation was investigated and it was found that (i) the initial coarse-grained material shows a clear one-step martensitic transformation on cooling process, (ii) the two-step transformations were found for all ultrafine-grained materials, (iii) the ultrafine-grained structure exhibited states, and (iv) the ultrafine-grained NiTi alloys showed a significantly higher stability by thermal cycling, indicating an excellent functional stability [5].

Figures 5.1 and 5.2 illustrate two-dimensional version of $[T, \sigma, \varepsilon]$-diagrams [6]. Typically, as the temperature increases (as seen in Figure 5.1), the transformation stress increases and the stress-hysteresis loop shifts upwardly. For SE, it is observed that the austenite deforms elastically until a loading transformation is reached. Further loading induces a transformation to the martensite phase with a large transformation strain. Upon unloading martensite, the transformation stain is recovered as the SMA returns to the original austenite. For the SME, the hysteresis is often observed when an SMA is subjected to a fixed load. In this case, an austenitic SMA will transform to martensite, when cooled, exhibiting a large transformation strain (which is proportion to the work output) in the process. When reheated to the austenite, the transformation strain is recovered. As stress increases, in the temperature–strain relationship, the transformation temperature increases [7].

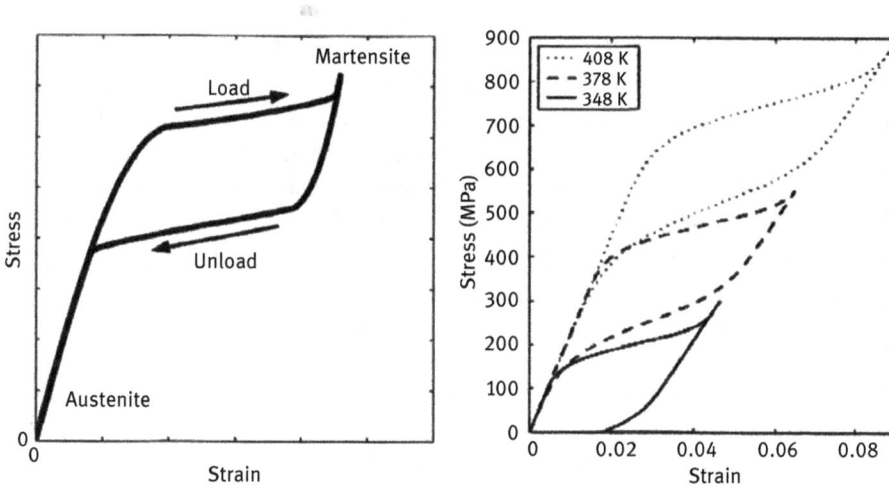

Figure 5.1: Stress–strain curves under influences of acting temperatures for superelastic phenomenon [6].

When SMAs become commercially viable, particularly in the medical device industry, R&D must be carried out on design-relevant properties including the cyclic stability,

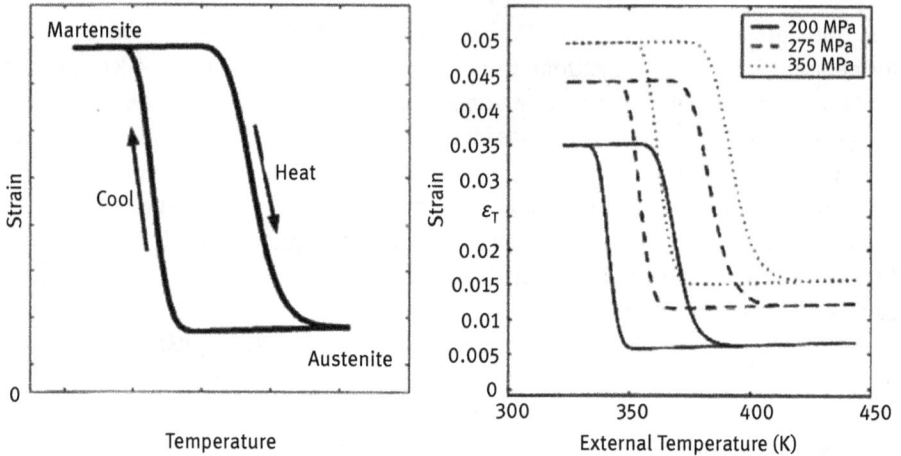

Figure 5.2: Temperature–strain curves under influences of acting stress levels for shape-memory phenomenon [6].

which is influenced by alloy type, thermal processing, heating/cooling rates, applied loads, prior cold work level, the structural integrity, and internal stress [8]. Strnadel et al. [9] examined the responses of three types of Ti–Ni and three types of Ti–Ni–Cu SMAs in a superelastic state to mechanical cycling in hard cycles with a constant ε_{max} and in soft cycles with a constant σ_{max}. It was found that (i) the transformation stress of the B2 parent phase into martensite and the hysteresis (or the amount of energy dissipated during one cycle) diminishes while the residual deformation increases as the number of cycles grows, and (ii) although the maximum deformation is greater in a soft cycle than in a hard cycle, the critical stress for inducing martensite, at least over the first 10 cycles, declines more slowly in soft cycles than in hard-loading cycles [9]. Stress and energy release during the phase transformation in NiTi alloy exhibit influences on orientations of the atomic structure, and if SMAs are exposed to repeated regular deforming and shape-recovery cyclings, some loses in their shape-memory recoverability (or amnesia) will be jeopardized, as Özkul et al. [10] observed with $Ni_{46.84}Ti_{53.16}$ (at%) alloy. Sedmák et al. [11] investigated the instability of the cyclic tensile superelastic behavior of NiTi polycrystal, being linked to its fatigue performance (number of cycles till failure), by high-resolution x-ray diffraction method. NiTi wires were cyclically deformed in tension at room temperature while x-ray diffraction patterns were recorded in three preselected states along the superelastic $[\sigma-\varepsilon]$ curve and analyzed and interpreted in terms of the gradual evolution of microstructural state during cycling. It is found that (i) the cyclic instability is due to the gradual redistribution of internal stresses originating from the accumulation of incremental plastic strains accompanying the stress induced martensitic transformation in constrained polycrystalline environment, and (ii) the degree of cyclic instability increases with the increasing involvement of slip in the hybrid slip/transformation process, which depends on initial microstructure (grain size, defects, precipitates), martensitic transformation (crystallographic incompatibility between transforming phases), temperature, and parameters of the cyclic loading (strain rate, amplitude, stress state, type of loading, etc.) [11]. Grossman et al. [12] using the binary NiTi and ternary NiTiCu (with 5, 7.5, and 10 at% Cu) shape-memory spring actuators, studied effects of the alloy composition and processing on the actuator stability during thermomechanical cycling. It was reported that (i) the cyclic actuator stability can be improved by using precycling, subjecting the material to cold work, and adding copper, (ii) adding copper is more attractive than cold work, because it improves cyclic stability without sacrificing the exploitable actuator stroke, (iii) Cu element reduces the width of the thermal hysteresis and improves geometrical and thermal actuator stability, because it results in a better crystallographic compatibility between the parent and the product phase; due to hypothesis that dislocations are created during the phase transformations that remain in the microstructure during subsequent cycling. These dislocations facilitate the formation of martensite (increasing M_S temperatures) and account for the accumulation of irreversible strain in martensite and austenite [12].

The left half of the diagram (Figure 5.3) illustrates a temperature hysteresis in terms of amount of austenitic phase [13] and the right one depicts the temperature hysteresis with regard to the amount of martensitic phase, respectively, where martensite finish temperature (M_S) and austenite finish temperature (A_F) are more frequently used to characterize the SME and SE phenomena. If the temperature is above A_F, the alloy is in austenitic state, that is, it is stiff, hard, and possesses superior superelastic properties [14]. If the temperature is below M_F, the NiTi alloy is in martensitic state, that is, it is soft, ductile, can easily be deformed, and possesses the SME [14, 15]. The most important message that we can learn from Figure 5.3 is that M_S and A_F points do not correspond to one point, rather show different temperatures, creating the temperature hysteresis, which is normally in a range of 10–30 °C in NiTi alloy while it will be several 100 degrees in steel. If the narrower temperature hysteresis exists in the martensitic transformation, this phenomenon is called pseudoelastic martensitic transformation. A change in these four transformation temperatures of the utilized NiTi alloy, which can be achieved by thermal and mechanical treatment or variation in the chemical composition, is the most important tool for manufacturers to alter the phase composition and consequently the mechanical properties.

During the phase transformation, stress is created which can particularly be measurable with thin-film sample. Since the lattice distortion during the martensitic transformation B2→R (ca. 1%) is much smaller than that during B2↔B19′ (ca. 10%) [19], Hou et al. [20], aiming a narrower hysteresis, prepared Ti–Ni$_{50.3}$ (at%) thin films with ~800 nm thickness deposited on silicon substrates by cosputtering Ti target and Ni target in a biased target ion beam deposition technique. The test temperature was cycled between 28 and 90 °C. The heating rate was exactly 1 °C/min, and cooling from 90 to 28 °C took about 2.5 h. Heating–cooling cycles were repeated four times. The results on the film stress versus temperature curves are shown in Figure 5.4 [20].

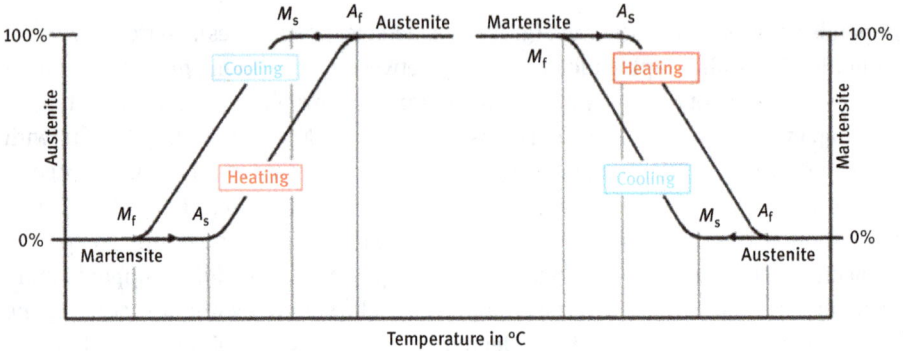

Figure 5.3: Amount of transformed phases versus treating temperature; left half for relationship between amount of austenitic phase versus temperature and right half for relationship between amount of martensitic phase versus temperature [13].

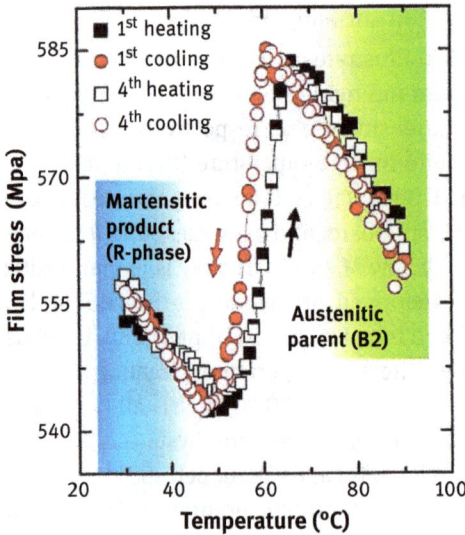

Figure 5.4: Stress–temperature curves; closed marks are from the first heating–cooling cycle; while open marks indicate data from the fourth heating–cooling cycle [20].

It was reported that (i) film stress increases from 543 MPa to 585 MPa as temperature increases from 50 to 65 °C, due to the phase transformation occurrence in the NiTi alloy thin films, (ii) A_S and A_F were 51 and 66 °C, respectively, and M_S and M_F were 62 °C and 49 °C, (iii) the thermal hysteresis was 4 °C ($A_F - M_S$), (iv) the negative slope can be attributed to differential coefficients of thermal expansion in thin film versus substrate, and the slope is proportional to the difference in their coefficients of thermal expansion, and (v) the stress associated with phase transformation dominates and can be much larger (>4 times) than the thermal stress due to thermal expansion; hence, the slope during phase transformation is positive [20].

5.2 Precipitation and [T-T-T] diagram

As described previously, the transformation temperature is altered by precipitating Ni-rich phases from the surrounding matrix. Precipitation processes in near-equiatomic $TiNi_{50}$ and Ni-rich $TiNi_{52}$ (at%) was metallographically investigated [21]. It was reported that (i) both did not show eutectoid and peritectoid reactions, and (ii) the precipitation sequence was: $\beta_0 \rightarrow Ti_{11}Ni1_4 \rightarrow Ti_2Ni_3 \rightarrow TiNi_3$, where β_0 is the initial matrix composition of the alloy. The precipitation process observed in NiTi resulting in an increase in A_F is well documented and commonly used in A_F tuning processes [22]. The precipitation process can be presented in the form of a [Temperature-Time-Transformation] diagram, as seen in Figure 5.5 [22]. In the figure, five curves represent different aging temperatures: A_F20 represents 300 °C, A_F25 for 350 °C, A_F30 for 400 °C, A_F35 for 450 °C, and A_F40 for 500 °C, respectively. The starting A_F was 11 °C. Each C-shaped curve represents contours of constant A_F [22, 23]. It was noted that a single precipitation reaction ($Ti_{11}Ni_{14}$) took place at about 400 °C [22].

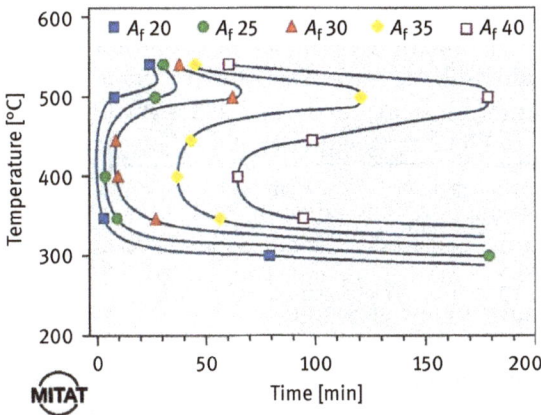

Figure 5.5: [Temperature-Time-Transformation] diagram of $TiNi_{50.8}$ wire as a function of different aging temperatures [22].

Zapoticla studied the effects of strain on [T-T-T] diagram, as seen in Figure 5.6 [23]. As seen clearly from Fig RRR, the nose of A_F contours shifts to left under the influence of prestrain condition, indicating shorter reaction duration. With Ni-rich NiTi alloys and under influence of external stresses, it was reported that (i) there found precipitates of Ni_4Ti_3 and Ni_3Ti_2 [24, 25] and (ii) these precipitates have been observed

(a)

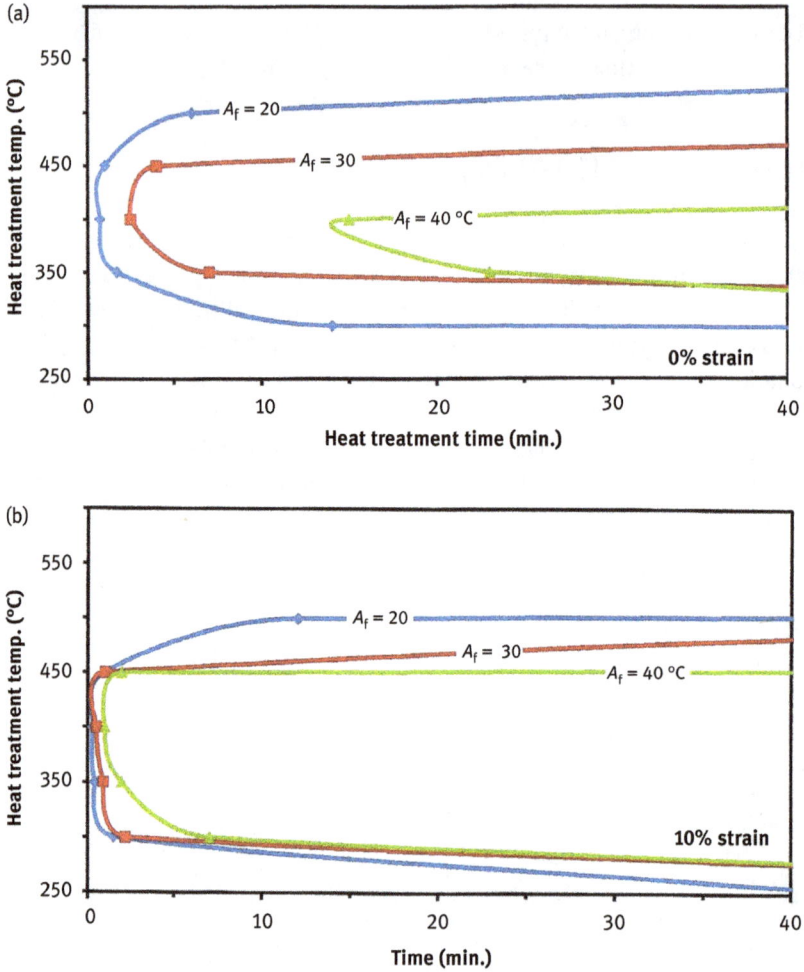

(b)

Figure 5.6: Comparison of [T-T-T] diagrams under influence of prestrain condition [23].

to be the source of a local stress field which influences both the nucleation and the growth of the R and martensitic phases; suggesting that the additional transformation step can be considered to be a locally stress-induced martensitic transformation [26]. Zarinejad et al. [27] investigated changes in the martensite start temperature (M_S) of NiTi-based SMAs as a result of second-phase precipitation, which would lead to a change in the matrix chemical composition and may alter the electron concentration of the matrix. It was reported that (i) when the electron concentration of the matrix increases, the transformation temperature decreases, whereas the M_S temperature rises when the electron concentration of the matrix decreases, and (ii) changes in the M_S temperature as a result of precipitation is mainly due to the change in electron concentration of the matrix.

Severe plastic deformation processes, such as equal channel angular pressing and high-pressure torsion, are successfully employed to produce ultrafine grain and nanocrystalline microstructures in a Ti–Ni$_{50.7}$ (at%) SMA. The effect of grain size on subsequent Ni-rich particle precipitation during annealing is investigated by transmission electron microscopy, selected-area electron diffraction and x-ray diffraction [28]. It was reported that (i) Ni$_4$Ti$_3$ precipitation is suppressed in grains of cross-sectional equivalent diameter below approximately 150 nm, and that particle coarsening is inhibited by very fine grain sizes, and (ii) the fine grain sizes impede precipitation processes by disrupting the formation of self-accommodating particle arrays and that the arrays locally compensate for coherency strains during nucleation and growth. Sji et al. [29] investigated the effect of grain size on the R-phase transformation of a nanocrystalline Ti–Ni$_{50.2}$ (at%) alloy. The nanometric grain size was created by severe cold deformation and low temperature anneal. It was found that (i) in the recrystallized state, achieving nanoscale grain sizes (<100 nm) was effective in suppressing the B2→B19′ martensitic transformation and revealing the B2↔R transformation, (ii) the B2↔R transformation temperature was found to increase with the decreasing grain size within the range of 22–155 nm and (iii) the suppression of the B19′ martensite in nanograins is attributed to the limited space within the grains to allow the formation of self-accommodation structures to contain the large lattice distortion of the martensite.

5.3 Thermocycling

For practical applications, thermal cycling effects in the alloys are an important factor to be considered [30]. Miyazaki et al. [31] investigated the influence of Ni content in Ti–Ni$_{50.2}$ (at%) alloy, which was aged after solution treatment and annealing after cold working on the thermal cycling effects. It was found that the shift of M_S, increases in electrical resistance prior to the martensitic transformation on cooling, and increases in $M_S - M_F$ (a measure for a scale of thermal hysteresis) observed in all of the solution-treated alloys irrespective of Ni content, but not in aged Ni-rich alloys and in cold-worked Ni-poor ones annealed at a temperature lower than the recrystallization temperature. Tadaki et al. [32] studied to examine thermal cycling effects in an aged Ti–Ni$_{51.0}$ (at%) alloy, extending the number of thermal cycles up to 104 times. It was reported that significant changes in transformation temperature and microstructure were observed even in the aged alloy when thermal cycling was extended to roughly more than 103 times.

Pelosin et al. [33] investigated the effect of thermal cycling on transformation temperature on a Ti-rich NiTi alloy and reported that (i) the martensite microstructure is modified by the successive cycling transformation, (ii) the premartensitic phase becomes stable following the diminishment of the beginning of the martensite formation, and (iii) the interrupted cooling has also shown that, contrary to the martensite, the R-phase exhibits no hysteretic behavior. The effects of thermal cycling through the martensite–austenite transformation were investigated in NiTi SMAs with DSC and TEM [34]. It was found that, after 100 thermal cycles, M_S decreased about 25 °C. Uchil et al. [35] investigated the stability of R-phase during thermal cycling for the near equiatomic NiTi wire samples which were 40% cold worked previously. Thermal cycle was in the temperature region of 340–620 °C. It was mentioned that (i) in the samples having only M↔A transformations after heat treatment, successive thermal cycles promote and result in the stabilization of intermediate R-phase, (ii) during the initial thermal cycles, M_S is found to decrease facilitating the development of intermediate R-phase as indicated by the emergence of peak in the cooling part of the transformation profile, (iii) once the R-phase is completely developed, the transformation parameters such as R_S, R_F, M_S, M_F and peak resistivity of the transformation profile remains unaffected during thermal cycling, and (iv) there exists a critical number of thermal cycles for the complete development of R-phase and this critical number is a function of heat treatment temperature. In the oral environment, orthodontic wires will be subject to thermal fluctuations. Berzins et al. [36] studied the effect of thermocycling on NiTi wire phase transformations. Straight sections of orthodontic wires were subjected to thermocycling between 5 and 55 °C for 1,000, 5,000, and 10,000 cycles. It was reported that repeated temperature fluctuations may contribute to qualitative and quantitative phase transformation changes in Cu-NiTi wires.

5.4 Heating rate and cooling rate

Wang et al. [37] investigated the effects of heating rate and cooling rate on the transformation characteristics in Cu-NiTi SMA by differential scanning calorimetry. It was reported that (i) the martensitic finish transformation temperature (M_F) and reverse end transformation temperature (A_F) depend strongly on the rate of the heating/cooling process, in such a manner that M_F decreased and A_F increased with increasing cooling/heating rate. while (ii) the martensitic start transformation temperature (M_S) and reverse start transformation temperature (A_S) are not so sensitive to the scanning rate. These findings were confirmed by Nurveren et al. [38].

Zhang et al. [39] and Motemani et al. [40] investigated the effect of cooling rate on the phase transformation in NiTi alloy. It was reported that the three-stage phase transformation can be induced at a very low cooling rate such as the furnace cooling [39]. Furthermore, it was found, commonly, that (i) the cooling rate has a great influence on the phase transformation temperatures; both M_S and M_F decrease with the decrease of the cooling rate, and decreasing the cooling rate contributes to enhancing the M→A austenite transformation temperature, and (ii) the phase transformation hysteresis ($A_F - M_F$) increases with the decrease of the cooling rate.

Adherapurapu et al. [41] studied the effects of aging and cooling rate on the transformation of nanostructured Ti–Ni$_{50.8}$ (at%) which was aged in a temperature range between 300 and 700 °C for 0.5, 1, and 5 h and cooling rates after aging was controlled by cooling media (water quench, air cooling, and furnace cooling). It was reported that the cooling rates indeed showed an impact not only on the transformation temperatures, but also on the transformation sequence of various phases involved in the martensitic transformations.

5.5 Cryogenic treatment

Kim et al. [42] investigated the effects of cryogenic treatment on $Ni_{56}Ti_{44}$ (wt%) alloy, which was subjected to a cryogenic treatment in liquid nitrogen (–196 °C). It was reported that (i) the cryogenically treated specimens had a significantly higher microhardness than the controls, and (ii) there was no measurable change in elemental or crystalline phase composition. Vinothkumar et al. [43, 44] studied the role of dry cryogenic treatment conditions on the microstructure, hardness, and wear resistance of martensitic shape-memory $Ni_{51}Ti_{49}$ (wt%) alloy. Samples were subjected to two groups: –185 and –80 °C. It was found that (i) microstructures of all specimens had equiaxed grains (approximately 25 μm) with well-defined boundaries and precipitates, (ii) the x-ray diffraction patterns of cryogenically treated specimens revealed accentuation of austenite and martensite peaks, (iii) the sample possessed A_F of 45.76 °C, (iii) reduction in Vickers hardness of specimens was high, and (iv) the weight loss due to wear was significantly high; concluding that dry cryogenic treatment for 24 h soaking period increases the martensite content of the shape-memory NiTi alloy without altering the grain size, and reduces the hardness and wear resistance of the alloy.

5.6 Aging

The heat treatment for NiTi-based alloys is conducted to train (or educate) the original shape that will be recovered later. For this specific aim, there are normally three major ways: (1) intermediate treatment, (2) low temperature treatment, and (3) aging treatment. By the intermediate treatment, the alloy was sufficiently cold worked by rolling or wire drawing to form a certain shape, followed by heating at 400–500 °C for up to several hours to memorize the shape. By increasing the treating temperature, larger shape recovery is expected. By the low-temperature treatment, the alloy is quenched from higher than 800 °C, followed by normalizing and subsequently the alloy is formed into a certain shape, followed by holding at a temperature range from 200 to 300 °C. This technique is suitable for memorizing relatively complicated shape, although shape recovery capability is somewhat less than the alloy treated by the intermediate treatment. By aging treatment, the alloy is first solution treated at 800–1,000 °C, followed by water quenching; thus, the solutionized alloy is subjected to aging treatment at around 400 °C and this aging treatment is applicable only for Ni-rich NiTi (with more than 50.5 at% Ni) alloy.

Aging heat treatment is normally performed on material that was previously solution-treated and quenched, so that aging the solutionized material allows the alloying elements to diffuse through the microstructure and form various types of intermetallic particles most of which can be found in the equilibrium phase diagram. Martensite stability is the result of aging in the martensitic phase. Ni-rich NiTi alloys are unstable in the sense that M_S shifts in annealed materials can occur during prolonged exposure at temperatures which could be met during service. This instability is the consequence of the formation of precipitates, and the precipitation sequence has been studied in detail. The most important macroscopic effect is the increase of the reverse transformation temperature of A_S and A_F. The precipitate Ti_3Ni_4 can be formed in NiTi alloys with more than 50.5 at% Ni, therefore it is clear that fine precipitate is effective in stabilizing the martensitic transformation temperature against thermal cycling. By aging (400 °C × 1 h) at higher temperatures, these precipitates grow and their density becomes lower, resulting in a decrease of the critical stress for slip. In such cases, M_S decreases with increasing thermal cycling, while T_R, the rhombohedral transition temperature increase in Ni-rich NiTi alloys decompose during aging through a series or more stable compounds: $Ni_{14}Ti_{11}$, Ni_4Ti_3, Ni_3Ti_2, and finally Ni_3Ti, as seen in the equilibrium phase diagram for the Ni–Ti binary alloy except the Ni_4Ti_3 which is not equilibrium and stable neither. These precipitates strengthen the NiTi matrix, which enhances SE but also causes a shift in the transformation temperatures [21, 45–47]. Hence, the martensitic transformation and related appearance of precipitates are, to some extent, related to aging heat treatment. In this section, we will discuss aging effects on these phenomena. There are numerous studies on the presence of Ni_4Ti_3 precipitates during aging in NiTi alloy, particularly Ni-rich NiTi alloys and the precipitation of Ni_4Ti_3 plays a critical role in determining the martensitic transformation path and temperature in NiTi SMAs [48–56]. The coherent Ni_4Ti_3 precipitates also give rise to the presence of an internal strain field [51, 57–60] around the precipitates as well as the depletion of Ni in the matrix [60–63].

In general, with respect to the effects of heat treatment to phase transformation, there are several phases involved. During and after these treatments, an intermediate rhombohedral phase (R-phase) between A-phase (B2) and M-phase (B19′) has been frequently observed, resulting in changing the transformation path from A→M into A→R→M [25, 35, 64, 65]. As the A↔M transformation, the A↔R also shows both shape recovery effect and pseudoelasticity [65, 66]. Moreover, compared to the M transformation, the R-phase transformation shows unique properties (e.g., narrow hysteresis, high fatigue life, etc.), which attracts widespread interest for practical applications. The R-phase transformation was characterized as a premartensitic transformation [67, 68]. It was moreover considered as a second-order transformation [69]. Now it is clear that the A→R is an independent martensitic transformation created by the rhombohedral distortion of the A lattice that occurs before the M transformation [66,

70–72], and it is mainly considered as a first-order transformation [73, 74]. However, the R-phase also shows second-order transformation behavior, as after the A→R transformation, the rhombohedral angle of the R-phase gradually decreases with decreasing temperature or increasing stress [63, 75, 76].

It was pointed out that, although the effect of intraoral environmental conditions on the superelastic properties of NiTi archwires and coil springs requires further research to establish the true spectrum of effects, it has been suggested that intraoral temperature variations might transiently affect their properties and that the fracture resistance of used NiTi wires is reduced when these orthodontic appliances were exposed to intraoral aging environment [77].

Aiming for more reliable applications of NiTi SMA in damping of civil engineering structures, a reliability study of the static and dynamic properties was demanded. Auguet et al. [78] studied the static (temperature and time effects) and dynamic actions in pseudoelastic NiTi SMA with a consideration of the long-time effects of temperature and time. It was reported that (i) the transformation temperatures change with time of aging even at moderate temperatures (i.e., near 100 °C), depending on time and temperature, (ii) it is possible to visualize the diffusion change in the R phase transformation via classical x-ray characterization, and (iii) for long times (10, 20 years) and direct sunny actions, more deep analysis is required. Furthermore, the same research group [79] investigated the metastable effects of aging on martensitic transformation of NiTi SMA at 90 °C for nearly 100 days and at 137 °C for 1.9 days. It was mentioned that the change of transformation temperatures under the action of the external temperature (or room temperature change) is close to 15 K.

In a temperature range from 200 to 300 °C aging temperature, Kim et al. [80] investigated the effects of low-temperature aging on the transformation behavior of a TiNi$_{50.9}$ (at%) alloy during the aging in the temperature range from 200 to 300 °C. It was reported that (i) the aging induced a two-stage R-phase transformation, which was followed by a single-stage martensitic transformation at a lower temperature, and (ii) the presence of fine coherent Ti$_3$Ni$_4$ precipitates was noticed; suggesting that the first R-phase transformation on cooling was associated with the formation of the precipitates whereas the second R-phase transformation at a lower temperature was from the matrix away from precipitates. It was furthermore mentioned that the martensitic transformation was associated with the second R-phase transformation at the lower temperature, that is, it was an incomplete transformation from the regions away from precipitates. The occurrence of the multiple-stage R-phase transformation is attributed to precipitation-induced inhomogeneity of the matrix, both in terms of composition and of internal stress fields. Chien et al. [81], characterizing 250 °C aged Ti$_{48.7}$Ni$_{51.3}$ SMA, concluded that (i) R-like phase and Ti$_3$Ni$_4$ precipitates coexisting in the strain glass of 250 °C early-aged specimens were observed and they also demonstrated the growth behavior of the real nanodomains exhibited in specimens during thermal aging, and (ii) increasing the aging time can elevate the T_G from about −60 °C to about −20 °C, which indicates the tunable characteristic of T_G for as-quenched $Ti_{48.7}Ni_{51.3}$ alloy.

The coherent Ni_4Ti_3 precipitates in a $TiNi_{50.8}$ (at%) after 250 °C aging treatment [82], in $TiNi_{50.9}$ (at%) after 300 °C aging treatment [83], and $TiNi_{50.7}$ (at%) after 300 °C × 100 h aging treatment [84] were reported. It was also mentioned that the transformation temperature of the A↔R transformation under both constant stress and constrained strain conditions can be controlled by aging time. This indicates that the R-phase transformation can be efficiently used in appropriate devices [83].

Wu et al. [85] investigated the $Ti_{49}Ni_{51}$ alloy aged at 400 °C and reported that (i) the transformation sequence of B2→r premartensite R-phase→r martensite was found, (ii) in the early aging stage, only the premartensitic transformation is observed due to the M_S point being deeply depressed by the coherent stress of $Ti_{11}Ni_{14}$ precipitates; while in the later aging stage, internal friction peaks associated with premartensitic and martensitic transformations are all observed on both heating and cooling, and (iii) the $Ti_{11}Ni_{14}$ precipitates can enhance the amount of martensite formed by unit of temperature or time during the martensitic transformation. Zheng et al. [83] investigated the effects of aging treatment on transformation behavior of $TiNi_{50.9}$ (at%) at between 300 and 540 °C. It was described that (i) although through the aging was at below 300 °C, the original A↔M transformation was suppressed, (ii) aging at between 300 and 560 °C resulted in complex multiple-stream transformations, and (iii) aging at above 560 °C led to a single-stage A↔M transformation, which indicates total dissolution of coherent Ti_3Ni_4. The effect of aging treatment of 300–550 °C for 2–180 min on $TiNi_{50.8}$ (at%) wire which was previously cold worked with 30% and 50% was investigated [86]. It was reported that (i) the maximum precipitation rate occurred at approximately 450 °C, (ii) the precipitation strengthening was evident in both 30% and 50% cold-worked wires, and (iii) recrystallization began at approximately 450 °C for both wires. Saedi et al. [87] investigated the effects of solution annealing and subsequent aging on shape-memory response of $TiNi_{50.8}$ (at%) alloys fabricated by the selective laser melting method. After solutionizing, samples were heat treated at selected times at 350 and 450 °C. It was found that (i) transformation temperatures, transformation behavior, strength, and recoverable strain are highly heat treatment dependent, (ii) samples aged at 350 °C showed better recovery in SE tests where 350 °C × 18 h aged samples exhibited almost perfect SE with 95% recovery ratio with 5.5% strain in the first cycle and stabilized SE with a recoverable strain of 4.2% after 10th cycle, and (iii) the 450 °C × 10 h aged sample exhibited 68% recovery with a recoverable strain of 4.2% in the first cycle and stabilized recoverable strain of 3.8% after 10th cycle. Radi et al. [88] studied the effect of a stress aging process on the microstructure and martensitic phase transformation of NiTi SMA, which was aged at 450 °C for 1 and 5 h under different levels of external tensile stress of 15, 60, and 150 MPa. It was reported that (i) the application of all stress levels restricts the formation of precipitates variants in the microstructure after 1 h stress aging process, (ii) the stress aging process resulted in changing the shape of precipitates in comparison with that of the stress-free aged samples, and (iii) the martensite start temperature on cooling shifts to higher temperatures with increasing the tensile stress during the aging process, being related to the change of austenite to martensite interface energy due to the different volume fractions and variants of precipitates [88, 89].

In a temperature range between 300 and 600 °C, the effects of aging treatment were investigated [49, 88, 90–102], the main findings are as follows: (i) phase transformation was B2↔R↔B19 in NiTi [90, 100] and B2↔B19′ in Cu–NiTi [90] and Hf–NiTi [98], (ii) the formation of fine Ni_4Ti_3 metastable precipitates [49, 91, 101], (iii) M_S and A_F temperatures for alloys aged at 400 and 450 °C were higher than those for others aged at 500 and 600 °C [92], (iv) aging at 550 °C for 3 h introduced coherent 10–20 nm precipitates in the matrix, which substantially improved the shape-memory and mechanical properties of the Hf–NiTi [97], (v) perfect SE with recoverable strain (if unrecovered strain exists, it should be a measure of dimensional instability) of >4% was observed for solutionized and 550 °C × 3 h aged single crystals [97], (vi) in the stress aging process, the M_s on cooling shifts to higher temperatures with increasing the tensile stress during the aging process; being related to the change of austenite to martensite interface energy due to the different volume fractions and variants of precipitates [88], (vii) precipitate crystal structure was identified as the H-phase that was recently reported in Ni-rich NiTiHf and NiTiZr alloys, and not the Ni_4Ti_3-type structure as reported in a few earlier studies [98], and (viii) aging at above 540 °C led to a single-stage A ↔ M transformation, which indicates total dissolution of coherent Ni_4Ti_3 [102].

Aging effects are studied at much higher temperatures. Karimzadeh et al. [103] investigated the effects of aging temperature ranging from 300 to 700 °C in $TiNi_{51.5}$ (at%). It was found that (i) aging at 300 and 400 °C led to the occurrence of the austenite to R phase (A↔R) transformation, (ii) aging at 500 and 600 °C led to the appearance of B19′ phase during cooling and showed the best mechanical properties, in comparison with the other aging temperatures, (iii) complete SE was observed by aging at 400 and 500 °C, and (iv) the alloy aged at 700 °C showed the transformation behavior and mechanical properties similar to the solution-annealed sample which showed no superelastic property. Aliasgarian et al. [104] heat-treated $TiNi_{52.5}$ (at%) and $TiNi_{50.5}$ (at%) and studied the tribological behavior. It was reported that (i) the alloys aged at 700 °C showed lower hardness comparing to the alloys aged at 400 °C, and (ii) under an applied load of 20 N, the samples aged at 700 °C showed better wear behavior in comparison with the samples aged at 400 °C with a higher hardness. The precipitation was noticed in $TiNi_{49}$ aged at 400, 500, 600, and 700 °C for 1 h [105]. Adharapurapu et al. [106] investigated effects of cooling rate and transformation behavior in $TiNi_{50.8}$ (at%) aged at between 300 and 700 °C for 0.5, 1, and 5 h. It was mentioned that the cooling rate effect becomes critical for structural applications of NiTi with thick sections since its influence on the multiple-step martensitic transformations.

There are a few scattered studies on the effects of cyclic aging. Zhang et al. [107] investigated the reversible change in transformation temperatures associated with alternating aging in $TiNi_{51}$ (at%). Since SME and SE are closely related to martensitic and R-phase transformations, transformation temperature control is very important for applications of SMAs. During alternating aging between 440 and 510 °C, it was mentioned that this phenomenon can be simply explained by considering the phase equilibrium between the TiNi matrix and the metastable Ti_3Ni_4 precipitates at different aging temperatures. Jiang et al. [108] studied the effects of cyclic ageing on the transformation behavior of $TiNi_{50.8}$ (at%) between 400 and 560 °C. It was reported that the alternating aging caused reversible changes in the transformation behavior and alternating increases and decreases in transformation temperatures.

5.7 Annealing

Generally, annealing is performed by heating a metal or alloy to a predetermined temperature, holding for a certain time, and then cooling to room temperature to improve ductility and reduce brittleness. Process annealing is carried out intermittently during the working of a piece of metal to restore ductility lost through repeated hammering or other working. Full annealing is done to give workability to such parts as forged blanks destined for use in the machine tool industry. Annealing is also done for relief of internal stresses. Annealing temperatures vary with metals and alloys and with properties desired but must be within a range that prevents the growth of crystals. Annealing at lower temperatures than the recrystallization temperature thermally rearranges dislocations which were introduced by the preceding cold work. Such thermally rearranged dislocations can also raise the critical stress for slip, resulting in stable transformation characteristics against thermal cycling.

Liu et al. [109] studied the damping capacity, tensile behavior, and the transformation characteristics of a NiTi SMA rolled sheet as a function of annealing conditions. It was reported that (i) both the martensite phase-damping capacity and the stress for martensite phase reorientation are a function of annealing temperature, and (ii) a higher damping of the martensite phase generally corresponds to a lower reorientation stress in the stress–strain curve, suggesting that the martensite phase-damping capacity and the stress for martensite phase reorientation are interrelated. In NiTi SMAs, it is known that both the annihilation of dislocations and the formation of Ni_4Ti_3 precipitates may occur during the postdeformation annealing, and the R-phase transformation temperatures exhibits different responses to the annealing conditions. Wang et al. [110], using $TiNi_{49.8}$ (at%) and $TiNi_{50.8}$ (at%) alloys, studied the main factor affecting the R-phase transformation temperatures during postdeformation annealing. It was found that the R-phase transformation temperatures are very stable in the $TiNi_{49.8}$ alloy; while a significant variation is observed in the $TiNi_{50.8}$ alloy with respect to the annealing and thermal cycling conditions: suggesting that the R-phase transformation temperatures are not susceptible to the change of dislocation density and depends mainly on the Ni concentration of the matrix, which can be modified by the formation of Ni_4Ti_3 precipitates. By appropriate annealing, Yang et al. [62] found a concentration gradient surrounding Ni_4Ti_3 precipitates grown by appropriate annealing in a $NiTi_{49.0}$ (at%) B2 austenite matrix, the concentration gradients of approximately 1.0–2.0 at% in Ni within the surrounding B2 matrix. Mahmud et al. [111] used $TiNi_{50.2}$ (at%) alloy that was subjected to cold work, followed by annealing treatment. It was found that the gradient-anneal resulted in varying thermal transformation behavior along the deformation behavior with a positive stress gradient.

Using Ti–50.6 at% Ni alloy, Miyazaki et al. [16] compared the mechanical behavior associated with (1) aging (aged at 400 °C for 1 h) after solution treatment and (2) annealing (annealed at 400 °C for 1 h, after cold working). It was found that (i) both aged and cold worked followed by annealed specimens show shape memory and SE associated with the R-phase transition, and (ii) when the specimen was aged after solution treatment, the critical stresses are low; while when it is annealed after the cold working, the critical stresses are higher, indicating that the dislocations have a stronger effect on the mechanical properties with the R-phase transition. Zinelis et al. [112] examined the effect of various treatments on the fatigue resistance of NiTi engine-driven endodontic file; in a range from 250 to 550 °C for 30 min. It was found that (i) the 430 and 440 °C groups showed the highest values, with fatigue resistance decreasing for thermal treatment at lower and higher temperatures. This may be the result of metallurgical changes during annealing. Ling et al. [113] studied the effects of annealing temperature (between 450 and 500 °C) on the shape recovery temperature T_F in the near-equiatomic NiTi alloy. It was reported that (i) varying the annealing temperature resulted in the largest change in these two parameters, causing an increase of 20 °C in T_F and a decrease of 13% in the shape recovery if the annealing temperature was increased much above 500 °C, and (ii) increasing the maximum strain (in the range between 0% and 8%) induced an increase of 10 °C in T_F at 500 °C with larger increases at other annealing temperatures but without any effect on the extent of shape recovery. The influence of cold work and heat treatment on the SME and plastic strain development was investigated for the thermally induced phase transformation of NiTi alloys under constant applied stress by cold rolling to reduce the specimen width 10, 20, 30 and 40% of the initial wire diameter and annealing treatment at 300, 400, or 500 °C for 15 min [114]. Kus et al. [115] studying the effect of annealing temperature (ranging from 400 to 600 °C) on the phase transformation in the near equiatomic NiTi alloy and mentioned that there was a critical temperature (about 600 °C), where the annealed specimens demonstrated remarkably different transformation courses, and also noticeable changes in transformation temperatures and heats depending on the annealing temperature.

There are some studies on annealing effect at relatively higher temperatures. Mitwally et al. [116] investigated the effects of cold work and annealing on the structure of $TiNi_{50.7}$ (at%), which was subjected to cold rolling with reductions of 20%, 30%, and 40%. It was mentioned that (i) increasing the extent of cold rolling causes an increase in the volume fraction of martensite, which results in an increase in recoverable strain, tensile strength and hardness; while this eliminates the SE plateau and results in a reduction in ductility and shape-memory recovery, (ii) after an initial sharp decrease in SE with cold rolling, further increase in cold rolling results in relative improvement in the SE (iii) annealing causes a significant increase in austenite volume fraction, leading to a decrease in hardness and SE in the temperature range from 400 to 500 °C, and (iv) recrystallization starts at 500 °C and grains grow with annealing at higher temperatures while hardness and SE increase. Huang et al. [117] studied the effect of annealing on the transformation behavior of NiTi alloy and found that (i) annealing temperature is found to significantly affect the SE of NiTi alloy, and (ii) the plateau strain of the forward stress-induced transformation increases with increasing annealing temperature, being 9% for specimens annealed at 600 °C. The influence of annealing temperatures on the thermomechanical behavior of NiTi alloy in terms of transformation temperatures, mechanical properties at ambient temperature and the recovery stress under constrained end conditions were investigated [118]. It was reported that (i) annealing the alloy above the recrystallization temperature (i.e., 600 °C ≈ 40% to 50% of melting point of 1,320 °C) reduces the maximum recovery stress significantly, even though the alloy still exhibits thermal transformations and a stress-induced martensite plateau at an annealing temperature above 600 °C [118, 119], and (ii) it is recommended that the alloy should be annealed at temperatures below 450 °C in order to produce the desired thermo-mechanical properties in the alloy for applications that exploit the SME. Khaleghi et al. [120] investigated the effect of short-time annealing treatment for 3, 9, and 18 s at 700 °C on the superelastic behavior of cold drawn Ni-rich NiTi wires. It was concluded that (i) for all cold drawn wires, annealing at 700 °C for only 3 s leads to the best superelastic behavior, and (ii) annealing at 700 °C for 3 s leads to production of an ultrafine-grained structure. Wang et al. [121] studied the effect of annealing on transformation behavior of TiNi$_{50.8}$ (at%) between 450 and 700 °C. It was found that (i) the transformation behavior and mechanical properties of both wires were sensitive to the annealing temperature, and (ii) multistage martensitic transformation was observed in the codrawn wire, compared with the one-stage A⟷M transformation in the commercial wire after annealing at 600 °C for 30 min.

5.8 Thermomechanical treatment

In passing decade, in order to improve mechanical properties of NiTi endodontic instruments, several ideas have been proposed, including thermal and mechanical treatments, machining procedures (such as twisting, electrical discharge machining, etc.), and surface modification [122], most of these items will be discussed later on in this book. Thermomechanical treatments for SMAs are found to be one of the more economical, simpler, and efficient methods adopted for manipulating the transformation properties [123]. The stability of phase transformation has been found to depend upon the thermomechanical treatments, such as hot- or cold working, heat treatment, and thermal cycling. While the fully solution annealed binary NiTi alloys exhibits only one-step B2↔B19′ transformation, precipitates or dislocations, which can be introduced by various thermomechanical treatments, are essential for inducing the R-phase transformation [122–125]. As stated previously, the phase transformations associated with SME in NiTi alloys can be one-stage, B19′ (martensite)↔B2 (austenite), two-stage including an intermediate R-phase stage, or multiple-stage depending on the thermal and/or mechanical history of the alloy [126]. Thermomechanical training is also an important process to develop a two-way memory effect in SMAs [127]. Figure 5.7 illustrates the x-ray diffractogram depicting the effects of temperature (from room temperature to 400 °C) and prestrain on phase changes [123]. It is clear shown that (i) as the temperature is increased, the peak corresponding to B2 structure starts to emerge around 190 °C and on further heating, the intensity of the peak increases [127, 128]. Broad and low intensity peaks are due to the deformation induced dislocations and vacancies which suppresses the martensitic transformation [123, 126, 130].

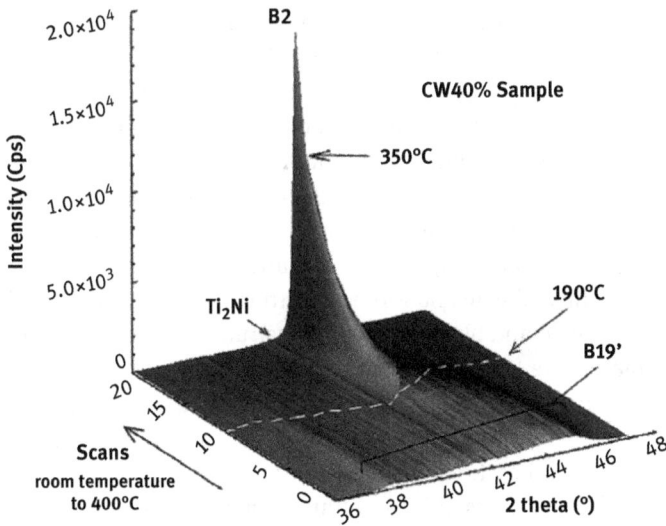

Figure 5.7: X-ray diffractogram showing effects of temperature and prestrain on transformation in NiTi alloy [123].

Bowers et al. [131] studied a near-equiatomic NiTi SMA which was subjected to a variety of thermomechanical treatments including pure thermal cycling and load-biased thermal cycling to investigate microstructural evolution of the material under actuating conditions. It was reported that (i) high temperature observations of the austenite phase show rapid accumulation of dislocations and moderate deformation twinning upon thermomechanical cycling, and (ii) there was an emergence of fine crystallites from the original coarse austenite grain structure. Pereira et al. [132] studied the mechanical properties of thermomechanically treated NiTi M-wire which was subjected to proprietary thermomechanical process; resulting in containing three crystalline phases (deformed and micro-twinned martensite, R-phase, and austenite). It was found that (i) B19′ martensite and the R-phase were detected, and (ii) the stress at the transformation plateau in the tensile load–unload curves was lower and more uniform and the M-wire showed the smallest stress hysteresis and apparent elastic modulus. The superiority of M-wire was demonstrated in terms of improved flexibility and fatigue resistance, and reduced microhardness while maintaining similar characteristics of the surface [133]. During the thermomechanical cycling, (i) the decrease in the transformation stresses, (ii) the rise in the transformation temperatures, and (iii) the variation in these phenomena with the number of cycles [134] were noted. Zupanc et al. [121] studied on endodontic files of the austenitic phase (conventional NiTi, M-wire, R-phase) and martensitic phase (CM wire, gold and blue heat-treated NiTi). Basically, the austenitic instruments possess superelastic properties and reveal high torque values at fracture. Thus, these files are appropriate to shape straight or slightly curved root canals. Additionally, the use of austenitic alloy in pathfinding instruments may compensate for the decreased torque resistance caused by the smaller diameter of these files. The martensitic instruments are more flexible with an enhanced resistance to cyclic fatigue and reveal a greater angle of rotation but lower torque at fracture. Cyclic fatigue is known to occur more likely in complex curvatures and root canal anatomies. Thus, martensitic instruments should be preferred in cases of severely curved root canals or those with a double curvature. It was reported that (i) the thermomechanical treatment of NiTi alloy allows a change in the phase composition leading to the appearance of martensite or R-phase under clinical conditions, and (ii) whilst M-wire and R-phase instruments maintain an austenitic state, CM wire and the gold and blue heat-treated instruments is composed of substantial amounts of martensite [121]. Seyyed et al. [135] treated NiTi and NiTiCu wires by thermomechanical treatments with final step of 20% cold drawing followed by annealing at different temperatures of 300 and 400 °C for varying times of 10, 30, and 60 min. It was concluded that the thermomechanical treatment at 300 °C for 30 min was the suitable process in terms of SE and transformation temperatures for orthodontic application. It was mentioned that, during the thermomechanical treatment on NiTi, (i) the hardening in stress–strain hysteresis loop and sharp change of strain in strain-temperature hysteresis loop were found, and (ii) phase transition took place by nucleation and propagation of transformation fronts [136].

Lekston et al. [137] investigated the phase transitions of TiNiCo and Ni-rich NiTi SMAs which are originally designed for medical applications. It was found that (i) aging after solution treatment and annealing below the recrystallization temperature after cold working in the alloys studied create separate reversible B2,↔R↔B19 transformations, (ii) during thermomechanical cycles characteristic temperatures of the reversible B2 ↔ R phase transition remain stable, and (iii) aging after solution treatment or recovery during annealing after cold working causes the precipitation process and the changes of the defect structure of the alloys promote transitions with the R-phase contribution. Tobushi et al. [138] investigated the thermomechanical properties of the SME and SE due to the martensitic transformation and the R-phase transformation of a TiNi SMA, and found that the thermomechanical properties due to the R-phase transformation were excellent for deformation with high cycles. Frenzel et al. [139], using as-cast TiNi$_{48}$Fe$_2$ (at%) SMAs, investigated their responses to the microstructure and the phase transformation behavior with different thermomechanical treatments; the applied procedures involve homogenization, swaging at different temperatures and recrystallization annealing. It was reported that (i) during repeated hot working in combination with recrystallization treatments, a fine-grained microstructure evolves in the surface volume of the swaged Fe–NiTi rod, (ii) in contrast, its center remains almost unaffected by recrystallization (because less dislocations are introduced during swaging and recovery processes are more intense), this microstructural heterogeneity can be avoided by cold work at the end of the thermomechanical procedure, (iii) swaging at low temperatures results in higher dislocation densities and thus in a more homogeneous microstructural evolution during recrystallization, and (iv) in the cold-worked state, the phase transition R $\rightarrow$ B19′ is strongly impeded by the presence of dislocations, whereas the transition characteristics of the recrystallized and of the nondeformed original material state almost coincide.

References

[1]   Van Humbeeck J, Stalmans R, Chandrasekaran M, Delaey L. On the Stability of Shape Memory Alloy. In: Duerig TW, et al. ed., Engineering Aspects of Shape Memory Alloys. Butterworth-Heinemann, London, UK, 1990, 96–105.

[2]   Mohouji HS, Hamedi M, Salehi M. Modeling, validation and testing of a Ti-49.8% Ni shape memory actuator. J. Intell. Mater. Syst. Struct. 2014, 26, 2196–204.

[3]   Santamarta R, Pons J, Cesari E, Segui C. Thermal martensite stabilization in Ni-Ti based alloys. J. Phys. IV (Proceed.) 2003, 112;doi: 10.1051/jp4:2003966.

[4]   Morgan NB, Friend CM. A review of shape memory stability in NiTi alloys. J. Phys. IV France 2001, 11, 325–32.

[5]   Burow J, Prokofiev E, Somsen C, Frenzel J, Valiev RZ, Eggeler G. Martensitic transformations and functional stability in ultra-fine grained NiTi shape memory alloys. Mater. Sci. Forum 2008, 584/586, 852–7.

[6]   Massa JE, Buchheit TE, McLaughkin JT. Characterization of shape memory alloys for safety mechanisms. Sandia Report 2008, SAND 2007–2008, 41 pages; https://prod-ng.sandia.gov/techlib-nonauth/access-control.cgi/2007/078000.pdf.

[7]   Sehitoglu H, Karaman I, Zhang X, Viswanath A, Chumlyakov Y, Maier HJ. Strain–temperature behavior of NiTiCu shape memory single crystals. Acta Mater. 2001, 49, 3621–34.

[8]   Morgan N. The stability of NiTi shape memory alloys and actuator applications. Thesis, Cranfield University, 1999; http://hdl.handle.net/1826/780.

[9]   Strnadel B, Ohashi S, Ohtsuka H, Ishihara T, Miyazaki S. Cyclic stress-strain characteristics of
        Ti-Ni and Ti-Ni-Cu shape memory alloys. Mater. Sci. Eng. A 1995, A202, 148–56.
[10]   Özkul I, Kalay E, Canbay CK. The investigation of shape memory recovery loss in NiTi alloy.
        Mater. Res. Exp. 2019, 6; doi: 10.1088/2053-1591/ab21c7.
[11]   Sedmák P, Šittner P, Pilch J, Curfs C. Instability of cyclic superelastic deformation of NiTi
        investigated by synchrotron X-ray diffraction. Acta Mater. 2015, 94, 257–70.
[12]   Grossmann C, Frenzel J, Sampath V, Depka T, Eggeler G. Elementary transformation and
        deformation processes and the cyclic stability of NiTi and NiTiCu shape memory spring
        actuators. Metall. Mater. Trans. A 2009, 40, 2530–44.
[13]   Zupanc J, Vahdat-Pajouh N, Schäfer E. New thermomechanically treated NiTi alloys – a review.
        Int'l Endod. J. 2018, 51, 1088–103.
[14]   Zhou H, Peng B, Zheng YF. An overview of the mechanical properties of nickel-titanium
        endodontic instruments. Endod. Topics 2013, 29, 42–54.
[15]   Ortín J, Delaey L. Hysteresis in shape-memory alloys. Int. J. Non-Lin. Mech. 2002, 37, 1275–81.
[16]   Miyazaki S, Otsuka K. Deformation and transition behaviour associated with the R Phase in
        Ti-Ni alloys. Metall. Mater. Trans. A 1986, 17A, 53–63.
[17]   Alapati SB, Brantley WA, Iijima M, Clark WA, Kovarik L, Buie C, Liu J, Ben Johnson
        W. Metallurgical characterization of a new nickel-titanium wire for rotary endodontic
        instruments. J. Endod. 2009, 35, 1589–93.
[18]   Kim HC, Yum J, Hur B, Cheung GS. Cyclic fatigue and fracture characteristics of ground and
        twisted nickel-titanium rotary files. J. Endod. 2010, 36, 147–52.
[19]   Tomozawa M, Kim HY, Miyazaki S. Microactuators using R-phase transformation of sputter-
        deposited Ti-47.3Ni shape memory alloy thin films. J. Intel. Mat. Syst. Str. 2006, 17, 1049–58.
[20]   Hou H, Hamilton RF, Horn MW. Narrow thermal hysteresis of NiTi shape memory alloy thin films
        with submicrometer thickness; https://avs.scitation.org/doi/am-pdf/10.1116/1.4959567.
[21]   Nishida M, Wayman CM, Honma T. Precipitation processes in near-equiatomic TiNi shape
        memory alloys. Precipitation processes in near-equiatomic TiNi shape memory alloys. Metall.
        Trans. A 1986, 9, 1505–15.
[22]   Pelton AR, DiCello J, Miyazaki S. Optimization of processing and properties of medical grade
        Nitinol wire. Minim. Invas. Therap. Allied Technol. 2000, 9, 107–18.
[23]   Zapoticla F. The effects of applied strain and heat treatment on the properties of NiTi wire
        during shape setting. Thesis, California Polytechnic State University, San Luis Obispo, 2010;
        https://pdfs.semanticscholar.org/3942/79567afaab2922f8bd7b3bdd9fcfc56b239b.pdf.
[24]   Michutta J, Carroll C, Yawny A, Somsen C, Neuking K, Eggeler G. Martensitic phase
        transformation in Ni-rich NiTi single crystals with one family of Ni4Ti3 precipitates. Mater. Sci.
        Eng. A 2004, 378, 152–6.
[25]   Bataillard L, Bidaux J-E, Gotthardt R. Interaction between microstructure and multi-step
        transformation in binary NiTi alloys using in-situ transmission and electron microscopy
        observations. J. Philos. Mag. A 1998, 78, 327–44.
[26]   Favier D, Liu Y, Orgeas L, Sandel A, Debove L, Comte-Gaz P. Influence of thermomechancial
        processing on the superelastic properties of a Ni-rich Nitinol shape memory alloy. Mater. Sci.
        Eng. A 2006, 429, 130–6.
[27]   Zarinejad M, Liu Y, Tong Y. Transformation temperature changes due to second phase precip-
        itation in NiTi-based shape memory alloys. Intermetallics 2009, 17, 914–9.
[28]   Prokofiev EA, Burow JA, Payton EJ, Zarnetta R, Frenzel J, Gunderov DV, Valiev RZ, Eggeler G.
        Suppression of Ni_4Ti_3 precipitation by grain size refinement in Ni-Rich NiTi shape memory
        alloys. Adv. Eng. Mater. 2010, 12, 747–53.
[29]   Shi X, Cui L, Jiang D, Yu C, Guo F, Yu M, Ren Y, Liu Y. Grain size effect on the R-phase
        transformation of nanocrystalline NiTi shape memory alloys. J. Mater. Sci. 2014, 49, 4643–7.

[30]   Jordan L, Masse M, Collier JY, Bouquet G. Effects of thermal and thermomechanical cycling on the phase transformations in NiTi and Ni-Ti-Co shape-memory alloys. *J. Alloys Compd.* 1994, 211, 204–7.

[31]   Miyazaki S, Igo Y, Otsuka K. Effect of thermal cycling on the transformation temperatures of Ti-Ni alloys. Acta Met. 1986, 34, 2045–51.

[32]   Tadaki T, Nakata Y, Shimizu K. Thermal cycling effects in an aged Ni-rich i-Ni shape memory alloy. Trans. JIM 1987, 28, 883–90.

[33]   Pelosin V, Riviere A. Effect of thermal cycling on the R-phase and martensitic transformations in a Ti-rich NiTi alloy. Metall. Mater. Trans. A 1998, 29, 1175–80.

[34]   Pelton AR, Huang GH, Moine P, Sinclair R. Effects of thermal cycling on microstructure and properties in Nitinol. Mater. Sci. Eng., A 2012, 532, 130–8.

[35]   Uchil J, Ganesh Kumara K, Mahesh KK. Effect of thermal cycling on R-phase stability in a NiTi shape memory alloy. Mater. Sci. Eng. A 2002, 332, 25–8.

[36]   Berzins DW, Roberts HW. Phase transformation changes in thermocycled nickel-titanium orthodontic wires. Dent. Mater. 2010, 26, 666–74.

[37]   Wang Z, Zu XT, Huo Y. Effect of heating/cooling rate on the transformation temperatures in TiNiCu shape memory alloys. Thermochim. Acta 2005, 436, 153–5.

[38]   Nurveren K, Akdoğan A, Huang WM. Evolution of transformation characteristics with heating/cooling rate in NiTi shape memory alloys. J. Mater. Process. Technol. 2008, 196, 129–34.

[39]   Zhang YQ, Jiang SY, Zhao YN, Tang M. Influence of cooling rate on phase transformation and microstructure of Ti-50.9%Ni shape memory alloy. Trans. Nonferrous Met. Soc. China 2012, 22, 2685–90.

[40]   Motemani Y, Nili-Ahmadabadi M, Tan MJ, Bornapour M, Rayagan S. Effect of cooling rate on the phase transformation behavior and mechanical properties of Ni-rich NiTi shape memory alloy, *J. Alloys Compd.* 2009, 469, 164–8.

[41]   Adharapurapu RR, Vecchio KS. Effects of aging and cooling rate on the transformation of nanostructured Ti-50.8Ni. J. Alloys Compd. 2017, 693, 150–63.

[42]   Kim JW, Griggs JA, Regan JD, Ellis RA, Cai Z. Effect of cryogenic treatment on nickel-titanium endodontic instruments. Intl. Endod. J. 2005, 38(6), 364–71.

[43]   Vinothkumar TS, Kandaswamy D, Prabhakaran G, Rajadurai A. Microstructure of cryogenically treated martensitic shape memory nickel-titanium alloy. J. Conserv. Dent. 2015, 18, 292–6.

[44]   Vinothkumar TS, Kandaswamy D, Prabhakaran G, Rajadurai A. Effect of dry cryogenic treatment on Vickers hardness and wear resistance of new martensitic shape memory nickel-titanium alloy. Eur. J. Dent. 2015, 9, 513–7.

[45]   Duerig TW, Zadno R. An Engineer's Perspective of Pseudoelasticity. In: Duerig TW, et al. ed., Engineering Aspects of Shape Memory Alloys. Butterworth-Heinemann, London, UK, 1990, 369–93.

[46]   Tadaki T, Nakata Y, Shimizu K. Thermal cycling effects in an aged Ni-rich i-Ni shape memory alloy. Trans. JIM 1987, 28, 883–90.

[47]   Nishida M, Wayman CM. Electron microscopy studies of precipitation processes in near-equiatomic TiNi shape memory alloys. Mater. Sci. Eng. 1987, 93, 191–203.

[48]   Zhou N, Shen C, Wagner F-X, Eggeler G, Mills MJ, Wang Y. Effect of Ni_4Ti_3 precipitation on martensitic transformation in Ti–Ni. Acta Mater. 2010, 58, 6685–94.

[49]   Khalil-Allafi J, Dlouhy A, Eggeler G. Ni_4Ti_3-precipitation during aging of NiTi shape memory alloys and its influence on martensitic phase transformations. Acta Mater. 2002, 50, 4255–74.

[50]   Zhang JX, Sato M, Ishida A. On the Ti_2Ni precipitates and Guinier–Preston zones in Ti-rich Ti–Ni thin films. Acta Mater. 2003, 51, 3121–30.

[51]   Tirry W, Schryvers D. Quantitative determination of strain fields around Ni_4Ti_3 precipitates in NiTi. Acta Mater. 2005, 53, 1041–9.

[52] Yang F, Coughlin DR, Phillips PJ, Yang L, Devaraj A, Kovarik L, Noebe RD, Mills MJ. Structure analysis of a precipitate phase in an Ni-rich high-temperature NiTiHf shape memory alloy. Acta Mater. 2013, 61, 3335–46.

[53] Santamarta R, Arróyave R, Pons J, Evirgen A, Karaman I, Karaca HE, Noebe RD. TEM study of structural and microstructural characteristics of a precipitate phase in Ni-rich Ni–Ti–Hf and Ni–Ti–Zr shape memory alloys. Acta Mater. 2013, 61, 6191–206.

[54] Karaca HE, Saghaian SM, Ded G, Tobe H, Basaran B, Maier HJ, Noebe RD, Chumlyakov YI. Effects of nanoprecipitation on the shape memory and material properties of an Ni-rich NiTiHf high temperature shape memory alloy. Acta Mater. 2013, 61, 7422–31.

[55] Wang X, Kustov S, Li K, Schryvers D, Verlinden B, Van Humbeeck J. Effect of nanoprecipitates on the transformation behavior and functional properties of a Ti–50.8 at.% Ni alloy with micron-sized grains. Acta Mater. 2015, 82, 224–33.

[56] Popoola O, Denanot MF, Moine P, Villair JP, Cahoreau M, Caisso J. Microstructural and analytical characterization of NiN_x precipitation in N^+-implanted equiatomic NiTi alloys. Acta Metall. 1989, 37, 867–76.

[57] Tadaki T, Nakata Y, Shimizu K, Otsuka K. Crystal structure, composition and morphology of a precipitate in an aged Ti-51 at%Ni shape memory alloy. Trans. Jpn. Inst. Met. 1986, 27, 731–40.

[58] Li DY, Wu XF, Ko T. The effect of stress on soft modes for the phase transformation in a Ti-Ni alloy. II. Effects of ageing and thermal cycling on the phase transformation. Philos. Mag. A 1991, 63A, 603–16.

[59] Zhang J, Cai W, Ren X, Otsuka K, Asai M. The nature of reversible change in M_s temperatures of Ti–Ni alloys with alternating aging. Mater. Trans. JIM, 1999, 40, 1367–75.

[60] Tirry W, Schryvers D. Linking a completely three-dimensional nanostrain to a structural transformation eigenstrain. Nat. Mater. 2009, 8, 752–7.

[61] Schryvers D, Tirry W, Yang ZQ. Measuring strain fields and concentration gradients around Ni_4Ti_3 precipitates. Mater. Sci. Eng. A 2006, A438/A440, 485–8.

[62] Yang Z, Schryvers D. Study of changes in composition and EELS ionization edges upon Ni_4Ti_3 precipitation in a NiTi alloy. Micron 2006, 37, 503–7.

[63] Yang Z, Tirry W, Schryvers D. Analytical TEM investigations on concentration gradients surrounding Ni_4Ti_3 precipitates in Ni–Ti shape memory material. Scr. Mater. 2005, 52, 1129–34.

[64] Fraj BB, Gahbiche A, Zghal S, Tourki Z. On the influence of the heat treatment temperature on the superelastic compressive behavior of the Ni-Rich NiTi shape memory alloy. J. Mater. Eng. Perform. 2017, 26, 5660–8.

[65] Miyazaki S, Otsuka K. Deformation and transition behavior associated with the R-phase.

[66] Miyazaki S, Kimura S, Otsuka K. Shape-memory effect and pseudoelasticity associated with the R-phase transition in Ti-50.5 at.% Ni single crystals. Philos. Mag. A 1988, 57A, 467–78.

[67] Duerig TW. Some unsolved aspects of Nitinol. Mater. Sci. Eng. A 2006, A438/A440, 69–74.

[68] Salamon MB, Meichle ME, Wayman CM. Premartensitic phases of Ti50Ni47Fe3. Phys. Rev. B 1985, 31B, 7306–15.

[69] Miyazaki S, Wayman CM. The R-phase transition and associated shape memory mechanism in Ti-Ni single crystals. Acta Metall. 1988, 36, 181–92.

[70] Otsuka K, Ren X. Physical metallurgy of Ti–Ni-based shape memory alloys. Prog. Mater. Sci. 2005, 50, 511–678.

[71] Sittner P, Lukás̆ P, Landa M, Novák V. Stress-strain-temperature behavior due to B2-R-B19′B2-R-B19′ transformation in NiTi polycrystals. J. Eng. Mater. Technol. 2006, 128, 268–78.

[72] Lukás̆ P, S̆ittner P, Lugovoy D, Neov D, Ceretti M. In situ neutron diffraction studies of the R-phase transformation in the NiTi shape memory alloy. Appl. Phys. A 2002, 74A, s1121–3.

[73] Goo E, Sinclair R. The B2 to R transformation in $Ti_{50}Ni_{47}Fe_3$ and $Ti_{49.5}Ni_{50.5}$ alloys. Acta Metall. 1985, 33, 1717–23.

[74] Fukuda T, Saburi T, Doi K, Nenno S. Nucleation and self-accommodation of the R-phase in Ti–Ni alloys. Mater. Trans. JIM 1992, 33, 271–7.

[75] Sittner P, Landa M, Lukás̆ P, Novák V. R-phase transformation phenomena in thermomechanically loaded NiTi polycrystals. Mech. Mater. 2006, 38, 475–92.

[76] Sittner P, Sedlák P, Landa M, Novák V, Lukás̆ P. *In situ* experimental evidence on R-phase related deformation processes in activated NiTi wires. Mater. Sci. Eng. A 2006, A438/A440, 579–84.

[77] Eliades T, Bourauel C. Intraoral aging of orthodontic materials: the picture we miss and its clinical relevance. Am. J. Orthod. Dentofacial Orthop. 2005, 127, 403–12.

[78] Auguet C, Isalgue A, Torra V, Lovey FC, Pelegrina JL. Metastable effects on martensitic transformation in SMA part VII. Aging problems in NiTi. J. Therm. Anal. Calorim. 2008, 92, 63–71.

[79] Auguet C, Isalgue A, Lovey FC, Pelegrina JL, Ruiz S, Torra V. Metastable effects on martensitic transformation in SMA. J. Therm. Anal. Cal. 2007, 89, 537–42.

[80] Kim JI, Liu Y, Miyazaki S. Ageing-induced two-stage R-phase transformation in Ti-50.9at.%Ni. Acta Mater. 2004, 52, 487–99.

[81] Chien C, Tsao C-S, Wu S-K, Chang C-Y, Chamg P-C, Kuo Y-K. Characteristics of the strain glass transition in as-quenched and 250 °C early-aged $Ti_{48.7}Ni_{51.3}$ shape memory alloy. Acta Mater. 2016, 120, 159–67.

[82] Wang X, Kustov S, Verlinden B, Van Humbeeck J. Fundamental development on utilizing the R-phase transformation in NiTi shape memory alloys. Shape Mem. Superelastic. 2015, 1, 231–9.

[83] Zheng Y, Jiang F, Li L, Yang H, Liu Y. Effect of ageing treatment on the transformation behaviour of Ti–50.9 at.% Ni alloy. Acta Mater. 2008, 56, 736–45.

[84] Chang SH, Lin KH, Wu SK. Effects of cold-rolling/aging treatments on the shape memory properties of Ti49.3Ni50.7 shape memory alloy. Materials (Basel) 2017, 10; doi: 10.3390/ma10070704.

[85] Wu SK, Lin HC, Chou TS. A study of electrical resistivity, internal friction and shear modulus on an aged $Ti_{49}Ni_{51}$ alloy. *Acta Metall. Mater.* 1990, 38, 95–102.

[86] Drexel M, Selvaduray G, Pelton A. The effects of cold work and heat treatment on the properties of nitinol wire. ASM International 2008, Proceedings of the International Conference on Shape Memory and Superelastic Technologies, 447–54.

[87] Saedi S, Turabi AS, Andani MT, Haberland C, Karaca H, Elahinia M. The influence of heat treatment on the thermomechanical response of Ni-rich NiTi alloys manufactured by selective laser melting. J. Alloys Compd. 2016, 677, 204–10.

[88] Radi A, Khalil-Allafi J, Etminanfar MR, Pourbabak S, Schryvers D, Amin-Ahmadi B. Influence of stress aging process on variants of nano-Ni_4Ti_3 precipitates and martensitic transformation temperatures in NiTi shape memory alloy. Mater. Des. 2018, 142, 93–100.

[89] Li DY, Chen LQ. Selective variant growth of coherent precipitate under external constraints. J. Phase Equilib. 1998, 19, 523–8.

[90] Seyyed Aghamiri SM, Ahmadabadi MN, Raygan Sh. Combined effects of different heat treatments and Cu element on transformation behavior of NiTi orthodontic wires. J. Mech. Behav. Biomed. Mater. 2011, 4, 298–302.

[91] Khoo ZX, An J, Chua CK, Shen YF, Kuo CN, Liu Y. Effect of heat treatment on repetitively scanned SLM NiTi shape memory alloy. Materials (Basel) 2018, 12(1), pii: E77; doi: 10.3390/ma12010077.

[92] Miyara K, Yahata Y, Hayashi Y, Tsutsumi Y, Ebihara A, Hanawa T, SUda H. The influence of heat treatment on the mechanical properties of Ni-Ti file materials. Dent. Mat. J. 2014, 33, 27–31.

[93]   Vojtěch D. Influence of heat treatment of shape memory NiTi alloy on its mechanical
         properties. Roznov pod Radhostem, Czech Republic, EU, 2010, 5, 18–20.

[94]   Yoneyama T, Doi H, Kobayashi E, Hamanaka H. Effect of heat treatment with the mould on
         the super-elastic property of Ti–Ni alloy castings for dental application. J. Mater. Sci.-Mater.
         M. 2002, 13, 947–51.

[95]   Mohammadi F, Kharaziha M, Ashrafi A. Role of heat treatment on the fabrication and electro-
         chemical property of ordered TiO_2 nanotubular layer on the as-cast NiTi. Met. Mater. Int.
         2019, 25, 617–26.

[96]   Bellini H, Moyano J, Gil FJ, Puigdollers A. Comparison of the superelasticity of different
         nickel–titanium orthodontic archwires and the loss of their properties by heat treatment. J.
         Mater. Sci.- Mater. M. 2016, 27; doi: 10.1007/s10856-016-5767-5.

[97]   Saghaian SM, Karaca HE, Tobe H, Souri M, Noebe RD, Chumlyakov YI. Effects of aging on the
         shape memory behavior of Ni-rich $Ni_{50.3}Ti_{29.7}Hf_{20}$ single crystals. Acta Mater. 2015, 87, 128–41.

[98]   Evirgen A, Karaman I, Santamarta R, Pons J, Noebe RD. Microstructural characterization and
         shape memory characteristics of the $Ni_{50.3}Ti_{34.7}Hf_{15}$ shape memory alloy. Acta Mater. 2015,
         83, 48–60.

[99]   Razali MF, Mahmud AS. Gradient deformation behavior of NiTi alloy by ageing treatment. J.
         Alloys Compd. 2015, 618, 182–6.

[100]   Eggeler G, Khalil-Allafi J, Gollerthan S, Somsen C, Schmahl W, Sheptyakov D. On the effect
         of aging on martensitic transformations in Ni-rich NiTi shape memory alloys. *Smart Mater.
         Struct.* 2005, 14, 186–91.

[101]   Zhou Y, Zhang J, Fan G, Ding X, Sun J, Ren X, Otsuka K. Origin of 2-stage R-phase
         transformation in low-temperature aged Ni-rich Ti–Ni alloys. Acta Mater. 2005, 53, 5365–77.

[102]   Zheng Y, Jiang F, Li L, Yang H, Liu Y. Effect of ageing treatment on the transformation
         behaviour of Ti–50.9 at.% Ni alloy. Acta Mater. 2008, 56, 736–45.

[103]   Karimzadeh M, Aboutalebi MR, Salehi MT, Abbasi SM, Morakabati M. Adjustment of aging
         temperature for reaching superelasticity in highly Ni-Rich Ti-51.5Ni NiTi shape memory alloy.
         Mater. Manuf. Process. 2016, 31, 1014–21.

[104]   Aliasgarian R, Ghasemi HM, Abedini M. Tribological behavior of heat treated Ni-rich NiTi
         alloys. J. Tribol. 2011, 133; doi: 10.1115/1.4004102.

[105]   Stroz D, Kwarciak J, Morawiec H. Effect of ageing on martensitic transformation in NiTi shape
         memory alloy. J. Mater. Sci. 1988, 23, 4127–31.

[106]   Adharapurapu RR, Vecchio KS. Effects of aging and cooling rate on the transformation of
         nanostructured Ti-50.8Ni. J. Alloys Compd. 2017, 693, 150–63.

[107]   Zhang J, Ren X, Otsuka K, Asai M. Reversible change in transformation temperatures of a
         Ti-51at% Ni alloy associated with alternating aging. Scr. Mater. 1999, 14, 1109–13.

[108]   Jiang F, Li L, Zheng Y, Yang H, Liu Y. Cyclic ageing of Ti–50.8 at.% Ni alloy. Intermetallics
         2008, 16, 394–8.

[109]   Liu Y, Van Humbeeck J, Stalmans R, Delaey L. Some aspects of the properties of NiTi shape
         memory alloy. J. Alloys Compd. 1997, 247, 115–21.

[110]   Wang X, Verlinden B, Van Humbeeck J. Effect of post-deformation annealing on the R-phase
         transformation temperatures in NiTi shape memory alloys. Intermetallics 2015, 62, 43–9.

[111]   Mahmud AS, Liu Y, Nam T-H. Gradient anneal of functionally graded NiTi. Smart Mater.
         Struct. 2008, 17; https://doi.org/10.1088/0964-1726/17/01/015031.

[112]   Zinelis S, Darabara M, Takase T, Ogane K, Papadimitriou GD. The effect of thermal treatment
         on the resistance of nickel-titanium rotary files in cyclic fatigue. Oral Surg. Oral Med. Oral
         Pathol. Oral Radiol. Endod. 2007, 103, 843–7.

[113]   Ling HC, Kaplow R. Variation in the shape recovery temperature in NiTi alloys. Mater. Sci.
         Eng. 1981, 48, 241–7.

[114]   Miller DA, Lagoudas DC. Influence of cold work and heat treatment on the shape memory effect and plastic strain development of NiTi. Mater. Sci. Eng. A, 2001, 308, 161–75.

[115]   Kus K, Breczko T. DSC investigations of the effect of annealing temperature on the phase transformation behaviour in Ni-Ti shape memory alloy. *Mater. Phys. Mech.* 2010, 9, 75–83.

[116]   Mitwally ME, Farag M. Effect of cold work and annealing on the structure and characteristics of NiTi alloy. Mater. Sci. Eng. A 2009, 519, 155–66.

[117]   Huang X, Liu Y. Effect of annealing on the transformation behavior and superelasticity of NiTi shape memory alloy. Scr. Mater. 2001, 45, 153–60.

[118]   Sadiq H, Wong MB, Al-Mahaidi R, Zhao XL. The effects of heat treatment on the recovery stresses of shape memory alloys. *Smart Mater. Struct.* 2010, 19; doi: 10.1088/0964-1726/19/3/035021.

[119]   Tadayyon G, Mazinani M, Guo Y, Zebarjad SM, Tofail SAM, Biggs MJ. The effect of annealing on the mechanical properties and microstructural evolution of Ti-rich NiTi shape memory alloy. Mater. Sci. Eng., A 2016, 662, 564–77.

[120]   Khaleghi F, Khalil-Allafi J, Abbasi-Chianeh V, Noori S. Effect of short-time annealing treatment on the superelastic behavior of cold drawn Ni-rich NiTi shape memory wires. *J. Alloys Compd.* 2013, 554, 32–8.

[121]   Wang X, Amin-Ahmadi B, Schryvers D, Verlinden B, Humbeeck JV. Effect of annealing on the transformation behavior and mechanical properties of two nanostructured Ti-50.8 at.%Ni thin wires produced by different methods. Mater. Sci. Forum 2013, 738/739, 306–10.

[122]   Zupanc J, Vahdat-Pajouh N, Schäfer E. New thermomechanically treated NiTi alloys – a review. Int. Endod. J. 2018, 51, 1088–103.

[123]   Braz Fernandes FM, Mahesh KK, dos Santos Paula A. Thermomechanical Treatments for Ni-Ti Alloys. In: Braz Fernandes FM, ed., Shape Memory Alloys – Processing, Characterization and Applications. Intechopen.com, Rijeka, Croatia, 2013, 23–6.

[124]   Favier D, Liu Y, Orgéas L, Sandel A, Debove L, Comte-Gaz P. Influence of thermomechanical processing on the superelastic properties of a Ni-rich Nitinol shape memory alloy. Mater. Sci. Eng., A 2006, 429, 130–6.

[125]   Wang X, Verlinden B, Van Humbeeck J. R-phase transformation in NiTi alloys. Mater. Sci. Technol. 2014, 30, 1517–29.

[126]   Paula AS, Mahesh KK, dos Santos CML, Braz Fernandes FM, da Costa Viana CS. Thermomechanical behavior of Ti-rich NiTi shape memory alloys. Mater. Sci. Eng. A 2008, 481/482, 146–50.

[127]   Wada K, Liu Y. Thermomechanical training and the shape recovery characteristics of NiTi alloys. Mater. Sci. Eng., A 2008, 481/482, 166–9.

[128]   Paula AS, Mahesh KK, Braz Fernandes FM, Martins RMS, Cardoso AMA, Schell N. In-situ high temperature texture characterisation in NiTi shape memory alloy using synchrotron radiation. Mater. Sci. Forum 2005, 495/497, 125–30.

[129]   Paula AS, Mahesh KK, Schell N, Braz Fernandes FM. Textural modifications during recovery in Ti-Rich Ni-Ti shape memory alloy subjected to low level of cold work reduction. Mater. Sci. Forum 2010, 636/637, 618–23.

[130]   Paula AS, Canejo JHPG, Mahesh KK, Silva RJC, Braz Fernandes FM, Martins RMS, Cardoso AMA, Schell N. Study of the textural evolution in Ti-rich NiTi using synchrotron radiation. Nucl. Instrum. Methods. Phys. Res. B 2006, 246, 206–10.

[131]   Bowers ML, Gao Y, Yang L, Gaydosh DJ, DeGraef M, Noebe RD, Wang Y, Milles MJ. Austenite grain refinement during load-biased thermal cycling of a $Ni_{49.9}Ti_{50.1}$ shape memory alloy. Acta Mater. 2015, 91, 318–29.

[132]   Pereira ESJ, Peixoto IFC, Viana ACD, Oliveira II, Gonzalez BM, Buono VTL, Bahia MGA. Physical and mechanical properties of a thermomechanically treated NiTi wire used in the manufacture of rotary endodontic instruments. Int. Endod. J. 2012, 45, 469–74.

[133]   De-Deus G, Silva EJ, Vieira VT, Belladonna FG, Elias CN, Plotino G, Grande NM. Blue thermo-mechanical treatment optimizes fatigue resistance and flexibility of the reciproc files. J. Endod. 2017, 43, 462–6.

[134]   Tobushi H, Iwanaga H, Tanaka K, Hori T, Sawada T. Stress-strain-temperature relationships of TiNi shape memory alloy suitable for thermomechanical cycling. JSME Int. J. 1-Solid M. 1992, 35, 271–7.

[135]   Seyyed Aghamiri SM, Nili Ahmadabadi M, Shahmir H, Naghdi F, Raygan S. Study of thermo-mechanical treatment on mechanical-induced phase transformation of NiTi and TiNiCu wires. J. Mech. Behav. Biomed. 2013, 21, 32–6.

[136]   Liu J-Y, Lu H, Chen J-M, Alain C, Wu T. Phenomenological description of thermomechanical behavior of shape memory alloy. J. Mater. Sci. 2008, 43, 4921–8.

[137]   Lekston Z, Lagiewka E. X-ray diffraction studies of NiTi shape memory alloys. Arch. Mater. Sci. Eng. 2007, 28, 665–72.

[138]   Tobushi H, Yamada S, Hachisuka T, Ikai A, Tanaka K. Thermomechanical properties due to martensitic and R-phase transformations of TiNi shape memory alloy subjected to cyclic loadings. Smart Mater. Struct. 1996, 5, 788–95.

[139]   Frenzel J, Pfetzing J, Neuking K, Eggeler G. On the influence of thermomechanical treatments on the microstructure and phase transformation behavior of Ni–Ti–Fe shape memory alloys. Mater. Sci. Eng. A 2008, 481/482, 635–8.

Chapter 6
Fabrication, synthesis, and product forms

In general, there are two major manufacturing technologies of NiTi-based alloys: casting and powder metallurgy (PM). Furthermore, depending on the heat source and detailed technology involved, casting can be divided into vacuum arc melting, vacuum induction melting (VIM), and electron beam melting (EBM). PM can be subdivided into conventional and additive manufacturing. Conventional processes can include conventional sintering, self-propagating high-temperature synthesis (SHS), combustion synthesis (CS), hot isostatic pressing (HIP), spark plasma sintering (SPS), and metal injection molding (MIM); while additive manufacturing includes selective laser sintering, selective laser melting (SLM), laser engineering net shaping, and EBM [1]. Due to unique properties of shape-memory effect (SME) and superelasticity (SE), NiTi material is an excellent candidate in many functional designs, but the manufacturing and processing complications of this alloy pose impediments to widespread applications [1]. In this chapter, we will discuss various methods to produce final shapes and forms which are used in both engineering and medical fields. Major topics are (1) liquid phase involved technique such as casting, melting, and welding; (2) solid-state technology such as PM, diffusion bonding, and mechanical alloying (MA); (3) additive manufacturing processes such as laser and electron beam techniques towards making 3D components; and (4) machining and forming technology. All these technologies are adequately and efficiently employed to produce useful various forms of wire, bar, thin film, plate, pipe and tube, joints, component with various controlled porosity, and foam structure in both macro- and microscales. Figure 6.1 illustrates generally practiced manufacturing flow route for titanium materials [2].

For these final products, mainly there are three ways to memorize the shape (shape-memory treatment): (1) intermediate temperature treatment, (2) low temperature treatment, and (3) aging treatment. By the intermediate treatment, materials which were severely deformed by cold rolling or cold drawing are fixed to hold the memorized shape during the heat treatment at 400–500 °C for several minutes to 1 h; resulting in exhibiting both SME and SE phenomena. By the low temperature treatment, the materials which were previously heated above 800 °C to normalize the microstructure are subjected to forming, followed by memorizing treatment at 200–300 °C; resulting in being suitable for intricate shape memorizing. By the aging treatment, materials which are solution treated at 800–1,000 °C and water-quenched, are then aged at about 400 °C while shape memorizing process; with a limited application for Ni-rich NiTi-based alloy (>50.5 at% Ni).

https://doi.org/10.1515/9783110666113-006

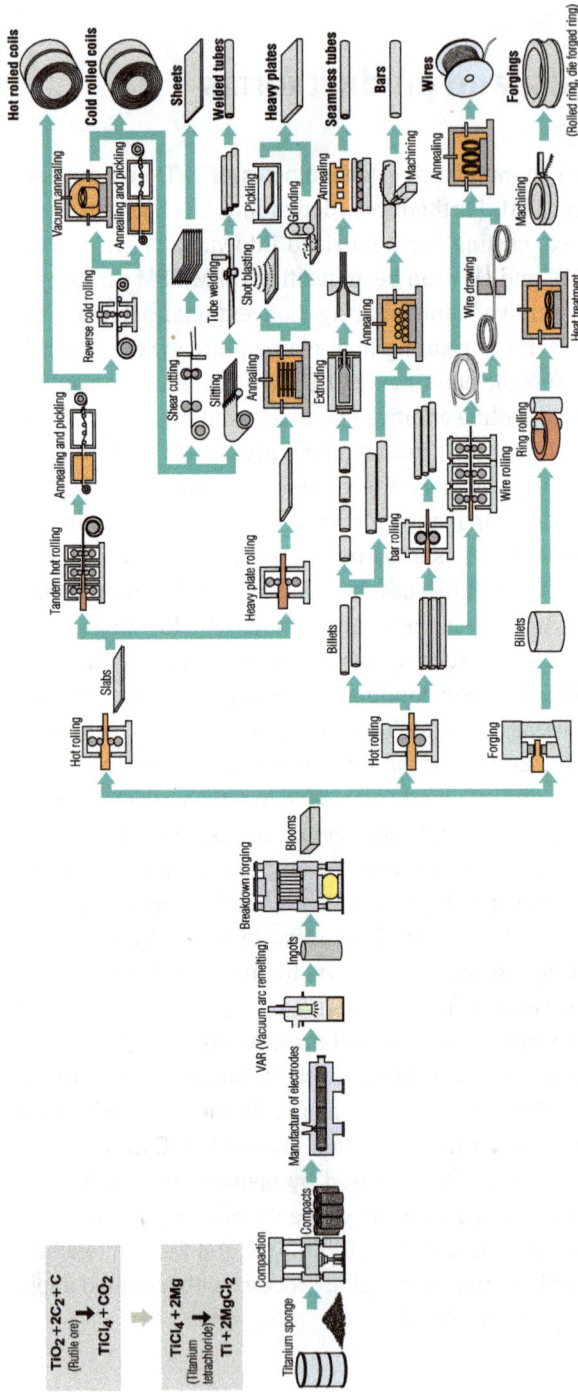

Figure 6.1: Manufacturing flow diagram for titanium materials [2].

6.1 Casting and other melting

Although, due to excellent corrosion resistance and biocompatibility and high specific strength (in other words, light weight and high mechanical strength), Ti materials (including commercially pure titanium (CPT) and Ti-based alloys) are utilized in versatile applications, Ti materials possess some engineering drawbacks. They should include; (1) the high melting point (ca. 1,700 ± 25 °C) would suppress a fabrication of dental prosthesis such as bridge and crown. (2) At higher temperature, particularly above 885 °C (for pure Ti) when low temperature HCP structured-Ti transforms to high temperature BCC structured-Ti, Ti shows a tremendous reactivity with oxygen, so that melting Ti should be done in argon gas or inert gas atmosphere. (3) It is so active at high temperature that crucible should be carefully selected. Sadrnezhaad et al. [3] investigated the possible contamination on $NiTi_{45}$ (wt%) with commercial ZrO_2-, Al_2O_3-, and SiC-base crucibles used for vacuum melting. The molten alloy was held under vacuum for 90 minutes at 1,450 °C to become homogenized. It was found that the relative degree of contamination declined in the following sequence: commercially pure SiC > SiC-5wt%Al_2O_3-5wt%SiO_2 > slurry cast alumina > recrystallized alumina > zircon type A > oxygen deficient high-purity zirconia. (4) It exhibits high solubility of oxygen and this reflects enhancement of strength while weakening ductility and fracture toughness as well [4, 5], as shown in Figure 6.2, where number 1 through 4 on the top x-axis indicates grade number for CPT. The improvements in mechanical strengths attribute to the interstitial oxygen atoms impeding dislocation motion and inhibiting low-temperature twinning [6]. (5) Ti possesses relatively low value of specific weight, which is about 4.5 gr/cm^3 at 20 °C and indicates some beneficial feature. However, since the casting pressure is proportional to melting metal type [7, 8], the smaller value of specific weight, the worse the castability, resulting in casting defects. In summary, metal titanium possesses several issues which are crucial and challenging to casting technique.

Takahashi et al. [9] cast an equiatomic NiTi alloy into molds of magnesia and silica investments by use of a dental argon-arc pressure casting machine with a copper crucible. It was reported that the shape recovery process was sharper in the specimens cast in magnesia investment molds than in those cast in silica (phosphate-bonded) investment molds; the latter casting had a hard region of the periphery, that is later called as the alpha case [10], suggesting that shape recovery process may be affected by reaction of molten metal with silica. Rods of $TiNi_{51.5}$ (at%) alloy were prepared under various casting conditions and it was reported that (i) specimens which were melted in a crucible and cast in a phosphate-bonded investment mold had less strength and elongation than those melted in a copper crucible and cast in a phosphate-bonded investment mold, and (ii) all castings exhibited SE at room temperature, with the upper limit of recoverable strain about 1.8% [11]. Yoneyama et al. [12] investigated the influence of composition and purity of titanium on the mechanical properties and the transformation temperatures of NiTi alloy dental castings (49.0–49.2 at%

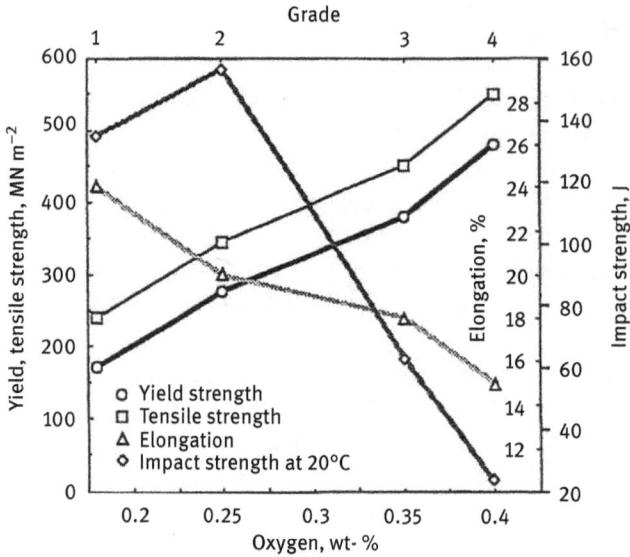

Figure 6.2: Mechanical properties as a function of level of oxygen [4].

of titanium content) by tensile testing and differential scanning calorimetry. It was shown that (i) NiTi$_{49.0}$ exhibited brittle in nature and NiTi$_{49.2}$ showed low apparent proof strength and large elongation, (ii) residual strain increased with increasing titanium content, and (iii) even small reductions of titanium purity influenced the tensile properties, and the transformation temperatures, causing high apparent proof strength, low residual strain, low elongation because of the reduction in transformation temperatures. Using the same alloy composition, Yoneyama et al., [13] investigated the influence of mold materials and heat treatment on the tensile properties and the transformation temperatures by tensile test and differential scanning calorimetry. A silica investment and a magnesia investment were used as the mold materials and heat treatment condition was 440 °C for 30 min. It was reported that (i) an apparent proof strength decreased in both compositions, and residual strain increased in NiTi$_{49.2}$ by the heat treatment, (ii) elongation increased in NiTi$_{49.0}$ with use of the magnesia mold or by the heat treatment, and (iii) the transformation temperatures of NiTi$_{49.2}$ increased with use of the magnesia mold; suggesting that the development of a suitable method for the casting of the alloy is expected to bring about the development of new devices and treatment plan in dentistry. Yoneyama et al. [14] studied tensile property of TiNi$_{50.85}$ (mol%) alloy castings in relation to the thermal behavior accompanied with phase transformation to examine the effect of heat treatment after casting with the mold in air. The heat treatment temperature was 440–500 °C for 0.9, 1.8, or 3.6 ks. It was found that (i) an apparent proof stress of the castings decreased with increasing period of heat treatment, and the decrease was larger with the treatment at 500 °C, (ii) residual strain also decreased by the heat treatment, however, it was low with the treatment for relatively short period, that is, 440 °C for 0.9 and 1.8 ks, and 500 °C for 0.9 ks treatments, (iii) from the thermal behavior measured by differential scanning calorimetry, the ascent in the transformation temperatures and the increase in the thermal peak height appeared to influence the changes in the tensile property, indicating that heat treatment was effective to utilize more flexibility, less stress, and less permanent deformation in dental castings.

Alloyed TiNi-based castings were also studied. Soga et al. [15] studied Pd-TiNi$_{50.8}$ (at%) the alloying effect of Pd to improve the SE of the alloy castings at body temperature for dental application. TiNi$_{50.8}$ (at%) alloy, which exhibited SE at 37 °C in castings, was used as a control material and variation of Pd content was 5.0, 7.5, 10.0, and 15.0 at% Pd, added to the equiatomic TiNi$_{50.0}$ alloy by the substitution for Ni. It was reported that (i) Ti-42.5Ni-7.5Pd alloy castings showed good SE among the examined alloys, (ii) an apparent proof stress could be changeable by the proportion of Ti and Ni with residual strain being kept low, and (iii) Ti-42.5Ni-7.5Pd alloy castings exhibited better superelastic flexibility than Ti-50.8Ni alloy, which is proven by lower apparent proof stress and larger elongation, due to its relatively high martensitic transformation starting temperature point. Alloying effect of Cu to NiTi alloy castings was studied on TiNi$_{50.8}$ and TiNi$_{40.8}$Cu$_{10.0}$ (mol%) by the three-point bending test in relation to the phase transformation. Castings were subject to heat treatment at 440, 500, and 520 °C for 30 min [16]. It was reported that (i) the difference between the load values in the loading and the unloading processes was relatively small for Ti-Ni-Cu alloy, (ii) with respect to the residual deflection, there was no significant difference between Ti-Ni and Ti-Ni-Cu alloys with the same treatment condition, (iii) the load values in the loading and the unloading processes decreased with each heat treatment for Ti-Ni alloy; however, the decrease in the load values for Ti-Ni-Cu alloy was not distinct; concluding that Ti-Ni-Cu alloy castings produce effective orthodontic force as well as stable low residual deflection, which is likely to be caused by the high and sharp thermal peaks during phase transformation. Goryczka et al. [17], using Ni$_{50}$Ti$_{50}$ and Ni$_{47}$Ti$_{50}$Co$_3$ shape-memory alloy (SMA) castings, studied the microtexture and reported that (i) relatively high cooling rate realized during solidification causes directional heat flow, causing that the high textured strips were produced, (ii) however, the total amount of the preferentially oriented grains differs between surface and cross section of the strip due to enhanced stiffness by Co addition, and (iii) at surface, where the crystallization of the grains is the most intensive amount of the grains can reach about 77%.

It is well known that NiTi-based alloy is exceedingly hard to produce because highly requirement for compositional control is demanded. Every titanium atom combining with oxygen or carbon is an atom that has been lost from the NiTi crystalline lattice; resulting in shifting the composition and making the transformation temperature that much lower. There are two primary melting methods used; vacuum arc remelting (VAR) and VIM. The VAR is done by striking an electrical arc between the raw material and a water-cooled copper strike plate. Melting is done in a high vacuum, and the mold itself is water-cooled copper; while VIM is done by using alternating magnetic fields to heat the raw materials in a crucible (generally carbon). This is also done in a high vacuum. While both methods have advantages, it has been demonstrated that an industrial state-of-the-art VIM melted material has smaller inclusions than an industrial state-of-the-art VAR one, leading to a higher fatigue resistance. Other researches report that VAR employing extreme high-purity raw materials may lead to a reduced number of inclusions and thus to an improved fatigue behavior.

Salvetr et al. [18] investigated $NiTi_{46}$ (wt%) and NiTiX (x = Nb, V, and Si with amount of 5 wt%) alloys prepared by arc remelting under argon protective atmosphere with repeated three times melting to obtain a homogenous chemical composition of samples. It was reported that the arc remelting produced a more homogeneous microstructure with a lower content of the Ti_2Ni phase than PM methods in case of Ni-Ti-X alloys, (ii) quite a high amount of the secondary phase (a mixture of the Ni_2Ti, NiTi, and Ti_2Ni phases) was created in the microstructure of $NiTi_{46}$ alloy, and (iii) the addition of silicon caused an increase of transformation temperatures and hardness. It is known that melting of Ti alloys in graphite crucibles is associated with a vigorous interface reaction, so that the carbon concentration of NiTi alloys needs to be kept below a certain minimum in order to assure that the functional properties of the alloys meet the required targets [19]. Zhang et al. [20, 21] studied the reaction between NiTi melts and electrographite crucibles during VIM processing. It was reported that the reaction results in the growth of a TiC layer and a simultaneous increase of carbon concentration in the melt [20]. Alloying with Cu, Fe, Hf, and Zr elements to NiTi-based alloys were studied in VIM method [21] and it was mentioned that VIM processing in graphite crucibles provides ternary NiTiX SMAs with good chemical homogeneity and acceptable impurity contents. Stambolič et al. [22] investigated the combined method with the vacuum-induction melting and continuous vertical casting of a NiTi alloy in order to produce the strand. It was found that (i) the microstructure is dendritic, where in the interdendritic region the eutectic is composed of a dark NiTi phase and a bright $TiNi_{3-x}$ phase, (ii) Ti carbides and Fe-rich phases were observed, and (iii) the microchemical analysis of the NiTi strand showed that the composition changed over the cross and longitudinal sections; indicating that the as-cast alloys are inhomogeneous.

The EBM and SLM are processes involving melting procedure, however these techniques are categorized into "additive manufacturing"; accordingly, we will discuss these methods later in this chapter.

6.2 Powder sintering technology

Figure 6.3 illustrates a standard classification of powder sintering methods; in which there are basically three major groups: conventional pressureless sintering, pressurized sintering, and powder injection molding.

Sintering method

- Pressure-less sintering
  - Liquid phase sintering
    - Conventional sintering
    - Thermal plasma sintering
  - Reaction sintering
    - Micro-, Milli wave sintering
  - Atmospheric sintering
    - Thermal plasma sintering
    - Micro-, Milli wave sintering
- Pressurized sintering
  - Solid compaction
    - Hot pressing (HP)
    - Spark plasma sintering (SPS)
  - Gas compaction
    - Ultra high-pressure sintering
- Powder injection molding
  - Metal injection molding (MIM)
    - Hot isostatic sintering (HIP)
    - High pressure gas reaction sintering
  - Ceramic injection molding (CIM)

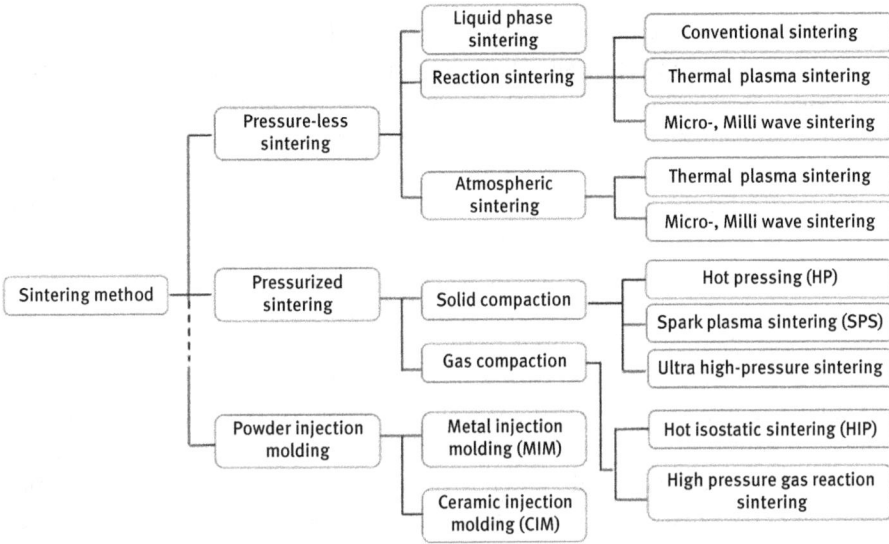

Figure 6.3: Classification of powder sintering methods.

6.2.1 Pressureless powder sintering

The pressureless sintering is a process where loose metal powders are poured into a metal die and vibrated until loosely compacted, followed by heating and sintering the thus-filled die with powders without applied pressure, indicating that the process does not require melting to its point of liquefaction. Sadrnezhaad et al. [23] fabricated porous NiTi SMA by using thermohydrogen process (THP), which enables a production of homogenous structures, appropriate pore-size distributions and short sintering times. It was mentioned that THPed NiTi alloy exhibited a low Young's modulus (19.8 GPa) and a high tensile strength of 255 MPa, indicating that these properties are close to those of the natural bone and can meet the mechanical property demands of the hard-tissue implants for heavy load-bearing applications. The samples produced exhibit sufficient thermoelastic effect distinguished by a 1.2% mean recoverable strain. To establish the biomechanical compatibility of implantable metallic material, the structure with controlled porosity is an ultimate answer [24, 25]. Andani et al. [26] mentioned that PM can provide an effective way of matching the stiffness of an implant with the surrounding tissue and the appropriate porosities and the overall geometry of the implant can be optimized for strain transduction and with a tailored stiffness profile. This discussion is directly related to the biomechanical compatibility that will be discussed later.

Li et al. [27] developed the elemental powder sintering (EPS) technique for the synthesis of porous NiTi alloy, in which Ni and Ti powders are used as the reactants and TiH_2 powder is added as a pore-forming agent and active agent, and investigated the effects of various experimental parameters (sintering temperature, sintering time, and TiH_2 content) on the porosity, pore size, and pore distribution as well as phase composition in experimental NiTi alloys. It was reported that (i) in order to avoid the formation of carcinogenic pure Ni phase, the porous NiTi alloy should be synthesized over a temperature of 950 °C; providing NiTi as the main phase without any elemental phase, and (ii) substitution of Ti by TiH_2 is more economic and more favorable to obtain homogeneous porous NiTi alloy.

Chen et al. [28] investigated the powder sintering behavior of NiTi from an elemental powder mixture of Ni and Ti. It was mentioned that (i) in the sintered alloys, the overall porosity ranges from 9.2% to 15.6%, while the open-to-overall porosity ratio is between 8.3% and 63.7% and largely depends on the sintering temperature, (ii) in comparison to powder compacts sintered at 950 °C and 1,100 °C, the powder compact sintered at 880 °C shows a much smaller pore size, a higher open-to-overall porosity ratio but smaller shrinkage and a lower density, (iii) direct evidence of eutectoid transformation in the binary Ni–Ti system during furnace cooling to ca. 617 °C is provided by in situ neutron diffraction, and (iv) the intensities of the B2-NiTi reflections decrease during the holding stage at 1,100 °C, which has been elaborated as an extinction effect according to the dynamical theory of neutron diffraction, when distorted crystallites gradually recover to perfect crystals. Chen et al. [29] also investigated NiTi powder sintering behavior from elemental powder mixtures of Ni and Ti, and Ni and TiH_2 using in situ neutron diffraction and in situ scanning electron microscopy. It was reported that (i) the sintered porous alloys have open porosities ranging from 2.7% to 36.0%, (ii) in comparison to the Ni and Ti compact, dehydrogenation occurring in the Ni and TiH_2 compact leads to less densification yet higher chemical homogenization only after high-temperature sintering, (iii) direct evidence of the eutectoid phase transformation of NiTi at 620 °C is reported by in situ neutron diffraction, and (iv) a comparative study of cyclic stress–strain behaviors of the porous NiTi alloys made from Ni and Ti, and Ni and TiH_2 compacts indicate that the samples sintered from the Ni and TiH_2 compact exhibited a much higher porosity, larger pore size, lower fracture strength, lower close-to-overall porosity ratio, and lower value of modulus of elasticity.

Taking an advantage of the brittle nature of $NiTi_2$ to promote a better efficiency of the mechanical activation process, Zhao et al. [30] studied an efficient PM method for the synthesis of NiTi alloys, involving mechanical activation of prealloyed $NiTi_2$ and elemental Ni powders $(NiTi_2–Ni)$ followed by a press-and-sinter step. A sintered specimen consisting mainly of NiTi phase was obtained after vacuum sintering at 1,050 °C for 0.5 h. It was reported that (i) using elemental Ti powders instead of prealloyed $NiTi_2$ powders, the structural homogenization of the synthesized NiTi alloys was delayed, (ii) only after vacuum sintering at 1,050 °C for 6 h, the NiTi phase was observed to be the predominant phase, (iii) the higher reactivity of the mechanically activated $(NiTi_2–Ni)$ powder particles can explain the different sintering behavior of those powders compared with the mechanically activated (Ti–Ni) powders; indicating that this innovative approach allows an effective time reduction in the mechanical activation and of the vacuum sintering step. Ma et al. [31] adopted the electroassisted powder metallurgical (EPM) route to prepare porous NiTi alloys with controllable porosities and pore sizes using nickel (Ni) and titanium (Ti) powders at a temperature ranging from 850 to 950 °C. Ammonium hydrogen carbonate (NH_4HCO_3) was used as a sacrificial space holder to tailor the porosity and pore size of the NiTi alloy and its properties (including modulus of elasticity, phase transformation temperature, SME, and SE). It was reported that by optimizing the synthesis temperature and content of the sacrificial space holder (0–30%), the porosity (41–75%), pore size (22–174 µm), elastic modulus (4.77–0.87 GPa), and the recovery strain (>2%) of the EPMed NiTi can meet or are close to the standard of natural bone (porosity: 30–80%, pore size: 100–600 µm, cancellous modulus: <3 GPa, recovery strain: >2%). Wang et al. [32] studied the phase transformation of the equiatomic NiTi alloys during vacuum sintering. It was found that (i) in the process of sintering $NiTi_2$ and Ni_3Ti phases are formed first and then transform into NiTi phase, (ii) the quantity of $NiTi_2$ and Ni_3Ti phases gradually decreased but not eliminated completely with increase of sintering time, (iii) the porosity of specimen sintering at 900 °C decreases slightly with increase of sintering time, and (iv) with increase of sintering time the porosity of specimen sintering at 1,050 °C decreased first and then increased because of generation of Ti-rich liquid in the process of sintering.

Shishkovsky et al. [33] investigated the microanalysis and histological studies of porous titanium and NiTi implants fabricated by selective laser sintering and emphasized that functional characteristics of the synthesized medical implants depend on the pore size distribution and their relative location as well as on the nanostructural morphology of the sintered surface. Deng et al. [34] utilized nanosecond pulsed laser sintering process to generate a nanoporous composite surface with NiTi alloy and hydroxyapatite (HA) by ultrafast laser heating and cooling of Ni, Ti, and HA nanoparticles mixtures precoated on the 3D NiTi substrates. It was found that the nanoscale porosity delivered by nanosecond pulsed laser sintering and the HA component positively contributed to osteoblast differentiation, as indicated by an increase in the expression of collagen and alkaline phosphatase, both of which are necessary for osteoblast mineralization; (ii) topological complexities that appeared to boost the activity of osteoblasts, including an increase in actin cytoskeletal structures and adhesion structures; (iii) concluding that the pulsed laser sintering method is an effective tool to generate biocompatible coatings in complex alloy-composite material systems with desired composition and topology; and (iv) these findings provide a better understanding of the osteoinductive behavior of the sintered nanocomposite coatings for use in orthopedic and bone regeneration applications.

Microwave sintering has emerged in recent years as a new method for sintering a variety of materials that have shown significant advantages against conventional sintering procedures [35–38]. Tang et al. [36] fabricated porous NiTi SMA by microwave sintering method, which enables formation of porous structures without using any pore-forming agents at a relatively low sintering temperature of 850 °C and a short sintering time of 15 min. It was reported that (i) the porous NiTi alloy exhibited porosity ratios from 27% to 48% and pore sizes range from 50 to 200 μm when using different sintering temperatures and holding times, (ii) the predominant B2 (NiTi) and B19′ (NiTi) phases were identified, (iii) a multistep phase transformation took place on heating and a two-step phase transformation took place on cooling of the porous NiTi alloy, and (iv) the irrecoverable strains decreased with increasing sintering temperature, but the holding time had little effect on the stress–strain behavior at 60 °C. Ibrahim et al. [37, 38] fabricated $TiNi_{51}$ (at%) NiTi alloy via a PM and microwave sintering technique. It was found that (i) varying the microwave temperature and holding time was found to strongly affect the density of porosity, presence of precipitates, transformation temperatures, and mechanical properties, and (ii) the lowest density and smallest pore size were observed in the NiTi sample sintered at 900 °C for 30 min, which exhibited the maximum strength and strain of 1,376 MPa and 29%, respectively [37]. The predominant martensite phases of B2 and B19′ were identified, which confirmed data reported by Tang et al. [36]. Ibrahim et al. [38] using the same processed NiTi SMA, investigated the effect of time and temperature on the microwave sintering process. It was reported that (i) the sintering condition of 700 °C for 15 min produced a part with coherent surface survey that does not exhibit gross defects, (ii) increasing the sintering time and temperature created defects on the outer surface; while reducing the temperature to 550 °C severely affected the mechanical properties, (iii) the microstructure of these samples showed two regions of Ni-rich region and Ti-rich region between them Ti_2Ni, NiTi, and Ni_3Ti phases, and (iv) an increase in the sintering temperature from 550 to 700 °C was found to increase the fracture strength significantly and decreased the fracture strain slightly; while reducing the sintering temperature from 700 to 550 °C severely affected the corrosion behaviors of NiTi alloy.

Liquid-phase sintering (LPS) is a process for forming high-performance, multiple-phase components from powders, involving sintering under conditions where solid grains coexist with a wetting liquid. Packed particles heated near their melting temperature bond together by sintering. As diffusion accelerates at higher temperatures, sintering is manifested by bonding between contacting particles. Sintering occurs over a range of temperature, but it is accelerated as the particles approach their melting range. It takes place faster as the particle size decreases, since diffusion distances are shorter and curvature stresses are larger [39]. Zhang et al. [40] studied the use of conventional press and sinter of titanium and nickel equiatomic blends, involving transient liquid phases and reported that (i) the employment of a double-stage sintering cycle is effective to sustain the geometrical integrity of the sintered sample, and (ii) the TiNi alloy prepared in this manner exhibits a recovery strain of 4.6% by generating a restoring stress of 230 MPa, with a martensitic transformation point M_S of 34 °C.

The FFC Cambridge process is a direct electrodeoxidation process used to reduce metal oxides to their constituent metals in a molten $CaCl_2$ salt bath. $NiTiO_3$ was used as a precursor (the first stable oxide to form upon blending and sintering NiO and TiO_2 powders) and was successfully reduced using the FFC Cambridge process at 900 °C and a constant cell voltage of –3.1 V to produce a NiTi alloy [41]. The main process mechanism is as follows: the first stage of the reaction involved the rapid formation of Ni and $CaTiO_3$. The reduction then proceeded via the formation of the intermediate compounds Ni_3Ti and Ni_2Ti_4O. All the $NiTiO_3$ and Ni were consumed after a period of 6 h, while the intermediate compounds remained until the reaction is completed [41]. Jackson et al. [42] investigated the effect of reduction temperature and current collector material on the microstructure of the final reduction product. It was reported that (i) the use of a Ni current collector caused Ni-enrichment at the surface of the product, stabilizing the high temperature B2 cubic form to room temperature, and (ii) Ni-rich phases Ni_4Ti_3 and Ni_3Ti were observed after 24 h reductions; the former is favored at higher reduction temperatures, and the latter at lower temperatures [43]. It was reported that (i) the reduction pathway consists of rapid initial reduction of $NiTiO_3$ to form $CaTiO_3$ and Ni, then the transformation of Ni to Ni_3Ti, and finally the consumption of $CaTiO_3$ and Ni_3Ti to produce NiTi.

6.2.2 Pressurized powder sintering

In this section, mainly conventional PM, HIP, and SPS will be discussed.

Li et al. [44] fabricated porous NiTi SMAs low-pressure sintering (LPS), and the pore features have been controlled by adjusting the processing parameters. It was reported that (i) the porous NiTi allots with high porosity (45%) and large pore size (200–350 μm) can be prepared by LPS using $TiH_{1.5}$ as pore-forming agent, (ii) these alloys exhibit isotropic pore structure with three-dimensional interconnected pores, and (iii) the porous NiTi SMA produced by LPS exhibits SE and mechanical properties superior to that by conventional sintering. Zhang et al. [45] and Chen et al. [46] adopted the conventional press-and-sinter process to fabricate porous NiTi SMAs. It was found that (i) employment of a double-stage sintering cycle is effective to sustain the geometrical integrity of the sintered sample, and (ii) the TiNi alloy exhibits a recovery strain of 4.6% by generating a restoring stress of 230 MPa, with a martensitic transformation point M_S of 34 °C [45]. Chen et al. [46] prepared two batches of powder mixture were prepared and compared: Ni/Ti – a mixture of titanium made from a hydrogenation–dehydrogenation process (HDH-Ti) and nickel, and Ni/TiH_2 – a mixture of TiH_2 and nickel, under three different compaction pressures and subsequently sintered in vacuum at three different temperatures (i.e., 1,000, 1,100, and 1,200 °C) for 2 h. It was reported that the porous alloys have open porosities from 10.2% to 33.8% and the largest pore size ranges from 3.5 to 27.4 μm, (ii) the predominant phase identified in all the produced porous NiTi alloys is B2 NiTi phase with the presence of other minor phases, (iii) in the Ni/Ti compacts sintered at 1,000 °C some Ti particles were left unreacted, and (iv) in comparison with the Ni/Ti sintered samples, the samples sintered from Ni/TiH_2 mixture exhibit a higher porosity, smaller pore size, higher fracture strength, smaller cyclic residual strain, higher recovery pseudoelastic strain, and lower secant modulus [46].

HIP is a manufacturing process for reducing the porosity of metals and increasing the density. HIP compresses materials by applying high temperature of several hundreds to 2,000 °C and isostatic pressure of several tens to 200 MPa at the same time under argon inert gas under medium. This improves the material's mechanical properties and workability. Hot pressing is very similar to HIP. Bram et al. [47] noticed that HIPed NiTi components showed reversible austenite⟷martensite transformations which are a prerequisite for SMEs [47]. The result was found by Bertheville et al. and also mentioned that a recovery strain of up to 1.6% is reached under a stress of 120 MPa [48]. Schullner et al. [49] compared the phase transformation temperatures of NiTi samples prepared by HIP and MIM and found that (i) the impurity content of the samples is influenced by the processing, and (ii) deviation in the phase transformation temperature was up to 50 K were found from values given in literature for samples with the same nominal Ni content; owing to the fact that preferentially the formation of Ti-consuming oxides was assumed to change the Ni content of the transforming matrix [49].

Yuan et al. [50] fabricated the porous $TiNi_{50.8}$ (at%) SMAs by capsule-free HIP method and investigated the phase transformation behaviors of the fabricated porous NiTi alloys with various porosity ratios. It was reported that (i) the porous alloys show an uniform pore distribution with the spherical pore size ranging mainly from 50 to 200 µm, (ii) their phase transformation behaviors are similar to that of conventional cast dense Ni-rich TiNi alloys; two steps, that is, R-phase transformation and B19′ martensitic transformation, (iii) the pores have a significant influence on B19′ martensitic transformation, while little influence on R-phase transformation, and (iv) the characteristic transformation temperatures and latent heats of B19′ martensitic transformation in porous $TiNi_{50.8}$ SMAs were found to decrease with increasing porosity ratio. Wu et al. [51] also investigated the phase transformation of porous HIPed NiTi alloys. It was found that (i) HIPed NiTi existed a multistage martensitic transformation on cooling while either a single or two-stage transformation on heating, (ii) the cold compaction pressure had great effect on the transformation temperatures, (iii) the transformation temperatures of NiTi with lower compaction pressure were higher than those of NiTi compacted at higher pressure, and (iv) with increase in the annealing time, the transformation temperatures quickly increased for those alloys compacted at 150 MPa; while there was only little change in the transformation temperatures for those compacted from 300 to 400 MPa. These changes in the transformation temperature can be attributed to the combined effect of high dislocations density and precipitation of the second phase [51]. Wu et al. [52] utilized titanium hydride powders to enhance the foaming process in the formation of orthopedic NiTi scaffolds during capsule-free HIP. It was found that (i) hydrogen is continuously released from titanium hydride as the temperature is gradually increased from 300 to 700 °C through two transitions: $TiH_{1.924} \rightarrow TiH_{1.5}/TiH_{1.7}$ between 300 and 400 °C and $TiH_{1.5}/TiH_{1.7} \rightarrow \alpha$-Ti between 400 and 600 °C, and (ii) holding processes at 425, 480, 500, 550, and 600 °C are found to significantly improve the porous structure in the

NiTi scaffolds due to the stepwise release of hydrogen; concluding that NiTi scaffolds foamed by stepwise release of hydrogen are conducive to the attachment and proliferation of osteoblasts and the resulting pore size also favor ingrowth of cells.

The conventional sintering of Ti and its alloys requires a high temperature (1,200–1,400 °C) in a high vacuum environment for a long time (24–48 h), causing adverse changes in the microstructure and mechanical properties of the sintered products. The SPS was developed to overcome these technical drawbacks associated with the conventional PM and SPS adopts the application of an external current to assist powder consolidation, have attracted much attention [50]. This technique sinters the powder under strong electric field and stress field and low sintering temperature, with advantages of fast heating/cooling rate and the combination of sintering and heat treatment. Therefore, SPS technology can efficiently save energy to sinter the powder used. Current studies on the preparation of porous Ti alloys by SPS mainly focus on the conditions of low temperature and low pressure that can easily sinter Ti and its alloys in powders, because high-current discharge, produces ionization in the plasma, can melt the local oxide film on the particles surface and allow the formation of junctions between particles. Ye et al. [53] prepared the densified samples with Ni_3Ti, NiTi, and $NiTi_2$ phases which were previously obtained from the MA technique. It was found that the formation of crystalline structure is controlled by solid-state reaction when the sample was sintered at 900 °C, but in the sample compacted at 1,100 °C, liquid was introduced for the low melting point phase $NiTi_2$ melt. Shearwood et al. [54] fabricated SPSed $Ti_{50}Ni_{50}$ at a temperature of 800 °C and found that below this temperature, specimens have high porosity but with apparent SME, while the specimens sintered at higher temperatures have bulk density, but experienced extensive oxidation with the resulting loss of the SME.

In order to use the excellent superelastic property of SMAs in many medical applications, the austenite transformation finish temperature (A_F) must be about 37 °C which is human body temperature. Due to the fact the Mo element exhibits an effective alloying influence on control in the transformation temperature, Kim et al. [55] fabricated $Ti_{50}Ni_{49.9}Mo_{0.1}$ and $Ti_{50}Ni_{49.7}Mo_{0.3}$ SMA powders by gas atomization and the transformation temperatures and microstructures of those powders were investigated as a function of powder size. It was reported that (i) the dependence of powder size on the martensitic transformation temperature is also very small in the powders ranging from 25 to 150 μm, (ii) the phase transformation temperatures of the porous specimens are almost the same as those of as-atomized powders and the A_F of $Ti_{50}Ni_{49.9}Mo_{0.1}$ and $Ti_{50}Ni_{49.7}Mo_{0.3}$ porous specimens is 40.4 and 34.4 °C, respectively, and (iii) the porous samples exhibit SME and the recovered stains of the $Ti_{50}Ni_{49.9}Mo_{0.1}$ and $Ti_{50}Ni_{49.7}Mo_{0.3}$ specimens are 1.5% and 2.0%, respectively. Alloying effect of Cu element can be found in modifications of their shape-memory properties by affecting their transformation behavior was studied [56].

Zhang et al. [57] investigated the effects of pore characteristics and microstructures on mechanical properties and superelastic behaviors of the porous NiTi alloys varying sintering temperatures (800–1,050 °C) and ammonium hydrogen carbonate (NH$_4$HCO$_3$) contents (0–20 wt%), which was used as space holder to manage the controlled porosity. It was reported that (i) porous NiTi alloys with 18–61% porosity and 21–415 μm average pore size all consisted of nearly single NiTi phase with few undesired phases Ti$_2$Ni and Ni$_3$Ti, (ii) the superelastic recovery strain ratio of the porous NiTi alloy could be improved up to 90% through training, (iii) this enhanced SE will greatly degrade the mechanical mismatch between bones and porous NiTi, (iv) furthermore, with increasing the porosity and pore size, elastic modulus and compressive strength of the porous NiTi alloys decreased; however, the compressive strength was higher or close to those of human bone and the elastic modulus was close to those of human bone; suggesting that the unique combination of interconnected pore characteristics, pure phase composition, low elastic modulus, high strength, and large superelastic recovery strain made this material a good candidate for ideal long-term load-bearing hard tissue implants [57]. In order to enhance the bioactivity of porous NiTi alloys as a candidate for orthopedic implant, Zhang et al. [58] fabricated composite structures of porous NiTi and HA with expectation of promoting the bone ingrowth and integration of the implant with the surrounding tissue. It was found that (i) interconnected pore characteristics and 29–37% porosity could be achieved by adding HA from 3 to 10 wt%; (ii) compression test revealed that porous NiTi–HA possessed not only low elastic modulus of 5.6–8.1 GPa (close to that of human bone) but also high compressive strength; (iii) furthermore, the addition of HA could improve the bioactivity of porous NiTi significantly, indicating that the combination of interconnected pore characteristics, low elastic modulus, high strength and good bioactivity might make this material a candidate for hard tissue implants. Velmurugan et al. [59] adopted a new approach of using the genetic algorithm and particle swarm optimization techniques with integrated artificial neural network to optimize the SPS process parameters to obtain better mechanical characteristics, for nickel titanium copper-SMAs. It was reported that (i) the density and microhardness can be enhanced by the reduction of particle size and increase in pressure and temperature, (ii) the maximum density of 6.21 g/cc and Vickers hardness of 766 Hv were obtained the optimal for process parameters of temperature, pressure, and particle size of ~800 °C, ~26 MPa and ~6 μm, respectively.

6.2.3 Metal injection molding

MIM is a metalworking process in which finely powdered metal is mixed with binder material to create a "feedstock" that is then shaped and solidified using injection molding. The molding process allows high volume, complex parts (which is called "green" MIM parts) to be shaped in a single step, being equivalent to the forming of plastic parts. The variety of part geometries that can be produced by this process is similar to the great variety of plastic components. After molding, the part undergoes conditioning operations to remove the binder (debinding) and densify the powders.

Finished products are small components used in many industries and applications. The behavior of MIM feedstock is governed by rheology.

Schöller et al. [60] applied MIM technique to produce NiTi shape-memory parts using prealloyed NiTi powders with different Ni contents as starting materials. It was found that (i) MIM process allows the production of near-net-shape components without the occurrence of rapid tool wear as found in the case of conventional machining operations, and (ii) with optimized manufacturing conditions, including feedstock preparation, injection parameters, and sintering conditions, densities of more than 98% of the theoretical value could be achieved. Siargos et al. [61] compared the galvanic coupling of conventional and MIMed brackets with commonly used orthodontic archwires in 37 °C lactic acid for 28 days while the potential differences between wires and brackets were recorded. It was reported that (i) the MIMed brackets exhibited potential differences similar to those seen for the conventional brackets, (ii) the greatest potential difference was found for MIMed brackets with nickel-titanium wires (512 mV), whereas MIMed brackets with Cu–NiTi wires had the smallest difference (115 mV), (iii) the MIMed bracket exhibited extensive internal porosity, whereas the conventional bracket was more solid internally. Alavi et al. [62] assessed the hardness of orthodontic brackets produced by MIM and conventional methods and different orthodontic wires (stainless steel, NiTi, and beta-titanium alloys) for better clinical results. It was reported that (i) the maximum mean hardness values of the wires were achieved for stainless steel (529.85 Vickers hardness [VHN]) versus the minimum values for beta-titanium (334.65 VHN), and (ii) among the brackets, Elite Opti-Mim exhibited significantly higher VHN values (262.66 VHN) compared to Ultratrimm (206.59 VHN). VHN values of wire alloys were significantly higher than those of the brackets, concluding that the MIMed orthodontic brackets exhibited hardness values much lower than those of SS orthodontic archwires and were more compatible with NiTi and beta-titanium archwires, and a wide range of microhardness values has been reported for conventional orthodontic brackets and it should be considered that the manufacturing method might be only one of the factors affecting the mechanical properties of orthodontic brackets including hardness. Using MIM technology, Deguchi et al. [63] fabricated pure titanium orthodontic brackets.

6.2.4 Space holder method

Space holder method is one of the PM techniques using space holder to sinter the metal powder, which can control the pores including volume fraction, size, shape, and distribution depending on the amount and characteristics of space holder used. Space holders should be strictly chosen since the left space holder may be detrimental to the human body. In general, NH_4HCO_3, NaCl, starch, Mg powder, polymethyl methacrylate, and urea are used as space holders, which are bioinert or biocompatible. Such space holders can be evaporated at a relatively low temperature or can be removed by water. Therefore, once the sintering and the spacer holder removal are finished, the porous metal materials are produced. Although this method to produce porous Ti parts is convenient and has a lot of advantages, the accuracy of pore size is not expected. Furthermore, the pore shape in the produced Ti parts is restrained by the shape of the space holder, which also influences the structure of the produced Ti parts. Therefore, unlike the porous structure produced by additive manufacturing technologies, the porous Ti alloys produced by the space holder method have limited porous structure choice [64–70].

Xu et al. [64] prepared porous NiTi alloys by microwave sintering using ammonium hydrogen carbonate (NH_4HCO_3) as the space holder agent to adjust the porosity in the range of 22–62%. It was reported that (i) the porosities and average pore sizes of the porous NiTi alloys increased with increasing the contents of NH_4HCO_3, (ii) the porous NiTi alloys consisted of nearly single NiTi phase, with a very small amount of two secondary phases (Ni_3Ti, $NiTi_2$) when the porosities are lower than 50%, (iii) the amount of Ni_3Ti and $NiTi_2$ phases increased with further increasing of the porosity proportion, (iv) the superelastic recovery strain of the trained porous NiTi alloys could reach between 3.1% and 4.7% at the prestrain of 5%, even if the porosity was up to 62%. Ma et al. [31] used ammonium hydrogen carbonate (NH_4HCO_3) was used as a sacrificial space holder to tailor the porosity and pore size of the NiTi alloy and its properties, that is, elastic modulus, phase transformation temperature, SME, and SE for fabricating porous NiTi alloys by an EPM route at a temperature ranging from 850 to 950 °C. Abidi et al. [65] utilized NaCl as a spacer while fabricating porous NiTi alloy by pressureless sintering under vacuum. It was found that (i) pore morphology and size were replica of the shape of the spacer powder, (ii) porous alloys with spherical pores showed better mechanical properties as compared to the cuboidal shape, and (iii) all the foams showed low modulus and almost complete recovery at 2% strain, which is critical to mimic the bone structure for implant applications. Li et al. [66] used TiH2 powder as a pore-forming agent and active agent during an EPS NiTi alloy. It was found that (i) in order to avoid the formation of carcinogenic pure Ni phase, the porous NiTi alloy should be synthesized over a temperature of 950 °C, and (ii) substitution of Ti by TiH2 is more economic and more favorable to obtain homogeneous porous NiTi alloy. Bansiddhi et al. [67] utilized niobium (Nb) wire as space holder for sintering the porous NiTi alloy. It was found that when NiTi powder compacts containing Nb wires are heated above 1,170 °C, each Nb wire reacts with adjacent NiTi powders to form a eutectic liquid which wicks into the space between the remaining NiTi powders, creating a macropore at the location of the Nb wire while eliminating the microporosity between NiTi powders which is filled with a NiTi/Nb eutectic phase after solidification.

Opposing what we have been discussing, there is a unique method for which spacer holders are not needed for producing porous Ti and Ti-based alloys based on the partial sintering of powders. Accordingly, it can be easily speculated that the possible contamination is minimized. For this method, the porosity increases with increasing the initial powder size or with decreasing the sintering pressure, while the sintering temperature has little effect on the densification of the produced scaffold. Correspondingly, the elastic modulus and yield strength of the scaffold linearly reduce with increasing their porosity. One of practical methods is the effective usage of TiH_2 admixed to NiTi powders. Li et al. [68] added TiH_2 powder as a pore-forming and active agent in synthesis of porous NiTi alloys by using a conventional powder sintering technique and investigated the influence of addition of TiH_2 on microstructure, SME and SE characteristics of porous NiTi alloys. It was reported that (i) porous NiTi alloys produced with or without addition of TiH_2 exhibit good SME and SE, and the addition of TiH_2 not only has a pronounced effect on the microstructure but also increases the SME and volume memory effect of porous NiTi alloys, and (ii) the addition of TiH_2 has a significant influence on the elastic behavior and Young's modulus of porous NiTi alloys. Chen et al. [69] presented an in situ observation of NiTi powder sintering from TiH_2 powder and reported that (i) hydrogen release during dehydrogenation significantly affects the sintering behavior and resultant microstructure, and (ii) in comparison to the blended Ni/Ti powders, dehydrogenation occurring in the Ni/TiH_2 blend leads to higher porosity, less densification, and a lower degree of chemical homogenization after being sintered at 900 °C for 1 h. Chen et al. [70] further investigated NiTi powder sintering behavior from elemental powder mixtures of Ni/Ti and Ni/TiH_2. It was found that (i) the sintered porous alloys have open porosities ranging from 2.7% to 36.0%, (ii) in comparison to the Ni/Ti compact, dehydrogenation occurring in the Ni/TiH_2 compact leads to less densification yet higher chemical homogenization only after high-temperature sintering, and (iii) a comparative study of cyclic stress–strain behavior of the porous NiTi alloys made from Ni/Ti and Ni/TiH_2 compacts indicate that the samples sintered from the Ni/TiH_2 compact exhibited a much higher porosity, larger pore size, lower fracture strength, lower close-to-overall porosity ratio, and lower modulus of elasticity. Wu et al. [51] utilized titanium hydride powders to enhance the foaming process in the formation of orthopedic NiTi scaffolds during capsule-free HIP. It was reported that (i) hydrogen is continuously released from titanium hydride as the temperature is gradually increased from 300 to 700 °C via two transitions: $TiH_{1.924} \rightarrow TiH_{1.5}/TiH_{1.7}$ between 300 and 400 °C and $TiH_{1.5}/TiH_{1.7} \rightarrow \alpha$-Ti between 400 and 600 °C, (ii) in the lower temperature range between 300 and 550 °C the rate of hydrogen release is slow, but the decomposition rate increases sharply above 550 °C, (iii) holding processes at 425, 480, 500, 550, and 600 °C are found to significantly improve the porous structure in the NiTi scaffolds due to the stepwise release of hydrogen, and (iv) NiTi scaffolds foamed by stepwise release of hydrogen are conducive to the attachment and proliferation of osteoblasts and the resulting pore size also favor ingrowth of cells.

6.2.5 Powder treatment

There are several techniques proposed to powders: crushing, grinding, chemical reactions, or electrolytic deposition. Powders of the elements titanium, vanadium, thorium, niobium, tantalum, calcium, and uranium are produced by high-temperature reduction of the corresponding nitrides and carbides.

Terayama et al. [71] and Sadrnezhaad et al. [72] prepared NiTi alloy powder by mechanically alloying (MA) technique. It was reported that (i) the sintered alloy of the MA powder showed more uniform phase of NiTi than that of the elementally mixed powders sintered in a same manner; however, the former showed a lower density than the latter due to a larger particle size of the MA powder of presintering, (ii) the alloy has shape-memory characteristics, and the transformation temperatures of the alloy are higher than those of the alloy of the elementally mixed powders due to waste of Ni powder [71], and (iii) the crystallite size of the MAed $Ni_{50}Ti_{50}$ samples decreased with MA duration and with the milling speed, (iv) metallographic studies proved the existence of martensitic B19′ after sintering of both the as-mixed and the mechanically alloyed samples, (v) sintering lowered the porosity of the samples; no matter what powder (as-mixed or mechanically alloyed) was used, and (vi) the porosity was greater, however, for the MA powders [72].

For powder compaction, Thadhani et al. [73] and Matsumoto et al. [74] employed the dynamic shock consolidation. It was reported that (i) the melted material at interparticle regions was observed to have rapidly solidified to largely amorphous and/or microcrystalline phases, (ii) particle interiors were also subjected to extensive plastic deformation which resulted in deformation twinning, grain elongation, and some recrystallization to defect-free grains, and (iii) unique microstructural modifications occurring due to inhomogeneous thermal and mechanical processing during the dynamic consolidation process, are reported [73], and (iv) with increasing flyer velocity, new pores are formed in the melted layer instead of the disappearance of initial interstices, and (v) effects of the mechanical deformation and the annealing brought by the shock treatment evidently appear in the temperature dependence of electrical resistivity accompanied with the martensitic transformation [74].

Although PM methods have been extensively used for producing near-net-shape components of NiTi SMAs in recent years, there are some drawback associated with this method of the undesirable formation of phases and intermetallics, such as $NiTi_2$, Ni_3Ti, and oxide phases during sintering process. Hosseini et al. [75] attempted to improve production of NiTi parts by preheat treatment of titanium powder under hydrogen atmosphere at temperatures between 100 and 300 °C. It was reported that the heat treatment could change the microstructure and facilitate martensitic transformation of sintered NiTi specimens and reduction of Ni_3Ti_3O oxide phase.

Whitney et al. [76] investigated the influence of Ni powder size on microstructural evolution. It was found that (i) the use of very fine Ni powders causes this transformation to occur at the eutectoid temperature (i.e., 765 °C); while the use of coarse Ni powders causes a gradual beta-Ti transformation from 765 to 882 °C, (ii) at 950 °C a large volume fraction of beta-Ti remains in coarse Ni/Ti mixtures whereas in fine Ni/Ti mixtures this phase is almost eliminated, (iii) further heating above 950 °C causes the beta-Ti to melt, initiating a large exothermic reaction in the coarse Ni/Ti mixtures at 980 °C, and (iv) the use of fine Ni significantly reduces this reaction, consequently, Ni powder size, and its influence over beta-Ti content can be used to control the reactive sintering behavior of Ni+Ti mixtures; indicating that a key microstructural feature that controls the sintering behavior of Ni+Ti powders was determined to be the transformation of alpha-Ti to beta-Ti during heating. Hryha et al. [77] evaluated the surface oxide state of the Ti, NiTi and Ti6Al4V powders in as-atomized state and after storage under air or Ar for up to 8 years. It was found that (i) powder in as-atomized state is covered by homogeneous Ti-oxide layer with the thickness of ~2.9 nm for Ti, ~3.2 nm and ~4.2 nm in case of Ti6Al4V and NiTi powders, respectively, (ii) exposure to the air results in oxide growth of about 30% in case of Ti and only about 10% in case of NiTi and Ti6Al4V, (iii) after the storage under the dry air for 2 years oxide growth of only about 3–4% was detected in case of both, Ti and NiTi powders, (iv) NiTi powder, stored under the dry air for 8 years, indicates oxide thickness of about 5.3 nm, which is about 30% thicker in comparison with the as-atomized powder, and (v) oxide thickness increase of only ~15% during the storage for 8 years in comparison with the powder, shortly exposed to the air after manufacturing, was detected.

6.3 Combustion syntheses

Since their mechanical properties are closer to those of cortical bones than stainless steels and titanium alloys [78], porous NiTi alloys show promising potential in the application of bone implantation. The porous structure also allows the ingrowths of new bone tissue along with the transport of body fluids, thus ensuring a harmonious bond between the implant and the body [78]. Porous NiTi alloys have been fabricated with PM processes such as SHS, MIM, HIP and SPS [78, 79–82]. These processes can avoid the problems associated with casting, like segregation or extensive grain growth and have the added advantages of precise control of composition and easy realization of complex part shapes [82]. CS can be accomplished in two modes: TE and SHS. CS systems typically have a non-monotonic dependence of reaction rate on temperature. For example, melting of the precursor is accompanied by a sharp increase in the reaction rate [83, 84]. CS is a highly exothermic reaction accompanying formation of compounds with high formation energies. The reaction is initiated by ignition and propagated at speeds of 0.1–20 cm/s. Usually, the reaction temperature reaches above approximately 1,800–3,200 °C.

Biswas et al. [81] prepared porous yet single phase NiTi alloy using the TE mode of SHS in combination with a controlled postreaction heat treatment under the variations of the heating rate and the green density. It was reported that (i) there was a threshold heating rate that varied with the green density, above which the product melted, (ii) the molten-cast product was single phase but devoid of the requisite porosity; below this threshold the product was porous but multiple phase, (iii) the postreaction heat treatment needed for the conversion to single phase porous NiTi also showed strong dependence on the heating rate, and (iv) a faster rate resulted in loss of both the porosity and the shape as the strong interaction with an intermediate liquid phase collapsed the porous structure; while this interaction could be controlled and heating up to 1,150 °C at a very low rate (1 °C/min) could produce single phase NiTi without sacrificing the porosity. Whitney et al. [76] investigated the TE synthesis of NiTi. It was reported that (i) the use of very fine Ni powders causes this transformation to occur at the eutectoid temperature (i.e., 765 °C), (ii) the use of coarse Ni powders causes a gradual beta-Ti transformation from 765 to 882 °C. At 950 °C a large volume fraction of β-Ti remains in coarse Ni/Ti mixtures whereas in fine Ni/Ti mixtures this phase is almost eliminated, and (iii) the use of fine Ni significantly reduces this reaction (i.e., 3 J/g); concluding that Ni powder size and its influence over β-Ti content can be used to control the reactive sintering behavior of admixture of Ni and Ti.

PM involves techniques which allow obtaining porous materials with adjustable porosity, pore size and pore distribution. Some of those techniques appear suitable for fabrication of porous metallic biomaterials. For successful implantology, there are at least three crucial requirements that should be melt and they include (1) biological compatibility, (2) biomechanical compatibility, and (3) macro- and micromorphological compatibility [85]. With regard to the morphological compatibility, Oshida et al. [86, 87] analyzed about 50 published articles on successful dental implant treatments in term of surface roughness (R_a) and pore size, and found that, independent of designs and materials for implants, (i) successful implants possess R_a value is in range from 5 to 50 μm, and (ii) pore size can be found in a range from 200 to 500 μm. The desired pore size ranges between 200–500 μm according to many authors; although, 100 μm is usually mentioned as the lower limit of pore size allowing bone tissue ingrowth. The presence of pores smaller than 100 μm is also advantageous because it increases roughness and enhances adhesion of proteins and cells to the surface and improves osteoconductivity [88]. Tosun and Tosun [89] studied the optimization of process parameters for the porosity in NiTi alloy produced as implant material by SHS method as a function of different process parameters (i.e., compaction pressure, preheating temperature, and heating rate). It was reported that (i) the highly effective parameter on production characteristics are the compaction pressure, whereas the preheating temperature are less effective factors, (ii) the optimal production parameters for the porosity were obtained at 50 MPa pressure, 60 °C/min heating rate, and 250 °C preheating temperature settings, and (iii) among the several functions tried, the power function is found to be the best-fit model.

Li et al. [90] prepared porous equiatomic NiTi SMAs by SHS using elemental nickel and titanium powders. It was found that (i) the porous NiTi alloys possess an open porous structure with about 60 vol% porosity, and the channel size is about 400 μm, (ii) the combustion temperature increases with increasing preheating temperature and results in melting of the NiTi compound above 450 °C, (iii) the preheating temperature has been shown to have a significant effect on the microstructure of the SHSed porous NiTi alloys, and (iv) the mechanism of anisotropy in pore structure is attributed to the convective flows of liquid and argon during combustion. SHSed porous NiTi SMA was prepared from elemental titanium and nickel powders and it was reported that (i) porous SHSed NiTi alloys had homogeneous pore distribution and proper mean pore size, and (ii) the mechanical property of pure product depended strongly on preheating temperature during synthesis process [91]. Porous NiTi (with an average porosity of 55 vol% and a general pore size of 100–600 µm) was SHSed with the addition of mechanically alloyed nanocrystalline NiTi as the reaction agent [92]. The SHS of porous NiTi using elemental powders was also performed for comparison. To enhance the bioactivity of the metal surface, porous NiTi synthesized by nanocrystalline NiTi was subjected to chemical treatment to form a layer of TiO_2 coating and the TiO_2 coated porous NiTi was subsequently immersed in a simulated body fluid to investigate its apatite forming ability. It was found [88] that (i) the main phases in porous NiTi synthesized by elemental powders were NiTi, Ti_2Ni, and unreacted free Ni, (ii) by using nanocrystalline Ni–Ti as reaction agent, the secondary intermetallic phase of Ti_2Ni was significantly reduced and the free Ni was eliminated, (iii) TiO_2 coating with anatase crystalline phase was formed on the surface of porous NiTi after the chemical treatment, and (iv) a layer consisting of nanocrystalline carbonate-containing apatite was formed on the surface of TiO_2 coating after soaking in the simulated body fluid. Tay et al. [93] fabricated porous NiTi alloy by SHS method to examine the effects of the green porosity and preheating temperature. It was found that (i) the porosity was in the range of 45–59 vol%, with the green porosity being the primary source of final porosity, and (ii) the effect of green porosity and preheating temperature was found to have an effect on the pore morphology and microstructure. Whitney et al. [94] investigated the mechanisms of reactive sintering in a Ni+Ti powder compact using differential scanning calorimetry and microstructural analysis. It was reported that (i) heating these mixtures up to 900 °C involves the slow growth of the three intermetallic compounds and the transformation of α-Ti to β-Ti followed by its rapid saturation with Ni, (ii) when samples were heated above 942 °C a TE mode of SHS was ignited by the melting of the (β-Ti) solid solution at 944 °C, (iii) increasing hold time at 900 °C prior to SHS decreased the volume fraction of (β-Ti) in the powder compact and reduced the magnitude of combustion, and (iv) the magnitude of the exothermic reaction occurring during SHS was found to decrease linearly with a decrease in the volume fraction of (β-Ti) developed at 900 °C.

Tosun et al. [95, 96] blended $Ni_{50.5}Ti$ (at%) 12 h and cold pressed in the different pressures (50, 75, and 100 MPa), followed by SHS at the different preheating temperatures (200, 250, and 300 °C) and heating rates (30, 60, and 90 °C/min). It was reported that (i) NiTi was seen as the dominant phase in the microstructure with other secondary intermetallic compounds, and (ii) the porosity of the synthesized products was in the range of from 50.7 to 59.7 vol% [95], (iii) the pressure, preheating temperature, and heating rate were found to have an important effect on the biocompatibility in vivo of the synthesized products, (iv) the fibrotic tissue within the porous implant was found in vivo periods (6 months), in which compacting pressure 100 MPa [96]. Resnina et al. [97] investigated the influence of the chemical composition and the preheating temperature of the NiTi powder mixture on kinetics of martensitic transformation and structure in the porous NiTi alloys produced by SHS. It was found that (i) a variation in the nickel concentration in the powder mixture influenced the volume fraction of the NiTi phase undergoing martensitic transformation, and (ii) an increase in the nickel concentration from 45 to 52 at% led to an increase in the volume fraction of the NiTi phase from 60% to 90% in the alloy and a decrease in the alloy volume undergoing martensitic transformation from 60% to 10%. It was further shown that (iii) in the porous NiTi alloys with the Ni concentration varied from 48 to 50 at% the martensitic transformations were found in two temperature intervals; the first interval was from 347 to 304 K and it was due to the B2 → B19′ transformation occurring in the volume of the TiNi phase with a nickel concentration close to 50.0 at%, and the second temperature interval was from 252 to 198 K and it was caused by the same transformation occurring in the volumes of the NiTi phase with a nickel concentration close to 51%, and (iv) the value of the preheating temperature did not influence the characteristics of the martensitic transformations, except in the $Ni_{48}Ti$ (at%) alloy where an increase in the preheating temperature resulted in an increase in the volume fraction of the NiTi phase undergoing martensitic transformation from 30% to 60%.

Bassani et al. [98, 99] mentioned that (i) process parameters, among which powder compaction degree and preheating temperature, strongly influence the reaction temperature and the resulting product: at low reaction temperatures, high quantity of secondary phases are formed, which are generally considered detrimental for biocompatibility; while, at higher reaction temperatures, the powders melt and crystallize in ingots, and (ii) changes in porous shape and size were observed especially for TiHx additions: the latter could be a promising route to obtain shaped porous products of improved quality [100]. It was further mentioned that (iii) in spite of toxicity of TiN_3 precipitate, phases with higher Ti content showed high biocompatibility, and (iv) a slightly reduced biocompatibility of porous NiTi was ascribed to combined effect of $TiNi_3$ presence and topography that requires higher effort for the cells to adapt to the surface [99].

Novák et al. investigated effect of SHS conditions on microstructure of NiTi alloy [100], and particle size effect of Tit and Ni on the TE-SHS synthesis of NiTi [103]. It was reported that (i) high heating rate (over 300 °C/min), temperature of 1,100 °C and process duration of 20 min were determined as the optimal parameters, and (ii) in order to prevent nickel depletion of the product and to minimize the porosity, the use of coarse titanium particles (200–600 μm) with fresh surface was recommended [100], and (iii) the optimum powder fraction of both nickel and titanium to achieve the most intensive SHS reaction was in a range from 25 to 45 μm [101]. Školáková et al. [102] investigated the effect of magnesium on the microstructure, phase composition, amount of undesirable Ti_2Ni phase, martensitic transformation, mechanical properties, and corrosion resistance of NiTi alloy which was fabricated by SHS process. It was mentioned that (i) a significant reduction of the content of undesirable Ti_2Ni phase by the addition of magnesium, and (ii) magnesium addition affects to increase corrosion resistance and enhance the yield strength. Biffi et al. [103] studied the SHS process on Ni and Ti elemental powders, ignited by a laser beam. It was mentioned that (i) samples with porosity in the range from 42% to 48% were produced, (ii) NiTi matrix and other secondary phases, like $NiTi_2$, Ni_3Ti, and Ni_4Ti_3 were detected, and

(iii) no dependence of microstructure and transformation temperatures with the process parameters was found. Aihara et al. [104] produced the porous NiTi alloys with a three-dimensional anisotropic interconnective open-pore structure has been successfully produced by the CS of elemental Ni and Ti powders. It was reported that (i) the average elastic modulus was approximately 1 GPa for a porosity of 60 vol% and the average pore size of 100–500 μm, (ii) the CSed NiTi alloy exhibits excellent corrosion resistance with breakdown potentials above 750 mV, and (iii) an ovine study in cortical sites of the tibia demonstrated rapid osseointegration into the porous structure as early as 2 weeks and complete bone growth across the implant at six weeks; indicating that the porous NiTi alloy is a promising biomaterial with proven biocompatibility and exceptional osseointegration performance which may enhance the healing process and promote long-term fixation, making it a strong candidate for a wide range of orthopedic implant applications.

6.4 Mechanical alloying and rapid solidification processing

6.4.1 Mechanical alloying

MA is a solid-state powder processing technique involving repeated welding, fracturing, and rewelding of powder particles in a high-energy ball mill. MA has now been shown to be capable of synthesizing a variety of equilibrium and nonequilibrium alloy phases starting from blended elemental or prealloyed powders. The nonequilibrium phases synthesized include supersaturated solid solutions, metastable crystalline and quasicrystalline phases, nanostructures, and amorphous alloys [105, 106]. Interdispersion of the ingredients occurs by repeated cold welding and fracture of free powder particles. Refinement of structure is approximately a logarithmic function of time and depends on the mechanical energy input to the process and work hardening of the materials being processed [107]. Takasaki et al. [108] and Saito et al. [109] prepared three kinds of NiTi elemental powders, including $Ti_{45}Ni_{55}$, $Ti_{50}Ni_{50}$ and $Ti_{55}Ni_{45}$ (at%) by MA using a planetary ball mill for alloying times up to 10 h, and the alloying process and the microstructures after heating at temperature of 1,000 °C. It was found that (i) the $Ti_{55}Ni_{45}$ powder formed an amorphous phase after MA for 10 h, while the $Ti_{50}Ni_{50}$ powders formed a disordered BCC-TiNi phase and the $Ti_{45}Ni_{55}$ powder also formed the disordered BCC-TiNi phase at an intermediate stage of MA but turned to an amorphous state with increasing alloying time, (ii) after heating at 1,000 °C, the Ti_2Ni phase has been formed in all powders, and the ordered B2-TiNi phase was observed in the $Ti_{50}Ni_{50}$ and $Ti_{45}Ni_{55}$ powders, but a monoclinic TiNi phase in the $Ti_{55}Ni_{45}$ powder, and (iii) in the $Ti_{45}Ni_{55}$ powder also the $TiNi_3$ phase has been formed. The amounts of these intermetallic phases are dependent on the chemical compositions of the starting powders.

Sadrnezhaad et al. [110] prepared commercially pure NiTi powders by mechanical milling in a vertical attritor mill under protective atmosphere for various times from

10 to 24 h, followed by compacting and sintering at different temperatures for different times. It was reported that (i) porosity, virtual density, transition temperatures, and the amount of Ni_3Ti first increased and then decreased with the milling time, and (ii) presence of oxygen in the milling atmosphere showed partial crystallization of NiTi intermetallic compound accompanied by titanium oxide formation. Mechanical alloyed NiTi powders were compacted and sintered and effects of milling time and milling speed on crystallite size, lattice strain, and XRD peak intensities were investigated by x-ray analysis of the alloy [72]. It was found that (i) the crystallite size of the MAed $Ni_{50}Ti_{50}$ samples decreased with MA duration and with the milling speed, (ii) depending on the crystal structure of the raw materials, the lattice strain increases with the milling duration, (iii) metallographic studies proved the existence of martensitic B19′ after sintering of both the as-mixed and the mechanically alloyed samples, and (iv) the porosity was greater, however, for the MA powders, owing to the sharper LPS effect of the as-mixed samples. Mousavi et al. [111] utilized MA to produce NiTi intermetallic with nanocrystalline structure from the elemental powders and it was reported that disordered B2–NiTi phase can be obtained with grain size of 25 nm, particle size of 15 μm, lattice distortion, and high microhardness of 1.2% and 922 HV, respectively. As to the mechanism of NiTi formation, in the early stages of MA, a composite lamellar structure of components is formed with the dissolution of Ti in Ni at the same time, and resulting solid solution finally leads to the formation of nanocrystalline disordered B2–NiTi phase. Annealing of the milled powder at 900 °C leads to grain growth, decrease of microhardness, and transformation of disordered structure to ordered NiTi. A small amount of $NiTi_2$ and Ni_3Ti phases was also detected.

Verdian [112] fabricated supersaturated NiTi (Al) alloys by MA and reported that (i) the diffraction lines corresponding to B2-NiTi shift toward higher angles with increasing milling time, which means that the lattice parameter of NiTi decreases and NiTi(Al) solid solution is formed, and (ii) the lattice parameter of NiTi reaches a constant value after 8 h of milling, indicating that alloying of NiTi and Al is completed. Ni and Ti elemental powders with a nominal composition $NiTi_{50}$ (at%) were mechanical alloyed in a planetary high-energy ball mill in different milling conditions (5, 10, 20, 40, and 60 h) [113], and it was found that increasing milling time leads to a reduction in crystallite size, and after 60 h of milling, the Ti dissolved in Ni lattice and NiTi (B2) phase was obtained. With milling time, morphology of prealloyed powders changed from lamella to globular. Annealing of as-milled powders at 900 °C for 900 s led to formation of nanocrystalline NiTi (B19′), grain growth, and release of internal strain. Alijani et al. [114] evaluated the effect of milling process on the chemical composition, structure, microhardness, and thermal behavior of $TiNi_{41}Cu_9$ compounds developed by MA. It was reported that at the initial stages of milling (up to 12 h), the structure consisted of a Ni-solid solution and amorphous phase, and by the milling evolution, nanocrystalline martensite (B19′) and austenite (B2) phases were initially formed from the initial materials and then from the amorphous phase. By the milling development, the composition uniformity is increased, the interlayer thickness is reduced, and the powders' microhardness is initially increased, then reduced, and afterward reincreased. The thermal behavior of the alloyed powders and the structure of heat-treated samples is considerably affected by the milling time. Khademzadeh et al. [115] studied the combination of high-energy MA and direct metal deposition techniques with the aim of assessing the feasibility of synthesizing near-single-phase NiTi from elemental nickel and titanium powders. It was found that the B2-NiTi phase with small quantity of $NiTi_2$ can be obtained by direct metal deposition of mechanically alloyed $Ni_{50.8}Ti_{49.2}$ powder. Akmal et al. [116] synthesized equiatomic NiTi alloy composites reinforced with 0, 2, 4, and 6 vol% nano-HA using pressureless sintering. It was reported that (i) mechanically alloyed NiTi and HA powders were blended, compacted, and then sintered for 3 h at 1,050 °C; (ii) the existence of multiple phases like NiTi, $NiTi_2$, Ni_3Ti, and Ni_4Ti_3 was detected; (iii) the 6 vol% HA reinforced composite showed Ni_3Ti as the major phase having the highest hardness value which can be attributed to the presence of relatively harder phases along with higher HA content as a reinforcement; (iv) the composite of NiTi with 2 vol% HA manifested the most desirable results in the form of better sintering density mainly due to the minute decomposition of NiTi into other phases, suggesting that the 2 vol% reinforced NiTi composite can be exploited as a novel material for manufacturing biomedical implants. The formulation of nanocrystalline NiTi SMAs has potential effects in mechanical stimulation and medical implantology. Arunkumar et al. [117] investigated the effect of milling time on the product's structural characteristics, chemical composition, and microhardness for NiTi synthesized by MA for different milling durations. It was found that (i) increasing the milling duration led to the formation of a nanocrystalline NiTi intermetallic at a higher level, (ii) particle size analysis revealed that the mean particle size was reduced to ~93 μm after 20 h of milling, and (iii) the mechanical strength was enhanced by the formation of a nanocrystalline intermetallic phase at longer milling time, which was confirmed by the results of Vickers hardness analyses.

There is another way to prepare alloyed powder – laser alloying. Laser alloying is a material surfacing process that utilizes the high density of laser beam power as the source of energy. Prior to the laser alloying process, a coating material is deposited onto the base material. The coating material and the base material are then targeted by the laser beam to melt metal coatings and a portion of the underlying substrate. Zhang et al. [118] utilized Mo surface-modified layer in NiTi alloy prepared by plasma surface alloying. It was reported that (i) the x-ray diffraction analysis revealed that the modified layers were composed of Mo, MoTi, MoNi, and Ti_2Ni; (ii) the microhardnesses of the Mo modified layers treated at 900 °C and 950 °C were 832.8 HV and 762.4 HV, respectively, which was about 3 times the microhardness of the TiNi substrate; (iii) scratch tests indicated that the modified layers possessed good adhesion with the substrate; and (iv) compared with as-received TiNi alloy, the modified alloys exhibited significant improvement of wear resistance against Si_3N_4 with low normal loads during the sliding tests. Mass spectrometry displayed that the Mo alloy layers had successfully inhibited the Ni release into the body.

6.4.2 Rapid solidification processing (RSP)

The term "rapid solidification" which is defined as a "rapid" quenching from the liquid state covers a broad range of material processes, covering a broad engineering regime that can be from the powder producing atomization techniques to the plasma spray process. Now the concept of rapid solidification is more general, and processes with lower cooling rates such as some powder producing techniques have been included under the rapid solidification technology (RST). RST accordingly incorporates a large number of different processes. These processes can be classified either as a function of their resulting products (e.g., powder producing techniques), as a function of the resulting microstructure (grain and particle size), or as a function of cooling rate (generally measured by secondary dendrite arm spacing) [119]. Common products obtained by rapid solidification techniques include powders, flakes, ribbons, wires, and foils. An essential factor associated with almost all current rapid solidification techniques is that the as-solidified product is very small (micron size) in at least one dimension. In other words, the surface-to-volume ratio for the product is very large. This is essential in obtaining very large cooling or solidification rates. Jiang et al. [120] fabricated a near-equiatomic NiTi SMA by rapid solidification process through vacuum arc melting followed by vacuum suction casting in water-cooled thick copper mold. It was mentioned that the rapidly solidified (or suction cast) NiTi alloy shows much finer grains and homogenous microstructure, in particular a uniform distribution of various fine precipitates, compared to the conventional cast one. It also exhibits the homogenous Ni distribution in the matrix of the alloy, allowing the martensitic transformation to take place throughout the NiTi alloy matrix simultaneously and resulting in sharper transformation peaks compared to the conventional cast alloy. The suction-cast NiTi alloy shows a significant improvement over the conventional cast one, in terms of possessing higher deformation recovery rates and displaying the increased compressive strength and damping capacity by 4% and 20%, respectively. Zadravec et al. [121] analyzed the influence of temperature boundary conditions on the solidification process of NiTi alloy. It was mentioned that (i) velocity and temperature field is strongly affected by the different cooling condition on the cylindrical wall and therefore the solidification process of the alloy and (ii) cooling on the cylindrical part is one of the major parameters for alloy solidification and, therefore, should not be neglected.

6.5 Additive manufacturing

American Society for Testing and Materials (ASTM) subcommittee defined the term additive manufacturing (AM) as "a process of joining materials to make objects from 3D model data, usually layer upon layer, as opposed to subtractive manufacturing methodologies." Additionally, synonyms are expressed as additive fabrication, additive processes, additive techniques, additive layer manufacturing, layer manufacturing, and freeform fabrication [122]. Opposing to structural components and devices, any functional structure and devices in either engineering or medical areas possess complex shapes and their sizes, depending on application protocols and users (or patients). Therefore, the production of, for example, biomedical implants using conventional technologies always accompanies with significant time, material, and energy costs. In recent decades, technology involved in the additive manufacturing such as SLM and EBM has been widely employed to solve the conventional manufacturing problems of NiTi-based materials [123]. Despite the fact that the type of the production process or raw material characteristics may be different in AM techniques, the fabrication consists of eight similar stages of all types throughout the development process as shown in Figure 6.4 [123].

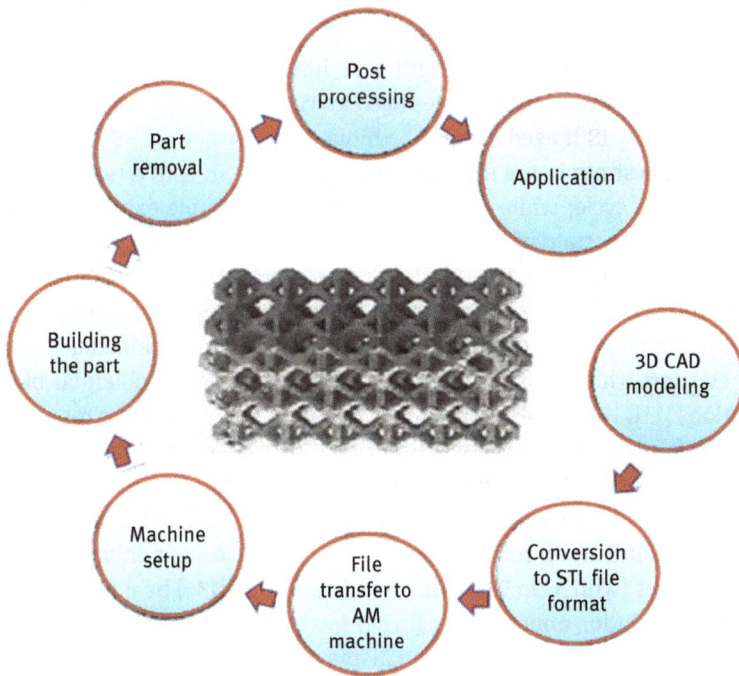

Figure 6.4: Process sequence of additive manufacturing [126].

6.5.1 Selective laser melting

SLM technology is a layer-wise process in accordance with a pre-designed 3D model under a protective atmosphere. Ti alloy parts are produced layer-by-layer by selectively melting, which consolidates the powder at a high rate by using a computer-controlled laser. SLM provides a wide range of advantages compared to the conventional manufacturing methods, such as lower production time, higher material utilization, almost no geometric constrictions, and further postprocessing. Due to the fast heating/cooling rate in SLM processing, SLM-produced alloys have distinctive microstructures in comparison with the conventional counterparts [124–127]. Particularly, for fabricating complex-shaped NiTi constructions which become more and more essential for biomedical applications especially for dental or craniomaxillofacial implants, the additive manufacturing method of SLM allows realizing complex-shaped elements with predefined porosity and three-dimensional microarchitecture directly out of the design data [128].

Saedi et al. [129–134] had extensively investigated the properties of SLMed NiTi alloys on various aspects. Using Ni-rich $Ni_{50.8}Ti_{49.2}$ (at%) alloys, effects of postfabrication heat treatment were studied [129, 130] and it was reported that (i) the SLM method and postheat treatments can be used to tailor the microstructure and shape-memory response, (ii) partial SE was observed after the SLM process, (iii) solution-treatment of fabricated samples increased the strength and improved the SE but slightly decreased the recoverable strain, (iv) samples aged at 350 °C showed better recovery in SE tests where 350 °C × 18 h aged samples exhibited almost perfect SE with 95% recovery ratio with 5.5% strain in the first cycle and stabilized SE with a recoverable strain of 4.2% after 10th cycle; while (v) 450 °C × 10 h aged sample exhibited 68% recovery with a recoverable strain of 4.2% in the first cycle and stabilized recoverable strain of 3.8% after 10th cycle [129, 130]. It was, furthermore, mentioned that transformation temperatures can be tailored and upon thermal treatments of 350 °C × 1 h and 600 °C × 1.5 h, superelastic recovery up to about 5.5% can be obtained at body and room temperatures which make these alloys very promising for potential biomedical applications [131, 132]. Studying the same NiTi alloy [133, 134], it was reported that (i) the sample fabricated with a laser power of 100 W and scanning speed of 125 mm/s exhibited almost perfect SE with a recovery ratio of 96% and strain recovery of 5.77% in the first cycle and (ii) the corresponding stabilized superelastic response demonstrated full strain recovery of 5.5% after 10 cycles. For enhancing SE of porous NiTi structures (with 32 to 58% porosity), Saedi et al. [134] heat treated the fabricated samples by solution annealing + aging at 350 °C for 15 min. It was reported that (i) SLMed NiTi with up to 58% porosity can display SME with full recovery under 100 MPa nominal stress, (ii) dense SLMed NiTi could show almost perfect SE with strain recovery of 5.65 after 6% deformation at body temperatures, (iii) the strain recoveries were 3.5%, 3.6%, and 2.7% for samples with porosity levels of 32%, 45%, and 58%, respectively, and (iv) the stiffness of NiTi parts can be tuned by adjusting the porosity levels to match the properties of the bones [134]. This last conclusion is related to the biomechanical compatibility [25] and stress shielding effect [135].

Mullen et al. [136], for developing a bone-ingrowth structure, investigated a novel porous titanium structure fabricated by SLM technique with specific requirements; in particular, functionally graded structures with bone-ingrowth surfaces exhibiting properties comparable to those of human bone. It was found that a porosity of a range from 10% to 95% and resultant compression strength of 0.5 to 350 MPa, which is comparable to the typical naturally occurring range, and (ii) optimized structures have been produced that possesses ideal qualities for bone in-growth applications and that these structures can be applied in the production of orthopedic devices. Research group of Shishkovskii et al. [137–139] investigated conditions of a layer-by-layer synthesis of 3D parts made of nitinol and HA additions using selective laser sintering/melting (SLS/SLM) since optimization for SLS/SLM parameters are needed for the synthesis of NiTi + HA to be used in tissue engineering and manufacture of medical devices (pins, nails, porous implants, drug delivery systems). It was reported that (i) no significant destruction of HA ceramics under laser treatment was observed, (ii) the amount of nickel released to the surface of 3D parts decreases owing to the additional oxidation of free titanium during SLS/SLM and the formation of a protective HA layer, and (iii) full-density 3D parts are produced from nitinol by SLM including preheating to 300 °C.

Moghaddam et al. [140] evaluated the anisotropic tensile properties of $Ni_{50.1}Ti_{49.9}$ (at%) components fabricated by SLM method. It was mentioned that (i) SLMed samples in the horizontal orientation had the highest ultimate tensile strength (606 MPa) and elongation (6.8%) with the strain recovery of 3.54% after four SME cycles, and (ii) at stress levels less than or equal to 200 MPa, these samples had the actuation strain greater than 3.8% without accumulation of noticeable residual strain. Chekotu et al. [141] reviewed the most recent publications related to the SLM processing of nitinol identify the various influential factors involved and process-related issues. It was described that (i) powder quality and material composition exhibit a significant effect on the produced microstructures and phase transformations, (ii) the effect of heat treatments after SLM fabrication on the functional and mechanical properties are noted, and (iii) optimization of several operating parameters were found to be critical in fabricating nitinol parts of high density. Zhou et al. [142] investigated the SLM to produce N-rich NiTi parts using the plasma rotating electrode processed prealloyed NiTi powder. It was reported that (i) the as-printed samples exhibit a dominant austenitic B2 phase and a minor in situ formed Ni_4Ti_3 precipitate, (ii) the asymmetric phase transformations of a two-step of B2 $\rightarrow$ R $\rightarrow$ B19′ phase transformation during cooling but the one-step B19′ $\rightarrow$ B2 transformation during heating was detected, and (iii) the as-printed NiTi samples achieve a fracture tensile stress of 1,411.0 ± 59.3 MPa and an elongation of 11.8 ± 0.9%.

6.5.2 Electron beam melting

In spite of unique applicability of NiTi due to SHE and SE phenomena, there are several drawbacks for fabrication including (1) the induced martensitic transformation and poor thermal conductivity properties, (2) the adhesion and the tool wear can easily occur during machining, (3) the grinding and laser cutting applications which involve crucial challenges create intermetallic phases and lead to crack formation in the heat-affected zone (HAZ), and (4) the intermetallic compounds cannot be removed with conventional processing methods and require postprocessing heat treatments [141, 143–146]. Additionally, there would be a technical difficulty for achieving fully dense materials even with PM processes owing to the capillary forces coming from these compounds [147, 148]. Accordingly, it is needed to establish a technology to accomplish the production of NiTi alloy with high chemical homogeneity and good shape-memory properties especially for complex shapes and geometries. EBM, as another additive manufacturing technique, is a technique to produce biomedical Ti alloys. Being similar to the SLM, EBM technology also adopts a layer-wise processing method using a high-energy electron beam as a heat source to fabricate alloy parts on a powder bed based on a computer aided designed model. In comparison to SLM technology, EBM process is conducted in a vacuum environment and can preheat the powder used, thereby producing alloy parts with higher densities. From an industrial point of view, EBM becomes a great alternative method for manufacturing biomedical Ti alloys [149–155].

Otubo et al. [156] utilized EBM with water-cooled copper crucible that that eliminates the carbon contamination and the oxygen contamination would be minimized due to operation in high vacuum to melt NiTi alloy and mentioned that the purity of the final product should be very important mainly in terms of biomedical applications and the contaminations by carbon and oxygen affect the direct and reverse martensitic transformation temperatures. Altug-Peduk et al. [157] studied the effectiveness of laser beam melting (LBM) and EBM and mentioned that the final manufactured material properties are highly affected by the properties of the powder particles in terms of particle size, density, distribution, and chemical properties.

6.6 Joining technology

Although the beneficial characteristics of SME and SE of NiTi alloy and its versatile applications in the automotive industry (e.g., diesel fuel injector), microelectromechanical devices, aerospace and power plants, civil engineering segment (actuators, passive energy dissipaters dampers etc.), and medical and dental applications are well documented, there should be recognized that a limiting factor in further application is their poor machinability, which often sets limits to the freedom in design. Another factor is the lack of available joining techniques, particularly for joining of NiTi alloys to other materials, because high temperatures can ruin SME and SE characteristics of NiTi alloys. At the same time, depending on heat involved of the joining method, it is basically believed that in joining NiTi alloys to other metals, formation of brittle intermetallic compounds is difficult to avoid.

As to joining technologies, there are the adhesion bonding, solid-state bonding, brazing or soldering, and fusion welding.

6.6.1 Adhesion bonding

Since the bonding strength by the adhesive bonding is relatively weak in comparison to other joining techniques to which some extent of heat is involved, joining surfaces of NiTi are needed to be modified to improve the wettability by mechanical, physical, or chemical treatment. Rossi et al. [158] studied mechanical and chemical surface treatments (immersion in acid and alkaline solutions), and different combinations of the above surface preparation procedures. It was reported that (i) all adopted treatments can remarkably influence the electrochemical properties of the wires, (ii) the acid treatments favor the formation of a surface passivation layer, while the alkaline treatments are effective in producing a rougher surface morphology, and (iii) the main effect of the mechanical surface treatment, consisting in abrading the alloy wires using an emery paper, was to increase the homogeneity of surface roughness. Merlin et al. [159] investigated the effect of different surface treatments on SMA wires embedded in polyester and vinylester polymeric matrices. It was mentioned that (i) considering the best results of the pull-out tests obtained with polyester resin, the debonding induced by strain recovery of 4%, 5%, and 6% prestrained NiTi wires has been evaluated with the wires being subjected to different surface treatment conditions and then being embedded in the polyester matrix and (ii) the debonding induced by strain recovery is strongly related to the propagation towards the radial direction of sharp cracks at the debonding region.

Jonnalagadda et al. [160] studied the effect of sand-blasting on adhesive strength of NiTi surface and conclude that sandblasting significantly increased the bond strength, whereas hand sanding and acid cleaning actually reduced interface strength. Surface coating of hard TiN dendrites provide large increase in effective surface area without process induced surface cracking which is processed by the laser gas nitriding [161, 162]. Man et al. [163] treated NiTi surface by the laser gas nitriding process to obtain NiTi surface consisting of TiN dendrites in NiTi matrix; followed by applying subsequent selective etching process, the matrix material in the alloyed layer can be selectively removed and a three-dimensional network of TiN dendrites is left on the surface protruding from the metal substrate. It was reported that (i) as to the adhesion jointing characteristics, a 150% increase in the lap-joint strength was achieved in the laser gas nitrided and etched specimen as compared with the sandblasted and etched ones. Sadrnezhaad et al. [164] investigated the strength of the NiTi/silicone bonds in a flexible composite piece and found that (i) greater adhesion strengths are obtained due to the presence of thin oxide layer, surface roughness, and frictional forces between the embedded-wires and the contacting phase, and (ii) the curing treatment shifts the transition points of the wires towards higher temperatures at the heating and lower temperatures at the cooling try-outs, respectively; affecting the shape-memory behavior of the NiTi wires embedded within the biocompatible flexible composite segments.

6.6.2 Solid-state joining

In this group, there are friction (stir) welding, resistance upset butt welding, explosion welding, ultrasonic spot welding, and diffusion boding. Friction welding (both conventional and friction stir welding) are solid-state processes, but the heavy deformation of the weld zone may cause degradation in the properties such as SE, shape memory, and phase transformation temperatures. Friction welding is normally an excellent process alternative when joining round bars since this joint fit-up satisfies the symmetry requirements in conventional and inertia friction welding [165]. Shinoda et al. [165] investigated characteristics of a TiNi-based SMA welded using friction welding It was reported that (i) friction welding could be performed in air and that the transformation temperatures, tensile properties, and shape-memory characteristics of the joints could be maintained similar to those of the base metal provided that proper heat treatment is carried out after friction welding, and (ii) an upsetting pressure of at least 127.8 MPa seems to be necessary for obtaining sound welds.

Fukumoto at el. [166] friction welded NiTi alloys to austenitic stainless steel with and without a Ni interlayer. It was found that, (i) without the use of Ni interlayer, a large amount of brittle Fe_2Ti intermetallic compound was formed at the weld interface, (ii) the formation of this brittle compound led to degradation of the joint strength, (iii) use of Ni interlayer changed the microstructures at the weld interface and improved the joint strength, (iv) fracture occurred at the interface between Ni and NiTi, and (v) the interface between Ni and NiTi was free from Fe_2Ti and consisted of mainly Ni_3Ti and NiTi. Barcellona et al. [167] and Mani et al. [168] investigates the feasibility of friction stir welding process to join NiTi SMAs. It was reported that (i) the weld showed significant grain refinement without formation of detrimental phases, (ii) the yield strength of the weld joint increased by 17% as compared to the base metal without substantial change in shape-memory behavior.

Resistance welding is based upon use of electrical current and mechanical pressure to produce a weld between two parts. Welding electrodes lead the current to the two parts that are squeezed together and subsequently welded. Usually, the weld cycle must first generate adequate heat to melt a small volume, with subsequent cooling under the influence of pressure until a weld is formed with sufficient strength to keep the parts together [169]. In conventional resistance spot and seam welding, the current density and the pressure are kept sufficiently high to form a weld, but not too high such that the material is squeezed out from the weld zone. However, in upset welding, there is an upset collar formed which usually needs to be removed by machining after welding [170]. Nishikawa et al. [170] welded thin NiTi shape-memory wires without fusion or case zone by the resistance butt welding. It was reported that (i) the weld current is fed to the butt welded parts with little offset during the time shorter than that in the forging process, and (ii) the tensile strength of the welded part is attained over 80% of that of the base metal. Li et al. [171] studied the effect of the impact speed on resistance butt weldments of NiTi. It was mentioned that (i) decreasing the impact speed can extend the effective heating time and escalate the heating rate, (ii) additional flash is generated near the NiTi/stainless steel interface due to the lengthening of the effective heating time, (iii) with decreasing impact speed, the plastic deformation zone of the SS is enlarged, and the microstructure in the HAZ is increasingly coarsened, but those of NiTi demonstrate the opposite trends, (iv) in all of the NiTi/SS joints, the weld consists of a diffusion layer with a thickness that increases slightly from 1 to 1.7 μm as the impact speed decreases from 40 to 27.5 mm/s, (v) the mechanical properties of the joint deteriorate with decreasing impact speed due to the increased remnant of semi-molten NiTi at the interface, and (vi) the joint welded at an impact speed of 40 mm/s has the highest strength of 522 ± 41 MPa with (7 ± 2)% rupture elongation, and it fractures via microvoid coalescence.

Although the joining techniques such as resistance spot welding, arc welding, and laser welding are some examples which are capable to produce defect-free joints [172], the formation of brittle intermetallic compounds in the weldment such as Ti_2Ni is a problematic issue since it might reduce the mechanical strength of joints [173, 174]. To control the formation of brittle precipitates and adjust the chemical compositions of the weldments, an insertion of Cu as an interlayer has been proposed [175, 176]. It is also mentioned that fusion-welding methods can also contribute to significant changes in the transformation temperatures which can impair the potential applications of the joints [174, 177]. The thermal history experienced by NiTi during welding can significantly affect its shape-memory and superelastic properties [178]. Thus, it is necessary to reduce the heat input of the welding process to restrict the thermophysical deterioration in the weldment [179]. Zhang et al. [172] using Cu interlayer for the ultrasonic jointing NiTi and reported that (i) no intermetallic compounds formed in the joints, (ii) the joining mechanisms involve a combination of shear plastic deformation, mechanical interlocking and formation of microwelds, (iii) a better bonding interface was obtained with higher welding energy levels, which contributed to a higher tensile load, (iii) an interfacial fracture mode occurred and the fracture surfaces exhibited both brittle and ductile-like characteristics with the existence of tear ridges and dimples, and (iv) the fracture initiated at the weak region of the joint border and then propagated through it, leading to tearing of Cu foil at the fracture interface.

Explosion welding is formed by the high velocity impact of the work pieces as a result of controlled detonation and the method consists of three main materials: (i) the base metal, (ii) the flyer, and (iii) the explosive. The main parameters in welding are the collision speed, collision angle, and the flyer speed. The interface or weld line between the base metal and the flyer is usually wavy. Zimmerly et al. [180] explosively welded thin layer (0.4–1.0 mm) of a near-equiatomic NiTi alloy to low-carbon steel substrates to fabricate high-strength, bimetallic tandems in which the NiTi provided resistance to cavitation damage and the low-carbon steel provided structural strength. It was reported that (i) tensile lap-shear tests on the welded material revealed bond strength of up to 387 MPa, and (ii) as-welded NiTi/steel tandems were less resistant to cavitation erosion than annealed, unwelded samples; however, a postweld heat treatment at 500 °C recovered most of the lost resistance. Belyaev et al. [181] also mentioned that the annealing at 500 °C recuperated partially kinetics of phase transitions. Yan et al. [182] prepared bulk NiTi/NiTi SMA laminates with a macroscopic heterogeneous composition by explosive welding and investigated their microstructures and martensitic transformation behaviors. It was mentioned that (i) an achievement of a fine granular structure and the maximum value of microhardness near the welding interface because of the excessive cold plastic deformation and the high impact velocity during the explosive welding was noticed, and (ii) the transformation temperatures of NiTi/NiTi alloys increase with the rise of the aging temperature, indicating that the shape-memory properties of NiTi/NiTi SMA fabricated by explosive welding can be improved by optimizing the aging technology. Sun et al. [183] fabricated NiTi SMA cladded on a copper foil by underwater explosive welding. It was reported that (i) the bond interface of the cladded metals shows a wavy morphology, (ii) the fracture mechanism of the composite showed predominantly quasicleavage and cleavage modes, and (iii) underwater explosive welding can solve the problem that materials with thin thickness and low formability difficult to bond by traditional explosive welding.

6.6.3 Diffusion bonding

For dissimilar metals joining, particularly in solid-state bonding, the thermal mismatching should be taken into considerations, which can lead to a harmful development of the residual stress at the boded interface layer, and it should also include metallurgical incompatibilities which could be prone to form brittle intermetallic compounds (such as Ni_3Ti, $NiTi$, and $NiTi_2$) at the interface [184–186]. There are mainly two distinct ways for diffusion bonding: with and without inserting interface layer(s). Velmurugan et al. [187] and others [188–191] conducted diffusion bonding without applying interlayers for bonding Ti-based alloy to stainless steel. Based on their findings along with discussions by Mo et al. [192], it can be mentioned that the diffusion bonding without interlayer can be determined by bonding temperature (in the range of 800–950 °C) and bonding time (in the range for a period of 60–120 min), although brittle intermetallic compound formation cannot be avoided.

For minimizing and controlling the content of brittle phases at the bonding interface, the interlayer is often employed. The interlayers can include Cu [172], Ni alloy [193], Ni/Al nanolayer [194], or Zr or Ni foil [186]. Appropriate selection for interlayers should be conducted based on how to control brittle intermetallic compound formation and to minimize the adverse residual interfacial stress.

The out-diffusion of toxic Ni ions from NiTi (nitinol), often used for biomedical applications, can be strongly reduced using oxygen plasma immersion ion implantation (PIII). Nonradioactive isotope markers, ^{60}Ni and ^{46}Ti, were implanted at 180 keV into NiTi prior to the oxygen implantation with fluences between 0.4×10^{16} and 4×10^{16} at/cm^2. Implanting oxygen ions by PIII in the temperature range of 400–550 °C leads to a surface oxide layer consisting of pure TiO_2. The results prove that Ni cations are the mobile species, while Ti is immobile during the oxygen insertion [195].

6.6.4 Soldering and brazing

Soldering and brazing are the only metal joining methods that can produce smooth and rounded fillets at the periphery of the joints. Both the operations involve heating the filler metal and joint surfaces above ambient temperature. Both are essentially the same metal joining technique, the difference being the temperature at which each method is performed. Solder filler metals melt below 450 °C. The soldering process may or may not require a fluxing agent. The filler metals melt at low temperatures so there is minimum part distortion and heat damage to sensitive parts. The filler metal is called solder which when solidifies, is then bonded to the metal parts to join them. The most commonly used solder is tin alloy and lead. Soldering is extensively used in the electronics industry for joining wires, capacitor, resistor, and so on with the joining plate. On the other hand, brazing filler metals melt above 450 °C and can be applied to a wide variety of materials, including metals, ceramics, glasses, plastic, and composite materials. Although it is not as strong as fusion welding, it is the strongest form of metal bonding without melting the parent metal of the components being joined together. Hence, this process requires more heat input than other soldering operations such as mechanical fastening, adhesive bonding, solid-state joining, welding and so on. This process can also be used to join dissimilar metals such as silver, gold, copper, aluminum, and so on. Joining NiTi with these solders has historically been difficult because fluxing agents which would enable good wetting of the solder to NiTi were not readily available. Although, in recently years, good flux has become available and soldering of NiTi is now easily achieved if proper cleaning and soldering techniques were used. These joints are still only acceptable for certain application, though since the solder metal is not very strong. Due to tits chemistry it cannot be used for extended periods in the body in medical applications [196].

In brazing processes, the base metals do not melt, and certain high temperature metallurgical phenomena (such as high temperature oxidation, elemental segregation, and grain growth) should be avoided; however there might be unwanted chemical reaction between the base metals and the filler alloy, so that vacuum atmosphere is recommended, and the use of filler alloys containing active elements may be an advantage. Shiue et al. [197] employed pure Au-based alloy as braze materials and Nb was also utilized [198] for brazing NiTi alloy. It is well known that brazing is more appropriate than welding for bonding dissimilar alloys. It is important to use an appropriate filler foil for brazing $Ti_{50}Ni_{50}$ alloy, and the wettability of braze melt on the base metal and reaction(s) at the interface between the braze melt and base metal must both be considered in filler metal selection. Accordingly, Shiue et al. [199, 200] selected Ag-based braze alloys. Lin et al. [205] employed Ag-based alloy (72Ag-28Cu, wt%) and Ag-based alloy foils (68Ag-26.7Cu-4.5Ti, wt%) braze alloy as the brazing fillers. The eutectic of the former fillers is 780 °C; while the latter has the liquidus of 850 °C, and the wetting and infrared brazing of $Ti_{50}Ni_{50}$ and $Ti-V_{15}Cr_3$ (Ti-15-3) alloys were investigated in terms of microstructures, dynamic wetting behaviors of the two fillers, as well as bonding strengths of joints [201]. It was reported that (i) a Ti-15-3/ Ag-based alloy/$Ti_{50}Ni_{50}$ joint brazed at 800 °C for 300 s consists of Cu_2Ti intermetallic compound in the Ag-rich matrix, and only the Ag-rich matrix remains in a joint infrared brazed at 850 °C for 300 s, (ii) a Ti-15-3/Ag-based foil/$Ti_{50}Ni_{50}$ joint brazed at 900 °C for 300 s consists of CuNiTi and $CuTi_2$, and (iii) the dissimilar infrared brazed joints show moderate average shear strengths of between 172 and 230 MPa.

6.6.5 Fusion welding

In general, laser welding, plasma welding, resistance welding, arc welding, and electron-beam welding are all proven ways of joining NiTi to itself. However, welding is generally not an acceptable method for joining NiTi to dissimilar materials such as stainless steel, because brittle intermetallic compounds will be formed at the weldment; hence to avoid such welding precipitates as well as oxides or nitrides, the welding should be done in a clean, inert atmosphere or in vacuum due to the reactivity of the titanium. In some cases, postwelding heat treatment is needed. It is normally believed that the HAZ will not exhibit superelastic properties in most cases, due to mismatching of chemical composition and microstructure as well as transformation temperature. Welding has been performed by gas tungsten or plasma arc welding without welding wire. During arc welding of NiTi SMA, embrittlement may occur due to the reactions with oxygen, nitrogen and hydrogen at high temperatures [202]. In addition, precipitation of brittle intermetallic compounds such as $NiTi_2$ and Ni_3Ti during solidification of NiTi SMA can have adverse effects on both strength and shape-memory characteristics of the material [203]. To recover the SME, post-welding heat treatment was recommended [204].

Table 6.1: Effects of fusion welding on four characteristic transformation temperatures [202].

		°C			
Welding method	Alloy	A_S	A_F	M_S	M_F
Tungsten inert gas welding	Base metal	63	98	53	35
	Fused metal	59	114	65	5
Laser beam welding	Base metal	67	93	44	10
	Fused metal	81	115	70	35

The phase transformation temperatures may change in welding, as given in Table 6.1, which may affect the applicability and functionality of NiTi SHMs. Among four temperatures related to austenite and martensite phases, M_S and A_F are more frequently used to characterize the SME and SE phenomena. It can be pointed out that (i) both tungsten inert gas (TIG) welding and laser beam welding affect raises in A_F and M_S temperatures, (ii) the laser beam welding influences more aggressive than TIG in these increments of transformation temperatures, which may be attributed to the fact that laser beam welding results in narrower weld zones [205]. These results indicate that certain post-welding heat treatment may be required to recover the initial transformation behavior. Oliveira et al. [206] investigated functional behavior of the TIG welding of 1.5 mm thick plates of Ni-rich NiTi. It was reported that (i) the SE was analyzed by cycling tests at maximum imposed strains of 4%, 8%, and 12% and for a total of 600 cycles, without rupture and the superelastic plateau was observed, in the stress–strain curves, 30 MPa below that of the base material, (ii) SME was evidenced by bending tests with full recovery of the initial shape of the welded joints, and (iii) uniaxial tensile tests of the joints showed a tensile strength of 700 MPa and an elongation to rupture of 20%, and the elongation is the highest reported for fusion welding of NiTi, including laser welding; indicating that these results can be of great interest for the wide-spread inclusion of NiTi in complex shaped components requiring welding, since TIG is not an expensive process and is simple to operate and implement in industrial environments.

As to the laser beam welding, both CO_2 gas laser [207] and more extensively Nd:YAG solid-state laser [208, 209] have been used. The Nd:YAG laser may give smaller welding spots and higher penetration depth than CO_2 laser since the plasma shielding effect is reduced due to the shorter wavelength [210–212]. Falvo et al. [213] studied the effect of Nd:YAG laser welding on the mechanical and functional properties of $NiTi_{51}$ (at%) alloy, which was prepared from thin sheets. Zhao et al. [214] investigated microstructure, phase transformation and stress–strain behaviors of a Ce-added laser welded TiNi alloy. The Nd:YAG laser was used for butt fusion of two $TiNi_{50.9}$ (at%) alloy plates with a TiNi–5Ce stripe preassembled between their joint. It was reported that unlike the typical monotonous SE of base metal and Ce-free welded TiNi, the Ce-added welded TiNi specimen shows a peculiar two-stage SE. Oliveira et al. [215–217] extensively studied laser weldments of NiTi SHMs to characterize various properties. The dissimilar laser welding of NiTi and CuAlMn SMAs was investigated [216] and reported that cycling tensile testing revealed that the joints preserved the superelastic behavior despite the unfavorable microstructure of the fusion zone which translates into an irrecoverable strain of 2% when cycled at 5% strain. With aged Ni-rich NiTiHf (at 500 °C for 3 h and air cooled), Oliveira et al. [217] fabricated defect-free joints with a conduction weld mode and transformation behavior was studied. It was mentioned that were obtained isothermal loading in both the martensite (at 30 °C) and austenite (at 200 °C) phases revealed equivalent strength and near-perfect SE in the welded and unwelded reference material.

Zeng et al. [218] investigated the joining NiTi and copper with Nd:YAG laser welding with an optimum peak power of 2.2 kW accommodated superelastic deformation of NiTi, proving their use with high strength actuators. It was mentioned that (i) fracture occurred through the cross section of these defect-free joints, (ii) a lower peak power of 1.8 kW created weak joints with limited weld penetration of the copper sheet, (iii) joints made with a higher peak power of 2.6 kW had significant cracking in the fusion zone, (iv) two regions of distinct Cu composition were found in the fusion zone, and cracking occurred at the interface between these regions because of their different physical properties, and (v) failure initiated at this cracking and propagated through the fusion zone that had been embrittled by mixing with over 20 at% Cu. Mehrpouya et al. [219] investigated the effect of welding parameters on functionality of dissimilar laser-welded NiTi SE to SME wires. It was reported that (i) a uniform and homogeneous welded joint without any crack or other significant defects was obtained, (ii) intermetallic compounds such as $NiTi_2$ and Ni_4Ti_3 were identified, compositional changes were noticed to alter the Ni/Ti balance in the weldment and significantly effects on the transformation temperatures, and (iii) welded joints showed an ultimate strength 20% less than that of base metal.

6.7 Controlled porosity

The specific aim for developing porous structured NiTi are three-fold; reflecting to several important demands including (1) increasing effective surface area, (2) interconnected pore structure, and (3) biomechanical compatibility. In particular, the requirement for the biomechanical compatibility exhibits is crucial to control the stress shielding; resulting in the success/failure of orthopedic and dental implants. Porous NiTi SMAs (TiNi SMA) can be fabricated using various engineering technique. It can be characterized by (i) the structure with interconnected pores, (ii) retaining the unique properties of solid NiTi alloys, and (iii) its application to orthopedic field such as the bone substitute and bone implant [220].

Zhu et al. [221] prepared an equiatomic porous TiNi alloy from elemental powders by Ar-sintering at 980 °C in argon for 8 hrs. It was reported that (i) the porous product showed the ultimate compressive and flexural strength was 677 and 246 MPa, respectively, (ii) the porosity ratio of porous TiNi alloy was up to 45%, and (iii) the differential scanning calorimetry measurements revealed that the A_S and A_F temperatures of the alloy was 32.5 and 45.6 °C, respectively. Tang et al. [36] fabricated porous NiTi SMA by microwave sintering method without using any pore-forming agents at 850 °C for short sintering time of 15 min. It was reported that (i) the porous NiTi SMA exhibited porosity ratios from 27% to 48% and pore sizes range from 50 to 200 μm at various combination of sintering temperatures and holding times, (ii) the predominant B2 (NiTi) and B19′ (NiTi) phases were identified in the porous NiTi SMA, (iii) a multistep phase transformation took place on heating and a two-step phase transformation took place on cooling of the porous NiTi SMA, and (iv) the irrecoverable strains decreased with increasing sintering temperature, but the holding time had little effect on the stress–strain behavior at 60 °C.

Zhang et al. [222] utilized the CS method to fabricate porous Ti-TiBx and NiTi composite materials. Aihara et al. [223] fabricated porous NiTi by the CS of elemental Ni and Ti powders. It was reported that (i) the average elastic modulus was approximately 1 GPa for a porosity of 60 vol% and the average pore size of 100–500 μm, (ii) the porous NiTi was composed of cubic (austenitic) and monoclinic (martensitic) NiTi compounds without the presence of Ni metal or Ni-rich phases, (iii) the product exhibits excellent corrosion resistance with passivation breakdown potentials above 750 mV, and (iv) an ovine study in cortical sites of the tibia demonstrated rapid osseointegration into the porous structure as early as two weeks and complete bone growth across the implant at six weeks; concluding that porous NiTi is a promising biomaterial with proven biocompatibility and exceptional osseointegration performance which may enhance the healing process and promote long-term fixation, making it a strong candidate for a wide range of orthopedic implant applications.

SHS has been extensively employed to fabricate porous NiTi alloy. Li et al. [224, 225] prepared the equiatomic porous NiTi SMAs, especially those with an unusual kind of linear-aligned elongated pore structure, by SHS using elemental nickel and titanium powders. It was mentioned that (i) the porous NiTi SMAs thus obtained have an open porous structure with about 60 vol% porosity, and the channel size is about 400 μm, and (ii) the preheating temperature has been shown to have a significant effect on the microstructure of the SHSed porous NiTi SMAs. Chung et al. [226] fabricated porous TiNi$_{51}$ (at%) SMA with high strength by the SHS method. It was found that (i) the compressive strength of the porous TiNi SMA with the general porosity of 52.8 vol% obtained under the preheating temperature of 400 °C could reach up to 500 MPa, and (ii) the transformation temperatures of the porous TiNi SMA with B2(TiNi) and B19′(TiNi) as the predominant phases could be easily adjusted to the suitable values for medical applications by heat treatment. Biswas [281] prepared porous yet single phase NiTi using the TE mode of SHS in combination with a controlled postreaction heat treatment. It was reported that (i) the postreaction heat treatment needed for the conversion to single phase porous NiTi also showed strong dependence on the heating rate, (ii) a faster rate resulted in loss of both the porosity and the shape as the strong interaction with an intermediate liquid phase collapsed the porous structure, and (iii) however, this interaction could be controlled and heating up to 1,150 °C at a very low rate (1 °C/min) could produce single phase NiTi without sacrificing the porosity. In order to decrease the A_F temperature to the human body temperature, 310 K (36 °C), Jiang et al. [227] investigated the effects of preheating temperature, thermal cycling and the third element Mo on the transformation temperatures of SHSed NiTi alloy. It was reported that (i) the preheating temperature has no obvious effect on the transformation temperatures; the transformation temperatures slightly decreased with increasing number of thermal cycles, but the A_F temperature was still above 310 K within 10 times, (ii) the addition of Mo decreased transformation temperatures considerably, (iii) the R-phase transformation was induced during cooling, and (iv) transformation temperatures of porous TiNi(Mo) SMAs with appropriate Mo contents fabricated by SHS meet the demands for human-body implants.

Tay et al. [93] produced porous NiTi SMA SHS process of compacts prepared from elemental nickel and titanium powders to examine the effects of the green porosity and preheating temperature. It was found that (i) the porosity in the SHSed products was in the range of 45–59 vol%, with the green porosity being the primary source of final porosity, (ii) NiTi was present as the dominant phase in the porous product with other secondary intermetallic compounds and element powders, and (iii) the effect of green porosity and preheating temperature was found to have an effect on the pore morphology and microstructure. Tosun et al. [89, 228] investigated the effects of varying cold compaction pressure, preheating temperature and heating rate for fabricating porous NiTi alloy and reported that (i) when the amount of liquid phase excessively increased, porosity decreased and the specimen deformed. This situation occurred with increase in heating rates ($\geq$60 °C/min), preheating temperatures (>250 °C) and cold compaction pressures (>7 MPa). Bassani et al. [99] produced an open-cell porous NiTi SHS, starting from Ni and Ti mixed powders. It was mentioned that (i) apart the well-known high toxicity of Ni, also toxicity of TiNi$_3$, while phases with higher Ti content showed high biocompatibility, (ii) a slightly reduced biocompatibility of porous NiTi was ascribed to combined effect of TiNi$_3$ presence and topography that requires higher effort for the cells to adapt to the surface. Saadati et al. [229] fabricated porous NiTi SMA by SMA process with 30 and 40 vol% green porosity from elemental Ni and Ti powders. It was found that (i) after synthesizing, the average porosity of specimens reached to 36.8% and 49.8% for green compacts with 30 and 40 vol% of green porosity, respectively, (ii) although desired B2 (NiTi) phase was the dominant phase, other phases like Ti$_2$Ni, Ni$_3$Ti, and Ni$_4$Ti$_3$ are found, and (iii) electrochemical polarization analysis in simulated body fluids shows that SHSed porous NiTi has better corrosion resistance than solid one and hydroxyapatite coating on porous NiTi worsen electrochemical corrosion resistance which is because of bioactive behavior of hydroxyapatite.

Laser technology is also applicable to fabricate porous NiTi alloy. Liang et al. [230] prepared porous microstructures on NiTi alloy surfaces by linearly polarized femtosecond lasers with moving focal point at a certain speed. It was found that (i) various novel microstructures from feather-like ripples to cluster-like porous textures could be formed with increasing laser energy, (ii) when the laser energy was 400 μJ, a periodic porous metal surface was generated, and (iii) analysis by x-ray photoelectron spectroscopy revealed that Ni/Ti on the sample surface was changed with an evident oxidization of titanium element under different laser energies. Krishna et al. [231] fabricated porous NiTi alloy 12–36% porosity from equiatomic NiTi alloy powder using laser-engineered net shaping and investigated the effects of processing parameters on density and properties of laser-processed NiTi alloy samples. It was reported that (i) the density increased rapidly with increasing the specific energy input up to 50 J/mm^3, (ii) further increase in the energy input had small effect on density, (iii) high cooling rates resulted in higher amount of cubic B2 phase, and increased the reverse transformation temperatures of porous NiTi samples due to thermally induced stresses and defects, (iv) transformation temperatures were found to be independent of pore volume, though higher pore volume in the samples decreased the maximum recoverable strain from 6% to 4%, (v) porous NiTi alloy samples with 12–36% porosity exhibited low value of modulus of elasticity between 2 and 18 GPa as well as high compressive strength and recoverable strain, and (vi) due to high open pore volume between 36% and 62% of total volume fraction porosity, these porous NiTi alloy samples can potentially accelerate the healing process and improve biological fixation when implanted in vivo.

Wen et al. [232] fabricated porous titanium with porosity gradients, that is, solid core with highly porous outer shell was successfully fabricated using a PM approach. It was reported that (i) satisfactory mechanical properties derived from the solid core and osseointegration capacity derived from the outer shell can be achieved simultaneously through the design of the porosity gradients of the porous titanium, (ii) the outer shell of porous titanium exhibited a porous architecture very close to that of natural bone, that is, a porosity of 70% and pore size distribution in the range of 200–500 μm, and (iii) the peak stress and the elastic modulus of the porous titanium with a porosity gradient (an overall porosity 63%) under compression were approximately 152 MPa and 4 GPa, respectively; being very close to those of natural bone. Zhang et al. [233] fabricated porous NiTi alloys with gradient porosity and large pore size by using a temporary space holder (NH_4HCO_3) and a conventional sintering method. It was found that (i) the pore characteristics of the porous alloys can be tailored effectively by changing the amount of NH_4HCO_3 added, (ii) the use of NH_4HCO_3 has no influence on the microstructure and phase transformation behavior of the porous alloys, and (iii) the fabricated samples with radial gradient porosity exhibit excellent SE (higher than 4%).

6.8 Amorphization and crystallization

During severe plastic deformation such as MA process, ball milling, or cold rolling, amorphization of NiTi can be achieved. Using MA procedure [234–236], an amorphous NiTi can be fabricated. Shen et al. [244] investigated the growth of amorphous phase and the formation of competing intermetallic compounds in mechanically deformed Ni/Ti multilayered composites. It was reported that the amorphous layers in the Ni/Ti multilayered composites continue to grow until they have attained a temperature-dependent critical thickness or a temperature-dependent critical annealing time. Kanchibhotla et al. [236] investigated the amorphous phase formation in the $(Ni_{51}Ti_{49})_{1-x}Ta_x$ system in the range of 0–20% Ta through MA. It was found that the crystallite sizes of both Ni and Ti were almost similar at any given percentage of Ta. Alijani et al. [237] evaluated the effect of milling process on the chemical composition, structure, microhardness, and thermal behavior of $TiNi_{41}Cu_9$ compounds developed by MA. It was reported that (i) at the initial stages of milling (typically up to 12 h), the structure consisted of a Ni solid solution and amorphous phase, and by the milling evolution, nanocrystalline martensite (B19′) and austenite (B2) phases were initially formed from the initial materials and then from the amorphous phase, (ii) by the milling development, the composition uniformity is increased, the interlayer thickness is reduced, and the powders microhardness is initially increased, then reduced, and afterward reincreased, and (iii) the thermal behavior of the alloyed powders and the structure of heat treated samples is considerably affected by the milling time.

Shen et al. [238] investigated the solid-state amorphizations in mechanically deformed NiTi multilayer composites. It was reported that the solid-state amorphization occurs first during heating, followed by the formation of intermetallic compounds through direct solid-state reaction of elemental nickel and titanium and by the crystallization of the amorphous alloy already formed by the solid-state amorphization reaction. Using the differential scanning calorimetry, Li et al. [239] determined the degree of crystallization and rate in TiNi alloys by severe plastic deformation. It was found that (i) the reverse martensitic transformation peak was not observed during the first heating at the rate of 40 K/min in the as-rolled samples, but one exothermic peak was observed at 620 K, which was associated with the amorphous crystallization process, (ii) during the second heating, reverse martensitic transformation was recovered, (iii) the onset crystallization temperature was low in the initial stage of crystallization with lower heating rates, but the crystallization fraction was found to increase with increasing temperature, (iv) however, the crystallization fraction was almost constant in the initial stage of crystallization with a relatively high heating rate, and (v) in all heating rates, the amorphous crystallization rates almost always reached maximum as the volumetric fraction of amorphous crystallization rose to 50%. Huang et al. [240] the initial stages of high-pressure-torsion-induced transformation from crystalline to amorphous TiNi. It is reported that (i) the deformation-induced amorphization initiated from dislocation core regions in the interior of grains and from grain boundaries, and (ii) both the energy stored in the dislocations and the energy stored in the grain boundaries contribute significantly to driving the crystalline-to-amorphous transformation. Resnina et al. [241] also utilized the high-pressure torsion and investigated the structure and martensitic transformations in $TiNi_{50.2}$ (at%) alloy. It was reported that (i) the high-pressure torsion under 8 GPa for 3.5 turns resulted in an amorphization of the alloy structure but the crystalline debris remained, (ii) further heating of the sample led to alloy crystallization and the bimodal grain size structure formed, and (iii) there found two distinct grain groups behaving differently due to the grain appeared during crystallization in two ways: from the amorphous phase and from the debris remained in the amorphous phase after the high-pressure torsion. Chellali et al. [242] prepared $Ni_{50}Ti_{50}$ amorphous particles using inert gas condensation followed by in situ compaction. It was reported that (i) the Ni-rich amorphous phase in the consolidated nanostructured material is responsible for the ferromagnetic behavior of the sample whereas the rapidly quenched amorphous and crystalline samples with the same chemical composition ($Ni_{50}Ti_{50}$) were found to be paramagnetic, and (ii) due to the high cooling rate obtained using the inert gas condensation technique, an exceptional control over the crystallization processes is possible, promoting the formation of various amorphous phases, which are not obtained by standard rapid quenching techniques; demonstrating that the potential of amorphous metallic nanostructures as advanced technological materials and useful magnetic compounds.

Nanocrystalline materials, because they have an advantage over coarse-grained materials in some physical properties, have aroused considerable interest in potential applications [243–246]. Ju et al. [247] investigated a series of NiTi nanocrystals with different annealing temperatures which were prepared by sputtering method. It was found that (i) the structure of nanophase powder is different from bulk NiTi alloy with BCC structure as target materials, (ii) when increasing the annealing temperature, a small fraction of the (Ni,Ti)-type nanocrystal with the hexagonal structure was presented except target materials and Ni, and it is atomic occupation in random, and (iii) there were four Ti and two Ni atoms around central Ni atoms, and the bond length of Ni–Ti and Ni–Ni were 0.2462 and 0.2585 nm at 800 °C annealed. Sadrnezhaad et al. [72] fabricated NiTi alloy from elemental powders by MA, followed sintering of the raw materials. It was reported that (i) the crystallite size of the MA $Ni_{50}Ti_{50}$ samples decreased with MA duration and with the milling speed, (ii) depending on the crystal structure of the raw materials, the lattice strain increases with the milling duration, and (iii) metallographic studies proved the existence of martensitic B19′ after sintering of both the as-mixed and the mechanically alloyed samples. Using equal-channel angular extrusion technique operated between 400 and 450 °C, Kockar et al. [248] fabricated $Ni_{49.7}Ti_{50.3}$ (at%) SMA and studied the effects of ultrafine grains on the thermomechanical cyclic stability of martensitic phase transformation. It was found that (i) ultrafine-grained NiTi alloy has average grain sizes of 100–300 nm, (ii) tensile failure experiments demonstrated that the strength differential between the onset of transformation and the macroscopic plastic yielding increases after the extrusion, (iii) such increase led to a notable improvement in the thermal cyclic stability under relatively high stresses, and (iv) the experimental observations are attributed to the increase in critical stress level for dislocation slip due to grain refinement, change in transformation twinning mode in submicron grains, the presence of R-phase, and multimartensite variants or a small fraction of untransforming grains due to grain boundary constraints. Singh et al. [249] fabricated Ni-rich nano-NiTi alloy, deformed by high-pressure torsion, which creates almost complete amorphization of initial B2 NiTi. It was reported that (i) the size of nanocrystals increases dramatically after annealing for 5 h, and (ii) the effective activation enthalpies for stress relaxation (along with crystallization) and grain growth were estimated at 115 and 289 kJ/mol, respectively.

Jiang et al. [250] fabricated amorphous and nanocrystalline NiTi SMAs by severe plastic deformation and local canning compression. It was reported that (i) the (001) martensite compound twins, the stacking faults, primary deformation bands, secondary deformation bands, and amorphous bands are observed, (ii) the (001) martensite compound twins belong to the deformation twins and thus lead to mechanical stabilization of stress-induced martensite, and (iii) deformation twinning and dislocation slip are of great importance in amorphization and nanocrystallization of NiTi SMA subjected to severe plastic deformation based on local canning compression. Park et al. [251] fabricated bulk nanocrystalline NiTi SMA by the cold marforming, followed by optimum postheat treatment. It was reported that (i) the total accumulative strain ($\varepsilon \sim 0.7$) during the process was much lower than that typically imposed during severe plastic deformation operations (i.e., $5 \le \varepsilon \le 12$), (ii) heating transmission electron microscopy revealed that the present process was effective in promoting static recrystallization and limiting grain growth, and (iii) the nanocrystalline alloy exhibited enhanced shape-memory behaviors.

6.9 Product forms

There are a variety of product forms in 1D (like a thin wire, fiber, and cable), 2D (like a sheet, film, or mesh), and 3D (like a tube, fiber, ribbon, stent, foam, scaffold, and truss).

6.9.1 One-dimensional product forms

6.9.1.1 Wire

Normally, there are two types of wires; round wire (0.02 mm to 2.5 mm or 0.0007 in to 0.1 in) and flat wire with minimum thickness of 0.02 mm (or 0.0007 in) and maximum thickness of 11.4 mm (or 0.5 in) [252–254]. Gall et al. [255] examined mechanical properties of cold drawn $TiNi_{50.1}$ and $TiNi_{50.9}$ (at%) SMA wires. As discussed in the previous chapters, it is confirmed that the more Ni rich wires contain fine second phase precipitates, while the wires with lower Ni content are relatively free of precipitates. It was found that the wire stress–strain response depends strongly on composition through operant deformation mechanisms. Delville et al. [256] also mentioned that mechanical properties depend on microstructural changes during the heat treatment. NiTi-based transcatheter endovascular devices, prior to medical usage, are normally subject to a complex thermomechanical prestrain associated with constraint onto a delivery catheter, device sterilization, and final deployment. As a result of such larger thermomechanical treatment should exhibit remarkable influences on the microstructural and mechanical properties [257]. Gupta et al. [257] investigated the effects of large thermomechanical prestrains on the fatigue of electropolished SE NiTi wire (which were bended to 0%, 8%, and 10%) using fully reversed rotary bend fatigue tests at 0.3–1.5% strain amplitudes for up to 10^8 cycles. It was reported that (i) 0% and 8% prestrain wires exhibited distinct low-cycle and high-cycle fatigue regions, reaching run out at 10^8 cycles at 0.6% and 0.4% strain amplitude, respectively, and (ii) over 70% fatigue cracks were found to initiate on the compressive prestrain surface in prestrained wires.

As to medical applications of NiTi-based alloy, there are two major areas; endodontic (rotary) file and reamers and orthodontic (arch)wires. Figures 6.5 illustrates typical straight endodontic files (a) and prebent endodontic files (b) [258, 259]. The same as for the ordinal wires, the rotary bending fatigue properties of medical-grade NiTi endodontic instruments are important and Pelton et al. [260] investigated under conditions of 0.5–10% strain amplitudes to a maximum of 10^7 cycles. It was reported that (i) for superelastic conditions there are four distinct regions of the strain cycle curves that are related to phases (austenite, stress-induced martensite, and R-Phase) and their respective strain accommodation mechanisms; while there are only two regions for the strain-cycle curves for thermal martensite, and (ii) the strain amplitude to achieve 107 cycles increases with both decreasing test temperature and increasing transformation temperature, (iii) fatigue behavior was not, however, strongly influenced by wire surface condition, and (iv) fractographically, fracture surfaces showed that the fatigue fracture area increased with decreasing strain amplitude. The interesting recommendation for the hybrid concept was introduced by Walsh [261], suggesting the idea of the hybrid concept to combine instruments of different file systems and use different instrumentation techniques to manage individual clinical situations to achieve the best biomechanical cleaning and shaping results and the least procedural

(a)   (b)

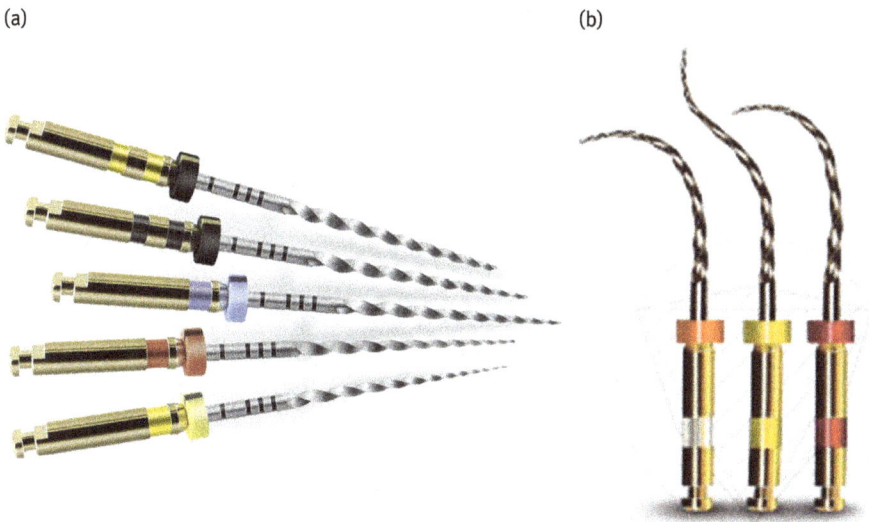

Figure 6.5: NiTi endodontic files [258, 259].

errors; hence, the hybrid concept combines the best features of different systems for safe, quick, and predictable results.

Figure 6.6 shows (a) colored (anodized) NiTi orthodontic archwires and (b) typical three cross sectional views of round, square and rectangular NiTi archwires (b) [262, 263]. Due to the nature of the usage of NiTi orthodontic archwires, there are several important properties that archwires should possess to accomplish satisfactory orthodontic mechanotherapy. They should include (1) excellent corrosion resistance in intraoral environment as well as daily foods and drinks in addition to tooth and oral care agents (particularly fluoride prophylactic agent) [264, 265], (2) good mechanical properties [266–271], (3) satisfactory weldability and strength of weldment [265, 272] and (4) excellent aesthetics and/or acceptable fashion [273, 274].

(a)   (b)

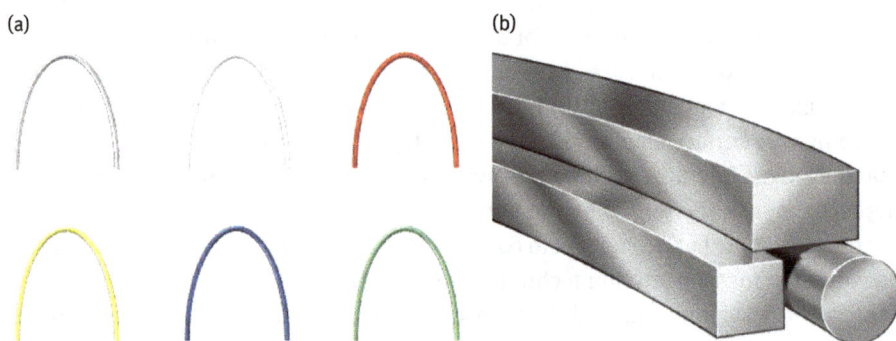

Figure 6.6: NiTi orthodontic archwires (a) colored wires and (b) three cross sections [262, 263].

Tian et al. [275] developed 55 μm diameter cold-worked NiTi microwire and studied the microstructure. It was found that (i) the surface consists of a few hundred nanometer thick oxide layer composed of TiO and TiO_2 with a small fraction of inhomogeneously distributed Ni, (ii) the interior of the wire has a core-shell structure with primarily B2 grains in the 1 μm thick shell, and heavily twinned B19' martensite in the core, and (iii) in between the B2 part of the metallic core-shell and the oxide layer, a Ni_3Ti interfacial layer is detected. Wu et al. [276] studied the deformation and shape recovery of NiTi nanowires. It was reported that (i) for a high loading rate, the occurrence of torsional buckling of a nanowire is faster and the buckling gradually develops near the location of the applied external loading, (ii) the critical torsional angle and critical buckling angle increase with aspect ratio of the nanowires, (iii) square nanowires have better mechanical strength than that of circular nanowires due to the effect of shape, and (iv) shape recovery naturally occurs before buckling.

6.9.1.2 Fiber

In order to alter the stiffness of a composite beam, Bidauex et al. [277] utilized NiTi fibers to actively change the stiffness on a model system composed of an epoxy matrix with a series of embedded pre-strained NiTi fibers. It was reported that (i) when electrically heated, the prestrained NiTi fibers undergo a phase transformation, (ii) it caused the shape recovery associated with the transformation restrained by the constraints of both the matrix and the clamping device, a force is generated, (iii) then, this force leads to an increase in the natural vibration frequency of the composite beam, and (iv) the R-phase or a mixture of the two can be stress-induced and the R-phase gives rise to the largest change in vibration frequency for a given temperature increase and the most reversible behavior. Budziak et al. [278–280] employed NiTi fiber coated with ZrO_2 as a new fiber to support the solid-phase microextraction. It was reported that (i) extraction efficiency of the NiTi-ZrO_2 fiber was similar to a commercial fiber, even though it had a lower coating thickness of 1.35 μm, and (iii) considering the amount extracted per unit volume, the NiTi-ZrO_2 fiber had a better extraction profile when compared to commercial fibers. For a similar solid-phase microextraction system, Wang et al. [281, 282] used uniform and dense ZnO nanosheets to support the surface of NiTi alloy wires.

Furthermore, NiTi fiber has been utilized as a reinforcing element in various types of composite; NiTi fiber for Al matric composite [283], NiTi fibers along with SiC particles for Al-based composites [284], and NiTi fiber coupled with Al_2O_3 nanoparticles for the basalt fiber reinforced polymer composite [285].

6.9.1.3 Strands

There could be double-strand or multistrand made of NiTi wire alone [286] or combination of other material [287]. Most of multistrand orthodontic wires are solely made of NiTi alloy wire [286]. Spriano et al. [287] developed a dual-material strand of NiTi wire and NiTiCr wire for mechanical applications with final diameter in the 1 to 2 mm range. Figure 6.7 depicts various types of multistrand NiTi wires for orthodontic applications [286].

6.9.1.4 Suture

Due to its easiness to handling, strength, and biocompatibility, Karjalainen et al. [288] evaluated the biomechanical behavior of 150 and 200 μm NiTi sutures in flexor tendon repair. Braided polyester (4-0 Ethibond) was used as control. It was reported that (i) NiTi wires were stiffer and reached higher tensile strength compared to braided polyester suture, (ii) repairs with 200 μm NiTi wire had a higher yield force, ultimate force and better resistance to gapping than 4-0 braided polyester repairs, and (iii) repairs made with 200 μm NiTi wire achieved higher stiffness and ultimate force than repairs made with 150 μm NiTi wire. Liu et al. [289], using NiTi open coil springs (50 gr)

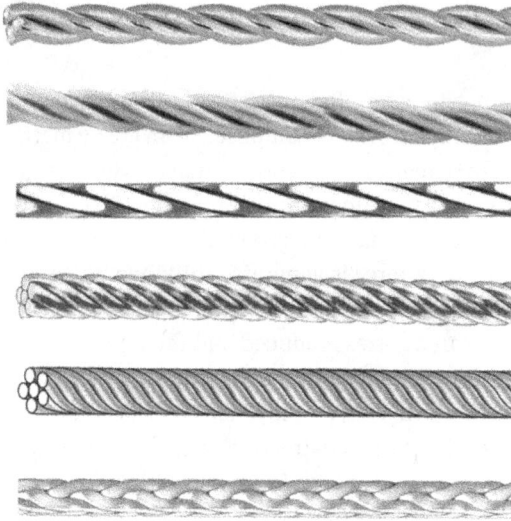

Figure 6.7: Various types of multistrand NiTi wire for orthodontic applications [286].

and 3 mm long miniscrew implants for skeletal anchorage, evaluated intermittent and continuous forces to expand the midsagittal sutures in 18 New Zealand white juvenile male rabbits. It was reported that (i) continuous forces produced significantly greater overall sutural separation (1.3 mm) than intermittent forces (0.8 mm), (ii) although they were delivered over a period of time 86% as long, intermittent forces produced only 61% of the sutural separation of continuous forces, and (iii) due to greater sutural separation and bone formation, continuous forces provide a more effective approach for separating sutures than intermittent forces.

6.9.1.5 Cable
Using SHE and SE characteristics, control approaches for buckling NiTi alloys were investigated [290]. It was mentioned that SE was effectively employed in a passive control approach, using the restraining cables to increase the postbuckling resistance and recentering capabilities of a compressed column while dissipating energy. Reed-lunn et al. [291, 292] conducted systematic studies with uniaxial tension tests on cables made from NiTi SMA wires to characterize their superelastic behavior in room temperature air. It was reported that (i) overall, the 7 × 7 construction has a mechanical response similar to that of straight wires with propagating transformation fronts and distinct stress plateaus during stress-induced transformations, however (ii) the 1 × 27 construction exhibits a more compliant and stable mechanical response, trading a decreased force for additional elongation, and does not exhibit transformation fronts due to the deeper helix angles of the layers, (iii) stress-induced phase transformations involved localized strain/temperature and front propagation in all of the tested 7 × 7 components

but none of the 1 × 27 components aside from the 1 × 27 core wire, and (iv) although the 1 × 27 multilayer strands exhibited temperature/strain localizations in a distributed pattern during transformations, the localizations did not propagate and their cause was traced back to contact indentations (stress concentrations) arising from the cable's fabrication. Figure 6.8 shows superelastic NiTi cable for construction usage [291].

Figure 6.8: Superelastic NiTi alloy cables for construction usage [291].

6.9.1.6 Staples
Staples are designed to facilitate fast and easy fixation in a variety of applications. Once they are implanted, return to their shape-memory compressed position. Dynamic and permanent compression is achieved at the arthrodesis or osteotomy site. There are several characteristics associated with NiTi staples; no heat activation required, fast and easy insertion, range of sizes available from 8 to 25 mm, symmetric or asymmetric leg lengths to fit anatomy, and simple intraoperative repositioning; if required [293]. Figure 6.9 shows NiTi compression staples [294] and Figure 6.10 illustrates fracture bone fixation. The staple offers internal fixation with residual compression. The wide variety of sizes and styles makes it ideal for most applications in the foot, ankle, hand, and wrist.

6.9.2 2-D product forms

6.9.2 1 Film and foil
Due to unique phenomena of SE and SME, excellent biocompatibility, and versatile capability of forming including thin film of NiTi-based alloys, NiTi alloy thin films

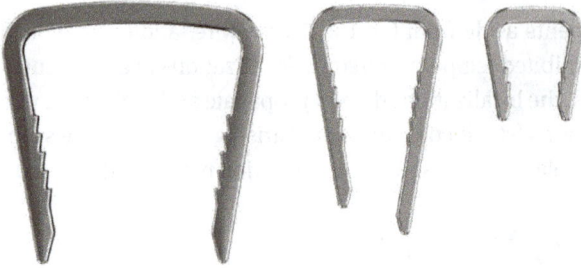

Figure 6.9: NiTi compression staples [293].

Figure 6.10: Fixation of fractured bones [294].

are very attractive material for industrial and medical applications such as microac-tuator, microsensors, and stents for blood vessels [295–297]. An important property besides SME in the application of SMA thin films is the adhesion between the film and the substrate, so that selection of substrate material is crucial [296]. NiTi thin films are prepared by the magnetron sputtering deposition on various types of substrate materials [296–309]; (1) glass substrates [295, 296] for 10 µm thick film [295], (2) Silicon substrates [300–304] 2 µm thick [200] and 5–15 µm [204], (3) MgO substrates [305], (4) Cu substrates [306], and (5) unknown substrates for 10–20 µm [307] and 3 µm [308]. A plasma sputtering can also produce NiTi thin film on Si substrates [304].

6.9.2.2 Strip

Dalle et al. [310] adopted the twin roll casting technique to produce $Ni_{49.8}Ti_{42.2}Hf_8$ SMA thin strips. Meng et al. [311] created two-dimensional functionally graded NiTi thin plates were created by laser surface scanning anneal. It was reported that (i) the microstructural gradient led to a unique SME, involving shape change in two oppo-site directions upon one heating, analogous to stingray motion, and (ii) such behavior also experiences enlarged temperature windows for both the forward and reverse martensitic transformations, rendering high controllability of actuation.

6.9.2.3 Sheet

Figure 6.11 shows a typical NiTi alloy sheet [312]. Daly et al. [313] investigated the stress-induced martensitic phase transformation in NiTi thin sheet using the in situ optical technique known as digital image correlation. It was mentioned that (i) the transformation initiates before the formation of localized bands, and the strain inside the bands does not saturate when they nucleate, (ii) the effect of rolling texture on the macroscopic stress–strain behavior was observed and it is shown that the resolved stress criterion or Clausius–Clapeyron relation does not hold for polycrystalline nitinol, and (iii) the effect of geometric defects on localization behavior was observed. Since NiTi SMAs are used in practice in their polycrystalline form and usually consist of strong textures, Liu et al. [314] studied the superelastic anisotropy in NiTi alloy thin sheet. It was found that (i) although the shape of the stress–strain curve is strongly orientation dependent, the shape recovery strain which is as high as 9% is nearly orientation independent, and (ii) the magnitude of the plateau-strain in the stress–strain curve is determined by the formation of the most favorably oriented martensite variants, while the magnitude of shape recovery strain which is independent of the plateau-strain is determined by the reverse transformation of both favorably and less-favorably oriented martensite variants. Hang et al. [315] investigated the hydrothermal synthesis of nanosheets on biomedical NiTi alloy in pure water. It was reported that (i) the rhombohedral $NiTiO_3$ nanosheets with thickness of 6 nm was grown at 200 °C, (ii) the nanosheets can well support cell growth, which suggests the ion release amount can be well tolerated, and (iii) good corrosion resistance and cytocompatibility combined with large specific surface area render the nanosheets promising as safe and efficient drug carriers of the biomedical NiTi alloy.

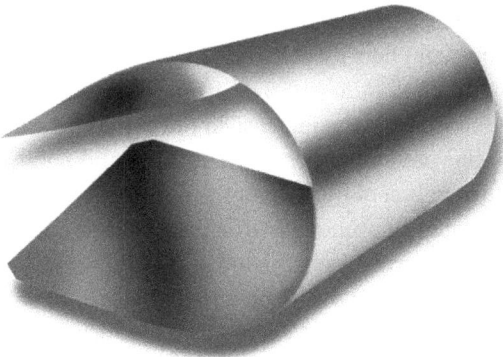

Figure 6.11: Typical NiTi alloy sheet [312].

6.9.2.4 Mesh

NiTi thin and fine mesh (as shown in Figure 6.12 [316]) has been used in medical field [317–319]. Typical examples of medical applications include the effective and supportive device for bone grafting procedure [317, 318] or constitutional materials for designing and creating scaffolds for tissue engineering [319].

Figure 6.12: Typical NiTi thin mesh [316].

6.9.3 3-D product forms

6.9.3.1 Tube

Poncet [320] mentioned that NiTi tubing is drawn in a seamless condition usually starting from gun-drilled barstock often referred to as hollows and typical tube drawing techniques conducted at most manufacturers can include (1) floating plug, (2) hard mandrel, and (3) sinking (drawn without internal mandrel or plug). Various conditions (as-received state, after chemical etching, aging in boiling water, and heat treatment) of NiTi tubes were evaluated by x-ray photoelectron spectroscopy, high-resolution Auger spectroscopy, electron backscattering, and scanning-electron microscopy [321]. It was reported that (i) NiTi tubing from various suppliers demonstrated great variability in Ni surface concentration (0.5–15 at%) and Ti/Ni ratio (0.4–35), (ii) shape-setting heat treatment at 500 °C for 15 min resulted in tremendous increase in the surface Ni concentration and complete Ni oxidation, (iii) preliminary chemical etching and boiling in water successfully prevented surface enrichment in Ni, initially resulting from heat treatment, and (iv) a stoichiometric uniformly amorphous TiO_2 oxide generated during chemical etching and aging in boiling water was reconstructed at 700 °C, revealing rutile structure [320] and findings about the surface oxidation and localized Ti depletion were confirmed by Potapov et al. [322]. There was a comparison in fatigue resistance of NiTi tubing samples collected from five different suppliers [323]. Figure 6.13 shows a typical flexible tube (6 mm inner diameter) by the slit processing [324].

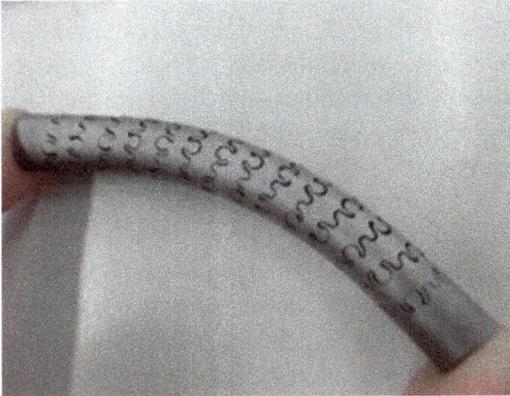

Figure 6.13: Flexible tube [324].

Favier et al. [325] deformed an initially austenitic polycrystalline $TiNi_{50.8}$ (at%) thin-walled tube with small grain sizes under tension in air at ambient temperature and moderate nominal axial strain rate to measure temperature and strain fields with visible-light and infrared digital cameras. It was found that (i) in the first apparently elastic deformation stage, both strain and temperature fields are homogeneous and increase, and (ii) during the first apparently elastic stage of the unloading, both strain and temperature fields are homogeneous and decrease. Ghassemi-Armaki et al. [326] studied compression-compression cyclic deformation of nanocrystalline NiTi tubes intended for medical stents and with an outer diameter of 1 mm and wall thickness of 70 μm using micropillars extracted from textured nanocrystalline NiTi thin-walled tubes. these micropillars were cycled in a displacement-controlled to strain levels of 4%, 6%, and 8% in each cycle and specimens were subjected to several hundred cycles. The cyclic response of two NiTi tubes, one with A_F of 17 °C and the other with an A_F of −5 °C is compared. It was found that (i) a reduction in the forward transformation stress, (ii) increase in maximum stress for a given displacement amplitude, and (iii) a reduction in the hysteresis loop area, all with increasing number of cycles; concluding that micropillar compression testing in a cyclic mode can enable characterizing the orientation-dependent response in such small dimension components that see complex loading in service, and additionally provide an opportunity for calibrating constitutive equations in micromechanical models.

Wu et al. [327] developed the enhanced photocatalytic activity of $NiO–TiO_2$ nanotube arrays which is mainly attributed to the p–n junction effect between p-type NiO and n-type TiO_2 heterojunction. Kim et al. [328] studied the synthesis and electrochemical properties of oriented $NiO–TiO_2$ nanotube arrays as electrodes for supercapacitors. It was reported that (i) annealing the as-grown nanotube arrays to a temperature of 600 °C transformed them from an amorphous phase to a mixture of crystalline rock salt NiO and rutile TiO_2, and (ii) changes in the morphology and crystal structure strongly influenced the electrochemical properties of the nanotube electrodes. Hang et al. [329] prepared Ni–Ti–O nanotube array on nearly equiatomic NiTi alloys which possess broad application potential such as for energy storage and biomedicine and investigated the influence of various anodization parameters on the formation and structure of Ni–Ti–O nanotube arrays and their potential applications. It was found that (i) well-controlled nanotube arrays can be fabricated during relatively wide ranges of the anodization voltage (5–90 V), electrolyte temperature (10–50 °C) and electrolyte NH_4F content (0.025–0.8 wt%) but within a narrow window of the electrolyte H_2O content (0.0–1.0 vol%), and (ii) through modulating these parameters, the Ni-Ti-O nanotube arrays with different diameter (15–70 nm) and length (45–1,320 nm) can be produced in a controlled manner. It is also suggested that the Ni-Ti-O nanotube arrays may be used as electrodes for electrochemical energy storage and nonenzymatic glucose detection, and may constitute nanoscaled biofunctional coating to improve the biological performance of NiTi-based biomedical implants. Figure 6.14 illustrates top and side views of iodine-doped TiO_2 nanotube arrays [330].

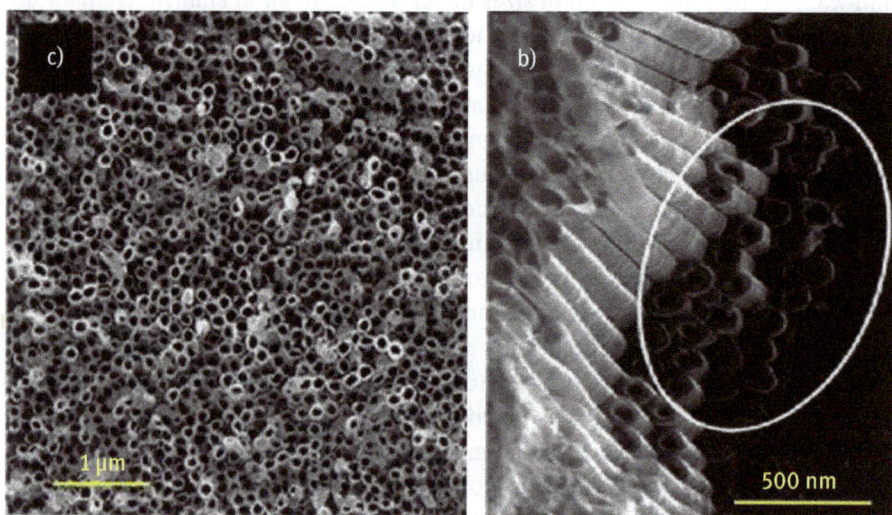

Figure 6.14: Top and side views of iodine-doped titania nanotube arrays [330].

Liu et al. [331] fabricated highly ordered Ni–Ti–O nanotube arrays on NiTi alloy substrates through pulse anodization in glycerol-based electrolytes and studied the effects of anodization parameters and the annealing process on the microstructures and surface morphology of Ni–Ti–O nanotube arrays. It was found that (i) the electrolyte type greatly affected the formation of nanotube arrays, (ii) anatase-type titania was formed on nanotube arrays annealed at 450 °C, (iii) the oxide nanotubes could be crystallized to rutile phase after annealing treatment at 650 °C, and (iv) the Ni–Ti–O nanotube arrays demonstrated an excellent thermal stability by keeping their nanotubular structures up to 650 °C. Mohammadi et al. [332] investigated the effect of various heat treatment processes on the formation and electrochemical properties of ordered TiO_2 nanotubes on NiTi. After solution treatment of as-cast NiTi samples at 900 °C for 1 h, various heat-treated groups were prepared; furnace-cooled sample, water-quenched sample, water-quenched, and aged. The heat-treated samples were anodized in ethylene glycol solution containing NH_4F. It was reported that (i) the microstructure, chemical composition and grain size of the NiTi samples depended on the heat treatment process, (ii) water-quenching and subsequent aging process provided fine precipitations distributed in the grain boundaries and reduced grain size, (iii) nanotube arrays with various distributions and microstructures could be developed depending on the heat-treatment process of NiTi samples as well as anodization voltage and time, and (iv) among various conditions, anodization of water-quenched and 500 °C aged treated NiTi at 50 V for 10 min could provide nanoscaled biofunctional coating to promote the biological applications of NiTi implants [332].

Although NiTi microtubes are versatile in medical applications including microcatheters and microstents, but it is difficult to fabricate them with dimensions of less than 100 μm by conventional tube-drawing; hence Buenconsejo et al. [333] fabricated high-strength NiTi microtubes by sputter deposition on Cu wire (50 μm diameter). It was reported that (i) all the microtubes exhibited shape-memory behavior after crystallization at 600 °C for 1 hr treatment, (ii) microtubes fabricated without rotating the Cu wire during deposition have low fracture strength due to the columnar grains and nonuniform tube wall thickness, (iii) microtubes fabricated by depositing NiTi on a rotating wire have a uniform wall thickness and the fracture strength increased with increasing rotation speed, and (iv) microtubes made by the rotating-wire method exhibited SE of 3% strain at room temperature with high fracture stress of 950 MPa, suggesting that they are suitable for practical applications. Lin et al. [334] developed radiopaque NiTi-Pt hypotube for potential applications as endovascular stents and other biomedical devices. It was reported that NiTi-Pt hypotube has very different anisotropic characteristics when compared to its NiTi counterpart including higher tensile strength and strain, higher stress–strain nonlinearity, smaller hysteresis loop, and sharper tails during loading–unloading.

6.9.3.2 Ribbon

Although the following picture shown in Figure 6.15 is nothing directly related to the main scheme of this book, it would provide a nice idea of a ring-formability of NiTi material.

Jenko et al. [336] characterized the surface microstructure of a melt-spun NiTi shape-memory ribbon. It was found that the surface consists of an oxide layer of 10 to 20 nm thickness, composed of Ti oxide and some Ni oxide with metallic Ni inhomogeneously distributed in the subsurface region. Tong et al. [337] investigated the relation between precipitation (volume fraction, morphology, type) and SME of

Figure 6.15: The NiTi ribbon ring [335].

$Ti_{50}Ni_{25}Cu_{25}$ melt-spun ribbon. It was reported that (i) the precipitation process takes as following sequence: TiCu →TiCu+Ti_2(Ni, Cu) → Ti_2(Ni, Cu) with increasing annealing temperature or duration, (ii) the SME is found to depend on both the volume fraction (which affects the shape recovery strain through reduction of the transformation volume participating the shape recovery) and the distribution of the precipitates (affects the shape recovery strain through strengthening the matrix thus reducing the martensite strain which is more predominant under low constraint stresses), (iii) precipitation strengthening, on the other hand, reduces the tendency of dislocation generation and/or movement, thus reducing the irreversible strain and improving shape recovery strain. Xie et al. [338] investigated microstructure and texture development in $Ti_{50}Ni_{25}Cu_{25}$ melt-spun ribbon and reported that the inherent brittleness of $Ti_{50}Ni_{25}Cu_{25}$ can be improved by annealing at 500 °C. Tomić et al. [339] prepared rapidly solidified NiTi shape-memory ribbons by melt-spinning and studied the biocompatibility and their immunomodulatory properties on human monocyte-derived dendritic cells. It was reported that (i) the melt-spinning of NiTi alloys can form a thin homogenous oxide layer, which improves their corrosion resistance and subsequent toxicity to human monocyte-derived dendritic cells, (ii) NiTi ribbons possess substantial immunomodulatory properties on monocyte cells, and (iii) these findings might be clinically relevant, because implanted Ni-Ti SMA devices can induce both desired and adverse effects on the immune system, depending on the microenvironmental stimuli.

6.9.3.3 Truss

Taylor et al. [340] fabricated NiTi–Nb microtrusses by extrusion-based 3D printing of powders and transient-LPS and investigated the effects of Nb concentration (0, 1.5, 3.1, 6.7 at.% Nb) on the porosity, microstructure, and phase transformations. It was found that (i) microtrusses with the highest Nb content exhibit long channels (from 3D-printing) and struts with smaller interconnected porosity (from partial sintering), resulting in overall porosities of ~75% and low compressive stiffnesses of 1–1.6 GPa, similar to those of trabecular bone and in agreement with analytical and finite element modeling predictions, (ii) diffusion of Nb into the NiTi particles from the bond regions results in a Ni-rich composition as the Nb replaces Ti atoms, leading to decreased martensite/austenite transformation temperatures, and (iii) adult human mesenchymal stem cells seeded on these microtrusses showed excellent viability, proliferation, and extracellular matrix deposition over 14 days in culture.

6.9.3.4 Spring

NiTi coil springs are designed to produce light continuous forces. The force delivery by NiTi open and closed coil springs during unloading (deactivation) has been compared to that provided by comparable stainless-steel springs [341]. It was found that (i) open-coil springs (0.010 × 0.035 inch) were compressed from their initial length of 15 mm to 6 mm and the forces generated with spring recovery recorded, (ii) closed-coil springs (0.009 × 0.035 inch) were distracted from their initial length of 3 mm to 9 mm and the force recorded as the spring recovered, (iii) the closed-coil NiTi springs produced light continuous forces of 75–90 g over the distraction range of 6 mm while the open-coil springs produced forces of 55–70 g within the 9 mm compression range, and (iv) stainless steel springs produced heavier forces, ca. 200 g, for an activation of 1 mm and the generated force increased rapidly as the activation was increased; indicating that NiTi coil springs deliver optimal forces for orthodontic tooth movement over a longer activation range than comparable stainless steel springs. Manhartsberger et al. [342] investigated NiTi springs of the open- and closed type. The closed coil springs were subjected to a tensile and the open coil springs to a compression test. After the first measurement, the springs were activated for a period of 4 weeks and then reinvestigated with the same procedure. It was reported that, (i) with the different coil springs, the force delivery given by the producer could be achieved only within certain limits, (ii) there was a distinct decrease in force delivery between the first and second measurement, and (iii) after considering the loading curves of all the NiTi coil springs and choosing the right activation range respective to the force delivery, it was found that the coil springs deliver a superior clinical behavior and open new treatment possibilities. Figure 6.16 depicts a typical orthodontic NiTi closed coil spring [343].

Barwart et al. [344], using differential scanning calorimetry, determined the transition temperature ranges of four types of superelastic orthodontic NiTi coil springs (Sentalloy). Data obtained from the transition temperature ranges provide valuable information on the temperature at which a NiTi wire or spring can assume superelastic properties and when this quality disappears. It was reported that (i) the hysteresis of the transition temperature, found between cooling and heating, was 3.4–5.2 K, (ii) depending on the spring type the austenite transformation started (A_S) at 9.7–17.1 °C and finished (A_F) at 29.2–37 °C, and (iii) the martensite transformation starting temperature (M_S) was evaluated at 32.6–25.4 °C, while M_F (martensite

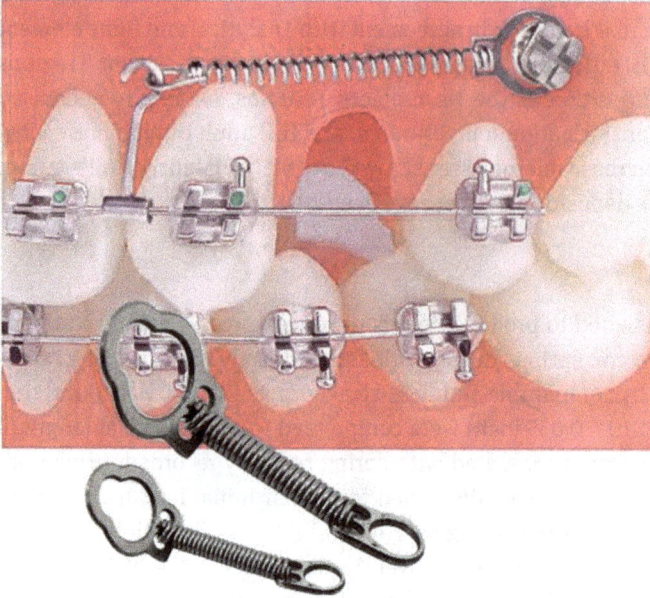

Figure 6.16: Typical view of orthodontic NiTi closed coil spring [343].

transformation finishing temperature) was 12.7–6.5 °C. It was also indicated that (iv) the springs become superelastic when the temperature increases and A_S is reached, and they undergo a loss of superelastic properties and a rapid decrease in force delivery when they are cooled to M_F, and (v) for the tested springs, M_F and A_S were found to be below room temperature; concluding that, at room temperature and some degrees lower, all the tested springs exert superelastic properties. For orthodontic treatment this means the maintenance of superelastic behavior, even when mouth temperature decreases to about room temperature as can occur, for example, during meals [342]. Keles [345], treating 15 patients, eight males, and seven females with a mean age of 13.32 years, selected for unilateral molar distalization, examined the clinical efficacy of NiTi coil spring. It was mentioned that the newly developed device achieved bodily distal molar movement with minimum anchorage loss.

6.9.3.5 Scaffold
Bone tissue engineering is one of the good examples for interdisciplinary science and engineering composed of medicine and dental science, material science, and biomechanics, and advanced development in scaffolds for tissue engineering can be recognized. Although, nowadays the majority of the research effort is in the development of scaffolds for nonload bearing applications, primarily using soft natural or synthetic polymers or natural scaffolds for soft tissue engineering; metallic scaffolds aimed for hard tissue engineering have been also the subject of in vitro and in vivo research and industrial development [346].

For developing bone scaffolds or bone grafting scaffolds, due to required morphological compatibility between scaffold surface texture and receiving vital hard/soft tissue [24], controlled porosity of scaffold structure is highly demanded. There are several studies on porous NiTi scaffolds [347–353]. Porous NiTi scaffolds were synthesized by selective laser sintering [347, 349]. The wear debris toxicity is a crucial issue on the interface between implants and bone tissues, thus inducing the subsequent mobilization of implants gradually and finally resulting in the failure of bone implants, which imposes restrictions on the applications of porous NiTi SMAs scaffolds for bone tissue engineering. Wu et al. [349] investigated the effects of the annealing temperature, applied load, and porosity on the tribological behavior and wear resistance of three-dimensional porous NiTi shape-memory bone scaffolds. It was reported that (i) the porous structure and phase transformation during the exothermic process affect the tribological properties and wear mechanism significantly, (ii) the porous NiTi phase during the exothermic reaction also plays an important role in the wear resistance, (iii) porous NiTi has smaller friction coefficients under high loads due to stress-induced SE, and (iv) the wear mechanism is discussed based on plastic deformation and microcrack propagation. Liu et al. [350] investigated the biomechanical characteristics of porous NiTi implants and bone ingrowth under actual load-bearing conditions by comparative study on porous NiTi, porous Ti, dense NiTi, and dense Ti which were implanted into 5 mm diameter holes in the distal part of the femur/tibia of rabbits for 15 weeks. It was reported that (i) the bone ingrowth and interfacial bonding strength are evaluated by histological analysis and push-out test, (ii) the porous NiTi materials bond very well to newly formed bone tissues and the highest average strength of 357 N and best ductility are achieved from the porous NiTi materials, and (iii) the bonding curve obtained from the NiTi scaffold shows similar SE as natural bones with a deflection of 0.30–0.85 mm thus shielding new bone tissues from large load stress; indicating that, in conjunction with the good cytocompatibility, the superelastic biomechanical properties of the porous NiTi scaffold bodes well for fast formation and ingrowth of new bones, and porous NiTi scaffolds are thus suitable for clinical applications under load-bearing conditions. Figure 6.17 shows scanning electron micrographs of cells cultured on 3D NiTi scaffolds [351].

SLM is one of the additive manufacturing techniques for producing complex functional parts via successively melting layers of metal powder. This process grants the freedom to design highly complex scaffold components to allow bone ingrowth and support mechanical anchorage [352]. The compression fatigue behavior of three different unit cells (octahedron, cellular gyroid, and sheet gyroid) of SLMed NiTi scaffolds were investigated. It was reported that (i) triply periodic minimal surfaces display superior static mechanical properties in comparison to conventional octahedron beam lattice structures at identical volume fractions, and (ii) oxygen analysis showed a large oxygen uptake during SLM processing which must be altered to meet ASTM medical grade standards and may significantly reduce fatigue life; concluding that fatigue properties of SLMed NiTi scaffolds provide sufficient mechanical

(a)
(b)

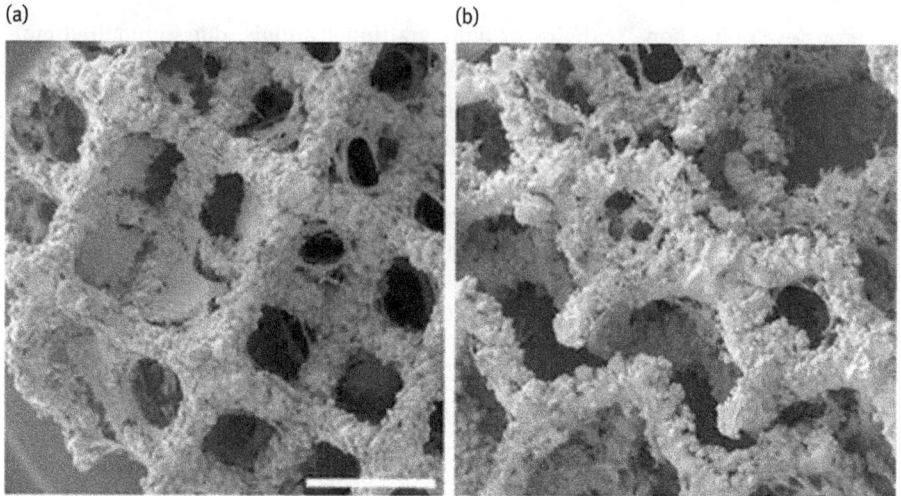

Figure 6.17: SEM images of cells cultured on 3D NiTi scaffolds, (a) periphery and (b) interior [351].

support over an implants lifetime within stress range values experienced in real life [352]. Wang et al. [353] mentioned about advantages of porous scaffold that (i) since their stiffness and porosity can be adjusted on demands, porous metals have found themselves to be suitable candidates for repairing or replacing the damaged bones since their stiffness and porosity can be adjusted on demands, and (ii) porous metals lies in their open space for the ingrowth of bone tissue, hence accelerating the osse-ointegration process. In order to elucidate forming mechanism of orthopedic NiTi scaffolds during capsule-free HIP, Wu et al. [52] investigated the thermal behavior of titanium hydride and hydrogen release. It was found that (i) hydrogen is continuously released from titanium hydride as the temperature is gradually increased from 300 to 700 °C, (ii) hydrogen is released in two transitions: $TiH1.924 \rightarrow TiH1.5/TiH1.7$ between 300 and 400 °C and $TiH1.5/TiH1.7 \rightarrow \alpha\text{-}Ti$ between 400 and 600 °C, (iii) in the lower temperature range between 300 and 550 °C the rate of hydrogen release is slow, but the decomposition rate increases sharply above 550 °C, and (iv) NiTi scaffolds foamed by stepwise release of hydrogen are conducive to the attachment and proliferation of osteoblasts and the resulting pore size also favor ingrowth of cells.

6.9.3.6 Foam

A metal foam is a cellular structure consisting of a solid metal (frequently aluminum) with gas-filled pores comprising a large portion of the volume. The pores can be sealed (closed-cell foam) or interconnected (open-cell foam). The defining characteristic of metal foams is a high porosity: typically only 5–25% of the volume is the base metal. Metal foams have great specific stiffness (ratio of stiffness to weight), and an almost reversible quasielastic zone, so they are applicable to the light structure. For metal foams with a 1/5 density, the specific stiffness is five times that of normally dense metal with the same weight. Metal foams typically retain some physical properties of their base material [354, 355].

Gotman et al. [356] fabricated highly porous NiTi scaffolds for bone ingrowth which were fabricated by reactive conversion of commercially available Ni foams. It was mentioned that these open cell trabecular NiTi scaffolds possess high strength and ductility and exhibit low-Ni ion release, and (ii) reactive conversion deposition of a thin titanium nitride (TiN) layer further improves the corrosion characteristics of trabecular NiTi and allows for material bioactivation by alkali treatment or biomimetic Ca phosphate deposition.

Grummon et al. [357] fabricated open-cell metallic foams having very low density which display martensite transformations required for shape-memory and superelastic behavior by powder-metallurgy technique. It was reported that (i) a polymeric precursor foam was coated with an equiatomic NiTi powder slurry and subsequently sintered to yield foams with relative densities as low as 0.039, (ii) although contaminated with interstitial impurities, they displayed unambiguous calorimetric signature of the B2→B19′ transformation, and (iii) the results are of considerable significance to potential applications requiring ultralight weight structures with the unusual dissipative and strain-recovery properties of NiTi shape-memory materials. Figure 6.18 shows NiTi alloy foam structure [357]. Barrabés et al. [358] examined NiTi foams that have been treated using a new oxidation treatment for obtaining Ni-free surfaces that could allow the ingrowth of living tissue, thereby increasing the mechanical anchorage of implants. It was reported that (i) a significant increase in the effective surface area of these materials can decrease corrosion resistance and favor the release of Ni which might induce allergic reactions or toxicity in the surrounding tissues, (ii) these foams have pores in an appropriate range of sizes and interconnectivity, and thus their morphology is similar to that of bone, and (iii) their mechanical properties are biomechanically compatible with bone. Young et al. [359] fabricated $Ni_{40}Ti_{50}Cu_{10}$ foams by

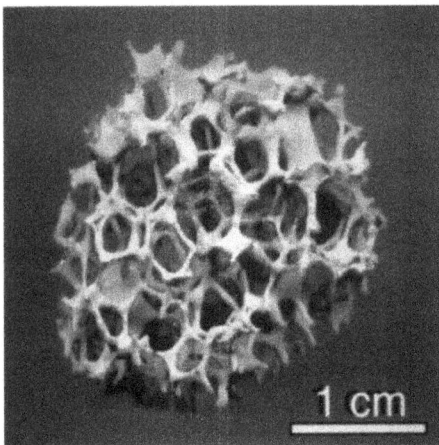

Figure 6.18: NiTi alloy foam structure [357].

the replication cast into a porous SrF_2 preform. This space holder is chemically stable in contact with liquid and solid $Ni_{40}Ti_{50}Cu_{10}$, but can be removed by dissolution in nitric acid. It was reported that (i) a $Ni_{40}Ti_{50}Cu_{10}$ foam with 60% porosity exhibits low stiffness (1–13 GPa) and large recoverable strains (~4%) during cyclical compression testing at 38 °C, within the superelastic range based on calorimetry results, and (ii) casting NiTi-based SMA foams enable the economical production of porous actuators, energy absorbers, and biomedical implants with complex shapes.

6.9.3.7 Stent

Nitinol (nickel-titanium) alloys exhibit a combination of properties which make these alloys particularly suited for self-expanding stents. It was mentioned that nitinol stents are manufactured to a size slightly larger than the target vessel size and delivered constrained in a delivery system; after deployment, they position themselves against the vessel wall with a low, chronic outward force. It then follows to resist outside forces with a significantly higher radial resistive force. Despite the high nickel content of nitinol, its corrosion resistance and biocompatibility are equal to that of other implant materials [360]. The most common nitinol stent is shown in Figure 6.19 [361].

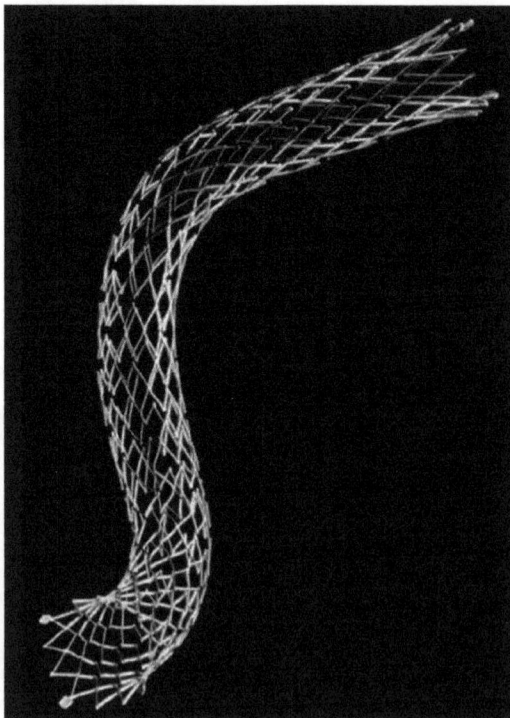

Figure 6.19: Typical NiTi alloy stent [361].

6.9.4 Others

In this section, we will discuss about hybrid structure using NiTi alloys and composites in which NiTi alloy is a main constituent. Accordingly, NiTi is dually utilized as (1) serving composite matrix material and (2) reinforcing filament or composite composition.

By hybridizing or incorporating shape-memory materials with other functional materials or structural materials, smart composites can be fabricated which may utilize the unique functions or properties of the individual bulk materials to achieve multiple responses and optimal properties, or, to tune their properties to adapt to environmental changes [362]. Maho et al. [363] investigated the surface characteristics of bare and modified NiTi samples with tantalum (Ta) coating by spectroscopic, microscopic, and electrochemical techniques. Frensemeier et al. [364] developed a switchable (reversible) dry adhesive based on a NiTi SMA with an adhesive silicone rubber surface. The developed system is based on the indentation-induced two-way shape-memory effect in NiTi alloys, which are trained by mechanical deformation through indentation and grinding to elicit a temperature-induced switchable topography with protrusions at high temperature and a flat surface at low temperature and the trained surfaces are coated with either a smooth or a patterned adhesive polydimethylsiloxane layer, resulting in a temperature-induced switchable surface, used for dry adhesion. It was reported that (i) adhesion tests show that the temperature-induced topographical change of the NiTi influences the adhesive performance of the hybrid system; for samples with a smooth polydimethylsiloxane layer the transition from flat to structured state reduces adhesion by 56%, and for samples with a micropatterned polydimethylsiloxane layer adhesion is switchable by nearly 100%, and (ii) both hybrid systems reveal strong reversibility related to the NiTi martensitic phase transformation, allowing repeated switching between an adhesive and a nonadhesive state [364]. SMA NiTi alloys with multiple superelastic plateaus, and hybrid components with both the SHE and SE effects were designed and developed utilizing a pulsed laser based vaporization process to precisely alter the local composition of NiTi-based alloys [365]. It was found that (i) each laser pulse decreased local Ni composition by 0.16 at%, leading to controlled changes in transformation temperatures, (ii) control of the local composition enabled strengthening by cold work and heat treatment, while maintaining distinct SMA properties throughout the material, and (iii) the combined cold rolling and annealing significantly increased the yield strength of the materials allowing for tunable performance offerings; concluding the synergistic behavior of adjacent locally processed regions demonstrated a significant increase in shape-memory functionality and their future application such as a multiforce orthodontic archwire capable of simultaneously applying biomechanically optimized tooth moving forces to each tooth along the dental arch. Witkowska et al. hybridized $CNH+TiO_2+TiN$-type surface layers on NiTi alloy [366] and $C(Ag) + TiO_2$ layer on NiTi alloy surface [367]. The hybrid process that combines oxidation under glow-discharge conditions with ion beam-assisted deposition has been applied to mechanically polished NiTi SMA in order to produce composite surface layers consisting of a TiO_2 layer and an external carbon coating with an addition of silver. It was found that the produced surface layers $C(Ag) + TiO_2$ type have shown increased surface roughness, improved corrosion resistance, altered wettability, and surface free energy, as well as reduced platelet adhesion, aggregation, and activation in comparison to NiTi alloy in initial state; indicating that such characteristics can be of great benefit for cardiac applications.

NiTi-TiC composites have been investigated by many researchers [368–373]. Ayers et al. [368] employed the SHS (or CS) technique to control the porosity and to engineer porous TiB–Ti, NiTi, NiTi–TiC, TiC–Ti, and multiphase/heterogeneous calcium phosphate for bone tissue engineering and drug delivery systems. The extent of the porosity and the size of the pores were controlled using certain gasifying agent (or space holder) such as B_2O_3, $CaCO_3$. Mari et al. [369] studied the transformation behavior of Ti-rich NiTi containing 0, 10, and 20 vol% equiaxed TiC particles. It was reported that (i) the thermoelastic phase transformation of the unreinforced matrix exhibits multiple steps, and (ii) the TiC particles inhibit the R phase and also lower some of the transformation temperatures. Using the same composites, the deformation behavior under uniaxial compression was investigated both below and above the matrix martensitic transformation temperature: (1) at room temperature, where the martensitic matrix deforms plastically by slip and/or twinning; and (2) at elevated temperature, where plastic deformation of the austenitic matrix takes place by slip and/or formation of stress-induced martensite [370]. It was found that the effect of TiC particles on the stress–strain curves of the composites depends upon which of these deformation mechanisms is dominant. There are three major mechanisms: (i) in the low-strain elastic region, the mismatch between the stiff, elastic particles and the elastic–plastic matrix is relaxed in the composites, (ii) in the moderate-strain plastic region where nonslip deformation mechanisms are dominant, mismatch dislocations stabilize the matrix for all samples, and (iii) in the high-strain region controlled by dislocation slip, weakening of the NiTi composites results [370]. Li et al. [371] developed a novel tribocomposites with TiNi SMA matrix and TiNi matrix. It was found that densities varied within 0.5% of average values. Luo et al. [372] developed the wear resistant NiTi enhanced by hard particle of TiC composites fabricated by the vacuum sintering process. It was reported that the nano-TiN/TiC/TiNi composite exhibited excellent wear resistance, superior to those of the TiC/TiNi composite and WC/NiCrBSi hardfacing overlay. Burkes et al. [373] employed the CS coupled with a quasiisostatic densification step to produce dense NiTi–TiC composites. The synthesis and characterization of five composites are presented, including ceramic-intermetallic (≥50% ceramic) composites and intermetallic-ceramic (≥50% intermetallic) composites. It was reported that (i) refractory TiC and NiTi intermetallic phases become more stoichiometric and the TiC particle size decreases with increasing intermetallic content, (ii) the Vickers hardness decreases as the matrix shifts from ceramic to intermetallic, and (iii) modulus of elasticity and compressive strength decreases with increasing amounts of Ni-Ti intermetallic.

There are three types of composites, depending on the type of matrix, which should include CMC (ceramic matrix composites), MMC (metal matrix composites) and PMC (polymer matrix composites). In the above, we have been looking at CMC. In the following, remaining MMC and PMC will be discussed. Kothalkar et al. [374] investigated NiTi/Ti$_3$SiC$_2$ interpenetrating composites using SPS that combine two unique material systems – a SMA and a MAX phase – demonstrating two different superelastic mechanisms. MAX phases are layered hexagonal carbides and nitrides which have the general formula: $M_{n+1}AX_n$ (MAX) where $n = 1$ to 3, and M is an early transition metallic element, A is an A-group (mostly IIIA and IVA or groups 13 and 14) element, and X is either carbon and/or nitrogen. Equiatomic NiTi and Ti$_3$SiC$_2$ were used. It was reported that the highest energy dissipation was observed for the TC composite followed by the as-sintered composite, pure NiTi, and pure Ti$_3$SiC$_2$ when compared at the same applied stress levels. Both the as-sintered and the composites showed higher damping up to 200 MPa stress than any of the metal-MAX phase composites reported in the literature. TiNi SMA and its composite using δ-Al$_2$O$_3$ nano-size particles were prepared by the PM method, and some mechanical properties like hardness, wear, and corrosion behavior were investigated [375]. It was found that (i) the lower wear rate was obtained for the nano-Al$_2$O$_3$-reinforced Ti alloy composite due to increased hardness, but the wear rate increased considerably with increasing the load over 25 N for Ti alloy, and (ii) among the tested samples, the best corrosion resistance was obtained for the Ti alloy because of lower porosity level. Zhang et al. [376] studied the seamlessly bridging the hard and the soft by fabricating hierarchically porous NiTi/hydrogels nanocomposites. It was reported that the nanocomposite surface can hold high-content water while keeping its hierarchical nanoscale topography, thus showing exceptional antibiofouling performance; leading to anti-biofouling alloy (e.g., NiTi)/hydrogel nanocomposites for improved stents and other blood-contacting implants and medical devices.

There are several studies on Al or Al-based alloy matrix composites with NiTi materials [284–287]. Porter et al. [284] prepared NiTi-reinforced Al composites by distributing NiTi throughout an aluminum matrix using powder-metallurgy processing, in the hope of using the shape-memory effect to achieve strengthening and improve the fatigue resistance, as compared to the aluminum matrix. It was found that (i) the shape-memory effect was activated by cold rolling the samples at –30 °C, (ii) upon reheating to the austenite phase, the NiTi was expected to return to its original shape while embedded in the aluminum matrix, due to the fact that this action created residual, internal stresses around each particle, which strengthened the material, and (iii) the yield and ultimate strengths, and the fatigue lives of the NiTi reinforced aluminum composites, have been improved considerably, as compared to the unreinforced material. Al6061 (precipitation-hardened aluminum alloy, containing mainly Mg and Si) matrix composites were investigated [284, 377–379]. Al6061–NiTi composites were fabricated via ultrasonic additive manufacturing for providing a light-weight solution for low thermal expansion applications. It was mentioned that the thermal expansion of

Al6061 can be reduced by over 50% by incorporating a 13% volume fraction of NiTi fibers, due to the contraction of the NiTi fiber during heating, thereby offsetting the thermal expansion of the Al matrix [378]. Chaudhury et al. [284] prepared NiTi short fiber and SiC particle reinforced 6061 Al alloy composites by pressure-assisted induction heating method in ambient atmosphere. Two different composites with significant difference in NiTi fiber diameter (127 and 51 μm) have been prepared. It was reported that (i) 51-μm NiTi fibers are better bonded with the Al matrix; debonding has seldom been observed, (ii) there is better microstructural compatibility, that is, fiber diameter is of the same order as the Al matrix grain size for this case; however, when 127-μm NiTi fibers are used, debonding and the pull-out mechanism can be frequently observed, and (iii) there is no evidence that the bonding of NiTi/Al is affected by either aging or size of NiTi fiber. Akalin et al. [379] studied wear characteristics of Al6061 composites, reinforced with short NiTi fibers using pressure-assisted sintering process in ambient air where the NiTi fibers are aligned unidirectional in the Al matrix. It was found that (i) transverse NiTi fibers improve the wear resistance significantly, (ii) samples with transverse fiber orientation show mostly abrasive wear, whereas, monolithic and parallel samples show adhesive wear mechanism, and (iii) since the Al6061 matrix material is smeared onto NiTi fibers in a short period, all composite samples show similar frictional characteristics after certain period of running in dry sliding.

Tin (Sn) matrix composites were investigated [380, 381]. Coughlin et al. [380] conducted mechanical tensile testing of NiTi–Sn3.5Ag single fiber composites. It was found that (i) this composite demonstrates superelastic behavior of the composite with 85% strain recovery, (ii) fatigue experiments show an evolution in damage over cycles, and an S-N (stress vs. numbers of cycles to failure) curve shows sharp transition between a nearly vertical low-cycle fatigue behavior and the high-cycle fatigue regime, and (iii) the solder composite exhibits constant fatigue strength over the superelastic range of the NiTi fiber. Poormir et al. [381] investigated the bending behavior of a self-healing MMC made from Sn–13 wt% Bi alloy as matrix and NiTi SMA strips as reinforcement. Specimens were fabricated in different reinforcement vol% (0.78, 1.55, and 2.33) and in various prestrains (0%, 2%, and 6%) and were heated at three temperatures (170, 180, and 190 °C). It was reported that (i) shape recovery was accomplished in all the specimens, but not all of them were able to withstand second loading after healing, and (ii) only specimens with 2.33 vol% of NiTi strips, 1.55 vol% of NiTi and 6% prestrain could endure bending force after healing, and they gained 35.31–51.83% of bending force self-healing efficiency.

Hao et al. [382] fabricated a novel composite by subtly combining two contrasting components: NiTi SMA and Ag. It was found that the composite exhibits simultaneously exceptional mechanical properties of high strength, good SE and high mechanical damping, and remarkable functional properties of high electric conductivity, high visibility under fluoroscopy and excellent thermal-driven ability; all of these result from the effective-synergy between the NiTi and Ag components, and place the composite in a unique position in the properties chart of all known structural-functional materials providing new opportunities for innovative electrical, mechanical, and biomedical applications; suggesting that this study may open new avenues for designing and fabricating advanced multifunctional materials by subtly combining contrasting multicomponents. Nanostructured nitinol with titanium- or tantalum-enriched surface layers was developed [383] to provide a decrease in the formation of reactive oxygen species and long-lived protein radicals in comparison to untreated nitinol. It was reported that (i) human peripheral vessel myofibroblasts and human bone marrow mesenchymal stromal cells grown on nitinol bases coated with titanium or tantalum-enriched surface layers exhibit a nearly two times higher mitotic index, (ii) response to implantation of pure nitinol, as well as nanostructure nitinol with titanium or tantalum-enriched surface layers, was expressed though formation of a mature uniform fibrous capsule peripherally to the fragment, (iii) no signs of calcinosis in the tissues surrounding implants with coatings were observed, and (iv) the nature and structure of the formed capsules testify bioinertia of the implanted samples. It was shown that the morphology and composition of the surface of metal samples does not alter following biological tests; indicating that nanostructure nitinol with titanium or tantalum enriched surface layers is a biocompatible material potentially suitable for medical applications [383].

There are several researches conducted on PMC with NiTi materials [384–387]. Hamada et al. [384] produced Ni-Ti SMA fiber embedded resin matrix composites for evaluation of smart denture (a newly developing denture with the function to close its own crack). It was found that (i) the crack closure of the composites was performed well simply by heating at 80 °C, and (ii) the embedded fibers could close the crack of the matrix with enough high accuracy for specimen repair, but they turned out to change the specimen shape after repair. Han et al. [385] prepared the polyurethane matrix composites consisting of carbon nanotube and NiTi springs to study the microstructure, mechanical property and damping property of the composites are investigated. It was found that (i) the carbon nanotube and NiTi spring exhibit a good interfacial bonding with the matrix, (ii) the elastic modulus and tensile strength have been enhanced by the addition of carbon nanotube and NiTi spring, (iii) the NiTi spring is helpful to increase the recovery strain and enhance the energy absorption of the composites by the superelastic deformation under the large deformation, and (iv) the NiTi spring cannot increase the damping property under low vibration amplitude because the superelasticity of the spring is not actuated when the deformation is small. Daghash et al. [386] investigated the cyclic behavior of composite materials that consists of a thermoset polymer matrix reinforced with superelastic NiTi shape-memory wires. It was reported that the SMA-FRP composites can recover relatively high strains upon unloading and exhibit very high failure strains. El-Tahan et al. [387] investigated the bond behavior between Nb–NiTi SMA wires and fiber reinforced polymer composites. It was found that the interfacial bond transfer consists of two components: cohesion (before the onset of debonding) and friction (after the onset of debonding).

6.10 Other important technologies for NiTi forming

Twohig et al. [388, 389] investigated the dieless drawing on NiTi alloy. Studying the effects of the dieless drawing process on commercial grade NiTi rods (5 mm diameter), it was reported [388] that (i) the rods were successfully dieless drawn with a maximum steady state reduction in cross-sectional area of 54%, (ii) uniform levels of stress and strain resulted in uniform reduction of the rod cross-sectional area, (iii) the grain structure was highly deformed in the drawing direction and increased porosity was observed as a result of the process, and (iv) the longitudinal section hardness of the rod was significantly reduced as a result of the dieless drawing process. Mechanical, microstructural, and thermal properties of an equiatomic NiTi alloy (which was subjected to the dieless drawing) were studied [389]. It was reported that (i) the transformation temperatures of the processed NiTi showed a significant reduction in the austenite finish transformation temperature of the processed material compared to that of the as-received material, (ii) tensile tests carried out to determine changes in the superelastic behavior of the material due to the dieless drawing process, and (iii) the process has resulted in a material with lower transformation temperatures, higher ductility, and reduced tensile flow stress.

Zhu et al. [390] cold-rolled NiTi alloys and subsequently electroplastic-rolled and studied the effect of rolling parameters on working plasticity. It was found that (i) NiTi alloy is easy to be embrittled and cracked in cold rolling, and the number of edge cracks is increased with higher thickness reduction, (ii) the electropulsing can enhance the ductility of NiTi alloy in rolling, and (iii) in rolling, the resistance of deformation can be decreased, and deformation degree at the same roll gap increases because NiTi alloy develops recrystallization and the hardness is decreased. Peterlechner et al. [391] evaluated an amorphous structure of NiTi which was obtained by repeated cold rolling in terms of relaxation and crystallization. It was reported that (i) upon deformation, a nanostructured mixture of crystalline and amorphous phase occurs, (ii) with increasing degree of repeated cold rolling, the fraction of the crystalline phase decreases; at an equivalent strain of 16.8%, an almost completely amorphous phase arises, and (iii) the crystallization kinetics of NiTi made amorphous by severe plastic deformation differ from those of amorphous NiTi obtained by melt spinning or sputter deposition.

The production NiTi tubes which are employed in medical area takes place usually via forging, gun-drilling, and hot/cold drawing procedures. Müller [392] produced solid and hollow semifinished NiTi materials by extrusion, which enable a substantially larger cross section of reduction in one step, compared to the above processes. Luo et al. [393] studied the hot extrusion of NiTi alloys. Since the second phase particles that are formed during melting process exhibits a significant effect on the hot extrusion and reextrusion processes of NiTi, the evolution of Ti_4Ni_2Ox (with $x \leq 1$) particles during hot extrusion and reextrusion of near equiatomic $TiNi_{50.6}$ (at%) billets, and their effects on the surrounding matrix microstructure were investigated. It was reported that (i) Ti_4Ni_2Ox (with $x \leq 1$) are the particles that precipitate in the austenite matrix, (ii) after hot extrusion and reextrusion, some particles are elongated and preferentially oriented along the elongation direction, and (iii) the presence of nondeforming second phase particles produce inhomogeneous deformation with a sharp increase in dislocation density in the surrounding matrix; suggesting to exert a profound effect on the processes of recovery and recrystallization for the surrounding matrix.

With regards to the machining, Mauer et al. [394] studied the electrochemical micromachining and Kong et al. [395] investigated the waterjet machining of NiTi alloys. Mauer et al. [394] utilized the electrochemical micromachining with ultrashort voltage pulses to fabricate microstructures on a NiTi SMA. It was mentioned that (i) because of its unique properties, NiTi is a desirable material for use in various applications including medical devices and actuators, (ii) electrochemical micromachining technique is a heat-free, strain-free, and mask-free method for microfabrication, and therefore well suited for use with SMAs, and (iii) microstructures were machined to a depth of 3 μm on NiTi surfaces, and the lateral resolution of the machining was found to be dependent on the duration of the voltage pulses used. Kong et al. [395] challenged the waterjet technology to NiTi alloys. Waterjet technology is well-known for cutting difficult-to-cut materials owing to its benefits of reduced mechanical and thermal damages to workpiece surfaces. By taking into consideration both of the waterjet temperatures at different material removal conditions (i.e., with and without abrasives in the focusing tube) and the transformation temperatures of NiTi, Kong et al. [395] tested three different working zones (100% martensite; mix of austenite and martensite; 100% austenite) under waterjet process. It was mentioned that the abrasive waterjetting is more viable than plain waterjetting for controlled-depth milling of NiTi SMAs.

Electrical discharge machining (EDM) is a well-known noncontact machining procedure that allows precise material removal via pulsed electrical discharge, for this technique, both the machining tool (electrode) and the workpiece should be electrically conductive. It is generally believed that EDM should harden the surface of the NiTi file, resulting in an improved fracture resistance and superior cutting efficiency (for particularly rotary endodontic NiTi files). Pirani et al. [396] evaluated the surface and microstructural alterations of new and used EDMed endodontic NiTi files. It was reported that (i) surface and microstructural characterization of EDMed files revealed the typical spark-machined surface, (ii) no fractures were registered during root canal instrumentation, nor evident surface alterations and minor degradation were observed between new and used instruments, (iii) the metallographic analysis of new and used files disclosed a homogeneous structure, mostly composed of lenticular martensite grains, and some residual austenite, and (v) the cyclic fatigue test showed an increase of fatigue resistance up to 700% on the EDM compared to conventional type of files. Feng et al. [397] analyzed EDMed NiTi alloy as stent material. It was found that (i) the composite structure of solidification bulge-crater-pore-particle can be prepared on the surface of the Ni–Ti alloy through magnetic mixed EDM using suitable processing parameters, and (ii) the contact angle of the surface reaches

138.2°. Takale et al. [398] examined the surface integrity of EDMed $Ti_{49.4}Ni_{50.6}$ (at%) as orthopedic implant material. It was reported that (i) although the conventional machining of NiTi yields poor surface finish and low dimensional accuracy of the machined components, EDM can achieve high dimensional accuracy, but its thermal nature causes great concern regarding surface integrity for biocompatibility application of NiTi material. In wire EDM process, to the machine, sharp corner without error is an extremely challenging issue which is vital for precision machining. Owing to the presence of huge wire deflection, the accuracy of the machined corner is undesirably affected. Bisaria et al. [399] investigated the effect of process parameters namely, spark on time, spark off time, wire tension, wire feed rate, and spark gap voltage on corner error for acute angle (60°), right angle (90°), and obtuse angle (120°) triangular profiles during EDM process of $Ni_{50.89}Ti_{49.11}$ (at%) SMA. It was mentioned that (i) owing to the effect of discharge concentration, corner error for acute angle profile is predominantly affected by spark on time, spark off time, and spark gap voltage, (ii) wire tension was the most influencing parameter for corner error of obtuse and right angle profile due to the effect of wire deflection and wire vibration, whereas wire feed rate has trivial effect for all type of profile, (iii) in order to reduce the corner error, pulse modification technique was utilized, and (v) at low pulse parameters, corner error for 60°, 90°, and 120° profile was reduced by 43.38%, 31.12%, and 29.04%, respectively, as compared to high pulse parameters.

Laser technology's versatility can be found in various applications. Khan et al. [400] mentioned that (i) the laser processing enables controlled augmentation of transformation temperatures, and (ii) the laser technique overcomes traditional fabrication challenges and promises to enhance SMA functionality and facilitate novel applications through producing a new class of smart materials; namely multiple memory materials. Marattukalam et al. [401] utilized laser engineered net shaping technology to fabricate dense equiatomic NiTi alloy components and evaluated the effect of laser power and scan speed on microstructure, phase constituents, hardness, and corrosion behavior of laser processed NiTi alloy. It was reported that (i) retention of large amount of high-temperature austenite phase at room temperature due to high cooling rates associated with laser processing, (ii) the high amount of austenite in these samples increased the hardness, (iii) the grain size and corrosion resistance were found to increase with laser power, and (iv) the decrease in surface energy shifted the corrosion potentials to nobler direction and decreased the corrosion current (i.e., better corrosion resistance). The similar effectiveness of the laser-engineered net shaping was reported [402]. Shiva et al. [403] determined the product's surface morphology, phase transformation temperature and crystalline nature of $Ti_{50}Ni_{25}Cu_{25}$ alloy using laser rapid manufacturing. It was reported that (i) the inclusion of Cu gives improved surface texture and multiple phase peaks, (ii) laser rapid manufacturing was a successful method for producing $Ti_{50}Ni_{25}Cu_{25}$ materials, without changing their properties of SME, and (iii) the properties of $NiTi_{50}$ and $Ti_{50}Ni_{25}Cu_{25}$ resembled each other. Pfeifer et al. [404] demonstrated the pulsed Nd:YAG laser cutting of 1 mm thick

NiTi SMAs for medical applications. It was mentioned that due to the local energy input only small HAZs occur and the shape-memory properties remain.

Equal-channel angular processing (ECAP) is one technique from the severe plastic deformation group to produce ultrafine grained material. This technique is cold work without reduction in the cross-sectional area. Shahmir et al. [405–407] had extensively studied the ECAP of NiTi alloys. It was reported that (i) near equiatomic NiTi SMAs may be successfully processed by ECAP at room temperature by containing the NiTi samples within Fe sheaths, (ii) processing was performed successfully for up to two passes of ECAP by controlling the processing variables and the initial structure of the NiTi alloys, and (iii) the results show that the austenitic transformation and cracking can be suppressed if adiabatic heating is minimized by reducing the processing ram speed and the specimen size [405]. Furthermore, it was pointed out that (iv) the imposed strain both decreases by comparison with the processing of conventional sheath-free NiTi billet and increases by increasing the dimensions of the core, and (v) the lower areas of the cores undergo less deformation than the upper areas after successful ECAP processing but the imposed strain, and thus the microhardness values, become homogeneously distributed on the longitudinal planes after two passes of ECAP [407]. About the annealing behavior of ECAPed NiTi alloy, it was mentioned that (vi) the SME improves by postdeformation annealing at 400 °C after ECAP processing, and (vii) annealing for 10 min gives a good SME which leads to a maximum in recoverable strain of 6.9% upon heating where this is more than a 25% improvement compared with the initial state.

References

[1] Elahinia MH, Hashemi M, Tabesh M, Bhaduri SB. Manufacturing and processing of NiTi implants: a review. Prog. Mater. Sci. 2012, 57, 911–46.
[2] https://www.kobelco.co.jp/english/titan/files/process.pdf.
[3] Sadrnezhaad SK, Badakhshan Raz S. Interaction between refractory crucible materials and the melted NiTi shape-memory alloy. Metall. Mater. Trans. B, 2005, 36B, 395–403.
[4] Wu H. Oxygen diffusion through titanium and other HCP metals. Thesis, University of Illinois, Urbana-Champaign, 2013.
[5] Wasz M, Brotzen F, McLellan R, Griffin A. Effect of oxygen and hydrogen on mechanical properties of commercial purity titanium. Int. Mater. Rev. 1996, 41, 1–12.
[6] Conrad H. Effect of interstitial solutes on the strength and ductility of titanium. Prog. Mat. Sci. 1981, 26, 123–403.
[7] Herø H, Syverud M, Waarli M. Mold filling and porosity in castings of titanium. Dent. Mater. 1993, 9, 15–8.
[8] Watanabe I, Watkins JH, Nakajima H, Atsuta M, Okabe T. Effect of pressure difference on the quality of titanium casting. J. Dent. Res. 1997, 76, 773–9.
[9] Takahashi J, Okazaki M, Kimura H, Furuta Y. Casting properties of Ni-Ti shape memory alloy. J. Biomed. Mater. Res. 1984, 18, 427–34.
[10] Koike M, Jacobson D, Chan K, Okabe T. Grindability of alpha-case formed on cast titanium. Dent. Mater. J. 2009, 28, 587–94.

[11] Takahashi J, Okazaki M, Kimura H, Furuta Y. Dental casting of superelastic Ni-Ti alloy. Dent. Mater. J. 1985, 4, 146–52.

[12] Yoneyama T, Doi H, Hamanaka H. Influence of composition and purity on tensile properties of Ni-Ti alloy castings. Dent. Mater. J. 1992, 11, 157–64.

[13] Yoneyama T, Kotake M, Kobayashi E, Doi H, Hamanaka H. Influence of mold materials and heat treatment on tensile properties of Ni-Ti alloy castings. Bull. Tokyo Med. Dent. Univ. 1993, 40, 167–72.

[14] Yoneyama T, Doi H, Kobayashi E, Hamanaka H. Effect of heat treatment with the mould on the super-elastic property of Ti–Ni alloy castings for dental application. J. Mater. Sci.-Mater. M. 2002, 13, 947–51.

[15] Soga Y, Doi H, Yoneyama T. Tensile properties and transformation temperatures of Pd added Ti-Ni alloy dental castings. J. Mater. Sci. Mater. Med. 2000, 11, 695–700.

[16] Yamamoto M, Kuroda T, Yoneyama T, Doi H. Bending property and phase transformation of Ti-Ni-Cu alloy dental castings for orthodontic application. J. Mater. Sci. Mater. Med. 2002, 13, 855–9.

[17] Goryczka T, Ochin P. Texture in NiTi-based shape memory alloys produced by twin roll casting. Sol. St. Phen. 2013, 203/204, 101–4.

[18] Salvetr P, Školáková A, Kopeček J, Novák P. Properties of Ni-Ti-X shape memory alloys produced by arc re-melting. Acta Metall. 2017, 23, 141–6.

[19] Frenzel J, Zhang Z, Neuking K, Eggeler G. High quality vacuum induction melting of small quantities of NiTi shape memory alloys in graphite crucibles. J. Alloys Compd. 2004, 385, 214–23.

[20] Zhang Z, Frenzel J, Neuking K, Eggeler G. On the reaction between NiTi melts and crucible graphite during vacuum induction melting of NiTi shape memory alloys. Acta Mater. 2005, 53, 3971–85.

[21] Zhang Z, Frenzel J, Neuking K, Eggeler G. Vacuum induction melting of ternary NiTiX (X=Cu, Fe, Hf, Zr) shape memory alloys using graphite crucibles. Mater. Trans. JIM, 2006, 47, 661–9.

[22] Stambolič A, Anzel I, Lojen G, Rudolf R. Continuous vertical casting of a NiTi alloy. Mater. Tehnol. 2016, 50, 981–8.

[23] Sadrnezhaad SK, Hosseini SA. Fabrication of porous NiTi-shape memory alloy objects by partially hydride titanium powder for biomedical applications. Mater. Design 2009, 30, 4483–7.

[24] Oshida Y. Bioscience and Bioengineering of Titanium Materials. Elsevier, Amsterdam, 1st edition, 2007.

[25] Oshida Y, Miyazaki T, Tominaga T. Some biomechanistic concerns on newly developed implantable materials. J. Dent. Oral Health 2018, 4, 5 pages; https://scientonline.org/open-access/some-biomechanistic-concerns-on-newly-developed-implantable-materials.pdf.

[26] Andani MT, Moghaddam NS, Haberland C, Dean D, Miller MJ, Elhinia M. Metals for bone implants. Part 1. Powder metallurgy and implant rendering. Acta Biomater. 2014, 10, 4058–70.

[27] Li B-Y, Rong L, Li Y-Y. Porous NiTi alloy prepared from elemental powder sintering. J. Mater. Res. 1998, 13, 2847–51.

[28] Chen G, Liss K-D, Cao P. An *in situ* study of sintering behavior and phase transformation kinetics in NiTi using neutron diffraction. Metall. Mater. Trans. A 2015, 46, 5887–99.

[29] Chen G, Liss K-D, Cao P. In situ observation and neutron diffraction of NiTi powder sintering. Acta Mater. 2014, 67, 32–44.

[30] Zhao X, Neves F, Correia JB, Liu K, Fernades FMB, Koledov V, von Gratowski S, Xu S, Huang J. Mechanical activation of pre-alloyed $NiTi_2$ and elemental Ni for the synthesis of NiTi alloys. J. Mater. Sci. 2018, 53, 13432–41.

[31] Ma X, Wang H, Xie H, Qu J, Chen X, Chen F, Song Q, Yin H. Engineering the porosity and superelastic behaviors of NiTi alloys prepared by an electro-assisted powder metallurgical route in molten salts. J. Alloys Compd. 2019, 794, 455–64.

[32] Wang J, Hu K. Phase transformation of NiTi alloys during vacuum sintering. IOP Conf. Ser.: Mater. Sci. Eng. 2017, 204, 012023; doi: 10.1088/1757-899X/204/1/012023.

[33] Shishkovsky I, Morozov Y, Smurov I. Nanofractal surface structure under laser sintering of titanium and nitinol for bone tissue engineering. Appl. Surf. Sci. 2007, 254, 1145–9.

[34] Deng B, Bruzzaniti A, Cheng GJ. Enhancement of osteoblast activity on nanostructured NiTi/hydroxyapatite coatings on additive manufactured NiTi metal implants by nanosecond pulsed laser sintering. Int. J. Nanomed. 2018, 13, 8217–30.

[35] Oghbaei M, Mirzaee O. Microwave versus conventional sintering: a review of fundamentals, advantages and applications. J. Alloys Compd. 2010, 494, 175–89.

[36] Tang CY, Zhang LN, Wong CT, Chan KC, Yue TM. Fabrication and characteristics of porous NiTi shape memory alloy synthesized by microwave sintering. Mater. Sci. Eng., A 2011, 528, 6006–11.

[37] Ibrahim MK, Hamzah E, Saud SN, Abu Bakar ENE, Bahador A. Microwave sintering effects on the microstructure and mechanical properties of Ti-51at%Ni shape memory alloys. Int. J. Min. Met. Mater. 2017, 24, 280–8.

[38] Ibrahim MK, Hamzah E, Saud SN, Nazim EM. Powder metallurgy fabrication of porous 51(at.%) Ni-Ti shape memory alloys for biomedical applications. Shape Mem. Superelast. 2018, 4, 327–36.

[39] German R, Suri P, Park S. Review: liquid phase sintering. J. Mat. Sci. 2009, 44, 1–39.

[40] Zhang N, Babayan Khosrovabadi P, Lindenhovius JH, Kolster BH. TiNi shape memory alloys prepared by normal sintering. Mater. Sci. Eng., A 1992, 150, 263–70.

[41] Jackson M, Dye D, Inman D, Dashwood R. Production of NiTi via the FFC Cambridge process. J. Electrochem. Soc. 2008, 155, E171–7.

[42] Jackson BK, Jackson M, Dye D, Inman D, Dashwood R. Optimization of the FFC Cambridge process for NiTi production. ECS Trans. 2009, 16, 211–9.

[43] Jackson BK, Dye D, Inman D, Bhagat R, Talling RJ, Raghunathan SL, Jackson M, Dashwood RJ. Characterization of the FFC Cambridge process for NiTi production using in situ x-ray synchrotron diffraction. J. Electrochem. Soc. 2010, 157, E57–63.

[44] Li H, Yuan B, Gao Y, Chung CY, Zhu M. High-porosity NiTi superelastic alloys fabricated by low-pressure sintering using titanium hydride as pore-forming agent. J. Mater. Sci. 2009, 44, 875–81.

[45] Zhang N, Khosrovabadi PB, Lindenhovius JH, Kolster BH. TiNi shape memory alloys prepared by normal sintering, Mater. Sci. Eng., A 1992, 150, 263–70.

[46] Chen G, Cao P, Edmonds N. Porous NiTi alloys produced by press-and-sinter from Ni/Ti and Ni/TiH$_2$ mixtures. Mater. Sci. Eng., A 2013, 582, 117–25.

[47] Bram M, Ahmad-Khanlou A, Heckmann A, Fuchs B, Buchkremer HP, Stöver D. Powder metallurgical fabrication processes for NiTi shape memory alloy parts. Mater. Sci. Eng., A 2002, 337, 254–63.

[48] Bertheville B, Neudenberger M, Bidaux JE. Powder sintering and shape-memory behaviour of NiTi compacts synthesized from Ni and TiH$_2$. Mater. Sci. Eng., A 2004, 384, 143–50.

[49] Schuller E, Bram M, Buchkremer HP, Stover D. Phase transformation temperatures for NiTi alloys prepared by powder metallurgical processes, Mater. Sci. Eng., A Struct. Mater. Prop. Microstruct. Process. 2004, 378, 165–9.

[50] Yuan B, Zhang XP, Chung CY, Zhu M. The effect of porosity on phase transformation behavior of porous Ti–50.8at%Ni shape memory alloys prepared by capsule-free hot isostatic pressing. Mater. Sci. Eng., A 2006, 438, 585–8.

[51] Wu SL, Liu XM, Chu PK, Chung CY, Chu CL, Yeung KWK. Phase transformation behavior of porous NiTi alloys fabricated by capsule-free hot isostatic pressing. J. Alloys Compd. 2008, 449, 139–43.

[52] Wu S, Liu X, Yeung KWK, Hub T, Xu Z, Chung JCY, Chu PK. Hydrogen release from titanium hydride in foaming of orthopedic NiTi scaffold. Acta Biomater. 2011, 7, 1387–97.
[53] Ye LL, Liu ZG, Raviprasad K, Quan MX, Umemoto M, Hu ZQ. Consolidation of MA amorphous NiTi powders by spark plasma sintering. Mater. Sci. Eng., A 1998, 241, 290–3.
[54] Shearwood C, Fu YQ, Yu L, Khor KA. Spark plasma sintering of TiNi nano-powder. Scr. Mater. 2005, 52, 455–60.
[55] Kim YW, Lee YJ, Nam TH. Shape memory characteristics of Ti-Ni-Mo alloys sintered by sparks plasma sintering. J. Alloys Compd. 2013, 577, S205–9.
[56] Cristea CD, Lungu M, Balagurov AM, Marinescu V, Culicov O, Sbarcea G, Cirstea V. Shape memory NiTi and NiTiCu alloys obtained by spark plasma sintering process. Adv. Eng. Forum 2015, 13, 83–90.
[57] Zhang L, Zhang YQ, Jiang YH, Zhou R. Superelastic behaviors of biomedical porous NiTi alloy with high porosity and large pore size prepared by spark plasma sintering. J. Alloys Compd. 2015, 644, 513–22.
[58] Zhang L, He ZY, Zhang YQ, Jiang YH, Zhou R. Enhanced in vitro bioactivity of porous NiTi–HA composites with interconnected pore characteristics prepared by spark plasma sintering. Mater. Design 2016, 101, 170–80.
[59] Velmurugan C, Senthilkumar V. Optimization of spark plasma sintering parameters for NiTiCu shape memory alloys. Mater. Manuf. Processes 2019, 34, 369–78.
[60] Schöller E, Krone L, Bram M, Buchkremer HP, Stäaver D. Metal injection molding of shape memory alloys using prealloyed NiTi powders. J. Mater. Sci. 2005, 40, 4231–8.
[61] Siargos B, Bradley TG, Darabara M, Papadimitriou G, Zinelis S. Galvanic corrosion of metal injection molded (MIM) and conventional brackets with nickel-titanium and copper-nickel-titanium archwires. Angle Orthod. 2007, 77, 355–60.
[62] Alavi S, Kachuie M. Assessment of the hardness of different orthodontic wires and brackets produced by metal injection molding and conventional methods. Dent. Res. J. (Isfahan) 2017, 14, 282–7.
[63] Deguchi T, Ito M, Obata A, Koh Y, Yamagishi T, Oshida Y. Trail production of titanium orthodontic brackets fabricated by sintering metal injection molding methods. J. Dent. Res. 1996, 75, 1491–6.
[64] Xu JL, Bao LZ, Liu AH, Jin XJ, Tong YX, Luo JM, Zhong ZC, Zheng YF. Microstructure, mechanical properties and superelasticity of biomedical porous NiTi alloy prepared by microwave sintering. Mater. Sci. Eng. C 2015, 46, 387–93.
[65] Abidi IH, Khalid FA, Farooq MU, Hussain MA, Maqbool A. Tailoring the pore morphology of porous nitinol with suitable mechanical properties for biomedical applications. Mater. Lett. 2015, 154, 17–20.
[66] Li BY, Rong LJ, Li YY. Porous NiTi alloy prepared from elemental powder sintering. J. Mater. Res. 1998, 13, 2847–51.
[67] Bansiddhi A, Dunand DC. Niobium wires as space holder and sintering aid for porous NiTi. Adv. Eng. Mater. 2011, 13, 301–5.
[68] Li B-Y, Rong L-J, Li Y-Y. The influence of addition of TiH$_2$ in elemental powder sintering porous Ni–Ti alloys. Mater. Sci. Eng., A 2000, 281, 169–75.
[69] Chen G, Cao P. NiTi powder sintering from TiH$_2$ powder: an *in situ* investigation. Metall. Mater. Trans., A 2013, 44, 5630–3.
[70] Chen G, Liss K-D, Cao P. In situ observation and neutron diffraction of NiTi powder sintering. Acta Mater. 2014, 67, 32–44.
[71] Terayama A, Kyogoku H, Sakamura M, Komatsu S. Fabrication of TiNi powder by mechanical alloying and shape memory characteristics of the sintered alloy. Mater. Trans. 2006, 47, 550–7.

[72] Sadrnezhaad SK, Arami H, Keivan H, Khalifehzadeh R. Powder metallurgical fabrication and characterization of nanostructured porous NiTi shape-memory alloy. Mater. Manuf. Processes 2006, 21, 727–35.

[73] Thadhani NN, Vreeland Jr T, Ahrens TJ. Microstructural modifications in a dynamically consolidated microcrystalline nickel titanium alloy powder. J. Mater. Sci. 1987, 22, 4446–52.

[74] Matsumoto H, Kondo K, Dohi S, Sawaoka A. Shock compaction of NiTi alloy powder. J. Mater. Sci. 1987, 22, 581–6.

[75] Hosseini SA, Yazdani-Rad R, Kazemzadeh A, Alizadeh M. Influence of thermal hydrogen treatment of titanium particles on powder metallurgical processing of NiTi-SMA. Mater. Manuf. Processes 2013, 28, 1179–83.

[76] Whitney M, Corbin SF, Gorbet RB. Investigation of the influence of Ni powder size on microstructural evolution and the thermal explosion combustion synthesis of NiTi. Intermetallics 2009, 17, 894–906.

[77] Hryha E, Shvab R, Bram M, Bitzer M, Nyborg L. Surface chemical state of Ti powders and its alloys: effect of storage conditions and alloy composition. Appl. Surf. Sci. 2016, 388, 294–303.

[78] Chu PK. Bioactivity of plasma implanted biomaterials. Nucl. Instrum. Meth. B 2006, 24, 1–7.

[79] Yeh CL, Sung WY. Synthesis of NiTi intermetallics by self-propagating combustion. J. Alloy Compd. 2004, 376, 79–88.

[80] Chu CL, Chung CY, Lin PH, Wang SD. Fabrication of porous NiTi shape memory alloy for hard tissue implants by combustion synthesis. Mat. Sci. Eng., A-Struct. 2004, 366, 114–9.

[81] Biswas A. Porous NiTi by thermal explosion mode of SHS: processing, mechanism and generation of single phase microstructure. Acta Mater. 2005, 53, 1415–25.

[82] Shearwood C, Fu YQ, Yu L, Khor KA. Spark plasma sintering of TiNi nano-powder. Scr. Mater. 2005, 52, 455–60.

[83] Morsi K, Moussa S, Wall JJ. Simultaneous combustion synthesis (thermal explosion mode) and extrusion of nickel aluminides. J. Mater. Sci. 2005, 40, 1027–30.

[84] Mukasyan AS, Rogachev AS. Thermal explosion mode of combustion synthesis. Con. Encyclop. Self-Prop. High-Temp. Synth. – Hist. Theor. Technol. Prod. 2017, 379–81.

[85] Oshida Y, Tuna EB. Science and Technology Integrated Titanium Dental Implant Systems. In: Basu, et al. ed., Advance Biomaterials. Wiley, 2009, 143–77.

[86] Oshida Y, Hashem A, Nishihara T, Yapchulay MV. Fractal dimension analysis of mandibular bones – toward a morphological compatibility of implants. J. Bio-Med. Mater. Eng. 1993, 4, 397–407.

[87] Elias CN, Oshida Y, Lima JHC, Muller CA. Relationship between surface properties (roughness, wettability and morphology) of titanium and dental implant torque. J. Mech. Behav. Biomed. Mater. 2008, 1, 234–42.

[88] Čapek J, Vojtech D, Novak P. Preparation of the NiTi alloy by a powder metallurgy technique. 2012, 6 pages; http://metal2013.tanger.cz/files/proceedings/02/reports/115.pdf.

[89] Tosun G, Tosun N. Analysis of process parameters for porosity in porous NiTi implants. Mater. Manuf. Process. 2012, 27, 1184–8.

[90] Li BY, Rong LJ, Li YY, Gjunter VE. Synthesis of porous Ni-Ti shape-memory alloys by self-propagating high-temperature synthesis: reaction mechanism and anisotropy in pore structure. Acta Mater. 2000, 48, 3895–904.

[91] Li Y-H, Rong L. Porous NiTi alloy prepared from combustion synthesis. Key Eng. Mater. 2002, 217, 137–42.

[92] Gu YW, Li H, Tay BY, Lim CS, Yong MS, Khor KA. In vitro bioactivity and osteoblast response of porous NiTi synthesized by SHS using nanocrystalline Ni-Ti reaction agent. J. Biomed. Mater. Res. A 2006, 78, 316–23.

[93]   Tay BY, Goh CW, Gu YW, Lim CS, Yong MS, Ho MK, Myint MH. Porous NiTi fabricated by
        self-propagating high-temperature synthesis of elemental powders. J. Mater. Process.
        Technol. 2008, 202, 359–64.

[94]   Whitney M, Corbin SF, Gorbet RB. Investigation of the mechanisms of reactive sintering and
        combustion synthesis of NiTi using differential scanning calorimetry and microstructural
        analysis. Acta Mater. 2008, 56, 559–70.

[95]   Tosun G, Özler L, Kaya M, Orhan N. A study on microstructure and porosity of NiTi alloy
        implants produced by SHS. J. Alloys Compd. 2009, 487, 605–11.

[96]   Tosun G, Ünsaldi E, Özler L, Orhan N, Durmuş AS, Eröksüz H. Biocompatibility of NiTi alloy
        implants in vivo. Int. J. Biomed. Biol. Eng. 2013, 7, 271–4.

[97]   Resnina N, Belayev S, Voronkov A. Influence of chemical composition and pre-heating
        temperature on the structure and martensitic transformation in porous TiNi-based shape
        memory alloys, produced by self-propagating high-temperature synthesis. Interme-
        tallics 2013, 32, 81–9.

[98]   Bassani P, Bassani E, Tuissi A, Giuliani P, Zanotti C. Nonequiatomic NiTi alloy produced by self
        propagating high temperature synthesis. J. Mater. Eng. Perform., 2014, 23, 2373–8.

[99]   Bassani P, Panseri S, Ruffini A, Montesi M, Ghetti M, Zanotti C, Tampieri A, Tuissi A. Porous
        NiTi shape memory alloys produced by SHS: microstructure and biocompatibility in
        comparison with Ti$_2$Ni and TiNi$_3$. J. Mater. Sci.: Mater. Med. 2014, 25, 2277–85.

[100]   Novák P, Mejzlíková L, Michalcová A, Čapek J, Beran P, Vojtěch D. Effect of SHS conditions on
        microstructure of NiTi shape memory alloy. Intermetallics 2013, 42, 85–91.

[101]   Novák P, Veselý T, Marek I, Dvořák P, Vojtěch V, Salvetr P, Karlík M, Haušild P, Kopeček J. Effect
        of particle size of titanium and nickel on the synthesis of NiTi by TE-SHS. Metall. Mater. Trans.
        B 2016, 47, 932–8.

[102]   Školáková A, Novák P, Salvetr P, Moravec H, Šefl V, Deduytsche D, Detavernier C. Investigation
        of the effect of magnesium on the microstructure and mechanical properties of NiTi shape
        memory alloy prepared by self-propagating high-temperature synthesis. Metall. Mater.
        Trans., A 2017, 48, 3559–69.

[103]   Biffi CA, Bassani P, Sajedi Z, Giuliani P, Tuissi A. Laser ignition in self-propagating high
        temperature synthesis of porous Nitinol shape memory alloy. Mater. Lett. 2017, 193, 54–7.

[104]   Aihara H, Zider J, Fanton G, Duerig T. Combustion synthesis porous nitinol for biomedical
        applications. Int. J. Biomat. 2019; https://doi.org/10.1155/2019/4307461.

[105]   Suryanarayana C. Mechanical alloying and milling. Prog. Mater. Sci. 2001, 46, 1–184.

[106]   Suryanarayana C, Ivanov E, Boldyrev VV. The science and technology of mechanical alloying.
        Mater. Sci. Eng. 2001, A304/A306, 151–8.

[107]   Benjamin JS, Volin TE. The mechanism of mechanical alloying, Metall. Trans. 1974, 5,
        1929–34.

[108]   Takasaki A. Mechanical alloying of the Ti–Ni system. Phys. Status Solidi A 1998, 169, 183–9.

[109]   Saito T, Takasaki A. The influence of chemical composition on shape memory effect of TiNi
        bulk alloy produced by mechanical alloying. Trans. Mater. Res. Soc. Jpn. 2009, 34, 403–6.

[110]   Sadrnezhaad SK, Selahi AR. Effect of mechanical alloying and sintering on Ni–Ti powders.
        Mater. Manuf. Processes 2004, 19, 475–86.

[111]   Mousavi T, Karimzadeh F, Abbasi MH. Synthesis and characterization of nanocrystalline NiTi
        intermetallic by mechanical alloying. Mater. Sci. Eng., A 2008, 487, 46–51.

[112]   Verdian MM. Fabrication of supersaturated NiTi (Al) alloys by mechanical alloying. Mater.
        Manuf. Processes 2010, 25, 1437–9.

[113]   Ghadimi M, Shokuhfar A, Rostami HR, Ghaffari M. Effects of milling and annealing on
        formation and structural characterization of nanocrystalline intermetallic compounds from
        Ni–Ti elemental powders, Mater. Lett. 2012, 80, 181–3.

[114] Alijani F, Amini R, Ghaffari M, Alizadeh M, Okyay AK. Effect of milling time on the structure, micro-hardness, and thermal behavior of amorphous/nanocrystalline TiNiCu shape memory alloys developed by mechanical alloying. Mater. Des. 2014, 55, 373–80.

[115] Khademzadeh S, Parvin N, Bariani PF. Production of NiTi alloy by direct metal deposition of mechanically alloyed powder mixtures. Int. J. Precis. Eng. Manuf. 2015, 16, 2333–8.

[116] Akmal M, Raza A, Khan MM, Khan MI, Hussain MA. Effect of nano-hydroxyapatite reinforcement in mechanically alloyed NiTi composites for biomedical implant. Mater. Sci. Eng. C Mater. Biol. Appl. 2016, 68, 30–6.

[117] Arunkumar S, Kumaravel P, Velmurugan C, Senthilkumar V. Microstructures and mechanical properties of nanocrystalline NiTi intermetallics formed by mechanosynthesis. Int. J. Min. Metall. Mater. 2018, 25, 80–7.

[118] Zhang H, Wang Z, Yang H, Shan X, Liu X, Yu S, He Z. Wear and corrosion properties of Mo surface-modified layer in TiNi alloy prepared by plasma surface alloying. J. Wuhan Univ. Technol.-Mater. Sci. Ed. 2016, 31, 910–7.

[119] Lavernia EJ, Srivatsan TS. The rapid solidification processing of materials: science, principles, technology, advances, and applications. J. Mater. Sci. 2010, 45, 287–325.

[120] Jiang H, Cao S, Ke C, Ma X, Zhang X. Fine-grained bulk NiTi shape memory alloy fabricated by rapid solidification process and its mechanical properties and damping performance. J. Mater. Sci. Technol. 2013, 29, 855–62.

[121] Zadravec M, Ternik P, Rudolf R, Svetec M. Numerical analysis of rapid solidification of NiTi alloy: influence of boundary conditions. Anali Pazu 2014, 4, 82–8.

[122] ASTM Committee F42 on Additive Manufacturing Technologies; http://www.astm.org/ COMMITTEE/F42.htm.

[123] Peduk GSA, Dilibal S, Harrysson O, Özbek S. Comparison of the production processes of nickel-titanium shape memory alloy through additive manufacturing. Int. Symp. 3D Print. (Addit. Manuf.) 2017, 3/4; https://nickel-titanium.com/wp-content/uploads/ NiTi_AM_Paper_Symposion-on-3D_AM_2017.pdf.

[124] Zhang L-C, Chen L-Y. A review on biomedical titanium alloys: recent progress and prospect. Adv. Eng. Mater. 2019, 21; https://doi.org/10.1002/adem.201801215.

[125] Yadroitsev I, Thivillon L, Bertrand P, Smurov I. Strategy of manufacturing components with designed internal structure by selective laser melting of metallic powder. Appl. Surf. Sci. 2007, 254, 980–3.

[126] Krishna BV, Bose S, Bandyopadhyay A. Laser processing of net-shape NiTi shape memory alloy. Metal. Mater. Trans. 2007, A38, 1096–130.

[127] Clare AT, Paul RC, Davies S, Sutcliffe JC, Tsopanos S. Selective laser melting of high aspect ratio 3D nickel–titanium structures two way trained for MEMS applications. Int. J. Mech. Mater. Des. 2008, 4, 181–7.

[128] Bormann T, Schumacher R, Müller B, Mertmann M, de Wild M. Tailoring selective laser melting process parameters for NiTi implants. J. Mater. Eng. Perform. 2012, 21, 2519–24.

[129] Saedi S, Turabi AS, Andani MT, Haberland C, Elahinia M, Karaca H. Thermomechanical characterization of Ni-rich NiTi fabricated by selective laser melting. Smart. Mater. Struct. 2016, 25, Article 035005; https://iopscience.iop.org/article/10.1088/0964-1726/25/3/035005/pdf.

[130] Saedi S, Turabi AS, Andani MT, Haberland C, Karaca H, Elahinia M. The influence of heat treatment on the thermomechanical response of Ni-rich NiTi alloys manufactured by selective laser melting. J. Alloys Compd. 2016, 677, 204–10.

[131] Saedi S, Turabi AS, Andani MT, Moghaddam NS, Elahinia M, Karaca HE. Texture, aging, and superelasticity of selective laser melting fabricated Ni-rich NiTi alloys. Mater. Sci. Eng., A, 2017, 686, 1–10.

[132]   Taheri Andani M, Saedi S, Turabi AS, Karamooz MR, Haberland C, Karaca HE, Elahinia M. Mechanical and shape memory properties of porous $Ni_{50.1}Ti_{49.9}$ alloys manufactured by selective laser melting. J. Mech. Behav. Biomed. Mater. 2017, 68, 224–31.

[133]   Saedi S, Moghaddam NS, Amerinatanzi A, Elahinia M, Karaca HE. On the effects of selective laser melting process parameters on microstructure and thermomechanical response of Ni-rich NiTi. Acta Mater. 2018, 144, 552–60.

[134]   Saedi S, Saghaian SE, Jahadakbar A, Shayesteh Moghaddam N, Taheri Andani M, Saghaian SM, Lu YC, Elahinia M, Karaca HE. Shape memory response of porous NiTi shape memory alloys fabricated by selective laser melting. J. Mater. Sci. Mater. Med. 2018, 29; doi: 10.1007/s10856-018-6044-6.

[135]   Engh CA, McGovern TF, Bobyn JO, Harris WH. A quantitative evaluation of periprosthetic bone remodeling after cementless total hip arthroplasty. J. Bone Joint Surg. 1992, 74-A, 1009–20.

[136]   Mullen L, Stamp RC, Brooks WK, Jones E, Sutcliffe CJ. Selective laser melting: a regular unit cell approach for the manufacture of porous, titanium, bone in-growth constructs, suitable for orthopedic applications. J. Biomed. Mater. Res. Part B Appl. Biomater. 2009, 89B, 325–34.

[137]   Shishkovskii IV, Yadroitsev IA, Smurov IY. Powder metal. Metal Ceram. 2011, 50, 275; https://doi.org/10/1007/s11106-011-9329-6.

[138]   Shishkovsky IV, Sherbakoff V, Yadroitsev I, Smurov I. Peculiar features of electrical resistivity and phase structure in 3-D porous nitinol after selective laser sintering/melting process. Proc. Inst. Mech. Eng. Part C: J. Mech. Eng. Sci. 2012, 226; https://doi.org/10.1177/0954406212440766.

[139]   Shishkovsky I, Yadroitsev I, Smurov I. Direct selective laser melting of nitinol powder. Phys. Procedia. 2012, 39, 447–54.

[140]   Moghaddam NS, Saghaian SE, Amerinatanzi A, Ibrahim H, Li P, Toker GP, Karaca HE, Elahinia M, Moghaddam NS. Anisotropic tensile and actuation properties of NiTi fabricated with selective laser melting. Mater. Sci. Eng., A 2018, 724, 220–30.

[141]   Chekotu JC, Groarke R, O'Toole K, Brabazon D. Advances in selective laser melting of nitinol shape memory alloy part production. Materials (Basel) 2019, 12, pii: E809; doi: 10.3390/ma12050809.

[142]   Zhou Q, Hayat MD, Chen G, Cai S, Qu X, Tang H, Cao P. Selective electron beam melting of NiTi: microstructure, phase transformation and mechanical properties. Mater. Sci. Eng. 2018, A 744; doi: 10.1016/j.msea.2018.12.023.[143] Haberland C, Elahinia M, Walker J, Meier H, Frenzel J, On the development of high quality NiTi shape memory and pseudoelastic parts by additive manufacturing. Smart Mater. Struct. 2014, 23; doi:10.1088/0964-1726/10/104002.

[144]   Dadbakhsh S, Speirs M, Van Humbeeck J, Kruth JP. Laser additive manufacturing of bulk and porous shape-memory NiTi alloys: from processes to potential biomedical applications. MRS Bull. 2016, 41, 765–74.

[145]   Guo Y, Klink A, Fu C, Snyder J. Machinability and surface integrity of nitinol shape memory alloy. CIRP Ann. – Manuf. Technol. 2013, 62, 83–6.

[146]   Moghaddam NS, Skoracki R, Miller M, Elahinia M, Dean D. Three dimensional printing of stiffness-tuned, nitinol skeletal fixation hardware with an example of mandibular segmental defect repair. Proc. CIRP 2016, 49, 45–50.

[147]   Andani MT, Moghaddam NS, Haberland C, Dean D, Miller M, Elahinia M. Metals for bone implants. Part 1. Powder metallurgy and implant rendering. Acta Biomater. 2014, 10, 4058–70.

[148]   Elahinia M, Moghaddam NS, Andani MT, Amerinatanzi A, Bimber BA, Hamilton RF. Fabrication of NiTi through additive manufacturing: a review. Prog. Mater. Sci. 2016, 83, 630–63.

[149]   Gibson I, Rosen D, Stucker B. Additive manufacturing technologies, 3D printing, rapid prototyping, and direct digital manufacturing. ISBN 978-1-4939-2113-3, Springer, 2015.

[150] Matsumoto H. Addition of an element to NiTi alloy by an electron-beam melting method. J. Mater. Sci. Lett. 1991, 10, 417–9.

[151] De Formanoir C, Michotte S, Rigo O, Germain L, Godet S. Electron beam melted Ti–6Al–4V: microstructure, texture and mechanical behavior of the as-built and heat-treated material. Mater. Sci. Eng. A 2016, 652, 105–19.

[152] Zhao MS, Li SJ, Hou WT, Hao YL, Yang R, Misra RDK. The influence of cell morphology on the compressive fatigue behavior of Ti-6Al-4V meshes fabricated by electron beam melting. J. Mech. Behav. Biomed. Mater. 2016, 59, 251–64.

[153] Hayat MD, Chen G, Liu N, Khan S, Tang HP, Cao P. Physical and tensile properties of NiTi alloy by selective electron beam melting. Key Eng. Mater. 2018, 770, 148–54.

[154] Bormann T, Müller B, Schinhammer M, Kessler A, Thalmann P, Wild M. Microstructure of selective laser melted nickel–titanium. Mater. Charact., 2014, 94, 189–202.

[155] Shishkovsky IV, Volova LT, Kuznetsov M, Morozov YG, Parkin IP. Porous biocompatible implants and tissue scaffolds synthesized by selective laser sintering from Ti and NiTi. J. Mater. Chem. 2008, 18, 1309–17.

[156] Otubo J, Rigo OD, Neto CM, Mei PR. The effects of vacuum induction melting and electron beam melting techniques on the purity of NiTi shape memory alloys. Mater. Sci. Eng. 2006, 438/440, 679–82.

[157] Altug-Peduk GS, Dilibal S, Harrysson O, Ozbek S, West H. Characterization of Ni–Ti alloy powders for use in additive manufacturing. Russ. J. Non-Ferr. Met. 2018, 59, 433–9.

[158] Rossi S, Deflorian F, Pegoretti A, D'Orazio D, Gialanella S. Chemical and mechanical treatments to improve the surface properties of shape memory alloy wires. Surf. Coat. Technol. 2008, 202, 2214–22.

[159] Merlin M, Scoponi M, Soffritti C, Fortini A, Rizzoni R, Garagnani GL. On the improved adhesion of NiTi wires embedded in polyester and vinylester resins. Frattura ed Integrità Strutturale 2015, 31, 127–37.

[160] Jonnalagadda K, Kline GE, Sottos NR. Local displacements and load transfer in SMA composites. Exp. Mech. 1997, 37, 78–86.

[161] Hwang TW, Woo YY, VanTyne CJ, Moon YH. Feasibility studies of laser surface nitriding on Ti-6Al-4V alloy using a nitric acid solution. J. Mech. Sci. Technol. 2017, 31, 4175–82.

[162] Man HC, Zhao NQ, Cui ZD. Surface morphology of a laser surface nitrided and etched Ti-6Al-4V alloy. Surf. Coat. Technol. 2005, 192, 341–6.

[163] Man HC, Zhao NQ. Enhancing the adhesive bonding strength of NiTi shape memory alloys by laser gas nitriding and selective etching. Appl. Surf. Sci. 2006, 253, 1595–600.

[164] Sadrnezhaad SK, Nermati NH, Bagheri R. Improved adhesion of NiTi wire to silicone matrix for smart composite medical applications. Mater. Des. 2009, 30, 3667–72.

[165] Shinoda T, Tsuchiya T, Takahashi H. Functional characteristics of friction welded near-equiatomic TiNi shape memory alloy. Trans. Jap. Weld. Soc. 1991, 22, 30–6.

[166] Fukumoto S, Inoue T, Mizuno S, Okita K, Tomita T, Yamamoto A. Friction welding of TiNi alloy to stainless steel using Ni interlayer. Sci. Technol. Weld. Join. 2010, 15, 124–30.

[167] Barcellona A, Fratini L, Palmeri D, Maletta C, Brandizzi M. Friction stir processing of Niti shape memory alloy: microstructural characterization. Int. J. Mater. Forming. 2010, 3, 1047–50.

[168] Mani Prabu SS, Madhu HC, Perugu CS, Akash K, Kumar PA, Kailas SV, Anbarasu M, Palani IA. Microstructure, mechanical properties and shape memory behaviour of friction stir welded nitinol. Mater. Sci. Eng. A. 2017, 693, 233–6.

[169] Tam B. Micro-welding of nitinol shape memory alloy. Thesis, University of Waterloo, 2010; https://pdfs.semanticscholar.org/cf4e/dd53c9af5fea8643945f4d92bf5ccd274ba1.pdf.

[170] Nishikawa N, Tanaka H, Kohda M, Nagaura T, Watanabe K. Behaviour of welded part of Ti-Ni shape memory alloys. J. de Phys. 1982, 43, C4, C4/839–44.

[171] Li Q, Zhu Y. Impact Butt welding of NiTi and stainless steel – an examination of impact speed effect. J. Mater. Process. Technol. 2018, 255; doi: 10.1016/j.jmatprotec. 2017.12.046.

[172] Zhang W, Ao S, Oliveira JP, Zeng Z, Huang Y, Luo Z. Microstructural characterization and mechanical behavior of NiTi shape memory alloys ultrasonic joints using Cu interlayer. Materials (Basel) 2018, 11, 1830; doi: 10.3390/ma11101830.

[173] Yang D, Jiang HC, Zhao MJ, Rong LJ. Microstructure and mechanical behaviors of electron beam welded NiTi shape memory alloys. Mater. Des. 2014, 57, 21–5.

[174] Tam B, Khan MI, Zhou Y. Mechanical and functional properties of laser-welded Ti-55.8 wt pct Ni nitinol wires. Metall. Mater. Trans. A. 2011, 42, 2166–75.

[175] Shojaei Zoeram A, Akbari Mousavi SAA. Effect of interlayer thickness on microstructure and mechanical properties of as welded Ti6Al4V/Cu/NiTi joints. Mater. Lett. 2014, 133, 5–8.

[176] Li HM, Sun DQ, Gu XY, Dong P, Lv ZP. Effects of the thickness of Cu filler metal on the microstructure and properties of laser-welded TiNi alloy and stainless steel joint. Mater. Des. 2013, 50, 342–50.

[177] Chan CW, Man HC, Yue TM. Effects of process parameters upon the shape memory and pseudo-elastic behaviors of laser-welded NiTi thin foil. Metall. Mater. Trans. A 2011, 42, 2264–70.

[178] Crăciunescu C, Ercuta A. Modulated interaction in double-layer shape memory-based micro-designed actuators. Sci. Technol. Adv. Mater. 2015, 16, 065003. doi: 10.1088/1468-6996/16/6/065003.

[179] Frick CP, Ortega AM, Tyber J, Maksound AEM, Maier HJ, Liu Y, Gall K. Thermal processing of polycrystalline NiTi shape memory alloys. Mater. Sci. Eng., A 2005, 405, 34–49.

[180] Zimmerly CA, Inal OT, Richman RH. Explosive welding of a near-equiatomic nickel-titanium alloy to low-carbon steel. Mater. Sci. Eng., A 1994, 188, 251–4.

[181] Belyaev S, Rubanik V, Resnina N, Rubanik Jr V, Rubanik O, Borisov V. Martensitic transformation and physical properties of 'steel – TiNi' bimetal composite, produced by explosion welding. Phase Transitions 2010, 83, 276–83.

[182] Yan Z, Cui L-S, Zheng Y-J. Microstructure and martensitic transformation behaviors of explosively welded NiTi/NiTi laminates. Chin. J. Aeronaut. 2007, 20, 168–71.

[183] Sun W, Li X, Yan H, Liu K. Underwater explosive welding of NiTi alloy/copper foil. Trans. China Weld. Inst. 2012, 33, 63–6.

[184] Song J, Kostka A, Veehmayer M, Raabe D. Hierarchical microstructure of explosive joints: example of titanium to steel cladding. Mater. Sci. Eng., A 2011, 528, 2641–7.

[185] Kundu S, Sam S, Chatterjee S. Interfacial reactions and strength properties in dissimilar titanium alloy/Ni alloy/microduplex stainless steel diffusion bonded joints. Mater. Sci. Eng., A 2013, 560, 288–95.

[186] Sam S, Kundu S, Chatterjee S. Diffusion bonding of titanium alloy to micro-duplex stainless steel using a nickel alloy interlayer: interface microstructure and strength properties. Mater. Des. 2012, 40, 237–44.

[187] Liu J, Cao J, Lin X, Chen H, Wang J, Feng J. Interfacial microstructure and joining properties of TiAl/Ti$_3$AlC$_2$ diffusion bonded joints using Zr and Ni foils as interlayer. Vacuum 2014, 102, 16–25.

[188] Velmurugan C, Senthilkumar V, Sarala S, Arivarasan J. Low temperature diffusion bonding of Ti-6Al-4V and duplex stainless steel. J. Mater. Process. Technol. 2016, 234, 272–9.

[189] Ghosh M, Chatterjee S. Characterization of transition joints of commercially pure titanium to 304 stainless steel. Mater. Charact. 2002, 48, 393–9.

[190] Vigraman T, Ravindran D, Narayanasamy R. Effect of phase transformation and intermetallic compounds on the microstructure and tensile strength properties of diffusion-bonded joints between Ti-6Al-4V and AISI 304L. Mater. Des. 2012, 36, 714–27.

[191]   Rizvia SA, Khanb TI. A novel fabrication method for nitinol shape memory alloys. Key Eng.
        Mater. 2010, 442, 309–15.
[192]   Mo D-F, Song T-F, Fang Y-J, Jiang X-S, Luo CQ, Simpson MD, Luo Z-P. A review on diffusion
        bonding between titanium alloys and stainless steels. Adv. Mater. Sci. Eng. 2018; https://doi.
        org/10.1155/2018/8701890.
[193]   Kundu S, Sam S, Mishra B, Chatterjee S. Diffusion bonding of microduplex stainless steel and
        Ti alloy with and without interlayer: interface microstructure and strength properties. Metall.
        Mater. Trans., A 2014, 45; doi: 10.1007/s11661-013-1977-3.
[194]   Simoes S, Viana F, Ramos AS, Vieira MT, Vieira MF. Reaction-assisted diffusion bonding of TiAl
        alloy to steel. Mater. Chem. Phys. 2016, 171, 73–82.
[195]   Lutz J, Lindner JKN, Mändl S. Marker experiments to determine diffusing species and
        diffusion path in medical Nitinol alloys. Appl. Surf. Sci. 2008, 255, 1107–9.
[196]   Hodgson DE. Fabrication, Heat Treatment, and Joining of Nitinol Components. In: Russell, et
        al. ed., SMST – 2000: Shape Memory and Superelasticity Technologies, SMAT 2001, CA, USA,
        10–24.
[197]   Shiue RK, Wu SK. Infrared brazing of Ti50Ni50 shape memory alloy using gold-based braze
        alloys. Gold Bull. 2006, 39, 200–4.
[198]   Grummon DS, Shaw JA, Foltz J. Fabrication of cellular shape memory alloy materials by
        reactive eutectic brazing using niobium. Mater. Sci. Eng., A. 2006, 438/440, 1113–8.
[199]   Shiue RK, Wu SK, Shiue JY. Infrared brazing of Ti-6Al-4V and 17-4 PH stainless steel with
        (Ni)/Cr barrier layer(s). Mater. Sci. Eng., A 2008, 488, 186–94.
[200]   Shiue R-K, Wu S-K, Yang S-H, Liu C-K. Infrared dissimilar joining of $Ti_{50}Ni_{50}$ and 316L stainless
        steel with copper barrier layer in between two silver-based fillers. Metals 2017, 7, 276; doi:
        10.3390/met7070276.
[201]   Lin C, Shiue R-K, Wu S-K, Yang T-E. Infrared brazed joints of $Ti_{50}Ni_{50}$ shape memory alloy
        and Ti-15-3 alloy using two Ag-based fillers. Materials (Basel) 2019, 12, 1603; doi: 10.3390/
        ma12101603.
[202]   Aksdelsen OM. Joining of Shape Memory Alloys. In: Cismasiu C., ed., Shape Memory Alloys.
        InTech, Croatia, 2010, 183–210.
[203]   Shinoda T, Tsuchiya T, Takahashi H. Functional characteristics of friction welded
        near-equiatomic TiNi shape memory alloy. Trans. Jap. Weld. Soc. 1991, 22, 30–6.
[204]   Ikai A, Kimura K, Tobushi H. TIG welding and shape memory effect of TiNi shape memory
        alloy. J. Int. Mat. Syst. Struct. 1999, 7, 646–55.
[205]   Pfeifer R, Herzog D, Meier O, Ostendorf A, Haferkamp H, Goesling T, Hurschler C, Mueller C.
        Laser welding of shape memory alloys for medical applications. J. Laser Appl. 2018; https://
        doi.org/10.2351/1.5061388.
[206]   Oliveira JP, Barbosa D, Fernandes FMB, Miranda RM. Tungsten inert gas (TIG) welding of
        Ni-rich NiTi plates: functional behavior. Smart Mater. Struct. 2016, 25; https://iopscience.iop.
        org/article/10.1088/0964-1726/25/3/03LT01/pdf.
[207]   Hsu YT, Wang YR, Wu SK, Chen C. Effect of CO_2 laser welding on the shape-memory and
        corrosion characteristics of TiNi alloys. Metall. Mater. Trans. 2001, 32A, 569–76.
[208]   Song YG, Li WS, Li L, Zheng YF. The influence of laser welding parameters on the microstructure
        and mechanical property of the as-jointed NiTi alloy wires. Mater. Lett. 2008, 62, 2325–8.
[209]   Maletta C, Falvo A, Furgiuele F, Barbieri G, Brandizzi M. Fracture behaviour of nickel-titanium
        laser welded joints. J. Mater. Eng. Perform. 2009, 18, 569–74.
[210]   Kim J-D. Prediction of the penetration depth in laser beam welding. KSME J. 1990, 4, 32–9.
[211]   Bharti A. Laser welding. Bull. Mater. Sci. 1988, 11, 191–212.
[212]   Tuissi A, Besseghini S, Ranucci T, Squatrito F, Pozzi M. Effect of Nd-YAG laser welding on the
        functional properties of the Ni-49.6 at.%Ti. Mater. Sci. Eng., A 1999, 273/275, 813–7.

[213] Falvo A, Furgiuele FM, Maletta C. Laser welding of a NiTi alloy: mechanical and shape memory behaviour. Mater. Sci. Eng., A 2005, 412, 235–40.

[214] Zhao X, Wang W, Chen L, Liu F, Chen G, Huang J, Zhang H. Two-stage superelasticity of a Ce-added laser-welded TiNi alloy. Mater. Lett. 2008, 62, 3539–41.

[215] Oliveira JP, Fernandes FMB, Miranda RM. Welding and joining of NiTi shape memory alloys: a review. Prog. Mater. Sci. 2017, 88, 412–66.

[216] Oliveira JP, Zeng Z, Andrei C, Fernandes FMB, Miranda RM, Ramirez AZ, Omori T, Zhou N. Dissimilar laser welding of superelastic NiTi and CuAlMn shape memory alloys. Mater. Des. 2017, 128, 166–75.

[217] Oliveira JP, Schell N, Zhou N, Wood L, Benafan O. Laser welding of precipitation strengthened Ni-rich NiTiHf high temperature shape memory alloys: microstructure and mechanical properties. Mater. Des. 2019, 162, 229–34.

[218] Zeng Z, Panton B, Oliveira JP, Han A, Zhou YN. Dissimilar laser welding of NiTi shape memory alloy and copper. Smart Mater. Struct. 2015, 24; https://iopscience.iop.org/article/10.1088/0964-1726/24/12/125036/pdf.

[219] Mehrpouya M, Gisario A, Broggiato GB, Puopolo M, Vesco S, Barletta M. Effect of welding parameters on functionality of dissimilar laser-welded NiTi superelastic (SE) to shape memory effect (SME) wires. Int. J. Adv. Manuf. Technol. 2019, 103, 1593–601.

[220] Kang SB, Yoon KS, Kim JS. In vivo result of porous TiNi shape memory alloy: bone response and growth. Mater. Trans. 2002, 43(5): SI1045–8.

[221] Zhu S, Yang X, Hu F, Deng S, Cui Z. Processing of porous TiNi shape memory alloy from elemental powders by Ar-sintering. Mater. Lett. 2004, 58, 2369–73.

[222] Zhang X, Ayers RA, Thorne K, Moore JJ, Schowengerdt F. Combustion synthesis of porous materials for bone replacement. Biomed. Sci. Instrum. 2001, 37, 463–8.

[223] Aihara H, Zider J, Fanton G, Duerig T. Combustion synthesis porous nitinol for biomedical applications. Int. J. Biomater. 2019; https://doi.org/10.1155/2019/4307461.

[224] Li BY, Rong LJ, Li YY, Gjunter VE. Synthesis of porous Ni–Ti shape-memory alloys by self-propagating high-temperature synthesis: reaction mechanism and anisotropy in pore structure. Acta Mater. 2000, 48, 3895–904.

[225] Li B-Y, Rong L-J, Li Y-Y, Gjunter V. A recent development in producing porous Ni–Ti shape memory alloys. Intermetallics 2000, 8, 881–4.

[226] Chung CY, Chu CL, Wang SD. Porous TiNi shape memory alloy with high strength fabricated by self-propagating high-temperature synthesis. Mater Lett. 2004, 58, 1683–6.

[227] Jiang HC, Rong LJ. Ways to lower transformation temperatures of porous NiTi shape memory alloy fabricated by self-propagating high-temperature synthesis. Mater. Sci. Eng. A. 2006, 438/440, 883–6.

[228] Tosun G, Orhan N, Özler L. Investigation of combustion channel in fabrication of porous NiTi alloy implants by SHS. Mater. Lett. 2012, 66, 138–40.

[229] Saadati A, Aghajani H. Fabrication of porous NiTi biomedical alloy by SHS method. J. Mater. Sci.: Mater. Med. 2019, 30, 92; https://doi.org/10.1007/s10856-019-6296-9.

[230] Liang C, Yang Y, Wang HS, Yang JJ, Yang XJ. Preparation of porous microstructures on NiTi alloy surface with femtosecond laser pulses. Chin. Sci. Bull. 2008, 53, 700–5.

[231] Krishna BV, Bose S, Bandyopadhyay A. Fabrication of porous NiTi shape memory alloy structures using laser engineered net shaping. J. Biomed. Mater. Res. Part B: Appl. Biomater. 2009, 89, 481–90.

[232] Wen CE, Yamada Y, Nouri A, Hodgson PD. Porous titanium with porosity gradients for biomedical applications, materials science forum. Mater. Sci. Forum 2007, 539/543, 720–5.

[233] Zhang YP, Li DS, Zhang XP. Gradient porosity and large pore size NiTi shape memory alloys. Scr. Mater. 2007, 57, 102–3.

[234] Weeber AW, Bakker H. Amorphization by ball milling, a review. Phys. B 1988, 153, 93–135.

[235] Shen TD, Quan MX, Wang JT, Hu ZQ. Amorphous phase growth by isothermal annealing-induced interdiffusion reactions in mechanically deformed Ni/Ti multilayered composites. J. Mater. Sci. 1994, 29, 2981–6.

[236] Kanchibhotla S, Munroe N, Kartikeyan T. Amorphization in Ni-Ti-Ta system through mechanical alloying. J. Mater. Sci. 2005, 40, 5003–6.

[237] Alijani F, Amini R, Ghaffari M, Alizadeh M, Okyay AK. Effect of milling time on the structure, micro-hardness, and thermal behavior of amorphous/nanocrystalline TiNiCu shape memory alloys developed by mechanical alloying. Mater. Des. 2014, 55, 373–80.

[238] Shen TD, Quan MX, Wang JT. Solid state amorphization reactions in Ni/Ti multilayer composites prepared by cold rolling. J. Mater. Sci. 1993, 28, 394–8.

[239] Li J-T, Miao W-D, Hu Y-L, Zheng Y-J, Cui L-S. Amorphization and crystallization characteristics of TiNi shape memory alloys by severe plastic deformation. Front. Mater. Sci. China 2009, 3, 325–8.

[240] Huang JY, Zhu YT, Liao XZ, Valiev RZ. Amorphization of TiNi induced by high-pressure torsion. Philos. Mag. Lett. 2004, 84, 183–90.

[241] Resnina N, Belyaev S, Zeldovich V, Pilyugin V, Frolova N, Glazova D. Variations in martensitic transformation parameters due to grains evolution during post-deformation heating of Ti-50.2 at.% Ni alloy amorphized by HPT. Thermochim. Acta, 2016, 627/629, 20–30.

[242] Chellali MR, Nandam SH, Li S, Fawey MH, Moreno-Pineda E, Velasco L, Boll T, Pastewka L, Kruk R, Gumbsh P, Hahn H. Amorphous nickel nanophases inducing ferromagnetism in equiatomic Ni-Ti alloy. Acta Mater. 2018, 161, 47–53.

[243] Karch J, Birringer R, Gleiter H. Nature 1987, 330, 536–8.

[244] Gleiter H. Nanocrystalline materials. Prog. Mater. Sci. 1989, 33, 223–315.

[245] Lu K. Nanocrystalline metals crystallized from amorphous solids: nanocrystallization, structure, and properties. Mater. Sci. Eng. 1996, R16, 161–221.

[246] Qin Y, Chen L, Zhu Y, Zhang L. Synthesis and structure of nanocrystalline NiTi alloy. J. Mater. Sci. Lett. 1996, 15, 1155–7.

[247] Ju X, Su Y. Local structure of NiTi nanocrystals studied by EXAFS and XRD. J. –Synchrotron. Radiat. 2001, 8, 520–1.

[248] Kockar B, Karaman I, Kim JI, Chumlyakov YI, Sharp J, Yu C-J. Thermomechanical cyclic response of an ultrafine-grained NiTi shape memory alloy. Acta Mater. 2008, 56, 3630–46.

[249] Singh RZ, Divinski SV, Rösner H, Prokofiev EA, Valiev RZ, Wilde G. Microstructure evolution in nanocrystalline NiTi alloy produced by HPT. J. Alloys Compd. 2011, 509, S290–3.

[250] Jiang S, Hu L, Zhang Y, Liang Y. Nanocrystallization and amorphization of NiTi shape memory alloy under severe plastic deformation based on local canning compression. J. Non-Cryst. Sol., 2013, 367, 23–9.

[251] Park CH, Han SH, Kim SW, Hong JK, Nam T, Yeom JT. An effective approach to produce a nanocrystalline Ni–Ti shape memory alloy without severe plastic deformation. J. Alloy Compd., 2016, 654, 379–83.

[252] Tuominen SM, Biermann RJ. Shape-memory wires. JOM 1988, 40, 32–5.

[253] Lau K-T, Tam W-Y, Meng X-L, Zhou L-M. Morphological study on twisted NiTi wires for smart composite systems. Mat. Lett. 2002, 57, 364–8.

[254] Shabalovskaya Rondelli SG, Anderegg J, Simpson B, Budko S. Effect of chemical etching and aging in boiling water on the corrosion resistance of nitinol wires with black oxide resulting from manufacturing process. J. Biomed. Mater. Res. Part B: Appl. Biomat. 2003, 66B, 331–40.

[255] Gall K, Tyber J, Brice V, Frick CP, Maier HJ, Morgan N. Tensile deformation of NiTi wires. J. Biomed. Mater. Res. Part A 2005, 15, 810–23.

[256] Delville R, Malard B, Pilch J, Sittner P, Schryvers D. Microstructure changes during non-conventional heat treatment of thin Ni–Ti wires by pulsed electric current studied by transmission electron microscopy. Acta Mater. 2010, 58, 4503–15.

[257] Gupta S, Pelton AR, Weaver JD, Gong X-Y, Nagaraja S. High compressive pre-strains reduce the bending fatigue life of nitinol wire. J. Mech. Behav. Biomed. Mater. 2015, 44, 96–108.

[258] https://www.medicalexpo.com/prod/dentsply-maillefer/product-72098-464720.html.

[259] https://www.pearsondental.com/catalog/products_bymfg_thumb1.asp?mfg_id=70&majcatid=940.

[260] Pelton AR, Fino-Decker J, Vien L, Bonsignore C, Saffari P, Launey M, Mitchell MR. Rotary-bending fatigue characteristics of medical-grade Nitinol wire. J. Mech. Behav. Biomed. Mater. 2013, 27, 19–32.

[261] Walsch H. The hybrid concept of nickel-titanium rotary instrumentation. Dent. Clin. North Am. 2004, 48, 183–202.

[262] https://protectmec.en.made-in-china.com/product/QvpmkSgdYEYX/China-Tapered-Shape-Orthodontic-Archwires-Niti-Arches.html.

[263] https://www.medicalexpo.com/prod/ormco/product-100426-834643.html.

[264] Zhang C, Sun X, Zhao S, Yu W, Sun D. Susceptibility to corrosion and *in vitro* biocompatibility of a laser-welded composite orthodontic arch wire. Ann. Biomed. Eng. 2014, 42, 222–30.

[265] Zhang C, Sun X. Susceptibility to stress corrosion of laser-welded composite arch wire in acid artificial saliva. Adv. Mater. Sci. Eng. 2013; http://dx.doi.org/10.1155/2013/738954.

[266] Seyyed Aghamiri SM, Nili Ahmadabadi M, Shahmir H, Naghdi F, Raygan S. Study of thermo-mechanical treatment on mechanical-induced phase transformation of NiTi and TiNiCu wires. J. Mech. Behav. Biomed. Mater. 2013, 21, 32–6.

[267] Paúl A, Ábalos C, Mendoza A, Solano E, Gil FJ. Relationship between the surface defects and the manufacturing process of orthodontic Ni–Ti archwires. Mater. Lett. 2011, 65, 3358–61.

[268] Rucker BK, Kusy RP. Elastic flexural properties of multistranded stainless steel versus conventional nickel titanium archwires. Angle Orthod. 2002, 72, 302–9.

[269] Rucker BK, Kusy RP. Elastic properties of alternative versus single-stranded leveling archwires. Am. J. Orthod. Dentofacial Orthop. 2002, 122, 528–41.

[270] Seyyed Aghamiri SM, Ahmadabadi MN, Raygan Sh. Combined effects of different heat treatments and Cu element on transformation behavior of NiTi orthodontic wires. J. Mech. Behav. Biomed. Mater. 2011, 4, 298–3–02.

[271] Shima Y, Otsubo K, Yoneyama T, Soma K. Bending properties of hollow super-elastic Ti-Ni alloy wires and compound wires with other wires inserted. J. Mater. Sci. Mater. Med. 2002, 13, 169–73.

[272] Iijima M, Brantley WA, Yuasa T, Muguruma T, Kawashima I, Mizoguchi I. Joining characteristics of orthodontic wires with laser welding. J. Biomed. Mater. Res. B Appl. Biomater. 2008, 84, 147–53.

[273] Hammad SM, Al-Wakeel EE, Gad el-S. Mechanical properties and surface characterization of translucent composite wire following topical fluoride treatment. Angle Orthod. 2012, 82, 8–13.

[274] Spendlove J, Berzins DW, Pruszynski JE, Ballard RW. Investigation of force decay in aesthetic, fibre-reinforced composite orthodontic archwires. Eur. J. Orthod. 2015, 37, 43–8.

[275] Tian H, Schryvers D, Shabalovskaya S, Van Humbeeck J. Microstructure of surface and subsurface layers of a Ni-Ti shape memory microwire. Microsc. Microanal. 2009, 15, 62–70.

[276] Wu CD, Sung PH, Fang TH. Study of deformation and shape recovery of NiTi nanowires under torsion. J. Mol. Model 2013, 19, 1883–90.

[277] Bidaux JE, Manson JA, Gotthardt R. Active Stiffening of Composite Materials by Embedded Shape-Memory-Alloy Fibres. Materials for Smart Systems II. Materials Research Society, Pittsburgh, PA, 1996, 107–17.

[278] Budziak D, Martendal E, Carasek E. Application of NiTi alloy coated with ZrO2 as a new fiber for solid-phase microextraction for determination of halophenols in water samples. Anal. Chim. Acta 2007, 598, 254–60.

[279] Budziak D, Martendal E, Carasek E. Preparation and application of NiTi alloy coated with ZrO(2) as a new fiber for solid-phase microextraction. J. Chromatogr. A 2007, 1164, 18–24.

[280] Budziak D, Martendal E, Carasek E. Preparation and characterization of new solid-phase microextraction fibers obtained by sol-gel technology and zirconium oxide electrodeposited on NiTi alloy. J. Chromatogr. A 2008, 1187, 34–9.

[281] Wang H, Du J, Zhen Q, Zhang R, Wang X, Du X. Selective solid-phase microextraction of ultraviolet filters in environmental water with oriented ZnO nanosheets coated nickel-titanium alloy fibers followed by high performance liquid chromatography with UV detection. Talanta 2019, 191, 193–201.

[282] Du J, Wang H, Zhang R, Wang X, Du X, Lu X. Oriented ZnO nanoflakes on nickel-titanium alloy fibers for solid-phase microextraction of polychlorinated biphenyls and polycyclic aromatic hydrocarbons. Mikrochim. Acta 2018, 185; doi: 10.1007/s00604-018-2971-7.

[283] Hu J, Wu G, Zhang Q, Kang P, Liu Y. Microstructure of multilayer interface in an Al matrix composite reinforced by TiNi fiber. Micron 2014, 64, 57–65.

[284] Chaudhury Z, Hailat M, Liu Y, Newaz G. Aluminum-based composites reinforced with SiC particles and NiTi fibers: influence of fiber dimensions and aging time on mechanical properties. J. Mater. Sci. 2011, 46, 1945–55.

[285] Liu Y, Wang Z, Li H, Sun M, Wang F, Chen B. Influence of embedding SMA fibres and SMA fibre surface modification on the mechanical performance of BFRP composite laminates. Materials (Basel) 2018, 11(1), pii: E70; doi: 10.3390/ma11010070.

[286] https://www.dentsplysirona.com/nb-no/explore/orthodontics/wires.html.

[287] Spriano S, Balagna C, Ferri A, Dotti F, Villa E, Nespoli A. Processing and surface treatments for pseudoelastic wires and strands. Mater. Manuf. Processes 2017, 32, 394–403.

[288] Karjalainen T, Göransson H, Viinikainen A, Jämsä T, Ryhänen J. Nickel-titanium wire as a flexor tendon suture material: an ex vivo study. J. Hand Surg. Eur. 2010, 35, 469–74.

[289] Liu SS, Kyung HM, Buschang PH. Continuous forces are more effective than intermittent forces in expanding sutures. Eur. J. Orthod. 2010, 32, 371–80.

[290] dos Santos FA. Buckling control using shape-memory alloy cables. J. Eng. Mech. 2016, 142; doi: 10.1061/(ASCE)EM.1943-7889.0001038.

[291] Reedlunn B, Daly S, Shaw J. Superelastic shape memory alloy cables: part I – isothermal tension experiments. Int. J. Sol. Struct. 2013, 50, 3009–26.

[292] Reedlunn B, Daly S, Shaw J. Superelastic shape memory alloy cables: part II – subcomponent isothermal responses. Int. J. Sol. Struct. 2013, 50, 3027–44.

[293] https://novastep.life/product/nitinol-compression-clips/.

[294] http://bioproimplants.com/portfolio-view/memory-staple.

[295] Nahdis S, Youngjae C. An overview of thin film nitinol endovascular devices. Acta Biomater. 2015, 21, 20–34.

[296] Kim D, Lee H, Bae J, Jeong H, Choi B, Nam T, Noh J. Effect of substrate roughness on adhesion and structural properties of Ti-Ni shape memory alloy thin film. J. Nanosci. Nanotechnol. 2018, 18, 6201–5.

[297] Grummon DS. Thin-film shape-memory materials for high-temperature applications. JOM 2003, 55, 24–32.

[298] Busch JD, Johnson AD, Lee CH, Stevenson DA. Shape-memory properties in Ni-Ti sputter-deposited film. J. Appl. Phys. 1990, 68, 6224–8.

[299] Mohri M, Nili-Ahmadabadi M. Evaluation of mechanical properties of Ni-Ti bi-layer thin film. Suppl. Proceed.: Mater. Process. Interfaces 2012; doi: 10.1002/9781118356074.[300] Sanjabi S, Barber ZH. The effect of film composition on the structure and mechanical properties of NiTi shape memory thin films. Surf. Coat. Tech. 2010, 204, 1299–304.

[301] Kumar A, Sharma SK, Bysakh S, Kamat SV, Mohan S. Effect of substrate and annealing temperatures on mechanical properties of Ti-rich NiTi films. J. Mater. Sci. Technol. 2010, 26, 961–6.

[302] Behera A, Aich S. Characterisation and properties of magnetron sputtered nanoscale bi-layered Ni/Ti thin films and effect of annealing. Surf. Interface Anal. 2015, 47, 805–14.

[303] Behera A, Suman R, Aich S, Mohapatra SS. Sputter-deposited Ni/Ti double-bilayer thin film and the effect of intermetallics during annealing. Surf. Interface Anal. 2017, 49, 620–9.

[304] Zamponi C, Rumpf H, Schmutz C, Quandt E. Structuring of sputtered superelastic NiTi thin films by photolithography and etching. Mater. Sci. Eng., A 2008, 481, 623–5.

[305] Kauffmann-Weiss S, Hahan S, Weigelt C, Schultz L, Wagner MF-X, Fähler S. Growth, microstructure and thermal transformation behaviour of epitaxial Ni-Ti films. Acta Mater. 2017, 132, 255–63.

[306] Meng F, Li Y, Wang Y, Zheng W. Flow stress of Ni-rich NiTi thin films. J. Mater. Sci. 2005, 40, 537–8.

[307] Loger K, de Miranda RL, Engel A, Marczynski-Bühlow M, Lutter G, Quandt E. Fabrication and evaluation of Nitinol thin film heart valves. Cardiovas. Eng. Technol. 2014, 5, 308–16.

[308] Tillmann W, Momeni S. Tribological performance of near equiatomic and Ti-rich NiTi shape memory alloy thin films. Acta Mater. 2015, 92, 189–96.

[309] Botterill NW, Grant DM. Novel micro-thermal characterisation of thin film NiTi shape memory alloys. Mater. Sci. Eng., A 2004, 378, 424–8.

[310] Dalle F, Despert G, Vermaut Ph, Portier R, Dezellus A, Plaindoux P, Ochin P. $Ni_{49.8}Ti_{42.2}Hf_8$ shape memory alloy strips production by the twin roll casting technique. Mater. Sci. Eng., A 2003, 346, 320–7.

[311] Meng Q, Liu Y, Yang H, Shariat BS, Nam T-H. Functionally graded NiTi strips prepared by laser surface anneal. Acta Mater. 2012, 60, 1658–68.

[312] http://www.smalloys.com/nitinol/.

[313] Daly S, Ravichandrank G, Bhattacharya K. Stress-induced martensitic phase transformation in thin sheets of Nitinol. Acta Mater. 2007, 55, 3593–600.

[314] Liu Y. The superelastic anisotropy in a NiTi shape memory alloy thin sheet. Acta Mater. 2015, 95, 411–27.

[315] Hang R, Liu S, Liu Y, Zhao Y, Bai L, Jin M, Zhang X, Huang X, Yao X, Tang B. Preparation, characterization, corrosion behavior and cytocompatibility of $NiTiO_3$ nanosheets hydrothermally synthesized on biomedical NiTi alloy. Mater. Sci. Eng. C Mater. Biol. Appl. 2019, 97, 715–22.

[316] https://www.aliexpress.com/item/32810707980.html.

[317] Yamauchi K, Nogami S, Tanaka K, Yokota S, Shimizu Y, Kanetaka H, Takahashi T. The effect of decortication for periosteal expansion osteogenesis using shape memory alloy mesh device. Clin. Implant Dent. Relat. Res. 2015, 17(Suppl 2), e376–84.

[318] Yamauchi K, Nogami S, Martinez-de la Cruz G, Hirayama B, Shimizu Y, Kumamoto H, Lethaus B, Kessler P, Takahashi T. Timed-release system for periosteal expansion osteogenesis using NiTi mesh and absorbable material in the rabbit calvaria. J. Craniomaxillofac Surg. 2016, 44, 1366–72.

[319] Loger K, Engel A, Haupt J, Li Q, Lima de Miranda R, Quandt E, Lutter G, Selhuber-Unkel C. Cell adhesion on NiTi thin film sputter-deposited meshes. Mater. Sci. Eng. C Mater. Biol. Appl. 2016, 59, 611–6.

[320] Poncet PP. Applications of Superelastic Nitinol Tubing; https://assets.website-files.com/59fcbaf103e295000131288b/5a318d4c0672a700015ad099_Applications_Superelastic-G%C3%87%C3%B4NiTl_tube.pdf.

[321] Shabalovskaya SA, Anderegg J, Laab F, Thiel PA, Rondelli G. Surface conditions of Nitinol wires, tubing, and as-cast alloys. The effect of chemical etching, aging in boiling water, and heat treatment. J. Biomed. Mater Res. Part B: Appl. Biomat. 2003, 65B, 193–203.

[322] Potapov PL, Tirry W, Schryvers D, Sivel VGM, Wu M-Y, Aslanidis D, Zandbergen H. Cross-section transmission electron microscopy characterization of the near-surface structure of medical Nitinol superelastic tubing. J. Mater. Sci.: Mater. Med. 2007, 18, 483–92.

[323] Robertson SW, Launey M, Shelley O, Ong O, Vien L, Senthilnathan K, Saffari P, Schlegel S, Pelton AR. A statistical approach to understand the role of inclusions on the fatigue resistance of superelastic Nitinol wire and tubing. J. Mech. Behav. Biomed. Mater. 2015, 51, 119–31.

[324] https://futaku.co.jp/en/tec0104en/.

[335] Favier D, Louche H, Schlosser P, Orgéas L, Vacher P, Debove L. Homogeneous and heterogeneous deformation mechanisms in an austenitic polycrystalline Ti–50.8 at.% Ni thin tube under tension. Investigation via temperature and strain fields measurements. Acta Mater. 2007, 55, 5310–22.

[326] Ghassemi-Armaki H, Leff AC, Taheri ML, Dahal J, Kamarajugadda M, Kumar KS. Cyclic compression response of micropillars extracted from textured nanocrystalline NiTi thin-walled tubes. Acta Mater. 2017, 136, 134–47.

[327] Wu Z, Wang Y, Sun L, Mao Y, Wang M, Lin C. An ultrasound-assisted deposition of NiO nanoparticles on TiO2 anotube arrays for enhanced photocatalytic activity. J. Mater. Chem. A 2014, 2, 8223–9.

[328] Kim JH, Zhu K, Yan Y, Perkins CL, Frank AJ. Microstructure and pseudocapacitive properties of electrodes constructed of oriented NiO-TiO2 nanotube arrays. Nano. Lett. 2010, 10, 4099–104.

[329] Hang R, Liu Y, Zhao L, Gao A, Bai L, Huang X, Zhang X, Tang B, Chu PK. Fabrication of Ni-Ti-O nanotube arrays by anodization of NiTi alloy and their potential applications. Sci. Rep. 2014, 18(4), 7547; doi: 10.1038/srep07547.

[330] Siuzdak K, Szkoda M, Sawczak M, Lisowska-Oleksiak A, Karczewski J, Ryl J. Enhanced photoelectrochemical and photocatalytic performance of iodine-doped titania nanotube arrays. RSC Adv. 2015, 5, 50379–91.

[331] Liu Q, Ding D, Ning C. Anodic fabrication of Ti-Ni-O nanotube arrays on shape memory alloy. Materials (Basel) 2014, 22, 3262–73.

[332] Mohammadi F, Kharaziha M, Ashrafi A. Role of heat treatment on the fabrication and electrochemical property of ordered TiO_2 nanotubular layer on the as-cast NiTi. Met. Mat. Int. 2019, 25, 617–26.

[333] Buenconsejo PJS, Ito K, Kim HY, Miyazaki S. High-strength superelastic Ti–Ni microtubes fabricated by sputter deposition. Acta Mater. 2008, 56, 2063–72.

[334] Lin Z, Hsiao H-M, Mackiewicz D, Anukhin B, Pike K. Anisotropic behavior of radiopaque NiTiPt hypotube for biomedical applications. Adv. Eng. Mater. 2009, 11, B189–93.

[335] https://www.sarvadajewels.com/the-niti-ribbon-ring.html.

[336] Jenko M, Grant JT, Rudolf R, Kokalj T, Mandrino D. Characterisation of the surface microtructure of a melt-spun Ni-Ti shape memory ribbon. Surf. Interface Anal. 2012, 44, 997–1000.

[337] Tong Y, Liu Y, Xie Z, Zarinejad M. Effect of precipitation on the shape memory effect of $Ti_{50}Ni_{25}Cu_{25}$ melt-spun ribbon. Acta Mater. 2008, 56, 1721–32.

[338] Xie ZL, Cheng GP, Liu Y. Microstructure and texture development in $Ti_{50}Ni_{25}Cu_{25}$ melt-spun ribbon. Acta Mater. 2007, 55, 361–9.

[339] Tomić S, Rudolf R, Brunčko M, Anžel I, Savić V, Colić M. Response of monocyte-derived dendritic cells to rapidly solidified nickel-titanium ribbons with shape memory properties. Eur. Cell Mater. 2012, 23, 58–80.

[340] Taylor SL, Iben AJ, Jakus AE, Shah RN, Dunand DC. NiTi-Nb micro-trusses fabricated via extrusion-based 3D-printing of powders and transient-liquid-phase sintering. Acta Biomater. 2018, 76, 359–70.

[341] von Fraunhofer JA, Bonds PW, Johnson BE. Force generation by orthodontic coil springs. Angle Orthod. 1993, 63, 145–8.

[342] Manhartsberger C, Seidenbusch W. Force delivery of Ni-Ti coil springs. Am. J. Orthod. Dentofacial Orthop. 1996, 109, 8–21.

[343] https://shopee.com.my/amp/5PCS-Sino-Dental-Orthodontic-Niti-Closed-Coil-Springs-0.010-0.012--i.126265908.1924043844.

[344] Barwart O, Rollinger JM, Burger A. An evaluation of the transition temperature range of super-elastic orthodontic NiTi springs using differential scanning calorimetry. Eur. J. Orthod. 1999, 21, 497–502.

[345] Keles A. Maxillary unilateral molar distalization with sliding mechanics: a preliminary investigation. Eur. J. Orthod. 2001, 23, 507–15.

[346] Alvarez K, Nakajima H. Metallic scaffold for bone regeneration. Materials 2009, 2, 790–832.

[347] Shishkovsky I. Stress–strain analysis of porous scaffolds made from titanium alloys synthesized via SLS method. Appl. Surface Sci. 2009, 255, 9902–5.

[348] Bormann T, Schumacher R, Müller B, de Wild M. Fabricating NiTi shape memory scaffolds by selective laser melting. Europ. Cells Mater. 2012, 21, 2519–25.

[349] Wu S, Liu X, Wu G, Yeung KW, Zheng D, Chung CY, Xu ZS, Chu PK. Wear mechanism and tribological characteristics of porous NiTi shape memory alloy for bone scaffold. J. Biomed. Mater. Res. A 2013, 101(9), 2586–601. doi: 10.1002/jbm.a.34568.

[350] Liu X, Wu S, Yeung KW, Chan YL, Hu T, Xu Z, Liu X, Chung JC, Cheung KM, Chu PK. Relationship between osseointegration and superelastic biomechanics in porous NiTi scaffolds. Biomaterials 2011, 32, 330–8.

[351] Hoffmann W, Wendt D, Bormann T, Rossi A, Müller B, Schumacher R, Martin I, de Wild M. Rapid prototyped porous nickel – titanium scaffolds as bone substitutes. J. TissueEng. 2014, 5, 1–14.

[352] Speirs M, Van Hooreweder B, Van Humbeeck J, Kruth J-P. Fatigue behaviour of NiTi shape memory alloy scaffolds produced by SLM, a unit cell design comparison. J. Mech. Behav. Biomed. Mater. 2017, 70, 53–5.

[353] Wang X, Xu S, Zhou S, Xu W, Leary M, Choong P, Qian M, Brandt M, Xie YM. Topological design and additive manufacturing of porous metals for bone scaffolds and orthopaedic implants: a review. Biomaterials 2016, 83, 127–41.

[354] Banhart J. Manufacture, characterization and application of cellular metals and metal foams. Prog. Mater. Sci. 2001, 46, 559–632.

[355] Strano M. A new FEM approach for simulation of metal foam filled tubes. J. Manuf. Sci. Eng. 2011, 133, 061003; doi: 10.1115/1.4005354.

[356] Gotman I. Fabrication of load-bearing NiTi scaffolds for bone ingrowth by Ni foam conversion. Adv. Eng. Mater. 2010, 12, B320–5.

[357] Grummon DS, Shaw JA, Gremillet A. Low-density open-cell foams in the NiTi system. Appl. Phys. Lett. 2003, 82, 2727–9.

[358] Barrabés M, Michiardi A, Aparicio C, Sevilla P, Planell JA, Gil FJ. Oxidized nickel – titanium foams for bone reconstructions: chemical and mechanical characterization. J. Mater. Sci.: Mater. Med. 2007, 18, 2123–9.

[359] Young ML, DeFouw JD, Frenzel J, Dunand DC. Cast-replicated NiTiCu foams with superelastic properties. Metall. Mater. Trans. A 2012, 43, 2939–44.

[360] Stöckel D, Pelton A, Duerig T. Self-expanding Nitinol stents: material and design considerations. Eur. Radiol. 2004, 14, 292–301.

[361] https://www.sciencedirect.com/topics/nursing-and-health-professions/nitinol-stent.

[362] Wei ZG, Sandstrom R, Miyazaki S. Shape-memory materials and hybrid composites for smart
systems – part I shape-memory materials. J. Mater. Sci. 1998, 33, 3743–62.

[363] Maho A, Kanoufi F, Combellas C, Delhalle J, Mekhalif Z. Electrochemical investigation of
Nitinol/tantalum hybrid surfaces modified by alkylphosphonic self-assembled monolayers.
Electrochim. Acta 2014, 116, 78–88.

[364] Frensemeier M, Kaiser JS, Frick CP, Schneider AS, Arzt E, Fertig RS, Kroner E. Temperature-
induced switchable adhesion using nickel-titanium-polydimethylsiloxane hybrid surfaces.
Adv. Funct. Mater. 2015, 25, 3013–21.

[365] Pequegnat A, Panton B, Zhou YN, Khan MI. Local composition and microstructure control for
multiple pseudoelastic plateau and hybrid self-biasing shape memory alloys. Mater. Des.
2016, 92, 802–13.

[366] Witkowska J, Sowińska A, Czarnowska E, Płociński T, Kamiński J, Wierzchoń T. Hybrid
a-CNH+TiO$_2$+TiN-type surface layers produced on NiTi shape memory alloy for cardiovascular
applications. Nanomedicine (Lond) 2017, 12, 2233–44.

[367] Witkowska J, Sowińska A, Czarnowska E, Płociński T, Rajchel B, Tarnowski M, Wierzchoń
T. Structure and properties of composite surface layers produced on NiTi shape
memory alloy by a hybrid method. J. Mater. Sci. Mater. Med. 2018, 29, 110; doi: 10.1007/
s10856-018-6118-5.

[368] Ayers R, Burkes D, Gottoli G, Yi HC, Moore JJ. Temperature synthesis of engineered porous
composite biomedical materials. Mater. Manuf. Processes 2007, 22, 481–8.

[369] Mari D, Dunand DC. NiTi and NiTi-TiC composites: part 1. Transformation and thermal cycling
behavior. Metall. Mater. Trans. A 1995, 26, 2833–47.

[370] Fukami-Ushiro KL, Mari D, Dunand DC. Niti and NiTi-TiC composites: part II. C1ompressive
mechanical properties. Metall. Mater. Trans. A 1996, 27, 183–91.

[371] Li DY, Luo YC. Effects of TiN nano-particles on porosity and wear behavior of TiC/TiNi tribo
composite. J. Mater. Sci. Lett. 2001, 20, 2249–52.

[372] Luo YC, Li DY. New wear-resistant material: nano-TiN/TiC/TiNi composite. J. Mater. Sci. 2001,
36, 4695–702.

[373] Burkes DE, Gottoli G, Moore JJ, Yi HC. Combustion synthesis and mechanical properties of
dense NiTi-TiC intermetallic-ceramic composites. Metall. Mater. Trans. A 2006, 37, 235–42.

[374] Kothalkar AD, Benitez R, Hu L, Radovic M, Karaman I. Thermo-mechanical response and
damping behavior of shape memory alloy – MAX phase composites. Metall. Mater. Trans. A
2014, 45, 2646–58.

[375] Şahin Y, Öksüz KE. Effects of Al$_2$O$_3$ nanopowders on the wear behavior of NiTi shape memory
alloys. JOM 2014, 66, 61–5.

[376] Zang D, Yi H, Gu Z, Chen L, Han D, Guo X, Wang S, Liu M, Jiang L. Interfacial engineering
of hierarchically porous NiTi/hydrogels nanocomposites with exceptional antibiofouling
surfaces. Adv. Mater. 2017, 29; doi: 10.1002/adma.201602869.

[377] Porter GA, Liaw PK, Tiegs TN, Wu KH. Ni-Ti SMA-reinforced Al composites. JOM 2000, 52,
52–6.

[378] Hehr A, Chen X, Pritchard J, Dapino MJ, Anderson PM. Al-NiTi metal matrix composites for zero
CTE materials: fabrication, design, and modeling. Adv. Comp. Aerospace, Marine, Land Appl.
II, 2015, 13–28.

[379] Akalin O, Ezirmik KV, Urgen M, Newaz GM. Wear characteristics of NiTi/Al6061 short
fiber metal matrix composite reinforced with SiC particulates. J. Tribol. 2010, 132(4); doi:
10.1115/1.4002332.

[380] Coughlin JP, Williams JJ, Chawla N. Mechanical behavior of NiTi shape memory alloy fiber reinforced Sn matrix "smart" composites. J. Mater. Sci. 2009, 44, 700–7.

[381] Poormir MA, Khalili SMR, Eslami-Farsani R. Investigation of the self-healing behavior of Sn-Bi metal matrix composite reinforced with NiTi shape memory alloy strips under flexural loading. JOM 2018, 70, 806–10.

[382] Hao S, Cui L, Jiang J, Guo F, Xiao X, Jiang D, Yu C, Chen Z, Zhou H, Wang Y, Liu Y, Brown DE, Ren Y. A novel multifunctional NiTi/Ag hierarchical composite. Sci. Rep. 2014, 4, 5267. doi: 10.1038/srep05267.

[383] Sevost'yanov MA, Nasakina EO, Baikin AS, Sergienko KV, Konushkin SV, Kaplan MK, Seregin AV, Leonov AV, Kozlov VA, Shkirinv AV, Bunkin NF, Kolmakov AG, Simakov SV, Gudkov SV. Biocompatibility of new materials based on nano-structured nitinol with titanium — and tantalum composite surface layers: experimental analysis in vitro and in vivo. J. Mater. Sci. Mater. Med. 2018, 29(3); doi: 10.1007/s10856-018-6039-3.

[384] Hamada K, Kawano F, Asaoka K. Shape recovery of shape memory alloy fiber embedded resin matrix smart composite after crack repair. Dent. Mater. J. 2003, 22, 160–7.

[385] Han R, Tian B, Wang YH, Chen F, Tong YX, Li L, Zheng YF. The preparation and characterization of NiTi/CNT/Polyurethane composite. Mater. Sci. Forum 2015, 813, 243–9.

[386] Daghash SM, Ozbulut OE. Characterization of superelastic shape memory alloy fiber-reinforced polymer composites under tensile cyclic loading. Mater. Des. 2016, 111, 504–12.

[387] El-Tahan M, Dawood M. Bond behavior of NiTiNb SMA wires embedded in CFRP composites. Polym. Comp. 2018, 39, 3780–91.

[388] Twohig E, Tiernan P, Tofail SAM. Experimental study on dieless drawing of Nickel–Titanium alloy. J. Mech. Behav.Biomed. Mater. 2012, 8, 8–20.

[389] Twohig E, Tiernan P, Butler J, Dickinson C, Tofail SAM. Mechanical, microstructural and thermal properties of a 50:50 at.% nickel–titanium alloy subjected to a dieless drawing process. Acta Mater. 2014, 68, 140–9.

[390] Zhu R, Tang G. The improved plasticity of NiTi alloy via electropulsing in rolling. Mater. Sci. Technol. 2016, 33(5), 1–6.

[391] Peterlechner M, Bokeloh J, Wilde G, Waitz T. Study of relaxation and crystallization kinetics of NiTi made amorphous by repeated cold rolling. Acta Mater. 2010, 58(20), 6637–48.

[392] Müller K. Extrusion of nickel–titanium alloys Nitinol to hollow shapes. J. Mater. Process. Technol. 2001, 111, 122–6.

[393] Luo J, Ye W-J, Ma X-X, Bobanga JO, Lewandowski JJ. The evolution and effects of second phase particles during hot extrusion and re-extrusion of a NiTi shape memory alloy. J. Alloys Compd. 2018, 735, 1145–51.

[394] Maurer J, Hudson JL, Fick SE, Moffat TP, Shaw GA. Electrochemical micromachining of NiTi shape memory alloys with ultrashort voltage pulses. Electrochem. Solid-State Lett. 2012, 15, D8–10.

[395] Kong MC, Axinte D, Voice W. Challenges in using waterjet machining of NiTi shape memory alloys: an analysis of controlled-depth milling. J. Mater. Process. Technol. 2011, 211, 959–71.

[396] Pirani C, Iacono F, Generali L, Sassatelli P, Nucci C, Lusvarghi L, Gandolfi MG, Prati C. HyFlex EDM: superficial features, metallurgical analysis and fatigue resistance of innovative electro discharge machined NiTi rotary instruments. Int. Endod. J. 2016, 49, 483–93.

[397] Feng CC, Li L, Zhang CS, Zheng GM, Bai X, Niu ZW. Surface characteristics and hydrophobicity of Ni-Ti alloy through magnetic mixed electrical discharge machining. Materials (Basel) 2019, 26, 12(3); doi: 10.3390/ma12030388.

[398] Takale AM, Chougule NK. Effect of wire electro discharge machining process parameters on surface integrity of $Ti_{49.4}Ni_{50.6}$ shape memory alloy for orthopedic implant application. Mater. Sci. Eng. C Mater. Biol. Appl. 2019, 97, 264–74.

[399] Bisaria H, Shandilya P. Processing of curved profiles on Ni-rich nickel–titanium shape memory alloy by WEDM. Mater. Manuf. Processes 2019; https://doi.org/10.1080/10426914.2019.1594264.

[400] Khan MI, Pequegnat A, Zhou YN. Multiple memory shape memory alloys. Adv. Eng. Mater. 2013, 15, 386–93.

[401] Marattukalam JJ, Singh AK, Datta S, Das M, Balla VK, Bontha S, Kalpathy SK. Microstructure and corrosion behavior of laser processed NiTi alloy. Mater. Sci. Eng., C 2015, 57, 309–13.

[402] Bandyopadhyay A, Krishna BV, Xue W, Bose S. Application of Laser Engineered Net Shaping (LENS) to manufacture porous and functionally graded structures for load bearing implants. J. Mater. Sci.: Mater. Med. 2009, 20, S29–34.

[403] Shiva S, Palani A, Mishra SK, Paul CP, Kukreja LM. Influence of Cu addition to improve shape memory properties in NiTi alloys developed by Laser Rapid Manufacturing. J. Laser Micro/Nanoeng. 2016, 11, 153–7.

[404] Pfeifer R, Herzog D, Hustedt M, Barcikowski S. Pulsed Nd:YAG laser cutting of NiTi shape memory alloys – influence of process parameters. J. Mater. Process. Technol. 2010, 210, 1918–25.

[405] Shahmir H, Nili-Ahmadabadi M, Mansouri-Arani M, Langdon TG. The processing of NiTi shape memory alloys by equal-channel angular pressing at room temperature. Mater. Sci. Eng. A 2013, 576, 178–84.

[406] Shahmir H, Nili-Ahmadabadi M, Arani MM, Ali Xajezade A, Langdon TG. Evaluating the room temperature ECAP processing of a NiTi alloy via simulation and experiments: room temperature ECAP processing of NiTi alloy. Adv. Eng. Mat. 2014, 17(4); doi: 10.1002/adem.201400248.

[407] Shahmir H, Nili-Ahmadabadi M, Wang CT, Jung JM, Kim HS, Langdon TG. Annealing behavior and shape memory effect in NiTi alloy processed by equal-channel angular pressing at room temperature. Mater. Sci. Eng. A 2015, 629, 16–22.

Chapter 7
Properties and their effects on biofunctionality

7.1 Basic mechanical and physical properties

In general, mechanical as well as physical properties are related to the crystalline structure. Basically, pure Ti has a hexagonal close-packed structure at room temperature (which is called α phase) and it will be transformed to β phase (body-centered cubic) at 883 °C (allotropic temperature). Many previous investigations show that the phase transformation temperature of pure Ti can be strongly influenced by alloying elements, and alloying elements are usually classified into α-stabilizers such as Al, C, and O which increase the temperature of β/α transus and β-stabilizers such as Mo, Ta, and Nb reduce the β/α transus temperature. Based on these metallurgical points of view, conventional Ti alloys can be primarily classified into four groups of α-, near-α, (α + β)-, and β-type alloys. Meanwhile, Ti-based shape-memory alloys (SMA) is another group of Ti alloy with superelasticity (SE) and shape-memory effect (SME), although the Ti-based SMA is concerned as an alloy, it actually behaves as an intermetallic compound. In addition, for exhibiting SME and/or SE, two distinct phases (austenitic parent phase and martensitic phase, respectively) play dominant roles. Normal metallic materials exhibit an elasticity prior to show yielding, followed by permanent displacement (or deformation). In regime of the elasticity, applied stress and resultant strain is linearly related which is known as Hooke's law. According to the Hooke's law, most metal alloys can be elastically deformed by up to 0.1% or 0.2%. NiTi (nickel–titanium) alloys, however, can be deformed up to 8% beyond their yield strength without showing any permanent strain. As the extent of elasticity is extraordinarily large, the phenomenon is called as SE, which can be defined as the ability of certain materials to recover their original shape once the load is removed even when they are deformed beyond their yield strength. In NiTi alloys, another uniqueness is associated with a solid-state martensitic transformation which can be induced by the application of stress or by a temperature reduction. During the diffusionless transformation, constituent atoms move coordinately by a shear-type mechanism and are rearranged into a new (into more thermodynamically stable crystalline structure) without any change in the chemical composition of the matrix. This martensitic transformation takes placed between austenitic (parent) phase and martensitic phase and the transformation is reversible [1–3]. Austenitic phase B2 shows a cubic crystalline structure, while the martensitic phase B19′ shows an orthorhombic structure, accordingly each of phases should show distinct features in mechanical and other properties, as will be discussed in this chapter.

Table 7.1 summarizes mainly currently employed NiTi endodontic instruments in terms of phase compositions/property along with trade name and manufacture's name, where A stands for austenitic phase, M for martensitic phase, R for R-phase,

https://doi.org/10.1515/9783110666113-007

Table 7.1: NiTi alloy for manufacturing various types of endodontic instruments [4–6].

Name	Phase/property	Makes
Con. NiTi	A/SE	Mtwo. Oneshape, ProFile, and Pro Taper Universal
Electropolishing		RaCe, BioRaCe, iRace, F360, and F6 Skytaper
R-phase	A/SE	Twisted File, Twisted File Adaptive, and K3XF
M-Wire	A(R)/SE	ProFile Vortex, ProFile GT Series X, and ProTaper Next; Reciproc
CM Wire	M(A&R)	Hyflex CM, TYPHOON Infinite Flex NiTi Files, V-Taper 2H, and Hyflex EDM
Gold heat-treated		ProTaper Gold
Blue heat-treated		Wave One Gold, ProFile Vortex Blue, and Reciproc Blue
MaxWire	M&A/SE&SME	XP-endo Finisher and XP-endo Shaper

and SE for superelasticity and SME for shape-memory effect [4–6]. As given in Table 7.1, NiTi endodontic instruments utilize mainly superelastic phenomenon with the austenite phase. It is well known that the austenite is transformed to martensite by stress; for a practical example, an insertion of the instrument into a curved root canal. This effect is called stress-induced martensite (SIM) transformation. The transformation of the austenitic phase to the martensitic phase allows a complete recovery of the deformation up to 8% strain, as described earlier. Because the stress-induced martensitic state is not stable at the intraoral temperature, unloading of the endodontic instrument (e.g., withdrawal of an instrument out of a curved root canal) leads to reverse transformation back to the austenite phase resulting in a spring-back of the endodontic instrument to its original shape.

Table 7.2 summarizes basic mechanical and physical properties for austenitic and martensitic phases. Generally, the martensitic phase of NiTi alloy is more ductile than the austenitic phase [7–13].

Table 7.2: Basic properties of NiTi alloy.

Property	Phase	Value
Mechanical properties		
Tensile strength	Austenite	700–750 MPa
	Martensite	900–960 MPa
Modulus of elasticity	Austenite	75–85 GPa
	Martensite (predeform)	28–30 GPa
	Martensite (postdeform)	45–48 GPa

Table 7.2 (continued)

Property	Phase	Value
Yield strength	Austenite	500–600 MPa
	Martensite (predeform)	100–105 MPa
	Martensite (postdeform)	530–560 MPa
Elongation	Austenite	~20%
	Martensite	~50%
Poisson's ratio		0.31–0.33
Springback in tension		$1.40 \times 10^{-2} (\sigma/E)$
Spring rate in bending		0.17 mm-N/degree
Spring rate in torsion		0.02 mm-N/degree
Physical and thermal properties		
Density		6.45–6.5 g/cm^3
Melting point		~1,310 °C
Electrical resistivity	Austenite	$80{-}100 \times 10^{-6}$ Ω·cm
	Martensite	$76{-}85 \times 10^{-6}$ Ω·cm
Thermal conductivity	Austenite	0.18 W/cm·K
	Martensite	0.086 W/cm·K
Coefficient of thermal expansion	Austenite	$10{-}12 \times 10^{-6}$/°C
	Martensite	$6.6{-}7 \times 10^{-6}$/°C
Magnetic susceptibility	Austenite	3.7×10^{-6} emu/g
	Martensite	2.4×10^{-6} emu/g
Magnetic permeability		<1.002
Specific heat		322 J/kg/°C
Latent heat		24,200 J/kg
Thermal hysteresis		30–50 °C

7.1.1 In general

Figure 7.1 shows a relationship diagram between the yield strength versus modulus of elasticity (E) of typical biomaterials. It is easy to find the engineering positioning of NiTi materials among other biomaterials. When any biomaterial is implanted in human being, the biomechanical compatibility should be taken into serious considerations. If any combination of implanted materials and receiving hard/soft tissue (e.g., bone) is not compatible to each other in a biomechanical environment, then

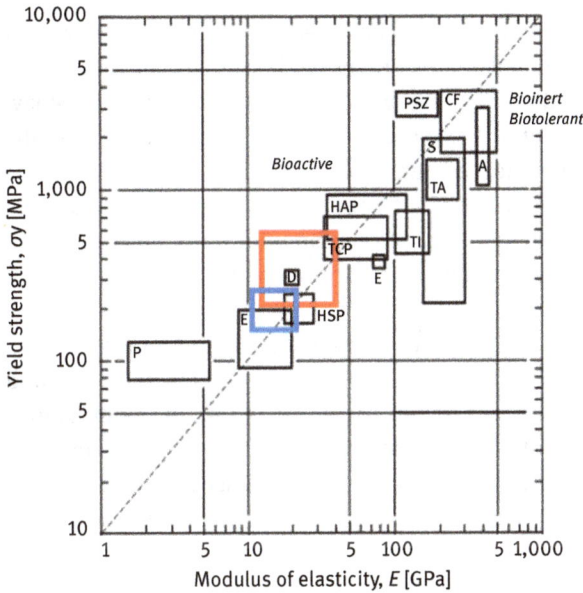

Figure 7.1: Relationship between yield strength and modulus of elasticity of typical biomaterials: P: polymeric materials, B: bone, D: dentin, HSP: high strength polymers, HAP: hydroxyapatite, TCP: tricalcium phosphate, E: enamel, TI: commercially pure Titanium, TA: Ti-Al based alloys, TZ: Ti−Zr alloy, S: 18-8 series stainless steel, CF: carbon fiber, A: alumina ceramics, PSZ: partially stabilized zirconia ceramics, and a red square for solid NiTi alloy (including both austenitic and martensitic phases) and a blue square for NiTi alloy with controlled porosity.

the interfacial stress will play a controlling role to maintain the osseointegrated fused bone/implant system [14–17]. If the strain under biofunctional action (e.g., occlusal force for the dental implant) is under the strain continuity, the attained osseointegration should be sustained; however, if the loading force overwhelms the interfacial stress, the osseointegrated couple will be failed due to the stress discontinuity [18]. Accordingly, it is necessitated to manipulate surface layer of NiTi alloy to exhibit mechanical matching to that of the bone.

Such mechanical matching is very important not only to maintain osseointegration but also for bone remodeling mechanism, directly related to the stress shielding. The implantation of solid titanium or stainless steel (SS) devices can lead to stress shielding due to mismatch of stiffness between the implant and the surrounding bone. "Stress shielding" refers to the slow healing and reduction of bone density that results when implants remove the stresses normally experienced by bone. Stress shielding is minimized by using implants that exhibit stiffness similar to that of the surrounding bone, that is, biomechanical compatibility. Stiffness, in turn, is influenced by both the design and by the elastic modulus of the implant material. The development of biocompatible materials that closely match the mechanical properties of bone is thus important to improve patient outcomes in a variety of orthopedic procedures [19–23].

In the following, there appear to be several characteristic trends in researches on mechanical properties of NiTi materials. One of them is related to orthodontic mechanotherapy, that is, concentrated on bending properties and some torsional behaviors; on the other hand, for endodontic files or reamers for endodontic treatment, the flexibility, torsional resistance, flexural resistance, and cyclic loading fatigue become crucial issue. Furthermore, since both orthodontic appliances and endodontic devices are normally subjected to sterilization under various conditions and agents between patients, the effects of such sterilization on mechanical properties have been investigated and reported.

Mechanical property, functional property, and porosity are tightly interrelated. Bram et al. [24] studied mechanical property of NiTi with a porosity of 51% and a pore size in the range of 300–500 μm. It was found that (i) at low deformations <6%, fully pronounced SE was found, and (ii) at higher strains, a shape recovery of maximum 6% took place, on which the onset of irreversible plastic deformation was superposed. Resnina et al. investigated mechanical behavior of porous $TiNi_{45}$ (at%) [25] and functional property of porous $NiTi_{48}$ (at%) [26]. It was reported that (i) the porous $TiNi_{45}$ is deformed by the same mechanisms as a cast $Ti_{50}Ni_{50}$ alloy, (ii) at low temperatures, the deformation of the porous alloy is realized via martensite reorientation at a low yield limit and by dislocation slip at a high yield limit, while at high temperatures (in the austenite B2 phase) the porous $TiNi_{45}$ alloy is deformed by the SIM at a low yield limit and by dislocation slip at a high yield limit, (iii) the temperatures of the martensitic transformation increase linearly when the stress rises up to 80 MPa, and (iv) the porous $TiNi_{45}$ alloy accumulates an irreversible strain on cooling and heating and demonstrates unstable functional behavior during thermal cycling [24]. On the other hand, for the porous $TiNi_{48}$ alloy, it was found that (i) a large unelastic strain recovered on unloading and it was not attributed to the SE effect, (ii) a decrease in deformation temperatures did not influence the value of strain that recovered on unloading, while the effective modulus decreased from 1.9 to 1.44 GPa, (iii) the porous $TiNi_{48}$ alloy demonstrated the transformation plasticity and the SMEs on cooling and heating under a stress, and (iv) the functional properties of the porous alloy were determined by the TiNi phase consisted of the two volumes $Ti_{49.3}Ni_{50.7}$ and $Ti_{50}Ni_{50}$ where the martensitic transformation occurred at different temperatures indicating that the existence of the $Ti_{49.3}Ni_{50.7}$ volumes in the porous $TiNi_{48}$ alloy improved the functional properties of the alloy [25]. Buchheit et al. [27] prepared a series of Ti-rich NiTiPt (with 13–18 at%Pt) by vacuum arc melting and found that (i) the transformation underwent between 175 and 225 °C and recoverable strain exceeding 2% was achieved, and (ii) alloys containing 15.5–16.5 at%Pt exhibited transformation temperatures in the vicinity of 200 °C and exhibited yield stresses of approximately 500 MPa and 3.5% elongation to failure.

For orthodontic mechanotherapy, one of four major material types is utilized in one of four different shapes of round, square, rectangular, or stranded shapes. Each material type exhibits distinct values of mechanical properties as seen in Table 7.3

Table 7.3: Mechanical properties and compositions of four major types of orthodontic wires [28].

Wire type	Composition (wt%)	Modulus of elasticity (GPa)	Yield strength (MPa)
Stainless steel	18Cr–8Ni–Fe	160–180	1,100–1,500
Co–Cr–Ni	40Co–20Cr–15Ni	160–190	830–1,000
β-Ti	78Ti–11Mo–7Zr	62–69	690–970
NiTi	55Ni–45Ti	34	210–410

[29]. Kusy et al. [29, 30] studied the elastic properties of NiTi orthodontic archwires and concluded that (i) NiTi makes the most active "leveling" archwire, and the β-Ti is a superior intermediate archwire where flexibility is required, (ii) SS and Co–Cr are the wires of choice for finishing and other applications where stability of form is required, (iii) beveling reduced the cross-sectional areas by 7–8% decreasing the wire stiffnesses by 15–19%, and (iv) the elastic limits of the superelastic NiTi wires were approximately 90% and 45% of their ultimate tensile strengths for the round and rectangular wires, respectively.

As to a deformation behavior of NiTi alloy, Favier et al. [31] deformed an austenitic polycrystalline $TiNi_{50.8}$ (at%) thin-walled tube with small grain sizes under tension in air at ambient temperature and moderate nominal axial strain rate. It was reported that (i) in a first apparently elastic deformation stage, both strain and temperature fields are homogeneous and increase in tandem, followed by initiation, propagation, and growth of localized helical bands inside which strain and temperature increases are markedly higher than the surrounding regions, and (ii) during the first apparently elastic stage of the unloading, both strain and temperature fields are homogeneous and decrease. Jiang et al. [32] investigated effects of aging treatment on the deformation behavior of $TiNi_{50.9}$ (at%) alloy. It was found that (i) the alloy exhibits SE within a certain temperature and stress window, which narrows with increasing aging temperature between 400 and 560 °C, with the sample aged at 400 °C for 1 h exhibiting the maximum temperature window of ~167 °C for SE, and (ii) the mechanically induced transformation sequence is different from the thermally induced transformation sequence, implying that the mechanical behavior of the alloy cannot be predicted from the knowledge of the thermally induced transformations based on the Clausius–Clapeyron correlation between stress and temperature. Furthermore, Hu et al. [33] studied deformation and degradation of superelastic NiTi under multiaxial loading. It was mentioned that (i) during the uniaxial loading, the R-phase as a transition between the austenite and the B19′ martensite was detected, (ii) degradation of the SE is found to depend strongly on the loading and unloading path followed, (iii) cycling biaxially leads to faster degradation than uniaxially due to a larger accumulation of dislocations, and (iv) if the deformation cycle contains a load path change, dislocation accumulation increases further and more martensite is retained.

It is well documented that the properties of NiTi alloys are significantly affected by thermomechanical treatments, which are essential to control the transformation temperatures, mechanical properties, and monotonic and cyclic responses [34, 35]. The common thermomechanical treatments should include ausforming [36–38] and marforming [35, 38], in which defects are introduced into the austenitic and martensitic structures, respectively. The specific aim for these treatments should include the enhancement of the characteristic SME and SE by suppressing the irreversible slip deformation during the martensite reorientation and stress-induced martensitic transformation by raising the critical shear stress for slip by controlling transformation temperatures with the aim of achieving the desired transformation path (thermally induced or stress-induced transformation at a given temperature), increasing the transformation stress of the stress-induced parent-to-martensite transformation while the SME and/or PE are expected to be preserved, and improving thermomechanical fatigue resistance. Pereira et al. [39] compared physical and mechanical properties of one conventional (C-Wire) and one thermomechanically treated NiTi wire (M-Wire) that are used to manufacture rotary endodontic instruments. It was reported that (i) B19′ martensite and the R-phase were found in M-Wire, in agreement with the higher transformation temperatures found in this wire compared with C-Wire, whose transformation temperatures were below room temperature, (ii) the stress at the transformation plateau in the tensile load–unload curves was lower and more uniform in the M-Wire, which also showed the smallest stress hysteresis and apparent elastic modulus, and (iii) the M-Wire had physical and mechanical properties that can render endodontic instruments more flexible and fatigue resistant than those made with conventionally processed NiTi wires.

7.1.2 Various factors affecting mechanical properties

Surface preparation is potentially important to the mechanical and biomedical properties of NiTi SMA. Miao et al. [40] studied the effect of surface preparation on mechanical properties of a $TiNi_{55.6}$ (wt%) alloy by preparation of chemical washing, mechanical polishing, and electropolishing. It was found that (i) electropolishing can significantly improve the surface smoothness of the NiTi alloy, (ii) the fractures of all tensile specimens showed dimple morphology, and (iii) electropolished samples tend to have better mechanical properties due to their smoother surface. Katić et al. [41] analyzed the effect of various coating formulations on the mechanical and corrosion properties of NiTi orthodontic wires. It was found that (i) uncoated and nitrified NiTi wires showed similar mechanical and anticorrosive properties, while rhodium-coated NiTi wires showed the highest surface roughness (SR) and significantly higher modulus of elasticity, yield strength, and delivery of forces during loading but not in unloading, (ii) rhodium-coated NiTi wires also had the highest corrosion current density and corrosion potential, and (iii) working properties of NiTi wires were unaffected by various coatings in unloading.

In order to study and define the mechanical characteristics of several currently available closed-coil retraction springs, Wichelhaus et al. [42] investigated mechanical behavior and clinical application of NiTi closed-coil springs under different stress levels and mechanical loading cycles. It was reported that among four groups according to the mechanical properties of the springs (strong SE without bias stress, weak SE without bias stress, strong SE with bias stress, and weak SE with bias stress), (i) the strongly superelastic closed-coil springs with preactivation are recommended, and (ii) the oral environment seems to have only a minor influence on their mechanical properties.

Khan et al. [43] studied effects of local phase conversion on the tensile loading of pulsed Nd:YAG laser processed $NiTi_{49.2}$ (at%) and found changes to room temperature phases of the processed metal when compared to the base material, which also affect the tensile properties of NiTi alloy. Nayan et al. [44] investigated the load-controlled tension–tension fatigue behavior of a martensitic NiTi SMA, in the ambient temperature. Fatigue life for several stress levels spanning the critical stress for detwinning was also determined. It was reported that (i) the fatigue life of the superplastic alloy is superior to superelastic SMA, (ii) the stress–strain hysteretic response, monitored throughout the fatigue loading, reveals progressive strain accumulation with the cyclic loading, and (iii) in addition, the area of hysteresis and recoverable and frictional energies were found to decrease with increasing number of fatigue cycles. Effects of R-phase on mechanical responses of three NiTi endodontic instruments (superelastic austenite NiTi, austenite + R-phase NiTi, and fully R-phased NiTi) were investigated [45]. It was mentioned that (i) the modeled instrument containing only R-phase demanded the lowest moment to be bent, followed by the one with mixed austenite + R-phase, and (ii) the superelastic instrument, containing essentially austenite, required the highest bending moment; during bending, the fully R-phased instrument reached the lowest stress values; however, it also experienced the highest angular deflection when subjected to torsion summarizing that NiTi endodontic instruments containing only R-phase in their microstructure would show higher flexibility without compromising their performance under torsion. Yoneyama et al. [46] investigated the influence of composition and purity of titanium on the mechanical properties and the transformation temperatures of $NiTi_{49.0\sim49.2}$ (at%) alloy dental castings. It was reported that (i) the $NiTi_{49.0}$ was a somewhat brittle property, and Ni-$49.2Ti_{49.2}$ had low apparent proof strength and large elongation, (ii) the R residual strain increased with increasing titanium content, and (iii) even small reductions of titanium purity influenced the tensile properties and the transformation temperatures causing high apparent proof strength, low residual strain, and low elongation because of the reduction in transformation temperatures.

7.1.3 Temperature effects

Due to their exceptional temperature sensitivity, superelastic NiTi archwires may be affected by temperature changes associated with ingestion of cold or hot food. Iijima et al. [47] investigated mechanical properties of superelastic NiTi orthodontic wires under controlled stress and temperature. The three-point bending test was carried out at constant temperature (23, 37 and 60°C) and stepwise temperature changes (37–60°C and to 37°C) and (37–2°C and to 37°C). Five specimens of each wire were tested. Micro-XRD spectra were measured at the tension side of the wire when the temperature changed from 37 to 60°C or 2°C. It was reported that (i) the load during the stepwise temperature changes (37–2°C and to 37°C) was consistent with that measured at a corresponding constant temperature, (ii) the austenite phase was transformed to martensite phase when the temperature is decreased from 37 to 2°C and (iii) in a stepwise temperature change (37–60°C and to 37°C), the load became higher than the original load at each corresponding constant temperature; however, there was no detectable change in the micro-XRD. Lombardo et al. [48] also studied the effect of temperature on the mechanical behavior of NiTi orthodontic archwire (cross-sectional diameter ranging from 0.010 to 0.016 inch) . A three-point wire-bending test was conducted on three analogous samples of each type of archwire at 55 and 5 °C, simulating an inserted archwire that is subjected to cold or hot drinks during a meal. It was obtained that (i) permanent strain was exhibited by all wires tested at 55 °C, (ii) statistically significant differences were found between almost all wires for the three considered parameters when tested at 55 and 5 °C. Loads were greater at 55 °C than at 5 °C, (iii) differences were also found between traditional and heat-activated archwires, the latter of which generated longer plateaus at 55 °C, shorter plateaus at 5 °C, and lighter mean forces at both temperatures, and (iv) the increase in average force seen with increasing diameter tended to be rather stable at both temperatures. Yanaru et al. [49] evaluated the temperature and deflection dependences of orthodontic force with NiTi wires with different A_F temperatures (namely, 35 and 40 °C). It was reported that (i) both wires showed typical superelastic hysteresis loops under the restraint condition at 40 °C, (ii) the force levels were significantly larger than those generally obtained by simple three-bending test, (iii) the recovery forces in the plateau region at 1.0 mm deflection were much larger than desired in the clinical guidelines around oral temperatures, and (iv) in the shape-memory wire with AF of 40 °C, the recovery force rapidly decreased to zero by a small reduction of the deflection from its maximum while the wire again exerted the force with the remaining permanent deflection by temperature rising. The effect of short-term temperature changes was investigated on the mechanical properties of rectangular NiTi archwires [50]. Eight rectangular superelastic wires were activated to 20 °C in longitudinal torsion at body temperature and subjected to cold (10 °C) or hot (80 °C) water with the strain held constant. It was reported that the effect of hot water disappeared quickly, but the wires remained at a level of reduced torsional stiffness (up to 85% less than baseline) after short applications of cold water, and (ii) the torsional stiffness remained low (up to 50% less than baseline) and showed no tendency to increase even after 2 h of postexposure restitution.

Barwart et al. [51] studied the effect of temperature change on the force delivery of NiTi springs in their superelastic range. Japanese NiTi closed coil springs were heated and cooled between 20 and 50 °C, while held in constant extension. It was found that (i) for all the springs examined, load values were found to increase with rising temperatures and decrease with a drop in temperature, (ii) this relationship between temperature change and load was more pronounced in the case of NiTi than in the steel springs, (iii) the force measured at 37 °C was about twice as high as at 20 °C for one type of NiTi spring, and (iv) on cooling, the superelastic springs showed unusual behavior; immediately after the temperature started to drop, a rapid decrease in force occurred to levels below those found at rising temperatures, this nonlinear decrease in load was not observed in the SS springs tested. Saedi et al. [52] fabricated Ni-rich $Ni_{50.8}Ti_{49.2}$ (at%) alloys by selective laser melting, followed by solutionizing and heat treating at 350 and 450 °C. It was found that (i) transformation temperatures, transformation behavior, strength, and recoverable strain are highly heat treatment dependent, (ii) samples aged at 350 °C showed better recovery in SE tests where 350 °C × 18 h aged samples exhibited almost perfect SE with 95% recovery ratio with 5.5% strain in the first cycle and stabilized SE with a recoverable strain of 4.2% after 10th cycle, and (iii) 450 °C × 10 h aged sample exhibited 68% recovery with a recoverable strain of 4.2% in the first cycle and stabilized recoverable strain of 3.8% after 10th cycle. Miyara et al. [53] investigated the influence of heat treatment on the mechanical properties of NiTi endodontic files (1.00 mm in diameter, a conical shape with 0.30-mm diameter tip and 0.06 taper). Specimens were heated for 30 min at 300, 400, 450, 500, or 600 °C. It was reported that (i) M_S and A_F temperatures for groups 400 and 450 °C were higher than those for others, (ii) the load/deflection ratios of groups 400, 450, and 500 °C were lower than that of group 600 °C, (iii) the bending load values at 2.0-mm deflection of groups 400, 450, and 500 °C were lower than those of group 300 °C and the control group, (iv) the cyclic fatigue strength of groups 400, 450, and 500 °C exceeded that of group 600 °C, and (v) changes in flexibility with heat treatment could improve the cyclic fatigue properties of NiTi instruments. Tadayyon et al. [54] investigated the influence of the annealing temperatures on the thermo-mechanical behavior of Ti-rich NiTi alloy with regard to transformation temperatures, mechanical properties at room temperature, and microstructure evolution under deformation. It was reported that (i) annealing above the recrystallization temperature (600 °C) modulated the mechanical behavior of the alloy significantly, (ii) by increasing the annealing temperature, the shape-memory behavior of the alloys improved, (iii) fractography on fractured surface revealed the brittle fracture area produced through the propagation of cleavage cracks; however, ductile fracture via nucleation growth and coalescence of microdimples in the martensitic phase at room temperature were also observed, and (iv) during plastic deformation, the NiTi alloy was also observed to undergo a detwinning process, dislocation slip and the formation of submicrocrystalline grains, nanocrystallization, and amorphous bands.

7.1.4 Tensile properties

The effect of casting mold materials (silica investment and magnesia investment) and subsequent heat treatment (at 440 °C × 30 min) on tensile properties of NiTi alloy with $Ti_{49.0}$ and $Ti_{49.2}$ (at%) casts was studied [55]. It was reported that (i) apparent proof strength decreased in both compositions, and residual strain increased in $NiTi_{49.2}$ by the heat treatment, (ii) elongation increased in $NiTi_{49.0}$ with use of the magnesia mold or by the heat treatment, and (iii) the transformation temperatures of $NiTi_{49.2}$ increased with use of the magnesia mold. Soga et al. [56] investigated the alloying effect of palladium (with 5.0, 7.5, 10.0, and 15.0 at%) to $Ni_{50}Ti_{50}$ alloy to evaluate tensile properties and transformation temperatures. It was found that (i) $TiNi_{42.5}Pd_{7.5}$ alloy castings showed good SE among the examined alloys from the viewpoint of residual strain and elongation, (ii) the apparent proof stress could be changeable by the proportion of Ti and Ni with residual strain being kept low, and (iii) $TiNi_{42.5}Pd_{7.5}$ alloy castings exhibited better superelastic flexibility than $TiNi_{50.8}$ alloy, which is proven by lower apparent proof stress and larger elongation, which might be caused by its relatively high martensitic transformation starting temperature point. Gall et al. [57] examined the structure and properties of cold drawn $TiNi_{50.1}$ and $TiNi_{50.9}$ (at%) SMA wires. It was found that (i) wires with both compositions possess a strong <111> fiber texture in the wire drawing direction, a grain size on the order of micrometers, and a high dislocation density, (ii) the more Ni-rich wires contain fine second phase precipitates, while the wires with lower Ni content are relatively free of precipitates, and (iii) the wire stress–strain response depends strongly on composition through operant deformation mechanisms and cannot be explained based solely on measured differences in the transformation temperatures. Kwon et al. [58] examined the effect of acidic fluoride solution on NiTi archwires by testing crystal structure, tensile strength, morphology after fracture, and element release from wire under four different test solutions after immersion for 1 or 3 days. It was reported that (i) three-day immersion in a 0.2% acid fluoride solution (pH 4) did not form any new crystal structure; however, tensile strength after immersion was changed compared to the as-received wires, (ii) the fractured wires showed dimple patterns in the inner part of the wire and ductile features on the outer part, (iii) element release in the test solution increased as NaF concentration and the period of immersion increased, and as pH valued decreased; that is, wires immersed in a 0.2% solution (pH 4) released several-fold greater amount of elements than wires in a 0.05% solution (pH 4), and (iv) tensile strength and element release were affected by acidic fluoride solution summarizing that NaF concentration, pH value, and the period of immersion were the factors affecting these properties.

7.1.5 Compression and hardness

Sehitoglu et al. [59] studied the deformation of NiTi shape-memory single crystals under compression loading for selected crystal orientations and two different Ti_3Ni_4 precipitate sizes. It was reported that (i) the transformation proceeds beyond the stress plateau region and extends until martensite yielding occurs resulting in a recoverable strain levels equivalent to the theoretical estimate of 6.4%, and (ii) since the austenite and martensite yield levels are reached at a smaller strain level in this case, the maximum recoverable strain was limited to 3.5% even though the theoretical estimates are near 5.1% suggesting that, in order to optimize the material performance, close attention must be paid to the selection of the crystallographic orientation and the precipitate size through heat treatment. Zhao et al. [60] processed porous NiTi alloy (with a potential applicability as a high-energy absorbing material) with several different porosities by spark plasma sintering. It was reported that the compression behavior of the porous NiTi showed a reasonably good agreement with those for porous NiTi with 13% porosity. The relationship between the heat treatment temperature and the thermomechanical behavior of the Ni-rich NiTi SMA was experimentally investigated [61]. It was reported that (i) the transformation temperatures and the compressive mechanical response are strongly affected by the heat treatment temperatures, (ii) the stress-induced phase transformation was characterized through the relationship between the heat treatment temperature and transformation stresses, elastic moduli, and the deformation level upon loading, and (iii) to better understand the global tendency of the various phase transformation stages, the critical stress–temperature diagram was provided for different proposed heat treatment temperatures. The deformation behavior under uniaxial compression of NiTi containing 0, 10, and 20 vol% TiC participates was studied both below and above the matrix martensitic transformation temperature: (1) at room temperature, where the martensitic matrix deforms plastically by slip and/or twinning and (2) at elevated temperature, where plastic deformation of the austenitic matrix takes place by slip and/or formation of SIM [62]. It was found that the effect of TiC particles on the stress–strain curves of the composites depends upon which of these deformation mechanisms is dominant (i) in the low-strain elastic region, the mismatch between the stiff, elastic particles, and the elastic-plastic matrix is relaxed in the composite, (ii) in the moderate-strain plastic region where nonslip deformation mechanisms are dominant, mismatch dislocations stabilize the matrix for all samples leading to strengthening of the composites, similar to the strain-hardening effect observed in metal matrix composites deforming solely by slip, and (iii) in the high-strain region controlled by dislocation slip, weakening of the NiTi composite results, because the matrix contains the untwinned martensite or retained austenite, which exhibit lower slip yield stress than twinned or stress-induced martensite, respectively.

The effect of various degrees of deformation was investigated at specific locations in the stress–strain curve under compression on the corrosion resistance of a wrought NiTi alloy with a martensite to austenite transformation peak of 110 °C [63]. Two metallurgical conditions were evaluated: 30% cold drawn and annealed at 900 °C for 1 h.

It was found that (i) the corrosion current density undergoes a significant reduction while the breakdown potential improves at increasing strains, and (ii) the alloy in the annealed condition exhibited breakdown potentials above 1,000 mV with current densities lower than 10 $\mu A/cm^2$ when it was strained to 24.4%. Qiu et al. [64] investigated the compressive response of martensitic NiTi SMA rods at various strain rates (400, 800, and 1,200 s^{-1}) and temperatures (room temperature in the martensite condition and 100 °C in the austenite condition). It was reported that (i) at room temperature, the critical stress increases slightly as strain rate increases, whereas the strain-hardening rate decreases; however, the critical stress under high strain rate compression at 100 °C increase first and then decrease due to competing strain hardening and thermal softening effects, (ii) after high rate compression, the microstructure of both martensitic and austenitic NiTi alloys changes as a function of increasing strain rate, while the phase transformation after deformation is independent of the strain rate at room temperature and 100 °C, and (iii) the dynamic recovery and recrystallization are also observed to occur after deformation of the austenitic NiTi alloy at 100 °C.

Xu et al. [65] studied the dynamic impact behavior of NiTi alloy at various impact velocities and temperatures. It was reported that (i) the maximum contact force increased with temperature, while the contact time decreased, and (ii) when temperature was above the reverse martensitic transformation finish temperature, the NiTi alloy specimen in the parent phase retained no permanent deformation but dissipated impact energy during the impact process. Wang et al. [66] conducted combined torsion–tension cycling experiments on thin-wall tubes (with thickness/radius ratio of 1:20, being similar to that found for stents) of nearly equiatomic NiTi SMAs. It was obtained that (i) the equivalent stress increases greatly with a small amount of applied axial strain, and the equivalent stress–strain curves have negative slopes in the phase transformation region, (ii) the shear stress drops when the torsional strain is maintained at its maximum value and the tensile strain is increased, and (iii) the shear stress increases with decreasing tensile strain, but it cannot recover to the original value after the complete unloading of the tensile strain. Lee et al. [67] examined the surface modification by boron implantation into NiTi alloy to improve NiTi root canal instruments with excellent cutting properties, without affecting their superelastic bulk-mechanical properties. With an implantation dose of 4.8 × 10^{17} boron/cm^2, a high concentration of boron (30 at%) was incorporated into NiTi alloy by 110 keV boron ions at room temperature (25 °C). It was reported that (i) boron-implanted and unimplanted NiTi alloys show surface hardness of 7.6 ± 0.2 and 3.2 ± 0.2 GPa, respectively, at the nanoindentation depth of 0.05 μm, and (ii) the ion-beam-modified NiTi alloy exceeds the surface hardness of SS.

7.1.6 Tension/compression

So far we have been seeing that mechanical behaviors of NiTi alloy are highly nonlinear and is strongly dependent on alloy composition, heat treatment, and (thermo) mechanical work. Orgéas et al. [68] conducted both uniaxial tension and compression tests which were performed with the same form for the specimens (sheet samples of gauge length 40 mm and cross section 5.6 mm × 2.7mm) in order to avoid any geometrical effect. The test NiTi sample was subjected to prior thermomechanical treatment, that is, a cold rolling leading to a thickness reduction of 18% followed by an annealing at 430 °C for 30 min. It was found that NiTi exhibited tension–compression asymmetry. Liu et al. [69] confirmed this tension–compression asymmetric behavior of NiTi alloy. This asymmetry was also examined by conducting the strain-controlled tension-compression test of NiTi shape-memory wires [70]. It was mentioned that the compression recovery forces were found to be markedly higher than tension forces. The effect of strain rate on tension–compression mechanical behavior was studied [71–73]. Nemat-Nasser et al. [72] and Pieczyska et al. [74] mentioned that there are potential of temperature rises during the high rate mechanical tests. This temperature rise was reflected to study the effect of temperature on mechanical behavior of NiTi alloy [75]. Siviour et al. [71] investigated the response of a superelastic NiTi alloy to mechanical deformation in tension and compression at strain rates of 10^{-3} and 10^3/s. The effect of loading direction has been understood through the nature of the phase transformation in the material, while comparison of quasi-static experiments at elevated temperatures to the response of the material at high strain rates provides better understanding of the importance of adiabatic heating in high strain rate loading. Chen et al. [73] determined the compressive stress–strain behavior of a NiTi SMA over strain rates of 10^{-3}–7.5×10^2/s. It was reported that (i) the plateau stress of the SMA is strain-rate dependent, (ii) under quasi-static deformation and low-amplitude impact loading, the stress–strain curve of the NiTi alloy forms a closed hysteresis loop, with the material returning to its original length upon unloading, and (iii) at higher dynamic loads, there is initially a residual deformation upon unloading, but the material slowly recovers its length. Since testing under relatively high strain rate (or deformation) might cause a temperature rise, the extent to which adiabatic heating affects rate behavior in these materials would be problematic. As discussed in early chapters, the origins of the SE of NiTi alloys arise from an austenitic phase to martensitic phase transformation. Superelastic response arises when the austenite finish temperature A_F lies a little below the operating temperature, so that the austenite phase is stable, but SIM can be produced. It is expected that raising the temperature (moving away from A_F) increases the transformation stress–strain.

7.1.7 Bending characteristics

Bending property is equally important for both endodontic instruments and orthodontic wires. Testarelli et al. [76] evaluated the bending properties of endodontic $Ni_{52}Ti$ (wt%) instruments to compare them with other commercially available NiTi rotary instruments, for both of which the tip size was 25, 0.06 taper. Tested instruments included (as trade name) Hyflex, EndoSequence, ProFile, Hero, and FlexMaster. It was found that (i) Hyflex files were found to be the most flexible instruments, with a significant difference in comparison with the other instruments, and (ii) among the other files, a significant difference has been reported for EndoSequence instruments compared with ProFile, Hero, and FlexMaster, whereas no significant differences have been reported among those three files. Tsao et al. [77] analyzed the stress distribution and flexibility of NiTi instruments which were subjected to bending forces. Two NiTi instruments (RaCe and Mani NRT) with different cross sections and geometries were evaluated. It was reported that (i) according to analytical results, the maximum curvature of the instrument occurs near the instrument tip, and (ii) results of the finite element analysis revealed that the position of maximum von Mises stress was near the instrument tip indicating that the model can be used to predict the position of maximum curvature in the instrument where fracture may occur. Schäfer et al. [78] compared the bending properties of different rotary NiTi instruments and investigated the correlation between their bending moments and their cross-sectional surface areas using ten files with different type, taper, and size. It was found that (i) bending moments were significantly lower for ProFile and RaCe files than for all other files, (ii) K3 files were significantly less flexible than all other instruments, and (iii) the correlation between stiffness and cross-sectional area was highly significant concluding that NiTi files with tapers greater than.04 should not be used for apical enlargement of curved canals because these files are considerably stiffer than those with 0.02 or 0.04 tapers. Yahata et al. [79] investigated the effect of heat treatment on the bending properties of NiTi endodontic instruments in relation to their transformation behavior. NiTi superelastic wire (1.00 mm Ø) was processed into a conical shape with a 0.30 mm diameter tip and 0.06 taper. The heat treatment temperature was set at 440 or 500 °C for a period of 10 or 30 min whereas specimens without heat treatment served as a control. It was reported that (i) the transformation temperature was higher for each heat treatment condition compared with the control, (ii) two clear thermal peaks were observed for the heat treatment at 440 °C; the specimen treated at 440 °C × 30 min exhibited the highest temperatures for M_S and A_F, with subsequently lower temperatures observed for specimens treated with other combinations as well as the control, and (iii) the specimen heated at 440 °C × 30 min had the lowest bending load values, both in the elastic range (0.5 mm deflection) and in the superelastic range (2.0 mm deflection) indicating that the influence of heat treatment time was less than that of heat treatment temperature. Silva et al. [80] evaluated the bending resistance and the cyclic fatigue life of a new heat-treated reciprocating instrument (ProDesign R (PDR)). Untreated PDR, Reciproc R25, and WaveOne Primary instruments were used as reference instruments for comparison. It was observed that (i) untreated PDR presented significantly higher bending resistance than the other tested systems, (ii) no differences were observed between PDR and Reciproc files regarding the bending resistance, (iii) PDR revealed a significantly longer cyclic fatigue life while untreated PDR and WaveOne instruments presented significantly lower cyclic fatigue life than

Reciproc, (iv) the new heat-treated reciprocating instrument PDR have higher cyclic fatigue resistance than untreated PDR, Reciproc, and WaveOne instruments, and (v) PDR and Reciproc were significantly more flexible than untreated PDR and WaveOne files.

Cantilever bending properties were evaluated for several clinically popular sizes of three superelastic and three nonsuperelastic brands of NiTi orthodontic wires in the as-received condition, and for 0.016-inch diameter wires after heat treatment at 500 and 600 °C for 10 min and for 2 h [81]. It was mentioned that (i) the bending properties were similar for the three brands of superelastic wires and for the three brands of nonsuperelastic wires, (ii) for the three brands of superelastic wires, heat treatment at 500 °C × 10 min had minimal effect on the bending plots, whereas heat treatment at 500 °C × 2 h caused decreases in the average superelastic bending moment during deactivation; heat treatment at 600 °C resulted in loss of SE, (iii) the bending properties for the three brands of nonsuperelastic wires were only slightly affected by these heat treatments, and (iv) the differences in the bending properties and heat treatment responses are attributed to the relative proportions of the austenitic and martensitic forms of NiTi in the microstructures of the wire alloys. Yamamoto et al. [82] investigated bending property of TiNiCu alloy castings in a three-point bending test for orthodontic application in relation to the phase transformation. The compositions of the alloys were $TiNi_{50.8}$ and $TiNi_{40.8}Cu_{10.0}$ (mol%), and four cross-sectional shapes of the specimens were selected and was heat-treated at 440, 480, and 520 °C for 30 min. It was reported that (i) the bending load changed by the cross-sectional size and shape mainly because of the difference in the moment of inertia of area, but the load-deflection relation did not differ proportionally in the unloading process, (ii) the difference between the load values in the loading and the unloading processes was relatively small for TiNiCu alloy, (iii) with respect to the residual deflection, there was no significant difference between TiNi and TiNiCu alloys with the same treatment condition, and (iv) the load values in the loading and the unloading processes decreased by each heat treatment for Ti–Ni alloy; however, the decrease in the load values for Ti–Ni–Cu alloy was not distinct concluding that TiNiCu alloy castings produce effective orthodontic force as well as stable low residual deflection, which is likely to be caused by the high and sharp thermal peaks during phase transformation. The heat-treated NiTi alloy wire (in a range from 400 to 540 °C for 30 min) were subjected to a three-point bending test and differential scanning calorimetry [83]. It was reported that (i) the transformation temperatures of the wires were lowered with increasing heat treatment temperature, (ii) the reverse transformation finishing temperature was below the body temperature with the treatment above 480 °C, (iii) residual deflection of the NiTi wire after bending was small with the secondary heat treatment above 460 °C, (iv) the load in the unloading process was less changeable and increased with the treatment temperature between 460 and 540 °C, and (v) secondary heat treatment in this range was suitable for using SE in expansion arch appliances. Moreover about the effects of heat treatment on bending–deflection behavior of NiTi wire was studied [84]. A total of 106 segments of NiTi wires (0.019 × 0.025 inch) and heat-activated NiTi wires (0.016 × 0.022 inch) from four commercial brands were tested. It was found that there were no statistically significant differences between the tested groups with the same size and brand of wire suggesting that heat treatment applied to the distal ends of rectangular NiTi archwires does not permanently change the elastic properties of the adjacent portions.

With respect to bending stiffness, Lim et al. [85] studied the transverse stiffness of two aesthetic orthodontic archwires (0·018-inch Teflon-coated SS and 0·017-inch Optiflex) in a simulated clinical setting and to assess the influence of deflection direction on the bending stiffness. The deflection of the archwires was measured with a travelling microscope and the load measured with a calibrated strain gauge ring transducer. It was reported that (i) the mean stiffnesses of the archwires in the lingual, labial, and occlusal deflection groups were found to be 2·9, 0·8, and 2·5 mN/mm, respectively, for 0·017-inch Optiflex, 13·2, 10·5, and 24·5 mN/mm, respectively, for 0·018-inch Teflon-coated SS and 26·6, 16·4, and 32·3 mN/mm, respectively, for the control, (ii) springback was found to be poor for Optiflex and the archwire remained bent upon deactivation, (iii) the influence of arch curvature on the bending stiffness was significantly different for Optiflex, Teflon-coated SS, and the control group, (iv) stiffness for the Teflon-coated archwire was found to be higher and more in line with the stiffness for the control, (v) both the Teflon-coated and SS archwires displayed good springback property; however, Optiflex was found to have low stiffness and resilience, and (vi) due to the poor springback, the clinical efficacy of Optiflex is probably limited. Garrec et al. [86] investigated the stiffness in bending of a superelastic NiTi orthodontic wire as a function of cross-sectional dimension. Fifteen NiTi archwires with three different cross-sectional dimensions were tested in three-point bending to determine the nature of forces in a loading and unloading cycle. It was mentioned that (i) the applied forces or stiffness dependence on cross-sectional size differs from the linear-elastic prediction because of the SE property, (ii) martensitic transformation is at the origin of nonlinear elasticity, (iii) the stiffness decreases with increasing deflection, and this phenomenon is emphasized in the unloading process, (iv) the value of stiffness appears to vary with wire size but depends on the ratio of volume of martensitic transformation, and (v) during martensitic transformation, the rigidity (elastic modulus) of the alloy is not constant suggesting that the obtained results should allow a different approach of biomechanical considerations, that is, a large-size square wire does not produce necessarily high forces.

Shima et al. [87] conducted feasibility study on hollow superelastic NiTi alloy wire and and hybrid wires with inserted different types of wires. It was obtained that (i) the hollow wire had lower load in the superelastic range, smaller load-deflection rate, and stress hysteresis in comparison with the conventional wire of the same diameter, (ii) the load of the hollow wire was controllable by heat treatment and the stress hysteresis was further decreased by a two-step heat treatment, (iii) the compound wire formed by inserting other types of wires into the hollow core exhibited changes in various bending properties such as increased load or load-deflection rate, according to the types and diameters of the inserted wire, and (iv) the hollow wire delivers much lighter and more continuous orthodontic force, and through heat treatment or deployment as a compound wire, it is possible to alter various bending properties indicating that the hollow wire was evaluated as a promising candidate for orthodontic application. Ballard et al. [88] evaluated the bending properties of fiber-reinforced polymeric composite archwires compared with similarly sized NiTi archwires. It was reported that (i) the 0.018-inch NiTi archwire demonstrated the highest force values at different deflection distances followed by Translucent Archwire II, 0.016-inch NiTi, and Translucent Archwire I, (ii) the 0.014-inch NiTi and 0.016-inch NiTi exhibited the highest modulus value, followed by 0.018-inch NiTi, 0.014-inch NiTi, Translucent Archwire II, and finally Translucent Archwire I, and (iii) during deactivation, the elastic recovery of 0.014-inch NiTi and 0.016-inch NiTi was significantly greater than Translucent Archwire II. Orthodontic mechanotherapy depends mainly on the loads developed by metal wires, and the load developed by a buckled orthodontic wire is of great concern for molar distalization and cannot be simply derived from mechanical properties measured through classical tests (i.e., tensile, torsion, and bending) [89]. The tested wires were activated and deactivated by loading and unloading. It was found that (i) the load due to buckling depends on material composition, wire length, the amount of activation, temperature, and deformation rate, (ii) at a temperature higher than the austenite finish transition temperature, superelastic wires were strongly dependent on temperature and deformation rate, (iii) the effect due to an increase of deformation rate was similar to that of a decrease of temperature, (iv) load variations due to temperature of a superelastic wire with a length of 20 mm were estimated to be approximately 4 g/ C, and (v) the high performance of an applied superelastic wire may be related to the high dynamics of the load in relation to temperature.

7.1.8 Load-deflection behavior

Wilkinson et al. [90] investigated the load-deflection characteristics of seven different 0.016-inch initial alignment archwires (Twistflex, NiTi, and five brands of heat-activated superelastic NiTi) with modified bending tests simulating a number of conditions encountered clinically and wire deflection was carried out at three temperatures (22.0, 35.5, and 44.0 °C) and to four deflection distances (1, 2, 3, and 4 mm). It was reported that (i) the effects of model, wire, and temperature variation were all statistically significant, (ii) Twistflex and the heat-activated superelastic wires produced a range of broadly comparable results, and NiTi gave the highest unloading values, and (iii) model rankings indicated that self-ligating Twin-Lock brackets produced lower friction than regular edgewise brackets. Force-deflection properties of initial orthodontic archwires (multistranded SS, conventional SS, superelastic NiTi, and thermoactivated NiTi archwires) were studied [91]. It was found that (i) significant

differences in deactivation forces were observed among the tested wires, (ii) the multistranded SS wire had the lowest mean deactivation force (1.94 N), while the conventional SS group had the highest value (4.70 N), and (iii) the superelastic and thermo-
activated NiTi groups were similar to the multistranded wire. Ahnadabadi et al. [92]
investigated the static and cyclic load-deflection behavior of NiTi archwires through
three-point bending test as a function of bending time, temperature, and number of
cycles which affects the energy dissipating capacity. It was mentioned that (i) NiTi
archwires are well suited for cyclic load–unload dental applications and (ii) reduction
in superelastic property for used archwires after long-time static bending was noted.
Watanabe et al. [93] compared the deflection load characteristics of homogeneous
and heterogeneous joints made by laser welding using various types of orthodontic
wires. Four kinds of straight orthodontic rectangular wires (0.017 inch × 0.025 inch)
included SS, Co–Cr–Ni, β-Ti, and NiTi. Homogeneous and heterogeneous end-to-end
joints (12 mm long each) were made by Nd:YAG laser welding. It was obtained that
(i) the deflection loads for control wires measured were as follows: SS: 21.7 ± 0.8 N;
Co–Cr–Ni: 20.0 ± 0.3 N; β-Ti: 13.9 ± 1.3 N; and NiTi: 6.6 ± 0.4 N, (ii) all of the homogeneously welded specimens showed lower deflection loads compared to corresponding control wires and exhibited higher deflection loads compared to heterogeneously
welded combinations, (iii) for homogeneous combinations, Co–Cr–Ni/Co–Cr–Ni
showed a significantly higher deflection load than those of the remaining homogeneously welded groups, (iv) in heterogeneous combinations, SS/Co–Cr–Ni and β-Ti/
NiTi showed higher deflection loads than those of the remaining heterogeneously
welded combinations (significantly higher for SS/Co–Cr–Ni), and (v) significance
was shown for the interaction between the two factors (materials combination and
welding method); however, no significant difference in deflection load was found
between four-point and two-point welding in each homogeneous or heterogeneous
combination. Lombardo et al. [94] investigated the characteristics of commonly
used types of traditional and heat-activated initial archwire by plotting their load/
deflection graphs and quantifying three suitable parameters describing the discharge
plateau phase. Forty-eight archwires (22 NiTi and 26 heat-activated) of cross-sectional
diameter ranging from 0.010 to 0.016 inch were obtained from seven different manufacturers. A modified three-point wire-bending test was performed on three analogous
samples of each type of archwire at a constant temperature (37.0 °C). It was obtained
that (i) statistically significant differences were found between almost all wires for
the three parameters considered, (ii) statistically significant differences were also
found between traditional and heat-activated archwires, the latter of which generated longer plateaus and lighter average forces, and (iii) the increase in average force
seen with increasing diameter tended to be rather stable, although some differences
were noted between traditional and heat-activated wires concluding that on average,
the increase in plateau force was roughly 50% when the diameter was increased by
0.002 inch (from 0.012 to 0.014 and from 0.014 to 0.016 inch) and about 150% when
the diameter was increased by 0.004 inch (from 0.012 to 0.016 inch), with differences

between traditional and heat-activated wires noted in this case. Aharari et al. [95] investigates the effect of a fluoride mouthwash on load-deflection characteristics of three types of NiTi-based orthodontic archwires. Twenty maxillary 0.016 inch round specimens from each of the single-strand NiTi (Rematitan "Lite"), multistrand NiTi (SPEED Supercable), and copper NiTi (Damon copper NiTi) wires were selected. The specimens were kept in either 0.2% NaF or artificial saliva solutions at 37 °C for 24 h. It was reported that (i) immersion in NaF solution affected the load-deflection properties of NiTi wires, (ii) the unloading forces at 0.5 and 1.0 mm deflections were significantly lower in fluoride-treated specimens compared with the control groups, and (iii) unloading forces at 1.5, 2.0, and 2.5 mm deflections were not statistically different between fluoride- and saliva-treated specimens suggesting that subjecting NiTi wires to fluoride agents decreased associated unloading forces, especially at lower deflections, and may result in delayed tooth alignment. Jamleh et al. [96] studied the dispersion of the lifetime of NiTi instruments, and their deflecting load changes during cyclic fatigue. A total of 120 ProFile NiTi rotary instruments (PRI) were tested using a specially designed cyclic fatigue testing apparatus with three pins. It was found that (i) the averages of number of cycles to failure and deflecting load of 120 samples were 584.3 ± 180.5 cycles and 6.44 ± 0.91 N, respectively, and (ii) all samples showed a sequential decrease in defecting load during rotation concluding that it is impossible to estimate the lifetime of a NiTi instrument from number of cycles to failure, thus, the change in deflecting load could be an alternative criterion to determine the remaining lifetime.

Gatto et al. [97] investigated the mechanical properties of superelastic and thermal NiTi archwires for correct selection of orthodontic wires. Seven different NiTi wires of two different sizes (0.014 and 0.016 inch), commonly used during the alignment phase, were tested. A three-point bending test was carried out to evaluate the load-deflection characteristics. The archwires were subjected to bending at a constant temperature of 37 °C and deflections were 2 and 4 mm. It was reported that (i) the thermal NiTi wires exerted significantly lower working forces than superelastic wires of the same size in all experimental tests, (ii) wire size had a significant effect on the forces produced: with an increase in archwire dimension, the released strength increased for both thermal and superelastic wires, (iii) superelastic wires showed, at a deflection of 2 mm, narrow and steep hysteresis curves in comparison with the corresponding thermal wires, which presented a wide interval between loading and unloading forces, (iv) during unloading at 4 mm of deflection, all wires showed curves with a wider plateau when compared with 2 mm deflection due to the fact that such a difference for the superelastic wires was caused by the martensite stress induced at higher deformation levels, (v) a comprehensive understanding of mechanical characteristics of orthodontic wires is essential and selection should be undertaken in accordance with the behavior of the different wires, (vi) in low-friction mechanics, thermal NiTi wires are to be preferred to superelastic wires, during the alignment phase due to their lower working forces, and (vii) in conventional straight-wire mechanics, a low force archwire would be unable to overcome the resistance to sliding. Arreghini et al. [98] determined the relative stiffness of a large selection of commonly-used square and rectangular steel, supertempered steel, NiTi, and TMA orthodontic archwires of various cross sections in order to provide the clinician with a useful, easy-to-consult guide to archwire sequence selection. Twenty-four archwires of different cross-sectional shape, size, and material were selected and they were subjected to a modified three-point bending test. It was obtained that (i) a considerable difference in resistance to deflection was revealed between all the tested archwires, (ii) as expected, the resistance to deflection of archwires of the same cross section was found to increase with increasing stiffness of their construction material, (iii) specifically, steel archwires can be as much as eight times stiffer than NiTi archwires of the same shape and cross section, and supertempered steel archwires are invariably stiffer than traditional steel versions, and (iv) marked differences in resistance to deflection were also found between NiTi archwires made of the same material but with different shape characteristics.

7.1.9 Torsional properties

Since a majority of accidental fractures of endodontic instruments is taken place during the treatment operation due to unwanted torsional situation, it is worth to review the researches on the torsional behavior of NiTi alloys.

Wolcott et al. [99] evaluated and compared the torsional properties of SS K-type 0.02 taper and NiTi U-type 0.02 and 0.04 taper instruments in terms of the maximum torque, torque at failure, and angular deflection. It was reported that (i) SS K-type 0.02 taper and NiTi U-type 0.02 and 0.04 taper instruments met or exceeded specification standards for maximum torque, and they also satisfied and far exceeded the standards for angular deflection at the failure point, and (ii) the SS instruments showed no significant difference between maximum torque and torque at failure, whereas both of the NiTi instruments showed a significant differential between maximum torque and torque at failure. Rowan et al. [100] also compared the torsional properties of SS endodontic files (Flex-O-File; Maillefer/L. D. Caulk Co., Milford, DE) and NiTi endodontic files (Quality Dental Products, Inc., Johnson City, TN). File sizes 15, 25, 35, 45, and 55 were subjected to torsional load in clockwise direction and counterclockwise direction independently. It was found that (i) SS files had a significantly greater rotation to failure in the clockwise direction, whereas the NiTi files had a significantly greater rotation to failure in the counterclockwise direction, and (ii) despite these differences in rotation to fracture, there was essentially no difference between both the instruments in the torque that it took to cause failure in both directions indicating that whereas the number of CW (clockwise) and CCW (counter-clockwise) rotations to failure differed for the two instruments, the actual force that it took to cause that failure was the same. Yum et al. [101] investigated the torsional strength, distortion angle, and toughness of five NiTi rotary files with different cross-sectional geometries (Twisted File (TF) and RaCe with equilateral triangle, ProTaper with convex-triangle, ProFile with U-shape, and Mtwo with S-shape). The size 25/.06 taper of TF, RaCe, ProFile, and Mtwo and the ProTaper F1 files were tested, all with the same diameter at D5. It was found that (i) TF and RaCe had significantly lower yield strength than other systems, (ii) TF had a significantly lower ultimate strength than other files, whereas Mtwo showed the greatest, (iii) ProFile showed the highest distortion angle at break, followed by TF, (iv) ProFile also showed the highest toughness value, whereas TF and RaCe both showed a lower toughness value than the others, and (v) fractographic examination revealed typical pattern of torsional fracture for all brands, characterized by circular abrasion marks and skewed dimples near the center of rotation. Based on these findings, it was concluded that the five tested NiTi rotary files showed a similar mechanical behavior under torsional load, with a period of plastic deformation before actual torsional breakage but with unequal strength and toughness value. Park et al. [102] studied the cyclic fatigue of five NiTi rotary instruments: TF (SybronEndo, Orange, CA, USA) and RaCe systems (FKG Dentaire, La Chaux-de-Fonds, Switzerland), both with an equilateral triangular cross section, and the ProTaper (Dentsply Maillefer, Ballaignes, Switzerland), Helix (DiaDent, Cheongju, Korea), and FlexMaster (VDW, Munchen, Germany), which had a convex triangular cross section. It was obtained that (i) TF had the lowest and FlexMaster the highest torsional resistance among the groups, (ii) scanning electron microscopy examination revealed a typical pattern of torsional fracture for TF, RaCe, and ProTaper that was characterized by circular abrasion marks and skewed dimples near the center of rotation, and (iii) Helix and FlexMaster presented a rough, torn-off appearance concluding that files of same cross-sectional design may exhibit different resistance to fracture probably as a result of the manufacturing process. The use of reciprocating movement has been claimed to increase the resistance of NiTi file to fatigue in comparison with continuous rotation. Kim et al. [103] compared the cyclic fatigue resistance and torsional resistance of relatively newly developed two NiTi files (Reciproc and WaveOne). It was reported that (i) Reciproc had a higher numbers of cycles to fatigue failure (NCF) and WaveOne had a higher torsional resistance than the others, (ii) both reciprocating files demonstrated significantly higher cyclic fatigue and torsional resistances than ProTaper, and (iii) the fractographic analysis showed typical features of cyclic fatigue and torsional failure for all instruments. Kaval et al. [104] also evaluated the cyclic fatigue related torsional resistance of NiTi files including Hyflex electrical discharge machining (HFEDM), ProTaper Gold (PTG), and ProTaper Universal (PTU) instruments. It was reported that (i) HFEDM instruments exhibited the highest cyclic fatigue resistance and were followed by PTG and PTU groups, respectively, (ii) the mean fragment length for PTU instruments was significantly shorter than that of HFEDM and PTG instruments, and (iii) PTG instruments demonstrated significantly higher torsional resistance than HFEDM files and the distortion angle of the instruments ranged as follows: PTU < PTG < HFEDM files. Lopes [105] compared the torsional resistance of two brands of rotary NiTi instruments (Mtwo retreatment instruments and PTU retreatment instruments D2 and D3), being subjected to a torsional assay in clockwise rotation. It was obtained that (i) the angular deflection at fracture decreased in the following order: Mtwo retreatment file #15 > Mtwo retreatment file #25 > PTU retreatment file D2 > PTU retreatment file D3, (ii) as for the maximum torque values, the results revealed the following descending order: PTU file D2 > Mtwo retreatment file #25 > PTU file D3 > Mtwo retreatment file #15, and (iii) scanning electron microscopic analysis revealed that plastic deformation occurred along the helical shaft of the fractured instruments and fractured surfaces were of the ductile type. It was hence concluded that instruments tested showed different torsional behavior depending on the parameter evaluated; if one considers that high angular deflection values may serve as a safety factor, then the Mtwo retreatment instruments showed significantly better results.

The influence of cross-sectional profile on the mechanical behaviors of six commercial NiTi instruments (ProTaper, Hero642, Mtwo, ProFile, Quantec, and NiTiflex) was studied [106]. Mathematical models for these instruments were constructed and their performances were analyzed under equal torque conditions. It was shown that (i) the ProTaper and Hero642 models achieved the lowest stress levels that made them the most torque-resistant while the NiTiflex model was the poorest, (ii) the maximum stress value and the stress distribution in a model were found strongly influenced by the cross-sectional profile, (iii) factors affecting the stress distribution include the cross-sectional inertia, depth of the flute, area of the inner core, radial land, and peripheral surface ground, and (iv) as the area of the inner core of the cross section increased, the model was more torque-resistant. Ninan et al. [107] investigated the torsion and bending properties of shape-memory files [controlled memory (CM) Wire, HyFlex CM (HFCM), and Phoenix Flex] and compared them with conventional (ProFile ISO and K3) and M-Wire (GT Series X (GTX) and ProFile Vortex) NiTi files. It was reported that (i) significant interactions were present among factors of size and file, (ii) variability in maximum torque values was noted among the shape-memory files' brands, sometimes exhibiting the greatest or least torque depending on brand, size, and taper, and (iii) in general, the shape-memory files showed a high angle of rotation before fracture but were not statistically different from some of the other files; however, the shape-memory files were more flexible, as evidenced by significantly lower bending moments. Yared et al. [108] compared the torque and rotation angle at fracture of new and used NiTi K3 0.04 rotary instruments. It was found that (i) the torque at fracture of the new instruments increased significantly with the diameter, (ii) the used instruments, sizes 25 to 40, had significantly lower torque at fracture values compared to the new ones and the used instruments, sizes 20 and 35, had significantly lower angle of rotation at fracture compared to the new ones, and (iii) a stronger relationship was found between the size of the file and the torque at fracture for the new instruments compared to the used ones, suggesting that the torque at fracture values of new instruments increased significantly with the diameter and the repeated use of 0.04 K3 instruments affected mainly the torque at fracture. Vieira et al. [109] investigated the influence of multiple clinical uses on the torsional behavior of PTU rotary NiTi instruments. It was obtained that (i) the use of the PTU rotary instruments by an experienced endodontist allowed for the cleaning and shaping of the root canal system of five molar teeth without fracture, (ii) the maximum torque for instruments S2, F1, and F2, and the angular deflection at fracture for instruments S2 and F1 were significantly lower following clinical use, (iii) the largest decrease in maximum torque was 18.6% for S2 instruments, and (iv) the same maximum percent decrease was found for angular deflection at fracture for F1 instruments concluding that torsional resistance and angular deflection of used instruments, as compared to that of new instruments, were reduced following clinical use. Camps et al. [110] compared stiffness and resistance to fracture of four brands of NiTi K files. Stiffness was determined by measuring the moment required to bend the instrument 45°. It was reported that (i) NiTi K files satisfied and far exceeded specification standards for stiffness, and they also satisfied and exceeded the standards for angular deflection at failure, and (ii) NiTi K files presented a null permanent deformation angle. Braga et al. [111] compared the flexibility, torsional resistance, and structural and dimensional characteristics of instruments produced by twisting with those of a geometrically similar NiTi produced by a grinding process, including TF and RaCe instruments. It was evaluated that (i) the two types of instruments had approximately the same chemical composition, phase constitution, and austenite finishing temperatures, and (ii) TF instruments had significantly lower Vickers microhardness values and were more flexible than RaCe instruments, but had similar (TF size 25, 0.08 taper and RaCe size 25, 0.06 taper) or significantly higher (TF size 25, 0.06 taper and RaCe size 25, 0.04 taper) torsional resistance.

Comparison of cyclic fatigue and torsional resistance was made in reciprocating single-file systems and continuous rotary instrumentation systems (80 instruments from the PTU, WaveOne, Mtwo, and Reciproc systems) being subjected to the dynamic bending testing in SS simulated curved canals [112]. It was found that (i) cyclic fatigue resistance was greater for reciprocating systems than for rotary systems, (ii) instruments from the Reciproc and WaveOne systems significantly differed only when axial displacement occurred, (iii) instruments of the PTU and Mtwo systems did not significantly differ, and (iv) cyclic fatigue and torsional resistance were greater for reciprocating systems than for continuous rotary systems, irrespective of axial displacement. Alcalde et al. [113] evaluated the cyclic and torsional fatigue resistance of the reciprocating single-file systems (Reciproc Blue (RB), ProDesign, and WaveOne). It was reported that (i) the cyclic fatigue resistance values of PDR 25.06 were significantly higher, (ii) RB 25.08 showed higher fatigue resistance than WaveOne Gold (WOG), (iii) the torsional test showed that PDR 25.06 had lower torsional strength, (iv) no differences were observed between RB 25.08 and WOG 25.07, (v) PDR 25.06 showed higher angular rotation values than RB 25.08 and WOG 25.07, and (vi) RB 25.08 presented higher angular rotation than WOG 25.07, concluding that PDR 25.06 presented the highest cyclic fatigue resistance and angular rotation until fracture compared to

RB 25.08 and WOG 25.07. In addition, RB 25.08 and WOG 25.07 had higher torsional strength than PDR 25.06. The authors mentioned as clinical relevance that in endodontic practice, thermally treated reciprocating instruments have been used for the root canal preparation of curved and constricted canals, therefore, these instruments should present high flexibility and suitable torsional strength to minimize the risk of instrument fracture [113].

There are several studies on effects of torsional actions on different mechanical properties as well as affecting factors on torsional behaviors of NiTi alloy. Oh et al. [114] studied the effect of torsional preloading on the torsional resistance of NiTi endodontic instruments. WaveOne Primary and PTU F2 files were used. In the phase 1 experiment, the ProTaper and WaveOne files were loaded to have a maximum load from 2.0 up to 2.7 or 2.8 Ncm, respectively. In the phase 2 experiment, the number of repetitions of preloading for each file was increased from 50 to 200, whereas the preloading torque was fixed at 2.4 Ncm. It was obtained that (i) all preloaded groups showed significantly higher ultimate strength than the unpreloaded groups, (ii) there was no significant difference among all groups for distortion angle and toughness, and (iii) although WaveOne had no significant difference between the repetition groups for ultimate strength, fracture angle, and toughness, ProTaper had a higher distortion angle and toughness in the 50-repetition group compared with the other repetition groups, concluding that torsional preloading within the ultimate values could enhance the torsional strength of NiTi instruments and the total energy until fracture was maintained constantly, regardless of the alloy type. Gambarini et al. [115] evaluated the effects of a final deoxidation process (Deox) on torsional resistance of the TF instruments. It was reported that TF instruments withstood 90% more torque (max. torque) than TF prototype instruments manufactured without the deoxidation process (88.7 vs. 53.3 g/cm) indicating that since design and dimensions of tested instruments were the same, the deoxidation process could be the only explanation of the improvement in torsional resistance. Miyai et al. [116] investigated the relationship between the functional properties and the phase transformation of NiTi endodontic instruments (EndoWave, HERO 642, K3, ProFile.06, and ProTaper). It was fund that (i) the maximum torsional torque values of HERO, K3, and ProTaper were significantly higher than those of EndoWave, ProFile, and K-file, (ii) the K-files had the lowest torque value, (iii) angular deflection at fracture was significantly higher for K-files than that of any NiTi instrument, (iv) the bending load values of HERO and K3 were significantly higher than those of EndoWave, ProFile, ProTaper, and K-file, (v) the K-files had the lowest load value, although residual deflection remained, and (vi) the transformation temperatures of HERO and K3 were significantly lower than those of EndoWave, ProFile, and ProTaper, concluding that the functional properties of NiTi endodontic instruments, especially their flexible bending load level, were closely related to the transformation behavior of the alloys. Acosta et al. [117] evaluated the influence of cyclic deformation on the torsional resistance of CM NiTi files (HyFlex, Typhoon (TYP), RaCe, and PTU F2 instruments) on comparison with superelastic instruments with similar geometric and dimensional characteristics. It was found that (i) new CM files had a significantly higher number of cycles to failure when compared with superelastic instruments; HF exhibited the highest value, (ii) the mean torque value for F2 was the highest, and (iii) the CM files precycled to 3/4 number of cycles to failure had a significantly lower torque than the new files, whereas the superelastic instruments displayed no significant differences, concluding that cyclic flexural loading significantly reduced the torsional resistance of the CM instruments. Silva et al. [118] evaluated the influence of CM thermal treatment on the torsional resistance and SR of NiTi instruments with identical geometric and dimensional characteristics. A sample of 25 conventional NiTi alloy and 25 CM thermal treated NiTi instruments with an identical geometric design and a nominal size of 0.25 mm at D_0 and a nominal taper of 0.06 mm/mm were selected. It was reported that (i) the torsional strength of thermal treated instruments was significantly lower than nontreated instruments; however, thermally treated instruments had significantly greater angular rotation to fracture than nontreated instruments, (ii) temperature did not influence the torsional strength and the angular rotation of the instruments, and (iii) regarding the roughness measurements, groove depth was lower at the surface of thermal-treated instruments when compared to nontreated instruments, concluding that thermal treatment influenced the torsional resistance and SR of NiTi instruments with identical geometric and dimensional characteristics, and temperature did not affect the torsional behavior. Bahia et al. [119] assessed the influence of cyclic deformation up to one half and three quarters of the fatigue life of NiTi rotary endodontic instruments (ProFile instruments in sizes #20, #25, and #30) on their behavior in torsion. It was mentioned that statistical analysis of the values obtained for the torque to failure and maximum angular deflection, using the Student t-test at the 95% confidence level, showed that the cyclic deformation of instruments of up to one half and three fourths of their fatigue life gives rise to a significant reduction in these parameters, and it was concluded that the simulated clinical use of ProFile instruments for shaping curved canals adversely affects their torsional properties. Campbell [120] studied the effects of fatigue damage on the torsional properties of both traditional NiTi and NiTi CM files (TYP). It was found that (i) TYP CM files had a number of cycles to failure 7 times higher than that of TYP files, (ii) no difference in torque between the CM files and the conventional NiTi files of either file size was detected, (iii) the torque of the size 40/.04 files was significantly higher than the torque of the size 25/.04 files, (iv) in the 40/.04 files group, TYP files in 75% precycling group had a significantly lower torque than files in the group with no precycling, whereas slight precycling (25%) significantly reduced the distortion angle on TYP CM files, (v) the CM files of both sizes had a significantly higher distortion angle than the corresponding NiTi files, and (vi) the fractured files in the precycling groups showed the typical pattern of torsional failure. Yared et al. [121] evaluated the influence of rotational speed, torque, and operator experience with a specific NiTi rotary instrumentation technique on the incidence of locking, deformation, and separation of instruments. PRI sizes 40–15 with a 6% taper were used in a crown-down technique. In one group of canals (n = 300) speeds of 150, 250, and 350 rpm (subgroups 1, 2, and 3) were used. Each one of the subgroups included 100 canals. In a second group (n = 300) torque was set at 20, 30, and 55 Ncm (subgroups 4, 5, and 6). In the third group (n = 300) three operators with varying experience (subgroups 7, 8, and 9) were also compared. It was reported that (i) in group 1 only one instrument was deformed in the 150-rpm group and no instruments separated or locked, (ii) in the 250-rpm group, instrument separation did not occur, however, a high incidence of locking, deformation, and separation was noted in the 350-rpm group, (iii) in general, instrument sizes 30–15 locked, deformed, and separated, (iv) overall, there was a trend toward a higher incidence of instrument deformation and separation in smaller instruments, (v) locking and separation occurred during the final passage of the instruments, in the last (tenth) canal in each subgroup. In the second group, neither separation nor deformation and locking occurred during the use of the ProFile instruments, at 150 rpm, and at the different torque values, (vi) in the third group, chi-squared analysis demonstrated that significantly more instruments separated with the least experienced operator, and (vii) instrument locking, deformation, and separation did not occur with the most experienced operator, concluding that preclinical training in the use of the PRI technique with crown-down at 150 rpm were crucial in avoiding instrument separation and reducing the incidence of instrument locking and deformation. Effect of high-pressure torsion deformation on surface properties and biocompatibility of $TiNi_{50.9}$ (mol%) alloys was investigated [122], which was subjected to high-pressure torsion deformation for different number of rotations (N) of 0.25, 0.5, 1, 5, and 10. The biocompatibility of the samples was evaluated based on a colony formation assay, nickel ion release, and protein adsorption behavior. It was found that (i) X-ray diffraction analysis revealed the occurrence of grain refinement, phase transformation, and amorphization in the TiNi samples by high-pressure torsion deformation due to high dislocation density, (ii) the changes in chemical composition and thickness of the passive film formed on the surface observed in X-ray photoelectron spectroscopy analysis reveals improvement in the stability of the passive film by torsional deformation, (iii) the microstructural change due to the deformation was found to influence the biocompatibility behaviors of TiNi, (iv) plating efficiency and protein adsorption were found to be higher when the samples are in stress-induced martensitic or amorphous state, and (v) the high-pressure torsion deformation was found to alter the surface behavior of the TiNi, which effectively reduced the Ni ion release and improved its biocompatibility.

In orthodontic treatment, there are several studies reported on comparison on torsion properties among different materials which are currently employed for mechanotherapy. Filleul et al. [123] examined the stress–strain behavior of two NiTi and two NiTiCu orthodontic wires in induced torsion under controlled conditions of moment and temperature. It was found that (i) the loading and unloading curves and plateau regions were found to be closely related to temperature with stiffness varying dramatically over mouth temperature range under identical stress, and (ii) diversity of reaction to stress is linked to the crystalline structure of the alloys. Kuroda et al. [124] measured the torque moment delivered by a novel elastic bendable wire, TiNb wire and compared with NiTi and TiMo alloy wires. The torque moment delivered by the various wire-bracket combinations was measured using a torque gauge at the temperature and humidity of 37 °C and 50%, respectively. It was reported that (i) as the degree of applied torque and the inserted wire size increased, the torque moment gradually increased, (ii) the torque moment of TiNb wires was smaller than those of NiTi wires and TiMo wires, at more than 20 degrees applied torque, and (iii) the torque moment with wire ligation was significantly larger than those with elastic ligation. Larson et al. [125] investigated the torsional elastic properties of orthodontic archwires using three different alloys: SS, β-Ti, and NiTi. It was reported that (i) after a clamping correction was applied, the values of the shear modulus (G) was 11.0–11.3, 4.3–4.4, and 2.3–3.0 Msi for SS, β-Ti, and NiTi wires, respectively, (ii) using a 10% energy loss criterion in a static test, the torsional yield strengths (τ_{ys}) were all nearly 100 ksi except for the 0.018″ SS (51 ksi) and rectangular NiTi (13–17 ksi); the latter exhibited some pseudoelastic behavior, and (iii) the NiTi wires displayed the greatest range and the least stiffness, whereas SS wires showed the greatest stiffness and the least range. Archambault et al. [126] compared the torque expression between wire types of SS, TiMo, and NiTiCu wire materials (0.019 × 0.0195 inch) with three 0.022-inch slot bracket combinations (Damon 3MX, In-Ovation-R, and SPEED). It was reported that (i) at low twist angles (<12 degrees), the differences in torque expression between wires were not statistically significant, (ii) at twist angles over 24 degrees, SS wire yielded 1.5 to 2 times the torque expression of TiMo and 2.5 to 3 times that of NiTi, and (iii) at high angles of torsion (over 40 degrees) with a stiff wire material, loss of linear torque expression sometimes occurred concluding that SS has the largest torque expression, followed by TiMo and NiTi.

7.1.10 Creep behaviors

If the SMA is subjected to the loading under the stress-controlled condition, creep (which is described as the transformation-induced creep [127]) and creep recovery can appear based on the martensitic transformation [127, 128]. During loading under constant stress rate, temperature increases due to the stress-induced martensitic transformation. If stress is held constant during the martensitic transformation stage in the loading process, temperature decreases and the condition for the progress of the martensitic transformation is satisfied, resulting in the transformation-induced creep deformation. If stress is held constant during the reverse transformation stage in the unloading process, creep recovery appears due to the reverse transformation [128–130]. Eggeler et al. [130] reported creep data of a Ni-rich NiTi alloy at stresses and temperatures in the range of 500 °C and 150 MPa. It was also mentioned that (i) the tempering and precreep result in a 20 °C increase of phase transformation temperatures, and (ii) the effect of an additional plastic deformation (as compared to a stress free aging treatment) is that phase transformation temperature-peaks broaden and overlap on cooling and broaden on heating. Russalian et al. [131] described the pseudo-creep phenomenon in such a condition that mentioned that the interruption of loading during reorientation and pseudoelasticity in SMAs with a strain arrest (namely, holding strain constant) results in a time-dependent evolution in stress or with a stress arrest (or holding stress constant) results in a time-dependent evolution in strain. Kobus et al. [132] studied the creep behavior of an Ni-rich NiTi SME alloy, which was exposed to temperatures in the range of about 500 °C. It was indicated that the effects of creep predeformation on the phase transition behavior of the material were determined.

7.1.11 Joint strengths

The bond behavior between NiTiNb SMA wires and fiber reinforced polymer composites was investigated [133]. A total of nine pull-out specimens were fabricated and tested. It was found that the interfacial bond transfer consists of two components: cohesion (before the onset of debonding) and friction (after the onset of debonding). Zhang et al. [134], utilizing the ultrasonic spot welding as a solid-state joining technique, fabricated a joining sheets of NiTi sheets using a Cu interlayer in between the two joining sheets to investigate the influence of the ultrasonic spot welding process on the microstructural characteristics and mechanical behavior of the NiTi joints. It was reported that (i) compared with conventional fusion welding techniques, no intermetallic compounds formed in the joints, which is of extreme importance for this particular class of alloys, (ii) the joining mechanisms involve a combination of shear plastic deformation, mechanical interlocking, and formation of microwelds, (iii) a better bonding interface was obtained with higher welding energy levels, which contributed to a higher tensile load, (iv) an interfacial fracture mode occurred and the fracture surfaces exhibited both brittle and ductilelike characteristics with the existence of tear ridges and dimples, and (v) the fracture initiated at the weak region of the joint border and then propagated through it leading to tearing of Cu foil at the fracture interface. Mesquita et al. [135] studied the appropriate power level for electric welding of three commercial brands of NiTi wires by electrical resistance welding. Three different wire products were tested, including GI (Orthometric), GII (3M), and GIII (GAC). It was reported that (i) the 2.5 power exhibited the lowest resistance to rupture in all tested groups while the 4.0 power provided the highest resistance in GI and GII (97.90 N and 99.61 N, respectively), while in GIII (79.28 N) the highest resistance was achieved with a 3.5 power welding, and (ii) the most appropriate power for welding varied for each brand, being 4.0 for Orthometric and 3M, and 3.5 for GAC NiTi wires. Vieira et al. [136] employed Nd:YAG continuous wave laser welding to superelastic

cold-rolled plates of NiTi 1 mm thick. It was reported that (i) the superelastic behavior of the welded joints was observed for applied stresses close to about 50 MPa below the ultimate tensile strength of the welds, (ii) the functionality was confirmed by analyzing the stabilization of the mechanical hysteretic response to strain levels up to 8%, and (iii) for tensile cycling involving strain levels larger than 6%, welded specimens were found to exhibit superior functional mechanical behavior presenting larger recoverable strain levels.

7.1.12 Sterilization effects

There are valuable information regarding the effects of various types of sterilization on NiTi materials in both endodontic instruments and orthodontic wires. Infection control is a major issue in dentistry, mainly, because of the concern over contagious diseases transmitted in health-care environment as well as professional oral heat providers. The frequency of recycling (or reusing) of endodontic instruments between patients is enormously higher than the cases in orthodontic wires/brackets or other medical instruments and devices. Hence, the cleaning and sterilization of endodontic instruments between treatment sessions are essential to prevent cross infection [137, 138]. As mentioned by Hurtt et al. [139], there are several different methods of file sterilization, including a glutaraldehyde immersion, steam autoclaving, and various techniques of salt sterilization. It was pointed out that (i) the only proper steam autoclaving reliably produced completely sterile instruments, and (ii) salt sterilization and glutaraldehyde solutions may not be adequate sterilization methods for endodontic hand files and should not be relied on to provide completely sterile instruments. Raju et al. [140] compared four different methods of sterilizing endodontic files in dental practice, including an autoclave, glass bead, glutaraldehyde, and CO_2 laser. It was found that (i) the files sterilized by autoclave and lasers were completely sterile, and (ii) those sterilized by glass bead were 90% sterile and those with glutaraldehyde were 80% sterile, concluding that autoclave or laser could be used as a method of sterilization in clinical practice and in advanced clinics; laser can be used also as a chair side method of sterilization.

There are many experimental evidences on sterilization effects on torsional properties of NiTi files. Iverson et al. [141] studied the effects of different sterilization methods on the torsional strength of two types of endodontic files (Kerr K-Flex files and Burns Unifiles) which were subjected to ten cycles of autoclaving, bead, dry heat, and cold chemical sterilization. It was reported that (i) repeated sterilization was found to have no effect on the torque resistance and degrees to failure for the Burns Unifile and no effect on degrees to failure for the K-Flex file, (ii) dry heat and cold sterilization, however, slightly increased the torque resistance of the K-Flex files as compared with the controls and the other sterilization methods, and (iii) no differences were found between the files in the degrees of rotation to failure but the

K-Flex files had a slightly greater torque resistance. Silvaggio et al. [142] determined whether heat sterilization adversely affects the torsional properties of nine hundred rotary NiTi ProFile files, making them more prone to fracture under torsional stress. It was reported that (i) 54 comparisons were made for torsional strength and 54 for rotational flexibility, (ii) 10 significant changes occurred for torsional strength and 10 for rotational flexibility, (iii) 8 of 10 changes in torsional strength were increased, and (iv) 52 of 54 (96.3%) comparisons for torsional strength and 47 of 54 (87%) for rotational flexibility showed a significant increase or no change concluding that heat sterilization of rotary NiTi files up to 10 times does not increase the likelihood of instrument fracture. The effect of sterilization on bending and torsional properties of K-files manufactured with different metallic alloys, including NiTi, Ti, and SS, which were subjected to dry heat and autoclave [143]. It was found that (i) the files made of Ti showed an increased flexibility after sterilization with autoclave (sizes 30 and 35) and dry heat (sizes 30, 35, and 40), (ii) resistance to fracture varied amongst the five groups of files tested as follows: it decreased in some sizes of SS instruments, decreased in all sizes of titanium files assessed by the torsional moment, and either increased or decreased in some sizes of NiTi files, and (iii) all files tested, however, satisfied relevant standards for angular deflection after being subjected to sterilization with an autoclave or dry heat. Hilt et al. [144] tested the hypothesis that multiple sterilizations of endodontic SS and NiTi K-type files will lead to a continuous decrease in the resistance of files to separation by torsion. It was indicated that neither the number of sterilization cycles nor the type of autoclave sterilization used affects the torsional properties, hardness, and microstructure of SS and NiTi files. King et al. [145] evaluated the effects of repeated autoclaving on torsional strength of two NiTi rotary endodontic files, including TFs and GTX files, which were subjected to 0, 1, 3, or 7 autoclave cycles. It was obtained that (i) there was no significant difference in torsional moment between the number of autoclave cycles for TFs; however, the mean torque at failure was significantly lower for GTX files after three and seven autoclave cycles, (ii) for TFs, there was a significant increase in mean number of degrees of rotation to failure with more autoclave cycles compared to nonautoclaved files, and (iii) for GTX files, there was no significant association between the number of autoclave cycles and the degrees of rotation to failure concluding that repeat autoclaving of unused GTX files between 3 and 7 times resulted in a significant reduction in torsional strength, while there was no effect observed for TFs. Casper et al. [146] compared the effects of multiple autoclaving cycles on the torsional load resistance of three new rotary endodontic files (Profile Vortex made from M-Wire (PV), TF, and CM Wire (CM), which were subjected to the steam autoclaving for 1, 2, 3, and 7 sterilization cycles. It was found that (i) autoclave cycles had no significant overall effect on file performance for any of the instrument systems tested, (ii) PV and CM displayed significantly greater resistance to torsional load than TF but were not different from each other, and (iii) angular deflection values for TF and CM were significantly higher than for PV, with TF demonstrating greater rotational distortion than CM concluding that (i) repeated steam autoclaving did not affect torsional resistance for unused files of the systems evaluated, and (ii) CM Wire files might have a combined advantage of greater torsional strength and high deformation before failure.

With respect to sterilization effects on fatigue behavior of NiTi files, Janardhanan et al. [147] investigated the fatigue resistance of NiTi endodontic files (ProTaper F2 size) before and after five cycles of autoclave sterilization. It was concluded that (i) autoclaving has relatively no negative impact on the fatigue failure of NiTi instruments; however, a slight increase in hardness is observed on repeated sterilization, and (ii) the most common reason for failure of the files was flexural strain, which almost always corresponded to the area of maximum curvature of the simulated root canal. Viana et al. [148] evaluated the effect of repeated sterilization cycles in dry oven or autoclave, on the mechanical behavior and fatigue resistance of rotary endodontic NiTi instruments which were subjected to five consecutive sterilization cycles in a dry oven or steam autoclave. It was obtained that (i) sterilization procedures resulted in no significant changes in Vickers microhardness, nor in the parameters describing the mechanical behavior of the wires, (ii) however, the number of cycles to failure was statistically higher for all instruments after dry heat or autoclave sterilization cycles, and (iii) in the instruments previously fatigued to one half of their fatigue life, autoclave sterilization gave rise to an increase of 39% in the remaining number of cycles to failure, concluding that changes in the mechanical properties of NiTi endodontic instruments after five cycles of commonly used sterilization procedures were insignificant, and the sterilization procedures are safe as they produced a significant increase in the fatigue resistance of the instruments. Oshida et al. [149] developed a novel technique to assess accumulative damages on ProFile files. Using extract tooth, three different degrees of root canal curvature (25°, 40°, and 55°) were used. Instrumentation of ProFile was conducted at every 50 rotational cycles' increments and between each instrumentation, the autoclave sterilization was performed at 121 °C in distilled water. It was found that there was no indication of adverse effects of autoclaving on subsequent instrumentation. This confirms results obtained by Hilt et al. [144]. Melo et al. [150] conducted a comparative study of the fatigue resistance of engine-driven NiTi endodontic ProFile and Quantec instruments in artificial canals with a 45-degree angle of curvature and 5-mm radius of curvature. It was observed that (i) the size of the instrument, which determines the maximum strain amplitude during cyclic deformation, is the most important factor controlling fatigue resistance, and (ii) the effect of heat sterilization on the fatigue resistance of the instruments was also examined indicating that the application of five sterilization procedures in dry heat increases the average number of cycles to failure of unused instruments by approximately 70%. Gambarini et al. [151] evaluated resistance to cyclic fatigue of new and used ProFile NiTi (which were operated in ten clinical cases indicating at least nine times sterilization done) rotary instruments. It was reported that (i) a significant reduction of rotation time to breakage (life span) was noted between new and used instruments, (ii) all sizes of new instruments were significantly more resistant than used ones, and (iii) no instrument underwent intracanal failure during clinical use, indicating that prolonged clinical use of NiTi engine-driven instruments significantly reduced their cyclic fatigue resistance and nevertheless, each rotary instrument was successfully operated in up to 10 clinical cases without any intracanal failure.

Aqueous sodium hypochlorite (NaOCl) solution is widely used in dental practice during root canal treatment. NaOCl is used as an endodontic irrigant as it is an effective antimicrobial and has tissue-dissolving capabilities. The antibacterial and tissue dissolution action of hypochlorite increases with its concentration, but this is accompanied by an increase in toxicity. Concentrations used vary down from 5.25% depending on the dilution and storage protocols of individual practitioners. Solution warmers are available to increase the temperature up to 60 °C [152]. Increasing the temperature of a solution of hypochlorite improves the bactericidal and pulp dissolution activity, although the effect of heat transfer to the adjacent tissues is uncertain [153]. However, recently Tada et al. [154] mentioned that (i) there should be the critical intraoral temperature and if the temperature exceeds this critical temperature, intraoral tissue could be subjected to the irreversible damage, and (ii) the critical temperature must be 42.5 °C. Yared et al. [155] evaluated cyclic fatigue of 0.06 ProFile NiTi rotary instruments after clinical use in molar teeth. In group 1, instruments sizes 40–15 were used in a crown-down technique using 2.5% NaOCl as an irrigant. Fifty-two molars were included and 13 sets of Profile NiTi rotary instruments were used. Each set of instruments was used in four molars, and was steam autoclaved before each use. Group 2 (10 sets of new ProFile NiTi rotary instruments) was the control group. Cyclic fatigue was tested by rotating the instruments in a 90 degrees metallic tube until they broke. It was concluded that sterilization and clinical use in the presence of NaOCl did not lead to a decrease in the number of rotations to breakage of the files. Mechanical properties of NiTi endodontic instruments (Maillefer, JS Dental, McSpadden, and Brasseler) and the effect of sodium hypochlorite treatment on mechanical properties were investigated [156]. It was reported that (i) all instruments evaluated complied with or exceeded ADA/ANSI Specification No. 28, with the sole exception of the Maillefer ISO size 40 for torsional moment. JS Dental and McSpadden NiTi files were the most resistant to torsional fracture, but all NiTi files were inferior when compared with SS files from a previous study, (ii) however, NiTi files were superior in flexibility, and Maillefer and Brasseler instruments were the best of the instruments tested, (iii) NiTi files also had negligible permanent deformation angles, and (iv) for all properties tested, NaOCl had no statistically significant effect. Peters et al. [157] investigated the effect of immersion in NaOCl on torque and fatigue resistance of two NiTi files (ProFile and RaCe) immersed in 5.25% NaOCl for 1 or 2 h at temperatures of 21 and 60 °C. It was reported that (i) torsional resistance of both rotaries was not significantly affected by immersion in NaOCl, except after 2 h of immersion at 60 °C, (ii) resistance to cyclic fatigue decreased significantly for ProFile and RaCe instruments after immersion in NaOCl, and (iii) spontaneous fractures occurred in 28 of 160 files during immersion in NaOCl concluding that NiTi rotaries have reduced resistance to cyclic fatigue after contact with heated NaOCl and may then be considered single-use instruments.

7.1.13 Bleaching effect

During the orthodontic mechanotherapy, maintenance and oral hygiene care are most important recommended practices for both patients and orthodontist as well as dental hygienist to help prevent dental caries and maintain healthy oral health. Normally, the use of fluoride-containing products, such as mouthwashes and gels, is recommended. Gupta et al. [158] evaluated the effects of fluoride prophylactic agents on mechanical properties of NiTi wires, using commercially available different mouthwash solutions: Phos-Flur gel (1.1% sodium acidulated phosphate fluoride, 0.5% w/v fluoride, pH = 5.1; Colgate Oral Pharmaceuticals) and Prevident 5000 (1.1% sodium fluoride neutral agent, 0.5% w/v fluoride, pH = 7; Colgate Oral Pharmaceuticals). It was found that (i) there was not much difference in the values of both modulus of elasticity and yield strength obtained after loading of stress on the wires in all the three experimental conditions, (ii) a significant difference in both modulus of elasticity and yield strength was observed during unloading of stress, and (iii) further, when the surface characteristics were observed for all the specimens using SEM images, it was observed that NiTi wires treated with Phos-Flur showed large surface defects which appeared as round, pitted areas depicting corrosion, numerous white inclusions, and overall damaged surface structure of the wire as compared with the control. Based on these findings, it was concluded fluoridated mouthwashes are essential to maintain good oral hygiene and decrease instance of caries in patients undergoing orthodontic treatment, and the prophylactic usage of topical fluoride agents on NiTi wire seems to diminish the mechanical properties of the orthodontic wire that could significantly affect future treatment outcomes. Srivastava [159] assessed the change in load deflection characteristics of NiTi, Cu–NiTi, SS (SS), and β-Ti wires by immersing in fluoride mouth rinses of two types – Phos-Flur and neutral NaF mouth rinse – utilizing a modified bending test and comparing it to control. It was obtained that there was no statistically significant reduction in load deflection characteristics of NiTi, Cu–NiTi, β-Ti, and SS wires on immersing in Phos-Flur mouth rinse and neutral sodium fluoride mouth rinses as compared to control at 2.5 and 1 mm of deflection in unloading phase. Walker et al. [160] investigated the effects of fluoride prophylactic agents on the mechanical properties of NiTi and Cu–NiTi orthodontic archwires. It was reported that (i) unloading mechanical properties of NiTi orthodontic wires were significantly decreased after exposure to both fluoride agents; however, Cu–NiTi wire mechanical properties were not significantly affected by either fluoride agent, (ii) corrosive changes in surface topography were observed for both wires, with

Cu–NiTi appearing to be more severely affected suggesting that using topical fluoride agents with NiTi wire could decrease the functional unloading mechanical properties of the wire and contribute to prolonged orthodontic treatment. Hammad et al. [161] determined the effects of a fluoride prophylactic agent on the mechanical properties and surface quality of a preformed round translucent composite archwire while comparing it with NiTi and multistranded SS wires. It was found that (i) fluoride treatment produced a statistically significant reduction in modulus of elasticity, yield strength, and yield strength/modulus of elasticity of the composite wire, (ii) a significant decrease in modulus of elasticity of NiTi wire was found after exposure to fluoride, upon comparison with distilled water control treatment, (iii) on the other hand, no significant effect of fluoride treatment was found on yield strength and its ratio with modulus of elasticity of NiTi wire and on studied properties of the multistranded SS wire, and (iv) corrosive changes in surface topography were observed after exposure to the fluoride agent and were more pronounced with the composite wire concluding that using a topical fluoride agent with translucent composite wire could decrease the mechanical properties and might damage the surface of the wire, potentially contributing to prolonged orthodontic treatment. Ramalingam et al. [162] determined the effect of topical fluoride agents on the mechanical properties of NiTi and copper NiTi archwires. Thirty patients with fixed appliances were randomly divided into three groups: Group I (control group) used no topical fluoride agents, Group II used a fluoride rinse, and Group III applied a fluoride gel to the teeth. After 30 days, the archwires were retrieved and the moduli of elasticity and yield strengths were measured. It was reported that (i) during unloading, the modulus of elasticity of the NiTi archwire fell significantly in the gel group, (ii) the modulus of elasticity and yield strengths of the NiTi archwires during loading, and the copper NiTi archwires during loading and unloading, were not affected by either the gel or the rinse, and (iii) scanning electron microscopic analysis revealed that the copper NiTi archwires in the gel group had the most pitting suggesting that topical fluoride agents alter the mechanical properties of NiTi wires and, hence, may prolong orthodontic treatment. Lin et al. [163] investigated the influence of fluoride and an acidic environment on the mechanical properties of NiTi-W orthodontic wires, which were immersed in 0% or 0.05% natrium-fluoride containing artificial saliva at a pH of 4 or 6 for 1 or 3 days. It was found that (i) a pH of 4 increased microhardness and decreased the three-point bending strength significantly, whereas immersion time and fluoride concentration had no significant influence on the microhardness or on the three-point bending strength, and (ii) when examining the test group NiTi-Ws after 3 days of immersion at a pH of 4, the SEM revealed a rough surface morphology, a damaged oxide layer, and signs of corrosion indicating that the most influential factor for decreasing the unloading force and increasing the hardness seems to be the pH value, whereas immersion time and NaF addition do not have a major influence.

7.2 Machinability, cutting, and shaping efficiency

Among versatile applications of NiTi materials in medical and dental fields, only endodontic instruments are devices which work for cutting and shaping. Hence, successful root canal therapy relies on the effective shaping and debridement of the root canal system without damaging the original configuration [164]. The increasing use of biomaterials in different areas of modern medicine raises the question of applicable technologies for their processing, as well as the ability to operate and control the parameters of such processes. The manufacturability of biomaterials primarily proceeds determined by their origins, mechanical properties, chemical composition, microstructure, the applicability of various technological and thermomechanical processes, dimensions, and the purpose of a finished part. All these analyses require application of the most advanced production methods in order to obtain a safe and reliable element that will replace the natural biological functions of humans. When it is considered that all of these complex functions and requirements need to be fully met, or at least to as great an extent as possible, merely in the initial reviewing process, the manufacturability of biomaterials can appear to be the most crucial mechanical property to be imposed [165]. In this section, machinability, machining, cutting efficiency, and shaping efficiency will be reviewed.

7.2.1 Machining and machinability

There are a quite numbers of researches on the machinability of NiTi alloy. Wu et al. [166] studied the machinability of a $Ti_{49.6}Ni_{50.4}$ SMA using a mechanical cutting test. It was observed that there is a wide hardened layer in front of the cutting edge of the $Ti_{49.6}Ni_{50.4}$ alloy which comes from the effects of strain hardening and cyclic hardening, (ii) meanwhile, $Ti_{49.6}Ni_{50.4}$ fragments can adhere to the diamond blade, (iii) the longer the cutting time, the more adhesion the $Ti_{49.6}Ni_{50.4}$ fragments have causing the $Ti_{49.6}Ni_{50.4}$ alloy to exhibit more difficult cutting characteristics than 18-8 SS and $Ti_{50}Al_{50}$ alloy, and (iv) from the viewpoint of cutting energy, the effect of applied load is more important than that of cutting speed and there is an optimal cutting load for the cutting of $Ti_{49.6}Ni_{50.4}$ SMA. Huang et al. [167] using the very similar alloy ($Ti_{49.4}Ni_{50.6}$ alloy), investigated the machinability by two machining methods: electrical discharge machining and femtosecond laser machining. It was reported that the electrical discharge wire cutting used resulted in an average SR of ~1.2 μm and a heat-affected layer of 150 μm depth, and (ii) in the laser machining, an ultrashort pulse laser with a width of 150 fs was used to minimize the effect of laser-generated heat on the surface integrity resulting in a much smaller SR of ~0.4 mm and a heat-affected layer of only 50 μm. One of the NiTi's drawbacks is that heat generated in NiTi during machining is not discharged smoothly and inner stress occurs when traditional machining methods are used. In order to overcome this difficulty, various nontraditional machining methods, including noncontact machining, have been investigated for use with NiTi. Lee et al. [168] studied the application of electrochemical polishing to the machining of NiTi alloy. In addition to the generated heat problem associated with NiTi alloy, Mehrpouya et al. [169] pointed out the high hardness, which requires a large amount of cutting force, resulting in high rate of tool wearing. Accordingly, as Shahedin et al. [170] pointed out, in order to make tool wear and cutting force are at minimum values, a particular range of cutting speeds should lead to diminishing machining barriers such as burr formation and chip-breaking, and also the lower cutting force and consequently lower temperature and stresses in the machining process improve the mechanical properties as well as reducing hardness, distortion, and residual stress.

In regard to machining efficiency, Camps et al. [171] compared the machining efficiency of four brands of NiTi and two brands of SS K-type files. It was reported that (i) the SS instruments with a triangular cross section were more efficient than the SS instruments with a square cross-section, (ii) there was a significant discrepancy between the machining ability of the nickel-titanium K files, and (iii) the Maillefer (SS) instruments were the most efficient. Lin et al. [172] studied the machining characteristics of TiNi SMAs. It was mentioned that (i) new cutting edges coated with SiC and Al_2O_3 powders exhibit a better cutting rate than a diamond blade to cut TiNi alloys, (ii) a high-speed steel drill coated with TiN film exhibits a better drilling ability than a high-speed steel drill for the TiNi alloys because the TiN film has high hardness and excellent wear resistance. Furthermore, a tungsten carbide drill exhibits the best drilling ability, (iii) plastic deformation occurs during drilling and hence the specimen's hardness near the drilled holes can reach 310 and 370 Hv for the $Ti_{50}Ni_{50}$ and $Ti_{49}Ni_{51}$ alloys, respectively, and (iv) at the same time $Ti_{50}Ni_{50}$ alloy exhibits better drilling characteristics than $Ti_{49}Ni_{51}$ alloy. Velmurugan et al. [173] overviewed the machining processes that can be used to machine the NiTi and its surface-induced characteristics such as microhardness, SR, topography, induced layer, residual stress, fatigue, and phase transformation. It was mentioned that (i) the conventional machining of NiTi alloys are quite complicated due to high toughness, severe strain hardening, fatigue hardening, and distinctive property of NiTi-SMAs such as PE and SME, and (ii) nontraditional process is significantly used to machine the NiTi alloys due to its better results on surface integrity characteristics. Therefore, the need for more effective and efficient manufacturing processes to transform the biocompatible materials into high standard artificial human body components (implants) is rapidly growing [174]. The nontraditional process should include the abrasive water jet machining, ultrasonic machining, ion beam machining, laser beam machining, electrical discharge machining, and electron beam machining[175].

The concept of miniaturizing machine tools has received a strong interest in the research community due to their ability to fabricate intricate components, characterized by the lower power consumption and higher productivity rate, and smaller sizes of work stations have enabled microscale machining operations to acquire an edge over other fabrication techniques in various applications [175]. Uppal et al. [176]

employed the femtosecond laser micromachining technique to machine NiTi SMA in air. Biermann et al. [177] employed a simulation-based approach to optimize five-axis micromilling processes with respect to the special material properties of NiTi alloy, especially, the influence of the various tool inclination angles is considered for introducing an intelligent tool inclination optimization algorithm. It was mentioned that the possible length-to-diameter ratio reaches up to 50 offering new possibilities in the manufacturing of microstents.

Kuppuswamy et al. [178] reviewed the high-speed micromachining as a possible way to process the NiTi medical components without compromising the productivity and quality of the machined surface textures. It was reported that machining behavior characterized in terms of low cutting forces and reduced burr size was achieved at 15 m/min of cutting speed when the NiTi alloy undergoes a transition from B2 phase to B19 phase. Weinert at al. [179], using solid carbide end milling cutters, examined the machinability of NiTi. It was reported that (i) despite the poor machinability of NiTi, good results concerning tool wear and shape accuracy of the milled slots were achieved, (ii) essential for a good machining result is the application of minimum quantity lubrication reducing NiTi adherences compared to dry machining, and (iii) work piece quality is improved and tool life is extended.

7.2.2 Cutting efficiency

As seen in the previous section, improvement of NiTi machinability and machining efficiency is not the only important issue, but the cutting and shaping efficiencies are more practically important when the machined NiTi files are employed to control root canal preparation.

Tepel et al. [180] studied the cutting efficiency of 24 different types of endodontic hand instruments. It was shown that (i) Nitinol K-files showed the least cutting efficiency, (ii) SS reamers and especially K-files showed better cutting efficiency than Nitinol K-files, (iii) flexible SS instruments displayed the best results, and (iv) with regard to cutting efficiency, flexible SS instruments were clearly superior to SS reamers and K-files, and especially to Nitinol K-files. Haïkel et al. [181] assessed the cutting efficiency of NiTi in the presence and absence of NaOCl treatment and compare them to a conventional SS K-type file. It was found that (i) without NaOCl, Brasseler (318 μg/J) and Maillefer (280 μg/J) NiTi files were most efficient, followed by JS Dental (71.4 μg/J) and McSpadden (40 μg/J), (ii) NaOCl treatment did not alter the cutting efficiency of any brand of instruments significantly, and (iii) when compared with conventional SS files, all NiTi files tested were less efficient. Schäfer et al. [182] investigated the cutting efficiency and the effects of instrumentation on curved canal shape of SS and NiTi ProFile Series 29 hand instruments and SS Flexoreamer. It was reported that (i) concerning cutting efficiency in rotary motion, the Flexoreamer had significantly greater cutting efficiency than SS ProFiles and NiTi ProFiles, (ii) changes in the canal shape differed significantly between the different instruments at all measuring points, and (iii) in this study, it seemed that flexible SS instruments with noncutting tips were superior to the nonstandardized ProFile Series 29 instruments with regard to cutting efficiency and instrumentation of curved canals. Rapisarda et al. [183] examined the surface nitridation treatment of the cutting surfaces with respect to surface or subsurface changes that produced an increase in the resistance to wear in NiTi endodontic files. It was observed that the experimental instruments showed in-depth distributions of chemical composition that were different from those seen in the control group; thermal-nitridated instruments demonstrated a surface ratio of nickel to titanium of 0.5. Implanted samples had a higher N/Ti ratio (1.2); this ratio may be due to the presence of a layer of titanium nitride, (ii) samples in the experimental groups showed an increase in cutting ability as compared with the controls, and (iii) thermal nitridation and nitrogen-ionic implantation treatment of NiTi files produced a higher wear resistance and an increased cutting capacity. Effect of sterilization on the cutting efficiency of physical vapor deposition (PVD)-coated NiTi endodontic instruments was studied [184] and it was concluded that repeated sterilization under autoclave or exposure to NaOCl prior to sterilization did not alter the cutting efficiency of physical vapor deposition (PVD)-coated NiTi K-files. Bui et al. [185] investigated the effect of electropolishing PRI on torque resistance, fatigue resistance, and cutting efficiency. It was reported that (i) electropolishing significantly reduced resistance to cyclic fatigue but did not affect torsional resistance, and (ii) however, electropolishing reduced the angle at failure and amount of unwinding, and (iii) electropolishing did not significantly affect cutting efficiency. Fayyad et al. [186] compared the efficacy of the cutting ability of two different instruments, concerning changes in the dentin thickness removed and root canal volume, with the multislice computed tomography (CT) scanning. It was obtained that (i) ProTaper removed significantly more dentin from the mesiodistal and buccolingual directions of the root canal than the TF, and (ii) no significant difference was recorded for the changes in root canal volume between the two systems concluding that the TF system was found to cut dentin efficiently with more uniform cutting than ProTaper system. Peters at al. [187] evaluated the cutting behavior of NiTi coronal flaring instruments, including BioRaCe BR0 (BR), HyFlex CM 1 (HY), ProFile OS No. 2 (PF), and ProTaper Sx (PT). It was found that (i) for all three methods, HY and PF were the most and the least cutting-efficient instruments, respectively, (ii) significant differences were detected between 250 and 500 rpm for HY and PT (area); for BR, HY and PT (depth); and for BR and HY (volume), and (iii) there were strong positive correlations between the results obtained with those three different methods with r-values ranging from 0.81 to 0.92. Based on these findings, it was concluded that HyFlex, manufactured with thermomechanically treated NiTi wire, was the most efficient instrument, and increased rotational speed was associated with increased cutting efficiency.

7.2.3 Shaping ability and canal preparation

The shaping ability of NT Engine and McXim NiTi rotary instruments in simulated root canals was evaluated [188]. In all, 40 canals consisting of four different shapes in terms of angle and position of curvature were prepared by a combination of NT Engine and McXim instruments using the technique recommended by the manufacturer. It was reported that (i) one instrument fractured and only four instruments deformed, with most of the failures occurring in canals with curves which began 12 mm from the orifice, that is, in short acute curves, (ii) none of the canals became blocked with debris, (iii) the canals were found to be smooth in the apical half of the canal in 33 specimens and in the coronal half of 39 specimens, and (iv) all canals had good taper characteristics and 35 had good flow characteristics concluding that NT Engine and McXim instruments prepared canals rapidly, with few deformations, no canal blockages and with minimal change in working length. Schäfer et al. [189] compared the shaping ability of K3 rotary NiTi instruments with SS K-Flexofiles manipulated by hand. It was mentioned that (i) in comparison with SS K-Flexofiles, rotary K3 instruments achieved better canal geometry and showed significantly less canal transportation, (ii) between both the canal types, K3 was significantly faster than K-Flexofiles, and (iii) both instruments maintained a good working distance concluding that K3 instruments prepared curved canals rapidly and with minimal transportation toward the outer aspect of the curve. Fractures occurred significantly more often with K3. Liu et al. [190] compared the shaping ability of engine-driven ProTaper and GT files, and manual preparation using K-Flexofile instruments in curved root canals of extracted human teeth. Irrigation was done with 2 mL 2.5% NaOCl after each instrument and, as the final rinse, 10 mL 2.5% NaOCl then 10 mL 17% ethylenediaminetetraacetic acid (EDTA) and finally 5 mL distilled water is used. It was mentioned that (i) two GT files but none of the K-Flexofile and ProTaper instruments separated, (ii) for debris removal, the ProTaper group achieved a better result than GT but not the K-Flexofile group at all three regions (apical, middle, and coronal), (iii) K-Flexofiles produced significantly less smear layer than ProTaper and GT files only in the middle third of the canal, and (iv) both NiTi rotary instruments maintained the original canal shape better than the K-Flexofiles ($P < 0.05$) and required significantly less time to complete the preparation. About the smear layer removal, the cleansing and smear layer removal capability of alternate canal irrigation with citric acid and NaOCl was evaluated [191]. The irrigation protocol was as follows: 5% NaOCl alone, NaOCl alternated with 1 mol L-1 citric acid solution or a combination of 15% EDTA and Cetrimide solution. It was reported that (i) qualitative SEM evaluation at x300 and x1000 showed no statistically significant differences in cleansing ability between citric acid, EDTA, and NaOCl groups, (ii) quantitative evaluation of smear layer removal, measured as open tubules/total dentinal surface ratio, showed that 1 mol L-1 citric acid solution was comparable to EDTA (11.97% vs. 10.36%); in samples treated with ProFile 0.04 taper instruments, citric acid was most effective (16.17%), while in the group treated with manual instrumentation EDTA and Cetrimide were the most effective (11.94%), and (iii) specimens irrigated with 5% NaOCl demonstrated significantly more cleansing than those obtained in the other two groups. Reddt et al. [192] compared the cleaning efficacy (debris and smear layer removal) of hand and two NiTi rotary instrumentation systems (K3 and ProTaper). It was found that ProTaper rotary instrumentation showed the maximum cleaning efficacy followed by K3 rotary instrumentation in the coronal, middle, and apical thirds of the root canal. Subtamaniam et al. [193] evaluated the effect of root canal instrumentation using both manual and rotary files in the root canals of primary anterior teeth. It was mentioned that (i) the rotary files cleaned the coronal and middle thirds of root canals more effectively, (ii) statistically there was no significant difference between the groups, and (iii) the lowest score of 2.6 in the apical third of root canals was seen with hand NiTi files concluding that rotary instrumentation was as effective as manual instrumentation in removal of smear layer in the root canals of primary anterior teeth.

Shaping abilities of two different engine-driven rotary NiTi systems (crown-down method) or SS hand files (balanced-force technique) in mesiobuccal canals of extracted mandibular molars were evaluated [194]. Group 1 was instrumented with Sequence (Brasseler USA, Savannah, GA, USA) rotary files, group 2 with Liberator (Miltex Inc., York, PA, USA) rotary files, and group 3 with Flex-R (Union Broach, New York, NY, USA) files. It was found that (i) sequence rotary files, Liberator rotary files, and Flex-R hand files had similar effects on apical canal transportation and changes in working length, with no significant differences detected among the three groups, and (ii) hand instrumentation times were longer than with either NiTi rotary group, whereas the rotary NiTi groups had a higher incidence of fracture. Vaudt et al. [195] investigated instrumentation time, working safety, and the shaping ability of two rotary NiTi systems (Alpha System and PTU) in comparison to SS hand instruments. It was reported that (i) active instrumentation time of the Alpha System was significantly reduced compared with PTU and hand instrumentation, (ii) no instrument fractures occurred in any of the groups, (iii) the Alpha System revealed significantly less apical straightening compared with the other instruments, and (iv) in the apical cross sections Alpha System resulted in significantly less uninstrumented canal walls compared with SS files concluding that despite the demonstrated differences between the systems, an apical straightening effect could not be prevented; areas of uninstrumented root canal wall were left in all regions using the various systems. Bonaccorso et al. [196] compared the shaping ability of ProTaper, Mtwo, BioRaCe, and BioRaCe + S-Apex instruments in simulated canals with an S-shaped curvature. It was found that (i) ProTaper instruments caused more pronounced canal transportation in the apical curvature than all other instruments, and (ii) the use of ProTaper, Mtwo, and BioRaCe instruments resulted in more canal aberrations compared with BioRaCe + S-Apex suggesting that NiTi systems including less tapered and more flexible instruments like S-Apex seem to be favorable when preparing S-shaped canals. Qunsi et al. [197] assessed the shaping efficacy of repeatedly used NiTi rotary instruments (PTU instruments). It was found that (i) two-way repeated-measures analysis of variance revealed significant differences between groups, (ii) regarding measurement type, there were no significant differences between buccolingual and mesiodistal measurements, but there were significant differences between micro-computed tomography (µCT) and buccolingual measurements and µCT and mesiodistal measurements, and (iii) significant differences were also noted between uses. Based on these findings, it was concluded that µCT scanning is more discriminative of the changes in canal space associated with repeated instrument use than photographic measurements. Burroughs et al. [198] determined the shaping ability of three NiTi endodontic file systems by measuring canal transportation: the Self-Adjusting File (ReDent Nova) group, the TYP group (TYP rotary files with CM Wire), and the Vortex group (ProFile Vortex rotary files with M-Wire NiTi). It was reported that (i) after adjusting for the level and canal wall side, the mean transportation was significantly higher for the TYP and Vortex groups compared with the Self-Adjusting File group, and (ii) the mean transportation was significantly higher for the TYP group versus the Vortex group indicating that the Self-Adjusting Files showed less canal transportation than ProFile Vortex and TYP files in simulated S-shaped root canals. About the canal transportation, Hartmann et al. [199] and Poly et al. [200] employed the CT technique to evaluate the canal transportation capability. According to the technique used for root canal instrumentation, they included (a) hand instrumentation with K-Flexofiles and K-Flexofiles which are activated by an oscillatory system and (b) ProTaper NiTi rotary system [199]. It was found that (i) in the buccal direction, the manual technique produced significantly less canal transportation than the oscillatory technique and both were similar to the rotary technique, (ii) in the distal and distopalatal directions, the oscillatory technique produced more canal transportation, and (iii) in the mesiopalatal direction, the oscillatory technique produced more canal transportation than the manual technique, and both were similar to the rotary technique concluding that all techniques produced canal transportation, and the oscillatory technique produced the greatest removal of root dentin toward the inner side of the root curvature. Poly et al. [200] compared canal transportation and centering ratio produced after instrumentation with a single heat-treated reciprocating system, WOG and a single heat-treated rotary instrument, XP-endo Shaper (XPS), using microcomputed tomographic imaging to detect canal transportation. It was reported that the microcomputed tomographic method showed that XPS's shaping ability regarding the centering ability, and canal transportation was significantly better than WOG only at the 7-mm level and concluded that root canal curvatures may lead to procedural errors during endodontic treatment.

Shaping ability of NiTi rotary instruments in curved root canals was investigated [201] with Mtwo, RaCe, and Medin rotary instruments. It was concluded that (i) Mtwo instruments maintained the canal curvature significantly better than Race and Medin instruments, and (ii) there was significant difference between the rotary instruments for iatrogenic transportation of the major foramen. Berutti et al. [202] compared the canal curvature and axis modification after instrumentation with WaveOne Primary reciprocating files and NiTi rotary ProTaper. It was obtained that the canal modifications are reduced when the new WaveOne NiTi single-file system is used. Hwang et al. [203] compared the shaping ability of Mtwo, a conventional NiTi file system, and Reciproc – a reciprocating file system morphologically similar to Mtwo. Root canal shaping was performed on the mesiobuccal and distobuccal canals of extracted maxillary molars. In the Reciproc file in reciprocating motion (RR) group, Reciproc was used in a reciprocating motion (150° counterclockwise/ 30° clockwise, 300 rpm); in the Mtwo file in reciprocating motion (MR) group, Mtwo was used in a reciprocating motion (150° clockwise/30° counterclockwise, 300 rpm); and in the Mtwo file in continuous rotating motion (MC) group, Mtwo was used in a continuous rotating motion (300 rpm). It was found that (i) no statistically significant differences were found among the three groups in the time for canal shaping or canal volume change, (ii) transportation values of the RR and MR groups were not significantly different at any level, (iii) however, the transportation value of the MC group was significantly higher than both the RR and MR groups at the cervical and apical levels, and (iv) in the scanning electron microscopic analysis, file deformation was observed for one file in group RR (1/15), three files in group MR (3/15), and five files in group MC (5/15). Based on these findings, it was concluded that in terms of shaping ability, Mtwo used in a reciprocating motion was not significantly different from the Reciproc system.

Arora et al. [204] compared the canal transportation, canal centering ability, and time taken for preparation of curved root canals after instrumentation with ProFile GTX files, Revo-S files, TFs, and Mtwo files by using cone-beam computed tomography (CBCT). It was found that (i) TF system showed significantly least canal transportation and highest canal centering ability values as compared to GTX, Revo-S, and Mtwo file systems, and (ii) overall, GTX, Revo-S, and Mtwo showed comparable results with respect to canal transportation and centering ability. Hence, it was concluded that the innovative method of manufacturing the TF system resulted in superior shaping ability in curved canals, with the instruments remaining more centered and producing less canal transportation than GTX, Revo-S, and Mtwo file systems. Saber et al. [205] conducted comparative evaluation of the shaping ability of ProTaper Next (PTN), iRaCe, and HFCM rotary NiTi files in severely curved root canals. It was mentioned that (i) during root canal preparation, no instruments fractured, (ii) the use of PTN resulted in significantly greater canal straightening than IR and HF, with no significant differences between IR and HF, (iii) there were no significant differences between the three groups with respect to apical transportation, and (iv) IR and HF were significantly faster than PTN, with no significant differences between IR and HF concluding that PTN, IR and HF respected original canal curvature well and were safe to use. Celikten et al. [206] performed comparative evaluation of shaping ability of two NiTi rotary systems (PTN and New One Shape) using CBCT. It was obtained that (i) significant differences were found between apical and coronal levels for both systems in canal transportation, (ii) in comparing the systems, similar values were found at each level, without significant difference in terms of canal curvature and volume, and (iii) voxel sizes did not affect the measurements on canal volume, curvature, or transportation; no significant difference was found between the 0.100- and 0.125-mm^3 voxel sizes, concluding that both instrumentation systems produced similar canal transportation and volume changes and the two voxel resolutions also showed similar results; however, a 0.125-mm^3 voxel size can be recommended for a flat panel CBCT scanner with lower exposure dose. Shaping ability of two NiTi instruments (TF Adaptive and Mtwo) was activated by continuous rotation or adaptive motion; a μCT observation was investigated [207]. It was reported that (i) volume and surface area increased less with TF file Adaptive type (TFA) files in continuous rotation than in other groups, (ii) TFA files had significantly less transportation and higher centering ability than Mtwo both in continuous and adaptive motion, and (iii) centering ratio, but not canal transportation, was improved by adaptive motion compared with continuous rotation for both instruments, however, no differences were found in canal transportation and centering ability in the apical third for both instruments and motions concluding that no difference between the devices and kinematics was found in the apical third; TFA performed significantly better in the middle and coronal parts of the root canal. Duque et al. [208] evaluated the influence of the NiTi wire in conventional NiTi (PTU) and CM NiTi (PTG) instrument systems on the quality of root canal preparation. It was found that (i) in the preoperative analysis, there were no statistically significant differences between the groups in terms of the area and volume of root canals, (ii) there was no statistical difference in the dentin thickness at the first apical level between, before, and after instrumentation for both systems, and (iii) the Conventional NiTi (PTU) and CM NiTi (PTG) instruments displayed comparable capabilities for shaping the straight mesial root canals of mandibular molars, although the PTG was better than the PTU at maintaining the centralization of the shape in the cervical portion. Wei et al. [209] evaluate the shaping ability of three new different NiTi rotary instruments in simulated root canals using μCT. It was reported that (i) Reciproc produced greater volume change in the apical part of the canals compared with PTU and K3XF, (ii) K3XF exhibited less transportation and better centering ability at the 2- and 3-mm levels from the apical foramen compared with PTU and Reciproc, and (iii) there were no significant differences in the centering ratio and transportation between PTU and Reciproc. Preparation time was significantly shorter in the Reciproc group. Shaping ability of Reciproc, WaveOne GOLD, and HFEDM single-file systems in simulated S-shaped canals was investigated [210]. It was obtained that (i) NiTi file fracture was not observed during shaping of the simulated canals although a danger zone formation in one sample and a ledge in one sample were observed in the Reciproc group, (ii) there was no statistically significant difference between the WaveOne and HyFlex groups' apical, medial, and coronal regions, and (iii) however, it was determined that the Reciproc group removed a statistically significantly higher amount of resin from all the canal regions when compared with the WaveOne and HyFlex groups. Espir et al. [211] evaluated the shaping ability and cleaning after oval root canal preparation using one or more instruments in reciprocating or rotary motion and concluded that a preparation that volumetrically increases the root canal is not necessarily associated with better cleaning, and shaping and hard-tissue debris removal depends on root canal anatomy, kinematics, number of instruments, and instrument design. Huang et al. [212] evaluated the shaping ability of three thermally treated rotary NiTi systems including PTN, HFCM, and HFEDM during root canal preparation in simulated root canals. It was reported that (i) HFEDM caused significantly greater volume increase than HFCM and PTN in the entire root canal and in the apical and middle thirds; (ii) HFCM removed the least amount of resin in the coronal third compared with HFEDM and PTN; and (iii) overall, HFCM caused significantly less transportation in the apical 2 mm and was better centered than PTN in the apical 3 mm. Drukteinis et al. [213] evaluated and compared the canal shaping ability of BioRace, PTN, and Genius engine-driven NiTi file systems in extracted mandibular first molars using μCT. It was observed that (i) there were no significant differences between the three groups in terms of dentine removed after preparation and determination of the root canal volume, or percentage of uninstrumented canal surface, and (ii) no significant differences were found between the systems for canal transportation in any canal third, concluding that the shaping ability of the BR, PTN, and GN NiTi file systems was equally effective and all instrumentation systems prepared curved root canal systems with no evidence of undesirable changes in 3D parameters or significant shaping errors.

With respect to canal preparation, there are still important information available as follows: Esposito et al. [214] compared the canal preparation with NiTi (Mac hand file) and SS (K-Flex) instruments. It was mentioned that (i) NiTi hand and engine-driven instruments maintained the original canal path in all cases, (ii) the incidence of deviation from the original canal path during instrumentation with SS files increased with file size, and (iii) the difference between NiTi groups and SS became statistically significant with instruments larger than size 30 concluding that NiTi files were more effective in maintaining the original canal path of curved root canals when the apical preparation was enlarged beyond size 30. Glossen et al. [215] compared the root canal preparations with NiTi (Mity files) hand, NiTi engine-driven (LightSpeed and NT Sensor file), and SS (K-Flex) hand endodontic instruments. It was obtained that (i) engine-driven NiTi instruments and hand instrumentation with the Canal Master "U" caused significantly less canal transportation, remained more centered in the canal, removed less dentin, and produced rounder canal preparations than K-Flex and Mity files, and (ii) engine instrumentation with LightSpeed and NT Sensor file was significantly faster than hand instrumentation. Guelzow et al. [216] conducted a comparative study of six rotary NiTi systems (FlexMaster, System GT, HERO 642, K3, ProTaper, and RaCe) and hand instrumentation for root canal preparation. It was found that (i) no significant differences were detected between the rotary NiTi instruments for alteration of working length, (ii) all NiTi systems maintained the original curvature well, with minor mean degrees of straightening ranging from 0.45 degrees (System GT) to 1.17 degrees (ProTaper), (iii) ProTaper had the lowest numbers of irregular postoperative root canal diameters; the results were comparable between the other systems, (iv) instrument fractures occurred with ProTaper in three root canals, while preparation with System GT, HERO 642, K3, and the manual technique resulted in one fracture each, (v) NiTi instruments prepared canals more rapidly than the manual technique, and (vi) the shortest time for instrumentation was achieved with System GT (11.7 s). Based on these findings, it was concluded that under the conditions of this ex vivo study, all NiTi systems maintained the canal curvature were associated with few instrument fractures and were more rapid than a standardized manual technique, and ProTaper instruments created more regular canal diameters. Taşdemir et al. [217] compared the root canal preparation with conventional SS K-files and Hero 642 rotary NiTi instruments. It was reported that (i) less transportation occurred with Hero 642 rotary instruments than SS K-files at the midroot and coronal levels, and (ii) Hero 642 rotary instruments had better centering ability than K-files at all three levels. Nordmeyer et al. [218] compared various parameters of root canal preparation using FlexMaster rotary NiTi and Endo-Eze AET SS instruments. Fifty curved mesial root canals of extracted mandibular molars were prepared to size 45 using FlexMaster or AET instruments. The following parameters were evaluated: straightening of root canal curvature, postoperative root canal cross sections, cleaning ability, safety issues, and working time. Statistical analyses were performed using Mann–Whitney U test and Wilcoxon test. It was described that (i) the mean degree of straightening was significantly less for FlexMaster than for AET, (ii) postoperative cross sections showed no significant differences between the systems, (iii) neither of the systems completely eliminated debris and smear layer, (iv) no procedural incidents occurred with the instruments, and (v) mean working time was significantly shorter for Flex-Master than for AET, concluding that AET cannot be recommended for preparation of curved root canals, owing to unacceptable straightening. Genç et al. [219] investigated the ability of two NiTi rotary apical preparation techniques used with an electronic apex locator-integrated endodontic motor and a manual technique to create an apical stop at a predetermined level (0.5 mm short of the apical foramen) in teeth with disrupted apical constriction, and to evaluate microleakage following obturation in such prepared teeth. It was reported that (i) all techniques performed slightly short of the predetermined level; (ii) closest preparation to the predetermined level was with the manual technique and the farthest was with S-Ape; (iii) a significant difference was found between the performances of these two techniques; (iv) LightSpeed ranked in between; and (v) leakage was similar for all techniques at either period; however, all groups leaked significantly more at 3 months compared to 1 week, concluding that despite statistically significant differences found among the techniques, deviations from the predetermined level were small and clinically acceptable for all techniques. Leakage following obturation was comparable in all groups.

There are still valuable and clinically important data relevant to canal dressing and preparation. Gluskin et al. [220] compared effects of preparation with conventional SS Flexofiles and Gates Glidden burs versus NiTi GT rotary files in the shaping of mesial root canals of extracted mandibular molars. It was found that (i) at the coronal and mid-root coronal one-third sections, the rotary GT files produced a significantly smaller postoperative canal area, (ii) in the midroot sections, there was significantly less transportation of the root canal toward the furcation, and less thinning of the root structure with GT files compared to the SS file, (iii) overall, there was greater conservation of structure coronally and more adequate shape in the midroot level, and (iv) the GT rotary technique was significantly faster than the SS hand-held file technique and the GT instruments fractured during the study concluding that novice dental students were able to prepare curved root canals with NiTi GT rotary files with less transportation and greater conservation of tooth structure, compared to canals prepared with hand instruments and the rotary technique was significantly faster. Ferreira et al. [221] evaluated histologically and morphometrically, the cleaning capacity of rotary instrumentation in root canals using the Profile system with ultrasonic irrigation. Twelve single-rooted mandibular incisors were divided randomly into three groups according to the irrigation method tested. The canals in the teeth were instrumented using Orifice Shapers, taper 0.6 and 0.4 in the cervical third, and up to a #35 file in the apical third. One per cent NaOCl was used as the irrigating solution. Group I: canals were irrigated with 5 ml of solution, using a Luer-Lok syringe, between each file. Group 2: canals were irrigated with the solution and instrumented using ultrasound for I min between each file. Group 3: canals were irrigated with 5 ml of solution with Luer-Lok syringe, between each file, and final irrigation was done using ultrasound for 3 min. After chemomechanical preparation, the apical thirds of the teeth were submitted for histological processing. It was reported that the rotary instrumentation with ProFile system NiTi files and ultrasonic irrigation for 3 min was more effective in cleaning root canals when the different methods were compared. Elayouti et al. [222] compared the preparation quality of two rotary systems and NiTi-hand files (Mtwo and ProTaper) in oval root canals, and to evaluate the effect of canal dimensions on the preparation. It was obtained that (i) with regards to the ratio of prepared canal outline, no statistical significant difference was found between Mtwo and ProTaper, but both systems performed significantly better than NiTi-hand files, (ii) in six root canals in Mtwo group (20%), and eight root canals in ProTaper group (27%), the minimal thickness of dentine wall after preparation was less than 0.5 mm, and (iii) in contrast to the maximum diameter of the root canal, the minimum diameter influenced the quality of the preparation concluding that (iv) no instrumentation technique was able to circumferentially prepare the oval outline of root canals, and (v) nevertheless, instruments with greater taper (ProTaper and Mtwo) were more efficient than NiTi-hand files, but this was, in some cases, at the expense of remaining dentine-wall thickness. Ajuz et al. [223] compared the incidence of deviation along S-shaped (double-curved) canals after glide path preparation with two NiTi rotary pathfinding instruments (PathFile and RaCe) and hand K-files. It was obtained that (i) intragroup analysis showed that all instruments promoted some deviation in virtually all levels, (ii) overall, regardless of the group, deviations were observed in the mesial wall at the canal terminus and at levels 4, 5, 6, and 7 mm and in the distal wall at levels 1, 2, and 3 mm, (iii) these levels corresponded to the inner walls of each curvature, (iv) both rotary NiTi instruments performed significantly better than hand K-files at all levels, except for PathFiles at the 0-mm level, and (v) RaCe instruments showed significantly better results than PathFiles at levels 0, 2, 3, 5, and 6 mm suggesting that rotary NiTi instruments are suitable for adequate glide path preparation because they promoted less deviation from the original canal anatomy when compared with hand-operated instruments, and of the two rotary pathfinding instruments, RaCe showed an overall significantly better performance. Paiva et al. [224] evaluated clinically the antibacterial effects of root canal treatment procedures using molecular microbiology analyses. Samples were taken from 14 necrotic root canals of teeth with apical periodontitis before (S1) and after instrumentation with NaOCl irrigation (S2), a final rinse with chlorhexidine (S3), and then 1-week interappointment medication with calcium hydroxide/chlorhexidine paste (S4). It was found that (i) all S1 samples were positive for bacteria in all tests, (ii) treatment procedures promoted a decrease in microbial diversity and significantly reduced the incidence of positive results and the bacterial counts, (iii) in general, each subsequent treatment step improved disinfection, and (iv) no specific taxon or community pattern was associated with posttreatment samples. Based on these findings, it was concluded that (i) supplementary steps consisting of a final rinse with chlorhexidine followed by calcium hydroxide interappointment medication promoted further decrease in the bacterial bioburden to levels significantly below those achieved by the chemomechanical procedures alone, and (ii) because the long-term outcome of root canal treatment is dependent upon maximal bacterial reduction, the present results are of clinical relevance.

7.2.4 Technical sensitivity of endodontic treatment

There are interesting evidences indicating the technical sensitivity of the endodontic treatment. We start with less experienced undergraduate dental students, then graduate students, and dental practitioners and experienced operators.

Al-Omari et al. [225] determined the influence of operator experience on the shaping ability of ProFile and K3 NiTi rotary instruments in simulated root canals. One hundred and sixty simulated canals consisting of four different shapes in terms of angle and position of curvature were prepared by experienced and inexperienced operators. One experienced operator prepared 80 canals and 2 inexperienced operators prepared 80 canals with a crown-down technique using either ProFile or K3 0.06 taper instruments. Images of the canals were taken, using a camera attached to a computer with image analysis software, before surgery and after preparation to sizes 20, 25, and 30 to working length. Postoperative images were combined with the preoperative image to highlight the amount and position of material removed during preparation as well as the shape of the prepared canal. It was observed that (i) overall, there was a highly statistically significant difference between the instruments for the time taken to prepare the canals, with K3 instruments taking a mean of 4.9 min and ProFile 6.0 min, (ii) six instruments fractured (three in each operator category); four were ProFile instruments, (iii) four instruments deformed, all with the inexperienced operators; three were K3, (iv) no perforations or zips were observed; however, one danger zone (created by the experienced operator using K3 instruments) and one ledge (created by an inexperienced operator using K3 instruments) were created, (v) twelve outer widenings were created with a highly significant difference between the operator and instrument used, and (vi) there was a highly significant difference by instrument, and experience regarding absolute transportation at the beginning of the curve and a statistically significant difference for the instrument used regarding absolute transportation half way to the orifice. It was hence concluded that the experienced operator prepared canals more quickly and safely than the inexperienced operators when using K3 instruments; both used ProFile instruments quickly and safely, and inexperienced operators would be advised to train using less aggressive instruments and when confident could progress to other instrument designs. Sonntag et al. [226] investigated root-canal shaping with manual and rotary NiTi files performed by students. Thirty undergraduate dental students prepared 150 simulated curved root canals in resin blocks with manual NiTi files with a stepback technique and 450 simulated curved canals with rotary NiTi files with a crown-down technique. Incidence of fracture, preparation length, canal shape, and preparation time were investigated. Questionnaires were then issued to the students for them to note their experience of the two preparation methods. It was reported that (i) zips and elbows occurred significantly less frequently with rotary than with manual preparation, (ii) the correct preparation length was achieved significantly more often with rotary files than with manual file, (iii) instrument fractures were recorded in only 1.3% of cases with both rotary and manual preparation, (iv) the mean time required for manual preparation was significantly longer than that required for rotary preparation, (v) prior experience with a hand preparation technique was not reflected in an improved quality of the subsequent rotary preparation, (vi) approximately 83% of the students claimed to have a greater sense of security in rotary than in manual preparation, and (vii) overall 50% felt that manual and engine-driven preparation should be given equal status in undergraduate dental education. It was concluded that (viii) inexperienced operators achieved better canal preparations with rotary instruments than with manual files, and (ix) no difference in fracture rate was recorded between the two systems. Pettiette et al. [227] compared the 1-year success rate of endodontic treatment of the same teeth used in our previous study. Twelve-month follow-up radiographs were compared with the immediate follow-up radiographs. Both sets of radiographs were taken with the same customized stent. Quantification of osseous changes using digital imaging was used. Thus, a reliable numerical estimation (densitometric ratio) of disease and healing processes could be established. Sixty-seven percent of the patients returned for the 12-month radiographs (19 NiTi vs 21 SS-K). It was found that (i) immediate postoperative periapical status was found to be similar, (ii) teeth instrumented with the NiTi files demonstrated a higher mean change in densitometric ratio, compared with SS-K files, and (iii) further tests of success and failure with the Fisher exact test showed more success (decreasing radiographic density) with NiTi files and more failures (increasing radiographic density) with SS-K type files indicating that maintaining the original canal shape after instrumentation leads to a better prognosis of endodontic treatment. Peru et al. [228] evaluated root canals instrumented by dental students using the modified double-flared technique, NiTi rotary System GT files and NiTi rotary ProTaper files by μCT. A total of 36 root canals from 18 mesial roots of mandibular molar teeth were prepared: 12 canals were prepared with the modified double-flared technique, using K-flexofiles and Gates-Glidden burs, 12 canals were prepared using System GT, and 12 using ProTaper rotary files. Each root was scanned using μCT preoperatively and postoperatively. It was found that (i) at the coronal and midroot sections, System GT and ProTaper files produced significantly less enlarged canal cross-sectional area, volume, and perimeter than the modified double-flared technique, (ii) in the midroot sections there was significantly less thinning of the root structure toward the furcation with System GT and ProTaper, (iii) the rotary techniques were both three times faster than the modified double-flared technique, and (iv) qualitative evaluation of the preparations showed that both ProTaper and System GT were able to prepare root canals with little or no procedural error compared with the modified double-flared technique. It was concluded that inexperienced dental students were able to prepare curved root canals with rotary files with greater preservation of tooth structure, low risk of procedural errors, and much quicker than with hand instruments. Shen et al. [229] analyzed the incidence and mode of ProFile instrument separation during a predefined schedule of clinical use by the undergraduate students in a dental school over 4 years. A total of 3,706 ProFile instruments discarded from the same undergraduate students program between 2003 and 2007 were analyzed. The lateral and fracture surfaces of 12 separated instruments were examined by scanning electron microscopy (SEM), and the location of the fractures was recorded. It was reported that (i) the overall proportion of instrument defects was 1.3%; deformation without fracture occurred in 1% and separation in 0.3%. The majority of instrument defects occurred in size 20 (34/48), (ii) the ProFile instruments (10/12) failed mostly because of shear stress, whereas only two failed because of fatigue fracture. The results of this study indicated that NiTi rotary instrument system was successfully introduced into an undergraduate endodontic program, and (iii) small size files should be considered as single-use, disposable instruments because of the higher possibility of torsional deformation

Arbab-Chirani et al. [230] evaluated the impact of rotary NiTi instruments on undergraduate teaching and clinical use in French dental schools and to evaluate the impressions of dental students when learning and using these techniques. A questionnaire was mailed to all 16 French undergraduate dental schools. Data were gathered on a range of issues concerning teaching and use of NiTi endodontic techniques. It was found that (i) the response rate was 100%, (ii) the need for teaching NiTi techniques to dental students was agreed by all schools, (iii) lectures and laboratory courses for rotary NiTi techniques were organized in all of the schools, (iv) in 13 of the 16 teaching hospitals, students could use rotary NiTi techniques for canal preparation under the supervision of teaching staff, (v) similarities were observed in the majority of responses, for example, type of rotary system taught and used clinically, and (vi) some differences were also observed, for example, the association of hand files to rotary NiTi instruments and at what stage in the undergraduate curriculum rotary instruments were introduced. It was concluded that (vii) there was a national consensus over the need for undergraduate teaching of rotary NiTi systems in France, and (viii) these techniques had made a substantial impact in endodontic teaching and were widely taught and used in French dental schools. Jungnickel et al. [231] evaluated factors associated with treatment quality of ex vivo root canal treatments performed by undergraduate dental students using different endodontic treatment systems. Four students performed root canal treatment on 80 extracted human teeth using four endodontic treatment systems in designated treatment order following a Latin square design. Lateral seal and length of root canal fillings was radiographically assessed; for lateral seal, a graded visual scale was used. Treatment time was measured separately for access preparation, biomechanical root canal preparation, obturation, and for the total procedure. Mishaps were registered. An ANOVA mirroring the Latin square design was performed. It was reported that (i) use of machine-driven NiTi systems resulted in overall better quality scores for lateral seal than use of the manual SS system, (ii) among systems with machine-driven files, scores did not significantly differ, (iii) use of machine-driven instruments resulted in shorter treatment time than manual instrumentation, (iv) machine-driven systems with few files achieved shorter treatment times, and (v) with increasing number of treatments, root canal-filling quality increased and treatment time decreased; a learning curve was plotted. No root canal shaping file separated. It was then concluded that (i) the use of endodontic treatment systems with machine-driven files led to higher quality lateral seal compared to the manual system, (ii) the three contemporary machine-driven systems delivered comparable results regarding quality of root canal fillings; they were safe to use and provided more efficient workflow than the manual technique, and (iii) increasing experience had a positive impact on the quality of root canal fillings while treatment time decreased.

Moving from less experienced operators to higher skill level operators, Kfir et al. [232] compared procedural errors that occur in patients during root canal preparation by senior dental students using a new "8-step method" versus the traditional "serial step-back technique." Senior dental students treated 221 root canals of maxillary and mandibular teeth. Instrumentation included coronal flaring with Gates-Glidden reamers and standardized SS K-files in all teeth. A new eight-step method was used to prepare 67 canals using standardized SS hand instruments (eight-step SS) and 69 canals using the rotary NiTi instruments (eight-step NiTi). The traditional serial step-back technique (step-back) was used for 85 root canals. In the apical third, reaming or filing motions were used up to sizes 25 and only reaming motion in sizes larger than 25 with the new eight-step method. A filing motion was used in the step-back for all sizes. Root canals of all groups were obturated with gutta-percha points and AH26 using a lateral condensation technique. Pre and postoperative radiographs were taken of each tooth. Procedural errors were recorded and statistically analyzed using a binomial test for comparison of proportion. It was reported that (i) procedural errors detected consisted of two canals with transportation (3%) with the eight-step SS, and three canals (4%) with transportation with eight-step NiTi, (ii) there were no canal obstructions or instrument separations, and (iii) with the step-back, 20 canals were transported (24%), seven canals had obstructions (8%), and in one canal instrument was separated (1%). It was concluded that the new eight-step method resulted in fewer procedural errors than the traditional serial step-back technique when senior students prepared root canals in patients either by hand with standardized K-files or by rotary NiTi instrumentation. Marending et al. [233] examined the factors affecting the outcome of root canal therapy in general dentistry hospital practice. Eighty-four patients were included. Of these, 66 (79%) were available for recall after > or = 30 months (mean = 46 months). Root canal treatments were performed using a standard protocol. At recall, teeth were scored by means of the periapical index (PAI), which was the dependent variable (dichotomized to sound/ unsound). Explanatory variables were patient age, integrity of the nonspecific immune system, smoking status, dichotomized PAI score before treatment, initial treatment versus retreatment, prior exposition of the root canal to saliva, SS hand versus NiTi rotary instrumentation, and quality of root filling. Unit of observation was the patient–individual. Data were analyzed using univariate tests and backward stepwise logistic regression analysis. It was found that (i) after five steps with elimination of the least significant independent variable, status of the immune system, initial PAI, and root filling quality were found to be the indispensable predictors for treatment outcome, (ii) using these three explanatory variables, the logistic regression model had a predictive value of 87%, compared to 91% with all eight variables, and (iii) success rate at recall (PAI < or = 2 without symptoms) was 88% (95% CI = 78, 94). It was concluded that the integrity of a patient's nonspecific immune system, which has been neglected in earlier investigations, is a significant predictor for endodontic treatment outcome and should receive more attention in future studies. Peciuliene et al. [234] gathered information about the various aspects, technical and biological, of endodontic treatment as performed by Lithuanian general dental practitioners and to compare their choices with established endodontic treatment standards of undergraduate education. Questionnaires were sent to all 2,850 Lithuanian dental practitioners. The structured questionnaire included inquiries about gender, duration of professional activity, working environment, and details about instruments and materials. It was mentioned that (i) from total 1,532 (53.8%) questionnaires were returned, (ii) only responses from general dental practitioners (1,431) were included, (iii) of the respondents 66% never used a rubber dam, (iv) most practitioners relied on conventional SS instruments, (v) the NiTi hand files were often and routinely used by 32.2% of the respondents, (vi) NaOCl was the most popular choice as a root canal irrigant, (vii) calcium hydroxide paste was used as an interappointment medicament, and (viii) cold-lateral condensation root filling method was used by 72.8% of the respondents while 15.6% used a paste for the root filling. It was then concluded that (i) the recently graduated dental practitioners were following the recommended standard of endodontic treatment better than those with a longer time from the graduation, (xi) it is important to improve the quality of existing courses of continuous education in endodontology in order to ensure the necessary competency in clinical practice, and (ii) the low use of a rubber dam and low adoption of new technologies in Lithuania is not acceptable and needs to be changed. Mesgouez et al. [235] determined the influence of operator experience on the time needed for canal preparation when using a rotary NiTi system. A total of 100 simulated curved canals in resin blocks were used. Four operators prepared a total of 25 canals each. The operators included practitioners with prior experience of the preparation technique and practitioners with no experience. The working length for each instrument was precisely predetermined. All canals were instrumented with rotary NiTi ProFile Variable Taper Series 29 engine-driven instruments using a high-torque handpiece (Maillefer, Ballaigues, Switzerland). The time taken to prepare each canal was recorded. Significant differences between the operators were analyzed using the Student's t-test and the Kruskal–Wallis and Dunn nonparametric tests. It was mentioned that (i) comparison of canal preparation times demonstrated a statistically significant difference between the four operators, and (ii) in the inexperienced group, a significant linear regression between canal number and preparation time occurred. It was concluded that time required for canal preparation was inversely related to operator experience.

7.3 Surface-related phenomena and other properties

7.3.1 Surface characterization

Surface is not just a free end of a substance, rather is a contact and boundary zone with other substances (either in gaseous, liquid, or solid). Surface and interface reactions include reactions with organic or inorganic materials, with vital or nonvital species, with hostile or friendly environments, and so on. Surface activities vary from mechanical actions (fatigue crack initiation and propagation, stress intensification, etc.), chemical action (discoloration, tarnishing, contamination, corrosion, oxidation, etc.), tribological action (wear, friction, lubricant, etc.) to physical action (surface contact, surface tension, diffusion, adsorption, absorption, wetting, etc.). The longevity, safety, reliability, and structural integrity of dental and medical materials and devices are governed by these surface phenomena. A physical system which comprises of a homogeneous component such as solid, liquid, or gas and is clearly distinguishable from each other is called as "phase," and a boundary at which two or three of these individual phases are in contact is called as "interface." Interfaces can be defined as a boundary face at which two phases are in contact. Accordingly, surface of solid or liquid can be interfaced with the gas. Although both phases interfacing to each other at the boundary face are separate phases, it can be considered that there would be an intermediate layer bridging between these distinct phases, and the intermediate layer possesses a continuous character having properties from both phases. Hence, it would be better to name the "interface layer" instead of "interface." However, since the thickness of the intermediate interface layer is only in order of several molecules, the layer can still be considered as a phase without macroscopic thickness [236–238].

Characterizing such surface and/or interface should provide more clear understanding various actions between two participating bodies (in this case, NiTi alloy and others). For dental and medical implant applications of NiTi biomaterials, osseointegration phenomenon is a predominant factor to control subsequent safety, longevity, and success rate. A combination of orthodontic archwire and bracket (particularly slot portion thereof) provides a crucial friction situation and successful mechanotherapy is totally relied on this biotribological condition. Furthermore, surface conditions of endodontic files should affect the machining efficiency, cleaning, and shaping capabilities. At the same time, various types of surface modifications have been conducted on biomedical NiTi alloy surface to promote the aforementioned target properties. Such surfaces regardless of modified or nonmodified conditions are subjected to nondestructive analytical equipment, including X-ray and electron diffractometer, optical and electron microscopy, and various spectroscopies for elemental analyses.

Lim et al. [239] investigated the surface characteristics and corrosion behavior of $Ni_{50.6}Ti$ (at%) SMA coated by a ceramiclike and highly biocompatible material, iridium oxide (IrO_2), which was prepared by thermal decomposition of $H_2IrCl_6·6H_2O$ precursor solution at the temperature of 300, 400, and 500 °C, respectively. The surface morphology and microstructure of the coatings were investigated by scanning electron microscope (SEM) and glancing angle X-ray diffraction. X-ray photoelectron spectroscopy was employed to determine the surface elemental composition. Corrosion resistance property of the coated samples was studied in a simulated body fluid at 37 ± 1 °C by electrochemical method. It was found that (i) the morphology and microstructure of the coatings were closely related to the oxidizing temperatures, (ii) a relatively smooth, intact, and amorphous coating was obtained when the $H_2IrCl_6·6H_2O$ precursor solution (0.03 mol/L) was thermally decomposed at 300 °C for 0.5 h, and (iii) compared with the bare NiTi alloy, IrO_2 coated samples exhibited better corrosion resistance behavior to some extent. Jenko et al. [240] studied the surface composition of a melt-spun NiTi SMA using different surface analytical techniques, namely, Auger electron spectroscopy and X-ray photoelectron spectroscopy, before and after testing its biocompatibility. It was reported that the surface consists of an oxide layer of 10–20 nm thickness, composed of Ti oxide and some Ni oxide with metallic Ni inhomogeneously distributed in the subsurface region. Lin et al. [241] prepared different surfaces of titanium by changing the etching temperature and time. It was reported that (i) surface topography and roughness were more proportional to etching temperature; however, diffusion of hydrogen and tensile strength are more time-related to titanium hydride formation on the surface, (ii) titanium becomes more hydrophilic after etching even though the micropits were not formed after etching, (iii) more and deeper cracks were found on the specimens with more hydrogen diffusion, and (iv) therefore, higher temperature and shorter time are an effective way to get a uniform surface and decrease the diffusion of hydrogen to prevent hydrogen embrittlement. Green et al. [242] amorphized NiTi alloy by N^+ ion implantation and by controlled shot peening in order to improve surface mechanical properties. It was obtained that (i) both the modified and unmodified NiTi surfaces were predominantly covered with TiO_2 and the underlying substrate crystallography determined both the affinity for surface OH^-/chemisorbed water and ultimately the wetting behavior of distilled water, (ii) N^+ ion implanted NiTi contained a TiN phase within the surface which reduced wetting, demonstrating a reduced interfacial energy, and (iii) the surface concentrations of Ni were unaffected by the surface modifications, with all samples containing less than 3 at% Ni suggesting that the surface TiO_2 oxide layer was maintained despite the surface amorphization treatments. Pequegnat et al. [243] investigated two $NiTi_{49.8}$ (at%) alloys which were subjected to the laser surface processing, including as-processed and polished, while comparing them to a chemically etched parent material. It was reported that (i) surface defects, including increased roughness, crystallinity, and presence of volatile oxide species, overshadowed any possible performance improvements from an increased Ti/Ni ratio or inclusion dissolution imparted by laser processing; and (ii) however, postlaser process mechanical polishing was shown to remove these defects and restore the performance, making it comparable to chemically etched NiTi material. Gu et al. [244] synthesized the surface oxide layers by heat treatment in air in the temperature range of 300–800 °C in order to enhance the bioactivity of NiTi alloy. The heat-treated sample was soaked in simulated body fluid to study the bioactivity of the thermally grown oxide layer. It was found that (i) a protective layer of TiO_2 was formed on the surface of NiTi alloy at heat treatment temperatures of 600 °C or higher with varying degree of anatase and rutile, (ii) small amount of nickel oxide was found on the surface of 300 and 400 °C treated samples, (iii) with further increase in the heat treatment temperature, the nickel concentration on the surface decreased and there was almost no nickel species on the surface after heat treatment at 600 or 800 °C, (iv) depth profiling revealed that the amount of TiO_2 (Ti^{4+}) decreased with depth with a concomitant increase of metallic Ti, (v) both TiO (Ti^{2+}) and Ti_2O_3 (Ti^{3+}) increased initially and then decreased gradually with depth, and (vi) Ni existed mainly in the oxidized state on the surface of heat-treated samples and it changed to metallic state with increasing depth. Based on these finding, it was concluded that the titanium oxide layer formed on the 600 and 800 °C heat-treated samples was bioactive, and a layer of apatite was formed on the surface of the titanium oxide layer after soaking in simulated body fluid. Wang et al. [245] modified surface structure of NiTi SMA by advanced oxidation processes in UV/ H_2O_2 photocatalytic system. It was obtained that (i) the advanced oxidation processes in UV/H_2O_2 photocatalytic system leads to formation of titanium oxides film on NiTi substrate, (ii) depth profiles of O, Ni, and Ti show such a film possesses a graded interface structure to NiTi substrate and there is no intermediate Ni-rich layer like that produced in conventional high temperature oxidation, (iii) except TiO_2 phase, some titanium suboxides (TiO, Ti_2O_3) may also exist in the titanium oxides film, (iv) oxygen is mainly present in metal oxides and some chemisorbed water and OH$^-$ are found in titanium oxides film, and (v) Ni nearly reaches zero on the upper surface and relatively depleted in the whole titanium oxides film indicating that the advanced oxidation processes in UV/H_2O_2 photocatalytic system is a promising way to favor the widespread application of biomedical NiTi SMA by improving its biocompatibility.

Due to crucial biotribological situation between slot portion of the bracket and archwire to achieve the best mechanotherapy as designed in advance, the friction behavior becomes very important. Oshida et al. [246] microanalyzed orthodontic archwires (equiatomic TiNi alloy) of both used (4 weeks) and unused conditions by optical and scanning electron microscopes, energy-dispersive X-ray spectroscopy, and electron diffraction to characterize the surface layers. They were also subjected to immersion and polarization corrosion tests in a 0.9% NaCl aqueous solution. It was obtained that (i) surface layers of used archwires were covered contaminants causing the discoloration, and the contaminants were identified as mainly KCl crystals, (ii) surfaces of both used and unused wires were observed to be irregular features characterized by lengthy islandlike structures, where nickel was selectively dissolved, (iii) corrosion tests in a 0.9% NaCl aqueous solution in immersion and polarization methods indicated that by increasing temperature from 3 to 60 °C and acidity from pH 11 to pH 3 calculated corrosion rates increased, and (iv) surface layers of TiNi archwires can be electrochemically modified to selectively etch nickel away, leaving a Ti-enriched surface layer and forming a uniformly distributed porous surface that may reduce the coefficient of friction against the orthodontic brackets. Eliades et al. [247] evaluated the structure and morphological condition of retrieved NiTi orthodontic archwires was evaluated and any possible alterations in the surface composition of the alloy following 1–6 months in vivo were characterized. It was reported that (i) scanning electron microscope and X-ray microanalysis showed that the elemental species precipitated on the material surface were Na, K, Cl, Ca, and P, forming NaCl, KCl, and Ca-P precipitates; (ii) increased intraoral exposure was consistently associated with the presence of a mature film, while evidence of alloy delamination, pitting, and crevice corrosion, as well as a notable reduction in the alloy grain size was observed; and (iii) intraoral exposure of NiTi wires alters the topography and structure of the alloy surface through surface attack in the form of pitting or crevice corrosion or formation of integuments suggesting that the in vivo research is required to resolve the implications of the described aging pattern in the corrosion resistance of the alloy, the potential for nickel leaching, as well as bracket-archwire friction variants. Huang et al. [248] investigated the surface characterizations and corrosion resistance of as-received commercial NiTi dental orthodontic archwires from different manufacturers using a cyclic potentiodynamic test in artificial saliva with various acidities. It was found that (i) the surface structure of the passive film on the tested NiTi wires were identical, containing mainly TiO_2 with small amounts of NiO; (ii) a different surface topography was observed on the NiTi wires from various manufacturers; (iii) the corrosion tests showed that both the wire manufacturer and solution pH had a statistically significant influence on the corrosion potential, corrosion rate, passive current, breakdown potential, and crevice-corrosion susceptibility; and (iv) the difference in the corrosion resistance among these NiTi dental orthodontic archwires did not correspond with the SR and preexisting defects. Huan et al. [249] also investigated the influence of a fluoride-containing environment on the surface topography variations of different NiTi orthodontic archwires. It was found that (i) both the archwire manufacturer and immersion environment had a significant influence on DeltaR(a), DeltaR(ms), and DeltaR(z), (ii) regardless of the archwire manufacturer, no statistically significant difference in DeltaR(a) (<70 nm), DeltaR(ms) (<90 nm), and DeltaR(z) (<450 nm) was observed on the tested NiTi archwires in lower fluoride-containing (<2,500 ppm) environments, including the various fluoride mouthwashes and the artificial saliva added with fluoride toothpastes, and (iii)in artificial saliva added with high fluoride prophylactic gel (around 17,000 ppm), a significant increase in DeltaR(a) (around 120–250 nm), DeltaR(ms) (around 140–320 nm), and DeltaR(z) (around 770–1,410 nm), that is, increasing the SR, was observed on the tested NiTi archwires. Zegan et al. [250] investigated the effects of oral fluids and archwire-bracket friction on the surface characteristics of NiTi alloy orthodontic archwires with/without aesthetic coating, in vivo for 2–3 months. It was reported that (i) initially, the orthodontic archwires showed microscopic manufacturing and coating defects in the physiognomic layer, (ii) after intraoral exposure, amorphous organic matter deposits were observed on the surface of the NiTi archwires and the wire coating presented exfoliation on the oral areas of friction with brackets, and (iii) X-ray microanalysis revealed changes in all atomic and mass percentages of chemical elements from the surface of all retrieved dental archwires, NiTi ion depletion and the occurrence of additional elements due to interactions with saliva, concluding that intraoral exposure of NiTi Archwires and the archwire-bracket friction of coated wire altered the morphology and changed the elemental composition of the surface due to the process of corrosion, adhesion of organic matters, and ionic exchange with oral fluids. Choi et al. [251] investigated the effects of sliding on the ultrastructure of three representative esthetic superelastic 0.014-inch NiTi archwires. It was reported that (i) epoxy resin-coated NiTi archwire and Teflon-coated NiTi archwire exhibited less SR than uncoated NiTi archwire, (ii) Ag/biopolymer-coated NiTi archwire showed the highest SR compared with the others because of its silver particles, (iii) in vitro sliding tests led to a significant increase in the SR of all 0.014-inch NiTi wires regardless of bracket type, (iv) the wire groups combined with SS brackets were rougher than those of ceramic brackets regardless of the coating materials because of exfoliation of the coating materials, and (v) the Teflon-coated NiTi archwire and SS bracket group showed the highest increase (5-fold) in SR compared to the others, while the epoxy resin-coated wire groups showed the lowest increase (1.4-fold) in SR compared with the others. Krishnan et al. [252] conducted surface characterization of NiTi orthodontic archwires. One conventional and five types of surface modified NiTi archwires were surface characterized with SEM, energy-dispersive analysis, Raman spectroscopy, atomic force microscopy, and 3D profilometry. It was found that (i) a considerable reduction in roughness values from conventional in a material specific pattern: Group I, conventional (578.56 nm) > Group V, Teflon (365.33 nm) > Group III, nitride (301.51 nm) > Group VI (i), rhodium (290.64 nm) > Group VI (ii), silver (252.22 nm) > Group IV, titanium (229.51 nm) > Group II, resin (158.60 nm), and (ii) the defects with aesthetic (resin/Teflon) and nitride surfaces and smooth topography achieved with metals; titanium/silver/rhodium were detected. It was, therefore, concluded that (i) resin, Teflon, titanium, silver, rhodium and nitrides were effective in decreasing SR of NiTi archwires albeit certain flaws, and (ii) findings have clinical implications, considering their potential in lessening biofilm adhesion, reducing friction, improving corrosion resistance, and preventing nickel leach and allergic reactions.

Bahia et al. [253] analyzed the influence of cyclic loading on the mechanical behavior of NiTi wires employed in the manufacture of ProFile rotary endodontic instruments. It was found that (i) the mechanical properties of the as-received wires, their chemical composition, the phases present and their transformation temperatures were consistent with their final application, (ii) only small changes, which decreased after the first few cycles, took place in the mechanical properties of the cycled wires, and (iii) the stress at maximum load and the plastic strain at breakage remained the same, while the critical stress for inducing the superelastic behavior, which is related to the restoring force of the endodontic instruments, decreased by approximately 27% concluding that the mechanical behavior of the NiTi wires was modified slightly by cyclic tensile loading in the superelastic plateau, and as the changes tended toward stabilization, the clinical use of rotary NiTi ProFile instruments does not compromise their superelastic properties until they fracture by fatigue or torsional overload or are otherwise discarded. Alapati et al. [254] examined numerous discarded ProFile GT, ProFile, and ProTaper NiTi rotary instruments obtained from two graduate endodontic clinics by SEM. It was reported that (i) the failure processes generally exhibited substantial ductile character, evidenced by a dimpled rupture fracture surface, (ii) crack propagation at grain boundaries and cleavage surfaces indicative of transgranular fracture were observed for some specimens, and (iii) it appeared that oxide particles from the manufacturing process served as nucleating sites for the microvoids, leading to dimpled rupture. Fatma et al. [255] evaluated the surface topography changes in three NiTi file systems using either rotary or reciprocal motion. It was found that (i) there were no significant differences preoperatively among the NiTi file systems in 1×1 or 5×5 μm^2 areas, and (ii) postoperatively, the WaveOne Primary had more surface irregularities, and (iii) three-dimensional Atomic Force Microscopy (AFM) images of instrument surfaces showed topographic irregularities preoperatively and postoperatively. Inan et al. [256] investigated the surface quality of new and used rotary NiTi instruments manufactured by the traditional grinding process and twisting methods. It was obtained that (i) mean root mean square values for new and used TF 25.06 files were 10.70 ± 2.80 nm and 21.58 ± 6.42 nm, respectively, and the difference between them was statistically significant, and (ii) mean root mean square values for new and used Mtwo 25.06 files were 24.16 ± 9.30 nm and 39.15 ± 16.20 nm, respectively, and the difference between them also was statistically significant. Ferreira et al. [257] assessed the assessment of nanoscale alterations in the surface topography of NiTi endodontic instruments using a high-resolution optical method and to verify the accuracy of the technique. It was mentioned that (i) the three-dimensional high-resolution image analysis showed clear alterations in the surface topography of the examined cutting blade and flute of the instrument, before and after use, with the presence of surface irregularities such as deformations, debris, grooves, cracks, steps, and microcavities, (ii) optical profilometry provided accurate qualitative nanoscale evaluation of similar surfaces before and after the fatigue test, and (iii) the stability and repeatability of the technique enables a more comprehensive understanding of the effects of wear on the surface of endodontic instruments.

7.3.2 Wettability and spreadability

The biological actions taking place at biomaterial/host tissue interface is governed by the surface wettability of biomaterials [258, 259]. Wettability is modulated by surface characteristics, such as surface physio-chemistry and surface morphology and topography, leading to success in implantology through well-established osseointegration [246, 260–262]. Interfaces between biomaterials, tissue and body fluids such as blood play a key role in determining the nature of the interaction between biomaterials and the living organism. The wettability of these biomaterials in relationship to their microenvironment is an important factor to consider when characterizing surface behavior. The measure of the contact angle between a fluid and material surface can be used to define wettability for that particular microenvironment. Oshida et al. [246] preoxidized using various commercially pure titanium grade III (CPT III), Ti-6AI-4V alloy, austenitic (M_S is close to 0 °C) and martensitic (M_S is ca. 50 °C) NiTi alloys, pure Ni, AISI Type 316L SS, and Co–Cr (27Cr-2Ni-5Mo) alloy at 300 °C for 30 min in pure oxygen. The initial contact angle (θ_o) and its changes ($\delta\theta/\delta t$) as a function of time in 1% NaCl solution drop were measured. The initial contact angle was measured by following equation:

$\theta = 2\tan^{-1}(2h/d)$, where h is the height of drop (1% NaCl) and d is its diameter, as shown in Figure 7.2 [246].

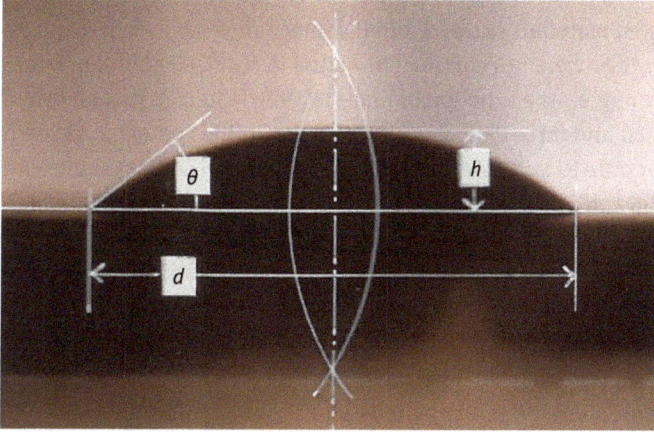

Figure 7.2: On photo print, the height (*h*) and diameter (*d*) of each drop were measured and the surface contact angle (*θ*) was calculated by the equation as mentioned in the text.

At the same time, we defined the spreading as a function of time, as follows [246]:

$$\delta\theta/\delta t = \{- (4h(\delta d/\delta t) + 4d(\delta h/\delta t)\}/(d^2 + 4h^2).$$

The obtained data for wetting θ (in degrees) and spread $\delta\theta/\delta t$ are shown in Table 7.4 [246, 260].

Table 7.4: Initial contact wetting angles and time-dependent spreading on various surfaces.

	mp → ox		bp → ox		mp → sp → ox	
	θ (deg)	δθ/δt	θ (deg)	δθ/δt	θ (deg)	δθ/δt
CPT III	54.24	−0.0046			54.82	−0.0045
Ti-6Al-4V	32.08	−0.0010	30.85	−0.0015	50.84	−0.0034
NiTi (M)	69.88	−0.0055	68.92	−0.0053	68.41	−0.0046
NiTi (A)	71.79	−0.0048			59.18	−0.0033
316L s.s.	56.46	−0.0024	55.73	−0.0025	50.17	−0.0013
Pure Ni	35.72	−0.0016			48.77	−0.0016
Co−Cr	62.04	−0.0023	61.85	−0.0021	62.04	−0.0018

mp: mechanical polishing with grit #600.
bp: buff polishing with alumina emulsion.
sp: shot-peened with alumina blast media.
ox: oxidation at 300 °C for 30 min in pure oxygen.

It was found that all tested materials showed initial contact angles between 0 and 90 degrees, indicating that they were not completely wetted in 1% NaCl solution. Moreover, changes in contact angles as a function of time (120 min) indicate that all tested materials can be classified into three distinct groups: (1) low initial contact angle with low spreading ($\delta\theta/\delta t$) for pure Ni and Ti6AL4V alloy, (2) high initial contact angle with low spreading for 316L SS and Co–Cr alloy, and (3) high initial contact angle with high spreading for CPT and NiTi alloys.

Cell attachment and spreading to titanium-based alloy surfaces is a major parameter in implant technology. Ponsonnet et al. [263] investigated substratum surface hydrophobicity, surface free energy, interfacial free energy, and SR to ascertain which of these parameters is predominant in human fibroblast spreading. Two methods for contact angle measurement were compared: the sessile drop method and the captive bubble two-probe method. Tested surface included CPT, Ti6Al4V, and NiTi. It was mentioned that (i) from captive bubble contact angle experiments (air or octane bubble under water), the interfacial free energy of the different surfaces in water was obtained, (ii) a relationship between cell spreading and the polar component of surface free energy was found, and (iii) interfacial free energy values were low for all the investigated surfaces indicating good biocompatibility for such alloys. Liu et al. [264] investigated the effects of plasma-implanted nitrogen in NiTi on the nanoscale surface morphology, structure, and wettability, as well as biocompatibility. It was reported that (i) X-ray photoelectron spectroscopy results show that the implantation depth of nitrogen increases with higher pulsing frequencies, (ii) atomic force microscopy discloses that the nanoscale SR increases and surface features are changed from islands to spiky cones with higher pulsing frequencies, (iii) the variation in the nanosurface structures leads to different surface free energy monitored by contact angle measurements, and (iv) the adhesion, spreading, and proliferation of osteoblasts on the implanted NiTi surface are assessed by cell culture tests indicating that the nanoscale surface morphology that is altered by the implantation frequencies impacts the surface free energy and wettability of the NiTi surfaces, and in turn affects the osteoblast adhesion behavior. Shabalovskaya et al. [265] studied the dependence of NiTi contact angles and surface energy on surface treatment in order to better understand the material hemocompatibility that was evaluated in our previous studies. It is found that (i) in the group of variously prepared surfaces of mechanically polished, additionally heat-treated, chemically etched, and additionally boiled in water, and further heat-treated, the contact angle could vary in the 50°–80° hydrophobic range and the total surface free energy in the 34–53 mN/m range, and (ii) the polar surface energy, varying from 5 to 29 mN/m, constitutes a decisive contribution to the total energy change, and it seems to be a direct function of the NiTi surface chemistry concluding that the alteration of the polar component of surface energy and thrombogenicity is due to changes of the electron-acceptor/electron-donor character of native NiTi surfaces during surface treatments. Tian et al. [266] fabricated micro/nanoscale structures on NiTi alloy to realize tunable anisotropic wetting and high adhesive capability. It was obtained that (i) the anisotropic wetting characterized by the difference between the water contact angles in the vertical and parallel directions ranges from 0° to 20.3°, which is far more than the value of natural rice leaves, (ii) the water sliding angles kept stable at 180°, successfully mimicking the adhesive ability of rose petals, (iii) the salinization process could strengthen the hydrophobicity but weaken anisotropic wetting, and (iv) these bioinspired NiTi surfaces have a tremendous potential applications such as microfluidic devices, biomimetic materials fabrication and lab on chip.

Active brazing is a commonly used method for joining ceramic materials. Siegmund et al. [267] investigated the wetting behavior of four Ti-rich ternary Ni–Ti–Zr alloys through sessile drop experiments on alumina disks of 96 and 99.9 % purity. It was obtained that (i) three of the analyzed alloys exhibited reactive wetting with final contact angles between 40° and 70°, (ii) the reaction phases at the metal/alumina interface had a thickness of about 1 µm and were of a similar composition for all alloys, (iii) dilatometer measurements showed thermal expansion coefficients between 13.2 and 15.8 × 10^{-6}/°C, and (iv) the lowest wetting angle of 40° was achieved with the alloy $Ti_{61}Zr_{20}Ni_{19}$ at temperatures above 980 °C.

7.3.3 Biofilm formation and cell adhesion

To enhance the safety and efficiency of metallic devices or instruments utilized in specific bioenvironments, numerous factors should be considered. The susceptibility to the fouling of the NiTi and other Ti-based alloys (such as Ti-6Al-4V or Ti-7Al-6Nb) due to the adhesion of microorganisms and the biofilm formation is very significant, especially in the context of an inflammatory state induced by implants contaminated by bacteria, and the implants corrosion stimulated by bacteria [268–272]. Cwalina et al. [269] examined the differences between the sulfur-oxidizing bacteria and sulfate-reducing bacteria strains in their affinity for NiTi and Ti-6Al-4V alloys. The biofilms formed on alloy surfaces by the cells of five bacterial strains (aerobic sulfur-oxidizing bacteria *Acidithiobacillus thiooxidans* and *Acidithiobacillus ferrooxidans*, and anaerobic sulfate-reducing bacteria *Desulfovibrio desulfuricans* – three strains) were studied using SEM and confocal laser scanning microscopy. It was indicated that (i) both alloys tested may be colonized by sulfur-oxidizing bacteria and sulfate-reducing bacteria strains, (ii) in the initial stage of the biofilm formation, the higher affinity of sulfate-reducing bacteria to both the alloys has been documented; however, the sulfur-oxidizing bacteria strains have indicated the higher (although differentiated) adaptability to changing environment as compared with sulfate-reducing bacteria, and (iii) stimulation of the sulfate-reducing bacteria growth on the alloys surface was observed during incubation in the liquid culture media supplemented with artificial saliva, especially of lower pH (imitated conditions under the inflammatory state, for example, in the periodontitis course); indicating that the results point to the possible threat to the human health resulting from the contamination of the titanium implant alloys surface by the sulfur-oxidizing bacteria (*A. thiooxidans* and *A. ferrooxidans*) and sulfate-reducing bacteria (*D. desulfuricans*).

Chezanowski et al. [272] attached the bacteria, *Staphylococcus aureus*, to the NiTi surface modified by a range of processes with and without light activation (used to elicit antimicrobial properties of materials). It was reported that (i) before the light activation, the number of bacterial colony forming units was the greatest for the samples thermally oxidized at 600 °C, (ii) this sample and the spark oxidized samples showed the highest photocatalytic activity but only the thermally oxidized samples at 600 °C showed a significant drop of *S. aureus* attachment, (iii) it was indicated that light activation and treating samples at 600 °C is a promising method for NiTi implant applications with inherent antimicrobial properties, (iv) light activation was shown to be an effective way to trigger photocatalytic reactions on samples covered with relatively thick titanium dioxide via accumulation of photons in the surface and a possible increase in defects which may result in free oxygen, and (v) light activation caused an increase in the total surface energy. Taha et al. [273] evaluated the in vitro ability of esthetic coated rectangular archwires to retain oral biofilms and in vivo biofilm formation on these wires after 4 and 8 weeks of clinical use and to correlate the findings with the SR of these wires. Three brands of esthetic coated NiTi archwires were selected. Arch wires retrieved after 4 and 8 weeks of intraoral use were obtained from 30 orthodontic patients. SR was assessed with an atomic force microscope. The in vitro adhesion assays were conducted by usage of *Streptococcus mutans* (MS), *Staphylococcus aureus*, and *Candida albicans*. The amount of bacterial adhesion was quantified using the colony-count method. It was obtained that (i) in vitro bacterial adhesion showed significant differences between wires in terms of *S. mutans* adhesion, (ii) all wires showed significant increases in SR and biofilm adhesion after intraoral exposure, and (iii) a significant positive correlation was observed between these two variables in vivo, but the correlation was not significant for in vitro bacterial adhesion concluding that SR and biofilm adhesion increased after intraoral use at all time intervals. Asiry et al. [274] investigated the effect of epoxy, polytetrafluoroethylene (PTFE) and rhodium surface coatings on SR, nanomechanical properties, and biofilm adhesion of NiTi archwires. Bacterial adhesion assays were performed using MS and *Streptococcus sobrinus* (SS) in an in vitro set up. It was found that (i) the highest roughness values (1.29 ± 0.49) were obtained for epoxy-coated wires and lowest values (0.29 ± 0.16) were obtained for the uncoated wires, (ii) no significant differences in the roughness values were observed between the rhodium coated and uncoated archwires, (iii) the adhesion of MS to the wires was significantly greater than that of SS, (iv) the epoxy coated wires demonstrated an increased adhesion of MS and SS and the uncoated wires demonstrated decreased biofilm adhesion, and (v) the Spearman correlation test showed that MS and SS adhesion was positively correlated with the SR of the wires. Based on these findings, it was concluded that (i) the different surface coatings significantly influence the roughness, nanomechanical properties, and biofilm adhesion parameters of the archwires, (ii) the evaluated parameters were most influenced by epoxy coating followed by PTFE and rhodium coating, and (iii) a positive correlation was detected between SR and biofilm adhesion. Musa et al. [275] determined if the presence of probiotic bacteria in an aging medium, that is, artificial saliva in this study, has relevant effects on the SR and the chemical composition of two main alloys used in dentistry (NiTi and SS). It was reported that (i) the probiotic bacteria *Lactobacillus reuteri* can induce processes that alter some features of the surface such as roughness and chemical composition, (ii) the effect is dependent on the type of alloy and coating, (iii) the bacteria increased roughness in the case of uncoated NiTi more than saliva alone (pH = 4.8), (iv) probiotic bacteria tend to decrease the corrosive influence of saliva on NiTi when the alloy is coated with rhodium or titanium nitride and this effect was also evidenced on stabiles steel, and (v) Raman spectroscopy confirmed that only SS samples are prone to oxidation processes, predominantly associated with exposure to saliva rather than probiotic bacteria.

As a countermeasure for biofilm formation, there are several studies for antiadherent and antimicrobial actions, Espinosa-Cristóbal et al. [276] prepared two sizes of AgNPs to determine the inhibitory effect and antiadherence activity of AgNPs on the adhesion of S. mutans on surfaces of brackets and wires for orthodontic therapies. It was found that the AgNPs exhibited good antimicrobial and antiadherence properties against *S. mutans* bacteria determining its high potential use for the control of white spot lesions in orthodontic treatments. A similar work was done by Mhaske et al. [277] He assessed the antiadherent and antibacterial properties of surface-modified SS and NiTi orthodontic wires with silver against *Lactobacillus acidophilus*. The study was done on 80 specimens of SS and NiTi orthodontic wires. The specimens were divided into eight test groups. Each group consisted of 10 specimens. Groups containing uncoated wires acted as a control group for their respective experimental group containing coated wires. Surface modification of wires was carried out by the thermal vacuum evaporation method with silver. Wires were then subjected to microbiological tests for assessment of the antiadherent and antibacterial properties of silver coating against *L. acidophilus*. It was obtained that (i) orthodontic wires coated with silver showed an antiadherent effect against *L. acidophilus* compared with uncoated wires, (ii) uncoated SS and NiTi wires, respectively, showed 35.4 and 20.5 % increase in weight which was statistically significant, whereas surface-modified wires showed only 4.08 and 4.4 % increase in weight, and (iii) the groups containing surface-modified wires showed statistically significant decrease in the survival rate of *L. acidophilus* expressed as colony-forming unit and as log of colony count when compared to groups containing uncoated wires, and it was 836.60 ± 48.97 colony-forming units in the case of uncoated SS whereas it was 220.90 ± 30.73 units for silver-modified SS, 748.90 ± 35.64 units for uncoated NiTi, and 203.20 ± 41.94 units for surface-modified NiTi, respectively.

There are several researches on endothelial cell attachment NiTi alloy. Ni-containing alloys (Pure nickel, titanium, and three biomedical alloys-18-8 SS, NiTi, and Rexillium III) were evaluated, which are commonly used in medical and dental applications that place them into long-term contact with soft tissues [278]. The release of Ni ions from these alloys is disturbing because of the toxic, immunologic, and carcinogenic effects that have been documented for some Ni compounds. Especially, Ni ions in solution recently have been shown to cause expression of inflammatory mediators, such as interleukin-1beta (IL-1beta), tumor necrosis factor-alpha (TNF-alpha), and intercellular adhesion molecules (ICAMs) from keratinocytes, monocytes, and endothelial cells [278]. First, it was determined whether or not the alloys or pure metals could cause cytotoxicity to THP-1 human monocytes or human microvascular endothelial cells (HMVECs) by measuring the succinic dehydrogenase (SDH) activity of the cells. Then, using identical conditions of exposure, the secretion of IL-1beta or TNF-alpha from monocytes or ICAM-1 expression on the HMVECs was determined. Only pure Ni suppressed (by 48% compared to Teflon controls) the SDH activity of the HMVECs or THP-1 monocytes. No alloy or metal caused the HMVECs to express ICAM-1, but the NiTi alloy caused a significant (ANOVA/Tukey) secretion of IL-1beta from the THP-1 monocytes. Secretion of TNF-alpha induced by NiTi was detectable but not statistically significant. The levels of IL-1beta secretion from monocytes were sufficient to induce ICAM-1 expression on HMVECs. The release of Ni from the NiTi was a logical suspect in causing the IL-1beta secretion by monocytes, but its role was not confirmed since other alloys, such as Rexillium III, released the same quantities of Ni yet did not activate the THP-1 monocytes. It was indicated that NiTi alloys pose a risk of promoting an inflammatory response in soft tissues by activating monocytes. Yang et al. [279] investigated the molecular mechanism of different effects of NiTi alloy surface and TiN coating on endothelial cell function. Release of Ni ion from uncoated and TiN-coated NiTi alloys and proliferation of endothelial cells on the two materials were evaluated, and then influence of the two materials on cellular protein expression profiles was investigated by proteomic technology. It was reported that (i) although the two materials did not affect cell proliferation, the Ni ions released from bare NiTi alloy generated inhibition on pathways associated with actin cytoskeleton, focal adhesion, energy metabolism, inflammation, and amino acid metabolism, and (ii) in comparison, TiN coating not only effectively prevented release of Ni ions from NiTi alloy but also promoted actin cytoskeleton and focal adhesion formation, increased energy metabolism, enhanced regulation of inflammation, and promoted amino acid metabolism. Furthermore, the two processes, "the initial mediation of adsorbed serum protein layer to endothelial cell adhesion and growth on the two materials" and "the following action of the two materials on cellular protein expression profile," were analyzed. It was found that (i) in stage of cell adhesion (within 4 h), release of Ni ions from bare NiTi alloy was very low, and the activation of adsorbed proteins to cell adhesion and growth related biological pathways (such as regulation of actin cytoskeleton, and focal adhesion pathways) was almost as same as TiN-coated NiTi alloy indicating that the released Ni ions did not affect the mediation of adsorbed proteins to endothelial cell adhesion, and (ii) however, in stage of cell growth and proliferation, the release of Ni ions from bare NiTi alloy increased with time and reached a higher level, which inhibited endothelial cell function at molecular level, whereas TiN coating improved endothelial cell function. NiTi medical device needs to be covered by endothelial cells after being placed in the blood vessel to reduce ischemic complications. Tateshima et al. [280] examined the impact of ultraviolet (UV) irradiation on the biocompatibility of NiTi surfaces with endothelial cells. NiTi sheets were treated with UV irradiation for 48 h and human aorta derived endothelial cells were used. It was reported that (i) UV irradiation converted the NiTi surface to hydrophilic state and increased albumin adsorption, and (ii) the number of endothelial cell migration, attachment, and proliferation as well as their metabolic activity were significantly increased on UV treated NiTi, suggesting that UV irradiation may promote endothelialization of NiTi devices in blood vessels. Lü et al. [281], using the signaling pathway polymerase chain reaction (PCR) arrays, investigated the activation of two important biological signaling pathways in endothelial cell adhesion and growth mediated by adsorbed serum protein on the surface of uncoated and TiN-coated NiTi alloys. First, the endothelial cells were cultured on the bare and TiN-coated NiTi alloys and chitosan films as control for 4 and 24 h, respectively. Then, the total RNA of the cells was collected and the PCR arrays were performed. After that, the differentially expressed genes in the transforming growth factor beta (TGF-β) signaling pathway and the regulation of actin cytoskeleton pathway were screened out and further bioinformatics analyses were performed. It was shown that (i) both TGF-β signaling pathway and regulation of actin cytoskeleton pathway were activated in the cells after 4 and 24 h culturing on the surface of bare and TiN-coated NiTi alloys compared to the chitosan group, (ii) the activated TGF-β signaling pathway promoted cell adhesion; the activated regulation of actin cytoskeleton pathway promoted cell adhesion, spreading, growth, and motility, and (iii) the activation of both pathways was much stronger in the cells cultured for 24 versus 4 h, which indicated that cell adhesion and growth became more favorable with longer time on the surface of two NiTi alloy materials.

There are several interesting works on cell adhesion. Scaffolds for tissue engineering enable the possibility to fabricate and form biomedical implants in vitro, which fulfill special functionality in vivo. Loger et al. [282] produced free-standing NiTi thin film meshes by means of magnetron sputter deposition. Meshes contained precisely defined rhombic holes in the size of 440 to 1,309 µm^2 and a strut width ranging from 5.3 to 9.2 µm. The effective mechanical properties of the microstructured superelastic NiTi thin film were examined by tensile testing. These results will be adapted for the design of the holes in the film. Using these meshes, the influence of hole and strut dimensions on the adhesion of sheep autologous cells (CD133+) was studied after 24 h and after 7 days of incubation. It was found that (i) optical analysis using fluorescence microscopy and SEM showed that cell adhesion depends on the structural parameters of the mesh, (ii) after 7 days in cell culture a large part of the mesh was covered with aligned fibrous material, and (iii) cell adhesion is particularly facilitated on meshes with small rhombic holes of 440 µm2 and a strut width of 5.3 µm. Based on these findings, it was concluded that free-standing NiTi thin film meshes have a promising potential for applications in cardiovascular tissue engineering, particularly for the fabrication of heart valves. Putters et al. [283] tested the effect of an increasing dose exposure to NiTi, pure Ni, or Ti on human fibroblasts in cell cultures in subgroups in comparison with a control group. It was shown that nickel induces a significant inhibition of mitosis in human fibroblasts, whereas no significant effects of this kind were found for titanium or NiTi, (ii) NiTi alloy is considered to be biocompatible and would be applied as a surgical implant as titanium material, which would seem to justify application as a surgical implant. Messer et al. [284] determined if release of elements from vascular stent alloys that contained nickel and cobalt was sufficient to activate expression of key cellular adhesion molecules (CAMs) by endothelial cells. Expression of these CAMs is a critical step in the long-term inflammatory response to stent materials and possibly to in-stent restenosis. SS, NiTi, CoCrNi, and NiCr alloys were placed in direct contact with primary human microvascular endothelial cells for 72 h after preparation at three roughnesses (120, 320, and 1,200 grit). Expression of three CAMs (ICAM1, VCAM1, and e-selectin) was assessed using a modified enzyme-linked immunosorbent assay (ELISA) procedure. Cytotoxicity of the alloys was assessed by measuring succinate dehydrogenase (SDH) activity and total protein content of the cells, and nickel release was measured by atomic absorption spectroscopy. It was obtained that (i) none of the alloys suppressed SDH activity or total cellular protein significantly at any SR, indicating little or no cytotoxicity, (ii) Ni release was measurable from all alloys, was greatest from the rougher surfaces, and was significantly different for the different alloy types, (iii) NiTi alloys exhibited the lowest nickel release; however, none of the alloys activated expression of the CAMs, regardless of SR or nickel release level, and (iv) supplemental experiments using nickel ions alone confirmed that ICAM1 was inducible on the endothelial cells by Ni(II) concentrations above 100 µM. Therefore, it was concluded that (i) nickel or other elemental release from several common types of stent alloys was not sufficient to activate expression of CAMs on endothelial surfaces, and (ii) although these results indicate a low risk for direct activation of endothelial cells by ions released from stent alloys, other mechanisms, such as modulation of CAM expression by monocytes or smooth muscle cells, must be considered before ion-mediated influence on CAM expression can be dismissed. Habijan et al. [285] determined whether cyclic dynamic strain, in a range relevant for orthopedic implants, diminishes the biocompatibility of NiTi-SMAs. In order to analyze the biocompatibility of NiTi-SMA surfaces subjected to cyclic loading, NiTi-SMA tensile specimens were preloaded with mesenchymal stem cells, transferred to a sterile cell culture system, and fixed to the pull rods of a tensile testing machine. It was reported that (i) dynamic loading of the tensile specimens did not influence the viability of adherent human mesenchymal stem cells (hMSCs) after 24 h or 7 days compared with the nonstrained control, (ii) dynamic cycles of loading and unloading did not affect nickel ion release from the tensile specimens, and (iii) the release of IL-6 from hMSCs cultured under dynamic conditions was significantly higher after mechanical load (873 pg/mL) compared with static conditions (323 pg/mL) indicating that a new type of mechanical in vitro cell culture experiment can provide information which previously could only be obtained in large animal experiments. Wang et al. [286] prepared a series of pH sensitive films composed of Ni(OH)2 and NiTi layered double hydroxide (NiTi LDH) with different Ni/Ti ratios on the surface of nitinol via hydrothermal treatment. It was mentioned that (i) the films with specific Ni/Ti ratios would release a large amount of nickel ions under acidic environments but were relatively stable in neutral or weak alkaline medium, (ii) cell viability tests showed that the films can effectively inhibit the growth of cancer cells but have little adverse effects to normal cells, (iii) extraordinarily high intracellular nickel content and reactive oxygen species level were found in cancer cells, indicating the death of cancer cells may be induced by the excessive intake of nickel ions, and (iv) such selective cancer cell inhibition effect of the films is supposed to relate with the reversed pH gradients of tumor cells. Tooth movement during orthodontic treatment is possible because of mechanical force-induced inflammation and remodeling in the periodontal ligament. Variation in the inflammatory response might be anticipated with initial aligning archwires of different composition. Myeloperoxidase (MPO) is an enzyme found in neutrophil granules that represents an important marker of inflammation. Based on this background, Fatima et al. [287] measured MPO enzyme activity during initial alignment with orthodontic archwires of different alloy types (superelastic NiTi, heat-activated NiTi, and multistranded SS). It was found that (i) MPO activity was significantly increased in GCF at 2 h, 7 and 14 days in all groups compared to baseline, and (ii) enzymatic activity was highest in the superelastic NiTi group followed by heat-activated NiTi and multistranded SS groups but with no significant difference between superelastic NiTi and heat-activated NiTi groups concluding that superelastic NiTi and heat-activated NiTi archwires produce an increased inflammatory response based upon MPO activity during initial levelling and alignment compared to multistranded SS.

7.4 Damping behavior

A high damping capacity is considered as one of the important functional properties of SMAs. Those properties are related to a thermoelastic martensitic transformation. As a consequence of this transformation, the internal friction or damping can be investigated for three different states: (1) during thermal transformation cycling, (2) during martensite induced strain cycling at constant temperature, (3) in the martensitic state [288–291]. The maxima of the damping capacity appearance during the phase transformation are attributed to the plastic strain and twin-interface movement, and stress-induced transformation formed by the applied external stress [288–290]. The net-shape NiTi parts layer-by-layer was fabricated using a laser beam that locally melted the NiTi powder [292]. It was found that the two cantilevers showed a damping ratio of about 0.03 at temperatures below austenite start, maximal values of up to 0.04 in the transformation regions, and low values of about 0.005 above austenite finish indicating that selective-laser-melted NiTi qualifies for the fabrication of shock-absorbing medical implants in the same manner than conventionally produced NiTi. Rong et al. [293] investigated damping behavior of Ti-rich TiNi SMA, porous TiNi SMA, and a novel TiNi/AlSi composite using dynamic mechanical analyzer. It was reported that (i) the internal friction of porous TiNi mainly originates from microplastic deformation and mobility of martensite interface and increases with the increase of the porosity, (ii) a novel TiNi/AlSi composite has been developed successfully by infiltrating AlSi alloy into the open pores of porous TiNi alloy with 60% porosity through compression casting, (iii) it shows the same phase transformation characteristics as the porous TiNi alloy, and (iv) the damping capacity of the composite has been increased and the compressive strength has been also promoted remarkably. Lammering et al. [294] investigated the dynamic behavior of superelastic SMAs in the austenitic phase at frequencies of up to 4 Hz. It was found that (i) the beginning of the transformation process is more difficult to observe and that the critical stress value for the transition to martensite becomes obviously smaller at higher strain rates, and (ii) the area of the hysteresis loop is reduced with increasing strain rate. Xing et al. [295] investigated transformation and damping characteristics of a $Ni_{51}Ti/Ni_{50.2}Ti$ alloy synthesized by explosive welding. It was reported that (i) the two endothermic peaks of the unprestrained specimen corresponded to the reverse transformation of each NiTi component, (ii) the reverse transformation temperature of $Ni_{50.2}Ti$ increased with increasing prestrain level, whereas the reverse transformation peak of $Ni_{51}Ti$ was split up into two independent endothermic peaks, and (iii) the internal friction results showed that the temperatures of the internal friction peaks of the $Ni_{51}Ti/Ni_{50.2}Ti$ alloy and the range of reverse transformation temperature increased with increasing prestrain level, which is consistent with the results indicating that the explosive welding is confirmed an effective method to fabricate chemical heterogeneous shape-memory materials.

Zhang et al. [296] prepared $Ni_{50.5}Ti_{49.5}$ foam with bimodal pore structure by elemental powder sintering and examined high damping capacity. NaCl particles of 500–600 and 75–90 µm diameters were adopted as space holder to create large and small pores with porosity of 40%. It was shown that the foam exhibits high damping capacities during phase transformation and at austenite region because of deformation, dislocation motion in thin nodes/walls, and SIM formation. $N_{50.5}Ti_{49.5}$ foams with porosity 25.5% were prepared by transient liquid sintering process [297]. The sintered $Ni_{50.5}Ti_{49.5}$ foams exhibited three-step martensite transformation (MT), that is, B2→ (R + B19′), B2→B19′, and R→B19′, due to the presence of heterogeneously distributed micron-sized Ni_4Ti_3 precipitates. By contrast, the aged $Ni_{50.5}Ti_{49.5}$ foam showed two-step B2→R→B19′ MT with homogeneous nanosized Ni_4Ti_3 phase. It was reported that (i) the sintered foam had a high damping capacity ($\tan\delta = 0.047$) in the martensite and low damping capacity ($\tan\delta = 0.002$) in the austenite, (ii) however, the aged foam exhibited improved damping capacity ($\tan\delta = 0.01$) in the austenite, (iii) the enhanced damping capacity of austenite was attributed to the localized plastic deformation in nodes/struts and stress-induced B2-R MT, (iv) a damping peak at a temperature of 130 °C related to the (R + B19')→B2 transition during heating was confirmed, and (v) the laminated pore architecture foams with anisotropic mechanical properties may be a promising candidate material for bio and mechanical applications.

The effect of hydrogen doping on the internal friction and the modulus of elasticity was investigated in the alloys $Ni_{50.8}Ti_{49.2}$, $Ni_{49}Ti_{51}$, and $Ni_{40}Ti_{50}Cu_{10}$ as a function of the hydrogen content nH ($nH = H/Me$ atomic) and frequency [298]. It was found that (i) hydrogen strongly affects the anelastic spectrum of these alloys in the temperature domain, (ii) the IF peak occurring at the austenite–martensite (A–M) transformation in the solubilized $Ni_{50.8}Ti_{49.2}$ alloy is enhanced by hydrogen at low contents ($nH < 1.3$) and reduced at high hydrogen contents ($nH > 1.3$), (iii) an additional IF peak is introduced in the same alloy by H for $nH > 1.3$, where this second peak, which appears at temperatures higher than the martensite start temperature Ms, is likely due to hydrogen redistributions over subsets of interstitial sites within a hydride, under the applied alternating stress fields associated with the sample vibrations, (iv) in the aged alloy $Ni_{49}Ti_{51}$ H suppresses the low temperature IF background and introduces a thermally activated relaxation, which is associated with H dragging processes by twin boundaries, and (v) a relaxation effect, again ascribable to H dragging processes by twin boundaries, has also been evidenced with the $Ni_{40}Ti_{50}Cu_{10}$ alloy. Similar research was conducted on hydrogen-doped NiTi and NiTiCu alloy [299]. The internal friction (Q^{-1}) and modulus of elasticity (E) were investigated as a function of temperature at frequencies in the $Ni_{30}Ti_{50}Cu_{20}$ alloy containing various amounts n_H ($n_H = H/Me$ at.) of H. It was reported that (i) several dissipation processes have been observed which are associated with stress-induced motions either of isolated H atoms or of twin boundaries interacting with H, (ii) values of Q^{-1} as high as 0.075 have been measured in the presence of H impurities over extended temperature regions at around the B2–B19 and B2–B19' martensitic transitions, (iii) the observed damping is not a transient effect as those usually reported at low frequencies in H-free materials, thus, it does not depend on the rate of temperature change, and (iv) no appreciable dependence of the damping on the frequency and strain-amplitude are observed between 0.48 and 1.5 kHz and between 1×10^{-7} mm/mm and 3×10^{-5} mm/mm, respectively.

7.5 Electric characteristics

Matsumoto conducted a precise measurement of the $Ni_{49.5}Ti_{50.5}$ by a four-probe potentiometric method for a better understanding of the transformation behavior [300]. It was mentioned that (i) the temperature dependence of the electrical resistivity is linear in the regions of the high- and low-temperature phases, and a drastic change in electrical resistivity is observed in the temperature range of the transformation, and (ii) the transformation remains of one stage between the high- and low-temperature phase even with thermal cycling, although with increasing number of thermal cycles the electrical resistivity versus temperature curve shifts to the low temperature side and the electrical resistivity increases in the single phase temperature region which is attributable to the generation of transformation-induced defects. A similar study was done to investigate the effects of the repetition of transformation with thermal cycling [301]. It was reported that (i) complete and incomplete thermal cycles induce an anomalous increase in electrical resistivity which is attributed to formation of the R-phase, (ii) the incomplete thermal cycle is effective in the formation of the R-phase in comparison with the complete thermal cycle, (iii) the behavior of the transformation to the R-phase with thermal cycles depends on the composition and it is more difficult to induce and stabilize the R-phase by thermal cycling in NiTi with higher M_S, and (iv) the characteristics on the effects of the thermal cycles are the shift of the M_S to the low-temperature side and the stabilization of the R-phase, which are taken to be attributable to the accumulation of transformation-induced defects concluding that the precise measurement of the electrical resistivity proved helpful in order to detect the trace of the R-phase and characterize the effects of the thermal cycles. Nagarajan et al. [302] compared eight different rectangular archwires for electrical resistivity values. It was found that (i) in tensile testing Ortho Organizers wires ranked first and GAC Lowland NiTi wires ranked last, and (ii) for resistivity tests Ormco A wires were found superior and Morelli remained last indicating that these rankings should be correlated clinically and need further studies. Sivarai et al. [303] evaluated the super elastic property of eight groups of austenite active NiTi wires through mechanical tensile testing and electrical resistivity methods. It was reported that (i) Ortho Organizers wires ranked first and superior, followed by American Orthodontics and Ormco A wires, and (ii) Morelli and GAC lowland NiTi wires were ranked last.

References

[1]   Gavini G, dos Santos M, Caldeira CL, de Lima Machado ME, Freire LG, Iglecias EF, Peters OA, de Miranda Candeiro GT. Nickel–titanium instruments in endodontics: a concise review of the state of the art. Brazilian Oral. Res. 2018, 32; http://dx.doi. org/10.1590/1803107bor-2018.vol32.0067.

[2]   Otsuka K, Wayman CM. Introduction. In: Ostuka, et al. ed., Shape Memory Materials. Cambridge University Press, 1998, 1–26.

[3]   Fernandes DJ, Peres RV, Mendes AM, Elias CN. Understanding the shape-memory alloys used in orthodontics. ISRN Dent. 2011; doi: 10.5402/2011/132408.

[4]   Zhou H, Peng B, Zheng Y-F. An overview of the mechanical properties of nickel–titanium endodontic instruments. Endod. Top. 2013, 29, 42–54.

[5]   Shim K-S, Oh S, Kum KY, Kim Y-C, Jee K-K, Chang SW. Mechanical and metallurgical properties of various nickel-titanium rotary instruments. Biomed. Res. Int. 2017; doi: 10.1155/2017/4528601.

[6]   Guelzow A, Stamm O, Martus P, Kielbassa AM. Comparative study of six rotary nickel-titanium systems and hand instrumentation for root canal preparation. Int. Endod. J. 2005, 38, 743–52.
[7]   https://www.furukawa-ftm.com/nt/li/furu-nt3.htm.
[8]   http://www.chemistrylearner.com/nitinol.html.
[9]   http://www.chemistrylearner.com/nitinol.html.
[10]   https://confluentmedical.com/wp-content/uploads/2016/01/Material-Data-Sheet-Shape-Memory.pdf.
[11]   https://matthey.com/markets/pharmaceutical-and-medical/medical-device-components/resource-library/nitinol-technical-properties.
[12]   https://commons.wikimedia.org/wiki/File:Nitinol_draht.jpg.
[13]   http://www.broadmoly.com/products/Nitinol_alloy_wire.html?gclid=CjwKCAjwtO7qBRBQEiwAl5WC27_Wq-_bozcNz5S3E8SZfvwyO7DYHQUnxhUpS4Tc-4yzc-IEcXfBQxoCTOwQAvD_BwE.
[14]   Chung CY, Chu CL, Wang SD. Porous TiNi shape memory alloy with high strength fabricated by self-propagating high-temperature synthesis. Mater. Lett. 2004, 58, 1683–6.
[15]   Chu CL, Chung CY, Lin PH, Wang SD. Fabrication of porous NiTi shape memory alloy for hard tissue implants by combustion synthesis. Mater. Sci. Eng., A Struct. Mater.: Prop. Microstruct. Process. 2004, 366, 114–9.
[16]   Zysset PK, Guo XE, Hoffler CE, Moore KE, Goldstein SA. Elastic modulus and hardness of cortical and trabecular bone lamellae measured by nanoindentation in the human femur. J. Biomech. 1999, 32, 1005–12.
[17]   Van Audekercke R, Martens M. Mechanical Properties of Cancellous Bone. In: Hastings GW, ed., Natural and Living Biomaterials. CRC Press, 2018, 89–98.
[18]   Oshida Y, Miyazaki T, Tominaga T. Some biomechanistic concerns on newly developed implantable materials. J. Dent. Oral Heal. 2018, 5 pages; http://scientonline.org/open-access/some-biomechanistic-concerns-on-newly-developed-implantable-materials.pdf.
[19]   Huiskes R, Weinans H, Van Rietbergen B. The relationship between stress shielding and bone resorption around total hip stems and the effects of flexible materials. Clin. Orthop. Rel. Res. 1992, 274, 124–34.
[20]   Nagels J, Stokdijk M, Rozing PM. Stress shielding and bone resorption in shoulder arthroplasty. J. Shoulder. Elbow Surg. 2003, 12, 35–9.
[21]   Orchard JW, Cook JL, Halpin N. Stress-shielding as a cause of insertional tendinopathy: the operative technique of limited adductor tenotomy supports this theory. J. Sci. Med. Sport 2004, 7, 424–8.
[22]   Niinomi M, Nakai M. Titanium-based biomaterials for preventing stress shielding between implant devices and bone. Int. J. Biomater. 2011; http://dx.doi.org/10.1155/2011/836587.
[23]   Sumner DR. Long-term implant fixation and stress-shielding in total hip replacement. J. Biomech. 2015, 48, 797–800.
[24]   Bram M, Köhl M, Buchkremer HP, Stöver D. Mechanical properties of highly porous NiTi alloys. J. Materi. Eng. Perform. 2011, 20, 522–8.
[25]   Resnina N, Belyaev S, Voronkov A, Gracheva A. Mechanical behaviour and functional properties of porous Ti-45 at. % Ni alloy produced by self-propagating high-temperature synthesis. Smart Mater. Struct. 2106, 25; doi: 10.1088/0964-1726/25/5/055018.
[26]   Resnina N, Belyaev SP, Voronkov A. Functional properties of porous Ti-48.0 at.% Ni shape memory alloy produced by self-propagating high-temperature synthesis. J. Mater. Eng. Perform. 2018, 27. doi: 10.1007/s11665-018-3231-z.
[27]   Buchheit TE, Susan DF, Massad JE, McElhanon JR, Noebe RD. Mechanical and functional behavior of high-temperature Ni-Ti-Pt shape memory alloys. Metall. Mater. Trans. A 2016, 47, 1587–99.

[28]   https://pocketdentistry.com/4-orthodontic-wires/.

[29]   Kusy RP, Greenberg AR. Effects of composition and cross section on the elastic properties of orthodontic wires. Angle Orthod. 1981, 51, 325–41.

[30]   Rucker BK, Kusy RP. Elastic properties of alternative versus single-stranded leveling archwires. Am. J. Orthod. Dentofacial Orthop. 2002, 122, 528–41.

[31]   Favier D, Louche H, Schlosser P, Orgéas L, Vacher P, Debove L. Homogeneous and hetero-geneous deformation mechanisms in an austenitic polycrystalline Ti–50.8 at.% Ni thin tube under tension: Investigation via temperature and strain fields measurements. Acta Mater. 2007, 55, 5310–22.

[32]   Jiang F, Liu Y, Yang H, Li L, Zheng Y. Effect of ageing treatment on the deformation behaviour of Ti–50.9 at.% Ni. Acta Mater. 2009, 57, 4773–81.

[33]   Hsu W-N, Polatidis E, Šmíd M, Petegem SV, Casati N, Swygenhoven HV. Deformation and degradation of superelastic NiTi under multiaxial loading. Acta Mater. 2019, 167, 149–58.

[34]   Paula AS, Mahesh KK, dos Santos CML, Braz Fernandes FM, da Costa Viana CS. Thermome-chanical behavior of Ti-rich NiTi shape memory alloys. Mater. Sci. Eng. A 2008, 481/482, 146–50.

[35]   Braz Fernandes FM, Mahesh KK, dos Santos Paula A. Thermomechanical Treatments for Ni-Ti Alloys. In: Braz Fernandes FM, ed., Shape Memory Alloys – Processing, Characterization and Applications. IntechOpen, 2013; doi: 10.5772/56087; https://www.intechopen.com/books/shape-memory-alloys-processing-characterization-and-applications/thermomechanical-treatments-for-ni-ti-alloys.

[36]   Hornbogen E. Ausforming of NiTi. J. Mater. Sci. 1999, 34, 599–606.

[37]   Kockar B, Karaman I, Kulkarni A, Chumlyakov Y, Kireeva IV. Effect of severe ausforming via equal channel angular extrusion on the shape memory response of a NiTi alloy. J. Nucl. Mater. 2007, 361, 298–305.

[38]   Paula AS, Mahesh KK, Santos CML, Braz Fernandes FM, Viana CSC. Mechanical behaviour of ausformed and marformed Ti-Rich Ni-Ti shape memory alloys. Mater. Sci. Forum 2008, 587/588, 635–9.

[39]   Pereira ESJ, Peixoto IFC, Viana ACD, Oliveira II, Gonzalez BM, Buono VTL, Bahia MGA. Physical and mechanical properties of a thermomechanically treated NiTi wire used in the manufacture of rotary endodontic instruments. Int. Endod. J. 2011, 45, 469–74.

[40]   Miao WD, Mi XJ, Zhu M, Guo JF, Kou YM. Effect of surface preparation on mechanical properties of a NiTi alloy. Mater. Sci. Forum 2001, 394/3, 173–6.

[41]   Katić V, Curković HO, Semenski D, Baršić G, Marušić K, Spalj S. Influence of surface layer on mechanical and corrosion properties of nickel – titanium orthodontic wires. Angle Orthod. 2014, 84(6), 1041–8.

[42]   Wichelhaus A, Brauchli L, Ball J, Mertmann M. Mechanical behavior and clinical application of nickel-titanium closed-coil springs under different stress levels and mechanical loading cycles. Am. J. Orthod. Dentofac. 2010, 137(5), 671–8.

[43]   Khan MI, Zhou Y. Effects of local phase conversion on the tensile loading of pulsed Nd:YAG laser processed nitinol. Mater. Sci. Eng., A 2010, 527, 6235–8.

[44]   Nayan N, Buravalla V, Ramamurty U. Effect of mechanical cycling on the stress–strain response of a martensitic nitinol shape memory alloy. Mater. Sci. Eng., A 2009, 525, 60–7.

[45]   Santos Lde A, Resende PD, Bahia MG, Buono VT. Effects of R-phase on mechanical responses of a nickel-titanium endodontic instrument: structural characterization and finite element analysis. Sci. World J. 2016, 2016, 7617493; doi: 10.1155/2016/7617493.

[46]   Yoneyama T, Doi H, Hamanaka H. Influence of composition and purity on tensile properties of Ni-Ti alloy castings. Dent. Mater. J. 1992, 11, 157–64.

[47]   Iijima M, Ohno H, Kawashima I, Endo K, Mizoguchi I. Mechanical behavior at different
         temperatures and stresses for superelastic nickel-titanium orthodontic wires having
         different transformation temperatures. Dent. Mater. 2002, 8, 88–93.
[48]   Lombardo L, Toni G, Stefanoni F, Mollica F, Guarneri MP, Siciliani G. The effect of
         temperature on the mechanical behavior of nickel-titanium orthodontic initial archwires.
         Angle Orthod. 2013, 83, 298–305.
[49]   Yanaru K, Yamaguchi K, Kakigawa H, Kozono Y. Temperature- and deflection- dependences
         of orthodontic force with Ni-Ti wires. Dent. Mater. J. 2003, 22, 146–59.
[50]   Meling TR, Odegaard J. The effect of short-term temperature changes on the mechanical
         properties of rectangular nickel titanium archwires tested in torsion. Angle Orthod. 1998,
         68, 369–76.
[51]   Barwart O. The effect of temperature change on the load value of Japanese NiTi coil springs
         in the superelastic range. Am. J. Orthod. Dentofacial Orthop. 1996, 110, 553–8.
[52]   Saedi S, Turabi AS, Andani MT, Haberland C, Karaca H, Elahinia M. The influence of heat
         treatment on the thermomechanical response of Ni-rich NiTi alloys manufactured by
         selective laser melting. J. Alloys Comp. 2016, 677, 204–10.
[53]   Miyara K, Yahata Y, Hayashi Y, Tsutsumi Y, Ebihara A, Hanawa T, Suda H. The influence of
         heat treatment on the mechanical properties of Ni-Ti file materials. Dent. Mater. J. 2014, 33,
         27–31.
[54]   Tadayyon G, Mazinani M, Guo Y, Zebarjad SM, Tofail SAM, Biggs MJ. The effect of annealing
         on the mechanical properties and microstructural evolution of Ti-rich NiTi shape memory
         alloy. Mater. Sci. Eng., A 2016, 662, 564–77.
[55]   Yoneyama T, Kotake M, Kobayashi E, Doi H, Hamanaka H. Influence of mold materials
         and heat treatment on tensile properties of Ni-Ti alloy castings. Bull. Tokyo Med. Dent.
         Univ. 1993, 40, 167–72.
[56]   Soga Y, Doi H, Yoneyama T. Tensile properties and transformation temperatures of Pd added
         Ti-Ni alloy dental castings. J. Mater. Sci. Mater. Med. 2000, 11, 695–700.
[57]   Gall K, Tyber J, Brice V, Frick CP, Maier HJ, Morgan N. Tensile deformation of NiTi wires. J.
         Biomed. Mater. Res. Part A 2005, 75(4), 810–23.
[58]   Kwon YH, Cheon YD, Seol HJ, Lee JH, Kim HI. Changes on NiTi orthodontic wired due to acidic
         fluoride solution. Dent Mater. J. 2004, 23, 557–65.
[59]   Sehitoglu H, Karaman I, Zhang XY, Chumlyakov Y, Maier HJ. Deformation of FeNiCoTi shape
         memory single crystals. Scr. Mater. 2001, 44, 779–84.
[60]   Zhao Y, Taya M, Kang Y, Kawasaki A. Compression behavior of porous NiTi shape memory
         alloy. Acta Mater. 2005, 53, 337–43.
[61]   Tadayyon G, Guo Y, Biggs MJP, Tofail SAM. Deformation behavior of heat-treated Ni-rich NiTi
         shape memory alloy. A. J. Mater. Sci. Appl. 2018, 6, 7–21.
[62]   Fukami-Ushiro KL, Mari D, Dunand DC. NiTi and NiTi-TiC composites: part II. Compressive
         mechanical properties. Metall. Mater. Trans. A1996, 27, 183–91.
[63]   Montero-Ocampo C, Lopez H, Salinas Rodriguez A. Effect of compressive straining on
         corrosion resistance of a shape memory Ni-Ti alloy in Ringer's solution. J. Biomed. Mater.
         Res. 1996, 32, 583–91.
[64]   Qiu Y, Young ML, Nie X. High strain rate compression of martensitic NiTi shape memory
         alloys. Shape Mem. Superelast. 2015, 1, 310–8.
[65]   Xu RB, Cui L, Zheng Y. The dynamic impact behavior of NiTi alloy. Mater. Sci. Eng. A 2006,
         438/440, 571–4.
[66]   Wang XM, Zhou QT, Hangzi Liu H, Deng CH, Yue ZF. Experimental study of the biaxial cyclic
         behavior of thin-wall tubes of NiTi shape memory alloys. Metall. Mater. Trans. A 2012, 43,
         4123–8.

[67]   Lee DH, Park B, Saxena A, Serene TP. Enhanced surface hardness by boron implantation in
         Nitinol alloy. J. Endod. 1996, 22, 543–6.
[68]   Orgéas L, Favier D. Non-symmetric tension-compression behaviour of NiTi alloy. J. Phys. IV
         (Proceedings) 1995, C8, 605–10.
[69]   Liu B, Dui G, Xie B, Xue L. A constitutive model of porous SMAs considering tensile-
         compressive asymmetry behaviors. J. Mech. Behav. Biomed. Mater. 2014, 32, 185–91.
[70]   Saigal A, Fonte M. Solid, shape recovered "bulk" Nitinol: part I – tension–compression
         asymmetry. Mater. Sci. Eng., A 2011, 528, 5536–50.
[71]   Siviour CR, Huber JE, Normanton T, Petrinic N. Strain rate dependence of a super-elastic NiTi
         alloy. DYMAT 2009, 1153–9; doi: 10.1051/dymat/2009161.
[72]   Nemat-Nasser S, Guo W-G. Superelastic and cyclic response of NiTi SMA at various strain
         rates and temperatures. Mech. Mater. 2006, 38, 463–74.
[73]   Chen WW, Wu QP, Kang JH, Winfree NA. Compressive superelastic behavior of a NiTi shape
         memory alloy at strain rates of 0.001–750 s⁻¹. Int. J. Sol. Struct. 2001, 38, 8989–98.
[74]   Pieczyska EA, Gadaj SP, Nowacki WK, Tobushi H. Transformation Induced Effects in TiNi
         Shape Memory Alloy Subjected to Tension. In: Gdoutos EE, ed., Experimental Analysis of
         Nano and Engineering Materials and Structures. Springer, Dordrecht, 2007, 727–8.
[75]   Costanza G, Tata ME, Libertini R. Effect of temperature on the mechanical behaviour of Ni-Ti
         shape memory sheets. TMS 2016 145th Annual Meeting & Exhibition 2016, 433–9; https://
         link.springer.com/chapter/10.1007/978-3-319-48254-5_52.
[76]   Testarelli L, Plotino G, Al-Sudani D, Vincenzi V, Giansiracusa A, Grande NM, Gambarini G.
         Bending properties of a new nickel-titanium alloy with a lower percent by weight of nickel. J.
         Endod. 2011, 37, 1293–5.
[77]   Tsao CC, Liou JU, Wen PH, Peng CC, Liu TS. Study on bending behaviour of nickel-titanium rotary
         endodontic instruments by analytical and numerical analyses Int. Endod. J. 2013, 46, 379–88.
[78]   Schäfer E, Dzepina A, Danesh G. Bending properties of rotary nickel-titanium instruments.
         Oral Surg. Oral. Med. Oral Pathol. Oral Radiol. Endod. 2003, 96, 757–63.
[79]   Yahata Y, Yoneyama T, Hayashi Y, Ebihara A, Doi H, Hanawa T, Suda H. Effect of heat
         treatment on transformation temperatures and bending properties of nickel-titanium
         endodontic instruments. Int. Endod. J. 2009, 42, 621–6.
[80]   Silva EJ, Rodrigues C, Vieira VT, Belladonna FG, De-Deus G, Lopes HP. Bending resistance and
         cyclic fatigue of a new heat-treated reciprocating instrument. Scanning 2016, 38, 837–41.
[81]   Khier SE, Brantley WA, Fournelle RA. Bending properties of superelastic and nonsuperelastic
         nickel-titanium orthodontic wires. Am. J. Orthod. Dentofacial. Orthop. 1991, 99, 310–8.
[82]   Yamamoto M, Kuroda T, Yoneyama T, Doi H. Bending property and phase transformation of
         Ti-Ni-Cu alloy dental castings for orthodontic application. J. Mater. Sci. Mater. Med. 2002,
         13, 855–9.
[83]   Yoneyama T, Doi H, Hamanaka H, Yamamoto M, Kuroda T. Bending properties and
         transformation temperatures of heat treated Ni-Ti alloy wire for orthodontic appliances. J.
         Biomed. Mater. Res. 1993, 27, 399–402.
[84]   Silva MF, Pinzan-Vercelino CR, Gurgel Jde A. The influence of distal-end heat treatment on
         deflection of nickel-titanium archwire. Dental. Press J. Orthod. 2016, 21, 83–8.
[85]   Lim KF, Lew KKK, Toh SL. Bending stiffness of two aesthetic orthodontic archwires: an in
         vitro comparative study. Clin. Mater. 1994, 16, 63–71.
[86]   Garrec P, Jordan L. Stiffness in bending of a superelastic Ni-Ti orthodontic wire as a function
         of cross-sectional dimension. Angle Orthod. 2004, 74, 691–6.
[87]   Shima Y, Otsubo K, Yoneyama T, Soma K. Bending properties of hollow super-elastic
         Ti-Ni alloy wires and compound wires with other wires inserted. J. Mater. Sci. Mater.
         Med. 2002, 13, 169–73.

[88]   Ballard RW, Sarkar NK, Irby MC, Armbruster PC, Berzins DW. Three-point bending test comparison of fiber-reinforced composite archwires to nickel-titanium archwires. Orthodontics (Chic.) 2012, 13, 46–51.

[89]   De Santis R, Dolci F, Laino A, Martina R, Ambrosio L, Nicolais L. The Eulerian buckling test for orthodontic wires. Eur. J. Orthod. 2008, 30, 190–8.

[90]   Wilkinson PD, Dysart PS, Hood JA, Herbison GP. Load-deflection characteristics of superelastic nickel-titanium orthodontic wires. Am. J. Orthod. Dentofacial Orthop. 2002, 121, 483–95.

[91]   Quintão CC, Cal-Neto JP, Menezes LM, Elias CN. Force-deflection properties of initial orthodontic archwires. World J. Orthod. 2009, 10, 29–32.

[92]   Ahmadabadi MN, Shahhoseini T, Habibi-Parsa M, Haj-Fathalian M, Hoseinzadeh-NikT, Ghadirian H. Static and cyclic load-deflection characteristics of NiTi orthodontic archwires using modified bending tests. J. Mater. Eng. Perfor. 2009, 18, 793–6.

[93]   Watanabe E, Stigall G, Elshahawy W, Watanabe I. Deflection load characteristics of laser-welded orthodontic wires. Angle Orthod. 2012, 82, 698–702.

[94]   Lombardo L, Marafioti M, Stefanoni F, Mollica F, Siciliani G. Load deflection characteristics and force level of nickel titanium initial archwires. Angle Orthod. 2012, 82, 507–21.

[95]   Ahrari F, Ramazanzadeh BA, Sabzevari B, Ahrari A. The effect of fluoride exposure on the load-deflection properties of superelastic nickel-titanium-based orthodontic archwires. Aust. Orthod. J. 2012, 28, 72–9.

[96]   Jamleh A, Kobayashi C, Yahata Y, Ebihara A, Suda H. Deflecting load of nickel titanium rotary instruments during cyclic fatigue. Dent. Mater. J. 2012, 31, 389–93.

[97]   Gatto E, Matarese G, Di Bella G, Nucera R, Borsellino C, Cordasco G. Load-deflection characteristics of superelastic and thermal nickel-titanium wires. Eur. J. Orthod. 2013, 35, 115–23.

[98]   Arreghini A, Lombardo L, Mollica F, Siciliani G. Load deflection characteristics of square and rectangular archwires. Int. Orthod. 2016, 14, 1–14.

[99]   Wolcott J, Himel VT. Torsional properties of nickel-titanium versus stainless steel endodontic files. J. Endod. 1997, 23, 217–20.

[100]   Rowan MB, Nicholls JI, Steiner J. Torsional properties of stainless steel and nickel-titanium endodontic files. J. Endod. 1996, 22, 341–5.

[101]   Yum J, Cheung GS, Park JK, Hur B, Kim HC. Torsional strength and toughness of nickel-titanium rotary files. J. Endod. 2011, 37, 382–6.

[102]   Park SY, Cheung GS, Yum J, Hur B, Park JK, Kim HC. Dynamic torsional resistance of nickel-titanium rotary instruments. J. Endod. 2010, 36, 1200–4.

[103]   Kim HC, Kwak SW, Cheung GS, Ko DH, Chung SM, Lee W. Cyclic fatigue and torsional resistance of two new nickel-titanium instruments used in reciprocation motion: reciproc versus WaveOne. J. Endod. 2012, 38, 541–4.

[104]   Kaval ME, Capar ID, Ertas H. Evaluation of the cyclic fatigue and torsional resistance of novel nickel titanium rotary files with various alloy properties. J. Endod. 2016, 42, 1840–3.

[105]   Lopes HP, Elias CN, Vedovello GA, Bueno CE, Mangelli M, Siqueira JF Jr. Torsional resistance of retreatment instruments. J. Endod. 2011, 37, 1442–5.

[106]   Xu X, Eng M, Zheng Y, Eng D. Comparative study of torsional and bending properties for six models of nickel-titanium root canal instruments with different cross-sections. J. Endod. 2006, 32, 372–5.

[107]   Ninan E, Berzins DW. Torsion and bending properties of shape memory and superelastic nickel-titanium rotary instruments. J. Endod. 2013, 39, 101–4.

[108]   Yared G, Kulkarni GK, Ghossayn F. Torsional properties of new and used rotary K3 NiTi files. Aust. Endod. J. 2003, 29, 75–8.

[109]   Vieira EP, Nakagawa RK, Buono VT, Bahia MG. Torsional behaviour of rotary NiTi ProTaper universal instruments after multiple clinical use. Int. Endod. J. 2009, 42, 947–53.

[110]   Camps JJ, Pertot WJ. Torsional and stiffness properties of nickel–titanium K files. Int. Endod. J. 1995, 28(5), 239–43.

[111]   Braga LC, Magalhães RR, Nakagawa RK, Puente CG, Buono VT, Bahia MG. Physical and mechanical properties of twisted or ground nickel-titanium instruments. Int. Endod. J. 2013, 46, 458–65.

[112]   da Frota MF, Espir CG, Berbert FL, Marques AA, Sponchiado-Junior EC, Tanomaru-Filho M, Garcia LF, Bonetti-Filho I. Comparison of cyclic fatigue and torsional resistance in reciprocating single-file systems and continuous rotary instrumentation systems. J. Oral Sci. 2014, 56, 269–75.

[113]   Alcalde MP, Duarte MAH, Bramante CM, de Vasconselos BC, Tanomaru-Filho M, Guerreiro-Tanomaru JM, Pinto JC, Só MVR, Vivan RR. Cyclic fatigue and torsional strength of three different thermally treated reciprocating nickel-titanium instruments. Clin. Oral Invest. 2018, 22, 1865–71.

[114]   Oh S-H, Ha J-H, Kwak SW, Ahn SW, Lee WC, Kim HC. The effects of torsional preloading on the torsional resistance of nickel-titanium instruments. J. Endod. 2017, 43, 157–62.

[115]   Gambarini G, Testarelli L, Galli M, Tucci E, De Luca M. The effect of a new finishing process on the torsional resistance of twisted nickel-titanium rotary instruments. Minerva Stomatol. 2010, 59, 401–6.

[116]   Miyai K, Ebihara A, Hayashi Y, Doi H, Suda H, Yoneyama T. Influence of phase transformation on the torsional and bending properties of nickel-titanium rotary endodontic instruments. Int. Endod. J. 2006, 39(2), 119–26.

[117]   Acosta EC, Resende PD, Peixoto IF, Pereira ÉS, Buono VT, Bahia MG. Influence of cyclic flexural deformation on the torsional resistance of controlled memory and conventional nickel-titanium instruments. J. Endod. 2017, 43, 613–8.

[118]   Silva EJNL, Giraldes JFN, de Lima CO, Vieira VTL, Elias CN, Antunes HS. Influence of heat-treatment on torsional resistance and surface roughness of nickel-titanium instruments. Int. Endod. J. 2019; doi: 10.1111/iej.13164.

[119]   Bahia MG, Melo MC, Buono VT. Influence of simulated clinical use on the torsional behavior of nickel-titanium rotary endodontic instruments. Oral Surg. Oral Med. Oral Pathol. Oral Radiol. Endod. 2006, 101, 675–80.

[120]   Campbell L, Shen Y, Zhou HM, Haapasalo M. Effect of fatigue on torsional failure of nickel-titanium controlled memory instruments. J. Endod. 2014, 40, 562–5.

[121]   Yared GM, Bou Daugher FE, Machtou P. Influence of rotational speed, torque and operator's proficiency on ProFile failures. Int. Endod. J. 2001, 34, 47–53.

[122]   Awang Shri DN, Tsuchiya K, Yamamoto A. Effect of high-pressure torsion deformation on surface properties and biocompatibility of Ti-50.9 mol. %Ni alloys. Biointerphases 2014, 9; doi: 10.1116/1.4867402.

[123]   Filleul MP, Jordan L. Torsional properties of Ni-Ti and copper Ni-Ti wires: the effect of temperature on physical properties. Eur. J. Orthod. 1997, 19, 637–46.

[124]   Kuroda S, Watanabe H, Nakajima A, Shimizu N, Tanaka E. Evaluation of torque moment in a novel elastic bendable orthodontic wire. Dent. Mater. J. 2014, 33, 363–7.

[125]   Larson BE, Kusy RP, Whitley JQ. Torsional elastic property measurements of selected orthodontic archwires. Clin. Mater. 1987, 2, 165–79.

[126]   Archambault A, Major TW, Carey JP, Heo G, Badawi H, Major PW. A comparison of torque expression between stainless steel, titanium molybdenum alloy, and copper nickel titanium wires in metallic self-ligating brackets. Angle Orthod. 2010, 80, 884–9.

[127]   Dunić V, Pieczyska EA, Kowalewski ZL, Matsui R, Slavković R. Experimental and numerical investigation of mechanical and thermal effects in TiNi SMA during transformation-induced creep phenomena. Materials (Basel) 2019, 12; doi: 10.3390/ma12060883.

[128]   Takeda K, Tobushi H, Pieczyska EA. Transformation-induced creep and creep recovery of shape memory alloy. Materials 2012, 5, 909–21.

[129]   Lagoudas DC, Chatzigeorgiou G, Kumar PK. Modeling and experimental study of simultaneous creep and transformation in polycrystalline high-temperature shape memory alloys. J. Intell. Mater. Syst. Struct. 2009, 20, 2257–67.

[130]   Eggeler G, Neuking K, Dlouhy A, Kobus E. Creep behavior of NiTi shape memory alloys and the effect of pre-creep on the martensitic phase transformation. Mater. Sci. Forum 2000, 327/328, 183–6.

[131]   Russalian VR, Bhattacharyya A. Pseudo-creep in shape memory alloy wires and sheets. Metall. Mater. Trans. A 2017, 48, 4511–24.

[132]   Kobus E, Neuking K, Eggeler G, Wittkamp I. The creep behaviour of a NiTi-alloy and the effect of creep deformation on its shape memory properties. Prakt. Metallogr. 2002, 39, 177–86.

[133]   El-Tahan M, Dawood M. Bond behavior of NiTiNb SMA wires embedded in CFRP composites. Polym. Comp. 2018, 39, 3780–91.

[134]   Zhang W, Ao S, Oliveira JP, Zeng Z, Huang Y, Luo Z. Microstructural characterization and mechanical behavior of NiTi shape memory alloys ultrasonic joints using Cu interlayer. Materials (Basel) 2018, 26; doi: 10.3390/ma11101830.

[135]   Mesquita TR, Martins LP, Martins RP. Welding strength of NiTi wires. Dental Press J. Orthod. 2018, 23, 58–62.

[136]   Vieira LA, Fernandes FMB, Miranda RM, Silva RJC, Quintino L, Cuesta LA, Ocaña JL. Mechanical behaviour of Nd:YAG laser welded superelastic NiTi. Mater. Sci. Eng., A 2011, 528, 5560–5.

[137]   Reams GJ, Baumgartner JC, Kulild JC. Practical application of infection control in endodontics. J. Endod. 1995, 21, 281–4.

[138]   Miller CH. Infection control. Dent. Clin. N. Am. 1996, 40, 437–56.

[139]   Hurtt CA, Rossman LE. The sterilization of endodontic hand files. J. Endod. 1996, 22, 321–2.

[140]   Raju TBVG, Garapati S, Agrawal R, Reddy S, Razdan A, Kumar SK. Sterilizing endodontic files by four different sterilization methods to prevent cross-infection – an in-vitro study. J. Int. Oral Health 2013, 5, 108–12.

[141]   Iverson GW, von Fraunhofer JA, Herrmann JW. The effects of various sterilization methods on the torsional strength of endodontic files. J. Endod. 1985, 11, 266–8.

[142]   Silvaggio J, Hicks ML. Effect of heat sterilization on the torsional properties of rotary nickel-titanium endodontic files. J. Endod. 1997, 23, 731–4.

[143]   Canalda-Sahli C, Brau-Aguadé E, Sentís-Vilalta J. The effect of sterilization on bending and torsional properties of K-files manufactured with different metallic alloys. Int. Endod. J. 1998, 31, 48–52.

[144]   Hilt BR, Cunningham CJ, Shen C, Richards N. Torsional properties of stainless-steel and nickel-titanium files after multiple autoclave sterilizations. J. Endod. 2000, 26, 76–80.

[145]   King JB, Roberts HW, Bergeron BE, Mayerchak MJ. The effect of autoclaving on torsional moment of two nickel-titanium endodontic files. Int. Endod. J. 2012, 45, 156–61.

[146]   Casper RB, Roberts HW, Roberts MD, Himel VT, Bergeron BE. Comparison of autoclaving effects on torsional deformation and fracture resistance of three innovative endodontic file systems. J. Endod. 2011, 37, 1572–5.

[147]   Janardhanan S, Kanisseri M, John MK. Influence of sterilization on mechanical properties and fatigue resistance of nickel-titanium rotary endodontic instruments: An *in vitro* study. Int. J. Oral Care Res. 2018, 6, 5–11.

[148]   Viana AC, Gonzalez BM, Buono VT, Bahia MG. Influence of sterilization on mechanical properties and fatigue resistance of nickel-titanium rotary endodontic instruments. Int. Endod. J. 2006, 39, 709–15.

[149] Oshida Y, Farzin-Nia F. Progressive Damage Assessment of TiNi Endodontic Files. In: Yahia L, ed., Shape Memory Implants. Springer-Verlag, Berlin, Germany, 2000, 236–49.

[150] Melo MCC, Bahia MGA, Buono VTL. Fatigue resistance of engine-driven rotary Ni-Ti endodontic instruments. Int. Endod. J. 2002, 28, 765–9.

[151] Gambarini G. Cyclic fatigue of ProFile rotary instruments after prolonged clinical use. Int. Endod. J. 2001, 34, 386–9.

[152] Spencer HR, Ike V, Brennan PA. Review: the use of sodium hypochlorite in endodontics – potential complications and their management. Brit. Dent. J 2007, 202, 555–9.

[153] Sirtes G, Waltimo T, Schaetzle M, Zehnder M. The effects of temperature on sodium hypochlorite short term stability, pulp dissolution capacity and antimicrobial efficacy. J. Endod. 2005, 31, 669–71.

[154] Tada E, Tominaga T, Yasukawa H, Oshida Y. Temperature increase during tooth whitening. Dent. Pract. 2019, 2; http://sciaeon.org/articles/Temperature-Increase-during-Tooth-Whitening.pdf.

[155] Yared GM, Bou Dagher FE, Machtou P. Cyclic fatigue of ProFile rotary instruments after clinical use. Int. Endod. J. 2000, 33, 204–7.

[156] Haïkel Y, Serfaty R, Wilson P, Speisser JM, Allemann C. Mechanical properties of nickel-titanium endodontic instruments and the effect of sodium hypochlorite treatment. J. Endod. 1998, 24, 731–5.

[157] Peters OA, Roehlike JO, Baumann MA. Effect of immersion in sodium hypochlorite on torque and fatigue resistance of nickel-titanium instruments. J. Endod. 2007, 33, 589–93.

[158] Gupta AK, Shukla G, Sharma P, Gupta AK, Kumar A, Gupta D. Evaluation of the effects of fluoride prophylactic agents on mechanical properties of nickel titanium wires using scanning electron microscope. J. Contemp. Dent. Pract. 2018, 19, 283–6.

[159] Srivastava K, Chandra PK, Kamat N. Effect of fluoride mouth rinses on various orthodontic archwire alloys tested by modified bending test: an in vitro study. Indian J. Dent. Res. 2012, 23, 433–4.

[160] Walker MP, White RJ, Kula KS. Effect of fluoride prophylactic agents on the mechanical properties of nickel-titanium-based orthodontic wires. Am. J. Orthod. Dentofacial Orthop. 2005, 127, 662–9.

[161] Hammad SM, Al-Wakeel EE, Gad el-S. Mechanical properties and surface characterization of translucent composite wire following topical fluoride treatment. Angle Orthod. 2012, 82, 8–13.

[162] Ramalingam A, Kailasam V, Padmanabhan S, Chitharanjan A. The effect of topical fluoride agents on the physical and mechanical properties of NiTi and copper NiTi archwires. An in vivo study. Aust. Orthod. J. 2008, 24, 26–31.

[163] Lin J, Han S, Zhu J, Wang X, Chen Y, Vollrath O, Wang H, Mehl C. Influence of fluoride-containing acidic artificial saliva on the mechanical properties of nickel-titanium orthodontics wires. Indian J. Dent. Res. 2012, 23, 591–5.

[164] Sjögren U, Hagglund B, Sundqvist G, Wing K. Factors affecting the long-term results of endodontic treatment. J. Endod. 1990, 16, 498–504.

[165] Ranđelović S. Manufacturability of Biomaterials. In: Zivic F, et al. ed., Biomaterials in Clinical Practice: Advances in Clinical Research and Medical Devices. Springer International Pub., 2018, 633–58.

[166] Wu SK, Lin HC, Chen CC. A study on the machinability of a Ti 49.6 Ni 50.4 shape memory alloy. Mater. Lett. 1999, 40, 27–32.

[167] Huang H, Zheng HY, Liu Y. Experimental investigations of the machinability of Ni50.6Ti49.4 alloy. Smart Mater. Struct. 2005, 14, 297–301.

[168] Lee ES, Shin TH. An evaluation of the machinability of nitinol shape memory alloy by electrochemical polishing. J. Mech. Sci. Technol. 2011, 25, 96–9.

[169] Mehrpouya M, Shahedin AM, Dawood SDS, Ariffin AK. An investigation on the optimum machinability of NiTi based shape memory alloy. Mater. Manuf. Processes. 2017, 32, 1497–504.

[170] Shahedin AM, Dawood SDS, Ariffin AK. An investigation on the optimum machinability of NiTi based shape memory alloy. Mater. Manuf. Process. 2017, 32(13), 1497–504.

[171] Camps JJ, Pertot WJ. Machining efficiency of nickel-titanium K-type files in a linear motion. Int. Endod. J. 1995, 28, 279–84.

[172] Lin HC, Lin KM, Chen YC. A study on the machining characteristics of TiNi shape memory alloys. J. Mater. Process Technol. 2000, 105, 327–32.

[173] Velmurugan C, Senthilkumar V, Dinesh S, Arulkirubakaran D. Machining of NiTi-shape memory alloys – a review. Mach. Sci. Technol. 2018, 22, 355–401.

[174] Saptaji K, Gebremariam MA, Azhari MABM. Machining of biocompatible materials: a review. Int. J. Adv. Manuf. Technol. 2018, 97, 2255–92.

[175] Venkatesh V, Swain N, Srinivas G, Kumar P, Barshilia HC. Review on the machining characteristics and research prospects of conventional microscale machining operations. Mater. Manuf. Process. 2017, 32, 235–62.

[176] Uppal N, Shiakolas PS. Micromachining characteristics of NiTi based shape memory alloy using femtosecond laser. J. Manuf. Sci. Eng. 2008, 130; https://doi.org/10.1115/1.2936380.

[177] Biermann D, Kahleyss F, Krebs E, Upmaier T. A study on micro-machining technology for the machining of NiTi: five-axis micro-milling and micro deep-hole drilling. J. Mater. Eng. Perform. 2010, 20, 745–51.

[178] Kuppuswamy R, Yui A. High-speed micromachining characteristics for the NiTi shape memory alloys. Int. J. Adv. Manuf. Technol. 2017, 93, 11–21.

[179] Weinert K, Petzoldt V. Machining NiTi micro-parts by micro-milling. Mater. Sci. Eng. 2008, 481/482, 672–5.

[180] Tepel J, Schäfer E, Hoppe W. Properties of endodontic hand instruments used in rotary motion. Part 1. Cutting efficiency. J. Endod. 1995, 21, 418–21.

[181] Haïkel Y, Serfaty R, Wilson P, Speisser JM, Allemann C. Cutting efficiency of nickel-titanium endodontic instruments and the effect of sodium hypochlorite treatment. J. Endod. 1998, 24, 736–9.

[182] Schäfer E, Lau R. Comparison of cutting efficiency and instrumentation of curved canals with nickel-titanium and stainless-steel instruments. J. Endod. 1999, 25, 427–30.

[183] Rapisarda E, Bonaccorso A, Tripi TR, Fragalk I, Condorelli GG. The effect of surface treatments of nickel-titanium files on wear and cutting efficiency. Oral Surg. Oral Med. Oral Pathol. Oral Radiol. Endod. 2000, 89, 363–8.

[184] Schäfer E. Effect of sterilization on the cutting efficiency of PVD-coated nickel–titanium endodontic instruments. Int. Endod. J. 2002, 35, 867–72.

[185] Bui TB, Mitchell JC, Baumgartner JC. Effect of electropolishing ProFile nickel-titanium rotary instruments on cyclic fatigue resistance, torsional resistance, and cutting efficiency. J. Endod. 2008, 34, 190–3.

[186] Fayyad DM, Elhakim Elgendy AA. Cutting efficiency of twisted versus machined nickel-titanium endodontic files. J. Endod. 2011, 37, 1143–6.

[187] Peters OA, Morgental RD, Schulze KA, Paqué F, Kopper PM, Vier-Pelisser FV. Determining cutting efficiency of nickel-titanium coronal flaring instruments used in lateral action. Int. Endod. J. 2014, 47, 505–13.

[188] Thompson SA, Dummer PMH. Shaping ability of NT engine and McXim rotary nickel-titanium instruments in simulated root canals. Part 1. Int. Endod. J. 1997, 30, 262–9.

[189] Schäfer E, Florek H. Efficiency of rotary nickel-titanium K3 instruments compared with stainless steel hand K-flexofile. Part 1. Shaping ability in simulated curved canals. Int. Endod. J. 2003, 36, 199–207.

[190]   Liu SB, Fan B, Cheung GS, Peng B, Fan MW, Gutmann JL, Song YL, Fu Q, Bian Z. Cleaning
        effectiveness and shaping ability of rotary ProTaper compared with rotary GT and manual
        K-flexofile. Am. J. Dent. 2006, 19, 353–8.

[191]   Di Lenarda R, Cadenaro M, Sbaizero O. Effectiveness of 1 mol L-1 citric acid and 15% EDTA
        irrigation on smear layer removal. Int. Endod. J. 2000, 33, 46–52.

[192]   Reddy KB, Dash S, Kallepalli S, Vallikanthan S, Chakrapani N, Kalepu V. A comparative
        evaluation of cleaning efficacy (debris and smear layer removal) of hand and two NiTi rotary
        instrumentation systems (K3 and ProTaper): a SEM study. J. Contemp. Dent. Pract. 2013, 14,
        1028–35.

[193]   Subramaniam P, Girish Babu KL, Tabrez TA. Effectiveness of rotary endodontic instruments
        on smear layer removal in root canals of primary teeth: a scanning electron microscopy
        study. J. Clin. Pediatr. Dent. 2016, 40, 141–6.

[194]   Matwychuk MJ, Bowles WR, McClanahan SB, Hodges JS, Pesun IJ. Shaping abilities of two
        different engine-driven rotary nickel titanium systems or stainless steel balanced-force
        technique in mandibular molars. J. Endod. 2007, 33, 868–71.

[195]   Vaudt J, Bitter K, Neumann K, Kielbassa AM. Ex vivo study on root canal instrumentation of
        two rotary nickel-titanium systems in comparison to stainless steel hand instruments. Int.
        Endod. J. 2009, 42, 22–33.

[196]   Bonaccorso A, Cantatore G, Condorelli GG, Schäfer E, Tripi TR. Shaping ability of four nickel-
        titanium rotary instruments in simulated S-shaped canals. J. Endod. 2009, 35, 883–6.

[197]   Ounsi HF, Franciosi G, Paragliola R, Al-Hezaimi K, Salameh Z, Tay FR, Ferrari M, Grandini S.
        Comparison of two techniques for assessing the shaping efficacy of repeatedly used nickel-
        titanium rotary instruments. J. Endod. 2011, 37, 847–50.

[198]   Burroughs JR, Bergeron BE, Roberts MD, Hagan JL, Himel VT. Shaping ability of three nickel-
        titanium endodontic file systems in simulated S-shaped root canals. J. Endod. 2012, 38,
        1618–21.

[199]   Hartmann MS, Fontanella VR, Vanni JR, Fornari VJ, Barletta FB. CT evaluation of apical canal
        transportation associated with stainless steel hand files, oscillatory technique and pro
        taper rotary system. Braz. Dent. J. 2011, 22, 288–93.

[200]   Poly A, AlMalki F, Marques F, Karabucak B. Canal transportation and centering ratio after
        preparation in severely curved canals: analysis by micro-computed tomography and
        double-digital radiography. Clin. Oral Invest. 2019, May, 1–8; doi: 10.1007/s00784-019-
        02870-8.

[201]   Talati A, Moradi S, Forghani M, Monajemzadeh A. Shaping ability of nickel-titanium rotary
        instruments in curved root canals. Iran Endod. J. 2013, 8, 55–8.

[202]   Berutti E, Chiandussi G, Paolino DS, Scotti N, Cantatore G, Castellucci A, Pasqualini D. Canal
        shaping with WaveOne primary reciprocating files and ProTaper system: a comparative
        study. J. Endod. 2012, 38, 505–9.

[203]   Hwang Y-H, Bae K-S, Baek SH, Kum K-T, Lee W-C, Shon W-J, Chang S-W. Shaping ability of the
        conventional nickel-titanium and reciprocating nickel-titanium file systems: a comparative
        study using micro-computed tomography. J. Endod. 2014, 40, 1186–9.

[204]   Arora A, Taneja S, Kumar M. Comparative evaluation of shaping ability of different rotary
        NiTi instruments in curved canals using CBCT. J. Conserv. Dent. 2014, 17, 35–9.

[205]   Saber SE, Nagy MM, Schafer E. Comparative evaluation of the shaping ability of ProTaper
        Next, iRaCe and Hyflex CM rotary NiTi files in severely curved root canals. Int. Endod. J.
        2015, 48, 131–6.

[206]   Celikten B, Uzuntas CF, Kursun S, Orhan AI, Tufekci P, Orhan K, Demiralp KÖ. Comparative
        evaluation of shaping ability of two nickel-titanium rotary systems using cone beam
        computed tomography. BMC Oral Health 2015, 15; doi: 10.1186/s12903-015-0019-5.

[207]   Pedullà E, Plotino G, Grande NM, Avarotti G, Gambarini G, Rapisarda E, Mannocci F. Shaping
         ability of two nickel–titanium instruments activated by continuous rotation or adaptive
         motion: a micro-computed tomography study. Clin. Oral Invest. 2016, 20, 2227–33.

[208]   Duque JA, Vivan RR, Cavenago BC, Amoroso-Silva PA, Bernardes RA, Vasconcelos BC, Duarte
         MA. Influence of NiTi alloy on the root canal shaping capabilities of the ProTaper universal
         and ProTaper gold rotary instrument systems. J. Appl. Oral Sci. 2017, 25, 27–33.

[209]   Wei Z, Cui Z, Yan P, Jiang H. A comparison of the shaping ability of three nickel-titanium
         rotary instruments: a micro-computed tomography study via a contrast radiopaque
         technique in vitro. BMC Oral Health 2017, 17, 39; doi: 10.1186/s12903-016-0326-5.

[210]   Özyürek T, Yılmaz K, Uslu G. Shaping ability of reciproc, WaveOne GOLD, and HyFlex EDM
         single-file systems in simulated S-shaped canals. J. Endod. 2017, 43, 805–9.

[211]   Espir CG, Nascimento-Mendes CA, Guerreiro-Tanomaru JM, Cavenago BC, Duarte MAH,
         Tanomaru-Filho M. Shaping ability of rotary or reciprocating systems for oval root canal
         preparation: a micro-computed tomography study. Clin. Oral Invest. 2018, 22, 3189–94.

[212]   Huang Z, Quan J, Liu J, Zhang W, Zhang X, Hu X. A microcomputed tomography evaluation of
         the shaping ability of three thermally-treated nickel-titanium rotary file systems in curved
         canals. J. Int. Med. Res. 2019, 47, 325–34.

[213]   Drukteinis S, Peciuliene V, Dummer PMH, Hupp J. Shaping ability of BioRace, ProTaper NEXT
         and genius nickel-titanium instruments in curved canals of mandibular molars: a MicroCT
         study. Int. Endod. J. 2019, 52, 86–93.

[214]   Esposito PT, Cunningham CJ. A comparison of canal preparation with nickel-titanium and
         stainless steel instruments. J. Endod. 1995, 21, 173–6.

[215]   Glossen CR, Haller RH, Dove SB, del Rio CE. A comparison of root canal preparations using Ni-Ti
         hand, Ni-Ti engine-driven, and K-Flex endodontic instruments. J. Endod. 1995, 21, 146–51.

[216]   Guelzow A, Stamm O, Martus P, Kielbassa AM. Comparative study of six rotary nickel-
         titanium systems and hand instrumentation for root canal preparation. Int. Endod. J. 2005,
         38, 743–52.

[217]   Taşdemir T, Aydemir H, Inan U, Unal O. Canal preparation with Hero 642 rotary Ni-Ti
         instruments compared with stainless steel hand K-file assessed using computed
         tomography. Int. Endod. J. 2005, 38, 402–8.

[218]   Nordmeyer S, Schnell V, Hülsmann M. Comparison of root canal preparation using Flex
         Master Ni-Ti and Endo-Eze AET stainless steel instruments. Oral Surg. Oral Med. Oral Pathol.
         Oral Radiol. Endod. 2011, 111, 251–9.

[219]   Genç Ö, Alaçam T, Kayaoglu G. Evaluation of three instrumentation techniques at the
         precision of apical stop and apical sealing of obturation. J. Appl. Oral Sci. 2011, 19, 350–4.

[220]   Gluskin AH, Brown DC, Buchanan LS. A reconstructed computerized tomographic
         comparison of Ni-Ti rotary GT files versus traditional instruments in canals shaped by novice
         operators. Int. Endod. J. 2001, 34, 476–84.

[221]   Ferreira RB, Alfredo E, Porto de Arruda M, Silva Sousa YT, Sousa-Neto MD. Histological
         analysis of the cleaning capacity of nickel-titanium rotary instrumentation with ultrasonic
         irrigation in root canals. Aust. Endod. J. 2004, 30, 56–8.

[222]   Elayouti A, Chu AL, Kimionis I, Klein C, Weiger R, Löst C. Efficacy of rotary instruments with
         greater taper in preparing oval root canals. Int. Endod. J. 2008, 41, 1088–92.

[223]   Ajuz NC, Armada L, Gonçalves LS, Debelian G, Siqueira JF Jr. Glide path preparation in S-shaped
         canals with rotary pathfinding nickel-titanium instruments. J. Endod. 2013, 39, 534–7.

[224]   Paiva SS, Siqueira JF Jr, Rôças IN, Carmo FL, Leite DC, Ferreira DC, Rachid CT, Rosado AS.
         Clinical antimicrobial efficacy of NiTi rotary instrumentation with NaOCl irrigation, final
         rinse with chlorhexidine and interappointment medication: a molecular study. Int. Endod.
         J. 2013, 46, 225–33.

[225]   Al-Omari MA, Aurich T, Wirtti SW, Jordan I. Shaping canals with ProFiles and K3 instruments: does operator experience matter? Oral Surg. Oral Med. Oral Pathol. Oral Radiol. Endod. 2010, 110, e50–e55.

[226]   Sonntag D, Delschen S, Stachniss V. Root-canal shaping with manual and rotary Ni-Ti files performed by students. Int. Endod. J. 2003, 36, 715–23.

[227]   Pettiette MT, Delano EO, Trope M. Evaluation of success rate of endodontic treatment performed by students with stainless-steel K-files and nickel-titanium hand files. J. Endod. 2001, 27, 124–7.

[228]   Peru M, Peru C, Mannocci F, Sherriff M, Buchanan LS, Pitt Ford TR. Hand and nickel-titanium root canal instrumentation performed by dental students: a micro-computed tomographic study. Eur. J. Dent. Educ. 2006, 10, 52–9.

[229]   Shen Y, Coil JM, Haapasalo M. Defects in nickel-titanium instruments after clinical use. Part 3: a 4-year retrospective study from an undergraduate clinic. J. Endod. 2009, 35, 193–6.

[230]   Arbab-Chirani R, Vulcain JM. Undergraduate teaching and clinical use of rotary nickel-titanium endodontic instruments: a survey of French dental schools. Int. Endod. J. 2004, 37, 320–4.

[231]   Jungnickel L, Kruse C, Vaeth M, Kirkevang LL. Quality aspects of ex vivo root canal treatments done by undergraduate dental students using four different endodontic treatment systems. Acta Odontol. Scand. 2018, 76, 169–74.

[232]   Kfir A, Rosenberg E, Zuckerman O, Tamse A, Fuss Z. Comparison of procedural errors resulting during root canal preparations completed by senior dental students in patients using an '8-step method' versus 'serial step-back technique'. Oral Surg. Oral Med. Oral Pathol. Oral Radiol. Endod. 2004, 97, 745–8.

[233]   Marending M, Peters OA, Zehnder M. Factors affecting the outcome of orthograde root canal therapy in a general dentistry hospital practice. Oral Surg. Oral Med. Oral Pathol. Oral Radiol. Endod. 2005, 99, 119–24.

[234]   Peciuliene V, Rimkuviene J, Aleksejuniene J, Haapasalo M, Drukteinis S, Maneliene R. Technical aspects of endodontic treatment procedures among lithuanian general dental practitioners. Stomatologija 2010, 12, 42–50.

[235]   Mesgouez C, Rilliard F, Matossian L, Nassiri K, Mandel E. Influence of operator experience on canal preparation time when using the rotary Ni-Ti ProFile system in simulated curved canals. Int. Endod. J. 2003, 36, 161–5.

[236]   Oshida Y. Bioscience and Bioengineering of Titanium Materials. Elsevier, Oxford, UK, 2007.

[237]   Oshida Y. Surface Engineering and Technology for Biomedical Implants. Momentum Press, New York, USA, 2014.

[238]   Tamerler C. Surfaces and their interfaces meet biology at the bio-interface. JOM 2015, 67, 2480–2.

[239]   Li M, Wang YB, Zhang X, Li QH, Liu Q, Cheng Y, Zheng YF, Xi TF, Wei SC. Surface characteristics and electrochemical corrosion behavior of NiTi alloy coated with IrO2. Mater. Sci. Eng. C Mater. Biol. Appl. 2013, 33, 15–20.

[240]   Jenko M, Grant JT, Rudolf R, Kokalj T, Mandrino D. Characterisation of the surface microstructure of a melt-spun Ni-Ti shape memory ribbon. Surf. Interface Anal. 2012, 44, 997–1000.

[241]   Lin X, Zhou L, Li S, Lu H, Ding X. Behavior of acid etching on titanium: topography, hydrophility and hydrogen concentration. Biomed. Mater. 2013, 9; doi: 10.1088/1748-6041/9/1/015002.

[242]   Green SM, Grant DM, Wood JV. XPS characterisation of surface modified Ni-Ti shape memory alloy. Mater. Sci. Eng., A 1997, 224, 21–6.

[243]   Pequegnat A, Michael A, Wang J, Lian K, Zhou Y, Khan MI. Surface characterizations of laser modified biomedical grade NiTi shape memory alloys. Mater. Sci. Eng., C 2015, 50, 36–78.

[244] Gu YW, Tay BY, Lim CS, Yong MS. Characterization of bioactive surface oxidation layer on NiTi alloy. Appl. Surf. Sci. 2005, 252, 2038–49.

[245] Wang RM, Chu CL, Hu T, Dong YS, Guo C, Sheng XB, Lin PH, Chung CY, Chu PK. Surface XPS characterization of NiTi shape memory alloy after advanced oxidation processes in UV/H2O2 photocatalytic system. Appl. Surf. Sci. 2007, 253, 8507–12.

[246] Oshida Y, Sachdeva RC, Miyazaki S. Microanalytical characterization and surface modification of TiNi orthodontic archwires. Biomed. Mater. Eng. 1992, 2, 51–69.

[247] Eliades T, Eliades G, Athanasiou AE, Bradley TG. Surface characterization of retrieved NiTi orthodontic archwires. Eur. J. Orthod. 2000, 22, 317–26.

[248] Huang HH. Surface characterizations and corrosion resistance of nickel-titanium orthodontic archwires in artificial saliva of various degrees of acidity. J. Biomed. Mater. Res. A 2005, 74, 629–39.

[249] Huang HH. Variation in surface topography of different NiTi orthodontic archwires in various commercial fluoride-containing environments. Dent. Mater. 2007, 23, 24–33.

[250] Zegan G, Sodor A, Munteanu C. Surface characteristics of retrieved coated and nickel-titanium orthodontic archwires. Rom. J. Morphol. Embryol. 2012, 53, 935–9.

[251] Choi S, Park DJ, Kim KA, Park KH, Park HK, Park YG. In vitro sliding-driven morphological changes in representative esthetic NiTi archwire surfaces. Microsc. Res. Tech. 2015, 78, 926–34.

[252] Krishnan M, Seema S, Tiwari B, Sharma HS, Londhe S, Arora V. Surface characterization of nickel titanium orthodontic arch wires. Med. J. Armed Forces India 2015, 71, S340–5.

[253] Bahia MG, Martins RC, Gonzalez BM, Buono VT. Physical and mechanical characterization and the influence of cyclic loading on the behaviour of nickel-titanium wires employed in the manufacture of rotary endodontic instruments. Int. Endod. J. 2005, 38, 795–801.

[254] Alapati SB, Brantley WA, Svec TA, Powers JM, Nusstein JM, Daehn GS. SEM observations of nickel-titanium rotary endodontic instruments that fractured during clinical Use. J. Endod. 2005, 31, 40–3.

[255] Fatma Y, Ozgur U. Evaluation of surface topography changes in three NiTi file systems using rotary and reciprocal motion: an atomic force microscopy study. Microsc. Res. Tech. 2014, 77, 177–82.

[256] Inan U, Gurel M. Evaluation of surface characteristics of rotary nickel-titanium instruments produced by different manufacturing methods. Niger. J. Clin. Pract. 2017, 20, 143–6.

[257] Ferreira F, Barbosa I, Scelza P, Russano D, Neff J, Montagnana M, Zaccaro Scelza M. A new method for the assessment of the surface topography of NiTi rotary instruments. Int. Endod. J. 2017, 50, 902–9.

[258] Rupp F, Gittens RA, Scheideler L, Marmur A, Boyan BD, Schwartz Z, Geis-Gerstorfer J. A review on the wettability of dental implant surfaces I: theoretical and experimental aspects. Acta Biomater. 2014, 10, 2894–906.

[259] Gittens RA, Scheideler L, Rupp F, Hyzy SL, Geis-Gerstorfer J, Schwartz Z, Boyan BD. A review on the wettability of dental implant surfaces II: biological and clinical aspects. Acta Biomater. 2014, 10, 2907–18.

[260] Oshida Y, Sachdeva R, Miyazaki S, Daly J. Effects of shot-peening on surface contact angles of biomaterials. J. Mater. Sci. Mater. Med. 1993, 4, 443–7.

[261] Lim YL, Oshida Y. Initial contact angle measurements on variously treated dental/medical titanium materials. J. Bio-Med. Mat. Eng. 2001, 11, 325–41.

[262] Elias CN, Oshida Y, Lima JHC, Muller CA. Relationship between surface properties (roughness, wettability and morphology) of titanium and dental implant torque. J. Mech. Behav. Biomed. Mater. 2008, 1, 234–42.

[263] Ponsonnet L, Reybier K, Jaffrezic N, Comte V, Lagneau C, Lissac M, Martelet C. Relationship between surface properties (roughness, wettability) of titanium and titanium alloys and cell behaviour. Mater. Sci. Eng., C 2003, 23, 551–60.

[264] Liu XM, Wu SL, Chu PK, Chung CY, Chu CL, Chan YL, Lam KO, Yeung KW, Lu WW, Cheung KM, Luk KD. Nano-scale surface morphology, wettability and osteoblast adhesion on nitrogen plasma-implanted NiTi shape memory alloy. J. Nanosci. Nanotechnol. 2009, 9, 3449–54.

[265] Shabalovskaya SA, Siegismund D, Heurich E, Rettenmayr M. Evaluation of wettability and surface energy of native Nitinol surfaces in relation to hemocompatibility. Mater. Sci. Eng. C 2013, 33, 127–32.

[266] Tian YL, Zhao YC, Yang CJ, Wang FJ, Liu XP, Jing XB. Fabrication of bio-inspired nitinol alloy surface with tunable anisotropic wetting and high adhesive ability. J. Colloid Interface Sci. 2018, 527, 328–38.

[267] Siegmund P, Guhl C, Schmidt E, Roßberg A, Rettenmayr M. Reactive wetting of alumina by Ti-rich Ni–Ti–Zr alloys. J. Mater. Sci. 2016, 51, 3693–700.

[268] Scarano A, Piattelli M, Caputi S, Favero GA, Piattelli A. Bacterial adhesion on commercially pure titanium and zirconium oxide disks: an *in vivo* human study. J. Periodontol. 2004, 75, 292–6.

[269] Cwalina B, Dec W, Michalska JK, Jaworska-Kik M, Student S. Initial stage of the biofilm formation on the NiTi and Ti6Al4V surface by the sulphur-oxidizing bacteria and sulphate-reducing bacteria. J. Mater. Sci. Mater. Med. 2017, 28, 173; doi: 10.1007/s10856-015988-2.

[270] Ramya S, George RP, Subba Rao RV, Dayal RK. Detection of algae and bacterial biofilms formed on titanium surfaces using micro-Raman analysis. Appl. Surf. Sci. 2010, 256, 5108–15.

[271] Jorand FPA, Debuy S, Kamagate SF, Engels-Deutsch M. Evaluation of a biofilm formation by *Desulfovibrio fairfieldensis* on titanium implants. Lett. Appl. Microbiol. 2014, 60; https://doi.org/10.1111/lam.12370.

[272] Chrzanowski W, Valappil SP, Dunnill CW, Abou Neel EA, Lee K, Parkin IP, Wilson M, Armitage DA, Knowles JC. Impaired bacterial attachment to light activated Ni-Ti alloy. Mater. Sci. Eng. C Mater. Biol. Appl. 2010, 30, 225–34.

[273] Taha M, El-Fallal A, Degla H. *In vitro* and *in vivo* biofilm adhesion to esthetic coated arch wires and its correlation with surface roughness. Angle Orthod. 2016, 86, 285–91.

[274] Asiry MA, Al Shahrani I, Almoammar S, Durgesh BH, Al Kheraif BA, Hashem MI. Influence of epoxy, polytetrafluoroethylene (PTFE) and rhodium surface coatings on surface roughness, nano-mechanical properties and biofilm adhesion of nickel titanium (Ni-Ti) archwires. Mater. Res. Express 2018, 5; doi: 10.1088/2053-1591/aaabe5.

[275] Musa Trolic I, Todoric Z, Pop Acev D, Makreski P, Pejova B, Spalj S. Effects of the presence of probiotic bacteria in the aging medium on the surface roughness and chemical composition of two dental alloys. Microsc. Res. Tech. 2019, 82, 1384–91.

[276] Espinosa-Cristóbal LF, López-Ruiz N, Cabada-Tarín D, Reyes-López SY, Zaragoza-Contreras A, Constandse-Cortéz D, Donohué-Cornejo A, Tovar-Carrillo K, Cuevas-González JC, Kobayashi T. Antiadherence and antimicrobial properties of silver nanoparticles against Streptococcus mutans on brackets and wires used for orthodontic treatments. J. Nanomater. 2018; https://doi.org/10.1155/2018/9248527.

[277] Mhaske AR, Shetty PC, Bhat NS, Ramachandra CS, Laxmikanth SM, Nagarahalli K, Tekale PD. Antiadherent and antibacterial properties of stainless steel and NiTi orthodontic wires coated with silver against Lactobacillus acidophilus – an in vitro study. Prog. Orthod. 2015, 16; doi:10.1186/s40510-015-0110-0.

[278] Wataha JC, Lockwood PE, Marek M, Ghazi M. Ability of Ni-containing biomedical alloys to activate monocytes and endothelial cells in vitro. J. Biomed. Mater. Res. 1999, 45, 251–7.

[279]   Yang D, Lü X, Hong Y, Xi T, Zhang D. The molecular mechanism for effects of TiN coating on NiTi alloy on endothelial cell function. Biomaterials 2014, 35, 6195–205.

[280]   Tateshima S, Kaneko N, Yamada M, Duckwiler G, Vinuela F, Ogawa T. Increased affinity of endothelial cells to NiTi using ultraviolet irradiation: an *in vitro* study. J. Biomed. Mater. Res. part A 2018, 106A, 1034–8.

[281]   Lü X, Qu Y, Hong Y, Huang Y, Zhang Y, Yang D, Zhang F, Xi T, Zhang D. A high-throughput study on endothelial cell adhesion and growth mediated by adsorbed serum protein *via* signaling pathway PCR array. Regen. Biomater. 2018, 5, 25–34.

[282]   Loger K, Engel A, Haupt J, Li Q, Lima de Miranda R, Quandt E, Lutter G, Selhuber-Unkel C. Cell adhesion on NiTi thin film sputter-deposited meshes. Mater. Sci. Eng. C Mater. Biol. Appl. 2016, 59, 611–6.

[283]   Putters JLM, Sukul DMKSK, Zeeuw GR, Bijma A, Besselink PA. Comparative cell culture effects of shape memory metal (Nitinol), nickel and titanium: a biocompatibility estimation. Eur. Surg. Res. 1992, 24, 378–82.

[284]   Messer RL, Wataha JC, Lewis JB, Lockwood PE, Caughman GB, Tseng WY. Effect of vascular stent alloys on expression of cellular adhesion molecules by endothelial cells. J. Long Term Eff. Med. Implants 2005, 15, 39–47.

[285]   Habijan T, Glogowski T, Kühn S, Pohl M, Wittsiepe J, Greulich C, Eggeler G, Schildhauer TA, Köller M. Can human mesenchymal stem cells survive on a NiTi implant material subjected to cyclic loading? Acta Biomater. 2011, 7, 2733–9.

[286]   Wang D, Ge N, Li J, Qiao Y, Zhu H, Liu X. Selective tumor cell inhibition effect of Ni-Ti layered double hydroxides thin films driven by the reversed pH gradients of tumor cells. ACS Appl. Mater. Interfaces 2015, 7, 7843–54.

[287]   Fatima A, Talapaneni AK, Saleh A, Sana S, Hussain A. An evaluation and comparison of myeloperoxidase enzymatic activity during initial orthodontic alignment: an in vivo study. J. Orthod. 2017, 44, 169–73.

[288]   van Humbeeck J. Damping properties of shape memory alloys during phase transformation. J. Phys. IV, Suppl. au J. de Physiq. III, 1996, 6, 371–80.

[289]   Lin HC, Wu SK, Yeh MT. Damping characteristics of TiNi shape memory alloys. Metal. Mater. Trans. A, 1993, 24, 2189–94.

[290]   Lin UC, Wu SK, Chang YC. Damping characteristics of $Ti_{50}Ni_{49.5}Fe_{0.5}$ and $Ti_{50}Ni_{40}Cu_{10}$ ternary shape memory alloys. Metall. Mater. Trans. A 1995, 26, 851–8.

[291]   van Humbeeck J, Stoiber J, Delay L, Goyhardt R. The high damping capacity of shape-memory alloys. Z. Metallkd. 1995, 86, 176–83.

[292]   de Wild M, Meier F, Bormann T, Howald CBC, Müller B. Damping of selective-laser-melted NiTi for medical implants. J. Mater. Eng. Perform. 2014, 23, 2614–9.

[293]   Rong L, Jiang H, Liu SW, Zhao X. Damping capacity of TiNi-based shape memory alloys. Proceed. SPIE – Int. Soc. Optical Eng. 2007, 6423; doi: 10.1117/12.780026.

[294]   Lammering R, Schmidt I. Experimental investigations on the damping capacity of NiTi components. Smart Mater. Struct. 2001, 10; https://iopscience.iop.org/article/10.1088/0964-1726/10/5/301/pdf.

[295]   Xing T, Zheng Y, Cui L. Transformation and damping characteristics of NiTi/NiTi alloys synthesized by explosive welding. Mater. Trans. 2006, 47, 658–60.

[296]   Zhang XX, Hou HW, Wei LS, Chen ZX, Wei WT, Geng L. High damping capacity in porous NiTi alloy with bimodal pore architecture. J. Alloys Compd. 2013, 550, 297–301.

[297]   Zhang X, Wei L. Processing and damping capacity of NiTi foams with laminated pore architecture. J. Mech. Behav. Biomed. Mater. 2019, 96, 108–17. doi: 10.1016/j.jmbbm.2019.04.036.

[298]   Coluzzi B, Biscarini A, Campanella R, Mazzolai G, Mazzolai FM. Effect of hydrogen on the damping properties of NiTi alloys. J. Phys. IV France 2001, 11, 8–87.

[299]   Biscarini A, Coluzzi B, Mazzolai G, Tuissi A, Mazzolai FM. Extraordinary high damping of hydrogen-doped NiTi and NiTiCu shape memory alloys. J. Alloys Compd. 2003, 355, 52–7.
[300]   Matsumoto H. Electrical resistivity of NiTi with a high transformation temperature. J. Mater. Sci. Lett. 1992, 11, 367–8.
[301]   Matsumoto H. Electrical resistivity study on transformation of NiTi alloys. J. Jpn. Inst. Met. 2002, 66, 1350–5.
[302]   Nagarajan D, Baskaranarayanan B, Usha K, Jayanthi MS, Vijjaykanth M. Comparison and ranking of superelasticity of different austenite active nickel-titanium orthodontic archwires using mechanical tensile testing and correlating with its electrical resistivity. J. Pharm. Bioallied Sci. 2016, 8, S71–6.
[303]   Sivarai A. Comparison of superelasticity of nickel titanium orthodontic arch wires using mechanical tensile testing and correlating with electrical resistivity. J. Int. Oral Health 2013, 5, 1–12.

Chapter 8
Corrosion and oxidation

The biolongevity and biofunctionality of materials are governed by their surface reactions in the in vivo environments. The intraoral environment is a hostile corrosive environment to nickel–titanium (NiTi) materials and mechanical actions inside the mouth produce significant stress on the materials [1]. The oral cavity is continuously filled with saliva, an aerated aqueous solution of chloride with varying amounts of Na, K, Ca, PO_4, CO_2, sulfur compounds, and mucin. The pH value is normally in the range of 6.5 to 7.5, but under plaque deposits it may be as low as 2.0. Temperatures can vary ±36.5 °C, and a variety of food and drink concentrations (with pH values ranging from 2.0 to 14.0) stay inside the mouth for short periods of time. Loads can go up to 1,000 Newtons (normally 200 N as a masticatory force) and sometimes it can be in an impact manner. Trapped food debris may decompose and release sulfur compounds, causing discoloration of natural teeth and restorative materials. Under these chemical and mechanochemical intraoral environments, materials in service in the mouth are expected to last for relatively long periods of time [2]. Such surface reactions can also influence physical, tribological, and biological processes. In Ti-based alloys, including CpTi (commercially pure titanium), Ti–6Al–4V, and NiTi, the aforementioned surface reactions are controlled by the stability of oxide(s) or oxide films formed on substrate layers of titanium.

8.1 Discoloration

Teeth that have been exposed to a long-term coffee and/or cigarette usage are normally stained. Viral bleaching of such teeth reflects patients' increasing desire to achieve an optimal esthetic appearance [3, 4]. Currently, mouth-rinses, toothpaste, dental prophylactic agents containing fluoride, and bleaching treatment agents, are popular for esthetic purposes and prevention of plaque and cavity formation. Normal orthodontic patients are referred to general dentists for fluoride treatments once every 6 months during the course of orthodontic mechanotherapy. Among various bleaching treatments for whitening, for stained teeth an overnight (normally for 8 h/day for two consecutive weeks) bleaching agents containing 10% carbamide peroxide are popular [5]. Some patients undergoing the nightguard bleaching (e.g., for 8-h treatment) are likely to have restorations containing metals (i.e., amalgam, crowns made of gold or porcelain fused to a base metal, fixed or removable prosthodontic bridge, or partial denture frameworks made of base alloys, and/or titanium implant fixtures). The corrosive effect of these agents (normally containing fluoride and hydrogen peroxide) on dental metallic materials has not been documented well, although it was reported to decrease the corrosion resistance of titanium in solutions containing fluoride [6–9].

https://doi.org/10.1515/9783110666113-008

The discoloration evaluation caused by fluoride treatment agents [2.0% NaF with pH 7.0, 0.4% SnF_2 with pH 7.0, and 1.23% acidulated phosphate fluoride (APF) with pH 3.5] and bleaching agents (10% carbamide peroxide) were conducted on CpTi (grade 2), NiTi, Ti–6Al–4V, 17–4 pH stainless steel, 70Ni–15Cr–5Mo, type IV gold alloy (70Au–10Ag–15Cu), and Dispersalloy amalgam [10]. The degree of discoloration on these treated alloys was examined by a colorimeter and naked eyes. After the baseline measurements of three, L^*, a^*, and b^*, comparisons were made with the Commission Internationale d'Eclairage (CIE-L^*a^*b) color system, and the value of ΔE^* can be calculated by $[(L_i - L_f)^2 + (a_i - a_f)^2 + (b_i - b_f)^2]^{1/2}$, where subscripts "i" and "f" indicate the initial value and final value, respectively. It was found that all tested metallic materials exhibit discoloration to various degrees, ranging from 10 to 18 in ΔE^*. Brushing of teeth between each treatment both fluoride and bleaching treatments indicate a remarkable reduction in the degree of discoloration (i.e., ΔE^* reduced from 2 to 8) of all tested materials. Although it is indicated to use beaching agent to natural tooth, it is clearly contraindicated to metallic restorations [10, 11]. Titanium implants were occasionally found to be strongly discolored after autoclaving. The discoloration is shown to be due to an accelerated growth of the surface oxide that covers the implants, and the formation of oxide thickness up to 650 A have been observed, that is, more than ten times thicker than on normal implants [12]. By applying surface sensitive spectroscopies [secondary ion mass spectrometry and X-ray photoelectron spectrometry (XPS) or electron spectroscopy for chemical analysis], it is also shown that these oxide films contain considerable amounts of fluorine, alkali metals, and silicon. Interestingly, it was mentioned that discoloration after autoclaving was observed for fluorine contaminations down to the ppm level. The source of fluorine was the textile cloths in which the titanium implant storage box had been wrapped during the autoclaving procedure, and the cloths contained residual Na_2SiF_6, which had been used as an additive to the rinsing water used in the last step of the cloth laundry procedure. Therefore, the biocompatibility of titanium implants is closely related to their surface oxides; hence, it is advisable to avoid all sources of fluorine in the implant preparation procedures.

8.2 Corrosion and electrochemical corrosion in various media

Internal fluids have chloride ion concentrations about seven times higher than that of oral fluids. A diet rich in sodium chloride, added to large volumes of acidulated beverages (phosphoric acid), provides a continuous source of corrosive agents despite the relatively short exposure. In addition, it has been calculated that an average urban mouth-breather inhales about a cubic meter of air every 2 h, with a potential intake of between 0.11 and 2.3 mg of sulfur dioxide [13]. Both sulfur dioxide and hydrogen sulfide have been found to accelerate the tarnishing and corrosion of metal implants.

Besides chlorine ion levels and pH of intraoral fluids, there is one more important factor involved in corrosion evaluation. It is an intraoral electrochemical potential.

Electrochemical corrosion studies are normally performed by means of potentiostatic or potentiodynamic measurements. Interpretation of the electrochemical corrosion data requires knowledge of expected intraoral potentials. Nilner et al. [14] reported that the intraoral potential ranging from −431 to −127 mV was measured on 407 amalgam restorations in the mouth of 28 patients. Corso et al. [15] reported that the intraoral potential is in a range from −300 to +300 mV. Reclaru et al. [16] reported a range from 0 to + 300 mV. Ewers et al. [17] reported that it ranges between −380 and +50 mV. Hence, the range of overlapping data is a very narrow window of potential zone from 0 to +50 mV. In addition to the previously listed intraoral chemistry and electrochemistry, it is important to know how to simulate the intraoral environments when the in vitro chemical or electrochemical corrosion test is prepared and conducted. Most of corrosion studies have been carried out using physiological isotonic electrolyte solutions such as 0.9% saline [18–20], Ringer's [21, 22], Tyrode's [23], Hank's [24] solutions, lactic acid to simulate the accumulated plaque [25, 26], and artificial saliva [27, 28] or simulated body fluid SBF, as mentioned later.

There are numerous studies on corrosion behavior of NiTi materials. One reason for this is the fact that almost half of atomic percentage of this alloy is Ni, which is considered as one of the three heavy toxic elements frequently used in biocompatible alloys. The other two elements are Cr and V. Therefore, the safety and biocompatibility should be examined thoroughly. Over the last decade, due to their unique shape-memory effect (SME) and superelasticity (SE) characteristics, NiTi alloys have been increasingly considered for use in external and internal biomedical devices, such as orthodontic wires, endodontic files, blade-type dental implants, self-expanding cardiovascular and urological stents, bone fracture fixation plates, and nails. For applications in human body, the corrosion resistance of NiTi becomes extremely important, as the amount and toxicity of corrosion products control the alloy biocompatibility.

8.2.1 Fluoride-containing solution

Noguti et al. [29] summarized the current data regarding the influence of fluoride on titanium corrosion process in the last 5 years, and it was reported that noxious effects induced by high-fluoride concentration as well as low pH in the oral cavity is found. Such conditions should be considered when prophylactic actions are administrated in patients containing titanium implants or other dental devices. In the followings, we will review various forms of fluorine-containing dental solutions as well as artificial saliva, physiological solution, and others.

Waler et al. [30] investigated the effect of fluoride prophylactic agents on the mechanical properties of Ti–Ni and Ti–Ni–Cu orthodontic archwires in an acidulated fluoride agent, a neutral fluoride agent, or distilled water (control) for 1.5 h at 37 °C. It was noted that unloading mechanical properties of Ti–Ni orthodontic wires were significantly decreased after exposure to both fluoride agents; however, Ti–Ni–Cu wire mechanical properties were not significantly affected by either fluoride agents, This suggests that using topical fluoride agents with Ti–Ni wire could decrease the functional unloading mechanical properties of the wire and contribute to prolonged orthodontic treatment. Yanisarapan et al. [31] determined the cytotoxicity, metal ion release and surface roughness of metal orthodontic appliances after immersion in different fluoride-containing product solutions. Twelve sets of twenty brackets and four tubes were ligated with three types of archwires: stainless steel, NiTi, and β-Ti, which were immersed (for 3 months) in solutions of fluoride toothpaste, 1.23% APF, or artificial saliva without fluoride as a control group. It was obtained that in the APF groups, the four metal ion levels and surface roughness of the brackets and archwires significantly increased, and the brackets and wires in the APF groups demonstrated more lines and grooves compared with the other groups; thereby, concluding that using APF gel during orthodontic treatment with fixed metal appliances should be avoided.

Schiff et al. [28, 32, 33] studied the effects of corrosion behavior of different Ti materials under various corrosive media. The four materials (TMA, TiNb, NiTi, and CuNiTi) were tested in three fluoride mouthwashes (Elmex, Meridol, and Acorea as well as in Fusayama–Meyer artificial saliva) [28]. Based on data, it was recommended that Elmex mouthwash for patients with titanium molybdenum alloy (TMA) and NiTi-based orthodontics wires are advised, but Acorea or Meridol mouthwash for patients with TiNb orthodontics wires was suggested. Moreover, three types of orthodontic brackets (CoCr, FeCrNi, and Ti) were electrochemically tested in test solutions (Elmex, Meridol, and Acorea fluoride mouthwashes) [32]. It was obtained that the bracket materials could be divided into two groups: Ti and FeCrNi in one, and CoCr, which has properties close to those of Pt, in the other. Similarly, two groups of electrolytes were identified: Elmex and Acorea mouthwashes in one group, and Meridol mouthwash in the second group. Based on these data, it was indicated that because of the risk of corrosion, Meridol mouthwash should not be prescribed for patients wearing Ti or FeCrNi-based orthodontic brackets. Schiff et al. [33] also investigated the influence of fluoride in certain mouthwashes on the risk of corrosion through galvanic coupling of orthodontic wires and brackets. Two titanium alloy wires (NiTi and Cu–NiTi), and the three most commonly used brackets (Ti, FeCrNi, and CoCr) were tested in a reference solution of Fusayama–Meyer artificial saliva and in two commercially available fluoride (250 ppm) mouthwashes (Elmex and Meridol) were subjected to the electrochemical corrosion tests. It was reported that Meridol mouthwash, which contains stannous fluoride, was the solution in which the NiTi wires coupled with the different brackets showed the highest corrosion risk, whereas in Elmex mouthwash, which contains sodium fluoride (NaF), the Cu–NiTi wires presented the highest corrosion risk; thereby, suggesting that mouthwashes should be prescribed according to the orthodontic materials used. Iijima et al. [34] investigated the corrosion properties of ion-implanted NiTi wire (Neo Sentalloy Ionguard) in artificial saliva and fluoride containing mouth rinse solutions (Butler F mouthrinse, Ora–Bliss). It was found that Neo

Sentalloy Ionguard in artificial saliva and Butler F mouthrinse (500 ppm) had a lower current density than Neo Sentalloy. A breakdown potential for passivation of Neo Sentalloy Ionguard in Ora–Bliss (900 ppm) was much higher than that of Neo Sentalloy, although both wires had similar corrosion potential (E_{CORR}) in Ora–Bliss (450 ppm and 900 ppm). The XPS results for Neo Sentalloy Ionguard suggested that the layers consisted of TiO_2, and TiN were present on the surface, and the layers may improve the corrosion properties. Srivastava et al. [35] assessed the change in load deflection characteristics of NiTi, Cu–NiTi, S.S, and β-Ti wires on immersing in fluoride mouth rinses of two types (phosflur and neutral NaF: S-Flo) mouth rinse utilizing a modified bending test for 1.5 h and incubated at 37 °C. It was obtained that there was no statistically significant reduction in load deflection characteristics of NiTi, copper NiTi, β-Ti, and S.S wires on immersing in phosflur mouth rinse and neutral NaF mouth rinses as compared to control at 2.5 and 1 mm of deflection in unloading phase.

Harzer et al. [36] investigated the corrosive properties of fluoride-containing toothpastes with different pH values on titanium brackets. Molar bands were placed on 18 orthodontic patients. In these patients, titanium brackets were bonded on the left quadrants and stainless steel brackets on the right quadrants of the upper and lower arches. Fifteen patients used gel Kam containing soluble tin fluoride (pH 3.2), whereas three used fluoride-free toothpaste. It was found that macroscopic evaluation showed the matte gray color of titanium brackets dominating over the silver gleam of the steel brackets. The plaque accumulation on titanium brackets is high because of the very rough surface. In addition, pitting and crevices were observed in only 3 of the 165 brackets tested; hence, confirming the results of in vitro studies, but the changes are so minor that titanium brackets can safely be used for up to 18 months. Schiff et al. [28] classified the different alloys (TMA, TiNb, NiTi, and Cu–NiTi) commonly used to make orthodontic wire according to their corrosion resistance in different media of three fluoride mouthwashes (Elmex®, Meridol®, and Acorea® as well as in Fusayama–Meyer artificial saliva). It was reported that the alloys could be divided into two groups: in one group, the NiTi-based alloys that were subjected to strong corrosion in the presence of monofluorophosphate were found in Acorea® solution; and in the other group, TiNb that was the most resistant to corrosion, and TMA, which corroded strongly with the stannous fluoride were found in Meridol® mouthwash. It was then advised that Elmex® mouthwash for patients with TMA and NiTi-based orthodontics wires, but Acorea® or Meridol® mouthwash for patients with TiNb orthodontics wires is suggested.

Kassab et al. [37] assessed the influence of fluoride concentration on the corrosion behavior of NiTi superelastic wire and compared the corrosion resistance of NiTi with that of β-Ti alloy in physiological solution (NaCl 0.15 M) with and without the addition of fluoride (NaF 0.02 M). It was obtained that polarization resistance decreased when NaF concentration was increased, and, depending on NaF concentration, NiTi can suffer localized or generalized corrosion. In NaCl solution with 0.02 M NaF, NiTi suffer localized corrosion, while β-Ti alloys remained passive.

Current values near zero were observed by galvanic coupling of NiTi and β-Ti. There was a decrease in NiTi corrosion resistance in the presence of fluoride, and the corrosion behavior of NiTi alloy depends on fluoride concentration. When 0.02 and 0.04 M of NaF were added to the NaCl solution, NiTi presented localized corrosion. When NaF concentration increased to 0.05, 0.07, and 0.12 M, the alloy presented general corrosion. NiTi corrosion resistance behavior is lower than that of β-Ti, and galvanic coupling of these alloys does not increase the corrosion rates. Kassab et al. [38] also assessed the influence of fluoride concentration on the corrosion behavior of NiTi superelastic wire and compared the corrosion resistance of NiTi with that of beta titanium alloy in physiological solution with and without addition of fluoride. It was found that polarization resistance decreased when NaF concentration was increased, and, depending on NaF concentration, NiTi can suffer localized or generalized corrosion. In NaCl solution with 0.02 M NaF, NiTi suffered localized corrosion, while beta titanium alloys remained passive. Current values near zero were observed by galvanic coupling of NiTi and beta titanium. Based on these findings, it was concluded that there is a decrease in NiTi corrosion resistance in the presence of fluoride. The corrosion behavior of NiTi alloy depends on the fluoride concentration. When 0.02 and 0.04 M of NaF were added to the NaCl solution, NiTi presented localized corrosion, and when NaF concentration increased to 0.05, 0.07, and 0.12 M, the alloy presented general corrosion. NiTi corrosion resistance behavior is lower than that of beta titanium, and galvanic coupling of these alloys does not increase the corrosion rates.

Kwon et al. evaluated changes in behavior of NiTi orthodontic wires in acidic fluoride solution [38] and fluoride-containing acetic acid solution [350]. It was found that a 3-day immersion in a 0.2%/pH 4 solution did not form any new crystal structure; however, tensile strength after immersion changed compared to the as-received wires. The 3M wires showed increased tensile strength whereas G&H and Ormco wires showed decreased strength. The release of the element in the test solution increased as NaF concentration and the period of immersion increased, and as pH valued decreased. The wires immersed in a 0.2%/pH 4 solution released several-fold greater amount of elements than wires in a 0.05%/pH 4 solution; and, in particular, NaF concentration, pH value, and the period of immersion were the factors affecting these properties [38]. The possibility of the formation of hydrofluoric acid (HF) by the reaction of fluoride with acetic acid seems natural in the oral cavity; hence, the effect of an acidic fluoride solution on NiTi wires was investigated by testing the microhardness and color changes on wires [39]. It was obtained that after immersion for 3 days, the microhardness of the tested-wires increased from 1.8% to 10.4% compared to that of the as-received state. The wires immersed in a higher NaF concentration, lower pH solution with longer immersion yielded a more corroded surface than those of the counter cases. The wires showed a different color after immersion; however, after 3 days in solutions of pH 4, wires showed an appreciable color change regardless of the products. In test solutions, 3M wires showed the highest volumetric and percentage (0.59 for 0.05%; 1.19 for 0.2% solution) weight loss whereas G&H wires showed the least volumetric and percentage (0.43 for 0.05%; 1.05 for 0.2% solution) weight loss among the tested-wires. In pH 6 solutions, wires lost weight and were under the detection limit of the testing machine. Ahn et al. [40] evaluated the effect of pH and temperature on orthodontic NiTi arch wires after immersion in an acidic fluoride solution, to which the acetic acid added for the adjustment of pH forms HF. More HF was formed in a solution of lower pH with higher temperature than that of the higher pH with lower temperature. It was reported that the reduction of microhardness (1.2–5.7%) occurred after immersion. However, within the same product, the pH and temperature had minor influence on the reduction. At pH 3.5 of 60 °C solution, the greatest weight loss, release of elements, and corrosion of surface occurred from the wires. On the other hand, at pH 6, no such loss or release occurred regardless of temperature. At 5 °C solution, the surface exhibited minor corrosion regardless of pH value.

8.2.2 Chlorine ion containing solution

Cheng et al. [41] studied the corrosion behavior of NiTi alloy in Cl^- ion concentration and pH value. It was found that the corrosion resistance decreased with increasing Cl^- ion concentration and pH value. The main form of corrosion was pitting corrosion. Sarkar et al. [42] conducted potentiodynamic cyclic polarization of four orthodontic wires (Permachrome, Elgiloy, β-Ti, NiTi) in a 1% NaCl solution within –500 and +300 mV (SCE). It was observed that the first three alloys to be passive whereas breakdown of passivity was observed on NiTi. The scanning electron microscopy (SEM) examination of the pre and postpolarized alloy surfaces provided evidence, which was consistent with the electrochemical measurements, in that the first three alloys exhibited no appreciable corrosion damage whereas pitting corrosion was observed on NiTi. The results obtained from X-ray analysis of the pitted surface indicated that this pitting could be due to selective dissolution of nickel from Nitinol. Cavalleri et al. [43] studied the corrosive effects of sodium hypochlorite (NaOCl) on NiTi endodontic instruments. Using a digital scanning microscope, the blade surfaces of three groups of ProTaper instruments were examined, particularly of the #1 shaping file. The blade's file (but not their handles) were soaked in NaOCl heated to 50 °C for three different times compatible with times in clinical practice, rather for 2, 5, and 10 min; the files were then compared with the control group. It was found that the micro-photographs did not reveal any appreciable signs of corrosion on the file blades. Therefore, it was evident that using NaOCl as an irrigating solution in root canals, where it comes into contact with NiTi rotary endodontic instruments, does not alter the surface structure of the files through corrosion. Hence, it is possible to conclude that, considering the length of time used in clinical practice, NaOCl does not cause any increase of risk of fracture to NiTi rotary instruments.

Berutti et al. [44] evaluated the influence of immersion in NaOCl on resistance to cyclic fatigue fracture and corrosion of ProTaper NiTi rotary instruments. A total of 120 new ProTaper NiTi Rotary files (F2) were randomly assigned to three different groups of 40 each. Group 1 was the control group; 20 mm (excluding the shaft) of group 2 instruments were immersed in 5% NaOCl at 50 °C for 5 min; instruments in group 3 were completely immersed in 5% NaOCl at 50 °C for 5 min. All instruments were then tested for cyclic fatigue by recording the time in seconds to fracture. It was found that instruments in group 3 had a significantly lower resistance to fracture because of cyclic fatigue than those in groups 1 and 2. In some instruments, in group 3, early fracture occurred after only a few seconds of fatigue testing, and SEM observations revealed evident signs of corrosion of the fractured instruments. Based on these findings, it was concluded that group 3 had significantly reduced resistance to cyclic fatigue compared with instruments in groups 1 and 2. The phenomenon of early fracture may be attributable to galvanic corrosion induced by the presence of dissimilar metals, where one acts as the cathode of a galvanic couple, established when the instrument is immersed in NaOCl solution. The NiTi alloy may acts as the anode and thus undergoes corrosion. Nóvoa et al. [45] evaluated the corrosion resistance of NiTi endodontic rotary instruments immersed in 5.25% NaOCl solution. It was reported that the corrosion potential (E_{CORR}) of the NiTi alloy reached the passive domain in approximately 20 seconds of immersion in the solution having a pH 10.1, and after this initial period the potential remained steady, indicating that stable passivation was achieved. However, at pH 12.3, no stationary state was achieved even after 6000 s of immersion time, concluding that the alloy was not stable in this medium from a corrosion point of view. The corrosion resistance of NiTi alloy was enhanced by lowering the pH of NaOCl solution to 10.1, which allows the system to reach the stability domain of the passivating species TiO_2 and NiO_2. Stokes et al. [46] evaluated and compared the corrosion susceptibility of stainless steel and NiTi endodontic files immersed in NaOCl. Each of the stainless-steel files (Kerr K-Flex, Caulk Flex-O, and Union Broach Flex-R) plus the NiTi files (Union Broach NiTi and Tulsa NiTi), was immersed into 5.25% NaOCl. The corresponding open-circuit potential (OCP) was recorded for 1 h on a strip chart with high impedance. It was found that the OCP classification of unstable and erratic files was evaluated as follows: K-Flex (16%), Flex-R (12%), Flex-O (75%), Union Broach NiTi (62%), and Tulsa NiTi (0%). There was a significant difference in corrosion frequency between brands when evaluated by OCP and light microscopy; however, there was no significant difference between stainless steel and NiTi.

The effects of deionized water, blood, phosphate-buffered saline (PBS) and a new anticorrosive solution based on methoxypropylamine (MOPA) on the cyclic fatigue resistance of endodontic NiTi rotary instruments was investigated [47]. Forty ProTaper F1 files were provided and divided to four groups (n = 10). Samples were first autoclaved and then stored in deionized water, blood, PBS, or MOPA for 24 h. Cyclic fatigue was tested with a custom-made stainless-steel block including artificial canals (curvature angle: 30°, radius of curvature: 5 mm). After immersion in test solutions, samples were rotated 300 rpm until fracture occurred. The number of cycles to failure (NCF) was calculated using recorded fracture time. It was reported that samples in blood group showed the lowest and samples in MOPA group showed the highest NCF values. A significant difference was observed between groups; NCF value of PBS group was significantly more than the NCF values of samples in blood and deionized water groups, indicating that the tested anticorrosive solution significantly increased the fracture resistance of the endodontic NiTi rotary instruments by reducing the cyclic fatigue. In contrast, blood and deionized water caused more corrosion and resulted in earlier file fracture.

8.2.3 Artificial saliva

The effects of multilayered Ti/TiN or single-layered TiN films deposited by pulse-biased arc ion plating on the corrosion behavior of NiTi orthodontic brackets in artificial saliva was studied [48]. It was reported that the multilayered Ti/TiN coating is found to exhibit a greater free corrosion potential, much lower passive current density, and no breakdown up to 1.5 V. Electrochemical impedance spectroscopy results indicate that the multilayered Ti/TiN coating has a larger impedance and lower porosity, which is believed to be responsible for the exceedingly low-metal ion release rate during 720 h exposure in the test solution. The visual inspection of the surfaces reveals different corrosion processes for the TiN and multilayered Ti/TiN coatings. Mareci et al. [49] investigated potentiodynamic polarization electrochemical corrosion behavior of the studied $Ni_{47.7}Ti_{37.8}Nb_{14.5}$ (NiTiNb) shape-memory alloy (SMA) in artificial saliva. It was found that very low-passive current densities were obtained from the anodic polarization curve, indicating a typical passive behavior for NiTiNb alloy over the surface of the NiTiNb alloy uniform corrosion appears, while in case of the NiTi alloy, surface pitting corrosion was developed. The role that Nb plays as an alloying element is by increasing the resistance of NiTi alloy to localized corrosion, and the electrochemical impedance spectroscopy (EIS) results exhibited high-impedance values (of $10^6 \, \Omega \, cm^2$) obtained from medium to low frequencies, which are indicative of the formation of a highly stable passive film on NiTiNb alloy in the artificial saliva.

Huang [50] investigated the surface characterizations and corrosion resistance of as-received commercial NiTi orthodontic archwires from different manufacturers using a cyclic potentiodynamic test in artificial saliva with various acidities. It was obtained that the surface structure of the passive film on the tested NiTi wires were identical, containing mainly TiO_2, with small amounts of NiO. A different surface topography was observed on the NiTi wires from various manufacturers. The corrosion tests showed that both the wire manufacturer and solution pH had a statistically significant influence on the corrosion potential, corrosion rate, passive current, breakdown potential, and crevice corrosion susceptibility. The difference in the corrosion resistance among these NiTi dental orthodontic archwires did not correspond with the surface roughness and preexisting defects. Huang also studied the corrosion resistance of stressed NiTi and stainless-steel orthodontic wires using cyclic potentiodynamic and potentiostatic tests in acid artificial saliva at 37 °C [51]. It was found that the pH had a significant influence on the corrosion parameters of the stressed NiTi and stainless steel wires. The pitting potential, protection potential, and passive range of stressed NiTi and stainless steel wires decreased on decreasing pH, whereas the passive current density increased on decreasing pH. The load had no significant influence on the above corrosion parameters. For all pH and load conditions, stainless steel wire showed higher pitting potential and wider passive range than NiTi wire, whereas NiTi wire had lower passive current density than stainless steel wire. The corrosion resistance of the stressed NiTi and stainless steel wires was related to the surface characterizations, including surface defect and passive film. Wang et al. [52] studied the cooperation of pH, temperature, and Cl⁻ concentration on electrochemical behavior of NiTi SMA in artificial saliva using orthogonal test method. It was reported that the pitting potential for NiTi in artificial saliva decreased at low and high pH. At 25 °C, the pitting potential was the lowest compared to those at 10, 37, and 50 °C; when the Cl⁻ concentration was not less than 0.05 mol/L, the pitting potential decreased with the increase of Cl⁻ concentration, and the free corrosion potential of austenitic NiTi was lower than that of mixture of austenite and martensite.

Trolić et al. [53] examined how probiotic supplements affect the corrosion stability of $Ni_{50.4}Ti_{49.6}$ orthodontic archwires, which were uncoated, nitrified, and rhodium-coated. It was obtained that rhodium-coated alloy in saliva has significantly higher general corrosion in saliva than nitrified alloy and uncoated alloy. In the presence of probiotics, the result was even more pronounced; probiotic supplement increases general and localized corrosion of rhodium-coated archwire and slightly decreases general corrosion and increases localized corrosion in uncoated archwire, whereas in the case of nitrified archwire, the probability of corrosion is very low. The differences in surface roughness between NiTi wires before corrosion are not significant. Exposure to saliva decreases roughness average in rhodium- coated wire; and media do not significantly influence surface microgeometry in nitrified and uncoated wires. Based on these findings, it was concluded that probiotic supplement affects corrosion depending on the type of coating of the NiTi archwire. It increases general corrosion of rhodium-coated wire and causes localized corrosion of uncoated and rhodium-coated archwire, and probiotic supplement does not have greater influence on surface roughness compared to that of saliva. Wang et al. [54] investigated the electrochemical behavior of $Ti_{50}Ni_{47.2}Co_{2.8}$ alloy in deaerated artificial saliva solutions (at 37 °C). The corrosion behavior of NiTiCo alloy was systematically studied by OCP, potentiodynamic, and potentiostatic measuring techniques. It was found that potentiodynamic and potentiostatic test results showed that the corrosion behavior of NiTiCo was similar to that of NiTi alloy. With the increase of pH value of the electrolytes, both corrosion potential (E_{CORR}) and pitting corrosion potential (E_B) decreased. The XPS results revealed that the outmost passive film consisted mainly of TiO_2, which were identical with that of NiTi alloy. The Ni ion released amount of NiTiCo was very close to that of NiTi alloy and neither Ti nor Co ion was detected due to the detection limitation.

Polarization resistance (R_P) was measured to test the hypothesis that different NiTi archwires may have dissimilar corrosion resistance in a fluoride-containing oral environment, which was an artificial saliva (pH 6.5) with various NaF concentrations (0%, 0.01%, 0.1%, 0.25%, and 0.5%) [55]. It was found that both archwire manufacturer and NaF concentration had a significant influence on R_P of NiTi archwires. Different surface topography was present on the test NiTi archwires that contained the similar surface chemical structure (TiO_2 and trace NiO). The surface topography did not correspond to the difference in corrosion resistance of the NiTi archwires. Increasing the NaF concentration in artificial saliva resulted in a decrease in R_P (or corrosion resistance) of all test NiTi archwires, and the NiTi archwires severely corroded and showed similar corrosion resistance in 0.5% NaF-containing environment. Based on these findings, it was concluded that different NiTi archwires had dissimilar corrosion resistance in acidic fluoride-containing artificial saliva, which did not correspond to the variation in the surface topography of the archwires. Moreover, the presence of fluoride in artificial saliva was detrimental to the corrosion resistance of the test NiTi archwires, especially at a 0.5% NaF concentration. Pulikkottil et al. [56] evaluated the corrosion resistance of four different orthodontic archwires and determined the effect of 0.5% NaF (simulating high fluoride-containing toothpaste of about 2250 ppm) on corrosion resistance of these archwires. Four different archwires (stainless steel [SS], NiTi, TMA, and ion-implanted TMA) were used. It was reported that the potentiostatic study reveals that the corrosion resistance of low-friction TMA (L-TMA) > TMA > NiTi > SS. Atomic force microscopy (AFM) analysis showed the surface roughness (R_a) of TMA > NiTi > L-TMA > SS; indicating that the chemical composition of the wire is the primary influential factor to have high-corrosion resistance and surface roughness is only secondary. The corrosion resistance of all wires had reduced significantly in 0.5% acidic fluoride-containing artificial saliva due to formation of fluoride complex compound. It was accordingly concluded that the presence of 0.5% NaF in artificial saliva was detrimental to the corrosion resistance of the orthodontic archwires, and complete removal of residual high-fluorinated toothpastes from the crevice between archwire and bracket during tooth brushing is mandatory. Ünal et al. [57] investigated the effect of fluoride added artificial saliva solution on NiTi and β-Ti alloys used for orthodontic wires. The orthodontic wires were immersed into fluoride added artificial saliva solution of pH 4.8 at 37 °C. It was concluded that the fluoride has an important effect on corrosion behavior of experimental wires that can be described as a type of wear; energy-dispersive spectroscopy (EDS) analysis denotes the existence of a thin oxide film layer in addition to the Ti-based oxide one on the surface of the metals. Kao et al. [58] assessed surface characteristics and compared the corrosion potential of metal brackets and wires in environments containing different media. Four brands of metal brackets and two types of orthodontic wires (stainless steel and NiTi) were investigated. The electrolyte was acidulated in NaF and pH 4 and 6 artificial saliva solutions at 37 °C. It was reported that the potentiodynamic curve showed that most brands of metal brackets were easily corroded in the NaF and pH 4 environments, while the NiTi and stainless steel wires were easily corroded in the pH 4 artificial saliva. The SEM observations showed that defects or pitting corrosion occurred on the surfaces of the brackets and wires in all tested media. The corrosion resistance of Ti–Ni, Ti–Ni–Co, and Ti–6Al–4V were compared with that of CpTi in Fusayama–Meyer artificial saliva with different pH and fluoride contents, using electrochemical polarization resistance measuring methods [27]. It was reported that the corrosion resistance of Ti–6Al–4V alloy is as good as that of CpTi in Fusayama–Meyer saliva and acid salivary solution, except Ti–Ni and Ti–Ni–Co alloys. A remarkable localized corrosion phenomenon of Ti–Ni, Ti–Ni–Co, and Ti–6Al–4V alloys occurred in fluoride and acid–fluoride salivary solutions. The fluoride ions could cause the breakdown of the protective passivation layer that normally exists on the titanium and its alloys, leading to pit corrosion.

8.2.4 Simulated body fluid

Anodic polarization measurements made in Hanks' physiological solution at 37 °C and pH of 7.4 were conducted on NiTi and Ti6Al4V alloys, CoCiCrMo, CoCrMo, 316L stainless steel and pure Ni [59]. It was reported that titanium materials to be the most passive. Among the influences caused by the amino acids, cysteine, and tryptophan on the corrosion behavior of Ti–Ni and Ti–6A1–4V, cysteine caused a lower breakdown potential for Ti–Ni, but it did not affect the breakdown of Ti–6A1–4V, although an increase in current density for Ti–6A1–4V was observed, and tryptophan produced no significant effects. Rondelli [60] evaluated the corrosion performances of NiTi SMAs in human body simulating fluids. It was obtained that as for the passivity current in potentiostatic conditions, taken as an index of ion release, the values are about three times higher for NiTi than for Ti6Al4V and austenitic stainless steels. Regarding the localized corrosion, while plain potentiodynamic scans indicated for NiTi alloy good resistance to pitting attack similar to Ti6Al4V, tests in which the passive film is abruptly damaged, pointed out that the characteristics of the passive film formed on NiTi alloy are not as good as those on Ti6Al4V but are comparable or inferior to those on austenitic stainless steels. Montero-Ocampo et al. [61] investigated the effect of various degrees of deformation at specific locations in the stress-strain curve under compression on the corrosion resistance of a wrought NiTi alloy with a martensite to austenite transformation peak of 110 °C. Two metallurgical conditions were evaluated: 30% cold drawn and annealed at 900 °C for 1 h. The cold drawn material was tested for corrosion resistance after 5.8%, 7.4%, 12.2%, and 24.5% applied strain. Similarly, the corrosion resistance of the annealed material condition was examined after deformation in compression to 11.9%, 22.3%, and 24.4% strain. Tafel extrapolation and cyclic polarization tests were used for corrosion characterization of each alloy condition. It was reported that the corrosion current density undergoes a significant reduction while the breakdown potential improves at increasing strains. In particular, the alloy in the annealed condition exhibited breakdown potentials above 1000 mV with current densities lower than 10 $\mu A/cm^2$ when it was strained to 24.4%.

The corrosion behaviors of various types of Ti materials (CpTi, NiTi, Ti6Al4V, Ti5Al2.5Fe, Ti5Al3Mo4Zr, and Ti4.5Al3V2Mo2Fe) through electrochemical polarization tests in 37 °C Ringer's solution were investigated [62]. It was found that CpTi and Ti–5Al–2.5Fe were the most resistant to corrosion; Ti–5Al–3Mo–4Zr, Ti–6Al–4V, and NiTi were the least resistant to corrosion; NiTi exhibited pitting corrosion along with transpassivation. Electron diffraction patterns indicated that all titanium alloys were covered mainly with rutile-type oxide (TiO_2) after corrosion tests; the oxides that formed on Ti–5Al–2.5Fe were identified as a mixture of TiO_2 and Ti_9O_{17}, and those that formed on NiTi were identified as a mixture of TiO_2 and Ni_2Ti_4O. Caroll et al. [63] measured the breakdown potentials for unpolished and mechanically polished NiTi wires in simulated body fluids. These wires are similar to those used in the manufacture of stents. It was mentioned that considerable scatter was observed in the results indicating a variable surface state. After appropriate heat treatments, the measured breakdown values were lower but more reproducible for the mechanically polished samples. Significantly, higher breakdown potentials were observed for cross-section wire samples. Energy-dispersive X-ray (EDX) analysis of the surface layers indicated that oxide thickening occurred after heat treatments. The oxide was predominantly made up of TiO_2 with a very thin layer of NiO at the outer surface, and in tests in which NiTi/gold couples were immersed in 0.9% NaCl for periods up to 12 months, only very small amounts of nickel (in the part per billion range) were released into solution and SEM examination revealed no corrosion. Yan et al. [64] investigated corrosion resistance of a laser spot-welded joint of NiTi alloy wires using potentiodynamic tests in Hank's solution at different pH values and the pH 7.4 NaCl solution for different Cl^- concentrations. It was reported that the corrosion resistance of a laser spot-welded joint of NiTi alloy wire was better than that of base metal, which exhibited a little higher breakdown potential and passive range, and a little lower passive current density. Corrosion resistances of a laser spot-welded joint and base metal decreased with increasing of the Cl^- concentration and pH value. The improvement of corrosion resistance of the laser spot-welded joint was due to the decrease of the surface defects and the increase of the Ti/Ni ratio. Liang et al. [65] investigated electrochemical behavior of TiNi SMA and Co–Cr alloys in dynamic Tyrode's simulated body fluid. It was indicated that for all alloys, the anodic dissolution and the pitting sensitivity increased with the flow rate of the Tyrode's solution increasing, while the open-circuit potentials and pitting corrosion potential E decreased with the Tyrode's solution increasing. The pitting corrosion of TiNi alloy was easier than Co–Cr alloys. Since the solution's flow enhanced oxygen transform and made it easy to reach the surface of electrodes, the plateau of oxygen diffusion control was diminished. All of these indicated that the cathodic reduction and the corrosion reaction, which was controlled by the electrochemical mass transport process, were all accelerated in dynamic Tyrode's simulated body fluid.

A NiTi$_{50.7}$ (at%) SMA was investigated by electrochemical tests in physiological environments of Ringer's solution and NaCl 0.9% solution [66]. It was found that the breakdown potential of the NiTi alloy in NaCl 0.9% solution is higher than that in Ringer's solution. The SEM reveal that low-pitting corrosion occurred in Ringer's solution compared with NaCl solution at potentiostatic tests, and the existence of hydride products in the X-ray diffraction (XRD) analysis confirms the decrease of the concentration of hydrogen ion in solutions. Furthermore, the biocompatibility tests were performed by reaction of mouse fibroblast cells (L929) and it was obtained that the figures of cells for different times showed natural growth of cells. The difference of the cell numbers between the test specimen and control specimen was negligible; therefore, it may be concluded that the NiTi SMA is not toxic in the physiological environments simulated with body fluids. Hansen et al. [67] evaluated the electrochemical behavior of polished NiTi surfaces exposed to different simulated body fluid solutions: Hanks solution, Hanks' balanced salt solution (HBSS), saline body fluid (SBF) solution, and Ringer's solution. It was found that the NiTi alloy shows the same corrosion mechanism (pitting) in all simulated body fluids that were studied; however, the corrosion potential E changes for each electrolyte, being HBSS, SBF, and Ringer's the most corrosive solutions. Furthermore, the Hanks' and HBSS solutions demonstrated good reproducibility of the electrochemical results. However, considering that the HBSS represents an extreme environment, this solution seems to be the most indicated to study the corrosion behavior of NiTi treated surfaces. Stergioudi et al. [68] studied the corrosion performance of two porous NiTi in physiological and Hank's solutions by potentiodynamic polarization, cyclic polarization, and impedance spectroscopy. It was obtained that both porous samples were susceptible to localized corrosion. The porosity increase (from 7% to 18%) resulted in larger and wider pore openings, thus favoring the corrosion resistance of 18% porous NiTi. Strengthening of corrosion resistance was observed in Hank's solution, and the pore morphology and microgalvanic corrosion phenomena were the determining factors affecting the corrosion resistance. Rosalbino et al. [69] studied the in vitro corrosion behavior of two Ti–Nb–Sn SMAs (TiNb$_{16}$Sn$_5$ and TiNb$_{18}$4Sn$_4$ in mass%), compared with that of Nitinol, in naturally aerated Ringer's physiological solution at 37 °C by corrosion potential E and EIS measurements as a function of exposure time, and potentiodynamic polarization curves. It was reported that corrosion potential E values indicated that both Ni–Ti and Ti-Nb-Sn alloys undergo spontaneous passivation due to spontaneously formed oxide film passivating the metallic surface, in the aggressive environment. It is also indicated that the tendency for the formation of a spontaneous oxide is greater for the Ti-18Nb-5Sn alloy. Significantly, low- anodic current density values were obtained from the polarization curves, indicating a typical passive behavior for all investigated alloys, but NiTi exhibited breakdown of passivity at potentials above approximately 450 mV (SCE), suggesting lower corrosion protection characteristics of its oxide film compared to the Ti-Nb-Sn alloys. The EIS studies showed high-impedance values for all samples, increasing with exposure time, indicating an improvement in corrosion resistance of the spontaneous oxide film. The obtained EIS spectra were analyzed using an equivalent electrical circuit representing a duplex structure oxide film, composed by an outer and porous layer (low resistance), and an inner barrier layer (high resistance) mainly responsible for the alloys' corrosion resistance. The resistance of passive film present on the metals' surface increases with exposure time displaying the highest values to Ti–18Nb–4Sn alloy; thereby suggesting that Ti–Nb–Sn alloys are promising materials for biomedical applications.

8.2.5 Sweat

Randin [70] measured the corrosion resistance of various nickel-containing alloys in artificial sweat (perspiration) using the Tafel extrapolation method. It was found that Ni, CuNi 25 (coin alloy), NiAl (colored intermetallic compounds), WC + Ni (hard metal), white gold (jewelry alloy), FN42 and Nilo Alby K (controlled expansion alloys), and NiP (electroless nickel coating) are in an active state and dissolve readily in oxygenated artificial sweat. By contrast, austenitic stainless steels, TiC + Mo_2C + Ni (hard metal), NiTi SMA, Hastelloy X (superalloy), Phydur (precipitation-hardening alloy), PdNi and SnNi (nickel-containing coatings) are in a passive state but may pit under certain conditions. Co and its alloys have poor corrosion resistance except for Stellite 20; while Cr and high-Cr ferritic stainless steels have a high pitting potential, but the latter are susceptible to crevice corrosion. Ti has a pitting potential greater than 3 V, and comparison between the in vitro measurements of the corrosion rate of nickel-based alloys and the clinical observation of the occurrence of contact dermatitis is discussed.

8.2.6 Others

There is scattered information on corrosion tests and results conducted in unique environment. Ahmed et al. [71] evaluated the corrosion performance of $Ni_{47}Ti_{49}Co_4$ SMA in artificial urine solution in comparison with $Ni_{51}Ti_{49}$ alloy as reference, at 37 °C and pH 5.6–6.4. It was reported that SEM results revealed less pitting attack for $Ni_{47}Ti_{49}Co_4$ SMA surface, after immersion in artificial urine solution. The XRD analysis demonstrated the formation of passive film on $Ni_{47}Ti_{49}Co_4$ SMA. The XPS analysis indicated that the film mainly consisted of O, Ti, Co, P, and a small amount of Ni. The concentration of Ni ions release was greatly reduced compared to that of the $Ni_{51}Ti_{49}$, SMA, and linear polarization results illustrated that corrosion potential (E_{CORR}), corrosion current density (i_{CORR}), and ac R_P were greatly affected by alloying

Co to NiTi alloy, indicating that the corrosion resistance of the ternary alloy, $Ni_{47}Ti_{49}Co_4$ SMA, offers superior corrosion resistance in artificial urine when compared to $Ni_{51}Ti_{49}$ SMA, which was suitable for medical applications. Hasegawa et al. [72] evaluated the effect of ethylenediaminetetraacetic acid (EDTA) solutions (3% and 10% EDTA $\cdot$ 2Na) on corrosion fatigue of three NiTi files with different shapes, in comparison with other solutions (6% NaClO, 3% H_2O_2, 0.9% NaCl, and distilled water). NiTi files were subjected to rotational bending in a bent glass tube (30° and 60° angles) filled with the solutions, and the number of rotations to failure was counted. It was obtained that at 30° bent angle, files in the two EDTA solutions showed significantly lower resistance than those in distilled water, but no significant difference was found between the two EDTA solutions. Fatigue resistance of two tested files in the two EDTA solutions was not significantly different from those in the other three solutions, whereas one file in EDTA solutions showed significantly lower resistance than that in 3% hydrogen peroxide (H_2O_2). At 60° bent angle, early failure within 1–2.5 min was observed for all tested files, and no significant difference was found among the six solutions, and at both angles, significant differences in fatigue resistance were observed among the three tested files, which could be related to the difference in the cross-sectional shapes of the files. Shabalovskaya et al. [73] investigated corrosion resistance, chemistry, and mechanical aspects of NiTi surfaces formed in H_2O_2 solutions (3% and 30%) for various periods of time. It was found that Ti oxides formed naturally on NiTi surfaces are only a few nanometers thick. The surface layers with variable Ti (6–15 at%) and Ni (5–13 at%) contents, which were oxidized in hydrogen peroxide solution showed the thickness up to 100 nm without Ni-enriched interfaces. The response of the surface oxides to stress in superelastic regime of deformations depended on oxide thickness. In the corrosion studies performed in both strained and strain-free states using potentiodynamic and potentiostatic polarizations, the surfaces treated in H_2O_2 showed no pitting in corrosive solution that was assigned to higher chemical homogeneity of the surfaces free of secondary phases and inclusions that assist better biocompatibility of Nitinol medical devices.

There are several studies on NiTi corrosive behavior when exposed to mucin, protein, and other biological substances. Chao et al. [74] studies NiTi alloy and stainless steel when they were exposed to an artificial saliva containing fibrinogen, IgG, or mucin. It was fund that fibrinogen, IgG, or mucin could have different influences on the susceptibility to corrosion of the same alloy. Adding protein leads to the decrease of corrosion resistance of stainless steel, whereas protein could slow down the corrosion process of NiTi. For NiTi, adding mucin could enhance the corrosion stability and repair capacity of passivation film, the susceptibility to pitting corrosion of NiTi and stainless steel in fibrinogen AS is not as high as mucin and IgG AS. There are different patterns of deposition formation on the metal surface by different types of protein, which is associated with their effects on the corrosion process of the alloys. When albumin was mixed with fluoride containing corrosion media, the corrosion behavior was noticed to change remarkably. Takemoto et al. [75] investigated the corrosion behavior and surface characterization of passive films on titanium immersed in a solution containing 2.0 g/L fluoride and albumin (either 0.1 or 1.0 g/L). It was found that fluorine was detected on the titanium surface immersed in the solution containing fluoride, and dissolution of the titanium was confirmed. The titanium immersed in a solution containing both fluoride and albumin had an albumin film regardless of the albumin concentration level. In addition, the amount of dissolved titanium from the titanium immersed in the solution was less than when the solution contained no albumin[7]. Therefore, it was suggested that the formation of adsorbed albumin films on the passive film acted to not only protect the titanium from attack by the fluoride but also suppressed dissolution of the titanium-fluoride compounds. Although this study was done for CpTi, the materials should include TiNi as well as Ti–6Al–4V since main oxide formed on these Ti-based alloys is TiO_2, which is same as the oxide formed on CpTi surface. Ide et al. [76] investigated the effect of albumin on the corrosion resistance of titanium in the presence of fluoride. The effects of the NaF concentration, albumin concentration, and pH on the corrosion characteristics of CPTi were examined by means of electrochemical techniques. It was reported that the corrosion resistance of titanium decreased as the NaF concentration increased and as pH decreased. The corrosion resistance of titanium in NaF solutions was improved in the presence of albumin. The natural electrode potential was elevated, and the passive current density was reduced by albumin at a concentration of 0.01%. The polarization resistance rose with increased concentrations of albumin in fluoride solution; thereby, indicating that the albumin in saliva and dental plaque affected the corrosion resistance of CPTi in fluoride solution. Hang et al. [77] studied the corrosion behavior of NiTi alloy in fetal bovine serum (FBS) at 37 °C using OCP, EIS, SEM, XPS, and inductively coupled plasma mass spectrometry (ICP-MS). It was obtained that the presence of FBS moves the OCP to the negative direction and makes the oxide film thinner and more porous than that in phosphate buffer saline (PBS, pH = 7.4). The impedance of the oxide film formed in FBS is smaller than that in PBS, but the total interface impedance is bigger in FBS because of the formation of a surface biofilm. Pits are formed on the NiTi alloy after immersing in FBS for 200 h, but they are not observed on the sample immersed in PBS. Further, XPS shows that the composition of the oxide film formed in FBS is similar to that formed in PBS and is composed of mainly Ti oxides with a small amount of Ti hydroxide; and hydrated Ti is observed on the outermost surface of the NiTi alloy. Ryhänen et al. [78] clarified the primary cytotoxicity and corrosion rate of NiTi material in human cell cultures. Comparisons were made with stainless steel (SS), titanium (Ti), composite material (C), and control cultures with no test discs. Human osteoblasts (OB) and fibroblasts (FB) were incubated for 10 days. It was reported that the proliferation of FB was 108% (NiTi), 134% (Ti) ($p < 0.02$), 107% (SS), and 48% (C) compared to the control cultures. The proliferation of OB was 101% (NiTi), 100% (Ti), 105% (SS), and 54% (C) compared to the controls. Initially, NiTi released more nickel (129–87 µg/L) into the cell culture media than SS (7 µg/L), but after 2 days the concentrations were approximately same (23–5 µg/L vs 11–1 µg/L). The titanium element concentrations from both NiTi and Ti samples were all <20 µg/L. Based on these results, it was concluded that NiTi has good in vitro biocompatibility with OB and FB. Despite the higher initial nickel dissolution, nitinol induced no toxic effects, decrease in cell proliferation, or inhibition on the growth of cells in contact with the metal surface. Rozali et al. [79] investigated the corrosion behavior for NiTi alloy under simulated biological conditions with the presence of proteins using electrochemical methods. The electrochemical tests were performed with the addition of bovine serum albumin with concentrations of 0%, 2%, 4%, 6%, 8%, and 10% at 37 °C. It was mentioned that the samples showed that they did not undergo any significant change in term of composition. NiTi alloys have a good corrosion resistant toward all concentrations; thereby, indicating that NiTi alloys can perform well with or without the presence of protein-containing solutions. Hence, they are compatible to use as implant materials in human body environment. Using NiTi alloy as a stent material, Marton et al. [80] examined the hypothesis that repetitive mechanical stresses interact synergistically with the atherosclerotic lesion environment to induce corrosive events that ultimately lead to nitinol stent material failure. Electropolished NiTi samples were subjected to cyclic defined stress regimens in the presence or absence of adherent acetylated lipoprotein-activated macrophage-like cells over a 5-day period. It was found that the combination of defined bending in the presence of adherent activated macrophages resulted in a significant increase in nitinol surface corrosion as compared to nitinol subjected to either treatment or media exposure alone. Deterioration of the protective oxide layer and an increase in the surface roughness were observed in nitinol specimens exposed to the combined influence of surface-adherent macrophages and repetitive loading. Based on these findings, it was suggested that there is a significant interaction between the biological and mechanical environment relative to nitinol corrosion potential E and provide a basis for design of future studies to improve the long-term performance of peripheral nitinol stents.

8.2.7 Influencing factors on NiTi corrosion behaviors

8.2.7.1 Surface condition

Surface conditions of NiTi materials possess variety of faces. The surface might be rough or smooth. It might be solid surface or porous one. It can be crystalline or amorphous. It possesses a thin or thick layer of Beilby's layer. Depending on the prefabrication conditions, it can exhibit a positive or negative surface residual stress. The surface can be hydrophobic or hydrophilic and others. All these surface conditions should affect subsequent chemical or electrochemical reactions. Li et al. [81] investigated corrosion characteristics of porous NiTi SMAs with different porosity prepared from combustion synthesis by potentiodynamic polarization measurements in physiological and Hank's solutions at 37 °C. It was indicated that (i) the corrosion resistance of porous NiTi SMA decreased with the increasing porosity, and porous NiTi

SMA was less corrosion resistant than the solid one. Es-Souni et al. [82] studied the effects of surface topography and surface finish residues on the potentiodynamic corrosion behavior on NiTi alloys. The cytotoxicity tests were performed on both alloys in fibroblast cell cultures from human gingiva using the MTT test. It was reported that the surface finish and the amounts of surface finish residues dramatically affect the corrosion resistance. Bad surface finish results in lower corrosion resistance. The in vitro biocompatibility, though not affected to the extent of corrosion resistance, is also reduced as the surface roughness and the amounts of residues increase, due to surface effects on corrosion and metallic ions release. Pereira et al. [83] investigated the surface morphology and electrochemical potential of superelastic, M-Wire (MW) and shape-memory technology (SMT) NiTi instruments before and after single clinical use in vivo. A total of 60 ProTaper Universal F2 (PTU-SE), ProTaper Next X2 (PTN-MW), Typhoon (TYP), Hyflex (HF), and Vortex Blue (VB), the last three SMT, and size 25, 0.06 taper ($n = 6$ of each type) files were examined. It was found that SEM observations of new instruments indicated the presence of marks left by the machining process during manufacturing and EDS revealed the existence of an oxide coating on shape-memory instruments. After clinical use, the five types were associated with propagation of transverse cracks 3 mm from the tip. The surface oxide layer of TYP, HF, and VB instruments had microcracks in multiple directions, while TYP and HF had fragmentation in chip form of the oxide layer. The EDS analysis demonstrated a significant reduction of the oxide layer in shape-memory instruments, except for VB. The electrochemical potentials were higher for shape-memory instruments than for MW and superelastic NiTi instruments. Based on these results, it was concluded that SMT NiTi instruments have a dysfunctional oxide layer after clinical use, and they featured higher electrochemical potential relative to NiTi instruments manufactured from MW and conventional superelastic NiTi alloy.

8.2.7.2 Coating

The effect of various coating formulations on the mechanical and corrosion properties of NiTi orthodontic wires was analyzed [84]. Uncoated, rhodium-coated, and nitrified NiTi wires were observed with a three-point bend test, surface roughness (R_a) measurement, SEM, EDS, and electrochemical testing OCP, EIS, and cyclic polarization scan). It was found that uncoated and nitrified NiTi wires showed similar mechanical and anticorrosive properties, while rhodium-coated NiTi wires showed the highest R_a and significantly higher modulus of elasticity, yield strength, and delivery of forces during loading but not in unloading. Rhodium-coated NiTi wires also had the highest corrosion current density and corrosion potential E, lowest impedance modulus, and two time constants on Bode plot, one related to the Rh/Au coating and the other to underlying NiTi. It was then concluded that working properties of NiTi wires were unaffected by various coatings in unloading. Nitrification improved corrosion resistance and rhodium coating reduced corrosion resistance and pronounced susceptibility to pitting corrosion in artificial saliva because of galvanic coupling between the noble coating and the base alloy. Viswanathan et al. [85] investigated the formation of nanostructured oxide layers by anodic oxidation on different surface finished (mirror finished, 600 and 400 grit polished) NiTi in electrolyte solution containing ethylene glycol and NH_4F. The anodized surface has been characterized by field emission SEM (FE-SEM), EDS, and XPS. The corrosion behaviors of the Ni–Ti substrate and anodized samples were tested in simulated body fluid (Hanks' solution). It was reported that the native oxide on the substrate is replaced by nanostructures through anodization process, and XPS of NiTi substrate shows the presence of Ni, NiO, Ti, and TiO_2 species, whereas Ni_2O_3, $Ni(OH)_2$, and TiO_2 are observed in the samples after anodization. Rokicki et al. [86] introduced a magneto electropolishing method to improve NiTi implantable devices, which exhibited better biocompatibility. It was also mentioned that the final NaOCl treatment should replace currently used NiTi implantable devices sterilization methods. Kim et al. [87] studied the influence of electropolishing on NiTi stents and its corrosion behavior. Electropolishing is an effective method for surface treatment, which not only controls the surface state but also helps to produce uniform surface layers. Therefore, to improve the surface quality of NiTi stents, an electropolishing (from 30–40 V and 10–30 s) was conducted as a postheat treatment for NiTi stent manufacturing process. It was reported that the electropolished NiTi stents under the condition of 40 V for 10 s exhibited the best corrosion performance as well as surface quality.

8.2.7.3 Polishing

About the chemical polishing effect, Shabalovskaya et al. [88] evaluated the effect of chemical etching in a HF/HNO_3 acid solution and aging in boiling water on the corrosion resistance of NiTi wires with black oxide with the use of potentiodynamic, potentiostatic equipment, and scratch tests. It was reported that after corrosion tests, as-received wires revealed uniformly cracked surfaces reminiscent of the stress corrosion cracking (SCC) phenomenon. These wires exhibited negative breakdown potentials in potentiostatic tests and variable breakdown potentials in potentiodynamic tests (–100 to + 400 mV vs SCE). Wires with treated surfaces did not reveal cracking or other traces of corrosion attacks in potentiodynamic tests up to +900 to 1,400 mV potentials and no pitting after stimulation at +800 mV in potentiostatic tests. They exhibited corrosion behavior satisfactory for medical applications. It can then be concluded that significant improvement of corrosion parameters was observed on the reverse scans in potentiodynamic tests after exposure of treated wires to potentials >1,000 mV. In scratch tests, the prepared surfaces repassivated only at low potentials, comparable to that of stainless steel. Tremendous improvement of the corrosion behavior of treated NiTi wires is associated with the removal of defect surface material and the growth of stable TiO_2 oxide. The corrosion behavior of NiTi in the body is of critical importance because of the known toxicological effects of nickel element. The stability of a NiTi alloy in the physiological environment is dependent primarily on the properties of the mostly TiO$_2$ oxide layer that is formed on the surface. Clarke et al. [89] prepared a range of NiTi wires using different drawing processes and a range of surface preparation procedures. It was mentioned that the wire samples with very thick oxides also contain a high nickel content in the oxide layer. The untreated samples with the thicker oxides show the lowest pitting potential values and greater nickel release in both long- and short-term experiments; and after long-term immersion tests, breakdown potentials increased for samples that exhibited lower values initially. This suggests that surface treatment is essential for the optimal bioperformance of nitinol.

8.2.7.4 Shot peeing

It is well documented that the advantages associated with shot peening exhibit two-fold: (1) beneficial development of surface compressive residual stress, and (2) enlargement of effective surface area. Olumi et al. [90] utilized shot pinning process to produce nanocrystalline NiTi alloy with increased corrosion resistance. The electrochemical behavior of NiTi alloy was evaluated using the potentiodynamic polarization scan and EIS tests in Ringer's solution after and prior to the shot pining process. It was obtained that the result of XRD analysis showed an average crystalline size of 23 nm. FE-SEM confirmed the development of a nanostructured alloy induced by shot-pinning process. Modification of NiTi alloy by shot-peening process resulted in corrosion resistance improvement and decrease in the corrosion rate, which consequently led to less release rate of the toxic nickel ions in the corrosive environment, compared to the nonmodified samples. Zhang et al. [91] characterized the effects of laser shock peening (LSP) on the biocompatibility, corrosion resistance, ion release rate, and hardness of NiTi alloy. It was reported that the cell culture study indicated that the LSPed NiTi samples had lower cytotoxicity and higher cell survival rate than the untreated samples; specifically, the cell survival rate increased from 88 ± 1.3% to 93 ± 1.1% due to LSP treatment. The treatment was shown to significantly decrease the initial Ni ion release rate compared with that of the untreated samples. Hence, electrochemical tests indicated that LSP improved the corrosion resistance of the NiTi alloy in simulated body fluid, with a decrease in the corrosion current density from 1.41 ± 0.20 μA/cm^2 to 0.67 ± 0.24 μA/cm^2. Immersion tests showed that calcium deposition was significantly enhanced by LSP, and the hardness of NiTi alloy increased from 226 ± 3 HV before LSP to 261 ± 3 HV after LSP. Based on these findings, titanium was indicated that the laser-shot peening is a promising surface modification method that can be used to improve the mechanical properties, corrosion resistance, and biocompatibility of NiTi alloy for biomedical applications.

8.2.7.5 Surface oxidation

Surface oxidation can modify the surface characteristics. Since the biocompatibility is strongly relied on the stability of the oxide film, which is formed on NiTi surface via electrochemical passivation or oxidation in air (or pure oxygen), Trépanier et al. [92] studied the effect on corrosion resistance and surface characteristics of electropolishing, heat treatment, and nitric acid passivation of NiTi stents. It was shown that all of these surface treatments improve the corrosion resistance of the alloy. This improvement is attributed to the plastically deformed native oxide layer removal and replacement by a newly grown, more uniform one. The uniformity of the oxide layer, rather than its thickness and composition, seems to be the predominant factor to explain the corrosion resistance improvement. Cissé et al. [93] determined the effect of surface modification of $Ni_{55.8}Ti$ (wt%) SMA on its corrosion behavior in Hanks' solution. determined. The surfaces of heat-treated samples were modified by mechanical polishing (MP), electropolishing (EP), and electropolishing followed by chemical passivation (CP). As-heat-treated samples with straw-colored oxide finishes (SCO) and blue-colored oxide finishes (BO) also were included in the study. It was found that surface roughness increased in the order CP < EP < SCO < BO < MP. The release of nickel within the five groups of NiTi samples, as determined by atomic absorption spectrophotometry, reduced in time over the measured period. The level of Ni ions released over a 25-day immersion period was highest in the SCO sample (0.002 µg/day). This Ni level is negligible compared with the daily intake of Ni in an ordinary diet. The auger electron spectroscopy (AES) analyses indicated that before immersion in Hanks' physiologic solution, the main surface composition of all the samples was titanium and nickel, with a small amount of oxygen, carbon, and sulfur as contaminants. The surface oxide thickness of the different samples increased in the order CP < EP < MP < BO < SCO. On the one hand, for the electrodes treated under the same conditions, the mean breakdown potential value decreased in the order BO > MP > CP > EP > SCO; on the other hand, the corrosion current density and rate increased in the order CP < SCO < EP < BO < MP. For the purpose of enhancement for the corrosion resistance of additively manufactured NiTi medical devices, a micro-arc oxidation was evaluated [94].

The corrosion behavior of NiTi alloys after mechanical polishing, electropolishing, and sterilization processes using cyclic polarization and atomic absorption was investigated [95]. It was mentioned that as a preparative surface treatment, electropolishing decreased the amount of nickel on the surface and remarkably improved the corrosion behavior of the alloy by increasing the mean breakdown potential value and the reproducibility of the results (0.99 +/– 0.05 V/SCE vs 0.53 +/– 0. 42). Ethylene oxide and Sterrad (R) sterilization techniques did not modify the corrosion resistance of electropolished NiTi, whereas a steam autoclave and, to a lesser extent, per acetic acid sterilization produced scattered breakdown potential. In comparing the corrosion resistance of common biomaterials, NiTi ranked between 316 L stainless steel and Ti6A14V even after sterilization. Electropolished NiTi and 316 L stainless steel alloys released similar amounts of nickel after a few days of immersion in Hank's solution. Measurements by atomic absorption have shown that the amount of released nickel from passive dissolution was below the expected toxic level in the human body. AES analyses indicated surface contamination by Ca and P on NiTi during immersion, but no significant modification in oxide thickness was observed. O'Hoy et al. [96] evaluated the effect of repeated cleaning procedures on fracture properties and corrosion of NiTi files. New NiTi instruments were subjected to 2, 5, and 10 cleaning cycles with the use of either diluted bleach (1% NaOCl) or Milton's solution (1% NaOCl plus 19% NaCl) as disinfectant. Each cleaning cycle consisted of scrubbing, rinsing, and immersing in NaOCl for 10 min followed by 5 min of ultrasonication. Files were then tested for torsional failure and flexural fatigue. It was found that up to 10 cleaning cycles did not significantly reduce the torque at fracture or number of revolutions to flexural fatigue, although decreasing values were noted with increasing number of cleaning cycles using Milton's solution. No corrosion was detected on the surface of these files. The files immersed in 1% NaOCl overnight displayed a variety of corrosion patterns. The extent of corrosion was variable among different brands of files and among files in each brand, and overall, Milton's solution was much more corrosive than diluted bleach. Based on these results, it was concluded that the files can be cleaned up to 10 times without affecting fracture susceptibility or corrosion, but should not be immersed in NaOCl overnight. Moreover, Milton's solution is much more corrosive than bleach with the same NaOCl concentration. Amorphous metallic alloy coatings of NiMo and NiTi substrates were fabricated by the ion beam mixing has been used to fabricate amorphous metallic alloy coatings of NiMo and NiTi substrates [97]. Ion beam mixing has been used to fabricate amorphous metallic alloy coatings of NiMo and NiTi substrates. Alternate thin layers (~100 Å) of Mo and Ni and Ti and Ni with different thicknesses were prepared by e-beam evaporation. The thicknesses of individual films were adjusted to obtain an overall composition of $Ni_{50}Mo_{50}$ and $Ni_{50}Ti_{50}$ after mixing. It was reported that the mixed layers were found to be amorphous and were tested for their corrosion behavior by potentiodynamic polarization techniques. Amorphous NiTi forms a better coating than amorphous NiMo when surveyed in 1 N HNO_3 and 0.1 N NaCl solutions. In general, amorphous coatings have demonstrated better corrosion resistance than their polycrystalline counterparts. Hu et al. [98] evaluated a nanocrystalline and partial amorphous structure in the surface layer of NiTi alloy by surface mechanical attrition treatment (SMAT). The corrosion behavior was systematically studied in a 0.9% NaCl physiological solution by electrochemical methods. It was reported that potentiodynamic polarization measurements indicate that the corrosion resistance of SMATed NiTi with the surface nanocrystalline and partial amorphous structure is significantly enhanced compared to the bare NiTi with coarse grains. Both corrosion potential (E_{CORR}) measurements and EIS reveal that a passive oxide layer is readily formed on the SMATed NiTi during early immersion in the 0.9% NaCl solution. When the passive oxide layer has stabilized after long exposure in the 0.9% NaCl solution, corrosion induced by Cl^- begins to degrade the passive oxide film. The observed corrosion behavior of SMATed NiTi is considered to be associated with the surface nanocrystalline and amorphous structure.

8.2.7.6 Heat treatment and welding

The temperature-dependent corrosion characteristics of shape-memory, superelastic, and nonsuperelastic orthodontic wires were investigated [99]. The following four orthodontic wires were tested: namely, Cu–NiTi (superelastic at 27 °C and shape memory at 40 °C), superelastic NiTi, and nonsuperelastic Nitinol Classic. Sectioned halves of as-received archwires were assessed electrochemically in artificial saliva at 5, 24, 37, and 45 °C. Open circuit potential was monitored for two h followed by polarization resistance and cyclic polarization tests. It was reported that DSC results showed Nitinol was primarily martensitic-stable whereas NiTi, 27 °C Cu–NiTi, and 40 °C Cu–NiTi possessed austenite-finish temperatures of approximately 19, 21, and 38 C, The OCP of the Cu–NiTi wires was significantly greater than NiTi and Nitinol, but no apparent trend in values was apparent with regard to temperature or phases present. Corrosion current density (i_{CORR}) increased with temperature for all wires, but not all were equally influenced. The two lowest austenite-finish temperature wires (27 °C Cu–NiTi and NiTi) approximately tripled in i_{CORR} from 37 to 45 °C, and greater incidence of pitting was observed in the Cu–NiTi wires. It was then concluded that the corrosion rate of various NiTi wires increases with temperature and different phases present may influence corrosion rate trends. Vojtěch et al. [100] heat-treated a $Ni_{50.9}Ti$ (at%) at temperature between 500 and 550 °C and durations from 5 to 10 min in air at normal and low pressure and in a salt bath. It was found that the heat treatments produced surface TiO_2 layers measuring 15–50 nm in thickness that were depleted in nickel, the sample covered by the 15-nm thick oxide that was treated at 500 °C × 5 min in a low pressure air showed the best corrosion performance in terms of Ni release. As the oxide thickness increased, due to either temperature or oxygen activity change, Ni release into the physiological solution accelerated. Ionita et al. [101] investigated effects of three different aging treatment on corrosion resistance of TiNi alloy in oral cavity environment. It was obtained that the best corrosion resistance in artificial saliva after aging treatment was found in the case of sample, which was heat treated at 700 °C × 20 min/water cooled, followed by aging at 400 °C × 1.5 h/water cooled. To such a subsequent heat treatment, the EDX surface analysis indicated an increase of Ti content of the matrix. This effect is in connection with stable and metastable precipitates that appear during the aging. Based on the greater affinity to oxygen of Ti compared to Ni, a uniform Ti-based oxide layer is predominant explaining the better electrochemical stability, and the obtained precipitates morphology are also an important factor in the corrosion behavior.

Welding causes complicated heat-affected features and their related microstructural alterations. Postweld heat-treatment (PWHT) was applied to NiTi weldments to improve the corrosion behavior by modifying the microstructure and surface composition [102]. It was mentioned that the surface oxide film on the weldments is principally TiO_2, together with some Ti, TiO, and Ti_2O_3. The surface Ti/Ni ratio of the weldments after PWHT is increased, the oxide film formed in Hanks' solution is thicker on the weldments after PWHT. The pitting resistance of the weldments is increased by PWHT as the galvanic effect in the weldments is very small, and the weldment with PWHT at 350 °C shows the best corrosion resistance among other heat-treated weldments in this study. Composite archwire is an archwire formed by solder connection of NiTi SMA and stainless steel wire. Zhao et al. [103] studied the biocompatibility of composite archwire as an important foundation for its clinical application. The electrochemical corrosion test was performed by immersion in solutions simulating oral cavity conditions. Murine L-929 cells were cocultured with composite archwire extract to evaluate the cytotoxicity of the corrosion products in vitro. It was reported that polarization tests indicated that composite archwire was resistant to corrosion in the tested artificial saliva-based solutions (chloric solution, simple artificial saliva, fluorinated artificial saliva, and protein-containing artificial saliva). The amount of toxic copper ions released after immersion was lower than average daily dietary intake levels. The cytotoxicity experiments demonstrated the in vitro biocompatibility of composite archwire. Based on the combined advantages of its base materials composite archwire, with its resistance to biocorrosion and in vitro cytocompatibility, it is a promising alternative material for use in orthodontic fixation applications. Mirshekari et al. [104] investigated the effects of Nd:YAG laser welding on the microstructure, phase transformation, cyclic deformation behavior, and corrosion resistance of $TiNi_{55}$ (wt%) wire. It was obtained that the laser welding altered the microstructure of the weld metal, which is mainly composed of columnar dendrites grown epitaxially from the fusion line. The DSC results indicated that the onset of the transformation temperatures of the weld metal differed from that of the base metal. The cyclic stress–strain behavior of laser-welded NiTi wire was comparable to the as-received material, while a little reduction in the pseudoelastic property was noted, the weld metal exhibited higher corrosion potential E, lower corrosion current density, higher breakdown potential, and wider passive region than the base metal. The weld metal was, therefore, more resistant to corrosion than the base metal. Michael et al. [105] conducted a systematic investigation of $Ni_{44.2}Ti$ (wt%) SMA with differing numbers of laser pulses and postprocess surface treatments. It was shown that fewer pulses were not significantly detrimental to the corrosion performance; however, increasing the number of pulses had a significant impact on oxide stability in the heat affected zone due to increased crystallinity. The PWHT restored corrosion performance to preprocessed capabilities; however, further optimization is required to achieve maximum corrosion resistance.

8.2.7.7 Surface modification

The surfaces of NiTi biomaterials play an important role in response to the artificial devices in a biological environment; for these materials to meet the clinical demands, it is necessary to modify their surface. The proper surface treatment expands the use of these materials in the biomedical field [106, 107]. Further improving the corrosion resistance of Ti biomaterials can enhance their stability during the long-term service in the human body. There are quite a number of volumes of proposed and exercised methods to improve the corrosion resistance of Ti biomaterials. These technologies should include passivation, laser surface modification, laser shot peening, metallic ion implantation, sol-gel, gaseous ion implantation, physical vapor deposition, electroplating, micro-arc oxidation, hybridization, and others [106, 108].

The effect of laser fluence on physical, mechanical, and chemical surface characteristics along with corrosion protection performance of the modified surface as compared to untreated NiTi surface was investigated [109]. Various phases were prominent on the top surface, namely nickel and titanium-rich phases, along with different nickel-titanium intermetallics and nanostructure of titanium oxide, based on varying melting pool recirculation time and cooling rate with laser fluence energy. It was reported that at low-laser fluence up to 4 J/mm^2, no significant melting pool were formed; and only transformation to the martensitic phase of NiTi took place on the top surface, which seemed to be highly too much corrosion-prone under simulated body fluid. At moderate laser fluence of 6–8 J/mm^2, mostly titanium-rich phases are prominent on the surfaces on account of optimum recirculation of melting pool and subsequent surfacing out of comparably light phase of titanium. Ti-rich phases on top surface exhibit superior corrosion resistance as compared to all other samples including bare nitinol; however, titanium oxide nanoparticles-reinforced martensitic structure is formed under high laser fluence due to over recirculation of molten pool. The modulus of elasticity also varied from 10 to 110 GPa based on top surface formation under different fluence levels, and thus this process can act as a tailor-made controllable pretreatment process over the traditional coating processes.

In order to enhance the corrosion resistance and to avoid Ni element leachout from the substrate surface into biological environment, nitrogen plasma immersion ion implantation (PIII) has been performed [110–112]. Poon et al. [111] produced barrier layers in NiTi by nitrogen PIII followed with vacuum annealing at 450 °C or 600 °C. It was shown that the barrier layer is not only mechanically stronger than the NiTi substrate, but also is effective in impeding the out-diffusion of Ni from the substrate. Among the samples, the 450 °C-annealed TiN barrier layer possesses the highest mechanical strength and best Ni out-diffusion impeding ability. The enhancement can be attributed to the consolidation of the Ti–N layer resulting from optimal diffusion at 450 °C. Liu et al. [112] modified the NiTi surface by nitrogen PIII in an effort to improve the corrosion resistance and mitigate nickel release from the materials. The implanted nitrogen depths and thicknesses of the surface TiN barrier layers were varied by changing the pulsing frequencies during PIII. It was reported that the nitride layer produced using a frequency of 50 Hz has the best stability under the OCP conditions, and the TiN layer produced using 200 Hz has the highest potentiodynamic stability after immersion in simulated body fluids for a long time. The TiN layer on the NiTi surface favors deposition of Ca–P composites thereby compensating for the instability of the TiN layer produced at a higher frequency.

Since titanium oxide films are deemed to be chemically inert and biocompatible, so that it can be expected to serve an effective barrier layers to impede the leaching of Ni from the NiTi substrate to biological tissues and fluids, same as nitrogen ion implantation. The oxygen plasma ion implantation has been conducted [113, 114]. Poon et al. [113] compared the anticorrosion efficacy of oxide films produced by atmospheric-pressure oxidation and oxygen plasma ion implantation. It was reported that the oxidized samples do not possess improved corrosion resistance and may even fare worse than the untreated samples. On the other hand, the plasma-implanted surfaces exhibit much improved corrosion resistance. The postimplantation annealing can further promote the anticorrosion capability of the samples. Mohan et al. [114] modified the surface of NiTi alloy substrates by the oxygen (PIII–O) technique at low temperature without affecting the substrate properties. It was found that polarization and electrochemical impedance spectroscopy studies reveal nearly ideal capacitor behavior with better passivation characteristics for the oxygen-implanted substrate. The sliding wear studies reveal lower friction of coefficient for the implanted layers as compared with the substrate. The bare and surface modified NiTi alloy samples are evaluated for biocompatibility using osteoblast-like cells (MG-63). According to the cellular behavior in terms of cell morphology, it was mentioned that oxygen implantation by PIII–O provides a better compatible surface for cell attachment and growth, and the modified surface exhibits a higher percentage of cell viability demonstrating the enhanced biocompatibility of the oxygen-implanted surface compared with bare NiTi alloy.

It is well documented the reason why hydroxyapatite (HA) is deposited onto implantable materials. First, its chemical composition is very close to a main constituent of receiving hard tissue (or bone), if the materials are to be implanted; meeting one of the most important requirements for successful implant treatment (i.e., a biological compatibility for excellent osseointegration) is essential. Second, HA is deposited to serve as if the deposited HA layer act as a periodontal membrane (which is lacking on metallic implant surface), so that the strain continuity can be maintained before externally loaded stress exceeds the interfacial stress developed at the implant/bone interface. This is due to the stress discrete situation caused by remarkable differences of modulus elasticity between bone and metallic implant. This is a sort of the hindsight (or overlooked) until Oshida [115, 116] pointed out. Maleki-Ghaleh [117] coated natural HA on NiTi alloy using electrophoretic deposition method to improve the corrosion resistance and biocompatibility. Coating process was performed at 120 s in various applied voltages of 40, 60, and 80 V, followed by sintering process at 800 °C under inert gas for 2 h. It was reported that the sample coated at 60 V reveals a uniform, dense coating accompanied with a higher corrosion resistance. After four weeks, nickel ions release was reduced to 0.205 $\mu g/cm^2$ for coated sample at 60 V. Marashi-Najafi [118] investigated corrosion resistance and in vitro evaluation of the pulsed current electrodeposited HA coatings on NiTi SMA. It was mentioned that the Ni release test showed that HA coating has the lowest amount of the released Ni in

Ringer's solution. The cell culture test revealed that the cell propagation and growth condition on HA-coated sample is better than bare and Ca–P-coated substrates. The corrosion behavior was examined in simulated body fluid (SBF) using Tafel polarization and EIS tests, which showed that HA-coated sample exhibited larger corrosion resistance in simulated body fluid; thereby, indicating that the stability of HA phase and the special morphology of HA coating are responsible for the better biological behavior of this coating. Qiu et al. [119] studied corrosion behavior of HA/zirconia composite coating on NiTi fabricated by electrochemical deposition. HA and HA/ZrO_2 composite coatings were directly electrodeposited on NiTi alloy surface. It was obtained that when ZrO_2 was added into the electrolyte, morphology of HAP was changed from thin flake–flower-like crystals to needle–flower-like crystals, and the coating was denser. The addition of ZrO_2 could improve the bonding strength between the coating and the substrate The corrosion resistance of NiTi in the SBF at 37 °C was significantly improved by almost 60 times by electrodeposition of the HA/ZrO_2 composite coating. Sheykholeslami et al. [120] investigated corrosion behavior of calcium phosphate coating electrodeposited on the modified nanoporous surface of NiTi alloy. The surface modification of the NiTi alloy was accomplished by anodizing in the ethylene glycol-based electrolyte and subsequent heat treatment at 550 °C for 30 min. It was reported that SEM images revealed that anodizing at 30 V led to the formation of a three-dimensional interconnected nanoporous TiO_2 layer on the surface of NiTi, and the existence of this oxide layer did not have a negative effect on the superelastic behavior of NiTi. Subsequently, this modified surface of samples was coated with Ca-P ceramic using the pulsed electrodeposition method. It was further reported that anodizing of the NiTi substrate before electrodeposition promoted the quality of the applied coating, electrodeposition at the higher current densities of 15 and 20 mA/cm^2 increased the possibility of the HA phase formation in the coating rather than the other less stable calcium phosphate phases. Both the TiO_2 layer and the Ca-P coating significantly improved the corrosion resistance of the NiTi alloy and suppressed the release of Ni ions from its surface. In comparison to the untreated NiTi, the mechanical locks between the nanoporous structure of the modified sample and the Ca–P coating increased the bonding strength.

Instead of forming TiN by chemical reaction between gaseous nitrogen ion and Ti element at the NiTi surface layer, there are several works for direct deposition of nitrides of Ti and Zr. Endo et al. [121] coated a titanium nitride film on a Ni–50Ti SMA by arc ion plating. The corrosion behavior of the titanium nitride-coated Ni–50Ti alloy was examined in 0.9% NaCl solution by potentiodynamic polarization measurements and a polarization resistance method. It was reported that XPS spectra showed that the titanium nitride film consisted of three layers: a top layer of TiO_2, a middle layer of TiN_x ($x > 1$), and an inner layer of TiN. The passive current density for the titanium nitride-coated alloy was approximately two orders of magnitude lower than that of the polished alloy in the potential range from the free corrosion potential E to +500 mV (vs Ag/AgCl). The pitting corrosion associated with breakdown of the coated film occurred above this potential, and the polarization resistance data also indicated that the corrosion rate of the titanium nitride-coated alloy at the free corrosion potential E (+50–+100 mV) was more than one order of magnitude lower than that for the polished alloy. Starosvetsky et al. [122] treated NiTi alloy to be nitrided using an original powder immersion reaction assisted coating (PIRAC) method to modify its surface properties. PIRAC nitriding method is based on annealing the samples in the atmosphere of highly reactive nitrogen supplied by decomposition of unstable nitride powders or, alternatively, by selective diffusion of the atmospheric nitrogen to the sample surface. Being a nonline-of-sight process, PIRAC nitriding allows uniform treatment of complex shape surgical implants. It was reported that hard two-layer titanium nitride (TiN)/Ti_2, Ni coatings were obtained on NiTi surface after PIRAC anneals at 900 and 1,000 °C, PIRAC-coating procedure was found to considerably improve the corrosion behavior of NiTi alloy in Ringer's solution, in contrast to untreated NiTi, no pitting was observed in the samples of PIRAC nitrided at 1,000 °C, 1 h up to 1.1 V. The coated samples were also characterized by very low-anodic currents in the passive region and by an exceedingly low-metal ion release rate; thereby, suggesting that PIRAC-nitriding procedure could improve the in vivo performance of NiTi alloys implanted into the human body. Krishnan et al. [123] compared the corrosion behavior of commercially available surface modified NiTi arch-wires with respect to a conventional NiTi and evaluated its association with surface characteristics. Five types of surface modified arch wires and a conventional NiTi arch wire (all from different manufacturers) were evaluated for their corrosion resistance from breakdown potential in an anodic polarization scan in Ringer's solution. It was reported that surface-modified NiTi wires showed significant improvement in corrosion resistance and reduction in surface roughness values, and breakdown potentials for passivation increased in the order of group 6 (conventional; 204 mV) < group 1 (nitride; 333 mV) < group 5 (epoxy resin; 346mV) < group 3 (oxide; 523 mV) < group 2 (gold; 872 mV) < group 4 (Teflon; 1181 mV), but root mean square (RMS) roughness values, which indicated surface roughness, followed a different pattern: group 3 (oxide; 74.12 nm) < group 1 (nitride; 221.651 nm) < group 4 (Teflon; 278.523 nm) < group 2 (gold; 317.894 nm) < group 5 (epoxy resin; 344.236 nm) < group 6 (conventional; 578.555 nm). It was therefore concluded that the surface modification of NiTi wires proved to be effective in improving its corrosion resistance and decreasing surface roughness; however, neither factor could maintain a direct, one-to-one relationship, indicating that the type and nature of coating material can effectively influence the anticorrosive features of NiTi wires, compared with its surface roughness values. Sugisawa et al. [124] evaluated the corrosion behaviors and mechanical properties of TiN coated SS and nickel titanium (Ni–Ti) orthodontic wires prepared by ion plating. It was obtained that TiN coating by ion plating improves the corrosion resistance of orthodontic wires. The corrosion pitting of the TiN coated wire surface become small. The tensile strength and stiffness of SS wire were increased after TiN coating, in contrast, its elastic force, which is a property for Ni–Ti wire, was decreased, TiN coating provided small friction forces and the low level of friction may increase tooth movement efficiently; hence, concluding that TiN-coated SS wire could be useful for orthodontics treatment.

Shao et al. [125] fabricated multilayer TiN/Ti coatings on the surface of NiTi alloy by filtered cathodic arc ion plating technique. It was mentioned that the coating consists of TiN/Ti multilayer with relatively wide transitional layer (Ti, N and O), and this coating can effectively improve the corrosion resistant property of NiTi alloy for the application of cardiac occluders. A titanium oxide–titanium (TiO$_2$–Ti) composite on NiTi alloy was coated using electrophoretic method, followed by heat-treating at 1,000 °C in two tube furnaces, the first one in argon atmosphere and the second one in nitrogen atmosphere at 1,000 °C [126]. The morphology and phase analysis of coatings were investigated using SEM and X-ray diffraction analysis, respectively. The electrochemical behavior of the NiTi and coated samples was examined using polarization and EIS tests. It was found that electrochemical tests in SBF demonstrated a considerable increase in corrosion resistance of composite-coated NiTi specimens compared to the non-coated one, and the heat-treated composite coating sample in nitrogen atmosphere had a higher level of corrosion resistance compared to the heat-treated sample in argon atmosphere, which is mainly due to having nitride phases. Roman et al. [127] deposited zirconium nitride (ZrN) thin films on NiTi and Si substrates in the 23–570 °C temperature range by direct current reactive magnetron sputtering using N$_2$/Ar gas mixture. It was found that the deposited films were composed mainly by the cubic ZrN phase, whose texture varies with substrate temperature, changing progressively from (111) to (200) texture as the temperature increases. The hardness of the films is influenced by the texture and has a linear relationship with the ratio of the texture coefficients P(111)/P(200). The higher hardness is obtained for ZrN thin films with (200) texture, and electrochemical tests show that NiTi coated with (200)-oriented ZrN films has higher tendency to passivation and greater stability of the passive film as compared to (111)-oriented ZrN films, despite no abrupt changes was observed when the texture changes from (111) to (200).

The anticorrosion property plays an important role in determining the effectiveness of the biocompatibility of metal implants. A functionally graded diamond-like carbon (DLC) coatings was deposited on the NiTi substrate by hybrid magnetron sputtering and plasma enhanced chemical vapor deposition [128]. A scratch test was adapted to study the adhesion strength of the coatings. It was reported that the Si/SiC graded layer (up to 150 nm thick) provided good adhesion between the coating and the substrate, up to approximately 47.17 ± 2.1 N. The effectiveness of corrosion protection for the coated specimen was investigated in Tyrode's SBF. Adhesion strength had a great influence on ness of corrosion protection, and the better adhesion strength, the better corrosion resistance. When compared to the others, the coating with a 150 nm thick Si/SiC graded layer provided better corrosion protection, and there existed no large amount of debonding and cracking of the coating around corrosion pits after the potentiodynamic polarization tests. Branzoi et al. [129] modified the surface of NiTi alloy by covering with DLC to improve the corrosion resistance and biocompatibility and prevented the toxic nickel ion release inside the human body. DLC films were deposited by d.c. magnetron sputtering using a graphite target and argon as the discharge gas. The electrochemical characteristics were investigated by electrochemical techniques (potentiodynamic polarization test and EIS) at 37 °C in the FBS. It was reported that in the case of uncoated samples, the release of metallic ions in bulk solution was much higher than in the case of DLC coated samples. The corrosion resistance and Vicker's microhardness of DLC-coated samples were improved in comparison with the uncoated samples. Huang et al. [130] coated DLC films on NiTi orthodontic archwires. The film protects against fluoride-induced corrosion and will improve orthodontic friction. The influence of a fluoride-containing environment on the surface topography and the friction force between the brackets and archwires was investigated. It was mentioned that the superior nature of the DLC coating, with less surface-roughness variation for DLC-coated archwires after immersion in a high-fluoride ion environment. We recognized that friction tests also showed that applying a DLC coating significantly decreased the fretting wear and the coefficient of friction, both in ambient air and artificial saliva; thereby, suggesting that DLC coatings are recommended to reduce fluoride-induced corrosion and improve orthodontic friction. Jamesh et al. [131] utilized hybrid cathodic arc/glow discharge plasma-assisted chemical vapor deposition to prepare graded films on Ti and NiTi. It was reported that XPS reveals the formation of a Ti and a carbon-graded interface between the substrate and carbon surface layer. The graded film significantly improves the corrosion resistance of Ti and NiTi in 154 mM NaCl as demonstrated by the smaller corrosion current density, larger resistance, and smaller capacitance. The graded film on NiTi shows near-capacitive impedance spectra and 14 times larger phase angle maxima at 10^5 Hz. The osteoblasts cells spread well on the graded film surface suggesting good cytocompatibility.

Wang et al. [132] employed two different surface modification techniques were used to change the surface morphology and roughness of stents at the micrometer level, and eventually improve their surface adhesion properties with respect to endothelial cells. One was chemical erosion followed by sol–gel TiO_2 coating, and the other was low temperature gas plasma deposition. It was reported that both techniques could change the surface morphology of the stents with microroughness. In comparison with the control, the treated NiTi alloy intravascular stents showed increased surface hydrophilicity and enhanced anticoagulation properties. However, the corrosion properties of the stents were not improved significantly. Sukuroglu et al. [133] deposited the TiO_2 coatings on the NiTi substrates and mentioned that the coated samples have higher corrosion resistance (against NaCl aqueous solution) than uncoated samples in the two different media.

Not only carbon element, but metallic Ti, Ni, Ta, and Zr elements have been implanted on NiTi surface. Li et al. [134] modified $TiNi_{49.8}$ (at%) alloy by Ti ion implantation to improve its corrosion resistance and biocompatibility. It was shown that Ti ion implantation caused the reduction of Ni concentration and the formation of a TiO_2 nanofilm on the TiNi alloy. The phase transformation temperatures of the Ti–TiNi alloy remained almost invariable after Ti ion implantation. The electrochemical tests indicated that the corrosion resistance of TiNi increased after Ti ion implantation. The Ni ion release rate in 0.9% NaCl solution for the TiNi alloy remarkably decreased due to the barrier effect of the TiO_2 nanofilm, and the cell proliferation behavior on Ti-implanted TiNi was better than that on the untreated TiNi after cell culture for 1 and 3 days. Fu et al. [135] prepared a Ni-coating on NiTi alloy by plasma nickelizing, at a low temperature of 450 °C, after a SMAT. It was mentioned that the coating consisted of Ni_3Ti and NiTi phases. The EIS measurements revealed the improved corrosion behavior of the samples after the nickelized SMAT Ti process. The oxide films formed on the coating were characterized via XPS and the complex titanium and nickel oxides provided excellent corrosion resistance to the 3.5% NaCl solution. Cheng et al. [136] coated element Ta on $TiNi_{50.6}$ (at%) alloy by an arc ion-plating method. It was mentioned that XPS survey and high- resolution spectra show that a thin oxide film with Ta_2O_5 in the outmost layer and tantalum suboxides in the inner layer are formed on the tantalum coating as a result of natural passivation of Ta in the atmosphere. The Ni ion release with time from uncoated and coated samples immersed in 0.9%NaCl solution was also investigated by atomic absorption spectrometry. When compared to the coated samples, the uncoated samples show a higher release rate that decreases slightly with time. The degree of dissolution for the coated sample is reduced from 0.28 to 0.74 $\mu g/cm^2$ after 49 days, implying the coating has a beneficial effect on the inhabitation of Ni ion release from the TiNi substrate. The atomic force microscope images and section analysis show that the root mean square roughness increases from 10.349 to 65.587 nm, and the maximum roughness (R_{MAX}) increases from 36.027 to 278.22 nm, confirming that immersion in the NaCl solution results in roughening of the coating surface. Zirconium ion implantation was performed on NiTi alloy to suppress Ni ion release as well as to improve corrosion resistance and cell-material interaction [137]. It was reported that a thicker Ni-depleted nanoscale composite layer formed after Zr implantation and the corrosion resistance was evidently increased in aspects of increased $E_B - E_{CORR}$ (difference between corrosion potential E and breakdown potential) and decreased corrosion current density. The coated NiTi sample possessed the highest $E_B - E_{CORR}$, more than 500 mV higher than that of untreated NiTi, suggesting a significant improvement on pitting corrosion resistance. Ni ion release rate of Zr–NiTi was decreased due to the depletion of Ni in the superficial surface layer and the diffusion resistance effect of the ZrO_2/TiO_2 nanofilm. The increased surface wettability induced by increased surface roughness was obtained after Zr implantation. The Zr–NiTi samples were found to be favorable to endothelial cells proliferation, especially after 5 and 7 days culture.

There are still surface modification using different materials. To improve the corrosion resistance and hemocompatibility of biomedical NiTi alloy, hydrophobic polymer coatings were deposited by plasma polymerization in the presence of a fluorine-containing precursor using an atmospheric-pressure plasma jet [138]. This process takes place at a low temperature in air and can be used to deposit fluoropolymer films using organic compounds that cannot be achieved by conventional polymerization techniques. It was found that the corrosion resistance of the coated NiTi alloy in physiological solutions including SBFs and Dulbecco's Modified Eagle's medium is evidently improved. The protein adsorption and platelet adhesion tests reveal that the adsorption ratio of albumin to fibrinogen is increased and the number of adherent platelets on the coating is greatly reduced. The plasma polymerized coating renders NiTi better in vitro hemocompatibility and is promising as a protective and hemocompatible coating on cardiovascular implants. Ahmed et al. [139] electrodeposited biocomposite consists of gold nanoparticles (AuNPs) and a natural polymer as chitosan (CS) on NiTi alloy to improve biocompatibility, biostability, surface corrosion resistance, and antibacterial effect for orthopedic implantation. It was mentioned that the nm-scale gold particles were embedded in the composite forming compact, thick, and smooth coat. The elemental analysis revealed significant less Ni ion release from the coated NiTi alloy compared with the uncoated one by 20-fold. The electrochemical corrosion measurements indicated that AuNPs/CS composite coat was effective for improving corrosion resistance in different immersion times and at all pH values, which suggests that the coated NiTi alloys have potential for orthopedic applications. The efficiencies of the biocomposite coats for inhibiting bacterial growth indicate high-antibacterial effect. Villermaux et al. [140] coated the plasma-polymerized tetra-fluoroethylene (PPFTE) coating to improve the corrosion resistance of NiTi plates and corresponding NiTi stables. It was reported that potentiodynamic tests in physiological Hank's solution show that PPTFE-coating improved the pitting corrosion resistance. The passivation range is increased from 35% to 96% compared to the untreated sample and the pit diameter is decreased from 100 microns to 10 microns, the uniformity of the deposited film is a very important parameter. When the film is damaged, the corrosion seems to increase in comparison to the untreated samples; otherwise, if the staple is carefully manipulated, the coating follows the large deformations induced by the memory effect of the alloy without cracking, and then, protects efficiently the staple from pitting.

There are several studies on effects of loading on corrosion behavior. Liu et al. [141] investigated the effect of bending stress on the corrosion of NiTi wires using potentiodynamic and potentiostatic tests in artificial saliva. It was indicated that bending stress induces a higher corrosion rate of NiTi wires in passive regions. The passive oxide film of specimens would be damaged under bending conditions. AES analysis showed a lower thickness of passive films on stressed NiTi wires compared with unstressed specimens in the passive region, and by SEM observation, localized corrosion was observed on stressed alloy specimens after a potentiodynamic test at pH 2; concluding that bending stress changed the corrosion properties and surface characteristics of NiTi wires in a simulated intra-oral environment. Liu et al. [142] investigated the influence of bending stress on the nickel release of commercial NiTi orthodontic wires in vitro, simulating the intraoral environment as realistically as possible. Two types of as-received orthodontic NiTi wires, free of performed internal stress, were immersed in artificial saliva. Half of the NiTi wires were exposed to continuous bending stress throughout the 14-day experimental period. It was found that the stressed NiTi wires exhibited substantial increases in the nickel release compared with the unstressed specimens during all experimental periods. The highest dissolution rate during the 0 to 1 day incubation period was observed for all stressed specimens; however, a slight increase of nickel released as a function of time was observed in the 3 groups of stressed specimens after 3 days of immersion. For the stressed specimens, it was hypothesized that the bending stress would induce buckling or cracking of the protective oxide film of the NiTi wires. It was then concluded that bending stress influences the nickel release of NiTi wires, and the factor of loading condition with respect to corrosion behavior and passive film should be considered in view of the widespread use of NiTi wires for dental devices.

There are important researches on the alloying effect on corrosion behavior of NiTi alloy. Iijima et al. [143] investigated alloying effects of Cr and Cu by potentiodynamic polarization measurement in 0.9% NaCl and 1% lactic acid solutions. It was reported that addition of 0.19 at% Cr had little effect on the structure of the oxide films and the corrosion resistance of the Ni–Ti alloys. For Ni–Ti–5Cu-0.3Cr alloy, the metallic Cu was enriched at the alloy/oxide film interface, resulting in increased susceptibility to pitting corrosion above +1,000 mV; however, the passive current density and the amount of released Ni were not significantly increased by the addition of Cu. Small amounts of Cr and Cu added to change the superelastic characteristics do not change the corrosion resistance of the NiTi alloy freely immersed in simulated physiological environments. Rondelli et al. [144] studied effect of Cu on localized corrosion resistance of Ni–Ti SMA. An electrochemical study using potentiodynamic and potentiostatic techniques was conducted in 0.9% NaCl on wires from equiatomic NiTi and ternary Ni44Ti51Cu5 superelastic alloys with Ti90Mo10 as a reference material. It was reported that the potentiostatic tests indicate that both NiTi and Cu–NiTi wires exhibit low corrosion potential E (approximately 50–150 mV vs SCE) inferior to that of TiMo alloy, which the latter proved to be immune to localized corrosion attacks up to 800 mV. Kassab et al. [145] examined effect of ternary element addition on the corrosion behavior of NiTi alloys in a physiological solution (0.9% NaCl) simulating a body temperature of 37 ± 1 °C and compared the corrosion behavior of selected ternary nickel titanium (NiTi)-based alloys ($Ni_{45}Ti_{50}Cu_5$, $Ni_{47}Ti_{50}Fe_3$, and $Ni_{39}Ti_{50}Pd_{11}$) with a binary, pseudoelastic $Ni_{50.7}Ti_{49.3}$ alloy. It was found that the localized corrosion resistance of these ternary alloys is lower than the binary NiTi alloy. By comparing the different NiTi-based alloys, the following relation has been proposed for their localized corrosion resistances: NiTiCu < NiTiFe < NiTiPd < NiTi, and depth profiling by XPS showed that the surface oxide film on all the investigated NiTi-based alloys is mainly of TiO_2; however, the NiTiPd and NiTiCu alloys showed metallic ternary element distributed within TiO_2 layer.

8.2.8 Galvanic corrosion

There are two types of galvanic cell: the first type is the microcells (or local cells) and the second type is the macrocells. As for microcells, within a single piece of metal or alloy there exist different regions of varying composition and hence different electrode potentials. This is true for different phases in heterogeneous alloys [146]. The positive way of applying these phenomena is by etching the polished surface of metal to reveal the microstructures for metallographic observations. A severely deformed portion within one piece of a material can serve as an anodic site [147]. For example, a portion of a fully annealed nail was bent, and the whole nail immersed into a NaCl aqueous solution. After several hours it was easily noticed that the localized plastic-deformed portion is rusted. For macrocells, which take place at the contact point of two dissimilar metals or alloys with different electrode potentials, the so-called galvanic corrosion (due to dissimilar metal coupling) is probably the best known of all the corrosion types [147]. Titanium is the metal choice for implant material due to excellent tissue compatibility; however, superstructures of such Ti implants are usually made of different alloys. This polymetallism leads to detectable (in macrolevel) galvanic corrosion. The intensity of the galvanism phenomenon is due to a number of factors such as the electrode potential, the extent of the polarization, the surface area ratio between anodic site and cathodic site, the distance between the electrodes, the surface state of the electrodes, the conductivity of the electrolyte, and the passage of a galvanic current on the electrolytic diffusion, stirring, aeration or deaeration, temperature, pH and the composition of the electrolytic milieu, coupling manner, and an accompanying crevice corrosion. When two solid bodies are in contact, it is fair to speculate that there could be marginal gaps between them because they are not subject to any bonding method like diffusion bonding or fusion welding. Therefore, when the galvanic corrosion behavior is studied in a dissimilar material couple, the crevice corrosion is another important localized corrosion measurement accompanied with galvanic corrosion [148–151].

Among these numerous influencing factors, the surface area ratio between anodic surface and cathodic surface appears to be the most important factor to be considered. If the ratio of surface cathode area/anode area (A_C/A_A) is large, the increased corrosion caused by coupling can be considerable. Conductivity of the electrolyte and geometry of the system enter the problem because only that part of the cathode area is effective for which resistance between anode and cathode is not a controlling factor [152].

Galvanic electrochemical corrosion was investigated in 0.9 wt% sodium chloride solution at 37 °C [153]. It was reported that when 316 L is coupled with NiTi, TMA, or stainless steel arch wire and was subjected to the immersion corrosion test, it was found that 316 L suffered from crevice corrosion. On the other hand, 2205 stainless steel did not show any localized crevice corrosion, although the surface of 2205 was covered with corrosion products, formed when coupled to NiTi and stainless steel wires; thereby, indicting that corrosion resistance. The 2205 duplex stainless steel is an improved alternative to 316 L for orthodontic bracket fabrication when used in conjunction with titanium, its alloys, or stainless steel arch wires. Iijima et al. [154] quantitatively assessed the galvanic corrosion behavior of orthodontic archwire alloys coupled to orthodontic bracket alloys in 0.9% NaCl solution, and studied the effect of surface area ratios. Two common bracket alloys, stainless steels and titanium, and four common wire alloys, NiTi alloy, β-Ti alloy, stainless steel, and Co-Cr-Ni were used. Three different surface area ratios, 1:1, 1:2.35, and 1:3.64, were used; two of them assumed that the multibracket appliances consists of 14 brackets and 0.016 inch of round archwire or 0.016 × 0.022 inch of rectangular archwire. It was reported that when the NiTi alloy was coupled with Ti (1:1, 1:2.35, and 1:3.64 of the surface area ratio) or β-Ti alloy was coupled with Ti (1:2.35 and 1:3.64 of the surface area ratio), Ti initially was the anode and corroded. However, the polarity reversed in 1 h, resulting in corrosion of the NiTi or β-Ti. The NiTi alloy coupled with SUS 304 or Ti exhibited a relatively large galvanic current density even after 72 h; thereby, suggesting that coupling SUS 304-NiTi and Ti–NiTi may remarkably accelerate the corrosion of NiTi alloy, which serves as the anode. The different anode-cathode area ratios used in this study had little effect on galvanic corrosion behavior, which does not agree with what was obtained by Oshida et al. [25]. Schiff et al. [33] investigated the influence of fluoride in certain mouthwashes on the risk of corrosion through galvanic coupling of orthodontic wires and brackets. Two titanium alloy wires, NiTi and Cu–NiTi, and the three most commonly used brackets, titanium (Ti), iron-chromium-nickel (FeCrNi) and cobalt-chromium (CoCr), were tested in a reference solution of Fusayama–Meyer artificial saliva and in two commercially available fluoride (250 ppm) mouthwashes, Elmex and Meridol. It was found that Meridol mouthwash, which contains stannous fluoride, was the solution in which the NiTi wires coupled with the different brackets showed the highest corrosion risk, whereas in Elmex mouthwash, which contains NaF, the CuNiTi wires presented the highest corrosion risk. Such corrosion has two consequences: deterioration in mechanical performance of the wire-bracket system, which would negatively affect the final esthetic result, and the risk of local allergic reactions caused by released Ni ions; suggesting that mouthwashes should be prescribed according to the orthodontic materials used. Siargos et al. [155] compared the galvanic coupling of conventional and metal injection molded (MIM) brackets with commonly used orthodontic archwires. Six of each type of brackets were suspended in lactic acid along with a sample of orthodontic wire (three NiTi and three Cu–NiTi) for 28 days at 37 °C. It was obtained that the MIMed brackets exhibited potential differences similar to those seen for the conventional brackets. The greatest potential difference was found for MIMed brackets with NiTi wires (512 mV), whereas MIMed brackets with Cu–NiTi wires had the smallest difference (115 mV). SEM-energy-dispersive spectroscopic analysis of the tie-wing area of each bracket type indicated similar elemental composition in both brackets, but in slightly different percentages by weight, and the MIMed bracket exhibited extensive internal porosity, whereas the conventional bracket was more solid internally. Based on these findings, it was concluded that the composition and manufacturing processes involved in fabricating MIMed brackets impart corrosive properties similar to those seen in the bracket-wing area of conventional brackets and may provide a measurable benefit when taking into account the increased corrosion between the bracket and brazing alloy of conventional brackets.

Darabara et al. [156] investigated the elemental composition, microstructure, hardness, corrosion properties, and ionic release of commercially available orthodontic brackets and Cu–NiTi archwires. Following the assessment of the elemental composition of the orthodontic wire (Copper Ni–Ti) and the six different brackets (Micro Loc, Equilibrium, OptiMESH(XRT), Gemini, Orthos2, and Rematitan), cyclic polarization curves were obtained for each material to estimate the susceptibility of each alloy to pitting corrosion in 1 M lactic acid. Galvanic corrosion between the orthodontic wire and each bracket took place in 1 M lactic acid for 28 days at 37 °C and then the ionic concentration of Ni and Cr was studied. It was reported that all tested wires and brackets with the exception of Gemini are not susceptible to pitting corrosion. In galvanic corrosion, following exposure for 28 days, the lowest potential difference (approximately 250 mV) appears for the orthodontic wire Cu–NiTi and the bracket is made up from pure titanium (Rematitan) or from the stainless steel AISI 316 grade (Micro Loc). Following completion of the galvanic corrosion experiments, measurable quantities of chromium and nickel ions were found in the residual lactic acid solution. Bakhtari et al. [157] compared the galvanic currents generated by different combinations of commonly used brackets and archwires. As-received stainless steel, NiTi, and β-Ti wires were coupled to stainless steel and titanium brackets in an artificial saliva medium. It was obtained that there was a significant difference in charge and galvanic currents when factored for type of bracket, but no significant difference between them when factored by type of wire. Specifically, a brazed stainless steel bracket was significantly greater in charge transferred and 10-h galvanic current than metal injection molded stainless steel and titanium brackets, which were not different from each other. It was therefore concluded that the method of bracket manufacturing might be of equal or more relevance to galvanic corrosion susceptibility than bracket composition. Heravi et al. [158] assessed the galvanic behavior of different bracket and archwire combinations that are commonly used in orthodontic treatments. Three types of orthodontic archwires with a diameter of 0.016 × 0.022 inch and 80 standard edgewise maxillary central incisor brackets were selected. Three groups consisted of different wire – bracket couples and one group was just brackets as a control group. Each group had five samples. Four brackets were then connected to each wire by elastic bands made from electrochemically neutral material. The samples were immersed into capped containers of Fusayama–Meyer artificial saliva. After six weeks, the released nickel ions were quantified via ion absorption technique. It was mentioned that the highest concentration of released nickel ions was for bracket+steel archwire and the least for the bracket without archwire. There were no significant differences among experimental groups, so it could be concluded that galvanic corrosion would not be a serious consideration through orthodontic treatment. Tahmasbi et al. [159] compared the galvanic corrosion of orthodontic wires and brackets from various manufacturers following exposure to a fluoride mouthwash. This study was conducted on 24 lower central incisor 0.022" Roth brackets of four different commercially available brands (Dentaurum, American Orthodontics, ORJ, Shinye). These brackets along with stainless steel (SS) or nickel-titanium (NiTi) orthodontic wires (0.016", round) were immersed in Oral-B mouthwash containing 0.05% NaF for 28 days. It was obtained that Cu ions released from specimens with NiTi wire were greater than those of samples containing SS wire. The ORJ brackets released more Cu ions than other samples. The Ni ions released from Shinye brackets were significantly more than those of other specimens. The corrosion rate of brackets coupled with NiTi wires was higher than that of brackets coupled with SS wires, and light and electron microscopic observations showed greater corrosion of ORJ brackets. Based on these findings, it was concluded that in fluoride mouthwash, Shinye and ORJ brackets exhibited greater corrosion than Dentaurum and American Orthodontics brackets. Stainless steel brackets used with NiTi wires showed greater corrosion and thus caution is recommended when using them. Furthermore, the galvanic corrosion of brackets manufactured by four different companies coupled with stainless steel (SS) or nickel-titanium (NiTi) wires in an artificial saliva solution was studied [160]. A total of 24 mandibular central incisor Roth brackets of four different manufacturers (American Orthodontics, Dentaurum, Shinye, ORJ) were used in this experimental study. These brackets were immersed in artificial saliva along with SS or NiTi orthodontic wires (0.016", round) for 28 days. The electric potential difference of each bracket/wire coupled with a saturated calomel reference electrode was measured via a voltmeter and recorded constantly. It was found that among ions evaluated, release of nickel ions from Shinye brackets was significantly higher than that of other brackets. The mean potential difference was significantly lower in specimens containing a couple of Shinye brackets and SS wire compared with other specimens, no significant difference was observed in the mean CR of various groups. Microscopic evaluation showed corrosion in two samples only: Shinye bracket coupled with SS wire and American Orthodontics bracket coupled with NiTi wire; thereby, indicating that Shinye brackets coupled with SS wire showed more susceptibility to galvanic corrosion. There were no significant differences among specimens in terms of the CR or released ions except the release of Ni ions, which was higher in Shinye brackets. Polychronis et al. [161] assessed galvanic behavior of lingual orthodontic brackets coupled with representative types of orthodontic wires. Three types of lingual brackets: Incognito (INC), In-Ovation L (IOV), and STb (STB) were combined with a stainless steel (SS) and a NiTi orthodontic archwire. All bracket-wire combinations were immersed in acidic 0.1 M NaCl 0.1 M lactic acid and neutral NaF 0.3% (wt) electrolyte, and the potential differences were continuously recorded for 48 h. It was found that SEM/EDX analysis revealed that INC is a single-unit bracket made of a high gold (Au) alloy while IOV and

STB are two-piece appliances, in which the base and wing are made of SS alloys. The SS wire demonstrated austenite and martensite iron phase, while NiTi wire illustrated an intense austenite crystallographic structure with limited martensite. All bracket wire combinations showed potential differences below the threshold of galvanic corrosion (200 mV) except for INC and STB coupled with NiTi wire in NaF media. Hence, it was concluded that the electrochemical results indicate that all brackets tested demonstrated galvanic compatibility with SS wire, but fluoride treatment should be used cautiously with NiTi wires coupled with Au and SS brackets.

8.2.9 Microbiology-induced corrosion

Microbiology-induced corrosion (MIC or biocorrosion) is defined as the deterioration and degradation of metallic structures and devices by corrosion processes that occur directly or indirectly as a result of the activity of living microorganisms, which either produce aggressive metabolites, or are able to participate directly in the electrochemical reactions occurring on the metal surface. MIC occurs in aquatic habitats varying in nutrient content, temperature, stress and pH. The oral environment of organisms, including humans, should be one of the most hospitable for MIC. Corrosion has long been thought to occur by one of three means: oxidation, dissolution, or electrochemical interaction. To this list the newly discovered microbiological corrosion must be added [162, 163]. Oxidation is somewhat beneficial in the case of stainless steel (mainly the Cr element) and titanium because they create a nonporous layer of passive oxide film that protects the substrate surface from further oxidation; Cr_2O_3 or $(Fe,Ni)O\cdot(Fe,Cr)_2O_3$ for stainless steel and TiO_2 for titanium materials. However, dissolution and electrochemical reactions are responsible for most of the detrimental corrosion. MIC has received increased attention by corrosion scientists and engineers in recent years. MIC is due to the presence of microorganisms on a metal surface, which leads to changes in the rates and, sometimes, also the types of the electrochemical reactions, which are, involved in the corrosion processes. It is, therefore, not surprising that many attempts have been made to use electrochemical techniques (corrosion potential E, the redox potential, the polarization resistance, the electrochemical impedance, electrochemical noise, and polarization curves including pitting) to study the details of MIC and determine its mechanism. Applications range from studies of the corrosion of steel pipes in the presence of sulfate-reducing bacteria to investigating the formation of biofilms and calcareous deposits on stainless steel in seawater, to the destruction of concrete pipes in sewers by microorganisms producing very low pH solutions, and to dental prostheses exposed to various types of bacteria intraorally [164, 165]. Study on MIC is a typical interdisciplinary subject that requires at least some understanding in the fields of chemistry, electrochemistry, metallurgy, microbiology, and biochemistry. In many cases, microbial corrosion is closely associated with biofouling phenomena, which are caused by the activity of organisms that produce deposits of gelatinous slime, or biogenically induced corrosion debris in aqueous systems. Familiar examples of this problem are the growth of algae in cooling towers, and barnacles, mussels and seaweed on marine structures. These growths either produce aggressive metabolites or create microhabitats suitable for the proliferation of other bacterial species, such as anaerobic conditions favoring the well-known sulfate-reducing bacteria. In addition, the presence of growths or deposits on a metal surface encourage the formation of differential aeration or concentration cells between the deposit and the surrounding environment, which might stimulate existing corrosion processes [164]. However, it is probable that microbiological corrosion rarely occurs as an isolated phenomenon, but is coupled with some type of electrochemical corrosion. For example, corrosion induced by sulfate-reducing bacteria usually is complicated by the chemical action of sulfides. Corrosion by an aerobic organism, by definition, always occurs in the presence of oxygen. To further compound the problem, microbiocidal chemicals may also have some conventional properties, such as corrosion inhibitors (film-forming). The materials on which a biofilm is developed have higher, or nobler, E_{CORR} (potential at corrosion current) values than those obtained in the absence of biofilms [166].

Microbes seem to accelerate the corrosion process. They locate susceptible areas, fix anodic sites, and produce or accumulate chemical species that promote corrosion. The microscopic heterogeneity of many materials – whether created intentionally or as an artifact – is quite clear on the scale of microbes and is an important and overlooked factor in microbiologically influenced corrosion. Weld regions are particularly attractive to microbes in many of the systems [167], since the weld line normally exhibits slight differences in chemical compositions as well as microstructures, resulting in a possibility for galvanic corrosion occurrence. Microbes are extremely tenacious. They can exist over a wide range of temperatures and chemical conditions, they are prolific, and they can exist in large colonies. They form synergistic communities with other microbes or higher life forms and accomplish remarkably complex chemical reactions in consort. Microbes can metabolize metals directly, and they require many of the chemical species produced by the physical chemistry of corrosion processes for their metabolisms. Microbes can also act to depolarize anodic or cathodic reactions and, in addition to catalyzing existing corrosion mechanisms, many microbes produce organic or mineral acids as metabolic products [167]. Accordingly, it is difficult to specify just what the important variables are when corrosion is influenced by microorganisms. In addition to the variables most important for the type of corrosion under consideration, variables such as dissolved oxygen, pH, temperature, and nutrients, which affect the life cycle of the bacteria, become important for both corrosion testing and corrosion mechanism. The critical situation involved in MIC is limited to the interfacial thin layer (usually with thickness in the 10 to 500 μm range) between the biofilm and metallic surface, since the chemistry of the electrolyte solution at the metal surface is ever-changing. Therefore, the bulk electrolyte properties may have little relevance to the corrosion as influenced by organisms within the film. The organism right at the metal surface influences corrosion activity, and those organisms multiply so rapidly on the surface that a low density of organisms in the bulk quickly becomes irrelevant [168].

Bacteria must adhere, spread, and proliferate on metallic surfaces in order to cause corrosive reactions. The process of bacterial adhesion is mediated by both an initial physiochemical interaction phase (i.e., electrostatic forces and hydrophobicity) and specific mechanisms (i.e., adhesion-receptor interactions, which allows bacteria to bind selectively to the surface), followed by a subsequent molecular and cellular interaction phase, which is normally followed by a plaque formation in dentistry [169]. The molecular and cellular interaction are complicated processes affected by many factors, including the characteristics of the bacteria themselves, the target material surface, and environmental factors such as the presence of serum proteins or bacterial substances. Microbial adhesion to surfaces is a common phenomenon. Many bacteria demonstrate a preference for the surface regions of solids, teeth, plant fibers, roots and so on, where nutrients may be concentrated by the adsorption of dissolved organic material. Since microorganism acts as colloidal particles, their interaction with solid surfaces can be anticipated, based on colloidal theory. However, microorganisms are more complex than typical colloids as they are capable of independent locomotion, growth in different shapes, and the production of extracellular polymeric materials that aid in anchoring the microbe to the surface [169]. An et al. [170] mentioned that adhesion is a situation where bacteria adhere firmly to a surface by complete physiochemical interactions between them, including an initial phase of reversible physical contact, and a time-dependent phase of irreversible chemical and cellular adherence. Bacteria and other microorganisms have a natural tendency to adhere to surfaces as a survival mechanism. This can occur in many environments including the living host, industrial systems, and natural waters. The general outcome of bacterial colonization of surfaces is the formation of an adherent layer (biofilm) composed of bacteria embedded in an organic matrix [171]. The starvation and growth phase influence bacterial cell surface hydrophobicity. Both the number and kind of microorganisms that colonized metal surfaces depend on the type of metal and the presence of an imposed electrical potential. No significant differences in attachment and growth of a pure culture were observed when metal surfaces were dipped in an exogenous energy source. The chemical composition of naturally occurring adsorbed organic films on metal surfaces was shown to be independent of surface composition and polarization [172].

Chang et al. [173] investigated electrochemical behaviors of CpTi, Ti–6Al–4V, Ti–Ni, Co–Cr–Mo, 316L stainless steel, 17-4 pH stainless steel, and Ni–Cr in (1) sterilized Ringer's solution, (2) *S. mutans* mixed with sterilized Ringer's solution, (3) sterilized tryptic soy broth, and (4) byproduct of *S. mutans* mixed with sterilized tryptic soy broth. It was concluded that among the four electrolytes, the byproduct mixed with sterilized tryptic soy broth was the most corrosive media, leading to an increase in I_{CORR} and reduction in E_{CORR}. The CpTi, Co–Cr–Mo, and stainless steel increased their I_{CORR} significantly, but Ti–6Al–4V and Ti–Ni did not show remarkable increase in I_{CORR} [173].

Koh et al. [149] studied the effect of surface area ratios on bacteria galvanic corrosion of commercially pure titanium coupled with other dental alloys. CpTi was coupled with a more noble metal (type IV dental gold alloy) and a less noble metal (Ni–Cr alloy) with different surface area ratios (ranging from 4:0 to 0:4 – in other words, 4:0 uncouple indicates that entire surface area of a noble metal was masked, while 0:4 uncouple indicates that only CpTi surface was masked, and area ratios between 4:0 and 0:4 indicate that both surfaces were partially masked to produce different surface areas). Galvanic couples were then electrochemically tested in bacteria culture media and culture media containing bacterial byproducts. It was found that I_{CORR} (and hence, corrosion rate) profiles as a function of surface area ratio showed a straight-line trend (meaning surface area ratio independence) for Ti/Au couples, whereas the representative curve for the Ti/Ni–Cr alloy was bowl-shaped. The latter curve indicates further higher corrosion rates at both ends (larger area of either CpTi or Ni–Cr), meaning that lowest corrosion rate was exhibited at the bottom of bowl (at equal surface area ratio) [149], supporting results obtained by Al-Ali et al. [25] and Garcia [148].

In the study [174], an inductively coupled plasma-optical emission spectrometer revealed elution of Fe, Cr, and Ni from stainless steel (SS) appliances incubated with oral bacteria. It was reported that three-dimensional laser confocal microscopy also revealed that oral bacterial culture promoted increased surface roughness and corrosion pits in SS appliances. The pH of the supernatant was lowered after co-culture of appliances and oral bacteria in any combinations, but not reached at the level of depassivation pH of their metallic materials, and *Streptococcus mutans* and *S. sanguinis*, which easily created biofilm on the surfaces of teeth and appliances, did corrode orthodontic SUS appliances. Rocher et al. [175] examined five alloys were examined by electrochemical assays and cell culture tests with different cell types and mentioned that all tests show the high biological and electrochemical performances of Ti6Al4V and NiTi, and in particular a significant influence of living cells on corrosion. Bahije et al. [176] investigated the electrochemical behavior of NiTi orthodontic wires in a solution containing *S. mutans* oral bacteria. The electrochemical behavior of the alloy (NiTi) was analyzed electrochemically in Ringer's sterile artificial saliva and in artificial saliva enriched with a sterile broth and modified by addition of bacteria. It was found that colonization of the metal surface by bacteria triggered a drop in the free corrosion potential E, the electrochemical impedance findings revealed no significant difference in NiTi behavior between the two media, and there was a slight difference between the two corrosion currents in favor of the bacteria-enriched solution, in which the NiTi underwent greater corrosion.

8.2.10 Localized corrosion and stress-assisted corrosion

General corrosion is the most common form of corrosion and is also called uniform corrosion where the electrochemical reactions (in aqueous or atmospheric media) proceed uniformly over the entire exposed metal surface over a large area. Hence, the corrosion margin can be reflected to design calculation (in particular, dimensions of crucial components or parts). So far, we have been discussing the corrosion in a sense of uniform attack. In reality, however there are numerous cases when the corrosive attack is localized after a short period of time for general and uniform attack. Localized corrosion (in contrast to general corrosion) is observed in small local areas and can be classified as one of types: pitting (pitting results when a small hole) or cavity (forms in the metal and usually as a result of depassivation of a small area). The occurrence of localized corrosion is a manifest proof that the anodic surface area can be much smaller than the cathodic; namely, the S_A/S_C ratio (or degree of localization) can be an important driving force of all localized corrosion problems since a corrosion situation corresponds to equal anodic and cathodic absolutes currents. As a result, there is an intense attack at localized sites on the surface of a component. Indeed, localized corrosion rates are often several orders of magnitude higher than the corrosion rates for general corrosion. Pitting corrosion and crevice corrosion can be distinguished as follows; if this event occurs under a deposit on the surface (perhaps a weld deposit or some solid debris from the environment) or at the joint of a bolted assembly and so on, the attack is termed, "crevice corrosion." If the attack initiates on the free surface of a component, it is termed "pitting." The resistance to these two types of localized corrosion varies greatly between different materials and is extremely dependent upon environmental factors.

Schulte et al. [177–179] combined two different electrochemical local probe techniques for examining the phenomena of localized (pitting) corrosion on the surface of passive metals and alloys: (1) alternating current scanning electrochemical microscopy (AC-SECM) for imaging, with high-spatial resolution, microscopically small heterogeneities in surface conductivity, and (2) SECM-integrated stripping voltammetry (SECM-SV) for quantifying/visualizing the local release of trace amount of metal ions at areas where passivity broke down and pitting corrosion started to appear. It was mentioned that alternating current AC-SECM has been used as an approach to detect inhomogeneities in the passive oxide layer of nitinol with high-spatial resolution. The appearance of microscopic corroding spots on the surface can be imaged due to the decreasing solution resistance between the sample and the used scanning tip. These spots bear a high reactivity toward the onset of pitting corrosion.

Nishimoto et al. [180] investigated the effects of the solution temperature, phase, elastic deformation, and the phase transformation on the inhibition of localized corrosion of a small amount of hydrogen-charged NiTi superelastic alloy in 0.9% NaCl solution by electrochemical measurements. It was mentioned that no pitting potentials are observed in the anodic polarization curves obtained above room temperature and under static applied stress in the presence of parent or stress-induced martensite phases. In a dynamic cyclic tensile test, in the elastic deformation region of the parent phase under various constant-applied anodic potentials, the critical applied potential for an increase in the current density, which corresponds to the occurrence of localized corrosion, considerably shifts in the noble direction at the stress plateau. For example, in the superelastic region, caused by stress-induced martensite and reverse transformations, the critical-applied potential for a marked increase in the current density slightly shifts in the noble direction. In the elastic deformation region of the stress-induced martensite phase, the critical applied potential for a marked increase in the current density shifts in the noble direction, thereby indicating that hydrogen charging is effective for inhibiting the localized corrosion of NiTi superelastic alloy in NaCl solution at various temperatures, irrespective of the phase itself, under dynamic elastic deformation and the phase transformation.

8.2.10.1 Pitting corrosion

The corrosion behavior of NiTi thin films fabricated by sputtering from Ni and Ti targets has been studied by cyclic potentiodynamic polarization tests in Hank's and Ringer's solution at 37 °C [181]. For comparison, bulk NiTi has also been studied to elucidate the different corrosion behavior of bulk and thin film material. It was found that the electrochemical experiments reveal that thin film NiTi has comparable corrosion current density, much higher pitting corrosion potential E and wider passive range than the bulk NiTi and NiTi vapor deposited thin films are less susceptible to pitting corrosion than the bulk. Bai et al. [182] evaluated the pitting corrosion property of nitinol, titanium, nickel, and 316 L stainless steel (316LSS) wires with different surface roughnesses in a saline solution at 37 °C. It was obtained that the cyclic potentiodynamic polarization results show that mechanically polished nitinol and Ti wires are highly resistant to pitting corrosion, while Ni and 316LSS wires are susceptible to pitting corrosion. EIS is used to study the interface of oxide film/solution and all mechanically polished nitinol wires are covered by 2–3 nm thick films formed under OCP. The electronic structures and semiconducting properties of passive films on nitinol, Ti, and Ni wires are studied by Mott-Schottky analysis and the passive films formed on nitinol and Ti exhibit n-type semiconducting characteristics, whereas films on Ni show p-type semiconducting characteristics.

Rondelli et al. [183] investigated localized corrosion behavior in simulated human body fluids of commercial NiTi orthodontic wires in artificial saliva at 40 °C. For comparison purposes, wires made of stainless steel and of cobalt-based alloy were also examined. It was reported that potentiodynamic tests in artificial saliva at 40 °C indicated a sufficient pitting resistance for the NiTi wires, similar to that of cobalt-based alloy wire; the stainless steel wire, instead, exhibited low-pitting potential. The potentiodynamic tests at 40 °C in isotonic saline solution (0.9% NaCl) showed that for NiTi and stainless steel wires, pitting potential values are in the range about 200 to 400 mV and about 350 mV versus SCE, respectively; consequently, these materials should be considered potentially susceptible to pitting; only the cobalt-based alloy should be immune from pitting. The localized corrosion potential E determined in the same environment (from 0 to 200 mV and 130 mV vs SCE for NiTi and stainless steel, respectively) pointed out that for these materials an even higher risk of localized corrosion, and slight differences in localized corrosion behavior among the various NiTi wires were detected. Kararia et al. [184] compared standard 3 M Unitek NiTi and stainless steel archwires with locally available JJ orthodontics wires. Ten archwires each of group 1-3 M 0.016" NiTi, group 2-JJ 0.016" NiTi, group 3-3 M 0.019" *0.025" SS and group 4-JJ SS contributed a 10 mm piece of wire for analysis prior to insertion in the patient and 6 weeks postinsertion. SEM images were recorded at ×2,000, ×4,000 and ×6,000 magnification. It was mentioned that the SEM images of all the archwires showed marked changes with deep scratches and grooves and dark pitting corrosion areas postintraoral use. 3M wires showed an uniform criss-cross pattern in as-received wires indicating a coating which was absent after intraoral use. There was a significant release of Ni and Cr elements from both group 3 and 4. Group 2 wires released ions significantly more than group.

Darabara et al. [185] evaluated the pitting and crevice corrosion characteristics of stainless steel (SS) and NiTi endodontic files in R-EDTA and NaOCl irrigating solutions. The corrosion behavior of two H-files produced from different SS alloys (Mani, AISI 303 SS, Dentsply Maillefer, AISI 304 SS) and one file produced from NiTi alloy (Maillefer) was determined in R-EDTA and NaOCl irrigating solutions at 37 °C by the cyclic potentiodynamic polarization method. Corrosion potential, corrosion current density (I_{CORR}) and pitting potential (E_P) were calculated from each curve. It was reported that cyclic polarization curves presented negative hysteresis implying that pitting or crevice corrosion are not likely to occur for all the materials examined in both irrigating solutions. In NaOCl, all materials showed significantly higher E_{CORR} as well as lower I_{CORR} compared with R-EDTA reagent, and all materials demonstrated equal E_P in NaOCl, which was to be found significantly lower than the value of E_P in R-EDTA. Based on these findings, it was concluded that none of the tested materials is susceptible to pitting or crevice corrosion in R-EDTA and NaOCl solutions and from this stand point are appropriate for the production of endodontic files.

8.2.10.2 Stress corrosion cracking

SCC is the cracking induced from the specific combinations of susceptible materials in suitable corrosive environments under adequate stress, specifically under constant tensile stress and a corrosive environment. Hence, SCC exhibits a synergistic effect of stress and corrosion. The impact of SCC on a material usually falls between dry cracking and the fatigue threshold of that material. The required tensile stresses may be in the form of directly applied stresses or in the form of residual stresses.

Wang et al. [186] studied the mechanism of the cracking of orthodontic NiTi wire. Two orthodontic NiTi wires were subjected: (1) optical and SEM to observe the fracture surface; (2) EDX determine the composition of the surface product; (3) anodic polarization to remove the surface product. Samples of NiTi alloy were subjected to the constant loading test to study the SCC behavior of NiTi SMA in artificial saliva. It was found that there were three typical areas at the fracture surface of NiTi orthodontic wire; Area "1" was a tool-made notch, crack initiated from the root of this notch and propagated to form Area "2", which was perpendicular to the wire axis and covered by surface film that consisted of Na, K, Cl, P, S, and O, except Ni and Ti. The cracking process of NiTi alloy under the constant loading test depended on the pH of saliva and applied stress, and the crack length was about 262 μm, the longest at 300 MPa and pH 3.0; indicating that a tool-made notch in orthodontic NiTi wires can cause SCC, and at high stress and low pH, this NiTi alloy was most sensitive to cracking. Zhang et al. [187] studied the corrosion resistance of laser-welded composite arch wire (CoAW) with Cu interlayer between NiTi SMA and stainless-steel wire in artificial saliva with different acidities and loads. It was reported that both the solution pH and the stress had a significant influence on the corrosion behaviors of the CoAW samples, decreasing the solution pH, or increasing the loading stress caused the increase of Cu release and weight loss, and the corroded morphology formed on the surfaces of the CoAW was the consequence under the combined effect of corrosion and stress. Liu et al. [188] developed titanium nitride (TiN)/titanium (Ti) coating on orthodontic NiTi wires and studied the stress corrosion of specimens in vitro, simulating the intraoral environment in as realistic a manner as possible. TiN/Ti coatings were formed on orthodontic NiTi wires by physical vapor deposition (PVD). The characteristics of untreated and TiN/Ti-coated NiTi wires were evaluated by measurement of corrosion potential E, corrosion current densities, breakdown potential, and surface morphology in artificial saliva with different pH and three-point bending conditions. It was reported that the untreated NiTi wires showed localized corrosion compared with the uniform corrosion observed in the TiN/Ti-coated specimen under both unstressed and stressed conditions. The bending stress influenced the corrosion current density and breakdown potential of untreated specimens at both pH 2 and pH 5.3. Although the bending stress influenced the corrosion current of the TiN/Ti-coated specimens, stable and passive corrosion behavior of the stressed specimen was observed even at 2.0 V (Ag/AgCl). It should be noted that the surface properties of the NiTi alloy could determine clinical performance. For orthodontic application, the mechanical damage destroys the protective oxide film of NiTi; however, the self-repairing capacity of the passive film of NiTi alloys is inferior to Ti in chloride-containing solutions. The TiN coating was found able to provide protection against mechanical damage, while the Ti interlayer improved the corrosion properties in an aggressive environment.

8.2.10.3 Reaction with hydrogen

It is generally agreed that the solubility of hydrogen in α-phase titanium and titanium alloys is quite low, namely approximately 20–2,000 ppm at room temperature [189]. However, severe problems can arise when Ti-based alloys pick up large amounts of hydrogen, especially at elevated temperatures [190–194]. The strong stabilizing effect of hydrogen on the β phase field results in a decrease of the (HCP crystalline structure) → β (BCC crystalline structure) transformation temperature from 885 °C to a eutectoid temperature of 300 °C [190]. Hydrogen absorption from the surrounding environment leads to degradation of the mechanical properties of the material [195]. This phenomenon has been referred to as hydrogen embrittlement (HE) over the past years. Hydrogen absorption often becomes a problem for high-strength steels even in air [196], and it also occurs for Ti in methanol solutions containing hydrochloric acid [197–201]. Hydrogen damage of Ti and Ti-based alloys is manifested as a loss of ductility (embrittlement) and/or reduction in the stress-intensity threshold for crack propagation [189]. CpTi is very resistant to HE, but it becomes susceptible to HE under an existence of notch, at low temperature, high-strain rates, or large grain sizes [202]. Eventually, all these thermal and mechanical situations make the material more brittle in nature. For α and near α Ti material, it is believed that such HE occurs due to precipitation of brittle hydride phases [203]. For α+β phase alloys, when a significant amount of β phase is present, hydrogen can be preferentially transported with the β lattice, and will react with α phase at α/β boundaries. Under this condition, degradation will generally become severe [204]. For β alloys, HE was observed to occur in the Ti–Mo–Nb–Al [205], Ti–V–Cr–Al–Sn [206], and Ti–V–Fe–Al alloys [207], due to δ-hydride phase formation, which is brittle at low temperatures [189].

There are some processes by which hydrogen could develop, and possibly be absorbed, under passive conditions: direct absorption of hydrogen produced by water hydrolysis and absorption of atomic hydrogen produced by the corrosion processes to produce an oxide. Under anodic conditions, when passive corrosion prevails, the corrosion potential E for passive titanium must rise at a value at which water reduction can cause titanium to oxidize, $Ti + 2H_2O \rightarrow TiO_2 + 2H_2$. Since Ti hydrides are thermodynamically stable at these potentials, the passive film can only be considered as a transport barrier, and not an absolute barrier. The rate of hydrogen absorption will be controlled by the rate of corrosion reaction, which dictates the rate of production of absorbable hydrogen. Since TiO_2 is extremely insoluble, the corrosion reaction will be effectively limited to an oxide film growth reaction [208]. Titanium corrosion processes are often accompanied with hydrogen production, introducing the possibility of the absorption of hydrogen into the material. Crevice corrosion, once initiated, is supported by both the reduction of oxygen on passive surfaces external to the crevice and the reduction of protons ($Ti + 4H^+ \rightarrow Ti^{4+} + 2H_2$) inside the crevice, leading to the absorption of atomic hydrogen in sufficient quantities to produce extensive hydride formation. For passive noncreviced or inert crevice conditions, corrosion could be sustained by reaction with water under neutral conditions ($Ti + 2H_2O \rightarrow TiO_2 + 2H_2$) and should proceed at an extremely slow rate. In the first step to possible failure by hydrogen-induced cracking, the hydrogen generated must pass through the TiO_2 film before absorption into the underlying titanium alloy. For absorption to proceed, redox transformation ($Ti^{4+} \rightarrow Ti^{3+}$) in the oxide film is necessary. This requires significant cathodic polarization of the metal, generally achievable by galvanic coupling to active materials (such as carbon steel), or application of cathodic protection potential [209].

8.2.10.4 Hydrogen embrittlement

Hydrogen embrittlement (HE) also known as hydrogen-assisted cracking (HAC) and hydrogen-induced cracking (HIC) describes the embrittling of metal after being exposed to hydrogen. It is a complex process that is not completely understood because of the variety and complexity of mechanisms that can lead to embrittlement. Mechanisms that have been proposed to explain embrittlement include the formation of brittle hydrides, the creation of voids that can lead to bubbles and pressure buildup within a material and enhanced decohesion or localized plasticity that assist in the propagation of cracks. Susceptibility to HE is often a result of the introduction of hydrogen during forming, coating, plating, cleaning, and finishing operations, often referred to as 'internal embrittlement'. Hydrogen also may be introduced over time ("external embrittlement") through environmental exposure (soils and chemicals, including water), corrosion processes (especially galvanic corrosion), cathodic protection, and/or from hydrogen generated by corrosion of a coating.

Superelastic Ti–Ni wire is widely used in orthodontic clinics, but delayed fracture in the oral cavity has been observed. Because HE is known to cause damage to Ti alloy systems, Yokoyama et al. [210] treated orthodontic wires cathodically in 0.9% NaCl solution for charging hydrogen, followed by conducting tensile tests and fractographic observations. It was reported that the strength of the Co–Cr alloy and stainless steel used in orthodontic mechanotherapy was not affected by the hydrogen charging; however, Ti–Ni wire showed significant decreases in its strength, and the fractured surface of the alloy with severe hydrogen charging exhibited dimple patterns, similar to those in the alloys from patients. In view of the galvanic current in the mouth, the fracture of the NiTi alloy might be attributed to the degradation of the mechanical properties due to hydrogen absorption [209]. HIC of NiTi was also found when it was tested in acidic solutions, and fluoride-containing solutions such as APF [211, 212]. However, it was suggested that the work-hardening Ti–Ni alloy was less sensitive to HE compared with Ti–Ni superelastic alloy [208]. Titanium reaction with hydrogen is not necessarily a useless phenomenon. The formation of titanium hydrides can provide an engineering advantage. It is well known that titanium is one of the most promising materials for hydrogen storage due to a stable formation of nonstoichiometric titanium hydrides (δ, ε, or γ type) over the composition range from $TiH_{1.53}$ to $TiH_{1.99}$ [197].

Asaoka et al. [213] treated NiTi alloys to hydrogen charging of 1 or 10 A/m^2 for 24 or 120 h, respectively. Theoretical distributions of the hydrogen concentration were computed for an infinite cylinder model using the differential equation of diffusion. It was reported that the diffusion constant of hydrogen through the alloy was estimated to be 9×10^{-15} m^2/s, assuming that the hardness is proportional to the concentration of hydride and/or hydrogen. The galvanic currents and fretting corrosion of the alloy might be effective factors in fracture formation during function. Li et al. [214] investigated the effect of substitution of Nb by Mo in $Nb_{40}Ti_{30}Ni_{30}$ with respect to microstructural features and hydrogen dissolution, diffusion and permeation.

As-cast $Nb_{40-x}Mo_xTi_{30}Ni_{30}$ ($x = 0$, 5, 10) alloys consist of primary BCC-Nb phase and binary eutectic (BCC-Nb + B2-TiNi). It was mentioned that the substitution of Nb by Mo reduces the hydrogen solubility in alloys, but may increase ($x = 5$) or decrease ($x = 10$) the apparent hydrogen diffusivity and permeability. As-cast $Nb_{35}Mo_5Ti_{30}Ni_{30}$ exhibits a combined enhancement of hydrogen permeability and embrittlement resistance as compared to $Nb_{40}Ti_{30}Ni_{30}$. Mo is a desirable alloying element in Nb that can contribute to a reduction in hydrogen absorption and an increase in intrinsic hydrogen diffusion, thus improving embrittlement resistance with minimal permeability penalty.

8.2.10.5 Corrosion fatigue

Corrosion fatigue is the result of the combined action of an alternating or cycling stresses and a corrosive environment. The fatigue process is thought to cause rupture of the protective passive film, upon which corrosion is accelerated. If the metal is simultaneously exposed to a corrosive environment, the failure can take place at even lower loads and after shorter time. In a corrosive environment the stress level at which it could be assumed a material has infinite life is lowered or removed completely. Contrary to a pure mechanical fatigue, there is no fatigue limit load in corrosion-assisted fatigue. Corrosion fatigue and fretting are both in this class. Much lower failure stresses and much shorter failure times can occur in a corrosive environment compared to the situation where the alternating stress is in a noncorrosive environment. Fractographically, the fatigue fracture is brittle nature and the cracks are most often transgranular, as in SCC.

Pedullà et al. [215] assessed resistance to cyclic fatigue of reciprocating NiTi files (Reciproc and WaveOne) after immersion in NaOCl solution over several time periods. A total of 90 new Reciproc R25 and WaveOne Primary files were tested, which were immersed in 5% NaOCl at 37 °C for 16 mm: no immersion (control), 1 or 5 min dynamically. Resistance to cyclic fatigue was determined by recording time to fracture in a stainless steel artificial canal with a 60° angle of curvature and 5 mm radius of curvature. It was obtained that resistance to cyclic fatigue of the same NiTi file was not significantly affected by immersion in NaOCl. Reciproc R25 was associated with a higher cyclic fatigue resistance in all groups compared to WaveOne Primary; hence, concluding that reciprocating dynamic immersion in NaOCl for 1 or 5 min did not reduce the cyclic fatigue resistance of NiTi files significantly; however, the type of reciprocating instrument influenced cyclic fatigue resistance with Reciproc R25 being more resistant than WaveOne Primary. Hasegawa et al. [72] evaluated the effect of EDTA solutions (3% and 10% EDTA · 2Na) on corrosion fatigue of three NiTi files with different shapes, in comparison with other solutions (6% NaClO, 3% H_2O_2, 0.9% NaCl and distilled water). Ni–Ti files were subjected to rotational bending in a bent glass tube (30° and 60° angles) filled with the solutions, and the number of rotations to failure was counted. It was reported that at 30° bent angle, files in the two EDTA solutions showed significantly lower resistance than those in distilled water, but no significant difference was found between the two EDTA solutions. Fatigue resistance of two tested files in the two EDTA solutions was not significantly different from those in the other three solutions, whereas one file in EDTA solutions showed significantly lower resistance than that in 3% H_2O_2. At 60° bent angle, early failure within 1–2.5 min was observed for all tested files, and no significant difference was found among the six solutions. At both angles, significant differences in fatigue resistance were observed among the three tested files, which could be related to the difference in the cross-sectional shapes of the files. Mohajeri et al. [216] investigated the corrosion behavior of a NiTi alloy. The NiTi alloy is subjected to cyclic thermal loading in corrosive environments (3.5 wt% NaCl and Ringer's solution). It was reported that the breakdown of passive oxide film on the surface with martensitic transformation. This breakdown would be an origin for fatigue crack initiation and an explanation for fatigue lifetime reduction of NiTi in corrosive environments. Huang et al. [217] examined a new fatigue test model that simulates the clinical situation for evaluating the corrosion effect of 5.25% NaOCl on NiTi files and to evaluate the effect of 3 different temperatures (22 °C, 37 °C, and 60 °C) on the cyclic fatigue of these files. Three NiTi files (size 25/.04), K3 (SybronEndo, Orange, CA), K3XF (SybronEndo), and Vortex (Dentsply Tulsa Dental Specialties, Tulsa, OK), were subjected to cyclic fatigue tests inside a novel artificial ceramic canal with a curvature of 60° and a 5-mm radius. It was obtained that the N_F of Vortex files was the highest followed by K3XF and K3 at all conditions. The N_F of all files was highest at 22 °C and lowest at 60 °C; however, no difference in N_F was detected in Vortex files between 22 °C and 37 °C. The N_F of all files in 5.25% NaOCl was shorter than that in water, although there was no statistically significant difference. No pitting or crevice corrosion was observed on the fracture surface. Based on these findings, it was concluded that NaOCl, 5.25%, does not significantly affect the fatigue behavior of NiTi files. The fatigue resistance should be tested under specific temperature conditions, and the austenite finish temperature of a file is important in determining the fracture risk at body temperature.

8.2.10.6 Tribocorrosion

Tribocorrosion describes the material degradation due to simultaneous mechanical and chemical/electrochemical interactions between surfaces in a tribological contact and relative motion. Tribocorrosion may occur under a variety of conditions (i.e., sliding, fretting, rolling, impingement) in a corrosive medium. In a tribocorrosion sliding system, two- or three-body contacts can be a cause of degradation and wear.

In order to improve the surface properties of NiTi such as corrosion resistance, Tan et al. [218] employed the plasma source ion implantation (PSII) technique with oxygen as incident ions at three levels of implantation dose (5×10^{16}, 1×10^{17}, and 3×10^{17} ions/cm^2). Pitting corrosion and wear-corrosion behavior of control and PSII-modified Ti-50.7 at% Ni alloy were evaluated by cyclic potentiodynamic polarization and wear-corrosion measurements. It was obtained that the corrosion behavior was influenced by both heat treatment and surface modification, the best pitting corrosion resistance was observed for samples with A_F (=21 °C) modified by oxygen implantation at a dose of 1×10^{17} ions/cm², and better wear-corrosion resistance was observed for oxygen-implanted samples. Kosec et al. [219] studied the electrochemical and tribocorrosion properties of superelastic NiTi sheet and orthodontic archwire, taking into account their microstructures and the effect of different surface finishes. It was mentioned that the microstructure of the investigated alloys had a significant effect on the measured electrochemical and tribocorrosion properties. Močnik et al. [220] analyzed the influence of environmental effects on the corrosion and tribocorrosion properties of NiTi and stainless-steel dental alloys. The effects of pH and fluorides on the electrochemical properties were studied using the cyclic potentiodynamic technique. It was found that lowering the pH preferentially affects the corrosion susceptibility of NiTi alloys, whereas stainless steel dental archwires are prone to local types of corrosion. The NiTi alloy is not affected by smaller increases of fluoride ions up to 0.024 M, while at 0.076 M (simulating the use of toothpaste) the properties are affected. A leaching test during wear-assisted corrosion showed that the concentrations of Ni ions released into the saliva exceeded the limit value of 0.5 µg/cm²/week. The oxide films on the NiTi and stainless steel alloys after the tribocorrosion experiment were thicker than those exposed to saliva only. Metallic medical devices such as intravascular stents can undergo fretting damage in vivo that might increase their susceptibility to pitting corrosion. As a result, the US Food and Drug Administration has recommended that such devices be evaluated for corrosion resistance after the devices have been fatigue tested in situations where significant can lead to fretting damage [221]. Three common alloys that cardiovascular implants are made from MP35N cobalt chromium (MP35N), electropolished nitinol (EP NiTi), and 316LVM stainless steel (316LVM) were selected to evaluate the effect of wire-fretting on the pitting corrosion susceptibility of these medical alloys, small- and large-fretting scar conditions of each alloy fretting against itself, and the other alloys in PBS at 37 °C were tested per ASTM F2129 and compared against as received or PBS immersed control specimens. It was observed that although the general trend observed was that fretting damage significantly lowered the rest potential (E_R) of these specimens, fretting damage had no significant effect on the breakdown potential (E_B) and hence did not affect the susceptibility to pitting corrosion. The fretting damage in PBS alone is not sufficient to cause increased susceptibility to pitting corrosion in the three common alloys investigated. Xue et al. [222] investigated the tribocorrosion behavior of a biomedical NiTi alloy in Ringer's simulated body fluid through friction experimentation, electrochemical testing, and abrasion mapping. The effects of applied load and particle concentration were considered. It was reported that due to the wear-accelerated corrosion, the corrosion potential E of the NiTi alloy shifted negatively and the corrosion current density increased by an order of magnitude, compared to the static electrochemical corrosion, the applied load effect on the wear rates of the NiTi alloy was higher than that of the particle concentration, the wear rates were decreased with the applied load increased, whereas the wear rates were decreased with the increase of abrasive particle concentrations, when the applied load was 1.5 N, the particle concentration was 0.03 g/cm^3, while the lowest wear rate of 9.47 × 10^{-5} mm^3/Nm was acquired under the corrosion-wear conditions. The wear contribution on the material loss exceeded the corrosion contribution, signifying the wear responsibility for the material loss, indicating that the corrosion-wear regime was mechanical abrasion dominant with corrosion.

8.3 Oxidation and oxides

8.3.1 Oxidation

ASTM F86 [223] standard recommends an appropriate chemical treatment of metallic implants to ensure a passive surface condition; such recommended treatment for stainless steel includes a nitric acid chemical passivation or electropolishing to modify the surface oxide characteristics increase their corrosion resistance and therefore improve their biocompatibility. NiTi is a passive alloy like CpTi and stainless steel and a stable surface oxide protects the substrate material from general corrosion. The surface oxide film or layer is the extreme outer layer on the titanium materials and such an extreme surface consists mainly of titanium oxide. Accordingly, the longevity and safety as well as biological reactions are all strongly related to the characteristics and crystal structures of formed oxides.

Titanium is a highly reactive metal and will react within microseconds to form an oxide layer when exposed to the atmosphere [224]. Although the standard electrode potential was reported in a range from –1.2 to –2.0 volts for the Ti $\leftrightarrow$ Ti^{+3} electrode reaction [225, 226], due to strong chemical affinity to oxygen, it easily produces a compact oxide film, ensuring high-corrosion resistance of the metal. This oxide, which is primarily TiO$_2$, forms readily because it has one of the highest heats of reaction known (ΔH= –915kJ/mol) (for 298.16–2,000 K) [227, 228]. It is also quite impenetrable by oxygen (since the atomic diameter of Ti is 0.29 nm, the primary protecting layer is only about 5–20 atoms thick) [229]. The formed oxide layer adheres strongly to the titanium substrate surface. The average single-bond strength of the TiO$_2$ to Ti substrate was reported to be about 300 kcal/mol, while it is 180 kcal/mol for Cr$_2$O$_3$/Cr, 320kcal/mol for Al$_2$O$_3$/Al and 420 kcal/mol for both Ta$_2$O$_5$/Ta and Nb$_2$O$_5$/Nb [230]. Adhesion and adhesive strength of Ti oxide to substrates are controlled by oxidation temperature and thickness of the oxide layer as well as the significant influence of nitrogen on oxidation in air. In addition, adhesion is greater for oxidation in air than in pure oxygen [231], suggesting that the influence of nitrogen on the oxidation process is significant.

Several studies have demonstrated that passivated Ti–Ni surface layers consist predominantly of a titanium oxide layer (TiO$_2$) [232–235], similar to that found on Ti alloys [236]. This is in agreement with theoretical thermodynamics, which specify that the free energy of formation of TiO_2 is favored over the other nickel or other titanium oxides. In addition to these oxidation conditions, it is well known that several parameters can modify the accommodation of the stresses developed during the oxidation of a metal and consequently play an important role in maintaining the protective properties of oxide layers. The results show that the adhesion of the oxide layers to the metal substrate decreases as the layer thickness increases. It is shown that the adhesion of the oxide layers decreases when the oxidation temperature increases, despite the increase in oxide plasticity [237]. Adhesive strength between Ti substrate and TiO_2 is also related to the thermal mismatching when the Ti/TiO_2 couple was subjected to the elevated temperatures. Although TiO_2 does not exhibit any phase transformation up to its melting point (i.e., 1,885 °C), its substrate Ti metal has an allotropic phase transformation at 885 °C (below which it is HCP – hexagonal close-packed crystalline structure and above which it becomes BCC (body-centered cubic crystalline structure). Accordingly, when Ti substrate is exposed beyond this phase transformation temperature, the bond strength between Ti and TiO_2 will be weakened due to the significant differences in coefficients of thermal expansion, particularly during the cooling stage passing through the 885°C transus temperature.

The performance of titanium and its alloys in surgical implant applications can be evaluated with respect to their biocompatibility and capability to withstand the corrosive species involved in fluids within the human body [238]. This may be considered as an electrolyte in an electrochemical reaction. It is well documented that the excellent corrosion resistance of titanium materials is due to the formation of a dense, protective, and strongly adhered film – which is called a passive film. Such a surface situation is referred to as passivity or a passivation state. The exact composition and structure of the passive film covering titanium and its alloys is controversial. This is the case not only for the "natural" air formed oxide, but also for films formed during exposure to various solutions, as well as those formed anodically. The "natural" oxide film on titanium ranges in thickness from 2 to 7 nm, depending on such parameters as the composition of the metal and surrounding medium, the maximum temperature reached during the working of the metal, the surface finish, . Chan et al. [233] studied the oxidation of the surface of an equiatomic NiTi alloy by XPS, AES and SEM. Samples of the NiTi alloy were oxidized at 23 °C and high temperatures at an oxygen pressure of 10^{-4} Torr and in atmosphere. It was reported that at 10^{-4} Torr of O_2 and 23 °C titanium was oxidized preferentially while nickel remained in the metallic form even after 40 min of exposure. At the same oxygen partial pressure and 400 °C, a mixture of TiOx, Ti_2O_3, and TiO_2 was formed, and nickel remained metallic. Eventually, after 40 min of O_2 exposure, the whole surface was covered with a layer of TiO_2, in atmosphere and at 23 °C, titanium again was oxidized preferentially, but NiO was detected on the surface after the first minute of air exposure. At a higher temperature of 450 °C, areas comprising a mixture of NiO and TiO_2, and areas consisting of TiO_2 only were found, and below the top surface layer, a complete layer of TiO_2 was detected, under which the TiO_2 layer, a nickel-rich layer was present.

Narushima et al. [239, 240] investigated the deoxidation behavior of NiTi alloy melts with metallic barium. Metallic barium (0.5–1.5 mass% with respect to NiTi alloy melt) was added to the NiTi alloy melts held in lime (CaO) crucibles at 800 °C, and the concentrations of oxygen, barium and calcium in the melts were quantitatively measured. The oxygen content in a raw NiTi alloy bar was 660 ppm, and the oxygen content in the NiTi alloy melts just before barium addition was around 1,100 ppm. Oxygen was introduced into the melts by the dissolution of lime during melting. It was reported that the addition of barium lowered the oxygen content in the melts with the formation of deoxidation product, which might be barium oxide. The oxygen content was measured to be around 350 ppm 120 s after barium addition. The barium and calcium contents were less than 10 ppm and 12 ppm, respectively, indicating that the deoxidation product was removed from the melts in a very short period of time; thereby, suggesting that barium is an effective deoxidation element for NiTi alloy melts. It was also mentioned that the nonmetallic inclusions observed in the NiTi alloys were the $Ti_4Ni_2O_X$ and $Ti(C,N,O)_X$ types, and decreased oxygen content through Ba deoxidation caused a change in the main phase of the nonmetallic inclusions from the $Ti_4Ni_2O_X$ type to the $Ti(C,N,O)_X$ type.

8.3.2 Air-formed oxide

When fresh titanium is exposed to the atmosphere by such cutting acts as lathing, milling, or sawing, an oxide layer begins to form within nanoseconds [229]. After only one second, a surface oxide (with 2 to 7 nanometers in thickness) will be formed. Air oxidation at room temperature produced titanium monoxide (TiO) with small quantities of titanium oxide, Ti_3O_5. Titanium has proved to be a highly successful material for implants inserted into human bone. Under certain conditions, titanium will establish and maintain a direct contact with the bone tissue in a process called osseointegration. The mechanisms underlying this behavior are not yet understood. However, it is clear that the properties of the implant surface are of vital importance for a successful osseointegration. The properties of titanium implant surfaces are determined by the thin (20–70 Å) oxide film that covers the metal. Thus, the biocompatibility of titanium implants is associated with the surface titanium oxide and not with the bulk titanium metal [224, 234, 235, 241, 242]. The surface oxide is also formed during the implant preparation procedures. During the machining of the implants, pure titanium metal is exposed to air and is rapidly oxidized. In the following cleaning and autoclaving procedures oxide film is then modified and grows. In order to obtain an oxide film that is reproducible with respect to the chemical composition, structure, and thickness, it is important that the different steps of the implant preparation are performed under carefully controlled conditions. Even small changes in the preparation procedures might lead to considerable changes of the implant surface [12].

As mentioned previously, because the surface air-formed oxide is nearly impenetrable, once this thin passivating film has formed, oxygen is prevented from reaching the metal beneath and further oxide layer thickening is quickly halted because the oxide film is dense and semiconductive (not like an electron-conductive metal substrate). The compositional structure and exact thickness of the passivating oxide layer is dependent on many factors associated with its formation. These include such factors as type of machining, surface roughness, coolants used during machining, and sterilization procedures. It should be no surprise that reported biocompatibility of devices is prepared by different investigators and therefore may vary to a significant degree [229].

It is well recognized that the excellent tissue-bone compatibility of titanium is mainly due to the properties of its stable surface oxide layer. Biocompatibility of implant materials relies on the chemical and electrochemical stability of this surface oxide layer, which interfaces with the soft and hard tissue and bone structure. Oxide layers formed on titanium change from lower to higher oxides as oxidation progresses and temperature increases. The following phases form in air: $Ti + O \rightarrow Ti(O) \rightarrow Ti_6O \rightarrow Ti_3O \rightarrow Ti_2O \rightarrow TiO \rightarrow Ti_2O_3 \rightarrow Ti_3O_5 \rightarrow TiO_2$. Among these oxides of different stoichiometry (e.g., TiO, Ti_2O_3, TiO_2), TiO_2 is the most common and stable thermodynamically. TiO_2 can have three different crystal structures – rutile, anatase, and brookite – but also can be amorphous. TiO_2 is very resistant against chemical attack, which makes titanium one of the most corrosion resistant metals. Another physical property that is unique to TiO_2 is its high-dielectric constant, which ranges from 14 to 110 (in 10^6 cycles; dielectric strength: 350 V/mil, volume resistivity: 10^{14}–10^{16} ohms-cm) depending on crystal structure [243]. This high dielectric constant would result in considerably stronger van der Waal's bonds on TiO_2 than on other oxides. TiO_2, like many other transition metal oxides, is catalytically active for a number of inorganic and organic chemical reactions, which also may influence the interface chemistry.

The surface and the "bulk" structure of Ti–Ni implants were characterized using SEM, transmission electron microscopy (TEM), XPS, and scanning Auger microprobe analysis (AES). Ti–Ni implants were compared with otherwise identically prepared nonimplanted specimens, and sputter-cleaned and reoxidized samples. Nonimplanted and implanted samples had essentially the same surface topography and microstructure. Ti, O, and C were the dominant elements detected on the surface. Trace amounts (1 at%) of Ni and Ca, N, Si, B, and S were also detected. Ti was present as TiO_2 on the surface, while nickel was present in metallic form. A significant difference in Ni peak intensity was observed when retrieved or nonimplanted control samples (very low-nickel content) were compared with sputter-cleaned and reoxidized samples (well-detected nickel). It is evident that the method of passivation is crucial for nickel loosening. No major changes occurred in the NiTi samples bulk structure or in the surface oxide during the implantation periods investigated [244]. The NiTi intermetallic alloy spontaneously forms a thin passive layer of TiO_2, which provides its biocompatibility. The oxide layer is thought to form as the Ti in the alloy surface reacts with oxygen, resulting in a depletion of Ti in the subsurface region – experimental evidence indicates formation of a Ni-rich layer below the oxide film. Nolan et al. [245] investigated the initial stages of oxide growth on the (110) surface of the NiTi alloy to understand the formation of alloy/oxide interface. Adsorption of atomic and molecular oxygen on the (110) surface was firstly achieved, followed by adding O_2 molecules, up to 2 monolayers of O_2. Oxygen adsorption always results in a large energy gain. It was mentioned that with atomic oxygen, Ti is pulled out of the surface layer leaving behind a Ni-rich subsurface region. Molecular O_2, on the other hand, adsorbs dissociatively and pulls a Ti atom farther out of the surface layer. The addition of further O_2 up to 1 monolayer is also dissociative and results in complete removal of Ti from the initial surface layer, when further O_2 is added up to 2 monolayer. Ti is pulled even further out of the surface and a single thin layer of composition O–Ti–O is formed, and the electronic structure shows that the metallic character of the alloy is unaffected by interaction with oxygen and formation of the oxide layer, consistent with the oxide layer being a passivant.

8.3.3 Passivation

Passivity is a property of a metal commonly defined in two ways. One of these is based on a change in the electrochemical behavior of the metal, and the other on its corrosion behavior [246]. Passivity is an unusual phenomenon observed during the corrosion of certain metals, indicating a loss of chemical or electrochemical reactivity under certain environmental conditions [247]. Such passivity appears on certain so-called passivating metals, many of which are 'transition metals', characterized by an unfilled d group of electrons in an inner electron shell. Passivating metals are, for example, Fe (covered with Fe_2O_3 and/or Fe_3O_4), Cr (with Cr_2O_3), Zr (with ZrO_2), and Ti (with TiO_2) [234, 235, 247]. This type of passivity is ascribed to an invisibly thin but dense and semiconducting oxide film on the metal substrate surface, displacing the electrode potential of the metal strongly in the positive (or more noble) direction (for example, Ti passive film exhibits its stable passivity up to 1.5–2.0 V). Hence, comparing the electrode potential (–1.2 to –1.5 V) to this stable passivity potential, surface nobility was remarkably improved. Passive films can be formed by either chemically or electrochemically (or anodic) treating titanium surfaces. Fraker et al. [248] found that more rigid oxidizing conditions produce higher oxides of titanium alloy (thin films) in saline water (3.5% NaCl), at a temperature range between 100 and 200 °C, using a transmission electron microscopy. They found that oxides ranged from Ti_2O to TiO_2 (anatase), with the higher oxides corresponding to the higher temperatures. Concerning oxide films formed anodically on titanium, most investigators agree that the film consists of TiO_2. All three crystalline forms of TiO_2 (rutile, anatase, and brookite) have been reported, as will be discussed later. It has also been reported that the anodic film is either oxygen excess or oxygen deficient. The controversy is widened by reports that the film is hydrated or contain mixed oxides. In addition to variations in oxygen stoichiometry, the films may contain various amounts of elements other than titanium and oxygen or incorporate such additional element(s) into TiO_2 structure. Thus, it is most probable that the anodic film is not necessarily stoichiometric TiO_2, and that the films studied by different authors had different compositions due to differences in the conditions of oxidation [248].

More recently, Trépanier et al. [249] performed an in vivo study on passivated Ti–Ni stents. Implantation of the material in rabbit paravertebral muscles and the study of the inflammatory reaction for periods ranging from 3 weeks to 12 weeks demonstrated good biological responses to Ti–Ni. Analysis of the fibrous capsule surrounding Ti–Ni stents revealed a decrease in thickness with time. A comparative 26-week follow-up study was conducted on rats to assess the effect of different materials on soft tissues [250]. In this study, short-time biocompatibility of polished Ti–Ni was similar to polished Ti–6Al–4V and electropolished stainless steel when in contact with muscle and perineutral tissue. These results indicate promising soft tissue biocompatibility of Ti–Ni material. When titanium alloys (Ti-6Al-7Nb and Ti-6Al-4V) were chemically treated, surface oxide films appear to be more complicated. Sittig et al. [251] treated these alloys along with CpTi in nitric acid hydrogen fluoride and identified formed oxides. Three types of oxides (TiO_2, Ti_2O_3, and TiO) were identified on the CpTi substrate, and it was found that the dissolution rate depends on grain orientations. On the other hand, for Ti–6Al–7Nb and Ti–6Al–4V alloys, Al_2O_3, Nb_2O_5 or V-oxides such as V_2O_5, were formed in addition to TiO_2 oxide, and it was found that the selective α-phase dissolution and enrichment of the β-phase appears to occur [251].

Studies on the nature and properties of thin films formed by the electrolytic oxidation on titanium, which, in turn, determine the behavior of the electrode in the open circuit, indicate the duplex nature of these films [252]. In both the acid and alkaline media, the outer layer of the anodic oxide film was found to be more susceptible to dissolution than the inner layer, thereby indicating a more defective structure of the outer layer. The surface reactivity was found to be higher in oxygen-saturated versus nitrogen-saturated solutions. Galvanostatic anodization of titanium has been studied in N_2-deaerated, 1 N solution of NaOH, Na_2SO_4, H_2SO_4, HCl, $HClO_4$, HNO_3, and H_2PO_4 at 25 °C at current densities ranging from 4 to 50×10^{-6} A/cm². From formation rate values, the following arrangement of passivation was obtained: (from the highest rate to the lowest rate) NaOH > Na_2SO_4 > $HClO_4$ > NHO_3 > HCl > H_3PO_4 > H_2SO_4 [252].

Metikoš-Huković et al. [253] employed structurally sensitive in situ methods, such as photopolarization and impedance, to examine the passivation process and the properties of the protective oxide layers on titanium. The thickness of anodic oxidation and the nonstoichiometry of the surface oxide were correlated. It was mentioned that the composition of the anodic film on titanium changes with the relative potential from lower to higher oxidation stages according to the equation: $TiH_2 + TiO \rightarrow nTi_2O_3 \cdot n\text{-}TiO_2 \rightarrow Ti_5O_9$ or $Ti_6O_{11} \rightarrow Ti_3O_5 \rightarrow TiO_2$. The characteristic behavior of titanium can easily be seen at higher anodic potential (approximately +1.5 V vs SCE in 5 mol H_2SO_4) when the electrode is covered with a nearly stoichiometric TiO_2 layer. The semiconducting properties of TiO_2 were investigated using an anodic film stabilized at +2 V (vs SCE: saturated calomel electrode with 0.2444 V vs NHE), and it was found that TiO_2, like the lower titanium oxides, is an n-type (metal-excess) semiconductor under anodic polarization [574]. Re-passivation capability [254] associated with titanium substrate can be considered as another reason for the high- corrosion resistance and biocompatibility of titanium materials. The repassivation capability can be evaluated by the physical scratch damage on the sample surface while being immersed in a certain corrosion media, and by studying the polarization behavior per ASTM F746 [255].

A recent in vitro study revealed no significant differences between cell growth behavior near the surfaces of different implant materials (mechanically polished CpTi and Ti–Ni, electropolished 316 L stainless steel). Passivated Ti–Ni showed no cytotoxic, allergic or genotoxic activity based on cytotoxicity tests, a guinea-pig sensitization test and genotoxicity testing, respectively [256]. In a different study that addressed only the genocompatibility of the material, TiNi exhibited a good biocompatibility behavior similar to CpTi and 316 L stainless steel on cellular chromatin [257].

Rondelli et al. [258] carried out potentiodynamic polarization scans, potentiostatic scratch tests, and modified ASTM F746 tests in simulated body fluids on commercial orthodontic wires made of different classes of materials, and on CpTi as a reference. The stability of passivating films, evaluated by electrochemical techniques that abruptly damage it, that is, potentiostatic scratch tests, increased in the following order: Ti–Ni<stainless steel < Cr–Co alloy < CpTi. Because satisfactory biocompatibility of implants relies on the presence of a stable and efficient self-healing passive film, this ranking should be considered in view of in vivo applications. Moreover, scratch tests show that it is not possible to enhance the performance of Ti–Ni samples by modifying surface-passive film by dipping it in an HF/HNO_3 mixture. Finally, straining of Ti–Ni wires under superelastic conditions and the consequent presence of stress-induced martensite does not substantially modify their localized corrosion resistance.

Cissé et al. [93] investigated the surface treatment effect on corrosion behavior of Ti–Ni alloy. The near-equiatomic superelastic Ti–Ni (Ni 55.8 wt%) alloy was used. The surface of heat-treated samples were modified by mechanical polishing (MP), electropolishing (EP), and electropolishing followed by chemical passivation (CP). As heat-treated samples with straw-colored oxide finishes (SCO) and blue-colored oxide finishes (BO) also were prepared. It was shown that surface roughness increased in the order of CP<EP<SCO<BO<MP. The Ni releases within the five groups of Ti–Ni samples, as determined by atomic absorption spectrophotometer, reduced in time over the measured period. The level of Ni ions released over a 25-day immersion period was highest in the SCO sample (0.002 µg/day). This Ni level is negligible compared with the dairy intake of Ni in an ordinary diet. The surface oxide thickness increased in the order CP<EP<MP<BO<SCO. On the other hand, for the electrodes treated under the same conditions, the mean breakdown potential value decreased in

the order BO>MP>CP>EP>SCO while the corrosion current density and rate increased in the order CP<SCO<EP<BO<MP.

The effect of surface roughness on the relative corrosion rates of wires of four alloys (stainless steel, Ti–Ni, Co–Cr alloy, and beta Ti) was investigated [259]. Batches of wire were divided into two groups; one group for polishing to provide a uniform surface finish, and the second group for as-received. The samples of as-received wires showed variations in surface finish, with beta Ti having the roughest appearance and Co-Cr the smoothest. Ti–Ni and stainless-steel surfaces were similar. Mechanical polishing provided a uniform finish, but significantly reduced the diameter of the wires. The relative corrosion rates (expressed in terms of corrosion current density) in a 0.9% NaCl solution were estimated using polarization resistance. Ti–Ni wires exhibited the greatest corrosion current density in the as-received group. Mechanical polishing significantly reduced the corrosion rate of Ti–Ni, such that comparison between the four alloys in the polished state revealed no significance in their relative corrosion rate/ corrosion current density [259].

Passive films formed on pure Ti and crystalline and amorphous Ti–Ni alloys containing from 30 to 60 at% Ti in a sulfuric acid solution were studied by XPS and photoelectrochemical methods. The photocurrent increased with applied anodic potential. The band-gap energies in the surface films were in the range of 3.2–3.4 eV which decreased with increasing applied potential. In the surface film, Ti ions were enriched. The Ti in the film consisted of Ti^{4+}, Ti^{3+}, and Ti^{2+} ions. The ratios between them on a specimen were independent of the applied potential, although the thickness of the surface film itself increased with the increasing potential. The surface films on pure Ti and Ti–Ni alloys contained OH-type oxygen [260]. Morawiec et al. [261] mentioned that the sterilization of the NiTi alloy in boiling water or steam causes passivation, which results in an amorphous 3.5 nm thick TiO_2 layer on the surface. It was also mentioned that between the surface and the matrix a transition layer of Ni_2O_3 and NiO was observed. The corrosion behavior of Nitinol-based medical implants is critical to their success in vivo. Contemporary Nitinol-based medical implants are typically chemically passivated or electrochemically polished to form a protective passive film. However, mechanically formed surfaces caused by handling damage, fretting, or fatigue fracture may also be present on a device in vivo. Based on this background, Schoerder et al. [262] mechanically polished surfaces to simulate mechanically damaged surfaces. These mechanically polished NiTi surfaces are compared with chemically passivated and electrochemically polished Nitinol surfaces and mechanically polished titanium surfaces in PBS solution. It was observed that the mechanically polished NiTi exhibits lower impedance at low frequencies, empirically modeled to a thinner film with lower film resistance than chemically passivated, electrochemically polished NiTi, and mechanically polished titanium. The passive film on mechanically polished NiTi continues to develop over time, increasing in its thickness and film resistance, and mechanically formed surfaces may be initially less protective than chemically passivated/electrochemically polished NiTi surfaces, but continue to become thicker and more resistant to electrochemical reactions with exposure to saline solution. Neelakantan et al. [263] studied the passivity behavior of NiTi SMAs with different levels of secondary phase titanium carbide TiC particles in an electrolyte of 0.9% sodium chloride at 37 °C. It was found that the influence of carbides and thermomechanical treatment/cold working on the passivity breakdown is highlighted. The polarization studies on the as-cast and cold-worked NiTi with high (0.05 wt%) and low (0.005 wt%) carbon levels show a significant difference in oxide stability. The alloy with extremely low carbon content shows a higher breakdown potential, while higher carbon levels result in higher density of larger TiC, and these carbide/matrix interfaces are more susceptible to pitting, the qualitative behavior of passive layer formed at 0.5 V on the cold-worked NiTi alloy with different carbon levels was ascertained by EIS. The oxide on the NiTi alloy with high (0.05 wt%) carbon levels showed lower resistance and poor stability at this condition.

Passive film can be formed by not only dry oxidation, but in electrochemical process, too. Speck et al. [264] measured anodic polarization in Hanks' physiological solution at 37 °C and a pH of 7.4 and listed from the most passive metals to least one as follows; titanium, Ti–6A1–4V, Ti–Ni (memory alloy), MP35N (Co–Ni–Cr–Mo), Co–Cr–Mo, 316L stainless steel, and nickel. The influence of the amino acids, cysteine, and tryptophan on the corrosion behavior of Ti–Ni and Ti–6A1–4V was studied. It was obtained that cysteine caused a lower breakdown potential for Ti–Ni, but it did not affect the breakdown of Ti–6A1–4V, although an increase in current density for Ti–6A1–4V was observed. Kawakita et al. [265] fabricated thick oxide films on the surface by an anodic oxidation process. It was noted that the films had a porous nature; and the structural analyses indicated that the films are amorphous, from the chemical analyses. The oxide films mainly contained titanium as metal component, the nickel content of the oxide films depends on pH of the electrolyte and anions could be incorporated from the electrolyte into the films. The asymmetrically rectangle waveform was effective to increase the density of the films, and the electrochemical technique showed that the corrosion resistance of the anodized films could improve the resistance against localized corrosion of NiTi.

8.3.4 Oxidation at elevated temperatures

Above 100 °C the surface is a lower oxide type, such as TiO. At elevated temperatures (above 200 °C); complex oxides are formed, such as Al_2TiO_5 on Ti–6Al–4V and $NiTiO_3$ on TiNi, in addition to rutile-type TiO_2 [234, 235]. Oxidation in air at 875 ° to 1,050 °C produced a flaky scale of TiO, Ti_2O_3, and rutile-type TiO_2. It was reported that the high oxide TiO_2 appears on the surface of titanium in the presence of the most rigid conditions of oxidation; in less rigid conditions a lower oxide ($3Ti_2O_3 \cdot 4TiO_2$) forms, while in still weaker oxidizing conditions the formation of even lower oxides (for example, TiO) are possible [266].

The characteristics of high-temperature oxidation of Ti are the main obstacle to strong Ti-ceramic bonding [267]. A thin oxide layer (approximately 32 nm) with good adherence to the substrate was formed when CpTi was oxidized at 750 °C in 0.1 atm, and a thick layer (approximately 1 μm) with poor adherence was formed when oxidized at 1,000 °C [268]. In another study [269] the thickness of TiO_2 on the CpTi surface increased with an increase in oxidation temperatures. Ti surface hardness also increases substantially after oxidation at and above 900 °C [269]. Therefore, the substrate is now in the BCC structure range. Lower Ti-ceramic bond strength was attributed to a thick TiO_2 zone on the metal surface when Ti was oxidized at higher temperatures [270]. Enhanced bond strength between titanium and ceramic was reported when porcelain was fired on cast Ti in a reduced argon atmosphere [271].

Modifying the Ti surfaces to control high-temperature oxidation has been examined. Published studies show that Ti surface nitridation [272] or a thin Cr coating [273] is an effective method in limiting Ti oxidation at high-temperatures. Improved Ti-ceramic bonding was found when a thin layer of Si_3N_4 was deposited on Ti surface [274].

Mechanically polished Ti–Ni alloy (50 at% Ni) was subjected to heat treatment in air in the temperature range 300–800 °C and characterized by SEM, XRD, XPS, and Raman spectroscopy. Thermogravimetry measurements were carried out to investigate the kinetics of oxidation. The results of thermodynamic calculations were compared to the experimental observations. It was found that Ti–Ni alloy exhibits different oxidation behaviors at temperatures below and above 500 °C. A Ni-free zone was found in the oxide layer for oxidation temperatures of 500 °C and 600 °C. The oxidation at 500 °C produces a smooth protective nickel-free oxide layer with a relatively small amount of Ni species at the air/oxide interface, which is in favor of good biocompatibility of Ti–Ni implants [275].

The corrosion of pure Ni and of binary Ti–Ni alloys containing 5, 10, and 15 wt% Ti respectively in molten $(0.62Li, 0.38K)_2 CO_3$ at 650 °C under air has been studied. The corrosion of the single-phase Ni-5Ti alloy was slower than that of pure Ni, forming an external scale composed of NiO and TiO_2. The two-phase Ni-10Ti and Ni-15Ti alloys underwent much faster corrosion than pure Ni, producing an external scale containing NiO and TiO_2, and a thick internal oxidation zone of titanium mainly involving the intermetallic compound $TiNi_3$ in the original alloys. The rates of growth of the external scales for the Ti–Ni alloys were reduced with the increase of their titanium content, while the internal oxidation was significantly enhanced. The corrosion mechanism of the alloys is also discussed [276].

Intravascular stents are being designed which utilize the shape-memory properties of Ti–Ni alloy. Despite the clinical advantages afforded by these stents their application has been limited by concerns about the large nickel ion content of the alloy. In this study, the surface chemistry of Ti–Ni alloy was modified by mechanical polishing and oxidizing heat treatments and subsequently characterized using X-ray photon spectroscopy (XPS). The effect of these surfaces on monolayer formation and the barrier integrity of human umbilical vein endothelial cells (HUVEC) were then assessed by confocal imaging of the adherent's junctional molecule VE-cadherin, perijunctional actin, and permeability to 42 kDa dextrans. Dichlorofluorescein assays were used to measure oxidative stress in the cells. XPS analysis of NiTi revealed its surface to be dominated by TiO_2. However, where oxidation had occurred after mechanical polishing or post polishing heat treatments at 300 and 400 °C in air, a significant amount of metallic nickel or nickel oxide species (10.5 and 18.5 at%) remained on the surface. Exposure of HUVECs to these surfaces resulted in increased oxidative stress within the cells, loss of VE-cadherin and F-actin and significantly increased paracellular permeability. These pathological phenomena were not found in cells grown on Ti–Ni, which had undergone heat treatment at 600 °C. At this temperature thickening of the TiO_2 layer had occurred due to diffusion of titanium ions from the bulk of the alloy, displacing nickel ions to sub-surface areas. This resulted in a significant reduction in nickel ions detectable on the sample surface (4.8 at%). This study proposes that the integrity of human endothelial monolayers on Ti–Ni is dependent upon the surface chemistry of the alloy and that this can be manipulated, using simple oxidizing heat treatments [277].

During implantation, titanium releases corrosion products into the surrounding tissue and fluids even though it is covered by a thermodynamically stable oxide film [278–280]. An increase in oxide thickness, as well as incorporation of elements from the extracellular fluid (P, Ca, and S) into the oxide, has been observed as a function of implantation time [281]. Moreover, changes in the oxide stoichiometry, composition, and thickness have been associated with the release of titanium corrosion products in vitro [282]. Properties of the oxide, such as stoichiometry, defect density, crystal structure and orientation, surface defects, and impurities were suggested as factors determining biological performance [283].

Chu et al. [284] studied the isothermal oxidation behavior of equiatomic TiNi SMA in dry air from 550 to 1,000 °C with thermogravimetric analysis. It was reported that a multi-layered scale is formed, consisting of an outer rutile layer, a porous intermediate layer of mixture of TiO_2 and Ni(Ti), and a thin inner $TiNi_3$ layer, the intermediate layer. Furthermore, it exhibits a stripe-like lamellar structure at the $TiNi_3$ interface, and the apparent activation energy of TiNi alloy oxidation is 226 kJ/mol, and the oxidation rate follows a parabolic law. Firstov et al. [285] heat-treated mechanically polished $Ni_{50}Ti$ alloy (at%) in air in the temperature range of 300–800 °C. It was found that NiTi alloy exhibits different oxidation behavior at temperatures below and above 500 °C. A Ni-free zone was found in the oxide layer for oxidation temperatures of 500 ° and 600 °C, and the oxidation at 500 °C produces a smooth protective nickel-free oxide layer with a relatively small amount of Ni species at the air/oxide interface, which is in favor of good biocompatibility of NiTi implants. Zeng et al. [286] studied the oxidation of pure Ni and three NiTi alloys containing 5, 10, and 15 wt% Ti over the temperature range 650–850 °C in air to examine the effect of titanium on the oxidation resistance of pure nickel. $NiTi_5$ is a single-phase solid solution, while the other two alloys consisted of nickel solid solution (α-Ni) and $TiNi_3$. It was reported that the oxidation of NiTi alloys at 650 °C follows an approximately parabolic rate law and produces a decrease in the oxidation rate of pure Ni by forming an almost pure TiO_2 scale, at higher temperatures. NiTi alloys also follow an approximately parabolic oxidation and their oxidation rates are close to or faster than those of pure Ni, duplex scales containing NiO, $NiTiO_3$, and TiO_2 formed, some internal oxides of titanium formed, especially at 850 °C, and (vi) the two-phase structure of $NiTi_{10}$ and $NiTi_{15}$ was transformed into a single-phase structure beneath the scales. Tian et al. [287, 288] investigated the oxidation behavior of both the ternary TiPdNi alloy and that containing the rare earth element Ce by using XRD, EPMA, and Nanoindenter. It was found that the oxidation scale exhibits different layers; the outmost layer is mainly composed of rutile TiO_2 while $TiNiO_3$ and Ti_4Pd_2O phases exist in the inner scale, and the alloying element addition (Ce) in the TiPdNi alloy effectively impedes the outward diffusion of Ti ions and obviously improves the oxidation resistance. The microhardness of the matrix changes significantly after oxidation.

Gu et al. [289] characterized the surface oxidation layers of NiTi which was oxidized in the temperature range of 300–800 °C. The heat-treated sample was soaked in simulated body fluid to study the bioactivity of the thermally grown oxide layer. It was shown that a protective layer of TiO_2 was formed on the surface of NiTi alloy at heat treatment temperatures of 600 °C or higher with varying degree of anatase and rutile, and small amount of nickel oxide was found on the surface of 300 and 400 °C treated samples by XPS. These results confirmed earlier reported data by Oshida et al. [234, 235]. It was also mentioned that with further increase in the heat treatment temperature, the nickel concentration on the surface decreased and there was almost no nickel species on the surface after heat treatment at 600 °C or 800 °C, depth profiling revealed that the amount of TiO_2 (Ti^{4+}) decreased with depth with a concomitant increase of metallic Ti. In addition, both TiO (Ti^{2+}) and Ti_2O_3 (Ti^{3+}) increased initially and then decreased gradually with depth. Ni existed mainly in the oxidized state on the surface of heat-treated samples, and it changed to metallic state with increasing depth. The in vitro test revealed that the titanium oxide layer formed on the 600 and 800 °C heat-treated samples was bioactive, and a layer of apatite was formed on the surface of the titanium oxide layer after soaking in simulated body fluid. Vojtěch et al. [290] studied both isothermal and cyclic oxidation behaviors of a commercial nitinol wire at 530–650 °C × 10–240 min in air. It was reported that the cyclic and isothermal oxidation rates are similar at the experimental temperatures because the scale spalling is not significant. The isothermal oxidation obeys the parabolic law with an activation energy of 158 kJ/mol, the main oxidation product even after early 10 min oxidation is rutile containing a few at.% Ni. The grain growth of rutile at 530–650 °C is characterized by the activation energy and grain growth exponent of 45.7 kJ/mol and 0.29–0.34, respectively. The preferential oxidation of titanium causes the formation of an Ni-enriched and Ti-depleted zone containing Ni_3Ti phase. XRD reveals that the presence of cubic B2 NiTi phase in the base alloy is not significantly affected by oxidation at 530–650 °C for 240 min. $NiTi_{49}$ and $NiPt_{30}Ti_{50}$ (at%) SMAs were isothermally oxidized in air over the temperature range of 500 °C to 900 °C for 100 h [291]. It was reported that a relatively pure TiO_2 rutile structure was identified as the predominant scale surface feature, typified by a distinct highly striated and faceted crystal morphology, with crystal size proportional to oxidation temperature. The complex-layered structure beneath these crystals was characterized by semiquantitative XRD of serial/taper polished sections, and SEM/EDS of cross-sections for samples oxidized at 700 °C.Eng. general, graded mixtures of TiO_2, $NiTiO_3$, NiO, $Ni(Ti)$, or $Pt(Ni)$ metallic dispersoids, and continuous Ni_3Ti or Pt-rich metal depletion zones, were observed from the gas surface to the substrate interior. Overall, substantial depletion of Ti occurred due to the formation of predominantly TiO_2 scales. Based on these finding, it was concluded that the $NiPt_{30}Ti_{50}$ alloy oxidized more slowly than the binary $NiTi_{49}$ alloy by decreasing oxygen and titanium diffusion through the thin Pt-rich layer. Several studies regarding superficial treatments of NiTi shape memory have been developed aiming to the improve corrosion resistance and to block the Ni release to adjacent tissues. The necessary heat treatment to achieve the SME normally occurs at temperatures between 500 and 600 °C. However, titanium oxide (TiO_2) is formed on the NiTi surface during the shape-memory process heat treatment; therefore, the effects of the heat treatment time on the surface characteristics of the formed NiTi oxide, at temperatures that promote the shape memory (530 and 570 °C) was evaluated [292]. It was found by increasing the exposure time at the temperature of 570 °C the formation of a thicker oxide is promoted, with less superficial roughness and of a hydrophobic nature, and these characteristics indicate that the obtained oxide layer has properties that promote the osseointegration process. Wu et al. [293] investigated the use of flowing argon as a protective gas for heat-treating NiTi at elevated temperatures. It was found that under flowing argon protection, complex oxidation reactions still take place within the surface layers of NiTi, including the formation of transient oxides of Ni_2Ti_4O and TiO at lower temperatures, followed by formation of $TiNi_3$ underneath the oxide layers, and conversion of TiO to TiO_2 at higher temperatures. The final product contained a rutile TiO_2 outer surface, a composite middle layer consisting of $Ni(Ti)$ + TiO_2, a $TiNi_3$ inner layer, and a Ti-depleted zone within the body of the alloy. For heating at 950 °C for 24 h, the surface layers have a combined thickness of ~15 µm and a Ti-depleted zone of ~400 µm within the body of the alloy. The Ti-depleted zone appeared to play a major role in affecting the transformation and mechanical properties of the alloy.

8.3.5 Crystal structures of Ti oxides

Nolan et al. [294] mentioned that the NiTi intermetallic alloy spontaneously forms a thin passive layer of TiO_2, which provides its biocompatibility. The oxide layer is thought to form as Ti in the alloy reacts with oxygen. It was investigated that the details of the oxide-alloy interface was investigated. The atomic model is the

(110) NiTi surface interfaced with the (100) rutile TiO_2 surface; this combination provides the best lattice match of alloy and oxide. When the interface forms, static minimizations and molecular dynamics, show that there is no migration of atoms between the alloy and the oxide. In the alloy, there are some notable structural relaxations. It was found that a columnar structure appears in which alternating long and short Ni–Ti bonds are present in each surface and subsurface plane into the fourth subsurface layer. The oxide undergoes some structural changes as a result of terminal oxygen coordinating to Ti in the NiTi surface. The electronic structure shows that Ti^{3+} species are present at the interface, with Ti^{4+} in the bulk of the oxide layer and that the metallic character of the alloy is unaffected by the interaction with oxygen, all of which is consistent with experiment. A thermodynamic analysis is used to examine the stability of different possible structures – a perfect interface and one with Ti and O vacancies. Under conditions typical of oxidation and shape-memory treatments, the most stable interface structure is that with Ti vacancies in the alloy surface, leaving an Ni-rich layer, consistent with the experimental findings for this interface. It should be clear that the proven high biocompatibility of titanium as implant material is connected with the properties of its surface oxide. In air or water, Ti quickly forms an oxide thickness of 2 to 7 nm at room temperature. TiO_2 possesses three crystalline structures: anatase, rutile, and brookite. Anatase-type TiO_2 is a tetragonal crystalline system with $a_o = 3.78$ Å and $c_o = 9.50$ Å; the rutile-type TiO_2 is also a tetragonal structure, but the lattice constants are quite different from those of anatase-type (i.e., $a_o = 4.58$ Å and $c_o = 2.98$ Å). The third type is brookite and has an orthorhombic crystalline structure with $a_o = 9.17$ Å, $b_o = 5.43$ Å, and $c_o = 5.13$ Å [295]. Among these oxides, rutile is known to be the most stable phase [296].

Identification of crystalline structures of formed oxides can be achieved in several ways. TFXRD (thin-film x-ray diffraction) or XRD if the oxide film is reasonably thick enough, so that diffracted intensities can be obtained which are high enough to be identified and differentiated from each other and background noise. The alternative way will be diffraction using accelerated electrons. If an appropriate sample holding device can be manufactured and installed inside the transmission electron microscopy, then the oxidized sample with surface oxide can be tilted almost parallel to the accelerated electron beam, resulting in a half ring of RED (reflection electron diffraction) [297, 298], from which one can identify the crystalline structure(s) of the oxide(s). Another way for crystalline identification of formed oxides is based on stripping (or isolating) formed oxide from the substrate by chemically etching the backside (the interface side of formed oxide and substrate) to be isolated from the metallic substrate. Normally for stainless steel, 10% bromine in anhydrated methanol solution is used [298, 299], and for titanium materials, a mixed acid of HNO_3 and HF is used or 10% bromine in ethyl acetate [296]. The stripped thin oxide film is carried on a copper mesh and TED (transmission electron diffraction) can be performed [300].

Information on titanium oxides of anatase-type, rutile-type, brookite-type, or a mixture of these, appears to be scattered. It is useful here to review and summarize these data to determine some common findings. The plasma assisted chemical vapor deposition process appears to form anatase type oxide [301–304]. By dry oxidation, the oxides are identified as rutile type oxide, oxidized at 300 °C for 0.5 h, 400 °C × 0.75 h, 500 °C × 3 h, 750 °C, and 875–1,050 ° [304, 305]. On the other hand, crystalline structures of oxides formed through wet oxidation are varied. Only rutile type of TiO_2 was identified by boiling in 5 wt% H_2SO_4 or 10 wt.% HCl [301], anodizing with 0.5 M H_2SO_4 at 5–10 V [306], boiling in 10% H_2SO_4, boiling in 10% HCl, and treating in 0.5–1 M HCl [307, 308]. A mixture of rutile and anatase type oxide was found when CpTi was treated in $HF/HNO_3/H_2O$, followed by heating in 5 mol NaOH at 75 °C, followed by air oxidation at 600 °C × 1 h, or boiling in 0.1 wt% H_2SO_4 [301, 302]. On the other hand, mainly anatase, with a small fraction of rutile type mixture, was identified by anodizing in 40% H_2SO_4 at 8 V [303] and boiling in 10% CrO_3 or 65% HNO_3 × 3 h [303]. When CpTi was treated in alkaline or mildly acid solution, formed oxides were identified to be solely anatase type. It can be formed by treating CpTi in an electrolyte containing $Ca(H_2PO_4)_2$, $Ca(COCH_3)_2$, and Na(EDTA) (pH: 14) [303], 1–10% H_2O_2, followed by in-air oxidation at 500 °C × 3 h [305], anodizing in HNO_3 solution [307], 0.1 M Na_2CO_3 or 0.01M HCl [308], anodizing in 0.1 M H_2SO_4 at 12.5 mA/cm^2 [309], anodizing in 0.1M H_2SO_4 at 5V [310], or anodizing in 1M H_2SO_4 at 155 V [311]. It was also reported that when CpTi was boiled in 0.2 wt% HCl for 24 h, a mixture of anatase and brookite was formed [312].

Table 8.1 summarizes the aforementioned information, and it was found that rutile type oxides are favored when the CpTi surface is exposed to in-air oxidation at elevated temperature. Processes involved in plasma spraying or vapor deposition make CpTi surfaces covered with anatase type oxide. In wet oxidation (by either simply boiling or electrochemically anodizing), if the electrolyte (or solution) is a strong acid, the rutile type oxide is favorable, whereas if the treating solution is mild or alkaline, anatase becomes dominant, with a small fraction of rutile type [313]. The surface electronic states on Ti–Ni ($Ti_{1-x}Nix$) electrodes, which have five types of intermetallic compounds (Ti, Ti_2Ni, TiNi, $TiNi_3$, Ni) and three types of oxides (TiO_2, $NiTiO_3$, NiO), were investigated by cyclic voltammeter for various nickel mole fraction x. An oxidation current peak attributed to the oxidation of the surface states appeared at 0.53 V versus SCE in the cyclic voltammogram [314].

Table 8.1: Summary of crystalline structures of titanium oxide.

Condition	Type of crystalline structure of formed titanium oxide
Physical deposition	Anatase
Dry oxidation	Rutile
Wet oxidation solution pH	Rutile R + A Anatase low pH ————————————→ high pH (acid) (alkaline)

Oshida [315] treated surface of NiTi in various methods as listed in Table 8.2. The oxide film that was isolated from NiTi substrate was subjected to TED (transmission electron diffractometer) to identify the crystalline structure of oxide films. The results of the diffraction analyses are also listed on right column of the table; indicating that, in general, oxide films formed on NiTi are very similar to those formed on CpTi surface. Figures 8.1 through 8.3 show typical electron diffraction patterns along with SEM images for three differently treated NiTi surfaces. Figure 8.1 shows rutile type TiO_2 only, Figure 8.2 depicts a mixture of rutile type TiO_2 + anatase type TiO_2 and Figure 8.4 also shows only rutile type TiO_2, while Figure 8.4 exhibits $NiTiO_3$ incorporated with rutile type TiO_2 + $NiTiO_3$.

8.3.6 Characterization of oxides

For understanding the complex interfacial phenomena between Ti and a biological system, Hanawa [316] prepared CpTi, Ti–6Al–4V, Ti–Ni, 316L stainless steel, Co–30Cr–5Ni, Ni–20Cr, Au–9Cu–6Ag, Ag–20Pd–15Cu–12Au, which were polished and immersed in Hank's electrolyte (pH 7.4) at 37 C for 1 h, 1 day, and 30 days. Such sample surfaces were microanalyzed. It was found that calcium phosphate layers are formed on the passive films of Ti and its alloys, stainless steel, Co–Cr, Ni–Cr alloy, in a neutral electrolyte solution. The calcium phosphate formed on Ti, more so than that formed on the other alloys, is similar to apatite. In all these materials, the formation of calcium phosphates is related to the biocompatibility of the material, while the passive oxide films are related to the corrosion resistance [316].

Table 8.2: Summary of structures of oxide film formed on variously treated TiNi surfaces.

	Treatment conditions	TED pattern analysis results
Chemically		
(a)	HF/HNO$_3$/H$_2$O for 30 s	Rutile type TiO_2
(b)	Boiling 3% H$_2$O$_2$ for 2 h	Rutile type TiO_2
(c)	75 °C 5M NaCl aq. solution for 24 h	Rutile type TiO_2 + anatase type TiO_2
Electrochemically		
(d)	0.9% NaCl aqueous solution electrolyte	Rutile type TiO_2 + anatase type Ti_{O2}
Thermally		
(e)	Autoclaved at 121 °C for 6 h	Rutile type TiO_2
(f)	In-air oxidation at 300 °C for 1 h	Rutile type TiO_2 + $NiTiO_3$
(g)	In-air oxidation of (a) at 300 °C for 1 h	Rutile type TiO_2

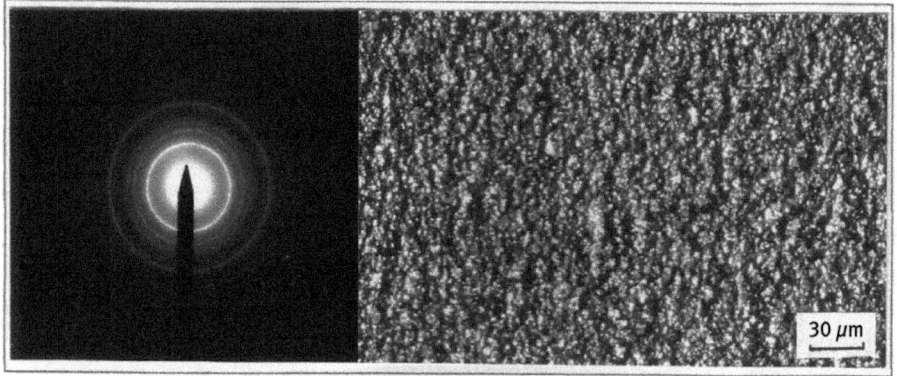

Figure 8.1: TED pattern and SEM image of oxide film formed on NiTi that was treated under condition (b) in Table 8.2.

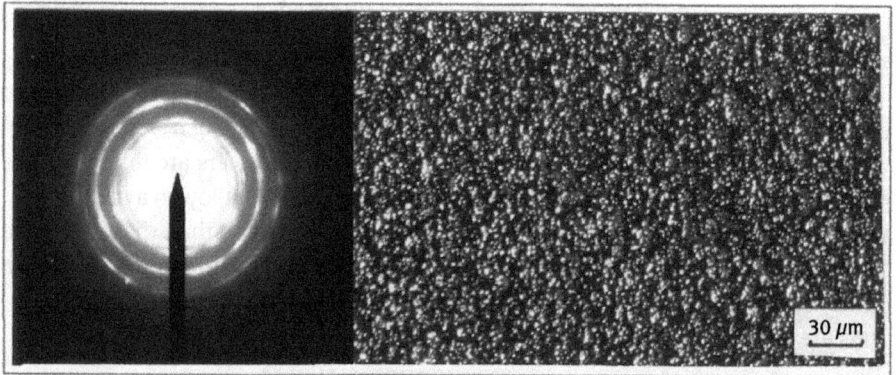

Figure 8.2: TED pattern and SEM image of oxide film formed on NiTi that was treated under condition (d) in Table 8.2.

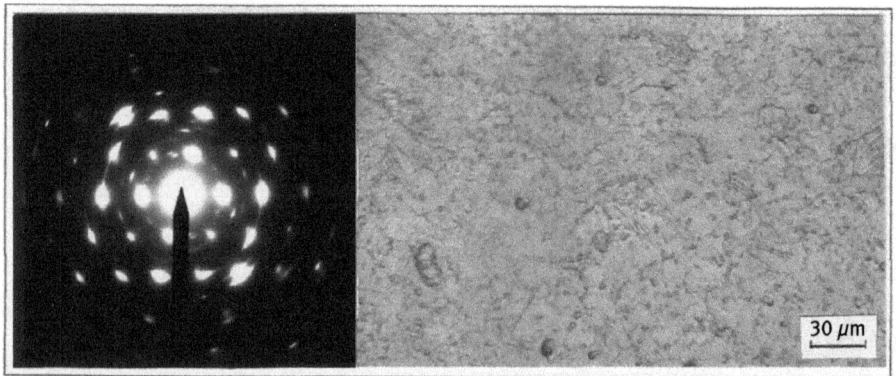

Figure 8.3: TED pattern and SEM image of oxide film formed on NiTi that was treated under condition (f) in Table 8.2.

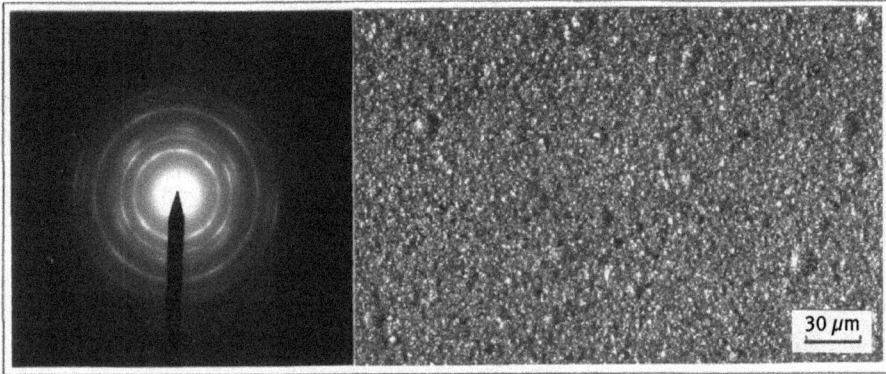

Figure 8.4: TED pattern and SEM image of oxide film formed on NiTi that was treated under condition (g) in Table 8.2.

Oshida et al. investigated the surface physics of oxides formed on various biomaterials, including Ti materials [234, 235]. Interfaces between biomaterials, tissue, and body fluids such as blood, play a key role in determining the nature of the interaction between biomaterials and the living organism. The wettability of these biomaterials in relationship to their microenvironment is an important factor to consider when characterizing surface behavior. The measure of the contact angle between a fluid and material surface can be used to define wettability for that particular microenvironment. In this study, pure Ti, Ti–6Al–4V alloy, austenitic and martensitic Ti–Ni alloys, pure Ni, AISI Type 316L stainless steel, Co–Cr alloy, and a-alumina were investigated. All metallic materials were mechanically polished and oxidized at 300 °C for 30 min in pure oxygen. Oxide films formed on the surfaces of these materials were examined under the electron microscope and their crystalline structures were identified by the electron diffraction method. The initial contact angle (q_0) and its changes (dq/dt) as a function of time in 1% NaCl solution drop was measured. The results of this study indicated that Ti and its alloys were covered with mainly TiO_2 (tetragonal structure). The NiO (cubic structure) was found on pure Ni, the spinel type oxide (cubic structure) was formed on both 316L stainless steel and Co–Cr alloy. TiO_2 (except for oxides formed on Ti–6Al–4V alloy) showed a rapid spreading characteristic in 1% NaCl solution, while a relatively slow spreading behavior was observed on the cubic structure oxides [234].

The effects of surface roughness on wettability were further investigated [235]. In addition, the spontaneous half-cell potential of all tested biomaterials was measured to correlate the wettability phenomenon to initial surface chemistry. Pure titanium and its alloys, including Ti-6Al-4V and Ti–Ni alloys, AISI Type 316L stainless, Co-Cr alloy, and pure nickel, were mechanically polished, shot-peened and pre-oxidized at 300 °C for 30 min in pure oxygen. It was found that shot-peening homogenized the surface conditions in terms of initial contact angles, TiO_2 oxide shows a higher spreading coefficient, while cubic structure oxides show a lower value, and the spreading coefficient was correlated to the magnitude of the spontaneous half-cell potential [235]. Trépanier et al. [317] studied the effect on corrosion resistance and surface characteristics of electropolishing, heat treatment, and nitric acid passivation of NiTi stents. It was shown that all of these surface treatments improve the corrosion resistance of the alloy. This improvement is attributed to the plastically deformed native oxide layer removal and replacement by a newly grown, more uniform one. The uniformity of the oxide layer, rather than its thickness and composition, seems to be the predominant factor to explain the corrosion resistance improvement. Tian et al. [287, 288] investigated the oxidation behavior of TiNi-Pd and alloying effect of Ce element. It was reported that the oxidation scale exhibits different layers, and the outmost layer is mainly composed of rutile TiO_2, while $TiNiO_2$ and Ti_4Pd_2O phases exist in the inner scale. The addition of alloying element in the TiPdNi alloy effectively impedes the outward diffusion of Ti ions and obviously improves the oxidation resistance. The addition of Ce in the TiNiPd alloy effectively impedes the Ti ions outward diffusion, causes a wide distribution of Ni and Pd over the scale depth, and obviously improves the oxidation resistance. Muhonen et al. [318] investigated the effect of different oxide layer thicknesses on the survival and attachment of osteoblastic cells (ROS-17/2.8). The AFM, XRD, and electrical resistance measurements were used to analyze the surface properties of oxidized NiTi samples and the effect of oxidation on material properties. It was shown clearly that straight correlation between oxide thickness and cellular well-being was not recognized, and the different thicknesses of oxide layer on NiTi had surprising impacts on cellular responses and also to the properties of the metal alloy. Zheng et al. [319] studied the electrochemical behavior of $Ti_{49.6}Ni_{45.1}Cu_5Cr_{0.3}$ alloy in artificial saliva solutions with a wide range of pH values and characterized the surface passive film after polarization tests. It was reported that potentiodynamic and potentiostatic test results showed that the corrosion behavior of TiNiCuCr was similar to that of NiTi alloy. Both corrosion potential E and pitting corrosion potential E showed a pH-dependent tendency that corrosion potential E and pitting corrosion potential E decreased with the increase of the pH value. The XPS results revealed the composition of the passive film consisted mainly of TiO_2 with a little amount of Ni oxides (NiO/Ni_2O_3) that was identical with NiTi alloy, besides Ni, a Cu enriched sub-layer was also found, and the nickel ion release rate showed a typical time-related decrease. Based on these findings, it was concluded that the addition of Cu and Cr had little effect on the corrosion behavior of NiTi or on the composition and the structure of the passive film. Vojtech et al. [320] oxidized NiTi nitinol wire at 480–530 °C/10 min in air. It was found that the main oxidation product at both temperatures is rutile containing a few at% Ni, beneath the rutile layer. There are titanium suboxides, showing characteristic maxima in depth profiles. Nickel in an oxidized state is present on the surface, whereas in a subsurface region of scales, there is only metallic nickel. Thickness of the total oxide layers is 70 and 140 nm after oxidation at 480 and 530 °C, respectively. The preferential oxidation of titanium causes the formation of a Ni-enriched and Ti-depleted zone, suggesting the presence of Ni_3Ti phase. The XRD studies reveal that the presence of cubic B2 NiTi phase in the base alloy is not affected by oxidation at 480–530 °C/10 min. Lin et al. [321] investigated the isothermal-oxidation behavior of $Ti_{50}Ni_{40}Cu_{10}$ SMA in 700–1,000 °C air. It was obtained that a multilayered oxide scale formed, consisting of an outermost $Cu_2O(Ni,Ti)$ layer, a layer of the mixture of TiO_2, $TiNiO_3$, and irregular small pores, a layer of the mixture of Ni(Ti,Cu), TiO_2, and irregular large pores, a $Ti(Ni,Cu)_3$ layer and an innermost $Ti_{30}Ni_{43-47}Cu_{27-23}$ layer. The apparent activation energy for the oxidation reaction of $Ti_{50}Ni_{40}Cu_{10}$ was determined to be 180 kJ/mol, and the oxidation rate follows a parabolic law. Smialek et al. [291] isothermally oxidized $NiTi_{49}$ and $NiPt_{30}Ti_{50}$ (at%) alloys (SMAs) in air over the temperature range of 500 °C to 900 °C for 100 h. It was reported that a relatively pure TiO_2 rutile structure was identified as the predominant scale surface feature, typified by a distinct highly striated and faceted crystal morphology, with crystal size proportional to oxidation temperature. In general, graded mixtures of TiO_2, $NiTiO_3$, NiO, Ni(Ti), or Pt(Ni) metallic dispersoids, and continuous Ni_3Ti or Pt-rich metal depletion zones, were observed from the gas surface to the substrate interior. Overall, substantial depletion of Ti occurred due to the formation of predominantly TiO_2 scales, and the $NiPt_{30}Ti_{50}$ alloy oxidized more slowly than the binary $NiTi_{49}$ alloy by decreasing oxygen and titanium diffusion through the thin Pt-rich layer.

The patterns of Ni release from Nitinol vary depending on the type of material (Ni–Ti alloys with low or no processing versus commercial wires or sheets). A thick TiO_2 layer generated on the wire surface during processing is often considered as a reliable barrier against Ni release. Shabalovskaya et al. [322] investigated NiTi wires with surface oxides resulting from production to identify the sources of Ni release and its distribution in the surface sublayers. Ni release was estimated using either ICPA or AAS. It was found that wire samples in the as-received state showed low breakdown potentials (200 mV). The improved corrosion resistance of these wires after treatment was not affected by strain and NiTi wires with the thickest surface oxide TiO_2 (up to 720 nm) showed the highest Ni release, attributed to the presence of particles of essentially pure Ni whose number and size increased while approaching the interface between the surface and the bulk. Esponar et al. [323] studied NiTi orthodontic archwires that have been treated using a new oxidation treatment for obtaining Ni-free surfaces. It was mentioned that the titanium oxide on the surface significantly improves corrosion resistance and decreases nickel ion release, while barely affecting transformation temperatures. This oxidation treatment avoids the allergic reactions or toxicity in the surrounding tissues produced by the chemical degradation of the NiTi. On the other hand, the lack of low friction coefficient for the NiTi superelastic archwires makes difficult the optimal use of these materials in orthodontic applications; thereby, suggesting that the decrease of this friction coefficient has been achieved by means of oxidation treatment, and transformation temperatures, friction coefficient, and ion release have been determined.

8.3.7 Oxide growth, stability, and breakdown

The oxide layer forms spontaneously in air, and normally has a thickness of between 2 and 7 nm, as mentioned previously. During machining, the relatively high-surface temperature can produce a much thicker oxide layer [324]. There is much evidence suggesting this layer grows steadily in vivo. The stoichiometry of the oxide is similar to titanium oxide (TiO_2) at the surface, and changes to a mixture of oxides at the metal/oxide interface [325]. TiO_2 is an n-type (metal-excess or donor type) semi-conductor, while TiO and Ti_2O_3 are listed as amphoteric conductors [326]. The surface is also known to have at least two types of hydroxyl groups attached to it [327]. TiO_2 is nonconducting, but electrons can tunnel through the layer. Thin oxide layers can allow the passage of electrons, leading to conformational changes and denaturing of proteins [328]. For implants located in cortical bone, the thickness of the interfacial oxide layer remains unaffected, while it increases by a factor of 3-4 on samples located in bone marrow. In general, when foreign agents, such as implant-material surface particles, are exposed to host tissue, circulating neutrophils and/or monocytes are recruited from the intravascular compartment to the location of the exposure [329]. Following recognition, the particles are encapsulated – literally engulfed by the phagocyte – and lysosomal granules, along with the particles, from a complex unit, the phagolysosome. Simultaneously, in the cytoplasmic vacuoles, enzyme releases occur and, as a result, degrading of several components takes place. Upon recognition, neutrophils and monocytes experience a "respiratory burst" and during the period, almost 20-fold oxygen consumption by the cells is observed [330]. There is convincing evidence that the consequent increase in oxygen secretion by these cells is mainly the result of this initial respiratory activity [331]. In addition, polymorphonuclear cells have been found to secrete superoxide anion (O_2^-) and H_2O_2 upon activation induced by several stimuli, including immunoglobulins and opsonized bacteria, among others [332]. O'Brien et al. [333] passivated Ti–Ni wires and vascular stent components in 10% nitric acid solution and evaluated electrochemical corrosion behaviors. It was reported that potentiodynamic polarization tests demonstrated a significant increase in breakdown potential for passivated samples (600–700 V vs SCE) compared to heat-treated surfaces (250–300 V vs SCE). Surface analysis indicated that the passivation reduces Ni and NiO content in the oxide and increases TiO_2 content. The long-term immersion tests demonstrated that Ni release from the surface of the material decreases with time and the quantity of Ni released is lower for passivated samples, and the improved corrosion resistance is maintained after prolonged immersion in 37 V 0.9% NaCl solution [333]. Tian et al. [334] investigated the stability of Ni in titanium oxide surface layers on NiTi wires known to release certain amounts of Ni by first principles density functional theory and transmission electron microscopy. It was reported that the oxides were identified as a combination of TiO and TiO_2 depending on the thickness of the layer, and the calculations indicate that free Ni atoms can exist in TiO at ambient temperature while Ni particles form in TiO_2, which was confirmed by the transmission electron microscopy observations.

8.3.8 Reaction with hydrogen peroxide

The insertion of an implant is inevitably associated with an inflammatory response due to the surgical trauma. Whether this reaction will subdue or persist may very well be dependent on the material selected, as well as the site of implantation and the loads put on it [335]. It is most probable that the interaction between material and inflammatory cells, either directly or via mediators of cell activation and migration such as complement factors [336], is of main importance for the biocompatibility of an implanted material. Further, experimental studies suggest that oxygen free radicals generated in large amounts during the respiratory burst by inflammatory cells are important for the tissue damage and persistence of the inflammatory reaction [337]. Further, it has been observed that an oxide-like layer grows on titanium as well as stainless steel implants in vivo during the implantation [338]. Model experiments have indicated that hydrogen peroxide in buffered saline solution in the millimolar concentration range produces an oxide-like layer on titanium surfaces in vitro [339], and at higher H_2O_2 concentrations, several oxidation intermediates are distinguished in the Ti-H_2O_2 system. It is well known that hydrogen peroxide is the one of the strongest oxidizing agents. These observations lead one to believe that in vivo the interaction of H_2O_2 and oxygen with the titanium would be the cause of the oxide growth observed. It is well known that Ti^{4+} ions will not produce hydroxyl radicals from H_2O_2; however, in the presence of the superoxide radical (O_2^-), both H_2O_2 and OH· may be produced through the action of superoxide dismutase and a cyclic Fenton reaction [340–343]. Such superoxide radical (O_2^-) generated during the metabolic activation is (in the body) enzymatically dismuted by hydrogen peroxide through the reaction: $O_2^- + O_2^- + 2H^+ \rightarrow H_2O_2 + O_2$. Superoxide or hydrogen peroxide alone is not potent enough in degrading adducts and, therefore, several possible mechanisms, starting from the reaction above have been suggested [343]. As a removal mechanism of bacteria inside the body, neutrophil activates such super-oxidants of the cells to generate hydrochlorous acid, which possesses the strongest antibacterial agent among various chlorine compounds. The most interesting is the Fenton type of reaction: $M^{n+} + H_2O_2 \rightarrow M^{(n+1)+} + OH· + OH^-$, where the potent hydroxyl radical, OH· is formed. In the presence of O_2^-, the oxidized metal ion, $M^{(n+1)+}$ is reduced again (to M^{n+}), and a cyclic continuous production of OH· would occur [344].

Hence, hydroxyl radicals formed from hydrogen peroxide during the inflammatory response are potent agents for cellular deterioration. The behavior of implanted material in terms of its ability to sustain or stop free radical formation may be therefore very important. In vitro studies of Ti, which is known to be biocompatible and osseointegrates into human bone, were carried out. The production of free radicals from H_2O_2 at Ti and TiO_2 surfaces was measured by spin trapping techniques. It was found that there is no sustained hydroxyl radical production at a titanium (oxide) surface, due to the quenching of the Fenton reaction through both trapping and oxidation of superoxide radicals in a TiO·OH adduct. Janzén et al. concluded that the degree of surface-induced hydroxyl radical formation from H_2O_2 through the Fenton reaction may be of importance for the behavior of implanted materials [344].

Tengvall et al. [345] investigated the role of Ti and TiO_2, as well as other metals, in the inflammatory response through the Fenton reaction. The TiOOH matrix formed traps the superoxide radical so that no or very small amounts of free hydroxyl radicals are produced. Ellipsometry and spin trapping with spectrophotometry and electron spin resonance were used to study the interaction between Ti and H_2O_2. Spectrophotometry results indicated that Ti, Zr, Au, and Al are low free OH-radical producers. It was suggested that a hydrated TiOOH matrix, after the inflammatory reaction, responded to good ion exchange properties, and extracellular components may interact with the Ti^{4+}-H_2O_2 compound before matrix formation. The TiOOH matrix is formed when the H_2O_2 coordinated to the Ti^{4+}-H_2O_2 complex is decomposed to water and oxygen [345].

It has been reported that Ti implants inserted in the human body for many years have shown high-oxidation rates at the areas relatively adjacent to the Ti implant, and the implants have shown a dark pigmentation due to the interaction with H_2O_2 [346]. It is well known that H_2O_2 is a strong oxidant that will increase the redox potential of the system. Gas evolution was observed on the Ti surface in the presence of H_2O_2 in the PBS (8.77NaCl g/L, 3.58 $Na_2HPO_4 \cdot 12H_2O$, 1.36KH_2PO_4, pH: 7.2–7.4) solution. This is due to the decomposition of H_2O_2 into water and molecular oxygen according to: $2H_2O_2 \rightarrow H_2O + O_2$, which is catalyzed by the presence of TiO_2 on the surface due to the combined effect of the electron-donating and electron-accepting properties of TiO_2 occurring simultaneously [345], and the formation of some Ti-complex, which can lead to a higher electrode potential [347]. Pan et al. [348] performed electrochemical measurements, XPS, and scanning tunneling microscopy analyses to study the effect of hydrogen peroxide on the passivity of CpTi (grade II) in a PBS solution. The results indicate that the passive film formed in the PBS solution – with and without addition of H_2O_2 – may be described with a two-layer structure model. The inner layer has a structure close to TiO_2, whereas the outer layer consists of hydroxylated compounds. The introduction of H_2O_2 in the PBS solution broadens the hydroxylate-rich region, probably due to the formation of a Ti^{4+}-H_2O_2 complex. Furthermore, the presence of H_2O_2 results in enhanced dissolution of Ti and a rougher surface on a microscopic scale. It was also observed that H_2O_2 addition furthermore seems to facilitate the incorporation of phosphate ions into the thicker porous layer [348].

The influence of H_2O_2 generated by an inflammatory response also has been suggested as an explanation for the high rate of Ti oxidation/corrosion in vivo [337, 345, 349]. The incorporation of mineral ions into the oxide film is believed to be important for the osseointegration behavior, which can lead to direct bone-titanium contact and strong bonding that in turn can contribute to the load bearing capacity of titanium implants [350]. To analyze titanium's response to representative surgical wound environments, Bearinger et al. [351] examined CpTi and Ti–6Al–4V, which were exposed to PBS with 30 mm with hydrogen peroxide (H_2O_2) addition. The study was characterized by simultaneous electrochemical AFM and step-polarization impedance spectroscopy. It was found that surfaces were covered with protective oxide domes that indicated topography changes with potential and time of immersion, and less oxide dome coarsening was noted on surfaces treated with PBS containing H_2O_2 than on surfaces exposed to pure PBS. In both types of solutions, oxide (early) resistances of CpTi samples were higher than Ti–6Al–4V oxide resistances, but CpTi oxide resistance was lower in the hydrogen peroxide solution compared to pure PBS. Differences in electrical properties between CpTi and Ti–6Al–4V surfaces suggest that CpTi, but not Ti–6Al–4V, has catalytic activity on H_2O_2, and that the catalytic activity of CpTi oxide affects its ability to grow TiO_2 [351]. Chu et al. [352] employed Fenton's oxidation to modify the surface of biomedical NiTi SMA. The influences of Fenton's oxidation on the surface microstructure, blood compatibility, leaching of harmful Ni ions, and corrosion resistance in simulated body fluids were also assessed. The mechanical stability of the surface titania film produced by Fenton's oxidation as well as their effects on the shape-memory behavior of the SMA are studied by bending tests. It was reported that Fenton's oxidation produces a novel nanostructured titania gel film with a graded structure on the NiTi substrate without an intermediate Ni-rich layer that is typical of high-temperature oxidation. There is a clear Ni-free zone near the top surface of the titania film. The surface structural changes introduced by Fenton's oxidation improve the electrochemical corrosion resistance and mitigate Ni release; aging in boiling water improves the crystallinity of the titania film and further reduces Ni leaching. Blood platelet adhesion is remarkably reduced after Fenton's oxidation, suggesting that the treated SMA has improved thromboresistance. The enhancement of blood compatibility is believed to stem from the improved hemolysis resistance, the surface wettability and the intrinsic electrical characteristics of the titania film. The titania film produced by Fenton's oxidation has good mechanical stability and does not adversely impact the shape-memory behavior of NiTi; suggesting that Fenton's oxidation is a promising low-temperature, low-cost surface modification method for improving the surface properties of biomedical NiTi SMA [352].

References

[1] Oshida Y, Farzin-Nia F. Response of Ti-Ni Alloys for Dental Biomaterials to Conditions in the Mouth. In: Yoneyama, et al. ed., Shape Memory Alloys for Biomedical Applications. CRC Woodhead Publishing Limited, Cambridge, England, 2009, 101–49.

[2] Brockhurst PJ. Dental Materials: New Territories for Materials Science. In: Metals Forum. Australian Inst Metals, 1980, 3, 200–10.

[3] Attin T. The security and use of carbamide peroxide gels in bleaching therapy. Dtsch. Zahnärztl Z. 1998, 53, 11–6.

[4] Rosentritt M, Lang R, Plein T, Behr M, Handel G. Discoloration of restorative materials after bleaching application. Quintessence Int. 2005, 36, 33–9.

[5] Goldstein GR, Kiremidjian-Schuhmacher L. Bleaching, 'Is it safe and effective?'. J. Prosthet. Dent. 1993, 69, 315–8.

[6]   Wilhelmsen W, Grande AP. The influence of hydrofluoric acid and fluoride ion on the corrosion and passive behavior of titanium. Electrochim. Acta 1987, 32, 1469–72.
[7]   Nakagawa M, Matsuya S, Udoh K. Effect of fluoride and dissolved oxygen concentration on the corrosion behavior of pure titanium and titanium alloys. Dent. Mater. J. 2002, 21, 83–92.
[8]   Takemoto S, Hatori M, Yoshinari M, Kawada E, Oda Y. Corrosion behavior and surface characterization of titanium in solution containing fluoride and albumin. Biomaterials 2005, 26, 829–37.
[9]   Dahl JE, Pallesen U. Tooth bleaching – a critical review of the biological aspects. Crit. Rev. Oral Biol. Med. 2003, 14, 292–304.
[10] Oshida Y, Sellers CB, Mirza K, Farzin-Nia F. Corrosion of dental materials by dental treatment agents. Mater. Sci. Eng. 2005, C25, 343–8.
[11] Oshida Y, Mirza K, Sellers CB, Panyayong W, Farzin-Nia F. Effects of Dental Treatment Agents on Discoloration/Corrosion Behavior of Metallic Dental Materials. In: Shrivastava, ed., Medical Device, Materials.ASM International, Park, OH, 2004, 467–70.
[12] Lausmaa J, Kasemo B, Hansson S. Accelerated oxide grown on titanium implants during autoclaving caused by fluorine contamination. Biomaterials 1985, 6, 23–7.
[13] Barton K. Protection Against Atmospheric Corrosion. J. Wiley & Sons, New York, 1973, 202.
[14] Nilner K, Holland RI. Electrochemical potentials of amalgam restorations in vivo. Scand. J. Dent. Res. 1985, 93, 357–9.
[15] Corso PP, German RM, Simmons HD. Corrosion evaluation of gold-based dental alloys. J. Dent. Res. 1985, 64, 854–9.
[16] Reclaru L, Meyer JM. Zonal coulometric analysis of the corrosion resistance of dental alloys. J. Dent. Res. 1995, 23, 301–11.
[17] Ewers GJ, Thornber MR. The effect of a simulated environment on dental alloys. J. Dent. Res. 1983, 62, 330 (Abstract No. 330).
[18] Brown SA, Simpson JP. Crevice and fretting corrosion of stainless steel plates and Screws. J. Biomed. Mater. Res. 1981, 15, 867–78.
[19] Man HC, Gabe DR. A study of pitting potentials for some austenitic stainless steels using a potentiodynamic technique. Corr. Sci. 1981, 21, 713–21.
[20] Oshida Y, Sachdeva R, Miyazaki S. Microanalytical characterization and surface modification of NiTi orthodontic arch wires. J. Biomed. Mater. Eng. 1992, 2, 51–69.
[21] Cahoon JR, Bandyopadhya R, Tennese L. The concept of protection potential applied to the corrosion of metallic orthopedic implants. J. Biomed. Mater. Res. 1975, 9, 259–64.
[22] Sutow EJ, Pollack SR, Korostoff A. An in vitro investigation of the anodic polarization and capacitance behavior of 316L stainless steel. J. Biomed. Mater. Res. 1976, 10, 671–93.
[23] Syrett BC, Wing SS. An electrochemical investigation of fretting corrosion of surgical implant materials. Corrosion 1978, 34, 379–86.
[24] Hoar TP, Mears DC. Corrosion-resistant alloys in chloride solutions: materials for surgical implants. Proc. Royal Soc. 1966, 294A, 486–510.
[25] Al-Ali S, Oshida Y, Andres CJ, Barco MT, Brown DT, Hovijitra S, Ito M, Nagasawa S, Yoshida T. Effects of coupling methods on galvanic corrosion behavior of commercially pure titanium with dental precious alloys. J. Biomed. Mater. Eng. 2005, 15, 307–16.
[26] Dunigan-Miller J. Electrochemical corrosion behavior of commercially pure titanium materials fabricated by different methods in three electrolytes. Indiana University Master Thesis, 2004.
[27] Schiff N, Grosgogeat B, Lissac M, Dalard F. Influence of fluoride content and pH on the corrosion resistance of titanium and its alloys. Biomaterials 2002, 23, 1995–2002.
[28] Schiff N, Grosgogeat B, Lissac M, Dalard F. Influence of fluoridated mouthwashes on corrosion resistance of orthodontics wires. Biomaterials 2004, 25, 4535–42.
[29] Noguti J, de Oliveira F, Peres RC, Renno ACM, Ribeiro DA. The role of fluoride on the process of titanium corrosion in oral cavity. BioMetals 2012, 25, 859–62.

[30]   Waler MP, White RJ, Kula KS. Effect of fluoride prophylactic agents on the mechanical properties of nickel-titanium-based orthodontic wires. Amer. J. Orthodont. Dentfac. Orthop. 2005, 127, 662–9.

[31]   Yanisarapan T, Thunyakitpisal P, Chantarawaratit P-O. Corrosion of metal orthodontic brackets and archwires caused by fluoride-containing products: cytotoxicity, metal ion release and surface roughness. Orthod. Waves 2018, 77, 79–89.

[32]   Schiff N, Dalard F, Lissac M, Morgon L, Grosgogeat B. Corrosion resistance of three orthodontic brackets: a comparative study of three fluoride mouthwashes. Eur. J. Orthod. 2005, 27, 541–9.

[33]   Schiff N, Boinet M, Morgon L, Lissac M, Dalard F, Grosgogeat B. Galvanic corrosion between orthodontic wires and brackets in fluoride mouthwashes. Eur. J. Orthod. 2006, 28, 298–304.

[34]   Iijima M, Yuasa T, Endo K, Muguruma T, Ohno H, Mizoguchi I. Corrosion behavior of ion implanted nickel-titanium orthodontic wire in fluoride mouth rinse solutions. Dent. Mater. J. 2010, 29, 53–8.

[35]   Srivastava K, Chandra PK, Kamat N. Effect of fluoride mouth rinses on various orthodontic archwire alloys tested by modified bending test: an in vitro study. Indian J. Dent. Res. 2012, 23, 433–4.

[36]   Harzer W, Schröter A, Gedrange T, Muschter F. Sensitivity of titanium brackets to the corrosive influence of fluoride-containing toothpaste and ea. Angle Orthod. 2001, 71, 318–23.

[37]   Kassab EJ, Gomes JP. Assessment of nickel titanium and beta titanium corrosion resistance behavior in fluoride and chloride environments. Angle Orthod. 2013, 83, 864–9.

[38]   Kwon YH, Cheon Y-D, Seol H-J, Lee J-H, Kim H-J. Changes on NiTi orthodontic wired due to acidic fluoride solution. Dent. Mater. J. 2004, 23, 557–65.

[39]   Kwon YH, Cho HS, Noh DJ, Kim HI, Kim KH. Evaluation of the effect of fluoride-containing acetic acid on NiTi wires. J. Biomed. Mater. Res. 2005, 72, 102–8.

[40]   Ahn H-S, Kim K-J, Seol H-J, Lee J-H, Kim H-I, Kwon YH. Effect of pH and temperature on orthodontic NiTi wires immersed in acidic fluoride solution. J. Biomed. Mater. Res. Part B: Appl. Biomater. 2006, 79B, 7–15.

[41]   Cheng Y, Cai W, Zhao LC. Effects of Cl⁻ ion concentration and pH on the corrosion properties of NiTi alloy in NaCl solution. J. Mater. Sci. Lett. 2003, 22, 239–40.

[42]   Sarkar NK, Redmond W, Schwaninger B, Goldberg AJ. The chloride corrosion behaviour of four orthodontic wires. J. Oral Rehabil. 1983, 10, 121–8.

[43]   Cavalleri G, Cantatore G, Costa A, Grillenzoni M, Comin Chiaramonti L, Gerosa R. The corrosive effects of sodium hypochlorite on nickel-titanium endodontic instruments: assessment by digital scanning microscope. Minerva Stomatol. 2009, 58, 225–31.

[44]   Berutti E, Angelini E, Rigolone M, Migliaretti G, Pasqualini D. Influence of sodium hypochlorite on fracture properties and corrosion of ProTaper Rotary instruments. Int. Endod. J. 2006, 39, 693–9.

[45]   Nóvoa XR, Martin-Biedma B, Varela-Patiño P, Collazo A, Macías-Luaces A, Cantatore G, Pérez MC, Magán-Muñoz F. The corrosion of nickel-titanium rotary endodontic instruments in sodium hypochlorite. Int. Endod. J. 2007, 40, 36–44.

[46]   Stokes OW, Di Fiore PM, Barss JT, Koerber A, Gilbert JL, Lautenschlager EP. Corrosion in stainless-steel and nickel-titanium files. J. Endod. 1999, 25, 17–20.

[47]   Ali Saghiri M, Asatourian A, Godoy FG, Sheibani N. Influence of an innovative anti-corrosive solution on resistance of endodontic NiTi rotary instruments: a preliminary study. Eur. Endod. J. 2018, 3, 55–60.

[48]   Liu C, Chu PK, Lin G, Yang D. Effects of Ti/TiN multilayer on corrosion resistance of nickel-titanium orthodontic brackets in artificial saliva. Corros. Sci. 2007, 49, 3783–96.

[49]   Mareci D, Chelariu R, Cailean A, Sutiman D. Electrochemical characterization of $Ni_{47.7}Ti_{37.8}Nb_{14.5}$ shape memory alloy in artificial saliva. Mater. Corros. 2011, 63, 807–12.

[50]   Huang H-H. Surface characterizations and corrosion resistance of nickel-titanium orthodontic archwires in artificial saliva of various degrees of acidity. J. Biomed. Mater. Res. A 2005, 74, 629–39.

[51]   Huang HH. Corrosion resistance of stressed NiTi and stainless steel orthodontic wires in acid artificial saliva. J. Biomed. Mater. Res. A 2003, 66, 829–39.

[52]   Wang J, Li N, Han EH, Ke W. Effect of pH, temperature and Cl- concentration on electrochemical behavior of NiTi shape memory alloy in artificial saliva. J. Mater. Sci. Mater. Med. 2006, 17, 885–90.

[53]   Trolić IM, Turco G, Contardo L, Serdarević NL, Otmačić H, Ćurković, Špalj S. Corrosion of nickel-titanium orthodontic archwires in saliva and oral probiotic supplements. Acta Stomatol. Croat. 2017, 51, 316–25.

[54]   Wang QY, Zheng YF. The electrochemical behavior and surface analysis of Ti50Ni47.2Co2.8 alloy for orthodontic use. Dent. Mater. 2008, 24, 1207–11.

[55]   Lee T-H, Huang T-K, Lin S-Y, Chen L-K, Chou M-Y, Huang H-H. Corrosion resistance of different nickel-titanium archwires in acidic fluoride-containing artificial saliva. Angle Orthod. 2010, 80, 547–53.

[56]   Pulikkottil VJ, Chidambaram S, Bejoy PU, Femin PK, Paul P, Rishad M. Corrosion resistance of stainless steel, nickel-titanium, titanium molybdenum alloy, and ion-implanted titanium molybdenum alloy archwires in acidic fluoride-containing artificial saliva: an *in vitro* study. J. Pharm. Bioallied. Sci. 2016, 8, S96–99.

[57]   Ünal Hİ. Effect of fluoride added artificial saliva solution on orthodontic wires. Prot. Met. Phys. Chem. S. 2012, 48, 367–70.

[58]   Kao CT, Huang TH. Variations in surface characteristics and corrosion behavior of metal brackets and wires in different electrolyte solutions. Eur. J. Orthod. 2010, 32, 555–60.

[59]   Speck KM, Fraker AC. Anodic polarization behavior of Ti-Ni and Ti-6A1-4V in simulated physiological solutions. J. Dent. Res. 1980, 59, 1590–5.

[60]   Rondelli G. Corrosion resistance tests on NiTi shape memory alloy. Biomaterials 1996, 17, 2003–8.

[61]   Montero-Ocampo C, Lopez H, Salinas Rodriguez A. Effect of compressive straining on corrosion resistance of a shape memory Ni-Ti alloy in Ringer's solution. J. Biomed. Mater. Res. 1996, 32, 583–91.

[62]   Kuphasuk C, Oshida Y, Andres CJ, Hovijitra ST, Barco MT, Brown DT. Electrochemical corrosion of titanium and titanium-based alloys. J. Prosthet. Dent. 2001, 85, 195–202.

[63]   Carroll WM, Kelly MJ. Corrosion behavior of nitinol wires in body fluid environments. J. Biomed. Mater. Res. Part A 2003, 67A, 1123–30.

[64]   Yan XJ, Yang DZ. Corrosion resistance of a laser spot-welded joint of NiTi wire in simulated human body fluids. J. Biomed. Mater. Res. A 2006, 77, 97–102.

[65]   Liang C, Zheng R, Huang N, Wu B. Electrochemical behaviour of Ti-Ni SMA and Co-Cr alloys in dynamic Tyrode's simulated body fluid. J. Mater. Sci. Mater. Med. 2010, 21, 1421–6.

[66]   Khalil-Allafi J, Amin-Ahmadi B, Zare M. Biocompatibility and corrosion behavior of the shape memory NiTi alloy in the physiological environments simulated with body fluids for medical applications. Mater. Sci. Eng. C 2010, 30, 1112–7.

[67]   Hansen AW, Führ LT, Antonini LM, Villarinho DJ, Marino CEB, de Fraga Malfatti C. The electrochemical behavior of the NiTi alloy in different simulated body fluids. Mater. Res. 2015, 18; http://dx.doi.org/10.1590/1516-1439.305614.

[68]   Stergioudi F, Vogiatzis C, Pavlidou E, Skolianos S, Michailidis N. Corrosion resistance of porous NiTi biomedical alloy in simulated body fluids. Smart Mater. Struct. 2016, 25(9), 095024; doi: 10.1088/0964-1726/25/9/095024.

[69]   Rosalbino F, Macciò D, Scavino G, Saccone A. In vitro corrosion behaviour of Ti-Nb-Sn shape memory alloys in Ringer's physiological solution. J. Mater. Sci. Mater. Med. 2012, 23, 865–71.

[70]   Randin JP. Corrosion behavior of nickel-containing alloys in artificial sweat. J. Biomed. Mater. Res. 1988, 22, 649–66.

[71]   Ahmed RA. Electrochemical properties of $Ni_{47}Ti_{49}Co_4$ shape memory alloy in artificial urine for urological implant. Ind. Eng. Chem. Res. 2015; https://doi.org/10.1021/acs.iecr.5b00838.

[72]   Hasegawa Y, Goto S, Ogura H. Effect of EDTA solution on corrosion fatigue of Ni-Ti files with different shapes. Dent. Mater. J. 2014, 33, 415–21.

[73]   Shabalovskaya SA, Anderegg JW, Undisz A, Rettenmayr M, Rondelli GC. Corrosion resistance, chemistry, and mechanical aspects of Nitinol surfaces formed in hydrogen peroxide solutions. J. Biomed. Mater. Res. Part B: Appl. Biomater. 2012, 100B, 1490–9.

[74]   Chao Z, Yaomu X, Chufeng L, Conghua L. The effect of mucin, fibrinogen and IgG on the corrosion behaviour of Ni–Ti alloy and stainless steel. BioMetals 2017, 30, 367–77.

[75]   Takemoto S, Hatori M, Yoshinari M, Kawada E, Oda Y. Corrosion behavior and surface characterization of titanium in solution containing fluoride and albumin. Biomaterials 2005, 26, 829–37.

[76]   Ide K, Hattori M, Yoshinari M, Kawada E, Oda Y. The influence of albumin on corrosion resistance of titanium in fluoride solution. Dent. Mater. J. 2003, 22, 359–70.

[77]   Hang R, Ma S, Ji V, Chu PK. Corrosion behavior of NiTi alloy in fetal bovine serum. Electrochim. Acta 2010, 55, 5551–60.

[78]   Ryhänen J, Niemi E, Serlo W, Niemala E, Sandvik P, Pernu H, Salo T. Biocompatibility of nickel-titanium shape memory metal and its corrosion behavior in human cell culture. J. Biomed. Mater. Res. 1997, 35, 451–7.

[79]   Rozali AA, Masdek NRN, Murad MC, Salleh Z, Koay MH. The effect of protein concentration on corrosion of nitinol alloy. Materialwiss. Werkst. 2018, 49, 489–93.

[80]   Marton D, Sprague EA. Synergistic effect of cyclic mechanical strain and activated macrophage cells on nitinol corrosion. Cardiovasc. Eng. Techn. 2010, 1, 216–24.

[81]   Li Y-H, Rao G-B, Rong L-J, Li Y-Y. The influence of porosity on corrosion characteristics of porous NiTi alloy in simulated body fluid. Mater. Lett. 2002, 57, 448–51.

[82]   Es-Souni M, Es-Souni M, Fischer-Brandies H. On the properties of two binary NiTi shape memory alloys. Effects of surface finish on the corrosion behaviour and in vitro biocompatibility. Biomaterials 2002, 23, 2887–94.

[83]   Pereira ESJ, Amaral CCF, Gomes JACP, Peters OA, Buono VTL, Bahia MGA. Influence of clinical use on physical-structural surface properties and electrochemical potential of NiTi endodontic instruments. Int. Endod. J. 2018, 51, 515–21.

[84]   Katić V, Curković HO, Semenski D, Baršić G, Marušić K, Spalj S. Influence of surface layer on mechanical and corrosion properties of nickel – titanium orthodontic wires. Angle Orthod. 2014, 84, 1041–8.

[85]   Viswanathan S, Mohan L, John S, Bera P, Anandan C. Effect of surface finishing on the formation of nanostructure and corrosion behavior of Ni–Ti alloy. Surf Interface Anal. 2017, 49, 450–6.

[86]   Rokicki R, Hryniewicz T, Pulletikurthi C, Rokosz K, Munroe N. Towards a better corrosion resistance and biocompatibility improvement of nitinol medical devices. J. Mater. Eng. Perform. 2015, 24, 1634–40.

[87]   Kim J, Park JK, Kim HK, Unnithan AR, Kim CS, Park CH. Optimization of electropolishing on NiTi alloy stents and its influence on corrosion behavior. J. Nanosci. Nanotechnol. 2017, 17, 2333–9.

[88]   Shabalovskaya S, Rondelli G, Anderegg J, Simpson B, Budko S. Effect of chemical etching and aging in boiling water on the corrosion resistance of nitinol wires with black oxide resulting from manufacturing process. J. Biomed. Mater. Res. Part B Appl. Biomater. 2003, 66B, 331–40.

[89]   Clarke B, Carroll W, Rochev Y, Hynes M, Bradley D, Plumley D. Influence of Nitinol wire surface treatment on oxide thickness and composition and its subsequent effect on corrosion resistance and nickel ion release. J. Biomed. Mater. Res. A 2006, 79, 61–70.

[90]   Olumi S, Sadrnezhaad SK, Atai M. The influence of surface nanocrystallization induced by shot peening on corrosion behavior of NiTi alloy. J. Mater. Eng. Perform. 2015, 24, 3093–9.

[91]   Zhang R, Mankoci S, Walters N, Gao H, Zhang H, Hou X, Qin H, Ren Z, Zhou X, Doll GL, Martini A, Sahai N, Dong Y, Ye C. Effects of laser shock peening on the corrosion behavior and biocompatibility of a nickel-titanium alloy. J. Biomed. Mater. Res. B Appl. Biomater. 2018; doi: 10.1002/jbm.b.34278.

[92]   Trépanier C, Tabrizian M, Yahia L'H, Bilodeau L, Piron DL. Effect of modification of oxide layer on NiTi stent corrosion resistance. J. Biomed. Mater. Res. 1998, 43, 433–40.

[93]   Cissé O, Savadogo O, Wu M, Yahia L. Effect of surface treatment of NiTi alloy on its corrosion behavior in Hanks' solution. J. Biomed. Mater. Res. 2002, 61, 339–45.

[94]   Dehghanghadikolaei A, Ibrahim H, Amerinatanzi A, Hashemi M, Moghaddam NS, Elahinia M. Improving corrosion resistance of additively manufactured nickel–titanium biomedical devices by micro-arc oxidation process. J. Mater. Sci. 2019, 54, 7333–55.

[95]   Thierry B, Tabrizian M, Trepanier C, Savadogo O, Yahia L. Effect of surface treatment and sterilization processes on the corrosion behavior of NiTi shape memory alloy. J. Biomed. Mater. Res. 2000, 51, 685–93.

[96]   O'Hoy PYZ, Messer HH, Palamara JEA. The effect of cleaning procedures on fracture properties and corrosion of NiTi files. Int. Endod. J. 2003, 36, 724–32.

[97]   Bhattacharya RS, Raffoul CN, Rai AK. A potentiodynamic evaluation of NiMo and NiTi amorphous thin films. Corrosion 1986, 42, 236–40.

[98]   Hu T, Xin YC, Wu SL, Chu CL, Lu J, Guan L, Chen HM, Hung TF, Yeung KWK, Chu PK. Corrosion behavior on orthopedic NiTi alloy with nanocrystalline/amorphous surface. Mater. Chem. Phys., 2011, 126, 102–7.

[99]   Pun DK, Berzins DW. Corrosion behavior of shape memory, superelastic, and nonsuperelastic nickel-titanium-based orthodontic wires at various temperatures. Dent. Mater. 2008, 24, 221–7.

[100]   Vojtěch D, Voděrová M, Fojt J, Novák P, Kubásek T. Surface structure and corrosion resistance of short-time heat-treated NiTi shape memory alloy. Appl. Surf. Sci. 2010, 257, 1573–82.

[101]   Ionita D, Caposi M, Demetrescu I, Ciuca S, Gherghescu IA. Effect of artificial aging conditions on corrosion resistance of a TiNi alloy. Mater. Corros. 2015, 66, 472–8.

[102]   Chan CW, Man HC, Yue TM. Effect of post-weld heat-treatment on the oxide film and corrosion behaviour of laser-welded shape memory NiTi wires. Corros. Sci. 2012, 56, 158–67.

[103]   Zhao S, Yu W, Sun D. Susceptibility to corrosion and *in vitro* biocompatibility of a laser-welded composite orthodontic arch wire. Ann. Biomed. Eng. 2014, 42, 222–30.

[104]   Mirshekari GR, Kermanpur A, Saatchi A, Sadrnezhaad SK, Soleymani AP. Microstructure, cyclic deformation and corrosion behavior of laser welded NiTi shape memory wires. J. Mater. Eng. Perform. 2015, 24, 3356–674.

[105]   Michael A, Pequegnat A, Wang J, Zhou YN, Khan MI. Corrosion performance of medical grade NiTi after laser processing. Surf. Coat. Technol. 2017, 324, 478–85.

[106]   Oshida Y. Surface Engineering and Technology for Biomedical Implants. Momentum Press, New York, USA, 2014.

[107]   Sasikumar Y, Indira K, Rajendran N. Surface modification methods for titanium and its alloys and their corrosion behavior in biological environment: a review. J. Bio- Tribo-Corros. 2019, 5, 36; doi: 10.1007/s40735-019-0229-5.

[108]   Mohammed MT, Khan ZA, Siddiquee AN. Surface modifications of titanium materials for developing corrosion behavior in human body environment: a review. Procedia Mater. Sci. 2014, 6, 1610–8.

[109] Chakraborty R, Datta S, Raza MS, Saha P. A comparative study of surface characterization and corrosion performance properties of laser surface modified biomedical grade nitinol. Appl. Surf. Sci. 2019, 469, 753–63.

[110] Green SM, Grant DM, Wood JV, Johanson A, Johanson E, Sarholt-Kristensen L. Effect of N⁺ implantation on the shape-memory behaviour and corrosion resistance of an equiatomic Ni Ti alloy. J. Mater. Sci. Lett. 1993, 12, 618–9.

[111] Poon RWY, Ho JPY, Liu X, Chung CY, Chu PK, Yeung KWK, Lu WW, Cheung KMC. Formation of titanium nitride barrier layer in nickel–titanium shape memory alloys by nitrogen plasma immersion ion implantation for better corrosion resistance. Thin Solid Films 2005, 488, 20–5.

[112] Liu XM, Wu SL, Chu PK, Chung CY, Chu CL, Chan YL, Yeung KWK, Lu WW, Cheung KMC, Luk KDK. In vitro corrosion behavior of TiN layer produced on orthopedic nickel–titanium shape memory alloy by nitrogen plasma immersion ion implantation using different frequencies. Surf. Coat. Technol. 2008, 202, 2463–6.

[113] Poon RWY, Ho JPY, Liu X, Chung CY, Chu PK, Yeung KWK, Lu WW, Cheung KMC. Anti-corrosion performance of oxidized and oxygen plasma-implanted NiTi alloys. Mater. Sci. Eng., A 2005, 390, 444–51.

[114] Mohan L, Chakraborty M, Viswanathan S, Mandal C, Bera P, Aruna ST, Anandan C. Corrosion, wear, and cell culture studies of oxygen ion implanted Ni–Ti alloy. Surf. Interface Anal. 2017, 49, 828–36.

[115] Oshida Y, Hashem A, Nishihara T, Yapchulay MV. Fractal dimension analysis of mandibular alveolar bone. The 3rd World Congress for Oral Implantology, 1994, 202 (Abstract No. F-124).

[116] Oshida Y. Bioscience and Bioengineering of Titanium Materials. Elsevier, Oxford, UK, 2007.

[117] Maleki-Ghaleh H, Khalil-Allafi J, Khalili V, Shakeri MS, Javidi M. Effect of hydroxyapatite coating fabricated by electrophoretic deposition method on corrosion behavior and nickel release of NiTi shape memory alloy. Mater. Corros. 2014, 65, 725–32.

[118] Marashi-Najafi F, Khalil-Allafi J, Etminanfar MR, Faezi-Alivand R. Corrosion resistance and in vitro evaluation of the pulsed current electrodeposited hydroxyapatite coatings on Nitinol shape memory alloy. Mater. Corros. 2017, 68, 1237–45.

[119] Qiu D, Wang A, Yin Y. Characterization and corrosion behavior of hydroxyapatite/zirconia composite coating on NiTi fabricated by electrochemical deposition. Appl. Surf. Sci. 2010, 257, 1774–8.

[120] Sheykholeslami SOR, Khalil-Allafi J, Fathyunes L. Preparation, characterization, and corrosion behavior of calcium phosphate coating electrodeposited on the modified nanoporous surface of NiTi alloy for biomedical applications. Metall. Mater. Trans. A 2018, 49, 5878–87.

[121] Endo K, Sachdeva R, Araki Y, Ohno H. Effects of titanium nitride coatings on surface and corrosion characteristics of NiTi alloy. Dent. Mater. J. 1994, 13, 228–39.

[122] Starosvetsky D, Gotman I. Corrosion behavior of titanium nitride coated Ni-Ti shape memory surgical alloy. Biomaterials 2001, 22, 1853–9.

[123] Krishnan M, Seema A, Kumar AV, Varthini NP, Sukumaran K, Pawar VR, Arora V. Corrosion resistance of surface modified nickel titanium archwires. Angle Orthod. 2014, 4, 358–67.

[124] Sugisawa H, Kitaura H, Ueda K, Kimura K, Ishida M, Ochi Y, Kishikawa A, Ogawa S, Takano-Yamamoto T. Corrosion resistance and mechanical properties of titanium nitride plating on orthodontic wires. Dent. Mater. J. 2018, 37, 286–92.

[125] Shao AL, Cheng Y, Zhou Y, Li M, Xi TF, Zheng YF, Wei SC, Zhang DY. Electrochemistry properties of multilayer TiN/Ti coatings on NiTi alloy for cardiac occluder application. Surf. Coat. Technol. 2013, 228, S257–61.

[126] Maleki-Ghaleh H, Khalil-Allafi J, Aghaie E, Siadati MH. Effect of TiO₂–Ti and TiO₂–TiN composite coatings on corrosion behavior of NiTi alloy. Surf. Interface Anal. 2015, 47, 99–104.

[127]   Roman D, Bernardi JC, Boeira CD, de Souza FS, Spinelli A, Figueroa CA, Basso RLO. Nanome-
        chanical and electrochemical properties of ZrN coated NiTi shape memory alloy. Surf. Coat.
        Technol. 2012, 206, 4645–50.

[128]   Liu CL, Hu D, Xu J, Yang D, Qi M. In vitro electrochemical corrosion behavior of functionally
        graded diamond-like carbon coatings on biomedical Nitinol alloy. Thin Solid Films 2006,
        496, 457–62.

[129]   Branzoi IV, Iordoc M, Branzoi F, Vasilescu-Mirea R, Sbarcea G. Influence of diamond-like
        carbon coating on the corrosion resistance of the NITINOL shape memory alloy. Surf.
        Interface Anal. 2010, 42, 502–9.

[130]   Huang SY, Huang JJ, Kang T, Diao DF, Duan YZ. Coating NiTi archwires with diamond-like
        carbon films: reducing fluoride-induced corrosion and improving frictional properties. J.
        Mater. Sci.: Mater. Med. 2013, 24, 2287–92.

[131]   Jamesh MI, Li P, Bilek MMM, Boxman RL, McKenzie DR, Chu PK. Evaluation of corrosion
        resistance and cytocompatibility of graded metal carbon film on Ti and NiTi prepared by
        hybrid cathodic arc/glow discharge plasma-assisted chemical vapor deposition. Corros. Sci.
        2015, 97, 126–38.

[132]   Wang GX, Shen Y, Zhang H, Quan XJ, Yu QS. Influence of surface microroughness by
        plasma deposition and chemical erosion followed by TiO2 coating upon anticoagulation,
        hydrophilicity, and corrosion resistance of NiTi alloy stent. J. Biomed. Mater. Res. A 2008, 85,
        1096–102.

[133]   Sukuroglu EE, Sukuroglu S, Akar K, Totik Y, Efeoglu I, Arslan E. The effect of TiO_2 coating on
        biological NiTi alloys after micro-arc oxidation treatment for corrosion resistance. Proc. Inst.
        Mech. Eng. H 2017, 231, 699–704.

[134]   Li Y, Zhou T, Luo P, Xu S-G. Surface modification of Ti–49.8at%Ni alloy by Ti ion implantation:
        phase transformation, corrosion, and cell behavior. Int. J. Miner. Metall. Mater. 2015, 22,
        868–75.

[135]   Fu T-L, Zhan Z-L, Zhang L, Yang Y-R, Liu Z, Liu J-X. Corrosion behaviors of low-temperature
        plasma nickelized coatings on titanium alloy. Mater. Corros. 2016, 67, 1321–8.

[136]   Cheng Y, Cai W, Zheng YF, Li HT, Zhao LC. Surface characterization and immersion tests of TiNi
        alloy coated with Ta. Surf. Coat. Technol. 2005, 10, 428–33.

[137]   Zhao T, Li Y, Xia Y, Venkatraman SS, Xiang Y, Zhao X. Formation of a nano-pattering NiTi surface
        with Ni-depleted superficial layer to promote corrosion resistance and endothelial
        cell-material interaction. J. Mater. Sci. Mater. Med. 2013, 24, 105–14.

[138]   Li P, Li L, Wang W, Jin W, Liu X, Yeung KWK, Chu PK. Enhanced corrosion resistance and
        hemocompatibility of biomedical NiTi alloy by atmospheric-pressure plasma polymerized
        fluorine-rich coating. Appl. Surf. Sci. 2014, 297, 109–15.

[139]   Ahmed RA, Fadl-allah SA, El-Bagoury N, El-Rab SMFG. Improvement of corrosion resistance
        and antibacterial effect of NiTi orthopedic materials by chitosan and gold nanoparticles.
        Appl. Surf. Sci. 2014, 292, 390–9.

[140]   Villermaux F, Tabrizian M, Yahia L, Czeremuszkin G, Piron DL. Corrosion resistance
        improvement of NiTi osteosynthesis staples by plasma polymerized tetrafluoroethylene
        coating. Biomed. Mater. Eng. 1996, 6, 241–54.

[141]   Liu IH, Lee TM, Chang CY, Liu CK. Effect of load deflection on corrosion behavior of NiTi wire.
        J. Dent. Res. 2007, 86, 539–43.

[142]   Liu J-K, Lee T-M, Liu I-H. Effect of loading force on the dissolution behavior and surface
        properties of nickel-titanium orthodontic archwires in artificial saliva. Am. J. Orthod.
        Dentofacial Orthop. 2011, 140, 166–76.

[143]   Iijima M, Endo K, Ohno H, Mizoguchi I. Effect of Cr and Cu addition on corrosion behavior of
        NiTi alloys. Dent. Mater. J. 1998, 17, 31–40.

[144]   Rondelli G, Vicentini B. Effect of copper on the localized corrosion resistance of Ni-Ti shape memory alloy. Biomaterials 2002, 23, 639–44.

[145]   Kassab E, Neelakantan L, Frotscher M, Swaminathan S, Maaß B, Rohwerder M, Gomes J, Eggeler G. Effect of ternary element addition on the corrosion behaviour of NiTi shape memory alloys. Mater. Corros. 2014, 65, 18–22.

[146]   Wranglén G. An Introduction to Corrosion and Protection of Metals. Chapman and Hill, New York, NY, 1985, 85–7.

[147]   Van Black L H. Elements of Materials Science and Engineering. Addison-Wesley Pub., Reading, MA, 1989, 509–11.

[148]   Garcia I. Galvanic corrosion of dental implant materials. Indiana University Master Degree Thesis, 2000.

[149]   Koh Il-W. Effects of bacteria-induced corrosion on galvanic couples of commercially pure titanium with other dental alloys. Indiana University Mater Degree Thesis, 2003.

[150]   Fontana MG, Greene ND. Corrosion Engineering. McGraw-Hill Book Co., New York, 1967, 20–5, 330–8.

[151]   Tomashov ND. Theory of Corrosion and Protection of Metals. MacMillan Co., New York, 1966, 212–8.

[152]   Uhlig HH, Revie RW. Corrosion and Corrosion Control. John Wiley & Sons, New York, 1985, 101–5.

[153]   Platt JA, Guzman A, Zuccari A, Thornburg DW, Rhodes BF, Oshida Y, Moore BK. Corrosion behavior of 2205 duplex stainless steel. Am. J. Orthod. Dentofacial Orthop. 1997, 112, 69–79.

[154]   Iijima M, Endo K, Yuasa T, Ohno H, Hayashi K, Kakizaki M, Mizoguchi I. Galvanic corrosion behavior of orthodontic archwire alloys coupled to bracket alloys. Angle Orthod. 2006, 76, 705–11.

[155]   Siargos B, Bradley TG, Darabara M, Papadimitriou G, Zinelis S. Galvanic corrosion of metal injection molded (MIM) and conventional brackets with nickel-titanium and copper-nickel-titanium archwires. Angle Orthod. 2007, 77, 355–60.

[156]   Darabara MS, Bourithis LI, Zinelis S, Papadimitriou GD. Metallurgical characterization, galvanic corrosion, and ionic release of orthodontic brackets coupled with Ni-Ti archwires. J. Biomed. Mater. Res. B Appl. Biomater. 2007, 81, 126–34.

[157]   Bakhtari A, Bradley TG, Lobb WK, Berzins DW. Galvanic corrosion between various combinations of orthodontic brackets and archwires. Am. J. Orthod. Dentofacial Orthop. 2011, 140, 25–31.

[158]   Heravi F, Mokhber N, Shayan E. Galvanic corrosion among different combination of orthodontic archwires and stainless steel brackets. J. Dent. Mater. Techniq. 2014, 3, 118–22.

[159]   Tahmasbi S, Ghorbani M, Masudrad M. Galvanic corrosion of and ion release from various orthodontic brackets and wires in a fluoride-containing mouthwash. J. Dent. Res., Dent. Clin., Dent. Prospects 2015, 9, 159–65.

[160]   Tahmasbi S, Sheikh T, Hemmati YB. Ion release and galvanic corrosion of different orthodontic brackets and wires in artificial saliva. J. Contemp. Dent. Pract. 2017, 18, 222–7.

[161]   Polychronis G, Al Jabbari YS, Eliades T, Zinelis S. Galvanic coupling of steel and gold alloy lingual brackets with orthodontic wires: is corrosion a concern? Angle Orthod. 2018, 88, 450–7.

[162]   Tiller AK. Aspects of Microbial Corrosion. In: Parkins RN, ed., Corrosion Processes. Applied Science Pub., London, 1982. 115–9.

[163]   Matasa CG. Stainless steels and direct-bonding brackets, III: microbiological properties. Inf. Orthod. Kierferorthop 1993, 25, 269–71.

[164]   Mansfeld F, Little BJ. A technical review of electrochemically techniques applied to microbiologically influenced corrosion. Corr. Sci. 1991, 32, 247–72.

[165] Dexter SC, Duquette DJ, Siebert OW, Videla HA. Use and limitations of electrochemical techniques for investigation microbiological corrosion. Corrosion 1991, 47, 308–18.

[166] Videla HA, Herrera LK. Microbiologically influenced corrosion: looking to the future. Int. Microbiol. 2005, 8, 169–80. [167] Walsh D, Pope D, Danford M, Huff T. The effect of microstructure on micro-biologically influenced corrosion. JOM 1993, 45, 22–30.

[168] Dexter SC. Microbiological Effects. In: Baboian R, ed., Corrosion Tests and Standards: Application and Interpretation. ASTM Manual Series MNL 20, 1995, 419–29.

[169] Gibbons RJ, Cohen L, Hay DI. Strains of streptococcus mutans and streptococcus sobrinus attach to different pellicle receptors. Infect. Immun. 1986, 52, 555–61.

[170] An YH, Friedman RJ. Concise review of mechanisms of bacterial adhesion to biomaterial surfaces. J. Biomed. Mater. Res. 1998, 43, 338–48.

[171] Tsibouklis J, Stone M, Thorpe AA, Graham P, Peters V, Heerlien R, Smith JR, Green KL, Nevell TG. Preventing bacterial adhesion onto surfaces: the low-surface-energy approach. Biomaterials 1999, 20, 1229–35.

[172] Little BJ, Wagner P. Factors influencing the adhesion of microorganism to surfaces. J. Adhesion 1986, 20, 187–210.

[173] Chang J-C, Oshida Y, Gregory RL, Andres CJ, Barco TM, Brown DT. Electro-chemical study on microbiology-related corrosion of metallic dental materials. J. Biomed. Mater. Eng. 2003, 13, 281–95.

[174] Kameda T, Oda H, Ohkuma K, Sano N, Batbayar N, Terashima Y, Sato S, Terada K. Microbio-logically influenced corrosion of orthodontic metallic appliances. Dent. Mater. J. 2014, 33, 187–95.

[175] Rocher P, Medawar L, Hornez J-C, Traisnel M, Breme J, Hildebrand HF. Biocorrosion and cytocompatibility assessment of NiTi shape memory alloys. Scr. Mater. 2004, 50, 255–60.

[176] Bahije L, Benyahia H, El Hamzaoui S, Ebn Touhami M, Bengueddour R, Rerhrhaye W, Abdallaoui F, Zaoui F. Behavior of NiTi in the presence of oral bacteria: corrosion by Streptococcus mutans. Int. Orthod. 2011, 9, 110–9.

[177] Schulte A, Ruhlig D, Erichsen T, Schuhmann WW. Imaging localized corrosion via AC-SECM and SECM-integrated stripping voltammetry. ECS Meeting Abstract MA2006-02 923.

[178] Schulte A, Belger S, Etienne M, Schuhmann W. Imaging localised corrosion of NiTi shape memory alloys by means of alternating current scanning electrochemical microscopy (AC-SECM). Mater. Sci. Eng., A 2004, 378, 523–6.

[179] Ruhlig D, Gugel H, Schulte A, Theisen W, Schuhmann W. Visualization of local electro-chemical activity and local nickel ion release on laser-welded NiTi/steel joints using combined alternating current mode and stripping mode SECM. Analyst 2008, 133, 1700–6.

[180] Nishimoto S, Yokoyama K, Inaba T, Mutoh K, Sakai J. Effects of temperature, phase, elastic deformation and transformation on inhibition of localized corrosion of hydrogen-charged Ni–Ti superelastic alloy in NaCl solution. J. Alloys Compd. 2016, 682, 22–8.

[181] Shahrabi T, Sanjabi S, Saebnoori E, Barber ZH. Extremely high pitting resistance of NiTi shape memory alloy thin film in simulated body fluids. Mater. Lett. 2008, 62, 2791–4.

[182] Bai Z, Rotermund HH. The intrinsically high pitting corrosion resistance of mechanically polished nitinol in simulated physiological solutions. J. Biomed. Mater. Res. Part B 2011, 99, 1–13.

[183] Rondelli G, Vicentitni B. Localized corrosion behavior in simulated human body fluids of commercial Ni–Ti orthodontic wires. Biomaterials 1999, 20, 785–92.

[184] Kararia V, Jain P, Chaudhary S, Kararia N. Estimation of changes in nickel and chromium content in nickel-titanium and stainless steel orthodontic wires used during orthodontic treatment: an analytical and scanning electron microscopic study. Contemp. Clin. Dent. 2015, 6, 44–50.

[185] Darabara M, Bourithis L, Zinelis S, Papadimitriou GD. Susceptibility to localized corrosion of stainless steel and NiTi endodontic instruments in irrigating solutions. Int. Endod. J. 2004, 37, 705–10.

[186] Wang J, Li N, Rao G, Han E-H, Ke W. Stress corrosion cracking of NiTi in artificial saliva. Dent. Mater. 2007, 23, 133–7.

[187] Zhang C, Sun X. Susceptibility to stress corrosion of laser-welded composite arch wire in acid artificial saliva. Adv. Mater. Sci. Eng. 2013; http://dx.doi.org/10.1155/2013/738954.

[188] Liu J-K, Liu I-H, Liu C, Chang C-J, Kung K-C, Liu Y-T, Lee T-M, Jou J-L. Effect of titanium nitride/titanium coatings on the stress corrosion of nickel–titanium orthodontic archwires in artificial saliva. Appl. Surf. Sci. 2014, 317, 974–81.

[189] Moodey NR, Costa JE. Microstructrue/Property Relationship in Titanium Alloys and Titanium Aluminides. Kim YW, Boyer RR, eds. TMS, Warrendlae, PA, , 1991, 587–604.

[190] Tal-Gutelmacher E, Eliezer D. The hydrogen embrittlement of titanium-based alloys. J. Metals, 2005, 57, 46–9.

[191] Williams DN, Jaffee RI. Relationships between impact and low-strain-rate hydrogen embrittlement of titanium alloys. J. Less-Common Metals 1960, 2, 42–8.

[191] Hardie D, Quyang S. Effect of hydrogen and strain rate upon the ductility of mill-annealed Ti6Al4V. Corr. Sci. 1999, 41, 155–77.

[192] Solovioff G, Eliezer D, Desch PB, Schwarz RB. Hydrogen-induced cracking in an Al-Al$_3$Ti-Al$_4$C$_3$ alloy. Scr. Metall. Mater. 1995, 33, 1315–20.

[194] Froes FH, Eliezer D, Nelson HG. Hydrogen Effects in Materials. Thompson AW, Moody NR, eds. TMS, Warrendale, PA, 1994, 719–33.

[195] Ogawa T, Yokoyama K, Asaoka K, Sakai J. Hydrogen absorption behavior of beta titanium alloy in acid fluoride solutions. Biomaterials 2004, 25, 2419–25.

[196] Hirth JP. Effects of hydrogen on the properties of iron and steel. Metall. Trans. A 1980, 11A, 861–90.

[197] Yokoyama K, Kaneko K, Ogwa T, Moriyama K, Asaoka K, Sakai J. Hydrogen embrittlement of work-hardened Ni-Ti alloy in fluoride solutions. Biomaterials 2005, 26, 101–8.

[198] Paton NE, Williams LC. Effect of Hydrogen on Titanium and Its Alloys. In: Hydrogen in Metals. American Society for Metals, 1974, 409–32.

[199] Ebtehaj K, Hardie D, Parkins RN. The stress corrosion and pre-exposure embrittlement of titanium in methanolic solutions of hydrochloric acid. Corr. Sci. 1985, 25, 415–29.

[200] Hollis AC, Scully JC. The stress corrosion cracking and hydrogen embrittlement of titanium in methanol-hydrochloric acid solutions. Corr. Sci. 1993, 34, 821–35.

[201] Nakahigashi J, Tsuru K, Yoshimura H. Application of hydrogen-treated titanium alloy to dental prosthesis – experimental manufacture of crown by superplastic forming. J. Jpn. Soc. Dental Mater. Dev. 2005, 24, 291–5.

[202] Gerard DA, Koss DA. The combined effect of stress state and grain size on hydrogen embrittlement of titanium. Scr. Met. Mater. 1985, 19, 1521–4.

[203] Shih DS, Robertson IM, Birnbaum HK. Hydrogen embrittlement of α titanium: *In situ* TEM studies. Acta Metall 1988, 36, 111–24.

[204] Nelson HG. Hydrogen embrittlement of α titanium: *in situ* tem studies. Met. Trans. 1973, 4, 364–7.

[205] Young GA, Scully JR. Effects of hydrogen on the mechanical properties of a Ti-Mo-Nb-Al alloy. Scr. Met. Mater. 1993, 28, 507–12.

[206] Kolman DG, Scully JR. Effects of the Environment on the Initiation of Crack Growth. ASTM, West Conshohocken, 1997, 61–73.

[207] Costa JE, Williams JC, Thompson AW. The effect of hydrogen on mechanical properties in Ti-10V-2Fe-3Al. Met. Trans. 1987, A18, 1421–30.

[208] Hua F, Mon K, Pasupathi P, Gordon G, Scoesmith D. Modeling the hydrogen-induced cracking of titanium alloys in nuclear waste repository environments. J. Metals. 2005, 57, 20–6.

[209] Murai T, Ishikawa M, Miwa C. The absorption of hydrogen into titanium under cathodic polarization. Corr. Eng. 1977, 26, 177–83.

[210] Yokoyama K, Hamada K, Moriyama K, Asaoka K. Degradation and fracture of Ni-Ti superelastic wire in an oral cavity. Biomaterials 2001, 22, 2257–62.

[211] Yokoyama K, Kaneko K, Moriyama K, Asaoka K, Sakai J, Nagumo M. Delayed fracture of Ni-Ti superelastic alloys in acidic and neutral fluoride solutions. J. Biomed. Mater. Res. 2004, 69A, 105–13.

[212] Yokoyama K, Kaneko K, Moriyama K, Asaoka K, Sakai J, Nagumo M. Hydrogen embrittlement of Ni-Ti superelastic alloy in fluoride solution. J. Biomed. Mater. Res. 2003, 65A, 182–7.

[213] Asaoka K, Yokoyama K, Nagumo M. Hydrogen embrittlement of nickel-titanium alloy in biological environment. Metall. Mater. Trans. A 2002, 33A, 495–501.

[214] Li X, Liang X, Liu D, Chen R, Huang F, Wang R, Rettenmayr M, Su Y, Guo J, Fu H. Design of (Nb, Mo)$_{40}$Ti$_{30}$Ni$_{30}$ alloy membranes for combined enhancement of hydrogen permeability and embrittlement resistance. Sci. Rep. 2017, 7; doi: 10.1038/s41598-017-00335-0.

[215] Pedullà E, Grande NM, Plotino G, Palermo F, Gambarini G, Rapisarda E. Cyclic fatigue resistance of two reciprocating nickel-titanium instruments after immersion in sodium hypochlorite. Int. Endod. J. 2013, 46, 155–9.

[216] Mohajeri M, Castaneda H, Lagoudas D. A study of NiTi alloy mechanical and corrosion behavior subjected to mechanical and cyclic thermal loading in corrosive environment. ECS Meeting Abstract 2017-02, 701.

[217] Huang X, Shen Y, Wei X, Haapasalo M. Fatigue resistance of nickel-titanium instruments exposed to high-concentration hypochlorite. J. Endod. 2017, 43, 1847–51.

[218] Tan L, Dodd RA, Crone WC. Corrosion and wear-corrosion behavior of NiTi modified by plasma source ion implantation. Biomaterials 2003, 24, 3931–9.

[219] Kosec T, Močnik P, Legat A. The tribocorrosion behaviour of NiTi alloy. Appl. Surf. Sci. 2014, 288, 727–35.

[220] Močnik P, Kosec T, Kovač J, Bizjak M. The effect of pH, fluoride and tribocorrosion on the surface properties of dental archwires. Mater. Sci. Eng. C Mater. Biol. Appl. 2017, 78, 682–9.

[221] Siddiqui DA, Sivan S, Weaver JD, Di Prima M. Effect of wire fretting on the corrosion resistance of common medical alloys. J. Biomed. Mater. Res. B Appl. Biomater. 2017, 105, 2487–94.

[222] Xue Y, Hu Y, Wang Z. Tribocorrosion behavior of NiTi alloy as orthopedic implants in Ringer's simulated body fluid. Biomed. Phys. Eng. Express 2019, 5; doi: 10.1088/2057-1976/ab1db0.

[223] American Society for Testing and Materials. ASTM F86 Standard practice for surface preparation and making of metallic surgical implants. ASTM, Philadelphia, PA, 1995, 6–8.

[224] Kasemo B. Biocompatibility of titanium implants: surface science aspects. J. Pros. Dent. 1983, 49, 832–7.

[225] Tomashov ND. Theory of corrosion and protection of metals: the science of corrosion. The MacMillan Co., New York, 1966, 144.

[226] CRC Handbook of Chemistry and Physics. The Chemical Rubber Co., Cleveland, 47th edition, 1966, D-54.

[227] CRC Handbook of Chemistry and Physics. The Chemical Rubber Co., Cleveland, 47th edition, 1966, D-27.

[228] West JM. Basic Corrosion and Oxidation. John Wiley and Sons, London, 2nd edition, 1986, 27–30.

[229] Lautenschlager EP, Monaghan P. Titanium and titanium alloys as dental materials. Intl.Dent. J. 1993, 43, 245–53.

[230] Oxidation of Metals and Alloys. American Society for Metals, Ohio, 1971, 46.

[231] Coddet C, Chaze AM, Beranger G. Measurements of the adhesion of thermal oxide films: application to the oxidation of titanium. J. Mater. Sci. 1987, 22, 2969–74.

[232] Trigwell S, Selvaduray G. Effects of surface finish on the corrosion of NiTi alloy for biomedical applications. In: Pleton AR, et al. ed., Shape Memory and Superelastic Technologies. Pacific Grove, San Francisco, CA, 1997, 383–8.

[233] Chan CM, Trigwell S, Duerig T. Oxidation of a NiTi alloy. Surface Interface Anal. 1990, 15, 349–54.

[234] Oshida Y, Sachdeva R, Miyazaki S. Changes in contact angles as a function of time on some pre-oxidized biomaterials. J. Mater. Sci.: Mater. Med. 1992, 3, 306–12.

[235] Oshida Y, Sachdeva R, Miyazaki S, Daly J. Effects of shot peening on surfaces contact angles of biomaterials. J. Mater. Sci.: Mater. Med. 1993, 4, 443–7.

[236] Lausmaa J, Mattsson L, Rolander U, Kasemo B. Chemical Composition and Morphology of Titanium Surface Oxides. In: Materials Research Society Symposium Proceeding Vol. 55. Materials Research Society, Pittsburgh, PA, 1986, 351–9.

[237] Oxidation of Metals and Alloys. American Society for Metals, Ohio, 1971, 142–3.

[238] Solar RJ. Corrosion Resistance of Titanium Surgical Implant Alloys: A Review. In: Syrett BD, ed., Corrosion and Degradation of Implant Materials, ASTM STP 684. American Society for Testing and Materials, Philadelphia, 1979, 259–73.

[239] Miyamoto S, Watanabe M, Narushima T, Iguchi Y. Deoxidation of NiTi alloy melts using metallic barium. Mater. Trans. 2008, 49(2), 289–93.

[240] Ito D, Nishiwaki N, Ueda K, Narushima T. Effect of Ba deoxidation on oxygen content in NiTi alloys and non-metallic inclusions. J. Mater. Sci. 2013, 48, 359–66.

[241] Wranglén G. An Introduction to Corrosion and Protection of Metals. Chapman and Hall, London, 1985, 65–6.

[242] Albrektsson T, Brånemark PI, Hansson H A, Kasemo B, Larsson K, Lundstroem I, McQueen DH, Skalak R. The interface zone of inorganic implants in vivo: titanium implants in bone. Ann. Biomed. Eng. 1983, 11, 1–27.

[243] CRC Handbook of Chemistry and Physics. The Chemical Rubber Co., Cleveland, 47th edition, 1966, E-58.

[244] Filip P, Lausmaa J, Musialek J, Mazanec K. Structure and surface of NiTi human implants. Biomaterials 2001, 22, 2131–8.

[245] Nolan M, Tofail SA. Density functional theory simulation of titanium migration and reaction with oxygen in the early stages of oxidation of equiatomic NiTi alloy. Biomaterials 2010, 31, 3439–48.

[246] Uhlig HH. The Corrosion Handbook. John Wiley & Sons, New York, 1966, 20.

[247] Fontana MG, Greene ND. Corrosion Engineering. McGraw-Hill Book, New York, 1967, 319–21.

[248] Fraker AC, Ruff AW. Corrosion of Titanium Alloys in Physiological Solutions. In: Titanium Science and Technology, Vol. 4. Plenum Press, New York, 1973, 2447–57.

[249] Trépenier C, Leung TK, Tabrizian M, Yahia L'H, Bienvenu J-G, Tanguay J-F, Piron DL, Bilodeau L. Preliminary investigation of the effects of surface treatments on the biological response to shape memory NiTi stents. J. Biomed. Mater. Res. 1999, 48, 165–71.

[250] Ryhänen J, Kallioinen M, Tuukkanen J, Junila J, Niemela E, Sandvik P, Serlo W. In vivo biocompatibility evaluation of nickel-titanium shape memory metal alloy: musle and perineural tissue responses and encapsule membrane thickness. J. Biomed. Mater. Res. 1998, 41, 481–8.

[251] Sittig C, Textor M, Spencer ND, Wieland M, Vallotton P-H. Surface characterization of implant materials c.p. Ti, Ti-6Al-7Nb and Ti-6Al-4V with different pretreatments. J. Mater. Sci.: Mater. Med. 1999, 10, 35–46.

[252] Mazhar AA, Heakal F, El-T, Gad-Allah AG. Anodic behavior of titanium in aqueous media. Corrosion, 1988, 44, 705–10.

[253] Metikoš-Huković M, Ceraj-Cerić M. Anodic oxidation of titanium: mechanism of non-stoichiometric oxide formation. Surf. Technol. 1985, 24, 273–83.

[254] Hanawa T, Asami K, Asaoka K. Repassivation of titanium and surface oxide film regenerated in stimulated bioliquid. J. Biomed. Mater. Res. 1998, 40, 530–8.

[255] ASTM F 746-87 Standard Test Method for Pitting or Crevice Corrosion of Metallic Surgical Implant Materials. In: Annual Book of ASTM Standards, Vol. 13.01. American Society for Testing and Materials, Philadelphia, PA, 1996, 80–5.

[256] Modjtahedi BS, Fortenbach CR, Marsano JG, Gandhi AM, Staab R, Maibach HI. Guinea pig sensitization assays: an experimental comparison of three methods. Cutan. Ocul. Toxicol. 2011, 30, 129–37.

[257] Assad M, Yahia L'H, Rivard CH, Lemieux N. In vitro biocompatibility assessment of a nickel-titanium alloy using electron microscopy in situ end labeling (EM-ISEL). J. Biomed. Mater. Res. 1998, 44, 154–61.

[258] Rondelli G, Vicentini B. Evaluation of electrochemical tests of the passive film stability of equiatomic Ni-Ti alloy also in presence of stress-induced martensite. J. Biomed. Mater. Res. 1999, 51, 47–54.

[259] Hunt NP, Cunningham SJ, Golden CG, Sheriff M. An investigation into the effects of polishing on surface hardness and corrosion of orthodontic archwires. Angle Orthodont. 1999, 69, 433–9.

[260] Asami K, Chen S-C, Habazaki H, Hashimoto K. The surface characterization of titanium and titanium-nickel alloys in sulfuric acid. Corrosion Sci. 1993, 35, 43–9.

[261] Morawiec H, Goryczka T, Lelątko J, Lekston Z, Winiarski A, Rówiński E, Stergioudis F. Surface structure of NiTi alloy passivated by autoclaving. Mater. Sci. Forum 2001, 636/637, 971–6.

[262] Schroeder V. Evolution of the passive film on mechanically damaged nitinol. J. Biomed. Mater. Res. Part A 2009, 90A, 1–17.

[263] Neelakantan L, Monchev B, Frotscher M, Eggeler G. The influence of secondary phase carbide particles on the passivity behaviour of NiTi shape memory alloys. Mater. Corros. 2012, 63, 979–84.

[264] Speck KM, Fraker AC. Anodic polarization behavior of Ti-Ni and Ti-6A1-4V in simulated physiological solutions. J. Dent. Res. 1980, 59, 1590–5.

[265] Kawakita J, Hassel A, Stratmann M. High voltage anodization of a NiTi shape memory alloy. ECS Trans. 2006, 1, 17–26.

[266] Coddet C, Chaze AM, Beranger G. Measurements of the adhesion of thermal oxide films: application to the oxidation of titanium. J. Mater. Sci. 1987, 22, 2969–74.

[267] Hruska AR, Borelli P. Quality criteria for pure titanium casting, laboratory soldering, intraoral welding, and a device to aid in making uncontaminated castings. J. Prosthet. Dent. 1999, 66, 561–5.

[268] Adachi M, Mackert JR Jr, Parry EE, Fairhurst CW. Oxide adherence and porcelain bonding to titanium and Ti-6Al-4V alloy. J. Dent. Res. 1990, 69, 1230–5.

[269] Kimura H, Horng CJ, Okazaki M, Takahashi J. Oxidation effects on porcelain-titanium interface reactions and bond strength. Dent. Mater. J. 1990, 9, 91–9.

[270] Hautaniemi JA, Herø H. Porcelain bonding on Ti: its dependence on surface roughness, firing and vacuum level. Surf. Interface Anal. 1993, 20, 421–36.

[271] Atsü S, Berksun B. Bond strength of three porcelain to two forms of titanium using two firing atmospheres. J. Prosthet. Dent. 2000, 84, 567–74.

[272] Oshida Y, Hashem A. Titanium-porcelain system. Part I: oxidation kinetics of nitrided pure titanium, simulated to porcelain firing process. J. Biomed. Mater. Eng. 1993, 3, 185–98.

[273] Wang RR, Fung KK. Oxidation behavior of surface-modified titanium for titanium-ceramic restorations. J. Prosthet. Dent. 1997, 77, 423–34.

[274] Wang RR, Welsch G, Monteiro O. Silicon nitride coating on titanium to enable titanium-ceramic bonding. J. Biomed. Mater. Res. 1999, 46, 262–70.

[275] Firstov GS, Vitchev RG, Kunmar H, Blanpain B, van Humbeeck J. Surface oxidation of NiTi shape memory alloy. Biomaterials 2002, 23, 4863–71.

[276] Zeng CL, We WT. Corrosion of Ni–Ti *alloys* in the molten $(Li,K)_2CO_3$ eutectic mixture. Corrosion Sci. 2002, 44, 1–12.

[277] Plant SD, Grant DM, Leach L. Behaviour of human endothelial cells on surface modified NiTi alloy. Biomaterials 2005, 26, 5359–67.

[278] Meachim G, Williams DF. Changes in nonosseous tissue adjacent to titanium implants. J. Biomed. Mater. Res. 1973, 7, 555–72.

[279] Ferguson AB, Akahoshi Y, Laing PG, Hodge ES. Characteristics of trace ions released from embedded metal implants in the rabbit. J. Bone Joint Surg. 1962, 44-A, 323–36.

[280] Ducheyne P, Williams G, Martens M, Helsen J. In vivo metal-ion release from porous titanium-silver material. J. Biomed Mater Res. 1984, 18, 293–308.

[281] Sundgren J-E, Bodö P, Lundström I. Auger electron spectroscopic studies of the interface between human tissue and implants of titanium and stainless steel. J. Colloid Interf. Sci. 1986, 110, 9–20.

[282] Ducheyne P, Healy KE. Surface Spectroscopy of Calcium Phosphate Ceramic and Titanium Implant Materials. In: Ratner BD, ed. Surface Characterization of Biomaterials. Elsevier Science, Amsterdam, 1988, 175–92.

[283] Albreksson T, Hansson HA. An ultrastructural characterization of the interface between bone and sputtered titanium or stainless steel surfaces. Biomaterials, 1986, 7, 201–5.

[284] Chu CL, Wu SK, Yen YC. Oxidation behavior of equiatomic TiNi alloy in high temperature air environment. Mater. Sci. Eng. A 1996, 216, 193–200.

[285] Firstov GS, Vitchev RG, Kumar H, Blanpain B, Van Humbeeck J. Surface oxidation of NiTi shape memory alloy. Biomaterials 2002, 23, 4863–71.

[286] Zeng CL, Li MC, Liu GQ, Wu WT. Air oxidation of Ni–Ti alloys at 650–850 °C. Oxid. Met. 2002, 58, 171–84.

[287] Tian Q, Wu J. Effect of the rare-earth element Ce on oxidation Behavior of TiPdNi alloys. Mater. Sci. Forum 2002, 394/395, 455–8.

[288] Tian Q, Chen J, Chen Y, Wu J. Oxidation behavior of TiNi–Pd shape memory alloys. Z. Metallkd. 2003, 94, 36–40.

[289] Gu YW, Tay BY, Lim CS, Yong M. Characterization of bioactive surface oxidation layer on NiTi alloy. Appl. Surf. Sci. 2005, 252, 2038–49.

[290] Vojtěch D, Novák P, Novák M, Jaska L, Fabián T, Maixner J, Machovič V. Cyclic and isothermal oxidations of nitinol wire at moderate temperatures. Intermetallics 2008, 16, 424–31.

[291] Smialek JL, Garg A, Rogers RB, Noebe RD. Oxide scales formed on NiTi and NiPtTi shape memory alloys. Metall. Mater. Trans. A 2012, 43, 2325–41.

[292] Hansen AW, Beltrami LVR, Antonini LM, Villarinho DJ, das Neves JCK, Marino CEB, de Fraga Malfatti C. Oxide formation on NiTi surface: influence of the heat treatment time to achieve the shape memory. Mater. Res. 2015, 18; http://dx.doi.org/10.1590/1516-1439.022415.

[293] Wu Z, Mahmud A, Zhang J, Liu Y, Yang H. Surface oxidation of NiTi during thermal exposure in flowing argon environment. Mater. Des. 2018, 140, 123–33.

[294] Nolan M, Tofail SA. The atomic level structure of the TiO(2)-NiTi interface. Phys. Chem. Chem. Phys. 2010, 12, 9742–50.

[295] Wyckoff RWG. The Structure of Crystals.Chemical Catalog Co., New York, 1931, 134, 239.

[296] Tang H, Prasad K, Sanjinès R, Schmid PE, Lévy F. Electrical and optical properties of TiO$_2$ anatase thin films. J. Appl. Phys. 1994, 75, 2042–7.

[297] Douglass DL, van Landuyt J. The structure and morphology of oxide films during the initial stages of titanium oxidation. Acta Met. 1966, 14, 491–503.

[298] Nakayama T, Oshida Y. Effects of surface working on the structure of oxide films by wet oxidation in austenitic stainless steels. Trans. Jpn. Inst. Metals, 1971, 12, 214–7.

[299] Mahla EM, Nielsen NA. A study of films isolated from passive stainless steels. Trans. Elctrochem. Soc. 1948, 81, 913–6.

[300] Nakayama T, Oshida Y. Identification of the initial oxide films on 18-8 stainless steel in high temperature water. Corrosion 1968, 24, 336–7.

[301] Lim YJ, Oshida Y, Andres CJ, Barco MT. Surface characterization of variously treated titanium materials. Int. J. Oral Maxillofac. Implants 2002, 16, 333–42.

[302] Głuszek J, Masalski J, Furman P, Nitsch K. Structural and electrcochemical examinations of PACVD TiO$_2$ films in Ringer solution. Biomaterials 1997, 18, 789–94.

[303] Liu X, Ding C. Plasma sprayed wollastonite/TiO$_2$ composite coatings on titanium alloys. Biomaterials 2002, 23, 4065–77.

[304] Browne M, Gregson PJ. Surface modification of titanium alloy implants. Biomaterials 1994, 15, 894–8.

[305] Takemoto S, Yamamoto T, Tsuru K, Hayakawa S, Osaka A, Takashima S. Platelet adhesion on titanium oxide gels: effect of surface oxidation. Biomaterials 2004, 25, 3845–92.

[306] Allen KW, Alsalim HS. Titanium and alloy surfaces for adhesive bonding. J. Adhesion 1974, 6, 229–37.

[307] Felske A, Plieth WJ. Raman spectroscopy of titanium dioxide layers. Electrochim. Acta 1989, 34, 75–7.

[308] Matthews A. The crystallization of anatase and rutile from amorphous titanium dioxide under hydrothermal conditions. Amer. Mineral 1976, 61, 419–24.

[309] Yahalom J, Zahavi J. Electrolytic breakdown crystallization of anodic oxide films on Al, Ta and Ti. Electrochim. Acta 1970, 15, 1429–35.

[310] Ohtsuka T. Structure of anodic oxide films on titanium. Surface Sci. 1998, 12, 799–804.

[311] Liang B, Fujibayashi S, Neo M, Tamura J, Kim H-M, Uchida M, Kikubo T, Nakamura T. Histological and mechanical investigation of the bone-bonding ability of anodically oxidized titanium in rabbits. Biomaterials 2003, 24, 4959–66.

[312] Koizumi T, Nakayama T. Structure of oxide films formed on Ti in boiling dilute H$_2$SO$_4$ and HCl. Corr. Sci. 1968, 8, 195–9.

[313] Oshida Y. Bioscience and Bioengineering of Titanium Materials. Elsevier, UK, 2006, 81–103.

[314] Kodama A, Ikeda H, Nakase C, Masuda T, Kitahara K, Arai A, Onuma Y, Tanaka K, Hayashi K, Yamashita J, Aikawa Y. Electrochemical characterization of electronic states at NiTi alloy/aqueous solution interface. J. Corr. Sci. Eng. 2005, 7, 23–31.

[315] Oshida Y. unpublished data, 2019.

[316] Hanawa T. Titanium and Its Oxide Film: A Substrate for Formation of Apatite. In: Davies JE, ed. The Bone-Biomaterial Interface. Univ. of Toronto Press, Toronto, 1991, 49–61.

[317] Trépanier C, Tabrizian M, Yahia LH, Bilodeau L, Piron DL. Effect of modification of oxide layer on NiTi stent corrosion resistance. J. Biomed. Mater. Res. 1998, 43, 433–40.

[318] Muhonen V, Heikkinen R, Danilov A, Jämsä T, Tuukkanen J. The effect of oxide thickness on osteoblast attachment and survival on NiTi alloy. J. Mater. Sci.: Mater. Med. 2007, 18, 959–67.

[319] Zheng YF, Wnag QY, Li L. The electrochemical behavior and surface analysis of Ti49.6Ni45.1Cu5Cr0.3 alloy for orthodontic usage. J. Biomed. Mater. Res. B Appl. Biomater. 2008, 86, 335–40.

[320] Vojtech D, Joska L, Leitner J. Influence of a controlled oxidation at moderate temperatures on the surface chemistry of nitinol wire. Appl. Surf. Sci. 2008, 254, 5664–69.

[321] Lin K-N, Wu S-K. Oxidation behavior of $Ti_{50}Ni_{40}Cu_{10}$ shape-memory alloy in 700–1,000 °C air. Oxid. Met. 2009, 71, 187–200.

[322] Shabalovskaya SA, Tian H, Anderegg JW, Schryvers DU, Carroll WU, Van Humbeeck J. The influence of surface oxides on the distribution and release of nickel from Nitinol wires. Biomaterials 2009, 30, 468–77.

[323] Espinar E, Llamas JM, Michiardi A, Ginebra MP, Gil FJ. Reduction of Ni release and improvement of the friction behaviour of NiTi orthodontic archwires by oxidation treatments. J. Mater. Sci. Mater. Med. 2011, 22, 1119–25.

[324] Sutherland DS, Forshaw PD, Allen GC, Brown IT, Williams KR. Surface analysis of titanium implants. Biomaterials 1993, 14, 893–9.

[325] Lausmaa GJ, Kasemo B. Surface spectroscopic characterization of titanium implant materials. Appl. Surf. Sci. 1990, 44, 133–46.

[326] Kubaschewski O, Hopkins BE. Oxidation of Metals and Alloys. Butterworths, London, 1962, 24.

[327] Healy KE, Ducheyne P. The mechanisms of passive dissolution of titanium in a model physiological environment. J. Biomed. Mater. Res. 1992, 26, 319–38.

[328] Tummler H, Thull R. Model of the metal/tissue connection of implants made of titanium or tantalum. Biol. Biochem. Perform. Biomater. Elsevier, Netherlands, 1986, 403–8.

[329] Eliades T. Passive film growth on titanium alloys: physiochemical and biologic considerations. Int. J. Oral Maxillofac. Implant 1997, 12, 621–7.

[330] Drath DB, Karnovsky ML. Superoxide production by phagocytic leukocytes. J. Exp. Med. 1975, 144, 257–61.

[331] Root BK, Metcalf JA. H_2O_2 release from human granulocytes during phagocytosis: relationship to superoxide anion formation and cellular catabolism of H_2O_2. Studies with normal and cytochalasin B-treated cells. J. Clin. Invest. 1977, 60, 1266–79.

[332] DeChatelet LR. Oxide bactericidal mechanisms of PMN. J. Infect. Dis. 1975, 131, 295–303.

[333] O'Brien B, Carroll WM, Kelly MJ. Passivation of nitinol wire for vascular implants – a demonstration of the benefits. Biomaterials 2002, 23, 1739–48.

[334] Tian H, Schryvers D, Liu D, Jiang Q, Van Humbeeck J. Stability of Ni in nitinol oxide surfaces. Acta Biomater. 2011, 7, 892–9.

[335] Tengvall P, Elwing H, Sjöqvist L, Lundström I. Interaction between hydrogen peroxide and titanium: a possible role in the biocompatibility of titanium. Biomaterials 1989, 10, 118–20.

[336] Jacob HS, Craddock PR, Hammerschmidt DE, Moldow CF. Induced granulocyte aggregation – unsuspected mechanism of disease. New Engl. J. Med. 1980, 302, 789–94.

[337] Del Masestro RF. An approach to free radicals in medicine and biology. Acta Physiol. Scand. (Suppl.) 1980, 492, 153–68.

[338] Sundgren J-E, Bodö P, Lundström I, Berggren A, Hellem S. Suger electron spectro-scopic studies of stainless steel implants. J. Biomed. Mater. Res. 1985, 19, 663–71.

[339] Tengvall P, Bjursten LM, Elwing H, Lunström I. Free radicals oxidation of metal implants and biocompatibility. 2nd Int'l Conf Biointercations '87, London, 1987.

[340] Jaeger CD, Bard AJ. Spin-trapping and electron spin resonance detection of radical intermediates in the photodecompositions of water at TiO_2 particulate systems. J. Phys. Chem. 1979, 83, 3146–51.

[341] Barb WG, Baxendale JH, George P, Hargrave KR. Reactions of ferrous and ferric ions with hydrogen peroxide. Trans. Farady Soc. 1951, 47, 462–500.

[342] Bielski BHJ. Fast kinetic studies of dioxygen-derived species and their metal-complexes. Phil. Trans. Roy. Soc. Lond. B 1985, 311, 473–82.

[343] Armstrong D. Free Radicals in Molecular Biology, Aging and Disease. Raven Press, New York, 1984.

[344] Janzén EG, Wang YY, Shetty RW. Spin-trapping with α-(4-pyridyl-1-oxide)-N-tert-nutylnitrone in aqueous solutions: a unique electron spin esponsonce spectrum for the hydroxyl radical adduct. J. Am. Chem. Soc. 1978, 100, 2923–5.

[345] Tengvall P, Lundström I, Sjöqvist L, Elwing H, Bjursten LM. Titanium-hydrogen peroxide intecation: model studies of the influence of the inflammatory response on titanium implants. Biomaterials 1989, 10, 166–75.

[346] Pan J, Thierry D, Leygraf C. Electrochemical and XPS studies of titanium for biomaterial applications with respect to the effect of hydrogen peroxide. J. Biomed. Mater. Res. 1994, 28, 113–22.

[347] Bianchi G, Malaguzzi S. Cathodic reduction of oxygen and hydrogen peroxide on titanium. Proc. 1st Intl Congr Metallic Corrosion London, April 10–15, 1961, 78–83.

[348] Pan J, Thierry D, Leygraf C. Hydrogen peroxide toward enhanced oxide growth on titanium in PBS solution: blue coloration and clinical relevance. J. Biomed. Mater. Res. 1996, 30, 393–402.

[349] Pan J, Liao H, Leygraf C, Thierry D, Li J. Variation of oxide films on titanium induced by osteoblast-like cell culture and the influence of an H_2O_2 pretreatment. J. Biomed. Mater. Res. 1998, 40, 244–56.

[350] Albrektsson T. The response of bone to titanium implants. CRC Crit. Rev. Bicompatibility 1984, 1, 53–84.

[351] Beariner JP, Orme CA, Gilbert JL. Effect of hydrogen peroxide on titanium surfaces: in situ imaging and step-polarization impedance spectroscopy of commercially pure titanium and titanium, 6-aluminum, 4-vanadium. J. Biomed. Mater. Res. 2003, 67A, 702–12.

[352] Chu CL, Hu T, Wu SL, Dong YS, Yin LH, Pu YP, Lin PH, Chung CY, Yeung KW, Chu PK. Surface structure and properties of biomedical NiTi shape memory alloy after Fenton's oxidation. Acta Biomater. 2007, 3, 795–806.

Chapter 9
Fatigue, fracture, and creep

9.1 Fatigue of NiTi, in general

For considering mechanical fatigue phenomena of ordinal materials, it is good enough to discuss the microstructure-related structural fatigue. However, for nickel–titanium (NiTi) materials, since they possess unique functional characteristics (shape-memory effect (SME) and superelasticity (SE)), it should include thermal cycling fatigue (to which phase transformation attributes) and functional (or degradation) fatigue (to which residual strain, hysteresis, or lower dissipated energy causing an increase in dislocation density can be attributed). Since the actuation is always provided by thermal activation of the alloy, the fatigue incurred by the material is called the thermo-mechanical or functional, implying that material performs mechanical work against the external applied load when heated during the actuation cycle. In short, the structural fatigue represents the physical failure (or fracture) of NiTi SME alloys with a lifetime of cyclic loading while the functional fatigue addresses the degradation of SE and SME occurred during the cyclic deformation of NiTi SME alloys [1–7]. Fatigue strength is one of the most important aspects to be considered and is controlled with two stages: crack initiation stage and crack propagation stage (which can be furthermore divided into a slow and gradual crack growth stage and the fast and final fracture). Depending on the number of cycles to failure (N_{CF}), fatigue life can be divided into low-cycle (or high-stress) fatigue (LCF) and high-cycle (or low-stress) fatigue (HCF).

Fatigue failures occur in many different forms. Mere fluctuations in externally applied stresses or strains result in mechanical fatigue. Cyclic loads acting in association with high temperatures cause creep fatigue; when the temperature of the cyclically loaded component also fluctuates, thermomechanical fatigue is induced. Recurring loads imposed in the presence of a chemically aggressive or embrittling environment give raise to corrosion fatigue. The repeated application of loads in conjunction with rolling contact between materials produces rolling contact fatigue, while fretting fatigue occurs as a result of pulsating stresses along with oscillatory relative motion and frictional sliding between surfaces. There are also two distinct terms on fatigue: static fatigue and dynamic fatigue. In the above, all are related to dynamic fatigue (which relates to materials that are cyclically stressed below levels of their ultimate tensile strengths). Static fatigue applies to ceramic materials and refers to stable crack propagation under sustained loads in the presence of an embrittling environment (in metallurgy and engineering is commonly known as stress corrosion cracking). Ceramics support a high static load for a long period of time and then fail abruptly. This type of failure occurs only when the materials are stored in a wet environment. The term fatigue means many things to many peoples in different disciplines. To the metal physicist: it is perhaps a study of how and why dislocations

https://doi.org/10.1515/9783110666113-009

lead to slip localization to nucleate cracks. To the metallographer: it is the characterization of surface changes leading to crack nucleation or the significance of fractographic features of the crack faces. To the corrosion chemist: it is the electrochemical reactions at the crack tip in the presence of an aqueous environment and the contaminant acceleration in crack growth. To the physical or process metallurgists: it is the development of fatigue resistance alloys for aircraft gear or turbine components. To the applied mechanist: it is the stress field of a predetermined crack shape in a given body acted on, by certain boundary conditions. To the statistician: it is the random nature of failure in a probability. To the design engineer: it is the development and application of design rules and codes for the prevention of failures in piping, pressure vessels, and so on.

There are two subprocesses involved with fatigue life in NiTi endodontic files: period consumed for the crack initiation (N_I) and the subsequent crack propagation period (N_P). The total fatigue life, N_F is the sum of these two periods, $N_F = N_I + N_P$. Relative proportion of these subperiods depends on whether LCF regime or HCF regimes were the dominant mode leading to failure. Maletta et al. [8] investigated the LCF of a commercial pseudoelastic Ni-rich NiTi alloy. Fatigue tests have been carried out within the stress-induced transformation regime, by using flat dog bone-shaped specimens obtained from as-received NiTi sheets. Both functional and structural fatigues were analyzed, namely, the evolution of the pseudoelastic capability and the cycles to failure. It was reported that (i) a degradation of the pseudoelastic recovery, during the first mechanical cycles, becomes more evident when increasing the strain amplitude, (ii) however, a stable functional response is always observed after the first stabilization cycles, which occurs between 100 and 150 cycles, (iii) structural fatigue data have been analyzed by a novel strain-life model, based on a modified Coffin–Manson approach, and (iv) fracture surfaces have been analyzed by scanning electron microscopic (SEM) observation in order to study the stable and unstable crack growth mechanisms. Cheung et al. [9] compared LCF behavior of commercial NiTi instruments subjected to rotational bending, a deformation mode similar to an engine-file rotating in a curved root canal, using a strain-life analysis, in water. A total of 286 NiTi rotary instruments from four manufacturers were constrained into a curvature by three rigid, stainless steel pins while rotating at a rate of 250 rpm in deionized water until broken. It was obtained that (i) a linear strain-life relationship, on logarithmic scales, was obtained for the LCF region with an apparent fatigue-ductility exponent ranging from −0.40 to −0.56, and (ii) the number of crack initiation sites, as observed on the fracture cross section, differed between brands, but not LCF life, concluding that the LCF life of NiTi instruments declines with an inverse power function dependence on surface strain amplitude, but is not affected by the cross-sectional shape of the instrument. Ye et al. [10] characterized microstructural changes of M-Wire throughout the cyclic fatigue (CF) process under controlled strain amplitude. The average fatigue life was calculated from 30 M-Wire samples that were subjected to a strain-controlled ($\approx 4\%$) rotating bend fatigue test at room temperature (RT) and rotational speed of 300 rpm. It was found that (i) during rotating bend fatigue test, no statistically significant difference was found on austenite finish temperatures between as-received M-Wire and fatigued samples, (ii) however, significant differences were observed on Vickers microhardness for samples with 60% and 90% fatigue life compared with as-received and 30% fatigue life, and (iii) substantial growth of martensite grains and martensite twins was observed in microstructure under transmission electron microscopy (TEM) after 60% fatigue life. Based on these findings, it was suggested that endodontic instruments manufactured with M-Wire are expected to have higher strength and wear resistance than similar instruments made of conventional superelastic NiTi wires because of its unique nanocrystalline martensitic microstructure.

HCF life is governed by the number of cycles required for crack initiation [11]. Kafka et al. [12] applying the computation of the fatigue-crack incubation life of a microstructure of interest under HCF examined the mechanistic modeling of the influence of microstructure on HCF of NiTi material. It was indicated that a larger void between halves of the inclusion increases fatigue life, while larger inclusion diameter reduces fatigue life. We will discuss more about the HCF of NiTi-made medical devices including stents later in this chapter.

During an entire fatigue life of materials, there are two stages involved: crack initiation stage and crack propagation stage, although its transition is never clearly detected. McKelvey et al. [13] studied the fatigue-crack propagation behavior in an equiatomic Nitinol, with particular emphasis on the effect of the stress-induced martensitic transformation on crack growth resistance (K_R). Specifically, fatigue-crack growth was characterized in stable austenite (at 120 °C), superelastic austenite (at 37 °C), and martensite (at −65 and − 196 °C). It was reported that (i) in general, fatigue-crack growth resistance was found to increase with decreasing temperature, such that fatigue thresholds were higher and crack-growth rates slower in martensite compared to stable austenite and superelastic austenite, and (ii) the stress-induced transformation of the superelastic austenite structure, which occurs readily at 37 °C during uniaxial tensile testing, could be suppressed during fatigue-crack propagation by the tensile hydrostatic stress state ahead of a crack tip in plane strain; this effect, however, was not seen in thinner specimens, where the constraint was relaxed due to prevailing plane-stress conditions. Chen et al. [14] investigated the fatigue-crack growth in cold-rolled and annealed polycrystalline superelastic NiTi alloys under cyclic stressing. The fatigue-crack growth morphologies of compact tension specimens are recorded. It was shown that (i) the main cracks of the annealed specimens grow along the original crack plane, while the crack paths of the cold-rolled specimens orient at 18° angle to the original horizontal crack plane, which might be due to the strong anisotropy of elastic modulus and hardness after the cold rolling, and (ii) the annealed specimens have higher fatigue-crack growth resistance than the cold-rolled specimens due to the stronger shielding mechanism caused by plastic deformation and reduced anisotropy by annealing. Barbosa et al. [15] evaluated the presence and propagation of defects and their effects on surfaces of NiTi endodontic instruments using noncontact and three-dimensional optical profilometry. The flute surface areas of instruments from two commercial instrumentation systems, namely Reciproc R25 ($n = 5$) and WaveOne Primary ($n = 5$), were assessed and compared before and after performing two instrumentation cycles in simulated root canals in clear resin blocks. It was obtained that (i) there was a significant increase in wear in both groups, especially between baseline and the second instrumentation cycle, with significantly higher wear values being observed on WaveOne instruments, (ii) a significant increase in surface roughness was observed in both groups from the first to the second instrumentation cycle, mostly in WaveOne specimens, and (iii) qualitative analysis revealed a greater number of defects on the flute topography of all the instruments after use. Hence, it was concluded that more defects were identified in WaveOne Primary instruments compared to Reciproc R25, irrespective of the evaluation stage.

During the entire fatigue life, the internal microstructure as well as surface/subsurface microstructures will progressively be affected and altered. These progressive changes can be monitored by means of mechanical property evaluation or physical property measurements [16]. Figure 9.1. [17] shows the damage evolution curves measured by eight different methods. Both damage fraction axis and life fraction axis are normalized to show unity as the 100% cumulative damage and onset of N_F, respectively.

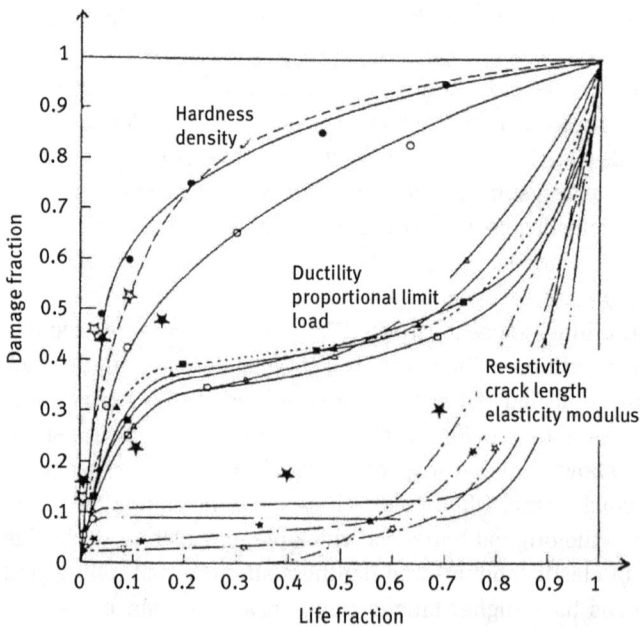

Figure 9.1: Nonlinear relationship between damage fraction and fatigue life fraction with respect to various damage parameters as damage indicators [17].

The above fatigue damage indicators in Figure 9.1 are nonlinear. Ideally, the fatigue damage indicator should change linearly as a function of life fraction. Oshida et al. [18–20] employed the X-ray diffraction parameters as a nondestructive fatigue damage indicator. Superficial layer and interior (which is undamaged, hence can be served as self-reference) were simultaneously subjected to X-ray diffraction with different wavelengths. From the analysis on progressive changes in line broadening, the dislocation densities at surface (D_S) and interior (D_I) were calculated. It was found that (i) the dislocation density ratio, D_I/D_S, is linearly related to the life fraction and (ii) if the ratio reaches over 85%, the target sample was subjected to severe fatigue damage.

Endodontic instrument damage may include one or a combination of the following: bending, stretching, or straightening of twist contour without bending, peeling, or tearing of metal at the cutting edges of the instrument without bending or straightening of the twist contour, partial revere twisting of the instrument, cracking along the file axis, and fracture of the instrument [21, 22]. The aforementioned types of file damage can be attributed to many factors, namely, stress concentrations [23], torsional forces [24–30], bending moments [24, 30, 31], wear [31, 32], inherent flaws [32], and metallurgical factors [33]. There should be other parameters which can be useful for fatigue damage assessments. There are still various materials-related indicators for damage assessment proposed: strain amplitude and volume under strain [34] and transformation strain [35], geometry and materials property [36, 37], stiffness [38], durability [39], or energy dissipation [40].

9.2 Fatigue on endodontic instruments

For considering mechanical fatigue phenomena of ordinal materials, it is good enough to discuss the microstructure-related structural fatigue. However, for NiTi materials, since they possess unique functional characteristics (SME and SE), it is needed to add functional fatigue such as the working displacement in a one way effect actuator or the dissipated energy in a loading–unloading cycle of a superplastic damping capacity decreases with increasing number of cycles. In short, the structural fatigue represents the physical failure (or fracture) of NiTi shape-memory alloys (SMAs) with a lifetime of cyclic loading while the functional fatigue addresses the degradation of SE and SME occurred during the cyclic deformation of NiTi SMAs. Fatigue phenomena in NiTi SMAs is complicated because it involves two-fold fatigue mechanisms of structural fatigue (which can be found in most ordinary materials and fatigue life can be controlled by the crack propagation), thermal cycling fatigue (to which phase transformation attributes, and the driving force of martensite transformation increases during the thermal cycling and the martensite transformation becomes more difficult), and functional (or degradation) fatigue (to which residual strain, hysteresis, or lower dissipated energy causing an increase in dislocation density can be attributed)

[1, 7, 41–43]. In the case of endodontic rotary NiTi files, it has been shown that the number of cycles to fracture decreases as the temperature increases from RT to the actual (in vivo) temperature [7]. For small-scale NiTi SMA members like microtubes used in endovascular stents, the critical crack size is minuscule and failure quickly follows crack development. Therefore, crack initiation monitoring is recommended instead of crack propagation control [44–46], which, in turn, is not easily detected. In any event, it is very important to nondestructively evaluate the cumulative fatigue damage, so that an early fatigue damage can be assessed.

For conducting fatigue tests on NiTi alloys, due to its uniqueness of SME and SE, two types of tests are normally performed: thermomechanical fatigue test and ordinal mechanical (cyclic, torsion, bending, and a combination of these) test.

9.2.1 Thermomechanical fatigue

In certain application field of NiTi alloys, a thermally activated cyclic transformation and a bias force are taking place and the degradation of memory effect (in other words, functional fatigue [47]) and the structural fatigue can also occur in thermal cycling. The thermomechanical fatigue life of NiTi SMAs strongly depends on the applied constant axial stress and temperature amplitude, and monotonically decreases with the increase in axial stress and the temperature amplitude (which determines whether a complete or partial transformation occurs during the thermal cycling) [47]. Saikrishna et al. [2] evaluated effect of intermittent overload cycles (OLCs) on fatigue behavior of NiTi SMA wire during thermomechanical cycling (TMC). It was shown that (i) fatigue life of NiTi is enhanced when the intermittent overload is above certain minimum level, (ii) an enhancement in fatigue life by ~50% is observed when the overload ratio is 2.0, and (iii) accumulation of plastic strain in the material under such TMC condition is found to be relatively high compared to that of TMC with no overload cycles. Tadaki et al. [48] examined thermal cycling effects in an aged Ni-rich Ti–Ni SMA by means of differential scanning calorimetry (DSC) and TEM, changing aging temperature and time, temperature range for thermal cycling, and the number of thermal cycles up to 10^4 times. Two kinds of thermal cycling termed "complete cycling" and "incomplete cycling" were adopted. It was reported that (i) the complete cycling caused the B2 matrix intermediate (R-phase) to martensite (M) transformations in the appropriately aged alloy and the incomplete cycling did only the first B2 to R transformations, (ii) TEM observation on the aged alloy subjected to the complete cycling of 10^4 times showed that the B2 matrix regions between the Ti_3Ni_4 precipitates were heavily roughened by some lattice defects, (iii) however, the TEM observation on the alloy subjected to the incomplete cycling showed no change. As SMAs have been recognized their importance by serving as high-energy density actuator, one characteristic that becomes particularly important is the thermomechanical transformation fatigue life, in addition to maximum transformation strain and stability of actuation cycles. Lagoudas et al. [49] investigated the thermally activated transformation fatigue characteristics of SMAs under various applied loads for both complete and partial phase transformation. A $Ni_{50}Ti_{40}Cu_{10}$ (at%) SMA was chosen to study the effects of various heat treatments on the transformation temperatures and the transformation fatigue lives of actuators were studied. For selected heat treatments, the evolution of recoverable and irrecoverable strains up to failure under different applied stress levels was studied in detail. It was mentioned that the irrecoverable strain accumulation as a function of the N_{CF} for different stress levels was related by a relationship similar to the Manson–Coffin law for both partial and complete transformations. Tan et al. [50] investigated the effects of heat treatment on transformation behavior, properties, and TMC fatigue of $Ti_{49.8}Ni_{50.2}$ alloy wire. It was found that (i) with increasing annealing temperature, R→B19′ transformation temperatures increase while B2→R transformation temperatures decrease; (ii) the two transformations temperatures will overlap upon annealing at 600 °C; (iii) recovery occurs in the specimen annealed below 600 °C, and σM decreases slightly with the increase of annealing temperature; (iv) the specimen annealed at 750 °C shows the largest stable recovery strain while upon annealing above 600 °C, the recrystallization of the specimen will occur; and (v) its σM and stable recovery strain decrease sharply, while its fatigue life increases from 103 to 104 magnitudes. Bertacchini et al. [51] characterized the thermally induced fatigue behavior of nickel-rich NiTi SMA actuators subjected to different constant applied stresses and found that fatigue limits of ~5,000 to ~60,000 cycles were found for applied stress levels ranging from 250 to 100 MPa. One of most debated aspects around Nitinol quality is microcleanliness, nowadays considered as the main factor affecting fatigue life. Recent results demonstrate that fatigue is undoubtedly associated with inclusions that can act as crack initiators. However, type, size, and distribution of such particles have been observed to strongly depend on Ni/Ti ratio as well as melting and thermomechanical processes. Coda et al. [52] characterized thermomechanical fatigue behavior of SMA wires (with diameter <100 µm), which were prepared by a peculiar, nonstandard combination of melting and thermomechanical processes (clean melt technology). It was reported that in the new NiTi clean melt alloy the maximum inclusion size and area fraction are significantly reduced compared to standard Nitinol, offering meaningful improvement in fatigue resistance over standard wires.

In mechanical fatigue testing for the NiTi SMAs, so many researches have been done to investigate their fatigue failure caused by the stress- or strain-controlled cyclic loading, which is denoted as the mechanical fatigue of NiTi SMAs. Such mechanical fatigue test should include the rotating–bending tests, cyclic-torsional tests, tension–compression tests, rolling-contact tests, compression tests, and so on.

9.2.2 Cyclic fatigue

Pruett et al. [53] studied CF of NiTi, engine-driven instruments by determining the effect of canal curvature and operating speed on the breakage of LightSpeed instruments.

It was mentioned that (i) for nickel–titanium, engine-driven rotary instruments, the radius of curvature, angle of curvature, and instrument size are more important than operating speed for predicting separation, and (ii) the effect of the radius of curvature as an independent variable should be considered when evaluating studies of root canal instrumentation. Yared et al. [54] evaluated CF of 0.06 ProFile NiTi rotary instruments after clinical use in molar teeth using 2.5% NaOCl as an irrigant. It was concluded that sterilization and clinical use in the presence of NaOCl did not lead to a decrease in the number of rotations to breakage of the files. After evaluating the CF of 0.04 ProFile NiTi rotary instruments operating at different rotational speeds and varied distances of pecking motion in metal blocks that simulated curved canals, Liu et al. [55] concluded that in order to prevent breakage of a NiTi rotary instrument, appropriate rotational speeds and continuous pecking motion in the root canals are recommended. Ullmann et al. [56] evaluated static fracture loads of PT (PT) NiTi instruments that had been subjected to various degrees of CF. It was concluded that (i) buildup of tension within NiTi rotary instruments depends on instrument diameter, and (ii) clinically, larger instruments that have been subjected to some CF should be used with great care or discarded. Tripi et al. [57] conducted a comparative study of the fatigue resistance of rotary NiTi endodontic instruments with the aim of assessing the influence of both instrument design and surface treatment on flexural fracture. A total of 120 instruments were tested; these came from different sources: ProFile, RaCe, K3, Hero, and Mtwo. To compare the effect of electropolishing procedures on fatigue resistance, a group of RaCe instruments (which are normally electropolished) without surface treatment was used. It was reported that (i) ProFile instruments gave the best values for fatigue resistance; (ii) for RaCe instruments, the surface treatment reduces the presence of microcracks, surface debris, and machining damage; (iii) the instrument design often proves to be an important factor in the fatigue resistance of NiTi rotary instruments; and (iv) in RaCe instruments the electropolishing surface treatment increases the fracture-related fatigue resistance.

Regarding nondestructive fatigue damage assessment as discussed earlier, Li et al. [58] interrupted the CF rotation periodically and removed and engaged instruments onto a device to monitor its stiffness by using two strain gauges in four different directions. It was mentioned that a potential nondestructive integrity assessment method for NiTi rotary instruments (as geometry change assessment) was developed.

Grande et al. [59] determined how instrument design affects the fatigue life of two NiTi rotary systems (Mtwo and PT) under CF stress in simulated root canals. A total of 260 instruments were rotated until fracture occurred and the N_{CF} were recorded. It was found that (i) cycles to failure significantly decreased as the instrument volume increased for both the radii of curvature tested, (ii) the radius of curvature had a statistically significant influence on the fatigue life of the instruments, and (iii) larger instruments underwent fracture in less time under cyclic stress than smaller ones concluding that (iv) the metal volume in the point of maximum stress during a CF test could affect the fatigue life of NiTi rotary instruments, and (v) the larger the metal volume, the lower the fatigue resistance. Larsen et al. [60] evaluated a new generation of NiTi rotary instruments including the Twisted File (TF) and ProFile GT Series X (GTX) and concluded that (i) TF was significantly more resistant to CF than ES (EndoSequence, from Brasseler, Savannah, GA), but not different from PF and (ii) the new developed manufacturing processes appeared to offer greater resistance to CF in a simulated canal model. Plotino et al. [61] evaluated the CF resistance of five NiTi rotary systems in an abrupt apical curvature. CF testing was performed in stainless steel artificial canals with a 2-mm radius of curvature and an angle of curvature of 90 degrees constructed to the dimensions of the instruments tested. Tested instruments were new and 25 mm in length, including ten PTU F2, FlexMaster, Mtwo, and ProFile, which rotated passively at 300 rpm until fracture occurred. It was found that (i) Mtwo had the highest fatigue resistance compared to the other instruments (N_{CF} 124 +/− 25); there was no statistical difference between ProFile from the two different brands, although ProFile from Maillefer had the higher fatigue life (N_{CF} 75 +/− 10) compared to ProFile from Tulsa (N_{CF} 66 +/− 10), and (ii) no difference was registered between FlexMaster (N_{CF} 53 +/− 5) and ProFile from Tulsa; PT F2 had a significantly ($P < 0.001$) lower fatigue life compared to the other instruments tested (N_{CF} 29 +/− 5) concluding that the lifespan registered for the instruments tested in an apical abrupt curvature was Mtwo > ProFile from Maillefer > ProFile from Tulsa > FlexMaster > PT. Pedullà et al. [62] assessed the resistance to CF of NiTi files (a total of 150 new Tiwsted Files and Mtwo files) after the immersion in 5% NaOCl solution at 37 °C for 16 mm, which was in similar conditions to those used in clinical practice. It was obtained that (i) resistance to CF of the same NiTi file was not significantly affected by immersion in NaOCl; (ii) the TF showed a higher resistance in all groups than Revo S SU; (iii) the comparison between the same groups of Twisted Files and Mtwo files or between Mtwo and Revo S files did not show significant differences except for two cases: group 2 of the Twisted Files and Mtwo files and group 5 of the Mtwo and Revo S SU files; (iv) static or dynamic immersion in NaOCl for 1 min or 5 min did not reduce the CF resistance of NiTi significantly; (v) however, the type of instrument influences CF resistance; and (vi) Twisted Files were more resistant followed by Mtwo and Revo S SU files. Rodrigues et al. [63] evaluated, by static and dynamic CF tests, N_{CF} of two types of rotary NiTi instruments including TF (which is manufactured by a proprietary twisting process) and RaCe files (which are manufactured by grinding). It was found that (i) measurements of the fractured fragments showed that fracture occurred at the point of maximum flexure in the midpoint of the curved segment; (ii) the N_{CF} was significantly lower for RaCe instruments compared with TFs; and (iii) the N_{CF} was also lower for instruments subjected to the static test compared with the dynamic model in both groups. Based on these findings, it was concluded that rotary NiTi endodontic instruments manufactured by twisting present greater resistance to CF compared with instruments manufactured by grinding.

Arias et al. [64] compared the CF resistance of new M-Wire reciprocating WaveOne and Reciproc files. Sixty Reciproc and 60 WaveOne new files were fixed to a specifically designed device and tested in tempered steel canals with a 3-mm radius and a 60°

angle of curvature. It was reported that (i) the probability of the mean life was higher for Reciproc than WaveOne files at both levels, with the probability of the Reciproc mean life being 62% higher than that of WaveOne at 5 mm from the tip and 100% higher at 13 mm (all statistically significant), and (ii) the probability of the mean life was higher at 5 mm than at 13 mm in both systems concluding that (i) Reciproc files were more resistant to CF than WaveOne files at both distances from the tip, and (ii) both systems had greater CF resistance at 5 mm than at 13 mm from the tip. Bouska et al. [65] compared CF resistance of TF, ProFile, GTX, and EndoSequence and obtained that significant differences were found between the various brands of files. The differences between file brands may be because of a different manufacturing process or differences in file design. Based on a simulated canal model, the ProFile Vortex (PV), TF, and GTX files appear to offer greater CF resistance than EndoSequence and ProFile files. Gambarini et al. [66] evaluated the CF fracture resistance of engine-driven twisted file instruments under reciprocating movement. It was reported that (i) reciprocating movement resulted in a significantly longer CF life when compared with continuous rotation, and (ii) no difference was found between reciprocation 150° clockwise/30° counterclockwise and 30° clockwise/150° counterclockwise. Inan et al. [67] compared the CF resistance of different rotary NiTi systems (R-Endo R3, PT D3, and Mtwo R) designed for root canal retreatment. It was obtained that (i) R-Endo R3 instruments showed better CF resistance than PT D3 and Mtwo R 25.05 instruments, and the difference was statistically significant, and there was no significant difference between PT D3 and Mtwo R 25.05 groups. The use of reciprocating movement was claimed to increase the resistance of NiTi file to fatigue in comparison with continuous rotation. Kim et al. [68] compared the CF resistance and torsional resistance of Reciproc and WaveOne files. It was shown that (i) Reciproc had a higher N_{CF} and WaveOne had a higher torsional resistance than the others, and (ii) both reciprocating files demonstrated significantly higher CF and torsional resistances than PT. Al-Sudani et al. [69] tested the fatigue resistance of NiTi rotary files in a double curvature (S-shaped) artificial root canal and compared those results with single curvature artificial root canals. ProFile instruments and Vortex instruments were tested. It was reported that (i) the N_{CF} value was always statistically lower in the double-curved artificial canal when compared with the single curve in both the apical and coronal curvatures, and (ii) statistically significant differences were noted between instruments of the same size of different brand only in the single curve: ProFile registered a mean of $633.5 \pm 75.1 \; N_{CF}$, whereas Vortex registered a mean of $548 \pm 48.9 \; N_{CF}$, indicating that regardless of the differences between the instruments used in the present study, the more complex is the root canal, the more adverse are the effects on the CF resistance of the instruments. Plontino et al. [70] evaluated the CF resistance of Reciproc and WaveOne instruments in simulated root canals. It was indicated that (i) a statistically significant difference was noted between Reciproc and WaveOne instruments, (ii) Reciproc R25 instruments were associated with a significant increase in the meantime to fracture when compared with primary WaveOne instruments (130.8±18.4 vs 97.8±

15.9s), and (iii) there was no significant difference in the mean length of the fractured fragments between the instruments concluding that Reciproc instruments were associated with a significantly higher CF resistance than WaveOne instruments.

Pedullà et al. [71] assessed resistance to CF of reciprocating NiTi files (Reciproc and WaveOne) after immersion in 5% NaOCl solution at 37 °C for 16 mm over several time periods. It was reported that (i) resistance to CF of the same NiTi file was not significantly affected by immersion in NaOCl and (ii) Reciproc R25 was associated with a higher CF resistance in all groups compared to WaveOne Primary concluding that reciprocating dynamic immersion in NaOCl for 1 or 5 min did not reduce the CF resistance of NiTi files significantly. However, the type of reciprocating instrument influenced CF resistance with Reciproc R25 being more resistant than WaveOne Primary. Duke et al. [72] determined the flexibility of PV and Vortex Blue (VB) files and compared their fatigue resistance in artificial single curvature (60° curvature, 5-mm radius) and two different artificial double curvature canals (first (coronal) curve of 60° curvature and 5-mm radius and the second one apical of 30° curvature and 2-mm radius and group 3: first curve of 60° curvature and 5-mm radius and the second one of 60° curvature and 2-mm radius). It was concluded that double curvature canals represent a much more stressful and challenging anatomy than single curvature canals, and, in them, fatigue resistance may be affected by the degrees and the radii of curvatures as well as by the bending properties of the files. Kaval et al. [73] evaluated the CF resistance of F6 SkyTaper, K3XF, new-generation OneShape, and TRUShape (TRS) 3D conforming files. Ten instruments from each group were selected and allowed to rotate using a low-torque motor in a stainless steel block with 1.5 mm diameter, 3 mm radius of 60° angle of curvature at the manufacturer's recommended speed, and N_{CF} from the beginning to the fracture was recorded. It was reported that the ranking of the groups from the highest to the lowest N_{CF} was as follows: F6 Sky-Taper (959 ± 92), K3XF (725 ± 71), TRS (575 ± 84), and OneShape (289 ± 58). It was concluded that (i) F6 SkyTaper instruments presented the highest CF resistance among the tested instruments and (ii) the S-shaped cross-sectional design of F6 SkyTaper instruments could be the most important factor on the superior cyclic life span of these instruments. Elnaghy et al. [74] assessed and compared the resistance to CF of XP-endo Shaper (XPS) instruments with TRS, HyFlex (HF) controlled memory (HCM), VB, and iRace (iR) NiTi rotary instruments, immersed in saline at 37 °C ± 1 °C during CF testing. It was found that (i) XPS had a significantly greater N_{CF} compared with the other instruments, (ii) the topographic appearance of the fracture surfaces of tested instruments revealed ductile fracture of CF failure, and (iii) XPS instruments exhibited greater CF resistance compared with the other tested instruments. Özyürek et al. [75] compared the CF resistance of Reciproc and Reciproc Blue (RB) files that were used to prepare root canals of mandibular molar teeth with or without a glide path. It was found that (i) the CF resistance of as-received condition of RB files was found to be higher than as-received condition of Reciproc files, and (ii) there was no statistically significant difference between Reciproc and RB files used with and without a glide path. Keskin et al. [76] compared the CF resistance of R-Pilot with ProGlider and WaveOne Gold Glider glide path instruments. It was found that (i) the CF resistance values of the WaveOne Gold Glider and R-Pilot were significantly higher than those of the ProGlider, with no significant difference between them, (ii) Weibull analysis revealed that WaveOne Gold Glider showed the highest predicted time-to-failure value for 99% survival rate, which was followed by R-Pilot and ProGlider, and (iii) regarding the length of the fractured tips, there were no significant differences among the instruments. Inan et al. [77] evaluated the CF resistance of Reciproc and RB by testing in a severe apical curvature at intracanal temperature. It was obtained that (i) the N_{CF} values of Reciproc R25 were significantly lower than RB R25, and (ii) there was no significant difference between the instruments regarding the length of fractured fragments indicating that (iii) RB R25 instruments displayed significantly higher N_{CF} than Reciproc R25, and (iv) novel reciprocating blue wire instruments exhibited higher CF resistance than its precedence M-Wire instrument when tested in severely curvatured canals. Ribio et al. [78] tested the null hypothesis that there were no significant differences between size 25 files F360, F6 SkyTaper, HF EDM, iR, Neoniti, One Shape Protaper Next, Reciproc, Revo-S, and Wave One Gold in terms of resistance to CF and length of broken fragments. It was reported that (i) the Levene's test showed no equal variances, (ii) the Welch's test showed significant differences in CF and separated fragment lengths, and (iii) the Games–Howell test exhibited significant differences in multiple comparisons concluding that (iv) the systems with controlled memory (CM)-Wire (HF EDM and Neoniti) were superior in resistance to the other systems for CF, and (v) for separated fragment lengths, F360 (conventional NiTi) and Reciproc (M-Wire) were significantly better in terms of resistance. Khalil et al. [79] evaluated the CF, bending resistance, and surface roughness of EdgeEvolve and PT Gold (PTG) NiTi rotary files. It was found that (i) EdgeEvolve files exhibited higher CF resistance than PTG files in single- and double-curved canals and both files were more resistant to CF in single-curved canals than double-curved canals, (ii) EdgeEvolve files exhibited significantly more flexibility than did PTG files, and (iii) both files had approximately similar Ni and Ti contents, and (iv) EdgeEvolve files showed significantly lower R_a values than PTG files ($p < 0.05$). Based on these findings, it was concluded that EdgeEvolve files exhibited significantly higher CF resistance than PTG files in both single- and double-curved canals.

Plotino et al. [80] reviewed the CF testing of NiTi rotary instruments and mentioned that (i) fractured rotary NiTi instruments can be classified into those that fail as a result of cyclic flexural fatigue or torsional failure or a combination of both, and (ii) clinically, NiTi rotary instruments are subjected to both torsional load and CF, and ongoing research aims to clarify the relative contributions of both factors to instrument separation suggesting that an international standard for CF testing of NiTi rotary instruments is required to ensure uniformity of methodology and comparable results.

9.2.3 Torsional fatigue

Within a curvature, a rotary NiTi instrument is subjected to constant changes in compression and tension, eventually leading to material fatigue and subsequently to fracture without previous signs of plastic deformation. The instrument part at the inside of the curvature is compressed, whereas tension occurs on the outside of the curvature. It was suggested that torsional failure takes place when the alloy exceeds the limits of plastic deformation in case a rotary instrument tip engages and binds within the canal yet the remainder of the file is continuing the rotation [81–83].

Best et al. [84] evaluated the torsional CF characteristics and specifically the endurance limit of a NiTi rotary instrument. It was reported that (i) instruments cycled at larger deflection angle consistently demonstrated fewer cycles to fracture than those cycled at smaller deflection angle, (ii) the differences among the mean log number of cycles of the different deflection angle were statistically significant, and (iii) cycles of 10^6 were completed without instrument fracture at 2.5 degrees. Bahia et al. [85] evaluated the influence of cyclic torsional loading on the flexural fatigue resistance and torsional properties of rotary NiTi instruments. Twelve sets of new K3 instruments, sizes 20, 25, and 30 with a 0.04 taper and sizes 20 and 25 with a 0.06 taper, were torsion tested until rupture to establish their mean values of maximum torque and angular deflection. It was found that (i) cyclic torsional loading caused no significant differences in maximum torque or in maximum angular deflection of the instruments analyzed, but comparative statistical analysis between measured N_{CF} values of new and previously cycled K3 instruments showed significant differences for all tested instrument, and (ii) longitudinal cracks, that is, cracks apparently parallel to the long axis of the instruments cycled in torsion was observed concluding that cyclic torsional loading experiments in new K3 rotary endodontic instruments showed that torsional fatigue decreased the resistance of these instruments to flexural fatigue, although it did not affect their torsional resistance. Kim et al. [86] evaluated the effect of CF on the torsional resistance of NiTi rotary instruments (ProFile and PT) using a fatigue testing machine for the N_{CF}. It was obtained that (i) in both ProFile and PT groups, the 75% preloading groups had significantly lower torsional strength than other preloaded files, (ii) in the ProFile group, the 50% and 75% preloading groups had a smaller distortion angle until fracture than the 25% and no preloading groups, (iii) the 75% preloading group showed a lower toughness value than the 25% and no preloading groups, (iv) in the PT group, all preloading groups had less distortion and toughness than the no preloading group, and (v) fractographic examinations revealed the 75% preloaded files showed less amount of reverse-wound flute than other preloading groups. Based on these findings, it was concluded that approximate 75% CF may reduce the torsional resistance of NiTi rotary instruments significantly. Setzer et al. [87] investigated the combined influence of CF and torsional stress on rotary NiTi instruments to determine possible differences in the fracture point of rotary NiTi instruments depending on the application of CF only or in combination with torsional stress. It was found that (i) all fractures, regardless if cyclic only or cyclic-torsion was used, occurred within the area of the curvature; (ii) the addition of a torsional load resulted in a mean 1.09-mm statistically significant difference between cyclic only and cyclic-torsion, relocating the fracture point toward the area where torsional load was applied; and (iii) there was a statistically significant difference between the three file systems when they were tested either in the cyclic only or the cyclic-torsion mode. Accordingly, it was concluded that (i) cyclic-torsion compared with cyclic only resulted in statistically significantly different mean fragment length (which is a distance from the shaft to the fracture point), and (ii) all fractures remained within the area of the curvature, but with the addition of a torsional load, the location of the fracture moved in the direction of the additionally applied torsional stress suggesting that stress was distributed from the area in which the torsional load was applied toward the area undergoing CF.

Campbell et al. [88] investigated the effects of the torsional properties on fatigue behaviors of both traditional NiTi and NiTi CM files. Typhoon (TYP) rotary files in both NiTi and CM were tested to obtain the mean N_{CF} using a three-point bending apparatus. It was obtained that (i) TYP CM files had an N_{CF} times higher than that of TYP files, (ii) no difference in torque between the CM files and the conventional NiTi files of either file size was detected, (iii) the torque of the size 40/0.04 files was significantly higher than the torque of the size 25/0.04 files, (iv) in the 40/0.04 files group, TYP files in the 75% precycling group had a significantly lower torque than files in the group with no precycling, whereas slight precycling (25%) significantly reduced the distortion angle on TYP CM files, (v) the CM files of both sizes had a significantly higher distortion angle than the corresponding NiTi files, and (vi) the fractured files in the precycling groups showed the typical pattern of torsional failure. Based on these results, it was concluded that (i) within the same amount of precycling (25%, 50%, and 75%), the CF life of TYP CM instruments was significantly higher than that of the TYP instruments; however, the torque value of TYP CM was similar to TYP files, and (ii) the larger instruments were not only less resistant to CF but also affected most by prestressing of both TYP and TYP CM files. Shen et al. [89] evaluated the effect of various degrees of CF on torsional failure and torsional preloading on the CF life of heat-treated K3XF NiTi instruments. The N_{CF} of K3XF and K3 NiTi instruments were examined in a three-point bending apparatus with a 7-mm radius and 45° curve. It was reported that (i) the fatigue resistance of K3XF instruments was two times higher than that of K3 instruments; (ii) the torque and angle of rotation at fracture of K3XF instruments were similar to those of K3 instruments; (iii) the 25%, 50%, and 75% torsional preloading significantly lowered the N_{CF} of both K3 and K3XF instruments; (iv) in the fatigue prestressed groups, K3 instruments with 75% preloading had significantly lower torque and distortion angles than unused K3 instruments; and (v) the fractographic patterns corresponded to the pattern defined by the last stage test. It was, therefore, concluded that (i) a low amount of torsional preloading reduced the fatigue resistance of K3 and K3XF instruments; (ii) a high amount of precycling fatigue significantly reduced the torsional resistance of K3 instruments; and (iii) the torsional resistance of K3XF instruments was less affected by previous load cycling even after extensive precycling. Alcalde et al. [90] evaluated the cyclic and torsional fatigue resistance of the reciprocating single-file systems RB 25.08, Prodesign R 25.06, and WaveOne Gold 25.07. It was obtained that (i) the CF resistance values of PDR 25.06 were significantly higher; (ii) RB 25.08 showed higher fatigue resistance than WOG 25.07; (iii) the torsional test showed that PDR 25.06 had lower torsional strength; (iv) no differences were observed between RB 25.08 and WOG 25.07; (v) PDR 25.06 showed higher angular rotation values than RB 25.08 and WOG 25.07; (vi) RB 25.08 presented higher angular rotation than WOG 25.07; and (vii) the cross-sectional area analysis showed that PDR 25.06 presented the smallest cross-sectional areas at 3 and 5 mm from the tip. Based on these results, it was concluded that PDR 25.06 presented the highest CF resistance and angular rotation until fracture compared to RB 25.08 and WOG 25.07, and RB 25.08 and WOG 25.07 had higher torsional strength than PDR 25.06.

9.2.4 Bending fatigue

Cheung et al. [91] conducted rotational bending fatigue tests to examine the fatigue behavior a NiTi rotary instrument using a strain-life approach and to determine the effect of water. A total of 212 instruments were tested. It was found that (i) a strain-life relationship typical of metals was found; (ii) N_{CF} declined with an inverse power function dependence on the effective surface strain amplitude; (iii) a fatigue limit was present at about 0.7% strain; (iv) the apparent fatigue-ductility exponent, a material constant for the LCF life of metals, was found to be between −0.45 and −0.55; and (v) there was a significant effect of the environmental condition on the LCF life, water being more detrimental than air. It was, hence, concluded that (i) fatigue behavior of NiTi rotary instrument is typical of most metals, provided that the analysis is based on the surface strain amplitude, and showed a HCF and a LCF region, and (ii) the LCF life is adversely affected by water. Kim et al. [92] compared the fatigue resistance of traditional, ground NiTi rotary instruments with the TF and to examine the fracture characteristics of the fatigued fragment. TF, RaCe, Helix, and PT F1 were examined with SEM for surface characteristics before subjected to a cyclic (rotational bending) fatigue test. It was found that TF showed a significantly higher resistance to CF than other NiTi files that were manufactured with a grinding process. The path of crack propagation appeared to be different for electropolished (TF and RaCe) versus non-electropolished (Helix and PT) instruments, concluding that although all specimens showed similar fractographic appearance, which indicated a similar fracture mechanism, instruments with abundant machining grooves seemed to have a higher risk of fatigue. Porous implants are known to promote cell adhesion and have low elastic modulus, a combination that can significantly increase the life of an implant. However, porosity can significantly reduce the fatigue life of porous implants. Bernard et al. [93]

reported on the fatigue behavior of bulk porous metals, specifically on porous NiTi alloy with regard to the high-cycle rotating bending fatigue response of porous NiTi alloys fabricated using laser-engineered net shaping. It was reported that rotating bending fatigue results showed that the presence of 10% porosity in NiTi alloys can decrease the actual fatigue failure stress, at 106 cycles, up to 54% and single reversal failure stress by ~30%. From fractographic analysis, it is clear that the effect of surface porosity dominates the rotating bending fatigue failure of porous NiTi samples. Shen et al. [94] examined the fatigue behavior of NiTi instruments from a novel CM NiTi wire (CM Wire) including ProFile, TYP, TYP CM, DS-SS0250425NEYY (NEYY), and DS-SS0250425NEYY CM (NEYY CM), which were subjected to rotational bending at the curvature of 35° and 45° in air at the temperature of 23 °C ± 2 °C, and the N_{CF} was recorded. It was obtained that (i) the alloy yielded an improvement of over three to eight times in N_{CF} of CM files than that of conventional NiTi files; (ii) the vast majority of CM instruments (50–92%) showed multiple crack origins, whereas most instruments made from conventional NiTi wire (58–100%) had one crack origin; (iii) the values of the fraction area occupied by the dimple region were significantly smaller on CM NiTi instruments compared with conventional NiTi instruments; and (iv) the square (NEYY CM) versus the triangular (TYP CM) configuration showed a significantly different lifetime on CM Wire at both curvatures. Based on these results, it was concluded that (i) the material property had a substantial impact on fatigue lifetime, (ii) instruments made from CM Wire had a significantly higher N_{CF} and lower surface strain amplitude than the conventional NiTi wire files with identical design. Cheung et al. [95] evaluated the bending fatigue lifetime of NiTi alloy and stainless steel endodontic files using finite element analysis. The strain-life approach was adopted and two theoretical geometry profiles, the triangular and the square cross sections, were considered. Both LCF lifetime and HCF lifetime were evaluated. It was reported that (i) the bending fatigue behavior was affected by the material property and the cross-sectional configuration of the instrument, (ii) both the cross-sectional factor and material property had a substantial impact on fatigue lifetime, (iii) the NiTi material and triangular geometry profiles were associated with better fatigue resistance than that of stainless steel and square cross sections, and (iv) finite element models were established for endodontic files to prejudge their fatigue lifetime, a tool that would be useful for dentist to prevent premature fatigue fracture of endodontic files. Hieawy et al. [96] compared the flexibility and CF resistance of PTU and PTG instruments in relation to their phase transformation behavior. It was found that (i) PTG had a CF resistance superior to PTU in all sizes, (ii) the N_{CF} of the NiTi files of sizes S1 and S2 was significantly higher than those of sizes F1 to F3, (iii) no significant difference in the N_{CF} of PTU instruments was detected between F1 and F2, (iv) the fractured files of both PTU and PTG showed the typical fracture pattern of fatigue failure, (v) the bending load values were significantly lower for PTG than for PTU, (vi) the DSC analyses showed that each segment of the PTG instruments had a higher austenite finish temperature (50.1 °C ± 1.7 °C) than the PTU instruments (21.2 °C ± 1.9 °C), (vii) PTG instruments had a two-stage transformation behavior, and (vii) there was no significant difference in the austenite finish between unused files and instruments subjected to the fatigue process. It was, therefore, concluded that PTG files were significantly more flexible and resistant to fatigue than PTU files, and PTG exhibited different phase transformation behavior than PTU, which may be attributed to the special heat treatment history of PTG instruments. PTG may be more suited for preparing canals with a more abrupt curvature. Silva et al. [97] evaluated the bending resistance and the CF life of a new heat-treated reciprocating instrument (ProDesign R). Untreated ProDesign R, Reciproc R25, and WaveOne Primary instruments were used as reference instruments for comparison. It was reported that untreated ProDesign R presented significantly higher bending resistance than the other tested systems. No differences were observed between ProDesign R and Reciproc files regarding the bending resistance. ProDesign R revealed a significantly longer CF life; in contrast, untreated ProDesign R and WaveOne instruments presented significantly lower CF life than Reciproc. The new heat-treated reciprocating instrument ProDesign R has higher CF resistance than Untreated ProDesign R, Reciproc, and WaveOne instruments. ProDesign R and Reciproc were significantly more flexible than Untreated ProDesign R and WaveOne files.

Figure 9.2 compares (S–N) (strain amplitude vs fatigue life) curves for cyclic torsion, bending, and tension, where R indicates a stress ratio ($R = \sigma_{MIN}/\sigma_{MAX}$). It can be said, in general, that torsional fatigue can be found in LCF regime, while bending fatigue in HCF regime.

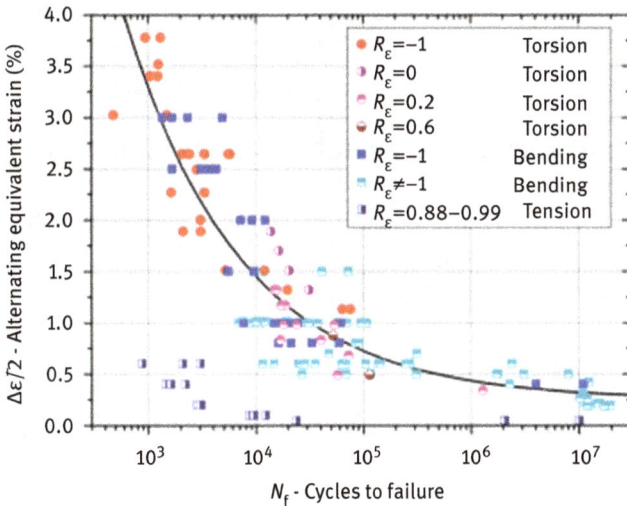

Figure 9.2: Strain amplitude versus fatigue life (S–N) curve [35, 47].

As other fatigue testing modes, a series of three ball-on-rod rolling contact fatigue tests were conducted using polished steel balls and NiTi rods prepared by vacuum casting and powder metallurgy techniques [98]. It was reported that alloyed NiTi rods containing small amounts of Hf as a microstructural processing aid generally endured higher stress levels than the baseline 60NiTi composition. Two predominant fatigue failure mechanisms were observed: intergranular (grain boundary) fracture and intragranular (through the grains) crack propagation. Further fatigue capability improvements could be obtained through process improvements, microstructural refinements, and alloying. Stress-induced martensite (SIM) currently available is recommended for mechanically benign applications involving modest stress levels and rates of stress cycle accumulation, and applications that include high continuous loads (stress) and high speeds for long durations should be avoided. Bernard et al. [99] conducted the compression fatigue test. Porous metals are being widely used in load bearing implant applications with an aim to increase osseointegration and also to reduce stress shielding. However, fatigue performance of porous metals is extremely important to ensure long-term implant stability, because porous metals are sensitive to crack propagation even at low stresses especially under cyclic loading conditions. The test was conducted for the high-cycle compression–compression fatigue behavior of laser processed NiTi alloy with varying porosities (between ~1% and 20%). It was shown that (i) compression fatigue of porous NiTi alloy samples is in part similar to metal foams, (ii) the applied stress amplitude is found to have strong influence on the accumulated strain and cyclic stability, and (iii) the critical stress amplitudes associated with rapid strain accumulation in porous NiTi alloy samples, with varying relative densities, were found to correspond to 140% of respective 0.2% proof strength indicating that these samples can sustain cyclic compression fatigue stresses up to 1.4 times their yield strength without failure.

9.3 Various factors influencing fatigue behaviors of NiTi alloys

There are various factors which influence fatigue strength and behaviors of NiTi alloys.

9.3.1 Surface conditions

Surface finish of any machined components plays a key role in their life performance, particularly the impact of surface roughness and residual stresses on the crack propagation should be noted, since both parameters can control the stress concentration factor at advancing crack tip. The surface roughness can also control the crack initiation site(s). Lopes et al. [100] evaluated the influence of surface grooves (peaks and valleys) resulting from machining during the manufacturing process of polished and unpolished NiTi BR4C endodontic files on the fatigue life of the instruments. It was reported that (i) analysis with the profilometer showed that surface grooves were deeper on the unpolished instruments compared with their electropolished counter-parts, and (ii) in the rotating bending fatigue test, the mean and standard deviation for the N_{CF} were greater for instruments with less pronounced grooves. Student's t-test revealed significant differences in all tests concluding that the depth of the surface grooves on the working part affected the N_{CF} of the instruments tested; the smaller the groove depth, the greater the N_{CF}. Cai et al. [101] investigated the effect of irrigation on the surface roughness and fatigue resistance of HF and M3 CM wire NiTi instruments. Two new files of each brand were dynamically immersed in either 5.25% sodium hypo-chlorite (NaOCl) or 17% ethylenediaminetetraacetic acid (EDTA) solution for 10 min, followed by atomic force microscopy (AFM) analysis. It was found that (i) for M3 files, the R_a and RMS values significantly increased after the immersion, (ii) for the HF file, the R_a and RMS values significantly increased only in EDTA, but not ($P > 0.05$) NaOCl, and (iii) the resistance to CF of both HF and M3 files did not significantly decrease by immersing in 5.25% NaOCl and 17% EDTA solutions indicating that except the HF files immersed in NaOCl, the surface roughness of other files exposed to irrigants increased; however, a change in the surface tomography of CM Wire instruments caused by contact with irrigants for 10 min did not trigger a decrease in CF resistance.

To compare the LCF behavior of electropolished and nonelectropolished NiTi instruments, Cheung et al. [102] used 45 electropolished and 62 nonelectropol-ished NiTi engine files that were subjected to rotational bending at various curvatures in 1.2% hypochlorite solution. It was reported that (i) a linear relationship was found between LCF life and surface-strain amplitude for both groups, with no discernible difference between the two, (ii) no electropolished instrument showed more than one crack origin, significantly fewer than for the nonelectropolished instruments, (iii) the square root of crack extension and strain amplitude were inversely related, and (iv) although surface smoothness is enhanced by electropolishing, this did not protect the instrument from LCF failure. Bui et al. [103] investigated the effect of electropolishing ProFile NiTi rotary instruments on torque resistance, fatigue resistance, and cutting efficiency. Size 25/0.04 ProFile files that were nonpolished for the control group and electropolished for the experimental group were used tests. It was reported that (i) electropolishing significantly reduced resistance to CF but did not affect torsional resistance; however, electropolishing reduced the angle at failure and amount of unwinding, and (ii) electropolishing did not significantly affect cutting efficiency. Praisarnti et al. [104] examined the fatigue behavior, especially at the LCF regime, of an experimentally electropolished FlexMaster and a commercial electropolished NiTi instrument (RaCe) in a 1.2% NaOCl solution. It was found that (i) the fatigue life of both instruments generally declined with increasing surface strain amplitude; there was a significant difference between the two instruments, and (ii) comparing the surface-treated FlexMaster with its commercially available nonelectropolished coun-terpart, an improved resistance to fatigue breakage as a result of electropolishing was noted concluding that (i) the LCF life of a NiTi instrument rotating with a curvature in a corrosive environment is enhanced by electropolishing, and (ii) the design, both cross-sectional and longitudinal, appears to have an effect on the fatigue behavior of NiTi rotary instruments. Sinan et al. [105] evaluated the effect of electropolishing time on the rupture resistance in flexion fatigue of the endodontics NiTi instruments. Forty-eight HeroShapers 6% 25/100 are divided into four groups of surface treatment (without electropolishing, with electropolishing of 70, 80, or 90 s). It was obtained that (i) the mean number of rotation to failure in each group of instrument decreases (407–355 h) with the increase of the electropolishing time, and (ii) there were no statistically significant difference between the different groups, concluding that electropolishing from 70 till 90 s improves the surface state without modifying the rupture resistance in flexion fatigue of triple helix endodontic instruments with 6% taper.

Condorelli et al. [106] assessed the failure mechanism of rotary NiTi instruments by chemical, structural, and morphological analyses to provide a rational explanation of the effects of surface and bulk treatments on their resistance to fatigue fracture. Thermal treatment (350–500 °C) was performed on electropolished and non-electropolished NiTi endodontic instruments. It was found that (i) before thermal treatment, significant differences in fatigue resistance between electropolished and nonelectropolished instruments (N_{CF} was 385 and 160, respectively) were attributed to differences in the surface morphology of the instruments, (ii) SEM analysis of the fracture surfaces indicated that flexural fatigue fractures occurred in two steps: first by a slow growth of initial cracks and then by rapid rupture of the remaining material, and (iii) thermal treatment did not affect the surface morphology but resulted in significant changes in the instrument bulk with the appearance of an R-phase and an improved fatigue resistance; indeed after treatment at 500 °C, N_{CF} increased up to 829 and 474 for electropolished and nonelectropolished instruments, respectively. It was, therefore, concluded that both thermal treatment and electropolishing improved the resistance of NiTi rotary instruments against fatigue fracture.

9.3.2 Heat treatment

Zinelis et al. [107] determined the effect of various thermal treatments on the fatigue resistance of a NiTi engine-driven endodontic file. Fifteen groups of 5 files each of ISO 30 and taper 0.04 were tested, which were heat treated for 30 min in temperatures 250, 300, 350, 375, 400, 410, 420, 425, 430, 440, 450, 475, 500, and 550 °C. It was obtained that the 430 and 440 °C groups showed the highest values, with fatigue resistance decreasing for thermal treatment at lower and higher temperatures. Cheung et al. [108] examined the LCF behavior of a NiTi engine-file under various environmental conditions. One brand of NiTi instrument was subjected to rotational-bending fatigue in air, deionized water, NaOCl, or silicone oil. It was reported that (i) a linear relationship, on logarithmic scales, between the LCF life and the surface strain amplitude, regression line slopes were significantly different between noncorrosive (air and silicone oil) and corrosive (water and hypochlorite) environments, as well as number of crack origins, and (ii) hypochlorite was more detrimental to fatigue life than water concluding that environmental conditions significantly affect the LCF behavior of NiTi rotary instruments. Shen et al. [109] examined the fatigue behavior of two types of NiTi instruments made from a novel CM Wire under various environment conditions. It was found that (i) two new CM Wire instruments yielded an improvement of >4 to 9 times in N_{CF} than conventional NiTi files with the same design under various environments, (ii) the fatigue life of three conventional superelastic NiTi instruments was similar under various environments, whereas the N_{CF} of two new CM Wire instruments was significantly longer in liquid media than in air, (iii) the vast majority of CM instruments showed multiple crack origins, whereas most instruments made from conventional NiTi wire had one crack origin, and (iv) the values of the area fraction occupied by the dimple region were significantly smaller on CM NiTi instruments than in conventional NiTi instruments under various environments. Based on these results, it was concluded that (i) the type of NiTi metal alloy (CM files vs conventional superelastic NiTi files) influences the CF resistance under various environments, and (ii) the fatigue life of CM instruments is longer in liquid media than in air. Chang et al. [110] investigated the effect of heat treatment on the CF resistance, thermal behavior and microstructural changes of K3 NiTi rotary instruments. It was fund that (i) there was a significant increase in the CF resistance between the heat-treated instruments and the as-received instruments, (ii) DSC showed that the as-received and heat-treated samples were different, with an increased A_F for the latter, and (iii) TEM analysis revealed that both as-received and heat-treated instruments were composed mainly of an austenite phase; however, the heat-treated samples had an increased appearance of larger grains, twinning martensite, TiO_2 surface layer, and a Ni-rich inner layer concluding that heat treatment increased the CF resistance of NiTi files and changed the thermal behavior of the instruments without marked changes in the constituting phases of NiTi alloy. Braga et al. [111] assessed the influence of M-Wire and CM technologies on the fatigue resistance of rotary NiTi files by comparing files, including EndoWave (EW), HF, PV and TYP systems together with PTU F2 instruments. It was obtained that (i) the PV, TYP, and HF files exhibited increased transformation temperatures, (ii) the PTU F2, PV, and TYP files had similar D3 values, which were less than those of the EW and HF files, and (iii) the average N_{CF} values were 150% higher for the TYP files compared with the PV files and 390% higher for the HF files compared with the EW files concluding that M-Wire and CM technologies increase the fatigue resistance of rotary NiTi files. Plotino et al. [112] evaluated the difference in CF resistance between VB and PV NiTi rotary instruments. It was found that when comparing the same size of the two different instruments, a statistically significant difference was noted between all sizes of VB and PV instruments except for tip size 15 and 0.04 taper, and concluded that VB showed a significant increase in CF resistance when compared with the same sizes of PV. The effect of two different temperatures (20 and 37 °C) on the CF life of rotary instruments was investigated [113] to correlate the results with martensitic transformation temperatures. It was reported that (i) for the tested size and at 20 °C, HF CM showed the highest resistance to fracture; no significant difference was found between TRS and VB, whereas PTU showed the lowest resistance to fracture, and (ii) at 37 °C, resistance to fatigue fracture was significantly reduced, up to 85%, for the tested instruments; at that temperature, HF CM and VB had similar and higher fatigue resistance compared with TRS and PTU concluding that using a novel testing design, immersion in water at simulated body temperature was associated with a marked decrease in the fatigue life of all rotary instruments tested.

Grande et al. [114] analyzed how a low environmental temperature can affect the fatigue life of instruments made by different types of heat-treated NiTi alloys. PTU F2, PTG F2, Twisted Files SM2, Mtwo #25.06, and VB #30.04 and #40.06 (Dentsply Tulsa Dental Specialties) were tested at two different environmental temperatures: 20 (±2 °C) for RT group and –20 (±2 °C) for the cooled environment (CE) group. It was obtained that (i) the mean N_{CF} values measured were significantly higher for the CE groups than the RT groups in all the systems tested, and (ii) the increase in CF resistance varied from 274%–854% concluding that a low environmental temperature determines a drastic increase in the flexural fatigue resistance of NiTi endodontic instruments manufactured with traditional alloy and different heat treatments. Dosanjh et al. [7] examined the effect of different temperatures on the CF of NiTi rotary files. EdgeFile group, VB group and ESX group were tested in a metal block that simulated a canal curvature of 60° and a 5-mm radius curvature. It was reported that (i) VB group showed a significant decrease in N_{CF} as the temperature increased from 3 to 60 °C, (ii) the ESX group showed a significant decrease in N_{CF} as the temperature increased from 3 to 37 °C, (iii) the EF group showed a significant increase in N_{CF} from 3 to 22 °C and a significant decrease in N_{CF} from 22 to 37 °C, and (iv) for each temperature, the EF group showed higher N_{CF} than the VB group, which showed higher N_{CF} than the ESX group. It was, accordingly, concluded that (i) the temperature was found to significantly affect the CF of NiTi rotary files, and (ii) at each tested temperature, N_{CF} was the highest for the EF group followed by the VB group and lowest for the ESX group. Klymus et al. [115] evaluated the impact of body temperature on the CF resistance of different NiTi alloys used for the manufacturing of RB R25, X1 Blue File 25, and WaveOne Gold Primary (WOG), which is subjected to the CF tests at RT (20 °C ± 1 °C) and at body temperature (37 °C ± 1 °C). It was found that (i) the CF test at 20 °C showed that RB 25.08 and X1 25.06 presented significantly higher time to fracture (TTF) and N_{CF} than WOG 25.07, (ii) at 37 °C, all groups presented significant reduction of TTF and N_{CF}, (iii) RB 25.08 presented significant higher TTF than WOG 25.07, (iv) regarding the N_{CF}, there was no significant difference among the groups, and (v) the WOG 25.07 presented the lowest percentage reduction of CF ($P < 0.05$). It was, then, concluded that (i) the body temperature treatment caused a marked reduction of the CF resistance for all reciprocating instruments tested, and (ii) the RB 25.08 and X1 25.06 systems presented similar results at both temperatures tested; however, WOG 25.07 presented the lowest percentage reduction in fatigue resistance at body temperature. Alfawaz et al. [116] tested the CF resistance of heat-treated instruments immersed in NaOCl solution under different concentrations (2.5% or 5.25%) and temperature conditions (25, 37, or 60 °C). PTG F2 instruments were tested. It was reported that (i) N_{CF} of the PTG F2 was highest in distilled water at 25 °C and lowest in 5.25% NaOCl at 60 °C, and (ii) changing the irrigating solution from distilled water to NaOCl and increasing the surrounding temperature reduced the fatigue resistance. Shen et al. [117] evaluated the effect of different temperatures (0, 10, 22, 37, and 60 °C) on the CF life of NiTi files, including EndoSequence, ProFile, K3 and three heat-treated (K3XF, Vortex, and HCM NiTi files). It was obtained that (i) when the temperature was reduced from 60 to 0 °C, the N_{CF} significantly increased from over 2 to 10 times for the NiTi file groups, (ii) K3XF had the highest fatigue resistance of all files at 0 °C, (iii) Vortex files had the highest N_{CF} at 60 °C, (iv) the N_{CF} of heat-treated files was significantly higher than superelastic NiTi files at 10 and 20 °C, (v) there was no significant difference in the N_{CF} of HCM at 0 and 22 °C, and (vii) there was little difference in the fractographic appearance among different temperatures, except that the fraction area occupied by the dimple region of some instruments at 0 °C was slightly smaller than at 60 °C concluding that cooling down to low temperatures may be an interesting strategy to improve the fatigue resistance of rotary NiTi files. Arias et al. [118] evaluated the effect of different ambient temperatures on CF life of two NiTi rotary systems and correlate the results with martensitic transformation temperatures. Heat-treated NiTi VB and EdgeSequel Sapphire (SP) instruments were tested for CF resistance at room and body temperature. It was found that (i) temperature had an effect on fatigue behavior: all instruments lasted significantly longer at room than at body temperature, and (ii) all VB significantly outlasted those of SP at body temperature; while smaller diameters of VB were also significantly more resistant than SP when tested at RT; SP with larger diameters (sizes no. 30, no. 35, and no. 40) lasted significantly longer than VB did and concluded that (i) immersion in water at body temperature was associated with a marked decrease in the fatigue life of all rotary instruments tested, and (ii) VB instruments were significantly more CF resistant at body temperature and showed the highest predictability in terms of fracture resistance.

9.3.3 Sterilization effect

Infection control is a major issue in dentistry, mainly, because of the concern over contagious diseases transmitted in health-care settings. The cleaning and sterilization of endodontic instruments between treatment sessions are essential to prevent cross infection. However, the effects of the heating and cooling cycles used during sterilization on the mechanical properties and resistance to fracture of endodontic instruments have not been well-stated. Mize et al. [119] evaluated the ability of heat treatment as a result of autoclave sterilization to extend the life of NiTi rotary endodontic instruments by reducing the effect of CF using 280 size 40 LightSpeed instruments.

In the first experimental protocol, instruments were cycled to either 25%, 50%, or 75% of the mean cycles-to-failure (M_{CF}) limit determined in the pilot study, then sterilized or not sterilized before being cycled to failure. In the second experimental protocol, instruments were cycled to 25% of the M_{CF} determined in the pilot study, and sterilized or not sterilized. The sequence of cycling to 25% of the predetermined cycles-to-failure limit followed by sterilization was repeated until the instruments failed. It was reported that (i) no significant increases in cycles to failure were observed between groups for either experimental protocol when instruments were evaluated at a similar radius, (ii) significant differences in cycles to failure were only observed when instruments cycled to failure in the artificial canal with 2 mm radius were compared with instruments cycled to failure in the artificial canal of 5 mm radius, and (iii) SEM images showed crack initiation and propagation in all instruments that were cycled to a percentage of the predetermined cycles-to-failure limit concluding that heat treatment as a result of autoclave sterilization does not extend the useful life of NiTi instruments. Martins et al. [120] evaluated the effect of 5.25% NaOCl (for 24 h) on the surface characteristics and fatigue resistance of ProFile. It was found that (i) surface characteristics showed no alteration after the immersion tests, and (ii) the fatigue resistance of instruments in EG2 and EG3 was significantly lower than in CG and EG1 indicating that (iii) immersion of ProFile in 5.25% NaOCl for 24 h had no influence on surface characteristics and fatigue resistance, and (iv) simulated clinical use was a decisive factor in the decrease of fatigue life. Viana et al. [121] studied the effect of repeated sterilization cycles in dry oven or autoclave, on the mechanical behavior and fatigue resistance of rotary endodontic NiTi instruments. It was mentioned that (i) sterilization procedures resulted in no significant changes in Vickers microhardness nor in the parameters describing the mechanical behavior of the wires, (ii) however, the N_{CF} was statistically higher for all instruments after dry heat or autoclave sterilization cycles, and (iii) in the instruments previously fatigued to one half of their fatigue life, autoclave sterilization gave rise to an increase of 39% in the remaining N_{CF}. It was, therefore, concluded that (iv) changes in the mechanical properties of NiTi endodontic instruments after five cycles of commonly used sterilization procedures were insignificant, and (v) the sterilization procedures are safe as they produced a significant increase in the fatigue resistance of the instruments. Peters et al. [122] investigated the effect of immersion in NaOCl on torque and fatigue resistance of two NiTi files, including ProFile and RaCe files, which were immersed in 5.25% NaOCl for 1 or 2 h at temperatures of 21 and 60 °C. It was mentioned that (i) torsional resistance of both rotaries was not significantly affected by immersion in NaOCl, except after 2 h of immersion at 60 °C, and (ii) resistance to CF decreased significantly for ProFile and RaCe instruments after immersion in NaOCl. Spontaneous fractures occurred in 28 of 160 files during immersion in NaOCl concluding that NiTi rotaries have reduced resistance to CF after contact with heated NaOCl and may then be considered single-use instruments. Cheung et al. [123] examined the LCF behavior of NiTi rotary endodontic instruments in aqueous 1.2% NaOCl solution. It was obtained that (i) a linear strain-life relationship was obtained for all groups; the apparent fatigue-ductility exponent was similar between various brands, but not for the number of crack origins, and (ii) there was an inverse, linear relationship between the square root of the extension of the fatigue-crack and the strain amplitude. Pedullà et al. [124] assessed the resistance to CF of three NiTi files after the immersion in 5% NaOCl (NaOCl) solution at 37 °C in conditions similar to those used in clinical practice. It was reported that (i) resistance to CF of the same NiTi file was not significantly affected by immersion in NaOCl, (ii) the TF showed a higher resistance in all groups than Revo S SU, and (iii) the comparison between the same groups of Twisted Files and Mtwo files or between Mtwo and Revo S files did not show significant differences except for two cases: group 2 of the Twisted Files and Mtwo files and group 5 of the Mtwo and Revo S SU files. It was, therefore, concluded that (i) static or dynamic immersion in NaOCl for 1 or 5 min did not reduce the CF resistance of NiTi significantly; however, the type of instrument influences CF resistance, and (ii) twisted Files were more resistant followed by Mtwo and Revo S SU files. Hasegawa et al. [125] evaluated the effect of EDTA solutions (3% and 10% EDTA · 2Na) on corrosion fatigue of three NiTi files with different shapes, in comparison with other solutions (6% NaClO, 3% H_2O_2, 0.9% NaCl, and distilled water). NiTi files were subjected to rotational bending in a bent glass tube (30° and 60° angles) filled with the solutions, and the number of rotations to failure was counted. It was indicated that (i) at 30° bent angle, files in the two EDTA solutions showed significantly lower resistance than those in distilled water, but no significant difference was found between the two EDTA solutions, (ii) fatigue resistance of two tested files in the two EDTA solutions was not significantly different from those in the other three solutions, whereas one file in EDTA solutions showed significantly lower resistance than that in 3% H_2O_2, (iii) at 60° bent angle, early failure within 1–2.5 min was observed for all tested files, and no significant difference was found among the six solutions, and (iv) at both angles, significant differences in fatigue resistance were observed among the three tested files, which could be related to the difference in the cross-sectional shapes of the files. The effect of NaOCl and EDTA solutions on the CF resistance of WaveOne and WaveOne Gold NiTi reciprocating files was evaluated [126]. A hundred WO (25/0.08), and 100 WOG (25/0.07) were randomly divided into five groups: group 1, no immersion; group 2, immersion in 5.25% NaOCl at 37 °C ± 1 °C for 5 min; group 3, immersion in 5.25% NaOCl at 37 °C ± 1 °C for 10 min; group 4, immersion in 17% EDTA at 37 °C ± 1 °C for 5 min; and group 5, immersion in 17 % EDTA at 37 °C ± 1 °C for 10 min. It was reported that (i) the CF resistance of the WOG was statistically higher than the WO in all the conditions tested, (ii) there was no statistically significant difference among the different conditions tested in terms of CF resistance for both WO and WOG files, and (iii) among the groups, there was no significant difference in the fracture lengths, concluding that NaOCl and EDTA solutions did not have any effect on the CF resistance of WO and WOG files.

For sterilization, chemical treatment as described earlier is not the only method, but autoclave sterilization has been also employed. Bergeron et al. [127] assessed multiple autoclave cycle effects on CF of GTX files and Twisted Files. A jig using a 5-mm radius curve with 90° of maximum file flexure was used to induce CF failure. Files representing each experimental group were first tested to establish baseline M_{CF}. Experimental groups were then cycled to 25% of the established baseline M_{CF} and then autoclaved. Additional autoclaving was accomplished at 50% and 75% of M_{CF} followed by continual testing until failure. It was found that (i) the GTX files showed no significant difference in M_{CF} for experimental versus control files, (ii) Twisted Files showed no significant difference in M_{CF} between experimental and control groups, however (iii) the Twisted Files experimental group showed a significantly lower M_{CF} compared with the controls. Based on these results, it was concluded that (i) autoclave sterilization significantly decreased CF resistance of one of the four file groups tested, and (ii) repeated autoclaving significantly reduced the M_{CF} of 25/0.06 Twisted Files; however, 25/0.04 Twisted Files and both GTX files tested were not significantly affected by the same conditions. Plotino et al. [128] evaluated the effect of autoclave sterilization on CF resistance of rotary endodontic instruments made of traditional and new NiTi alloys. Four NiTi rotary endodontic instruments of the same size (tip diameter 0.40 mm and constant 0.04 taper) were selected: K3, Mtwo, Vortex, and K3 XF prototypes. Each group was then divided into two subgroups: unsterilized instruments and sterilized instruments. The sterilized instruments were subjected to 10 cycles of autoclave sterilization. It was observed that (i) comparing the results between unsterilized and sterilized instruments for each type of file, differences were statistically significant only between sterilized and unsterilized K3XF files (762 vs 651 N_{CF}), (ii) the other instruments did not show significant differences in the mean N_{CF} as a result of sterilization cycles (K3, 424 vs 439 N_{CF}; Mtwo, 409 vs. 419 N_{CF}; and Vortex, 454 vs 480 N_{CF}), and (iii) comparing the results among the different groups, K3 XF (either sterilized or not) showed a mean N_{CF} significantly higher than all other files concluding that repeated cycles of autoclave sterilization do not seem to influence the mechanical properties of NiTi endodontic instruments except for the K3 XF prototypes of rotary instruments that demonstrated a significant increase of CF resistance. Zhao et al. [129] compared the CF resistance of HCM, Twisted Files (TF), K3XF, Race, and K3 and evaluated the effect of autoclave sterilization on the CF resistance of these instruments both before and after the files were cycled. There were four groups: group 1 for unsterilized instruments, group 2 for presterilized instruments subjected to 10 cycles of autoclave sterilization, group 3 for instruments sterilized at 25%, 50%, and 75% of the M_{CF} as determined in group 1, and then cycled to failure, and group 4 for instruments cycled in the same manner as group 3 but without sterilization. It was found that (i) HCM, TF, and K3XF had significantly higher CF resistance than Race and K3 in the unsterilized group 1, and (ii) autoclave sterilization significantly increased the M_{CF} of HCM and K3XF both before and after the files were cycled indicating that HCM, TF, and K3XF instruments composed of new thermal-treated alloy were more resistant to fatigue failure than Race and K3. Autoclaving extended the CF life of HCM and K3XF. Özyürek et al. [130] compared the CF resistances of PTU, PTN, and PTG and the effects of sterilization by autoclave on the CF life of NiTi instruments. Each brand of the NiTi files were divided into four subgroups: group 1, as-received condition; group 2, presterilized instruments exposed to 10 times sterilization by autoclave; group 3, instruments tested were sterilized after being exposed to 25%, 50%, and 75% of the M_{CF}, then cycled fatigue test was performed; and group 4, instruments exposed to the same experiment with group 3 without sterilization. It was obtained that (i) PTG showed significantly higher N_{CF} than PTU and PTN in group 1, (ii) sterilization significantly increased the N_{CF} of PTN and PTG in group 2, (iii) PTN in group 3 had significantly higher CF resistance than PTN group 4, and (iv) also, significantly higher N_{CF} was observed for PTG in group 2 than in groups 3 and 4. Accordingly, it was concluded that (i) PTG instrument made of new gold alloy was more resistant to fatigue failure than PTN and PTU, and (ii) autoclaving increased the CF resistances of PTN and PTG.

9.3.4 Material's parameters

To evaluate two NiTi wires with different carbon and oxygen contents in terms of mechanical resistance to rotary bending fatigue under varied parameters of strain amplitude and rotational speed, the wires produced from two vacuum induction melting (VIM) processed NiTi ingots were tested, $TiNi_{49.81}$ and $TiNi_{50.33}$ (at%) , named VIM 1 and VIM 2. The VIM 1 wire had a high carbon content of 0.188 wt% and a low oxygen content of 0.036 wt%. The oxygen and carbon contents of wire VIM 2 did not exceed their maximum [131]. It was found that (i) the wire with lower carbon content performed better when compared to the one with higher carbon content, withstanding 29,441 and 12,895 cycles, respectively, to fracture, (ii) the surface quality of the wire was associated with resistance to CF, and (iii) overall, the N_{CF} was higher for VIM 2 wires with lower carbon content. In fatigue experiments of NiTi SMAs, TiC inclusions have been found to cause cracks. Based on bending–rotation fatigue experiments, which have evolved as one standard method to study the structural fatigue of superelastic NiTi wires, the influence of TiC inclusions on the fatigue behavior of NiTi SMAs has been analyzed quantitatively [132]. It was mentioned that (i) the stress distributions at the cross sections are nonlinear, and there is a stress plateau in the cross section when the phase transformation occurs, (ii) the stress distribution in the cross section of the specimen without inclusion is not only dependent on the load but also dependent on the loading path and loading history, (iii) on the other hand, the maximum stress of the specimen without inclusion is not always at the surface, which is due to the phase transformation behavior of NiTi alloys, (iv) the existence of the inclusions changes the stress distributions in the cross section, (v) the maximum stress is dependent on the position of the inclusions, the load, and the loading path, (vi) the maximum stresses increase as the distance from the inclusion to the neutral axis increases, (vii) when the inclusion is at the specimen surface, the maximum stress is the highest among all the studied cases, and (viii) such high stresses caused by the inclusions can easily induce fatigue cracks. Rahim et al. [11] studied the effects of different oxygen (O) and carbon (C) levels on fatigue lives of pseudoelastic NiTi SMAs. It was mentioned that (i) HCF life is governed by the number of cycles required for crack initiation, (ii) in the LCF regime, the high-purity alloy outperforms the materials with higher number densities of carbides and oxides, (iii) in the HCF regime, on the other hand, the high-purity and C-containing alloys show higher fatigue lives than the alloy with oxide particles, (iv) there is high experimental scatter in the HCF regime where fatigue cracks preferentially nucleate at particle/void assemblies which form during processing, and (v) cyclic crack growth follows the Paris law and does not depend on impurity levels.

A complex influence of various structural parameters (including an increased dislocation density, size of Ni-rich (Ni_4Ti_3 and Ni_3Ti_2) particles, and volume fraction of Ti-rich (Ti_4Ni_2Ox) particles as well as content of Ni in the alloy) on the strain-controlled fatigue behavior of NiTi alloy was investigated [133]. It was found that (i) in low-cycle conditions strain-controlled fatigue resistance of NiTi alloy may be improved by the creation of a microstructure which increases the part of deformation that is realized by martensitic mechanism, and (ii) no correlation between NiTi alloy strain-controlled fatigue resistance and critical strain parameter was observed for the high-cycle conditions. Gall et al. [134] investigated a relationship between material microstructure to monotonic and fatigue properties of NiTi SMAs. Ni-rich NiTi materials were either (1) hot rolled or (2) hot rolled and cold drawn. In addition to the two material processing routes, heat treatments are used to systematically alter material microstructure giving rise to a broad range of thermal, monotonic, and cyclic properties. The strength and hardness of the austenite and martensite phases initially increase with mild heat treatment (300 °C) and subsequently decrease with increased aging temperature above 300 °C. It was reported that (i) the low-cycle pseudoelastic fatigue properties of the NiTi materials generally improve with increasing material strength, although comparison across the two product forms demonstrates that higher measured flow strength does not assure superior resistance to pseudoelastic cyclic degradation, (ii) fatigue-crack growth rates in the hot-rolled material are relatively independent of heat treatment and demonstrate similar fatigue-crack growth rates to other NiTi product forms; however, the cold-drawn material demonstrates fatigue threshold values 5 times smaller than the hot-rolled material, and (iii) the difference in the fatigue performance of hot-rolled and cold-drawn NiTi bars is attributed to significant residual stresses in the cold-drawn material, which amplify fatigue susceptibility despite superior measured monotonic properties.

9.4 Fatigue on orthodontic archwires

The fracture of NiTi alloy orthodontic wires is caused by fatigue damage due to multiple rebending, degradation under stress in the oral environment, and hydrogen embrittlement using fluoride-containing dental products [135–138]. The fatigue phenomenon is an important mechanical performance characteristic of biomaterials, because they are typically used under cyclic loading [139, 140]. Among various important parameters which should determine fatigue behavior, the stress ratio ($R = \sigma_{MIN}/\sigma_{MAX}$) plays a key role and takes between $-\infty$ and $+1$. When the ratio enters a positive zone, the applied stress is in tension side, so that it indicates a crack opening; the more negative value of the stress ratio the more tendency for the crack closure. Dubey et al. [141] examined the effects of positive stress ratios on the fatigue-crack growth behavior of a forged, mill-annealed Ti-6Al-4V alloy. It was reported that (i) differences between fatigue-crack growth rates at low and high stress ratios are shown to be due largely to crack closure, (ii) 1% offset procedure is shown to collapse "closure corrected" low stress ratio data with the "closure-free" high stress ratio data, and (iii) for prediction of the fatigue-crack growth, stress ratio, crack closure, and stress intensify factor should be considered. At the same time, the mean stress ($\sigma_{MEAN} = (\sigma_{MIN} + \sigma_{MAX})/2$) is also important, in particular, for fatigue life prediction [142]. Since a small cyclic stress can trigger martensitic transformation in NiTi alloy when the mean stress is close to a critical stress, it is crucial to understand the effect of mean stress. The effects of the mean stress on fatigue properties of a NiTi SMA through systematic experiments were studied and preliminary results indicate that the cycles-to-failure has a nonmonotone dependence on the mean strain at fixed strain amplitude [143].

Intraoral temperature variation at archwire sites adjacent to the maxillary right central incisor and first premolar and its correlation with ambient temperature was investigated [144]. It was reported that (i) temperatures ranged from 5.6 to 58.5 °C at the incisor and from 7.9 to 54 °C at the premolar, with medians of 34.9 and 35.6 °C, respectively, (ii) ambient temperature correlated poorly with the intraoral temperatures, (iii) on average during the 24-h period, temperatures at the incisor site were in the range of 33–37 °C for 79% of the time, below it for 20%, and above it for only 1% of the time, (iv) corresponding figures for the premolar site were 92, 6, and 2%, and (v) at both archwire sites the most frequent temperatures were in the range of 35–36 °C concluding that the temperature at sites on an archwire in site varies considerably over a 24-h period.

Their use is due to their properties of shape memory and SE [2]. Both SME and SE are related to the ability of NiTi archwire to easily transform to and from a martensitic phase and the transformation can occur by means of stress or temperature changes. Temperature sensitivity of NiTi alloy was reported [145–150]. Mullins et al. [145] investigated the mechanical behavior of thermoresponsive nitinol archwires in flexure at 5 and 37 °C. Four same-sized (but different force level) rectangular archwires were examined using a three-point bend test. Samples were tested at 5 and 37 °C. It was found that (i) superelastic behavior was exhibited by all wires tested at 37 °C but not at 5 °C, (ii) permanent deformation was greater at 5 °C than 37 °C, (iii) the initial slope of the load-deflection data averaged 1,230 g/mm at 37 °C, which was significantly different from the average at 5 °C (500 g/mm), (iv) loads at the apparent yield point and the loads at 1, 2, and 3 mm of deflection were greater at 37 °C than at 5 °C, (v) while the slope and length of the superelastic region were not judged to be clinically significantly different, the average load of the superelastic region was significantly different: F300 (340 g) > F200 (250 g) > F100 (180 g) and Bioforce (180 g), and (vi) when loaded at 5 °C and then unloaded at 37 °C, the mechanical hysteresis of the wires tested at 37 °C and the permanent deformation of the wires tested at 5 °C were reduced for all wires indicating that (vii) Nitinol wires are available with a variety of mechanical properties, (viii) the different mechanical properties of thermoresponsive wires at 5 and 37 °C result in clinically useful shape-memory behavior, and (ix) utilizing the superelastic and shape-memory features of thermoresponsive wires has clinical advantages. Meling et al. [147] compared the effect of short-term cooling or heating on the bending force exerted by NiTi archwires. Two rectangular superelastic and one conventional NiTi wire were tested in bending at 37 °C. The specimens were tested during the activation phase and during the deactivation phase. The wires were kept at constant strain and the bending force was measured continually while the activated specimens were subjected to cold (10 °C) or hot (80 °C) water. The test situation simulates a patient's archwire that is subjected to cold or hot drinks or food during a meal. It was mentioned that (i) although the conventional NiTi wire was marginally affected by brief cooling or heating, regardless of activation phase, the superelastic wires were strongly affected by short-term application of cold or hot water, (ii) when tested in activation phase, the effect of heating was transient whereas the wires continued to exert sub-baseline bending forces after short-term application of cold water, (iii) when tested in deactivation phase, the effect of cooling was transient whereas the wires exerted supra-baseline bending forces after a short-term application of hot water, (iv) the effect of short-term temperature changes on the bending stiffness of superelastic NiTi archwires is dependent upon the bending phase, and (v) cooling induced transient effects on a wire when the wire was tested in deactivation phase, but the effects were prolonged when the wire was tested in the activation phase. Iijima et al. [148] studied the phase transformation behavior in three commercial NiTi orthodontic wires having different transformation temperatures (25, 37, and 60 °C) by micro-X-ray diffraction. The X-ray diffraction intensity ratio (M002/A110) between the (002) peak for martensitic NiTi and the (110) peak for austenitic NiTi was employed as the index to the proportions of the martensite and austenite phases. It was reported that the ratio of martensite to austenite increased in all three NiTi wires with decreasing included bending angle (greater permanent bending deformation) and was lower within the compression area for all wires at all bending angles than within the tension area. Berzins et al. [149] investigated the effect of thermocycling on NiTi wire phase transformations. Straight segments from single 27 and 35 °C copper NiTi (Ormco), Sentalloy (GAC), and Nitinol Heat Activated (3 M Unitek) archwires were sectioned into 5 mm segments. A control group consisted of five randomly selected nonthermocycled segments. The remaining segments were thermocycled between 5 and 55 °C with five randomly selected segments analyzed with DSC (−100 ↔ 150 °C at 10 °C/min) after 1000, 5000, and 10,000 cycles. It was found that (i) Nitinol HA and Sentalloy did not demonstrate qualitative or quantitative phase transformation behavior differences, (ii) significant differences were observed in some of the copper NiTi transformation temperatures, as well as the heating enthalpy with the 27 °C copper NiTi wires, and (iii) qualitatively, with increased thermocycling the extent of R-phase in the heating peaks decreased in the 35 °C copper NiTi, and an austenite to martensite peak shoulder developed during cooling in the 27 °C copper NiTi indicating that repeated temperature fluctuations may contribute to qualitative and quantitative phase transformation changes in some NiTi wires. Lombardo et al. [150] investigated and compared the characteristics of commonly used types of traditional and heat-activated initial archwires at different temperatures by plotting their load/deflection graphs and quantifying three parameters describing the discharge plateau phase. Forty-eight archwires of cross-sectional diameters ranging from 0.010 inches to 0.016 inches were tested. A modified three-point wire-bending test was performed on three analogous samples of each type of archwire at 55 and 5 °C, simulating an inserted archwire that is subjected to cold or hot drinks during a meal. For each resulting load/deflection curve, the plateau section was isolated and the mean value of each parameter for each type of wire was obtained. It was obtained that (i) permanent strain was exhibited by all wires tested at 55 °C, (ii) statistically significant differences were found between almost all wires for the three considered parameters when tested at 55 and 5 °C, (iii) loads were greater at 55 °C than at 5 °C, (iv) differences were also found between traditional and heat-activated archwires, the latter of which generated longer plateaus at 55 °C, shorter plateaus at 5 °C, and lighter mean forces at both temperatures, and (v) the increase in average force seen with increasing diameter tended to be rather stable at both temperatures. Based on these results, it was concluded that (i) all NiTi wires tested showed a significant change related to temperature in terms of behavior and force for both traditional and heat-activated wires, and (ii) stress under high temperatures can induce permanent strain, whereas the residual strain detected at low temperatures can be recovered as temperature increases.

Eliades e al. [151] evaluated the structure and morphological condition of retrieved NiTi orthodontic archwires (GAC, German Orthodontics and ORMCO). It was observed that (i) optical microscopy revealed islands of amorphous precipitants and accumulated microcrystalline particle, (ii) micro-MIR-FTIR investigation of the retrieved samples demonstrated the presence of a proteinaceous biofilm, the organic constituents of which were mainly amide, alcohol, and carbonate, (iii) SEM and X-ray microanalysis showed that the elemental species precipitated on the material surface were Na, K, Cl, Ca, and P, forming NaCl, KCl, and Ca-P precipitates, (iv) increased intraoral exposure was consistently associated with the presence of a mature film, while evidence of alloy delamination, pitting, and crevice corrosion, as well as a notable reduction in the alloy grain size was observed (v) intraoral exposure of NiTi wires alters the topography and structure of the alloy surface through surface attack in the form of pitting or crevice corrosion or formation of integuments, and (vi) the in vivo research is required to resolve the implications of the described ageing pattern in the corrosion resistance of the alloy, the potential for nickel leaching, as well as bracket-archwire friction variants. Wilkinson et al. [152] investigated the load-deflection characteristics of seven different 0.016-in initial alignment archwires (Twistflex, NiTi, and five brands of heat-activated superelastic nickel-titanium (HASN)) with modified bending tests simulating a number of conditions encountered clinically. Wire deflection was carried out at 22.0, 35.5, and 44.0 °C and to four deflection distances (1 mm, 2 mm, 3 mm, and 4 mm). It was reported that (i) the effects of model, wire, and temperature variation were all statistically significant, (ii) Twistflex and the 5 HASN wires produced a range of broadly comparable results, and NiTi gave the highest unloading values, and (ii) model rankings indicated that self-ligating Twin-Lock brackets produced lower friction than regular edgewise brackets suggesting using the rankings from the mechanical test simulations to predict possible clinical performance of archwires. Prymak et al. [153] analyzed in vitro the fatigue resistance of NiTi and Cu-NiTi orthodontic wires when subjected to forces and fluids which are present intraorally. Stainless steel wires were used for comparison. It was found that (i) in general, NiTi wires fractured earlier than the stainless steel specimens, (ii) survival times were lower for the NiTi wires when immersed in fluids (water, citric acid, NaCl solution, artificial saliva, and fluoridated artificial saliva) than in air, (iii) SEM surface analysis showed that the NiTi and Cu–NiTi wires had a rougher surface than steel wires, (iv) the fracture occurred within a short number of loading cycles, and (v) until fracture occurred, the mechanical properties remained mostly constant.

Lombardo et al. [154] investigated and compared the characteristics of as-received and retrieved NiTi archwires at a constant temperature by plotting their load/deflection graphs and quantifying three parameters describing the discharge plateau phase. Two hundred four NiTi archwires, traditional and heat-activated, of various cross sections, were obtained from five different manufacturers. Specimens prepared from the selected wires were subjected to a three-point bending test where 92 were retrieved through an in vivo retrieval protocol (crowding group C1 and group C2), 56 went through an in vitro retrieval protocol, and 56 were as received. The in vitro retrieval protocol was performed by a gear motor connected to a stainless steel support that performed fatigue cycles to the bent wires in artificial saliva. The load/deflection graphs of as-received and retrieved wires were described through three parameters. It was found that statistically significant differences between as-received and retrieved wires were found only for the parameter plateau slope which represents the constancy of force expressed by the wire. It was concluded that (i) the aging of NiTi archwires influences the force constancy expressed, (ii) the behavior of the wires changes depending on the size, brand, and type of retrieval protocol, and (iii) in terms of performance, the poorest is represented by all wires retrieved in vitro and in vivo group C2 (moderate to severe crowding). Bourauel et al. [155] investigated the fracture resistance of as-received and retrieved NiTi archwires (German Orthodontics, CA, USA) of various cross sections, which were retrieved from orthodontic patients and brand-, type-, and size-matched wires were included as controls. It was reported that (i) retrieved wires fractured at a significantly lower number of cycles compared to their as-received matches, (ii) the size of the wire played a role in determining fracture with larger cross sections showing reduced fatigue failure properties, and 0.30 and 0.035 mm specimens showing no fracture at the selected strain and number of cycles, (iii) the SEM investigations revealed evidence of smoother fractured surfaces in large-sized, rectangular cross-sectional retrieved wires, and (iv) Kaplan-Meier survival analysis with log rank test demonstrated the effect of aging and size in predicting survival of NiTi wires. It was further mentioned that (i) the extended life expectancy of NiTi archwires proposed in current treatment trends is associated with higher probability of fatigue fracture of wires, (ii) large diameter and square or rectangular cross sections possess an increased chance of failure relative to smaller diameter wires, and (iii) a necessity may emerge to monitor patients to identify wire failures despite the commonly employed practice of increased time intervals between appointments in patients treated with NiTi wires.

Hernández et al. [156] examined the number of cycles that NiTi archwires withstand in flexural fatigue. Sixty circular 0.016″ archwires, ORMCO (20), GAC (20), and 3M (20), tested at 37 °C. It was found that ORMCO archwires showed more elasticity than 3M and GAC archwires. Murakami et al. [136] investigated HCF behavior in three β-Ti wires (TMA, Resolve, and Gummetal). Fatigue was evaluated using a static three-point bending test and a HCF test with a three-point bending mode. It was reported that (i) the Gummetal wire exhibited the lowest elastic modulus, bending strength, and fatigue limit and exhibited the highest resilience of the three types of wire studied; however, no difference in the N_{CF} was observed among the three types of wire, (ii) the fatigue-crack propagation and rapid propagation regions of all wires contained single-phase β-Ti, (iii) the elastic modulus and bending strength influenced the fatigue limit, although these properties did not affect the N_{CF}, and (iv) the three types of β-Ti wires exhibited similar risks of wire fracture. During the orthodontic mechanotreatment, some orthodontic wires may break in the oral cavity and the susceptibility of these alloys to cyclic loadings and to hydrogen embrittlement is supposed to be main causes of these unexpected failures. Rokbani et al. [157] investigated the effect of hydrogen, obtained after cathodically charging in 0.9% NaCl solution, on the fatigue behavior of NiTi commercial orthodontic wires subjected to HCF. Fatigue tests were analyzed using self-heating method based on observing thermal effects under mechanical cyclic loading. It was obtained that (i) the increase in hydrogen charging time is connected with an increase in the mean stabilized temperature and a decrease in the fatigue life, (ii) self-heating method allows a rapid prediction of the endurance limit with a good reproducibility of fatigue tests at high number of cycles, and (iii) cyclic stress-induced transformations and conventional fatigue tests under strain control are considered in this work to investigate the effect of hydrogen on cyclic loading type and to acquire a better understanding of the interaction between hydrogen and thermomechanical mechanisms in NiTi alloys.

9.5 Fatigue on other applications

Applications, devices, and equipment which utilize NiTi alloy are not limited to previously discussed endodontic instruments and orthodontic archwire, but it should extend to devices using high temperature shape memory alloys (HTSMAs), medical stent, and wire and tubing as well.

Karakoc et al. [158] studied the effects of upper cycle temperature (UCT) on the actuation fatigue response of nanoprecipitation hardened $Ni_{50.3}Ti_{29.7}Hf_{20}$ HTSMA. A series of actuation fatigue experiments were conducted under different constant tensile stresses while the test temperature was cycled between two extreme martensitic transformation temperatures. It was reported that (i) the samples subjected to 300 °C UCT exhibit fatigue lives twice that of the samples with 350 °C UCT, and those tested under 300 MPa with 300 °C UCT withstand more than 10,000 cycles with actuation strains of 2–3%, (ii) actuation strains remained constant or increased with thermal cycling in the 350 °C UCT experiments, while those subjected to 300 °C UCT exhibited decreasing actuation strains, at all stress levels, (ii) in the 300 °C UCT experiments, partial martensitic transformation becomes operative, resulting in a reduction of actuation strain in each cycle, and postpones damage accumulation, and (iii) in the 350 °C UCT cases, increase of actuation strain is attributed to the partial recovery of cyclically induced remnant deformation at higher temperatures and as a result, larger volume of transforming material; this, in turn, accelerates the formation of cracks, that open and close during thermal cycling and reversible martensitic transformation, and manifests itself as additional recoverable strain in each cycle. Using the same HTSMA, the actuation fatigue response was studied undergoing thermal cycling between martensite and austenite under various tensile stress levels up to 500 MPa [159]. It was reported that (i) the data revealed a consistent increase in actuation strain concomitant with the applied load at the expense of fatigue life, (ii) significantly high N_{CF} were observed for this class of materials: specimens tested under 200 MPa achieved ~21,000 cycles with the average actuation strain of ~2.15% while those tested under 500 MPa experienced ~2,100 cycles to failure with the average actuation strain of 3.22%, (iii) fracture surface and crack density analyses revealed notable crack formation prior to failure at all stress levels; however, the rate of crack formation during repeated transformation increased with the applied stress, (iv) the actuation fatigue lives of the present HTSMAs exhibit an almost perfect power law correlation with average actuation work output, (v) the same work-based power law was shown to successfully capture actuation fatigue lives of several low temperature SMAs, and (vi) the power law exponents for many SMAs were shown here to be either ~ −0.5 or ~ −0.8, which points out the likelihood of the existence of a universal empirical rule for actuation fatigue response of SMAs.

Speirs et al. [160] investigates the compression fatigue behavior of three different unit cells (octahedron, cellular gyroid, and sheet gyroid) of selective laser melting (SLM) Nitinol scaffolds. It was reported that (i) triply periodic minimal surfaces display superior static mechanical properties in comparison to conventional octahedron beam lattice structures at identical volume fractions, (ii) fatigue resistance was also found to be highly geometry dependent due to the effects of additive manufacturing processing techniques on the surface topography and notch sensitivity, (iii) geometries minimizing nodal points and the staircase effect displayed the greatest fatigue resistance when normalized to yield strength, (iv) oxygen analysis showed a large oxygen uptake during SLM processing which must be altered to meet ASTM medical grade standards and may significantly reduce fatigue life, and (v) these achieved fatigue properties indicate that NiTi scaffolds produced via SLM can provide sufficient mechanical support over an implants lifetime within stress range values experienced in real life.

Intravascular stenting has emerged as the primary treatment for vascular diseases and has received great attention from the medical community since its introduction two decades ago [161]. Inherent risks of stenting include restenosis and thrombosis. Recently, stent fractures have been recognized as a complication that may result in thrombosis, perforation, restenosis, and migration of the stent resulting in morbidity and mortality. Stent fractures were originally seen in the superficial femoral arteries but have since then been reported in almost all vascular sites including the coronary, renal, carotid, iliac, and femoropopliteal arteries. Fractures are the result of the complex interplay between stent manufacturing, the stented segment, pulsatile and nonpulsatile biomechanical forces, and plaque morphology at a particular vascular site. [162]. In particular, the patient daily activities expose the peripheral arteries to large and cyclic deformations which may cause long-term failure of the device and consequently reocclusion of the artery. Accordingly, the assessment of the stent fatigue rupture is of primary importance to assure the effectiveness of stenting procedure. The fatigue behavior characterization of Nitinol for peripheral stent is a quite difficult problem because of the complexity of the in vivo solicitations the stent is subjected to and the strong nonlinearity in the material response [163]. McKelvey et al. [164] studied the fatigue-crack propagation behavior in the superelastic alloy Nitinol, which is specifically an equiatomic NiTi alloy and for endovascular stents. Characterization of fatigue-crack growth rates was performed at 37 °C on disk-shaped compact-tension samples in environments of air, aerated deionized water, and aerated Hank's solution (a simulated body fluid). The effect of cyclic loading on the uniaxial constitutive behavior was investigated at a strain range of 6.4%, and results indicated that the magnitude of available superelastic strain (~5.0%) is maintained even after cyclic softening. It was further mentioned that, however, despite the persistence of nucleating the stress-induced martensitic phase after cycling with a maximum strain slightly below the plastic yield point, Nitinol was found to have the lowest fatigue-crack growth resistance of the principal metallic alloys currently used for implant applications. Pelton et al. [165] evaluated combined effects of cardiac pulsatile fatigue and stent-vessel oversizing for application to both stents and stent subcomponents. In particular, displacement-controlled fatigue tests were performed on stent-like specimens processed from Nitinol microtubing. It was demonstrated that (i) the nonlinear constant fatigue-life response of Nitinol stents is unusual and contrary to conventional engineering materials and (ii) as a result, the fatigue life of Nitinol is observed to increase with increasing mean strain.

Figueiredo et al. [166] analyzed LCF life under strain control (εa–N_F curve) of NiTi wires in bending–rotation tests. These were carried out on stable austenite, superelastic, and stable martensite wires, with strain amplitudes from 0.6% to 12%. It was reported that (i) for strain amplitudes up to 4%, εa–N_F curves of superelastic wires are close to those values reported in the literature, and close to that of the stable austenite wire, (ii) for higher strain amplitudes, the fatigue life of superelastic wires increases with strain and approaches the fatigue life of stable martensite wire, and (iii) fatigue-crack characteristics were studied by SEM suggesting that the abnormal shape of the superelastic wire curve is associated to the changes in fatigue properties that occurs when the superelastic material transform to martensite. Wang et al. [167] conducted torsion–tension cycling experiments on thin-wall tubes (with thickness/radius ratio of 1:20, similar to that found for stents) of nearly equiatomic NiTi SMAs. Experiments were controlled by axial displacement and torsional angle with step loading involving torsional loading to a maximum strain, followed by tensile loading and reverse-order unloading. It was mentioned that (i) the SE of the material is confirmed by pure torsion and tension experiments at the test temperature, (ii) the evolution of equivalent stress–strain curves as well as the separated tensile and torsional stress–strain curves during cycling is analyzed, (iii) the equivalent stress increases greatly with a small amount of applied axial strain, and the equivalent stress–strain curves have negative slopes in the phase transformation region, (iv) the shear stress drops when the torsional strain is maintained at its maximum value and the tensile strain is increased, and (v) the shear stress increases with decreasing tensile strain, but it cannot recover to the original value after the complete unloading of the tensile strain. Pelton et al. [168] investigated the rotary bending fatigue properties of medical-grade Nitinol wires under conditions of 0.5–10% strain amplitudes to a maximum of 10^7cycles. It was reported that (i) for pseudoelastic conditions there are four distinct regions of the strain-cycle curves that are related to phases (austenite, stress-induced martensite, and R-phase) and their respective strain accommodation mechanisms; in contrast, there are only two regions for the strain-cycle curves for thermal martensite, (ii) the strain amplitude to achieve 10^7-cycles increases with both decreasing test temperature and increasing transformation temperature, and (iii) fatigue behavior was not, however, strongly influenced by wire surface condition. Robertson et al. [169]

manufactured superelastic wires and diamond-shaped stent surrogates from Nitinol rods and tubing, respectively, from five different mill product suppliers – Standard VAR, Standard VIM, Standard VIM+VAR, Process-Optimized VIM+VAR, and High-Purity VAR. HCF tests up to 10^7 cycles were conducted under tension–tension conditions for wires and bending conditions for diamonds. These materials were compared under both testing methods at 37 °C with 6% prestrain and 3% mean strain (unloading plateau) with a range of alternating strains. It was found that (i) the High-Purity VAR material outperformed all alloys tested with a measured 10^7-fatigue alternating strain limit of 0.32% for wire and 1.75% for diamonds, (ii) Process-Optimized VIM+VAR material was only slightly inferior to the High Purity VAR with a diamond alternating bending strain limit of 1.5%, (iii) these two "second generation" Nitinol alloys demonstrated approximately a 2× increase in 10^7-cycle fatigue strain limit compared to all of the standard-grade Nitinol alloys (VAR, VIM, and VIM+VAR) that demonstrated virtually indistinguishable fatigue performance, (iv) this statistically significant increase in fatigue resistance in the contemporary alloys is ascribed to smaller inclusions in the Process-Optimized VIM+VAR material and both smaller and fewer inclusions in the High-Purity VAR Nitinol. Prior to implantation, Nitinol-based transcatheter endovascular devices are subject to a complex thermomechanical prestrain associated with constraint onto a delivery catheter, device sterilization, and final deployment. Though such large thermomechanical excursions are known to impact the microstructural and mechanical properties of Nitinol, their effect on fatigue properties is still not well understood. Gupta et al. [170] investigated the effects of large thermomechanical prestrains on the fatigue of pseudoelastic Nitinol wire using fully reversed rotary bend fatigue experiments. Electropolished Nitinol wires were subjected to a 0%, 8% or 10% bending prestrain and rotary bend fatigue testing at 0.3–1.5% strain amplitudes for up to 10^8 cycles. It was obtained that (i) the imposition of 8% or 10% bending prestrain resulted in residual set in the wire, (ii) large prestrains also significantly reduced the fatigue life of Nitinol wires below 0.8% strain amplitude, (iii) while 0% and 8% prestrain wires exhibited distinct LCF and HCF regions, reaching run out at 10^8 cycles at 0.6% and 0.4% strain amplitude, respectively, 10% pre-strain wires continued to fracture at less than 10^5 cycles, even at 0.3% strain amplitude, (iv) over 70% fatigue cracks were found to initiate on the compressive prestrain surface in prestrained wires, and (v) in light of the texture-dependent tension–compression asymmetry in Nitinol, this reduction in fatigue life and preferential crack initiation in prestrained wires is thought to be attributed to compressive prestrain-induced plasticity and tensile residual stresses as well as the formation of martensite variants. Saikrishnak et al. [171] investigated the effect of intermittent OLCs on the fatigue life of NiTi wires. It was shown that (i) the OLCs were effective in enhancing the fatigue life of the NiTi wires undergoing TMC, (ii) three distinct regions were identified in the N_{CF} versus overload ratio plot; region I (OLR ≤ 1.5) corresponded to threshold OLR, at and below which the OLCs did not have any effect on the fatigue behavior of the NiTi wires; in region II (1.5 < OLR < 2.5) the fatigue life was found to increase monotonically with the increase in OLR; in region III (OLR ≥ 2.5), the increase in fatigue life was quite substantial, (iii) the actuator wire with OLCs of OLR 2.5 showed fatigue life of three times of that of the wire with OLR of 2.25 and 9 times of that without OLCs, (iv) the effect of OLCs toward enhancement in fatigue life was cumulative, and its withdrawal at any stage during the TMC resulted in a life lower than that of the wire subjected to OLCs till fracture, and (v) the presence of stabilized martensite phase in the microstructure was found beneficial toward enhancement in fatigue life and its restoration by healing treatment had a detrimental effect on the life of the actuator wires.

Porous metals are being widely used in load bearing implant applications with an aim to increase osseointegration and also to reduce stress shielding. However, fatigue performance of porous metals is extremely important to ensure long-term implant stability, because porous metals are sensitive to crack propagation even at low stresses especially under cyclic loading conditions. Bernard et al. [172] conducted high-cycle compression–compression fatigue behavior of laser processed NiTi alloy with varying porosities between ~1% and 20%. It was found that (i) compression fatigue of porous NiTi alloy samples is in part similar to metal foams, (ii) the applied stress amplitude is found to have strong influence on the accumulated strain and cyclic stability, and (iii) the critical stress amplitudes associated with rapid strain accumulation in porous NiTi alloy samples, with varying relative densities, were found to correspond to 140% of respective 0.2% proof strength indicating that these samples can sustain cyclic compression fatigue stresses up to 1.4 times their yield strength without failure.

9.6 Fracture, fracture toughness (K_{IC}), and fractography

9.6.1 Fracture mechanisms

For products with geometrically small sizes such as endovascular stents, the technological importance of fracture mechanics is limited, with the emphasis being placed on preventing crack nucleation rather than controlling crack growth. On the other hand, devices characterized with actuation, energy absorption, and vibration damping applications requires understanding and practice of fracture mechanics concepts in SMAs. The fracture response of SMAs is rather complex owing to the reversibility of phase transformation, detwinning and reorientation of martensitic variants, the possibility of dislocation and transformation-induced plasticity, and the strong thermomechanical coupling. Large-scale phase transformation under actuation loading paths, that is, combined thermomechanical loading, and the associated configuration dependence complicate the phenomenon even further and question the applicability of single parameter fracture mechanics theories [173]. The reversible stress-induced and thermally phase transition mechanisms are occurring in the crack tip region as a consequence of the highly localized stresses [174]. The high values of local stresses arising near the crack tip in NiTi-based SMAs causes a stress-induced martensitic transformation, as it significantly changes the crack tip stress distribution with respect to common engineering metals [175]. Hence there is the proposed approach that uses a trilinear stress–strain constitutive behavior, that is, it is able to analyze SMAs with nonconstant transformation stress [175]. To understand the localized microstructural changes including several thermomechanical parameters and loading conditions on the crack tip transformation region, the manner of the stress distribution becomes very important [176–181]. It should be also noted that temperature plays an important role on the stress intensity factor and on critical fast fracture conditions. As a consequence, linear elastic fracture mechanics approaches are not suitable to predict fracture properties of SMAs, as they do not consider the effects of temperature [182, 183]. Therefore, it can be stated that SMAs exhibit very large recoverable deformation capabilities either by temperature or stress variations, which can be obtained through a thermal or a mechanical induced reversible phase transformation between a parent austenite phase (B2) and a product martensite one (B19′), the so-called thermally induced martensite and SIM. The evidence that makes the situation more complicated is that there are two different stress intensity factors that have been defined to describe the stress distribution in both transformed and untransformed regions, that is, in the martensitic and austenitic phases, respectively. Since the certain thermal-related phenomenon is involved in the fracture process in SMAs, nonlinear fracture mechanics should be considered [184].

For these applications such as endovascular stents and vena filters and for endodontic instruments, the stress-induced transformation (or SE) has been used extensively for these self-expanding implantable devices. Most of these applications involve cyclically varying biomechanical stresses or strains that drive the need to fully understand the fatigue and fracture resistance of this alloy [185]. Several key fracture-mechanics parameters associated with the onset of subcritical and critical cracking, specifically the K_{IC}, crack-resistance curve, and fatigue threshold, have recently been reported for the superelastic alloy Nitinol, in the product form of the thin-walled tube that is used to manufacture several biomedical devices, most notably endovascular stents [186]. For fully understanding such fracture behavior of this material, the criterions including crack-initiation K_{IC} and fatigue threshold stress-intensity range together with the general yield strength and fatigue endurance strength should be considered [186, 187].

Gall et al. [188, 189] using the Ti_3Ni_4 precipitates containing Ni50.8Ti (at%) SMAs, studied deformation and fracture behaviors. It was mentioned that (i) the mechanisms of material fracture identified in the single crystal NiTi are (1) nucleation, growth, and coalescence of voids from the Ti_3Ni_4 precipitates, (2) cleavage fracture on {100} and {110} crystallographic planes, and (3) nucleation, growth, and coalescence of voids from fractured Ti-C inclusions, (ii) as the Ti_3Ni_4 precipitate size increases to about 400 nm, the overall fracture is dominated by failure mechanism and the cleavage markings become diffused. Chen et al. [190] investigated the fracture behavior of shape memory $Ni_{50.7}Ti$ (at%) alloy at RT. It was reported that (i) specimens with different thicknesses show various SMEs and superelasticities, (ii) the main crack with a quasi-cleavage mode that combines cleavage with ductile tearing is initiated at the notch tip and is stress-control-propagated in line with the direction of the maximum normal stress, (iii) the microstructure has little effect on the direction of crack propagation, but coarser substructures show lower resistance to the crack propagation, and (iv) in specimens with various types of notches, various notch acuities present different effects on the crack initiation and propagation and result in different fracture behaviors. Kasiri et al. [191] examined whether the theory of critical distances (TCDs) can predict the failures caused by stress concentrations and defects such as notches and cracks. The TCD uses the stress at a certain distance ahead of the notch to predict the failure of the material due to the stress concentration. The critical distance is believed to be a material property which is related to the microstructure of the material. The TCD is simply applied to a linear model of the material without the need to model the complication of its nonlinear behavior. The nonlinear behavior of the material at fracture is represented in the critical stress. Hence the effect of notches and short cracks on the fracture of SMA NiTi was studied by analyzing experimental data from the literature. Using a finite element model with elastic material behavior, it was shown that (i) the TCD can predict the effect of crack length and notch geometry on the critical stress and stress intensity for fracture, with prediction errors of less than 5%, and (ii) the value of the critical distance obtained for this material was $L = 90\ \mu m$; this may be related to its grain size. Hsu et al. [192] studied the degradation of the superelastic properties of a commercial NiTi alloy during uniaxial, biaxial, and load-path-change cycling performed in situ with synchrotron X-ray diffraction. It was reported that (i) careful examination of the diffraction pattern during uniaxial loading shows the R-phase as a transition between the austenite and the B19′ martensite, (ii) degradation of the SE is found to depend strongly on the loading and unloading path followed, and it is discussed in terms of the B19′ martensitic variant selection, the accumulation of dislocations, and the residual R-phase and B19′ martensite, (iii) cycling biaxially leads to faster degradation than uniaxially due to a larger accumulation of dislocations, and (iv) if the deformation cycle contains a load path change, dislocation accumulation increases further and more martensite is retained.

9.6.2 Fracture toughness

The crack is assumed to propagate at a critical level of the crack-tip energy release rate. Results pertaining to the influence of forward and reverse phase transformation on the near-tip mechanical fields and K_{IC} are presented for a range of thermomechanical parameters and temperature. The K_{IC} is obtained as the ratio of the far-field applied energy release rate to the crack-tip energy release rate. A substantial fracture toughening is observed, in accordance with experimental observations, and associated with the energy dissipated by the transformed material in the wake of the growing crack.

Reverse phase transformation, being a dissipative process itself, is found to increase the levels of toughness enhancement [193]. The K_{IC} and K_R are essential properties for successful application of metals and alloys [193]. Jape et al. [194] analyzed the effect of transformation-induced plasticity (TRIP) on the fracture response of polycrystalline SMA in the prototype infinite center-cracked plate subjected to thermal cycling under constant mechanical loading in plain strain. It was mentioned that (i) similar to phase transformation, TRIP is found to affect both the driving force for crack growth and the crack growth kinetics by promoting crack advance when occurring in a fan in front of the crack tip and providing a "shielding" effect when occurring behind that fan, (ii) accumulation of TRIP strains over the cycles results in higher energy release rates from one cycle to another and may result in crack growth if the crack-tip energy release rate reaches a material "specific" critical value after a sufficient number of cycles, and (iii) during crack advance, the shielding effect of the TRIP strains left in the wake of the growing crack dominates and therefore TRIP is found to both promote the initiation of crack growth and extend the stable crack growth regime. Owing to a stress-induced martensitic phase transformation, the most distinctive feature in the fracture process of SMAs is the formation of a kidney-like crack tip phase transformation zone which can transform back at the crack wake [194–198].

Yi et al. [199, 200] studied the fracture toughening mechanism of SMAs. It was reported that (i) the martensite transformation reduces the crack tip stress intensity factor and increases the toughness, (ii) the toughness of SMAs is enhanced by the transformed strain fields which tend to limit or prevent crack opening and advancing, and (iii) the toughness of SMAs is enhanced by the transformation strain, which tends to limit or prevents crack advancing. Yan et al. [201] investigated the effect of phase-transformation-induced volume contraction on the fracture properties of superelastic SMAs. It was mentioned that (i) during steady-state crack propagation, the transformation zone extends ahead of the crack tip due to forward transformation while partial reverse transformation occurs in the wake, (ii) as a result of the volume contraction associated with the austenite-to-martensite transformation, the induced stress-intensity factor is positive, (iii) the reverse transformation has been found to have a negligible effect on the induced stress-intensity factor, and (iv) an important implication of the present results is that the phase transformation with volume contraction in SMAs tends to reduce their fracture resistance and increase the brittleness. Kashelf et al. [202] characterized the fracture behavior of titanium open foam and examined and the R-curves of crack propagation from precracks are measured. It was reported that (i) the K_{IC} was found to be dependent on the expanding crack bridging zone at the back of the crack tip, (ii) the compact tension specimens also have some plastic collapse along the ligaments and it has shown that the titanium foam with a higher relative density is tougher, and (iii) the nonuniform stressing within the plastic zone at the crack tip and the plastic collapse of cell topology behind the tip was found to be the primary cause of the R-curve behavior in low relative density titanium foams. Ahadi et al. [203] investigated the grain size dependence of K_{IC} and static K_R of superelastic NiTi with average grain size (GS) from 10 to 1500 nm. The measurements of strain and temperature fields at the crack-tip region are synchronized with the force–displacement curves under mode-I crack opening tests. It was found that (i) with GS reduction down to nanoscale, the K_{IC} and the size of crack-tip phase transformation zone monotonically decrease and the K_R changes from a rising to a flat R-curve, and (ii) the roles of intrinsic and extrinsic toughening mechanisms in the GS dependence of the fracture process are discussed. Haghgouyan et al. [204] proposed a new test methodology for measuring the K_{IC} of SMAs using the critical value of J-integral as the fracture criterion is proposed, which relies on the ASTM standard method for measuring the K_{IC} of conventional ductile materials extended to account for the martensitic transformation/martensite orientation-induced changes in the apparent elastic properties. A comprehensive set of nominally isothermal fracture experiments was carried out on near-equiatomic NiTi compact tension specimens at three distinct temperatures: (1) below the martensite finish temperature, MF; (2) between the martensite-start temperature, MS, and the martensite desist temperature, MD, above which the stress-induced martensitic transformation is suppressed; and (3) above MD. It was found that (i) at these temperatures, the material either remains in the martensite state throughout the loading (martensitic material, case (1)) or transforms from austenite to martensite close to the crack tip (transforming material, case (2)) or remains always in the austenite state (austenitic material, case (3)), respectively, (ii) the critical J-values for crack growth, namely, the K_{IC}, reported in all three cases, result in extrapolated stress intensity factors that are much higher than the corresponding values reported in literature on the basis of linear elastic fracture mechanics and (iii) the K_{IC} of martensitic and transforming materials is found to be approximately the same while the K_{IC} of stable austenite is considerably higher.

9.6.3 Fractography

Fractography is the study of the fracture surfaces of materials in order to determine the relation between the microstructure and the mechanism(s) of crack initiation and propagation and, eventually, the root cause of the fracture. Different types of crack growth (e.g., fatigue, stress corrosion cracking, and hydrogen embrittlement) produce characteristic features on the surface, which can be used to help identify the failure mode. Close examination of the topography and fracture features can help to determine the fracture mode as well as determine the fracture origin and crack direction. The overall pattern of cracking can be more important than a single crack, however, especially in the case of brittle materials like ceramics and glasses. Fractographic methods are routinely used to determine the cause of failure in engineering structures, especially in product failure and the practice of failure analysis. In material science research, fractography is used to develop and evaluate theoretical models of crack growth behavior.

Vaidyanathan et al. [205] examined the fatigue-crack growth characteristics of shape memory NiTi matrix composites reinforced with 10 and 20 vol.% of TiC particles. It was observed that (i) microstructural characterization of these hot-isostatically pressed materials shows that the TiC particles do not react with the NiTi matrix and that they lack any texture, (ii) overall fatigue crack-growth characteristics were found to be similar for the unreinforced and reinforced materials; however, a slight increase in the threshold for fatigue-crack initiation was noted for the composites, (iii) the KIC, as indicated by the failure stress intensity factor range, was found to be similar for all materials. Neutron diffraction studies near the crack tip of the loaded fracture NiTi specimen detected no significant development of texture at the crack tip, and (iv) a comparison is made between the micromechanisms of fracture of metal matrix composites, which deform by dislocation plasticity, and those of the present NiTi–TiC composites, which deform additionally by twinning. Alapati et al. [206] examined numerous discarded ProFile GT, ProFile, and PT NiTi rotary instruments obtained from two graduate endodontic clinics by SEM. These instruments had an unknown history of clinical use and had fractured or experienced considerable permanent torsional deformation without complete separation. It was observed that (i) the failure processes generally exhibited substantial ductile character, evidenced by a dimpled rupture fracture surface, and (ii) crack propagation at grain boundaries and cleavage surfaces indicative of transgranular fracture were observed for some specimens. Spanaki-Voreadi et al. [207] evaluated the failure mechanism of PT NiTi rotary instruments fractured under clinical conditions. A total of 46 PT instruments that failed (fractured and/or plastically deformed) during the clinical use were collected from various dental clinics, whereas a new set of PT instruments served as control. It was reported that after inspection under stereomicroscopy the instruments were classified into three categories: (1) plastically deformed but not fractured, (2) fractured with plastic deformation, and (3) fractured without plastic deformation. It was observed that (ii) stereomicroscopic inspection showed that 17.4% of the discarded instruments were only plastically deformed, 8.7% were fractured with plastic deformation, and 73.9% were fractured without plastic deformation, (iii) micro-XCT revealed instruments without any surface or bulk defects along with a few files with crack development below the fracture surface, and (iv) no defects were identified in the unused instruments. SEM examination of fractured surfaces demonstrated the presence of dimples and cones, a typical pattern of dimple rupture developed because of ductile failure, suggesting that a single overloading event causing ductile fracture of PT instruments is the most common fracture mechanism encountered under the clinical conditions. Shen et al. [208] investigated the mode of failure of three brands of NiTi instruments, including PT, PT for hand use, and K3 that separated during clinical use. A total of 79 fractured instruments were collected from three endodontic clinics over 16 months. The fracture surface of each fragment was examined by SEM. It was reported that (i) most of the rotary instruments (78% of K3 and 66% of PT) failed because of fatigue fracture, whereas 91% of NiTi hand instruments failed as a result of shear, (ii) the fracture mode of shaping files in rotary PT was different between two different clinics, (iii) all surfaces with fatigue fracture ($n = 47$) revealed the presence of either one or two crack origins, and (iv) the vast majority (86%) of K3 fatigue failure had two crack origins that could be found not only at the cutting edge but also at various places along the flute compared with only 28% of PT showing multiple crack origins; the latter showed one crack origin in 81% of the fatigued shaping files but only 37% for finishing files indicating that the failure mode of NiTi instruments is related to preparation technique.

9.6.4 Fracture of endodontic instruments

The dynamic fracture of endodontic hand instruments, including the K file and H file, as reference traditional files, the K-Flex, Flexofile, Unifile, and Helifile were tested [209]. It was reported that (i) the life span of instruments, the distance between the instrument tip and the rupture point, and the resistance to fracture were dependent on the size and design of the instruments, and (ii) there were two types of breakage patterns: the K file, H file, Unifile, and Helifile showed a distinct fracture starting point with crack striations and ductile fractures, while the K-Flex and Flexofile showed only plastic deformations and axial fissures. Tepel et al. [210] investigated bending and torsional properties of 24 different types of NiTi K-files, TiAl K-files and reamers, conventional stainless steel K-files and reamers, and flexible stainless steel instruments. It was found that (i) in ascending order of bending moment the instruments ranked: NiTi K-files, TiAl K-files and reamers, flexible stainless steel instruments, conventional stainless steel K-files, and reamers, and (ii) NiTi, TiAl, and flexible stainless steel instruments displayed lower torque values than conventional stainless steel K-files and reamers. Endodontic file fracture has traditionally been considered an uncommon event; however, a recent perception of increased fracture incidence with rotary NiTi instruments has emerged [211, 212]. Over the last ten years, a range of NiTi alloy modifications have been made by instrument manufacturers, with varying reports of success, in an attempt to reduce the likelihood of file separation. McGuigan et al. [211] investigated the incidence and etiology of file fracture as well as analyzing recommended prevention protocols. It was mentioned that (i) the bulk of the literature relating to instrument fracture is in vitro evidence, which limits its clinical relevance, (ii) the reported incidence of NiTi instrument fracture is similar to stainless steel files; however, inconsistent methodologies hamper accurate comparison, (iii) NiTi instruments are reported to fail by torsional overload and/or flexural fatigue, with file fracture occurring principally in the apical third of the canal or with inappropriate use, and (iv) operator skill, manufacturer modifications, and limiting file reuse have been demonstrated to be significant in reducing fracture incidence indicating the importance of a prevention strategy. Gil et al. [213] investigated the premature catastrophic fracture produced for different periods during clinical endodontic treatment of two brands of NiTi endodontic rotary instruments. It was found that (i) calorimetric studies have shown an increase of the M_S and A_S transformation temperatures with time of use as well as a decrease of their stress transformations, (ii) reverse transformation enthalpies decreased along the time, (iii) the enthalpies of transformation decreased because martensitic plates were anchored, which prevented their transformation to austenite thus losing its superelastic effect, and (iv) the stabilization of the martensitic plates induced the collapse of the structure, causing the fracture. Adoption of rotary NiTi instruments has renewed concerns regarding instrument fracture and its consequences. Spili et al. [214] determined the frequency of instrument fracture and its impact on treatment outcome from an analysis of specialist endodontic practice records involving 8460 cases. It was reported that (i) overall prevalence of retained fractured instruments was 3.3% of treated teeth, (ii) in the case-control study, overall healing rates were 91.8% for cases with a fractured instrument and 94.5% for matched controls, and (iii) healing in both groups was lower in teeth with a preoperative periapical radiolucency (86.7% vs 92.9%), and (iv) in the hands of skilled endodontists prognosis was not significantly affected by the presence of a retained fractured instrument.

Kottoor et al. [215] evaluated the effect of multiple root canal usage on the surface topography and fracture of TF and PT (PT) rotary NiTi file systems. Ten sets of PT and TF instruments were used to prepare the mesial canals of mandibular first molars. This sequence was repeated for both the TF and PT groups until 12 uses. It was found that (i) fresh TF instruments showed no surface wear when compared to PT instruments, (ii) spiral distortion scores remained the same for both the groups till the 6th usage, while at the 9th usage TF showed a steep increase in the spiral distortion score when compared to PT, (iii) PT instruments fractured at a mean root canal usage of 17.4, while TF instruments showed a mean root canal usage of 11.8, and (iv) fractographically, all the TF instruments failed due to torsion, while all the PT instruments failed because of CF concluding that PT instruments showed more resistance to fracture than TF instruments. Practically, cleaning should be done between each usage of endodontic instrument. O'Hoy et al. [216] evaluated the effect of repeated cleaning procedures on fracture properties and corrosion of NiTi files. New NiTi instruments were subjected to 2, 5, and 10 cleaning cycles with the use of either diluted bleach (1% NaOCl) or Milton's solution (1% NaOCl plus 19% NaCl) as disinfectant. Each cleaning cycle consisted of scrubbing, rinsing, and immersing in NaOCl for 10 min followed by 5 min of ultrasonication. Files were then tested for torsional failure and flexural fatigue. It was found that (i) up to 10 cleaning cycles did not significantly reduce the torque at fracture or number of revolutions to flexural fatigue, although decreasing values were noted with increasing number of cleaning cycles using Milton's solution, (ii) no corrosion was detected on the surface of these files, (iii) files immersed in 1% NaOCl overnight displayed a variety of corrosion patterns, (iv) the extent of corrosion was variable amongst different brands of files and amongst files in each brand, and (v) overall, Milton's solution was much more corrosive than diluted bleach. It was, therefore, concluded that (i) files can be cleaned up to 10 times without affecting fracture susceptibility or corrosion, but should not be immersed in NaOCl overnight, and (ii) Milton's solution is much more corrosive than bleach with the same NaOCl concentration.

9.6.4.1 Rotary file

The endodontic engine-driven instrumentation of root canal system aims to decrease the preparation time and simplification of root canal instrumentation. Nonetheless, in the early era of engine-driven instruments there was a high risk of instrument separation when compared to hand instruments. Instrument separation of NiTi Rotary instruments might occur due to torsion, CF or a combination of both forces [217]. The clinical approach to discard rotary instruments is aiming to avoid the separation is the number of uses. However, the number of times an instrument can be used with minimum risk of separation is still unclear: different studies show controversial data, as will be seen.

Sattapan et al. [218] analyzed the type and frequency of defects in NiTi rotary endodontic files after routine clinical use. All of the files (total: 378, Quantec Series 2000) discarded after normal use from a specialist endodontic practice over 6 months were analyzed. It was observed that (i) almost 50% of the files showed some visible defect; 21% were fractured and 28% showed other defects without fracture, (ii) fractured files could be divided into two groups according to the characteristics of the defects observed, (iii) torsional fracture occurred in 55.7% of all fractured files, whereas flexural fatigue occurred in 44.3% indicating that (iv) torsional failure, which may be caused by using too much apical force during instrumentation, occurred more frequently than flexural fatigue, which may result from use in curved canals. Arens et al. [219] analyzed the number and types of defects observed in single-use, rotary NiTi instruments. Every ProFile Series 29.04 taper NiTi instrument used during a 4-week period in an endodontic specialty practice was collected. All instruments were new and were used by experienced clinicians during a single patient visit. It was reported that (i) a total of 786 ProFile Series 29 NiTi rotary instruments were evaluated; 115 (14.63%) showed some type of defect after one clinical use. Size 3 instruments had the highest defect rate (22.66%) followed by size 5 (17.30%), size 2 (17.24%), and size 4 instruments (16.10%), (ii) there was no statistically significant difference, (iii) the size 6 and size 7 instruments showed minimal defects (2.38% and 4.76%, respectively), (iii) seven of 786 files had fractured (0.891%), and (iv) there was no statistically significant difference in the type of failure seen within each file size indicating that defects can occur even with new files in the hands of experienced endodontists, and for absolute safety a single-use approach should be followed. Similar results on the technical sensitivity were reported by Parashos et al. [220, 221].

Berutti et al. [222] studied the influence of immersion in NaOCl on resistance to CF fracture and corrosion of PT rotary instruments. A total of 120 new PT Rotary files (F2) were randomized and assigned to three different groups of 40 each. Group 1 was the control group; 20 mm (excluding the shaft) of group 2 instruments were immersed in 5% NaOCl at 50 degrees C for 5 min; instruments in group 3 were completely immersed in 5% NaOCl at 50 °C for 5 min. All instruments were then tested for CF, recording the time in seconds to fracture. It was obtained that (i) instruments in group 3 had a significantly lower resistance to fracture because of CF than those in groups 1 and 2, (ii) in some instruments in group 3, early fracture occurred after only a few seconds of fatigue testing, and (iii) SEM observations revealed evident signs of corrosion of the fractured instruments. It was then concluded that (iv) the group 3 had significantly reduced resistance to CF compared with instruments in groups 1 and 2, and (v) the phenomenon of early fracture may be attributed to galvanic corrosion induced by the presence of dissimilar metals, where one acts as the cathode of a galvanic couple, established when the instrument is immersed in NaOCl solution. The NiTi alloy may acts as the anode and thus undergoes corrosion. Wolcott et al. [223] determined if the number of uses affects the separation incidence of PT rotary instruments. The 4,652 consecutively treated root canals were performed in an endodontic group practice over a 17-month period. Both the separation incidence and the number of uses were tracked for each file. It was found that (i) the overall rate of instrument fracture in this study was 2.4% with no significant differences over the first four uses, (ii) PT rotary files may be safely reused at least four times, and (iii) the size of the rotary file, among other factors, will determine how many times a particular file should be used. Shen et al. [224] compared the incidence and mode of instrument separation of the two NiTi rotary systems that were used according to a predefined schedule of clinical use by the same group of operators. A total of 166 ProFile and 325 PT instruments, discarded from this endodontic clinic over 17 months, was analyzed. It was reported that (i) the incidences of instrument separation were 7% for ProFile and 14% for PT, (ii) the proportion of unwinding defects was 5% in ProFile and 0.3% in PT instruments, and (iii) flexural fatigue was implicated in the majority of separations in both groups indicating that while PT was more likely to separate without warning, ProFile tended to exhibit unwinding of flutes more frequently.

Spanaki-Voreadi et al. [225] evaluated the failure mechanism of PT NiTi rotary instruments fractured under clinical conditions. A total of 46 PT instruments that failed (fractured and/or plastically deformed) during the clinical use were collected from various dental clinics, whereas a new set of PT instruments served as control. After inspection under stereomicroscopy the instruments were classified into three categories: (1) plastically deformed but not fractured, (2) fractured with plastic deformation, and (3) fractured without plastic deformation. It was found that (i) stereomicroscopic inspection showed that 17.4% of the discarded instruments were only plastically deformed, 8.7% were fractured with plastic deformation, and 73.9% were fractured without plastic deformation, (ii) micro-XCT revealed instruments without any surface or bulk defects along with a few files with crack development below the fracture surface, and (iii) no defects were identified in the unused instruments.

SEM examination of fractured surfaces demonstrated the presence of dimples and cones, a typical pattern of dimple rupture developed because of ductile failure suggesting that a single overloading event causing ductile fracture of PT instruments is the most common fracture mechanism encountered under the clinical conditions. Di Fiore [226] listed several recommended measures that practitioners can take to prevent NiTi rotary instrument fracture during root canal preparation: (1) avoid subjecting NiTi rotary instruments to excessive stress, (2) use instruments that are less prone to fracture, (3) follow protocol for instrument use, (4) assess root canal curvatures radiographically and instrument them carefully, (5) ensure that the endodontic access preparation is adequate, (6) open orifices before negotiating canals, (7) enlarge root canals with fine hand instruments, (8) set rotational speed and torque at low levels, (9) use the crown-down technique, (10) irrigate and lubricate root canals during preparation, (11) manipulate rotary instruments with a pecking or pumping motion, and (12) if inexperienced, engage in preclinical training in the use of rotary instruments. In recent years, a number of rotary NiTi systems have been developed to provide better, faster, and easier cleaning and shaping of the root canal system. Although the NiTi instruments are more flexible than the stainless steel files, the main problem with the rotary NiTi instruments is the failure of the instruments. Inan et al. [227] evaluated the deformation and fracture rate of Mtwo rotary NiTi instruments discarded after routine clinical use. A total of 593 Mtwo rotary NiTi instruments were collected after clinical use from the clinic of endodontics over 12 months. It was reported that (i) a percentage of all files (25.80%) showed defects, and the major defect was fracture (16.02%), (ii) The most frequently fractured file was #10.04 (30.39%), and (iii) deformations without fracture were mostly observed on #15.05 files (25.47%). Based on these results, it was concluded that (i) a higher rate of deformation was observed for #10.04 and #15.05 files; therefore, these files should be considered as single-use instruments, and (ii) because CF was the cause of 71.58% of the instrument fractures, it is also important not to exceed the maximum number of usage recommended by the manufacturer and discard the instruments on a regular basis. Varela-Patiño et al. [228] determined the influence of the type of instrument rotation on the frequency of fractures or deformation. Instrumentation was performed on 120 molar root canals with an angle of curvature greater than 30 degrees using alternating rotation (group A: 60 degrees clockwise, 45 degrees counterclockwise) and continuous rotation (group B). It was indicated that instruments used with alternating rotation have a higher mean number of uses (13.0) compared with the continuous rotation group (10.05); this difference was statistically significant. It was then concluded that the PT shaping instruments (S1 and S2) are those that achieved the greatest difference in use with alternating rotation, with S2 being the most resistant to fracture or deformation with the two types of movement used. Stojanac et al. [229] examined the lifespan or N_{CF} of tapered rotary NiTi endodontic instruments. Simulated root canals with different curvatures were used to determine a relation between canal curvature and instrument lifespan. It was reported that (i) using a novel mathematical model for the deformation of pseudoelastic Ni–Ti alloy shows that maximum stress need not necessarily occur at the outer layer, (ii) on the basis of this observation, the Coffin–Manson relation was modified with parameters determined from this experiment, (iii) the N_{CF} was influenced by the angle and radius of canal curvature and the size of instrument at the beginning of canal curvature, and (iv) the resulting quantitative mathematical relation could be used to predict the lifespan of rotary NiTi endodontic instruments under clinical conditions and thereby reducing the incidence of instrument failure in vivo.

9.6.4.2 Reciprocating file

When conventional NiTi instruments are rotated in root canals, they are subjected to structural fatigue that, if continued, will eventually lead to fracture [230–232]. As Serene et al. [233] also mentioned that torsion and fatigue through flexure are the two main reasons to the fracture. Torsional fracture occurs when the tip or any other part of the rotating instrument binds to the root canal walls, while the rest of the file keeps turning. Fracture due to flexural fatigue (bending stress) occurs when an instrument that has already been weakened by metal fatigue is placed under further stress. Yared [234] studied an alternating movement, which used the PT F2 instrument (Dentsply/Maillefer) in a reciprocating movement. It was claimed that (i) diminishing the necessary steps for root canal preparation, and (ii) the single use of these instruments would decrease the risk of cross-contamination and instrument separation. Therefore, minimizing the reciprocating rotating motion on the instrument compared to the instruments in a consistent rotating motion indicates a promise for the reduction in the number of instruments required for the cleaning and shaping sequence in minimizing possible contamination and alleviating operator anxiety of the possibility of instrument failure [228,235]. It is recognized that reciprocation of NiTi instruments possess advantages over continuous rotation [232]: (1) binding of the instruments into the root canal dentine walls is less frequent, reducing torsional stress [236] and (2) the reduction of the number of cycles within the root canal during preparation results in less flexural stress on the instrument [237]. As De-Deus et al. [238] pointed out that (i) a cutting angle is larger than the relief angle, and (ii) the shift of the kinematics increases the resistance to the CF when the reciprocating asymmetrical movement is applied; the critical areas of stress move progressively to new locations during the periodical change of the angle, thus distributing effectively the areas of stress to different points of the instrument decreasing the damage and increasing the life span of the instrument.

Troian et al. [239] evaluated the deformation and fracture of NiTi RaCe and K3 size 25, 0.04 taper instruments. Ten sets of instruments from RaCe and K3 NiTi rotary systems were used to prepare 100 simulated canals in epoxy resin blocks with 20 or 40-degree curvatures beginning 8 or 12 mm from the orifice. Each instrument set was used to prepare five simulated canals using a crown-down technique. It was found that (i) no fractures occurred with K3 instruments, whereas six RaCe instruments fractured, (ii) a statistically significant difference occurred between RaCe and K3 instruments in terms of distortion of spirals and surface wear, (iii) distortion of spirals and wear increased with progressive use of RaCe instruments, whereas K3 instruments remained relatively undamaged after their fifth use, and (iv) the simulated canals with smaller radii of curvature were positively associated with fracture of RaCe instruments concluding that a significant difference was found between RaCe and K3 in terms of deformation and fracture of size 25, 0.04 taper instruments; K3 instruments had more favorable results. Beruttie et al. evaluated the influence of glide path on canal curvature and axis modification after instrumentation with WaveOne Primary reciprocating files [240] and compared the canal curvature and axis modification after instrumentation with WaveOne Primary reciprocating files and NiTi rotary PT [241]. It was reported that (i) glide path was found to be extremely significant for both curvature radius ratio parameter and the relative axis error parameter, (ii) canal modifications seem to be significantly reduced when previous glide path is performed by using the new WaveOne NiTi single-file system, (iii) an instrument factor was extremely significant for both the curvature ratio parameter and the relative axis error parameter, and (iv) canal modifications are reduced when the new WaveOne NiTi single-file system is used. Plotino et al. [70] evaluated the CF resistance of Reciproc and WaveOne instruments in simulated root canals. Two groups of 15 NiTi endodontic instruments of identical tip size of 0.25mm were tested, group A: Reciproc R25 and group B WaveOne primary. CF testing was performed in a stainless steel artificial canal manufactured by reproducing the instrument's size and taper. A simulated root canal with a 60° angle of curvature and 5-mm radius of curvature was constructed for both the instruments tested. It was found that (i) a statistically significant difference was noted between Reciproc and WaveOne instruments., (ii) Reciproc R25 instruments were associated with a significant increase in the meantime to fracture when compared with primary WaveOne instruments, and (iii) there was no significant difference in the mean length of the fractured fragments between the instruments concluding that Reciproc instruments were associated with a significantly higher CF resistance than WaveOne instruments. Reciprocating instruments were developed to improve and simplify the preparation of the root canal system by allowing greater centralization of the canal and requiring a shorter learning curve. Despite the risk of instrument separation, using a reciprocating instrument in more than one case is a relatively common clinical practice. Bueno et al. [242] evaluated the fracture resistance of Reciproc R25 and WaveOne Primary instruments according to the number of uses during the preparation of root canals in up to three posterior teeth. It was obtained that (i) none of the instruments showed any signs of deformation, but three instruments fractured (0.26% of the number of canals and 0.84% of the number of teeth). All fractures occurred in mandibular molars (1 WaveOne Primary file during the third use and 2 Reciproc R25 files, 1 during the first use and the other during the third use), and (ii) there was a low incidence of fracture when reciprocating files were used up to three cases of endodontic treatment in posterior teeth.

Klymus et al. [243] evaluated the impact of body temperature on the CF resistance of different NiTi alloys used for the manufacturing of RB R25 (RB 25.08), X1 Blue File 25 (X1 25.06), and WOG (WOG 25.07). Sixty instruments of the RB 25.08, X1 25.06, and WOG 25.07 systems were used ($n = 20$). CF tests were performed at RT (20 °C ± 1 °C) and at body temperature (37 °C ± 1 °C). The instruments were reciprocated until fracture occurred in an artificial stainless steel canal with a 60° angle and a 5-mm radius of curvature. It was found that (i) the CF test at 20 °C showed that RB 25.08 and X1 25.06 presented significantly higher TTF and N_{CF} than WOG 25.07, (ii) at 37 °C, all groups presented significant reduction of TTF and N_{CF}, (iii) RB 25.08 presented significant higher TTF than WOG 25.07, (iv) regarding the N_{CF}, there was no significant difference among the groups, and (v) the WOG 25.07 presented the lowest percentage reduction of CF. Based on these findings, it was concluded that (i) the body temperature treatment caused a marked reduction of the CF resistance for all reciprocating instruments tested, and (ii) the RB 25.08 and X1 25.06 systems presented similar results at both temperatures tested; however, WOG 25.07 presented the lowest percentage reduction in fatigue resistance at body temperature.

9.6.5 Removal of fractured or separated endodontic files

Complications can occur during many dental procedures. The prepared clinician responds by either correcting the problem during treatment or, ideally, preventing the problem from occurring in the first place [244]. In endodontic treatment, separation of rotary NiTi files is problematic. The separations might be taken placed by one or complicated reasons of CF, torsional stress, elongated flute portion and others by creating a straight-line (glide path) access into a canal, and preflaring the coronal portion before using rotary files in the apical third of the canal. If a file breaks, successful removal primarily depends on the location of the file in the canal rather than the specific technique employed for removal [244]. It is worthwhile to notice that instrument separation is not directly associated to failure. A follow-up of eight patients with irretrievable instruments has shown that after 5 years, 100% of these patients presented functional teeth [245]. Only 12.5% of these patients were presented with radiographic characteristics of no healing. That is likely to happen if the fragment prevents a proper cleaning of the apical third in a necrotic case [246]. It is generally accepted understanding that an attempt of instrument removal after the curvature is too risky to overtake the benefits. Although the separated file cannot be removed, the prognosis when file separation occurs can still be favorable, especially if care was taken to reduce the critical concentration of canal debris with hand instrumentation and chemical irrigation prior to rotary file insertion, and it should be noticed that if NiTi file was utilized, it contains about 50 wt%Ni, which is believed to cause Ni element hyper-sensitivity or even exhibit an oncogenicity. Shen et al. [247] evaluated the influence of various factors on the success or failure of attempts to remove fragments of separated NiTi instruments from root canals. The possible influencing factors should include type of tooth, degree of root canal curvature, location of fragment in relation to the root canal curvature, and radiographic length of fragment. A success of treatment was defined as removal or complete bypassing of the fragments. It was reported that (i) the overall success rate was 53%, in which the success rate for ProFile fragments was 41% and for the NiTi K-file 60%, (ii) the success rate in maxillary teeth was higher than that of mandibular teeth, (iii) of 52 instruments in molars, 28 were successfully removed or bypassed, while of the 12 fragments in premolars, only 2 were removed, (iv) all 8 cases in anterior teeth were retrieved completely, (v) when the fragment was localized before the curvature, complete removal was achieved, while when the fragments were located at and beyond the curvature, the success rates were 60% and 31%, respectively, (vi) in canals with a slight, moderate, and severe curvature, the success rates were 100%, 83%, and 43%, respectively, and (vii) in general, the longer the fragment, the greater the chance for successful removal or bypass concluding that favorable factors for removal of separated NiTi fragments are straight root canals, anterior teeth, localization before the curvature, fragments longer than 5 mm, and hand NiTi K-file. Hülsmann et al. [248] evaluated postoperatively the influence of several factors on the success rate of removal procedures of fractured endodontic instruments. In 105 teeth with 113 fragments removal attempts were undertaken using a wide range of techniques and instruments. It was mentioned that (i) anatomical factors favorable for removal were straight canals, incisors and canines, localization before the curvature, length of fragment more than 5 mm, localization in the coronal or mesial third of the root canal, or reamer, and (ii) in molars removal procedures were most successful in the palatal canals of maxillary molars.

About the probability of removing fractured instruments from root canals, Suter et al. [249] evaluated in a clinical case series the location of fractured instruments, how many of them could be removed, and to compare these findings with the results of a similar study. Within an 18-month period all referred endodontic cases involving fractured instruments within root canals were analyzed. The protocol for removal of fractured instruments is as follows: create straight-line access to the coronal portion of the fractured instrument, attempt to create a ditched groove around the coronal aspect of the instrument using ultrasonic files and/or to bypass it with K-Files. Subsequently, the fractured instrument was vibrated ultrasonically and flushed out of the root canal or an attempt was made to remove the instrument with the tube and Hedström file method or similar techniques. The location of the fractured instrument and the time required for removal were recorded. Successful removal was defined as complete removal from the root canal without creating a clinically detectable perforation [249]. It was found that (i) in total, 97 consecutive cases of instrument fracture were included in the time period. In all, 84 instruments (87%) were removed successfully, (ii) there was a significant correlation between the time needed to remove fractured instruments and a decrease in success rate, (iii) curved canals had significantly more fractured instruments than straight canals, (iv) rotary instruments fractured significantly more often in curved canals compared with other instruments, (v) half of all instrument fractures occurred in mesial roots of lower molars and most often when using rotating instruments, and (vi) there was no statistically significant difference in the success rate with respect to the location of the fractured instrument (tooth/root type), the type of fractured instrument, or the different methods of instrument removal. Based on these findings, it was concluded that (vii) curved canals are a higher risk for instrument fracture than straight canals, and in curved canals, rotary instruments fractured more often than other instruments, (viii) in all, 87% of the fractured instruments were removed successfully, (ix) a decrease in success rate was evident with increasing treatment time, and (x) the use of an operating microscope was a prerequisite for the techniques used to remove the fractured instruments. Rosen et al. [250] compared the diagnostic ability to radiographically detect separated stainless steel versus NiTi instruments located at the apical third of filled root canals with either AH 26 or Roth sealer. Sixty single-rooted extracted human teeth with one straight root canal were instrumented to a size 25 apical diameter. In 40 teeth, apical 2-mm segments of stainless steel (n = 20) or NiTi (n = 20) files were intentionally fractured in the apical part of the root canal. The remaining 20 teeth without fractured files served as a control group. Subsequently, the root canals were filled using laterally condensed gutta-percha and either AH 26 sealer or Roth sealer. It was obtained that (i) the kappa values were 0.76 and 0.615 for the first and second observers, respectively, and 0.584 between the observers, (ii) there were no significant differences in the diagnostic ability between digital and conventional radiography or the different root canal sealers (AH vs, Roth), and (iii) the sensitivity to detect fractured stainless steel was significantly higher than NiTi. Based on these findings, it was concluded that (i) it may be difficult to radiographically detect a retained separated instrument, and (ii) it is easier to radiographically detect fractured stainless steel than NiTi instruments retained at the apical third of the root canal.

Among various techniques to remove the fractured file, the ultrasonic technique (US) has been widely employed. Ward et al. [251] evaluated an US to remove fractured rotary NiTi endodontic instruments from root canals. It was mentioned that an US using tips combined with the creation of a "staging platform" using Gates Glidden instruments and the use of the dental operating microscope was consistently successful at removing fractured rotary NiTi instruments from narrow, curved root canals when some part of the fractured instrument segment was located in the straight portion of the canal. When the fractured instrument segment was located entirely around the curve, care must be taken because the success rate significantly decreased and major canal damage may ensue. Souter et al. [252] demonstrated a technique utilizing modified Gates Glidden burs and ultrasonics to remove fractured instruments from root canals. Varying extents of tooth structure are removed during this procedure, potentially leading to complications. It was reported that (i) fractured instrument fragments were removed from three different levels (coronal, middle, or apical third) of mesiolingual canals of extracted human mandibular molars, (ii) the success rate, frequency of perforations, and root strength were recorded for each group, (iii) perforations and unsuccessful file removal occurred only with fragments lodged in the apical third, (iv) fracture resistance declined significantly with more apically located file fragments, (v) a review of 60 clinical cases showed similar rates of successful file removal and rate of perforations, and (vi) removal of a fractured file fragment from the apical third of curved canals should not be routinely attempted. Alomairy et al. [253] evaluated the US and instrument removal system (iRS) in removing fractured rotary NiTi file from root canals. A total of 30 extracted human molars with closed apices were collected. Angles and radii of curvature were measured for all the canals with the fractured fragments, and then canals were classified into slight, moderate, and severe subgroups according to their curvature angles. It was reported that (i) within each subgroup, teeth were randomly distributed into US or iRS according to measured angles and radii; a total of 15 canals were assigned to both US and iRS techniques, (ii) the overall success rate was 70% ($n = 21$), (iii) less curved canals and longer radii of curvature showed more success, (iv) the median time used for retrieval was 40 min using the US technique, whereas it was 55 min for iRS, and (v) the removal of fractured rotary NiTi endodontic instruments was more successful with less curved canals and longer radius of curvature above 4.4 mm. Wohlgemuth et al. [254] evaluated the effectiveness of the minimally invasive GentleWave System in removing separated stainless steel endodontic files from the apical and midroot regions of molar root canals. Thirty-six extracted human molars were accessed, and the glide path was confirmed to the apex. ISO #10, #15, and #20 K-file fragments of 2.5-mm length were separated at the apical ($n = 18$) or midroot ($n = 18$) region of the molars by engaging a weakened file with downward pressure. During analysis, the teeth were divided into two curved groups based on the curvature of the root (<30° and >30°). It was reported that (i) the overall success rate of instrument removal when the separated files were engaged in the apical region was 61%, and for the midroot region, it was 83%, (ii) less curved canals (<30°) showed a 91% success rate ($n = 24$), whereas canals with an angle of curvature greater than 30° showed a 42% success rate ($n = 12$), and (iii) the median treatment time for instrument retrieval was 10 min 44 s concluding that the GentleWave System is effective in retrieving separated instruments while conserving the dentinal structure. Hirayama et al. [255] evaluated the effect of removing broken H-file pieces and removal times for various conditions of the ultrasonic device or broken file pieces. Cut H-files as broken files were pushed into transparent resin blocks at apical parts, and models of intracanal broken instruments were reproduced. The relationships among the output strength of the ultrasonic device, lengths and sizes of broken file pieces, presence or absence of water coolant, and the parts contacted by the tip of the ultrasonic device were examined. It was found that (i) the times taken to remove broken files were compared and discussed, (ii) a stronger output power of the ultrasonic device was more effective for removing broken files, (iii) the relationship with the length of broken files was not regular, but the time was short for the shortest length of 3 mm; the larger pieces of broken files were more difficult to remove, and (iv) the use of water coolant was effective. Regarding the contact parts, the time for removal was short when contacting the cut surface suggesting that the time taken to remove broken files closely corresponded with the conditions of the ultrasonic device or status of broken file pieces using transparent resin blocks. Madarati et al. [256] investigated the temperature rise induced by ultrasonic tips when activated against two types of fragments at two power-settings. Twenty-four F2-PTU rotary files and 36 stainless steel K-files, size 50, were sectioned at a point 5.5 mm from their tips. At specific power-settings, ET40D ultrasonic tips were activated against the peripheral surface of the fragment coronal part (1 mm) for 30 s with/without air-active function according to each group: (i) without air-active function at power-setting 1.5 against NiTi fragments, (ii) without air-active function at power-setting 1.5 against SS fragments, (iii) without air-active function at power-setting 3 against SS fragments, (iv) with air-active function at power-setting 3 against stainless steel fragments, and (v) with Air-Active at power-setting 3 against NiTi fragments. Temperature rises were inspected at 15 and 30 s using the thermocouples. It was found that (i) with no air-active function and power-setting 1.5, the temperature rises induced on NiTi fragments at 15 and 30 s (36.33 and 55.44 °C, respectively) were significantly greater than those induced on SS fragments at 15 and 30 s (26.08 and 35.27 °C, respectively), (ii) when ultrasonic tips were activated against SS fragments without Air-Active, the temperature rises induced by power-setting 3 at 15 and 30 s (35.25 and 45.32 °C, respectively) were significantly greater than those induced by power-setting 1.5 at the same intervals (26.08 and 35.27 °C, respectively), and (iii) at 30 s, the overall temperature rise induced with Air-Active (25.56 °C) was significantly lower than that induced without Air-Active (45.34 °C). Based on these findings, it was concluded that (iv) lower power-settings and shorter application times are recommended when using ultrasonics for removal of NiTi fragments compared with SS ones, and (vi) air-active function, as a coolant, is recommended when dealing with both types of fragments. Arslan et al. [257] determined the effect of taper (0.08, 0.06, and 0.04) of separated K3XF instruments on duration taken for the secondary fracture formation during ultrasonic activation. Ten 25/0.08 K3XF, ten 25/0.06 K3XF, and ten 25/0.04 K3XF instruments were used for the study. The apical 5 mm of the instruments was cut to simulate the fragments in root canals. Fragments of the instruments were sandwiched between two straight dentin blocks. An ultrasonic tip was used to cause a secondary fracture of the fragment. The time needed for the secondary fracture was recorded for each instrument. It was obtained that (i) secondary fractures occurred in all instruments, (ii) in the 0.08 taper group, secondary fractures took longer than in the case of the 0.06 and the 0.04 taper groups, and (iii) there were no significant differences between the 0.06 and the 0.04 taper groups in terms of the time required for the occurrence of a secondary fracture. Typically, when removing separated instruments, a much lower power setting is chosen. The purpose of this in vitro study was to determine which tapered files were more resilient to secondary fracture, thus allowing a higher power setting to be chosen. Thus, the results of the present study cannot be used in clinical practice. If the clinician knows the taper of the broken file, the clinician should be very careful with regard to secondary fractures when using ultrasonics to remove the separated smaller tapered instruments [257].

There are still several different techniques proposed for removal of fractured endodontic instruments. Yang [258] evaluated the effects on root dentin of two trephining techniques using an ultrasonic tip or a trepan bur in the mesial canals of mandibular molars during attempts to remove fractured file fragments using microcomputed tomographic imaging. Twenty-one teeth with a similar anatomic configuration in mesial (buccal and lingual) canals were selected. A 4-mm apical segment of K3 file size 25/0.06 was fractured in each mesiobuccal and mesiolingual canal 5 mm apically from the canal orifice. A staging platform was prepared at the coronal aspect of the broken instrument followed by either ultrasonics or a new trepan bur technique to expose a 1- to 1.5-mm length of the fragment. If the broken instrument could not be removed by exposing it either by ultrasound or the trepan bur, a microtube device was used to attach to and withdraw the fragment. Micro-CT scanning was performed before and after removing the broken instrument. Canal volume, diameter, and furcal root dentin thickness were measured by using image analysis software. The time required for the removal of the instrument fragments was recorded. It was obtained that (i) the trepan bur technique had significantly less impact on canal volume, diameter, and furcal root dentin thickness change than the US, and (ii) the time consumed for successful removal of the fragments was significantly less in the trepan bur group (8.9 ± 3.5 min) than in the ultrasonic group concluding that a new small-sized trepan bur technique was superior to the use of ultrasound with regard to the amount of dentin removed and the speed in the removal of fractured instruments from root canals. Ormiga et al. [259] introduced a new concept of retrieval of fractured instruments from root canals based on an electrochemical process. Current register tests were used to evaluate the dissolution process of 25.04 NiTi K3 rotary files. A constant anodic potential was applied to the NiTi files, whereas the potentiostat registered the anodic current. After the tests, all files were observed by using an optical microscope. It was reported that (i) the current attained initial values of approximately 55 mA that declined during the entire test, (ii) a good reproducibility of results was observed, and (iii) the optical microscopy analysis evidenced a progressive consumption of the files with increasing polarization time indicating that (iv) the concept of fractured file retrieval by an electrochemical process is feasible, and (v) the concept resulted in a consistent basis for the development of a method to remove fractured instruments from root canals.

In spite of great efforts for retrieving fractured instruments at the dental chairside, there could be uncoverable accidental ingestion and aspiration could happen [260, 261]. Susini et al. [261] determined the incidence of aspiration and ingestion of endodontic instruments in France during root canal treatment without using rubber dam. Data was provided by two insurance companies representing 24,651 French general dentists over 11 years. The type and number of accidents per year, the number of dental items involved, and the percentage of occurrence of either aspiration or ingestion were reported. The incidence of accidental aspiration or ingestion was calculated. The need for hospitalization to remove the endodontic instruments and other dental items was reported and compared using chi square tests. It was obtained that (i) one endodontic instrument was aspirated and 57 were ingested, (ii) forty-three other dental items were aspirated and 409 were ingested, (iii) for the endodontic instruments, the incidence of aspiration was 0.001 per 100,000 root canal treatments and the incidence of ingestion was 0.12 per 100,000 root canal treatments, (iv) the aspirated endodontic instruments and dental items required statistically more frequent hospitalization than the ingested items, and (v) the endodontic instruments did not require more frequent hospitalization than other dental items when aspirated (ns) and when ingested (ns). No fatal outcome was reported. It was, then, concluded that (i) the incidence of ingestion or aspiration of endodontic instruments was low even though most general practitioners do not routinely use rubber dam, and (ii) use of rubber dam by general practitioners for endodontic procedures should be encouraged by stressing its advantages rather than the fear factor of accidents. Nevares et al. [262] evaluated the success rates of standardized techniques for removing or bypassing fractured instruments from root canals and determine whether visualization of the 112 fractured instruments with the aid of an operating microscope has any impact on the success rates. It was obtained that (i) the overall success rate (removal and bypassing) was 70.5% ($n = 79$), (ii) in the visible fragment group, the success rate was 85.3% ($n = 58$), and in the nonvisible fragment group it was 47.7% ($n = 21$), and (ii) success rates were significantly higher when the fragment was visible concluding the standardized techniques used in this study for removing or bypassing fractured instruments were effective, and approximately two times greater success rate was obtained when the fragment was visible inside the root canal compared with when it was nonvisible.

To minimize the incident of separation of endodontic instrument, there are some recommended consideration [251]: (1) appropriate training before using files of new design, (2) understanding of root canal anatomy and establishing of glide path before cleaning and shaping, (3) examination of new files as some defects can occur while being manufactured, (4) examination of files during treatment regularly even with a single use, (5) use of magnification for file examination as some defects cannot be seen by naked eye, and (6) adherence to manufacturer's instructions. Intracanal separation of endodontic instruments may hinder cleaning and shaping procedures within the root canal system, with a potential impact on the outcome of treatment. Madarati et al. [263] mentioned that guidelines for management of intracanal separated instruments have not been formulated. Decisions on management should consider the following: (1) the constraints of the root canal accommodating the fragment, (2) the stage of root canal preparation at which the instrument separated, (3) the expertise of the clinician, (4) the armamentaria available, (5) the potential complications of the treatment approach adopted, and (6) the strategic importance of the tooth involved and the presence or absence of periapical pathosis. For managing fractured instruments, Lambrianidis [264] mentioned that (i) the therapeutic options for the management of fractured instruments include no intervention, nonsurgical (orthograde and conservative) management, surgical management, and tooth extraction, (ii) the optimal management of instrument fragments during root canal treatment is their retrieval, (iii) informing the patient, localization, and identification of the fragment are essential steps prior to any decision and particularly prior to initiation of efforts to retrieve the fragment, (iv) a plethora of different means and techniques have been proposed for nonsurgical management of instrument fragments, (v) regardless of how sophisticated they may be, the use of an operating microscope and the appropriate armamentarium are required along with knowledge, training, and experience, (vi) any clinical procedure for the management of fractured instruments includes the risk of creating additional error(s) that may eventually jeopardize the prognosis of the tooth, and (vii) therefore, continuous evaluation of the progress is needed in order to consider alternative options if required. It is further described that (viii) as a general rule of thumb, surgical management follows nonsurgical management of the error and is performed either immediately after the completion of the nonsurgical management or at a later time based on future evaluation, (ix) there are some cases, though, where weight of risks versus benefits changes this rule and surgical procedures are performed without any nonsurgical attempt, and (x) the timing and type of surgical procedure are determined by several interdependent factors that require evaluation and careful consideration on an individual case-by-case basis.

9.6.6 Fracture of orthodontic archwires

Zinelis et al. [265] characterized intraorally fractured NiTi archwires and determined the type of fracture to assess changes in the alloy's hardness and structure. Eleven NiTi SE 200 and 19 copper-NiTi intraorally fractured archwires were collected. The location of fracture (anterior or posterior), wire type, cross section, and period of service before fracture were recorded. The retrieved wires and brand-, type-, and size-matched specimens of unused wires were subjected to SEM to assess the fracture type and morphological variation of fracture site of retrieved specimens, and to Vickers hardness (HV200) testing to investigate the hardness of as-received and in vivo fractured specimens. It was reported that (i) the fracture site distribution showed a preferential location at the midspan between the premolar and the molar, suggesting that masticatory forces and complex loading during engagement of the wire to the bracket slot and potential intraoral aging might account for fracture incidence, and (ii) all retrieved wires had the distinct features of brittle fracture without plastic deformation or crack propagation, whereas no increase in hardness was observed for the retrieved specimens. It was then concluded that (iii) most fractures sites were in the posterior region of the arch, probably because of the high-magnitude masticatory forces, (iv) brittle fracture without plastic deformation was observed in most NiTi wires regardless of archwire composition, (v) there was no increase in the hardness of the intraorally exposed specimens regardless of wire type, and (vi) this rules out hydrogen embrittlement as the cause of fracture. Guzman et al. [266] determined the incidence and location of fracture in round NiTi and round stainless steel orthodontic archwires, both commonly used in orthodontics. One thousand orthodontic patients (1,434 archwires) were evaluated during regular treatment visits to assess archwire fracture and location. The patient's gender, age, type of archwire (round NiTi and round stainless steel), diameter of the archwire, arch type, location of fracture (anterior or posterior), and period of service before fracture were recorded. It was reported that (i) twenty-five archwire failures were reported (1.7%) of the total sample size, (ii) all fractured archwires were NiTi, and 76% of the fractures were located in the posterior region, (iii) no statistical significance was found between archwire fracture and gender, arch type (maxillary/mandibular), archwire diameter, or bracket type, (iv) the frequency of archwire fracture during regular orthodontic visits is very low, (v) the most common archwire fracture site is the posterior region, (vi) NiTi wires are the most commonly fractured archwire and (vii) no statistically significant correlation exists between archwire fracture and gender, arch type, bracket type, or diameter of archwire. Superelastic NiTi wire is widely used in orthodontic clinics, but delayed fracture in the oral cavity has been observed. Because hydrogen embrittlement is known to cause damage to Ti alloy systems, orthodontic wires were charged with hydrogen using an electro-chemical system in saline. Yokoyama et al. [267] investigated the degradation and fracture of NiTi superelastic wire in an oral cavity. It was found that (i) fracture surfaces were observed after hydrogen charging, (ii) the strength of the Co–Cr alloy and stainless steel used in orthodontic treatment was not affected by the hydrogen charging; however, NiTi wire showed significant decreases in strength, (iii) the critical stress of martensite transformation was increased with increasing hydrogen charging, and the alloy was embrittled, (iv) the fractured surface of the alloys with severe hydrogen charging exhibited dimple patterns similar to those in the alloys from patients, and (v) in view of the galvanic current in the mouth, the fracture of the NiTi alloys might be attributed to the degradation of the mechanical properties due to hydrogen absorption.

Perinetti et al. [268] evaluated the surface corrosion and fracture resistance of two commercially available NiTi-based archwires, as induced by a combination of fluoride, pH, and thermocycling. One hundred and ten rectangular section NiTi-based archwires were used, 55 of each of the following: thermally activated Thermaloy and superelastic NeoSentalloy 100 g. Each of these was divided into five equal subgroups. One of these five subgroups did not undergo any treatment and served as the control, while the other four were subjected to 30 days of incubation at 37 °C under fluoridated artificial saliva (FS) at 1500 ppm fluoride treatment alone (two subgroups) or combined with a session of thermocycling (FS + Th) treatment at the end of incubation (two subgroups). Within each of the Thermaloy® and NeoSentalloy® groups, the FS and FS + Th treatments were performed under two different pH conditions: 5.5 and 3.5 (each with one subgroup per treatment). It was observed that (i) significant effects in terms of surface corrosion, but not fracture resistance, were seen mainly for the Thermaloy group at the lowest pH, with no effects of Th irrespective of the group or pH condition, and (ii) different NiTi-based archwires can have different corrosion resistance, even though the effects of surface corrosion and fracture resistance appear not to be significant in clinical situations, especially considering that thermocycling had no effect on these parameters. Rerhrhaye et al. [269] investigated, in vitro, the impact of this acidic and fluoridated environment on the electrochemical behavior and the mechanical properties of orthodontic alloys in NiTi and in stainless steel (controls) for the following parameters: Young's modulus (E), elastic limit (σ_e) and the maximum tensile load (σ_m). It was reported that (i) for the NiTi archwires, immersion in the fluoridated and acidic medium showed a statistically significant reduction of the Young's modulus (E), the elastic limit (σ_e), and the maximum tensile load (σ_m), (ii) similarly, a higher level of released nickel proportionate to the increase in the fluoride concentration and acidity was observed in the immersion solutions, and (iii) ESM observations revealed the status of the surface of the different alloys and the presence of corrosive pitting. Bock et al. [270] compared the mechanical strength of different joints made by conventional brazing, tungsten inert gas (TIG) and laser welding with and without filling material. Five standardized joining configurations of orthodontic wire in spring hard quality were used: round, cross, 3 mm length, 9 mm length, and 7 mm to orthodontic band. The joints were made by five different methods: brazing, TIG and laser welding with and without filling material. It was reported that (i) in all cases, brazing joints were ruptured on a low level of fracture strength (186–407 N), (ii) significant differences between brazing and TIG or laser welding were found in each joint configuration, (iii) the highest fracture strength means were observed for laser welding with filling material and 3 mm joint length (998 N), and (iv) using filling materials, there was a clear tendency to higher mean values of fracture strength in TIG and laser welding; however, statistically significant differences were found only in the 9-mm long joints. Based on these results, it was concluded that (i) the fracture strength of welded joints was positively influenced by the additional use of filling material, and (ii) TIG welding was comparable to laser welding except for the impossibility of joining orthodontic wire with orthodontic band.

9.7 Creep

Among smart materials, SMA actuators undergo a dimensional change due to the martensite (M) to austenite (A) transformation on heating. Since these dimensional changes can be large, SMA actuators are very suitable for applications requiring large displacements and a high work output [271, 272]. SMAs with high transformation temperatures can enable simplifications and improvements in operating efficiency of many mechanical components designed to operate at temperatures above 100 °C, potentially impacting the automotive, aerospace, manufacturing, and energy exploration industries [271].

There is a large body of evidence which suggests that thermal cycling of many of these SMAs under an applied stress results in an offset strain at the end of each cycle, where the start and end points of the hysteresis loop are displaced by a finite amount of strain [271]. This offset strain increases cumulatively during each subsequent thermal cycle, which is likely to decrease the effectiveness of the actuator in actual applications. Raj et al. [272] investigated the effect of stress on the long-term low-temperature tensile creep behavior of nominally stoichiometric NiTi deformed at 27, 100, and 200 °C under initial applied stresses between 200 and 350 MPa corresponding to the martensitic, two-phase and austenitic phase fields, respectively. The creep tests were conducted for several months to as long as 15 months. It was found that (i) the observed creep limit for the martensitic phase was determined to be 200 MPa at 27 °C for creep times less than 2000 h, (ii) similarly, the austenitic phase did not exhibit measurable creep below 220 MPa at 200 °C for creep times less than 1500 h, (iii) the martensitic phase exhibits a very large strain increase on loading followed by normal primary creep above 200 MPa, and (iv) as expected, steady-state creep was not observed even after several months of testing. Based on these finding, it was concluded that (i) the creep of the austenitic phase occurs by a dislocation glide-controlled creep mechanism accommodated by the nucleation and growth of deformation twins, (ii) normal primary creep was also observed at 100 °C, and (iii) however, sudden increases in the creep strain were observed within the first 1200 s after loading due to microstructural instability and due to stress-induced martensitic transformation of the austenite present in the material.

Eggeler et al. [273] reported on creep data of a Ni-rich NiTi alloy at stresses and temperatures in the range of 500°C and 150 MPa that (i) the apparent activation energy for creep was 421 kJ/mole and the creep stress exponent was 5, (ii) tempering and precreep result in a 20 °C increase of phase transformation temperatures (PTT), and (iii) the effect of an additional plastic deformation (as compared to a stress free aging treatment) is that PTT-peaks broaden and overlap on cooling and broaden on heating. If the SMA is subjected to the subloop loading under the stress-controlled condition, creep and creep recovery can appear based on the martensitic transformation and in the design of SMA elements, these deformation properties are important since the deflection of SMA elements can change under constant stress [274]. Takeda et al. [274] reported that (i) during loading under constant stress rate, temperature increases due to the stress-induced martensitic transformation, (ii) if stress is held constant during the martensitic transformation stage in the loading process, temperature decreases and the condition for the progress of the martensitic transformation is satisfied, resulting in the transformation-induced creep deformation, (iii) if stress is held constant during the reverse transformation stage in the unloading process, creep recovery appears due to the reverse transformation, and (iv) the volume fraction of the martensitic phase increases in proportion to an increase in creep strain. Dunić et al. [275] reported on the TiNi SMA which was subjected to a modified program of force-controlled tensile loading to investigate the time-dependent development of transformation strain under the constant-force conditions in order to describe transformation-induced creep phenomena. It was found that (i) mechanical characteristics of the TiNi SMA were derived using a testing machine, whereas the SMA temperature changes accompanying its deformation were obtained in a contactless manner with an infrared camera, (ii) the stress and related temperature changes demonstrated how the transformation-induced creep process started and evolved at various stages of the SMA loading, and (iii) it was demonstrated how the transformation-induced creep process occurring in the SMA under such conditions was involved in thermomechanical couplings and the related temperature changes.

9.8 Cavitation erosion

Cavitation erosion is the process of surface deterioration and surface material loss due to the generation of vapor or gas pockets inside the flow of liquid. These pockets are formed due to a low pressure well below the saturation vapor pressure of the liquid and erosion caused by the bombardment of vapor bubbles on the surface. It occurs under conditions of rapid pressure changes. This can occur in what can be a rather explosive and dramatic fashion. Additionally, this condition can form an airlock, which prevents any incoming fluid from offering cooling effects, further exacerbating the problem. The locations where this is most likely to occur is at other geometry-affected flow areas, such as pipe elbows and expansions, and so on. The harder the material, the greater its resistance to cavitation erosion. Richman et al. [276] tested two near-equiatomic NiTi alloys (austenite parent phase and martensite product phase) for resistance to cavitation erosion. It was reported that (i) although both compositions show very low mass-loss rates, the parent-phase alloy is clearly superior, and (ii) correlations based on LCF properties correctly predict the relative erosion resistances of these unusual materials, but do not account quantitatively for the recoverable deformation.

NiTi alloy has much higher cavitation erosion resistance than conventional metallic materials [277–280]. Superelasticity (in which the deformation can be recovered automatically while unloading) and SME (in which the deformation can be recovered through heating while unloading) of NiTi alloy can effectively accommodate the cavitation impact stress and deformation strain, which contributes to excellent cavitation erosion resistance of NiTi alloy [281]. Liu et al. [281] prepared NiTi alloy by melting Ti (99.7%) and Ni (99.98%) in a middle frequency vacuum induction furnace to cast 25-mm diameter rod followed by solution treatment at 800 °C for 1 h, which was further annealed at 700 °C for 30 min followed by quenching water, and then was aged at

400 °C for 10 min followed by air cooling. The cavitation erosion test was conducted in distilled water at RT conforming to ASTM standard G32-92 and as interrupted for measuring the weight loss until 20 h. It was reported that (i) the quantitative results indicated that the amount of pits increased with the increase of cavitation erosion time, but the mean area of erosion pits is about 3 μm^2 until 11 h of cavitation erosion; hence, the main damage characteristic of NiTi alloy was the appearance of many tiny erosion pits with nearly equal area during the initial 11 h, and (ii) after CE for 11-h cavitation tests, the amount and mean area of erosion pits decrease and increase sharply, respectively, which was the result of many erosion pits connecting and merging. Yang et al. [282] fabricated NiTi thin films by using the filtered arc deposition system to investigate the effect of substrate temperature (130–600 °C) on the structures and properties of NiTi thin films. It was found that (i) the fabricated thin NiTi film was dense and homogenous, (ii) higher substrate temperatures resulted in crystalline thin films dominated by the parent phase (B2), (iii) the cavitation erosion resistance of the films assessed by using ASTM test method G32 indicated that NiTi thin films showed superior cavitation erosion resistance compared with 316 austenitic stainless steel, and (iv) the improved cavitation erosion resistance was attributed to the recoverable deformation associated with pseudoelasticity of the NiTi thin film.

References

[1]   Eggeler G, Hornbogen E, Yawny A, Heckmann A, Wagner M. Structural and functional fatigue of NiTi shape memory alloys. Mater. Sci. Eng., A 2004, 378, 24–33.
[2]   Saikrishna CN, Ramaiah KV, Vidyashankar B, Bhaumik SK. Effect of intermittent overload cycles on thermomechanical fatigue life of NiTi shape memory alloy wire. Metall. Mater. Trans. A 2013, 44, 5–8.
[3]   Ataalla T, Leary M, Subic A. Functional fatigue of shape memory alloys. Sus. Automot. Technol. 2012, 39–43.
[4]   Mammano GS, Dragoni E. Functional fatigue of NiTi shape memory wires for a range of end loadings and constraints. Frattura ed Integrità Strutturale, 2013, 23, 25–33.
[5]   Hou H, Tang Y, Hamilton RF, Horn MW. Functional fatigue of submicrometer NiTi shape memory alloy thin films. J. Vac. Sci. Technol. A 2017, 35, 040601; https://doi.org/10.1116/1.4983011.
[6]   LePage WS, Ahadi A, Lenthe WC, Sun QP, Pollock TM, Shaw JA, Daly SH. Grain size effects on NiTi shape memory alloy fatigue crack growth. J. Mater. Res. 2018, 33, 91–107.
[7]   Dosanjh A, Paurazas S, Askar M. The effect of temperature on cyclic fatigue of nickel-titanium rotary endodontic instruments. J. Endod. 2017, 43, 823–6.
[8]   Maletta C, Sgambitterra E, Furgiuele F, Casati R, Tuissi A. Low cycle fatigue of pseudoelastic NiTi alloys; https://scholar.google.co.jp/scholar?q=167+Low+cycle+fatigue+of+pseudoelastic+NiTi+alloys&hl=en&as_sdt=0&as_vis=1&oi=scholart.
[9]   Cheung GS, Darvell BW. Low-cycle fatigue of NiTi rotary instruments of various cross-sectional shapes. Int. Endod. J. 2007, 40, 626–32.
[10] Ye J, Gao Y. Metallurgical characterization of M-wire nickel-titanium shape memory alloy used for endodontic rotary instruments during low-cycle fatigue. J. Endod. 2012, 38, 105–7.

[11]   Rahim M, Frenzel J, Frotscher M, Pfetzing-Micklich F, Steegmüller R, Wohlschlögel M, Mughrabi H, Eggeler G. Impurity levels and fatigue lives of pseudoelastic NiTi shape memory alloys. Acta Mater. 2013, 61, 3667–86.

[12]   Kafka OL, Yu C, Shakoor M, Liu Z, Wagner GL, Liu WK. Data-driven mechanistic modeling of influence of microstructure on high-cycle fatigue life of nickel titanium. JOM 2018, 70, 1154–8.

[13]   McKelvey AL, Ritchie RO. Fatigue-crack growth behavior in the superelastic and shape-memory alloy nitinol. Metall. Mater. Trans. A 2001, 32, 731–43.

[14]   Chen J, Yin H, Kang G, Sun Q. Fatigue crack growth in cold-rolled and annealed polycrystalline superelastic NiTi alloys. Acta Mech. Solida Sin. 2018, 31, 599–607.

[15]   Barbosa I, Ferreira F, Scelza P, Neff J, Russano D, Montagnana M, Zaccaro Scelza M. Defect propagation in NiTi rotary instruments: a noncontact optical profilometry analysis. Int. Endod. J. 2018, 51, 1271–8.

[16]   Oshida Y, Farzin-Nia K. Progressive Damage Assessment of TiNi Endodontic Files. In: Yahia L'H, ed. Shape Memory Implants. Springer-Verlag, Berlin, Germany, 2000, 236–49.

[17]   Pluvinage GC, Raquet MN. Physical and Mechanical Measurements of Damage in Low-Cycle Fatigue: Application for Two-Level Tests. In: Lankford, et al. ed., ASTA STP811. Fatigue Mechanisms: Advances in Quantitative Measurement of Physical Damage. ASTM International, West Conshohocken, PA, 1983, 139–150.

[18]   Oshida Y, Daly J. Fatigue Damage Evaluation of Shot-Peened High Strength Aluminum Alloy. In: Meguid SA, ed., Surface Engineering. Elsevier Applied Science, London, 1990, 404–16.

[19]   Oshida Y, Chen PC. Non-destructive low-cycle fatigue characterization of multi-layer thin film structures. J. Non-Destruct. Eval. 1990, 8, 235–5.

[20]   Oshida Y, Chen PC. High and low-cycle fatigue damage evaluation of multi-layer thin film structure. Trans. ASME J. Electron. Packag. 1991, 113, 58–62.

[21]   Sotokawa T. An analysis of clinical breakage of root canal instruments. J. Endod. 1988, 14, 75–82.

[22]   Zuolo M, Waloton R, Murgel C. Canal mater files: scanning electron microscopic evaluation of new instruments and their wear with clinical usage. J. Endod. 1992, 18, 336–9.

[23]   Lilley J, Smith D. An investigation of the fracture of root canal reamers. Br. Dent. J. 1996, 19, 364–72.

[24]   Lautenschlager E, Jacobs J, Marshall G, Heuer M. Brittle and ductile torsional failures of endodontic instruments. J. Endod. 1977, 3, 175–8.

[25]   Craig R, Mcliwain E, Peyton F. Bending and tension properties of endodontic instruments. Oral Surg. Oral Med. Oral Pathol. Oral Radiol. Endod. 1968, 25, 239–54.

[26]   Dolan D, Craig R. Bending and torsion of endodontic files with rhombus cross-sections. J. Endod. 1982, 8, 260–4.

[27]   Luebke N, Brantley W. Physical dimensions and torsional properties of rotary endodontic instruments. Part 1. Gates Glidden drills. J. Endod. 1990, 16, 438–41.

[28]   Lausten L, Luebke N, Brantley W. Bending and metallurgical properties of rotary endodntic instruments. Part 4. Gates Glidden and Peeso drills. J. Endod. 1993, 19, 440–7.

[29]   Brantley W, Luebke N, Luebke F, Mittchell J. Performance of engine-driven rotary endodntic instruments with a superimposed bending deflection. Part 5: Gates Glidden and Peeso drills. J. Endod. 1994, 20, 241–5.

[30]   Wolcott J, Himel V. Torsional properties of nickel-titanium versus stainless steel endodontic files. J. Endod. 1997, 23, 217–20.

[31]   Campus J, Pertot W. Machining efficiency of Ni-Ti K-type files in a linear motion. Int. Endod. J. 1995, 28, 239–43.

[32]   Scott G, Walton R. Ultrasonic endodontics: the wear of instruments with usage. J. Endod. 1986, 12, 279–83.

[33] Chernick L, Jacobs J, Lautenschlager E, Heuer M. Torsional failure of endodontic files. J. Endod. 1976, 2, 94–7.

[34] Young JM, Van Vliet KJ. Predicting in vivo failure of pseudoelastic NiTi devices under low cycle, high amplitude fatigue. J. Biomed. Mater. Res. B Appl. Biomater. 2005, 72, 17–26.

[35] Runciman A, Xu D, Pelton AR, Ritchie RO. An equivalent strain/Coffin–Manson approach to multiaxial fatigue and life prediction in superelastic Nitinol medical devices. Biomaterials 2011, 32, 4987–93.

[36] Cheung GS, Zhang EW, Zheng YF. A numerical method for predicting the bending fatigue life of NiTi and stainless steel root canal instruments. Int. Endod. J. 2011, 44, 357–61.

[37] Dordoni E, Petrini L, Wu W, Migliavacca F, Dubini G, Pennati G. Computational modeling to predict fatigue behavior of NiTi stents: what do we need? J. Funct. Biomater. 2015, 6, 299–317.

[38] Petrini L, Trotta A, Dordoni E, Migliavacca F, Dubini G, Lawford PV, Gosai JN, Ryan DM, Testi D, Pennati G. A computational approach for the prediction of fatigue behaviour in peripheral stents: application to a clinical case. Ann. Biomed. Eng. 2016, 44, 536–47.

[39] Bonsignore C. Present and future approaches to lifetime prediction of superelastic nitinol. Theor. Appl. Fract. Mech. 2017, 92, 298–305.

[40] Peigney M. A Direct Method for Predicting the High-Cycle Fatigue Regime of Shape-Memory Alloys Structures. In: Barrera O, et al. ed., Advances in Direct Methods for Materials and Structures. Springer, 2018, 13–28.

[41] Ashbli S, Menzemer CC. On the fatigue behavior of nanocrystalline NiTi shape memory alloys: a review. J. Nanomed. Nanotechnol. 2019, 10, 2, 7 pages; doi: 10.4172/2157-7439.1000529.

[42] LePage WS, Ahadi A, Lenthe WC, Sun QP, Pollock TM, Shaw JA, Daly SH. Grain size effects on NiTi shape memory alloy fatigue crack growth. J. Mater. Res. 2018, 33, 91–107.

[43] van Humbeeck J. Cycling effects, fatigue and degradation of shape memory alloys. J. Phys. IV 1991, 1, C4-189–C4-197.

[44] Mahtabi MJ, Shamsaei N, Mitchell MR. Fatigue of nitinol: the state-of-the-art and ongoing challenges. J. Mech. Behav. Biomed. Mater. 2015, 50, 228–54.

[45] Kang G, Song D. Review on structural fatigue of NiTi shape memory alloys: pure mechanical and thermo-mechanical ones. Theor. Appl. Mech. Lett. 2015, 5, 245–54.

[46] Bonsignore C. Present and future approaches to lifetime prediction of superelastic nitinol. Theor. Appl. Fract. Mech. 2017, 92, 298–305.

[47] Barrera N, Biscari P, Urbano MF. Macroscopic modeling of functional fatigue in shape memory alloys. Eur. J. Mech. A 2014, 45, 101–9.

[48] Tadaki T, Nakata Y, Shimizu K. Thermal cycling effects in an aged Ni-rich Ti–Ni shape memory alloy. Trans. Japan Inst. Met. 1987, 28, 883–90.

[49] Lagoudas DC, Miller AD, Rong L, Kumar PK. Thermomechanical fatigue of shape memory alloys. Smart Mater. Struct. 2009, 18; https://iopscience.iop.org/article/10.1088/0964-1726/18/8/085021.

[50] Tan J, Li Y, Gao B, Mi X. Thermo-mechanical cycling fatigue and transformation behavior of Ti49.8N50.2 alloy wire. Rare Metal Mat. Eng. 2009, 38, 784–7.

[51] Bertacchini OW, Lagoudas DC, Calkins FT, Mabe JH. Thermomechanical cyclic loading and fatigue life characterization of nickel rich NiTi shape-memory alloy actuators. SPIE Smart Struct. Mater. + Nondestruct. Evaluat. Health Monitor. 2008; doi: 10.1117/12.776502.

[52] Coda A, Cadelli A, Zanella M, Fumagalli L. Straightforward downsizing of inclusions in NiTi alloys: a new generation of SMA wires with outstanding fatigue life. SMST 2018, 4, 41–7.

[53] Pruett JP, Clement DJ, Carnes DL Jr. Cyclic fatigue testing of nickel-titanium endodontic instruments. J. Endod. 1997, 23, 77–85.

[54] Yared GM, Bou Dagher FE, Machtou P. Cyclic fatigue of ProFile rotary instruments after clinical use. Int. Endod. J. 2000, 33, 204–7.

[55] Li UM, Lee BS, Shih CT, Lan WH, Lin CP. Cyclic fatigue of endodontic nickel titanium rotary instruments: static and dynamic tests. J. Endod. 2002, 28, 448–51.

[56] Ullmann CJ, Peters OA. Effect of cyclic fatigue on static fracture loads in PT nickel-titanium rotary instruments. J. Endod. 2005, 31, 183–6.

[57] Tripi TR, Bonaccorso A, Condorelli GG. Cyclic fatigue of different nickel-titanium endodontic rotary instruments. Oral Surg. Oral Med. Oral Pathol. Oral Radiol. Endod. 2006, 102, e106–14.

[58] Li U-M, Shin C-S, Lan W-H, Lin C-P. Application of nondestructive testing in cyclic fatigue evaluation of endodontic Ni-Ti rotary instruments. Dent. Mater. J. 2006, 25, 247–52.

[59] Grande NM, Plotino G, Pecci R, Bedini R, Malagnino VA, Somma F. Cyclic fatigue resistance and three-dimensional analysis of instruments from two nickel-titanium rotary systems. Int. Endod. J. 2006, 39, 755–63.

[60] Larsen CM, Watanabe I, Glickman GN, He J. Cyclic fatigue analysis of a new generation of nickel titanium rotary instruments. J. Endod. 2009, 35, 401–3.

[61] Plotino G, Grande NM, Melo MC, Bahia MG, Testarelli L, Gambarini G. Cyclic fatigue of NiTi rotary instruments in a simulated apical abrupt curvature. Int. Endod. J. 2010, 43, 226–30.

[62] Pedullà E, Grande NM, Plotino G, Pappalardo A, Rapisarda E. Cyclic fatigue resistance of three different nickel-titanium instruments after immersion in sodium hypochlorite. J. Endod. 2011, 37, 1139–42.

[63] Rodrigues RC, Lopes HP, Elias CN, Amaral G, Vieira VT, De Martin AS. Influence of different manufacturing methods on the cyclic fatigue of rotary nickel-titanium endodontic instruments. J. Endod. 2011, 37, 1553–7.

[64] Arias A, Perez-Higueras JJ, de la Macorra JC. Differences in cyclic fatigue resistance at apical and coronal levels of reciproc and WaveOne new files. J. Endod. 2012, 38, 1244–8.

[65] Bouska J, Justman B, Williamson A, DeLong C, Qian F. Resistance to cyclic fatigue failure of a new endodontic rotary file. J. Endod. 2012, 38, 667–9.

[66] Gambarini G, Gergi R, Naaman A, Osta N, Al Sudani D. Cyclic fatigue analysis of twisted file rotary NiTi instruments used in reciprocating motion. Int. Endod. J. 2012, 45, 802–6.

[67] Inan U, Aydin C. Comparison of cyclic fatigue resistance of three different rotary nickel-titanium instruments designed for retreatment. J. Endod. 2012, 38, 108–11.

[68] Kim HC, Kwak SW, Cheung GS, Ko DH, Chung SM, Lee W. Cyclic fatigue and torsional resistance of two new nickel-titanium instruments used in reciprocation motion: reciproc versus WaveOne. J. Endod. 2012, 38, 541–4.

[69] Al-Sudani D, Grande NM, Plotino G, Pompa G, Di Carlo S, Testarelli L, Gambarini G. Cyclic fatigue of nickel-titanium rotary instruments in a double (S-shaped) simulated curvature. J. Endod. 2012, 38, 987–9.

[70] Plotino G, Grande NM, Testarelli L, Gambarini G. Cyclic fatigue of reciproc and WaveOne reciprocating instruments. Int. Endod. J. 2012, 45, 614–8.

[71] Pedullà E, Grande NM, Plotino G, Palermo F, Gambarini G, Rapisarda E. Cyclic fatigue resistance of two reciprocating nickel-titanium instruments after immersion in sodium hypochlorite. Int. Endod. J. 2013, 46, 155–9.

[72] Duke F, Shen Y, Zhou H, Ruse ND, Wang ZJ, Hieawy A, Haapasalo M. Cyclic fatigue of ProFile vortex and vortex blue nickel-titanium files in single and double curvatures. J. Endod. 2015, 41, 1686–90.

[73] Kaval ME, Capar ID, Ertas H, Sen BH. Comparative evaluation of cyclic fatigue resistance of four different nickel-titanium rotary files with different cross-sectional designs and alloy properties. Clin. Oral Invest. 2017, 21, 1527–30.

[74] Elnaghy A, Elsaka S. Cyclic fatigue resistance of XP-endo shaper compared with different nickel-titanium alloy instruments. Clin. Oral Invest. 2018, 22, 1433–7.

[75]   Özyürek T, Uslu G, Yılmaz K, Gündoğar M. Effect of glide path creating on cyclic fatigue
       resistance of reciproc and reciproc blue nickel-titanium files: a laboratory study. J. Endod.
       2018, 44, 1033–7.
[76]   Keskin C, İnan U, Demiral M, Keleş A. Cyclic fatigue resistance of R-pilot, WaveOne gold glider,
       and ProGlider glide path instruments. Clin. Oral Invest. 2018, 22, 3007–12.
[77]   Inan U, Keskin C, Yilmaz ÖS, Baş G. Cyclic fatigue of reciproc blue and reciproc instruments
       exposed to intracanal temperature in simulated severe apical curvature. Clin. Oral Invest.
       2019, 23, 2077–82.
[78]   Rubio J, Zarzosa JI, Pallarés A. A comparative study of cyclic fatigue of 10 different types of
       endodontic instruments: an in vitro study. Acta Stomatol Croat 2019, 53, 28–36.
[79]   Khalil WA, Natto ZS. Cyclic fatigue, bending resistance, and surface roughness of PT gold
       and EdgeEvolve files in canals with single- and double-curvature. Restor. Dent. Endod. 2019,
       44(2), 1–9.
[80]   Plotino G, Grande NM, Cordaro M, Testarelli L, Gambarini G. A review of cyclic fatigue testing
       of nickel-titanium rotary instruments. J. Endod. 2009, 35, 1469–76.
[81]   Sattapan B, Nervo GJ, Palamara JEA, Messer HH. Defects in rotary nickel-titanium files after
       clinical use. J. Endod. 2000, 26, 161–5.
[82]   Martín B, Zelada G, Varela P, Bahillo JG, Magán F, Ahn S, Rodríguez C. Factors influencing the
       fracture of nickel titanium rotary instruments. Int. Endod. J. 2003, 36, 262–6.
[83]   Barbosa FO, Gomes JA, de Araújo MC. Influence of electrochemical polishing on the
       mechanical properties of K3 nickel-titanium rotary instruments. J. Endod. 2008, 34,
       1533–6.
[84]   Best S, Watson P, Pilliar R, Kulkarni GG, Yared G. Torsional fatigue and endurance limit of a
       size 30.06 ProFile rotary instrument. Int. Endod. J. 2004, 37(6), 370–3.
[85]   Bahia MG, Melo MC, Buono VT. Influence of cyclic torsional loading on the fatigue resistance
       of K3 instruments. Int. Endod. J. 2008, 41, 883–91.
[86]   Kim JY, Cheung GS, Park SH, Ko DC, Kim JW, Kim HC. Effect from cyclic fatigue of nickel-
       titanium rotary files on torsional resistance. J. Endod. 2012, 38, 527–30.
[87]   Setzer FC, Böhme CP. Influence of combined cyclic fatigue and torsional stress on the fracture
       point of nickel-titanium rotary instruments. J. Endod. 2013, 39, 133–7.
[88]   Campbell L, Shen Y, Zhou HM, Haapasalo M. Effect of fatigue on torsional failure of nickel-
       titanium controlled memory instruments. J. Endod. 2014, 40, 562–5.
[89]   Shen Y, Riyahi AM, Campbell L, Zhou H, Du T, Wang Z, Qian W, Haapasalo M. Effect of a
       combination of torsional and cyclic fatigue preloading on the fracture behavior of K3 and K3XF
       instruments. J. Endod. 2015, 41, 526–30.
[90]   Alcalde MP, Duarte MAH, Bramante CM, de Vasconselos BC, Tanomaru-Filho M, Guerreiro-
       Tanomaru JM, Pinto JC, Só MVR, Vivan RR. Cyclic fatigue and torsional strength of three
       different thermally treated reciprocating nickel-titanium instruments. Clin. Oral Investig.
       2018, 22, 1865–71.
[91]   Cheung GS, Darvell BW. Fatigue testing of a NiTi rotary instrument. Part 1: strain-life
       relationship. Int. Endod. J. 2007, 40, 612–8.
[92]   Kim HC, Yum J, Hur B, Cheung GS. Cyclic fatigue and fracture characteristics of ground and
       twisted nickel-titanium rotary files. J. Endod. 2010, 36, 147–52.
[93]   Bernard S, Balla VK, Bose S, Bandyopadhyay A. Rotating bending fatigue response of laser
       processed porous NiTi alloy. Mater. Sci. Eng. C. 2011, 31, 815–20.
[94]   Shen Y, Qian W, Abtin H, Gao Y, Haapasalo M. Fatigue testing of controlled memory wire nickel-
       titanium rotary instruments. J. Endod. 2011, 37, 997–1001.
[95]   Cheung GSP, Zhang EW, Zheng YF. A numerical method for predicting the bending fatigue life
       of NiTi and stainless steel root canal instruments. Int. Endod. J. 2011, 44, 357–61.

[96]   Hieawy A, Haapasalo M, Zhou H, Wang ZJ, Shen Y. Phase transformation behavior and resistance to bending and cyclic fatigue of PT gold and PT universal instruments. J. Endod. 2015, 41, 1134–8.

[97]   Silva EJ, Rodrigues C, Vieira VT, Belladonna FG, De-Deus G, Lopes HP. Bending resistance and cyclic fatigue of a new heat-treated reciprocating instrument. Scanning 2016, 38, 837–41.

[98]   Della Corte C, Stanford MK, Jett TR. Rolling contact fatigue of superelastic intermetallic materials (SIM) for use as resilient corrosion resistant bearings. Tribol. Lett. 2015, 57; doi: 10.1007/s11249-014-0456-3.

[99]   Bernard S, Balla VK, Bose S, Bandyopadhyay A. Compression fatigue behavior of laser processed porous NiTi alloy. J. Mech. Behav. Biomed. Mater. 2012, 13, 62–8.

[100]   Lopes HP, Elias CN, Vieira MV, Vieira VT, de Souza LC, Dos Santos AL. Influence of surface roughness on the fatigue life of nickel-titanium rotary endodontic instruments. J. Endod. 2016, 42, 965–8.

[101]   Cai JJ, Tang XN, Ge JY. Effect of irrigation on surface roughness and fatigue resistance of controlled memory wire nickel-titanium instruments. Int. Endod. J. 2017, 50, 718–24.

[102]   Cheung GS, Shen Y, Darvell BW. Does electropolishing improve the low-cycle fatigue behavior of a nickel-titanium rotary instrument in hypochlorite? J. Endod. 2007, 33, 1217–21.

[103]   Bui TB, Mitchell JC, Baumgartner JC. Effect of electropolishing ProFile nickel-titanium rotary instruments on cyclic fatigue resistance, torsional resistance, and cutting efficiency. J. Endod. 2008, 34, 190–3.

[104]   Praisarnti C, Chang JW, Cheung GS. Electropolishing enhances the resistance of nickel-titanium rotary files to corrosion-fatigue failure in hypochlorite. J. Endod. 2010, 36, 1354–7.

[105]   Sinan AA, Thiémélé-Yacé SE, Abouattier-Mansilla E, Vallaeys K, Diemer F. Effects of electropolishing on rupture resistance in cyclic fatigue and on the surface condition of endodontic instruments. Odontostomatol Trop. 2010, 33, 23–8.

[106]   Condorelli GG, Bonaccorso A, Smecca E, Schäfer E, Cantatore G, Tripi TR. Improvement of the fatigue resistance of NiTi endodontic files by surface and bulk modifications. Int. Endod. J. 2010, 43, 866–73.

[107]   Zinelis S, Darabara M, Takase T, Ogane K, Papadimitriou GD. The effect of thermal treatment on the resistance of nickel-titanium rotary files in cyclic fatigue. Oral Surg. Oral Med. Oral Pathol. Oral Radiol. Endod. 2007, 103, 843–7.

[108]   Cheung GS, Shen Y, Darvell BW. Effect of environment on low-cycle fatigue of a nickel-titanium instrument. J. Endod. 2007, 33, 1433–7.

[109]   Shen Y, Qian W, Abtin H, Gao Y, Haapasalo M. Effect of environment on fatigue failure of controlled memory wire nickel-titanium rotary instruments. J. Endod. 2012, 38, 376–80.

[110]   Chang SW, Kim YC, Chang H, Jee KK, Zhu Q, Safavi K, Shon WJ, Bae KS, Spangberg LS, Kum KY. Effect of heat treatment on cyclic fatigue resistance, thermal behavior and microstructures of K3 NiTi rotary instruments. Acta Odontol. Scand. 2013, 71, 1656–62.

[111]   Braga LC, Faria Silva AC, Buono VT, de Azevedo Bahia MG. Impact of heat treatments on the fatigue resistance of different rotary nickel-titanium instruments. J. Endod. 2014, 40, 1494–7.

[112]   Plotino G, Grande NM, Cotti E, Testarelli L, Gambarini G. Blue treatment enhances cyclic fatigue resistance of vortex nickel-titanium rotary files. J. Endod. 2014, 40, 1451–3.

[113]   de Vasconcelos RA, Murphy S, Carvalho CA, Govindjee RG, Govindjee S, Peters OA. Evidence for reduced fatigue resistance of contemporary rotary instruments exposed to body temperature. J. Endod. 2016, 42, 782–7.

[114]   Grande NM, Plotino G, Silla E, Pedullà E, DeDeus G, Gambarini G, Somma F. Environmental temperature drastically affects flexural fatigue resistance of nickel-titanium rotary files. J. Endod. 2017, 43, 1157–60.

[115]   Klymus ME, Alcalde MP, Vivan RR, Só MVR, de Vasconselos BC, Duarte MAH. Effect of
       temperature on the cyclic fatigue resistance of thermally treated reciprocating instruments.
       Clin. Oral Investig. 2018; doi: 10.1007/s00784-018-2718-1.

[116]   Alfawaz H, Alqedairi A, Alsharekh H, Almuzaini E, Alzahrani S, Jamleh A. Effects of sodium
       hypochlorite concentration and temperature on the cyclic fatigue resistance of heat-treated
       nickel-titanium rotary instruments. J. Endod. 2018, 44, 1563–6.

[117]   Shen Y, Huang X, Wang Z, Wei X, Haapasalo M. Low environmental temperature influences
       the fatigue resistance of nickel-titanium files. J. Endod. 2018, 44, 626–9.

[118]   Arias A, Hejlawy S, Murphy S, de la Macorra JC, Govindjee S, Peters OA. Variable impact by
       ambient temperature on fatigue resistance of heat-treated nickel titanium instruments. Clin.
       Oral Investig. 2019, 23, 1101–9.

[119]   Mize SB, Clement DJ, Pruett JP, Carnes DL Jr. Effect of sterilization on cyclic fatigue of rotary
       nickel-titanium endodontic instruments. J. Endod. 1998, 24, 843–7.

[120]   Martins RC, Bahia MGA, Buono VTL. The effect of sodium hypochlorite on the surface
       characteristics and fatigue resistance of ProFile nickel-titanium instruments. Oral Pathol.
       Oral Radiol. Endod. 2006, 102, e99–105.

[121]   Viana AC, Gonzalez BM, Buono VT, Bahia MG. Influence of sterilization on mechanical
       properties and fatigue resistance of nickel-titanium rotary endodontic instruments. Int.
       Endod. J. 2006, 39, 709–15.

[122]   Peters OA, Roehlike JO, Baumann MA. Effect of immersion in sodium hypochlorite on torque
       and fatigue resistance of nickel-titanium instruments. J. Endod. 2007, 33, 589–93.

[123]   Cheung GS, Darvell BW. Low-cycle fatigue of rotary NiTi endodontic instruments in
       hypochlorite solution. Dent. Mater. 2008, 24, 753–9.

[124]   Pedullà E, Grande NM, Plotino G, Pappalardo A, Rapisarda E. Cyclic fatigue resistance of
       three different nickel-titanium instruments after immersion in sodium hypochlorite. J.
       Endod. 2011, 37, 1139–42.

[125]   Hasegawa Y, Goto S, Ogura H. Effect of EDTA solution on corrosion fatigue of Ni-Ti files with
       different shapes. Dent. Mater. J. 2014, 33, 415–21.

[126]   Gülşah Uslu G, Özyürek T, Yılmaz K, Plotino G. Effect of dynamic immersion in sodium
       hypochlorite and EDTA solutions on cyclic fatigue resistance of WaveOne and WaveOne gold
       reciprocating nickel-titanium files. J. Endod. 2018, 44, 834–7.

[127]   Bergeron BE, Mayerchak MJ, Roberts MJ, Jeansonne BG. Multiple autoclave cycle effects
       on cyclic fatigue of nickel-titanium rotary files produced by new manufacturing methods. J.
       Endod. 2011, 37, 72–4.

[128]   Plotino G, Costanzo A, Grande NM, Petrovic R, Testarelli L, Gambarini G. Experimental
       evaluation on the influence of autoclave sterilization on the cyclic fatigue of new nickel-
       titanium rotary instruments. J. Endod. 2012, 38, 222–5.

[129]   Zhao D, Shen Y, Peng B, Haapasalo M. Effect of autoclave sterilization on the cyclic fatigue
       resistance of thermally treated nickel-titanium instruments. Int. Endod. J. 2016, 49, 990–5.

[130]   Özyürek T, Yılmaz K, Uslu G. The effects of autoclave sterilization on the cyclic fatigue
       resistance of PT universal, PT next, and PT gold nickel-titanium instruments. Restor. Dent.
       Endod. 2017, 42, 301–8.

[131]   Matheus TC, Menezes WM, Rigo OD, Kabayama LK, Viana CS, Otubo J. The influence of
       carbon content on cyclic fatigue of NiTi SMA wires. Int. Endod. J. 2011, 44, 567–73.

[132]   Wang XM, Wang YF, Yue ZF. Finite element simulation of the influence of TiC inclusions on the
       fatigue behavior of NiTi shape-memory alloys. Metall. Mater. Trans. A 2005, 36, 2615–20.

[133]   Kollerov M, Lukina E, Gusev D, Mason P, Wagstaff P. Impact of material structure on the
       fatigue behaviour of NiTi leading to a modified Coffin–Manson equation. Mater. Sci. Eng., A
       2013, 585, 356–62.

[134]   Gall K, Tyber J, Wilkesanders G, Robertson SW, Ritchie RO, Maier JH. Effect of microstructure on the fatigue of hot-rolled and cold-drawn NiTi shape memory alloys. Mater. Sci. Eng., A 2008, 486, 389–403.

[135]   Bian X, Gazder AA, Saleh AA, Pereloma EV. A comparative study of a NiTi alloy subjected to uniaxial monotonic and cyclic loading-unloading in tension using digital image correlation: the grain size effect. J. Alloys Compd. 2019, 777, 723–35.

[136]   Murakami T, Iijima M, Muguruma T, Yano F, Kawashima I, Mizoguchi I. High-cycle fatigue behavior of beta-titanium orthodontic wires. Dent. Mater. J. 2015, 34, 189–95.

[137]   Brantley WA. Orthodontic Wires. In: Brantley WA, et al. ed., Orthodontic Materials: Scientific and Clinical Aspects.Thieme, Stuttgart, 2001, 77–103.

[138]   Kaneko K, Yokoyama K, Moriyama K, Asaoka K, Sakai J, Nagumo M. Delayed fracture of beta titanium orthodontic wire in fluoride aqueous solutions. Biomaterials 2003, 24, 2113–20.

[139]   Kappert PF, Kelly JR. Cyclic fatigue testing of denture teeth for bulk. Dent. Mater. 2013, 29, 1012–9.

[140]   Lin CW, Ju CP, Chern Lin JH. A comparison of the fatigue behavior of cast Ti-7.5Mo with c.p. titanium, Ti-6Al-4V and Ti-13Nb-13Zr alloys. Biomaterials 2005, 26, 2899–907.

[141]   Dubey S, Soboyejo ABO, Soboyejo WO. An investigation of the effects of stress ratio and crack closure on the micromechanism of fatigue-crack growth in Ti-6Al-4V. Acta Mater. 1997, 45, 2777–87.

[142]   Meng W, Xie L, Zhang Y, Wang Y, Sun X, Zhang S. Effect of mean stress on the fatigue life prediction of notched fiber-reinforced 2060 Al-Li alloy laminates under spectrum loading. Adv. Mater. Sci. Eng. 2018; https://doi.org/10.1155/2018/5728174.

[143]   Tabanlí R, Simha N, Berg B. Mean stress on fatigue of NiTi. Mater. Sci. Eng. A, 1999, 273/275, 644–8.

[144]   Moore R, Watts JTF, Hood JAA, Burritt DJ. Intra-oral temperature variation over 24 hours. Eur. J. Orthod., 1999, 21, 1–13.

[145]   Mullins W, Bagby M, Norman T. Mechanical behavior of thermoresponsive orthodontic archwires. Dent. Mat., 1996, 12, 308–14.

[146]   Filleul M, Jordan L. Torsional properties of Ni-Ti and cooper Ni-Ti wires: the effect of temperature on physical properties. Eur. J. Orthod., 1997, 19, 637–46.

[147]   Meling TR, Ødegaard J. The effect of short-term temperature changes on superelastic nickel-titanium archwires activated in orthodontic bending. Am. J. Orthod. Dentofac. Orthop. 2001, 19, 263–73.

[148]   Iijima M, Ohno H, Kawashima I, Endo K, Brantley WA, Mizoguchi I. Micro x-ray diffraction study of superelastic nickel-titanium orthodontic wires at different temperatures and stresses. Biomaterials 2002, 23, 1769–74.

[149]   Berzins DW, Roberts HW. Phase transformation changes in thermocycled nickel-titanium orthodontic wires. Dent. Mater. 2010, 26, 666–74.

[150]   Lombardo L, Toni G, Stefanoni F, Mollica F, Siciliani G. The effect of temperature on the mechanical behaviour of nickel-titanium orthodontic initial archwires. Angle Orthod. 2013, 83, 298–305.

[151]   Eliades T, Eliades G, Athanasiou AE, Bradley TG. Surface characterization of retrieved NiTi orthodontic archwires. Eur. J. Orthod. 2000, 22, 317–26.

[152]   Wilkinson P, Dysart P, Hood J, Herbison P. Load-deflection characteristics of superelastic nickel-titanium orthodontic wires. Am. J. Orthod. Dentofacial Orthop. 2002, 121, 483–95.

[153]   Prymak O, Klocke A, Kahl-Nieke B, Epple M. Fatigue of orthodontic nickel-titanium (NiTi) wires in different fluids under constant mechanical stress. Mater. Sci. Eng. A 2004, 378, 110–4.

[154]   Lombardo L, Toni G, Mazzanti V, Mollica F, Spedicato A, Siciliani G. The mechanical behavior of as received and retrieved nickel titanium orthodontic archwires. Prog. Orthod. 2019, 20; doi: 10.1186/s40510-018-0251-z.

[155]   Bourauel C, Scharold W, Jäger A, Eliades T. Fatigue failure of as-received and retrieved NiTi orthodontic archwires. Dent. Mater. 2008, 24, 1095–101.

[156]   Hernández GS, Espínola GS, Gayosso CA, Furuki HK. A comparative study of fatigue resistance of NiTi archwires from three commercial brands. Revista Mexicana de Orthodoncia 2014, 2, e247–50.

[157]   Rokbani M, Saint-Sulpice L, Chirani SA, Bouraoui T. Fatigue properties by "self-heating" method: application to orthodontic Ni-Ti wires after hydrogen charging. J. Intell. Mater. Syst. Struct. 2018; https://doi.org/10.1177/1045389X18778371.

[158]   Karakoc O, Hayrettin C, Bass M, Wang SJ, Canadinc D, Mabe JH, Lagoudas DC, Karaman I. Effects of upper cycle temperature on the actuation fatigue response of NiTiHf high temperature shape memory alloys. Acta Mater. 2017, 138, 185–97.

[159]   Karakoc O, Hayrettin C, Canadinc D, Karaman I. Role of applied stress level on the actuation fatigue behavior of NiTiHf high temperature shape memory alloys. Acta Mater. 2018, 153, 156–68.

[160]   Speirs M, Van Hooreweder B, Van Humbeeck J, Kruth JP. Fatigue behaviour of NiTi shape memory alloy scaffolds produced by SLM, a unit cell design comparison. J. Mech. Behav. Biomed. Mater. 2017, 70, 53–9.

[161]   Hsiao HM, Yin MT. An intriguing design concept to enhance the pulsatile fatigue life of self-expanding stents. Biomed. Microdevices 2014, 16, 133–41.

[162]   Adlakha S, Sheikh M, Wu J, Burket MW, Pandya U, Colyer W, Eltahawy E, Cooper CJ. Stent fracture in the coronary and peripheral arteries. J. Interv. Cardiol. 2010, 23, 411–9.

[163]   Petrini L, Wu W, Dordoni E, Meoli A, Migliavacca F, Pennati G. Fatigue behavior characterization of nitinol for peripheral stents. Funct. Mater. Lett. 2012, 5, 216–23.

[164]   McKelvey AL, Ritchie RO. Fatigue-crack propagation in Nitinol, a shape-memory and superelastic endovascular stent material. J. Biomed. Mater. Res. 1999, 47, 301–8.

[165]   Pelton A, Schroeder V, Mitchell M, Gong XY, Barney M, Robertson S. Fatigue and durability of Nitinol stents. J. Mech. Behav. Biomed. Mater. 2008, 1, 153–64.

[166]   Figueiredo AM, Modenesi P, Buono V. Low-cycle fatigue life of superelastic NiTi wires. Int. J. Fatigue 2009, 31, 751–8.

[167]   Wang XM, Zhou QT, Liu H, Deng CH, Yue ZF. Experimental study of the biaxial cyclic behavior of thin-wall tubes of NiTi shape memory alloys. Metall. Mater. Trans. A 2012, 43, 4123–8.

[168]   Pelton AR, Fino-Decker J, Vien L, Bonsignore C, Saffari P, Launey M, Mitchell MR. Rotary-bending fatigue characteristics of medical-grade Nitinol wire. J. Mech. Behav. Biomed. Mater. 2013, 27, 19–32.

[169]   Robertson SW, Launey M, Shelley O, Ong O, Vien L, Senthilnathan K, Saffari P, Schlegel S, Pelton AR. A statistical approach to understand the role of inclusions on the fatigue resistance of superelastic Nitinol wire and tubing. J. Mech. Behav. Biomed. Mater. 2015, 51, 119–31.

[170]   Gupta S, Pelton AR, Weaver JD, Gong X-Y, Nagaraja S. High compressive pre-strains reduce the bending fatigue life of nitinol wire. J. Mech. Behav. Biomed. Mater. 2015, 44, 96–108.

[171]   Saikrishnak CN, Ramaiah KV, Paul D, Bhaumik SK. Enhancement in fatigue life of NiTi shape memory alloy thermal actuator wire. Acta Mater. 2016, 102, 385–96.

[172]   Bernard S, Krishna Balla V, Bose S, Bandyopadhyay A. Compression fatigue behavior of laser processed porous NiTi alloy. J. Mech. Behav. Biomed. Mater. 2012, 13, 62–8.

[173]   Baxevanis T, Lagoudas DC. Fracture mechanics of shape memory alloys: review and perspectives. Int. J. Fract. 2015, 191;doi: 10.1007/s10704-015-9999-z.

[174]   Maletta C, Furgiuele FM, Sgambitterra E. Fracture mechanics of pseudoelastic NiTi alloys: review of the research activities carried out at University of Calabria. Frattura ed Integrità Strutturale, 2012, 23, 13–24; doi: 10.3221/IGF-ESIS.23.02.

[175] Maletta C. A novel fracture mechanics approach for shape memory alloys with trilinear stress–strain behavior. Int. J. Fract. 2012, 177, 39–51.

[176] Wang GZ, Xuan FZ, Tu ST, Wang ZD. Effects of triaxial stress on martensite transformation, stress–strain and failure behavior in front of crack tips in shape memory alloy NiTi. Mater. Sci. Eng. A 2010, 527, 1529–36.

[177] Hazar S, Anlas G, Moumni Z. Evaluation of transformation region around crack tip in shape memory alloys. Int. J. Fract. 2016, 197, 99–110.

[178] Lexcellent C, Thiebaud F. Determination of the phase transformation zone at a crack tip in a shape memory alloy exhibiting asymmetry between tension and compression. Scr. Mater. 2008, 59, 321–3.

[179] Lexcellent C, Laydi MR, Taillebot V. Analytical prediction of the phase transformation onset zone at a crack tip of a shape memory alloy exhibiting asymmetry between tension and compression. Int. J. Fract. 2011, 169, 1–13.

[180] Maletta C, Falvo A, Furgiuele F, Leonardi A. Stress-induced martensitic transformation in the crack tip region of a NiTi alloy. J. Mater. Eng. Perform 2009, 18, 679–85.

[181] Birman V. On mode I fracture of shape memory alloy plates. Smart Mater. Struct. 1998, 7, 433–7.

[182] Maletta C, Sgambitterra E, Niccoli F. Temperature dependent fracture properties of shape memory alloys: novel findings and a comprehensive model. Sci. Rep. 2016, 6; doi: 10.1038/s41598-016-0024-1.

[183] Sgambitterra E, Maletta C, Furgiuele F. Temperature dependent local phase transformation in shape memory alloys by nanoindentation. Scr. Mater. 2015, 101, 64–7.

[184] Maletta C, Furgiuele F. Fracture control parameters for NiTi based shape memory alloys. Int. J. Solids Struct. 2011, 48, 1658–64.

[185] Robertson SW, Pelton AR, Ritchie RO. Mechanical fatigue and fracture of Nitinol. J. Int. Mater. Rev. 2012, 57, 1–37; doi: 10.1179/1743280411y.0000000009.

[186] Robertson SW, Ritchie RO. A fracture-mechanics-based approach to fracture control in biomedical devices manufactured from superelastic nitinol tube. J. Biomed Mater. Res. Part B Appl. Biomater. 2008, 84, 26–33.

[187] McKelvey AL, Ritchie RO. Fatigue-crack propagation in Nitinol, a shape-memory and superelastic endovascular stent material. J. Biomed. Mater. Res. 1999, 47, 301–8.

[188] Gall K, Yang N, Sehitoglu H, Chumlyakov YI. Fracture of precipitated NiTi shape memory alloys. Int. J. Fract. 2001, 109, 189–207.

[189] Gall K, Maier HJ. Cyclic deformation mechanisms in precipitated NiTi shape memory alloys. Acta Mater. 2002, 50, 4643–57.

[190] Chen JH, Wang GZ, Sun W. Investigation on the fracture behavior of shape memory alloy NiTi. Metall. Mater. Trans. A 2005, 36, 941–55.

[191] Kasiri S, Kelly DJ, Taylor D. Can the theory of critical distances predict the failure of shape memory alloys? Comput. Methods Biomech. Biomed. Eng. 2011, 14, 491–6.

[192] Hsu W-N, Polatidis E, Šmíd M, Petegem SV, Casati N, Swygenhoven HV. Deformation and degradation of superelastic NiTi under multiaxial loading. Acta Mater. 2019, 167, 149–58.

[193] Baxevanis T, Landis CM, Lagoudas DC. On the fracture toughness of pseudoelastic shape memory alloys. J. Appl. Mech. 2013, 81; https://doi.org/10.1115/1.4025139.

[194] Jape S, Baxevanis T, Lagoudas DC. On the fracture toughness and stable crack growth in shape memory alloy actuators in the presence of transformation-induced plasticity. Int. J. Fract. 2018, 209, 117–30.

[195] Boulbitch AA, Korzhenevskii AL, Self-oscillating regime of crack propagation induced by a local phase transition at its tip. Phys. Rev. Lett. 2011, 107; https://www.ncbi.nlm.nih.gov/pubmed/21929176.

[196] Daly S, Miller A, Ravichandran G, Bhattacharya K. An experimental investigation of crack initiation in thin sheets of nitinol. Acta Mater. 2007, 55, 6322–30.

[197] Daymond MR, Young ML, Almer JD, Dunand DC. Strain and texture evolution during mechanical loading of a crack tip in martensitic shape-memory NiTi. Acta Mater. 2007, 55, 3929–42.

[198] Maletta C, Furgiuele F. Analytical modeling of stress-induced martensitic transformation in the crack tip region of nickel–titanium alloys. Acta Mater. 2010, 58, 92–101.

[199] Yi S, Gao S. Fracture toughening mechanism of shape memory alloys due to martensite transformation. Int. J. Solids Struct. 2000, 37, 5315–27.

[200] Yi S, Gao S, Shen L. Fracture toughening mechanism of shape memory alloys under mixed-mode loading due to martensite transformation. Int. J. Solids Struct. 2001, 38, 4463–76.

[201] Yan WY, Wang CH, Zhang XP, Mai YW. Effect of transformation volume contraction on the toughness of superelastic shape memory alloys. Smart Mater. Struct. 2002, 11, 947–55.

[202] Kashef S, Asgari A, Hilditch TB, Yan W, Goel VK, Hodgson PD. Fracture toughness of titanium foams for medical applications. Mater. Sci. Eng. A 2010, 527, 7689–93.

[203] Ahadi A, Sun Q. Grain size dependence of fracture toughness and crack-growth resistance of superelastic NiTi. Scr. Mater. 2016, 113, 171–5.

[204] Haghgouyan B, Hayrettin C, Baxevanis T, Karaman I, Lagoudas DC. Fracture toughness of NiTi – towards establishing standard test methods for phase transforming materials. Acta Mater. 2019, 162, 226–38.

[205] Vaidyanathan DC, Dunand DC, Ramamurty U. Fatigue crack-growth in shape-memory NiTi and NiTi–TiC composites. Mater. Sci. Eng., A 2000, 289, 208–16.

[206] Alapati SB, Brantley WA, Svec TA, Powers JM, Nusstein JM, Daehn GS. SEM observations of nickel-titanium rotary endodontic instruments that fractured during clinical use. J. Endod. 2005, 31, 40–3.

[207] Spanaki-Voreadi AP, Kerezoudis NP, Zinelis S. Failure mechanism of PT Ni–Ti rotary instruments during clinical use: fractographic analysis. Int. Endod. J. 2006, 39, 171–8.

[208] Shen Y, Cheung GS, Peng B. Haapasalo M. Defects in nickel-titanium instruments after clinical use. Part 2: fractographic analysis of fractured surface in a cohort study. J. Endod. 2009, 35, 133–6.

[209] Haikel Y, Gasser P, Allemann C. Dynamic fracture of hybrid endodontic hand instruments compared with traditional files. J. Endod. 1991, 17, 217–20.

[210] Tepel J, Schäfer E, Hoppe W. Properties of endodontic hand instruments used in rotary motion. Part 3. Resistance to bending and fracture. J. Endod. 1997, 23, 141–5.

[211] McGuigan MB, Louca C, Duncan HF. Endodontic instrument fracture: causes and prevention. Br. Dent. J. 2013, 214, 341–8.

[212] Zupanc J, Vahdat-Pajouh N, Schäfer E. New thermomechanically treated NiTi alloys – a review. Int. Endod. J. 2018, 51, 1088–103.

[213] Gil J, Rupérez E, Velasco E, Aparicio C, Manero JM. Mechanism of fracture of NiTi superelastic endodontic rotary instruments. J. Mater. Sci. Mater. Med. 2018, 29; doi: 10.1007/s10856-018-6140-7.

[214] Spili P, Parashos P, Messer HH. The impact of instrument fracture on outcome of endodontic treatment. J. Endod. 2005, 31, 845–50.

[215] Kottoor J, Velmurugan N, Gopikrishna V, Krithikadatta J. Effects of multiple root canal usage on the surface topography and fracture of two different Ni-Ti rotary file systems. Indian J. Dent. Res. 2013, 24, 42–7.

[216] O'Hoy PY, Messer HH, Palamara JE. The effect of cleaning procedures on fracture properties and corrosion of NiTi files. Int. Endod. J. 2003, 36, 724–32.

[217]   Coelho MS, de Azevêdo Rios M, da Silveira Bueno CE. Separation of nickel-titanium rotary and reciprocating instruments: a mini-review of clinical studies. Open Dent. J. 2018, 12, 864–72.

[218]   Sattapan B, Nervo GJ, Palamara JE, Messer HH. Defects in rotary nickel-titanium files after clinical use. J. Endod. 2000, 26, 161–5.

[219]   Arens FC, Hoen MM, Steiman HR, Dietz GC Jr. Evaluation of single-use rotary nickel-titanium instruments. J. Endod. 2003, 29, 664–6.

[220]   Parashos P, Gordon I, Messer HH. Factors influencing defects of rotary nickel-titanium endodontic instruments after clinical use. J. Endod. 2004, 30, 722–5.

[221]   Parashos P, Messer HH. Rotary NiTi instrument fracture and its consequences. J. Endod. 2006, 32, 1031–43.

[222]   Berutti E, Angelini E, Rigolone M, Migliaretti G, Pasqualini D. Influence of sodium hypochlorite on fracture properties and corrosion of PT rotary instruments. Int. Endod. J. 2006, 39, 693–9.

[223]   Wolcott S, Wolcott J, Ishley D, Kennedy W, Johnson S, Minnich S, Meyers J. Separation incidence of protaper rotary instruments: a large cohort clinical evaluation. J. Endod. 2006, 32, 1139–41.

[224]   Shen Y, Cheung GS, Bian Z, Peng B. Comparison of defects in ProFile and PT systems after clinical use. J. Endod. 2006, 32, 61–5.

[225]   Spanaki-Voreadi AP, Kerezoudis NP, Zinelis S. Failure mechanism of PT Ni-Ti rotary instruments during clinical use: fractographic analysis. Int. Endod. J. 2006, 39, 171–8.

[226]   Di Fiore PM. A dozen way to prevent nickel-titanium rotary instrument fracture. J. Am. Dent. Assoc. 2007, 138, 196–201.

[227]   Inan U, Gonulol N. Deformation and fracture of Mtwo rotary nickel-titanium instruments after clinical use. J. Endod. 2009, 35, 1396–9.

[228]   Varela-Patiño P, Ibañez-Párraga A, Rivas-Mundiña B, Cantatore G, Otero XL, Martin-Biedma B. Alternating versus continuous rotation: a comparative study of the effect on instrument life. J. Endod. 2010, 36, 157–9.

[229]   Stojanac I, Drobac M, Petrovic L, Atanackovic T. Predicting in vivo failure of rotary nickel-titanium endodontic instruments under cyclic fatigue. Dent. Mater. J. 2012, 31, 650–5.

[230]   Sotokawa T. An analysis of clinical breakage of root canal instruments. J. Endod. 1998, 14, 75–82.

[231]   Pruett JP, Clement DJ, Carnes DL Jr. Cyclic fatigue testing of nickel-titanium endodontic instruments. J. Endod. 1997, 23, 77–85.

[232]   van der Vyver PJ, Jonker C. Reciprocating instruments in endodontics: a review of the literature. S. Afr. Dent. J. 2014, 69; http://www.scielo.org.za/scielo.php?script=sci_arttext&pid=S0011-85162014000900008.

[233]   Serene TP, Adams JD, Saxena A. Nickel-titanium instruments: application in endodontics. J. Endod. 1995, 2, 92–4.

[234]   Yared G. Canal preparation using only one Ni-Ti rotary instrument: preliminary observations. Int. Endod. J. 2008, 41, 339–44.

[235]   You SY, Bae KS, Baek SH, Kum KY, Shon WJ, Lee W. Lifespan of one nickel-titanium rotary file with reciprocating motion in curved root canals. J. Endod. 2010, 36, 1991–4.

[236]   Varela-Patiño P, Martin Biedma B, Rodriguez N, Cantatore G, Malentaca A, Ruiz-Pinon M. Fracture rate of nickel-titanium instruments using continuous versus alternating rotation. Endod. Pract. Today 2008, 2, 193–7.

[237]   Sattapan B, Palmara JE, Messer HH. Torque during canal instrumentation using rotary nickel-titanium files. J. Endod. 2000, 26, 156–60.

[238]   De-Deus G, Moreira EJ, Lopes HP, Elias CN. Extended cyclic fatigue life of F2 PT instruments used in reciprocating movement. Int. Endod. J. 2010, 43, 1063–8.

[239] Troian CH, Só MV, Figueiredo JA, Oliveira EP. Deformation and fracture of RaCe and K3 endodontic instruments according to the number of uses. Int. Endod. J. 2006, 39, 616–25.
[240] Berutti E, Paolino DS, Chiandussi G, Alovisi M, Cantatore G, Castellucci A, Pasquali D. Root canal anatomy preservation of WaveOne reciprocating files with or without glide path. J. Endod. 2012, 38, 101–4.
[241] Berutti E, Chiandussi G, Paolino DS, Scott N, Cantatore G, Castellucci A, Pasqualini D. Canal shaping with WaveOne primary reciprocating files and PT system: a comparative study. J. Endod. 2012, 38, 505–9.
[242] Bueno CSP, de Oliveira DP, Pelegrine RA, Fontana CE, Rocha DGP, Bueno CEDS. Fracture incidence of WaveOne and reciproc files during root canal preparation of up to 3 posterior teeth: a prospective clinical study. J. Endod. 2017, 43, 705–8.
[243] Klymus ME, Alcalde MP, Vivan RR, Só MVR, de Vasconselos BC, Duarte MAH. Effect of temperature on the cyclic fatigue resistance of thermally treated reciprocating instruments. Clin. Oral Invest. 2019, 23, 3047–52.
[244] Bahcall JK, Carp S, Miner M, Skidmore L. The causes, prevention, and clinical management of broken endodontic rotary files. Dent. Today 2005, 24, 74, 76, 78–80.
[245] Hansen JR, Beeson TJ, Ibarrola JL. Case series: tooth retention 5 years after irretrievable separation of Lightspeed LSX instruments. J. Endod. 2013, 39, 1467–70.
[246] Panitvisai P, Parunnit P, Sathorn C, Messer HH. Impact of a retained instrument on treatment outcome: a systematic review and meta-analysis. J. Endod. 2010, 36, 775–80.
[247] Shen Y, Peng B, Cheung GS. Factors associated with the removal of fractured NiTi instruments from root canal systems. Oral Surg. Oral Med. Oral Pathol. Oral Radiol. Endod. 2004, 98, 605–10.
[248] Hülsmann M, Schinkel I. Influence of several factors on the success or failure of removal of fractured instruments from the root canal. Endod. Dent. Traumatol. 1999, 15, 252–8.
[249] Suter B, Lussi A, Sequeira A. Probability of removing fractured instruments from root canals. Int. Endod. J. 2005, 38, 112–23.
[250] Rosen E, Azizi H, Friedlander C, Taschieri S, Tsesis I. Radiographic identification of separated instruments retained in the apical third of root canal-filled teeth. J. Endod. 2014, 40, 1549–52.
[251] Ward JR, Parashos P, Messer HH. Evaluation of an ultrasonic technique to remove fractured rotary nickel-titanium endodontic instruments from root canals: an experimental study. J. Endod. 2003, 29, 756–63.
[252] Souter NJ, Messer HH. Complications associated with fractured file removal using an ultrasonic technique. J. Endod. 2005, 31, 450–2.
[253] Alomairy KH. Evaluating two techniques on removal of fractured rotary nickel-titanium endodontic instruments from root canals: an in vitro study. J. Endod. 2009, 35, 559–62.
[254] Wohlgemuth P, Cuocolo D, Vandrangi P, Sigurdsson A. Effectiveness of the gentlewave system in removing separated instruments. J. Endod. 2015, 41, 1895–8.
[255] Hirayama S, Kimura Y, Amano Y. Basic research on removal of H-file fragments by ultrasonic equipment. J. Jpn. Soc. Endod. 2010, 31, 194–9.
[256] Madarati AA. Temperature rise on the surface of NiTi and stainless steel fractured instruments during ultrasonic removal. Int. Endod. J. 2015, 48, 872–7.
[257] Arslan H, Yıldız ED, Taş G, Akbıyık N, Topçuoğlu HS. Duration of ultrasonic activation causing secondary fractures during the removal of the separated instruments with different tapers. Clin. Oral Invest. 2019, May, 5 pages; doi: 10.1007/s00784-019-02936-7.
[258] Yang Q, Shen Y, Huang D, Zhou X, Gao Y, Haapasalo M. Evaluation of two trephine techniques for removal of fractured rotary nickel-titanium instruments from root canals. J. Endod. 2017, 43, 116–20.

[259]   Ormiga F, da Cunha Ponciano Gomes JA, de Araújo MC. Dissolution of nickel-titanium endodontic files via an electrochemical process: a new concept for future retrieval of fractured files in root canals. J. Endod. 2010, 36, 717–20.

[260]   Sumitake K, Yoshioka T, Kitajima K, Igarashi M. Consideration from the case of accidental ingestion of instruments during root canal treatment. J. Jpn. Soc. Endod. 2015, 36, 126–9.

[261]   Susini G, Pommel L, Camps J. Accidental ingestion and aspiration of root canal instruments and other dental foreign bodies in a French population. Int. Endod. J. 2007, 40, 585–9.

[262]   Nevares G, Cunha RS, Zuolo ML, Bueno CE. Success rates for removing or bypassing fractured instruments: a prospective clinical study. J. Endod. 2012, 38, 442–4.

[263]   Madarati AA, Hunter MJ, Dummer PM. Management of intracanal separated instruments. J. Endod. 2013, 39, 569–81.

[264]   Lambrianidis T. Therapeutic Options for the Management of Fractured Instruments. In: Lambrianidis, ed., Management of Fractured Endodontic Instruments, 2018, 75–195.

[265]   Zinelis S, Eliades T, Pandis N, Eliades G, Bourauel C. Why do nickel-titanium archwires fracture intraorally? Fractographic analysis and failure mechanism of in-vivo fractured wires. Am. J. Orthod. Dentofacial Orthop. 2007, 132, 84–9.

[266]   Guzman U, Jerrold L, Abdelkarim A. An in vivo study on the incidence and location of fracture in round orthodontic archwires. J. Orthod. 2013, 40, 307–12.

[267]   Yokoyama K, Hamada K, Moriyama K, Asaoka K. Degradation and fracture of Ni-Ti superelastic wire in an oral cavity. Biomaterials 2001, 22, 2257–62.

[268]   Perinetti G, Contardo L, Ceschi M, Antoniolli F, Franchi L, Baccetti T, Di Lenarda R. Surface corrosion and fracture resistance of two nickel-titanium-based archwires induced by fluoride, pH, and thermocycling. An in vitro comparative study. Eur. J. Orthod. 2012, 34, 1–9.

[269]   Rerhrhaye W, Bahije L, El Mabrouk K, Zaoui F, Marzouk N. Degradation of the mechanical properties of orthodontic NiTi alloys in the oral environment: an in vitro study. Int. Orthod. 2014, 12; doi:10.1016/j.ortho.2014.06.006.

[270]   Bock JJ, Bailly J, Gernhardt CR, Fuhrmann AW. Fracture strength of different soldered and welded orthodontic joining configurations with and without filling material. J. Appl. Oral Sci. 2008, 16, 328–35.

[271]   Ma J, Karaman I, Noebe RD. High temperature shape memory alloys. Intern. Mater. Rev. 2010, 55, 257–315.

[272]   Raj SV, Noebe RD. Low temperature creep of hot-extruded near-stoichiometric NiTi shape memory alloy part I: isothermal creep. NASA/TM – 2013-217888/PART1, 2013, 16 pages; https://ntrs.nasa.gov/archive/nasa/casi.ntrs.nasa.gov/20130014001.pdf.

[273]   Eggeler GF, Neuking K, Dlouhy A, Kobus E. Creep behavior of NiTi shape memory alloys and the effect of pre-creep on the martensitic phase transformation. Mater. Sci. Forum 1999, 327/328, 183–6.

[274]   Takeda K, Tobushi H, Pieczyska EA. Transformation-induced creep and creep recovery of shape memory alloy. Materials 2012, 5, 909–21.

[275]   Dunić V, Pieczyska EA, Kowalewski ZL, Matsui R, Slavković R. Experimental and numerical investigation of mechanical and thermal effects in TiNi SMA during transformation-induced creep phenomena. Materials (Basel) 2019, 12, pii: E883; doi: 10.3390/ma12060883.

[276]   Richman RH, Rao AS, Hodgson DE. Cavitation erosion of two NiTi alloys. Wear 1992, 157, 401–7.

[277]   Wu SK, Lin HC, Yeh CH. A comparison of the cavitation erosion resistance of TiNi alloys, SUS304 stainless steel and Ni-based self-fluxing alloy. Wear 2000, 244, 85–93.

[278]   Cheng FT, Shi P, Man HC. NiTi cladding on stainless steel by TIG surfacing process: part I. Cavitation erosion behavior. Scr. Mater. 2001, 45, 1083–9.

[279]   Richman RH, Rao AS, Kung D. Cavitation erosion characteristics of a NiTi alloy. Wear 1995, 181/183, 80–5.

[280]   He JL, Won KW, Chang CT, Chen KC, Lin HC. Cavitation erosion characteristics of a NiTi alloy.
        Wear 1999, 233/235, 104–10.
[281]   Liu W, Zheng Y, Yao ZM, Liu CS. Cavitation erosion characteristics of a NiTi alloy. Metall.
        Mater. Trans. A 2004, 35(1), 356–62.
[282]   Yang LM, Tieu AK, Dunne DP, Huang SW, Li HJ, Wexler D, Jiang ZY. Cavitation erosion
        resistance of NiTi thin films produced by filtered arc deposition. Wear 2009, 267, 233–43.

Chapter 10
Nonmedical applications

For medical applications of NiTi alloys, the functional and/or operational temperature should be around human body's temperature (37 °C) or, for a short time of period, it should not exceed the critical temperature of 42.5 °C (above which the intracanal tissue could be subjected to the irreversible damage during the endodontic treatment [1]). If NiTi alloy is constrained to shape change upon phase transformation, then 700 MPa stresses can be generated, which are too much as compared to shape-memory polymers where the amount of stress is much smaller. Due to the possibility of large recoverable strain of about 8% without force generation and 700 MPa stress without recoverable strain, there is a high possibility to use NiTi shape-memory alloy (SMA) for the design of components with different strain outputs and different amounts of external work output [2]. Because of their remarkable properties, SMA can be used in a large number of nonmedical applications [3, 4]. SMA can solve problems in the aerospace industry, especially those related to vibration control of slender structures and solar panels, and nonexplosive release devices [5, 6]. Micromanipulators and robotic actuators have been employed in order to mimic the smooth movement of human muscles [7]. SMA are commonly used as external actuators or as SMA fibers embedded in a composite matrix so that they can alter the mechanical properties of slender structures for the control of buckling and vibration [8].

10.1 High-temperature shape-memory alloys

SMAs with stable reverse transformation temperatures above 120 °C can be considered as the high-temperature shape-memory alloys (HTSMAs. Due to the lack of minimum quality standards for stability, ductility, functional behavior, and reliability, although there are numerous proposed applications reported, not many successful applications have been realized so far. Such potential nonmedical applications include household electric appliances, engine component for aircrafts, aerospace industries, automotive, nuclear power plant, manufacturing and energy exploration industries, and so on. There are several notable remarks mentioned on the HTSMAs: (i) after suitable deformation and processing, a shape change is observed while heating the alloy through the temperature interval from 175 to 190 °C, (ii) this shape change can be completely reversed during subsequent cooling from 155 to 125°C, (iii) the magnitude of the reversible strain produced by this alloy is 1.5%; somewhat higher strains can be achieved if lower memory temperatures can be accepted, and conversely, better high-temperature capabilities can be achieved by accepting smaller reversible strains, and (iv) the memory effects in this alloy have been found to be unaffected

https://doi.org/10.1515/9783110666113-010

by short over heating to temperatures as high as 300 °C [9]. In Recent years, R&D in the HTSMAs materials has been remarkably advanced. Ma et al. [10] categorized all HTSMAs into three groups based on their martensitic transformation temperatures: group I, transformation temperatures in the range of 100–400 °C; group II, in the range of 400–700 °C; and group III, above 700 °C.

Atli et al. [11] treated a $Ti_{49.5}Ni_{25}Pd_{25}Sc_{0.5}$ HTSMA thermo-mechanically to improve the dimensional stability upon repeated thermal cycles under constant loads. This is accomplished using severe plastic deformation via equal channel angular extrusion and postprocessing annealing heat treatments. It was reported that (i) the results of the thermomechanical experiments reveal that the processed materials display enhanced shape-memory response, exhibiting higher recoverable transformation and reduced irrecoverable strain levels upon thermal cycling compared with the unprocessed material, and (ii) this improvement is attributed to the increased strength and resistance of the material against defect generation upon phase transformation as a result of the microstructural refinement due to the equal channel angular extrusion process. Karaca et al. [12] characterized shape-memory properties of a $Ni_{50.3}Ti_{29.7}Hf_{20}$ (at%) polycrystalline alloy. It was reported that (i) precipitation was found to alter the martensite morphology and significantly improve the shape-memory properties of the Ni-rich NiTiHf alloy, (ii) for the peak aged condition shape-memory strains of up to 3.6%, the lowest hysteresis, and a fully reversible superelastic (SE) response were observed at temperatures up to 240 °C, and (iii) in general, the nickel-rich NiTiHf polycrystalline alloy exhibited a higher work output ($\approx$16.5 J/cm^3) than other NiTi-based high-temperature alloys. Karakoc et al. [13] investigated the effects of upper cycle temperature (UCT) on the actuation fatigue response of nanoprecipitation-hardened $Ni_{50.3}Ti_{29.7}Hf_{20}$ HTSMA. It was found that (i) the samples subjected to 300 °C UCT exhibit fatigue lives twice that of the samples with 350 °C UCT, and those tested under 300 MPa with 300 °C UCT withstand more than 10,000 cycles with actuation strains of 2–3%, (ii) actuation strains remained constant or increased with thermal cycling in the 350 °C UCT experiments, while those subjected to 300 °C UCT exhibited decreasing actuation strains, at all stress levels, (iii) in the 300 °C UCT experiments, partial martensitic transformation becomes operative, resulting in a reduction of actuation strain in each cycle, and postpones damage accumulation, and (iv) in the 350 °C UCT cases, increase of actuation strain is attributed to the partial recovery of cyclically induced remnant deformation at higher temperatures and as a result, larger volume of transforming material which, in turn, accelerates the formation of cracks that open and close during thermal cycling and reversible martensitic transformation, and manifests itself as additional recoverable strain in each cycle.

Evirgen et al. [14] studied the relationship between the crystallographic compatibility of austenite and martensite phases and the transformation thermal hysteresis of Ni-rich $Ni_{50.3}Ti_{29.7}Hf_{20}$ and $Ni_{50.3}Ti_{29.7}Zr_{20}$ alloys undergoing B2–B19' martensitic transformation as a function of microstructure. It was reported that an experimental linear relationship of ΔT versus $\lambda 2$ (the second eigenvalue of the transformation stretch matrix) was observed for these NiTi(Hf/Zr) alloys, but with a shallower slope as compared to the universal behavior followed by alloys showing B2–B19 martensitic phase transformation. Elahinia et al. [15] fabricated a NiTiHf$_{20}$ HTSMA by selective laser melting technique using NiTiHf powder. It was mentioned that (i) transformation temperature was found to be above 200 °C and slightly lower due to the additional oxygen pick up from the gas atomization and melting process, and (ii) the shape-memory response in compression was measured for stresses up to 500 MPa, and transformation strains were found to be very comparable (up to 1.26% for as-extruded; up to 1.52% for selective laser melting). Casalena et al. [16] studied the Ni$_{54}$Ti$_{45}$Hf$_1$ HTSMA and reported that (i) it exhibits strengths more than 40% greater than those of conventional NiTi-based SMAs – 2.5 GPa in compression and 1.9 GPa in torsion – and retains those strengths during cycling, (ii) the SE hysteresis is very small and stable with cycling, (iii) aging treatments are used to induce a very high density of Ni$_4$Ti$_3$precipitates, which impede plasticity during cycling yet do not impart substantial dissipation to the reversibility of the phase transformation, (iv) a combination of small, untwinned retained martensite laths, and dislocations on the austenite–martensite interfaces primarily strengthen the alloy as opposed to dislocation networks, and (v) some combination of nanoprecipitation and interface dislocations is responsible for the remarkably low mechanical hysteresis exhibited by this material. Babacan et al. [17] investigated the effects of cold and warm rolling on the shape-memory response and thermomechanical cyclic stability of Ni$_{50}$Ti$_{30}$Hf$_{20}$ HTSMA. Cold rolling without intermediate annealing was performed up to the maximum possible thickness reduction before surface cracks appeared. These samples were subsequently annealed at various temperatures. 550 °C was determined to be the optimum annealing temperature for 30 min durations based on the differential scanning calorimetry and microhardness test results. It was found that (i) the rolling led to an increase in the resistance against defect generation accompanying martensitic transformation in this alloy, resulting in a significant improvement in dimensional stability during thermal cycling, (ii) 15% cold rolling followed by the 550 °C 30 min annealing condition showed the best actuation response, exhibiting comparable recoverable transformation levels to the hot extruded samples, with much lower residual strains, (iii) all warm rolled samples exhibited notable two-way SME with compressive two-way shape-memory strains, pointing out the existence of compressive internal stress storage following warm rolling, and (iv) while transformation temperatures of all rolled samples were lower than those of the starting hot extruded sample, thermal hysteresis was notably higher in the rolled samples being attributed to the increase in dislocation density and the change in the martensite microstructure with rolling. It was also mentioned that NiTiHf HTSMAs that have attracted recent interest in high-temperature applications can be processed by using conventional rolling methods while preserving desired cyclic shape-memory response. Hayrettin et al. [18] investigated the two-way shape-memory effect (SME) in nanoprecipitation hardened, Ni$_{50.3}$Ti$_{29.7}$Hf$_{20}$ thin walled tubes and its thermal stability. It was reported that (i) under 200 MPa, 600 thermal cycles were sufficient to reach a two-way shape-memory strain as high as 2.95%, which was shown to be stable upon annealing up to 400 °C for 30 min, (ii) this two-way shape memory strain was 85% of the maximum actuation strain measured under 200 MPa (iii) the microstructure after thermo-mechanical training was investigated using transmission electron microscopy, which did not indicate a significant change in precipitate structure and size after the training; however, small amount of remnant austenite was revealed at 100 °C below the martensite finish temperature, with notable amount of dislocations, (iv) nanoprecipitation-hardened $Ni_{50.3}Ti_{29.7}Hf_{20}$ shows relatively high two-way shape-memory strain and stable actuation response after much less number of training cycles as compared to binary NiTi and nickel lean NiTiHf compositions, and (v) tube wall thickness and training stress levels have been found to have negligible effect on shape-memory strains and number of cycles to reach the desired training level, for the ranges studied.

10.2 Applications

There are two typical types of materials: needs-oriented material (such as stainless steel for searching of steels that exhibit excellent resistance against staining) and seeds-oriented material and technology, to the latter nickel–titanium materials are most typical materials. When looking at functionality and applicability associated with NiTi materials, it can be classified into application using (1) a single SME, (2) repeatable SME, (3) reversible SME, (4) omnidirectional SME, and (5) transformation pseudoelasticity. Among these, application using reversible SME is the most popular for actuators and energy-conversion materials from heat energy to mechanical energy.

In nonmedical applications, it is convenient to classify them into two groups, depending on either using SME phenomenon or SE one as follows. Application examples utilizing SME should include (1) antenna or pipe connector or pipe lock ring (see Figure 10.1) through the shape recovery upon heating, (2) actuator through stress induced by heating, (3) wind direction control for air conditioner through shape change by temperature, and (4) steam valve through the stress changes by temperature. Application examples using SE can include (1) stent or fishing string through the shape maintaining mechanism, (2) guide wire, core wire for retractable cellular phone's whip antenna (see Figure 10.2) through flexibility, and (3) eye glass frame or bra through constant stress level. In Figure 10.1, the change in shape is prevented so that the shape-memory tubular element can be used as a mechanical coupling to join two pipes together securely. Such couplings have replaced welding in hydraulic lines in aircraft and ships and in the repair of undersea pipelines.

Actually, the SE behavior, reversible strains of several percent during heating or cooling over a limited temperature range, generation of high recovery stresses, and a work output with a high power/weight ratio will bring versatile applications in industry and engineering field [21].

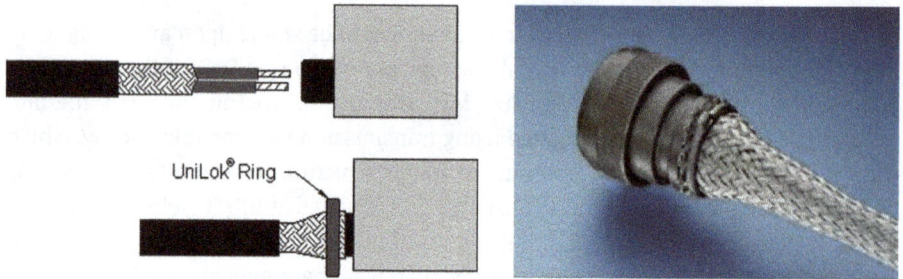

Figure 10.1: Pipe lock ring utilizing shape-memory effect of NiTi alloy [19, 20].

Figure 10.2: Core wire of NiTi alloy for cellular phone's antenna [20].

10.2.1 Actuators

An actuator is a device to convert energy into motion and depending on the energy source; there are generally two types of actuators: thermal actuator and electrical actuator. The actuator typically is a mechanical device that takes energy, which is usually created by air, electricity, or liquid, and converts it into some kinds of motion. Thermal actuators are mechanical systems that use the thermally induced expansion and contraction of materials as a mechanism for the creation of motion. These devices are compliant structures, using elastic deformation and mechanical constraints, which frequently are designed to amplify the motion generated by thermal expansion or contraction. Temperature changes that result in thermal actuation are most commonly provided by environmental changes or by Joule heating from electrical current flow. The electrical actuator is an electromechanical device that converts electrical energy into mechanical energy. Most electric actuators operate through the interaction of magnetic fields and current-carrying conductors to generate force. The reverse process, producing electrical energy from mechanical energy, is done by generators such as an alternator or a dynamo.

Pozzi et al. [22] investigated the martensitic transformation and its fundamental properties in the newly developed high-temperature alloys (e.g., NiTiHf) and mentioned that an equiatomic NiTi wires, which show poor sensing features and melt spun TiNiCu ribbons, with remarkably promising sensing/actuating characteristics. Tomozawa et al. [23] prepared TiNiCu/SiO$_2$ two-layer diaphragm-type microactuators by sputter deposition and micromachining. The influence of heat treatment temperature on the actuation behavior was investigated under quasi-static conditions. It was reported that (i) the reaction layer formed between the TiNiCu and SiO$_2$ layers and, preferentially grew into the SiO$_2$ side, (ii) the reaction layer formed at 750 °C mainly consisted of Ti$_4$(Ni,Cu)$_2$O, (iii) the maximum height of the diaphragm decreased with increasing heat treatment temperature, (iv) the growth of the reaction layer also affected the microstructure of the TiNiCu layer, (v) the density of fine platelets and Ti$_2$Ni precipitates decreased with increasing heat treatment temperature from 600 to 650 °C, and they disappeared at 700 °C due to the fact that the reaction layer mainly consisted of a Ti-rich phase, and (vi) the microactuator heat treated at 700 °C showed the highest transformation temperature with the lowest transformation temperature hysteresis, which is attractive for high speed actuation. Krulevitch et al. [24] mentioned that thin-film SMAs have the potential to become a primary actuating mechanism for mechanical devices with dimensions in the micron-to-millimeter range requiring large forces over long displacements. The work output per volume of thin-film SMA microactuators exceeds that of other microactuation mechanisms such as electrostatic, magnetic, thermal bimorph, piezoelectric, and thermopneumatic, and it is possible to achieve cycling frequencies on the order of 100 Hz due to the rapid heat transfer rates associated with thin-film devices [17]. Skrobanek et al. [25] developed devices consisted of stress-optimized microfabricated NiTi beam-cantilever with 100 μm thickness. It was mentioned that (i) for stress optimization, the lateral widths of the beams have been designed for homogeneous stress distributions without local stress maxima, allowing a maximum use of active material for bending actuation, (ii) as a result, higher work outputs for a given stress level and smaller hysteresis are obtained compared to nonoptimized actuators with parallel beam-cantilever devices, (iii) thus, improved hysteresis control is possible and better lifetime characteristics are expected, and (iv) for a strain limit of 2% the achieved work output is about 24 (mu) Nm. Karakoc et al. [26] studied the actuation fatigue behavior of a precipitation-hardened HTSMA Ni$_{50.3}$Ti$_{29.7}$Hf$_{20}$ undergoing thermal cycling between martensite phase and austenite phase under various tensile stress levels up to 500 MPa. It was reported that (i) a consistent increase in actuation strain concomitant with the applied load at the expense of fatigue life, (ii) significantly high number of cycles to failure were observed for this class of materials: specimens tested under 200 MPa achieved ~21,000 cycles with the average actuation strain of ~2.15% while those tested under 500 MPa experienced ~2,100 cycles to failure with the average actuation strain of 3.22%, and (iii) the actuation fatigue lives of the present HTSMAs exhibit an almost perfect power law correlation with average actuation work output. In the microscale the most suitable conformation of a SMA actuator is given by a planar wavy formed arrangement, that is, the snake-like shape, which allows high strokes, considerable forces, and devices with very low sizes. Nespoli et al. [27] laser-machined micro-snake-like actuators using a nanosecond pulsed fiber laser, starting from a 120-μm-thick NiTi sheet and mentioned laser machining has to be followed by some postprocesses in order to obtain a microactuator with good thermo-mechanical properties.

10.2.2 Heat engine

Nitinol material has been utilized as a rotary drive element in the heat engine. During the shape recovering process of NiTi alloy, geometrical change causes a rotational movement which can, as a result, generate energy. Heat source can be solar energy or waste heat, which the latter could be originated from back of a refrigerator. A way of utilizing shape recovery of NiTi wire as a rotary drive element of a solid state heat engine can be by (1) revering force of U-shaped wire upon heating, (2) contracting force of a stretched wire upon heating, or (3) contracting force of coil upon heating to what a coil-shape was previously trained and the coil was stretched at room temperature. It is shown that efficiencies on the order of 20% may be expected and can be improved with alloys having certain transformation hysteresis loop characteristics [28]. Measurements of engine power cycled through 12 and 80 °C water baths were as high as 3 W/kg, which compare favorably with more complex SME heat engine designs [29].

Figure 10.3 explains a solar hot water Nitinol heat engine [30]. In this design the springs are heated by sunlight and cooled by dipping into a tank of water as the wheel spins. This engine should put out somewhere around the equivalent of 40 W of power.

10.2.3 Seismic and vibration application

SMAs exhibit peculiar thermomechanical, thermoelectrical, and thermochemical behaviors under mechanical, thermal, electrical, and chemical conditions. A stress-induced micromechanical phase transition occurs in SMAs that causes inelastic deformation and gives rise to a large energy-absorbing capacity. Because it is possible to achieve large hysteretic deformation in SMAs without incurring plastic deformation, SMAs have potential for use in earthquake-engineering passive damping schemes [31]. In particular, NiTi SMA possesses unique thermomechanical behaviors such as SME and SE, which have made them attractive candidates for structural vibration control applications [32, 33] SMAs can be also applied in bridges and buildings for seismic resistant design and retrofit [34–36]. NiTi wires are very

Front view   Side view

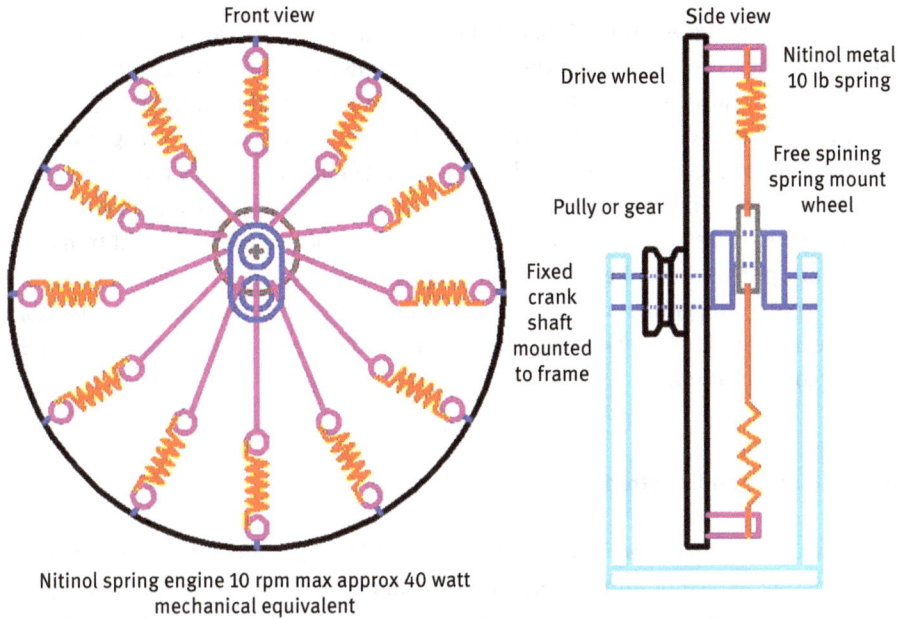

Drive wheel   Nitinol metal
                               10 lb spring

Pully or gear   Free spining
                               spring mount
                                   wheel

Fixed
crank
shaft
mounted
to frame

Nitinol spring engine 10 rpm max approx 40 watt
mechanical equivalent

Figure 10.3: Solar hot water Nitinol heat engine [30].

attractive for passive vibration control as they have a pseudoelastic property and can sustain large amounts (up to 10% strain) of inelastic deformation. Parulekar et al. [37] employed an SMA damper device made up of austenite NiTi wires as a passive energy absorber. Qian et al. [38] investigated the evaluation of an innovative energy dissipation system with SMAs for structural seismic protection. A recentering (or self-centering) SMA damper as SE nitinol wires as energy dissipation components [38–40]. It was reported that (i) the hysteretic behaviors of the damper can be modified to best fit the needs for passive structural control applications by adjusting the pretension of the nitinol wires, and the damper performance is not sensitive to frequencies greater than 0.5 Hz, and (ii) the simulation results indicate that SE SMA dampers are effective in mitigating the structural response of building structures subjected to strong earthquakes.

The adaptive-passive vibration absorber using SMAs shows promise for combining the stability and low complexity of passive tuned absorbers with the robust performance of active vibration control schemes [41]. NiTi alloy is commonly used for passive damping applications, in which the energy may be dissipated by the conversion from mechanical to thermal energy [42]. Mirzaeifar et al. [43] studied the pseudoelastic response of SMA helical springs under axial force and concluded that NiTi SMA helical springs can be used as energy dissipating devices, for example, for seismic applications. The mechanical vibrations/oscillations may become undesirable and may cause temporary and even irreversible damage to the system. There are several techniques to minimizing these vibration effects ranging from passive methods to the use of controllers with smart materials [44]. Moraes et al. [44] analyzed a passive vibration control system installed in a structure that simulates two-floor buildings and the system based on the incorporation of one SMA-SE coil springs configuration for energy dissipation and the addition of damping. It was reported that (i) as compared with the structure configuration with steel spring, the forced vibrations frequency response function analysis showed a reduction in displacement transmissibility of up to 51% for the first modal shape and 73% for the second mode in the SMA-SE coil spring configuration, and (ii) as for damping, there was a considerable increase in the order of 59% in the first mode and 119% in the second, for the SMA-SE springs configuration.

10.2.4 Battery and electrode

Luo et al. [45] mentioned that an equiatomic NiTi alloy has shown exceptional properties as a fixed potential liquid chromatography with electrochemical detector for carbohydrates and related substances and it exhibited excellent sensitivity and superior long-term stability compared to pure Ni, and investigated the role of Ti and the respective surface oxides of Ni and Ti in the catalytic stability of the detector. It was reported that (i) cyclic voltammetry results showed that Ti is initially oxidized, most likely to TiO_2 in 0.1 M NaOH solution, (ii) the oxidation of Ni to Ni^{2+} oxide also occurs at potentials close to that of Ti, (iii) at higher potentials in the range of +0.4 to +0.5 V versus Ag/AgCl reference, Ni^{3+} oxide undergoes further oxidation to the Ni^{3+} oxidation state, which is responsible for the catalysis of carbohydrates, amino acids, and other biosubstances, (iv) when NiTi and Ni are, repetitively, cyclic voltammetry cycled in the potential range of 0.0 to +0.6 V, a second wave appears at more negative potentials during the reverse cathodic scan for Ni but not for NiTi, and (v) XPS results for the nature of the surface oxides are consistent with oxidized Ti as TiO_2, Ni^{2+} predominantly as $Ni(OH)_2$, and Ni^{3+} possibly as NiOOH. Sato et al. [46] used NiTi alloy electrode as an electrochemical detector for the analysis of underivatized amino acids in flow systems. It was mentioned that (i) cyclic voltammetry experiments confirmed that electrogenerated $Ni^{3+}O(OH)$ functioned as the key redox mediator associated with the oxidation of the amine group in amino acids, (ii) the electrochemical behavior of the Ni–Ti electrode in alkaline medium was very similar to the Ni electrode; however, the oxide film was found to be much stable on NiTi than on Ni, and (iii) consequently, the Ni–Ti alloy electrode exhibited an excellent stability for constant-potential amperometric detection of amino acids in flow system, demonstrating that the NiTi alloy electrodes are evaluated to be very suitable for the amperometric detection of underivatized amino acids in anion-exchange chromatography. Kim et al. [47] studied the synthesis and electrochemical properties of oriented $NiO–TiO_2$

nanotube arrays as electrodes for supercapacitors. It was reported that (i) annealing the as-grown NT arrays to a temperature of 600 °C transformed them from an amorphous phase to a mixture of crystalline rock salt NiO and rutile TiO_2, (ii) changes in the morphology and crystal structure strongly influenced the electrochemical properties of the nanotube electrodes, (iii) electrodes composed of NT films annealed at 600 °C displayed pseudocapacitor (redox-capacitor) behavior, including rapid charge/discharge kinetics and stable long-term cycling performance, (iv) at similar film thicknesses and surface areas, the nanotube-based electrodes showed a higher rate capability than the randomly packed nanoparticle-based electrodes, (v) even at the highest scan rate (500 mV/s), the capacitance of the nanotube electrodes was not much smaller (within 12%) than the capacitance measured at the slowest scan rate (5 mV/s), and (vi) the faster charge/discharge kinetics of NT electrodes at high scan rates is attributed to the more ordered NT film architecture, which is expected to facilitate electron and ion transport during the charge–discharge reactions. Kwon et al. [48] fabricated the rapidly solidified Si–xTiNi ($x = 0.2$–0.45) alloy ribbons via melt spinning process. The thickness of the melt-spun ribbons was about 12.5 μm, and the sound section was selected for the experiment. It was found that (i) the charge/discharge energy capacity and electrochemical properties were significantly influenced by the relative ratio of NiTi to silicon, (ii) with increasing the total amount of Ni and Ti content up to 45 at%, the amount of $Si_7Ni_4Ti_4$ phase increased and the cycle performance was improved, and (iii) the $Si_7Ni_4Ti_4$ phase acted as a buffer for the volume expansion/contraction of Si occurring during the alloying and dealloying, and it could prevent a significant deterioration in cycle performance of the battery.

10.2.5 Others

Lah et al. [49] developed a prototype of a new shape-memory nitinol knitted fabric intended for use as an active thermal insulating interlining in firefighting protective clothing. Weft knitted fabrics were made from commercially available cold-worked nickel titanium alloy monofils. Knits were made on a manual knitting machine from a monofil measuring 0.1 mm in diameter, while a hand-made knit was prepared from a monofil measuring 0.2 mm in diameter. Nitinol fabrics were annealed at 500 °C to achieve an austenite transition temperature of 75 °C. A special constructed mold made of a steel frame and aluminum domes measuring 30 and 20 mm in height was used to give the nitinol fabrics a new temporary shape. It was reported that (i) a two-way, shape-memory effect (SME) of the nitinol fabrics was achieved using a 15-cycle training process, (ii) the achieved SME was tested in a heated chamber at 100 °C, where bulges measuring 12–25 mm in height occurred, and (iii) NiTi knits made from finer monofil were the most successful shape-memory knits. It was further mentioned that (i) when it was exposed to environmental temperatures of 75 °C and higher, it instantly changed its form from a two-dimensional shape to a three-dimensional shape, while increasing the air gap in the pocket, and (ii) a quilted fabric made from such a smart textile system could be used in firefighting protective clothing to locally improve thermal insulation and protect the human skin from overheating or burns.

Akalin et al. [50] investigated the wear characteristics of Al6061 composites, reinforced with short NiTi fibers, which were fabricated using pressure-assisted sintering process in ambient air where the NiTi fibers are aligned unidirectionally in the Al matrix. The wear tests were performed using a reciprocating tribometer in ball-on-flat configuration where the counter-body material was martensitic steel. It was reported that (i) transverse NiTi fibers improve the wear resistance significantly, (ii) samples with transverse fiber orientation show mostly abrasive wear, whereas, monolithic and parallel samples show adhesive wear mechanism, (iii) SiC reinforcements improve the wear resistance of the composite and the monolithic samples, and (iv) since the Al6061 matrix material is smeared onto NiTi fibers in a short period, all composite samples show similar frictional characteristics after certain period of running in dry sliding.

Sutapun et al. [51] introduced the application of NiTi SMA thin films in optical devices. Physical and optical properties of NiTi SMA thin films change as these films undergo phase transformation on heating. An optical beam can be modulated either mechanically with a NiTi actuator or by the changes that occur in NiTi optical properties upon heating and phase transformation. Reflection coefficients of NiTi films were measured in their so-called martensitic (room-temperature) and austenitic (elevated-temperature) phases. It was reported that (i) the reflection coefficients of the austenitic phase were higher than those of the martensitic phase by more than 45% in the wavelength range between 550 and 850 nm, (ii) a microfabricated NiTi diaphragm with a 0.26-mm-diameter hole was used as a prototype light valve, and (iii) the intensity of the transmitted light through the hole was reduced by 10–17% when the diaphragm was heated.

Among other versatile applications of NiTi alloys, the following applications can be found: as sports goods, due to the high damping capacity of SE NiTi, it is also used as a golf club insert or fishing line (<0.045 mm diameter); as household appliances, in a coffee maker, a shape-memory NiTi spring will open when the water boils to pour hot water to brew the coffee; and as spectacle eyeglass frames made of NiTi alloy can be found, which will return to their original shape if deformed through accidental damage (see Figure 10.4 [52]).

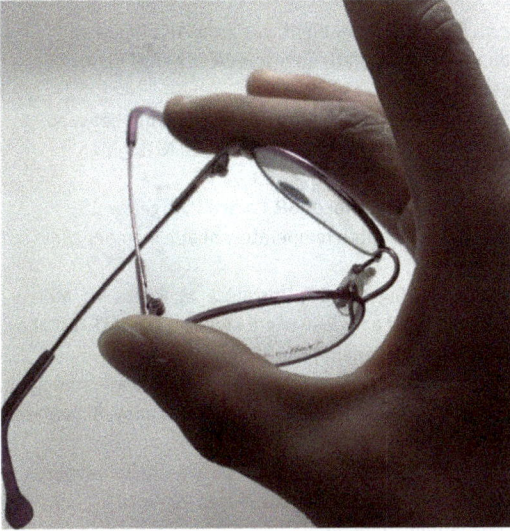

Figure 10.4: Flexible NiTi frames for eyeglasses [52].

References

[1] Tada E, Tominaga T, Yasukawa H, Oshida Y. Temperature increase during tooth whitening. Dentistry Pract. 2019, 2, 1–8.

[2] Otsuka SK, Wayman CM. Shape Memory Materials. Cambridge University Press, 1999.

[3] van Humbeeck J. Non-medical applications of shape memory alloys. Mater. Sci. Eng. A 1999, 273/275, 134–48.

[4] Schetky LMcD. The industrial applications of shape memory alloys in North America. Mater. Sci. Forum 2000, 327/328, 9–16.

[5] Denoyer KK, Erwin RS, Ninneman RR. Advanced smart structures flight experiments for precision spacecraft. Acta Astronaut. 2000, 47, 389–97.

[6] Webb G, Wilson L, Lagoudas DC, Rediniotis O. Adaptive control of shape memory alloy actuators for underwater biomimetic applications. AIAA J. 2000, 38, 325–34.

[7] Rogers CA. Intelligent materials. Sci. Am. 1995, 9, 122–7.

[8] Birman V. Theory and comparison of the effect of composite and shape memory alloy stiffeners on stability of composite shells and plates. Int. J. Mech. Sci., 1997, 39, 1139–49.

[9] Duerig TW, Albrecht J, Gessinger GH. A shape-memory alloy for high-temperature applications. JOM 1982, 34, 14–20.

[10] Ma J, Karaman I, Noebe RD. High temperature shape memory alloys. Intern. Mater. Rev. 2010, 55, 257–315.

[11] Atli KC, Karaman I, Noebe RD, Garg A, Chumlyakov YI, Kireeva IV. Shape memory charac-teristics of $Ti_{49.5}Ni_{25}Pd_{25}Sc_{0.5}$ high-temperature shape memory alloy after severe plastic deformation. Acta Mater. 2011, 59, 4747–60.

[12] Karaca HE, Saghaian SM, Ded G, Tobe H, Basaran B, Maier HJ, Noebe RD, Chumlyakov YI. Effects of nanoprecipitation on the shape memory and material properties of an Ni-rich NiTiHf high temperature shape memory alloy. Acta Mater. 2013, 61, 7422–31.

[13] Karakoc O, Hayrettin C, Bass M, Wang SJ, Canadinc D, Mabe JH, Lagoudas DC, Karaman I. Effects of upper cycle temperature on the actuation fatigue response of NiTiHf high temperature shape memory alloys. Acta Mater. 2017, 138, 185–97.

[14] Evirgen A, Karaman I, Santamarta R, Pons J, Hayrettin C, Noebe RD. Relationship between crystallographic compatibility and thermal hysteresis in Ni-rich NiTiHf and NiTiZr high temperature shape memory alloys. Acta Mater. 2016, 121, 374–83.

[15] Elahinia M, Moghaddam NS, Amerinatanzi A, Saedi S, Toker GP, Karaca H, Bigelow GS, Benafan O. (2018) Additive manufacturing of NiTiHf high temperature shape memory alloy. Scr. Mater. 2018, 145, 90–4.

[16] Casalena L, Bucsek AN, Pagan DC, Hommer GM, Bigelow GS, Obstalecki M, Noebe RD, Mills MJ, Stebner AP. Structure-property relationships of a high strength superelastic NiTi–1Hf alloy. Adv. Eng. Mater. 2018, 20; https://doi.org/10.1002/adem.201800046.

[17] Babacan N, Bilal M, Hayrettin C, Liu J, Benafan O, Karamn I. Effects of cold and warm rolling on the shape memory response of $Ni_{50}Ti_{30}Hf_{20}$ high-temperature shape memory alloy. Acta Mater. 2018, 157, 228–44.

[18] Hayrettin C, Karakoc O, Karaman I, Mabe JH, Santamarta R, Pons J. Two way shape memory effect in NiTiHf high temperature shape memory alloy tubes. Acta Mater. 2019, 163, 1–13.

[19] https://www.intrinsicdevices.com/unilok_applications.html.

[20] https://www.is-rayfast.com/wp/wp-content/uploads/2017/03/TXR_brochure.pdf.

[21] Van Humbeeck J. Shape memory alloys: a material and a technology. Adv. Eng. Mater. 2001, 3, 837–50.

[22] Pozzi M, Airoldi G. The electrical transport properties of shape memory alloys. Mater. Sci. Eng. A 1999, 273, 300–4.

[23] Tomozawa M, Kim HY, Yamamoto A, Hiromoto S, Miyazaki S. Effect of heat treatment temperature on the microstructure and actuation behavior of a Ti–Ni–Cu thin film microactuator. Acta Mater. 2010, 58, 6064–71.

[24] Krulevitch P, Lee AP, Ramsey PB, Trevino JC, Hamilton J, Northrup MA. Thin film shape memory alloy microactuators. J. Microelectromech. Sys. 1996, 5, 270–82.

[25] Skrobanek KD, Kohl M, Miyazaki S. Stress-optimised shape memory microactuator. Proc. SPIE 1996, 2779, 499–504.

[26] Karakoc O, Hayrettin C, Canadinc D, Karaman I. Role of applied stress level on the actuation fatigue behavior of NiTiHf high temperature shape memory alloys. Acta Mater. 2018, 153, 156–68.

[27] Nespoli A, Biffi CA, Previtali B, Villa E, Tuissi A. Laser and surface processes of NiTi shape memory elements for micro-actuation. Metall. Mater. Trans. A 2014, 45, 2242–9.

[28] Tong HC, Wayman CM. Thermodynamic considerations of 'solid state engines' based on thermoelastic martensitic transformations and the shape memory effect. Metall. Trans. 6A 1975, 6, 29–32.

[29] Jardine AP. A shape memory effect swashplate heat engine. Mater. Lett. 1988, 7, 102–5.

[30] http://scholar.lib.vt.edu/theses/available/etd-02102001-172947/unrestricted/ETD.pdf.

[31] Graesser EJ, Cozzarelli FA. Shape-memory alloys as new materials for aseismic isolation. J. Eng. Mech. 1991, 117, 2590–608.

[32] Saadat S, Salichs J, Noori M, Hou Z, Davoodi H, Bar-On I, Suzuki Y, Masuda A. An overview of vibration and seismic applications of NiTi shape memory alloy. Smart Mater. Struct. 2002, 11; https://iopscience.iop.org/article/10.1088/0964-1726/11/2/305/pdf.

[33] Pines DJ. Real-time seismic damping and frequency control of steel structures using nitinol wire. Proc. SPIE –Int. Soc. Optical Eng. 2002, 4696, 176–85.

[34] DesRoches R, Smith B. Shape memory alloys in seismic resistant design and retrofit: a critical review of their potential and limitations. J. Earthq. Eng. 2004, 8, 415–29.

[35]   Wilson J, Wesolowsky M, Shape memory alloys for seismic response modification: a state-of-the-art review. Earthq. Spectra 2005, 21, 569–601.

[36]   Speicher M, Hodgson D, DesRoches R, Leon R. (2009) Shape memory alloy tension/compression device for seismic retrofit of buildings. J. Mater. Eng. Perform. 2009, 18, 746–53.

[37]   Parulekar YM, Reddy GR, Vaze KK, Guha S, Gupta C, Muthumani K, Sreekala R. Seismic response attenuation of structures using shape memory alloy dampers. Struct. Cont. Health Monit. 2012, 19, 102–19.

[38]   Qian H, Li H, Song G, Guo W. Recentering shape memory alloy passive damper for structural vibration control. Math. Probl. Eng. 2013, 3, 1–13.

[39]   Tang W, Lui EM. Hybrid recentering energy dissipative device for seismic protection. J. Structures 2014, 17 pages; http://dx.doi.org/10.1155/2014/262409.

[40]   Xu X, Cheng G, Zheng J. Tests on pretrained superelastic NiTi shape memory alloy rods: towards application in self-centering link beams. Adv. Civil Eng. 2018, 13 pages; https://doi.org/10.1155/2018/2037376.

[41]   Williams K, Chiu G, Bernhard R. Adaptive-passive absorbers using shape-memory alloys. J. Sound Vib. 2002, 249, 835–48.

[42]   Pan Q, Cho C. The investigation of a shape memory alloy micro-damper for MEMS applications. Sensors (Basel) 2007, 7, 1887–900.

[43]   Mirzaeifar R, DesRoches R, Yavari A. A combined analytical, numerical, and experimental study of shape-memory-alloy helical springs. Int. J. Sol. Struct. 2011, 48, 611–24.

[44]   Moraes YJO, Silva AA, Rodrigues MC, de Lima AGB, dos Reis RPB, da Silva PCS. Dynamical analysis applied to passive control of vibrations in a structural model incorporating SMA-SE coil springs. Adv. Mater. Sci. Eng. 2018, 15 pages; https://doi.org/10.1155/2018/2025839.

[45]   Luo PF, Kuwana T, Paul DK, Sherwood PM. Electrochemical and XPS study of the nickel-titanium electrode surface. Anal. Chem. 1996, 68, 3330–7.

[46]   Sato K, Jin JY, Takeuchi T, Miwa T, Takekoshi Y, Kanno S, Kawase S. Nickel-titanium alloy electrodes for stable amperometric detection of underivatized amino acids in anion-exchange chromatography. Talanta 2001, 53, 1037–44.

[47]   Kim JH, Zhu K, Yan Y, Perkins CL, Frank AJ. Microstructure and pseudocapacitive properties of electrodes constructed of oriented NiO-TiO2 nanotube arrays. Nano. Lett. 2010, 10, 4099–104.

[48]   Kwon HJ, Song JJ, Ahn DK, Hong SH, Cho JS, Moon JT, Sohn KY, Park WW. Microstructures and electrochemical properties of Si-xTiNi alloys for lithium secondary batteries. J. Nanosci. Nanotechnol. 2013, 13, 3417–21.

[49] Lah AŠ, Fajfar P, Kugler G, Rijavec T. A NiTi alloy weft knitted fabric for smart firefighting clothing. Smart Mater. Struct. 2019, 28; doi: 10.1088/1361-665X/ab18b9.

[50]   Akalin O, Ezirmik KV, Urgen M, Newaz GM. Wear characteristics of NiTi/Al6061 short fiber metal matrix composite reinforced with SiC particulates. J. Tribol. 2010, 132; doi: 10.1115/1.4002332.

[51]   Sutapun B, Tabib-Azar M, Huff MA. Applications of shape memory alloys in optics. Appl. Opt. 1998, 37, 6811–5.

[52]   http://resource.download.wjec.co.uk.s3.amazonaws.com/vtc/2016-17/16-17_1-4/website/category/2/shape-memory-alloys-sma/index.html.

Chapter 11
Properties in biological environment

In this chapter, several important issues will be discussed in terms of ion release, allergic reaction, several compatibilities [biological, hemo, cyto, and magnetic resonance imaging (MRI)], toxicity, their related osseointegration, and others.

11.1 Ion release and dissolution

Dental materials contain a wide range of Ni element contents from about 8% in type 304 stainless steel (Fe–18Cr–8Ni) for orthodontic archwire as well as a reinforcing wire for denture base, about 27Ni in Co–Cr alloy [45Co-20Cr-27Ni-12(Fe, Mo, W, Mn)] for clasp wire, about 50Ni in NiTi-based alloy for orthodontic archwire as well as endodontic file and reamer, 73Ni–5Cr–12Cu alloy for crown, bridge, or denture base, to 86% in 86Ni–8Cr–4Cu as a reinforcing filament for resin denture base [1]. Nickel is a known allergen and it might also have carcinogen properties. Due to this reason, the nickel-containing alloys have raised great concerns about its usefulness in the medical industry. Excessive release of Ni ions may cause an allergic response, leading to severe health problems. Therefore, real applications of NiTi-based shape-memory alloys (SMA) should consider the problem of Ni ions release. Hence, there are extensively large number of researches reported on this issue. The increasing utilization of heavy metals in modern industries leads to an increase in the environmental burden. Nickel represents a good example of a metal whose use is widening in modern technologies. As a result of accelerated consumption of nickel-containing products, nickel compounds are released to the environment at all stages of production and utilization. Their accumulation in the environment may represent a serious hazard to human health. Among the known health-related effects of nickel are skin allergies, lung fibrosis, variable degrees of kidney and cardiovascular system poisoning, and stimulation of neoplastic transformation [2]. Nickel is one of the relatively nontoxic trace metals found in the tissues of man, ranking in this respect with the essential elements, iron, cobalt, copper, and zinc. Its physiological role, if any, has not been established and there have been few biological studies of this transitional metal. Nickel has been found in soils, in a variety of plants, in sea foods, and in many organs and tissues of animals. It may also enter foods during processing. Few analyses on the sources of nickel in man have been made, and little is known of this interesting but relatively neglected element [3]. On the other hand, it should be recognized that there are essential elements (or sometimes called bioelements) for the human body to maintain homeostasis; such bioelements include Ni, which is contained in human tissue with approximately 0.1 ppm [4]. The potential for higher nickel concentration release from NiTi material may generate harmful allergic, toxic, or carcinogenic

https://doi.org/10.1515/9783110666113-011

reactions [5, 6], as mentioned earlier. The atomic bonding forces between Ni and Ti in intermetallic Ti–Ni are considerably higher than in a Ti alloy with a small amount of Ni [7] and will not produce the same reactions as pure metals. Thus, it is important to recognize the synergistic effect of alloying elements when evaluating the biocompatibility for any alloy [4]. Although unique properties such as the shape-memory effect and superelasticity can enhance the performance of medical implants, the biocompatibility of the materials remains a concern. There are two main factors determining the biocompatibility of materials, namely, the host reaction induced by the materials and degradation of the materials in the body environment. Nitinol consists of 50% of Ni and dissolution of Ni ions can induce allergic, toxic, and carcinogenic effects. The corrosion performance of nitinol in vivo determines the release of Ni ions. Studies have shown that the corrosion performance can range from excellent to poor, indicating the lack of complete understanding of the chemistry of the nitinol surface. For small diameter devices such as fine wires and caliber vascular stents, a small surface defect may be sufficient to increase the leaching of Ni. Implants in the body are usually under stress/strain because of loading/unloading conditions and such actions can aggravate Ni release. In addition, sterilization procedures may modify the materials surface and accelerate Ni release and a multitude of factors must be considered simultaneously.

11.1.1 Metallic ion release

Doi et al. [8] examined dissolution of components from pure Ni, pure Ti, NiTi alloy, and 316 L stainless steel in Eagle's minimum essential medium to evaluate the effects of dynamic condition, pH, extraction period, and filtration on the corrosion behavior. Extraction medium was pH 3.5, 7.0, and 9.5. Extraction was done under static and gyrating conditions of 160, 200, and 230 rpm for 3 and 7 days. It was reported that under static conditions, Ni was eluted from pure Ni and NiTi alloy at pH 3.5. Ni and Fe were slightly eluted from 316 L stainless steel at pH 3.5; however, dissolution of components from alloys was greater under dynamic conditions and with longer extraction periods. Dissolution of components, particularly of nickel at pH 3.5, was the highest from pure Ni, NiTi alloy, and 316 L stainless steel. Ti was not detected in the filtrate under the conditions tested, indicating that a considerable amount of nickel was detected in the extract and filtrate indicated that the element was present in both a soluble and particulate form, especially at pH 3.5, and that extraction under dynamic conditions is useful for investigating the degradation of metallic materials for medical and dental use. Watarai et al. [9] measured the current density of $Ti_{50}Ni$ (at%) alloy after abrasion in simulated bioliquids using a potentiostat to estimate the amount of metallic ions released from the alloy during repassivation and maturation. The current density in saline, saline with and without N_2 bubbling, and Hanks' solutions with and without proteins after abrasion was measured, and the amount of released ion was calculated from the integrated current density with time, assuming that Ti^{4+} and Ni^{2+} are equivalently released. It was reported that (i) no difference in the amount of released ion was observed between saline with and without N_2 bubbling; (ii) no difference was observed between saline and pH 7.4 Hanks' solution; (iii) more Ti^{4+} and Ni^{2+} were released in bioliquids with proteins than in saline with and without N_2 bubbling, namely, dissolved oxygen and inorganic ions in Hanks' solution did not influence the amount of released ion, but proteins influenced it; and (iv) the release of metallic ions from metals and alloys in biological systems can be estimated by the methodology employed in this study. Ryhänen et al. [10] determined the corrosion of NiTi in vivo and to evaluate the possible deleterious effects of NiTi on osteotomy healing, bone mineralization, and the remodeling response. Femoral osteotomies of 40 rats were fixed with either NiTi or stainless steel intramedullary nails. It was mentioned that there were no statistically significant differences in nickel concentration between the NiTi and stainless steel groups in any of the organs. NiTi appears to be an appropriate material for further intramedullary use because it has good biocompatibility in bone tissue.

Cortizo et al. [11] evaluated the biocompatibility of two of the most labile components of metallic dental alloys on osteoblastlike cells. The influence of protein and ions on metal dissolution properties was also investigated using different electrolyte solutions. Morphological alterations, cell growth, and differentiation of osteoblasts were assessed after exposure to pure metals (Ag, Cu, Pd, Au) and Ni–Ti alloy and correlated with the kinetics of elements released into the culture media. It was shown that (i) Cu and Ag were the most cytotoxic elements and the other metals were biocompatible with the osteoblasts; (ii) metal ions induced cell death through early mitosis arrest, apoptotic phenomena, and necrotic processes; (iii) voltammograms showed that anions and proteins interfered in the corrosion process; and (iv) fetal bovine serum strongly affected the electrochemical process, decreasing the oxidation rate of the metals, concluding that Cu and Ag ions showed a time-dependent low biocompatibility, which correlated with the concentration of released ions, and the dissolution of the metallic materials was dependent on the composition of the simulated biological media. Copper-based coatings can reduce infections for Ti implants. However, Cu is also cytotoxic. To examine the balance of antibacterial versus adverse tissue effects, Hoene et al. [12] evaluated a Cu coating regarding in vivo Cu release and local inflammatory reactions for 72 h. TiAl6V4 plates received either plasma electrolytic oxidation only Ti, or an additional galvanic Cu deposition (TiCu). It was found that no *Staphylococcus aureus* were found in vitro on TiCu after 24 h. Total and tissue macrophages around implants increased until 72 h for both series and were increased for TiCu. As numbers of total and tissue macrophages were comparable, macrophages were probably tissue-derived. MHC-class-II-positive cells increased for TiCu only. T-lymphocytes had considerably lower numbers than macrophages, did not increase or differ between both series, and thus had minor importance. It was also mentioned that tissue reactions increased beyond Cu release, indicating effects of either surface-bound Cu or more likely the implants themselves. TiCu samples possessed antibacterial effectiveness in vitro, released measurable Cu amounts in vivo, and caused a moderately increased local inflammatory response, demonstrating anti-infective potential of Cu coatings.

Okazaki et al. [13] conducted a 7-day immersion test using several solutions on stainless steel, Co-based alloy, and Ni–Ti alloy, which are used for stents and stent grafts. The quantitative data on the release of each metal ion and the correlation between metal ion release rate and pH were obtained. It was reported that the quantities of Fe and Ni released from stainless steel gradually decreased with increasing solution pH (pH 2–7.5). For Co-Cr-Mo-Ni-Fe alloy, the quantity of Cr released steadily increased as pH decreased (pH ≤ 6) and reached nearly zero at pH higher than 6 (pH 6–7.5). Co release was slightly affected by a variation in pH. The quantities of Ni and Ti released from NiTi alloy markedly increased with decreasing pH (pH ≤ 4) and they leveled off from pH 4 (pH 4–7.5). Although the rapid increases were observed at approximately pH 2, the quantities were even higher than that of Co released from the Co–Cr–Mo and Co-Cr-Mo-Ni-Fe alloys. For further investigation of the rapid increase in the quantities of metals released at pH 2, an anodic polarization test was employed to study the passive and transpassive behaviors of NiTi alloy. It was shown that the critical current density for the passivation of NiTi alloy markedly increased as pH decreased (pH ≤ 4) and was low (1.4 μA/cm^2) at pH higher than 4 (pH 4–7.5), and the potential at a current density of 10 μA/cm^2, by contrast, markedly rose with decreasing pH (pH ≤ 2), and was 1.2 V from pH 2 (pH 2–7.5). Intra-arterial stenosis due to atherosclerosis is often treated with endovascular balloon dilatation with a metal stent. Observations of stent fractures, stent compression, accumulation of immunocompetent cells around stents have suggested the possibility of immunologic reactions to substances released from stents. An accelerated corrosion model was developed to study corrosion behavior of commonly used surgical peripheral stents. Høl et al. [14] investigated single nitinol stents, connected stents of the same material (stent in stent, both nitinol), and connected stents of dissimilar alloys (Nitinol with stainless steel stent inside). The stents were subjected to mechanical pulsatile radial strain (up to 8% strain at 1 Hz) and electrochemical stress (+112 mV vs SCE). The release of nickel and titanium ions was compared. It was found that there was a higher release of nickel when combining two similar (range: 1,382–8,018 μg/L) and dissimilar (range: 170–2,497 μg/L) stents compared to single stents (range: 0.4–216 μg/L). The concentration of titanium was low (range: 1.6–98.4 μg/L) with only a difference between the single and two similar stents. Deposits of corrosion products were clearly visible after fretting and pitting corrosion mainly on the Nitinol stents. Several mesh wires were fractured, indicating that (i) mechanical strain combined with weak electric potential resulted in pronounced corrosion and fracture of stents, especially with overlapping stents; (ii) single stents after pulsatile load released the lowest amount of ions; and (iii) the combination of stents of the same material (Nitinol) had the highest release of metal ions.

11.1.2 Ni release, in general

Metal ion release is strongly related to stability of passive film. If it suffers from pitting corrosion, the metal ion release takes place from the substrate surface. The formation of pits is related to the equilibrium potentials of (passive) film formation given as a function of the activities of the components of the film substance. The solubility of the product, with respect to the ions in the electrolyte, depends on the electrode potential, since oxidation states of the metal in the film and in the electrolyte are often different. Heusler [15] discussed the kinetics of uniform film formation and dissolution with respect to (1) equilibrium of all components across both interfaces, (2) partial equilibrium of one component at the outer film/electrolyte interface, and (3) irreversible ion transfer reactions at the outer interface. Examples are oxide films on iron (i.e., F_2O_3 and/or F_3O_4), titanium (TiO_2) and aluminum (Al_2O_3). It was mentioned that the processes during the incubation time (which should include chlorine ion adhesion and absorption) of pitting corrosion corresponded to nonuniform dissolution and formation of the passivating film.

Since nickel release during the biodegradation of TiNi is an important concern for its use as an implant or other prostheses, the common finding among numerous studies is that nickel element released from TiNi alloy was at a noticeable level for a brief period of time, but the release amount decreased rapidly after 24 to 48 h in Hank's solution. Wever et al. [16] studied the nickel element release in Hank's solution. For clinical implantation purposes in shape-memory metals, the nearly equiatomic TiNi alloy is generally used. The corrosion properties and surface characteristics of this alloy were investigated and compared with two reference controls (316L stainless steel and Ti–6Al–4V). The anodic polarization curves, performed in Hank's solution at 37 °C, demonstrated a passive behavior for the TiNi alloy. A more pronounced difference between the corrosion and breakdown potential, that is, a better resistance to chemical breakdown of passivity was found for the TiNi alloy compared to 316L stainless steel. The passive film on the TiNi consists of mainly TiO_2-based oxide with minimal amounts of nickel in the outermost surface layers. After immersion in Hank's solution, the growth of a calcium phosphate layer was observed. The passive diffusion of Ni from the TiNi alloy measured by atomic absorption spectrophotometer reduced significantly in time from an initial release rate of 14.5×10^{-7} µg/cm^2/s to a nickel release that could not detect more after 10 days. It is suggested that the good corrosion properties of TiNi alloy and the related promising biological response, as reported in literature, may be ascribed to the presence of mainly a TiO_2-based surface layer and its specific properties, including the formation of a calcium phosphate later after exposure to a bioenvironment. Dong et al. [17] treated NiTi alloy with ceramic conversion at 400 and 650 °C, and investigated the effect of the surface treatment on the fretting corrosion behavior of NiTi alloy using fretting corrosion tests in the Ringer's solution. It was reported that the experimental results have shown that the ceramic conversion treatment can convert the surface of NiTi into a TiO_2 layer, which can effectively improve the fretting corrosion resistance of NiTi alloy and significantly reduce Ni ion release into the Ringer's solution. Scanning electron microscopy (SEM) observations revealed that the untreated samples were severely damaged by adhesion and delamination; the high temperature (650 °C/1 h) treated samples were damaged mainly by spallation and adhesion; and the low temperature (400 °C/50 h) treated samples were characterized by mild abrasion. Mild oxidation and corrosion were also observed for all three types of samples tested under fretting corrosion conditions. Ni-ion release rates from NiTi surfaces exposed in the cell culture media, and human vascular endothelial cell culture environments were investigated [18]. The NiTi surface layers situated in the depth of 70 μm below a NiTi oxide scale are affected by interactions between the NiTi alloys and the bioenvironments. The finding was proved with the use of inductively coupled plasma mass spectrometry (ICP-MS) and electron microscopy experiments. As the exclusive factor controlling the Ni-ion release rates was not only thicknesses of the oxide scale, but also the passivation depth, which was twofold larger. It was suggested that some other factors, in addition to the Ni concentration in the oxide scale, admittedly hydrogen soaking deep below the oxide scale, must be taken into account in order to rationalize the concentrations of Ni ions released into the bioenvironments and the suggested role of hydrogen as the surface passivation agent is also in line with the fact that the Ni ion release rates considerably decrease in NiTi samples that were annealed in controlled hydrogen atmospheres prior to bioenvironmental exposures.

11.1.3 Nickel ion release from orthodontic appliances

Since only orthodontic appliances (archwires and brackets) will be exposed to intraoral environment for a certain period of time (basically varying from 6 months to 2–3 years), risk of metallic element release would be much higher than other NiTi devices. It should be noted that the intraoral environment is not necessarily friendly to such appliances, rather it contains noxious elements to metallic materials, a wide range fluctuation of pH value, depending on drinks and foods, chemistry of saliva, or potential occurrence of the galvanic corrosion (if dissimilar metal is located adjacent to NiTi devices). A numerous number of research reports have been published in this regard.

11.1.3.1 Appliances, in general
Park et al. [19] measured the amounts of nickel and chromium released from a simulated orthodontic appliance incubated in 0.05% sodium chloride solution. It was found that the average release of metals was 40 μg nickel and 36 μg chromium per day for a full-mouth appliance, which was well below the average dietary intake of nickel and chromium consumed by Americans; however, the clinician should be aware that

release of nickel and chromium from orthodontic bands might sensitize patients to nickel and chromium and may cause hypersensitivity reactions in patients with a prior history of hypersensitivity of these metals. Kerosuo et al. [20] conducted tests by immersing five identical samples, each consisting of a fixed appliance, a head-gear, and a quad-helix for one-half of a dental arch in 0.9% sodium chloride for 2 h, 24 h, and 7 days. It was reported that a significant release of nickel was detected from the quad-helix during the first 2 h in static conditions, whereas during the following two periods significantly less nickel was released from the quad-helix than from the other appliances. The fixed appliance with simulated function showed a significantly higher cumulative release of nickel than the similar appliance in static conditions, 44.2 and 17.1 µg. The total amounts of chromium released from the fixed appliance were significantly lower than those of nickel. No difference in the release of chromium was seen between the static and dynamic conditions, indicating that there was certain differences in the amount and pattern of nickel release from different stainless steel orthodontic appliances in vitro. The release rate of nickel from dynamically loaded fixed appliances was found to be accelerated compared with that released under static conditions. Barrett and Bishara et al. [21, 22] compared in vitro the corrosion rate of a standard orthodontic appliance consisting of bands, brackets, and either stainless steel or NiTi archwires. Evaluation was conducted with the appliances immersed for 4 weeks in a prepared artificial saliva (AS) medium at 37 °C. It was reported that ortho-dontic appliances release measurable amounts of nickel and chromium when placed in an AS medium. The nickel release reaches a maximum after approximately 1 week, then the rate of release diminishes with time; on the other hand, chromium release increases during the first 2 weeks and levels off during the subsequent 2 weeks. The release rates of nickel or chromium from stainless steel and NiTi archwires are not sig-nificantly different. For both arch wire types, the release for nickel averaged 37 times greater than that for chromium [21]. It was also determined whether orthodontic patients accumulate measurable concentrations of nickel in their blood during their initial course of orthodontic therapy. Blood samples were collected at three different time periods: before the placement of orthodontic appliances, 2 months after their placement, and 4–5 months after their placement. The study involved 31 subjects, 18 females and 13 males, who had malocclusions that required the use of a fully banded and bonded edgewise appliance. The age of the subjects in the study ranged between 12 and 38 years. It was obtained that patients with fully banded and bonded orthodon-tic appliances did not show either a significant or consistent increase in nickel blood levels during the first 4–5 months of orthodontic therapy. Orthodontic therapy using appliances made of alloys containing NiTi did not result in a significant or consistent increase in the blood levels of nickel, indicating that orthodontic appliances used, in their "as-received" condition, corrode in the oral environment releasing both nickel and chromium, in amounts significantly below the average dietary intake [22].

Ousehal et al. [23] evaluated levels of nickel released into saliva by fixed ortho-dontic appliances by performing an in vivo study on 16 patients (eight boys and eight girls). It was shown that there was a significant increase in nickel levels just after NiTi archwire insertion; however, the difference was nonsignificant 8 weeks later. Orthodontic appliances release nickel ions mainly at the start of orthodontic treatment. In order metal released from the fixed orthodontic appliances currently in use, Hwang et al. [24] fabricated simulated fixed orthodontic appliances that corresponded to half of the maxillary arch and soaked them in 50 mL of AS (pH 6.75 ± 0.15, 37 °C) for 3 months. Four groups were established according to the appliance manufacturer and the type of metal in the 0.016 × 0.022-in archwires. Groups A and B were stainless steel archwires from Ormco and Dentaurum, respectively, and groups C and D were both NiTi archwires with Ormco's copper NiTi and Tomy's Bioforce sentalloy, respectively. Stainless steel archwires were heat treated in an electric furnace at 500 °C for 1 min and quenched in water. The amount of metal released from each group by immersion time was measured. It was reported that there was no increase in the amount of chromium released after 4 weeks in group A, 2 weeks in group B, 3 weeks in group C, and 8 weeks in group D. There was no increase in the amount of nickel released after 2 weeks in group A, 3 days in group B, 7 days in group C, and 3 weeks in group D. There was no increase in the amount of iron released after 2 weeks in group A, 3 days in group B, and 1 day in groups C and D. In our 3-month-long investigation, we saw a decrease in metal released as immersion time increased. Eliades et al. [25] characterized qualitatively and quantitatively the substances released from orthodontic brackets and NiTi wires and to comparatively assess the cytotoxicity of the ions released from these orthodontic alloys. Two full sets of stainless steel brackets of 20 brackets each (weight 2.1 g) and two groups of 0.018 × 0.025 NiTi archwires of 10 wires each (weight 2.0 g) were immersed in 0.9% saline solution for a month. It was obtained that there was no indications of ionic release for the NiTi alloy aging solution, whereas measurable nickel and traces of chromium were found in the stainless steel bracket-aging medium. Concentrations of the nickel chloride solution greater than 2 mM were found to reduce by more than 50% viability. Ortiz et al. [26] determined the amounts of metallic ions that stainless steel, nickel-free, and titanium alloys release to a culture medium, and to evaluate the cellular viability and DNA damage of cultivated human fibroblasts with those mediums. The metals were extracted from 10 samples (each consisting of four buccal tubes and 20 brackets) of the three orthodontic alloys that were submerged for 30 days in minimum essential medium. Next, the determination of metals was performed by using ICP-MS, cellular viability was assessed by using the tetrazolium reduction assay (MTT assay) (3-[4,5-dimethylthiazol-2-yl]-2, 5-diphenyltetrazolium bromide), and DNA damage was determined with the Comet assay. The metals measured in all the samples were Ti(47), Cr(52), Mn(55), Co(59), Ni(60), Mo(92), Fe(56), Cu(63), Zn(66), As(75), Se(78), Cd(111), and Pb(208). It was found that the cellular viability of the cultured fibroblasts incubated for 7 days with minimum essential medium, with the stainless steel alloy submerged, was close to 0%. High concentrations of titanium, chromium, manganese, cobalt, nickel, molybdenum, iron, copper, and zinc were detected. The nickel-free alloy released lower amounts of ions to the medium. The greatest damage in the cellular DNA, measured as the olive moment, was also produced by the stainless steel alloy followed by the nickel-free alloy. Conversely, the titanium alloy had an increased cellular viability and did not damage the cellular DNA, when compared with the control values. Based on these findings, it was concluded that titanium brackets and tubes are the most biocompatible of the three alloys studied.

Mikulewicz et al. [27–29] extensively investigated trace metal release from orthodontic appliances. It was pointed out that for in vivo studies, nickel concentrations in blood and urine, long-term metal release was detected and significant differences were found and it leads to the conclusion that nickel ions are released from orthodontic appliances in measurable amounts to human organism [27]. For in vitro studies, it must be underlined that the main disadvantage of in vitro tests was that the experimental setup did not reflect in vivo conditions, for example, the presence of biofilm, which grows on the surface of the materials in oral cavity, and the presence and activity of microflora to a large extent is responsible for the process of corrosion, in particular, biodeterioration. It was also mentioned that the further scheme of in vitro research should incorporate changeable conditions of oral cavity environment (pH, dynamic conditions – saliva flow) and the presence of microbiological flora (microbiological attack) in the experimental design and, first of all, the real proportions of appliance elements [28]. The results of an in vitro experiment were reported on the release of metal ions from orthodontic appliances composed of alloys containing iron, chromium, nickel, silicon, and molybdenum into AS [29]. The concentrations of magnesium, aluminum, silicon, phosphorus, sulfur, potassium, calcium, titanium, vanadium, manganese, iron, cobalt, copper, zinc, nickel, and chromium were significantly higher in AS in which metal brackets, bands, and wires used in orthodontics were incubated. In relation to the maximum acceptable concentrations of metal ions in drinking water and to recommended daily doses, two elements of concern were nickel (573 vs 15 µg/L in the controls) and chromium (101 vs 8 µg/L in the controls). Three ion release coefficients were defined: α, a dimensionless multiplication factor; β, the difference in concentrations (in µg/L); and y, the ion release coefficient (in percent). The elevated levels of metals in saliva are thought to occur by corrosion of the chemical elements in the alloys or welding materials. The concentrations of some groups of dissolved elements appear to be interrelated. Amini et al. [30] evaluated hair nickel and chromium levels in fixed orthodontic patients. Scalp hair nickel/chromium concentrations of 12 female and 12 male fixed orthodontic patients were measured before treatment and 6 months later, using atomic absorption spectrophotometry. The effects of treatment, gender, and age on hair ions were analyzed statistically. The patients' mean age was 18.38 ± 3.98 years. It was found that the mean nickel levels were 0.1380 ± 0.0570 and 0.6715 ± 0.1785 µg/g dry hair mass, respectively, in the baseline and sixth month of treatment. Chromium concentrations were 0.1455 ± 0.0769 and 0.1683 ± 0.0707 µg/g dry hair mass, respectively. After 6 months, nickel increased for 387% and chromium increased for 16%, and no significant correlations were observed between any ion levels with age or gender, indicating that it seems that 6 months of fixed orthodontic treatment might increase levels of hair nickel and chromium.

There are some studies on a coupling of bracket and wire. Yanisarapan et al. [31] determined the cytotoxicity, metal ion release, and surface roughness of metal orthodontic appliances after immersion in different fluoride product solutions. Twelve sets of 20 brackets and four tubes were ligated with three types of archwires: stainless steel, NiTi, and β-Ti. The samples in each archwire group were divided into three subgroups and immersed in solutions of fluoride toothpaste, 1.23% acidulated phosphate fluoride (APF), or AS without fluoride as a control group. The immersion times were estimated from the recommended time of using each fluoride product for 3 months. The samples were immersed in cell culture medium for 7 days. Primary gingival fibroblast cell viability was determined by an MTT assay. Metal ion (Ni, Cr, Fe, and Mo) release and surface roughness were measured by ICP-MS and a noncontact optical three-dimensional (3D) surface characterization and roughness measuring device, respectively. It was reported that in the APF groups, the four metal ion levels and surface roughness of the brackets and archwires significantly increased, while cell viability significantly decreased, especially in the TMA subgroup. The SEM results showed that the brackets and wires in the APF groups demonstrated more lines and grooves compared with the other groups. Using APF gel during orthodontic treatment with fixed metal appliances should be avoided. Tahmasbi et al. [32, 33] investigated the galvanic corrosion and ion release from a combination of brackets and wires in a fluoride-containing mouthwash. The test was conducted on 24 lower central incisor 0.022" Roth brackets of four different commercially available brands. These brackets along with stainless steel or NiTi orthodontic wires were immersed in Oral-B mouthwash-containing 0.05% sodium fluoride for 28 days. It was reported that the copper ions released from specimens with NiTi wire were greater than those of sample-containing stainless steel wire. Corrosion rate of brackets coupled with NiTi wires was higher than that of brackets coupled with stainless steel wires, and stainless steel brackets used with NiTi wires showed greater corrosion and thus caution is recommended when using them [32]. A similar test was conducted in an AS solution [33]. A total of 24 mandibular central incisor Roth brackets of four different manufacturers were used in this experimental study. These brackets were immersed in AS along with stainless steel or NiTi orthodontic wires for 28 days. It was found that among ions evaluated, release of nickel ions from brackets was significantly higher than that of other brackets. The mean potential difference was significantly lower in specimens containing a couple of brackets and stainless steel wire compared with other specimens. No significant difference was observed in the mean corrosion rate of various groups, concluding that Shinye brackets coupled with stainless steel wire showed more susceptibility to galvanic corrosion, and there were no significant differences among specimens in terms of the CR or released ions except the release of Ni ions, which was higher in Shinye brackets. Darabara et al. [34] investigated the elemental composition, microstructure, hardness, corrosion properties, and ionic release of commercially available orthodontic brackets and Cu–NiTi archwires. Galvanic corrosion between the orthodontic wire and each bracket took place in 1 M lactic acid for 28 days at 37 °C and then the ionic concentration of nickel and chromium was studied. The orthodontic wire is made up from a NiTi alloy with copper additions, while the orthodontic brackets are manufactured by different stainless steel grades or titanium alloys. It was reported that all tested wires and brackets with the exception of Gemini are not susceptible to pitting corrosion. In galvanic corrosion, following exposure for 28 days, the lowest potential difference (~250 mV) appears for the orthodontic wire Cu–NiTi and the bracket made up from pure titanium or from the stainless steel AISI 316 grade. Following completion of the galvanic corrosion experiments, measurable quantities of chromium and nickel ions were found in the residual lactic acid solution.

11.1.3.2 Archwire

The corrosion resistance of different NiTi orthodontic wires in AS with various acidities, in terms of ion release, was evaluated [35]. Four types of as-received commercial NiTi orthodontic wires were immersed in 37 °C AS at pH 2.5–6.25 for different periods from 1 to 28 days. It was shown that the manufacturer, pH value, and immersion period, respectively, had a significantly statistical influence on the release amount of Ni and Ti ions, (ii) the amount of Ni ions released in all test solutions was well below the critical value necessary to induce allergy and below daily dietary intake level, (iii) the amount of Ti ions released in pH > 3.75 solution was mostly not detectable, representing that the TiO_2 film on NiTi wires exhibited a good protection against corrosion, (iv) preexisted surface defects on NiTi wires might be the preferred locations for corrosion, (v) the NiTi wire with the highest release amount of metal ions had the maximal increase in surface roughness after immersion test, while a rougher surface did not correspond to a higher metal ion release, and (vi) the average amount of Ni ions released per day from the tested TiNi wires was well below the critical concentration necessary to introduce allergy (600–2,500 µg) [36] and under daily dietary intake level (300–500 µg) [3]. Cioffi et al. [37] investigated the combined effect of strain and fluoridated media: the wires were examined both under strained (5% tensile strain) and unstrained conditions, in fluoridated AS at 37 °C. It was reported that the corrosion behavior of NiTi alloy is highly affected by the fluoride content, showing a release of 4.79 ± 0.10 µg/cm²/day, but, differently from other biomaterials, it does not seem to be affected by elastic tensile strain. Petoumeno et al. [38] determined whether the clinical application of NiTi wires would lead to corrosion defects on the wire surfaces, and whether an influence on the patients' salivary Ni ion concentration would become apparent. A total of 115 wires of different manufacturers was retrieved after intraoral application lasting 1–12 months. The wires were examined after cleaning with a scanning electron microscope. The salivary Ni ion concentration in 18 patients at predefined intervals was analyzed, following a detailed orthodontic treatment protocol during the initial phase of orthodontic therapy. The intervals were: (1) a saliva sample before treatment commenced; (2) after bonding of brackets and bands; (3) 2 weeks after bonding, immediately before; (4) immediately after fitting the archwires; and (5) 4 and 8 weeks after placing the archwires. It was reported that (i) surface analysis revealed no differences in the degree of corrosion of the different products; (ii) no statistically significant differences were noted in the Ni ion concentration at time points 1 (reference value), 3, 5, and 6 (34, 34, 28, and 30 µg/L, respectively); (iii) the samples taken immediately after bracket bonding or the NiTi wire application, however, displayed a significant increase in the salivary Ni ion concentration (2: 78 µg/L and 4: 56 µg/L); and (iv) it was significantly higher after bonding of the steel brackets than after NiTi wire application. Based on these results, it was concluded that (i) increased Ni ions are released initially after the orthodontic devices have been fitted, but they decay quickly; (ii) this is reflected in miniscule corrosion defects as pitting; and (iii) it is unlikely that orthodontic NiTi wires are a relevant additional Ni load for the patient. Poosti et al. [39] compared the nickel ion concentrations released from recycled NiTi wires after sterilization by either dry heat or steam autoclave. Eighty preformed NiTi wires were assigned to four equal groups. In Groups 1, 2 and 3 the archwires were used intraorally for 4 weeks. The Group 4 archwires were not used. Group 1 archwires were sterilized by dry heat, the Group 2 archwires were sterilized by steam autoclave, the Group 3 archwires were not sterilized and the Group 4 archwires were as-received condition. A 2 cm length, cut from each archwire, was immersed in AS for 4 weeks and the nickel ion concentrations in the AS measured with an atomic absorption spectrophotometer. It was obtained that there were no significant differences in the nickel ion concentrations released into the AS by each group of archwires, concluding that sterilization of used NiTi wires by either dry heat and steam autoclave does not affect the concentrations of nickel ions released into AS.

It was examined whether NiTi archwires cause an increase of nickel concentration in the saliva of 18 orthodontic patients to estimate the possible risk of these archwires in patients who have nickel hypersensitivity [40]. Saliva samples were collected before orthodontic treatment, after placement of the bands and brackets, 2 weeks later and before placing the NiTi archwires, immediately after placing the NiTi archwires, 4 weeks after placing the wires, and 8 weeks after placing the wires. It was found that by using mass spectrometry, no statistically significant differences were found in the nickel concentrations in the samples taken without appliances, in those obtained 2 weeks after placement of the bands and brackets, and 4 and 8 weeks after placement of the archwires. Samples taken immediately after placement of the bands and brackets and the NiTi archwires showed slight but significant increases in nickel concentration of 78 and 56 µg/L, respectively, compared with the pretreatment value of 34 µg/L. It was, therefore, concluded that (i) nickel leaching occurred after placement of the bands and brackets and after placement of the NiTi archwires, associated with an increase of the nickel ion concentration in the patient's saliva; and (ii) this effect decreased within 10 weeks. Kuhta et al. [41] examined the effects of three different parameters (pH value, type of archwire, and length of immersion) on release of metal ions from orthodontic appliances. Simulated fixed orthodontic appliances that corresponded to one-half of the maxillary arch were immersed in AS of different pH values (6.75 ± 0.15 and 3.5 ± 0.15) during a 28-day period. Three types of archwires were used: stainless steel, NiTi, and thermo-NiTi. The quantity of metal ions was determined with the use of a high-resolution ICP-MS (HR-ICP-MS). It was reported that (i) the release of six different metal ions was observed: Ti, Cr, Ni, Fe, Cu, and Zn; (ii) the appliances released measurable quantities of all ions examined; (iii) the change in pH had a very strong effect (up to 100-fold) on the release of ions; (iv) the release of ions was dependent on wire composition, but it was not proportional to the content of metal in the wire; and (v) the largest number of ions was released during the first week of appliance immersion. It was then concluded that (i) levels of released ions are sufficient to cause delayed allergic reactions, and (ii) this must be taken into account when type of archwire is selected, especially in patients with hypersensitivity or compromised oral hygiene. Suárez et al. [42] studied the surface topographic changes and nickel release in lingual orthodontic archwires in vitro. Stainless steel, NiTi and Cu-NiTi lingual orthodontic archwires were studied using atomic absorption spectrometry for nickel release after immersion in a saline solution. It was mentioned that (i) statistically significant changes in roughness were seen in all archwires except NiTi; (ii) surface changes were most severe in the Cu-NiTi alloy; and (iii) stainless steel archwires released the highest amount of nickel, concluding that only roughness changes in Cu-NiTi archwires seemed to be clinically significant, and the amount of nickel released for all archwires tested is below the levels known to cause cell damage. Liu et al. [43] investigated the influence of bending stress on the nickel release of commercial NiTi orthodontic wires in vitro, simulating the intra-oral environment as realistically as possible. Two types of as-received orthodontic NiTi wires, free of performed internal stress, were immersed in AS. Half of the NiTi wires were exposed to continuous bending stress throughout the 14-day experimental period. It was obtained that the stressed NiTi wires exhibited substantial increases in the nickel release compared with the unstressed specimens during all experimental periods. The highest dissolution rate during the 0 to 1 day incubation period was observed for all stressed specimens; however, a slight increase of nickel released as a function of time was observed in the three groups of stressed specimens after 3 days of immersion. For the stressed specimens, it was hypothesized that the bending stress would induce buckling or cracking of the protective oxide film of the NiTi wires. The mechanism of nickel release was the underlying metal surface reacting with the surrounding environment. Based on these findings, it was concluded that bending stress influences the nickel release of NiTi wires, and the factor of loading condition with respect to corrosion behavior and passive film should be considered in view of the widespread use of NiTi wires for dental devices.

The amount of nickel, chromium, copper, cobalt, and iron ions released from simulated orthodontic appliance made of new archwires and brackets was determined

[44]. Sixty sets of new archwire, band material, brackets and ligature wires were prepared simulating fixed orthodontic appliance. These sets were divided into four groups of 15 samples each. Group 1: Stainless steel rectangular archwires. Group 2: Rectangular NiTi archwires. Group 3: Rectangular copper NiTi archwires. Group 4: Rectangular elgiloy archwires. These appliances were immersed in 50 mL of AS solution and stored in polypropylene bottles in the incubator to simulate oral conditions. After 90 days the solution was subjected to detect Ni, Cr, Cu, Co, and Fe ions using atomic absorption spectrophotometer. It was found that high levels of nickel ions were released from all four groups, compared to all other ions, followed by release of iron ion levels. There is no significant difference in the levels of all metal ions released in the different groups, concluding that the use of newer brackets and newer archwires confirms the negligible release of metal ions from the orthodontic appliance. Senkutvan et al. [45] investigated the rate of Ni ion release from different types arch wires used in orthodontics. Four groups of arch wires (NiTi, stainless steel, Cu–NiTi, and ion-implanted NiTi) with 12 samples were stored in AS with a pH 5.6–7.0 thermostated at (36.5 °C) and tested at different intervals, that is, 7th day, 14th day, and 21st day. It was obtained that results showed significantly statistical influence on the release amount of Ni and Ti ions. Large variation in concentration of Ni released from brackets and bands combined; however, the amount of Ni ions released in all test solutions diminished with time and was below the critical value necessary to induce allergy and below daily dietary intake level, indicating that the daily release of NiTi, stainless steel, Cu–NiTi, and ion-implanted NiTi by an orthodontic appliance in acid pH, particularly favorable to corrosion, was well below that ingested with a normal daily diet. The quantities of metal ions released in our experimental conditions should not be a cause for concern in utilizing the appliance. Azizi et al. [46] evaluated the amount of nickel and titanium ions released from two wires with different shapes and a similar surface area. Forty round NiTi archwires with the diameter of 0.020 in. and 40 rectangular NiTi arch wires with the diameter of 0.016 × 0.016 in. were immersed in AS during a 21-day period. The surface area of both wires was 0.44 in^2. Wires were separately dipped into polypropylene tubes containing 50 mL of buffer solution and were incubated and maintained at 37 °C. It was reported that the amount of nickel and titanium concentrations was significantly higher in the rectangular wire group. The most significant release of all metals was measured after the first hour of immersion. In the rectangular wire group, 243 ± 4.2 ng/mL of nickel was released after 1 h, while 221.4 ± 1.7 ng/mL of nickel was released in the round wire group, and similarly, 243.3 ± 2.8 ng/mL of titanium was released in the rectangular wire group and a significantly lower amount of 211.9 ± 2.3 ng/mL of titanium was released in the round wire group, concluding that release of metal ions was influenced by the shape of the wire and increase of time. Mirhashemi et al. [47] assessed the release of nickel and chromium ions from NiTi and stainless steel orthodontic wires following the use of four common mouthwashes available on the market. The in vitro experimental study was conducted on 120 orthodontic appliances for one maxillary quadrant including five

brackets, one band and half of the required length of stainless steel, and NiTi wires. The samples were immersed in Oral B, Oral B 3D White Luxe, Listerine, and Listerine Advance White for 1, 6, 24, and 168 h. It was found that Ni ions were released from both wires at all time points; the highest amount was in Listerine and the lowest in Oral B mouthwashes. The remaining two solutions were in between this range. The process of release of chromium from the stainless steel wire was the same as that of nickel; however, the release trend in NiTi wires was not uniform. It was concluded that (i) Listerine caused the highest release of ions; (ii) Listerine Advance White, Oral B 3D White Luxe, and distilled water were the same in terms of ion release; and (iii) Oral B showed the lowest amount of ion release.

11.1.4 NiTi dissolution from endodontic instruments

It is difficult to find published articles on nickel release phenomenon of NiTi alloy endodontic instruments; possible reason for that might be the fact that the length of time when the NiTi instrument is exposed to body fluid (including saliva) is so limited and eventually not long enough for a chemical dissolution reaction to take place. However, there are scattered reports on dissolution of NiTi alloy to retrieve fractured endodontic instruments. Ormiga et al. [48, 49] introduced a new concept of retrieval of fractured instruments from root canals based on an electrochemical process and evaluated current register tests for dissolution process of 25.04 NiTi K3 rotary files. A constant anodic potential was applied to the NiTi files, whereas the Potentiostat registered the anodic current [48]. It was reported that (i) the current attained initial values of approximately 55 mA that declined during the entire test; (ii) a good reproducibility of results was observed; and (iii) the optical microscopy analysis evidenced a progressive consumption of the files with increasing polarization time, concluding that the concept of fractured file retrieval by an electrochemical process is feasible, and this concept resulted in a consistent basis for the development of a method to remove fractured instruments from root canals. Ormiga et al. [49] treated NiTi K3 file anodically and mentioned that (i) the total electrical charge values generated during the tests evidence a statistical difference among the three groups of fragments, (ii) the larger is the diameter of the exposed surface cross section, the higher is the total value of electrical charge, and (iii) the radiographic images obtained before and after the tests showed a significant reduction of the fragment length as a result of the polarization imposed, suggesting that it is possible to obtain a significant dissolution of K3 NiTi endodontic instrument fragments, and the diameter of the surface of fragment exposed to the medium affects the current levels used to promote the dissolution. Mitchell et al. [50] investigated the effect of file dissolution products on the periodontal ligament fibroblasts. Endodontic files were dissolved in sodium fluoride (NaF) by passing a 50 mA current through the NiTi files while immersed in the NaF solution. NaF/NiTi solutions were diluted with minimal essential medium-α containing 10% serum. Periodontal ligament (PDL) cells were treated for up to 24 h, and cell viability was quantified by using calcein AM to label live cells and ethidium homodimer to label dead cells. This was repeated by using AS as an alternative to NaF. It was found that (i) NaF solution reduced PDL cell survival, and the NaF/NiTi solution further reduced PDL cell survival; (ii) AS alone did not reduce cell survival, whereas AS/NiTi solution reduced PDL cell survival; and (iii) particles that resulted from the electrochemical dissolution of NiTi files were highly cytotoxic, indicating that electrochemically dissolving NiTi files in NaF results in solutions that are cytotoxic to PDL fibroblasts, and AS may be a less toxic alternative for dissolving NiTi files.

11.1.5 Prevention of nickel ion release

There are basically two ways to prevent the release of Ni ions: selectively leach out Ni ions chemically or electrochemically, or entirely cover the Ni-containing surface with a noble species [51, 52]. The actual methods used to achieve these ends include etching in acids and alkaline solutions, electropolishing, heat and ion beam treatments, boiling in water and autoclaving, conventional and ion plasma implantations, laser melting and bioactive coating deposition, as will be discussed in detail. It should be noted that electrochemical treatments run the risk of causing hydrogen embrittlement, depending on the extent of hydrogen overpotential [53–55].

11.1.5.1 Oxidation

Various oxidation treatments were applied to nearly equiatomic NiTi alloys so as to form a Ni-free protective oxide on the surface. An ion release experiment was carried out up to 1 month of immersion in the simulated body fluid (SBF) for both oxidized and untreated surfaces [56]. It was reported that (i) oxidation treatment in a low-oxygen pressure atmosphere leads to a high surface Ti/Ni ratio, a very low Ni surface concentration and a thick oxide layer, (ii) the oxidation treatment does not significantly affect the shape memory properties of the alloy, (iii) the oxide formed significantly decreases Ni release into exterior medium comparing with untreated surfaces, and (iv) as a result, this new oxidation treatment could be of great interest for biomedical applications, as it could minimize sensitization and allergies and improve biocompatibility and corrosion resistance of NiTi SMA. Chu et al. [57] developed a new surface modification protocol encompassing an electropolishing pretreatment and subsequent photoelectrocatalytic oxidation (PEO) to improve the surface properties of biomedical NiTi SMA. It was indicated that electropolishing is a good way to improve the resistance to localized breakdown of NiTi alloy whereas PEO offers the synergistic effects of advanced oxidation and electrochemical oxidation. PEO leads to the formation of a sturdy titania film on the electropolishing treated NiTi substrate. There is an Ni-free zone near the top surface and a graded interface between the titania layer and NiTi substrate, which bodes well for both biocompatibility and mechanical stability. Ni ion release from the NiTi substrate is suppressed, as confirmed by the 10-week immersion test. In comparison, after undergoing only electropolishing, the mechanical properties of NiTi exhibit an inverse change with depth. Surface modification by dual electropolishing and PEO can notably suppress Ni ion release and improve the biocompatibility of NiTi SMA while the surface mechanical properties are not compromised, making the treated materials suitable for hard tissue replacements. The patterns of Ni release from Nitinol vary depending on the type of material (Ni–Ti alloys with low or no processing versus commercial wires or sheets). A thick TiO_2 layer generated on the wire surface during processing is often considered as a reliable barrier against Ni release. Shabalovskaya et al. [58] conducted NiTi wires with surface oxides resulting from production to identify the sources of Ni release and its distribution in the surface sublayers. It was shown that wire samples in the as-received state showed low breakdown potentials (200 mV); the improved corrosion resistance of these wires after treatment was not affected by strain. NiTi wires with the thickest surface oxide TiO_2 (up to 720 nM) showed the highest Ni release, attributed to the presence of particles of essentially pure Ni whose number and size increased while approaching the interface between the surface and the bulk. The biological implications of high and lasting Ni release are also discussed. Espinar et al. [59] investigated NiTi orthodontic archwires that have been treated using a new oxidation treatment for obtaining Ni-free surfaces. It was mentioned that (i) the titanium oxide on the surface significantly improves corrosion resistance and decreases nickel ion release, while barely affecting transformation temperatures, and (ii) this oxidation treatment avoids the allergic reactions or toxicity in the surrounding tissues produced by the chemical degradation of the NiTi.

11.1.5.2 Chemical or electrochemical treatment

Despite the suggested surface treatments, the use of electrochemical techniques makes possible a controlled oxidation through a low-temperature route, which manly depends on the relation between electrode and electrolyte. Dick et al. [60] studied the NiTi behavior in different aggressive electrolytes, as to develop surface treatments for the selective dissolution of Ni, and the dealloying of a passive alloy, as opposed to the known dealloying behavior of noble metal alloys. It was reported that electrolytes containing anions that complex Ti such as Cl^- or F^- cause localized corrosion of the NiTi alloy, nucleated mainly at the grain boundaries between cracked Ti_2Ni precipitates and the NiTi matrix. In concentrated H_3PO_4, Na_2SO_4, and H_2SO_4, the NiTi alloy is passive up to 1.25 V, where transpassivity starts caused by Ni dissolution and formation of less protective higher oxidation number oxides. Longer polarizations at potentials in the passive region result in the slow dissolution of Ni, while in the transpassive region, the intense dissolution of Ni and Ti at similar rates takes place. At these high potentials, the formation of a porous oxide layer is promoted, while in the passive region the affected layer is thinner than 50 nm. In Na$_2$SO$_4$, the dissolution rate is approximately one order of magnitude higher than in H$_3$PO$_4$. A previous 2 h polarization in H$_3$PO$_4$ at transpassive potentials increases the stability of NiTi in Hank's solution comparing to untreated samples or to samples treated in Na$_2$SO$_4$ under similar conditions. Chu et al. [61] treated chemically polished NiTi SMA substrate with a boiling aqueous solution containing hydrogen peroxide to form titania film in situ at low temperature. It was reported that (i) titania film is successfully fabricated in situ on NiTi SMA by this surface oxidation method, and (ii) it is mainly composed of rutile and anatase, whose surface compositions and morphologies are sensitive to H$_2$O$_2$ content. Chu et al. [62] studied oxidation behavior and surface characterizations of NiTi SMA which as treated with a boiling H$_2$O$_2$ aqueous solution. It was obtained that (i) the low-temperature oxidation of NiTi alloy in H$_2$O$_2$ solution resulted in the formation of a titania scale enriched with Ti–OH groups; (ii) the titania scale is mainly composed of rutile and anatase, and is relatively depleted in Ni, which can improve the biocompatibility of NiTi; and (iii) depth profiles of O, Ni, and Ti show the titania scale possesses a smooth graded interface structure to NiTi substrate, which is in favor of high bonding strength of the titania scale with NiTi substrate. Oshida [55, 63] identified the crystalline structure of H$_2$O$_2$-oxidized film as a rutile-type TiO$_2$, which does not agree with data by Chu [62, 63], although it was found that Ni element was not incorporated into this rutile oxide [64], so that this dealloying causes the surface later of NiTi as a Ti-rich surface, which, in turn, is more safe and beneficial in biocompatibility.

11.1.5.3 Surface modification

Villermaux et al. [64] employed the excimer laser surface treatment to improve corrosion resistance of NiTi SMA plates, so that Ni release can be prevented. It was mentioned that, due to a combination of the homogenization of the surface by melting, the hardening owing to N incorporation and the thickening of the oxide layer, the laser treatment improved the corrosion resistance of NiTi SMA. Similarly, NiTi SMA was nitrided using an original powder immersion reaction-assisted coating (PIRAC) method in order to modify its surface properties [65]. PIRAC nitriding method is based on annealing the samples in the atmosphere of highly reactive nitrogen supplied by decomposition of unstable nitride powders or, alternatively, by selective diffusion of the atmospheric nitrogen to the sample surface. Being a nonline-of-sight process, PIRAC nitriding allows uniform treatment of complex shape surgical implants. It was reported that hard two-layer titanium nitride (TiN)/Ti$_2$, Ni coatings were obtained on NiTi surface after PIRAC anneals at 900 and 1,000 °C. PIRAC coating procedure was found to considerably improve the corrosion behavior of NiTi alloy in Ringer's solution; in contrast to untreated nitinol, no pitting was observed in the samples PIRAC nitrided at 1,000 °C, 1 h up to 1.1 V. The coated samples were also characterized by very low anodic currents in the passive region and by an exceedingly low metal ion release rate, suggesting that PIRAC nitriding procedure could improve the in vivo performance of NiTi alloys implanted into the human body. Ju et al. [66] treated $Ti_{49}Ni_{51}$ samples by plasma surface alloying in air at 400 and 500 °C. It was shown that (i) plasma surface alloying treatment can convert the original surface into a TiO_2 layer, thus reducing the Ni content in the surface, (ii) the surface hardness of the $Ti_{49}Ni_{51}$ alloy increased from 4.6 to 23.0 GPa following the plasma surface alloying treatment, and the sliding wear resistance of all the treated samples has been effectively increased. The influence of surface modification in NiTi alloy was evaluated by nitrogen plasma immersion ion implantation (N-PIII; varying temperatures, and exposure time as follows: <250 °C/2 h, 290 °C/2 h, and 560 °C/1 h) in the amount of nickel released using immersion test in SBF [67]. It was found that (i) the depth of the nitrogen implanted layer increased as the implantation temperature increased resulting in the decrease of nickel release, and (ii) the sample implanted in high implantation temperature presented 35% of nickel release reduction compared to the reference sample.

11.1.5.4 Coating

The release of Ni from biomedical NiTi materials has stimulated a research on its surface modifications and coatings [68–70]. Kimura et al. [71] coated NiTi implants with oxide film to suppress dissolution of Ni and estimated the corrosion resistance in 1% NaCl solution by means of anodic polarization measurement. It was found that (i) by coating, dissolution at low potential was suppressed and the dissolute current density decreased, (ii) a further decrease in current density, which results from stabilization of the passive state on the surface, was observed by the repeated polarization, and (iii) the oxide film showed close adhesion with the matrix, and did not form cracks or peel off by plastic deformation associated with the shape-memory effect. To alleviate the effects of Ni allergy from NiTi alloy implants, Ozeki et al. [72] coated NiTi alloy plates with hydroxyapatite ($Ca_{10}(PO_4)_6(OH)_2$; HA), alumina (Al_2O_3), or titanium (Ti) to form 1-μm-thick films using radio frequency magnetron sputtering. After the plates had been immersed in physiological saline for periods of 1, 4, or 8 weeks, the concentration of Ni ions released in each solution was detected using a microwave-induced plasma mass spectrometer. It was reported that (i) after 8 weeks, the concentration of Ni ions released from the noncoated, the Ti-coated, the HA-coated, and the alumina-coated plates were 238, 19.7, 183, and 106 ppb, respectively. The bonding strength of the Ti film, the HA film, and the alumina film to the NiTi substrate were 3.8 ± 1.2, 2.6 ± 0.7, and 3.1 ± 1.2 MPa, respectively. The noncoated, the HA-coated, the alumina-coated, and the Ti-coated plates were implanted into the femurs of a dog for 4 weeks for histological observation. In case of the noncoated plates, connective tissue more than 300 μm thick was observed, whereas for the coated plates the thickness of the connective tissue was around 100 μm. Yeung et al. [73] investigated the use of N-PIII, or oxygen PIII (O-PIII) to mitigate the deleterious effect of Ni element. It was found that (i) the near-surface Ni concentration in all the treated samples is significantly suppressed, (ii) the plasma-treated surfaces are cytologically compatible allowing the attachment and proliferation of osteoblasts, and (iii) among the two types of samples, the best biological effects are found on the samples with nitrogen implantation. Surmenev et al. [74] deposited thin calcium phosphate coatings on NiTi substrates by rf-magnetron sputtering. The release of nickel upon immersion in water or in saline solution (0.9% NaCl in water) was measured by atomic absorption spectroscopy (AAS) for 42 days. It was observed that (i) after an initial burst during the first 7 days that was observed for all samples, the rate of nickel release decreased 0.4–0.5 ng/cm^2/day for a 0.5-μm-thick calcium phosphate coating (deposited at 290 W), which was much less than the release from uncoated NiTi (3.4–4.4 ng/cm^2/day), and (ii) notably, the nickel release rate was not significantly different in pure water and in aqueous saline solution. Natural HA (nHA) was coated on NiTi alloy using electrophoretic deposition (EPD) method to improve the corrosion resistance and biocompatibility [75]. Coating process was performed at 120 s in various applied voltages of 40, 60, and 80 V. Sintering process was done at 800 °C under inert gas for 2 h. Electrochemical behavior of the coated samples was investigated in SBF by using electrochemical impedance spectroscopy and polarization tests. It was reported that (i) the sample coated at 60 V reveals a uniform, dense coating accompanied with a higher corrosion resistance; and (ii) after 4 weeks, nickel ion release was reduced to 0.205 μg/cm^2 for coated sample at 60 V. Anuradha et al. [76] developed surface coatings on NiTi archwires capable of protection against nickel release and investigated the stability, mechanical performance, and prevention of nickel release of titanium sputter-coated NiTi arch wires. Coated and uncoated specimens immersed in AS were subjected to critical evaluation of parameters such as surface analysis, mechanical testing, element release, friction coefficient, and adhesion of the coating. It was obtained that (i) titanium coatings exhibited high reliability on exposure even for a prolonged period of 30 days in AS, (ii) the coatings were found to be relatively stable on linear scratch test with reduced frictional coefficient compared to uncoated samples, and (iii) titanium sputtering coatings seem to be promising for nickel-sensitive patients, indicating that the superior nature of the coating, evident as reduced surface roughness, friction coefficient, good adhesion, and minimal hardness and elastic modulus variations in AS over a given time period. Katic et al. [77] determined the effect of pH, fluoride (F$^-$) and hydrofluoric acid concentration (HF) on dynamic of nickel (Ni^{2+}) and titanium (Ti^{4+}) ions release. NiTi wires with untreated surface (NiTi), rhodium (RhNiTi), and nitride (NNiTi) coating were immersed once a week for 5 min in remineralizing agents, followed by immersion to AS. Ion release was recorded after 3, 7, 14, 21, and 28 days. It was reported that release of Ni^{2+} from NiTi and NNiTi wires correlated highly linearly positively with HF ($r = 0.948$ and 0.940, respectively); for RhNiTi the correlation was lower and negative ($r = -0.605$). The prediction of Ti^{4+} release was significant for NiTi ($r = 0.797$) and NNiTi ($r = 0.788$) wire. Association with F$^-$was lower; for pH it was not significant. HF predicts the release of ions from the NiTi wires better than the pH and F$^-$ of the prophylactic agents.

Porous NiTi SMAs are one of the promising biomaterials for surgical implants because of their unique shape-memory effects and porous structure with open pores. Wu et al. [78–80] extensively studied nickel leaching issue from porous NiTi alloy. Oxidation in conjunction with postreaction heat treatment was used to modify the surfaces of porous single-phase NiTi prepared by capsule-free hot isostatic pressing to mitigate Ni leaching and enhance the surface properties. It was mentioned that differential scanning calorimetry thermal analysis, uniaxial compression tests, ICP-MS, and cell cultures reveal that porous NiTi alloys oxidized at 450 °C for 1 h have an austenite transition temperature below 37 °C, excellent superelasticity, lower nickel release, and no cytotoxicity [78, 79]. Three different chemical processes are used to treat porous NiTi alloys, including H_2O_2 treatment, NaOH treatment, and H_2O_2 pretreatment plus subsequent NaOH treatment to mitigate leaching of nickel from the alloy. The porous NiTi samples modified by the two latter processes favor deposition of a layer composed of Ca and P due to the formation of bioactive Na_2TiO_3 on the surface. It was reported that (i) among the three processes, H_2O_2 pretreatment plus subsequent NaOH modification is the most effective in suppressing nickel release; (ii) the sample modified by the H_2O_2 treatment is composed of rough TiO_2 on the outer surface and an oxide transition layer underneath whereas the sample treated by NaOH comprises a surface layer of titanium oxide and Na_2TiO_3 together with a transition layer; and (iii) the sample processed by the $H_2(2)O_2(2)$ and NaOH treatment has a pure Na_2TiO_3 layer on the surface and a transition layer underneath, indicating that the different nickel release behavior and bioactivity of porous NiTi alloys processed by different methods [80].

11.2 Allergic reaction, in general

Allergy is the opposite of AIDS (acquired immune deficiency syndrome). Allergy symptoms are the result of too much immunity. The immune system produces antibodies to fight infections. AIDS is too little immunity, and allergy is too much. Metal sensitivity is thought to be a very important factor in the overall biocompatibility of implants. Although titanium has not been found to show sensitivity, other materials such as nickel from stainless steel, other high nickel alloys (e.g., TiNi alloy), and cobalt from cobalt-based alloys have been shown to be sensitizers in skin dermatitis and might, therefore, show an allergic response upon implantation [81]. Allergic contact dermatitis to metals is a common skin disease in many countries of the world. Ni allergy is most frequent, and the prevalence is reported to be 10% in females and 1% in men [82]. The incidence of Ni allergy has increased especially in the young female generation where the cause is probably due to the increased habit of ear piercing in females in recent years [83, 84]. In this section, allergic reaction in general, Ni element allergy and Ti element allergy will be discussed. The response of the host to the presence of foreign substances can trigger four types of hypersensitivity reactions: immediate, cytotoxic, immune complex, and cell mediated. Type I: immediate hypersensitivity (anaphylactic reaction), in which the allergic reactions are systemic or localized, as in allergic dermatitis (e.g., hives, wheal, and erythema reactions). The reaction is the result of an antigen cross-linking with membrane-bound IgE antibody of a mast cell or basophil. Histamine, serotonin, bradykinin, and lipid mediators (e.g., platelet activating factor, prostaglandins, and leukotrienes) are released during the anaphylactic reaction. These released substances have the potential to cause tissue damage. Type II: cytotoxic reaction (antibody-dependent), in which, the antibody reacts directly with the antigen that is bound to the cell membrane to induce cell lysis through complement activation. These antigens may be intrinsic or "self" as in autoimmune reactions or extrinsic or "nonself." Cytotoxic reactions are mediated by IgG and IgM. Examples of cytotoxic reaction are the Rh incompatibility of a newborn, blood transfusion reactions, and autoimmune diseases such as pemphigus vulgaris, bullous pemphigoid, autoimmune hemolytic anemia, and Goodpasture's syndrome to name a few. Type III: immune complex reaction, in which IgG and IgM bind antigen, forming antigen–antibody (immune) complexes. These activate complement, which results in polymorphonuclear (PMN) chemotaxis and activation. PMNs then release tissue damaging enzymes. Tissue damage present in autoimmune diseases (e.g., systemic lupus erythematosus), and chronic infectious diseases (e.g., *leprosy*) can be attributed, in part, to immune complex reactions. Type IV: cell-mediated (delayed hypersensitivity), which are initiated by T-lymphocytes and mediated by effector T cells and macrophages. This response involves the interaction of antigens with the surface of lymphocytes. Sensitized lymphocytes can produce cytokines, which are biologically active substances that affect the functions of other cells. This type of reaction takes 48–72 h, or longer, after contact with the antigen to fully develop. Many chronic infectious diseases, including tuberculosis and fungal infections, exhibit delayed hypersensitivity. Evidence suggests that hypersensitivity reactions, particularly type III and IV, may be involved in the pathogenesis of periodontal disease [85]. Honari et al. [86] mentioned about the hypersensitivity reactions associated with endovascular devices that (i) allergic reactions to endoprostheses are uncommon and reported in association with orthopedic, dental, endovascular, and other implanted devices, and (ii) hypersensitivity reactions to the biomaterials used in endovascular prostheses are among the infrequent reactions that may lead to local or systemic complications following cardiovascular therapeutic interventions.

11.2.1 Ni allergy

Nickel is one of the most common causes of allergic contact dermatitis and produces more allergic reactions than all other metals combined. Currently several brands of orthodontic wires are made of NiTi alloy and potentially have a high enough nickel content to provoke manifestations of allergic reactions in the oral cavity. Ni is capable of eliciting toxic and allergic responses. Ni can produce more allergic reactions than all other metal elements. Nickel is a strong biological sensitizer and consequently may induce a delayed hypersensitivity reaction (type IV immune response). The incidence of nickel sensitivity in a population above the age of 10 was examined through epicutaneous tests with 5% nickel sulfate performed on certain school and occupational test subjects and on subjects at a home for elderly people [87]. It was reported that (i) Ni sensitivity was observed in 4.5% (in 44 cases of 980 tested subjects), in 8% of the females and in 0.8% of the males; (ii) in 42 of the 44 nickel-sensitive subjects there was a history of dermatitis from metal contact; (iii) at the time of testing, 16 (34%) of the nickel-sensitive subjects revealed eczema; (iv) a manifest nickel sensitivity was thus found in 1.6% of all tested subjects, in 2.8% of females and in 0.4% of males; (v) Ni sensitivity and a simultaneous hand eczema was noted in 0.9% of the tested population, in 1.6% of females and in 0.2% of males; (vi) hand eczemas were rarer (20.5%) in the nickel-sensitive subjects in the population study than in the nickel-sensitive patients tested at the same time in the clinic (56.6%); and (vii) no case of nickel sensitivity was occupationally related. Nickel is a metal of widespread distribution in the environment: there are almost 100 minerals of which it is an essential constituent element and which have many industrial and commercial uses. Nickel and nickel compounds belong to the classic noxious agents encountered in industry but are also known to affect non-occupationally exposed individuals. The general population may be exposed to nickel in the air, water, and food. Inhalation is an important route of occupational exposure to nickel in relation to health risks. Most nickel in the human body originates from drinking water and food; however, the gastrointestinal route is of lesser importance, due to its limited intestinal absorption. The toxicity and carcinogenicity of some nickel compounds (in the nasal cavity, larynx, and lungs) in experimental animals, as well as in the occupationally exposed population, are well documented [88].

11.2.1.1 Orthodontic appliances
Nickel is one of the most common causes of allergic contact dermatitis and produces more allergic reactions than all other metals combined. Currently, several brands of orthodontic wires are made of NiTi alloy and potentially have a high enough nickel content to provoke manifestations of allergic reactions in the oral cavity. Bass et al. [5] determined if standard orthodontic therapy can sensitize patients to nickel and assessed gingival response to nickel-containing orthodontic appliances in patients who are nickel sensitive before treatment. Twenty-nine patients from the Division of Orthodontics, Albert Einstein/Montefiore Medical Center were tested, ranging in age from 12 to 48 years. Of the 29 patients, there were 18 female and 11 males. It was found that (i) five of the patients had a positive nickel patch test, a rate of 18.5%; (ii) the five patients that tested positive were all female, meaning that the overall rate for females was 27.7% (5:18); (iii) the five female patients sensitive to nickel were followed monthly by intraoral photos and gingival and plaque index scores; (iv) the remaining patients began routine orthodontic therapy and were retested 3 months

into treatment to see whether sensitization occurred; (v) two patients converted from an initial negative patch test to a positive test; and (vi) there may be a risk of sensitizing patients to nickel with long-term exposure to nickel-containing appliances as occurs in routine orthodontic therapy. Kerosuo et al. [89] investigated the frequency of nickel hypersensitivity in adolescents in relation to sex, onset, duration and type of orthodontic treatment, and the age at which ears were pierced. The subjects were 700 Finnish adolescents, from 14 to 18 years of age, of which 476 (68%) had a history of orthodontic treatment with metallic appliances. The study consisted of patch testing for a nickel allergy and a patient history obtained by a questionnaire and from patient record. It was found that the frequency of nickel sensitization in the whole group was 19%. Nickel allergy was significantly more often found in girls (30%) than in boys (3%) and in subjects with pierced ears (31%) than in those with no piercing of ears (2%). Orthodontic treatment did not seem to affect the prevalence of nickel sensitization. None of the girls who were treated with fixed orthodontic appliances before ear piercing showed hypersensitivity to nickel, whereas 35% of the girls who had experienced ear piercing before the onset of orthodontic treatment were sensitized to nickel, suggesting that orthodontic treatment does not seem to increase the risk for nickel hypersensitivity, and rather, the data suggests that treatment with nickel-containing metallic orthodontic appliances before sensitization to nickel (ear piercing) may have reduced the frequency of nickel hypersensitivity. Nickel is a strong biological sensitizer and consequently may induce a delayed hypersensitivity reaction (type IV immune response). Janson et al. [90] conducted the cross-sectional study to determine the prevalence of nickel hypersensitivity reaction before, during, and after orthodontic therapy with conventional stainless steel brackets and wires; to evidence the induction of this reaction by the orthodontic appliances; and to characterize the nickel hypersensitive persons. Nickel patch tests and a questionnaire were used to evaluate the hypersensitivity to this metal. The total sample consisted of 170 patients, 105 females and 65 males, from the orthodontic department at Bauru Dental School, University of São Paulo. They were divided into three groups as follows: A ($n = 60$), patients before the beginning of orthodontic therapy; B ($n = 66$), patients currently undergoing orthodontic treatment, and C ($n = 44$), patients who had undergone orthodontic treatment previously. It was reported that the chi-square test (χ^2) showed an allergic reaction in 28.3% of the total sample with 23% female and 5.3% male, indicating a gender difference ($\chi^2 = 10.75$). There was a positive association between nickel hypersensitivity and previous personal allergic history to metals ($\chi^2 = 34.88$) as well as with the daily use of metal objects ($\chi^2 = 11.95$). There was no statistically significant difference in the prevalence of contact dermatitis among the three groups ($\chi^2 = 0.39$), suggesting that orthodontic therapy with conventional stainless steel appliances does not initiate or aggravate a nickel hypersensitivity reaction. Fixed orthodontic appliances usually include brackets, bands, and archwires made of stainless steel, NiTi, or nickel–cobalt alloys, and these can release metal ions. Faccioni et al. [91] investigated the biocompatibility in vivo of fixed orthodontic appliances, evaluating the presence of metal ions in oral mucosa cells, their cytotoxicity, and their possible genotoxic effects. Mucosa samples were collected by gentle brushing of the internal part of the right and left cheeks of 55 orthodontic patients and 30 control subjects who were not receiving orthodontic treatment. It was reported that nickel and cobalt concentrations were 3.4-fold and 2.8-fold higher, respectively, in the patients than in the controls; cellular viability was significantly lower in the patients than in the controls, and there was a significant negative correlation with metal levels. The biologic effects, evaluated by alkaline comet assay, indicated that both metals induced DNA damage (more cells with comets and apoptotic cells). There were significant positive correlations between (1) cobalt levels and the number of comets and apoptotic cells, (2) nickel levels and number of comet cells, and (3) cobalt levels and comet tails. Schuster et al. [92] determined the incidence of suspected allergic reactions during fixed appliance therapy in 68 orthodontic offices in the German State of Hesse by questionnaire at approximately 0.3% of the 60,000 patients covered. It was found that (i) more extraoral (45%) than intraoral (17%) skin changes were registered, with both intraoral and extraoral changes being observed in 38%; (ii) in 53% of the affected cases the therapy was adapted to nickel-free materials, whereas it was continued as planned after a brief recovery period in 33%; (iii) the treatment was discontinued in 14% of the affected patients, corresponding to one in every 3,150; and (iv) the individual tolerance can often be tested by inserting one bracket or one band. In addition, early orthodontic treatment seems to promote a certain immune tolerance, especially toward extraoral nickel contacts; however, if a patient is known to have a nickel allergy, materials containing nickel should be renounced on principle in the orthodontic appliances. Based on these findings, it was concluded that (i) skin changes occurring in the course of orthodontic treatment should be examined and verified if necessary by a dermatologist, (ii) gold plating and other coatings (titanium nitride) of the metal elements even encourage corrosion after a brief protection period, and (iii) soldering should be avoided.

In some orthodontic patients, an oral inflammatory response is induced by corrosion of orthodontic appliances and subsequent nickel release and this inflammatory response is manifested as stomatitis (nickel-induced allergic contact stomatitis: NiACS) [93]. Genelhu et al. [93] investigated the roles of age, sex, previous allergic history, and time of exposure to fixed orthodontic appliances in the etiopathogeny of NiACS. Forty-four orthodontic patients (range: 10–44 years) were divided into 2 groups, depending on their NiACS clinical manifestations. It was reported that young patients, especially females with a history of allergic reactions, had a greater predisposition to NiACS clinical manifestations; time of exposure to orthodontic appliances was not a significant factor and thus concluded that a previous allergic reaction should be considered a predictive factor of NiACS clinical manifestations and should be noted in the patient's medical history. Pazzini et al. [94] determined the prevalence of nickel allergy in a sample of orthodontic patients and longitudinally compare the clinical periodontal status of these individuals with that of a group of nonallergic patients. The initial sample consisted of 96 patients selected randomly from a databank of patients who sought orthodontic care at a teaching institution. Following the selection and beginning of treatment, periodontal status was assessed over a 12-month period (one evaluation every 3 months – T(1), T(2,) T(3,) T(4)). The evaluations were performed blindly by a single, calibrated examiner and were followed by prophylaxis and orientation regarding oral hygiene. The prevalence of nickel allergy was determined by the patch test 9 months after the beginning of treatment and occurred in 16 individuals (17.2%). Two groups were then established: the allergic group (AG, n = 16) and the age-paired nonallergic control group (NAG, n = 16). It was found that (i) significant differences were present between groups at the T(3) and T(4) evaluations for the loss of effectiveness (LOE) index, with allergic individuals showing higher mean values than nonallergic individuals (hyperplasia, change in color, and bleeding); and (ii) no significant differences were found in the intragroup evaluations between the four evaluations, suggesting a cumulative effect from nickel throughout orthodontic treatment associated with clinically significant periodontal abnormalities in allergic individuals over time. Amini et al. extensively investigated effects of fixed orthodontic treatment on salivary Ni and Cr levels [95], urinary nickel level [96] and hair Ni and Cr levels [97]. Saliva samples were collected from 20 orthodontic patients, before treatment (control) and 6 and 12 months later. It was reported that (i) average nickel level changed from 9.75 ± 5.02 to 10.37 ± 6.94 and then to 8.32 ± 4.36 µg/L in 1 year and (ii) average chromium concentration changed from 3.86 ± 1.34 to 4.6 ± 6.11 and then to 2.04 ± 1.66 µg/L [95]. Urinary nickel concentrations in 20 female and 10 male patients being treated with stainless steel appliances were measured using atomic absorption spectrophotometry. It was found that (i) the mean nickel concentrations in male and female patients were 9.67 ± 3.25 and 9.9 ± 3.83 µg/L, respectively; (ii) these statistics for male and female control subjects were 6.65 ± 2.57 and 8.43 ± 2.94 µg/L, respectively; and (iii) orthodontic therapy for longer durations with stainless-steel archwires might elevate slightly, but significantly, urinary nickel levels [99]. Scalp hair nickel and chromium concentrations of 12 female and 12 male fixed orthodontic patients were measured before treatment and 6 months later, using atomic absorption spectrophotometry. It was observed that (i) the mean nickel levels were 0.1380 ± 0.0570 and 0.6715 ± 0.1785 µg/g dry hair mass, respectively, in the baseline and sixth month of treatment; (ii) chromium concentrations were 0.1455 ± 0.0769 and 0.1683 ± 0.0707 µg/g dry hair mass, respectively; (iii) after 6 months, nickel increased for 387% and chromium increased for 16%; and (iv) no significant correlations were observed between any ion levels with age or gender, indicating that it seems that 6 months of fixed orthodontic treatment might increase levels of hair nickel and chromium [97].

11.2.1.2 Ear pierce

Piercing the earlobes has increased in popularity among males in recent years. This habit would be expected to increase the incidence of Ni/Co sensitization. Meijer et al.

[84] performed patch testing with nickel sulfate and cobalt chloride in 520 young Swedish men doing compulsory military service. It was reported that (i) the overall frequency of Ni/Co positive tests was 4.2%; (ii) the prevalence of nickel and cobalt positive tests was significantly higher in 152 men with pierced earlobes than in those 368 with unpierced earlobes (2.7%); (iii) a history of hand eczema (7/152 = 4.6%) or other types of eczema (22/152 = 14.5%) in individuals with pierced earlobes was no more common than in those with unpierced earlobes: 24/368 = 6.5% and 51/386 = 13.9%, respectively, and (iv) hand eczema was no more common in sensitized (1/22 = 4.5%) than in nonsensitized individuals (32/498 = 6.4%). Fors et al. [98] conducted a cross-sectional study to investigate the association between nickel sensitization and exposure to orthodontic appliances and piercings. A total 4376 adolescents were patch tested following a questionnaire asking for earlier piercing and orthodontic treatment. It was mentioned that (i) questionnaire data demonstrated a reduced risk of nickel sensitization when orthodontic treatment preceded piercing (OR 0.46; CI 0.27–0.78). Data from dental records demonstrated similar results (OR 0.61, CI 0.36–1.02), but statistical significance was lost when adjusting for background factors, and (ii) exposure to full, fixed appliances with NiTi-containing alloys (OR 0.31, CI 0.10–0.98) as well as a pooled "high nickel-releasing" appliance group (OR 0.56, CI 0.32–0.97) prior to piercing was associated with a significantly reduced risk of nickel sensitization. It was, therefore, concluded that (i) high nickel-containing orthodontic appliances preceding piercing reduces the risk of nickel sensitization by a factor of 1.5–2, (ii) the risk reduction is associated with estimated nickel release of the appliance and length of treatment, and (iii) sex, age at piercing, and number of piercings are also important risk indicators.

11.2.1.3 Contact dermatitis

Romaguera et al. [83] conducted patch tests with the GEIDC (Spanish Contact Dermatitis Research Group; originally in Spanish language) standard series of allergens, and with eight washers made of copper, nickel, nickel–palladium, palladium, brass, bronze, gold, and iron, were carried out in 964 consecutive patients who complained of intolerance to metals and in 200 controls who did not. It was reported that all subjects were also questioned as to personal and family history of atopy, occupational contact, and intolerance to gold, suggesting the substitution of nickel in imitation jewelry with metals such as palladium or bronze. Jensen et al. [99] investigated a dose–response dependency of oral exposure to nickel. In a double-blind, placebo-controlled oral exposure trial, 40 nickel-sensitive persons and 20 healthy (nonnickel-sensitive) controls were given nickel sulfate hexahydrate in doses similar to and greater than the amount of nickel ingested in the normal Danish daily diet. The nickel content in urine and serum before and after oral exposure was measured to determine nickel uptake and excretion. The influence of the amount of nickel ingested on the clinical reactions to oral exposure and on nickel concentrations in serum and urine was evaluated. It was reported that (i) among nickel-sensitive individuals, a definite dose–response dependency was seen, following oral exposure to nickel, (ii) seven of 10 nickel-sensitive individuals had cutaneous reactions to oral exposure to 4.0 mg nickel, an amount approximately 10 times greater than the estimated normal daily dietary intake of nickel, (iii) four of 10 nickel-sensitive individuals had cutaneous reactions to 1.0 mg nickel, a dose which is close to the estimated maximum amount of nickel contained in the daily diet, (iv) four of 10 nickel-sensitive individuals reacted to 0.3 mg nickel or to the amount equivalent to that contained in a normal daily diet, and one of 10 reacted to a placebo, (iv) none of the 20 healthy controls had cutaneous reactions to 4.0 mg nickel or to a placebo, (v) prior to oral exposure, there was no measurable difference in the amount of nickel in the urine or serum of nickel-sensitive persons and healthy controls, and (vi) following the oral challenge, the nickel content in the urine and serum of both nickel-sensitive and healthy control individuals was directly related to the dose of nickel ingested.

11.2.2 Ti allergy

Titanium is considered to be biocompatible because of a number of favorable theoretical properties: resistance to corrosion, bioinertness, capacity for osseointegration, high fatigue limit, and nonferromagnetism. Consequently, it has already been widely used in medical and dental implants, and projections suggest that its use in such bioapplications will only increase. Other routes of exposure include food and personal care products. Because few diagnosed cases of type IV reactions to titanium have been *reported*, and its allergenic status has been controversial. There are reasons, however, to expect that cases are being underreported. Not only are cases of hypersensitivity likely underrecognized, in vivo diagnostic testing (patch testing) for titanium hypersensitivity has been conducted with water-insoluble compounds that do not penetrate the stratum corneum. Not surprisingly, the results of such tests have overwhelmingly been negative. None of the issues surrounding the allergenicity of titanium will likely be resolved until diagnostic testing for type IV hypersensitivity is conducted with a stable, solvent-soluble, protein-reactive titanium salt that penetrates the skin [55, 100–103].

11.2.2.1 Dental implant and orthopedic implant

It has been shown that titanium has had dramatic success in many surgical procedures as a result of its excellent mechanical properties and resistance to corrosion. There is still concern, however, about the release of metal and controversy surrounding whether or not the plates should be removed after bone healing. Meningaud et al. [104] investigated whether or not there is a relationship between duration of plating and metal release from Ti miniplates in maxillofacial surgery. A prospective cohort study design was used. The concentration of Ti, in the soft tissues covering the plates, was examined in all patients who underwent removal of Ti miniplates from January 1998 to April 1999 (51 cases). It was reported that the study did not support the existence of a relationship between duration of plating and total Ti nor soluble Ti in the soft tissue surrounding the plates. The only independent factor of Ti release found was associated with mechanical constraints during surgery. Almost 100% of Ti is released during the osteosynthesis; then Ti levels remain constant in the surrounding tissues. Most of the time, Ti seems to be clinically inert. It was then concluded that when compared to the possible risks of a second operation, removal of Ti miniplates should not be a routine procedure except in the case of complaints from patients, particularly in the case of infection, hypersensitivity, dehiscence, or screw loosening. Hallab et al. [105] mentioned that (i) if cutaneous signs of an allergic response appear after implantation of a metal device, metal sensitivity should be considered; (ii) currently, there is no generally accepted test for the clinical determination of metal hypersensitivity to implanted devices; (iii) the prevalence of dermal sensitivity in patients with a joint replacement device, particularly those with a failed implant, is substantially higher than that in the general population; and (iv) until the roles of delayed hypersensitivity and humoral immune responses to metallic orthopedic implants are more clearly defined, and the risk to patients may be considered minimal. Langford et al. [106] evaluated histomorphologically the soft tissues adjacent to titanium maxillofacial miniplates and screws in patients and determined the nature of pigmented, particulate debris found in the tissues. Thirty-five soft tissue specimens were excised from the tissues adjacent to titanium miniplates that had been in situ for between 1 month and 13 years. It was obtained that (i) all of the soft tissues showed fibrosis, (ii) pigmented debris was present in 70% of the specimens and titanium was identified by energy dispersive X-ray analysis (EDX) analysis, (iii) the debris was predominantly extracellular and was not associated with any inflammatory response or giant cell reaction, and (iv) fibroblasts were the predominant cell with small aggregates of lymphocytes and scattered macrophages, concluding that titanium is apparently well tolerated for up to 13 years. Mylanus et al. [107] determined whether histologic features or histomorphometric outcomes of retrieved craniofacial percutaneous titanium implants could be found that could be related to chronic pain at the implant site in the temporal bone. Four patients who had previously received percutaneous titanium implants for auricular prostheses (one patient) and a bone anchored hearing aid (three patients) had chronic pain at the implant site despite conservative treatment. The implants were retrieved, sectioned, and ground for qualitative light microscopic inspection and quantitative measurement of the bone-to-metal contact and bone area. It was found that (i) qualitative inspection of the sections of the implants demonstrated soft tissue zones at the interface, especially under the flange; (ii) inflammatory cells (IC) were seen in the interface in all seven implants; (iii) in the implants with good bone-to-metal contact, the percentage of bone-to-metal contact rose with time; and (iv) in the two implants with poor bone-to-metal contact, the soft tissue zones were more extensive, and slightly more skin reactions were observed than in the other implants while still in situ. Based on these findings, it was concluded that a clear explanation for the chronic pain is not at hand. The only common histologic findings were the IC present in the interface, but with varying density, and the presence of soft tissue zones, and a variety of bone-to-metal was found among the retrieved implants. Titanium is generally considered a safe metal to use in implantation but some studies have suggested that particulate titanium may cause health problems either at the site overlying the implant or in distant organs, particularly after frictional wear of a medical prosthesis, causing the so-called wear debris (toxicity). Frisken et al. [108] studied the levels of dissemination of titanium from threaded screw type implants following placement of single implants in sheep mandibles. Twelve sheep were implanted with a single 10 × 3.75 mm self-tapping implant for time intervals of 1, 4, and 8–12 weeks. Four unoperated sheep served as controls. It was obtained that (i) there was no statistically significant different levels of titanium in any organ compared to controls, although some minor elevations in titanium levels within the lungs and regional lymph nodes were noted, and (ii) two implants failed to integrate and these showed higher levels of titanium in the lungs (2.2–3.8 times the mean of the controls) and regional lymph nodes (7–9.4 times the levels in controls). In conclusion, it was also mentioned that (i) debris from a single implant insertion is at such a low level that it is unlikely to pose a health problem, and (ii) even though the number of failed implants was low, multiple failed implants may result in considerably more titanium release which can track through the regional lymph nodes; suggesting that sheep would be an excellent model for following biological changes associated with successful and failed implants and the effect this may have on titanium release.

The incidence and possible causes of abnormal concentrations of metallic elements (Ti, Al, and V) were investigated [109]. Titanium, aluminum, and vanadium concentrations in serum and hair were measured after surgery in 46 patients with titanium alloy spinal implants (12 patients in the implant failure group and 34 patients in the no implant failure group) and 20 patients without spinal implants (control group). All the subjects were examined again 1 year after the first examination or implant removal. It was obtained that (i) of the 46 patients with titanium alloy spinal implants, 16 patients (34.8%) exhibited abnormal serum metal concentrations and 11 patients (23.9%) exhibited abnormal hair metal concentrations; (ii) in the control group, three patients (15%) exhibited only abnormal serum and metal aluminum concentrations at the first examination; (iii) in both of the two patients who exhibited abnormal serum titanium concentrations and then had their spinal implants removed, the serum and hair titanium levels decreased to beneath the reference value limit in 1 year after the removal; (iv) comparison of the implant failure and no implant failure groups showed no significant differences in the incidence of abnormal serum concentrations of titanium, aluminum, or both metals; and (v) serum metal concentrations did not seem to be a useful indicator of hardware loosening or implant failure. Based on these results, it was concluded that (i) approximately one third of patients with titanium alloy spinal implants exhibited abnormal serum or

hair metal concentrations at a mean time of mean 5.1 years after surgery, and (ii) titanium or aluminum may travel to distant organs after dissolution of metals from the
spinal implants. The nonperforated mucosa covering submerged maxillary titanium
implants with regard to induced tissue reactions was histologically evaluated [110].
Thirteen patients, 21–69 years of age, without previous implants were included. After
initial examination, the bone crest areas destined for dental implant placement were
exposed and threaded external hex dental implants were inserted. Prior to wound
closure, a full mucosal tissue slice was biopsied from the edge of the mucoperiosteal
flap (baseline). The patients were monitored monthly for 6 months. At the abutment
connection, biopsies were taken by a 6-mm punch, altogether yielding 26 specimens.
Tissue reactions were analyzed by coded histometric analysis at four defined areas at
increasing distance from the oral epithelium, including ratios of IC/epithelial cells,
IC/fibroblasts, and number of dense particles. It was mentioned that (i) the stained
sections portrayed gingival tissue with intact oral epithelium and connective tissue
with variable accumulation of IC; (ii) experimental biopsies demonstrated mineralized areas and dense particles of different sizes. Analysis of variance revealed a
higher IC/fibroblast ratio for level 3 at baseline compared to level 3 at 6 months; (iii)
a significant decrease in IC/fibroblast ratio was observed between levels 2 and 3, and
2 and 4 at 6 months; and (iv) the connective tissue level facing the cover screw contained the highest number of dense particles, concluding that tissue sensitivity reactions to titanium implants were not disclosed. All 6-month biopsies contained dense
particles that were most likely metals. Olmedo et al. [111] evaluated histologically the
biological effect of pitting corrosion and to contribute clinically relevant data on the
permanence of titanium metal structures used in osteosynthesis in the body. Commercially pure titanium (CPT) laminar implants (control) and CPT laminar implants
with pitting corrosion (experimental) were implanted in the tibiae of rats. It was
found that the histological study of the titanium implants submitted to pitting corrosion showed scarce bone–implant contact, it was only present in the areas with no
pitting and/or surface alterations. There was a statistically significant lower percentage of bone–implant contact in the experimental group (6% ± 4) than in the control
group (26% ± 6). Products of corrosion in the peri-implant bed, especially around the
blood vessels and areas of bone marrow in the metal–tissue interface, were observed.
The adverse local effects caused by pitting corrosion suggest that titanium plates
and grids should be used with caution as permanent fixation structures. Richardson
et al. [112] determined serum titanium levels in patients after instrumented spinal
arthrodesis with implants composed of titanium alloy and to identify potential factors
responsible for any increase in ion levels. Serum titanium concentrations were measured in 30 patients with titanium spinal instrumentation at a mean of 26 months
after surgery and compared with a control group without metallic implants. Comparisons were made regarding serum titanium levels with respect to specific instrumentation characteristics such as number of pedicle screws used, and the presence of cross
connectors or titanium interbody devices. It was reported that serum titanium levels

were significantly higher in patients with titanium spinal implants (mean: 2.6 µg/L) when compared with controls (mean: 0.71 µg/L). Subjects who underwent an instrumented arthrodesis of only one spinal segment had decreased serum titanium levels when compared with those who were fused at two or more spinal segments (mean: 2.3 vs 3.1 µg/L) and patients with four or less pedicle screws also had decreased serum titanium levels when compared with constructs of six to eight pedicle screws (mean: 2.3 vs 3.35 µg/L); however, both of these findings were not statistically significant. Patients without cross connectors had a slightly increased serum titanium level when compared with those with connectors (mean: 2.7 vs 2.44 µg/L); however, this finding was also not statistically significant. Patients with titanium interbody devices had a statistically significant elevation in serum titanium levels when compared with those without (mean: 3.3 vs 1.98 µg/L). It was, therefore, concluded that significantly higher serum titanium concentrations were observed in subjects with titanium spinal instrumentation when compared with controls.

Sicilia et al. [113] evaluated the presence of titanium allergy by the anamnesis and examination of patients, together with the selective use of cutaneous and epicutaneous testing, in patients treated with or intending to receive dental implants of such material. Thirty-five subjects out of 1,500 implant patients treated and/or examined (2002–2004) were selected for Ti allergy analysis. Sixteen presented allergic symptoms after implant placement or unexplained implant failures [allergy-compatible response group (ACRG)], while 19 had a history of other allergies, or were heavily Ti exposed during implant surgeries or had explained implant failures [predisposing factors group (PFG)]. Thirty-five controls were randomly selected (CG) in the allergy center. Cutaneous and epicutaneous tests were carried out. It was reported that (i) nine out of 1,500 patients displayed positive (+) reactions to Ti allergy tests (0.6%): eight in the ACRG (50%), one in the PFG (5.3%)($P = 0.009$) and zero in the control group, and (ii) five positives were unexplained implant failures (five out of eight), concluding that Ti allergy can be detected in dental implant patients, even though its estimated prevalence is low (0.6%), and (iv) a significantly higher risk of positive allergic reaction was found in patients showing postoperative ACRG, in which cases allergy tests could be recommended. Egusa et al. [114] demonstrated the emergence of facial eczema in association with a titanium dental implant placed for a mandibular overdenture supported by two implants and reported that (i) complete remission was achieved by the removal of the titanium material, and (ii) the possibility that in rare circumstances, for some patients, the use of titanium dental implants may induce an allergic reaction. Olmedo et al. [115] reported two novel clinical cases of reactive lesions of the peri-implant mucosa associated with titanium dental implants where metal-like particles which were observed histologically. In both cases, the lesions were diagnosed as epulis, based on clinical evidence. Extirpation biopsies were carried out. It was reported that (i) case 1 was diagnosed as pyogenic granuloma and case 2 as peripheral giant cell granuloma., (ii) the presence of metal-like particles in the tissues suggests that the etiology of the lesions might be related to the corrosion process of the metal structure, and (iii) all clinical cases of soft tissue lesions associated with implants should be reported to contribute to the understanding of the etiology and pathogeny of these lesions. Degradation products of metallic biomaterials including titanium may result in metal hypersensitivity reaction. Hypersensitivity to biomaterials is often described in terms of vague pain, skin rashes, fatigue and malaise and in some cases implant loss. Recently, titanium hypersensitivity has been suggested as one of the factors responsible for implant failure. Siddiqi et al. [116] studied epidemiological data on incidence of titanium-related allergic reactions. A computer search of electronic databases primarily MEDLINE and PUBMED was performed with the following key words: "titanium hypersensitivity," "titanium allergy," "titanium release" without any language restriction. Manual searches of the bibliographies of all the retrieved articles were also performed. In addition, a complementary hand search was also conducted to identify recent articles and case reports. It was reported that most of the literature comprised case reports and prospective in vivo/in vitro trials. About 127 publications were selected for full text reading. The bulk of the literature originated from the orthopedic discipline, reporting wear debris following knee/hip arthroplasties. The rest comprised osteosynthesis (plates/screws), oral implant/ dental materials, dermatology/cardiac-pacemaker, pathology/cancer, biomaterials, and general reports, indicating that (i) titanium can induce hypersensitivity in susceptible patients and could play a critical role in implant failure, (ii) there is a need for long-term clinical and radiographic follow-up of all implant patients who are sensitive to metals, and (iii) at present, a little is known about titanium hypersensitivity, but it cannot be excluded as a reason for implant failure. Basko-Plluska et al. [117] mentioned about the hypersensitivity reactions to metallic implants cutaneous reactions to metal implants, orthopedic or otherwise, which are well documented in the literature. Complex immune reactions may develop around the implants, resulting in pain, inflammation, and loosening. Nickel, cobalt, and chromium are the three most common metals that elicit both cutaneous and extracutaneous allergic reactions from chronic internal exposure. However, other metal ions as well as bone cement components (such as PMMA) can cause such hypersensitivity reactions [117]. To complicate things, patients may also develop delayed-type hypersensitivity reactions to metals (namely, in-stent restenosis, prosthesis loosening, inflammation, pain, or allergic contact dermatitis) following the insertion of intravascular stents, dental implants, cardiac pacemakers, or implanted gynecologic devices. Chaturvedi [118] mentioned that (i) the long-term presence of corrosion reaction products and ongoing corrosion lead to fractures of the alloy-abutment interface, abutment, or implant body; and (ii) the combination of stress, corrosion, and bacteria contribute to implant and/or suprastructure failure. Cundy et al. [119] determined serum titanium, niobium, and aluminum levels in pediatric patients within the first postoperative year after instrumented spinal arthrodesis. The pattern of systemic metal release over time was evaluated by measuring serum titanium, niobium, and aluminum levels preoperatively and 1 week, 1 month, 6 months, and 12 months after instrumented spinal arthrodesis using a titanium alloy. Serum metal levels were measured using HR-ICP-MS. It was found that (i) 32 patients were included in the study group, (ii) mean age at surgery was 14.7 years, (iii) preoperative and postoperative concentrations of serum titanium and niobium were significantly different, (iv) median postoperative serum concentrations of titanium and niobium were elevated 2.4- and 5.9-fold above the normal range respectively with 95% and 99% of samples elevated postoperatively, and (v) a significant and rapid rise in serum titanium and niobium levels was observed within the first postoperative week, after which elevated serum levels persisted up to 12 months. Based on these results, it was concluded that (i) abnormally elevated serum titanium and niobium levels in patients with titanium-based spinal instrumentation up to 12 months was reported, and (ii) the long-term systemic consequences of debris generated by wear and corrosion of spinal instrumentation is unclear but concerning, particularly as these implants inserted into the pediatric population may remain in situ for beyond 6 decades. Javed et al. [120] assessed whether or not Ti sensitivity is associated with allergic reactions in patients with dental implants. To address the focused question "Can Ti cause allergic reactions in patients with dental implants?," databases were explored from 1977 until May 2010 using a combination of the following keywords: allergy, dental, hypersensitivity, implant, oral, and titanium. Letters to the editor and unpublished data were excluded. It was reported that (i) seven studies (six clinical and one experimental) were included; (ii) the participants were aged between 14.3 and 84.1 years; (iii) in five clinical studies, Ti implants were inserted in the mandible; (iv) five studies reported dermal inflammatory conditions and gingival hyperplasia as allergic reactions in patients with Ti dental implants; (v) a case report presented swelling in submental and labial sulcus and hyperemia of soft tissues in a patient with Ti dental implants; (vi) two studies reported that Ti implants are well tolerated in host tissues; and (vii) the patch test was performed in two clinical studies for the diagnosis of allergic reactions, indicating that the significance of Ti as a cause of allergic reactions in patients with dental implants remains unproven.

There have been some concerns that titanium might evoke an unwelcome host reaction and judgments are based on case studies [115] and isolated clinical reports [114]. There is a possible association between surface corrosion of titanium and hypersensitivity reactions or others [110, 113, 116, 118, 120]. In summary, the clinical relevance of allergic reactions in patients with titanium dental implants remains debatable. The results of two recent reviews on the topic reported different conclusions [116, 120]. In the first review, it was concluded that the significance of titanium as a cause of allergic reactions in patients with dental implants remains unproven [120]. On the other hand, the results of the other review indicated that titanium could induce hypersensitivity in susceptible patients and might play a critical role in causing implant failure [116].

Recently, zirconia (ZrO_2) implant have been used in some cases. Oliva et al. [121] introduced the case of the full-mouth oral rehabilitation of a titanium allergic patient. The patient was a young female with amelogenesis imperfecta who had generalized massive tooth destruction. All teeth in the mouth were extracted and 15 CeraRoot acid-etched (ICE surface) implants were placed (seven implants in the maxilla and eight implants in the mandible). No immediate temporaries were placed. Temporaries were placed 3 months after surgery and left in function for 2 months. The case was finally restored with zirconium oxide bridges and ceramic veneering (three bridges in the maxilla and another three in the mandible). It was reported that (i) a good prognosis about the 3-year follow-up showing good stability of soft tissues and bone level was reported and (ii) zirconium oxide implants and restorations might be an alternative for the oral rehabilitation of titanium allergic patients. Osman et al. [122] made a review of dental implant materials in terms of metallic titanium versus ceramic zirconia. It was mentioned that metallic titanium remains the gold standard for the fabrication of oral implants, even though sensitivity does occur, though its clinical relevance is not yet clear; ceramic zirconia implants may prove to be promising in the future; however, further in vitro and well-designed in vivo clinical studies are needed before such a recommendation can be made; and special considerations and technical experience are needed when dealing with zirconia implants to minimize the incidence of mechanical failure.

There are scattered reports on potential allergic reaction against Ti pacemaker [123–126].

11.3 Biocompatibility

11.3.1 In general

Biological compatibility (or biocompatibility) is one of three major requirements for achieving successful implant treatment; the other two are biomechanical compatibility and macro- and micro-morphological compatibility [52, 55]. Ideal implants can serve in the human body for a long-term without revision surgery. After being implanted in the human body, implants induce a lot of reactions in the biological environment with body fluid, proteins and cells, determining the ultimate success of implantation. Therefore, evaluation on the biocompatibility is sine quo non for potential and implantable Ti and Ti alloy implants. As well known, Ti and Ti alloys are inert materials in the human body environment, which can spontaneously form a stable passive film of TiO_2 on the surface of the implants. Even if the passive film is damaged, it is rebuilt rapidly, due to the so-called self-healing ability. This is an important reason that Ti and Ti alloys possess good biocompatibility. In fact, the stable formation of surface passive film on Ti materials is crucial in terms of biological inertness. However, although Ti and Ti alloys have these advantages, the concerns on the biocompatibility of Ti and Ti alloys have been widely discussed and studied since the 1970s [127, 128]. As discussed later in this section, some other properties like hemocompatibility, osseointegration, cytocompatibility, and antibacterial function should be considered.

Compared with stainless steel and Co–Cr-based alloys, Ti and its alloys are widely used as biomedical implants due to many fascinating properties, such as superior mechanical properties, strong corrosion resistance, and excellent biocompatibility. One of the most important things for designing and selecting the biomedical materials is their medical intentions. Once a biomedical implant is placed, a considerable number of reactions would take place at the interface of the host tissue and the implant. These reactions determine the biocompatibility of an implant, which also determines the success of the implantation. Zhang et al. [127] mentioned that, to be biocompatible, toxic elements are not expected to be used in biomedical materials; hence, elements such as Ti, Nb, Mo, Ta, Zr, Au, W, and Sn are highly biocompatible, while Al, V, Cr, Ni, and so on are considered as hazardous elements for the human body [129]. Kuroda et al. [129] reported the detailed information on pure metallic elements and their biocompatibility (see Figure 11.1).

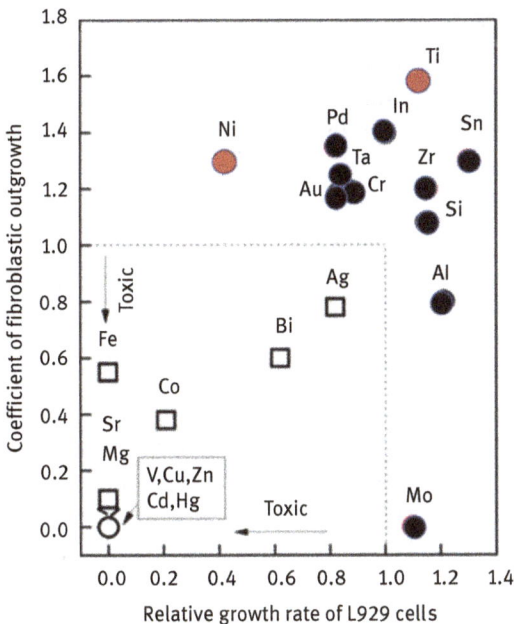

Figure 11.1: Pure metallic elements and their biocompatibilities [129].

11.3.2 Corrosion resistance and biocompatibility

Among various methods to evaluate the biocompatibility of NiTi materials, the evaluation on corrosion resistance has been most frequently employed. Ryhänen et al. [130] clarified the primary cytotoxicity and corrosion rate of NiTi in human cell cultures.

Comparisons were made with stainless steel, Ti (Ti), composite material (CM), and control cultures with no test disks. Human osteoblasts and fibroblasts were incubated for 10 days with test disks of equal size, 6 × 7 mm. It was reported that (i) the proliferation of fibroblasts was 108% (NiTi), 134% (Ti), 107% (Stst), and 48% (C) compared to the control cultures; (ii) the proliferation of osteoblasts was 101% (Nitinol), 100% (Ti), 105% (stainless steel), and 54% (C) compared to the controls; (iii) initially, NiTi released more nickel (129–87 μg/L) into the cell culture media than stainless steel (7 μg/L), but after 2 days the concentrations were about equal (23–5 μg/L vs 11–1 μg/L); and (iv) the titanium concentrations from both NiTi and Ti samples were all <20 μg/L, concluding that NiTi has good in vitro biocompatibility with human osteoblasts and fibroblasts, and despite the higher initial nickel dissolution, NiTi induced no toxic effects, decrease in cell proliferation, or inhibition on the growth of cells in contact with the metal surface. Wen et al. [131] investigated the corrosion resistance and tissue compatibility of $Ti_{50}Ni_{50}$ and $Ti_{50}Ni_{50-x}CuX$ (x = 1, 2, 4, 6, 8). It was found that (i) the repassivation potential of $Ti_{50}Ni_{50-x}CuX$ (x = 2, 4, 6, 8) alloys is about 200 mV higher than that of $Ti_{50}Ni_{50}$ alloy, namely, the addition of Cu raises the repassivation potential of TiNi SMAs and improves their corrosion resistance; (ii) pitting potentials of $Ti_{50}Ni_{50}$ and $Ti_{50}Ni_{50-x}CuX$ (x = 1, 2, 4, 6, 8) alloys increase with solution pH value, but the repassivation potentials keep constant; (iii) the adding of Cu has no obvious influence on pitting potential of TiNi alloys, meanwhile, the corrosion potential and corrosion rate of $Ti_{50}Ni_{50-x}CuX$ (x = 1, 2, 4, 6, 8) alloys are irrelevant to its Cu content and the values are almost the same as those of TiNi alloys; (iv) the connective tissue layer covering the plates is statistically significantly thicker for $Ti_{50}Ni_{42}Cu_8$ plates than that of $Ti_{50}Ni_{50}$, $Ti_{50}Ni_{48}Cu_2$, $Ti_{50}Ni_{44}Cu_6$ plates after 1 month; (v) the numbers of connective tissue cells, polynucleated cells, macrophages and round cells are higher for $Ti_{50}Ni_{42}Cu_8$ plates than those of the other three types of plates, but no statistically significant differences are detected; and (vi) there are no significant differences on tissue reaction parameters after 2 and 3 months among four alloys, concluding that $Ti_{50}Ni_{50-x}CuX$ (x = 2, 6, 8) SMAs have good biocompatibility. Khalil-Allafi et al. [132] investigated $Ni_{50.7}Ti$ (at%) by electrochemical corrosion evaluation in two physiological environments of Ringer solution and NaCl 0.9% solution. It was reported that (i) the breakdown potential of the NiTi alloy in NaCl 0.9% solution is higher than that in Ringer solution, (ii) the pH value of the solutions increases after the electrochemical tests, and (iii) the existence of hydride products in the X-ray diffraction analysis confirms the decrease of the concentration of hydrogen ion in solutions. Zhang et al. [133] studied the biocompatibility of composite archwire (CAW) as an important foundation for its clinical application. The electrochemical corrosion and ion release behavior of CAW upon immersion in solutions simulating oral cavity conditions were measured to evaluate the corrosion behavior of CAW and found that (i) CAW is resistant to corrosion in the tested AS-based solutions (chloric solution, simple AS, fluorinated AS, and protein-containing AS), and (ii) the amount of toxic copper ions released after immersion was lower than average daily dietary intake levels.

Figure 11.2 depicts a relationship between corrosion resistance (in terms of polarization resistance) against biological compatibility [134].

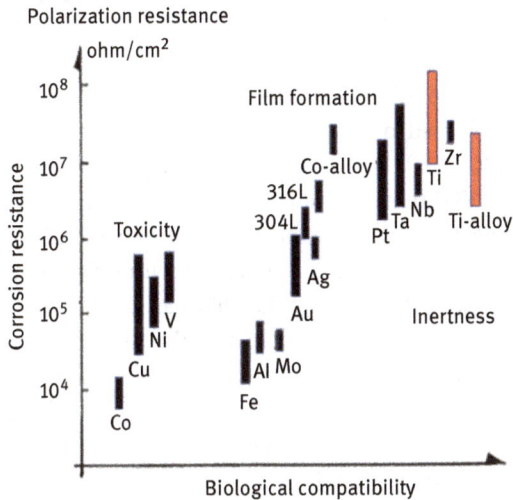

Figure 11.2: Corrosion resistance versus biological compatibility [134].

A stable and passive oxide film of TiO_2 formed on NiTi can control an excellent biocompatibility [135, 136] and such passive oxide film can control undesired ion leaching out to make a good corrosion resistance to NiTi substrate as pointed out in several works [11, 137, 138].

11.3.3 Evaluation

A considerable number of studies have been done and reported about the biological performance of NiTi alloys, indicating that the alloys exhibits excellent biocompatible behavior both in vitro [139], in vivo [140] or in vitro and in vivo [141].

11.3.3.1 In vitro evaluation
The biocompatibility of NiTi screws was evaluated using immunohistochemistry to observe the distribution of bone proteins during bone remodeling process around NiTi implant [139]. It was reported that biocompatibility results of the NiTi screws compared with the other screws (including Vitallium, CpTi, duplex austenite-ferrite stainless steel, 316L type stainless steel) showed a slower osteogenesis process characterized by no close contact between implant and bone, disorganized migration of osteoblasts around the implant, and a lower activity of osteonectin synthesis. Ryhänen et al. [130]

clarified the primary cytotoxicity and corrosion rate of NiTi in human cell cultures, by comparisons with stainless steel, titanium (Ti), CM, and control cultures with no test disks. Human osteoblasts and fibroblasts were incubated for 10 days with test disks of equal size, 6×7 mm. The cultures were photographed and the cells counted. Samples from culture media were collected on days 2, 4, 6, and 8, and the analysis of metals in the media was done using flameless atomic absorption spectrophotometry. It was reported that (i) the proliferation of fibroblasts was 108% (NiTi), 134% (Ti), 107% (stainless steel), and 48% (CM) compared to the control cultures; (ii) the proliferation of osteoblasts was 101% (NiTi), 100% (Ti), 105% (stainless steel), and 54% (CM) compared to the controls; (iii) initially, NiTi released more nickel (129–87 µg/L) into the cell culture media than stainless steel (7 µg/L), but after 2 days the concentrations were about equal (23–5 µg/L vs 11–1 µg/L); and (iv) the titanium concentrations from both NiTi and Ti samples were all <20 µg/L, concluding that NiTi has good in vitro biocompatibility with human osteoblasts and fibroblasts, despite the higher initial nickel dissolution, Nitinol induced no toxic effects, decrease in cell proliferation, or inhibition on the growth of cells in contact with the metal surface. Assad et al. [142] evaluated the relative in vitro genotoxicity of NiTi and compared to CPT, 316L stainless steel, and positive and negative controls. Human peripheral blood lymphocytes were cultured in semiphysiological medium that previously had been exposed to the biomaterials. It was mentioned that (i) cellular chromatin exposed to the positive control demonstrated a significantly stronger immunogold labeling than when it was exposed to NiTi, CPT, stainless steel extracts, or the untreated control; (ii) gold particle counts, whether in the presence of NiTi, CPT, or the negative control medium, were not statistically different; and (iii) NiTi genocompatibility therefore presents promising prescreening results toward its biocompatibility approval. Rhalmi et al. [143] evaluated the general muscle and bone reaction to porous NiTi alloy. The latter material was implanted in rabbit tibias and back muscle, and assessed after 3, 6, and 12 weeks of implantation. Porous NiTi specimens did not cause any adverse effect regardless of both implantation site and postsurgery recovery time. Muscle tissue exhibited thin tightly adherent fibrous capsules with fibers penetrating into implant pores. It was obtained that (i) attachment strength of the soft tissue to the porous implant seemed to increase with postimplantation time; (ii) bone tissue demonstrated good healing of the osteotomy; (iii) there was bone remodeling characterized by osteoclastic and osteoblastic activity in the cortex; (iv) this general good in vivo biocompatibility with muscle and bone tissue corresponded very well with the in vitro cell culture results we obtained; and (v) fibroblasts seeded on porous NiTi sheets managed to grow into the pores and all around specimen edges showing an another interesting cytocompatibility behavior, indicating a good biocompatibility acceptance of porous NiTi and very promising toward eventual NiTi medical device approbation. The biocompatibility of an orthopedic implant depends on the effect of the implant on bone-forming cells, osteoblasts. Changes in osteoblastic proliferation, maturation, and differentiation are important events in ossification that enable monitoring the effect of the implant.

Kapanen et al. [144] measured the levels of transforming growth factor (TGF-beta) with enzyme-linked immunosorbent assay (ELISA) from a ROS-17/2.8 osteosarcoma cell line cultured on different metal alloy disks, including NiTi, stainless steel, pure titanium (Ti) and pure nickel (Ni). It was found that (i) the TGF-beta1/DNA value in the NiTi group (0.0007 ± 0.0003) was comparable with those seen in the stainless steel (0.0008 ± 0.0001) and Ti (0.0007 ± 0.0001) groups; (ii) the concentration in the Ni group was lower (0.0006 ± 0.0003), though not statistically significantly; (iii) increasing roughness of the NiTi surface increased the TGF-beta1 concentration; on the other hand, all roughness groups of TiII showed low levels of TGF-beta1, while a rough TiI surface induced similar TGF-beta1, expression as rough NiTi; and (iv) these same measurements made with interleukin 6 (IL-6) were found to be under the detection limit in these cultures, concluding that a rough NiTi surface promotes TGF-beta1 expression in ROS-17/2.8 cells. Es-Souni et al. [145] analyzed research work published on the biocompatibility of NiTi alloys, considering aspects related to: (1) corrosion properties and the different methods used to test them, as well as specimen surface states; (2) biocompatibility tests in vitro and in vivo; (3) the release of Ni ions. It was mentioned that (i) NiTi SMAs are generally characterized by good corrosion properties, in most cases superior to those of conventional stainless steel or CoCrMo-based biomedical materials, (ii) the majority of biocompatibility studies suggest that these alloys have low cytotoxicity (both in vitro and in vivo) as well as low genotoxicity, (iii) the release of Ni ions depends on the surface state and the surface chemistry, and (iv) smooth surfaces with well-controlled structures and chemistries of the outermost protective TiO_2 layer lead to negligible release of Ni ions, with concentrations below the normal human daily intake. Bogdanski et al. [146] investigated the biocompatibility of NiTi alloys by single-culture experiments on functionally graded samples with a stepwise change in composition from pure nickel to pure titanium. It was reported that a good biocompatibility for a nickel content up to about 50% was found. User et al. [147] investigated the biocompatibility of NiTi archwires by simulating actual contact state of archwires around brackets, which enabled incorporation of realistic mechanical conditions into *ex situ* experiments, by immersing in AS for 31 days. Following the immersion, the archwires and the immersion solutions were analyzed with the aid of various electron-optical techniques. It was reported that (i) carbon-rich corrosion products was formed on both archwire sets upon immersion; (ii) the corrosion products preferentially formed at the archwire–bracket contact zones, which is promoted by the high energy of these regions and the microcracks brought about by stress assisted corrosion; and (iii) these corrosion products prevented significant Ni or Ti ion release by blocking the microcracks, which, otherwise, would have led to enhanced ion release during immersion, indicating the need for incorporating both realistic chemical and mechanical conditions into the ex situ biocompatibility experiments of orthodontic archwires, including the archwire–bracket contact. Toker et al. [148] compared to companion wires retrieved from patients in terms of chemical changes and formation of new structures on the surface. It was mentioned that (i) the acidic erosion effective at the earlier stages of immersion led to the formation of new structures as the immersion period approached 30 days, (ii) comparison of these results with the analysis of wires utilized in clinical treatment evidenced that ex situ experiments are reliable in terms predicting C-rich structure formation on the wire surfaces; however, the formation of C pileups at the contact sites of arch wires and brackets could not be simulated with the aid of static immersion experiments, warranting the simulation of the intraoral environment in terms of both chemical and physical conditions, including mechanical loading, when evaluating the biocompatibility of NiTi orthodontic arch wires.

11.3.3.2 In vitro and in vivo evaluation

Morita et al. [149] evaluated the biocompatibility of NiTi SMA used as a medical implant material, by electrochemical corrosion test and in vivo and in vitro biological tests for the alloy. The tests included (1) anodic polarization test for the alloy in a quasibody fluid; (2) cell proliferation tests for pure titanium, pure nickel, SUS316L stainless steel, Ti6Al4V, TiNi$_{55}$ by using of L929 fibroblastic cells; (3) lactate dehydrogenase (LDH), human interleukin-1 (hIL-1), and human tumor necrosis factor (hTNF) biochemical assays by using of U937 human macrophages administered the corrosion products of these alloys for the cells; (4) measurement of the mount of excretions of the metallic corrosion products of TiNi$_{55}$, SUS316L stainless steel, and Ti6Al4V with urine and feces injected into the abdomen cavity of Wistar rats; and (5) tissue reaction observations for SUS316L, TiNi$_{55}$, and Ni wires implanted along the femoral bone axis of the rats. It was found that (i) the pitting corrosion potentials of TiNi$_{55}$ alloy was drastically improved by the aging treatment; (ii) in the case of TiNi$_{55}$ alloy, the inflammatory cytokines, hIL-6 and hTNF, were suppressed to lower levels compared with Ti6Al4V alloy; (iii) corrosion products prepared from the titanium alloys were stable in the body; (iv) it was very hard to eliminate the titanium ions with urine and feces; (v) TiNi$_{55}$ alloy was shown an excellent biocompatibility evaluated by the in vivo implantation test, because of the stable passive film formed on the surface and protected the metal ion release to the surrounding tissue. Hodorenko et al. [141] studied the applicability of a porous NiTi-based SMA scaffold as an incubator for bone marrow mesenchymal cells, hepatocytes, and pancreatic islet cells. The porous NiTi-based alloy was fabricated with a self-propagating high-temperature synthesis (SHS) technique, in which scaffold blocks measuring 4 × 4 × 10 mm were prepared. The in vitro tests were done using mesenchymal stem cells (MSC) isolated from mature bone marrow of CBA/j inbred mice and cultured in three different culture media – control medium, osteogenic medium, and chondrogenic medium. Hepatocytes and islet cells were isolated from the livers and pancreatic glands of Wistar rats, respectively, seeded on porous TiNi-based SMA scaffolds, and cultured. The scaffolds were then implanted into the abdominal cavity of Wistar rats and later harvested, at days 7, 14, 21, and 28, postimplantation. SEM imaging was performed with preimplanted scaffolds at day

0 and harvested scaffolds at days 7, 14, 21, and 28, postimplantation. It was reported that (i) based on weight increase percentages, the *in vitro* study revealed that the osteogenic group showed a twofold increase, and the chondrogenic group showed a 1.33-fold increase, compared to the control group; and (ii) the in vivo study, on the other hand, showed that from day 7 postimplantation, the cellular ingrowth gradually invaded the inner porous structure from the periphery toward the center, and at day 28 postimplantation, all pores were closed and completely filled with cells and the extracellular matrix; suggesting that porous NiTi-based SMA is a unique biocompatible incubator for cell cultures and can be successfully used for tissue bioengineering and artificial organs.

11.3.3.3 In vivo evaluation

General soft tissue response and biocompatibility to NiTi alloy in vivo was investigated and clarified neural and perineural responses [150]. Seventy-five rats were randomized into three groups. Test specimens were implanted into paravertebral muscle and near the sciatic nerve. A comparison was made between Nitinol, stainless steel, and Ti6Al4V. The animals were euthanized at 2, 4, 8, 12, and 26 weeks after implantation. General morphologic and histologic observations were made under light microscopy. Semiautomatic computerized image analysis was used to measure the encapsule membrane thickness around the implants. It was found that (i) the muscular tissue response to NiTi was clearly nontoxic, regardless of the time period; (ii) the overall inflammatory response to NiTi was very similar to that of stainless steel and Ti6Al4V alloy; (iii) the immune cell response to NiTi remained low; (iv) only a few foreign-body giant cells were present; (v) the detected neural and perineural responses were also clearly nontoxic and nonirritating with NiTi alloy; (vi) no qualitative differences in histology between the different test materials could be seen; (vii) at 8 weeks, the encapsule membrane of Nitinol was thicker than that of stainless steel (mean 62 ± 25 µm vs 41 ± 8 µm); and (vii) at the end of the study, the encapsule thickness was equal to all the materials tested, concluding that NiTi exhibited good in vivo biocompatibility after intramuscular and perineural implantation in rats in the 26-week follow-up. Kang et al. [140] fabricated porous NiTi SMAs with interconnected pores and evaluated bone tissue response and histocompatibility of porous NiTi SMA in vivo. Thirty block implants (5 mm × 5 mm × 7 mm) were prepared. Analysis of pore structure of the implant was performed using Hg-porosimetry and scanning electron microscope. Fifteen New Zealand white rabbits were used. Sterile porous TiNi SMA implant was implanted in the defects of proximal tibia metaphysis. Limbs of five rabbits were harvested respectively at 2, 4, and 6 weeks post implantation. It was obtained that (i) the pore sizes of porous NiTi alloy were 323 ± 89 µm, with porosity of $55.3 \pm 6.7\%$, (ii) no apparent adverse reactions such as inflammation and foreign body reaction were noted on or around all implanted porous NiTi blocks, (iii) bone ingrowth was found in the pore space of all implanted blocks and the percent bone ingrowth into the pore space of porous NiTi alloy increased over time, and (iv) at 6-week postimplantation, bone ingrowth into pore in the block was very excellent (at 6 weeks, 78.3 ± 9.7%) and this percent bone ingrowth was much higher than that of other porous materials; indicating the possibility that porous TiNi SMA could be used as an ideal bone substitute. Tosun et al. [151] blended the powders of Ni and Ti with 50.5 at% Ni for 12 h and cold-pressed at the different pressures (50, 75, and 100 MPa). The thus obtained porous NiTi compacts was further synthesized by the SHS in the different preheating temperatures (200, 250, and 300 °C) and heating rates (30, 60, and 90 °C/min). The effects of the pressure, preheating temperature and heating rate were investigated on biocompatibility in vivo. It was reported that (i) the porosity in the synthesized products was in the range of 50.7–59.7 vol%, (ii) the pressure, preheating temperature and heating rate were found to have an important effect on the biocompatibility in vivo of the synthesized products, and (iii) maximum fibrotic tissue within the porous implant was found in vivo periods (6 months), in which compacting pressure 100 MPa.

11.3.4 Improvement

Although a basic concept to improve the biocompatibility of NiTi alloy is to establish a stable and passive surface oxide film, there can be alternative methods.

11.3.4.1 Oxidation treatment

Various oxidation treatments were applied to nearly equiatomic NiTi alloys so as to form a Ni-free protective oxide on the surface. Michiardi et al. [152] analyzed surfaces of NiTi by the x-ray photoelectron spectroscopy (XPS). It was reported that (i) oxidation treatment in a low-oxygen pressure atmosphere leads to a high surface Ti/Ni ratio, a very low Ni surface concentration and a thick oxide layer; (ii) this oxidation treatment does not significantly affect the shape memory properties of the alloy; and (iii) the oxide formed significantly decreases Ni release into exterior medium comparing with untreated surfaces. As a consequence, this new oxidation treatment could be of great interest for biomedical applications, as it could minimize sensitization and allergies and improve biocompatibility and corrosion resistance of NiTi SMAs. Chu et al. [153] fabricated a dense titania film in situ on NiTi by anodic oxidation in a Na_2SO_4 electrolyte. The microstructure of the titania film and its influence on the biocompatibility of NiTi SMA are investigated by SEM, XPS, ICP-MS, hemolysis analysis, and platelet adhesion test. It was indicated that (i) the titania film has a Ni-free zone near the surface and can effectively block the release of harmful Ni ions from the NiTi substrate in SBF, and (ii) the wettability, hemolysis resistance, and thromboresistance of the NiTi sample are improved by this anodic oxidation method. Toker et al. [154] carried out a systematic set of ex situ experiments on NiTi SMA in order to identify the dependence of its biocompatibility on sample geometry and body location. NiTi samples with three different geometries were immersed into three different fluids simulating different body parts. The changes observed in alloy surface and chemical content of fluids upon immersion experiments designed for four different time periods were analyzed in terms of ion release, oxide layer formation, and chemical composition of the surface layer. It was reported that both sample geometry and immersion fluid significantly affect the alloy biocompatibility, as evidenced by the passive oxide layer formation on the alloy surface and ion release from the samples. Upon a 30-day immersion period, all three types of NiTi samples exhibited lower ion release than the critical value for clinic applications; however; a significant amount of ion release was detected in the case of gastric fluid, warranting a thorough investigation prior to utility of NiTi in gastrointestinal treatments involving long-time contact with tissue. Certain geometries appear to be safer than the others for each fluid, providing a new set of guidelines to follow while designing implants making use of NiTi SMAs to be employed in treatments targeting specific body parts. Chembath et al. [155] anodized NiTi alloy on a biomedical grade NiTi alloy in phosphoric acid-based electrolyte and subjected to posttreatment by annealing. It was obtained that (i) the corrosion potential shifted to more noble direction after annealing and there was one order decrease in corrosion current density compared to as-anodized surface; (ii) cyclic potentiodynamic polarization behavior of mechanically polished (MP) and anodized surfaces was almost similar and both the surfaces exhibited positive hysteresis loop on reversing the scan direction; (iii) the formation of negative hysteresis loop in the case of annealed anodized alloy showed that the surface was capable of retaining its passive behavior even at higher potentials; (iv) dynamic impedance spectroscopy studies carried out on annealed anodized surfaces showed that on increasing the potential from open circuit potential to higher potentials (0.8 V), the thickness of the oxide layer also increased, but behaved as a leaky capacitor; and (v) in terms of bioactivity, the entire surface of the annealed sample was covered with HA after 12 days of immersion in Hanks' solution and the nickel released (370 ppb) was comparatively less than bare NiTi surface (633 ppb).

11.3.4.2 HA coating

Microstructural characteristics and biocompatibility of a Type-B carbonated HA coating prepared on NiTi SMA by biomimetic deposition were characterized using X-ray diffraction (XRD), SEM, XPS, Fourier transform infrared spectroscopy (FTIR) and in vitro studies including hemolysis test, MTT cytotoxicity test and fibroblasts cytocompatibility test [156]. It was reported that (i) CO_3^{2-} groups were present as substitution of PO_4^{3-} anions in HA crystal lattice due to Type-B carbonate; (ii) the growth of Type-B carbonated HA coating in SBF-containing HCO_3^{1-} ions is stable during all periods of biomimetic deposition; (iii) the carbonated HA coating has better blood compatibility than the chemically polished NiTi SMA; (iv) there was a good cell adhesion to this HA coating surface and cell proliferation in the vicinity of the coating was better than that for the chemically polished NiTi SMA; and (v) thus biomimetic deposition of this carbonated HA coating is a promising way to improve the biocompatibility of NiTi SMA for implant applications. Marashi-Najafi et al. [157] studied the pulse electrochemical deposition of HA on NiTi alloy and in vitro evaluation of coatings. At first step, a thermochemical surface modification process was applied to control the Ni release of the alloy. The electrochemical deposition of CaP coatings was examined at both dilute and concentrated solutions. The morphology and the composition of coatings were studied using SEM and FTIR. It was found that (i) plate-like and needle-like morphologies were formed for dilute and concentrated solutions, respectively, and HA phase was formed by increasing the pulse current density for both electrolyte; (ii) the thickness of the samples was measured using cross-sectioning technique; (iii) fibroblast cell culture test on the coated samples revealed that the HA coating obtained by dilute solution shows the best biocompatibility; (iv) MTT assay showed the highest cell density and cell proliferation after 5 days for the HA coating of dilute solution; (vi) the contact angle of samples was measured and the coated samples showed a hydrophilic surface; (vii) soaking the sample in SBF revealed that the crystallization rate of calcium phosphate compounds is higher on the plate-like HA coating as compared to the needle-like morphology; (viii) the P release of the HA-coated samples was measured in a physiological saline solution and the results show that the ions releasing in the plate-like coating are less than the needle-like coating; and (ix) it seems that the stability of the plate like coating in biological environments is responsible for the better biocompatibility of the coating. Kocijan et al. [158] employed the electrodeposition method to apply an HA coating on the surface of a biocompatible NiTi alloy with the aim to enhance the corrosion resistance of the substrate and therefore decrease the release of nickel ions from the surface of the alloy. It was mentioned that (i) a uniform inner layer of HA was formed and overlaid by clusters of spherical and flake-like morphological structures, (ii) the HA coating exhibited superhydrophilic wetting properties compared to the moderate hydrophilicity of the NiTi alloy, (iii) the barrier properties of the HA coating were investigated by using potentiodynamic measurements, (iv) the HA coating significantly enhanced the corrosion resistance of the NiTi alloy and confirmed the effective barrier properties of the coating, which can be used to minimize the nickel release in biomedical applications.

11.3.4.3 TiN coating

TiN coating has been demonstrated to improve the biocompatibility of bare NiTi alloys; however, essential biocompatibility differences between NiTi alloys before and after TiN coating are not known so far. Lifeng et al. [159] explored the underlying biological mechanisms of biocompatibility differences between bare and TiN-coated NiTi substrate. The changes of bare and TiN-coated NiTi alloys in surface chemical composition, morphology, hydrophilicity, Ni ions release, cytotoxicity, apoptosis, and gene expression profiles were also compared. It was reported that compared with the bare NiTi alloys, TiN coating significantly decreased Ni ions content on the surfaces of the NiTi alloys and reduced the release of Ni ions from the alloys, attenuated the inhibition of Ni ions to the expression of genes associated with anti-inflammatory, and also suppressed the promotion of Ni ions to the expression of apoptosis-related genes. TiN coating distinctly improved the hydrophilicity and uniformity of the surfaces of the NiTi alloys, and contributed to the expression of genes participating in cell adhesion and other physiological activities, indicating that the TiN-coated NiTi alloys will help overcome the shortcomings of NiTi alloys used in clinical application currently, and can be expected to be a replacement of biomaterials for a medical device field. Jin et al. [160] evaluated the biocompatibility of TiN-coated NiTi SMA to compare with that of the uncoated NiTi SMA. Based on the orthodontic clinical application, the surface properties and biocompatibility were characterized by SEM, XRD, wettability test, mechanical test, and in vitro tests including MTT, cell apoptosis, and cell adhesion tests. It was found that the bonding between the substrate and TiN coating is excellent. The roughness and wettability increased as for the TiN coating compared with the uncoated NiTi SMA. MTT test showed no significant difference between the coated and uncoated NiTi SMA, however, the percentage of early cell apoptosis was significantly higher as for the uncoated NiTi alloy. SEM results showed that TiN coating could enhance the cell attachment, spreading, and proliferation on NiTi SMA, suggesting that TiN coating bonded with the substrate well and could lead to a better biocompatibility.

11.3.4.4 Carbon coating

The biocompatibility of diamond-like carbon (DLC)-coated NiTi SMA in vitro and in vivo was investigated [161]. The in vitro study was carried out by coculturing the DLC-coated and DLC-uncoated NiTi SMA with bone marrow MSCs, respectively, and the in vivo study was carried out by fixing the rabbits' femoral fracture model by DLC-coated and DLC-uncoated NiTi SMA embracing fixator for 4 weeks, respectively. It was reported that the concentration of the cells, alkaline phosphatase, and nickel ion in culture media were detected, respectively, at the first to fifth day after coculturing. The inorganic substance, osteocalcin, alkaline phosphatase, and tumor necrosis factor (TNF) in callus surrounding fracture and the Ni^+ in muscles surrounding fracture site, liver, and brain were detected 4 weeks postoperatively. The in vitro study showed that the proliferation of MSCs and the expression of alkaline phosphatase in the DLC-coated group were higher than the uncoated group, while the uncoated group released more Ni^{2+} into the culture media than that in the coated group. The in vivo study revealed that the inorganic substance and alkaline phosphatase, osteocalcin, and TNF expression were significantly higher in the DLC coated NiTi SMA embracing fixator than that in the uncoated group. Ni^{2+} in the liver, brain, and muscles surrounding the fracture were significantly lower in the DLC-coated groups than that in the uncoated group. Based on the results, it was concluded that NiTi SMA coated by DLC appears to have better biocompatibility in vitro and in vivo compared to the uncoated one. Markhoff et al. [162] conducted an in vitro study by culturing the osteoblastic cell line MG-63 as well as primary human osteoblasts, fibroblasts, and macrophages on titanium alloys (forged Ti6Al4V, additive manufactured Ti6Al4V, NiTi, and DLC-coated NiTi) to verify their specific biocompatibility and inflammatory potential. It was mentioned that (i) additive manufactured Ti6Al4V and NiTi revealed the highest levels of metabolic cell activity; (ii) DLC-coated NiTi appeared as a suitable surface for cell growth, showing the highest collagen production; (iii) none of the implant materials caused a strong inflammatory response; and (iv) in general, no distinct cell-specific response could be observed for the materials and surface coating used, summarizing that all tested titanium alloys seem to be biologically appropriate for application in orthopedic surgery.

11.3.4.5 Coating of composites

Surface treatment of NiTi alloys to improve their biocompatibility is of interest. Kei et al. [163] deposited TiO_2 and/or Al_2O_3 layer on the surface of NiTi thin plates as a protection layer by using atomic layer deposition. Huang et al. [164] developed the new TiO_2/heparin coatings on NiTi SMA, for the purpose of reducing possible release of toxic Ni ions and hence improve hemocompatibility. Sevost'yanov et al. [165] coated NiTi substrate with titanium or tantalum-enriched surface layers exhibiting a nearly two times higher mitotic index. Thangavel et al. [166] employed RF magnetron sputtering to coat NiTi/Ag compounds on NiTi substrate. Sheiko et al. [167] coated high quality biocompatible poly-ether-ether-ketone (PEEK) on NiTi shape-memory wires using dipping deposition from colloidal aqueous PEEK dispersions after substrate surface treatment.

11.3.4.6 Plasma treatment

Wang et al. [168] assessed the surface modification effects of plasma coatings on biocompatibility of nitinol intravascular stent in terms of anticoagulation, hemocytolysis rate, hydrophilicity, and cytotoxicity. In order to improve their surface-adhesive properties to endothelial cells, NiTi alloy intravascular stents were treated and coated using a low-temperature plasma deposition technique. It was reported that (i) plasma coating changed the surface morphology of the stents to a micron-level surface roughness in the range of 1–5 μm, and (ii) in comparison with the untreated control, the plasma-treated NiTi alloy intravascular stents showed increased surface hydrophilicity and enhanced anticoagulation property. Liu et al. [169] modified the surface of NiTi alloy by N-PIII at various voltages. It was mentioned that (i) the XPS results reveal that near-surface Ni concentration is significantly reduced by PIII and the surface TiN layer suppresses nickel release and favors osteoblast proliferation, especially for samples implanted at higher voltages; (ii) the surfaces produced at higher voltages of 30 and 40 kV show better adhesion ability to osteoblasts compared to the unimplanted and 20 kV PIII samples; and (iii) the effects of heating during PIII on the phase transformation behavior and cyclic deformation response of the materials were investigated by differential scanning calorimetry and three-point bending tests, indicating that N-PIII conducted using the proper conditions improves the biocompatibility and mechanical properties of the NiTi alloy significantly. Chrzanowski et al. [170] investigated three types of surface modifications: thermal oxidation, alkali treatment, and plasma sputtering, and compared with smooth, ground surface. It was reported that thermal oxidation caused a drop in surface nickel content, while negligible chemistry changes were observed for plasma-modified samples when compared with control ground samples. In contrast, alkali treatment caused significant increase in surface nickel concentration and accelerated nickel release. Nickel release was also accelerated in thermally oxidized samples at 600 °C, while in other samples it remained at low level. Both thermal oxidation and alkali treatment increased the roughness of the surface, but mean roughness R(a) was significantly greater for the alkali-treated ones. Ground and plasma-modified samples had "smooth" surfaces with R(a) = 4 nm. Deformability tests showed that the adhesion of the surface layers on samples oxidized at 600 °C and alkali treatment samples was not sufficient; the layer delaminated upon deformation, and the cell cytoskeletons on the samples with a high nickel content or release were less developed, suggesting some negative effects of nickel on cell growth, indicating that smooth, plasma-modified surfaces provide sufficient properties for cells to grow. Gudimova et al. [171] reported the results of the chemical, topographic and structural properties of the NiTi alloy surface and their changes after surface treatments by ion implantation techniques with use of ions such as Ta* and Si*. The influence of physicochemical properties of the surface ion-modified NiTi alloy was studied on in vitro cultured MSCs of the rats' bone marrow. It was observed that (i) the ion surface modification improves histocompatibility of the NiTi alloy and leads to increase of proliferative activity of MSCs on its surface, and (ii) a major contribution to viability improvement MSCs of rat marrow has the chemical composition and the microstructure of the surface area. In order to improve the biocompatibility and prevent the release of nickel element, Jenko et al. [172] oxidized NiTi alloy using different procedures. It was reported that the only procedure that led to the formation of nickel-free oxide films involved a pretreatment with hydrogen plasma for 10 s followed by a treatment with a plasma composed of 90% H_2 and 10% O_2. The extreme chemical reactivity of such a plasma resulted in the formation of an oxide film in about 15 s, meaning that external oxidation took place. The biocompatibility investigations, performed according to the ISO standard protocol using L929 cells, showed the absence of any cytotoxic effects that might be due to a contact between the biological materials and nickel, and the investigation of nickel release of samples exposed to Hank's solution, measured by ICP optical emission spectroscopy showed negligible Ni concentrations.

11.3.4.7 Porous structure

Porous NiTi alloys have demonstrated bone attachment as well as tissue ingrowth in the past. Rhalmi et al. [143] evaluated the general muscle and bone reaction to porous NiTi, which was implanted in rabbit tibias and back muscle, and assessed after 3, 6, and 12 weeks of implantation. It was found that (i) porous NiTi specimens did not cause any adverse effect regardless of both implantation site and postsurgery recovery time; (ii) muscle tissue exhibited thin tightly adherent fibrous capsules with fibers penetrating into implant pores; (iii) attachment strength of the soft tissue to the porous implant seemed to increase with postimplantation time; (iv) bone tissue demonstrated good healing of the osteotomy and there was bone remodeling characterized by osteoclastic and osteoblastic activity in the cortex; and (v) fibroblasts seeded on porous NiTi sheets managed to grow into the pores and all around specimen edges showing an another interesting cytocompatibility behavior, indicating good biocompatibility acceptance of porous NiTi and very promising toward eventual NiTi medical device approbation. Prymak et al. [173] utilized disks consisting of macroporous NiTi alloy as implants in clinical surgery, for example, for fixation of spinal dysfunctions. Studies on the biocompatibility were performed by coincubation of porous NiTi samples with isolated peripheral blood leukocyte fractions (PMN neutrophil granulocytes; peripheral blood mononuclear leukocytes, PBMC) in comparison with control cultures without NiTi samples. It was reported that the cytokine response of PMN (analyzed by the release of IL-1ra and IL-8) was not significantly different between cell cultures with or without NiTi. There was a significant increase in the release of IL-1ra, IL-6, and IL-8 from PBMC in the presence of NiTi samples, In contrast, the release of TNF-alpha by PBMC was not significantly elevated in the presence of NiTi. IL-2 was released from PBMC only in the range of the lower detection limit in all cell cultures. The material, clearly macroporous with an interconnecting porosity, consists of NiTi (martensite; monoclinic, and austenite; cubic) with small impurities of $NiTi_2$ and possibly NiC_X, and the material is not superelastic upon manual compression and shows a good biocompatibility. Li et al. [174] developed an in situ nitriding method to modify the outer surface and the pore walls of both open and closed pores of porous NiTi SMA as part of their sintering process. It was reported that (i) the in situ nitrided porous NiTi SMAs exhibit much better corrosion resistance, cell adherence, and bone tissue-induced capability than the porous NiTi alloys without surface modification, (ii) the released Ni ion content in the blood of rabbit is reduced greatly by the in situ nitriding, and (iii) the excellent biocompatibility of in situ nitrided sample is attributed to the formation of the TiN layer on all the pore walls including both open and closed pores. Zheng et al. [175] fabricated bulk ultrafine-grained $Ni_{50.8}Ti_{49.2}$ alloy (UFG-NiTi) by equal-channel angular pressing technique to improve its surface biocompatibility. It was mentioned that (i) the pitting corrosion potential was increased from 393 mV (SCE) to 704 mV (SCE) with sandblasting and further increased to 1,539 mV (SCE) with following acid etching in HF/HNO_3 solution, (ii) all the above surface treatment increased the apatite forming ability of UFG-NiTi in varying degrees when soaked them in SBF, (iii) both sandblasting and acid etching could promote the cytocompatibility for osteoblasts: sandblasting enhanced cell attachment and acid etching increased cell proliferation, and (iv) the different corrosion behavior, apatite forming ability and cellular response of UFG-NiTi after different surface modifications are attributed to the topography and wettability of the resulting surface oxide layer. Porous NiTi can exhibit an engineering elastic modulus comparable to that of cortical bone (12–17 GPa), so that the biomechanical compatibility can be achieved. Bassani et al. [176] produced open cell porous NiTi by SHS, starting from Ni and Ti mixed powders. The biocompatibility of such material was investigated by single culture experiment and ionic release on small specimen. It was noted that in particular, NiTi and porous NiTi were evaluated together with elemental Ti and Ni reference metals and the two intermetallic $TiNi_3$, Ti2Ni phases, so that the influence of secondary phases in porous NiTi materials and relation with Ni-ion release can be identified, indicating (i) apart from the well-known high toxicity of Ni, also toxicity of $TiNi_3$, while phases with higher Ti content showed high biocompatibility, and (ii) a slightly reduced biocompatibility of porous NiTi was ascribed to combined effect of $TiNi_3$ presence and topography that requires higher effort for the cells to adapt to the surface.

11.4 Hemocompatibility

Hemocompatibility (blood compatibility) is the most important aspect and a key property for biomaterial devices contacting with blood including cardiovascular stents. Hemocompatibility of biomaterials is mainly dependent on its surface characteristics, including composition, roughness wettability, surface free energy, and morphology will affect an implant material's hemocompatibility.

Thierry et al. [177] compared the relative thrombogenicity of nitinol versus stainless steel stents. Nitinol stents were laser cut to reproduce the exact geometry of the stainless steel Palmaz® stents and tested in an ex vivo Aeteriovenous (AV) shunt porcine model under controlled conditions. It was reported that Nitinol stents presented only small amounts of white and/or red thrombus principally located at the strut intersections while Palmaz® stents clearly exhibited more thrombus. Along with the unique mechanical properties of nitinol, its promising hemocompatibility may promote their increasing use for both peripheral and coronary revascularization procedures. Huang et al. [178] prepared Ti–O thin films to improve hemocompatibility by PIII and deposition and by sputtering. The behavior of fibrinogen adsorption was investigated by ^{125}I radioactive isotope labeling. It was mentioned that the systematic evaluation of hemocompatibility, including in vitro clotting time, thrombin time, prethrombin time, platelet adhesion, and in vivo implantation into dog's ventral aorta or right auricle from 17 to 90 days, proved that Ti–O films have excellent hemocompatibility, suggesting that the significantly lower interface tension between Ti–O films and blood and plasma proteins and the semiconducting nature of Ti–O films give them their improved hemocompatibility. Gao et al. [179] coated a thin TiO_2 film on the surface of NiTi SMA by activated sputter method. The blood platelet adherence and antithrombogenicity of the TiO_2-coated NiTi alloy were evaluated. It was shown that (i) the platelets on the TiO_2-coated NiTi alloy were fewer than those on 316L stainless steel, and no agglomeration or distortion for the platelets on the coated alloy was found, which means less probability of blood coagulation for the alloy, (ii) the coagulation time on the coated NiTi SMA was longer than that on the 316L, and (iii) compared with that on the 316L stainless steel, the TiO_2-coated NiTi SMA showed better blood compatibility, indicating that the NiTi alloy with TiO_2 coating is a kind of ideal biomedical materials with high clinical value. Liu et al. [180] established the micromagnetic field on the surfaces of 316L stainless steel and NiTi alloy through the magnetization process of sol–gel prepared TiO_2 thin film with the powder of $SrFe_{12}O_{19}$. It was reported that (i) the deposited thin film can decrease the etching of body fluid as well as prevent the hazardous Ni ions released from the metal, and (ii) with evaluation of dynamic cruor time test and blood platelets adhesion test, the micromagnetic field of the thin film can improve the blood compatibility. Zhao et al. [181] formed a composite TiO_2/Ta_2O_5 nanofilm the NiTi SMA by Ta implantation. The wettability, protein adsorption, platelets adhesion, and hemolysis tests are conducted to evaluate the hemocompatibility. It was found that the contact angle measurements showed that the surface of the NiTi alloy kept hydrophilic before and after Ta implantation, although the water contact angle increased with the increasing implantation current. Both of the surface energy and the interfacial tension decreased after Ta implantation. The protein adsorption behavior was investigated by ^{125}I isotope labeling. The fibrinogen adsorption was enhanced by a high surface roughness or a large interfacial tension, while the albumin adsorption was insensitive to the surface modification. Platelet adhesion and activation were weakened and the hemolysis rate was reduced at least 46% after Ta implantation due to the decreased surface energy and improved corrosion resistance ability, respectively.

Yang et al. [182] and Cheng et al. [183] studied DLC. Yang et al. [194] fabricated hydrogenated amorphous carbon (a-C:H) films fabricated on silicon wafers (1 0 0) using PIII deposition. It was mentioned that (i) film graphitization is promoted at higher substrate bias, and (ii) the film deposited at a lower substrate bias of –75 V possesses better blood compatibility than the films at higher bias and stainless steel, suggesting that there are two possible paths to improve the blood compatibility, suppression of the endogenic clotting system, and reduction of platelet activation. Cheng et al. [183] deposited DLC films TiNi50.8 (at%) alloys using plasma-based ion implantation technique. The influence of the pulsed negative bias voltage applied to the substrate from 12 to 40 eV on the surface characteristics and corrosion resistant property as well as hemocompatibility was investigated. It was reported that the uncoated TiNi alloy shows severe pitting corrosion, which could be due to the presence of Cl^- ions in the solution. On the contrary, the coated sample shows very little pitting corrosion and behaves with better corrosion-resistant property especially for the specimens deposited at 20 kV bias voltages. The platelet adhesion test shows that the hemocompatibility of DLC-coated TiNi alloy is much better than that of the bare TiNi alloy, and the hemocompatibility performance of DLC-coated TiNi alloy deposited at 20 kV is superior to that of the other coated specimens. Lee et al. [184] coated the surface of NiTi stent with DLC and then subsequently grafted by using zwitterion (N^+ and SO_3^-)-linked poly(ethylene glycol) (PEG). It was shown that (i) the surface grafting of zwitterionic PEG derivatives could substantially enhance the blood compatibility of TiNi-DLC stent, and (ii) antifouling properties of PEG and zwitterions are expected to be very useful in advancing overall stent performance. Liang et al. [185] investigated corrosion resistance and hemocompatibility of ion implanted biomaterials of NiTi SMA and Co-based alloys with electrochemical method, dynamic clotting time, and hemolysis rate tests. It was obtained that (i) the electrochemical stability and anodic polarization behavior of the materials were improved significantly after ion implantation; (ii) when TiNi, Co-based alloys were implanted Mo + C and Ti + C, respectively, the corrosion potentials were enhanced more than 200 mV, passive current densities decreased, and passive ranges were broadened; (iii) dynamic clotting time of the ion implanted substances was prolonged and hemolysis rate decreased; (iv) corrosion resistance and hemocompatibility of the alloys were improved by ion implantation, and effects of dual implantation was better than that of C single implantation. X-ray diffraction analysis of the alloys after dual implantation revealed that TiC, Mo_2C, Mo_9Ti_4, and Mo appeared on the surface of TiNi alloy, and CoCX, Co_3Ti, TiC, and TiO on the surface of Co-based alloys. It was also mentioned that these phases dispersing on the alloy surface formed amorphous film, prevented dissolving of alloy elements, and improved the corrosion resistance and hemocompatibility of the alloys. Plant et al. [186] studied the effect of surface nickel species on NiTi alloy thrombogeneity by assessment of platelet activation and whether oxidative modification of the alloy would affect platelet response. Tests were conducted under static conditions and arterial levels of shear stress. It was observed that (i) heat treatment of the alloy at 600 °C significantly reduced surface nickel species due to passive film formation of titanium dioxide, and (ii) under both static and flow conditions platelet activation on the heat-treated alloy was comparable to that on pure titanium and was significantly lower than that on polished NiTi, demonstrating that the risk of thrombotic complications associated with NiTi in vivo can be reduced through heat modification of the alloy surface to reduce surface nickel.

Zhao et al. [187] applied hafnium ion implantation to NiTi alloy for suppression of Ni ion release and enhancement of osteoblast–material interactions and hemocompatibility. The auger electron spectroscopy, XPS, and atomic force microscope results showed that a composite TiO_2/HfO_2 nanofilm with increased surface roughness was formed on the surface of NiTi, and Ni concentration was reduced in the superficial surface layer. It was reported that (i) potentiodynamic polarization tests displayed that 4 mA NiTi sample possessed the highest $E(br) - E(corr)$, 470 mV higher than that of untreated NiTi, suggesting a significant improvement on pitting corrosion resistance; (ii) ICP-MS tests during 60 days immersion demonstrated that Ni ion release rate was remarkably decreased, for example, a reduction of 67% in the first day; (iii) the water contact angle increased and surface energy decreased after Hf implantation; (iv) cell culture and methyl-thiazol-tetrazolium indicated that Hf-implanted NiTi expressed enhanced osteoblasts adhesion and proliferation, especially after 7 days culture. (v) Hf implantation decreased fibrinogen adsorption, but had almost no effect on albumin adsorption; (vi) platelets adhesion and activation were suppressed significantly (97% for 4 mA NiTi) and hemolysis rate was decreased by at least 57% after Hf implantation; and (vii) modified surface composition and morphology and decreased surface energy should be responsible for the improvement of cytocompatibility and hemocompatibility. Shabalovskaya et al. [188] investigated the dependence of NiTi contact angles and surface energy on surface treatment in order to understand the material hemocompatibility. In the group of surfaces: (1) MP, (2) additionally heat treated, (3) chemically etched, and (4) additionally boiled in water, and (5) further heat treated, it was obtained that (i) the contact angle could vary in the 50–80° hydrophobic range and the total surface free energy in the 34–53 mN/m range, and (ii) the polar surface energy, varying from 5 to 29 mN/m, constitutes a decisive contribution to the total energy change, and it seems to be a direct function of the NiTi surface chemistry. Based on the complex analysis of surface energy on electrochemistry and hemocompatibility it was concluded that the alteration of the polar component of surface energy and thrombogenicity is due to changes of the electron-acceptor/electron-donor character of native NiTi surfaces during surface treatments. In order to improve the corrosion resistance and hemocompatibility of NiTi alloy, Li et al. [189] deposited hydrophobic polymer coatings by plasma polymerization in the presence of a fluorine-containing precursor using an atmospheric-pressure plasma jet at a low temperature in air. It was reported that (i) the corrosion resistance of the coated NiTi alloy was evidently improved, (ii) protein adsorption and platelet adhesion tests reveal that the adsorption ratio of albumin to fibrinogen is increased and the number of adherent platelets on the coating is greatly reduced, and (iii) the plasma-polymerized coating renders NiTi better *in vitro* hemocompatibility and is promising as a protective and hemocompatible coating on cardiovascular implants. Pulletikurthi et al. [190] investigated the utility of magnetoelectropolished (MEP) ternary NiTi alloys, NiTiTa, and NiTiCr as blood contacting materials. The hemocompatibility of these alloys were compared to MP metallic biomaterial counterparts. It was obtained that (i) the in vitro thrombogenicity tests revealed significantly less platelet adherence on ternary MEP Nitinol, especially MEP NiTiTa$_{10}$ as compared to the MP metals, (ii) the enhanced anti-platelet-adhesive property of MEP NiTiTa$_{10}$ was in part, attributed to the Ta$_2$O$_5$ component of the alloy, and (iii) the formation of a dense and mixed hydrophobic oxide layer during MEP is believed to have inhibited the adhesion of negatively charged platelets, concluding that MEP ternary Nitinol alloys can potentially be utilized for blood-contacting devices, where complications resulting from thrombogenicity can be minimized.

11.5 Cytocompatibility and cytotoxicity

For the prolonged use of Ti–Ni material in a human body, deterioration of the corrosion resistance of the materials becomes a critical issue because of the increasing possibility of deleterious ions released from the substrate to living tissues. The safety, success rate, and longevity of placed dental or orthopedic implants are governed by various biofunctional factors, including three major compatibilities (biological, biomechanical, and macro- and micromorphological), hemocompatibility, cytocompatibility, and osseointegration. Some of these factors were already discussed in previous portions of this book and, in this section, the cytocompatibility and cytotoxicity will be reviewed.

11.5.1 Cytocompatibility

Mikulewicz et al. [191] performed an extensive review on the cytocompatibility of medical biomaterials containing nickel, as assessed by cell culture of human and animal osteoblasts or osteoblast-like cells, based on a survey of 21 studies on the biomaterials including stainless steel, NiTi alloys, pure Ni, Ti, and other pure metals. It was mentioned that the observation that the layers significantly reduced the initial release of metal ions and increased cytocompatibility was confirmed in cell culture experiments. Physical and chemical characterization of the materials was performed, including surface characterization (roughness, wettability, corrosion behavior, quantity of released ions, microhardness, and characterization of passivation layer). Cytocompatibility tests of the materials were conducted in the cultures of human or animal osteoblasts and osteoblast-like cells in terms of cell proliferation and viability test, adhesion test, morphology (by fluorescent microscopy or SEM). In the majority of works, it was found that the most cytocompatible materials were stainless steel and NiTi alloy; while pure Ni was rendered and less cytocompatible. All the papers confirmed that the consequence of the formation of protective layers was in significant increase of cytocompatibility of the materials, indicating the possible further modifications of the manufacturing process (formation of the passivation layer). Nickel–titanium SMAs are increasingly being used in orthopedic applications. However, there is a concern that Ni is harmful to the human body. Yeung et al. [73, 192, 193] treated NiTi surface by PIII and mentioned that (i) the near-surface Ni concentration in all the treated samples was significantly suppressed, (ii) the plasma-treated surfaces are cytologically compatible allowing the attachment and proliferation of osteoblasts, and (iii) among the two types of samples, the best biological effects are found on the samples with nitrogen implantation. Wu et al. [78, 194] utilized the O-PIII in order to reduce the amount of nickel leached from porous NiTi alloys with a porosity of 42% prepared by capsule-free hot isostatic pressing. It was reported that (i) the O-PIII porous NiTi SMAs have good mechanical properties and excellent superelasticity, and the amount of nickel leached from the O-PIII porous NiTi is much less than that from the untreated samples, (ii) XPS results indicate that a nickel-depleted surface layer predominantly composed of TiO_2 is produced by O-PIII and acts as a barrier against out-diffusion of nickel, and (iii) in the cell culturing tests, both the O-PIII and untreated porous NiTi alloys have good biocompatibility. Zhao et al. [195] modified NiTi alloy with tantalum (Ta) PIII. It was reported that (i) the surface became rougher and was covered by ordered and uniform grains after Ta implantation, (ii) Ni release was averagely inhibited by Ta-NiTi to up to 60% of that in unmodified NiTi alloy within 30 days, (iii) MTT assays demonstrated that Ta-NiTi sample allowed greater degree of cell proliferation for both smooth muscle cell and osteoblasts, indicating excellent protection and cytocompatibility, (iv) a negative correlation was observed between Ni release and cell proliferation, and (v) analysis of the cell morphology revealed healthy cells extending on the alloy surface, which indicated that NiTi alloy had good cytocompatibility despite the initial Ni dissolution, but the implanted Ta would endow traditional NiTi alloy much lower Ni release, improved cytocompatibility, and other potential merits.

Jian et al. [196] fabricated five types of porous NiTi alloy samples of different porosities and pore sizes. According to compressive and fracture strengths, three groups of porous NiTi alloy samples underwent further cytocompatibility experiments. It was observed that (i) porous NiTi alloys exhibited a lower Young's modulus (2.0–0.8 GPa), (ii) both compressive strength (108.8–56.2 MPa) and fracture strength (64.6–41.6 MPa) decreased gradually with increasing mean pore size, (iii) cells grew and spread well on all porous NiTi alloy samples, (iv) cells attached more strongly on control group and blank group than on all porous NiTi alloy samples, (vi) cell adhesion on porous NiTi alloys was correlated negatively to mean pore size (277.2–566.5 µm), (vii) cells proliferated on control group and blank group than on all porous NiTi alloy samples, (viii) cellular alkaline phosphatase activity on all porous NiTi alloy samples was higher than on control group and blank group, and (ix) the porous NiTi alloys with optimized pore size could be a potential orthopedic material. Wu et al. [79] employed the oxidation in conjunction with postreaction heat treatment to modify the surfaces of porous single-phase NiTi prepared by capsule-free hot isostatic pressing to mitigate Ni leaching and enhance the surface properties. It was mentioned that cell cultures reveal that porous NiTi alloys oxidized at 450 °C for 1 h have an austenite transition temperature below 37 °C, excellent superelasticity, lower nickel release, and no cytotoxicity. The NiTi archwires were studied as-received and after conditioning for 24 h or 35 days in a cell culture medium under static conditions [197]. It was mentioned that (i) all of the tested archwires, including their conditioned medium, were noncytotoxic for L929 cells, but Rematitan SE (both as received and conditioned) induced the apoptosis of rat thymocytes in a direct contact, (ii) in contrast, TruFlex SE and Equire TA increased the proliferation of thymocytes, and (iii) the cytotoxic effect of Rematitan SE correlated with the higher release of Ni ions in conditioned medium, higher concentration of surface Ni and an increased oxygen layer thickness after the conditioning. Based on these findings it was concluded that the apoptosis assay on rat thymocytes, in contrast to the less sensitive standard assay on L929 cells, revealed that Rematitan SE was less cytocompatible compared to other archwires and the effect was most probably associated with a higher exposition of the cells to Ni on the surface of the archwire, due to the formation of unstable oxide layer. Hang et al. [198] fabricated NiTiO (nickel–titanium–oxygen) nanopores with different length (0.55–114 μm) which were anodically grown on nearly equiatomic NiTi alloy. Length-dependent corrosion behavior, nickel ion (Ni^{2+}) release, cytocompatibility, and antibacterial ability were investigated by electrochemical, analytical chemistry, and biological methods. It was reported that constructing nanoporous structure on the NiTi alloy improve its corrosion resistance; however, the anodized samples release more Ni^{2+} than that of the bare NiTi alloy, suggesting chemical dissolution of the nanopores rather than electrochemical corrosion governs the Ni^{2+} release. In addition, the Ni^{2+} release amount increases with nanopore length. The anodized samples show good cytocompatibility when the nanopore length is <11 μm. Encouragingly, the length scale covers the one (1–11 μm) that the nanopores showing favorable antibacterial ability, concluding that the nanopores with length in the range of 1–11 μm are promising as coatings of biomedical NiTi alloy for anti-infection, drug delivery, and other desirable applications. Hang et al. [199] also studied the hydrothermal synthesis of nanosheets on biomedical NiTi alloy in pure water. It was shown that (i) rhombohedral $NiTiO_3$ nanosheets with thickness of 6 nm can be grown at 200 °C, (ii) hydrothermal treatment enhances the corrosion resistance of the NiTi alloy, (iii) the treatment for 30 min significantly reduces Ni ion release, while prolonged hydrothermal time results in increased Ni ion release because of the growth of the nanosheets with large specific surface area, and (iv) excitingly, the nanosheets can well support cell growth, which suggests the release amount can be well tolerated, suggesting that good corrosion resistance and cytocompatibility combined with large specific surface area render the nanosheets promising as safe and efficient drug carriers of the biomedical NiTi alloy.

11.5.2 Cytotoxicity

Cytotoxicity of Ni ions on fibroblasts was examined by cell count and neutral red assay. It was found that (i) Ni ions had dose-dependent cytotoxicity and (ii) the dissolution of Ni ions from Ni-containing metallic restorations must be lower than these concentration levels so that body tissues might not be severely damaged [200]. A study was done to qualitatively and quantitatively characterize the substances released from orthodontic brackets and NiTi archwires and to comparatively assess the cytotoxicity of the ions released from these orthodontic alloys [25]. Two full sets of stainless-steel brackets and NiTi archwires were immersed in 0.9% saline solution for a month. Human periodontal ligament fibroblasts and gingival fibroblasts were exposed to various concentrations of the two immersion media; nickel chloride was used as a positive control for comparison purposes. It was reported that there was no ionic release for the NiTi alloy aging solution, whereas measurable nickel and traces of chromium were found in the stainless steel bracket-aging medium. Concentrations of the nickel chloride solution greater then 2 mM were found to reduce by more than 50% the viability and DNA synthesis of fibroblasts; however, neither orthodontic material-derived media had any effect on the survival and DNA synthesis of either cell. Assad et al. [201] determined the biological response that NiTi elicits compared to other orthopedic metals currently used in orthopedic surgery. Confluent L-929 fibroblasts culture plates were incubated (directly or under an agar bed) in presence of NiTi, titanium (Ti), vitallium (Co-Cr-Mo) and 316L stainless steel disks. It was reported that all cultures were evaluated for cytotoxic reactions, under light microscopy. Direct contact and agar diffusion assays indicated that all metals tested induced a mild biological reaction. Specimens were ranked according to an index of biological response, and they are enumerated here in decreasing order of cytotoxicity: NiTi approximately Co-Cr-Mo $\gg$ pure grade 4 Ti approximately pure grade 1 Ti approximately Ti-6A1-4V approximately 316L stainless steel. Plasma surface modification increased the cytocompatibility of NiTi. Porous titanium–nickel alloys represent new biomaterials for long-term implantation. Their porosity properties might confer them the capacity to trigger fluid capillarity, tissue ingrowth, as well as good tissue-implant apposition and fixation. Porous Ti–Ni was therefore extracted in a saline semiphysiological solution and materials were evaluated for potential cytotoxicity and genotoxicity reactions. Based on results, it was concluded that porous Ti–Ni can be considered completely cytocompatible and genocompatible, and therefore represents a good candidate for long-term implantation [202]. The transformation behavior, mechanical properties and cytotoxicity of a binary $NiTi_{42}$ and a ternary $NiTi_{42}Cu_7$ (at%) alloy have been investigated [203]. The cytotoxicity tests were performed on both alloys in cultured epithelial cells from human gingiva, including both MTT (colorimetric assay for assessing cell metabolic activity) tests and morphological observations. It was found that (i) although the ternary alloy is characterized by a narrower hysteresis and superior mechanical properties, including fatigue resistance, its cytotoxicity is higher than that of the binary alloy, and (ii) this is thought to arise from the release of copper ions in the medium, which upon AAS measurements amount to approximately 2.8 $\mu g/cm^2$ for an incubation period of 7 days. Manceur et al. [204] conducted in vitro tests to evaluate the biocompatibility of $Ni_{50.8}Ti$ (at%) single crystals in the orientation <001> after four different heat treatments in a helium atmosphere followed by mechanical polishing. The study was performed on the material extracts after immersion of the specimens in cell culture medium (DMEM) for 7 days at 37 °C. Cytotoxicity studies were performed on L-929 mouse fibroblasts using the MTT assay. J-774 macrophages were used to assess the potential inflammatory effect of the extracts by IL1-beta and TNF-alpha dosages (sandwich ELISA method). It was reported that (i) exposure of L-929 to material extracts did not affect cell viability;

(ii) in addition, IL1-beta and TNF-alpha secretion was not stimulated after incubation with NiTi extracts compared to the negative controls; and (iii) these results were predictable since AAS did not detect nickel ions in the extracts with a resolution of 1 ppm, concluding that the TiNi$_{50.8}$Ti (at%) single crystals do not trigger a cytotoxic reaction. David et al. [205] used murine cortical cell cultures to examine the in vitro neurotoxicity of commonly used orthodontic archwire alloys, including 0.016 inch NiTi, Cu–NiTi, TiMo, Elgiloy, and stainless steel archwire alloys. Standard sized samples of each material were placed on tissue culture inserts suspended above the cell cultures. Neuronal death was determined using the LDH release assay 24 h after exposure to the archwires. It was shown that NiTi, Cu–NiTi, and TiMo alloys were not neurotoxic, while stainless steel and Elgiloy were significantly toxic. Washing the archwires for 7 days in a saline solution did not alter the toxicity; however, the free radical scavenger, Trolox, blocked the toxicity of both stainless steel and Elgiloy, indicating that the death was free radical mediated. The caspase inhibitor, Z-VAl-Ala-Asp-fluoromethylketone (zVAD-FMK), blocked the toxicity of stainless steel, but not Elgiloy, suggesting that stainless steel induced apoptosis. Stainless steel induced apoptosis was provided by propidium staining which showed nuclear chromatin condensation and fragmentation into discrete spherical or irregular shapes, characteristic of apoptosis. The specific metal responsible for the toxicity was not determined; the metals common to each of the toxic archwires were nickel, iron, and chromium. Dinca et al. [206] conducted cytocompatibility tests on NiTi thin films through in vitro assays using microorganisms culture cells such as yeasts (*Saccharomyces cerevisiae*) and bacteria (*Escherichia coli*), in order to determine the thin film's toxic potential at the in vitro cellular level. Microorganism's adhesion on the nitinol surface was observed and the biofilm formation has been analyzed and quantified. It was shown that (i) there was no reactivity detected in cell culture exposed to NiTi films in comparison with the negative controls and a low adherence of the microorganisms on the nitinol surface that is an important factor for biofilm prevention, concluding that (ii) NiTi is a good candidate material to be used for implants and medical devices.

The toxicity of fluoride corrosion extracts of stainless steel and NiTi archwires on a human osteosarcoma cell line (U2OS) was investigated [207]. The stainless steel and NiTi wires were corroded by an electrochemical method with the application of three kinds of electrolytes: 0.2% pH 3.5 APF (NaF) in AS, and pH 4 and pH 6.75 AS solutions. The extracts were analyzed for nickel, chromium, and titanium ions by the atomic absorption method. It was obtained that (i) the release of ionic nickel was different in different extract groups, (ii) the stainless steel and NiTi wires in the 0.2% pH 3.5 NaF AS group caused a dose-dependent decrease in the survival rate, and (iii) survival rates of cells in the groups exposed to extracts of stainless steel and NiTi wires in pH 4 and pH 6.75 AS solutions showed no statistical differences; concluding that orthodontic wires in acidulated fluoride saliva solution can cause U2OS cell toxicity. McMahon et al. [208] investigated the cytotoxicity behaviors of NiTi$_{49.2}$ (at%) and TiNb$_{26}$ (at%) SMAs. It was reported that (i) TiNb SMAs are less cytotoxic than NiTi

SMAs, at least under static culture conditions, (ii) this increased TiNb cytocompatibility was correlated with reduced ion release as well as with increased corrosion resistance according to potentiodynamic tests, (iii) measurements of the surface composition of samples exposed to cell culture medium further supported the reduced ion release observed from TiNb relative to NiTi SMAs, (iv) alloy composition depth profiles also suggested the formation of calcium phosphate deposits within the surface oxide layers of medium-exposed NiTi but not of TiNb, and (v) collectively, the present results indicate that TiNb SMAs may be promising alternatives to NiTi for certain biomedical applications. Sodor et al. [209] evaluated the biocompatibility of some orthodontic biomaterials like stainless steel archwires, brackets, and NiTi alloy coil springs. The studies were performed in vitro using human fibroblasts cultures on which the orthodontic materials were applied. The positive control was the copper amalgam. Readings of the cell reactions were performed at 3 and 6 days. It was found that (i) the materials used in the study cause cell alterations of variable intensity, (ii) the metallic brackets represent an important cell stress factor causing shape changes, (iii) for the metallic brackets, a preferential tropism for different areas of the bracket was also obvious, (iv) a preferential tropism for the areas between the NiTi coil spring spirals was observed, and (v) for the stainless steel archwires, it was found that at 6 days a decay of cell density and also a higher amount of cells near the archwire areas on which bends were performed. It was then concluded that all biomaterials analyzed in our study cause cellular changes of varying intensity without necessarily showing a cytotoxic character. McNamara et al. [210] mentioned, on surface chemistry and cytotoxicity of reactively sputtered tantalum oxide films on NiTi plates, that (i) despite the widely perceived biocompatibility there remain some concerns about the sustainability of the alloy's biocompatibility due to the defects in the TiO_2 protective layer and the presence of high amount of sub-surface Ni, which can give allergic reactions, and (ii) many surface treatments have been investigated to try to improve both the corrosion resistance and biocompatibility of this layer. Tantalum (Ta) oxide thin films was sputter deposited onto the surface of the NiTi alloy [210]. It was reported that (i) in general, reactive sputtering especially in the presence of a low oxygen mixture yields a thicker film with better control of the film quality, (ii) the sputtering power influenced the surface oxidation states of Ta, and (iii) both microscopic and quantitative cytotoxicity measurements show that Ta films on NiTi are biocompatible with little to no variation in cytotoxic response when the surface oxidation state of Ta changes. Yanisarapan et al. [211] determined the cytotoxicity, metal ion release and surface roughness of metal orthodontic appliances after immersion in different fluoride product solutions. Twelve sets of 20 brackets and four tubes were ligated with three types of archwires: stainless steel, NiTi, and β-Ti. The samples in each archwire group were divided into three subgroups and immersed in solutions of fluoride toothpaste, 1.23% APF, or AS without fluoride as a control group. It was obtained that in the APF groups, the four metal ion levels and surface roughness of the brackets and archwires significantly increased, while cell viability significantly decreased, especially in the TMA subgroup. The SEM results showed that the brackets and wires in the APF groups demonstrated more lines and grooves compared with the other groups, concluding that using APF gel during orthodontic treatment with fixed metal appliances should be avoided. The different contents of tantalum pentoxide (Ta_2O_5: 10, 15, 20, and 30 wt%) nanoparticles were introduced into the nHA coating structure on NiTi substrate through EPD method [212]. The phase compositions of coatings were perused before and after the sintering at 800 °C for 1 h by XRD. It was mentioned that (i) the incorporation of 30 wt%Ta_2O_5 into nHA matrix induced the formation of undesirable soluble $Ca_3(PO_4)_2$ phase in composite coating, (ii) the SEM images showed that the density of continuous nHA coating increased by compositing with Ta_2O_5, (iii) the maximum adhesion strength of 28.3 ± 0.7 MPa accomplished from the nHA-20 wt%Ta_2O_5 composite coating, (iv) the Ni ions concentration measurement results from the passivated-NiTi with nHA and nHA-(10, 15, and 20)wt%Ta_2O_5 coatings during 30 days of immersion in PBS clarified the positive role of Ta_2O_5 in decreasing the Ni leaching due to the lowering the open porosities of nHA structure, (v) the biological response of the coating surfaces was assessed in vitro by cell culturing and MTS assay, (vi) by considering the morphology and density of adsorbed cells on each coating, the improved biocompatibility of nHA coating in the presence of Ta_2O_5 was justified by scrutinizing the surface roughness, wettability and charge, (vii) the highest cell attachment and proliferation on nHA-20 wt%Ta_2O_5 coating was related to owning the lowest roughness, wetting angle of 34° ± 0.5 and the highest negative surface charge density, and (viii) the concentration of the highest negative charge density on nHA-20 wt%Ta_2O_5 coating surface in the SBF solution caused to the enhancement of the amount of the apatite nuclei through providing more sites to calcium absorption.

11.6 Magnetic resonance imaging compatibility

11.6.1 In general

MRI is probably the most innovative and revolutionary imaging technology, with the exception of compound tomography. MRI is a 3D imaging technique used to image the protons of the body by employing magnetic fields, radio frequencies, electromagnetic detectors, and computers [213]. For millions of patients worldwide, MRI examinations provide essential and potentially life-saving information. Some devices such as pacemakers and neurostimulator have limitations related to MRI safety and may be contraindicated for use with MRI. Other internal devices such as stents, vena cava filters, and some types of catheters and guidewires, are safe for use with MRI but have limited MRI image compatibility. Some of these devices are simply not well imaged under MRI. Others have properties that interfere with the MRI image by causing an image artifact (distortion) in the area and around the device, limiting the effectiveness of MRI for assisting placement or diagnostic follow-up on these implants. It may be contraindicated in certain situations because the magnetic field present in the MRI environment may, under certain circumstances, result in movement or heating of a metallic orthopedic implant device. Further, metals that exhibit magnetic attraction in the MRI setting may be subject to movement (deflection) during the procedure. Both magnetic and nonmagnetic metallic devices of certain geometries may be subjected to heating caused by interactions with the magnetic field. There are currently several researchers and an American Society for Testing and Materials committee exploring methods for accurately assessing the MRI compatibility of implant devices. The primary focus of their research has been the measurement of implant movement in response to a magnetic field. Shellock et al. [214–216] conducted several studies in which the movement/deflection of various orthopedic implants was measured in the high magnetic field (0.3–1.5 Tesla) region of MRI units. The results of these studies show no measurable movement of implants fabricated from cobalt, titanium, and stainless steel alloys [217]. Ferromagnetic metals will cause a magnetic field inhomogeneity, which, in turn, causes a local signal void, often accompanied by an area of high signal intensity, as well as a distortion of the image. They create their own magnetic fields and dramatically alter precession frequencies of protons in the adjacent tissues. Tissues adjacent of ferromagnetic components become influenced by the induced magnetic field of the metal hardware rather than the parent field and, therefore, either fail to process or do so at a different frequency and hence do not generate useful signal. Two components contribute to the susceptibility artifact: induced magnetism in the ferromagnetic component itself and induced magnetism in protons adjacent to the component. Artifacts from metal may have varied appearance on MRI scans due to different types of metal or configuration of the piece of metal. In relation to MRI, titanium alloys are less ferromagnetic than both cobalt and stainless steel, induce less susceptibility to artifacts, and result in less marked image degradation [217–219].

There are several research reports on MRI application on orthodontic appliances [220–223]. Okano et al. [220] investigated the accuracy of MRI diagnosis of the temporomandibular joint in those with orthodontic appliances by comparing magnetic resonance images of the temporomandibular joints with and without orthodontic appliances. MRI was performed in a total of 20 temporomandibular joints before and after the insertion of six kinds of orthodontic appliances (types 1–6), and the magnetic resonance images were compared. It was reported that with respect to disk position, the diagnostic accuracy was 80%, 75%, 70%, 70%, 65%, and 60% in order from type 1 through type 6. The distribution of stages for the assessment of condylar configurations was 80%, 55%, 40%, 40%, 20%, and 10% in order from type 1 through type 6. No significant changes were found in the condylar head marrow signals, concluding that MRI is performed in orthodontic patients preferably by using ceramic brackets in the front teeth and direct bonding tubes in the molar teeth while removing arch wires. Klocke et al. measured forces on orthodontic wires caused by the static magnetic field of a 1.5-Tesla MRI system, and to assess the safety hazards associated with these forces [221] and assessed forces on orthodontic wires in a high field strength

MRI system at 3 Tesla [222]. Thirty-two different orthodontic wires (21 arch wires, eight ligature wires and three retainer wires) were investigated. It was reported that (i) steel ligature wires and arch wires made of cobalt chromium, titanium molybdenum, NiTi, and brass alloys showed no or negligible forces within the magnetic field, (ii) the translational and rotational forces within the MRI magnetic field should pose no risk to carefully ligated arch wires, (iii) steel retainer wire bonds should be checked to ensure secure attachment prior to an MRI investigation. Anterior cervical discoidectomy with or without fusion is well-established surgical remedy for cervical prolapsed intervertebral disk (PIVD) disease. If fusion is done by an iliac bone graft then internal fixation is commonly used to keep the graft in position. Singh et al. [223] determined the efficacy and tolerability of SMAs, especially NiTi clips in the stabilization of grafts following anterior cervical discoidectomy. About 133 NiTi clips were applied in 119 patients between January 2002 and December 2008. The patients' age ranged from 38 to 60 years. There were 66 males and 53 females. Various indications for fixation of the spine included degenerated cervical spondylosis with single-level PIVD (105) and two-level PIVD in 14 patients. The cine mode fluoroscopy confirmed the perioperative correct placement of grafts and clips in all the patients. Follow-up ranged from 2 to 8 years (mean: 4.6 years). It was found that (i) single-level discoidectomy was performed in 105 patients and two-level disk removal was done in 14 patients; (ii) a single NiTi clip was applied in all the cases except for 14 cases of two-level PIVD; (iii) no procedural complication or adverse reaction to the clip was noted; (iv) there was no movement at the operated level in dynamic lateral view X-ray of cervical spine at the first postoperative day as well as on follow-up; (v) bony fusion occurred in all patients after 9–12 months of surgery; and (vi) there was no incidence of breakage or dislodgement of the clip from the site where it was inserted. No artifact was noted in cervical MRI done in 33 patients. Based on these findings, it was concluded that (vii) NiTi clips are a simple alternative for cervical spine stabilization after discoidectomy, and (viii) their insertion is simple, minimally invasive, does not require any special set of instruments, and they are much more economical than other established methods of treatment. These clips are accepted well by human tissue and do not interfere with MRI. With the increased usage of MRI as a diagnostic tool in clinic, the currently used metals for vascular stents, such as 316L stainless steel, Co–Cr alloys, and Ni–Ti alloys, are challenged by their unsatisfactory MRI compatibility, due to their constituents containing ferromagnetic elements. Li et al. [224] selected the Nb-xTa-2Zr ($30 \leq x \leq 70$) series alloys to provide more MRI compatible vascular stents. It was reported that increasing the Ta content gave rise to the decreased volume magnetic susceptibility and the increased modulus of elasticity, together with the elevated yield strength but less changed elongation. From multiple requirements for the stents, the Nb-60Ta-2Zr alloy exhibits an optimal properties, including the volume magnetic susceptibility of about 3% of the 316L stainless steel, the modulus of elasticity of 142 GPa superior to pure niobium, high mass density of 12.03 g/cm^3 favored to the X-ray visibility, yield strength of ~330 MPa comparable to the 316L stainless steel, and an elongation of ~24%.

11.6.2 Artifacts

MRI is widely used as an important diagnostic tool, especially for orthopedic and brain surgery. This method has remarkable advantages for obtaining various cross-sectional views and for diagnosis of the human body with no invasion and no exposure of the human body to x-ray radiation. However, MRI diagnosis is inhibited when metals are implanted in the body, since metallic implants, such as stainless steels, Co–Cr alloys, and Ti alloys become magnetized in the intense magnetic field of the MRI instrument, and artifacts occur in the image [225–228]. Metallic objects produce artifacts on magnetic resonance images. Abbaszadeh et al. [225] studied the magnitude of such artifacts. Samples of various dental materials (dental gold, amalgam, stainless steel, titanium, silver–palladium, and vitallium) were embedded in bovine muscle and then subjected to Tl-weighted MRI. It was reported that (i) all metallic objects were found to produce artifacts and to interfere with the interpretation of magnetic resonance (MR) images; (ii) artifacts were most pronounced in the central plane of the object; and (iii) gold produced the greatest artifact, and amalgam produced the least, concluding that because metals commonly used in the maxillofacial region all produce artifacts on MR images, avoidance measures should be used to minimize the effect of these artifacts. Susceptibility artifacts due to metallic prostheses are a major problem in clinical MRI. We theoretically and experimentally analyze slice distortion arising from susceptibility differences in a phantom consisting of a stainless steel ball bearing embedded in agarose gel. To relate the observed image artifacts to slice distortion, we simulate images produced by 2D and 3D spin-echo (SE) and a view angle tilting (VAT) sequence. Two-dimensional SE sequences suffer from extreme slice distortion when a metal prosthesis is present, unlike 3D SE sequences for which – since slices are phase-encoded – distortion of the slice profile is minimized, provided the selected slab is larger than the region of interest. In a VAT sequence, artifacts are reduced by the application of a gradient along the slice direction during readout. However, VAT does not correct for the excitation slice profile, which results in the excitation of spins outside the desired slice location and can lead to incorrect anatomical information in MR images. We propose that the best sequences for imaging in the presence of a metal prosthesis utilize 3D acquisition, with phase encoding replacing slice selection to minimize slice distortion, combined with excitation and readout gradient strengths at their maximum values. Elison et al. [227] evaluated cranial MRI distortion caused by various orthodontic brackets. Ten subjects received five consecutive cranial MR scans. A control scan was conducted with Essix trays (GAC International, Bohemia, NY, USA) fitted over the maxillary and mandibular teeth. Four experimental MR scans of the head were conducted with plastic, ceramic, titanium, and stainless steel brackets incorporated into the Essix tray material. It was found that (i) there is a statistically significant difference between the mean distortion scores of stainless steel brackets and the mean distortion scores of the other experimental MR scans; (ii) interrater and intrarater agreement was high; (iii) the study showed that plastic, ceramic, and titanium brackets cause minimal distortion of cranial MR images (similar to the control); on the other hand, stainless steel brackets cause significant distortion, rendering several cranial regions nondiagnostic; (iv) areas with the most distortion were the body of the mandible, the hard palate, the base of the tongue, the globes, the nasopharynx, and the frontal lobes; (v) in general, the closer the stainless steel appliance was to a specific anatomic region, the greater the distortion of the MR image. Orthodontic appliances are often prophylactically removed prior to MRI examinations, although they are sometimes left in situ (out of ignorance). Blankenstein et al. [229] measured the size of experimental artifacts created by orthodontic devices and to develop criteria using sound material-science research for making MRI more compatible, thereby supporting radiologists and orthodontists in their efforts. Sixteen orthodontic small device and wire specimens made of different steel and titanium or CoCr alloys were placed in a chambered water-filled phantom for MRI. Each was subjected to SE and gradient-echo sequences at 1.5 and 3 Tesla. It was observed that (i) artifact formation depends on the material properties (specimen size, crystalline structure, manufacture-related processing) and on the specifications of the MRI system used (main field strength, sequence type), (ii) artifact radii ranged from 14 mm (SE at 1.5 Tesla) to 51 mm (gradient echo at 3 Tesla), and (iii) no artifacts occurred at 1.5 Tesla around the titanium and Co–Cr specimens; the same observation was made with one of the steel grades. Based on these results, it was concluded that (iv) artifact size cannot be predicted merely from the designation steel, (v) nor did the crystalline structure of the baseline material from which a steel device had been produced have major implications for artifact size; (vi) relevant, however, was the magnetic permeability (or susceptibility) of the final products, which is not disclosed by the manufacturers, and it cannot be measured on fixed intraoral appliances; and (vii) the present investigation reveals that some steel devices can remain in situ without triggering adverse consequences.

11.6.3 Indications and contraindications

There are different types of contraindications that would prevent a person from being examined with an MRI scanner. MRI systems use strong magnetic fields that attract any ferromagnetic objects with enormous force. Caused by the potential risk of heating, produced from the radio frequency pulses during the MRI procedure, metallic objects like wires, foreign bodies, and other implants need to be checked for compatibility. High-field MRI requires particular safety precautions. In addition, any device or MRI equipment that enters the magnet room has to be MR compatible. MRI examinations are safe and harmless, if these MRI risks are observed. Safety concerns in MRI include the magnetic field strength, possible missile effects caused by magnetic forces, the potential for heating of body tissue due to the application of radio frequency energy, the effects on implanted active devices such as cardiac pacemakers or insulin pumps, magnetic torque effects on indwelling metal (clips, etc.), the audible acoustic noise, danger due to cryogenic liquids, and the application of contrast medium.

MRI use is contraindicated for the following devices (except where noted) [52, 230]:

- *Cardiac pacemakers*: are absolutely contraindicated; however, MRI-compatible pacemakers are being developed. A few pacemaker patients have been scanned for life-threatening situations or inadvertently. Subsequently, their pacemakers must be meticulously checked for function. Some deaths have been reported.
- *Intracranial aneurysm clips*: are contraindicated, unless the specific type of MRI-compatible clip can be absolutely documented.
- *Neurostimulators/spinal-fusion stimulators*: are generally contraindicated. There are two manufacturers that are seeking FDA approval for usage of their product in MRI, but under strict guidelines.
- *Drug fusion pumps*: are generally contraindicated. There are two models Synchro Med and Synchro Med EL that are currently FDA approved with specific guidelines.
- *Metallic foreign bodies*: may or may not preclude MRI scanning depending on type, size and location. These patients must be screened by X-ray. Detection should also target possible orbital metallic foreign bodies.
- *Cochlear implants*: contraindicated.
- *Other otologic implants*: generally, these are nonferromagnetic and MRI compatible. The exception is the McGee stapedectomy piston prosthesis – it is not MRI compatible.
- *Dental implants*: are generally compatible, except for those that contain magnetically activated components.
- *Ocular implants*: some are MRI compatible.
- *Intravascular stents, filters, and coils*: most of these are compatible 6–8 weeks following placement, unless they are made of nonferrous metal, for example, titanium, in which case they can be imaged right after the placement. Drug-eluting stents must be cleared by the implanting physician if they have been in less than 3 months.
- *Vascular access ports and catheters*: are compatible, excluding the Swan Ganz Catheter.
- *Penile implants*: these are of mixed compatibility and should be checked.
- *Orthopedic implants/prosthesis*: are compatible, though MRI may cause local heating and MRI will cause local image artifact.
- *Heart valve prosthesis*: is compatible, but there is a prototype electromagnetically controlled heart valve that is being developed, which is contraindicated.
- *Pacer wires (without pacemaker)*: to have an MRI compatibility is controversial, but they are probably fine at low MRI fields but questionable at high-field imaging. All patients must have a chest X-ray prior to having an MRI to ensure that the wires are not looped or crossed.
- *Holter monitor*: contraindicated.

- *Ventricular-peritoneal shunts*: compatible, except SOPHY adjustable pressure valve type.
- *Swan Ganz catheter*: contraindicated. The thermal dilutor may melt.
- *Dermal patches*: can cause burns during MRI study and should be removed.
- *Pessary and diaphragm*: compatible, but it will produce local image artifact.
- *Hearing aids*: Must be removed.
- *Permanent eyeliner/tattoos*: can cause local burns.
- *Body rings/spikes*: can cause local burns, dislodgment, or artifacts. It is suggested to remove these devices under normal MRI or the use under low field magnetic resonance.
- *Breast tissue expanders*: not compatible.
- *Bullets, shrapnel, and pellets*: depends on the foreign body's location and duration of the MRI. An X-ray is required before an MRI can be done.

11.7 Osseointegration and its evaluation

11.7.1 Osseointegration

Broadly speaking, two types of anchorage mechanisms have been described: biomechanical and biochemical. Biomechanical binding is when bone ingrowth occurs into micrometer sized surface irregularities. The term osseointegration is probably, realistically, the biomechanical phenomenon. Biochemical bonding may occur with certain bioactive materials where there is primarily a chemical bonding, with possible supplemental biomechanical interlocking. The distinct advantage with the biochemical bonding is that the anchorage is accomplished within a relatively short period of time, while biomechanical anchorage takes weeks to develop. This would clinically translate into the possibility of earlier restorative loading of implants. Most commercially available implants depend on biomechanical interlocking for anchorage. All implants must exhibit biomechanical as well as morphological compatibility. As we have seen previously, Ti implants coated with CaP, or inorganic or organic bone-like apatite, possess both anchorage mechanisms, because coated surfaces are normally rough which can facilitate the biomechanical anchorage (morphological compatibility), coated materials per se accommodate biochemical bonding (biological compatibility). In orthodontic mechanotherapy, the anchorage mechanism is slightly different from ordinary implants. Saito et al. [231] investigated the anchorage potential of titanium implants with the use of a sectional arch wire technique for orthodontic mesiodistal tooth movement, as assessed by the osseointegration of implants and tooth movement. Two implants were surgically placed in healed mandibular extraction sites of the second and third premolars on each side in four adult male beagle dogs. It was reported that (i) there was no statistical difference in the percentage of peri-implant bone volume between the loaded and unloaded sides, and (ii) there was also no statistical difference between the compression and tension sides in both loaded and unloaded implants, suggesting that the implants maintained rigid osseointegration. It was, hence, concluded that endosseous titanium implants can function as anchors for long-term orthodontic mesiodistal movement [248]. Bone fusing to titanium was first reported in 1940 by Bothe et al. [232]. Brånemark began extensive experimental studies in 1952 on the microscopic circulation of bone marrow healing. These studies led to dental implant application in early 1960, 10-year implant integration was established in dogs without significant adverse reactions to hard or soft tissues. Studies in humans began in 1965, were followed for 10 years, and reported in 1977 [233].

Osseointegration, as first defined by Brånemark, denotes at least some direct contact between vital bone with the surface of an implant at the light microscopic level of magnification [234]. The percentage of direct bone–implant contact is variable. To determine osseointegration, the implant must be removed and evaluated under a microscope. Rigid fixation defines the clinical aspect of this microscopic bone contact with an implant and is the absence of mobility with 1–500 g force applied in a vertical or horizontal direction. Rigid fixation is the clinical result of a direct bone interface but has also been reported with fibrous tissue interfaces [235]. Osseointegration was originally defined as a direct structural and functional connection between ordered living bone and the surface of a load-carrying artificial implant (which is typically made of titanium materials). It is now said that an implant is regarded as osseointegrated when there is no progressive relative movement between the implant and the bone with which it has direct contact. In practice, this means that in osseointegration there is an anchorage mechanism, whereby nonvital components can be reliably and predictably incorporated into living bone and that this anchorage can persist under all normal conditions of loading. Bioactive (or biochemical) retention can be achieved in cases where the implant is coated with bioactive materials, like HA. These bioactive materials stimulate bone formation leading, to a physio-chemical bond. If it is recognized that the implant is ankylosed with the bone, it is sometimes called a biointegration instead of osseointegration [235]. For profound understanding of the osseointegration mechanism, there are several works reported. Fourteen titanium dental implants (TioblastTM) were implanted singly in the proximal tibia of New Zealand rabbits for 120 days. A bone defect was surgically produced and filled with Bio-Oss® (which is a natural, osteoconductive bone substitute that promotes bone growth in periodontal and maxillofacial osseous defects) around six of these implants. After the animals were sacrificed and their organs harvested, bone segments were fixed, and methacrylate embedded after the push-in test had been performed. The results showed that (i) the implants were apically and coronally surrounded by bone, whether Bio-Oss® was used or not, (ii) fractures were evident through the newly formed bone and between the preexisting and newly formed bone, and (iii) detachment between the implant and the bone occurred at the coronal extremity of the implants and along its cervical region. Based on these findings, it was concluded that the bone–titanium interface has a high resistance to loading, exhibiting a greater resistance than the newly formed bone [236]. The effect of a dual treatment of titanium implants and the subsequent bone response after implantation were investigated by Rønold et al. [237]. Coin-shaped CpTi implants were placed into the tibias of 12 rabbits. The implant was dually blasted with TiO_2 particles of two different sizes (i.e., finer particles of 22–28 μm in size, and coarser particles of 180–220 μm in size). It was found that (i) the Ti surface blasted with coarse particles showed a significantly better functional attachment than the fine surface, and (ii) the Ti surface blasted with fine particles showed lower retention in bone, indicating that there is a positive correlation between the topographical and mechanical evaluation of the surfaces [237].

In cementless fixation systems, surface character is an important factor. Alkali and heat treatments of titanium have been shown to produce strong bonding to bone and a higher ongrowth rate. With regard to the cementless hip stem, Nishiguchi et al. [238] examined the effect of alkali and heat treatments on titanium rods in an intramedullary rabbit femur model. Half of the implants were immersed in the 5 mol/L sodium hydroxide solution and heated at 600 °C for 1 h, and the other half were untreated. The bone–implant interfaces were evaluated at 3, 6, and 12 weeks after implantation. Pull-out tests showed that the treated implants had a significantly higher bonding strength to bone than the untreated implants at each time point. As postoperative time elapsed, histological examination revealed that new bone formed on the surface of both types of implants, but significantly more bone made direct contact with the surface of the treated implants. At 12 weeks, ~56% of the whole surface of the treated implants was covered with the bone. Based on aforementioned results, it was concluded that (i) alkali-treated and heat-treated titanium create strong bone bonding and a high affinity to bone, as opposed to a conventional mechanical interlocking mechanism, and (ii) alkali and heat treatments of titanium may be suitable surface treatments for cementless joint replacement implants [238]. Several factors influence the healing process and the long-term mechanical stability of cementless fixed implants, such as bone remodeling and mineralization processes [239]. Histomorphometric and bone hardness measurements were taken in implants inserted in sheep femoral cortical bone at different times to compare the in vivo osseointegration of titanium screws with the following surface treatments: machined (Ti-MA); acid-etched with 25 vol% of HNO_3 solution for 1 h (Ti-HF); HA vacuum plasma spray (Ti-HA); and Ca-P anodization (with 0.06 ml/L β-glycerophosphate plus 0.3 mL/L calcium acetate at 275 V and 50 mA/cm^2), followed by a 300°C autoclave-hydrothermal treatment for 2 h (Ti-AM/HA). It was reported that (i) the bone-to-implant contact (BIC) of Ti-HF was lower than that of the other surface treatments at both experimental times, (ii) significant differences in MAR (mineral apposition rate) were also found between the different experimental times for Ti-MA and Ti-HF, demonstrating that bone growth had slowed inside the screw threads of Ti-HA and Ti-AM/HA after 12 weeks, (iii) no bone microhardness changes in preexisting host bone were found, while Ti–MA showed the lowest value for the inner thread area at 8 weeks. Based on these findings, it was confirmed that osteointegration may be accelerated by adequate surface roughness and a bioactive ceramic coating, such as Ca-P anodization followed by a hydrothermal treatment, which enhances bone interlocking and mineralization [239]. Fujibayashi et al. [240] demonstrated that bone formation took place after 12-month implantation in the muscles of beagle dogs and reported that chemically and thermally treated plasma-sprayed porous titanium surfaces exhibited an intrinsic osteoinductivity. Takemoto et al. [241] tested on porous plasma-sprayed Ti blocks with the following characteristics: porosity of 41%, pore size of a ranging between 300 and 500 μm, and yield compressive strength of 85.2 MPa. Three types of surface treatments were applied (i) alkali (in the NaOH solution at 60 °C for 24 h) and heat treatment (600 °C for 1 h, followed by a furnace cooling), (ii) alkali, hot water, and heat treatment, and (iii) alkali, dilute 40 °C HCl for 24 h, hot water, and heat treatment. The osteoinductivity of the materials implanted in the back muscles of adult beagle dogs was examined at 3, 6, and 12 months. It was found that (i) the alkali–acid/heat-treated porous bioactive titanium implant had the highest osteoinductivity, with induction of a large amount of bone formation within 3 months, (ii) the dilute HCl treatment was considered to give both chemical (titania formation and sodium removal) and topographic (etching) effects on the titanium surface, and (iii) adding the dilute HCl treatment to the conventional chemical and thermal treatments is a promising candidate for advanced surface treatment of porous titanium implants [241, 242]. A number of experimental and clinical data on so-called oxidized implants have reported promising outcomes, as we have previously reviewed. However, little has been investigated on the role of the surface oxide properties and osseointegration mechanism of the oxidized implant. Sul et al. [243] recently proposed two action mechanisms for osseointegration of oxidized implants, that is, mechanical interlocking through bone growth in pores/other surface irregularities, and biochemical bonding. Two groups of oxidized implants were prepared using a micro arc oxidation process and were then inserted into rabbit bone. One group consisted of magnesium ion incorporated implants, and the other consisted of TiO_2 stoichiometry implants. It was reported that after 6 weeks of follow-up, the mean peak values for removal torque of the Mg-incorporated Ti implants dominated significantly over the TiO_2/Ti implants. Bonding failure generally occurred in the bone away from the bone to implant interface for the Mg-incorporated Ti implant and mainly occurred at the bone to implant interface for the TiO_2/Ti implant which consisted mainly of TiO_2 chemistry and significantly rougher surfaces, as compared to the Mg-incorporated Ti implant. Between bone and the Mg-incorporated Ti implant surface, ionic movements and ion concentrations gradient were detected. It was therefore concluded that the surface chemistry-mediated biochemical bonding can be theorized for explaining the oxidized bioactive implants [260]. Lim and Oshida [244] investigated the surface characteristics of acid-treated and alkali-treated CpTi. It was reported that acid-treated CpTi was covered with rutile-type TiO_2 with high hydrophobicity (in other words, less surface activity), whereas oxide film formed on the alkali-treated CPTi surface was identified to be an anatase-type dominantly with a trace amount of rutile, which exhibited high hydrophilicity (i.e., high surface energy).

For the last 15 years, orthopedic implants have been coated with HA to improve implant fixation. The osteoconductive effect of HA coatings has been demonstrated in experimental and clinical studies. However, there are ongoing developments to improve the quality of HA coatings. Aebli et al. [245] investigated whether a rough and highly crystalline HA coating applied by vacuum plasma spraying had a positive effect on the osteointegration of special, high-grade titanium (Ti) implants with the same surface roughness. Ti alloy implants were vacuum plasma spray-coated with high-grade Ti or HA. The osseointegration of the implants was evaluated by light microscopy or pullout tests after 1, 2, and 4 weeks of unloaded implantation in the cancellous bone of 18 sheep. It was found that (i) the interface shear strength increased significantly over all time intervals, (ii) by 4 weeks, values had reached ~10 N/mm^2, but (iii) the difference between the coatings was not significant at any time interval, and (iv) direct bone–implant contact was significantly different between the coatings after 2 and 4 weeks, and reached 46% for Ti and 68% for HA implants by 4 weeks. Therefore, it was indicated that the use of a rough and highly crystalline HA coating, applied by the vacuum plasma spray method, enhances early osteointegration [245]. Stewart et al. [246] investigated the effects of a plasma-sprayed HA/tricalcium phosphate (TCP) coating on osteointegration of plasma-sprayed titanium alloy implants in a lapine, distal femoral intramedullary model. The effects of the HA/TCP coating were assessed at 1, 3, and 6 months after implant placement. The HA/TCP coating significantly increased new bone apposition onto the implant surfaces at all-time points. The ceramic coating also stimulated intramedullary bone formation at the middle and distal levels of the implants. There was no associated increase in pullout strength at either 3 or 6 months; however, postpullout evaluation of the implants indicated that the HA/TCP coating itself was not the primary site of construct failure. Rather, failure was most commonly observed through the peri-prosthetic osseous struts that bridged the medullary cavity. It was therefore concluded that the osteoconductive activity of HA/TCP coating on plasma-sprayed titanium alloy implant surfaces may have considerable clinical relevance to early host–implant interactions, by accelerating the establishment of a stable prosthesis–bone interface [246].

The in vivo behavior of a porous Ti-6Al-4V material that was produced by a positive replica technique, with and without an octacalcium phosphate coating, has been studied both in the back muscle and femur of goats by Habibovic et al. [247]. Macro- and microporous biphasic CaP ceramic, known to be both osteoconductive and able to induce ectopic bone formation, was used for comparison. The three groups of materials (Ti-6Al-4V, octacalcium-phosphate–coated Ti-6Al-4V and biphasic CaP ceramic) were implanted transcortically and intramuscularly for 6 and 12 weeks in 10 adults. Dutch milk goats in order to study their osteointegration and osteoinductive potential. It was found that (i), in femoral defects, both octacalcium-phosphate–

coated Ti-6Al-4V and biphasic CaP ceramic performed better than the uncoated Ti-6Al-4V at both time points, (ii) biphasic CaP ceramic showed a higher bone amount than octacalcium-phosphate–coated Ti-6Al-4V after 6 weeks of implantation, while after 12 weeks, this difference was no longer significant, (iii) ectopic bone formation was found in both octacalcium-phosphate–coated Ti-6Al-4V and biphasic CaP ceramic implants after 6 and 12 weeks, and (iv) the ectopic bone formation was not found in uncoated titanium alloy implants, suggesting that the presence of CaP is important for bone induction [264]. The osseointegration of the copper vapor laser-superfinished titanium alloy (Ti-6Al-4V) implants with pore sizes of 25, 50, and 200 μm was evaluated in a rabbit intramedullary model by Stangl et al. [248]. Control implants were prepared by corundum blasting. Each animal received all four different implants in both femora and humeri. Using static and dynamic histomorphometry, the bone–implant interface and the peri-implant bone tissue were examined 3, 6, and 12 weeks postimplantation. Among the laser-superfinished implants, it was reported that total bone–implant contact was the smallest for the 25-μm pores and was similar for 50 and 200-μm pore sizes at all-time points, and however, all laser-superfinished surfaces were inferior to corundum-blasted control implants in terms of bone–implant contact. Implants with 25-μm pores showed the highest amount of peri-implant bone volume at all-time points, indicating that the amount of peri-implant bone was not correlated with the quality of the bone–implant interface. Accordingly, it was concluded that although laser-superfinished implants were not superior to the corundum-blasted control implants in terms of osteointegration, understanding of the mechanisms of bone remodeling within pores of various sizes was advanced, and optimal implant surfaces can be further developed with the help of modern laser technology [248]. There is another work using the laser technology for surface modification. Götz et al. [249] examined the osteointegration of laser-textured Ti-6Al-4V implants with pore sizes of 100, 200, and 300 μm, specifically comparing 200-μm implants with polished and corundum-blasted surfaces in a rabbit transcortical model. Using a distal and proximal implantation site in the distal femoral cortex, each animal received all four different implants in both femora. The bone/implant interface and the newly formed bone tissue within the pores and in peri-implant bone tissue were examined 3, 6, and 12 weeks postimplantation by static and dynamic histomorphometry. It was shown that (i) additional surface blasting of laser-textured Ti-6Al-4V implants with 200-μm pores resulted in a profound improvement in osteointegration at 12-week postimplantation, (ii) although lamellar bone formation was found in pores of all sizes, the amount of lamellar bone within the pores was linearly related to the pore size, (iii) in 100-μm pores, bone remodeling occurred with a pronounced time lag relative to larger pores, and (iv) implants with 300-μm pores showed a delayed osteointegration compared with 200-μm pores. Therefore, it was concluded that 200 μm may be the optimal pore size for laser-textured Ti-6Al-4V implants, and that laser treating in combination with surface blasting may be a very interesting technology for the structuring of implant surfaces [249]. There are some researches done on method or technique to evaluate/examine the osseointegration [250–253]. Since bone should be observed to grow into the porous structure of the coating, yielding direct evidence of a mechanism of the bone/implant interface, it was proposed to employ the advanced various analytical devices including dual-beam focused ion beam microscopy, SEM, transmission electron microscopy (TEM), and x-ray energy-dispersive spectrometer [250]. Stenport et al. [251] conducted comparison tests of tissue integration to CpTi and Ti-6Al-4V implants using various 3D biomechanical and two-dimensional histomorphometrical techniques and mentioned the loosening torque during in vivo removal torque in rabbits. It was found that significantly higher mean value of removal torque and shear strength test were observed for the CpTi implants compared to Ti-6Al-4V implants. In order to chronologically examine the titanium–bone interfaces and to clarify the process of osteointegration, Morinaga et al. [252] utilized light microcopy, TEM, quantitative tomography, and microcomputed tomography. Experimental implants (Ti-coating plastic implants) were placed into tibiae of 8-week-old rats. It was reported that (i) bone formation in osteointegration of Ti implants did not occur from the surfaces of the implant or preexisting bone, but it was likely that bone formation progressed at a site a small distance away from the surface, (ii) small bone fragments adhered to each other and transformed into reticular-shaped bone, and finally these bones became lamellar bone [251]. Kerner et al. [253] assessed *in vivo* new bone formation around Ti-6Al-4V alloy implants chemically grafted with macromolecules bearing ionic sulfonate or carboxylate groups, or both. Unmodified and grafted Ti-6Al-4V implants were placed bilaterally into rabbit femoral condyle. It was found that (i) the BIC was the lowest (13.4%) for the implants modified with grafted carboxylate only, (ii) the value of BIC on the implants with 20% sulfonate was significantly lower than that observed on 100% sulfonate surfaces, (iii) after both 4 and 12 weeks postimplantation, the BIC value for implants with more than 50% sulfonate was similar to that obtained with the unmodified Ti-6Al-4V, and (iv) the grafted Ti-6Al-4V alloy exhibiting either 100% sulfonate or carboxylate and sulfonate (50% each) groups promoted bone formation. Such materials are of clinical interest because, they do not promote bacteria adhesion but, they support new bone formation, a condition which can lead to osseointegration of bone implants while preventing peri-implant infections [253]. The design, surface characteristics and strength of metallic implants are dependent on their intended use and clinical application. Surface modifications of materials may enable reduction of the time taken for osseointegration and improve the biological response of biomechanically favorable metals and alloys. Titanium aluminum nitride (TAN) coated and plain rods of stainless steel and CpTi were implanted into the mid-shaft of the femur of Wistar rats. The femurs were harvested at 4 weeks and processed for scanning electron and light microscopy. It was reported that all implants exhibited a favorable response in bone with no evidence of fibrous encapsulation. Although there was no significant difference in the amount of new bone formed around the different rods (osseoconduction), there was a greater degree of shrinkage separation of bone from the coated rods than from the plain rods, indicating that the TAN coating may result in reduced osseointegration between bone and implant [254].

Kapanen et al. [255] determined the biocompatibility of NiTi alloy on bone formation in vivo, using ectopic bone formation assay which goes through all the events of bone formation and calcification. Comparisons were made between NiTi, stainless steel and Ti-6Al-4V, which were implanted for 8 weeks under the fascia of the latissimus dorsi muscle in 3-month-old rats. It was found that (i) the total bone mineral density values were nearly equal between the control and the NiTi samples, the stainless steel samples and the Ti-6Al-4V samples had lower bone mineral density values (ii) there were no significant differences between the implanted materials, although Ti-6Al-4V showed the largest matrix powder areas, (iii) the same method was used for measurements of proportional cartilage and new bone areas in the ossicles, (iv) NiTi showed the largest cartilage area, and (v) between implant groups the new bone area was largest in NiTi. Based on these findings, it was concluded that (i) NiTi has good biocompatibility, as its effects on ectopic bone formation are similar to those of stainless steel, and (ii) the ectopic bone formation assay developed here can be used for biocompatibility studies. Kujala et al. [256] evaluated the effect of porosity on the osteointegration of NiTi implants in rat bone. It was found that (i) the porosities (average void volume) and the mean pore size were 66.1% and 259 ± 30 µm (group 1, $n = 14$), 59.2% and 272 ± 17 µm (group 2, $n = 4$) and 46.6% and 505 ± 136 µm (group 3, $n = 15$), respectively, (ii) the implants were implanted in the distal femoral metaphysis of the rats for 30 weeks and the proportional bone–implant contact was best in group 1 (51%) without a significant difference compared to group 3 (39%), (iii) group 2 had lower contact values (29%) than group 1, and (iv) fibrotic tissue within the porous implant was found more often in group 1 than in group 3, in which 12 samples out of 15 showed no signs of fibrosis; concluding that (v) porosity of 66.1% (MPS 259 ± 30 µm) showed best bone contact (51%) of the porosities tested here; however, the porosity of 46.6% (mean pore size: 505 ± 136 µm) with bone contact of 39% was not significantly inferior in this respect and showed lower incidence of fibrosis within the porous implant. Complex physical and chemical interactions take place in the interface between the implant surface and bone. Various descriptions of the ultrastructural arrangement to various implant design features, ranging from solid and macroporous geometries to surface modifications on the micron level, submicron level, and nanolevel, have been put forward. Shah et al. [257] reviewed the current knowledge regarding structural organization of the bone–implant interface with a focus on solid devices, mainly metal (or alloy) intended for permanent anchorage in bone. It was mentioned that (i) the bone–implant interface is a heterogeneous zone consisting of mineralized, partially mineralized, and unmineralized areas, (ii) within the meso-micro-nano-continuum, mineralized collagen fibrils form the structural basis of the bone–implant interface, in addition to accumulation of noncollagenous macromolecules such as osteopontin, bone sialoprotein, and osteocalcin, (iii) the interpretation is influenced by the in vivo model and species-specific characteristics, healing

time point(s), physicochemical properties of the implant surface, implant geometry, sample preparation route(s) and associated artefacts, analytical technique(s) and
their limitations, and noncompromised versus compromised local tissue conditions,
(iv) the understanding of the ultrastructure of the interface under experimental conditions is rapidly evolving due to the introduction of novel techniques for sample
preparation and analysis, (v) nevertheless, the current understanding of the interface
zone in humans in relation to clinical implant performance is still hampered by the
shortcomings of clinical methods for resolving the finer details of the bone–implant
interface, and (vi) within the meso-micro-nano-continuum, mineralized collagen
fibrils form the structural basis of the bone–implant interface, in addition to accumulation of noncollagenous macromolecules such as osteopontin, bone sialoprotein,
and osteocalcin.

11.7.2 Stability and evaluation

Implant stability is a prerequisite characteristic of osseointegration. Continuous monitoring in a quantitative and objective manner is important to determine the status of
implant stability [258, 259]. Osseointegration is also a measure of implant stability
which can occur in two stages: primary and secondary [260]. Primary stability mostly
occurs from mechanical engagement with cortical bone. A key factor for the implant
primary stability is the BIC [261], therefore, the primary stability is affected by bone
quality and quantity, surgical technique and implant geometry (length, diameter,
surface characteristics). Secondary stability offers biological stability through bone
regeneration and remodeling [262–265]. Secondary stability is affected by primary stability [264, 266]. During the transition period from the primary to the secondary stability, the implant faces the risk of micromotion, possibly leading to implant failure.
It is estimated that this period in humans occurs roughly 2–3 weeks after implant
placement when osteoclastic activity decreases the initial mechanical stability of
the implant but not enough new bone has been produced to provide an equivalent
or greater amount of compensatory biological stability [261, 267]. This is related to
the biologic reaction of the bone to surgical trauma during the initial bone remodeling phase; bone and necrotic materials resorbed by osteoclastic activity is reflected
by a reduction in implant stability quotient value. This process is followed by new
bone apposition initiated by osteoblastic activity, therefore leading to adaptive bone
remodeling around the implant [268]. Summarizing, it is true that implant stability
is one of the most important factors for the success of implant treatments. Although
most studies showed a correlation between bone density and implant stability, some
studies suggest the opposite due to the differences in the methods used. Recent
studies suggest that implant stability during the healing process only increases for
implants with low initial stabilities; meanwhile, loss of stability during the healing
can be observed in implants with high initial stabilities [269].

11.7.2.1 In vitro evaluation of stability

In vitro cell culture models are routinely used to study the response of osteoblastic cells in contact with different substrates for implantation in bone tissue. Cell cultures focused on the morphological aspect, growth capacity and the state of differentiation of the cells on materials with various chemical, composition and topography [270, 271]. It is well documented that the biochemistry and topography of biomaterials' surfaces play a key role on success or failure upon placement in a biological environment [272]. Wettability, texture, chemical composition, and surface topography are properties of the biomaterials that directly influence their interaction with cells [273–275]. Historically, the gold standard method used to evaluate the degree of osseointegration was microscopic or histologic analysis. And biological responses study can include cell morphology and cell activity (cell adhesion, differentiation, and proliferation) [276–280]. Since cell interactions with extracellular matrix affect directly the cellular processes of adhesion, proliferation, and differentiation [281], the surface properties of biomaterials are essential to the response of cells at biomaterial interface, affecting the growth and quality of newly formed bone tissue [282]. In general, the common finding of the aforementioned references can be found as a fact that the cell activity is strongly related to implant surface morphology and topology as related to the process of osseointegration [272]. Actually, this surface characteristic is one of the three major requirements for placed implant to exhibit the subsequent retaining in the bone (in other words, osseointegration) – known as morphological compatibility [52, 283]. Two other requirements are biological compatibility and biomechanical compatibility [52, 284]. To manipulate surface structure as being morphologically compatible to receiving hard tissue surface configuration, there have been various methods and techniques proposed; including as-machined, blasted surface, acid or alkaline etching, chemical treating on blasted surface, HA coating, recently the biomimetic calcium phosphate coating [52].

11.7.2.2 In vivo animal evaluation of stability

It is generally believed that outcomes on the initial biological behavior of implantable materials obtained in vitro cannot be fully correlated to in vivo performance. Cell cultures cannot reproduce the dynamic environment that involves the in vivo bone/implant interaction, and their results can only be confirmed in animal models and subsequently in clinical trials [271, 285, 286]. Irrespective of the different animal models or surgical sites, valuable information can be retrieved from properly designed animal studies. Static and dynamic histomorphometric parameters plus biomechanical testing are recommended as measurable indicators of the host/implant response where different surface designs are compared. BIC, that is, the most often evaluated parameter in in vivo studies together with bone density and amount and type of cellular content are examples of static parameters [286]. In most majority of publications, canine [287, 288], sheep/goat [289], pig [290], rabbit [291–294], rat and mice [295–298] are popular kinds of animals as nonhuman primates which are mostly sacrificed for research purpose [299]. Although animal tests appear to be a well-established technique to provide valuable and applicable information to humans, there are still several concerns. One of the main problems associated with the in vivo tests using animal models is the test duration and its validity for application. Dziuba et al. [293], using rabbit's model, investigated the biological behavior of Mg-implantable alloy for the longest of 12 months. Amerstorfer et al. [297] had conducted the in vivo with Sprague-Dawley rats for 12 months to investigate the biodegradability for Mg alloy as a potential implant material. Furthermore, Akens et al. [300], using sheep for 18 months at the longest test duration, studied the efficacy of photo-oxidized bovine osteochondral transplant. Although the conclusive remarks common over these tests indicate that the result are promising, but still a longer period of testing time is required before applying the materials to human subjects. The relevancy of results obtained from animal models has been subject to a debate with great extent. The use of animal models in the study of dental implants has contributed greatly to understand many different devices in use. Animal testing plays a major role in assessing the safety and efficacy of dental implants. To date, animal testing has shown the nature of soft tissue attachment to implants and the types of interfacial tissues within bone sites. There have been increased studies correlating animal tests with in vitro analysis and human studies.

There is an important aspect to the question of the relevance of animal research to humans, namely, the way of the observations and results are evaluated and adopted [301, 302]. Since some results from in vitro studies can be difficult to extrapolate to the in vivo situation, the use of animal models is often an essential step in the testing of orthopedic and dental implants prior to clinical use in humans [303]. As mentioned earlier, variety of animals such as the dog, sheep, goat, pig, and rabbit models has been used for the evaluation of bone–implant interactions. There are differences in bone composition between the various species and humans. While no species fulfils all of the requirements of an ideal model, an understanding of the differences in bone architecture and remodeling between the species is likely to assist in the selection of a suitable species for a defined research question [303]. One other important issue was raised from the veterinarians. Let's listen to what they say: "Animal research is useless because disease mechanisms are very different between animals and humans, so drugs that work an animal do not work on humans." Animals have been used repeatedly throughout the history of biomedical research. In recent years, the practice of using animals for biomedical research has come under severe criticism by animal protection and animal rights groups. Those against the animal tests contend that the benefit to humans does not justify the harm to animals. Many people also believe that animals are inferior to humans and very different from them, hence results from animals can't be applied to humans. Those in favor of animal testing argue that experiments on animals are necessary to advance medical and biological knowledge and it was recognized that the practice of using animals in biomedical research has led to significant advances in the treatment of various diseases [304, 305]. Veterinarian adopted technology and treatment technique developed for humans to their patients [306]. Veterinary dentistry has made great strides providing treatments that were once available only to people. When performed for the medical benefit of animals, procedures such as professional dental cleaning, root canal treatment, and even, in selected cases, periodontal surgery and orthodontic correction can prevent and treat disease, improve quality of life, and enhance and sustain the human–animal bond. On the contrary, even quality of life of animals have been emphasized among veterinarians and animal rights proponents [307]. Issues such as cruelty to animals and the humane treatment of animals are valid concerns, and hence, the use of animals in experimentation is greatly regulated. This has led to the 3Rs campaign, which advocates the search (1) for the *replacement* of animals with nonliving models; (2) *reduction* in the use of animals; and (3) *refinement* of animal use practices. However, total elimination of animal testing will significantly set back the development of essential medical devices, medicines, and treatment. By employing the 3Rs when continuing to use animals for scientific research, the scientific community can affirm its moral conscience as well as uphold its obligation to humanity to further the advancement of science for civilization and humanity [308]. Now back to the stability evaluation, besides above-mentioned histological analyses, there are the biomechanical tests (torque, push-out, pullout, etc.) to measure the amount of force that a torque needs to fail the bone–implant interface surrounding different implant surfaces [309, 310]. Considering the several factors that influence the osseointegration, the evaluation of the largest possible number of host/implant response parameters is desirable for better understanding the bone healing around different implant surface, clarifying their indications of use and supporting their immediate/early loading. These tests check clinically for mobility with the help of blunt-ended instruments, radiographs, cutting torque resistance, reverse torque, and resonance frequency analysis [311, 312].

References

[1]   Setcos JC, Babaei-Mahani A, Silvio LD, Mjör IA, Wilson NHF. The safety of nickel containing dental alloys. Dent. Mater. 2006, 22, 1163–8.
[2]   Denkhaus E, Salnikow K. Nickel essentiality, toxicity, and carcinogenicity. CRC Cr. Rev. Oncol. Hem. 2002, 42, 35–56.
[3]   Schroeder HA, Balassa JJ, Tipton IH. Abnormal trace elements in man – nickel. J. Chron. Dis. 1962, 15, 51–65.
[4]   Trépanier C, VenugopalanR, Pelton AR. Corrosion Resistance and Biocompatibility of Passivated NiTi. In L'H Yahia, ed., Shape Memory Implants. Springer, New York, 2000, 35–45.
[5]   Bass JK, Fine H, Cisneros GJ. Nickel hypersensitivity in the orthodontics patent. Am. J. Orthod. Dentofac. Orthop. 1993, 103, 280–5.
[6]   Takamura K, Hayashi K, Ishinishi N, Sugioka Y. Evaluation of carcinogenicity and chronic toxicity associated with orthopedic implants in mice. J. Biomed. Mater. Res. 1994, 28, 583–9.

[7]   Hultgren R, Desai PD, Hawkins T, Gleiser M, Kelley KK. Selected value of the thermodynamic properties of binary alloys. American Society for Metals, Metals Park, OH, 1973, 1244–6.

[8]   Doi H, Takeda S. In vitro metal dissolution under various extraction conditions. Shika Zairyo Kikai 1990, 9, 375–86.

[9]   Watarai M, Hanawa T, Moriyama K, Asaoka K. Amount of metallic ions released from Ti-Ni alloy by abrasion in simulated bioliquids. Biomed. Mater. Eng. 1999, 9, 73–9.

[10]   Ryhänen J, Kallioinen M, Serlo W, Perämäki P, Junila J, Sandvik P, Niemelä E, Tuukkanen J. Bone healing and mineralization, implant corrosion, and trace metals after nickel-titanium shape memory metal intramedullary fixation. J Biomed. Mater. Res. 1999, 47, 472–80.

[11]   Cortizo MC, De Mele MF, Cortizo AM. Metallic dental material biocompatibility in osteoblastlike cells: correlation with metal ion release. Biol. Trace Elem. Res. 2004, 100, 151–68.

[12]   Hoene A, Prinz C, Walschus U, Lucke S, Patrzyk M, Wilhelm L, Neumann H-G, Schlosser M. In vivo evaluation of copper release and acute local tissue reactions after implantation of copper-coated titanium implants in rats. Biomed. Mater. 2013, 8; doi: 10.1088/1748-6041/8/3/035009.

[13]   Okazaki Y, Gotoh E. Metal release from stainless steel, Co-Cr-Mo-Ni-Fe and Ni-Ti alloys in vascular implants. Corros. Sci. 2008, 50, 3429–38.

[14]   Høl PJ, Gjerdet NR, Jonung T. Corrosion and metal release from overlapping arterial stents under mechanical and electrochemical stress – an experimental study. J. Mech. Behav. Biomed. Mater. 2019, 93, 31–5.

[15]   Heusler KE. The influence of electrolyte composition on the formation and dissolution of passivating films. Corr. Sci. 1989, 29, 131–47.

[16]   Wever DJ, Veldhuizen AG, deVries J, Busscher HJ, Uges DRA, van Horn JR. Electrochemical and surface characterization of a nickel-titanium alloy. Biomaterials 1998, 19, 761–9.

[17]   Dong H, Ju X, Yang H, Qian L, Zhou Z. Effect of ceramic conversion treatments on the surface damage and nickel ion release of NiTi alloys under fretting corrosion conditions. J. Mater. Sci.: Mater. Med. 2008, 19, 937–46.

[18]   Ševčíková J, Bártková D, Goldbergová M, Kuběnová M, Čermák J, Frenzel J, Weiser A, Dlouhý A. On the Ni-ion release rate from surfaces of binary NiTi shape memory alloys. Appl. Surf. Sci. 2018, 427, 434–43.

[19]   Park HY, Shearer TR. In vitro release of nickel and chromium from simulated orthodontic appliances. Am. J. Orthod. 1983, 84, 156–9.

[20]   Kerosuo H, Moe G, Kleven E. In vitro release of nickel and chromium from different types of simulated orthodontic appliances. Angle Orthod. 1995, 65, 111–6.

[21]   Barrett RD, Bishara SE, Quinn JK. Biodegradation of orthodontic appliances. Part I. Biodegradation of nickel and chromium in vitro. Am. J. Orthod. Dentofac. Orthop. 1993, 103, 8–14.

[22]   Bishara SE, Barrett RD, Selim MI. Biodegradation of orthodontic appliances. Part II. Changes in the blood level of nickel. Am. J. Orthod. Dentofac. Orthop. 1993, 103, 115–9.

[23]   Ousehal L, Lazrak L. Change in nickel levels in the saliva of patients with fixed orthodontic appliances. Int. Orthod. 2012, 10, 190–7.

[24]   Hwang C-J, Shin J-S, Cha J-Y. Metal release from simulated fixed orthodontic appliances. Am. J. Orthod. Dentofac. Orthop. 2001, 120, 383–91.

[25]   Eliades T, Pratsinis H, Kletsas D, Eliades G, Makou M. Characterization and cytotoxicity of ions released from stainless steel and nickel-titanium orthodontic alloys. Am. J. Orthod. Dentofac. Orthop. 2004, 125, 24–9.

[26]   Ortiz AJ, Fernández E, Vicente A, Calvo JL, Ortiz C. Metallic ions released from stainless steel, nickel-free, and titanium orthodontic alloys: toxicity and DNA damage. Am. J. Orthod. Dentofac. Orthop. 2011, 140, e115–22.

[27]   Mikulewicz M, Chojnacka K. Trace metal release from orthodontic appliances by in vivo studies: a systematic literature review. Biol. Trace Elem. Res. 2010, 137, 127–38.

[28]   Mikulewicz M, Chojnacka K. Trace metal release from orthodontic appliances by in vitro studies: a systematic literature review. Biol. Trace Elem. Res. 2010, 139, 241–56.

[29]   Mikulewicz M, Chojnacka K, Woźniak B, Downarowicz P. Release of metal ions from orthodontic appliances: an in vitro study. Biol. Trace Elem. Res. 2012, 146, 272–80.

[30]   Amini F, Mollaei M, Harandi S, Rakhshan V. Effects of fixed orthodontic treatment on hair nickel and chromium levels: a 6-month prospective preliminary study. Biol. Trace Elem. Res. 2015, 164, 12–7.

[31]   Yanisarapan T, Thunyakitpisal P, Chantarawaratit P-O. Corrosion of metal orthodontic brackets and archwires caused by fluoride-containing products: cytotoxicity, metal ion release and surface roughness. Orthod. Waves 2018, 77, 79–89.

[32]   Tahmasbi S, Ghorbani M, Masudrad M. Galvanic corrosion of and ion release from various orthodontic brackets and wires in a fluoride-containing mouthwash. J. Dent. Res., Dent. Clin., Dent. Prospects 2015, 9, 159–65.

[33]   Tahmasbi S, Sheikh T, Hemmati YB. Ion release and galvanic corrosion of different orthodontic brackets and wires in artificial saliva. J. Contemp. Dent. Pract. 2017, 18, 222–7.

[34]   Darabara MS, Bourithis LI, Zinelis S, Papadimitriou GD. Metallurgical characterization, galvanic corrosion, and ionic release of orthodontic brackets coupled with Ni-Ti archwires. J. Biomed. Mater. Res. B Appl. Biomater. 2007, 81, 126–34.

[35]   Huang HH, Chiu YH, Lee TH, Wu SC, Yang HW, Su KH, Hsu CC. Ion release from NiTi orthodontic wires in artificial saliva with various acidities. Biomaterials 2003, 24, 3585–92.

[36]   Kaaber K, Veien NK, Tjell JC. Low nickel diet in the treatment of patients with chronic nickel dermatitis. Br. J. Dermatol. 1978, 98, 197–201.

[37]   Cioffi M, Gilliland D, Ceccone G, Chiesa R, Cigada A. Electrochemical release testing of nickel-titanium orthodontic wires in artificial saliva using thin layer activation. Acta Biomater. 2005, 1, 717–24.

[38]   Petoumeno E, Kislyuk M, Hoederath H, Keilig L, Bourauel C, Jäger A. Corrosion susceptibility and nickel release of nickel titanium wires during clinical application. J. Orofac. Orthop. / Forts 2008, 69, 411–23.

[39]   Poosti M, Rad HP, Kianoush K, Hadizadeh B. Are more nickel ions released from NiTi wires after sterilisation? Aust. Orthod. J. 2009, 25, 30–3.

[40]   Petoumenou E, Arndt M, Keilig L, Reimann S, Hoederath H, Eliades T, Jäger A, Bourauel C. Nickel concentration in the saliva of patients with nickel-titanium orthodontic appliances. Am. J. Orthod. Dentofac. Orthop. 2009, 135, 59–65.

[41]   Kuhta M, Pavlin D, Slaj M, Varga S, Lapter-Varga M, Slaj M. Type of Archwire and level of acidity: effects on the release of metal ions from orthodontic appliances. Angle Orthod. 2009, 79, 102–10.

[42]   Suárez C, Vilar T, Gil FJ, Sevilla P. In vitro evaluation of surface topographic changes and nickel release of lingual orthodontic archwires. J. Mater. Sci. Mater. Med. 2010, 21, 675–83.

[43]   Liu J-K, Lee T-M, Liu I-H. Effect of loading force on the dissolution behavior and surface properties of nickel-titanium orthodontic archwires in artificial saliva. Am. J. Orthod. Dentofac. Orthop. 2011, 140, 166–76.

[44]   Karnam SK, Reddy AN, Manjith CM. Comparison of metal ion release from different bracket archwire combinations: an in vitro study. J. Contemp. Dent. Pract. 2012, 13, 376–81.

[45]   Senkutvan RS, Jacob S, Charles A, Vadgaonkar V, Jatol-Tekade S, Gangurde P. Evaluation of nickel ion release from various orthodontic arch wires: an in vitro study. J. Int. Soc. Prev. Community Dent. 2014, 4, 12–6.

[46]   Azizi A, Jamilian A, Nucci F, Kamali Z, Hosseinikhoo N, Perillo L. Release of metal ions from round and rectangular NiTi wires. Prog. Orthod. 2016, 17; doi: 10.1186/s40510-016-0123-3.

[47] Mirhashemi A, Jahangiri S, Kharrazifard M. Release of nickel and chromium ions from orthodontic wires following the use of teeth whitening mouthwashes. Prog. Orthod. 2018, 19; doi: 10.1186/s40510-018-0203-7.

[48] Ormiga F, da Cunha Ponciano Gomes JA, de Araújo MC. Dissolution of nickel-titanium endodontic files via an electrochemical process: a new concept for future retrieval of fractured files in root canals. J. Endod. 2010, 36, 717–20.

[49] Ormiga F, da Cunha Ponciano Gomes JA, de Araújo MC, Barbosa AO. An initial investigation of the electrochemical dissolution of fragments of nickel-titanium endodontic files. J. Endod. 2011, 37, 526–30.

[50] Mitchell Q, Jeansonne BG, Stoute D, Lallier TE. Electrochemical dissolution of nickel-titanium endodontic files induces periodontal ligament cell death. J. Endod. 2013, 39, 679–84.

[51] Shabalovskaya SA, Anderegg J, Van Humbeeck J. Critical overview of Nitinol surfaces and their modifications for medical applications. Acta Biomater. 2008, 4, 447–67.

[52] Oshida Y. Surface Engineering and Technology for Biomedical Implants. Moment Press, New York, NY, USA, 2014.

[53] Steegmueller R, Wagner C, Fleckenstein T, Schuessler A. Gold coating of nitinol devices for medical applications. Mater. Sci. Forum 2002, 394/395, 161–4.

[54] Cheng Y, Cai W, Li HT, Zheng YF. Surface modification of NiTi alloy with tantalum to improve its biocompatibility and radiopacity. J. Mater. Sci. 2006, 41, 4961–4.

[55] Oshida Y. Bioscience and Bioengineering of Titanium Materials. Elsevier, London, UK, 2007.

[56] Michiardi A, Aparicio C, Planell JA, Gil FJ. New oxidation treatment of NiTi shape memory alloys to obtain Ni-free surfaces and to improve biocompatibility. J. Biomed. Mater. Res. B Appl. Biomater. 2006, 77, 249–56.

[57] Chu CL, Guo C, Sheng XB, Dong YS, Lin PH, Yeung KWK, Chu PK. Microstructure, nickel suppression and mechanical characteristics of electropolished and photo-electrocatalytically oxidized biomedical nickel titanium shape memory alloy. Acta Biomater. 2009, 5, 2238–45.

[58] Shabalovskaya SA, Tian H, Anderegg JW, Schryvers DU, Carroll WU, van Humbeeck J. The influence of surface oxides on the distribution and release of nickel from Nitinol wires. Biomaterials 2009, 30, 468–77.

[59] Espinar E, Llamas JM, Michiardi A, Ginebra MP, Gil FJ. Reduction of Ni release and improvement of the friction behaviour of NiTi orthodontic archwires by oxidation treatments. J. Mater. Sci. Mater. Med. 2011, 22, 1119–25.

[60] Dick LFP, Corte DD. Selective dissolution of Ni from Nitinol for increasing the biocompatibility. ECS Trans. 2008, 11, 29–38.

[61] Chu C-L, Zhou J, Chung JCY, Pu Y-P, Lin P-H. In situ formation of titania film on NiTi alloy treated with hydrogen peroxide solution at low temperature. Trans. Nonferr. Metal. Soc. China 2005, 15, 834–8.

[62] Chu CL, Chung JCY, Chu PK. Surface oxidation of NiTi shape memory alloy in boiling aqueous solution containing hydrogen peroxide. Mater. Sci. Eng. 2006, A 417, 104–9.

[63] Oshida Y. unpublished data, 2019.

[64] Villermaux F, Tabrizian M, Yahia LH, Meunier M, Piron DL. Excimer laser treatment of NiTi shape memory alloy biomaterials. Appl. Surf. Sci. 1997, 109/110, 62–6.

[65] Starosvetsky D, Gotman I. Corrosion behavior of titanium nitride coated Ni-Ti shape memory surgical alloy. Biomaterials 2001, 22, 1853–9.

[66] Ju X, Dong H. Plasma surface modification of NiTi shape memory alloy. Surf. Coat. Technol. 2006, 201, 1542–7.

[67] de Camargo EN, Lobo AO, da Silva MM, Ueda M, Garcia EE, Pichon L, Reuther H, Otubo J. Determination of Ni release in NiTi SMA with surface modification by nitrogen plasma immersion ion implantation. J. Mater. Eng. Perform. 2011, 20, 798–801.

[68] Duerig T, Pelton A, Stöckel D. An overview of nitinol medical applications. Mater. Sci. Eng. (A) 1999, 273/275, 149–60.
[69] Thompson A. An overview of nickel-titanium alloys used in dentistry. Inter. Endo. J. 2000, 33, 297–310.
[70] Feninat FE, Laroche G, Fiset M. Mantovani D. Shape memory materials for biomedical applications. Adv. Eng. Mater. 2000, 4, 91–104.
[71] Kimura H, Sohmura T. Improvement in corrosion resistance of Ti-Ni shape memory alloy by oxide film coating. Jpn. J. Dent. Mater. 1988, 7, 106–10.
[72] Ozeki K, Yuhta T, Aoki H, Fukui Y. Inhibition of Ni release from NiTi alloy by hydroxyapatite, alumina, and titanium sputtered coatings. Biomed. Mater. Eng. 2003, 13, 271–9.
[73] Yeung KWK, Poon RWY, Liu XY, Ho JPY, Chung CY, Chu PK, Lu WW, Chan D, Cheung KMC. Investigation of nickel suppression and cytocompatibility of surface-treated nickel-titanium shape memory alloys by using plasma immersion ion implantation. J. Biomed. Mater. Res. Part A 2005, 72, 238–45.
[74] Surmenev RA, Ryabtseva MA, Shesterikov EV, Pichugin VF, Peitsch T, Epple M. The release of nickel from nickel–titanium (NiTi) is strongly reduced by a sub-micrometer thin layer of calcium phosphate deposited by rf-magnetron sputtering. J. Mater. Sci. Mater. Med. 2010, 21, 1233–9.
[75] Maleki-Ghaleh H, Khalil-Allafi J, Khalili V, Shakeri MS, Javidi M. Effect of hydroxyapatite coating fabricated by electrophoretic deposition method on corrosion behavior and nickel release of NiTi shape memory alloy. Mater. Corros. 2014, 65, 725–32.
[76] Anuradha P, Varma NK, Balakrishnan A. Reliability performance of titanium sputter coated Ni-Ti arch wires: mechanical performance and nickel release evaluation. Biomed. Mater. Eng. 2015, 26, 67–77.
[77] Katic V, Curkovic L, Ujevic Bosnjak M, Peros K, Mandic D, Spalj S. Effect of pH, fluoride and hydrofluoric acid concentration on ion release from NiTi wires with various coatings. Dent. Mater. J. 2017, 36, 149–56.
[78] Wu S, Liu X, Chan YL, Ho JP, Chung CY, Chu PK, Chu CL, Yeung KW, Lu WW, Cheung KM, Luk KD. Nickel release behavior, cytocompatibility, and superelasticity of oxidized porous single-phase NiTi. J. Biomed. Mater. Res. A 2007, 81, 948–55.
[79] Wu S, Liu X, Chan YL, Ho JPY, Chung CY, Chu PK, Chu CL, Yeung KWK, Lu WW, Cheung KMC, Luk KDK. Nickel release behavior, cytocompatibility, and superelasticity of oxidized porous single-phase NiTi. J. Biomed. Mater. Res. 2016; https://doi.org/10.1002/jbm.a.31115.
[80] Wu S, Liu X, Chan YL, Chu PK, Chung CY, Chu C, Yeung KW, Lu WW, Cheung KM, Luk KD. Nickel release behavior and surface characteristics of porous NiTi shape memory alloy modified by different chemical processes. J. Biomed. Mater. Res. A. 2009, 89, 483–9.
[81] Wood JFL. Patient sensitivity to alloy used in partial denture. Br. Dent. J. 1974, 136, 423–4.
[82] Menne T, Christophersen J, Green A. Epidemiology if Nickel Dermatitis. In: Maibach HI, Menne T, ed., Nickel and the Skin: Immunology and Toxicology. CRC Press, Boca Raton, FL, 1989, 109–16.
[83] Romaguera C, Grimalt F, Vilaplana J. Contact dermatitis from nickel: an investigation of its source. Cont. Dermatitis, 1988, 19, 52–7.
[84] Meijer C, Bredberg M, Fisher T, Widstrom L. Ear piercing and nickel and cobalt sensitization, in 520 young Swedish men doing compulsory military service. Cont. Dermatitis, 1995, 32, 147–79.
[85] Mills MP. Immunological and inflammatory aspects of periodontal disease; https://www.dentalcare.com/en-us/professional-education/ce-courses/ce1/types-of-hypersensitivity-reactions, 14–23.
[86] Honari G, Ellis SG, Wilkoff BL, Aronica MA, Svensson LG, Taylor JS. Hypersensitivity reactions associated with endovascular devices. Cont. Dermatitis 2008, 59, 7–22.

[87]   Peltonen L. Nickel sensitivity in the general population. Cont. Dermatitis 1979, 5, 27–32.

[88]   Cempel M, Nikel G. Nickel: a review of its sources and environmental toxicology. Pol. J. Environ. Stud. 2006, 15, 375–82.

[89]   Kerosuo H, Kullaa A, Kerosuo E, Kanerva L, Hensten-Pettersen A. Nickel allergy in adolescents in relation to orthodontic treatment and piercing of ears. Am. J. Orthod. Dentofac. Orthop. 1996, 109, 148–54.

[90]   Janson GRP, Dainesi EA, Consolaro A, Woodside DG, Freitas MR. Nickel hypersensitivity reaction before, during, and after orthodontic therapy. Am. J. Orthod. Dentofac. Orthop. 1998, 113, 655–60.

[91]   Faccioni F, Franceschetti P, Cerpelloni M, Fracasso ME. In vivo study on metal release from fixed orthodontic appliances and DNA damage in oral mucosa cells. Am. J. Orthod. Dentofac. Orthop. 2003, 124, 687–93.

[92]   Schuster G, Reichle R, Bauer RR, Schopf PM. Allergies induced by orthodontic alloys: incidence and impact on treatment. J. Orofac. Orthop. Der Kieferorthopädie 2004, 65, 48–59.

[93]   Genelhu MC, Marigo M, Alves-Oliveira LF, Malaquias LC, Gomez RS. Characterization of nickel-induced allergic contact stomatitis associated with fixed orthodontic appliances. Am. J. Orthod. Dentofac. Orthop. 2005, 128, 378–81.

[94]   Pazzini CA, Júnior GO, Marques LS, Pereira CV, Pereira LJ. Prevalence of nickel allergy and longitudinal evaluation of periodontal abnormalities in orthodontic allergic patients. Angle Orthod. 2009, 79(5), 922–7.

[95]   Amini F, Rakhshan V, Mesgarzadeh N. Effects of long-term fixed orthodontic treatment on salivary nickel and chromium levels: a 1-year prospective cohort study. Biol. Trace Elem. Res. 2012, 150, 15–20.

[96]   Amini F, Rakhshan V, Sadeghi P. Effect of fixed orthodontic therapy on urinary nickel levels: a long-term retrospective cohort study. Biol. Trace Elem. Res. 2012, 150, 31–6.

[97]   Amini F, Mollaei M, Harandi S, Rakhshan V. Effects of fixed orthodontic treatment on hair nickel and chromium levels: a 6-month prospective preliminary study. Biol. Trace Elem. Res. 2015, 164, 12–7.

[98]   Fors R, Stenberg B, Stenlund H, Persson M. Nickel allergy in relation to piercing and orthodontic appliances – a population study. Cont. Dermat. 2012, 67, 342–50.

[99]   Jensen CS, Menné T, Lisby S, Kristiansen J, Veien NK. Experimental systemic contact dermatitis from nickel: a dose response study. Cont. Dermat. 2003, 49, 124–32.

[100]   Hamann C. Metal allergy: titanium. Met. Aller. 2018, 443–66; doi: 10.1007/978-3-319-58503-1_34.

[101]   Steinemann SG. Titanium –the material of choice? Periodontol 2000, 1998, 17, 7–21.

[102]   Maspero C, Giannini L, Galbiati G, Nolet F, Esposito L, Farronato G. Titanium orthodontic appliances for allergic patients. Minerva Stomatol 2014, 63, 403–10.

[103]   Wood MM, Warshaw EM. Hypersensitivity reactions to titanium: diagnosis and management. Dermatitis 2015, 26, 7–25.

[104]   Meningaud J-P, Poupon J, Bertrand J-C, Chenevier C, Galliot-Guilley M, Guilbert F. Dynamic study about metal release from titanium miniplates in maxillofacial surgery. Int. J. Oral Maxillofac. Surg. 2001, 30, 185–8.

[105]   Hallab N, Merritt K, Jacobs JJ. Metal sensitivity in patients with orthopaedic implants. J. Bone Joint Surg. Am. 2001, 83, 428–36.

[106]   Langford R, Frame J. Tissue changes adjacent to titanium plates in patients. J. Craniomaxillfac. Surg. 2002, 30, 103–7.

[107]   Mylanus EA, Johansson CB, Cremers CW. Craniofacial titanium implants and chronic pain: histologic findings. Otol. Neurotol. 2002, 23, 920–5.

[108]   Frisken KW, Dandie GW, Lugowski S, Jordan G. A study of titanium release into body organs following the insertion of single threaded screw implants into the mandibles of sheep. Aust. Dent. J. 2002, 47, 214–7.

[109]   Kasai Y, Lida R, Uchida A. Metal concentrations in the serum and hair of patients with titanium alloy spinal implants. Spine 2003, 25, 1320–6.

[110]   Flatebø RS, Johannessen AC, Grønningsæter AG, Bøe OE, Gjerdet NR, Grung B, Leknes KN. Host response to titanium dental implant placement evaluated in a human oral model. J. Periodontol. 2006, 77, 1201–10.

[111]   Olmedo DG, Duffo G, Cabrini RL, Guglielmotti MB. Local effect of titanium implant corrosion: an experimental study in rats. Int. J. Oral Maxillofac. Surg. 2008, 37, 1032–8.

[112]   Richardson TD, Pineda SJ, Strenge KB, van Fleet TA, Macgregor M, Milbrandt JC, Espinosa JA, Freitag P. Serum titanium levels after instrumented spinal arthrodesis. Spine 2008, 33, 792–6.

[113]   Sicilia A, Cuesta S, Coma G, Arregui I, Guisasola C, Ruiz E, Maestro A. Titanium allergy in dental implant patients: a clinical study on 1500 consecutive patients. Clin. Oral Implants Res. 2008, 19, 823–35.

[114]   Egusa H, Ko, H, Shimazu T, Yatani H. Suspected association of an allergic reaction with titanium dental implants: a clinical report. J. Prosthet. Dent. 2008, 100, 344–7.

[115]   Olmedo DG, Paparella ML, Brandizzi D, Cabrini RL. Reactive lesions of peri-implant mucosa associated with titanium dental implants: a report of 2 cases. Int. J. Oral Maxillofac. Surg. 2010, 39, 503–7.

[116]   Siddiqi A, Payne AGT, de Silva RK, Duncan WJ. Titanium allergy: could it affect dental implant integration? Clin. Oral Implants Res. 2011, 22, 673–80.

[117]   Basko-Plluska JL, Thyssen JP, Schalock PC. Cutaneous and systemic hypersensitivity reactions to metallic implants – an update. Dermatitis 2011, 22, 65–79.

[118]   Chaturvedi TP. An overview of the corrosion aspect of dental implants (titanium and its alloys) Indian J. Dent. Res. 2009, 20, 91–8.

[119]   Cundy TP, Antoniou G, Sutherland LM, Freeman BJC, Cundy PJ. Serum titanium, niobium, and aluminum levels after instrumented spinal arthrodesis in children. Spine 2013, 38, 564–70.

[120]   Javed F, Al-Hezaimi K, Almas K, Romanos GE. Is titanium sensitivity associated with allergic reactions in patients with dental implants? A systematic review. Clin. Implant Dent. Relat. Res. 2013, 15, 47–52.

[121]   Oliva X, Oliva J, Oliva JD. Full-mouth oral rehabilitation in a titanium allergy patient using zirconium oxide dental implants and zirconium oxide restorations. A case report from an ongoing clinical study. Eur. J. Esthet. Dent. 2010, 5, 190–203.

[122]   Osman RB, Swain MV. A critical review of dental implant materials with an emphasis on titanium *versus* zirconia. Materials (Basel) 2015, 8, 932–58.

[123]   Viraben R, Boulinguez S, Alba C. Granulomatous dermatitis after implantation of a titanium-containing pacemaker. Cont. Dermat. 1995, 33, 437–41.

[124]   Freeman S. Allergic contact dermatitis to titanium in a pacemaker. Cont. Dermat. 2006, 55, 41–4.

[125]   Syburra T, Schurr U, Rahn M, Graves K, Genoni M. Gold-coated pacemaker implantation after allergic reactions to pacemaker compounds. EP Europace 2010, 12, 749–50.

[126]   Kypta A, Blessberger H, Lichtenauer M, Lambert T, Kammler J, Steinwender C. Gold-coated pacemaker implantation for a patient with type IV allergy to titanium. Indian Pacing Electrophysiol. J. 2015, 15, 291–2.

[127]   Zhang L-C, Chen L-Y. A review on biomedical titanium alloys: recent progress and prospect. Adv. Eng. Mater. 2019; https://doi.org/10.1002/adem.201801215.

[128]   Castleman LS, Motzkin SM, Alicandri FP, Bonawit VL, Johnson AA. Biocompatibility of nitinol alloy as an implant material. J. Biomed. Mater. Res. 1976, 10, 695–731.

[129]   Kuroda D, Niinomi M, Morinaga M, Kato Y, Yashiro T. Design and mechanical properties of new β type titanium alloys for implant materials. Mater. Sci. Eng. A 1998, 243, 244–9.

[130]   Ryhänen J, Niemi E, Serlo W, Niemala E, Sandvik P, Pernu H, Salo T. Biocompatibility of nickel-titanium shape memory metal and its corrosion behavior in human cell culture. J. Biomed. Mater. Res. 1997, 35, 451–7.

[131]   Wen X, Zhang N, Li X, Cao Z. Electrochemical and histomorphometric evaluation of the TiNiCu shape memory alloy. Bio-med. Mater. Eng. 1997, 7, 1–11.

[132]   Khalil-Allafi J, Amin-Ahmadi B, Zare M. Biocompatibility and corrosion behavior of the shape memory NiTi alloy in the physiological environments simulated with body fluids for medical applications. Mater. Sci. Eng. C 2010, 30, 1112–7.

[133]   Zhang C, Sun X, Zhao S, Yu W, Sun D. Susceptibility to corrosion and *in vitro* biocompatibility of a laser-welded composite orthodontic arch wire. Ann. Biomed. Eng. 2014, 42, 222–30.

[134]   Oshida Y. Lecture note, 2010.

[135]   Ponsonnet L, Treheux D, Lissac M, Jaffrezic-Renault N, Grosgogeat B. Review of in vitro studies on the biocompatibility of NiTi alloys. Int. J. Appl. Electromag. Mech. 2006, 23, 147–51.

[136]   Shabalovskaya SA. On the nature of the biocompatibility and on medical applications of NiTi shape memory and superelastic alloys. Bio-Med. Mater. Eng. 1996, 6, 267–89.

[137]   Shabalovskaya SA. Surface, corrosion and biocompatibility aspects of Nitinol as an implant material. Biomed. Mater. Eng. 2002, 12, 69–109.

[138]   Es-Souni M, Es-Souni M, Brandies HF. On the properties of two binary NiTi shape memory alloys. Effects of surface finish on the corrosion behaviour and in vitro biocompatibility. Biomater. 2002, 23, 2887–94.

[139]   Berger-Gorbet M, Broxup B, Rivard C, Yahia LH. Biocompatibility testing of NiTi screws using immunohistochemistry on sections containing metallic implants. J. Biomed. Mater. Res. 1996, 32, 243–8.

[140]   Kang S-B, Yoon K-S, Kim J-S, Nam T-H, Gjunter VE. In vivo result of porous TiNi shape memory alloy: bone response and growth. Mater. Trans. 2002, 43, 1045–8.

[141]   Hodorenko VN, Chekalkin TL, Kim J-S, Kang S-B, Dambaev GT, Gunther VE. *In vitro* and *in vivo* evaluation of porous TiNi-based alloy as a scaffold for cell tissue engineering. Artif. Cells Nanomed. Biotechnol. 2016, 44, 704–9.

[142]   Assad M, Yahia LH, Rivard CH, Lemieux N. In vitro biocompatibility assessment of a nickel-titanium alloy using electron microscopy in situ end-labeling (EM-ISEL). J. Biomed. Mater. Res. 1998, 41, 154–61.

[143]   Rhalmi S, Odin M, Assad M, Tabrizian M, Rivard CH, Yahia L. Hard, soft tissue and in vitro cell response to porous nickel-titanium: a biocompatibility evaluation. Biomed. Mater. Eng. 1999, 9, 151–62.

[144]   Kapanen A, Kinnunen A, Ryhänen J, Tuukkanen J. TGF-beta1 secretion of ROS-17/2.8 cultures on NiTi implant material. Biomaterials 2002, 23, 3341–6.

[145]   Es-Souni M, Es-Souni M, Fischer-Brandies H. Assessing the biocompatibility of NiTi shape memory alloys used for medical applications. Anal. Bioanal. Chem. 2005, 381, 557–67.

[146]   Bogdanski D, Köller M, Müller D, Muhr G, Bram M, Buchkremer HP, Stöver D, Choi J, Epple M. Easy assessment of the biocompatibility of Ni-Ti alloys by in vitro cell culture experiments on a functionally graded Ni-NiTi-Ti material. Biomaterials 2002, 23, 4549–55.

[147]   Uzer B, Gumus B, Toker SM, Sahbazoglu D, Saher D, Yildirim C, Polat-Altintas S, Canadinc D. A critical approach to the biocompatibility testing of NiTi orthodontic archwires. Int. J. Metall. Metal Phys. 2016, 1; doi: 10.35840/2631-5076/9203.

[148]   Toker SM, Canadinc D. Evaluation of the biocompatibility of NiTi dental wires: a comparison of laboratory experiments and clinical conditions. Mater. Sci. Eng. C Mater. Biol. Appl. 2014, 40, 142–7.

[149] Morita M, Hashimoto T, Yamauchi K, Suto Y, Homma T, Kimura Y. Evaluation of biocompatibility for titanium-nickel shape memory alloy in vivo and in vitro environments. Mater. Trans. 2007, 48, 352–60.

[150] Ryhänen J, Kallioinen M, Tuukkanen J, Junila J, Niemelä E, Sandvik P, Serlo W. In vivo biocompatibility evaluation of nickel-titanium shape memory metal alloy: muscle and perineural tissue responses and encapsule membrane thickness. J. Biomed. Mater. Res. 1998, 5, 481–8.

[151] Tosun G, Özler EÜL, Orhan N, Durmuş AS, Eröksüz H. Biocompatibility of NiTi alloy implants in vivo. Int. J. Biomed. Biol. Eng. 2013, 7; https://publications.waset.org/299/pdf.

[152] Michiardi A, Aparicio C, Planell JA, Gil FJ. New oxidation treatment of NiTi shape memory alloys to obtain Ni-free surfaces and to improve biocompatibility. J. Biomed. Mater. Res. B Appl. Biomater. 2006, 77, 249–56.

[153] Chu CL, Wang RM, Hu T, Yin LH, Pu YP, Lin PH, Dong YS, Guo C, Chung CY, Yeung KW, Chu PK. XPS and biocompatibility studies of titania film on anodized NiTi shape memory alloy. J. Mater. Sci. Mater. Med. 2009, 20, 223–8.

[154] Toker SM, Canadinc D, Maier HJ, Birer O. Evaluation of passive oxide layer formation-biocompatibility relationship in NiTi shape memory alloys: geometry and body location dependency. Mater. Sci. Eng. C Mater. Biol. Appl. 2014, 36, 118–29; doi: 10.1016/j.msec.2013.11.040.

[155] Chembath M, Balaraju JN, Sujata M. Effect of anodization and annealing on corrosion and biocompatibility of NiTi alloy. Surf. Coat. Technol. 2016, 302, 302–9.

[156] Chu C, Hu T, Yin LH, Pu YP, Dong YS, Lin PH, Chung CY, Yeung KW, Chu PK. Microstructural characteristics and biocompatibility of a type-B carbonated hydroxyapatite coating deposited on NiTi shape memory alloy. Biomed. Mater. Eng. 2009, 19, 401–8.

[157] Marashi-Najafi F, Khalil-Allafi J, Etminanfar MR. Biocompatibility of hydroxyapatite coatings deposited by pulse electrodeposition technique on the Nitinol superelastic alloy. Mater. Sci. Eng. C Mater. Biol. Appl. 2017, 76, 278–86.

[158] Kocijan A. Electrodeposition of a hydroxyapatite coating on a biocompatible NiTi alloy. Materiali Tehnologije 2018, 52, 67–70.

[159] Lifeng Z, Yan H, Dayun Y, Xiaoying L, Tingfei X, Deyuan Z, Ying H, Jinfeng Y. The underlying biological mechanisms of biocompatibility differences between bare and TiN-coated NiTi alloys. Biomed. Mater. 2011, 6; doi: 10.1088/1748-6041/6/2/025012.

[160] Jin S, Zhang Y, Wang Q, Zhang D, Zhang S. Influence of TiN coating on the biocompatibility of medical NiTi alloy. Colloids Surf. B 2013, 101, 343–9

[161] Li Q, Xia YY, Tang JC, Wang RY, Bei CY, Zeng Y. In vitro and in vivo biocompatibility investigation of diamond-like carbon coated nickel-titanium shape memory alloy. Artif. Cells Blood Substit. Immobil. Biotechnol. 2011, 39, 137–42.

[162] Markhoff J, Krogull M, Schulze C, Rotsch C, Hunger S, Bader R. Biocompatibility and inflammatory potential of titanium alloys cultivated with human osteoblasts, fibroblasts and macrophages. Materials (Basel) 2017, 10; doi: 10.3390/ma10010052.

[163] Kei CC, Vokoun D, Racek J. Atomic layer deposited Al2O3 coatings on NiTi alloy. J. Mater. Eng. Perform. 2014, 23, 2641–9.

[164] Huang J, Dong P, Hao W, Wang T, Xia Y, Da G, Fan Y. Biocompatibility of TiO2 and TiO2/heparin coatings on NiTi alloy. Appl. Surf. Sci. 2014, 313, 172–82.

[165] Sevost'yanov MA, Nasakina EO, Baikin AS, Sergienko KV, Konushkin SV, Kaplan MK, Seregin AV, Leonov AV, Kozlov VA, Shkirinv AV, Bunkin NF, Kolmakov AG, Simakov SV, Gudkov SV. Biocompatibility of new materials based on nano-structured nitinol with titanium and tantalum composite surface layers: experimental analysis in vitro and in vivo. J. Mater. Sci. Mater. Med. 2018, 29(3); doi: 10.1007/s10856-018-6039-3.

[166] Thangavel E, Dhandapani VS, Dharmalingam K, Marimuthu M, Veerapandian M, Arumugam MK, Kim S, Kim B, Ramasundaram S, Kim DE. RF magnetron sputtering mediated NiTi/Ag

coating on Ti-alloy substrate with enhanced biocompatibility and durability. Mater. Sci. Eng. C Mater. Biol. Appl. 2019; doi: 10.1016/j.msec.2019.01.099.

[167] Sheiko N, Kékicheff P, Marie P, Schmutz M, Jacomine L, Perrin-Schmitt F. PEEK (polyether-ether-ketone)-coated nitinol wire: film stability for biocompatibility applications. Appl. Surf. Sci. 2016, 389, 651–65.

[168] Wang G, Shen Y, Cao Y, Yu Q, Guidoin R. Biocompatibility study of plasma-coated nitinol (NiTi alloy) stents. IET Nanobiotechnol. 2007, 1, 102–6.

[169] Liu XM, Wu SL, Chan YL, Chu PK, Chung CY, Chu CL, Yeung KW, Lu WW, Cheung KM, Luk KD. Surface characteristics, biocompatibility, and mechanical properties of nickel-titanium plasma-implanted with nitrogen at different implantation voltages. J. Biomed. Mater. Res. A 2007, 82, 469–78.

[170] Chrzanowski W, Szade J, Hart AD, Knowles JC, Dalby MJ. Biocompatible, smooth, plasma-treated nickel-titanium surface – an adequate platform for cell growth. J. Biomater. Appl. 2012, 26, 707–31.

[171] Gudimova EY, Meisner L, Lotkov AI, Matveeva VA, Meisner S, Matveev AL, Shabalina OI. In vitro biocompatibility of the surface ion modified NiTi alloy. Conference: Advanced Materials with Hierarchical Structure for New Technologies and Reliable Structures 2016: Proceedings of the International Conference on Advanced Materials with Hierarchical Structure for New Technologies and Reliable Structures 2016; doi: 10.1063/1.4966364.

[172] Jenko M, Godec M, Kocijan A, Rudolf R, Dolinar D, Ovsenik M, Gorenšek M, Zaplotnik R, Mozetic M. A new route to biocompatible Nitinol based on a rapid treatment with H_2/O_2 gaseous plasma. Appl. Surf. Sci. 2019, 473, 976–84.

[173] Prymak O, Bogdanski D, Köller M, Esenwein SA, Muhr G, Beckmann F, Donath T, Assad M, Epple M. Morphological characterization and in vitro biocompatibility of a porous nickel-titanium alloy. Biomaterials 2005, 26, 5801–7.

[174] Li H, Yuan B, Gao Y, Chung CY, Zhu M. Remarkable biocompatibility enhancement of porous NiTi alloys by a new surface modification approach: in-situ nitriding and in vitro and in vivo evaluation. J. Biomed. Mater. Res. A 2011, 99, 544–53.

[175] Zheng CY, Nie FL, Zheng YF, Cheng Y, Wei SC, Valiev RZ. Enhanced in vitro biocompatibility of ultrafine-grained biomedical NiTi alloy with microporous surface. Appl. Surf. Sci. 2011, 257, 9086–93.

[176] Bassani P, Panseri S, Ruffini A, Montesi M, Ghetti M, Zanotti C, Tampieri A, Tuissi A. Porous NiTi shape memory alloys produced by SHS: microstructure and biocompatibility in comparison with Ti2Ni and TiNi3. J. Mater. Sci. Mater. Med. 2014, 25, 2277–85.

[177] Thierry B, Merhi Y, Bilodeau L, Trepanier C, Tabrizian M. Nitinol versus stainless steel stents: acute thrombogenicity study in an ex vivo porcine model. Biomaterials 2002, 23, 2997–3005.

[178] Huang N, Yang P, Leng YX, Chen JY, Sun H, Wang J, Wang GJ, Ding PD, Xi TF, Leng Y. Hemocom-patibility of titanium oxide films. Biomaterials 2003, 24, 2177–87.

[179] Gao S, Zhai Y, Hu J. Study of blood compatibility on TiO2 coated biomedical Ni-Ti shape memory alloy. Sheng Wu Yi Xue Gong Cheng Xue Za Zhi 2011, 28, 968–71.

[180] Liu Q, Cheng XN, Fei HX. Effects of micro-magnetic field at the surface of 316L and NiTi alloy on blood compatibility. Med. Biol. Eng. Comput. 2011, 49, 359–64.

[181] Zhao T, Li Y, Gao Y, Xiang Y, Chen H, Zhang T. Hemocompatibility investigation of the NiTi alloy implanted with tantalum. J. Mater. Sci. Mater. Med. 2011, 22, 2311–8.

[182] Yang P, Kwok SCH, Chu PK, Leng YX, Wang J, Huang N. Haemocompatibility of hydrogenated amorphous carbon (a-C:H) films synthesized by plasma immersion ion implantation-deposition. Nucl. Instrum. Methods Phys. Res. Sect. B 2003, 206, 721–5.

[183] Cheng Y, Zheng YF. The corrosion behavior and hemocompatibility of TiNi alloys coated with DLC by plasma based ion implantation. Surf. Coat. Technol. 2006, 200, 4543–8.

[184] Lee BS, Shin H-S, Park K, Han DK. Surface grafting of blood compatible zwitterionic poly(ethylene glycol) on diamond-like carbon-coated stent. J. Mater. Sci. Mater. Med. 2011, 22, 507–14.

[185] Liang CH, Huang NB. Study on hemocompatibility and corrosion behavior of ion implanted TiNi shape memory alloy and Co-based alloys. J. Biomed. Mater. Res. A 2007, 83, 235–40.

[186] Plant SD, Grant DM, Leach L. Surface modification of NiTi alloy and human platelet activation under static and flow conditions. Mater. Lett. 2007, 61, 2864–7.

[187] Zhao T, Li Y, Zhao X, Chen H, Zhang T. Ni ion release, osteoblast-material interactions, and hemocompatibility of hafnium-implanted NiTi alloy. J. Biomed. Mater. Res. B Appl. Biomater. 2012, 100, 646–59.

[188] Shabalovskaya SA, Siegismund D, Heurich E, Rettenmayr M. Evaluation of wettability and surface energy of native Nitinol surfaces in relation to hemocompatibility. Mater. Sci. Eng. C 2013, 33, 127–32.

[189] Li P, Li L, Wang W, Jin W, Liu X, Yeung KWK, Chu PK. Enhanced corrosion resistance and hemocompatibility of biomedical NiTi alloy by atmospheric-pressure plasma polymerized fluorine-rich coating. Appl. Surf. Sci. 2014, 297, 109–15.

[190] Pulletikurthi C, Munroe N, Stewart D, Haider W, Amruthaluri S, Rokicki R, Dugrot M, Ramaswamy S. Utility of magneto-electropolished ternary nitinol alloys for blood contacting applications. J. Biomed. Mater. Res. Part B Appl. Biomater. 2015, 103B, 1366–74.

[191] Mikulewicz M, Chojnacka K. Cytocompatibility of medical biomaterials containing nickel by osteoblasts: a systematic literature review. Biol. Trace Elem. Res. 2011, 142, 865–89.

[192] Yeung KWK, Poon RWY, Liu XY, Ho JPY, Chung CY, Chu PK, Lu WW, Chan D, Cheung KMC. Corrosion resistance, surface mechanical properties, and cytocompatibility of plasma immersion ion implantation – treated nickel-titanium shape memory alloys. J. Biomed. Mater. Res. Part A 2005, 75, 256–67.

[193] Yeung KW, Poon RW, Chu PK, Chung CY, Liu XY, Lu WW, Chan D, Chan SC, Luk KD, Cheung KM. Surface mechanical properties, corrosion resistance, and cytocompatibility of nitrogen plasma-implanted nickel-titanium alloys: a comparative study with commonly used medical grade materials. J. Biomed. Mater. Res. A 2007, 82, 403–14.

[194] Wu SL, Chu PK, Liu XM, Chung CY, Ho JP, Chu CL, Tjong SC, Yeung KW, Lu WW, Cheung KM, Luk KD. Surface characteristics, mechanical properties, and cytocompatibility of oxygen plasma-implanted porous nickel titanium shape memory alloy. J. Biomed. Mater. Res. A 2006, 79, 139–46.

[195] Zhao T, Yang R, Zhong C, Li Y, Xiang Y. Effective inhibition of nickel release by tantalum-implanted TiNi alloy and its cyto-compatibility evaluation in vitro. J. Mater. Sci. 2011, 46, 2529–35.

[196] Jian YT, Yang Y, Tian T, Stanford C, Zhang XP, Zhao K. Effect of pore size and porosity on the biomechanical properties and cytocompatibility of porous NiTi alloys. PLoS One 2015, 10; doi: 10.1371/journal.pone.0128138.

[197] Čolić M, Tomić S, Rudolf R, Marković E, Šćepan I. Differences in cytocompatibility, dynamics of the oxide layers' formation, and nickel release between superelastic and thermo-activated nickel – titanium archwires. J. Mater. Sci. Mater. Med. 2016, 27; doi: 10.1007/s10856-016-5742-1.

[198] Hang R, Liu Y, Bai L, Zhang X, Huang X, Jia H, Tang B. Length-dependent corrosion behavior, Ni^{2+} release, cytocompatibility, and antibacterial ability of Ni-Ti-O nanopores anodically grown on biomedical NiTi alloy. Mater. Sci. Eng. C Mater. Biol. Appl. 2018, 89, 1–7.

[199] Hang R, Liu S, Liu Y, Zhao Y, Bai L, Jin M, Zhang X, Huang X, Yao X, Tang B. Preparation, characterization, corrosion behavior and cytocompatibility of $NiTiO_3$ nanosheets hydrothermally synthesized on biomedical NiTi alloy. Mater. Sci. Eng. C Mater. Biol. Appl. 2019, 97, 715–22.

[200] Taira M, Toguchi MS, Hamada Y, Takahashi J, Itou R, Toyosawa S, Ijyuin N, Okazaki M. Studies on cytotoxic effect of nickel ions on three cultured fibroblasts. J.Mater. Sci. Mater. Med. 2001, 12, 373–76.

[201] Assad M, Lombardi S, Bernèche S, Desrosiers EA, Yahia LH, Rivard CH. Assays of cytotoxicity of the nickel-titanium shape memory alloy. Ann. Chir. 1994, 48, 731–6.

[202] Assad M, Chernyshov A, Leroux MA, Rivard C-H. A new porous titanium-nickel alloy: part 1. Cytotoxicity and genotoxicity evaluation. Bio-Med. Mater. Eng. 2002, 12, 225–37.

[203] Es-Souni M, Es-Souni M, Brandies HF. On the transformation behaviour, mechanical properties and biocompatibility of two niti-based shape memory alloys: NiTi42 and NiTi42Cu7. Biomaterials 2001, 22, 2153–61.

[204] Manceur A, Chellat F, Merhi Y, Chumlyakov Y, Yahia L. In vitro cytotoxicity evaluation of a 50.8% NiTi single crystal. J. Biomed. Mater. Res. A 2003, 67, 641–6.

[205] David A, Lobner D. In vitro cytotoxicity of orthodontic archwires in cortical cell cultures. Eur. J. Orthod. 2004, 26, 421–6.

[206] Dinca VC, Soare S, Barbalat A, Dinu CZ, Moldovan A, Stoica I, Vassu T, Purice A, Scarisoareanu N, Birjega R, Craciun V, De Stefano VF, Dinescu M. Nickel–titanium alloy: cytotoxicity evaluation on microorganism culture. Appl. Surf. Sci. 2006, 252, 4619–24.

[207] Kao CT, Ding SJ, He H, Chou MY, Huang TH. Cytotoxicity of orthodontic wire corroded in fluoride solution in vitro. Angle Orthod. 2007, 77, 349–54.

[208] McMahon RE, Ma J, Verkhoturov SV, Munoz-Pinto D, Karaman I, Rubitschek F, Maier HJ, Hahn MS. A comparative study of the cytotoxicity and corrosion resistance of nickel-titanium and titanium-niobium shape memory alloys. Acta Biomater. 2012, 8, 2863–70.

[209] Sodor A, Ogodescu AS, Petreuş T, Şişu AM, Zetu IN. Assessment of orthodontic biomaterials' cytotoxicity: an in vitro study on cell culture. Rom. J. Morphol. Embryol. 2015, 56, 1119–25.

[210] McNamara K, Kolaj-Robin O, Belochapkine S, Laffir F, Gandhi AA, Tofail SAM. Surface chemistry and cytotoxicity of reactively sputtered tantalum oxide films on NiTi plates. Thin Solid Films 2015, 589, 1–7.

[211] Yanisarapan T, Thunyakitpisal P, Chantarawaratit P-O. Corrosion of metal orthodontic brackets and archwires caused by fluoride-containing products: cytotoxicity, metal ion release and surface roughness. Orthod. Waves 2018, 77, 79–89.

[212] Horandghadim N, Khalil-Allafi J, Urgen M. Effect of Ta_2O_5 content on the osseointe-gration and cytotoxicity behaviors in hydroxyapatite-Ta_2O_5 coatings applied by EPD on superelastic NiTi alloys. Mater. Sci. Eng. C Mater. Biol. Appl. 2019, 102, 683–95.

[213] Lauterbur PC. Image formation by induced local interactions: example employing nuclear magnetic resonance. Nature 1973, 242, 190–1.

[214] Shellock FG, Crues JV. High field strength MR imaging and metallic biomedical implants: an ex vivo evaluation of deflection forces. Am. J. Roentgenol 1988, 151, 389–92.

[215] Shellock FG, Morisoli S, Kanal E. MR procedures and biomedical implants, materials, and devices, update. Radiology 1993, 189, 587–99.

[216] Shellock FG, Mink JH, Curtin S, Friesman MJ. MR imaging and metallic implants for anterior cruciate ligament reconstruction: assessment of ferromagnetism and artifact. J. Magn. Reason Imaging 1992, 2, 225–8.

[217] Shellock FG. Biomedical implants and devices: assessment of magenta field interactions with a 3.0 tesla MR system. J. Magn. Reason Imaging 2002, 16, 721–32.

[218] Shellock FG, Fieno DS, Thompson LJ, Talavage TM, Merman DS. Cardiac pacemaker: in vitro assessment at 1.5 T. Am. Heart. J. 2006, 151, 436–43.

[219] Oshida Y, Tuna EB, Aktören O, Gençay K. Dental implant systems. Int. J. Mol. Sci. 2010, 11, 1580–678.

[220] Okano Y, Yamashiro M, Kaneda T, Kasai K. Magnetic resonance imaging diagnosis of the temporomandibular joint in patients with orthodontic appliances. Oral Surg. Oral Med. Oral Pathol. 2003, 95, 255–63.

[221] Klocke A, Kemper J, Schulze D, Adam G, Kahl-Nieke B. Magnetic field interactions of orthodontic wires during magnetic resonance imaging (MRI) at 1,5 T. J. Orofac. Orthop. 2005, 66, 279–87.

[222] Klocke A, Kahl-Nieke B, Adam G, Kemper J. Magnetic forces on orthodontic wires in high field magnetic resonance imaging (MRI) at 3 T. J. Orofac. Orthop. 2006, 67, 424–9.

[223] Singh D, Sinha S, Singh H, Jagetia A, Gupta S, Gangoo P, Tandon M. Use of nitinol shape memory alloy staples (NiTi clips) after cervical discoidectomy: minimally invasive instrumentation and long-term results. Minim. Invasive Neurosurg. 2011, 54, 172–8.

[224] Li HZ, Xu J. MRI compatible Nb-Ta-Zr alloys used for vascular stents: optimization for mechanical properties. J. Mech. Behav. Biomed. Mater. 2014, 32, 166–76.

[225] Abbaszadeh K, Heffez LB, Mafee MF. Effect of interference of metallic objects on interpretation of T1-weighted magnetic resonance images in the maxillofacial region. Oral Surg. Oral Med. Oral Pathol. Oral Radiol. Endod. 2000, 89, 759–65.

[226] Hopper TAJ, Vasilić B, Pope JM, Jones CE, Epstein C, Song HK, Wehrli FW. Experimental and computational analyses of the effects of slice distortion from a metallic sphere in an MRI phantom. Magn. Reson. Imaging 2006, 24, 1077–85.

[227] Elison JM, Leggitt VL, Thomson M, Oyoyo U, Wycliffe ND. Influence of common orthodontic appliances on the diagnostic quality of cranial magnetic resonance images. Am. J. Orthod. Dentofac. Orthop. 2008, 134, 563–72.

[228] Costa ALF, Appenzeller S, Yasuda CL, Pereira FR, Zanardi VA, Cendes F. Artifacts in brain magnetic resonance imaging due to metallic dental objects. Med. Oral. Pathol. Oral Cir. Buccal. 2009, 14, E278–82.

[229] Blankenstein F, Truong BT, Thomas A, Thieme N, Zachriat C. Predictability of magnetic susceptibility artifacts from metallic orthodontic appliances in magnetic resonance imaging. J. Orofac. Orthop. 2015, 76, 14–29.

[230] www.stmri.com/ht_docs/safety-2.html; www.mri.tju.edu/Policies-contraindications.html.

[231] Saito S, Sugimoto N, Morohashi T, Kurabayashi H, Shimizu H, Yamasaki K, Shiba A, Yamada S, Shibasaki Y. Endosseous titanium implants as anchors for mesiodistal tooth movement in the beagle dog. Am. J. Orthod. Dentofac. Orthop. 2000, 118, 601–7.

[232] Bothe RT, Beaton LE, Davenport HA. Reaction of bone to multiple metallic implants. Surg. Gynecol. Obstet. 1940, 71, 598–602.

[233] Brånemark P-I, Hansson BO, Adell R, Briene U, Lindstrom J, Hallen O, Ohman. Osseointegrated implants in the treatment of the edentulous jaw, experience from a 10-year period. Scand. J. Plast. Reconstr. Surg. Suppl. 1977, 16, 1–132.

[234] Misch CE, Misch CM. Generic Root Form Component Terminology. In: Misch CE, ed., Implant Dentistry, Mosby: An Affiliate of Elsevier, 1999, 13–15.

[235] Brånemark R, Brånemark P-I, Rydevik B, Myers RR. Osseointegration in skeletal reconstruction and rehabilitation. J. Rehabil. Res. Dev. 2001, 38, 175–82.

[236] Zaffe D, Baena RR, Rizzo S, Brusotti C, Soncini M, Pietrabissa R, Cavani F, Quaglini V. Behavior of the bone-titanium interface after push-in testing: a morphological study. J. Biomed. Mater. Res. A 2003, 64, 365–71.

[237] Rønold HJ, Lyngstadaas SP, Ellingsen JE. A study on the effect of dual blasting with TiO_2 on titanium implant surfaces on functional attachment in bon. J. Biomed. Mater. Res. A 2003, 67, 524–30.

[238] Nishiguchi S, Fujibayashi S, Kim H-M, Kokubo T, Nakamura T. Biology of alkali- and heat-treated titanium implants. J. Biomed. Mater. Res. A 2003, 67, 26–35.

[239] Giavaresi G, Fini M, Cigada A, Chiesa R, Rondelli G, Rimondini L, Aldini NN, Martini L, Giardino R. Histomorphometric and microhardness assessments of sheep cortical bone surrounding titanium implants with different surface treatments. J. Biomed. Mater. Res. A 2003, 67, 112–20.

[240] Fujibayashi S, Neo M, Kim HM, Kokubo T, Namamura T. Osteoinduction of porous bioactive titanium metal. Biomaterials 2004, 25, 443–50.

[241] Takemoto M, Fujibayashi S, Neo M, Suzuki J, Kokubo T, Nakamura T. Mechanical properties and osteoinductivity of porous bioactive titanium. Biomaterials 2005, 26, 6014–23.

[242] Takemoto M, Fujibayashi S, Neo M, Suzuki J, Matsushita T, Kokubo T, Nakamura T. Osteoinductive porous titanium implants: effect of sodium removal by dilute HCl treatment. Biomaterials 2006, 27, 2682–91.

[244] Sul Y-T, Johansson C, Byon E, Albrektsson T. The bone response of oxidized bioactive and non-bioactive titanium implants. Biomaterials 2005, 26, 6720–30.

[244] Lim YJ, Oshida Y. Initial contact angle measurements on variously treated dental/medical titanium materials. J. Bio-med. Mater. Eng. 2001, 11, 325–41.

[245] Aebli N, Krebs J, Stich H, Schawalder P, Walton M, Schwenke D, Gruner H, Gasser B, Theis J-C. *In vivo* comparison of the osseointegration of vacuum plasma sprayed titanium- and hydroxyapatite-coated implants. J. Biomed. Mater. Res. A 2003, 66, 356–63.

[246] Stewart M, Welter JF, Goldberg VM. Effect of hydroxyapatite/tricalcium-phosphate coating on osseointegration of plasma-sprayed titanium alloy implants. J. Biomed. Mater. Res. A 2004, 69, 1–10.

[247] Habibovic P, Li J, van der Valk CM, Meijer G, Layrolle P, van Blitterswijk CA, de Groot K. Biological performance of uncoated and octacalcium phosphate-coated Ti6Al4V. Biomaterials 2005, 26, 23–36.

[248] Stangl R, Pries A, Loos B, Müller M, Erben RG. Influence of pores created by laser superfinishing on osseointegration of titanium alloy implants. J. Biomed. Mater. Res. A 2004, 69, 444–53.

[249] Götz HE, Müller M, Emmel A, Holzwarth U, Erben RG, Stangl R. Effect of surface finish on the osseointegration of laser-treated titanium alloy implants. Biomaterials 2004, 25, 4057–64.

[250] Giannuzzi LA, Phifer D, Giannuzzi NJ, Capuano MJ. Two-dimensional analysis of bone/dental implant interfaces with the use of focused ion beam and electron microscopy. J. Oral Maxillofac. Surg. 2007, 65, 737–47.

[251] Stenport VF, Johansson CB. Evaluations of bone tissue integration to pure and alloyed titanium implants. Clin. Implant Dent. Relat. Res. 2008, 10, 191–9.

[252] Morinaga K, Kido H, Sato A, Watazu A, Matsuura M. Chronological changes in the ultrastructure of titanium-bone interfaces: analysis by light microscopy, transmission electron microscopy, and micro-computed tomography. Clin. Implant Dent. Relat. Res. 2009, 11, 59–68.

[253] Kerner S, Mogonney V, Pavon-Djavid G, Helay G, Sedel L, Anagnostou F. Bone tissue response to titanium implant surfaces modified with carboxylate and sulfonate groups. J. Mater. Sci. Mater. Med. 2010, 21, 707–15.

[254] Freeman CO, Brook IM. Bone response to a titanium aluminium nitride coating on metallic implants. J. Mater. Sci. Mater. Med. 2006, 17, 465–70.

[255] Kapanen A, Ryhänen J, Danilov A, Tuukkanen J. Effect of nickel-titanium shape memory metal alloy on bone formation. Biomaterials 2001, 22, 2475–80.

[256] Kujala S, Ryhänen J, Danilov A, Tuukkanen J. Effect of porosity on the osteointegration and bone in growth of a weight-bearing nickel-titanium bone graft substitute. Biomaterials 2003, 24, 4691–7.

[257] Shah FA, Thomsen P, Palmquist A. Osseointegration and current interpretations of the bone-implant interface. Acta Biomater. 2019, 84, 1–15.

[258] Albrektsson T, Zarb G, Worthington P, Eriksson AR. The long-term efficacy of currently used dental implants: a review and proposed criteria of success. Int. J. Oral Maxillofac. Implants 1986, 1, 11–25.

[259] Swami V, Vijayaraghavan V, Swami V. Current trends to measure implant stability. J. Indian Prosthodont Soc. 2016, 16, 124–30.

[260] Meredith N. Assessment of implant stability as a prognostic determinant. Int. J. Prosthodont. 1998, 11, 491–501.

[261] Barikani H, Rashtak S, Akbari S, Fard MK, Rokn A. The effect of shape, length and diameter of implants on primary stability based on resonance frequency analysis. Dent. Res. J. (Isfahan) 2014, 11, 87–91.

[262] Brunski JB. Biomechanical factors affecting the bone-dental implant interface. Clin. Mater. 1992, 10, 153–201.

[263] Sennerby L, Roos J. Surgical determinants of clinical success of osseointegrated oral implants: a review of the literature. Int. J. Prosthodont. 1998, 11, 408–20.

[264] Cochran DL, Schenk RK, Lussi A, Higginbottom FL, Buser D. Bone response to unloaded and loaded titanium implants with a sandblasted and acid-etched surface: a histometric study in the canine mandible. J. Biomed. Mater. Res. 1998, 40, 1–11.

[265] Mistry G, Shetty O, Shetty S, Raghuwar D, Singh R. Measuring implant stability: a review of different methods. J. Dent. Implants 2014, 4, 165–9.

[266] Patil R, Bharadwaj D. Is primary stability a predictable parameter for loading implant? J. Innovations Dent. 2016, 8, 84–8.

[267] Norton M. Primary stability versus viable constraint a need to redefine. Int. J. Oral Maxillofac. Implants. 2013, 28, 19–21.

[268] Dos Santos MV, Elias CN, Cavalcanti Lima JH. The effects of superficial roughness and design on the primary stability of dental implants. Clin. Implant Dent. Relat. Res. 2011, 13, 215–23.

[269] Bajaj G, Bathiya A, Gade J, Mahale Y, Ulemale M, Atulkar M. Primary versus secondary implant stability in immediate and early loaded implants. Int. J. Oral Health. Med. Res. 2017, 3, 49–54.

[270] Anselme K. Osteoblast adhesion on biomaterials. Biomaterials 2000, 21, 667–81.

[271] Novaes AB, de Souza SLS, de Barros RRM, Pereira KKY, Iezzi G, Piattelli A. Influence of implant surfaces on osseointegration. Braz. Dent. J. 2010, 21, 471–81.

[272] Kasemo B. Biological surface science. Surf. Sci. 2002, 500, 656–77.

[274] Deligianni DD, Katsala N, Ladas S, Sotiropoulou D, Amedee J, Missirlis YF. Effect of surface roughness of the titanium alloy Ti6Al-4V on human broadcast marrow cell response and on protein adsorption. Biomaterials 2001, 22, 1241–51.

[274] Lamers E, Walboomers XF, Domanski M, Riet J, van Delft FC, Luttge R, Winnubst LA, Gardeniers HJ, Jansen JA. The influence of nanoscale grooved substrates on osteoblast behavior and extracellular matrix deposition. Biomaterials 2010, 31, 3307–16.

[275] Mendonça G, Mendonça DB, Aragão FJ, Cooper LF. The combination of micron and nanoto-pography by H(2)SO(4)/H(2) O(2) treatment and its effects on osteoblast-specific gene expression of hMSCs. J. Biomed. Mater. Res. A 2010, 94, 169–79.

[276] Sandrini E, Morris C, Chiesa R, Cigada A, Santin M. In vitro assessment of the osteoonte-grative potential of a novel multiphase anodic spark deposition coating for orthopaedic and dental implants. J. Biomed. Mater. Res. Part B Appl. Biomater. 2005, 73B, 392–9.

[277] Sandrini E, Giordano C, Busini V, Signorelli E, Cigada A. Apatite formation and cellular response of a novel bioactive titanium. J. Mater. Sci. Mater. Med. 2007, 18, 1225–37.

[278] Guéhennec LLe, Soueidan A, Layrolle P, Amouriq Y. Surface treatments of titanium dental implants for rapid osseointegration. Dent. Mater. 2007, 23, 844–54.

[279] Alves SF, Wassall T. In vitro evaluation of osteoblastic cell adhesion on machined osseointegrated implants. Brz. Oral Res. 2009, 23, 131–6.

[280] Thalji G, Cooper LF. Molecular assessment of osseointegration in vitro: a review of current literature. Intl. J. Oral Maxillofac. Implants. 2014, 29, e171–99.

[281] Lincks J, Boyan BD, Blanchard CR, Lohmann CH, Liu Y, Cochran DL, Dean DD, Schwartz Z. Response of MG63 osteoblast-like cells to titanium and titanium alloy is dependent on surface roughness and composition. Biomaterials 1998, 19, 2219–32.

[282] von der Mark K, Park J, Bauer S, Schmuki P. Nanoscale engineering of biomimetic surfaces: cues from the extracellular matrix. Cell Tissue Res. 2010, 339, 131–53.

[283] Oshida Y, Hashem A, Nishihara T, Yapchulay MV. Fractal dimension analysis of mandibular bones – toward a morphological compatibility of implants. J. BioMed. Mater. Eng. 1994, 4, 397–407.

[284] Oshida Y, Miyazaki T, Tominaga T. Some biomechanistic concerns of newly developed implantable materials. J. Dent. Oral Health. 2018, 4; http://scientonline.org/open-access/some-biomechanistic-concerns-on-newly-developed-implantable-materials.pdf.

[285] Lemons JE. Biomaterials, biomechanics, tissue healing, and immediate-function dental implants. J. Oral Implantol. 2004, 30, 318–24.

[286] Babuska V, Moztarzadeh O, Kubikova T, Moztarzadeh A, Hrusak D, Tonar Z. Evaluating the osseointegration of nanostructured titanium implants in animal models: current experimental methods and perspectives (review). Biointerphases 2016, 11; https://doi.org/10.1116/1.4958793.

[287] van Oirschot BAJA, Alghamdi HS, Näˉrhi TO, Anil S, Aldosari AAF, van den Beucken JJP, Lansen JA. In vivo evaluation of bioactive glass-based coatings on dental implants in a dog implantation model. Clin. Oral Impl. Res. 2012; doi: 10.1111/clr.12060.

[288] Im JH, Kim SG, Oh JS, Lim SC. A comparative study of stability after the installation of 2 different surface types of implants in the maxillae of dogs. Implant Dent. 2015, 24, 586–91.

[289] Yoo D, Marin C, Freitas G, Tovar N, Bonfante E, Teixeira HS, et al. Surface characterization and *in vivo* evaluation of dual acid-etched and grit-blasted/acid-etched implants in sheep. Implant Dent. 2015, 24, 256–62.

[290] Pettersson M, Pettersson J, Thorén MM, Johansson A. Release of titanium after insertion of dental implants with different surface characteristics – an *ex vivo* animal study. Acta Biomater. Odontol. Scand. 2017, 3, 63–73.

[291] Hermida JC, Bergula A, Dimaano F, Hawkins M, Colwell C, D'Lime DD. An in vivo evaluation of bone response to three implant surfaces using a rabbit intramedullary rod model. J. Orthop. Surg. Res. 2010, 5, 57–64.

[292] Tsetsenekou E, Papadopoulos T, Kalyvas D, Papaioannou N, Tangl S, Watzek G. The influence of alendronate on osseointegration of nanotreated dental implants in New Zealand rabbits. Clin. Oral Implants Res. 2012, 23, 659–66.

[293] Dziuba D, Meyer-Lindenberg A, Seitz JM, Waizy H, Angrisani N, Reifenrath J. Long-term in vivo degradation behaviour and biocompatibility of the magnesium alloy ZEK100 for use as a biodegradable bone implant. Acta Biomater. 2013, 9, 8548–60.

[294] Liu Y, Zhou Y, Jiang T, Liang YD, Zhang Z, Wang YN. Evaluation of the osseointegration of dental implants coated with calcium carbonate: an animal study. Intl. J. Oral Sci. 2017, 9, 133–8.

[295] Ikuta K, Urakawa H, Kozawa E, Hamada S, Ota T, Kato R, et al. *In vivo* heat-stimulus-triggered osteogenesis. Intl. J. Hyperthermia 2015, 31, 58–66.

[296] Dang Y, Zhang L, Song W, Chang B, Han T, Zhang Y, et al. In vivo osseointegration of Ti implants with a strontium-containing nanotubular coating. Int. J. Nanomed. 2016, 11, 1003–11.

[297]   Amerstorfer F, Fischer L, Eichler J, Draxler J, Zitek A, Meischel M, et al. Long-term *in
        vivo* degradation behavior and near-implant distribution of resorbed elements for
        magnesium alloys WZ21 and ZX50. Acta Biomater. 2016, 42, 440–50.
[298]   Ota T, Nishida Y, Ikuta K, Kato R, Kozawa E, Hamada S, Sakai T, Ishiguro N. Heat-stimuli-
        enhanced osteogenesis using clinically available biomaterials. PLoS One 2017, 12; doi:
        10.1371/journal.pone.0181404.
[299]   Turner AS. Animal models of osteoporosis – necessity and limitations. Eur. Cells Mater. 2001,
        1, 66–81.
[300]   Akens MK, von Rechenberg B, Bittmann P, Nadler D, Zlinszky K, Auer JA. Long
        term in-vivo studies of a photo-oxidized bovine osteochondral transplant in sheep. BMC
        Musculoskeletal Disorders 2001, 2; doi: 10.1186/1471-2474-2-9.
[301]   Natiella JR. The use of animal models in research on dental implants. J. Dent. Educ. 1988, 52,
        792–7.
[302]   Pound P, Bracken MB, Bliss SD. Is animal research sufficiently evidence based to be a
        cornerstone of biomedical research? BMJ (formerly the Brit. Med. J.). 2014, 348, g3387–9.
[303]   Pearce AI, Richards RG, Milz S, Schneider E, Pearce SG. Animal models for implant
        biomaterial research in bone: a review. Eur. Cells Mater. 2007, 13, 1–10.
[304]   Ringach DL. The use of nonhuman animals in biomedical research. Am. J. Med. Sci. 2011,
        342, 305–13.
[305]   Marvizon JC, Walwyn W, Minasyan A, Chen W, Taylor BK. Latent sensitization: a model for
        stress-sensitive chronic pain. Curr. Protoc. Neurosci. 2015, 71; doi: 10.1002/0471142301.
        ns0950s71.
[306]   Rowan AN. The benefits and ethics of animal research. Trends in Anim. Res.. Sci. Amer. 1997,
        276, 79–93.
[307]   Tannenbaum J, Arzi B, Reiter AM, Peralta S, Snyder CJ, Lommer MJ, et al. The case against the
        use of dental implants in dogs and cats. J. Amer. Vet. Med. Assoc. 2013, 243, 1680–5.
[308]   Hajar R. Animal testing and medicine. Heart Views 2011, 12, 42. 73. Atsumi M, Park SH, Wang
        HL. Methods used to assess implant stability: current status. Int. J. Oral Maxillofac. Implants.
        2007, 22, 743–54.
[309]   Sachdeva A, Dhawan P, Sindwani S. Assessment of implant stability: methods and recent
        advances. Brit. J. Med. Med. Res. 2016, 12, 1–10.
[310]   Kose OD, Karatasli B, Demircan S, Kose TE, Cene E, Aya SA, et al. In vitro evaluation of manual
        torque values applied to implant-abutment complex by different clinicians and abutment
        screw loosening. BioMed. Res. Intl. 2017, article ID 7376261, 9 pages.
[311]   De Oliveira GJP, Barros-Filho LAB, Barros LAB, Queiroz TP, Marcantonio E. In vivo evaluation of
        the primary stability of short and conventional implants. J. Oral Implant. 2016, 42, 458–63.
[312]   Kanth KL, Swamy DN, Mohan TK, Swarna C, Sanivarapu S, Pasupuleti M. Determination of
        implant stability by resonance frequency analysis device during early healing period. J. Dr.
        NTR Univ. Health Serv. 2014, 3, 169–75.

Chapter 12
Surface modification and engineering

12.1 Introduction

The materials used in medical or dental applications may be referred to as biomaterials. The metals commonly used in dental/medical applications are stainless steels (e.g., 316L Mo-containing stainless steel), cobalt-based alloys (e.g., Cr–Co alloy), titanium alloys (e.g., Ti-6Al-4V and Ti-6Al-7Nb), NiTi shape-memory alloy (SMA), zirconium-based alloys, and commercially pure titanium, zirconium, and tantalum. They are found mainly in orthopedic devices (e.g., joints, plates, screws, rods, and bars), dental devices (e.g., periodontal implants, orthodontic appliances, and endodontic instruments), neurological implants (e.g., cochlear and pacemaker), stents for coronary angioplasty, and general surgery tools. The mechanical properties of metallic materials (modulus of elasticity, ultimate tensile and yield strengths, and ductility) have enabled the creation of devices with long-term in vivo stability. In addition, the passive surface oxide layers have provided chemical inertness and stability within biologic environments, providing excellent biocompatibility. Porous metallic biomaterials have recently been developed. The hardness of some porous biomaterials has been reduced to magnitudes less than 50% of the nonporous alloys, which can be biomechanically compatible with adjacent hard tissue. An increase in effective surface area of such porous structure (3- to 10-fold) can be useful for bone ingrowth processes, though it has been reported that such biomaterials are prone to biocorrosion, resulting in elemental leaching into the tissues, which has biologic consequences [1–4]. In the field of biomaterials, many novel processing techniques and methods have been developed to improve the desirable characteristics of biomaterials. Today, the use of biomaterials has been widely spread. Estimates of the numbers of biomedical devices incorporating biomaterials used in the United States in 2002 include total hip joint replacements (448,000), knee joint replacements (452,000), shoulder joint replacements (24,000), dental implants (854,000), coronary stents (1,204,000), and coronary catheters (1,328,000) [5]. The clinical application of these devices is still a critical issue because of the tissue trauma that occurs during implantation and the presence of the device in the body. The in vivo functionality and durability of any implantable device can be compromised by the body's response to the foreign material. Numerous strategies to overcome negative body reactions have been reported. Some key issues of biomaterial/tissue interactions should include foreign body response and biocompatibility and biocompatibility assessment. General approaches used to overcome the in vivo instability of implantable devices can include the use of biocompatible biomaterial coatings [6, 7].

Coatings, claddings, thin-film deposition, and other surface additive modifications may be categorized in many ways. One of these is principally by the form of the material deposited at any given instant on a small area of the surface. Deposition

https://doi.org/10.1515/9783110666113-012

may be categorized by (1) atomic size [including electrolytic deposition by electroplating, physical vapor deposition (PVD) by ion implantation, vacuum deposition, vacuum evaporation, ion beam, plasma deposition by sputter deposition, chemical vapor deposition (CVD) by spray pyrolysis, etc.], (2) particulate size (including flame thermal spray, electric wire arc spray, plasma spray, fuse thermal spray, etc.), (3) bulk coating or cladding (including wetting processes by dip coating, spin coating, cladding by explosive roll bonding, overlay by laser cladding, weld overlay, etc.), and (4) surface modification (including chemical conversion by electrolytic, anodization, chemical by thermal vapor or plasma vapor, mechanical by shot peening; thermal surface enhancement by diffusion from bulk, sputtering, or ion implantation) [8, 9]. In many cases, a coating is applied to a substrate to improve its appearance or surface properties such as surface roughness, adhesion, wettability, corrosion resistance, and wear resistance. Surface treatments serve several purposes: (1) to control tribological action (friction, lubrication, and wear), (2) to improve corrosion resistance (or passivation), (3) to change physical property, for example, conductivity, resistivity, and reflection, (4) to alter surface mechanical properties (hardness, modulus of elasticity, etc.) (5) to change dimension (flatten, smooth, etc.), (6) to vary appearance, for example, color and roughness, (7) to reduce cost (replace bulk material), and (8) to provide an anti-allergenic surface [10, 11]. It follows that coatings can be considered a "key technology" [12] to promote biocompatibility and adhesion of foreign materials within a biological host.

In this chapter, the technology and methods for surface coating and additive modification, the materials for such surface engineering, and some advanced technologies which might be applicable in medical/dental fields will be discussed.

12.2 Nature of surface and interface

Before discussing surface coating and additive modification in details, it is worth reviewing briefly the nature of surface and interface. The surface of a material defines the boundary of an object and is not just a free end side of a substance, but it is a contact and boundary zone with other substances (either in gaseous, liquid, or solid state). A physical system which comprises of a homogeneous component such as solid, liquid, or gas and is clearly distinguishable from each other is called a phase, and a boundary at which two or three of these individual phases are in contact is called an interface. Surface and interface reactions include reactions with organic or inorganic materials, vital or nonvital species, hostile or friendly environments, and so on. Surface activities may vary from mechanical actions (fatigue crack initiation and propagation, stress intensification, dislocation movement, etc.), chemical action (discoloration, tarnishing, contamination, corrosion, oxidation, etc.), mechanochemical action (corrosion fatigue, stress–corrosion cracking, etc.), thermomechanical action (thermal fatigue and creep), tribological and biotribological actions (wear and wear debris toxicity, friction, etc.) to physical and biophysical actions (surface contact and adhesion, adsorption, absorption, diffusion, cellular attachment, cell proliferation and differentiation, etc.). Consequently, the properties of the surface determine the interaction of the second material [13, 14]. The surface has a certain characteristic thickness: (1) In cases where interatomic reactions such as wetting or adhesion are dominant, it is important to target and analyze atoms up to a depth of 100 nm, (2) when mechanical interactions such as tribology and surface hardening come into play, a thickness of about 0.1–10 μm will be important since the elasticity due to surface contact and the plastically deformed layer will be a governing area, and (3) for cases when mass transfer or corrosion is involved, the effective layer for preventing the diffusion will be within 1–100 μm. Since the thickness of the desired coating layer is accordingly selected, it becomes important to control the thickness of coated or surface-modified film/layer. Coating is just a small portion of surface science and engineering. Structural elements beneath the surface function principally to support the surface and its characteristics determine not only esthetic appearance but also interaction with the environment, including mechanical interactions, chemical interactions, optical and thermal interactions [8], and of course biological interactions. Accordingly, the longevity, safety, reliability, and structural integrity of dental and medical materials and devices are greatly governed by these surface phenomena, which can be detected, observed, characterized, and analyzed by virtue of various means of equipment and technologies [14–16].

12.3 Surface modifications

There are various reasons for surface layer to receive alteration: for changing surface topology, for controlling surface energy, for managing hydrophilicity, for varying appearance, for enhancing surface chemistry and physics, for enriching surface elements or substances and so on. Methods for achieving such surface modification can be found in mechanical, chemical and electrochemical, physical, thermal, or combined techniques. As a result, modified surface can be (i) rough surface, (ii) one extreme end of the functionally graded structure, (iii) thin layer of noble, bioactive or immobilization or biomolecule, as depicted in Figure 12.1 [17].

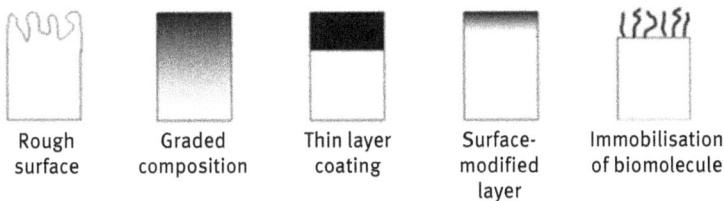

| Rough surface | Graded composition | Thin layer coating | Surface-modified layer | Immobilisation of biomolecule |

Figure 12.1: Several features of modified surface [17].

According to Hanawa [17], surface modification technologies can be divided into dry process and wet process, in which various modification techniques are listed as seen in Figure 12.2.

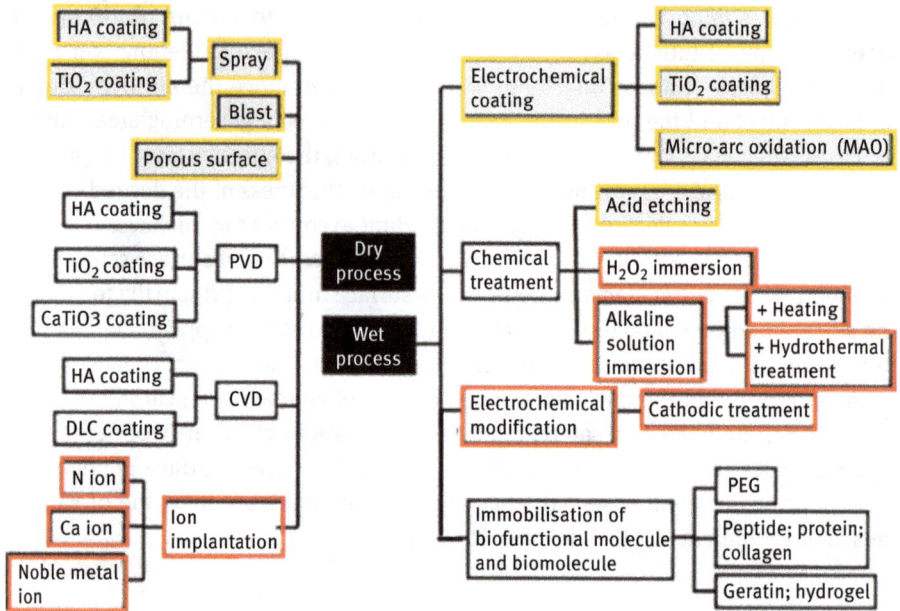

Figure 12.2: Surface modification techniques by both dry and wet processes used in research and industry, where red blocks indicate the surface layer formation and yellow blocks represent the mechanical anchoring [17].

12.3.1 Mechanical modification

Although sand blasting, shot peening, and laser peening has been employed for mainly (1) cleaning surface contaminants, (2) roughening surfaces to increase effective surface area, and (3) producing beneficial surface compressive residual stress. They are not directly used for coating purposes. However, there are two unique solid state processes which can be utilized as a cladding method: the mechanical alloying process (a new method for producing composite metal powders with controlled microstructures) [18], and friction stir (or inertia) welding [19, 20]. These additive technologies can be used for modifying surface layers. Zhang et al. [21] studied the effects of laser shock peening (LSP) on the biocompatibility, corrosion resistance, ion release rate, and hardness of NiTi. It was reported that (i) the cell culture study indicated that the LSPed NiTi samples had lower cytotoxicity and higher cell survival rate than the untreated samples, (ii) specifically, the cell survival rate increased from 88 ± 1.3% to 93 ± 1.1% due to LSP treatment, (iii) LSP treatment was shown to significantly decrease the initial Ni ion release rate compared with that of the untreated samples, (iv) electrochemical tests indicated that LSP improved the corrosion resistance of the NiTi alloy in simulated body fluid, with a decrease in the corrosion current density from 1.41 ± 0.20 $\mu A/cm^2$ to 0.67 ± 0.24 $\mu A/cm^2$, (v) immersion tests showed that calcium deposition was significantly enhanced by LSP, and (vi) the hardness of NiTi alloy increased from 226 ± 3 HV before LSP to 261 ± 3 HV after LSP indicating that LSP is a promising surface modification method that can be used to improve the mechanical properties, corrosion resistance, and biocompatibility of NiTi alloy for biomedical applications. Zhao et al. [22] investigated the efficient powder metallurgy method for the synthesis of NiTi alloys, involving mechanical activation of prealloyed $NiTi_2$ and elemental Ni powders ($NiTi_2$–Ni) followed by a press-and-sinter step. The idea is to take advantage of the brittle nature of $NiTi_2$ to promote a better efficiency of the mechanical activation process. The conventional mechanical activation route using elemental Ti and Ni powders (NiTi) was also used for comparative purposes. It was found that (i) starting with ($NiTi_2$–Ni) powder mixtures resulted in the formation of a predominant amorphous structure after mechanical activation at 300 rpm for 2 h, (ii) a sintered specimen consisting mainly of NiTi phase was obtained after vacuum sintering at 1,050 °C for 0.5 h, (iii) the produced NiTi phase exhibited the martensitic transformation behavior, (iv) using elemental Ti powders instead of prealloyed $NiTi_2$ powders, the structural homogenization of the synthesized NiTi alloys was delayed, (v) performing the mechanical activation at 300 rpm for the (NiTi) powder mixtures gave rise to the formation of composite particles consisting in dense areas of alternate fine layers of Ni and Ti; however, no significant structural modification was observed even after 16 h of mechanical activation, (vi) only after vacuum sintering at 1,050 °C for 6 h, the NiTi phase was observed to be the predominant phase, and (vii) the higher reactivity of the mechanically activated ($NiTi_2$–Ni) powder particles can explain the different sintering behavior of those powders compared with the mechanically activated (NiTi) powders; indicating that this innovative approach allows an effective time reduction in the mechanical activation and of the vacuum sintering step.

12.3.2 Polishing

Electropolishing (EP) is an established final surface finishing process for metal workpieces that allows for a controlled electrochemical removal of surface material leading to a smoother surface. Lopes et al. studied effect of surface roughness on fatigue life [23] and on the torsional resistance [24] of NiTi rotary endodontic instruments. It was concluded that (i) the depth of the surface grooves on the working part affected the number of cycles until fracture (N_{CF}) of the instruments tested; the smaller the groove depth, the greater the N_{CF}, (ii) the depth of the surface grooves resulting from machining during the manufacturing process did not affect the maximum torque and angular deflection of the instruments tested. During manufacturing of NiTi endodontic instruments, EP is used to remove surface irregularities, cracks, and residual stress that are caused by the previous grinding process [25]. This is supposed to improve fracture resistance (due to minimized local stress raisers), cutting efficiency, and resistance to corrosion (minimized local anodic sites) [26–28]. Anderson et al. [29] investigated the effect of EP on cyclic flexural fatigue and torsional strength of rotary nickel–titanium endodontic instruments (EndoWave, ProFile, and RaCe). The number of rotations to fracture and torque at fracture were determined and compared among the instruments tested. It was found that (i) overall, electropolished instruments performed significantly better than nonelectropolished instruments in cyclic fatigue testing and, to a lesser extent, in static torsional loading, (ii) when viewing electropolished instruments with the scanning electron microscopy (SEM), milling grooves, cracks, pits, and areas of metal rollover were observed, although they were more evident in the nonelectropolished instruments, (iii) EP may have beneficial effects in prolonging the fatigue life of rotary NiTi endodontic instruments, and (iv) the benefits of EP are likely to be caused by a reduction in surface irregularities that serve as points for stress concentration and crack initiation. Similar conclusion was obtained [30], conducting a comparative study of the fatigue resistance of rotary NiTi endodontic instruments in order to assess the influence of both instrument design and surface treatment on flexural fracture. Hence, it is indicated that the presence of microcracks, surface debris, and milling grooves can be reduced by EP. However, EP can't inhibit the development of microfractures [31].

Condorelli et al. [32] assessed the failure mechanism of rotary NiTi instruments by chemical, structural, and morphological analyses to provide a rational explanation of the effects of surface and bulk treatments on their resistance to fatigue fracture. Thermal treatment (350–500 °C) was performed on electropolished and nonelectropolished NiTi endodontic instruments. It was reported that (i) before thermal treatment, significant differences in fatigue resistance between electropolished and nonelectropolished instruments (the number of revolutions to failure, N(f), was 385 and 160, respectively) were attributed to differences in the surface morphology of the instruments, (ii) SEM analysis of the fracture surfaces indicated that flexural fatigue fractures occurred in two steps: first by a slow growth of initial cracks and then rapid rupture of the remaining material, and (iii) thermal treatment did not affect the surface morphology but resulted in significant changes in the instrument bulk with the appearance of an R-phase and an improved fatigue resistance; indeed after treatment at 500 °C, N(f) increased up to 829 and 474 for electropolished and nonelectropolished instruments, respectively concluding that both thermal treatment and EP improved the resistance of NiTi rotary instruments against fatigue fracture. Praisarnti et al. [33] examined the fatigue behavior, especially at the low-cycle fatigue (LCF) region, of an experimentally electropolished FlexMaster and a commercial electropolished NiTi instrument (RaCe) in a corrosive environment. A total of 90 NiTi rotary instruments were subjected to rotational bending at various degrees of curvatures while immersed in 1.2% sodium hypochlorite solution until broken. The maximum surface strain amplitude, calculated from the curvature of the instrument and the diameter of the cross section at break, was plotted against the LCF life. The results were compared with data for a nonelectropolished commercial product tested by using the same methodology. It was obtained that (i) the fatigue life of both instruments generally declined with increasing surface strain amplitude; there was a significant difference between the two instruments, and (ii) comparing the surface-treated FlexMaster with its commercially available nonelectropolished counterpart, an improved resistance to fatigue breakage as a result of EP was noted indicating that (iii) the LCF life of a NiTi instrument rotating with a curvature in a corrosive environment is enhanced by EP, (iv) the design, both cross-sectional and longitudinal, appears to have an effect on the fatigue behavior of NiTi rotary instruments. Data [32, 33] along with results [23, 24, 29] indicates an advanced resistance to cyclic fatigue of electropolished versus nonelectropolished instruments. In accordance with these findings, SEM analysis of fractured surfaces revealed that the cracks of nonelectropolished instruments run along the machining grooves, whilst electropolished instruments exhibited a fine irregular zigzag crack pattern [34].

12.3.3 Chemical and electrochemical modification

In general, chemical method has been adopted for (i) forming a stable oxide film and (ii) resulting surface having an appropriate level of surface roughness, on the other hand, electrochemical method possesses also has 2-fold purposes: (i) forming a stable oxide film, and (ii) through an electrochemical reaction to deposit target substance(s) onto substrate surface [35]. Electrochemistry deals with charge transfer across interfaces. One of the first applications of electrochemistry was for coating conducting surfaces with either metals (e.g., electroplating) or by inorganic or organic substances (such as oxides and polymers). In either method, enhanced biocompatibility of implants (in particular, titanium implants) highly depends on the possibility of achieving high degrees of surface functionalization for an enhanced mineralization of bioactive minerals [e.g., hydroxyapatite (HA)] [36]. Detailed information about HA coating will be described later in this section. Electrochemical coatings ranging from monoatomic layers to microns thick have been carried out, which affected the physical and chemical properties of the coated surface [37]. Using electrochemical reaction, deposition of biocompatible platinum foil (Pt) onto TiN substrate [38] was reported. Moreover, (Ti,Al)N was coated electrochemically onto Ni–Cr dental base alloy [39]. There are several techniques for the deposition of coatings on metals, including PVD, CVD, electrochemical deposition, plasma spraying, and sol–gel process. The sol–gel coating process possesses unique advantages as follows: (1) excellent chemical stability, oxidation control, and enhanced corrosion resistance for metal substrates

[40, 41], (2) it is an environmentally friendly technique of surface protection and had showed the potential for the replacement of toxic pretreatments and coatings which have traditionally been used for increasing corrosion resistance of metals [42, 43], (3) the processing temperature generally is low, frequently close to room temperature, thus thermal volatilization and degradation of entrapped species, such as organic inhibitors, is minimized, (4) since liquid precursors are used, it is possible to cast coatings in complex shapes and to produce thin films without the need for machining or melting, and (5) the sol–gel films are formed by "green" coating technologies [43]. Kado et al. [44] developed a simple chemical modification method to immobilize cell-adhesive molecules on a titanium surface to improve its biocompatibility with human periodontal ligament cells (HPDLCs). It was concluded that cell-adhesive molecules successfully immobilized on the titanium surface and improved the compatibility of the surface with HPDLCs, suggesting that the collagen-immobilized titanium surface could be used for forming ligament-like tissues around titanium dental implants. Lee et al. [45] examined the gelatin nanogold (GnG) composite for surface dipping modification of titanium in addition to insure biocompatibility on dental implants or biomaterials. The GnG composite was constructed by gelatin and hydrogen tetrachloroaurate in presence of reducing agent, sodium borohydrate ($NaBH_4$). It was reported that (i) the GnG composite showed well dispersed character, the strong absorption at 530 nm, roughness, regular crystal and clear C, Na, Cl, P, and Au signals onto titanium, and (ii) further, this composite allowed MC-3T3 E1 cell growth and viability compared to gelatin and pure titanium, indicating that GnG composite coated titanium surfaces have a good biocompatibility for osteoblast growth and attachment than by simple and versatile dipping method.

There are several research works that reported on anodization for which NiTi is connected to anodic electrode in a certain type of electrolyte against a cathode (which is normally carbon). Speck et al. [46] reported that anodic polarization measurements made in Hanks' physiological solution at 37 °C and a pH of 7.4 show titanium materials to be the most passive of the following metals: titanium, Ti-6A1-4V, Ti–Ni, MP35N (Co–Ni–Cr–Mo), Co–Cr–Mo, 316L stainless steel, and nickel. The influence of the amino acids, cysteine, and tryptophan on the corrosion behavior of Ti–Ni and Ti-6A1-4V was studied. It was also mentioned that cysteine caused a lower breakdown potential for Ti–Ni, but it did not affect the breakdown of Ti-6A1-4V, although an increase in current density for Ti-6A1-4V was observed, and tryptophan produced no significant effects. A dense titania film is fabricated in situ on NiTi anodic oxidation in a Na_2SO_4 electrolyte [47]. The microstructure of the titania film and its influence on the biocompatibility of NiTi SMA are investigated by SEM, X-ray photoelectron spectroscopy (XPS), inductively coupled plasma mass spectrometry, hemolysis analysis, and platelet adhesion test. It was reported that (i) the titania film has a Ni-free zone near the surface and can effectively block the release of harmful Ni ions from the NiTi substrate in simulated body fluids, and (ii) the wettability, hemolysis resistance, and thromboresistance of the NiTi sample are improved by this anodic oxidation method.

Bernard et al. [48] anodized a laser processed NiTi alloy for different times in H_2SO_4 electrolyte with varying pH to create biocompatible surfaces with low Ni ion release as well as bioactive surfaces to enhance biocompatibility and bone cell-material interactions. The anodized surfaces were assessed for their in vitro cell-material interactions using human fetal osteoblast cells for 3, 7, and 11 days, and Ni ion release up to 8 weeks in simulated body fluids. The results were correlated with the surface morphologies of anodized surfaces characterized using field-emission scanning electron microscopy. It was obtained that (i) anodization creates a surface with nano/microroughness depending on the anodization conditions, (ii) the hydrophilicity of the NiTi surface was found to improve after anodization, as shown by the lower contact angles in cell medium, which dropped from 32° to <5°, (iii) the improved wettability of anodized surfaces is further corroborated by their high surface energy, comparable with that of commercially pure Ti, (iv) relatively high surface energies, especially the polar component, and nano/microsurface features of anodized surfaces significantly increased the number of living cells and their adherence and growth on these surfaces, and (v) a significant drop in Ni ion release from 268 ± 11 to 136 ± 15 ppb was observed for NiTi surfaces after anodization indicating that anodization of a NiTi alloy has a positive influence on the surface energy and surface morphology, which in turn improves bone cell-material interactions and reduces Ni ion release in vitro. Anodization of NiTi alloy was carried out on a biomedical grade NiTi alloy in phosphoric acid-based electrolyte and subjected to posttreatment by annealing [49]. The morphology of anodized surface was porous with irregularly distributed pores and annealing resulted in the formation of nodules of titanium oxides on the porous structure. It was found that (i) the corrosion potential shifted to more noble direction after annealing and there was one order decrease in corrosion current density compared to as-anodized surface, (ii) cyclic potentiodynamic polarization behavior of mechanically polished and anodized surfaces was almost similar and both the surfaces exhibited positive hysteresis loop on reversing the scan direction, and (iii) the formation of negative hysteresis loop in the case of annealed anodized alloy showed that the surface was capable of retaining its passive behavior even at higher potentials. Dynamic impedance spectroscopy studies carried out on annealed anodized surfaces showed that on increasing the potential from open circuit potential to higher potentials (0.8 V), the thickness of the oxide layer also increased, but behaved as a leaky capacitor, and in terms of bioactivity, the entire surface of the annealed sample was covered with HA after 12 days of immersion in Hanks' solution and the nickel released (370 ppb) was comparatively less than bare NiTi surface (633 ppb). Ni–Ti–O nanotube arrays (NTAs) was prepared on nearly equiatomic NiTi alloy, which will show broad application potential such as for energy storage and biomedicine, but their precise structure control is a great challenge because of the high content of alloying element of Ni, a nonvalve metal that cannot form a compact electronic insulating passive layer when anodized. Hang et al. [50] investigated the influence of various anodization parameters on the formation and structure of Ni–Ti–O NTAs and their potential applications. It was shown that

(i) excellent controlled NTAs can be fabricated during relatively wide ranges of the anodization voltage (5–90 V), electrolyte temperature (10–50 °C) and electrolyte NH_4F content (0.025–0.8 wt%) but within a narrow window of the electrolyte H_2O content (0.0–1.0 vol%), (ii) through modulating these parameters, the Ni–Ti–O NTAs with different diameter (15–70 nm) and length (45–1,320 nm) can be produced in a controlled manner, (iii) regarding potential applications, the Ni–Ti–O NTAs may be used as electrodes for electrochemical energy storage and nonenzymic glucose detection and may constitute nanoscaled biofunctional coating to improve the biological performance of NiTi-based biomedical implants. Huan et al. [51] fabricated hybrid micro/nanostructures on biomedical NiTi, by first electrochemically etched and then anodized in fluoride-containing electrolyte. It was reported that (i) with the etching process, the NiTi surface was microroughened through the formation of micropits uniformly distributed over the entire surface, (ii) following the subsequent anodizing process, self-organized nanotube structures enriched in TiO_2 could be superimposed on the etched surface under specific conditions, (iii) the anodizing treatment significantly reduced water contact angles and increased the surface free energy compared to the surfaces prior to anodizing, and (iv) it is possible to create hybrid micro/nanostructures on biomedical NiTi alloys by combining electrochemical etching and anodizing under controlled conditions concluding that these novel structures are expected to significantly enhance the surface biofunctionality of the material when compared to conventional implant devices with either micro- or nanostructured surfaces. Ni–Ti–O nanopores with different length (0.55–114 µm) were anodically grown on nearly equiatomic NiTi alloy [52]. Length-dependent corrosion behavior, nickel ion (Ni^{2+}) release, cytocompatibility, and antibacterial ability were investigated by electrochemical, analytical chemistry, and biological methods. It was obtained that (i) constructing nanoporous structure on the NiTi alloy improve its corrosion resistance; however, the anodized samples release more Ni^{2+} than that of the bare NiTi alloy, suggesting chemical dissolution of the nanopores rather than electrochemical corrosion that governs the Ni^{2+} release, (ii) the Ni^{2+} release amount increases with nanopore length, (iii) the anodized samples show good cytocompatibility when the nanopore length is <11 µm, and (iv) the length scale covers the nanopores (1–11 µm) showing favorable antibacterial ability indicating that the nanopores with length in the range of 1–11 µm are promising as coatings of biomedical NiTi alloy for antiinfection, drug delivery, and other desirable applications.

Endo et al. [53] studied the corrosion resistance of a NiTi alloy, chemically modified with a human plasma fibronectin using an aminosilane and a glutaraldehyde, by electrochemical techniques in a 0.9% NaCl solution and a cell culture medium containing serum. The role of serum proteins in influencing the passive behavior of the alloy was extensively studied by anodic polarization and XPS. It was found that (i) the passive current density increased in the presence of serum proteins, (ii) the enhanced dissolution of the passive film appears to be a consequence of quick adsorption of the serum proteins and the subsequent formation of metal ion–protein complexes on the film surface, (iii) with the chemical modification, the corrosion rate was reduced by approximately 50% in both solutions due to a highly cross-linked siloxane network formed on the alloy surface, and (iv) this network layer limited the diffusion of dissolved oxygen, metal ions, and biological molecules involved in the corrosion reactions. Fukushima et al. [54] improved the corrosion resistance of Ti–Ni alloy by electrolytic treatment, using different acidic electrolyte compositions. It was reported that (i) specimens electrolyzed with lactic acid, water, and glycerol were found to show higher corrosion potential and release lower amount of titanium and nickel ions than mechanical-polished specimens, (ii) with the electrolytic treatment, nickel concentration in the surface oxide layer of Ti–Ni alloy decreased, and the thickness of the surface oxide layer increased, and (iii) electrolytic treatment with suitable electrolyte could improve the corrosion resistance of Ti–Ni alloy, which is effective to produce medical and dental devices that utilize shape-memory effect or superelasticity (SE) with better biocompatibility. Wu et al. [55] treated porous NiTi alloys with three different chemical processes. It was found that (i) H_2O_2 treatment, NaOH treatment, and H_2O_2 pretreatment plus subsequent NaOH treatment can mitigate leaching of nickel from the alloy, (ii) the porous NiTi samples modified by the two latter processes favor deposition of a layer composed of Ca and P due to the formation of bioactive Na_2TiO_3 on the surface, (iii) among the three processes, H_2O_2 pretreatment plus subsequent NaOH modification is the most effective in suppressing nickel release, (iv) small area XPS reveals that the surfaces treated by different chemical processes have different structures and compositions, (v) the sample modified by the H_2O_2 treatment is composed of rough TiO_2 on the outer surface and an oxide transition layer underneath whereas the sample treated by NaOH comprises a surface layer of titanium oxide and Na_2TiO_3 together with a transition layer, and (vi) the sample processed by the H_2O_2 and NaOH treatment has a pure Na_2TiO_3 layer on the surface and a transition layer underneath.

The effect of surface modification of $Ni_{55.8}Ti$ (wt%) alloy on its corrosion behavior in Hanks' solution was determined [56]. The surfaces of heat-treated samples were modified by mechanical polishing (MP), EP, and EP followed by chemical passivation (CP). As-heat-treated samples with straw-colored oxide (SCO) finishes and blue-colored oxide (BO) finishes also were included. It was reported that (i) surface roughness increased in the order CP<EP<SCO<BO<MP, (ii) the nickel release within the five groups of NiTi samples, as determined by atomic absorption spectrophotometry, reduced in time over the measured period, (iii) the level of Ni ions released over a 25-day immersion period was highest in the SCO sample ($0.002\ \mu g/day$), (iv) this Ni level is negligible compared with the daily intake of Ni in an ordinary diet. The auger electron spectroscopy analyses indicated that before immersion in Hanks' physiologic solution, the main surface composition of all the samples was titanium and nickel, with a small amount of oxygen, carbon, and sulfur as contaminants, (v) the surface oxide thickness of the different samples increased in the order CP<EP<MP<BO<SCO; whilst, for the electrodes treated under the same conditions, the mean breakdown

potential value decreased in the order BO>MP>CP>EP>SCO while the corrosion current density and rate increased in the order CP<SCO<EP<BO<MP.

12.3.4 Physical modification

Mändl et al. [57] treated NiTi and CpTi in oxygen–plasma immersion ion implantation method and found that a hard, dense, and thick rutile type TiO_2 layer formed with excellent biocompatibility. Surface passivation can be achieved by plasma immersion ion implantation (PIII) processing in oxygen atmosphere on Ti [58] and NiTi [59]. PIII nitridization were conducted on Ti and/or NiTi [60]. The surface layers of titanium and titanium alloys were modified by ion beam implantation [61]. A similar work was reported on NiTi [62]. Both negative ion and positive ion are implanted for surface modification of materials. Tateshima et al. [63] examined the impact of ultraviolet (UV) irradiation on the biocompatibility of NiTi surfaces with endothelial cells. NiTi sheets were treated with UV irradiation for 48 h and human aorta-derived endothelial cells were used in this study. UV irradiation converted the NiTi surface to hydrophilic state and increased albumin adsorption. The number of endothelial cell migration, attachment, and proliferation as well as their metabolic activity were significantly increased on UV-treated NiTi. It was found that the photoactivation of NiTi surfaces by UV irradiation demonstrates improved biocompatibility of UV-treated NiTi surfaces with vascular endothelial cells and it was suggested that the UV irradiation may promote endothelialization of NiTi devices in blood vessels. Chakraborty et al. [64] investigated the effect of laser fluence on physical, mechanical, and chemical surface characteristics along with corrosion protection performance of the modified surface as compared to bare NiTi alloy surface. Various phases were prominent on the top surface, namely nickel and titanium-rich phase along with different NiTi intermetallics and nanostructure of titanium oxide, based on varying melting pool recirculation time and cooling rate with laser fluence energy. It was mentioned that (i) at low laser fluence up to 4 J/mm^2, no significant melting pool were formed and only transformation to the martensitic phase of Nitinol took place on the top surface, which seemed to be highly corrosion-prone under simulated body fluid, (ii) at moderate laser fluence of 6–8 J/mm^2, mostly Ti-rich phases are prominent on the surfaces on account of optimum recirculation of melting pool and subsequent surfacing out of comparably light phase of titanium, (iii) Ti-rich phases on top surface exhibit superior corrosion resistance as compared to all other samples including bare nitinol; however, titanium oxide nanoparticles-reinforced martensitic structure is formed under high laser fluence due to over recirculation of molten pool, and (iv) the modulus of elasticity also varied from 10 to 110 GPa based on top surface formation under different fluence levels, and thus this process can act as a tailor-made controllable pretreatment process over the traditional coating processes. Thangavel et al. [65] formed mechanically robust, biocompatible, and corrosion resistant Ag doped NiTi (NiTi/Ag) coatings on implant grade commercially pure titanium substrates by RF magnetron sputtering. Five samples with varying silver content (0, 1, 3, 7, and 10 at%) were prepared by controlling the power applied to Ag and NiTi targets. It was reported that (i) the soft Ag crystallites decreased the roughness and crystallinity of NiTi/Ag, (ii) among all compositions, NiTi/Ag coating with 3 at% Ag exhibited lowest friction coefficient (0.1) and wear rate (6.9×10^{-8} mm^3/N $*$ mm), (iii) electrochemical corrosion measurements indicated that Ag incorporation increased the corrosion resistance of NiTi, (iv) increase in Ag content shifted E_{corr} values in the anodic direction and reduced the current density by one order of magnitude. When cultured on NiTi/Ag coating with 3 at% Ag, human dermal fibroblast neonatal cells demonstrated highest cell viability, and (v) overall, NiTi/Ag-coated titanium substrates were found to be a promising orthopedic implant material.

Huan et al. [66] modified surface of NiTi SMA by plasma electrolytic oxidation in Na$_3$PO$_4$ with an aim to produce porous NiTi surfaces for biomedical applications. It was concluded that the wettability and surface free energy of NiTi increased significantly, indicating that the plasma oxidation process shows potential for expanding the biofunctionality of NiTi. Chrzanowski et al. [67] plasma treated NiTi since high nickel content such as in NiTi is believed to reduce the number of biomedical applications due to the reported toxicity of nickel. The reduction in nickel release and minimized exposure of the cell to nickel can optimize the biocompatibility of the alloy. Among many ideas proposed to prevent Ni element release from NiTi, it was hypothesized that the native surface of NiTi supports cell differentiation and insures good biocompatibility. Three types of surface modifications were investigated: thermal oxidation, alkali treatment, and plasma sputtering, and compared with smooth, ground surface. It was concluded that the most favorable cell responses were observed for ground and plasma-sputtered surfaces, suggesting that smooth, plasma-modified surfaces provide sufficient properties for cells to grow. Samples of Nitinol were oxidized using different procedures in order to improve their biocompatibility and prevent the Ni release [68]. It was reported that (i) the extreme chemical reactivity of such a plasma resulted in the formation of an oxide film in about 15 s, meaning that external oxidation took place, (ii) the biocompatibility investigations, performed according to the ISO standard protocol using L929 cells, showed the absence of any cytotoxic effects that might be due to the contact between the biological materials and nickel, and (iii) the investigation of nickel release of samples exposed to Hank's solution, measured by ICP-OES, showed negligible Ni concentrations. Jabłoński et al. [69] studied the functional coatings based on biopolymers (chitosan) as prospective and attractive approach in surface functionalization of SMA. Alloy substrates were pre-modified in two steps: (1) piranha solution treatment, and (2) plasmochemical etching (Ar$^+$) and activation in RF CVD reactor in argon–ammonia atmosphere. Then, the activated substrates were immersed in biopolymer solution(s) resulting in formation of biopolymer layer(s). Furthermore, the modified alloy limits transfer of the metals (Ni, Ti) to Hank's solution. It was mentioned that (i) the obtained coatings with prefunctionalization in RF reactor and polymer-based layers inhibit corrosion of NiTi alloy, (ii) the obtained chitosan coatings on modified surface after plasma treatment in Ar/NH$_3$ mixture are homogeneous, possess good adhesive properties and improved mechanical properties as well as are characterized by noncytotoxicity, and (iii) the modified alloy limits transfer of the metals (Ni, Ti) to Hank's solution.

Various types of compounds have been deposited physically onto NiTi alloy substrate. Sun et al. [70] fabricated (Ti,O)/Ti or (Ti,O,N)/Ti coatings on a Ni$_{50.8}$Ti (at%) SMA by the plasma immersion ion implantation and deposition (PIIID) technique in order to improve its corrosion, wear resistance, and bioactivity. It was found that (i) SEM examination of coating surfaces and cross sections showed that (Ti,O)/Ti and (Ti,O,N)/Ti composite coatings were dense and uniform, having thickness values of 1.16 ± 0.08 µm and 0.95 ± 0.06 µm, respectively, (ii) X-ray diffraction (XRD) revealed that there were no diffraction peaks corresponding to TiO$_2$ or TiN for (Ti,O)/Ti and (Ti,O,N)/Ti composite coatings, suggesting that after the PIIID treatment, TiO$_2$ and TiN were amorphous or nanosized in the coatings, (iii) the width of wear tracks on (Ti,O,N)/Ti coated NiTi SMA samples was reduced 6.5-fold, in comparison with that of uncoated samples, (iv) E$_{CORR}$ was improved from −466.20 ± 37.82 mV for uncoated samples to 125.50 ± 21.49 mV and −185.40 ± 37.05 mV for (Ti,O)/Ti coated and (Ti,O,N)/Ti coated samples, respectively, and (v) both types of coatings facilitated bonelike apatite formation on the surface of NiTi SMA in simulated body fluid, indicating their in vitro bioactivity. Mirak et al. [71] electrodeposited zinc-hydroxyapatite (Zn-HA) and zinc-hydroxyapatite-titania (Zn-HA-TiO$_2$) nanocomposite coatings onto a NiTi SMA, using a chloride zinc plating bath. It was reported that (i) the Zn-HA-TiO$_2$ coating exhibited a platelike surface morphology, where the addition of the nanoparticles caused an increase in roughness, (ii) due to applying a proper stirring procedure during codeposition, a homogenous dispersion of the nanoparticles in the coatings was achieved, and (iii) the addition of the TiO$_2$ nanoparticles to the Zn-HA-TiO$_2$ coating enhanced the microhardness and wear resistance. Two kinds of biocompatible coatings (TiO$_2$-Ti and TiO$_2$-TiN) were produced in order to improve the corrosion resistance of NiTi alloy [72]. A titanium oxide–titanium (TiO$_2$-Ti) composite was coated on NiTi alloy using electrophoretic method. After the coating process, the samples were heat-treated at 1,000 °C in two tube furnaces: the first one in argon atmosphere and the second one in nitrogen atmosphere at 1,000 °C. It was mentioned that (i) electrochemical tests in simulated body fluid demonstrated a considerable increase in corrosion resistance of composite-coated NiTi specimens compared to the noncoated one, and (ii) the heat-treated composite coating sample in nitrogen atmosphere had a higher level of corrosion resistance compared to the heat-treated sample in argon atmosphere, which is mainly due to having nitride phases. Chakraborty et al. [73] utilized a high power fiber laser to synthesize titanium HA composite coating over biomedical-grade NiTi alloy surface through laser in situ formation, cladding, and alloying processes. It was obtained that (i) Ni-free titanium–calcium phosphate coating developed through laser cladding, (ii) layer composition and microstructure are directly dependent on input laser fluence, (iii) migration of titanium from base layer to clad layer is observed, (iv) clad surface exhibits high corrosion resistance compared to bare NiTi allot, and (v) surface modulus of elasticity is found in 6–30 GPa range, similar to natural bone.

There are some researches on nitirization and nitride coating onto NiTi alloy surface. Chu et al. [74] studied surface nanomechanical behavior under nanoindentation of ZrN and ZrCN film on NiTi substrate. It was reported that (i) the surface hardness and modulus of the films increase initially with larger nanoindentation depths and then reach their maximum values, (ii) afterwards, they diminish gradually and finally reaching plateau values which are the composite modulus and composite hardness derived from the ZrN/ZrCN film and NiTi substrate, (iii) they are higher than those of electropolished NiTi SMA due to the properties of ZrN and ZrCN, and (iv) in comparison, the surface nanomechanical properties of electropolished NiTi exhibit a different change with depths. Jin et al. [75] evaluated the biocompatibility of TiN coated NiTi-SMA to compare with that of the uncoated NiTi-SMA. Based on the orthodontic clinical application, the surface properties and biocompatibility were characterized by SEM, XRD), wettability test, mechanical test, and in vitro tests including MTT, cell apoptosis, and cell adhesion tests. It was found that (i) the bonding between the substrate and TiN coating is excellent, (ii) the roughness and wettability increased as for the TiN coating compared with the uncoated NiTi-SMA, (iii) MTT test (evaluation of cell respiration by the mitochondria) showed no significant difference between the coated and uncoated NiTi-SMA, however the percentage of early cell apoptosis was significantly higher for the uncoated NiTi alloy, and (iv) SEM results showed that TiN coating could enhance the cell attachment, spreading, and proliferation on NiTi-SMA; indicating that TiN coating bonded with the substrate well and could lead to a better biocompatibility. Czarnowska et al. [76] described the microstructure, topography, and morphology of TiN surface layer on NiTi alloy, and corrosion resistance, both before and after nitriding in low-temperature plasma at 290 °C. It was mentioned that (i) the surface titanium nitride layer improved the adhesion of platelets and the proliferation of osteoblasts, which was investigated in in vitro experiments with human cells, and (ii) experimental data revealed that nitriding NiTi-SMA under low-temperature plasma improves its properties for bone implant applications. Maleki-Ghaleh et al. [77] studied the biological behavior of NiTi-SMA while preserving its superelastic behavior in order to facilitate its compatibility for application in human body. The surfaces of NiTi samples were bombarded by three different nitrogen doses. It was reported that (i) the improvement of both corrosion and biological behaviors of the modified NiTi samples was observed; however, no significant change in the superelasticity was observed, and (ii) samples modified at 1.4×10^{18} ion/cm^2 showed the highest corrosion resistance and the lowest Ni ion release.

Some types of metal elements were coated on substrates of NiTi alloy. Ng et al. [78] used Mo and ZrO$_2$ to enhance the wear resistance, Hf was ion-implanted for improvement of wear resistance [79], Bell et al. [80] developed a radiopaque nanoporous Ta coating for Nitinol stents, Ta was sputter-coated to enhance the biocompatibility [81], and the bioactive TiO_2 was fabricated by plasma electrolytic oxidation [82]. Diamond-like carbon film was deposited using plasma-based ion implantation technique [83], using a radiofrequency plasma chemical vapor deposition method [84] or by sputter coating [85].

12.3.5 Hydroxyapatite coating

Due to its attractive bioactive properties and chemical similarity to hard tissues including bone and tooth structures, either HAs alone or HA-based biocomposites have been coated on orthopedic prostheses and dental implants. Critical factors for long-term stability of fixation of these prostheses include biocompatibility, material selection, implant design, and surface modification and texturing. Various coating technologies have been developed and employed. Among them, the thermal spray techniques (plasma spray and high velocity oxy-fuel) are mostly employed to deposit HA because of their high thermal efficiency and relative economy. However, some of the bioactive properties of HA are lost during thermal spraying. HA is thermally decomposed to less biocompatible products, such as tricalcium phosphate, tetracalcium phosphate, calcium oxide, and amorphous calcium phosphates, causing the weakness of their interface with the metallic substrate, so that the longevity of prostheses will be limited. Ideally, any biomaterial coatings must have high degrees of crystallinity and phase purity, good adhesive and cohesive integrity, and adequate porosity to promote bone ingrowth. And the adhesion of bone cells, strength, and durability of biomaterials are highly relied on characteristics of the substrate and the methods utilized in the HA coating process [86].

Bioceramic coatings can be applied on substrates by a number of methods, including dip coating-sintering, immersion coating, electrophoretic deposition, ion-beam sputter coating and dynamic mixing, hot isostatic pressing (HIP), and thermal spraying techniques such as plasma spraying, flame spraying, and high-velocity oxy-fuel (HVOF) combustion spraying. The currently popular method of deposition of HA on Ti implants is by plasma spraying or arc plasma spraying, with three environmental variations: air plasma spraying, controlled atmosphere plasma spraying, and vacuum plasma spraying, which has proved to be one of the leading technologies in the medical practice for implant production [87]. The outcomes of coating products are normally controlled by several parameters, including heat power of the plasma jet; type, pressure, and flow rate of the plasma and transporting gas; type, mean size and size distribution, flow rate, and shape of the injected powder particles; location, angle, and type of powder injection into the plasma jet; roughness, purity, and temperature of the coated surface; spraying-off distance between the spraying gun and substrate surface; speed relation between the spraying gun and coated surface; and type of the ambient atmosphere in the plasmatron [88–94]. Using this technology, coating materials are usually formed from powder particles injected into a high temperature field created in a plasma torch (plasmatron), where they are accelerated, molten, and propelled toward the coated substrate surface. However, the initial powder particles usually differ in shape, dimensions, temperature, and velocity that can result in evaporation, partial melting, or even absence of melting of some particles [95], causing formation of inhomogeneous and porous structure. This unwanted porosity along with presence of a small amount of amorphous phase with nonstoichiometric composition and nonuniformity affects adversely mechanical strength and bond strength as well. Moreover, the partially molten particles solidify on the surface as spherical grains that can get depleted or become sources of residual stress fields in the coating, influencing substantially its properties and performance [95, 96]. There is another drawback associated with high temperature plasma spraying technique; that is, decomposition of coated HA through the following reaction: $Ca_5(PO_4)_3(OH) \rightarrow CaO + Ca_3P_2O_8$ [97] and the HA decomposition is more pronounced when the HA powder particles are smaller and the spraying voltages are higher [98]. Although plasma spraying of HA coatings is currently the only commercial process in use, due to these technical drawbacks, the long-term stability of plasma sprayed coatings could be a problem.

By the bipolar pulsed current electrochemical deposition technique, HA was coated on NiTi substrate [99, 100]. It was reported that (i) reducing the electrocrystallization rate by the pulse deposition technique can promote HA formation on both abraded and modified substrates (ii) comparing the corrosion resistance of the bare substrates revealed that the modified alloy has a higher corrosion resistance than the abraded alloy and the modified surface is well passivized during anodic polarization in Ringer's solution; however, this condition is reversed after the deposition of HA film, (iii) because of the lower crystallization sites on the abraded alloy, the produced HA film is denser and more protective against the corrosive mediums as compared to the coating on the modified alloy, and (iv) although the HA coating can improve the bioactivity of both substrates, the resulted film on the oxidized alloy is porous and deteriorates the implant permanence in the vicinity of body fluids. Similarly, Kocijan et al. [101] elecrodeposited HA on NiTi with the aim to enhance the corrosion resistance of the substrate and therefore decrease the release of nickel ions from the surface of the alloy. It was reported that (i) a uniform inner layer of HAP was formed and overlaid by clusters of spherical and flakelike morphological structures, (ii) HA coating exhibited super-hydrophilic wetting properties compared to the moderate hydrophilicity of the NiTi alloy, and (iii) the HA coating significantly enhanced the corrosion resistance of the NiTi alloy and confirmed the effective barrier properties of the coating, which can be used to minimize the nickel release in biomedical applications.

Deng et al. [102] enhanced the osteoblast activity on NiTi/HA coatings on additive manufactured NiTi metal implants by nanosecond pulsed laser sintering. It was obtained that (i) the nanoscale porosity delivered by nanosecond pulsed laser sintering and the HA component positively contributed to osteoblast differentiation, as indicated by an increase in the expression of collagen and alkaline phosphatase, both of which are necessary for osteoblast mineralization, and (ii) topological complexities were observed which appeared to boost the activity of osteoblasts, including an increase in actin cytoskeletal structures and adhesion structures. Based on these findings, it was concluded that (iii) the pulsed laser sintering method is an effective tool to generate biocompatible coatings in complex alloy-composite material systems with desired composition and topology. Sattar et al. [103] investigated the electrochemical behavior of HA-coated and thermally oxidized near equiatomic NiTi SMAs for biomedical applications. NiTi samples were spin-coated with laboratory-synthesized HA followed by drying and calcination at 120 ± 5 °C and 450 ± 5 °C, respectively. Similarly, uncoated NiTi samples were thermally oxidized by heating at 450 ± 5 °C in air. The comparison was made among HA-coated, thermally oxidized, and uncoated samples. It was found that (i) SEM, atomic force microscope, and X-ray diffraction confirm the presence of titania and HA coating on the NiTi substrates, (ii) electrochemical testing results showed better resistance against corrosion for the HA-coated samples because of the protective coatings of HA and TiO_2, as compared to the rest, and (iii) Fourier transform infrared spectroscopy and SEM reveal that the HA-coated sample will provide bioactive cushion to the host structure for better adhesion during the implanted period, while the implant will do its job.

12.3.6 Thermal modification

Various oxidation treatments were applied to nearly equiatomic NiTi alloys to form a Ni-free protective oxide on the surface. Michiardi et al. [104] studied the oxidation treatment of NiTi alloy to obtain Ni-free surfaces and to improve biocompatibility. It was reported (i) oxidation treatment in a low-oxygen pressure atmosphere leads to a high surface Ti/Ni ratio, a very low Ni surface concentration, and a thick oxide layer, (ii) the oxidation treatment does not significantly affect the shape-memory properties of the alloy, (iii) the oxide formed significantly decreases Ni release into exterior medium when compared untreated surfaces, and (iv) as a result, the present new oxidation treatment could be of great interest for biomedical applications, as it could minimize sensitization and allergies and improve biocompatibility and corrosion resistance of NiTi SMAs. Danilov et al. [105] examined the effect of NiTi oxidation on material surface characteristics related to biocompatibility. Correspondence between electron work function (EWF) and adhesive force predicted by electron theory of adsorption as well as the effect of surface mechanical stress on the adhesive force were studied on the nonoxidized and oxidized at 350, 450, and 600 degrees C NiTi alloy for medical application. The adhesive force generated by the material surface towards the drops of alpha-minimal essential medium was used as a characteristic of NiTi adsorption properties. It was reported that (i) variations in EWF and mechanical stress caused by surface treatment were accompanied by variations in adhesive force, (ii) NiTi oxidation at all temperatures used gave rise to decrease in adhesive force and surface stress values in comparison to the nonoxidized state, (iii) in contrary, the EWF value revealed an increase under the same condition. Variations in surface oxide layer thickness and its phase composition were also followed, (iv) the important role of oxide crystallite size in EWF values within the range of crystallite dimensions typical for NiTi surface oxide as an instrument for the fine regulation of NiTi adsorption properties was demonstrated, and (v) the comparative oxidation of pure titanium and NiTi showed that the effect of Ni on the EWF value of NiTi surface oxide is negligible. Porous NiTi SMAs are one of the promising biomaterials for surgical implants because of porous structure with open pores which can accommodate for ingrowth of boney cells. However, the complex surface morphology and larger area of porous NiTi compared to dense NiTi make it more vulnerable from the viewpoint of release of nickel, which can cause deleterious effects in the human body. Wu et al. [106] modified surface of porous single-phase NiTi (which was prepared by capsule-free HIP) by oxidation in conjunction with postreaction heat treatment to mitigate Ni leaching and enhance the surface properties. It was reported that differential scanning calorimetry thermal analysis, uniaxial compression tests, inductively coupled plasma mass spectrometry, and cell cultures reveal that porous NiTi alloys oxidized at 450 °C for 1 h have an austenite transition temperature below 37 °C, excellent SE, lower nickel release, and no cytotoxicity. Chrzanowski et al. [107] compared different surface treatments used for bioactivation of pure titanium surfaces – thermal, alkali treatment and spark oxidation – and to assess their suitability as treatments for NiTi alloys. This was considered by examining the surface properties, calcium phosphate precipitation from a physiological solution, and nickel ion release. It was found that (i) the native surface of Ni-Ti alloy is highly bioactive when assessing the precipitation of calcium phosphates from Hank's solution, (ii) low temperature heat treatments also produced promising surfaces while high temperature treatment resulted in a very low rate of Ca and P precipitation, (iii) alkali treatment and spark oxidation resulted in some bioactivity, (iv) Ni ion release was greatest for alkali treated and sparks oxidized samples, and the rate of its release from these two samples was on the verge of daily safe dose for adolescent human, and (v) heat treatment at 400 °C resulted in significant increase in the transformation temperatures, and a further increase of the treatment temperature up to 600 °C caused a drop of the transformation temperature. TiN and TiO_2 coatings, which are known from their low chemical reactivity, high hardness, and wear and corrosion resistance, are used for protecting the NiTi surface. Lelatko et al. [108] coated nearly equiatomic $Ni_{50.6}Ti$ (at%) SMA with the layers obtained by nitriding under glow discharge at 800 °C. Additionally, at the end of the process some amount of oxygen was added. It was mentioned that (i) the surface is formed from nanocrystalline and columnar grains of the TiN phase, (ii) between the top layer and β-NiTi substrate, the interface Ti_2Ni layer was formed. Addition of oxygen at the end of the process created a thin layer of TiO_2 phase nanograins at the surface of the TiN phase, (iii) in the same areas, small amount of amorphous phase was identified, and (iv) the combination of nitriding and oxidation formed layers that reveal relatively high corrosion resistance.

A $Ni_{50.9}Ti$ (at%) SMA was heat treated by several regimes simulating the shape setting procedure, the last step in the manufacture of implants [109]. Heating temperatures were between 500 and 550 °C and durations from 5 to 10 min. Heat treatments were performed in air at normal and low pressure and in a salt bath. The purpose of the treatments was to obtain and compare different surface states of the NiTi alloy. It was reported that (i) the heat treatments produced surface TiO_2 layers measuring 15–50 nm in thickness that were depleted in nickel, (ii) the sample covered by the 15-nm thick oxide that was treated at 500 °C/5 min in a low pressure air showed the best corrosion performance in terms of Ni release, and (iii) as the oxide thickness increased, due to either temperature or oxygen activity change, Ni release into the physiological solution accelerated. Xu et al. [110] investigated the effects of micro-arc oxidation (MAO) surface modification (alumina coatings) on the phase transformation behavior, shape-memory characteristics, in vitro hemocopatibility, and cytocompatibility of the biomedical NiTi alloy by differential scanning calorimetry, bending test, hemolysis ratio test, dynamic blood clotting test, platelet adhesion test, and cytotoxicity testing by human osteoblasts (Hobs). It was obtained that (i) there were no obvious changes of the phase transformation temperatures and shape-memory characteristics of the NiTi alloy after the MAO surface modification and the coating could withstand the thermal shock and volume change caused by martensite–austenite phase transformation, and (ii) compared to the uncoated NiTi alloys, the MAO surface modification could effectively improve the haemocopatibility of the coated NiTi alloys by the reduced hemolysis ratio, the prolonged dynamic clotting time, and the decreased number of platelet adhesion; and the rough and porous alumina coatings could obviously promote the adherence, spread, and proliferation of the Hobs with the significant increase of proliferation number of Hobs adhered on the surface of the coated NiTi alloys. Aun et al. [111] developed a procedure to coat superelastic NiTi alloys with flexible TiO_2 protective nanocomposite films using sol–gel technology to improve the metal biocompatibility without deteriorating its superelastic mechanical properties. It was mentioned that (i) a short densification by thermal treatment at 500 °C for 10 min yielded a bilayer film consisting of a 50 nm-thick crystallized TiO_2 at the inner interface with another 50-nm-thick amorphous oxide film at the outer interface, and (ii) this bilayer could sustain over 6.4% strain without cracking and could thus be used to coat biomedical instruments as well as other devices made with superelastic NiTi alloys. Using the same MAO method, Sukuroglu et al. [112] deposited the TiO_2 coatings on the NiTi substrates. The surface topography, morphology, crystallographic structure, and thickness of the coatings were determined using scanning electron microscopy and X-ray diffraction. It was shown that the coated samples have higher corrosion resistance than uncoated samples in the two different media. Viswanathan et al. [113] investigated the formation of nanostructured oxide layers by anodic oxidation on different surface finished (mirror finished, 600 and 400 grit polished) NiTi alloy in electrolyte solution containing ethylene glycol and NH_4F. It was reported that (i) the native oxide on the substrate is replaced by nanostructures through anodization process, (ii) XPS of NiTi substrate shows the presence of Ni, NiO, Ti, and TiO_2 species, whereas Ni_2O_3 and $Ni(OH)_2$ and TiO_2 are observed in the samples after anodization, and (iii) corrosion resistance of the anodized sample is comparable with that of the untreated sample. The MAO coating was evaluated in order to enhance the corrosion resistance of additively manufactured NiTi medical devices [114]. It was mentioned that the corrosion characteristics of the MAO-coated specimens revealed that the proposed coating methodology significantly improves the corrosion resistance of NiTi parts produced using additive manufacturing process.

12.4 Surface modifications for dental and medical devices

For summarizing the above, we selected endodontic instruments, orthodontic archwires, implant (for both dental and orthopedic), orthopedic stents, and orthopedic devices as typical medical application examples of NiTi alloys and listed surface modification methods for each of these applications in mechanical, chemical, physical, and thermal means. Although some of listed references might be duplicated in the main text, we present all references in the table at the bottom of Table 12.1.

Table 12.1: Typical medical applications using NiTi alloys and surface modification methods.

	Endodontic instruments	Orthodontic archwires	Implants	Orthopedic Stents	Orthopedic devices
Polishing chemical treatment	[28, 29, 33, 115, 116]	[129, 130]	[146–149]	[172–178]	
Nitridation	[117–119, 109, 120]	[131–135]	[79, 150–154]	[179–181]	[192]
Oxidation heating	[110–114]	[136–138]	[155–157]	[146]	[193]
Plasma ion implantation	[57, 61, 121–123]	[139, 140]	[158–161]	[182, 183]	[194]
Cryogenic treatment	[124–127]				
Carbon deposition	[128]	[141, 142]	[162–165]	[184–187]	[195, 196]
Metal deposition		Rd: [143–145]		Silicone: [188, 189]	Au: [197]
HA coating			[166–171]	[190, 191]	

References

[1]   Oshida Y, Guven Y. Biocompatible Coatings for Metallic Biomaterials. In: Wen C, ed., Surface Coating and Modification of Metallic Biomaterials. Woodhead Pub., Elsevier, 2015, 287–343; https://doi.org/10.1016/B978-1-78242-303-4.00010-7.

[2]   Corina B, Felicia S. About the tribological behavior of ceramic materials. Tribol. Ind. 2008, 30, 10–4.

[3]   Pike M. Medical device materials. Ad. Mater. Process. 2009, 167, 50–6.

[4]   Pilliar RM. Metallic Biomaterials. In: Narayan R, ed., Biomedical Materials, Springer Science and Business Media, 2009, 41–2.

[5]   Zivić F, Babić M, Grujović N, Mitrović S. Tribometry of materials for bioengineering applications. Tribol. Ind. 2010, 32, 25–32.

[6]   Sandberg T, Carlsson J, Ott MK. Interactions between human neutrophils and mucin-coated surfaces. J. Mater. Sci. Mater. Med. 2009, 20, 621–31.

[7]   Morais JM, Papadimitrakopoulos F, Burgess DJ. Biomaterials/tissue interactions: possible solutions to overcome foreign body response. Am. Ass. Pharm. Sci. J. 2010, 12, 188–96.

[8]   Tucker RC. Surface engineering. J. Met. 2002, 160, 1–3.

[9]   Söderholm KJM. Review: coatings in dentistry – a review of some basic principles. Coatings 2012, 2, 138–59.

[10]   Rieu J. Ceramic formation on metallic surfaces (ceramization) for medical applications. Clin. Mater. 1993, 12, 227–35.

[11]   Hallab N, Merritt K, Jacobs JJ. Metal sensitivity in patients with orthopaedic implants. J. Bone Joint Surg. Am. 2001, 83, 428–38.

[12] Fuchs T, Schmidmaier G, Raschke MJ, Stange R. Bioactive-coated implants in trauma surgery. Eur. J. Trauma Emer. Surg. 2008, 34, 60–8.

[13]   Oshida Y. Bioscience and Bioengineering of Titanium Materials. Elsevier, London, UK, 1st edition, 2007.

[14]   Oshida Y. Bioscience and Bioengineering of Titanium Materials. Elsevier, London, UK, 2nd edition, 2013.

[15]   Oshida Y, Hashem A, Nishihara T, Yapchulay MV. Fractal dimension analysis of mandibular bones – toward a morphological compatibility of implants. J. Bio-Med. Mater. Eng. 1994, 4, 397–407.

[16]   Allen P. Titanium alloy development. Adv. Mater. Processes. 1996, 154, 35–7.

[17]   Hanawa T. Functionalisation of Metallic Surfaces for Biomedical Applications. In: Wen C, ed., Surface Coating and Modification of Metallic Biomaterials, 2015, 275–286; https://doi.org/10.1016/B978-1-78242-303-4.00009-0.

[18]   Suryanarayana C. Mechanical alloying and milling. Prog. Mater. Sci. 2001, 46, 1–184.

[19]   Mishra RS, Ma ZY. Friction stir welding and processing. Mater. Sci. Eng. 2005, R50, 1–78.

[20]   Li B, Shen Y, Hu W. Surface nitriding on Ti-6Al-4V alloy via friction stir processing method under nitrogen atmosphere. Appl. Surf. Sci. 2013, 274, 356–64.

[21]   Zhang R, Mankoci S, Walters N, Gao H, Zhang H, Hou X, Qin H, Ren Z, Zhou X, Doll GL, Martini A, Sahai N, Dong Y, Ye C. Effects of laser shock peening on the corrosion behavior and biocompatibility of a nickel-titanium alloy. J. Biomed. Mater. Res. B Appl. Biomatter. 2018; doi: 10.1002/jbm.b.34278.

[22]   Zhao X, Neves F, Correia JB, Liu K, Fernades FMB, Koledov V, von Gratowski S, Xu S, Huang J. Mechanical activation of pre-alloyed NiTi2 and elemental Ni for the synthesis of NiTi alloys. J. Mater. Sci. 2018, 53, 13432–41.

[23]   Lopes HP, Elias CN, Vieira MV, Vieira VT, de Souza LC, Dos Santos AL. Influence of surface roughness on the fatigue life of nickel-titanium rotary endodontic instruments. J. Endod. 2016, 42, 965–8.

[24]   Lopes H, Elias C, Vieira M, Vieira V, Inojosa I, Ferreira L. Influence of the surface roughness on the torsional resistance of nickel-titanium rotary endodontic instruments. ENDO – Endod. Pract. Today 2017, 11, 51–5.

[25]   Zupanc J, Vahdat-Pajouh N, Schäfer E. New thermomechanically treated NiTi alloys – a review. Intl. Endod. J. 2018; https://doi.org/10.1111/iej.12924.

[26]   Bonaccorso A, Tripi TR, Rondelli G, Condorelli GG, Cantatore G, Schäfer E. Pitting corrosion resistance of nickel-titanium rotary instruments with different surface treatments in seventeen percent ethylenediaminetetraacetic acid and sodium chloride solutions. J. Endod. 2008, 34, 208–11.

[27]   Kuhn G, Tavernier B, Jordan L. Influence of structure on nickel-titanium endodontic instruments failure. J. Endod. 2001, 27, 516–20.

[28]   Bui TB, Mitchell JC, Baumgartner JC. Effect of electropolishing ProFile nickel-titanium rotary instruments on cyclic fatigue resistance, torsional resistance, and cutting efficiency. J. Endod. 2008, 34, 190–3.

[29]   Anderson ME, Price JW, Parashos P. Fracture resistance of electropolished rotary nickel-titanium endodontic instruments. J. Endod. 2007, 33, 1212–6.

[30]   Tripi TR, Bonaccorso A, Condorelli GG. Cyclic fatigue of different nickel-titanium endodontic rotary instruments. Oral Surg. Oral Med. Oral Pathol. Oral Radiol. Endod. 2006, 102, e106–14.

[31]   Herold KS, Johnson BR, Wenckus CS. A scanning electron microscopy evaluation of microfractures, deformation and separation in EndoSequence and profile nickel-titanium rotary files using an extracted molar tooth model. J. Endod. 2007, 33, 712–4.

[32]   Condorelli GG, Bonaccorso A, Smecca E, Schäfer E, Cantatore G, Tripi TR. Improvement of the fatigue resistance of NiTi endodontic files by surface and bulk modifications. Int. Endod. J. 2010, 43, 866–73.

[33]   Praisarnti C, Chang JW, Cheung GS. Electropolishing enhances the resistance of nickel-titanium rotary files to corrosion-fatigue failure in hypochlorite. J. Endod. 2010, 36, 1354–7.

[34]   Lopes HP, Elias CN, Vieira VT, Moreira EJL, Marques RVL, de Oliveira JCM, Debelian G, Siqueira JF Jr. Effects of electropolishing surface treatment on the cyclic fatigue resistance of BioRace nickel-titanium rotary instruments. J. Endod. 2010, 36, 1653–7.

[35]   Elias CN, Oshida Y, Lima JHC, Muller CA. Relationship between surface properties (roughness, wettability and morphology) of titanium and dental implant removal torque. J. Mech. Behav. Biomed. Mat. 2008, 1, 234–42.

[36]   Ajami E. Aguey-Zinsou K-F. Formation of OTS self-assembled monolayers at chemically treated titanium surfaces. J. Mat. Sci. Mat. Med. 2011, 22, 1813–24.

[37]   Guslitzer-Okner R, Mandler D. Applications of electrochemistry and nanotechnology in biology and medicine, I: electrochemical coating of medical implants. Mod. Aspect. Electroc. 2011, 52, 291–342.

[38]   Morcos BM, O'Callaghan JM, Amira MF, Van Hoof C, de Beeck MO. Electrodeposition of platinum thin films as interconnects material for implantable medical applications electrochemical/electroless deposition. *J. Electrochem. Soc. 2013, 160,* D300–6.

[39]   Chung KH, Liu GT, Duh JG, Wang JH. Biocompatibility of a titanium-aluminum nitride film coating on a dental alloy. Surf. Coat. Tech. 2004, 188/189, 745–9.

[40]   Brinker CJ, Hurd AJ, Shunrk PR, Frye GC, Asley CS. Review of sol-gel thin film formation. J. Non-Cryst. Sol. 1992, 147, 424–36.

[41]   Wright JD, Sommerdijk NAJ. Sol-Gel Materials Chemistry and Applications. CRC Press, OPA Overseas Publishers Association, 2001.

[42]   Durán A, Castro Y, Aparicio M, Conde A, de Damborenea JJ. Protection and surface modification of metals with sol-gel coatings. Int. Mater. Rev. 2007, 52, 175–92.

[43] Wang D, Bierwagen GP. Sol-gel coatings on metals for corrosion protection. Progress. Org. Coat. 2009, 64, 327–38.

[44] Kado T, Hidaka T, Aita H, Endo K, Furuichi Y. Enhanced compatibility of chemically modified titanium surface with periodontal ligament cells. Appl. Surf. Sci. 2012, 262, 240–7.

[45] Lee Y-H, Bhattarai G, Aryal S, Lee, N-H, Lee M-H, Kim T-G, Jhee E-C, Kim H-Y, Yi H-K. Modified titanium surface with gelatin nano gold composite increases osteoblast cell biocompatibility. Appl. Surf. Sci. 2010, 256, 5882–7.

[46] Speck KM, Fraker AC. Anodic polarization behavior of Ti-Ni and Ti-6A1-4V in simulated physiological solutions. J. Dent. Res. 1980, 59, 1590–5.

[47] Chu CL, Wang RM, Hu T, Yin LH, Pu YP, Lin PH, Dong YS, Guo C, Chung CY, Yeung KW, Chu PK. XPS and biocompatibility studies of titania film on anodized NiTi shape memory alloy. J. Mater. Sci. Mater. Med. 2009, 20, 223–8.

[48] Bernard SA, Balla VK, Davies NM, Bose S, Bandyopadhyay A. Bone cell-materials interactions and Ni ion release of anodized equiatomic NiTi alloy. Acta Biomater. 2011, 7, 1902–12.

[49] Chembath M, Balaraju JN, Sujata M. Effect of anodization and annealing on corrosion and biocompatibility of NiTi alloy. Surf. Coat. Technol. 2016, 302, 302–9.

[50] Hang R, Liu Y, Zhao L, Gao A, Bai L, Huang X, Zhang X, Tang B, Chu PK. Fabrication of Ni-Ti-O nanotube arrays by anodization of NiTi alloy and their potential applications. Sci. Rep. 2014, 18(4), 7547; doi: 10.1038/srep07547.

[51] Huan Z, Fratila-Apachitei LE, Apachitei I, Duszczyk J. Synthesis and characterization of hybrid micro/nano-structured NiTi surfaces by a combination of etching and anodizing. Nanotechnology 2014, 25; doi: 10.1088/0957-4484/25/5/055602.

[52] Hang R, Liu Y, Bai L, Zhang X, Huang X, Jia H, Tang B. Length-dependent corrosion behavior, Ni^{2+} release, cytocompatibility, and antibacterial ability of Ni-Ti-O nanopores anodically grown on biomedical NiTi alloy. Mater. Sci. Eng. C Mater. Biol. Appl. 2018, 89, 1–7.

[53] Endo K. Chemical modification of metallic implant surfaces with biofunctional proteins (Part 2). Corrosion resistance of a chemically modified NiTi alloy. Dent. Mater. J. 1995, 14, 199–210.

[54] Fukushima O, Yoneyama T, Doi H, Hanawa T. Corrosion resistance and surface characterization of electrolyzed Ti-Ni alloy. Dent. Mater. J. 2006, 25, 151–60.

[55] Wu S, Liu X, Chan YL, Chu PK, Chung CY, Chu C, Yeung KW, Lu WW, Cheung KM, Luk KD. Nickel release behavior and surface characteristics of porous NiTi shape memory alloy modified by different chemical processes. J. Biomed. Mater. Res. A 2009, 89, 483–9.

[56] Cissé O, Savadogo O, Wu M, Yahia L. Effect of surface treatment of NiTi alloy on its corrosion behavior in Hanks' solution. J. Biomed. Mater. Res. 2002, 61, 339–45.

[57] Mändl S. PIII treatment of Ti alloys and NiTi for medical applications. Surf. Coat. Tech. 2007, 201, 6833–8.

[58] Yankov RA, Shevchenko N, Rogozin A, Maitz MF, Richter E, Möller W, Donchev A, Schütze M. Reactive plasma immersion ion implantation for surface passivation. Surf. Coatings Tech. 2004, 201, 6752–8.

[59] Qi H, Wu HY. Effect of surface modification of pure Ti on tribological and biological properties of bone tissue. Surf. Eng. 2013, 29, 300–5.

[60] Firouzi-Arani M, Savaloni H, Ghoranneviss M. Dependence of surface nano-structural modifications of Ti implanted by N^+ ions on temperature. Appl. Surf. Sci. 2010, 256, 4502–11.

[61] Rautray TR, Narayanan R, Kwon T-Y, Kim K-H. Surface modification of titanium and titanium alloys by ion implantation. J. Biomed. Mater. Res. Part B 2010, 93B, 581–91.

[62] Liu XM, Wu SL, Chu PK, Chung CY, Chu CL, Yeung KWK, Lu WW, Cheung KMC, Luk KDK. Effects of water plasma immersion ion implantation on surface electrochemical behavior of NiTi shape memory alloys in simulated body fluids. Appl. Surf. Sci. 2007, 253, 3154–9.

[63] Tateshima S, Kaneko N, Yamada M, Duckwiler G, Vinuela F, Ogawa T. Increased affinity of endothelial cells to NiTi using ultraviolet irradiation: an *in vitro* study. J. Biomed. Mater. Res. Part A 2108, 106A, 1034–8.

[64] Chakraborty R, Datta S, Raza MS, Saha P. A comparative study of surface characterization and corrosion performance properties of laser surface modified biomedical grade nitinol. Appl. Surf. Sci. 2019, 469, 753–63.

[65] Thangavel E, Dhandapani VS, Dharmalingam K, Marimuthu M, Veerapandian M, Arumugam MK, Kim S, Kim B, Ramasundaram S, Kim DE. RF magnetron sputtering mediated NiTi/Ag coating on Ti-alloy substrate with enhanced biocompatibility and durability. Mater. Sci. Eng. C Mater. Biol. Appl. 2019; doi: 10.1016/j.msec.2019.01.099.

[66] Huan Z, Fratila-Apachitei LE, Apachitei I, Duszczyk J. Porous NiTi surfaces for biomedical applications. Appl. Surf. Sci. 2012, 258, 5244–9.

[67] Chrzanowski W, Szade J, Hart AD, Knowles JC, Dalby MJ. Biocompatible, smooth, plasma-treated nickel–titanium surface – an adequate platform for cell growth. J. Biomater. 2012, 26, 707–31.

[68] Jenko M, Godec M, Kocijan A, Rudolf R, Dolinar D, Ovsenik M, Gorenšek M, Zaplotnik R, Mozetic M. A new route to biocompatible Nitinol based on a rapid treatment with H_2/O_2 gaseous plasma. Appl. Surf. Sci. 2019, 473, 976–84.

[69] Jabłoński P, Niemiec W, Kaczmarek L, Hebda M, Krawiec H, Kyzioł K. Biopolymer-based coatings on the plasmochemical activated surface of NiTi alloy. The Int. Surfaces, Coatings and Interfaces Conference SurfCoat Korea, 2019; https://www.researchgate.net/profile/Piotr_Jablonski6/publication/332142603_Biopolymer-based_coatings_on_the_plasmochemical_activated_surface_of_NiTi_alloy/links/5ca31420a6fdccab2f67d3f6/Biopolymer-based-coatings-on-the-plasmochemical-activated-surface-of-NiTi-alloy.pdf.

[70] Sun T, Wang LP, Wang M. (Ti, O)/Ti and (Ti, O, N)/Ti composite coatings fabricated via PIIID for the medical application of NiTi shape memory alloy. J. Biomed. Mater. Res. B Appl. Biomater. 2011, 96, 249–60.

[71] Mirak M, ALizadeh M, Salahinejad E, Amini R. Zn–HA–TiO_2 nanocomposite coatings electrodeposited on a NiTi shape memory alloy. Surf. Interface Anal. 2015, 47, 176–83.

[72] Maleki-Ghaleh H, Khalil-Allafi J, Aghaie E, Siadati MH. Effect of TiO_2–Ti and TiO_2–TiN composite coatings on corrosion behavior of NiTi alloy. Surf. Interface Anal. 2015, 47, 99–104.

[73] Chakraborty R, Raza MS, Datta S, Saha P. Synthesis and characterization of nickel free titanium–hydroxyapatite composite coating over Nitinol surface through in-situ laser cladding and alloying. Surf. Coat. Technol. 2019, 358, 539–50.

[74] Chu CL, Ji HL, Guo C, Sheng XB, Dong YS, Lin PH, Hu T, Chu PK. Surface nanomechanical behavior of ZrN and ZrCN films deposited on NiTi shape memory alloy by magnetron sputtering. J. Nanosci. Nanotechnol. 2011, 11, 11176–80.

[75] Jin S, Zhang Y, Wang Q, Zhang D, Zhang S. Influence of TiN coating on the biocompatibility of medical NiTi alloy. Colloids Surf. B 2013, 101, 343–9.

[76] Czarnowska E, Borowski T, Sowińska A, Lelątko J, Oleksiak J, Kamiński J, Tarnowski M, Wierzchoń T. Structure and properties of nitrided surface layer produced on NiTi shape memory alloy by low temperature plasma nitriding. Appl. Surf. Sci. 2015, 334, 24–31.

[77] Maleki-Ghaleh H, Khalil-Allafi J, Sadeghpour-Motlagh M, Shakeri MS, Masoudfar S, Farrokhi A, Beygi Khosrowshahi Y, Nadernezhad A, Siadati MH, Javidi M, Shakiba M, Aghaie E. Effect of surface modification by nitrogen ion implantation on the electrochemical and cellular behaviors of super-elastic NiTi shape memory alloy. J. Mater. Sci. Mater. Med. 2014, 25, 2605–17.

[78] Ng KW, Man HC, Yue TM. Corrosion and wear properties of laser surface modified NiTi with Mo and ZrO_2. Appl. Surf. Sci. 2008, 254, 6725–30.

[79] Zhao T, Li Y, Liu Y, Zhao X. Nano-hardness, wear resistance and pseudoelasticity of hafnium implanted NiTi shape memory alloy. J. Mech. Behav. Biomed. Mater. 2012, 13, 174–84.

[80] Bell BC, Glocker DA. Radiopaque nano-porous Ta coatings for nitinol medical devices. J. Med. Dev. 2008, 2; doi: 10.1115/1.2936118.

[81] McNamara K, Kolaj-Robin O, Belochapkine S, Laffir F, Gandhi AA, Tofail SAM. Surface chemistry and cytotoxicity of reactively sputtered tantalum oxide films on NiTi plates. Thin Solid Films 2015, 589, 1–7.

[82] Siu HT, Man HC. Fabrication of bioactive titania coating on nitinol by plasma electrolytic oxidation. Appl. Surf. Sci. 2013, 274, 181–7.

[83] Cheng Y, Zheng YF. Influence of negative voltage on the structure and properties of DLC films deposited on NiTi alloys by PBII. J. Mater. Sci. 2006, 41, 4179–83.

[84] Ozeki K, Masuzawa T, Hirakuri KK. The wear properties and adhesion strength of the diamond-like carbon film coated on SUS, Ti and Ni-Ti with plasma pre-treatment. Biomed. Mater. Eng. 2010, 20, 21–35.

[85] Li Q, Xia YY, Tang JC, Wang RY, Bei CY, Zeng Y. In vitro and in vivo biocompatibility investigation of diamond-like carbon coated nickel-titanium shape memory alloy. Artif. Cells Blood Substit. Immobil. Biotechnol. 2011, 39, 137–42.

[86] Oshida Y. Hydroxyapatite – Synthesis and Applications. Momentum Press, New York, NY, USA, 2015.

[87] Iordanova I, Antonov V, Sprecher CM, Skulev HK, Gueorguiev B. Chapter 5: Plasma Sprayed Bioceramic Coatings on Ti-Based Substrates: Methods for Investigation of Their Crystallographic Structures and Mechanical Properties. In: Jazi H, ed., Advanced Plasma Spray Applications. In Tech Pub., Croatia, 2012, 99–120.

[88] Azarmi F, Coyle T, Mostaghimi J. Optimization of atmospheric plasma spray process parameters using a design of experiment for alloy 625 coatings. J. Thermal Spray Technol. 2008, 17, 144–55.

[89] Harsha S, Dwivedi D, Agarwal A. Performance of flame sprayed Ni-WC coating under abrasive wear conditions. J. Mater. Eng. Perform. 2008, 17, 104–10.

[90] Sarikaya O. Effect of some parameters on microstructure and hardness of alumina coatings prepared by the air plasma spraying process. Surf. Coat. Technol. 2005, 190, 388–93.

[91] Skulev H, Malinov S, Sha W, Basheer P. Microstructural and mechanical properties of nickel-base plasma sprayed coatings on steel and cast iron substrates. Surf. Coat Technol. 2005, 197, 177–84.

[92] Vilotijević M, Marković P, Zec S, Marinković S, Jokanović V. Hydroxyapatite coatings prepared by a high power laminar plasma jet. J. Mater. Process. Technol. 2011, 211, 996–1004.

[93] Quek CH, Khor KA, Cheang P. Influence of processing parameters in the plasma spraying of hydroxyapatite/Ti–6Al–4V composite coatings. J. Mater. Process. Technol. 1999, 89/90, 550–5.

[94] Altuncu E, Erdogan G, Ozturk S, Ustel E. Characterisation of plasma sprayed HAP coatings on Ti alloys, thermal spray 2011: Proceedings of the International Thermal Spray Conference (DVS-ASM), 2011, 2011, 1225–8.

[95] Brossa F, Lang E. Plasma Spraying – A Versatile Coating Technique. In: Gissler W, et al. ed., Mechanical and Materials Science: Advanced Techniques for Surface Engineering. Kluwer Academic Publishers, London, UK, 1992, 199–252.

[96] Iordanova I, Surtchev M, Forcey K. Metallographic and SEM investigation of the microstructure of thermally sprayed coatings on steel substrates. Surf. Coat Technol. 2001, 139, 118–26.

[97] Morris H, Ochi K. Hydroxyapatite-coated implants: a case for their use. J. Oral Maxillofac. Surg. 1998, 56, 1303–11.

[98] Frayssinet P, Primout I, Mathon D, Tourenne F, Rouquet N. Ca/P ratio and biocompatibility of hydroxyapatite coatings. J. Therm. Spray Technol. 2004, 13, 190–9.

[99] Marashi-Najafi F, Khalil-Allafi J, Etminanfar MR. Biocompatibility of hydroxyapatite coatings deposited by pulse electrodeposition technique on the Nitinol superelastic alloy. Mater. Sci. Eng. C Mater. Biol. Appl. 2017, 76, 278–86.

[100]   Etminanfar MR, Khalil-Allafi J, Sheykholeslami SOR. The effect of hydroxyapatite coatings on the passivation behavior of oxidized and unoxidized superelastic nitinol alloys. J. Mater. Eng. Perform. 2018, 27, 501–9.

[101]   Kocijan A. Electrodeposition of a hydroxyapatite coating on a biocompatible NiTi alloy. Materiali in Tehnologije 2018, 52, 67–70.

[102]   Deng B, Bruzzaniti A, Cheng GJ. Enhancement of osteoblast activity on nanostructured NiTi/hydroxyapatite coatings on additive manufactured NiTi metal implants by nanosecond pulsed laser sintering. Int. J. Nanomed. 2018, 13, 8217–30.

[103]   Sattar T, Manzoor T, Khalid FA, Akmal M, Saeed G. Improved in vitro bioactivity and electrochemical behavior of hydroxyapatite-coated NiTi shape memory alloy. J. Mater. Sci. 2019, 54, 7300–6.

[104]   Michiardi A, Aparicio C, Planell JA, Gil FJ. New oxidation treatment of NiTi shape memory alloys to obtain Ni-free surfaces and to improve biocompatibility. J. Biomed. Mater. Res. B Appl. Biomater. 2006, 77, 249–56.

[105]   Danilov A, Tuukkanen T, Tuukkanen J, Jämsä T. Biocompatilibity-related surface characteristics of oxidized NiTi. J. Biomed. Mater. Res. A 2007, 82, 810–9.

[106]   Wu S, Liu X, Chan YL, Ho JP, Chung CY, Chu PK, Chu CL, Yeung KW, Lu WW, Cheung KM, Luk KD. Nickel release behavior, cytocompatibility, and superelasticity of oxidized porous single-phase NiTi. J. Biomed. Mater. Res. A 2007, 81, 948–55.

[107]   Chrzanowski W, Abou Neel EA, Armitage DA, Knowles JC. Surface preparation of bioactive Ni-Ti alloy using alkali, thermal treatments and spark oxidation. J. Mater. Sci. Mater. Med. 2008, 19, 1553–7.

[108]   Lelatko J, Goryczka T, Paczkowski P, Wierzchoń T, Morawiec H. TEM studies of the nitrided/oxided Ni-Ti surface layer. J. Microsc. 2010, 237, 435–8.

[109]   Vojtěch D, Voděrová M, Fojt J, Novák P, Kubásek T. Surface structure and corrosion resistance of short-time heat-treated NiTi shape memory alloy. Appl. Surf. Sci. 2010, 257, 1573–82.

[110]   Xu JL, Zhong ZC, Yu DZ, Liu F, Luo JM. Effect of micro-arc oxidation surface modification on the properties of the NiTi shape memory alloy. J. Mater. Sci. Mater. Med. 2012, 23, 2839–46.

[111]   Aun DP, Houmard M, Mermoux M, Latu-Romain L, Joud J-C, Berthomé G, Buono VTL. Development of a flexible nanocomposite TiO$_2$ film as a protective coating for bioapplications of superelastic NiTi alloys. Appl. Surf. Sci. 2016, 375, 41–9.

[112]   Sukuroglu EE, Sukuroglu S, Akar K, Totik Y, Efeoglu I, Arslan E. The effect of TiO$_2$ coating on biological NiTi alloys after micro-arc oxidation treatment for corrosion resistance. Proc. Inst. Mech. Eng. H 2017, 231, 699–704.

[113]   Viswanathan S, Mohan L, John S, Bera P, Anandan C. Effect of surface finishing on the formation of nanostructure and corrosion behavior of Ni–Ti alloy. Surf. Interface Anal. 2017, 49, 450–6.

[114]   Dehghanghadikolaei A, Ibrahim H, Amerinatanzi A, Hashemi M, Moghaddam NS, Elahinia M. Improving corrosion resistance of additively manufactured nickel–titanium biomedical devices by micro-arc oxidation process. J. Mater. Sci. 2019, 54, 7333–55.

[115]   Boessler C et al., The effect of electropolishing on torque and force during simulated root canal preparation with ProTaper shaping files. J. Endod. 2009, 35, 102–6.

[116]   Lopes HP et al., Effects of electropolishing surface treatment on the cyclic fatigue resistance of BioRace nickel-titanium rotary instruments. J. Endod. 2010, 36, 1653–7.

[117]   Shenhar A et al., Microstructure and fretting behavior of hard TiN-based coatings on surgical titanium alloys. Ceramics Int. 2000, 26, 709–13.

[118]   Huang H-H et al., Corrosion and cell adhesion behavior of TiN-coated and ion-nitrided titanium for dental applications. Appl. Surf. Sci. 2005, 244, 252–6.

[119]   Li U-M et al., Study of the effects of thermal nitriding surface modification of nickel titanium
         rotary instruments on the wear resistance and cutting efficiency. J. Dental Sci. 2006, 1, 53–8.
[120]   Lelatko J et al., TEM studies of the nitrided/oxided Ni-Ti surface layer. J. Microsc. 2010, 237,
         435–8.
[121]   Gavini G et al., Cyclic fatigue resistance of rotary nickel-titanium instruments submitted to
         nitrogen ion implantation. J. Endod. 2010, 36, 1183–6.
[122]   Wolle CF e al., The effect of argon and nitrogen ion implantation on nickel-titanium rotary
         instruments. J. Endod. 2009, 35, 1558–62.
[123]   Shevchenko N et al., Studies of surface modified NiTi alloy. Appl. Surf. Sci. 2004, 235,
         126–31.
[124]   Kim JW et al., Effect of cryogenic treatment on nickel-titanium endodontic instruments.
         Int. Endod. J. 2005, 38, 364–71.
[125]   Vinothkumar TS et al., Influence of deep dry cryogenic treatment on cutting efficiency
         and wear resistance of nickel-titanium rotary endodontic instruments. J. Endod. 2007, 33,
         1355–8.
[126]   George GK et al., An in vitro evaluation of the effect of deep dry cryotreatment on the cutting
         efficiency of three rotary nickel titanium instruments. J. Conserv. Dent. 2011, 14, 169–72.
[127]   Vijaykumar S et al., Effect of deep cryogenic treatment on machinability of NiTi shape
         memory alloys in electro discharge machining. Appl. Mech. Mater. 2014, 592/594, 197–201.
[128]   Caetano DAS. Corrosion resistance of diamond-like carbon coated rotary endodontic
         instruments – pilot study. 2018; https://docplayer.com.br/157467182-Corrosion-resistance-
         of-diamond-like-carbon-coated-rotary-endodontic-instruments-pilot-study.html.
[129]   Oshida Y et al., Microanalytical characterization and surface modification of TiNi orthodontic
         archwires. Biomed. Mater. Eng. 1992, 2, 51–69.
[130]   Iijima M et al., Corrosion behavior of ion implanted nickel-titanium orthodontic wire in
         fluoride mouth rinse solutions. Dent. Mater. J. 2010, 29, 53–8.
[131]   Liu J-K et al., Effect of titanium nitride/titanium coatings on the stress corrosion of nickel–
         titanium orthodontic archwires in artificial saliva. Appl. Surf. Sci. 2014, 317, 974–81.
[132]   Sugisawa H et al., Corrosion resistance and mechanical properties of titanium nitride plating
         on orthodontic wires. Dent. Mater. J. 2018, 37, 286–92.
[133]   Gil FJ et al., Improvement of the friction behaviour of NiTi orthodontic archwires by nitrogen
         diffusion. Biomed. Mater. Eng. 1998, 8, 335–42.
[134]   Gil FJ et al., Inhibition of Ni release from NiTi and NiTiCu orthodontic archwires by nitrogen
         diffusion treatment. J. Appl. Biomater. Biomech. 2004, 2, 151–5.
[135]   Katic V et al., Effect of pH, fluoride and hydrofluoric acid concentration on ion release from
         NiTi wires with various coatings. Dent. Mater. J. 2017, 36, 149–56.
[136]   Wichelhaus A et al., The effect of surface treatment and clinical use on friction
         in NiTi orthodontic wires. Dent. Mater. 2005, 21, 938–45.
[137]   Espinar E et al., Reduction of Ni release and improvement of the friction behaviour of NiTi
         orthodontic archwires by oxidation treatments. J. Mater. Sci. Mater. Med. 2011, 22, 1119–25.
[138]   Čolić M, Tomić S, Rudolf R, Marković E, Šćepan I. Differences in cytocompatibility, dynamics
         of the oxide layers' formation, and nickel release between superelastic and thermo-
         activated nickel–titanium archwires. J. Mater. Sci. Mater. Med. 2016, 128; doi: 10.1007/
         s10856-016-5742-1.
[139]   Cobb NW III et al., Efficiency of multi-strand steel, superelastic Ni-Ti and ion-implanted Ni-Ti
         archwires for initial alignment. Clin. Orthod. Res. 1998, 1, 12–9.
[140]   Iijima M et al., Corrosion behavior of ion implanted nickel-titanium orthodontic wire in
         fluoride mouth rinse solutions. Dent. Mater. J. 2010, 29, 53–8.

[141] Kobayashi S et al., Dissolution effect and cytotoxicity of diamond-like carbon coatings on orthodontic archwires. J. Mater. Sci. Mater. Med. 2007, 18, 2263–8.

[142] Huang SY et al., Coating NiTi archwires with diamond-like carbon films: reducing fluoride-induced corrosion and improving frictional properties. J. Mater. Sci. Mater. Med. 2013, 24, 2287–92.

[143] Katić V et al., Influence of surface layer on mechanical and corrosion properties of nickel – titanium orthodontic wires. Angle Orthod. 2014, 84, 1041–8.

[144] Katić V et al., Influence of various fluoride agents on working properties and surface characteristics of uncoated, rhodium coated and nitrified nickel-titanium orthodontic wires. Acta Odontol. Scand. 2015, 73, 241–9.

[145] Asiry MA et al., Influence of epoxy, polytetrafluoroethylene (PTFE) and rhodium surface coatings on surface roughness, nano-mechanical properties and biofilm adhesion of nickel titanium (Ni-Ti) archwires. Mater. Res. Express 2018, 5: doi: 10.1088/2053-1591/aaabe5.

[146] Trépanier C et al., Effect of modification of oxide layer on NiTi stent corrosion resistance. J. Biomed. Mater. Res. 1998, 43, 433–40.

[147] Thierry B et al., Effect of surface treatment and sterilization processes on the corrosion behaviour of NiTi shape memory alloy. J. Biomed. Mater. Res. 2000, 51, 685–93.

[148] Standard practice for surface preparation and making of metallic surgical implants, ASTM Standard F86–91. ASTM, Conshohocken, PA, 1991.

[149] Es-Souni M et al., Human gingival fibroblast response to electropolished NiTi surfaces. J. Biomed. Mater. Res. A 2007, 80, 159–66.

[150] Liu XM et al., Nano-scale surface morphology, wettability and osteoblast adhesion on nitrogen plasma-implanted NiTi shape memory alloy. J. Nanosci. Nanotechnol. 2009, 9, 3449–54.

[151] Yang D et al., The molecular mechanism for effects of TiN coating on NiTi alloy on endothelial cell function. Biomaterials 2014, 35, 6195–205.

[152] Subramanian B et al., A comparative study of titanium nitride (TiN), titanium oxy nitride (TiON) and titanium aluminum nitride (TiAlN), as surface coatings for bio implants. Surf. Coat. Technol. 2011, 205, 5014–20.

[153] Zhou X et al., Mechanical property and microstructure evolution of nitrogen-modified Ti-6Al-4V alloy with core-shell structure by hot compression. Mater. Charact. 2018, 142, 270–5.

[154] Horandghadim N et al., Effect of Ta_2O_5 content on the osseointegration and cytotoxicity behaviors in hydroxyapatite-Ta_2O_5 coatings applied by EPD on superelastic NiTi alloys. Mater. Sci. Eng. C Mater. Biol. Appl. 2019, 102, 683–95.

[155] Filip P et al., Structure and surface of NiTi human implants. Biomaterials 2001, 22, 2131–8.

[156] Firstov GS et al., Surface oxidation of NiTi shape memory alloy. Biomaterials 2002, 23, 4863–71.

[157] Michiardi A, Engel E, Aparicio C, Planell JA, Gil FJ. Oxidized NiTi surfaces enhance differentiation of osteoblast-like cells. J. Biomed. Mater. Res. A 2008, 85, 108–14.

[158] Jin G et al., Synergistic effects of dual Zn/Ag ion implantation in osteogenic activity and antibacterial ability of titanium. Biomaterials 2014, 35, 7699–713.

[159] Mändl S et al., Plasma immersion ion implantation treatment of medical implants. Surf. Coat. Technol. 2001, 142/144, 1046–50.

[160] Bansiddhi A et al., Porous NiTi for bone implants: a review. Acta Biomater. 2008, 4, 773–82.

[161] Kwok DTK et al., Surface Treatments of Nearly Equiatomic NiTi Alloy (Nitinol) for Surgical Implants. In: Laskovski A, ed., Biomedical Engineering, Trends in Materials Science, IntechOpen, 2011; doi: 10.5772/13212, 269-282.

[162]   Erdemir A et al., Tribology of diamond-like carbon films: recent progress and future prospects. J. Phys. D Appl. Phys. 2006, 39; https://iopscience.iop.org/article/10.1088/0022-3727/39/18/R01/pdf.

[163]   Donnet C, Erdemir A. Tribology of Diamond-Like Carbon Films – Fundamentals and Applications. Springer Pub., New York, NY, USA, 2008.

[164]   Marin E et al., Diffusive thermal treatments combined with PVD coatings for tribological protection of titanium alloys. Mater. Des. 2016, 89, 314–22.

[165]   Hang R et al., Biological response of endothelial cells to diamond-like carbon-coated NiTi alloy. J. Biomed. Mater. Res. A 2012, 100, 496–506.

[166]   Chen MF et al., Bioactive NiTi shape memory alloy used as bone bonding implants. Mater. Sci. Eng. C 2004, 24, 497–502.

[167]   Shi P et al., Preparation of titania–hydroxyapatite coating on NiTi via a low-temperature route. Mater. Lett. 2006, 60, 1996–9.

[168]   Qiu D et al., Characterization and corrosion behavior of hydroxyapatite/zirconia composite coating on NiTi fabricated by electrochemical deposition. Appl. Surf. Sci. 2010, 257, 1774–8.

[169]   Qiu D et al. Preparation and characterization of hydroxyapatite/titania composite coating on NiTi alloy by electrochemical deposition. Surf. Coat. Technol. 2011, 205, 3280–4.

[170]   Maleki-Ghaleh H et al., Effect of hydroxyapatite coating fabricated by electrophoretic deposition method on corrosion behavior and nickel release of NiTi shape memory alloy. Mater. Corros. 2014, 65, 725–32.

[171]   Karimi E et al., Electrophoretic deposition of double-layer HA/Al composite coating on NiTi. Mater. Sci. Eng. C 2016, 58, 882–90.

[172]   Trépanier C et al., Preliminary investigation of the effects of surface treatments on biological response to shape memory NiTi stents. J. Biomed. Mater. Res. 1999, 48, 165–71.

[173]   Chu CL et al., Microstructure, nickel suppression and mechanical characteristics of electropolished and photoelectrocatalytically oxidized biomedical nickel titanium shape memory alloy. Acta Biomater. 2009, 5, 2238–45.

[174]   Zhang K et al., Surface modification of implanted cardiovascular metal stents: from anti-thrombosis and anti-restenosis to endothelialization. J. Biomed. Mater. Res. A 2013, 102, 588–609.

[175]   Gill P et al., Surface modification of Ni–Ti alloys for stent application after magnetoelectropolishing. Mater. Sci. Eng. C Mater. Biol. 2015, 50, 37–44.

[176]   Kim J et al., Optimization of electropolishing on NiTi alloy stents and its influence on corrosion behavior. J. Nanosci. Nanotechnol. 2017, 17, 2333–9.

[177]   Pérez LM et al., Effect of Nitinol surface treatments on its physico-chemical properties. J. Biomed. Mater. Res. 2009, 91B, 337–47.

[178]   Rokicki R et al., Modifying metallic implants with magnetoelectropolishing. Med. Dev. Diagn. Ind. 2008; http://www.mddionline.com/article/modifying-metallic-implants-magnetoelectropolishing.

[179]   Neelakantan L et al., Selective surface oxidation and nitridation of NiTi shape memory alloys by reduction annealing. Corros. Sci. 2009, 51, 635–41.

[180]   Zorn G et al., Tailoring the surface of NiTi alloy using PIRAC nitriding followed by anodization and phosphonate monolayer deposition. Chem. Mater. 2008, 20, 5368–73.

[181]   Ion R et al., Nitride coating enhances endothelialization on biomedical NiTi shape memory alloy. Mater. Sci. Eng. C, 2016, 62, 686–91.

[182]   Shen Y et al., Investigation of surface end othelialization on biomedical nitinol (NiTi) alloy: effects of surface micropatterning combined with plasma nanocoatings. Acta Biomater. 2009, 5, 3593–604.

[183]   Haider W et al., Surface modifications of nitinol. J. Long Term Eff. Med. Implants. 2009, 19, 113–22.

[184]   Roy RK et al., Biomedical applications of diamond-like carbon coatings: a review. J. Biomed. Mater. Res. Part B Appl. Biomater. 2007, 83B, 72–84.

[185]   Shin H-S et al., Biocompatible PEG grafting on DLC-coated nitinol alloy for vascular stents. J. Bioact. Compat. Polym. 2009, 24, 316–28.

[186]   Lee BS et al., Surface grafting of blood compatible zwitterionic poly(ethylene glycol) on diamond-like carbon-coated stent. J. Mater. Sci. Mater. Med. 2011, 22, 507–14.

[187]   Kim JH et al., Comparison of diamond-like carbon-coated nitinol stents with or without polyethylene glycol grafting and uncoated nitinol stents in a canine iliac artery model. Br. J. Radiol. 2011, 84, 210–5.

[188]   Nakamura T, Shimizu Y, Ito Y, Matsui T, Okumura N, Takimoto Y, Ariyasu T, Kiyotani T. A new thermal shape memory Ti-Ni alloy stent covered with silicone. ASAIO J. 1992, 38, M347–50.

[189]   Chepeleva E et al., Distribution of nickel after modified nitinol stent implantation in animals. AIP Conf. Proceed. 1882, 020013, 2017; https://doi.org/10.1063/1.5001592.

[190]   Maleki-Ghaleh H et al., Hydroxyapatite coating on NiTi shape memory alloy by electrophoretic deposition process. Surf. Coat. Technol. 2012, 208, 57–63.

[191]   Dunne CF et al., Deposition of hydroxyapatite onto shape memory NiTi wire. Mater. Lett. 2016, 176, 185–8.

[192]   Shao AL et al., Electrochemistry properties of multilayer TiN/Ti coatings on NiTi alloy for cardiac occluder application. Surf. Coat. Technol. 2013, 228, S257–261.

[193]   Müller CW et al., Electromagnetic induction heating of an orthopaedic nickel-titanium shape memory device. J. Orthop. Res. 2010, 28, 1671–6.

[194]   Wu S et al., Electrochemical stability of orthopedic porous NiTi shape memory alloys treated by different surface modification techniques. J. Electrochem. Soc. 2009, 156, C187–94.

[195]   Hedayat A et al., Phase transformation in carbon-coated nitinol, with application to the design of a prosthesis for the reconstruction of the anterior cruciate ligament. J. Mater. Sci. Mater. Med. 1992, 3, 65–74.

[196]   Witkowska J et al., NiTi shape-memory alloy oxidized in low-temperature plasma with carbon coating: characteristic and a potential for cardiovascular applications. Appl. Surf. Sci. 2017, 421, 89–96.

[197]   Ahmed RA et al., Improvement of corrosion resistance and antibacterial effect of NiTi orthopedic materials by chitosan and gold nanoparticles. Appl. Surf. Sci. 2014, 292, 390–9.

Chapter 13
Sterilization

Although sterilization is performed for various types of dental devices and medical tools, the devices most frequently subjected to the routine sterilization is, without any exceptions, endodontic instruments. Endodontics is the aspect of dentistry involved in the treatment or precautions taken to maintain the vital pulp, moribund tooth, or nonvital tooth in the dental arch [1]. Reuse of instruments in dentistry is common, and endodontic treatment involves the use of instruments which are usually reused. During endodontic instrumentation, vital tissue, dentin shavings, necrotic tissue, bacteria, blood, blood by-products, and other potential irritants are encountered with accumulation of the debris on the flutes of the instruments [2]. Transfer of these debris from patient to patient and to dental staff is highly undesirable, as these debris can act as antigens and infecting agents capable of transmitting diseases such as Creutzfeldt–Jakob disease [3, 4]. The presence of debris has been reported to interfere with sterilization by forming a protective barrier that prevents the complete sterilization of the surface beneath [5]. If this debris is not removed, sterilization procedure may be pointless [6]. In the absence of adequate infection control procedures, there is a high probability of transmitting pathogenic microorganism through endodontic instruments [5]. Sterilization is a process to render an object free from viable microorganisms including bacterial spores and viruses. Sterilization of instruments in dentistry is required to protect patients and oral health care staff from cross contamination through instruments [7]. Resterilization (conducted between reuse occasions) is the repeated application of the terminal process designed to remove or destroy all viable forms of microbial life, including bacterial spores, to an acceptable sterility assurance level [8, 9]. Resterilization of instruments used on one patient for reuse on another has been common practice in dentistry. Some instruments used in endodontic procedures, such as files, reamers, and nickel–titanium files, are class I instruments as defined by the United States Food and Drug Administration and can be reused if sterility can be guaranteed [10]; however, there is now evidence that the sterilization process is complex and that if strict adherence to an effective protocol is not followed, contamination of instruments may result, as will be seen later in this chapter. It is very important for endodontic instruments to be cleaned, disinfected, and sterilized effectively. Various methods for cleaning and sterilizing endodontic instruments have been proposed. These include use of brushes and sponges, ultrasonic cleaning, glutaraldehyde solution, sodium hypochlorite (NaOCl), glass bead sterilization, dry heat, or steam sterilization. The miniature complex architecture of endodontic instruments makes precleaning and sterilization difficult [1, 8].

https://doi.org/10.1515/9783110666113-013

13.1 Process

The following methods are used to sterilize endodontic instruments normally: auto-claving, carbon dioxide laser sterilization, chemical sterilization (with glutaralde-hyde), glass bead sterilization, and gamma sterilization.

Hurtt et al. [11] analyzed to determine the best method of providing complete file sterility, including the metal shaft and plastic handle. Six test groups of 15 files were studied using *Bacillus stearothermophilus* as the test organism. Groups were "steril-ized" by glutaraldehyde immersion, steam autoclaving, and various techniques of salt sterilization. It was reported that (i) only proper steam autoclaving reliably pro-duced completely sterile instruments and (ii) salt sterilization and glutaraldehyde solutions may not be the adequate sterilization methods for endodontic hand files and should not be relied on to provide completely sterile instruments. Ferreira et al. [4] examined both qualitatively and quantitatively the presence of organic debris on endodontic files decontaminated and sterilized after use. Thirty K files #30 were used, 10 of which served as positive and negative control. Ten pig molars were instrumented using the crown-down technique by inserting one file in each root canal, totaling 25 files. The files in group 1 (n = 10) were sterilized by autoclave. Files in group 2 (n = 10) were placed in an ultrasonic bath with enzyme solution and then sterilized by autoclave. Those in group 3 (n = 5) were used but not sterilized, and finally the files in group 4 (n = 5) were neither used nor sterilized. The experimental and control files were subsequently stained with Van Gieson's solution and observed by optical micros-copy. A value representative of the amount of organic material still present on the file was then assigned in accordance with a previously established scale. The same mea-surement was carried out in the apical, middle, and cervical thirds of each file body while tables were formulated comparing the different groups. It was reported that (i) both experimental groups produced significantly inferior results compared to the files in the positive control group, and (ii) in comparing the experimental groups, the files immersed in ultrasonic bath with enzyme solution exhibited values that were inferior to those of the files which had not been subjected to this procedure. Raju et al. [12] compared four methods of (autoclaving, carbon dioxide laser sterilization, chemical sterilization with glutaraldehyde, and glass bead sterilization) endodontic files in dental practice and recommend the effective method among these. The present study was carried out on 100 K-files, 21 mm long, size 25. 20 K-files were taken as control group, and the remaining 80 files were divided into 4 groups of 20 files each in 4 dif-ferent modes of sterilization: Autoclave, glass bead, glutaraldehyde, and CO_2 laser. All the 100 files were pre-sterilized in an endodontic instrument box by autoclaving for 30 min at 121 °C at a pressure of 15 pounds, for standardization to eliminate any bias. Later the test files were divided into 5 groups of 20 files each and labeled as A (auto-clave), B (glass bead), G (glutaraldehyde), L (CO_2 laser), and C (control) and were numbered 1 to 20. The spore suspension was prepared by immersing the commer-cially available *Bacillus stearothermophillus* strips into thioglycollate medium broth and incubating at 55 °C for 48 h. Growth that occurred in the test tube was confirmed by doing Gram's stain that showed the presence of gram-positive *Bacillus stearothermophillus*. All the pre-sterilized files were contaminated with *Bacillus stearothermophillus* in a sterile Petri dish for 5 min. After 5 min of immersion, the files were transferred to another sterile Petri dish under vacuum hood safety with the help of a sterile tweezer, following which the files were dried in an incubator for 10 min at 37 °C and stored in an endodontic instrument box till they were sterilized by different methods. The 20 contaminated files in group A were placed in an endodontic instrument box and subjected to autoclave at 121 °C for 15 min at a pressure of 15 pounds. The 20 contaminated files in group B were taken in 4 batches of 5 files each and wiped for 10 s with a gauze soaked with surgical spirit and placed in the periphery of the glass bead sterilizer and sterilized for 45 s at 240 °C. The 20 contaminated files in group G were placed in a sterile plastic container containing 2.4% glutaraldehyde solution and were left in it for 12 h for cold (chemical sterilization). The 20 contaminated files in group L were irradiated for 3 s per surface at 10 watts using CO_2 laser system. The laser beam was moved along the length of the instrument during the 3-s period. A sterile tweezer was used to hold the handle of the file and change the surface for exposure. After completion of sterilization, the shaft of the instrument was removed from the handle by means of a sterile autoclaved wire cutter and each file was placed in separate tubes containing thioglycollate medium with the help of a sterile tweezer to check for any microbial growth and the efficacy of sterilization method. The 20 contaminated files in group C (control group) were put in separate tubes containing thioglycollate medium by the method described above without doing any sterilization. It was found that (i) the endodontic files sterilized by autoclaving in an instrument box at 121 °C for 15 min at a pressure of 15 pounds (group A) showed total sterility, (ii) the files subjected to sterilization by glass bead sterilizer after wiping for 10 s with a gauze soaked with surgical spirit and sterilized for 45 s at 240 °C (group B) showed presence of turbidity in 2 test tubes, (iii) incomplete sterilization to the range of 10% was observed when the files were sterilized in glass bead sterilizer, (iv) the endodontic files sterilized by immersing in glutaraldehyde for 24 h (group G) showed sterilization up to only 80%. This method showed contamination of 4 files after incubation, (v) the files on sterilization by CO_2 laser for 3 s per surface at 10 watts (group L) showed 100% sterility. There was total sterility seen by this method of sterilization, (vi) statistical analysis of the four different sterilization methods showed a statistically significant difference between groups with regard to their efficacies in sterilization, and (vii) autoclave and laser sterilization showed 100% sterilization as compared to chemical and glass bead sterilization. Based on these findings, it was concluded that (viii) autoclaving and exposing to laser give complete sterilization, whereas glass bead sterilizer can be used as an alternative when these two methods are not available, though autoclave is an effective method for sterilizing endodontic files, the time taken by it to sterilize is more. Kumar et al. [13] determined the effectiveness of various disinfectants and sterilization techniques for disinfection and resterilization of dental burs and endodontic files. Disinfectants used were Quitanet plus, glutaraldehyde, glass bead sterilizer, and autoclave. It was reported that (i) the endodontic files and burs sterilized by autoclaving and glutaraldehyde showed complete sterilization, (ii) burs and files immersed in glutaraldehyde (2.4%) for 12 h showed complete sterilization, whereas Quitanet plus solution and glass bead sterilizer showed incomplete sterilization, concluding that (iii) autoclaving and glutaraldehyde (2.4%) showed complete sterilization, and (iv) other methods cannot be relied upon for sterilization. The reuse of instruments in the field of endodontics leads to cross infection due to contamination with microbes as the instruments come into direct contact with saliva, blood, and infected tissues. Yenni et al. [14] compared the effectiveness of four different methods of sterilizing contaminated endodontic files. A total of 48 stainless-steel K files were divided into four groups based on the sterilization method followed – group A: autoclave, group B: glass bead sterilization, group C: glutaraldehyde, and group D: Quitanet plus (aldehyde-free solution). In all the tested groups, half of the files were contaminated with *Escherichia coli* and remaining with *Enterococcus faecalis*. Then, presterilization colony counts were recorded, followed by sterilization through respective methods. Later, the sterilized files were rinsed with distilled water and 100 ul of the diluted concentration was transferred and cultured onto the respective agar plates to determine the total microbial reduction. It was found that the (i) autoclave showed complete effectiveness in reducing the microbial count followed by Quitanet Plus, glass bead sterilizer, and glutaraldehyde, and (ii) autoclave is considered to be the best sterilization technique to prevent cross-infection in endodontic therapy.

13.2 Influences

13.2.1 Surface characteristics

There are studies concerning possible changes after the sterilization. Thierry et al. [15] studied the effect of dry heat, steam autoclaving, ethylene oxide, peracetic acid, and plasma-based sterilization techniques on the surface properties of NiTi. After processing electropolished NiTi disks with these techniques, surface analyses were performed by Auger electron spectroscopy (AES), atomic force microscopy (AFM), and contact angle measurements. It was reported that (i) AES analyses revealed a higher Ni concentration (6–7 vs 1%) and a slightly thicker oxide layer on the surface for heat and ethylene oxide processed materials, (ii) studies of surface topography by AFM showed up to a threefold increase of the surface roughness when disks were dry heat sterilized, and (iii) an increase of the surface energy of up to 100% was calculated for plasma treated surfaces indicating that (iv) some surface modifications are induced by sterilization procedures. Morawiec et al. [16] briefly mentioned that (i) sterilization of the NiTi alloy in boiling water or steam caused passivation, which results in an amorphous 3.5 nm thick TiO_2 layer on the surface, (ii) between the surface and the matrix a transition layer of Ni_2O_3 and NiO was observed, using the X-ray photoelectron spectroscopy, and (iii) differences in sterilization conditions affect the amount of metallic nickel on the surface. Valois et al. [17] evaluated the surface of rotary NiTi files after multiple autoclave cycles. Two different types of rotary NiTi (Greater Taper and ProFile) were attached to a glass base. After 1, 5, and 10 autoclave cycles the files were positioned in the atomic force microscope. The analyses were performed on 15 different points. The same files were used as control before any autoclave cycle. The following vertical topographic parameters were measured: arithmetic mean roughness, maximum height, and root mean square (RMS). It was reported that (i) all topographic parameters were higher for both Greater Taper and ProFile after 10 cycles compared with the control, and (ii) ProFile also showed higher topographic parameters after 5 cycles compared with the control indicating that multiple autoclave cycles increase the depth of surface irregularities located on rotary NiTi files. Spagnulo et al. [18] evaluated the effects of repeated autoclave sterilization cycles on surface topography of conventional NiTi and titanium nitride (TiN)-coated rotary instruments. A total of 60 NiTi rotary instruments, 30 ProTaper (Dentsply Maillefer) and 30 TiN-coated AlphaKite (Komet/Gebr. Brasseler) were analyzed. Instruments were evaluated in the as-received condition and after 1, 5, and 10 sterilization cycles. After sterilization, the samples were observed using scanning electron microscope (SEM), and surface chemical analysis was performed on each instrument with energy dispersive X-ray spectroscopy (EDS). Moreover, the samples were analyzed by AFM, and roughness average (Ra) and the root mean square value of the scanned surface profiles were recorded. It was found that the (i) SEM observations revealed the presence of pitting and deep milling marks in all instruments, (ii) EDS analysis confirmed that both types of instruments were composed mainly of nickel and titanium, whilst AlphaKite had additional nitride, (iii) after multiple autoclave sterilization cycles, SEM examinations revealed an increase in surface alterations, and EDS values indicated changes in chemical surface composition in all instruments, and (iv) Ra and RMS values of ProTaper significantly increased after 5 and 10 cycles with respect to the as-received instruments, whilst AlphaKite showed significant differences compared with the controls after 10 cycles. Based on these findings, it was concluded that multiple autoclave sterilization cycles modified the surface topography and chemical composition of conventional and TiN-coated NiTi rotary instruments.

Cai [19] investigated the effect of irrigation on the surface roughness and fatigue resistance of HyFlex and M3 controlled memory (CM) wire NiTi instruments. Two new files of each brand were analyzed by AFM. Then, the instruments were dynamically immersed in either 5.25% NaOCl or 17% ethylene diamine tetra-acetic acid (EDTA) solution for 10 min, followed by AFM analysis. The Ra and RMS values were analyzed. Then, 36 files of each brand were randomly assigned to 3 groups ($n = 12$). Group 1 (the control group) was composed of new instruments. Groups 2 and 3 were dynamically immersed in 5.25% NaOCl and 17% EDTA solutions for 10 min, respectively. The number of rotations to failure for various groups was analyzed using the one-way analysis of variance software. It was mentioned that (i) for M3 files, the Ra and RMS values significantly increased after the immersion, (ii) for the HyFlex file, the Ra and RMS values significantly increased only in EDTA, but not NaOCl, and (iii) the resistance to cyclic fatigue of both HyFlex and M3 files did not significantly decrease by immersing in 5.25% NaOCl and 17% EDTA solutions concluding that except the HyFlex files immersed in NaOCl, the surface roughness of other files exposed to irrigants increased; however, a change in the surface tomography of CM wire instruments caused by contact with irrigants for 10 min did not trigger a decrease in cyclic fatigue resistance. Yılmaz et al. [20] compared the effect of autoclave cycles on the surface topography and roughness of HyFlex CM and HyFlex EDM instruments using AFM analysis. Eight new files of each brand were subdivided into four subgroups ($n = 2$/ each subgroup). One group was allocated as the control group and not subjected to autoclave sterilization. The other 3 groups were subjected to different numbers (1, 5, and 10) of autoclave sterilization cycles. After the cycle instruments were subjected to AFM analysis, Ra and the RMS values were chosen to investigate the surface features of endodontic files. It was obtained that (i) the lowest Ra and RMS values were observed in the HyFlex EDM files that served as the control and in those subjected to a single cycle of autoclave sterilization, (ii) the highest Ra and RMS values were observed in the HyFlex CM and HyFlex EDM files that were subjected to 10 cycles of autoclave sterilization, and (iii) the surface roughness values of the HyFlex CM group showed a significant increase after 10 autoclave cycles, whereas those of the HyFlex EDM group exhibited a significant change after 5 autoclave cycles. It was then concluded that (iv) although the initial surface roughness values of the HyFlex EDM files were lower than those of the HyFlex CM files, the surface roughness values of the EDM files showed a statistically significant increase after 5 cycles of autoclave sterilization, and (v) in contrast, the surface roughness values of the HyFlex CM files did not increase until 10 cycles of autoclave sterilization. Baldin et al. [21] evaluated the influence of sterilization on a hybrid coating obtained from a sol composed of alkoxysilane tetraethoxysilane and organoalkoxysilane methyltriethoxysilane containing 10% (mass) of hydroxyapatite particles. The coating was obtained by dip coating, by applying two layers (protective/bioactive), which were cured at different temperatures (450 °C and 60 °C). The effects of sterilization on the superficial, electrochemical, bioactive, and mechanical properties of the coating were evaluated by performing different sterilization processes, namely, steam autoclave, hydrogen peroxide plasma, and ethylene oxide. The main results were the following: (i) all sterilization processes caused significant morphological changes in the hybrid coating, (ii) the autoclaved sample presented the highest structural chemical changes, and, consequently, the highest degradability, even though it had a superior bioactive behavior in relation to the other samples, and (iii) in addition, the sterilization processes influenced the electrochemical behavior of the hybrid coating and altered the mechanical resistance to abrasion, thus presenting lower wear performance in relation to the un-sterilized sample.

13.2.2 Fatigue

As seen in the previous section, surface roughness increases by dry heat sterilization [15], autoclave sterilization [18, 20] or irrigation [19]. It is, in general, believed
that fatigue crack is initiated at a very localized surface irregular site, so that roughening surface might be harmful for fatigue behavior. Mize at el., [22] evaluated the
ability of heat treatment as a result of autoclave sterilization to extend the life of NiTi
rotary endodontic instruments by reducing the effect of cyclic fatigue using 280 size
40 LightSpeed instruments. Instruments were cycled in artificial canals with angles
of curvature of 30 degrees and either 2 or 5 mm radii of curvature. The sequence of
cycling to 25% of the predetermined cycles-to-failure limit followed by sterilization
was repeated until the instruments failed. It was reported that (i) no significant
increases in cycles to failure were observed between groups for either experimental
protocol when instruments were evaluated at a similar radius, (ii) significant differences in cycles to failure were only observed when instruments cycled to failure in the
artificial canal with 2 mm radius were compared with instruments cycled to failure
in the artificial canal of 5 mm radius, and (iii) scanning electron microscopic photos
showed crack initiation and propagation in all instruments that were cycled to a percentage of the predetermined cycles-to-failure limit concluding that heat treatment
as a result of autoclave sterilization does not extend the useful life of nickel–titanium
instruments. Viana et al. [23] evaluated the effect of repeated sterilization cycles in dry
oven or autoclave on the mechanical behaviors and fatigue resistance of rotary endodontic NiTi instruments. New NiTi instruments were subjected to five consecutive
sterilization cycles in a dry oven or steam autoclave. It was found that (i) sterilization
procedures resulted in no significant changes in Vickers microhardness, nor in the
parameters describing the mechanical behavior of the wires; however, the number of
cycles to failure (N_{CF}) was statistically higher for all instruments after dry heat or autoclave sterilization cycles, and (ii) in the instruments previously fatigued to one half of
their fatigue life, autoclave sterilization gave rise to an increase of 39% in the remaining N_{CF}. Hence, it was concluded that (i) changes in the mechanical properties of NiTi
endodontic instruments after five cycles of commonly used sterilization procedures
were insignificant, and (ii) the sterilization procedures are safe as they produced a
significant increase in the fatigue resistance of the instruments. Multiple autoclave
cycle effects on cyclic fatigue of GT Series X files (Dentsply Tulsa Dental Specialties,
Tulsa, OK) and Twisted Files (TFs) (SybronEndo, Orange, CA) were investigated [24].
A jig using a 5-mm radius curve with 90° of maximum file flexure was used to induce
cyclic fatigue failure. Files ($n = 10$) representing each experimental group (GT Series
X 20/0.04 and 20/0.06; TFs 25/0.04 and 25/0.06) were first tested to establish baseline mean cycles to failure (M_{CF}). Experimental groups ($n = 20$) were then cycled to
25% of the established baseline M_{CF} and then autoclaved. Additional autoclaving was
accomplished at 50% and 75% of M_{CF} followed by continual testing until failure. It
was obtained that (i) the GT Series X (20/0.04 and 20/0.06) files showed no significant

difference in M_{CF} for experimental versus control files, (ii) the TFs (25/0.04) showed no significant difference in M_{CF} between experimental and control groups, and (iii) however, the TFs (25/0.06) experimental group showed a significantly lower M_{CF} compared with the controls. Based on these results, it was suggested that (i) under the conditions of this evaluation, autoclave sterilization significantly decreased cyclic fatigue resistance of one of the four file groups tested, and (ii) repeated autoclaving significantly reduced the M_{CF} of 25/0.06 TFs; however, 25/0.04 TFs and both GT Series X files tested were not significantly affected by the same conditions.

Plotino et al. [25] evaluated the effect of autoclave sterilization on cyclic fatigue resistance of rotary endodontic instruments made of traditional and new NiTi alloys. Four NiTi rotary endodontic instruments of the same size (tip diameter 0.40 mm and constant .04 taper) were selected: K3, Mtwo, Vortex, and K3 XF prototypes. Each group was then divided into two subgroups, unsterilized instruments and sterilized instruments. The sterilized instruments were subjected to 10 cycles of autoclave sterilization. Twelve files from each different subgroup were tested for cyclic fatigue resistance. Means and standard deviations of N_{CF} and fragment length of the fractured tip were calculated for each group. It was reported that (i) comparing the results between unsterilized and sterilized instruments for each type of file, differences were statistically significant only between sterilized and unsterilized K3XF files (762 vs 651 N_{CF}), (ii) the other instruments did not show significant differences in the mean N_{CF} as a result of sterilization cycles (K3, 424 vs 439 N_{CF}; Mtwo, 409 vs 419 N_{CF}; Vortex, 454 vs 480 N_{CF}), and (iii) comparing the results among the different groups, K3 XF (either sterilized or not) showed a mean N_{CF} significantly higher than all other files concluding that repeated cycles of autoclave sterilization do not seem to influence the mechanical properties of NiTi endodontic instruments except for the K3 XF prototypes of rotary instruments that demonstrated a significant increase of cyclic fatigue resistance. Özyürek et al. [26] compared the cyclic fatigue resistances of ProTaper Universal (PTU), ProTaper Next (PTN), and ProTaper Gold (PTG) and the effects of sterilization by autoclave on the cyclic fatigue life of NiTi instruments. Eighty PTU, 80 PTN, and 80 PTG were included to the present study. Files were tested in a simulated canal. Each brand of the NiTi files were divided into 4 subgroups: group 1, as-received condition; group 2, presterilized instruments exposed to 10 times sterilization by autoclave; group 3, instruments tested were sterilized after being exposed to 25%, 50%, and 75% of the M_{CF}, then cycled fatigue test was performed; group 4, instruments exposed to the same experiment with group 3 without sterilization. The N_{CF} was calculated. It was found that (i) PTG showed significantly higher N_{CF} than PTU and PTN in group 1, (ii) sterilization significantly increased the N_{CF} of PTN and PTG in group 2, (iii) PTN in group 3 had significantly higher cyclic fatigue resistance than PTN group 4, and (iv) significantly higher N_{CF} was observed for PTG in group 2 than in groups 3 and 4 concluding that PTG instrument made of new gold alloy was more resistant to fatigue failure than PTN and PTU. Autoclaving increased the cyclic fatigue resistances of PTN and PTG. Janardhanan et al. [27] evaluated the fatigue resistance of NiTi endodontic files before and after sterilization and the effects of sterilization on NiTi rotary files that were fatigued to half of their average life. Thirty-six NiTi files (ProTaper F2 size) were evaluated for fatigue test and hardness without sterilization and after five cycles of autoclaving. It was mentioned that (i) autoclaving has relatively no negative impact on the fatigue failure of NiTi instruments; however, a slight increase in hardness is observed on repeated sterilization, and (ii) the most common reason for failure of the files was flexural strain, which almost always corresponded to the area of maximum curvature of the simulated root canal. Pedullà et al. [28] assessed the effects of NaOCl immersion and sterilization on the cyclic fatigue resistance of heat-treated NiTi rotary instruments. Two hundred and ten new 25/0.06 TFs and Hyflex CM (Coltene Whaledent, Cuyahoga Falls, OH) files were divided into seven groups (n = 15) for each brand. Group 1 (control group) included new instruments that were not immersed in NaOCl or subjected to autoclave sterilization. Groups 2 and 3 were composed of instruments dynamically immersed for 3 min in 5% NaOCl solution 1 and 3 times, respectively. Groups 4 and 5 consisted of instruments only autoclaved 1 and 3 times, respectively. Groups 6 and 7 recruited instruments that received a cycle of both immersion in NaOCl and sterilization 1 and 3 times, respectively. Instruments were subsequently subjected to a fatigue test. It was reported that (i) comparison among groups indicated no significant difference of N_{CF} except for the groups of TFs sterilized 3 times without and with immersion in NaOCl, and (ii) HyFlex CM files exhibited higher cyclic fatigue resistance than TFs when files were sterilized 3 times, independently from immersion in NaOCl. It was concluded that (i) repeated cycles of sterilization did not influence the cyclic fatigue of NiTi files except for TFs, which showed a significant decrease of flexural resistance after 3 cycles of sterilization, and (ii) immersion in NaOCl did not reduce the cyclic fatigue resistance of all heat-treated NiTi files tested significantly.

13.2.3 Torsion

Silvaggio et al. [29] determined whether heat sterilization adversely affects the torsional properties of rotary NiTi files, making them more prone to fracture under torsional stress. Nine hundred sizes 2 through 10 Profile Series 29.04 taper files were divided into groups of 10 files each and sterilized 0, 1, 5, or 10 times in the steam autoclave, Statim autoclave, or dry heat sterilizer. Complete data were collected for sizes 2 through 7, but not for sizes 8 through 10 because their torque resistance exceeded the testing limits of the Torquemeter Memocouple. It was reported that (i) fifty-four comparisons were made for torsional strength and 54 for rotational flexibility and 10 significant changes occurred for torsional strength and 10 for rotational flexibility, (ii) eight of 10 changes in torsional strength were increases, and (iii) fifty-two of 54 (96.3%) comparisons for torsional strength and 47 of 54 (87%) for rotational flexibility showed a significant increase or no change indicating that heat sterilization of rotary NiTi files up to 10 times does not increase the likelihood of instrument fracture. Mitchell et al. [30] evaluated the effect

of cyclic autoclave sterilization and simulated clinical usage on a mechanical property of one brand of stainless-steel endodontic files. The angular deflection moments
were measured by a torque apparatus approved by the American Dental Association
for such purposes. Comparisons of values for sterilized and nonsterilized files were
made. It was reported that (i) a significant decrease in angular deflection values exists
for stainless-steel endodontic files having undergone 10 cycles of autoclave sterilization
versus files having undergone only two or five similar cycles, (ii) all file sizes (15, 20, 25,
30, 35, and 40) tested in torsion were detrimentally affected by the autoclave sterilization, (iii) of the files investigated, sizes 35 and 40 were the most adversely affected by
the steam-under-pressure sterilization, and (iv) the angular deflection values of those
files subjected to repeated autoclaving did not decrease below the minimum value
accepted by the American Dental Association (ADA) for resistance to torque forces concluding that (v) the repeated sterilization of a stainless steel endodontic file does result
in a significant reduction in the torque resistance of that file, which was not significant
clinically. Iverson et al. [31] studied the effects of different sterilization methods on the
torsional strength of two types of endodontic files (Kerr K-Flex files and Burns Unifiles),
which were subjected to 10 cycles of autoclaving, bead, dry heat, and cold chemical
sterilization. It was reported that (i) repeated sterilization was found to have no effect
on the torque resistance and degrees to failure for the Burns Unifile and no effect on
degrees to failure for the K-Flex file, (ii) dry heat and cold sterilization, however, slightly
increased the torque resistance of the K-Flex files as compared with the controls and the
other sterilization methods, and (iii) no differences were found between the files in the
degrees of rotation to failure but the K-Flex files had a slightly greater torque resistance.
Canalda-Sahli et al. [32] investigated the effect of dry-heat or autoclave sterilization on
the resistance to fracture in torque and angular deflection and the resistance to bending
of K-type files made of NiTi (Nitiflex, Naviflex), Ti (Microtitane) or stainless steel (Flexofile, Flex-R). Ten K-type files of each sort, from size 25 to 40, were tested, according to
The American National Standards Institute (ANSI)/ADA specification 28 (1988) and ISO
specification 3630 (1992). It was mentioned that (i) sterilization with dry heat and autoclave slightly decreased the flexibility of files made of stainless steel and NiTi for most
of the sizes, although the values obtained satisfied International Organization for Standardization (ISO) specifications, (ii) the files made of titanium showed an increased
flexibility after sterilization with autoclave (sizes 30 and 35) and dry heat (sizes 30,
35, and 40), (iii) resistance to fracture varied amongst the five groups of files tested as
follows: it decreased in some sizes of stainless-steel instruments, decreased in all sizes
of titanium files assessed by the torsional moment, and either increased or decreased in
some sizes of NiTi files, and (iv) all files tested, however, satisfied relevant standards for
angular deflection after being subjected to sterilization with an autoclave or dry heat.
Hilt et al. [33] tested the hypothesis that multiple sterilizations of endodontic stainless
steel and NiTi files will lead to a continuous decrease in the resistance of files to separation by torsion. One hundred stainless steel and 100 nickel–titanium #30 K-type
files were divided into 20 groups of 10 and sterilized in increments of 10 cycles, using

a full cycle and a fast cycle autoclave. These files were tested by twisting each of them in a clockwise direction until fracture (torque g-cm). It was indicated that neither the number of sterilization cycles nor the type of autoclave sterilization used affects the torsional properties, hardness, and microstructure of stainless steel and NiTi files.

13.2.4 Cutting efficiency

The effects of steam sterilization and usage on sharpness were evaluated on #25 endodontic files [34]. Files were used to instrument 1, 5, and 10 molars. Control groups determined the effect of steam sterilization alone on cutting efficiency of unused files. A cutting efficiency test was performed on an apparatus that compares sharpness of files when used in linear motion. Scanning electron microscopic analysis was performed in each group. It was mentioned that (i) significant differences were found between experimental files used for instrument 1 molar and those used for 5 or 10 molars, (ii) the difference in cutting efficiency between the second and third experimental groups was not significant, indicating that most of the decrease in sharpness occurred with use between one and five molars, and (iii) no significant difference was found between the control groups, indicating no decrease in cutting efficiency by sterilization alone. Schäfer et al. [35] tested effects of sterilization on the outline efficiency of PVD-coated NiTi endodontic instruments and concluded that repeated sterilization under autoclave or exposure to NaOCl prior to sterilization did not alter the cutting efficiency of PVD-coated NiTi files. The efficacy of four methods of sterilizing endodontic instruments (including autoclaving, carbon dioxide laser sterilization, chemical sterilization with glutaraldehyde and glass bead sterilization) was investigated [36]. The endodontic file was sterilized by four different methods after contaminating it with *Bacillus stearothermophillus* and then checked for sterility by incubating after putting it in test tubes containing thioglycollate medium. It was indicated that (i) the files sterilized by autoclave and lasers were completely sterile, and (ii) those sterilized by glass bead were 90% sterile and those with glutaraldehyde were 80% sterile concluding that autoclave or laser could be used as a method of sterilization in clinical practice and in advanced clinics; laser can be used also as a chair side method of sterilization. Kommmineni et al. [8] investigated the sterilization efficacy using two methods of sterilization using rotary endodontic files, including autoclave, chemical sterilization using two solutions (chlorhexidine and glutaraldehyde). Sixty new presterilized rotary NiTi files contaminated with commercially available *Bacillus stearothermophillus*. The files were sterilized by the two methods and checked for sterility by incubating the files in test tubes containing thioglycollate medium. It was reported that the files sterilized by autoclave were 100% sterile, and those sterilized by chlorhexidine showed 87% sterility and with glutaraldehyde showed only 60% sterility indicating that the autoclave could be used as best reliable method of sterilization.

13.2.5 Corrosion

Thierry et al. [37] studied the corrosion behavior of NiTi alloys after mechanical polishing, electropolishing, and sterilization processes using cyclic polarization and atomic absorption. It was reported that (i) as a preparative surface treatment, electropolishing decreased the amount of nickel on the surface and remarkably improved the corrosion behavior of the alloy by increasing the mean breakdown potential value and the reproducibility of the results (0.99 ± 0.05 V/SCE vs 0.53 ±0. 42), (ii) ethylene oxide and Sterrad sterilization techniques (which utilizes hydrogen peroxide vapor and low-temperature gas plasma) did not modify the corrosion resistance of electropolished NiTi, whereas a steam autoclave and, to a lesser extent, peracetic acid sterilization produced scattered breakdown potential, (iii) in comparing the corrosion resistance of common biomaterials, NiTi ranked between 316L stainless steel and Ti6A14V even after sterilization, (iv) electropolished NiTi and 316L stainless-steel alloys released similar amounts of nickel after a few days of immersion in Hank's solution, (v) measurements by atomic absorption have shown that the amount of released nickel from passive dissolution was below the expected toxic level in the human body, and (vi) AES analyses indicated surface contamination by Ca and P on NiTi during immersion, but no significant modification in oxide thickness was observed.

13.2.6 Fracture

The effect of repeated cleaning procedures on fracture properties and corrosion of NiTi files was evaluated [38]. New NiTi instruments were subjected to 2, 5, and 10 cleaning cycles with the use of either diluted bleach (1% NaOCl) or Milton's solution (1% NaOCl plus 19% NaCl) as disinfectant. Each cleaning cycle consisted of scrubbing, rinsing, and immersing in NaOCl for 10 min followed by 5 min of ultrasonication. Files were then tested for torsional failure and flexural fatigue and four brands of NiTi files were immersed in either Milton's solution or diluted bleach overnight and evaluated for corrosion. It was found that (i) results up to 10 cleaning cycles did not significantly reduce the torque at fracture or number of revolutions to flexural fatigue, although decreasing values were noted with increasing number of cleaning cycles using Milton's solution, (ii) no corrosion was detected on the surface of these files, (iii) files immersed in 1% NaOCl overnight displayed a variety of corrosion patterns, (iv) the extent of corrosion was variable amongst different brands of files and amongst files in each brand, (v) overall, Milton's solution was much more corrosive than diluted bleach, and (vi) corrosion of file handles was often extreme. It was then concluded that (i) files can be cleaned up to 10 times without affecting fracture susceptibility or corrosion, but should not be immersed in NaOCl overnight, and (ii) Milton's solution is much more corrosive than bleach with the same NaOCl concentration. Medical device fractures during gamma and electron beam (eBeam) sterilization have been reported.

Two common factors in these device fractures were a constraining force and the presence of fluorinated ethylene propylene (FEP). Smith et al. [39] investigated the effects of eBeam sterilization on constrained light-oxide NiTi alloy wires in FEP to recreate these fractures and determine their root cause. NiTi superelastic wires were placed inside FEP tubes and constrained with nominal outer fiber strains of 10%, 15%, and 20%. These samples were then subjected to a range of eBeam sterilization doses up to 400 kGy and compared with unconstrained wires also subjected to sterilization. It was reported that (i) factures were observed at doses of >100 kGy, and (ii) analysis of the fracture surfaces indicated that the samples failed due to irradiation-assisted stress-corrosion cracking indicating that it was also observed to occur with polytetrafluoroethylene (or simply Teflon) (PTFE) at 400 kGy. It was then suggested that NiTi alloy was susceptible to irradiation-assisted stress corrosion cracking when in the presence of a constraining stress, fluorinated polymers, and irradiation. Norwich et al. [40] determined the susceptibility of NiTi alloy to this type of failure. The variables studied included wire diameter, wire surface finish, wire oxide layer, quantity of wires encased, type of tubing, and strain level during gamma sterilization. The greatest susceptibility to fracture occurred to single wire samples with a light oxide layer held under high strain in FEP shrink tube. Gamma sterilization experiments were conducted to isolate and confirm this failure mechanism. Scanning electron microscopy was used to analyze the fractured samples. Chemical analysis was performed in an attempt to detect trace elements to determine the root cause of the failures. Stress corrosion cracking caused by the liberation of fluorine due to the degradation of the polymer during gamma sterilization is suspected.

13.2.7 Cases in orthodontic archwire

There are a few studies on sterilization on orthodontic NiTi archwires. Smith et al. [41] evaluated the effect of clinical use and various sterilization/disinfection protocols on three types of nickel–titanium, and one type each of beta-titanium and stainless-steel arch wire. The sterilization/disinfection procedures included disinfection alone or in concert with steam autoclave, dry heat, or cold solution sterilization. It was reported that (i) no clinically significant differences were found between new and used arch wires, and (ii) the direction of load application to the arch wire and the particular segment of arch wire tested was found to cause substantial differences in generated loads for certain arch wire types. Poosti et al. [42] compared the concentration of Ni release from recycled NiTi wires after sterilization by either dry heat or steam autoclave. Eighty preformed NiTi wires were assigned to four equal groups. In groups 1, 2, and 3 the archwires were used intraorally for 4 weeks. Group 4 archwires were not used. Group 1 archwires were sterilized by dry heat, group 2 archwires were sterilized by steam autoclave, group 3 archwires were not sterilized, and group 4 archwires were as-received wires. A 2 cm length, cut from each archwire, was immersed in artificial saliva for 4 weeks and the nickel ion concentrations in the artificial saliva were measured with an atomic absorption spectrophotometer. It was obtained that there were no significant differences in the nickel ion concentrations released into the artificial saliva by each group of archwires concluding that sterilization of used NiTi wires by either dry heat and steam autoclave does not affect the concentrations of nickel ions released into artificial saliva.

References

[1]   Enabulele JE, Omo JO. Sterilization in endodontics: knowledge, attitude, and practice of dental assistants in training in Nigeria – a cross-sectional study. Saudi Endod. J. 2018, 8, 106–10.

[2]   Mustafa M. Knowledge, attitude and practice of general dentists towards sterilization of endodontic files: a cross – sectional study. Indian J. Sci. Tech. 2016, 9, 11–6.

[3]   Limbhore M, Saraf A, Medha A, Jain D, Mattigatti S, Mahaparale R. Endodontic hand instrument sterilization procedures followed by dental practitioners. Unique J. Med. Dent. Sci. 2014, 2, 106–11.

[4]   Ferreira MM, Michelotto AL, Alexandre AR, Morgantio R, Carnillo EV. Endodontic files: sterilize or discard. Dent. Press Endod. 2012, 2, 46–51.

[5]   Aslam A, Panuganti V, Nanjundarethy JK, Halappa M, Krishna VH. Knowledge and attitude of endodontic post graduate students towards sterilization of endodontic files: a cross – sectional study. Saudi Endod. J. 2016, 4, 18–22.

[6]   Linsuwamont P, Parashos P, Messor HH. Cleaning of rotary nickel titanium endodontic instruments. Int. Endod. J. 2004, 37, 19–28.

[7]   Takkar H, Kumar SA, Kumar MS, Takkar S. Contribution of endodontic field in clean India campaign by the dentists – survey in Sri Ganganagar District, Rajanthan. J. Adv. Med. Dent. Sci. Res. 2015, 3, 23–8

[8]   Kommmineni NK, Dappili SRR, Prathyusha P, Vanaja P, Reddy KVKK, Vasanthi D. Comparative evaluation of sterilization efficacy using two methods of sterilization for rotary endodontic files: an in vitro study. J. Dr. NTR Univ. Health Sci. 2016, 5, 142–6.

[9]   Dunn D. Reprocessing single-use devices – the ethical dilemma. AORN J. 2002, 75, 989–99.

[10]   Dunn D. Reprocessing single-use devices – regulatory roles. AORN J 2002, 76, 100–8.

[11]   Hurtt CA, Rossman LE. The sterilization of endodontic hand files. J. Endod. 1996, 22, 321–2.

[12]   Raju TBVG, Garapati S, Agrawal R, Reddy S, Razdan A, Kumar SK. Sterilizing endodontic files by four different sterilization methods to prevent cross-infection – an in-vitro study. J. Int. Oral Health 2013, 5, 108–12.

[13]   Kumar KV, Kumar KSK, Supreetha S, Raghu KN, Veerabhadrappa AC, Deepthi S. Pathological evaluation for sterilization of routinely used prosthodontic and endodontic instruments. Thermal Anal. Polym. 2015, 5, 232–6.

[14]   Yenni M, Bandi S, Avula SS, Margana PG, Kakarla P, Amrutavalli A. Comparative evaluation of four different sterilization methods on contaminated endodontic files. CHRISMED J. Health. Res. 2017, 4, 194–7.

[15]   Thierry B, Tabrizian M, Savadogo O, Yahia L'H. Effects of sterilization processes on NiTi alloy: surface characterization. J. Biomed. Mater. Res. 2000, 49, 88–98.

[16]   Morawiec H, Goryczka T, Lelątko J, Lekston Z, Winiarski A, Równiński E, Stergioudis F. Surface structure of NiTi alloy passivated by autoclaving. Mater. Sci. Forum 2001, 636/637, 971–6.

[17]   Valois CR, Silva LP, Azevedo RB. Multiple autoclave cycles affect the surface of rotary nickel-titanium files: an atomic force microscopy study. J. Endod. 2008, 34, 859–62.

[18] Spagnuolo G, Ametrano G, D'Antò V, Rengo C, Simeone M, Riccitiello F, Amato M. Effect of autoclaving on the surfaces of TiN -coated and conventional nickel-titanium rotary instruments. Int. Endod. J. 2012, 45, 1148–55.

[19] Cai JJ, Tang XN, Ge JY. Effect of irrigation on surface roughness and fatigue resistance of controlled memory wire nickel-titanium instruments. Int. Endod. J. 2017, 50, 718–24.

[20] Yılmaz K, Uslu G, Özyürek T. Effect of multiple autoclave cycles on the surface roughness of HyFlex CM and HyFlex EDM files: an atomic force microscopy study. Clin. Oral Invest. 2018, 22, 2975–80.

[21] Baldin EKK, Malfatti CF, Rodói V, Brandalise RN. Effect of sterilization on the properties of a bioactive hybrid coating containing hydroxyapatite. Adv. Mater. Sci. Eng. 2019; https://doi.org/10.1155/2019/8593193.

[22] Mize SB, Clement DJ, Pruett JP, Carnes DL Jr. Effect of sterilization on cyclic fatigue of rotary nickel-titanium endodontic instruments. J. Endod. 1998, 24, 843–7.

[23] Viana AC, Gonzalez BM, Buono VT, Bahia MG. Influence of sterilization on mechanical properties and fatigue resistance of nickel-titanium rotary endodontic instruments. Int. Endod. J. 2006, 39, 709–15.

[24] Bergeron BE, Mayerchak MJ, Roberts MJ, Jeansonne BG. Multiple autoclave cycle effects on cyclic fatigue of nickel-titanium rotary files produced by new manufacturing methods. J. Endod. 2011, 37, 72–4.

[25] Plotino G, Costanzo A, Grande NM, Petrovic R, Testarelli L, Gambarini G. Experimental evaluation on the influence of autoclave sterilization on the cyclic fatigue of new nickel-titanium rotary instruments. J. Endod. 2012, 38, 222–5.

[26] Özyürek T, Yılmaz K, Uslu G. The effects of autoclave sterilization on the cyclic fatigue resistance of ProTaper Universal, ProTaper Next, and ProTaper Gold nickel-titanium instruments. Restor. Dent. Endod. 2017, 42, 301–8.

[27] Janardhanan S, Kanisseri M, John MK. Influence of sterilization on mechanical properties and fatigue resistance of nickel-titanium rotary endodontic instruments: an *in vitro* study. Int. J. Oral Care Res. 2018, 6, 5–11.

[28] Pedullà E, Benites A, La Rosa GM, Plotino G, Grande NM, Rapisarda E, Generali L. Cyclic fatigue resistance of heat-treated nickel-titanium instruments after immersion in sodium hypochlorite and/or sterilization. J. Endod. 2018, 44, 648–53.

[29] Silvaggio J, Hicks ML. Effect of heat sterilization on the torsional properties of rotary nickel-titanium endodontic files. J. Endod. 1997, 23, 731–4.

[30] Mitchell BF, James GA, Nelson RC. The effect of autoclave sterilization on endodontic files. Oral Surg. Oral Med. Oral Pathol. 1983, 55, 204–7.

[31] Iverson GW, von Fraunhofer JA, Herrmann JW. The effects of various sterilization methods on the torsional strength of endodontic files. J. Endod. 1985, 11, 266–8.

[32] Canalda-Sahli C, Brau-Aguadé E, Sentís-Vilalta J. The effect of sterilization on bending and torsional properties of K-files manufactured with different metallic alloys. Int. Endod. J. 1998, 31, 48–52.

[33] Hilt BR, Cunningham CJ, Shen C, Richards N. Torsional properties of stainless-steel and nickel-titanium files after multiple autoclave sterilizations. J. Endod. 2000, 26, 76–80.

[34] Morrison SW, Newton CW, Brown CE Jr. The effects of steam sterilization and usage on cutting efficiency of endodontic instruments. J. Endod. 1989, 15, 427–31.

[35] Schäfer E. Effect of sterilization on the cutting efficiency of PVD-coated nickel–titanium endodontic instruments. Int. Endod. J. 2002, 35, 867–72.

[36] Venkatasubramanian R, Jayanthi, Das UM, Bhatnagar S. Comparison of the effectiveness of sterilizing endodontic files by 4 different methods: an *in vitro* study. J. Indian Soc. Pedod. Prev. Dent. 2010, 28, 2–5.

[37] Thierry B, Tabrizian M, Trepanier C, Savadogo O, Yahia L. Effect of surface treatment and sterilization processes on the corrosion behavior of NiTi shape memory alloy. J. Biomed. Mater. Res. 2000, 51, 685–93.

[38] O'Hoy PYZ, Messer HH, Palamara JEA. The effect of cleaning procedures on fracture properties and corrosion of NiTi files. Int. Endod. J. 2003, 36, 724–32.

[39] Smith SA, Gause B, Plumley DL, Drexel MJ. Irradiation-assisted stress-corrosion cracking of Nitinol during eBeam sterilization. J. Mater. Eng. Perform. 2012, 21, 2638–42.

[40] Norwich DW. Fracture of polymer-coated nitinol during gamma sterilization. J. Mater. Eng. Perform. 2012, 21, 2618–21.

[41] Smith GA, von Fraunhofer JA, Casey GR. The effect of clinical use and sterilization on selected orthodontic arch wires. Am. J. Orthod. Dentofacial Orthop. 1992, 102, 153–9.

[42] Poosti M, Rad HP, Kianoush K, Hadizadeh B. Are more nickel ions released from NiTi wires after sterilisation? Aust. Orthod. J. 2009, 25, 30–3.

Chapter 14
Biotribology and wear debris toxicity

The longevity, safety, reliability, functionality, and structural integrity of biomedical materials are governed by surface and interface phenomena. Surface is not just a free end of the solid or liquid substance but is a contact zone with other substances (gaseous, liquid, or solid state). Surface and interface reactions include reactions with organic or inorganic materials, with vital or nonvital species, with hostile or friendly environments, and so on. Surface activities vary from mechanical actions (fatigue crack initiation and propagation, stress intensification, etc.), chemical action (discoloration, tarnishing, contamination, corrosion, and oxidation), tribological action (wear, friction, lubricant, and wear debris toxicity) to physical action (surface contact, surface tension, diffusion, absorption and desorption, etc.). The term biotribology is a compound word of biology and tribology (tribology in biological environment). The term tribology* was introduced from an origin of Greek word tribw (to rub). Tribology was defined as a science and technology dealing with the friction, wear, and lubricant.

> Note: The prefix "tri" normally refers to "three" in those words such as "tripot," "trigonometry," "trilingual," "trifocal". At the same time, we know that tribology is a science about three phenomena (friction, wear, and lubricant), so, it is easily misinterpreted that tribology is a science dealing with the aforementioned three phenomena. But it is not true. The Greek word tribos does not have a prefix meaning "three." This is just a matter of coincidence.

14.1 In general

Tribology is the science dealing with the interaction of surfaces in tangential motion. Hence, tribology includes the nature of surfaces from both a chemical and physical point of view, including topography, the interaction of surfaces under load and the changes in the interaction when tangential motion is introduced. Macroscopically, the interactions are manifested in the phenomena of friction and wear. Modification of the interaction through the interposition of liquid, gaseous, or solid films is known as the lubrication process. Hence, from a macroscopic point of view, tribology includes lubrication, friction, and wear [1]. There are four major clinical reasons to remove the implants: (1) fracture, (2) infection, (3) wear, and (4) loosening. Among these, the removal due to the infection generally occurs in relatively early stages after implantation, while the other three incidents typically increase gradually over years, because they are somewhat related to biotribological reactions. Human joints are an example of natural joints and show low wear and exceedingly low friction through efficient lubrication. Disease or accident can impair the function at a joint, and this

https://doi.org/10.1515/9783110666113-014

can lead to the necessity for joint replacement [1]. Sivasankar et al. [2] mentioned that the hip is one of the largest weight-bearing joints in our body. It consists of two parts namely, a ball (femoral head) at the top of our thighbone (femur) and it fits into a rounded

socket (acetabulum) in our pelvis. A band of tissues called ligaments connect the ball to the socket and provide stability to the joint. The hip joint may get damaged due to diseases like rheumatoid arthritis, osteoarthritis, fractures, and dislocations and sometimes due to accidents too. This may cause the fracture in hip and will permanently handicap the person. There are several types of hip fractures, including (1) femoral neck fracture: pins (surgical screws) are used if the person is younger and more active, and if the broken bone is not removed much out of place. If the person is older and less active, a high strength metal device that fits into hip socket replacing the head of the femur (hemiarthroplasty) is needed; (2) intertrochanteric fracture: a metallic device (compression screw and side plate) holds the broken bone in place while it lets the head of the femur move normally in the hip socket. The total hip replacement (THR) is the latest technology, which has been implemented as a boon for the humanity. By the implementation of the THR, a person who is facing a problem in the hip or met with an accident will be benefited and it will enable the person to regain normal work. Special tribological element problems arise in many mechanical devices used in various areas, for example, computer devices and medical implants (prostheses, assists, and artificial organs). Medical implants, such as prostheses, should demonstrate very good tribological properties, as well as biological inertness. Their longevity depends significantly on the wear of the rubbing elements and must be extended as much as possible. The developments of total replacement of hip, knee, ankle, elbow, shoulder, and hand joints (and the appropriate surgical techniques) has been the major success of orthopedic surgery, and would not have been possible without extensive in vitro and in vivo studies of the tribological problems, especially wear, associated with such artificial joints.

There are a number of important biotribological testing parameters that can greatly affect the outcome of a wear study in addition to the implant design and material selection. The current ASTM and ISO wear testing standards for spine arthroplasty leave many choices as testing parameters, including but not limited to sequence of kinematics and load, phasing, type of lubricant, and specimen preparation (sterilization and artificial aging) [3]. In situ techniques of optical microscopy and Raman spectroscopy were used to observe interfacial sliding dynamics and identify near-surface structural and chemical changes. Third-body physical and chemical processes, such as thickening, thinning, loss of transfer films, generation of wear debris, and sliding-induced chemical changes, were identified for sapphire against Ti–Si–C, nanocrystalline diamond, and titanium- and tungsten-doped diamond-like carbon (DLC) coatings. These processes observed by in situ methods were also used to explain why friction and wear behavior changed with coating composition, properties, or test conditions [4].

14.2 Friction

The most important incidence involving friction in dentistry is frictional behavior between the orthodontic bracket (slot portions) and archwires. While orthodontic mechanotherapy is active, sliding of bonded brackets along archwires occur in all orthodontic cases. When these sliding movements occur, frictional forces between the archwire and the bracket are produced. Different combinations of materials have varying coefficients of friction, leading to different amounts of frictional force created within the bracket/archwire system. This force either retracts or protracts between brackets. It is thought that static, rather than kinetic, frictional forces may be more relevant in orthodontic tooth movement. This is due to the bracket/archwire system having stopping and starting movements as the teeth are moved. In vitro testing of orthodontic archwires and brackets can be performed in either a wet or dry environment. Wet testing is usually done by submerging the archwire/bracket system in saliva, salivary substitute, or glycerin solution. Surface coating is also an influencing factor on frictional behavior of bracket/archwire system.

14.2.1 General characterization

Rucker et al. [5] compared the sliding mechanics of multistranded stainless steel (SS) wires with single-stranded leveling wires in the passive and active regions when dominated by classical friction and elastic binding, respectively. Tests were done under both dry and wet (human saliva) conditions. The round multistranded wires had 3- (triple) and 6-stranded (coax) configurations in nominal sizes of 15.5, 17.5, 19.5, and 21.5 mil; the rectangular wires had 3- (rect3) and 8-stranded (rect8) configurations in nominal sizes of 16×16, 16×22, 17×25, and 19×25 mil. It was reported that (i) in the passive region, the kinetic coefficients of friction μ(k-FR) in the wet state were the same as, lower than, and higher than in the dry state for single-stranded SS, single-stranded nickel–titanium (NiTi), and multistranded SS wires, respectively, (ii) because the kinetic coefficients of friction were similar for multistranded and single-stranded SS wires, μ(k-FR) is a material property for SS and perhaps also for NiTi, (iii) in the active region, the frictional behaviors of multistranded SS wires compared with other leveling archwires are as follows: (1) coax wires had low friction, (2) triple and rect8 wires had midrange friction, and (3) rect3 wires had high friction, (iv) the coefficients of binding (μ(BI)) were not affected by saliva and were proportional to the wire stiffnesses, and (v) the resistance to sliding (RS) depended on wire stiffnesses to the extent that the differences in the μ(k-FR)'s of SS versus NiTi became unimportant shortly after binding occurred. Stress-induced martensite formation with stress hysteresis that changes the elasticity and stiffness of NiTi wire influences the sliding mechanics of archwire-guided tooth movement. Liaw et al. [6] investigated the frictional behavior of an improved superelastic NiTi wire with low-stress hysteresis.

Improved superelastic NiTi alloy wires (L & H Titan) with low-stress hysteresis were examined by using 3-point bending and frictional resistance tests with a universal test machine at a constant temperature of 35 °C and compared with the former conventional austenitic-active superelastic NiTi wires (Sentalloy). Wire stiffness levels were derived from differentiation of the polynomial regression of the unloading curves, and values for kinetic friction were measured at constant bending deflection distances of 0, 2, 3, and 4 mm, respectively. It was mentioned that (i) compared with conventional Sentalloy wires, the L & H Titan wire had a narrower stress hysteresis including a lower loading plateau and a higher unloading plateau, (ii) L & H Titan wires were less stiff than the Sentalloy wires during most unloading stages. Values of friction measured at deflections of 0, 2, and 3 mm were significantly increased in both types of wire, (iii) however, they showed a significant decrease in friction from 3 to 4 mm of deflection. L & H Titan wires had less friction than Sentalloy wires at all bending deflections; concluding that (iv) stress-induced martensite formation significantly reduced the stiffness and, thus, could be beneficial to decrease the binding friction of superelastic NiTi wires during sliding with large bending deflections, and (v) austenitic-active alloy wires with low-stress hysteresis and lower stiffness and friction offer significant potential. Regis et al. [7] assessed the effect of clinical exposure on the surface morphology, dimensions, and frictional behavior of metallic orthodontic brackets. Ninety-five brackets, of three commercial brands, were retrieved from patients who had finished orthodontic treatment. As-received brackets, matched by type and brand, were used for comparisons. Surface morphology and precipitated material were analyzed by optical and scanning electron microscopy and x-ray microanalysis. Bracket dimensions were measured with a measuring microscope. RS on a stainless-steel wire (SSW) was assessed. It was found that (i) retrieved brackets showed surface alterations from corrosion, wear, and plastic deformation, especially in the external slot edges, (ii) film deposition over the alloy surface was observed to a variable extent, (iii) the main elements in the film were carbon, oxygen, calcium, and phosphorus, (iv) the as-received brackets showed differences in the slot sizes among brands, and one brand showed a 3% increase in the retrieved brackets' slots, and (v) the frictional behavior differed among brands. Retrieved brackets of two brands showed 10% to 20% increase in RS. Based on these findings, it was concluded that metallic brackets undergo significant degradation during orthodontic treatment, possibly with increased friction. Kumar et al. [8] evaluated the frictional forces generated by five different orthodontic brackets when used in combination with SS, titanium molybdenum alloy (TMA or β-titanium (β-Ti)), and NiTi archwires in dry conditions at physiological temperature. Five different types of maxillary upper right side self-ligating brackets (SLBs) (Damon 3MX) and conventional SS brackets (Mini 2000) with a slot size 0.022 inch were coupled with 0.016" NiTi and 0.019 × 0.025" SS/TMA archwires. It was reported that (i) SLB showed lower fictional values in comparison with elastic ligatures, and (ii) frictional force increased proportionally to the wire size, (iii) TMA and NiTi archwires presented higher frictional resistance than SS archwires;

concluding that SS brackets tied with conventional ligatures produced high and low friction when ligated with SLBs with passive clip.

14.2.2 Frictional behavior between brackets and archwires

SS, cobalt chromium (Co–Cr), NiTi, and β-Ti wires were tested in narrow single (0.050 in), medium (0.130 in), and wide twin (0.180 in) SS brackets in both 0.018 and 0.022 in slots. It was reported that (i) β-Ti and NiTi wires generated greater amounts of frictional forces than stainless steel or Co–Cr wires did for most wire size, and (ii) increase in wire size generally resulted in increased bracket-wire friction [9]. However, it was reported that wires in ceramic brackets generated significantly stronger frictional forces than the wires in SS brackets [10]. Prososki et al. [11] measured surface roughness and static frictional force resistance of orthodontic archwires. Nine NiTi alloy archwires were studied. One β-Ti alloy wire, one SS alloy wire, and one cobalt–chromium alloy wire were included for comparison. Frictional force resistance was quantified by pushing wire segments through the SS SLBs of a four-tooth clinical model. It was found that (i) the cobalt–chromium alloy and the NiTi alloywires, with the exception of Sentalloy and Orthonol, exhibited the lowest frictional resistance, (ii) the SS alloy and the β-Ti alloy wires showed the highest frictional resistance, (iii) the SS alloy wire was the smoothest wire tested, whereas NiTi, Marsenol, and Orthonol were the roughest, and (iv) no significant correlation was found between arithmetic average roughness and frictional force values. Vaughan et al. [12] measured the level of kinetic frictional forces generated during in vitro translation at the bracket-wire interface for two sintered SS brackets as a function of two slot sizes, four wire alloys, and five to eight wire sizes. The two types of sintered SS brackets were tested in both 0.018-inch and 0.022-inch slots. Wires of four different alloy types, SS, Co–Cr, NiTi, and β-Ti were tested. There were five wire sizes for the 0.018-inch slot and eight wire sizes for the 0.022-inch slot. It was reported that (i) for most wire sizes, lower frictional forces were generated with the SS of Co–Cr wires than with the β-Ti or NiTi wires, (ii) increase in wire size generally resulted in increased bracket-wire friction, and (iii) there were no significant differences between manufacturer for the sintered SS brackets. Schumacher et al. [13] conducted the in vitro study to investigate the influence of different bracket designs on sliding mechanics. Five differently shaped SS brackets (Discovery: Dentaurum, Damon SL: A-Company, Synergy: Rocky Mountain Orthodontics, Viazis bracket and Omni Arch appliance: GAC) were compared in the 0.022"-slot system. The orthodontic measurement and simulation system (OMSS) was used to quantify the difference between applied force (NiTi coil spring, 1.0 N) and orthodontically effective force and to determine leveling losses occurring during the sliding process in arch-guided tooth movement. It was mentioned that (i) comparison of the brackets revealed friction-induced losses ranging from 20% to 70%, with clear-cut advantages resulting from the newly developed bracket types, (ii) however, an increased tendency towards leveling losses in terms of distal rotation (maximum 15 degrees) or buccal root torque (maximum 20 degrees) was recorded, especially with those brackets giving the archwire increased mobility due to their shaping or lack of ligature wire. Cacciafesta et al. evaluated the friction of SS and esthetic SLBs [14] and conventional and metal-insert ceramic brackets [15] in various bracket–archwire combinations. It was reported that (i) SS SLBs generated significantly lower static and kinetic frictional forces than both conventional SS and polycarbonate SLBs, which showed no significant differences between them, (ii) β-Ti archwires had higher frictional resistances than SS and NiTi archwires, (iii) no significant differences were found between SS and NiTi archwires, and (iv) all brackets showed higher static and kinetic frictional forces as the wire size increased. For metal-insert ceramic brackets [15], it was mentioned that (v) metal-insert ceramic brackets generated significantly lower frictional forces than did conventional ceramic brackets, but higher values than SS brackets, (vi) β-Ti archwires had higher frictional resistances than did SS and NiTi archwires, (vii) no significant differences were found between SS and NiTi archwires, (viii) all the brackets showed higher static and kinetic frictional forces as the wire size increased, and (ix) metal-insert ceramic brackets are not only visually pleasing, but also a valuable alternative to conventional SS brackets in patients with esthetic demands. Clocheret et al. [16] evaluated the frictional behavior of 15 different archwires and 16 different brackets using small oscillating displacements when opposed to a standard SS bracket or a standard SS wire. Tests were run according to a pilot study at a frequency of 1 Hz and with a reciprocating tangential displacement of 200 μm, while the wire remained centered in the bracket slot under a load of 2 N. It was mentioned that (i) there was a significant difference between the evaluated wires and brackets, and (ii) the mean coefficient of friction (COF) of the wires varied from 0.16 for Imagination NiTi tooth-colored wire to 0.69 for the True Chrome Resilient Purple wire, while for the brackets it ranged from 0.39 for Ultratrimm to 0.72 for the Master Series, indicating that a large number of different commercially available archwires and brackets were evaluated with the same apparatus according to the same protocol, allows a direct comparison of the different archwire and bracket combinations, and can assist in the choice of the optimal bracket-wire combination with regard to friction. It has been suggested that the frictional resistance of ceramic brackets can be reduced by either lining the slots with SS or by contouring the base of the slot. Kapur et al. [17] compared in vitro the static and kinetic frictional resistances of ceramic brackets with metal lined slots ("Clarity," CL), SS brackets ("Miniature Twin," MT), and two ceramic brackets with different slot designs ("Contour," CO; "Transcend," TR). Two sizes (0.018 × 0.025 inch; 0.021 × 0.025 inch) of SS, NiTi, and β-Ti wires were drawn through the brackets. All brackets had 0.022 inch slots, and the brackets and wires were used once. The brackets were of different widths: CL, 0.180 inch; CO, 0.114 inch; MT, 0.118 inch; TR, 0.138 inch. It was reported that (i) there were no significant static or kinetic frictional differences when the smaller 0.018 × 0.025 inch wires were drawn through the brackets; (ii) there were no statistically significant static or kinetic frictional differences between the CL-CO, CL-MT, and CO-MT bracket pairs when the 0.021 × 0.025 inch wires (SS, NiTi, β-Ti) were used; (iii) there were no significant kinetic frictional resistance differences between the CL-TR and MT-TR when the SS wires were used; (iv) in general, the static and kinetic resistances of the 0.021 × 0.025 inch wires of NiTi wire < SS wire < β-Ti wire; (v) regardless of wire type some of the lowest kinetic resistances were found with the narrow CO brackets with the rounded slot bases; and (vi) the highest static and kinetic frictional resistances were found with the wide TR bracket, and with SS and β-Ti wires, concluding that the high static and kinetic frictional resistances of ceramic brackets can be reduced either by lining the slots with SS or by reducing the bracket width and rounding the slot base.

SLBs are claimed to eliminate or minimize the force of ligation at the bracket-wire interface. Krishnan et al. [18] evaluated the frictional features of contemporary SLBs with different archwire alloys (SS, NiTi, and β-Ti archwires) on frictional forces of passive and active SLBs with a conventional bracket (CB). All brackets had 0.022-in slots, and the wires were 0.019 × 0.025 in. Friction was evaluated in a simulated half-arch fixed appliance on a testing machine. It was found that (i) static and kinetic frictional forces were lower for both the passive and active designs than for the CBs, (ii) maximum values were seen with the β-Ti archwires and significant differences were observed between NiTi and SS archwires, and (iii) with the passive or active SLBs, SS wire did not produce a significant difference, but differences were significant with NiTi and β-Ti wires. Based on these results, it was concluded that (iv) when NiTi and β-Ti wires are used for guided tooth movement, passive appliances can minimize frictional resistance. Tecco et al. [19] tested the null hypothesis that no statistically significant difference in frictional resistance is noted when round or rectangular archwires are used in conjunction with low-friction ligatures (small, medium, or large) or conventional ligatures. A total of 10 SS brackets, a 0.022-in slot, and various orthodontic archwires, ligated with low-friction ligatures or conventional ligatures, were tested to compare frictional resistance. The archwires employed were 0.014-in and 0.016-in NiTi, 0.018-in SS, 0.016 × 0.022-in NiTi, 0.016 × 0.022-in SS, 0.017 × 0.025-in TMA, 0.017 × 0.025-in NiTi, 0.017 × 0.025-in SS, 0.019 × 0.025-in SS, and 0.019 × 0.025-in NiTi. Each bracket/archwire combination was tested 10 times in the dry state at an ambient temperature of 34 °C. It was reported that (i) low-friction ligatures with round archwires showed statistically significantly lower frictional resistance than did conventional ligatures, (ii) when coupled with 0.016 × 0.022-in NiTi and SS, no statistically significant difference was observed among the four groups, (iii) when coupled with 0.017 × 0.025-in archwires, low-friction ligatures showed statistically significantly greater frictional resistance than was seen with conventional ligatures, and (iv) when coupled with 0.019 × 0.025-in NiTi, low-friction ligatures showed statistically significantly greater frictional resistance than did conventional ligatures, but no difference among the four groups was observed with the 0.019 × 0.025-in SS. No significant difference was assessed among low-friction ligatures of different sizes. It was then concluded that (v) low-friction ligatures show lower friction when compared with conventional ligatures when coupled with round archwires, but not when coupled with rectangular ones. Tecco et al. [20] also compared friction (F) of conventional and ceramic brackets (0.022-inch slot) using a model that tests the sliding of the archwire through 10 aligned brackets. Polycrystalline alumina brackets (PCAs), PCA brackets with a SS slot (PCA-M), and monocrystalline sapphire brackets (MCS) were tested under elastic ligatures using various archwires in dry and wet (saliva) states. Conventional SS brackets were used as controls. It was obtained that (i) in both dry and wet states, PCA and MCS brackets expressed a statistically significant higher friction value with respect to SS and PCA-M brackets when combined with the rectangular archwires, (ii) PCA brackets showed significantly higher friction than MCS brackets when coupled with 0.014 × 0.025-inch NiTi archwire, and (iii) in the wet state, the mean friction values were generally higher than in the dry state. Based on these findings, it was concluded that (iv) PCA brackets showed significantly higher friction than MCS brackets only when combined with 0.014 × 0.025-inch NiTi archwires, and (v) a 10 aligned-brackets study model showed similar results when compared to a single bracket system except for friction level with 0.014 × 0.025-inch NiTi archwires. Hsu et al. [21] evaluated the static and kinetic frictional forces produced between different combination of orthodontic archwires and brackets. Three types of archwires were examined: (1) SS, (2) conventional NiTi alloy, and (3) improved superelastic NiTi alloy. Two types of brackets were tested: (1) SS and (2) plastic. It was observed that (i) the static frictional force was significantly higher than the kinetic frictional force in all archwire–bracket combinations, (ii) the frictional force was lower for the SS bracket than for the plastic bracket with SS wire and the improved superelastic NiTi-alloy wire, (iii) the frictional force was lower for the improved superelastic NiTi-alloy wire than for NiTi wire with the SS bracket, but higher for NiTi wire with the plastic bracket, and (iv) the frictional force was lowest for SS wire for both two types of bracket; indicating that (v) the frictional forces of brackets are influenced by different combinations of bracket and archwire, and (vi) the improved superelastic NiTi-alloy wire does not exhibit "low friction" (as claimed by the manufacturers) in all cases. Heo et al. [22] compared frictional properties according to the amounts of vertical displacement (VD) and horizontal displacement (HD) of teeth and bracket types during the initial leveling/alignment stage. Combinations of SLBs (SLBs; two active type: In-Ovation-R and In-Ovation-C; four passive type: Damon-3Mx, Damon-Q, SmartClip-SL3, and Clarity-SL) and 0.014-inch NiTi archwires (austenitic type, A-NiTi, and copper type, Cu-NiTi) were tested in a stereolithographically made typodont system that could simulate malocclusion status and periodontal ligament space. The upper canines (UCs) were displaced in the gingival direction and the upper lateral incisors (ULIs) in the lingual direction from their ideal positions by up to 3 mm, with 1-mm intervals, respectively. It was reported that (i) in the gingival displacement of UCs, Clarity-SL produced significantly lower frictional force, while Damon-3Mx, In-Ovation-R, and SmartClip-SL3 produced higher frictional force among SLBs, (ii) in the lingual displacement of ULIs, Damon-Q and

Damon-3Mx produced significantly lower frictional force, while Clarity-SL produced the highest frictional force among SLBs, (iii) Clarity-SL combined with A-NiTi and C-NiTi, Damon-3Mx combined with A-NiTi, and (iv) In-Ovation-C combined with Cu-NiTi showed differences in frictional properties between VD and HD. It was then concluded that (v) since the frictional properties of SLBs would be different between VD and HD of teeth, it is necessary to develop SLBs with low friction in both VD and HD of teeth.

In order to investigate the tribological behavior of medical devices in contact with tissue, friction tests for four kinds of medical metallic alloys (316L SS, CoCr, NiTi, and TiMoSn) on soft tissue–mimicking poly(vinyl alcohol) hydrogel (PVA-H) biomodel were carried out at low normal load [23]. XPS analysis and wettability tests for them were prepared to understand the difference in friction. According to the surface oxide compositions, these alloys can be divided into two groups: "Fe/Cr-oxide-surface alloys" for 316L and CoCr, and "Ti-oxide-surface alloys" for NiTi and TiMoSn. It was reported that (i) from the wettability test, Fe/Cr-oxide-surface alloys show lower polar components of surface free energy than Ti-oxide-surface alloys, and (ii) Fe/Cr-oxide-surface alloys show higher friction coefficients in the elastic friction domain than those of Ti-oxide-surface alloys, while there was no significant difference in the hydrodynamic lubrication. These results confirmed the major results on research investigating changes in surface contact angles as a function of time (moving from initial wettability to time-dependent spreadability) on some preoxidized biomaterials [24]. The static and kinetic friction forces of the contact bracket–archwire with different dental material compositions was evaluated in order to select those materials with lower RS [25] using an artificial saliva solution at 36.5 °C. The bracket–archwire pairs studied were: SS–SS; SS–glass fiber composite; SS–Nitinol 60; sapphire–SS; sapphire–glass fiber composite; and sapphire–Nitinol 60. It was found that the best performance is obtained for Nitinol 60 archwire sliding against a stainless steel bracket, both under dry and lubricated conditions; these results are in agreement with the low-surface roughness of Nitinol 60 with respect to the glass fiber composite archwire. Sridharan et al. [26] compared the frictional attributes of SS CBs and self-ligating SS brackets with different dimensions of archwires, using two sets of maxillary brackets: (1) conventional SS (Victory Series), (2) SS self-ligating (SmartClip) without first premolar brackets. SS, NiTi, and β-Ti wires are the types of orthodontic wire alloys. It was obtained that for Victory Series in static friction, p-value was 0.946 and for kinetic friction, it was 0.944; at the same time for SmartClip, the p-value for static and kinetic frictional resistance was 0.497 and 0.518 respectively; hence, there was no statistically significant difference between the NiTi and SS archwires. It was then concluded that (i) when compared with CBs with SS ligatures, SLBs can produce significantly less friction during sliding, and (ii) β-Ti wires archwires expressed high amount of frictional resistance and the SS archwires comprise low-frictional resistance among all the archwire materials. Of the variables used by in vitro studies of RS in orthodontics, sliding velocity (SV) of the wire is often the one farthest from its clinical counterpart. Savoldi et al. [27] investigated whether velocity influences the RS at values approximating the orthodontic movement. A SS SLB with a NiTi clip was fixed onto a custom-made model. Different shaped orthodontic SS wires of four sizes and two types (round, 0.020″ and 0.022″; rectangular, 0.016″ × 0.022″ and 0.017″ × 0.025″) were tested using an Instron® testing machine. Wires were pulled at four velocities $(1 \times 10^{-2}$ mm/s, 1×10^{-3} mm/s, 1×10^{-4} mm/s, 1×10^{-5} mm/s). It was found that (i) RS was higher for rectangular wires, and for those with larger diameters, (ii) lower SV was associated with lower RS, with wire type and size having an interaction effect, and (iii) the RS relatively to SV can be represented as: RS $\propto \alpha[\ln(\text{SV})]+\beta$, where α and β are constants. It was, then, concluded that (iv) at very low SV and low normal forces, SV influences the RS of SS archwires in orthodontic brackets, and the proportionality is logarithmic, (v) although respecting these parameters in vitro is challenging, quantitative evaluations of RS should be carried out at clinically relevant velocities if aiming at translational application in the clinical scenario.

14.2.3 Effect of lubricant

Effects of lubricant in the frictional behavior are one of the most important elements in the biotribological situation in the archwire/bracket system and salivary flow is playing a crucial role in this action. Saliva is clear viscous fluid secreted by the salivary and mucous glands in the mouth. Saliva contains (1) organic substances such as salivary protein, mucin and albumin, salivary enzymes, amylase (ptyalin) maltase, lipids, lysozyme, phosphatase, lactoferrin, sialoperoxidase and carbonic anhydrase, kallikrein, blood component and blood derivatives, antigens, serum cells, gingival cervicular fluid, immunoglobulins, IgA, IgG, IgM, nonprotein nitrogenous substances, urea, uric acid, creatine, xanthine, hypoxanthin, free amino acids, glycoproteins, and proteoglycans, (2) organic ionic substances such as sodium, potassium, calcium, chloride, bicarbonate, fluoride, bromide, phosphate, thiocyanate, and (3) gases present in saliva including oxygen (1 mL), carbon dioxide (50 mL/10 mL) and nitrogen (2.5 mL) [28]. About the salivary secretion; (1) under spontaneous situation, it occurs all the time, without any known stimulus to keep mouth moist all the time, and (2) under stimulated condition, it occurs because of known stimulus; may be psychological, visual, taste, and others (like, during vomiting). Regarding the salivary flow, (1) under resting condition, it is anything above 0.1 mL/min and it is a slow flow of saliva; keeping mouth moist and lubricates mucosa, (2) under stimulated condition, it is above 0.2 mL/min with max 7 mL/min, (3) during the sleep, it is normally nearly zero. Bongaerts et al. [29] demonstrated the efficient boundary lubricating properties of human whole saliva (HWS) in a soft hydrophobic rubbing contact, consisting of a poly(dimethylsiloxane) (PDMS) ball and a PDMS disk. The influence of applied load, entrainment speed and surface roughness were investigated for mechanically stimulated HWS. It was mentioned that (i) lubrication by HWS results in a boundary friction coefficient of $\mu \approx 0.02$, two orders of magnitude lower than that obtained for water, (ii) dried saliva on the other hand results in $\mu \approx 2$–3, illustrating the importance of hydration for efficient salivary lubrication. Increasing the surface roughness increases the friction coefficient for HWS, while it decreases that for water, (iii) the boundary lubricating properties of HWS are less sensitive to saliva treatment than are its bulk viscoelastic properties, (iv) centrifugation and aging of HWS almost completely removes the shear thinning and elastic nature observed for fresh HWS, (v) in contrast, the boundary friction coefficients are hardly affected, which indicates that the high-molecular-weight (supra)molecular structures in saliva, which are expected to be responsible for its rheology, are not responsible for its boundary lubricating properties, and (vi) saliva-coated PDMS surfaces form an ideal model system for ex-vivo investigations into oral lubrication and how the lubricating properties of saliva are influenced by other components like food, beverages, oral care products, and pharmaceuticals. Kusy et al. [30] measured the RS in vitro for various archwires against SS brackets. Using SS ligatures, a constant normal force (300 g) was maintained while second-order angulation (straight θ) was varied from –12 degrees to +12 degrees. Using miniature bearings to simulate contiguous teeth, five experiments each were run in the dry or wet states with human saliva at 34 °C as a function of four archwire alloys, five interbracket distances (IBDs), and two bracket engagements. It was reported that in the active configuration couples comprised of titanium alloys (NiTi and β-Ti) had higher RS values in the wet versus the dry state. Sarkar et al. [31] mentioned that oral lubrication deals with one of the most intricate examples of biotribology, where surfaces under sliding conditions span from the hardest enamel to soft oral tissues in human physiology. Complexity further arises with surfaces being covered by an endogenous biolubricant saliva before exogenous food particles can wet, stick, or slip at the surfaces. Salivary lubrication has been extensively studied ex vivo in enamel contacts, and the effect of its components has also been assessed separately [32–34]. Between hard contact surfaces, human saliva has shown to reduce the friction coefficient by a factor of 20 [35], having friction coefficient of $\mu \approx 0.02$, that is, two orders of magnitude lower than that of water [36].

14.2.3.1 Human saliva
The lubricating features and viscosity of human saliva and five commercially available saliva substitutes were compared [37] and it was indicated that (i) there was a little correlation existing between these parameters, (ii) saliva substitutes based on carboxymethylcellulose do not appear to lubricate biocompatible hard interfaces well and, therefore, might not protect against the rapid attrition observed in xerostomic individuals. In contrast, a mucin-based substitute proved to be a better lubricant with values comparable to whole human saliva. Due to the increase in life expectancy, new treatments have emerged which, although palliative, provide individuals with a better quality of life. Artificial saliva is a solution that contains

substances that moisten a dry mouth, thus mimicking the role of saliva in lubricating the oral cavity and controlling the existing normal oral microbiota. Silva et al. [38] assessed the influence of commercially available artificial saliva on biofilm formation by *Candida albicans*. Artificial saliva I consists of carboxymethylcellulose, while artificial saliva II is composed of glucose oxidase, lactoferrin, lysozyme, and lactoperoxidase. A control group used sterile distilled water. It was reported that statistically significant reduction of 29.89% (1.45 CFU/mL) of *Candida albicans* was observed in saliva I when compared to saliva II. Apart from the original purpose of citing articles regarding biofunction of the artificial salivary lubrication in frictional behavior in the orthodontic archwire/bracket system, it is worthy to know about the other function of the artificial saliva, in particular, against the xerostomia symptom. Advances in medical research has resulted in successful treatment of many life-threatening infectious diseases as well as autoimmune and lifestyle-related diseases, increasing life-expectancy of both the developed and developing world. As a result of a growing aging population, the focus has also turned on chronic diseases which seriously affect the quality of older patient life. Xerostomia (dry mouth) is one such condition, which leads to bad oral health and difficulty in consumption of dry foods and speech. Saliva substitutes are used to ease symptoms. However, they often don't work properly and objective comparison of saliva substitutes to mimic natural salivary functions does not exist [39]. Reduced saliva secretion (hyposalivation) and changed saliva composition are associated with xerostomia, that is, a dry mouth feeling [40, 41], and a variety of additional dryness-related complaints [42]. Approximately 25% of all elderly and 50–60% of hospitalized elderly suffer from xerostomia [43, 44]. Major causes are the use of medication [45, 46], Sjögren's syndrome [41, 47], and radiotherapy in the maxillofacial region [48, 49]. Although xerostomia is not a life-threatening condition, it certainly reduces the patient's quality of life, with a significant negative impact on healthy aging [50].

14.2.3.2 Artificial saliva

Stannard et al. [51] measured kinetic coefficients of friction for SS, β-Ti, NiTi, and Co–Cr archwires on a smooth SS or Teflon surface. A universal material-testing instrument was used to pull rectangular archwire (0.17 × 0.025 in) through pneumatically controlled binding surfaces. Coefficients of friction were determined under dry and wet (artificial saliva) conditions. It was found that (i) frictional force values (and thus, coefficients of friction) were found to increase with increasing normal force for all materials, (ii) β-Ti and SS were sliding against SS, and SS wire on Teflon consistently exhibited the lowest dry friction values, (iii) artificial saliva increased friction for SS, β-Ti, and NiTi wires sliding against SS, but (iv) artificial saliva did not increase friction of Co–Cr, SS sliding against SS, or SS wire on Teflon compared to the dry condition, and (v) SS and β-Ti wires sliding against SS, and SS wire on Teflon, showed the lowest friction values for the wet conditions [51]. Kusy et al. [52] evaluated coefficients of friction in the dry and wet (saliva) states of SS, Co–Cr, NiTi, and β-Ti wires against either SS or polycrystalline alumina brackets. It was found that (i) in the dry state and regardless of slot size, the mean kinetic coefficients of friction were smallest for the all SS combinations (0.14) and largest for the β-Ti wire combination (0.46), (ii) the coefficients of polycrystalline alumina combinations were generally greater than the corresponding combinations that included SS brackets, (iii) in the wet state, the kinetic coefficients of the all SS combinations increased to 0.05 over the dry state, but (iv) all β-Ti wire combinations in the wet state decreased to 50% of the values in the dry state. Alfonso et al. [53] analyzed the influence of the nature of the orthodontic archwires on the friction coefficient and wear rate against materials used commonly as brackets (Ti–6Al–4V and 316L SS). The materials selected as orthodontic archwires were ASI304 SS, NiTi, Ti, TiMo (or TMA or β-Ti), and Cu-NiTi. The array archwire's materials selected presented very similar roughness but different hardness. Materials were chosen from lower and higher hardness degrees than that of the brackets. Wear tests were carried out at in artificial saliva at 37 °C. It was shown that (i) there was a linear relationship between the hardness of the materials and the friction coefficients, (ii) the material that showed lower wear rate was the ASI304 (18Cr-8Ni-Fe) SS, and (iii) to prevent wear, the wire and the brackets have high hardness values and in the same order of magnitude. In sliding mechanics, frictional force is an important counter-balancing element to orthodontic tooth movement, which must be controlled in order to allow application of light continuous forces. Phukaoluan et al. [54] compared the frictional forces between a SS bracket and five different wire alloys under dry and wet (artificial saliva) conditions. TiNi, Cu-TiNi, Co-TiNi, commercial wires A, and commercial wires B with equal dimensions of 0.016 × 0.022″ were tested. The SS bracket was chosen with a slot dimension of 0.022″. Static and kinetic friction forces were measured using a custom-designed apparatus, with a 3-mm stretch of wire alloy at a crosshead speed of 1 mm/min. It was reported that (i) the static and dynamic frictions in the wet condition tended to decrease more slowly than those in the dry condition, (ii) the friction of Cu-TiNi and commercial wires B would increase, and (iii) these results were associated with scarred surfaces, that is the increase in friction would result in a larger bracket microfracture; suggesting that (iv) it is seen that copper addition resulted in an increase in friction under both wet and dry conditions, and (v) the friction in the wet condition was less than that in dry condition due to the lubricating effect of artificial saliva.

14.2.3.3 Other lubricant media

So far, the effects of human saliva or artificial saliva on frictional behavior of orthodontic archwire coupled with bracket (slot portion). There are a few studies on effects metallic dental devices on natural saliva chemistry and biology. Lara-Carrillo et al. [55] identified changes in the oral environment with clinical, salivary, and bacterial risk markers after placement of fixed orthodontic appliances on permanent dentition, and, with ethical approval, utilized different techniques to analyzed clinical, salivary, and bacterial risk markers in 34 patients (mean age, 16.7 ± 5.2 years), 14 males, and 20 females; before starting orthodontic treatment and a month after. Clinical risk markers (decayed, missing, and filled surfaces [DMFS], O'Leary's plaque index, and plaque pH); salivary markers (unstimulated and stimulated saliva flow rate, buffer capacity, pH, and occult blood in saliva) and bacterial counts (*Streptococcus mutans* and *Lactobacillus*). It was reported that (i) orthodontic appliances increased the stimulated salivary flow rate, buffer capacity, salivary pH and occult blood in saliva, (ii) bacterial levels increased slightly after a month of treatment, without statistical significance, and (iii) between genders, initially it was observed differences in: stimulated saliva, buffer capacity, and plaque pH; after treatment the unstimulated saliva showed differences. Based on these results, it was concluded that (iv) orthodontic treatment changes the oral environmental factors, promotes an increase in stimulated flow rate, buffer capacity and salivary pH, which augment the anticaries activity of saliva, and (v) in contrast, increased occult blood indicated more gingival inflammation, apparently because it augmented the retentive plaque surfaces and is difficult to maintain a good oral hygiene, and rinsed the bleeding in saliva by periodontal damage. The quality (defined as salivary protein content, viscosity, pH, and buffer capacity) and the quantity of saliva (mostly related to flow rate) play a crucial role in the equilibrium between demineralization and remineralization of enamel in a cariogenic environment [56]. Specific changes, such as increased pH, buffer capacity, and flow rate, may contribute to decreased susceptibility to dental caries [57, 58]. All these salivary properties become of utmost importance during orthodontic treatment with fixed appliances, when an increased chance of plaque retention and a greater difficulty in optimal oral hygiene maintenance are thought to predispose to enamel demineralization and white spot formation [59, 60]. There is still no consensus on the way the quality and the quantity of saliva change during orthodontic treatment [61]. So far, a few studies have tried to investigate the relationship between the placement of fixed orthodontic appliances and the change of nonmicrobial salivary properties, mostly with conflicting outcomes and short-term assessment (up to 6 months from bracket placement). Bonetti et al. [62] evaluated the salivary flow rate, pH and buffer capacity prior to the beginning of therapy and after one year from bracket placement using a simple and commercially available chairside saliva check kit. The study population consisted of 20 healthy patients (mean age, 16.5 ± 4 years) scheduled for fixed orthodontic treatment. Salivary samples were taken just before bracket bonding (T0; baseline assessment) and after one year of treatment (T1; half-treatment assessment) using the GC Saliva-Check Kit (GC Corp., Leuven, Belgium). It was reported that (i) no statistically significant difference was detected between T0 and T1 for the salivary parameters examined in the present study, and (ii) the placement of fixed orthodontic appliances did not change the salivary pH, buffer capacity, and flow rate after a year of treatment if compared with the baseline assessment.

14.2.3.4 Effects of metallic orthodontic devices on saliva

There are several studies on frictional behavior under influences of lubricant other than human saliva or artificial saliva. Berradja et al. [63] evaluated the frictional behavior of orthodontic archwires (SS and NiTi) in dry and wet conditions in vitro in three different environments: ambient air with 50% relative humidity, 0.9 wt% sodium chloride solution, and deionized water at 23 °C. It was found that (i) NiTi archwires sliding against alumina exhibited high coefficients of friction (about 0.6) in the three environments, (ii) SS archwires sliding against alumina had relatively low coefficients of friction (0.3) in the solutions, but high coefficients (0.8) in air, (iii) the low-frictional forces of the SS wires sliding against alumina in the solutions were due to a lubricating effect of the solutions and corrosion-wear debris, and (iv) the high frictional forces between the NiTi wires and alumina are attributed to an abrasive interfacial transfer film between the wires and alumina. Hosseinzadeh et al. [64] assessed the surface characterization and frictional resistance between SS brackets and two types of orthodontic wires made of SS and NiTi alloys after immersion in a chlorhexidine-containing prophylactic agent. SS orthodontic brackets with either SS or heat-activated NiTi wires were immersed in a 0.2% chlorhexidine and an artificial saliva environment for 1.5 h. The frictional force was measured on a universal testing machine with a crosshead speed of 10 mm/min over a 5-mm of archwire. It was obtained that (i) there was no significant difference in the frictional-resistance values between SS and NiTi wires immersed in either chlorhexidine or artificial saliva, (ii) the frictional resistance values for the SS and NiTi wires immersed in 0.2% chlorhexidine solution were not significantly different from that in artificial saliva, and (iii) no significant difference in the average surface roughness for both wires before (as-received) and after immersion in either chlorhexidine or artificial saliva was observed. It was, then, concluded that (iv) one-and-half-hour immersion in 0.2% chlorhexidine mouthrinse did not have significant influence on the archwires surface roughness or the frictional resistance between SS orthodontic brackets and archwires made of SS and NiTi, (v) chlorhexidine-containing mouth-rinses may be prescribed as nondestructive prophylactic agents on materials evaluated in the present study for orthodontic patients. Kao et al. [65] investigated and compared the levels of frictional resistance between metal brackets and orthodontic wires after immersion in an acidified phosphate fluoride (APF) agent. Three types of mandibular incisor SS metal brackets with beta-titanium alloy wire (TMA), heat-activated NiTi wire , and 2 sizes of SSWs were immersed in 0.2% APF and pH 6.75 artificial saliva solutions for 24 hours. The study included 480 bracket-wire specimens. It was reported that (i) in the APF-immersed group, the static frictional force was greater than the kinetic frictional force, (ii) the frictional forces of the orthodontic wires had statistically significant differences in this progressive order: TMA, Ni-Ti, and SSW. Similar frictional force results were obtained in the pH 6.75 saliva group, and (iii) the frictional force values of the APF group were higher than those of the pH 6.75 saliva group; indicating that (iv) the frictional forces of orthodontic brackets and wires are influenced by contact with fluoride-containing solutions. Geramy et al. [66] assessed the effect of 0.05% sodium fluoride mouthwash on the friction between orthodontic brackets and wire. Four types of orthodontic wires including rectangular standard SS, TMA, NiTi, and Cu-NiTi were selected. In each group, half of the samples were immersed in 0.05% sodium fluoride mouthwash and the others were immersed in artificial saliva for 10 hours. It was found that (i) the friction rate was significantly higher after immersion in 0.05% sodium fluoride mouthwash in comparison with artificial saliva, and (ii) Cu-NiTi wire showed the highest friction value followed by TMA, NiTi, and SS wires; concluding that (iii) the 0.05% sodium fluoride mouthwash increased the frictional characteristics of all the evaluated orthodontic wires. Zeng et al. [67] investigated the influence of load and sliding speed on super-low friction of Nitinol 60 alloy using Nitinol 60 alloy pin sliding over GCr15 steel disk under castor oil lubrication. It was mentioned that (i) super-low COF of Nitinol 60 alloy was achieved at the stable state, corresponding to so-called super-lubricity regime in the presence of castor oil, (ii) sliding speed employed in the friction tests plays great roles in the lubrication behaviors of Nitinol 60 alloy, (iii) when the friction tests are executed from low sliding speed to high sliding speed, COF decreases at the initial stage and on the contrary, COF increases at the stable stage, (iv) however, with the increase in load, COF firstly increases and then decreases at the initial stage, (v) COF is unstable but still super low and remains almost the same value at the stable stage, and (vi) the influence of sliding speed on the lubrication behaviors of Nitinol 60 alloy under castor oil lubrication is more obvious than that of load due to the characteristics structure of castor oil, therefore maintaining appropriate level of sliding speed to achieve super-low friction of Nitinol 60 alloy.

Dridi et al. [68] studied the influence of biolubricants (human saliva, olive oil, aloe vera oil, sesame oil, and sunflower oil) on the friction force of SS and NiTi rectangular archwires against SS brackets. Two types of brackets were used namely: SLBs and CBs. It was reported that (i) under oil lubrication, the friction behavior in the archwire/bracket assembly were the best, (ii) the SLB ligation was better than the conventional ligation system, and (iii) the enhancement of the frictional behavior with natural oils was linked to their main components: fatty acids.

14.2.4 Influencing parameters

All materials have two coefficients of friction: static coefficient and kinetic coefficient [69]. Suppose a box is sitting on the tabletop, the static friction is what keeps the box from moving without being pushed, and it must be overcome with a sufficient opposing force before the box will move; while the kinetic friction (also referred to as dynamic friction) is the force that resists the relative movement of the surfaces once they're in sliding motion. Sliding between bracket and wire in the oral cavity occurs at a low velocity as a sequence of short steps rather than as a continuous motion [70]. In such conditions, the distinction between static and kinetic frictional resistance is arbitrary because these two forms of friction are dynamically related [71]. In orthodontics, a tooth undergoing a sliding movement along an archwire goes through many tipping and uprighting cycles, moving in small increments. Therefore, orthodontic space closure depends more on static friction than on kinetic friction [72].

14.2.4.1 Surface roughness

Surface roughness and static frictional force resistance of orthodontic NiTi archwires along with β-Ti alloy wire, SS, and Co–Cr alloy wires were measured. It was found that (i) the Co–Cr alloy and the NiTi alloy wires exhibited the lowest frictional resistance, (ii) the SS alloy and the β-Ti wires showed the highest frictional resistance, and (iii) no significant correlation was found between arithmetic average roughness and frictional force values [11].

To investigate whether the surface roughnesses of opposing materials influence the coefficients of friction and ultimately the movement of teeth, archwires were slid between contact flats to simulate orthodontic archwire–bracket appliances [73]. From laser specular reflectance measurements, the RMS surface roughness of these archwires varied from 0.04 μm for SS to 0.23 μm for NiTi. Using the same technique, the roughness of the contact flats varied from 0.03 μm for the 1 μm lapped SS, to 0.26 μm for the as-received alumina. After each of the archwire-contact flat couples was placed in a friction tester, fifteen normal forces were systematically applied at 34 °C. It was reported that (i) from plots of the static and kinetic frictional forces versus the normal forces, dry coefficients of friction were obtained that were greater than those reported in the dental literature, (ii) the all SS couples had lower kinetic coefficients (0.120–0.148) than the SS-polycrystalline alumina couple (0.187), and (iii) when pressed against the various flats, the β-Ti archwire (RMS = 0.14 μm) had the highest coefficients of friction (0.445–0.658), although the NiTi archwire was the roughest (RMS = 0.23 μm).

14.2.4.2 Surface treatment

Oshida et al. [74] conducted friction tests to compare the wet static frictional forces of low friction "colors" (surfaces of wires are colored by anodizing process) TMA (Ti–Mo alloy) archwires with archwires of other materials (SS, NiTi, and uncoated TMA), and to test the effects of repetitive sliding. The results showed that uncoated TMA wires produce the highest wet static frictional forces in all cases. In most cases, NiTi wires produced the next highest force levels followed by the colored TMA wires, and then SS. Repetition seemed to have little to no effect on all of the archwires with the exception of NiTi and uncoated TMA. NiTi wires showed a decrease in force values as the wire was subjected to more repetitions. Uncoated TMA, to the contrary, showed that more friction developed between runs and a trend toward increasing friction as the wire had repeated use [57]. It is then recommended that further studies are required to correlate types of oxide(s) formed on Ti surfaces to frictional behaviors. Espinar et al. [75] studied NiTi orthodontic archwires that have been treated using a new oxidation treatment for obtaining Ni-free surfaces. It was mentioned that (i) the titanium oxide on the surface significantly improves corrosion resistance and decreases nickel ion release, while barely affecting transformation temperatures, (ii) this oxidation treatment avoids the allergic reactions or toxicity in the surrounding tissues produced by the chemical degradation of the NiTi; on the other hand, the lack of low friction coefficient for the NiTi superelastic archwires makes the optimal use of these materials in orthodontic applications difficult, and (iii) the decrease of this friction coefficient has been achieved by means of oxidation treatment. Meier et al. [76] investigated whether electrochemical surface treatment of NiTi and TiMo archwires (OptoTherm™ and Beta-Titan™) reduces friction inside the bracket–archwire complex. The material properties of the surface-treated wires (Optotherm/LoFrix™ and BetaTitan/LoFrix™) were compared to untreated wires made by the same manufacturer (see above) and by another manufacturer (Neo Sentalloy®; GAC, Bohemia, NY, USA). It was reported that (i) force losses due to friction were reduced by 10% points (from 36% to 26%) in the NiTi and by 12% points (from 59% to 47%) in the TiMo wire specimens. Most of the other material properties exhibited no significant changes after surface treatment, while the three-point bending tests revealed mildly reduced force levels in the TiMo specimens due to diameter losses of roughly 2%, these force levels remained almost unchanged in the NiTi specimens; concluding that compared to untreated NiTi and TiMo archwire specimens, the surface-treated specimens demonstrated reductions in friction loss by 10% and 12% points, respectively. Since the low friction of NiTi wires allows a rapid and efficient orthodontic tooth movement, Wichelhaus et al. [77] investigated the friction and surface roughness of different commercially available superelastic NiTi wires before and after clinical use. The surface of all wires had been pretreated by the manufacturer. Forty superelastic wires (Titanol Low Force, Titanol Low Force River Finish Gold, Neo Sentalloy, and Neo Sentalloy Ionguard) of diameter 0.016 × 0.022 in. were tested. The friction for each type of NiTi archwire ligated into a commercial SS bracket was determined with a universal testing machine. Having ligated the wire into the bracket, it could then be moved forward and backwards along a fixed archwire while a torquing moment was applied. It was found that (i) initially, the surface treated wires demonstrated significantly less friction than the nontreated wires, (ii) the surface roughness showed no significant difference between the treated and the nontreated surfaces of the wires, and (iii) all 40 wires however showed a significant increase in friction and surface roughness during clinical use, suggesting that (iv) while the Titanol Low Force River Finish Gold (Forestadent, Pforzheim, Germany) wires showed the least friction of all the samples and consequently should be more conservative on anchorage, the increase in friction of all the surface treated wires during orthodontic treatment almost cancels out this initial effect on friction, (v) it is therefore recommended that surface treated NiTi orthodontic archwires should only be used once. The lack of low friction coefficient for the NiTi superelastic archwires makes the optimal use of these materials in orthodontic applications difficult. Gil et al. [78] achieved the decrease of this friction coefficient by means of nitrogen diffusion heat treatments.

14.2.4.3 Coating

Huang et al. [79] coated DLC films onto NiTi orthodontic archwires. The film protects against fluoride-induced corrosion and will improve orthodontic friction. The influence of a fluoride-containing environment on the surface topography and the friction force between the brackets and archwires were investigated. It was obtained that (i) the superior nature of the DLC coating was noticed, with less surface roughness variation for DLC-coated archwires after immersion in a high fluoride ion environment, and (ii) friction tests also showed that applying a DLC coating significantly decreased the fretting wear and the COF, both in ambient air and artificial saliva; concluding that DLC coatings are recommended to reduce fluoride-induced corrosion and improve orthodontic friction. Rokaya et al. [80] evaluated the mechanical and tribological properties of graphene oxide/silver nanoparticle (GO/AgNP)-coated medical-grade NiTi alloy. It was reported that (i) the coating thickness ranged from 0.46–1.34 μm and the mean surface roughness (Ra) ranged from 50.72–69.93 nm, (ii) increasing the coating time from 1–10 min increased the roughness, thickness, and elastic modulus of surface coating, (iii) the friction coefficients of the coated NiTi alloy were significantly lower compared with that of the uncoated NiTi alloy, and (iv) the GO/AgNP nanocomposite coated NiTi alloy demonstrated improved mechanical strength and a reduced friction coefficient that would be more favorable for biomedical applications. Bandeira et al. [81] evaluated the friction force in sliding systems composed of coated NiTi archwires, coated NiTi subjected to thermal cycling, and coated NiTi subjected to acid solution immersion, and compare them to NiTi and polymeric wires. Samples of NiTi ($n = 05$), coated NiTi ($n = 15$), and OPTIS ($n = 05$) 0.016 inches in diameter and 50 mm long, in conjunction with Metafasix ligatures and saliva in InVu brackets, were submitted to friction testing. It was found that (i) the mean (± standard deviation) maximum friction force for NiTi, coated NiTi, and OPTIS was 105.20 ± (2.63); 99.65 ± (0.64); 59.76 ± (4.93), respectively, (ii) there was no significant difference in NiTi, coated NiTi, and acid-immersed coated NiTi, (iii) among the thermal-cycled or acid-immersed coated NiTi wires there was lower friction force in those undergoing thermal cycling, (iv) the coated NiTi and the OPTIS presented homogeneous surfaces, whereas NiTi wires presented a heterogeneous surface, and (v) fractures were observed in the coated NiTi wires that underwent thermal cycling; concluding that (vi) OPTIS, thermal-cycled coated NiTi, coated NiTi, NiTi, and acid-immersed coated NiTi presented, respectively, increasing values of maximum friction force. Kachoei et al. [82] fabricated a friction-reducing and antibacterial coating with zinc oxide (ZnO) nanoparticles on NiTi wire. NiTi orthodontic wires were coated with ZnO nanoparticles using the chemical deposition method. Characteristics of the coating as well as the physical, mechanical, and antibacterial properties of the wires were investigated. It was mentioned that (i) a stable and well-adhered ZnO coating on the NiTi wires was obtained, (ii) the hardness and elastic modulus of the ZnO nanocoating were 2.3 ± 0.2 and 61.0 ± 3.6 GPa, respectively, (iii) the coated wires presented up to 21% reduction in the frictional forces and antibacterial activity against *Streptococcus mutans*, (iv) ZnO nanocoating significantly improved the surface quality of NiTi wires, and (v) the modulus of elasticity, unloading forces and austenite finish temperature were not significantly different after coating. Based on these results, it was concluded that (vi) the coating could be implemented into practice for safer and faster treatment to the benefit of both patient and clinician. A friction-reducing coating was applied onto different NiTi substrates using inorganic fullerene-like tungsten disulfide (IF-WS$_2$) nanoparticles to estimate in vitro friction reducing extent of the coating [83]. Different NiTi substrates were coated with cobalt and IF-WS$_2$ nanoparticles film by the electrodeposition procedure. It was reported that (i) stable and well-adhered cobalt + IF-WS$_2$ coating of the NiTi substrates was obtained, and (ii) friction tests presented up to 66% reduction of the friction coefficient, indicating that NiTi alloy is widely used for many medical appliances; hence, this unique friction-reducing coating could be implemented to provide better manipulation and lower piercing rates.

14.2.4.4 Type and setting of bracket

The interaction between the bracket of an axially rotated tooth and archwire produces a moment. This moment influences tooth movement and rotational control and is itself influenced by bracket width and bracket ligation. Bednar et al. [84] mentioned that (i) empirically both bracket width and ligation technique significantly affect the moment produced during axial rotation, (ii) for the range of bracket widths and types evaluated, *ligation technique* was found to have a greater influence on moment production than did bracket width, and (iii) the self-ligated spring clip bracket delivered the least force over the greatest range of axial rotation. Kusy et al. [30] and Whitley et al. [85] investigated the influence of IBD on the RS) and found that the RS was inversely proportional to the total IBD regardless of archwire alloy or bracket slot. Angolkar et al. [86] determined the frictional resistance offered by ceramic brackets used in combination with wires of different alloys and sizes during in vitro translatory displacement of brackets. It was reported that (i) wire friction in the ceramic brackets increased as wire size increased, and rectangular wires produced greater friction than round wires, and (ii) the β-Ti and NiTi wires were associated with higher frictional forces than SS or cobalt-chromium wires.

14.3 Wear and wear debris toxicity

14.3.1 Wear phenomenon

Titanium–nickel alloys based on the unusual intermetallic compound NiTi exhibit a ductility comparable with metallic alloys. The primary interest in these materials has focused on their shape-memory alloy effects and superelasticity. Near-equiatomic TiNi alloy has been found to exhibit high resistance to wear, especially to erosion.

The high wear resistance of the alloy may largely benefit from its superelasticity [87]. Wear behavior of NiTi SMA is closely corresponds to deformation mechanisms associated with different plastic strain accumulation process. In general, high wear resistance can prevent the loosening of biomedical implants. Although Ti alloy has high specific strength and high corrosion resistance, their wear resistance is not desirable. Among versatile medical applications of NiTi alloys, the wear phenomenon becomes a crucial issue when they are used as endodontic instruments and orthopedic implants. Unlike dental implants, most orthopedic implants will be subjected to biotribological actions. Metal ions released from the implant surface are suspected of playing some contributing role in the loosening of hip and knee prostheses, which are substantially subjected to biotribological environments. Because of its biocompatibility, superelasticity and shape-memory characteristics, NiTi alloys have been gaining immense interest in the medical field. However, there is still concern on the corrosion resistance of this alloy if it is going to be implanted in the human body for a long time. Titanium is not toxic but nickel is carcinogenic and is implicated in various reactions including allergic response and degeneration of muscle tissue. Debris from wear and the subsequent release of Ni^+ ions due to corrosion in the body system are fatal issues for long-term application of this alloy in the human body [88].

The wear behavior can be most easily understood when we consider the human hip joint, in which wear can be designated as reciprocating sliding wear, because the contact area is smaller than the stroke of the wear path. Furthermore, the wear paths of the back and forth section of the cycle do not lie on the same geometrical lines, which lead to sliding wear. Even though, in sliding as well as in reciprocating sliding wear, all the other wear process – adhesion, abrasion, surface fatigue, and tribochemical (or tribocorrosion) reactions [89]. If such healthy hip joint is deteriorated due to various reasons, the total hip arthroplasty (THA) will be surgically operated. The hip is one of the most important joints that support our body, having the task of joining the femurs with the pelvis. The smooth and spherical head of the femur fits perfectly into the natural seat of the acetabulum, which is a cup-shaped cavity; the whole joint is wrapped in very resistant ligaments that make the joint stable. The hip joint is subjected to high daily stresses, having to bear the weight of the upper part of the body. Thus, especially with advancing age, these stresses can jeopardize its functioning. Among the bearing surfaces involved in THA, the biomaterials are submitted to sliding friction, producing particle debris, which, in turn, initiate an inflammatory reaction ultimately leading to osteolysis [90]. Wear is defined as a cumulative surface damage phenomenon in which material is removed from a body in the form of small particles, primarily by mechanical processes [91]. The wear mechanism is the transfer of energy with removal or displacement of material and in that follows an explanation of the mechanisms of wear observed with different biomaterials.

Liang et al. [92] studied the wear behavior of a TiNi alloy after various heat treatments in three conditions: sliding wear, impact abrasion and sand-blasting erosion. It was shown that (i) for all three conditions, the TiNi in a pseudoplastic state shows much better wear resistance as compared with that in a pseudoelastic state, and (ii) the wear resistance of a TiNi alloy is mainly dependent on the recoverable strain limit, that is, the sum of the pseudoelastic and pseudoplastic strain limits. Eggert et al. [93] evaluated defects of Lightspeed cutting tips before and after usage. Instrument sizes 20 to 32.5, 35 to 60, and 65 to 100 were used in 9, 18, and 36 canals, respectively, and autoclaved after shaping every third root canal. It was reported that (i) no instruments fractured during the test, but all the cutting heads had one or more imperfections, even before usage, (ii) the presence of debris, pitting, and metal strips changed significantly. Imperfections were found on new and used Lightspeed cutting heads, indicating the general difficulty in machining defect-free NiTi rotary instruments, and (iii) however, high quality should remain a goal to improve instrument efficiency. Liu et al. [94] studied the deformation and crack initiation of a wearing surface during sliding abrasion, using a finite element model. As finished surfaces consist of asperities, when two surfaces come into contact; normal and tangential loads are transmitted through interacting asperities. The damage and removal of materials during the wear process are therefore a consequence of the asperity interaction. A finite element model was employed to investigate such a situation of asperity interaction, with the emphasis on the plastic deformation in the contact region and its accumulation, which resulted in eventual cracking. The finite element model was applied to the contact deformation in superelastic TiNi alloy, with a comparison to 304 SS. The plastic strain and thus the life of an asperity before cracking were analyzed. It was demonstrated that the superelasticity is greatly beneficial to the wear resistance of TiNi alloy due to the reduction of accumulated plastic strain. Gialanella et al. [95] investigated the wear behavior of two NiTi shape-memory alloys, one of them being martensitic, the other one austenitic at room temperature. Wear tests have been conducted with a disk-on-block geometry. The block was made of the NiTi alloy, whereas counter-face disk materials were AISI M2 high-speed steel and a WC–Co hard metal. It was reported that (i) in the wear tests involving the M2 steel disk, both NiTi alloys display a transition, as a function of the applied load, from a mainly oxidation regime to a more complex situation, in which oxidation wear is accompanied by delamination of metallic alloy fragments, and (ii) higher wear rates of the shape-memory alloys have been observed for the NiTi/WC–Co coupling, in which a transition from a mainly delamination wear to a regime featuring a mixture of delamination and oxidation wear has been observed. Reciprocating instruments made from M-wire alloy have been proposed to reduce the risk of fracture. Two reciprocating NiTi instruments were used on extracted teeth up to three times. ESEM/EDS analysis was conducted to determine defects, alterations, and wear features of the apical third of instruments and metallographic analysis was performed on the cross-section of new and used instruments to compare alloy properties [96]. It was reported that (i) no instrument fractured and no spiral distortions were observed under optical microscope even when the number of uses increased, (ii) no significant differences were found for WaveOne and Reciproc, (iii) blades presented a wrapped portion in WaveOne group and a more symmetrical feature in Reciproc group, (iv) metallographic analysis revealed in both groups the presence of twinned martensitic grains with isolated flat austenitic areas, and (v) both instruments demonstrated limited alteration, such as tip deformation and wear, confirming the safe clinical use of both instruments for shaping multirooted teeth.

14.3.2 Influencing factors on wear

There are several studies on effect of phase composition and transformation on wear behavior. Abedini et al. [97, 98] studied the wear behavior of $TiNi_{50.3}$ (at%) alloy in martensitic and austenitic states was studied. The alloy was prepared in a vacuum induction melting furnace, forged at 800 °C, annealed at 1,000 °C for 12 h, quenched in water, then aged at 400 °C for 1 h and followed by water quenching. It was mentioned that (i) the highest deflection recovery due to the superelasticity was observed at temperature of 50 °C, and (ii) the wear tests were conducted using a pin-on-disk tribometer in a water media at temperatures ranging from 0 to 50 °C, resulting that the wear rate of NiTi alloy was decreased as the wear testing temperature increased; due to mainly attributing to the superelasticity effect and higher strength of the alloy in the austenitic state at temperature of 50 °C, indicating a lower COF in the austenitic state compared to the martensitic state. Yan et al. [99] studied the deformation mechanisms of NiTi shape-memory alloy during the wear process at different temperatures when different microstructures are present. Three temperature regimes were selected namely, $T < M_F$, $A_S < T < A_F$, and $T > A_F$, where fully martensitic, martensite coexisting with austenite, and fully austenitic microstructures were formed, respectively. It was reported that (i) when $T < M_F$, it was observed that the COF had decreased initially and thereafter stabilized at a lower value with increasing wear cycles, (ii) more decrease was found when the temperature was near to A_S, (iii) furthermore, when tested above A_F, the COF had decreased more significantly under higher load, and (iv) difference in the trend of COF at different temperatures is originated from the different deformation mechanisms involved in the wear process, particularly the martensite detwinning process, the stress-induced phase transformation process, and the plastic deformation of martensite. Furthermore, Yan et al. [100] conducted the ball-on-disk sliding wear tests on austenitic NiTi alloy with alumina counter ball were conducted at different temperatures and under different loads. Based on the coefficients of friction, the surface wear features, temperature-dependent stress–strain curves and the estimated contact stresses, the deformation mechanisms involved in the wear process were examined. It was reported that (i) there were two wear modes, (ii) mode I is temperature-sensitive and occurred when $A_F < T < M_D$ and wear process was dominated by the interplay among contact stress (elastic deformation of austenite or stress-induced martensitic transformation), temperature, (by increasing temperature, the wear resistance is improved) and shape recovery property, and (iii) mode II occurs when $T > M_D$ and it is less temperature-sensitive within the testing range, in which the austenitic NiTi loses its superelasticity and obeys a conventional deformation sequence, and the key factor dominating the wear process is the magnitude of contact stress. Aliasgarian et al. [101] investigated effects of heat treatments and applied loads on the tribological behavior of $TiNi_{56.5}$ and $TiNi_{57.5}$ (wt%) alloys. Wear tests were performed on a pin-on-disk tribometer under normal loads of 20 and 60 N at a sliding speed of 0.3 m/s. It was found that (i) the alloys aged at 700 °C showed lower hardness comparing to the alloys aged at 400 °C, (ii) under an applied load of 20 N, the samples aged at 700 °C showed better wear behavior in comparison with the samples aged at 400 °C with a higher hardness; attributing to higher toughness of the samples aged at 700 °C, (iii) the wear of the samples aged at 400 °C decreased with the increase in normal load; however, there was an increase in the wear of the samples aged at 700 °C with the increase of load, and (iv) formation and stability of tribological layers on the contacting surfaces could be the main reason for the reduction of the wear of the samples aged at 400 °C with the increase in normal load.

Liu et al. [102] investigated the wear resistance of nanocrystalline NiTi sheets with different grain sizes of 10, 42, and 80 nm by nanowear tests under different normal forces of 100, 300, and 500 µN. It was indicated that (i) the minimization of grain size remarkably improves the nanohardness of alloy since the martensitic phase transformation is restrained, and affects the competition between the hardness and the martensitic phase transformation, consequently makes direct influence upon the wear resistance, and (ii) the nanohardness is a dominant factor of nanocrystalline NiTi sheets against wear until the martensitic phase transformation gains an advantage under a heavy load condition. Wear behavior of NiTi SMA is closely corresponds to deformation mechanisms associated with different plastic strain accumulation process, as has been described in the above. Plastic strain accumulation is achieved by dislocation motion; however, grain boundary acts as a strong barrier. Yan et al. [103] studied the wear behavior of single-crystalline and polycrystalline NiTi SMAs to understand the effect of grain boundary on the plastic strain accumulation in the wear process. Wear tests were conducted at $M_F < T < A_F$, where phase boundary exists between martensitic and austenitic phases. Tests were conducted under ball-on-disk sliding wear mode, and alumina (Al_2O_3) counterbody was used. For single-crystalline NiTi SMA, transition wear occurred even when the applied load was relatively low (i.e., 100 mN). It was mentioned that (i) for polycrystalline NiTi SMA, with increasing applied load and wear cycles, the wear has shifted from near-zero wear stage to severe wear stage; no transition behavior was observed, and (ii) significant differences in the wear process were discussed with respect to deformation mechanisms associated with dislocation motion in the single-crystalline and polycrystalline NiTi SMAs.

14.3.3 Improvement of wear resistance

In general, high wear resistance can prevent the loosening of biomedical implants. Although Ti alloys possess relatively high specific strength (strength/specific weight) and high corrosion resistance, their wear resistance is not at satisfactory level [104, 105]. To enhance the wear resistance, a variety of surface modification and technologies that was discussed in Chapter 12 would be applicable. Although physical deposition methods and thermochemical surface treatments can improve the wear resistance of Ti alloy implants to some extent, some problems still exist. The physical deposition methods are prone to delaminate between the coating and substrate under the condition of repeated loading due to mismatching the interfacial stress field. The thermochemical surface treatments are usually conducted at high temperatures (HT), leading to a twist of the substrate. Therefore, surface severe plastic deformation methods have been proposed and developed to overcome these problems associated with a certain types of surface modifications [106–109].

14.3.3.1 Improvement by metal addition

Ng et al. [88] reported the corrosion and wear properties of laser surface modified NiTi using Mo and ZrO_2 as surface alloying elements, respectively. It was mentioned that the modified layers which are free from microcracks and porosity, act as both physical barrier to nickel release and enhance the bulk properties, such as hardness, wear resistance, and corrosion resistance. Zhang et al. [110] studied the effects of Nb element alloying to NiTi alloy and found that (i) it is significant for practical applications because impact energy increases while the wear rate decreases, and (ii) the abnormal wear behavior of the TiNiNb alloy can be attributed to the excellent wear behavior of amorphous structure and the consumption of impact energy during amorphous structure production. Zhao et al. [111] modified NiTi shape-memory alloy by Hf ion implantation to improve its wear resistance and surface integrity against deformation. It was reported that (i) the Auger electron spectroscopy and x-ray photoelectron spectroscopy results indicated that the oxide thickness of NiTi alloy was increased by the formation of TiO_2/HfO_2 nanofilm on the surface, (ii) the nanohardness measured by nanoindentation was decreased even at the depth larger than the maximum reach of the implanted Hf ion, (iii) the lower COF with much longer fretting time indicated the remarkable improvement of wear resistance of Hf implanted NiTi, especially for the sample with a moderate incident dose, (iv) the formation of TiO_2/HfO_2 nanofilm with larger thickness and decrease of the nanohardness played important roles in the improvement of wear resistance, and (v) Hf implanted NiTi exhibited larger superelastic recovery strain and retained better surface integrity even after being strained to 10% as demonstrated by in situ scanning electron microscope observation. Ahmadi et al. [112] demonstrated that by addition of yttrium, hardness properties and resistance to wear and corrosive wear of TiNi alloy were improved. New yttrium-rich regions were formed in microstructure of TiNi alloy. The improved properties of this alloy by the yttrium addition could be attributed to the formation of these regions. It was mentioned that there was an optimum content for addition of yttrium between 2% and 5% (in wt%), and above this content the improvement in properties of TiNi became minor.

Alhumdany et al. [113] investigated the effects of added yttrium and Tantalum on the wear behaviors of NiTi shape-memory alloys to examine the effects of compacting pressure levels between 400 and 650 MPa and to investigate wear parameters such as load, sliding distance, and time to determine the wear rate for the prepared alloys. Yttrium and tantalum were then added at weight percentages 1, 2, and 3 wt% of each element (at the expense of nickel) to basic chemical composition of 55 wt% Nickel and 45 wt% titanium. The wear behavior was then studied using pin-on-disk tests at variable loads and times (2, 5, 10, and 15 N and 5, 10, 15, and 20 min, respectively). It was mentioned that (i) the wear volume loss decreases with the addition of tantalum by 0.52% with 3% tantalum addition at 650 MPa compacting stress for a 15 N load, (ii) it also decreases by 0.48% with 2% yttrium addition at 650 MPa for a 10 N load, (iii) the wear volume loss further decreases as the compacting stresses increase, and (iv) the wear rate increased as the load and time increased for all tested specimens.

14.3.3.2 Ion implantation modification

In order to improve the surface properties of NiTi such as corrosion resistance, Tan et al. [114] employed plasma source ion implantation (PSII) technique with oxygen as incident ions at three levels of implantation dose (5×10^{16}, 1×10^{17}, and 3×10^{17} ions/cm^2). It was shown that (i) corrosion behavior was influenced by both heat treatment and surface modification, (ii) the best pitting corrosion resistance was observed for samples with $A_F = 21\ °C$ modified by oxygen implantation at a dose of 1×10^{17} ions/cm^2, and (iii) better wear-corrosion resistance was observed for oxygen-implanted samples. Using suitable surface modifications, it is possible to form pure titania (rutile) surface layers, thus prohibiting the outdiffusion of toxic nickel cations. Mändl et al. [115] studied the wear behavior and the lifetime of such surface layers, produced by oxygen plasma immersion ion implantation (PIII) at 25 kV and 250–550 °C to compare to that of untreated NiTi. It was reported that (i) ball-on-disk tests with intermittent impact loading revealed a specific wear volume of the treated samples comparable to that of the base material, slightly lower and higher depending on the process conditions, (ii) an increased fatigue lifetime was found for lower temperatures and higher oxygen fluences, indicating that the layer thickness is not the decisive factor, and (iii) instead, internal stress relaxation and atomic rearrangement are proposed as the dominant mechanism. Similarly, Mohan et al. [116] implanted oxygen ions by the PIII (PIII–O) technique at low temperature (LT) without affecting the substrate properties. It was found that (i) polarization and electrochemical impedance spectroscopy studies reveal nearly ideal capacitor behavior with better passivation characteristics for the oxygen-implanted substrate, and (ii) sliding wear studies reveal lower friction of coefficient for the implanted layers as compared with the substrate. The bare and surface modified NiTi alloy samples are evaluated for biocompatibility using osteoblast-like cells (MG-63). It was also mentioned that (iii) oxygen implantation by PIII–O provides a better compatible surface for cell attachment and growth, and (iv) the modified surface exhibits a higher percentage of cell viability demonstrating the enhanced biocompatibility of the oxygen-implanted surface compared with bare NiTi alloy. NiTi shape-memory alloy samples were plasma-implanted with nitrogen at voltages ranging from –10 to –40 kV [117]. It was reported that (i) x-ray photoelectron spectroscopy results disclose the formation of gradient TiN layers which thicknesses and elemental in-depth distributions depend on the applied voltages, and (ii) the wear resistance of the plasma-implanted NiTi samples increases with implantation voltages and decreases with the applied loads; indicating that (iii) the wear mechanism of the implanted samples is adhesive-dominant under low applied loads but becomes abrasive-dominant at high applied loads. Alves-Claro et al. [118] treated NiTi files with PIII to enhance the wear resistance. It was reported that (i) the hardness values found for the treated NiTi files were significantly lower than the hardness values measured before the implantation process, and (ii) the comparison of commercially available instruments shows that the wear resistance of the SS file is higher than the resistance of the NiTi, concluding that (iii) the surface treatment significantly increased the NiTi files wear resistance.

14.3.3.3 Composite reinforcement

Luo et al. [119] developed a tribocomposite such as TiC as a reinforcing phase. Attempt was made to develop such a tribocomposite, using nano-TiN powder to strengthen the matrix of the TiC/TiNi composite. The composite (TiC/TiNi) was made using a vacuum sintering process. Sliding wear behavior of this material was evaluated. It was reported that the nano-TiN/TiC/TiNi composite exhibited excellent wear resistance, superior to those of the TiC/TiNi composite and WC/NiCrBSi hardfacing overlay. Akalin et al. [120] investigated the wear characteristics of Al6061 composites, reinforced with short NiTi fibers. The NiTi/Al6061 composite samples were fabricated using pressure-assisted sintering process in ambient air where the NiTi fibers are aligned unidirectional in the Al matrix. In addition, NiTi/Al6061 composite with 5 wt% SiC5 wt% SiC particulates and monolithic Al6061 and Al6061 with 5 wt% SiC5 wt% SiC particulates were processed in similar conditions. The wear tests were performed using a reciprocating tribometer in ball-on-flat configuration where the counterbody material was martensitic steel. It was mentioned that (i) transverse NiTi fibers improve the wear resistance significantly, (ii) samples with transverse fiber orientation show mostly abrasive wear, whereas, monolithic and parallel samples show adhesive wear mechanism, (iii) in addition, SiC reinforcements improve the wear resistance of the composite and the monolithic samples, and (iv) since the Al6061 matrix material is smeared onto NiTi fibers in a short period, all composite samples show similar frictional characteristics after certain period of running in dry sliding.

14.3.3.4 Surface modification

Zhang [121] studied the wear property of Mo surface-modified layer in TiNi alloy prepared by the double glow plasma surface alloying technique. It was obtained that

(i) the x-ray diffraction analysis revealed that the modified layers were composed of Mo, MoTi, MoNi, and Ti_2Ni, (ii) the microhardness of the Mo modified layers treated at 900 °C and 950 °C were 832.8 HV and 762.4 HV, respectively, which was about three times the microhardness of the TiNi substrate, (iii) compared with as-received TiNi alloy, the modified alloys exhibited significant improvement of wear resistance against Si_3N_4 with low normal loads during the sliding tests, and (iv) mass spectrometry displayed that the Mo alloy layers had successfully inhibited the Ni release into the body. Ozeki et al. [122] deposited DLC films on SS, Ti, and NiTi substrates using a radiofrequency plasma chemical vapor deposition method. It was reported that (i) the XPS results showed that the N_2 and O_2 plasma pretreatment produced nitride and oxide on the substrate surfaces, such as TiO_2, TiO, Fe_2O_3, CrN, and TiNO, (ii) in the pull-out test, the adhesion strengths of the DLC film to the SS, Ti, and NiTi substrates were improved with the plasma pretreatment, (iii) in the ball-on-disk test, the DLC-coated SS, Ti, and NiTi substrates without the plasma pretreatment showed severe film failure following the test, (iii) the DLC coated SUS and NiTi substrates with the N_2 plasma pretreatment showed good wear resistance, compared with that with the O_2 plasma pretreatment. Vinothkumar et al. [123] investigated the role of dry cryogenic treatment (CT) temperature and time on the Vickers hardness and wear resistance of new martensitic NiTi SME alloy. Fifteen cylindrical specimens and 50 sheet specimens were subjected to different CT conditions: Deep cryogenic treatment (DCT) 24 group: –185 °C, 24 h; DCT six group: -185 °C, 6 h; shallow cryogenic treatment (SCT) 24 group: –80 °C, 24 h; SCT six group: –80 °C, 6 h; and control group. Wear resistance was assessed from weight loss before and after reciprocatory wet sliding wear. It was reported that (i) the as-received NiTi alloy contained 50.8 wt% nickel and possessed austenite finish temperature (A_F) of 45.76 °C, (ii) reduction in Vickers hardness of specimens in DCT 24 group was highly significant, and (iii) the weight loss was significantly higher in DCT 24 group; concluding that (iv) deep dry CT with 24 h soaking period significantly reduces the hardness and wear resistance of NiTi SME alloy. TiNi shape-memory alloy and its composite using δ-Al_2O_3 nanosize particles were prepared by the powder metallurgy method, and some mechanical properties like hardness, wear, and corrosion behavior were investigated [124]. It was mentioned that (i) the lower wear rate was obtained for the nano-Al_2O_3-reinforced Ti alloy composite due to increased hardness, but the wear rate increased considerably with increasing the load over 25 N for Ti alloy, and (ii) however, the best corrosion resistance was obtained for the base alloy, which is very important for implant applications. It was, then, concluded that (iii) the microhardness of NiTi alloy improved with the introduction of Al_2O_3particles.

14.3.4 Wear debris toxicity

Unlike dental implants, most orthopedic implants will be subjected to biotribological actions. Metal ions released from the implant surface are suspected of playing some contributing role in the loosening of hip and knee prostheses, which are substantially subjected to biotribological environments. The fresh surface will be revealed due to the formation of fraction/wear products, and there may be a chemical reaction taking place between the freshly revealed surface of implants, and the surrounding environments, and the selective dissolution of the alloy constituents into the surrounding tissues. Besides, it is generally believed that the solid powder such as wear/friction products particles are allergic reaction to the living tissue, so that the biological effects of wear products (debris) should be considered separately from the biocompatibility of the implants. Particulate wear debris is detected in histocytes/macrophages of granulomatous tissues adjacent to loose joint prostheses. Such cell–particle interactions have been simulated in vitro by challenging macrophages with particles doses according to weight percent, volume percent, and number of particles. Macrophages stimulated by wear particles are expected to synthesize numerous factors affecting events in the bone–implant interface. Wear debris from orthopedic joint implants initiates a cascade of complex cellular events that can result in aseptic loosening of the prosthesis [125]. Macrophases are cells which are mainly involved in phagocytosis and signaling and maintaining inflammation, which leads to cell damage ion soft tissues and bone resorption [126, 127]. The presence of large particles provides the major stimulus for cell recruitment and granuloma formation, whereas the small particles are likely to be the main stimulus for the activation of cells which release proinflammatory products. Apoptosis can be morphologically recognized by a number of features, such as loss of specialized membrane structures, blebbing, condensation of the cytoplasm, condensation of nuclear chromatin, and splitting of the cell into a cluster of membrane-bound bodies [126, 127]. Recent advances in the understanding of the mechanisms of osteoclastogenesis and osteoclast activation at the cellular and molecular levels have indicated that bone marrow-derived macrophages may play a dual role in osteolsyis associated with the total joint replacement: (1) the major cell in host defense responds to UHMWPE particles via the production of cytokines, and (2) the precursors for the osteoclasts responsible for the ensuing bone resorption [128]. The long-term effects of metal-on-metal arthroplasty are currently under scrutiny because of the potential biological effects of metal wear debris [129]. Information regarding metal-induced toxicity is based on a limited amount of epidemiological and experimental studies involving in vitro and in vivo models. Unfortunately, there are few data available on the systemic effects of metal in arthroplasty patients as follows [129]; the blood [130, 131], the immune system [132, 133], the liver [130, 134, 135], the kidney [136, 137], the respiratory system [138, 139], the nervous system [130, 140, 141], the heart and vascular systems [142, 143], the musculoskeletal system [144, 145], the endocrine system [144, 146], the visual and auditory systems.[147, 148], the skin [149, 150], and the reproductive system [151–153].

Although there are numerous studies on wear debris toxicity on Ti-based alloy because it is a major metallic material for component of orthopedic implants such as total hip or knee replacements [128, 129], there is no research works on the same topics for NiTi alloy; however there are several studies on cytotoxicity and genotoxicity of solid NiTi alloy, for example [154, 155]. Since the chemical composition of the wear debris should not differ from that of the parent material, it is worthy to study effects of the NiTi powder. Metallic particles are clearly cytotoxic in vitro, whether being produced from titanium-based [156–159] or cobalt-chromium alloys [156, 160, 161]. This cytotoxicity is probably secondary to intracellular dissolution at low pH levels, since it can be affected by blocking H^+ release [162]. Discussions of biological response to wear debris have tended to emphasize the number of particles produced by a device in a period of time or, occasionally, the total weight or rate of production (by weight) of the debris. The in vitro investigations of macrophage response to small particles suggest that the total surface area of the debris may be an important parameter [163]. Wear debris particle size, produced by adhesion, tends to vary inversely with material rigidity (i.e., modulus of elasticity), thus metallic and ceramic materials can be expected to produce smaller particles than the polyethylene debris released by conventional bearings [164].

Cunningham et al. [165] determined if the presence of spinal instrumentation wear particulate debris deleteriously influences early osseointegration of posterolateral bone graft or disrupts an established posterolateral fusion mass. Thirty-four New Zealand White rabbits were randomized into two groups based on postoperative time periods of 2 months (Group 1) and 4 months (Group II). Group I underwent a posterolateral arthrodesis (PLF) at L5-L6 using tricortical iliac autograft or tricortical iliac autograft plus titanium particulate. Group 2 all received iliac autograft at the initial surgery and were reoperated on after 8 weeks and treated with PLF exposure alone or titanium particulate. Postoperative analysis included serological quantification of systemic cytokines. Postmortem microradiographic, immunocytochemical, and histopathological assessment of the intertransverse fusion mass quantified the extent of osteolysis, local proinflammatory cytokines, osteoclasts, and inflammatory infiltrates. It was reported that (i) basic science phase: serological analysis of systemic cytokines indicated no significant differences in cytokine levels between the titanium or autograft treatments, (ii) immunocytochemistry indicated increased levels of local cytokines: TNF-alpha at the titanium-treated PLF sites at both time periods, (iii) osteoclast cell counts and regions of osteolytic resorption lacunae were higher in the titanium-treated versus autograft-alone groups, and the extent of cellular apoptosis was markedly higher in the titanium-treated sites at both time intervals, (iv) electron microscopy indicated definitive evidence of phagocytized titanium particles and foci of local, chronic inflammatory changes in the titanium-treated sites. Based on these findings, it was concluded that (v) titanium particulate debris introduced at the level of a spinal arthrodesis elicits a cytokine-mediated particulate-induced response favoring proinflammatory infiltrates, increased expression of intracellular TNF-alpha, increased osteoclastic activity and cellular apoptosis, (vi) this is the first basic scientific study and the first clinical study demonstrating associations of spinal instrumentation particulates wear debris and increased cytokines and increased osteoclastic activity, and (vii) osteolysis is the number one cause of failure of ortho-pedic implants in the appendicular skeleton. Spinal surgeons need to increase their awareness of this destructive process.

Rhalmi et al. [166, 167] extensively investigated the effects of porous or particles of NiTi material on spinal cord dura mater reaction. New porous metals have been developed for medical applications in order to improve the interface between the implant and the adjacent biological structures; due to macro- and micromorphological compatibility [168]. As opposed to coated materials, porous NiTi represents a clear advantage for biological tissues, which can fully integrate throughout its complete network of interconnected fenestration, and also a requirement for the biomechanical compatibility is satisfied since the modulus of elasticity of the porous NiTi is very close value to that of receiving hard tissue (bone) [168]. Potential biomedical applications of porous NiTi include soft tissue attachment and bone carrier devices for guided bone regeneration. For example, porous NiTi intervertebral fusion devices for degenerative disc disease conditions represent a clinical application that may become an alternative to traditional interbody cages, which otherwise require autologous bone grafting and longer surgery time. Since porous NiTi fusion devices are intended for implantation at the spinal level, potential wear debris (if any) must be evaluated for the possibility of developing adverse reactions in adjacent spinal nervous tissues. A spinal subchronic assay was performed on porous NiTi particles using a rabbit model for 1-, 4-, and 12-week postsurgical recovery periods [166]. It was reported that (i) the local subchronic implantation study is one of its kind since NiTi particles have never been tested in vivo on the nervous tissue, especially the spinal cord: certainly the most adjacent soft tissue to an intervertebral device and one of the most sensitive and fragile tissue of the human body, (ii) of course, this in vivo assay represented a worst-case scenario since porous NiTi intervertebral fusion device debris are not expected to normally occur following interbody implantation, (iii) nevertheless, this macroscopic and histologic evaluation confirmed the local compatibility of porous NiTi with the spinal cord and its nerve roots in a rabbit model. Indeed, the spinal cord, dura mater and nerve root tissues did not show adverse reactions such as severe inflammation and necrosis, (iv) conversely, the porous NiTi particles were kept entrapped in a soft tissue capsule at the posterior side of the spinal cord: a normal biological tissue encapsulation process that takes place against foreign bodies, (v) these local subchronic results further support previous reports showing high biocompatibility of nitinol alloys, and (vi) long-term implantation of porous NiTi interbody fusion devices is currently under development in order to further characterize the biofunctionality and biocompatibility of porous nitinol as a spinal implant. Simulating in an animal model the particles released from a porous nitinol interbody fusion device and to evaluate its consequences on the dura mater, spinal cord and nerve roots, lymph nodes (abdominal para-aortic), and organs (kidneys, spleen, pancreas, liver, and lungs), Rhalmi et al. [167] evaluated the compatibility of the NiTi particles with the dura mater in comparison with titanium alloy. In spite of the great use of

metallic devices in spine surgery, the proximity of the spinal cord to the devices raised concerns about the effect of the metal debris that might be released onto the neural tissue. Forty-five New Zealand white female rabbits were divided into three groups: nitinol (treated: $N = 4$ per implantation period), titanium (treated: $N = 4$ per implantation period), and sham rabbits (control: $N = 1$ per observation period). The nitinol and titanium alloy particles were implanted in the spinal canal on the dura mater at the lumbar level L2–L3. The rabbits were sacrificed at 1, 4, 12, 26, and 52 weeks. Histologic sections from the regional lymph nodes, organs, from remote and implantation sites, were analyzed for any abnormalities and inflammation. It was reported that (i) regardless of the implantation time, both nitinol and titanium particles remained at the implantation site and clung to the spinal cord lining soft tissue of the dura mater, (ii) the inflammation was limited to the epidural space around the particles and then reduced from acute to mild chronic during the follow-up, (iii) the dura mater, subdural space, nerve roots, and the spinal cord were free of reaction, (iv) no particles or abnormalities were found either in the lymph nodes or in the organs, (v) in contact with the dura, the nitinol elicits an inflammatory response similar to that of titanium, and (vi) the tolerance of nitinol by a sensitive tissue such as the dura mater during the span of 1 year of implantation demonstrated the safety of nitinol and its potential use as an intervertebral fusion device.

14.4 Fretting

Although a combination of shape-memory characteristics, pseudoelasticity, and good damping properties make near-equiatomic NiTi alloy a desirable candidate material for certain biomedical device applications and the alloy has moderately good wear resistance, further improvements in this regard would be beneficial from the perspective of reducing wear debris generation, improving biocompatibility, and preventing failure during service. Tan et al. [169] conducted fretting wear tests of NiTi in both austenitic and martensitic microstructural conditions with the goal of simulating wear which medical devices such as stents may experience during surgical implantation or service. The tests were performed using a SS stylus counter-wearing surface under dry conditions and also with artificial plasma containing 80 g/L albumen protein as lubricant. Additionally, the research explores the feasibility of surface modification by sequential ion implantation with argon and oxygen to enhance the wear characteristics of the NiTi alloy. Each of these implantations was performed to a dose of 3×10^{17} atom/cm^2 and an energy of 50 kV, using the PSII process. It was reported that (i) improvements in wear resistance were observed for the austenitic samples implanted with argon and oxygen, (ii) ion implantation with argon also reduced the surface Ni content with respect to Ti due to differential sputtering rates of the two elements, an effect that points toward improved biocompatibility. Qian et al. [170] studied the fretting behavior of superelastic NiTi shape-memory alloy at various displacement amplitudes on a serve-hydraulic dynamic test machine. It was shown that (i) the superelastic properties of the material played a key role in the observed excellent fretting behavior of NiTi alloy, (ii) due to the low phase transition stress (only 1/4 the value of its plastic yield stress) and the large recoverable phase transition strain (5%) of NiTi, (iii) the friction force of NiTi/GCr15 SS pair is smaller than the value of GCr15/GCr15 pair and at the same time the Rabinowicz wear coefficient of NiTi plate is about 1/9 the value of GCr15 plate under the same fretting conditions, (iv) for NiTi/GCr15 pair, even NiTi has a much lower hardness than GCr15, the superelastic NiTi alloy exhibits superior fretting wear property than GCr15 steel, (v) the weak ploughing was the main wear mechanism of NiTi alloy in the partial slip regime, (vi) while in the mixed regime and gross slip regime, the wear of NiTi was mainly caused by the abrasive wear of the GCr15 debris in the three-body wear mode. Yang et al. [171] prepared three types of surface-treated NiTi samples, M-1 (700 °C/0.5 h), M-2 (650 °C/1 h) and M-3 (400 °C/50 h) by ceramic conversion (CC) treatment under different conditions to investigated the effect of the surface treatment on the fretting behavior of NiTi alloy in the Ringer's solution by using a horizontal servohydraulic fretting apparatus. It was found that (i) the surface layer of the LT (400 °C) treated samples M-3 was dominated by a single TiO_2 layer, while the HT (650 °C and 700 °C) treated samples M-1 and M-2 consisted of surface TiO_2 layer followed by a $TiNi_3$ layer, (ii) these surface layers were found to have a strong effect on the fretting behavior of the NiTi alloy in terms of changes in the shape of the curves of the tangential force (F_T) versus displacement (d), the fretting regimes and the damage mechanisms involved, (iii) the stress-induced reorientation of martensite bands in the NiTi alloy could decrease the slope of the F_T–d curve and thus increase the elastic accommodation ability of the NiTi plate against 1Cr13 steel ball pair, (iv) however, since the surface-treated layers could suppress the martensite reorientation in the NiTi substrate and thus decrease the elastic accommodation ability of NiTi, the gross slip started at a smaller displacement amplitude for the surface-treated NiTi samples than for the untreated one, (v) the main wear mechanism of the as-received NiTi alloy in slip regime was adhesion and delamination, while the major damage to the HT-treated NiTi samples M-1 and M-2 was determined as the spallation of surface-treated layers, (vi) due to the high bonding strength of the surface-treated layer with NiTi substrate, the LT-treated NiTi samples M-3 showed the best fretting wear resistance in all samples tested. It has been demonstrated that the surface modification can improve the fretting wear resistance of NiTi alloys in air or enhance their aqueous corrosion resistance without fretting. However, little is known about the behavior of surface engineered NiTi under fretting corrosion conditions. This is important for such body implants as orthodontic archwires and orthopedic bone fixation devices because they need to withstand the combined attack of corrosion from body fluid and mechanical fretting. Hence, Dong et al. [172] treated a NiTi alloy by CC method at 400 and 650 °C. The effect of the surface treatment on the fretting corrosion behavior of NiTi alloy was investigated using fretting corrosion tests in the Ringer's solution. It was shown that (i) the experimental results have shown that the CC treatment can convert the surface of NiTi into a TiO_2 layer, which can effectively improve the fretting corrosion resistance of NiTi alloy and significantly reduce Ni ion release into the Ringer's solution, (ii) detailed SEM observations revealed that the untreated samples were severely damaged by adhesion and delamination; the HT (650 °C/1 h) treated samples were damaged mainly by spallation and adhesion; and the LT (400 °C/50 h) treated samples were characterized by mild abrasion, and (iii) mild oxidation and corrosion were also observed for all three types of samples tested under fretting corrosion conditions. Metallic medical devices such as intravascular stents can undergo fretting damage in vivo that might increase their susceptibility to pitting corrosion. As a result, the US Food and Drug Administration has recommended that such devices be evaluated for corrosion resistance after the devices have been fatigue tested in situations where significant micromotion can lead to fretting damage. Siddiqui et al. [173] selected three common alloys that cardiovascular implants are made from [MP35N cobalt chromium (MP35N), electropolished nitinol, and 316LVM SS (316LVM)] study. In order to evaluate the effect of wire fretting on the pitting corrosion susceptibility of these medical alloys, small and large fretting scar conditions of each alloy fretting against itself, and the other alloys in phosphate buffered saline (PBS) at 37 °C were tested per ASTM F2129 and compared against as received or PBS immersed control specimens. It was reported that (i) although the general trend observed was that fretting damage significantly lowered the rest potential (E_R) of these specimens, fretting damage had no significant effect on the breakdown potential (E_B) and hence did not affect the susceptibility to pitting corrosion, and (ii) fretting damage in PBS alone is not sufficient to cause increased susceptibility to pitting corrosion in the three common alloys investigated.

14.5 Tribocorrosion

More frequently, the wear of metal bearings can be distinguished in three main processes and their combinations: abrasive wear, due to either two or three bodies, adhesive wear and fatigue wear. However, other types of wear such as corrosive can occur. The dynamic loading these implants undergo, together with the corrosiveness of physiological fluids can accelerate the degradation processes. The synergistic adverse effect of wear and corrosion does not consist of a simple sum of the two but more as a synergy realized between them called tribocorrosion. Tribocorrosion is defined as an "irreversible transformation of material in tribological contact caused by simultaneous physicochemical and mechanical surface interactions" [174]. In the last decades, a scaring occurrence of inflammatory reactions has been seen in patients with large head metal-on-metal , often with signs of tribocorrosion at the head-neck interface. Tribocorrosion arises not only at MoM bearing surfaces, but also at metal/metal modular junctions where micro-motions between the two components are possible. In many biomedical applications of NiTi alloys, the tribocorrosion properties of these alloys can be of critical concern. Kosec et al. [175] studied the electrochemical and tribocorrosion properties of superelastic NiTi sheet and orthodontic archwire, taking into account their microstructures and the effect of different surface finishes. In the case of the electrochemical tests, samples were tested in artificial saliva, whereas in the tribocorrosion tests the experiments were performed in ambient air, distilled water, and artificial saliva, the latter as a corrosive medium. In these tests, the total wear rate of the alloy samples was determined, together with the corresponding chemical and tribological contributions. It was mentioned that the microstructure of the investigated alloys had a significant effect on the measured electrochemical and tribocorrosion properties. Even though NiTi and SS are the most commonly used alloys for orthodontic treatments and both are known to be resistant to corrosion, there are circumstances that can lead to undesired situations, like localized types of corrosion attack, wear during sliding of an archwire though brackets and breakdowns due to iatrogenic causes. Močnik et al. [176] analyzed the influence of environmental effects on the corrosion and tribocorrosion properties of NiTi and SS dental alloys. The effects of pH and fluorides on the electrochemical properties were studied using the cyclic potentiodynamic technique. The migration of ions from the alloy into saliva during exposure to saliva with and without the presence of wear was analyzed using ICP-MS analyses. Auger spectroscopy was used to study the formation of a passive oxide layer on different dental alloys. It was reported that (i) lowering the pH preferentially affects the corrosion susceptibility of NiTi alloys, whereas SS dental archwires are prone to local types of corrosion, (ii) the NiTi alloy is not affected by smaller increases of fluoride ions up to 0.024 M, while at 0.076 M (simulating the use of toothpaste) the properties are affected, (iii) a leaching test during wear-assisted corrosion showed that the concentrations of Ni ions released into the saliva exceeded the limit value of 0.5 μg/cm^2/week, (iv) the oxide films on the NiTi and SS alloys after the tribocorrosion experiment were thicker than those exposed to saliva only.

References

[1]   Dumbleton JH. Tribology Series 3: Tribology of Natural and Artificial Joints.Elsevier, New York, 1981.
[2]   Sivasankar M, Arunkumar S, Bakkiyaraj V, Muruganandam A, Sathishkumar S. A review on total hip replacement. Int'l. Res. J. Adv. Eng. Technol. 2016, 9, 589–642.
[3]   Harper LH, Dooris A, Paré PE. The fundamentals of biotribology and its application to spine arthroplasty. SAS J. 2009, 3, 125–32.
[4]   Chromik RR, Strauss HW, Scharf TW. Materials phenomena revealed by in situ tribometry. J. Met. 2012, 31, 35–43.
[5]   Rucker BK, Kusy RP. Resistance to sliding of stainless steel multistranded archwires and comparison with single-stranded leveling wires. Am. J. Orthod. Dentofac. Orthop. 2002, 122, 73–83.
[6]   Liaw YC, Su YY, Lai YL, Lee SY. Stiffness and frictional resistance of a superelastic nickel-titanium orthodontic wire with low-stress hysteresis. Am. J. Orthod. Dentofac. Orthop. 2007, 131, e12–8.

[7]   Regis S Jr, Soares P, Camargo ES, Guariza Filho O, Tanaka O, Maruo H. Biodegradation
      of orthodontic metallic brackets and associated implications for friction. Am. J. Orthod.
      Dentofac. Orthop. 2011, 140, 501–9.

[8]   Kumar D, Dua V, Mangla R, Solanki R, Solanki M, Sharma R. Frictional force released during
      sliding mechanics in nonconventional elastomerics and self-ligation: an in vitro comparative
      study. Ind. J. Dent. 2016, 7, 60–5.

[9]   Kapila S, Angolkar PV, Duncanson MG, Nanda RS. Evaluation of friction between edgewise
      stainless steel brackets and orthodontic wires of four alloys. Am. J. Orthod. Dentofac. Othop.
      1990, 98, 117–26.

[10] Angolkar PV, Kapila S, Duncanson MG, Nanda RS. Evaluation of friction between ceramic
      brackets and orthodontic wires of four alloys. Am. J. Orthod. Dentofac. Orthop. 1990, 98,
      499–506.

[11] Prososki RR, Bagby MD, Erickson LC. Static frictional force and surface roughness of nickel-
      titanium arch wires. Am. J. Orthod. Dentofac. Orthop. 1991, 100, 341–8.

[12] Vaughan JL, Duncanson MG Jr, Nanda RS, Currier GF. Relative kinetic frictional forces
      between sintered stainless steel brackets and orthodontic wires. Am. J. Orthod. Dentofac.
      Orthop. 1995, 107, 20–7.

[13] Schumacher HA, Bourauel C, Drescher D. The influence of bracket design on frictional losses
      in the bracket/arch wire system. J. Orofac. Orthop. 1999, 60, 335–47.

[14] Cacciafesta V, Sfondrini MF, Ricciardi A, Scribante A, Klersy C, Auricchio F. Evaluation of friction
      of stainless steel and esthetic self-ligating brackets in various bracket-archwire combinations.
      Am. J. Orthod. Dentofac. Orthop. 2003, 124, 395–402.

[15] Cacciafesta V, Sfondrini MF, Scribante A, Klersy C, Auricchio F. Evaluation of friction of
      conventional and metal-insert ceramic brackets in various bracket-archwire combinations.
      Am. J. Orthod. Dentofac. Orthop. 2003, 124, 403–9.

[16] Clocheret K, Willems G, Carels C, Celis JP. Dynamic frictional behaviour of orthodontic
      archwires and brackets. Eur. J. Orthod. 2004, 26, 163–70.

[17] Kapur WR, Kwon HK, Close JM. Frictional resistances of different bracket-wire combinations.
      Aust. Orthod. J. 2004, 20, 25–30.

[18] Krishnan M, Kalathil S, Abraham KM. Comparative evaluation of frictional forces in active and
      passive self-ligating brackets with various archwire alloys. A. J. Orthod. Dentofac. Orthop.
      2009, 136, 675–82.

[19] Tecco S, Tetè S, Festa F. Friction between archwires of different sizes, cross-section and alloy
      and brackets ligated with low-friction or conventional ligatures. Angle Orthod. 2009, 79,
      111–6.

[20] Tecco S, Teté S, Festa M, Festa F. An in vitro investigation on friction generated by ceramic
      brackets. World J. Orthod. 2010, 11, e133–44.

[21] Hsu J-T, Wu L-C, Chang Y-Y, Weng T-N, Huang H-L, Yu C-H. Frictional forces of conventional and
      improved superelastic NiTi-alloy orthodontic archwires in stainless steel and plastic brackets.
      IFMBE Proc. 2009, 312–5.

[22] Heo W, Baek SH. Friction properties according to vertical and horizontal tooth displacement
      and bracket type during initial leveling and alignment. Angle Orthod. 2011, 81, 653–61.

[23] Kosukegawa H, Fridrici V, Kapsa P, Sutou Y, Adachi K, Ohta M. Friction properties of medical
      metallic alloys on soft tissue–mimicking Poly(Vinyl Alcohol) hydrogel biomodel. Tribol. Lett.
      2013, 51, 311–21.

[24] Oshida Y, Sachdeva R, Miyazaki S. Changes in contact angles as a function of time on some
      pre-oxidized biomaterials. J. Mater. Sci. Mater. Med. 1992, 3, 306–12.

[25] Carrion-Vilches FJ, Bermudez M-D, Fructuoso P. Static and kinetic friction force and surface
      roughness of different archwire bracket sliding contacts. Dent. Mater. J. 2015, 34, 648–53.

[26] Sridharan K, Sandbhor S, Rajasekaran UB, Sam G, Ramees MM, Abraham EA. An in vitro evaluation of friction characteristics of conventional stainless steel and self-ligating stainless steel brackets with different dimensions of archwires in various bracket-archwire combination. J. Contemp. Dent. Pract. 2017, 18, 660–4.

[27] Savoldi F, Visconti L, Dalessandri D, Bonetti S, Tsoi JKH, Matinlinna JP, Paganelli C. In vitro evaluation of the influence of velocity on sliding resistance of stainless steel arch wires in a self-ligating orthodontic bracket. Orthod. Craniofac. Res. 2017, 20, 119–25.

[28] Gupta A, Gupta A, Agarwal L. Role of saliva: an orthodontic perspective. Int. J. Oral Health Dent. 2016, 2, 126–31.

[29] Bongaerts JHH, Rossetti D, Stokes JR. The lubricating properties of human whole saliva. Tribol. Lett. 2007, 27, 277–87.

[30] Kusy RP, Whitley JQ. Resistance to sliding of orthodontic appliances in the dry and wet states: influence of archwire alloy, interbracket distance, and bracket engagement. J. Biomed. Mater. Res. 2000, 52, 797–811.

[31] Sarkar A, Andablo-Reyes E, Bryant M, Dowson D, Neville A. Lubrication of soft oral surfaces. Curr. Opin. Colloid In. 2019, 39, 61–75.

[32] Aguirre A, Mendoza B, Reddy MS, Scannapieco FA, Levine MJ, Hatton MN. Lubrication of selected salivary molecules and artificial salivas. Dysphagia 1989, 4, 95–100.

[33] Douglas WH, Reeh ES, Ramasubbu N, Raj PA, Bhandary KK, Levine MJ. Statherin: a major boundary lubricant of human saliva. Biochem. Biophys. Res. Commun. 1991, 180, 91–7.

[34] Reeh ES, Douglas WH, Levine MJ. Lubrication of saliva substitutes at enamel-to-enamel contacts in an artificial mouth. J. Prosthet. Dent. 1996, 75, 649–56.

[35] Berg ICH, Rutland MW, Arnebrant T. Lubricating properties of the initial salivary pellicle – an AFM study. Biofouling 2003, 19, 365–9.

[36] Bongaerts JHH, Rossetti D, Stokes JR. The lubricating properties of human whole saliva. Tribol. Lett. 2007, 27, 277–87.

[37] Hatton MN, Levine MJ, Margarone JE, Aguirre A. Lubrication and viscosity features of human saliva and commercially available saliva substitutes. Oral Maxil. Surg. 1987, 45, 496–9.

[38] Silva MP, Chibebe Jr J, Jorjão AL, da Silva Machado AK, de Oliveira LD, Junqueira JC, Jorge AOC. Influence of artificial saliva in biofilm formation of Candida albicans in vitro. Braz. Oral. Res. 2012, 26, 24–8.

[39] Vinke J, Kaper HJ, Vissink A, Sharma PK. An ex vivo salivary lubrication system to mimic xerostomic conditions and to predict the lubricating properties of xerostomia relieving agents. Sci. Rep. 2018, 8; doi: 10.1038/s41598-018-27380-7.

[40] Ben-Aryeh H, Miron D, Berdicevski I, Szargel R, Gutman D. Xerostomia in the elderly: prevalence, diagnosis, complications and treatment. Gerodontology 1985, 4, 77–82.

[41] Von Bültzingslöwen I, Sollecito TP, Fox PC, Daniels T, Jonsson R, Lockhart PB, Wray D, Brennan MT, Carrozzo M, Gandera B, Fujibayashi T, Navazesh M, Rhodus NL, Schiødt M. Salivary dysfunction associated with systemic diseases: systematic review and clinical management recommendations. Oral Surg. Oral Med. Oral Pathol. Oral Radiol. Endod. 2007, 103, e1–15.

[42] Plemons JM, Al-Hashimi I, Marek CL. Managing xerostomia and salivary gland hypofunction. J. Am. Dent. Assoc. 2014, 145, 867–73.

[43] Pajukoski H, Meurman JH, Halonen P, Sulkava R. Prevalence of subjective dry mouth and burning mouth in hospitalized elderly patients and outpatients in relation to saliva, medication, and systemic diseases. Oral Surg. Oral Med. Oral Pathol. Oral Radiol. Endod. 2001, 92, 641–9.

[44] Smidt D, Torpet LA, Nauntofte B, Heegaard KM, Pedersen AML. Associations between labial and whole salivary flow rates, systemic diseases and medications in a sample of older people. Comm. Dent. Oral Epidemiol. 2010, 38, 422–35.

[45] Porter SR, Scully C, Hegarty AM. An update of the etiology and management of xerostomia. Oral Surg. Oral Med. Oral Pathol. Oral Radiol. Endodontology 2004, 97, 28–46.

[46] Aliko A, Wolff A, Dawes C, Aframian D, Proctor G, Ekström J, Narayana N, Villa A, Sia YW, Joshi RK, McGowan R, Jensen SB, Kerr AR, Pedersen AML, Vissink A. World workshop on oral medicine VI: clinical implications of medication-induced salivary gland dysfunction. Oral Surg. Oral Med. Oral Pathol. Oral Radiol. 2015, 120, 185–206.

[47] Peri Y, Agmon-Levin N, Theodor E, Shoenfeld Y. Sjögren's syndrome, the old and the new. Best Pract. Res. Clin. Rheumatol. 2012, 26, 105–17.

[48] Vissink A, Mitchell JB, Baum BJ, Limesand KH, Jensen SB, Fox PC, Elting LS, Langendijk JA, Coppes RP, Reyland ME. Clinical management of salivary gland hypofunction and xerostomia in head-and-neck cancer patients: successes and barriers. Int. J. Radiat. Oncol. Biol. Phys. 2010, 78, 983–91.

[49] Klein Hesselink EN, Brouwers AH, de Jong JR, van der Horst-Schrivers AN, Coppes RP, Lefrandt JD, Jager PL, Vissink A, Links TP. Effects of radioiodine treatment on salivary gland function in patients with differentiated thyroid carcinoma: a prospective study. J. Nucl. Med. 2016, 57, 1685–91.

[50] Jellema AP, Slotman BJ, Doornaert P, Leemans CR, Langendijk JA. Impact of radiation-induced xerostomia on quality of life after primary radiotherapy among patients with head and neck cancer. Int. J. Radiat. Oncol. Biol. Phys. 2007, 69, 751–60.

[51] Stannard JG, Gau JM, Hanna MA. Comparative friction of orthodontic wires under dry and wet conditions. Am. J. Orthod. 1986, 89, 485–91.

[52] Kusy RP, Whitley JQ, Prewitt MJ. Comparison of the frictional coefficients for selected archwire-bracket slot combinations in the dry and wet state. Angle Orthod. 1991, 61, 293–301.

[53] Alfonso MV, Espinar E, Llamas JM, Rupérez E, Manero JM, Barrera JM, Solano E, Gil FJ. Friction coefficients and wear rates of different orthodontic archwires in artificial saliva. J. Mater. Sci. Mater. Med. 2013, 24, 1327–32.

[54] Phukaoluan A, Khantachawana A, Kaewtatip P, Dechkunakorn S, Anuwongnukroh N, Santiwong P, Kajornchaiyakul J. Comparison of friction forces between stainless orthodontic steel brackets and TiNi wires in wet and dry conditions. Int. Orthod. 2017, 15, 13–24.

[55] Lara-Carrillo E, Montiel-Bastida N-M, Sánchez-Pérez L, Alanís-Tavira J. Effect of orthodontic treatment on saliva, plaque and the levels of Streptococcus mutans and Lactobacillus. Med. Oral Patol. Oral Cir. Bucal. 2010, 15, e924–9.

[56] Leone CW, Oppenheim FG. Physical and chemical aspects of saliva as indicators of risk for dental caries in humans. J. Dent. Educ. 2001, 65, 1154–62.

[57] Edgar WM, Higham SM. Role of saliva in caries models. Adv. Dent. Res. 1995, 9, 235–8.

[58] Bardow A, Nyvad B, Nauntofte B. Relationships between medication intake, complaints of dry mouth, salivary flow rate and composition, and the rate of tooth demineralization in situ. Arch. Oral Biol. 2001, 46, 413–23.

[59] Richter AE, Arruda AO, Peters MC, Sohn W. Incidence of caries lesions among patients treated with comprehensive orthodontics. Am. J. Orthod. Dentofac. Orthop. 2011, 139, 657–64.

[60] Zanarini M, Pazzi E, Bonetti S, Ruggeri O, Bonetti GA, Prati C. In vitro evaluation of the effects of a fluoride-releasing composite on enamel demineralization around brackets. Prog. Orthod. 2012, 13, 10–6.

[61] Peros K, Mestrovic S, Milosevic SA, Slaj M. Salivary microbial and nonmicrobial parameters in children with fixed orthodontic appliances. Angle Orthod. 2011, 81, 901–6.

[62] Bonetti GA, Parenti SI, Garulli G, Gatto MR, Checchi L. Effect of fixed orthodontic appliances on salivary properties. Prog. Orthod. 2013, 14; doi: 10.1186/2196-1042-14-13.

[63] Berradja A, Willems G, Celis JP. Tribological behaviour of orthodontic archwires under dry and wet sliding conditions in-vitro. I – frictional behaviour. Aust. Orthod. J. 2006, 22, 11–9.

[64]   Hosseinzadeh NT, Hooshmand T, Farazdaghi H, Mehrabi A, Razavi ES. Effect of chlorhexidine-containing prophylactic agent on the surface characterization and frictional resistance between orthodontic brackets and archwires: an in vitro study. Prog. Orthod. 2013, 14, 1–8.

[65]   Kao CT, Ding SJ, Wang CK, He H, Chou MY, Huang TH. Comparison of frictional resistance after immersion of metal brackets and orthodontic wires in a fluoride-containing prophylactic agent. Am. J. Orthod. Dentofac. Orthop. 2006, 130, e1–9.

[66]   Geramy A, Hooshmand T, Etezadi T. Effect of sodium fluoride mouthwash on the frictional resistance of orthodontic wires. J. Dent. (Tehran) 2017, 14, 254–8.

[67]   Zeng Q, Dong G. Influence of load and sliding speed on super-low friction of Nitinol 60 alloy under castor oil lubrication. Tribol. Lett. 2013, 52, 47–55.

[68]   Dridi A, Bensalah W, Mezlini S, Tobji S, Zidi M. Influence of bio-lubricants on the orthodontic friction. J. Mech. Behav. Biomed. Mater. 2016, 60, 1–7.

[69]   Prashant PS, Nandan H, Gopalakrishnan M. Friction in orthodontics. J. Pharm. Bioall. Sci. 2015, 7, S334–8.

[70]   Frank CA, Nikolai RJ. A comparative study of frictional resistances between orthodontic bracket and arch wire. Am. J. Orthod. 1980, 78, 593–609.

[71]   Rossouw PE, Kamelchuk LS, Kusy RP. A fundamental review of variables associated with low velocity frictional dynamics. Semin. Orthod. 2003, 9, 223–35.

[72]   Omana HM, Moore RN, Bagby MD. Frictional properties of metal and ceramic brackets. J. Clin. Orthod. 1992, 26, 425–32.

[73]   Kusy RP, Whitley JQ. Effects of surface roughness on the coefficients of friction in model orthodontic systems. J. Biomech. 1990, 23, 913–25.

[74]   Oshida Y, Kotona TR, Farzin-Nia F, Rosenthall MR. Comparison of Frictional Forces Between Three Grades of Low Friction "Colors" TMA. In: Shrivastava S, ed., Medical Device Materials. ASM International, Materials Park, 2004, 455–8.

[75]   Espinar E, Llamas JM, Michiardi A, Ginebra MP, Gil FJ. Reduction of Ni release and improvement of the friction behaviour of NiTi orthodontic archwires by oxidation treatments. J. Mater. Sci. Mater. Med. 2011, 22, 1119–25.

[76]   Meier MJ, Bourauel C, Roehlike J, Reimann S, Keilig L, Braumann B. Friction behavior and other material properties of nickel–titanium and titanium–molybdenum archwires following electro-chemical surface refinement. J. Orofac. Orthop. 2014, 75, 308–19.

[77]   Wichelhaus A, Geserick M, Hibst R, Sander FG. The effect of surface treatment and clinical use on friction in NiTi orthodontic wires. Dent. Mater. 2005, 21, 938–45.

[78]   Gil FJ, Solano E, Campos A, Boccio F, Sáez I, Alfonso MV, Planell JA. Improvement of the friction behaviour of NiTi orthodontic archwires by nitrogen diffusion. Biomed. Mater. Eng. 1998, 8, 335–42.

[79]   Huang SY, Huang JJ, Kang T, Diao DF, Duan YZ. Coating NiTi archwires with diamond-like carbon films: reducing fluoride-induced corrosion and improving frictional properties. J. Mater. Sci. Mater. Med. 2013, 24, 2287–92.

[80]   Rokaya D, Srimaneepong V, Qin J, Siraleartmukul K, Siriwongrungson V. Graphene oxide/silver nanoparticle coating produced by electrophoretic deposition improved the mechanical and tribological properties of NiTi alloy for biomedical applications. J. Nanosci. Nanotechnol. 2019, 19, 3804–10.

[81]   Bandeira AMB, Alves dos Santos MP, Pulitini G, Elias CN, Ferreira da Costa M. Influence of thermal or chemical degradation on the frictional force of an experimental coated NiTi wire. Angle Orthod. 2011, 81, 484–9.

[82]   Kachoei M, Nourian A, Divband B, Kachoei Z, Shirazi S. Zinc-oxide nanocoating for improvement of the antibacterial and frictional behavior of nickel-titanium alloy. Nanomedicine (Lond) 2016, 11, 2511–27.

[83]   Samorodnitzky-Naveh GR, Redlich M, Rapoport L, Feldman Y, Tenne R. Inorganic fullerene-like tungsten disulfide nanocoating for friction reduction of nickel-titanium alloys. Nanomedicine (Lond) 2009, 4, 943–50.

[84]   Bednar JR, Gruendeman GW. The influence of bracket design on moment production during axial rotation. Am. J. Orthod. Dentof. Orthop. 1993, 104, 254–61.

[85]   Whitley JQ, Kusy RP. Influence of interbracket distances on the resistance to sliding of orthodontic appliances. Am. J. Orthod. Dentofac. Orthop. 2007, 132, 360–72.

[86]   Angolkar PV, Kapila S, Duncanson MG Jr, Nanda RS. Evaluation of friction between ceramic brackets and orthodontic wires of four alloys. Am. J. Orthod. Dentofac. Orthop. 1990, 98, 499–506.

[87]   Clayton P. Tribological behavior of a titanium-nickel alloy. Wear 1993, 162/164, 202–10.

[88]   Ng KW, Man HC, Yue TM. Corrosion and wear properties of laser surface modified NiTi with Mo and ZrO_2. Appl. Surf. Sci. 2008, 254, 6725–30.

[89]   Merola M, Affatato S. Materials for hip prostheses: a review of wear and loading considerations. Materials (Basel) 2019, 12; doi: 10.3390/ma12030495.

[90]   Massin P. Achour S. Wear products of total hip arthroplasty: the case of polyethylene Produits d'usure des arthroplasties totales de hanche: le cas du polyéthylène. Morphologie 2017, 101, 1–8.

[91]   Guy R, Nockolds C, Phillips M, Roques-Carmes C. Implications of polishing techniques in quantitative x-ray microanalysis. J. Res. Natl. Inst. Stand. Technol. 2002, 107, 639–62.

[92]   Liang YN, Li SZ, Jin YB, Jin W, Li S. Wear behavior of a TiNi alloy. Wear, 1996, 198, 236–41.

[93]   Eggert C, Peters O, Barbakow F. Wear of nickel-titanium lightspeed instruments evaluated by scanning electron microscopy. J. Endod. 1999, 25, 494–7.

[94]   Liu R, Li DY. A finite element model study on wear resistance of pseudoelastic TiNi alloy. Mater. Sci. Eng. 2000, A277, 169–75.

[95]   Gialanella S, Ischia G, Straffelini G. Phase composition and wear behavior of NiTi alloys. J. Mater. Sci. 2008, 43, 1701–10.

[96]   Pirani C, Paolucci A, Ruggeri O, Bossù M, Polimeni A, Gatto MRA, Gandolfi MG, Prati C. Wear and metallographic analysis of WaveOne and reciproc NiTi instruments before and after three uses in root canals. Scanning 2014, 36, 517–25.

[97]   Abedini M, Ghasemi HM, Nili-Ahmadabadi M. Tribological behavior of NiTi alloy in martensitic and austenitic states. Mater. Des. 2009, 30, 4493–7.

[98]   Abedini M, Ghasemi HM, Nili-Ahmadabadi M, Mahmudi R. Effect of phase transformation on the wear behavior of NiTi alloy. J. Eng. Mater. Technol. 2010, 132; doi: 10.1115/1.4001593.

[99]   Yan L, Liu Y. Effect of temperature on the wear behavior of NiTi shape memory alloy. J. Mater. Res. 2014, 30, 186–96.

[100]   Yan L, Liu Y. Wear behavior of austenitic NiTi shape memory alloy. Shape Mem. Superelast. 2015, 1, 58–68.

[101]   Aliasgarian R, Ghasemi HM, Abedini M. Tribological behavior of heat treated Ni-rich NiTi alloys. J. Tribol. 2011, 133; doi: 10.1115/1.4004102.

[102]   Liu P, Kan Q, Yin H. Effect of grain size on wear resistance of nanocrystalline NiTi shape memory alloy. Mater. Lett. 2019, 241, 43–6.

[103]   Yan L, Liu Y, O'Neill G, Wang W. Effect of grain boundary on the wear behaviour of NiTi shape memory alloys when $M_f < T < A_f$. Tribol. Lett. 2018, 66; https://www.springerprofessional.de/en/effect-of-grain-boundary-on-the-wear-behaviour-of-niti-shape-mem/15447716.

[104]   Zhang L-C, Chen L-Y. A review on biomedical titanium alloys: recent progress and prospect. Adv. Eng. Mater. 2019, 21; https://doi.org/10.1002/adem.201801215.

[105]   Geetha M, Singh AK, Asokamani R, Gogia AK. Ti based biomaterials, the ultimate choice for orthopaedic implants – a review. Prog. Mater. Sci. 2009, 54, 397–425.

[106] Wang L, Qu J, Chen L, Meng Q, Zhang LC, Qin J, Zhang D, Lu W. Investigation of deformation mechanisms in β-type Ti-35Nb-2Ta-3Zr alloy *via* FSP leading to surface strengthening. Metall. Mater. Trans. 2015, 46, 4813–8.

[107] Wen M, Liu G, Gu J, Guan W, Lu J. Dislocation evolution in titanium during surface severe plastic deformation. Appl. Surf. Sci. 2009, 255, 6097–102.

[108] Lv Y, Ding Z, Xue J, Sha G, Lu E, Wang L, Lu W, Su C, Zhang L-C. Deformation mechanisms in surface nano-crystallization of low elastic modulus Ti6Al4V/Zn composite during severe plastic deformation. Scr. Mater. 2018, 157, 142–7.

[109] Xie L, Wang L, Wang K, Yin G, Fu Y, Zhang D, Lu W, Hua L, Zhang L-C. TEM characterization on microstructure of Ti–6Al–4V/Ag nanocomposite formed by friction stir processing. Materialia 2018, 3, 139–44.

[110] Zhang J, Zhu J. Surface structure evolution and abnormal wear behavior of the TiNiNb alloy under impact load. Metall. Mater. Trans. A 2009, 40, 1126–30.

[111] Zhao T, Li Y, Liu Y, Zhao X. Nano-hardness, wear resistance and pseudoelasticity of hafnium implanted NiTi shape memory alloy. J. Mech. Behav. Biomed. Mater. 2012, 13, 174–84.

[112] Ahmadi H, Nouri M. Effects of yttrium addition on microstructure, hardness and resistance to wear and corrosive wear of TiNi alloy. J. Mater. Sci. Technol. 2011, 27, 851–5.

[113] Alhumdany AA, Abidali ALK, Abdulredha HJ. Investigation of wear behaviour for NiTi alloys with yttrium and tantalum additions. IOP Conf. Ser. Mater. Sci. Eng. 2018, 433; doi: 10.1088/1757-899X/433/1/012072.

[114] Tan L, Dodd RA, Crone WC. Corrosion and wear-corrosion behavior of NiTi modified by plasma source ion implantation. Biomaterials 2003, 24, 3931–9.

[115] Mändl S, Fleischer A, Manova D, Rauschenbach B. Wear behaviour of NiTi shape memory alloy after oxygen-PIII treatment. Surf. Coat. Technol. 2006, 200, 6225–9.

[116] Mohan L. Chakraborty M, Viswanathan S, Mandal C, Bera P, Aruna ST, Anandan C. Corrosion, wear, and cell culture studies of oxygen ion implanted Ni–Ti alloy. Surf. Interface Anal. 2017, 49, 828–36.

[117] Liu X, Wu S, Chan YL, Chu PK, Chung CY, Chu CL, Yeung KWK, Lu WW, Cheung KMC, Luk KDK. Structure and wear properties of NiTi modified by nitrogen plasma immersion ion implantation. Mater. Sci. Eng. A 2007, 444, 192–7.

[118] Alves-Claro APR, Claro FAE, Uzumaki ET. Wear resistance of nickel–titanium endodontic files after surface treatment. J. Mater. Sci. Mater. Med. 2008, 19, 3273–7.

[119] Luo YC, Li DY. New wear-resistant material: Nano-TiN/TiC/TiNi composite. J. Mater. Sci. 2001, 36, 4695–702.

[120] Akalin O, Ezirmik KV, Urgen M, Newaz GM. Wear characteristics of NiTi/Al6061 short fiber metal matrix composite reinforced with SiC particulates. J. Tribol. 2010, 132(4); doi: 10.1115/1.4002332.

[121] Zhang H, Wang Z, Yang H, Shan X, Liu X, Yu S, He Z. Wear and corrosion properties of Mo surface-modified layer in TiNi alloy prepared by plasma surface alloying. J. Wuhan Univ. Technol. -Mater. Sci. Ed. 2016, 31, 910–7.

[122] Ozeki K, Masuzawa T, Hirakuri KK. The wear properties and adhesion strength of the diamond-like carbon film coated on SUS, Ti and Ni-Ti with plasma pre-treatment. Biomed. Mater. Eng. 2010, 20, 21–35.

[123] Vinothkumar TS, Kandaswamy D, Prabhakaran G, Rajadurai A. Effect of dry cryogenic treatment on Vickers hardness and wear resistance of new martensitic shape memory nickel-titanium alloy. Eur. J. Dent. 2015, 9, 513–7.

[124] Şahin Y, Öksüz KE. Effects of Al₂O₃ nanopowders on the wear behavior of NiTi shape memory alloys. JOM 2014, 66, 61–5.

[125] Stea S, Visentin M, Granchi D, Cenni E, Ciapetti G, Sudanese A, Toni A. Apotosis in peri-implant tissue. Biomaterials 2000, 21, 1393–8.

[126]   Chiba J, Rubash H, Kim KJ, Iwaki Y. The characterization of cytokines in the interface tissue obtained from failed cementless total hip arthroplasty with and without femoral osteolysis. Clin. Orthop. Rel. Res. 1994, 300, 304–12.

[127]   Perry M, Murtuza FY, Ponsford FM, Elson CJ, Atkins RM, Learmonth D. Properties of tissue from around cemented joint implants with erosive and/or linear osteolysis. Arthroplasty 1997, 12, 670–6.

[128]   Ingham E, Fisher J. The role of macrophages in osteolysis of total hip joint replacement. Biomaterials 2005, 26, 1271–86.

[129]   Keegan GM, Learmonth ID, Case CP. Orthopaedic metals and their potential toxicity in the arthroplasty patient: a review of current knowledge and future strategies. J. Bone Joint Surg. Br. 2007, 89B; https://doi.org/10.1302/0301-620X.89B5.18903.

[130]   Nayak P. Aluminium: impacts and disease. Environ. Res. 2002, 89, 101–15.

[131]   Tkeshekashvili L, Tsakadze K, Khulusauri O. Effect of some nickel compounds on red blood cell characteristics. Biol. Trace Elem. Res. 1989, 21, 337–42.

[132]   Hart AJ, Hester T, Sinclair K, Powell JJ, Goodship AE, Pele L, Fersht NL, Skinner J. The association between metal ions from his resurfacing and reduced T-cell counts. J. Bone Joint Surg. Br. 2006, 88B, 449–54.

[133]   Ferreira M, de Lourdes Pereira M, Garcia e Costa F, Sousa JP, de Carvalho GS. Comparative study of metallic biomaterials toxicity: a histochemical and immunohistochemical demonstration in mouse spleen. J. Trace Elem. Med. Biol. 2003, 17, 45–9.

[134]   Kurosaki K, Nakamura T, Mukai T, Endo T. Unusual findings in a fatal case of poisoning with chromate compounds. Forensic. Sci. Int. 1995, 75, 57–65.

[135]   Kametani K, Nagata T. Quantitative elemental analysis on aluminium accumulation by HVTEM-EDX in liver tissues of mice orally administered with aluminium chloride. Med. Mol. Morphol. 2006, 39, 97–105.

[136]   Oliveira H, Santos TM, Ramalho-Santos J, de Lourdes Pereira M. Histopathological effects of hexavalent chromium in mouse kidney. Bull. Environ. Contam. Toxicol. 2006, 76, 977–83.

[137]   Barceloux DG. Chromium. Clin. Toxicol. 1999, 37, 173–94.

[138]   Nemery B. Metal toxicity and the respiratory tract. Eur. Respir. J. 1990, 3, 202–19.

[139]   Antonini J, Lewis AB, Roberts JR, Whaley DA. Pulmonary effects of welding fumes: review of worker and experimental animal studies. Am. J. Indust. Med. 2003, 43, 350–60.

[140]   Yokel R. The toxicology of aluminium in the brain: a review. Neurotoxicology 2000, 21, 813–28.

[141]   Garcia GB, Biancardi M, Quiroga A. Vanadium (V)-induced neurotoxicity in the rat central nervous system: a histo-immunohistochemical study. Drug. Chem. Toxicol. 2005, 28, 329–44.

[142]   Linna A, Oksa P, Groundstroem K, Halkosaari M, Palmroos P, Huikko S, Uitti J. Exposure to cobalt in the production of cobalt and cobalt compounds and its effects on the heart. Occup. Environ. Med. 2004, 61, 877–85.

[143]   Lippmann M, Ito K, Hwang JS, Maciejczyk P, Chen LC. Cardiovascular effects of nickel in ambient air. Environ. Health. Perspect. 2006, 114, 1662–9.

[144]   Jeffery E, Ebero K, Burgess E, Cannata J, Greger JL. Systemic aluminium toxicity: effects on bone, hematopoietic tissue, and kidney. J. Toxicol. Environ. Health 1996, 48, 649–65.

[145]   Vermes C, Glant TT, Hallab NJ, Fritz EA, Roebuck KA, Jacobs JJ. The potential role of the osteoblast in the development of periprosthetic osteolysis: review of in vitro osteoblast responses to wear debris, corrosion products, and cytokines and growth factors. J. Arthroplasty 2001, 16, 95–100.

[146]   Darbre PD. Metalloestrogens: an emerging class of inorganic xenoestrogens with potential to add to the oestrogenic burden of the human breast. J. Appl. Toxicol. 2006, 26, 191–7.

[147]   Lu Z-Y, Gong H, Ameniya J. Aluminum chloride induces retinal changes in the rat. Toxicol. Sci. 2002, 66, 253–60.

[148]   Steens W, von Foerster G, Katzer A. Severe cobtalt poisoning with loss of sight after ceramic-metal pairing in a hip: a case report. Acta Orthop. 2006, 77, 830–2.

[149]   Hallab N, Jacobs J, Black J. Hypersensitivity to metallic biomaterials: a review of leukocyte migration inhibition assays. Biomaterials 2000, 21, 1301–14.

[150]   Hallab N, Mikecz K, Jacobs J. Metal sensitivity in patients with orthopaedic implants. J. Bone Joint Surg. Am. 2001, 83A, 428–36.

[151]   Aruldhas M, Subramaniam S, Sekar P, Vengatesh G, Chandrahasan G, Govindarajulu P, Akbasha MA. Chronic chromium exposure-induced changes in testicular histoarchitecture are associated with oxidative stress: study in a non-human primate (Macaca radiata Geoffroy). Hum. Reprod. 2005, 20, 2801–13.

[152]   Pandey R, Kumar R, Singh SP, Saxena DK, Srivastava SP. Male reproductive effect of nickel sulphate in mice. Biometals 1999, 12, 339–46.

[153]   Domingo JL. Vanadium: a review of the reproductive and developmental toxicity. Reprod. Toxicol. 1996, 10, 175–82.

[154]   Assad M, Chernyshov AV, Leroux MA, Rivard C-H. A new porous titanium-nickel alloy: part 1. Cytotoxicity and genotoxicity evaluation. Biomed. Mater. Eng. 2002, 12, 225–37.

[155]   Wever DJ, Veldhuizen AG, Sanders MM. Cytotoxic and genotoxic activity of a nickel-titanium alloy. Biomaterials. 1997, 18, 1115–20.

[156]   Haynes DR, Boyle SJ, Rogers SD, Susan D, Howie DW, Barrie V-R. Variation in cytokines induced by particles from different prosthetic materials. Clin. Orthop. 1998, 352, 223–30.

[157]   Nakashima Y, Sun DH, Trindale MC, Trindale MCD, Chun LE, Song Y, Goodman SB, Schuman DJ, Maloney WJ, Smith RL. Induction of macrophase C-C chemokine expression by titanium alloy and bone cement particles. J. Bone Joint Surg. Br. 1999, 81, 155–62.

[158]   Shanbhag AS, Jacobs JJ, Black J. Human monocyte response to particulate biomaterials generated in vivo and in vitor. J. Orthop. Res. 1995, 13, 792–801.

[159]   Shanbhag AS, Jacobs JJ, Black J, Galante JO, Glant TT. Macrophage/particle interactions: effect of size, composition and surface area. J. Biomed. Mater. Res. 1994, 28, 81–90.

[160]   Howie DW, Rogers SD, McGee MA, Haynes DR. Biologic effects of cobalt chrome in cell and animal models. Clin. Orthop. 1996, 329S, S217–32.

[161]   Maloney WJ, Smith RL, Castro F, Schurman DJ. Fibroblast response to metallic debris in vitro. Enzyme induction cell proliferation, and toxicity. J. Bone Joint Surg. Am. 1993, 75, 835–44.

[162]   Haynes DR, Rogers SD, Howie DW, Pearcy MJ, Barrie V-R. Drug inhibition of the macrophage response to metal wear particles in vitro. Clin. Orthop. 1996, 323, 316–26.

[163]   Shanbhag AS, Jacobs JJ, Black J, Galante JO, Glant TT. Effects of particles on fibroblast proliferation and bone resorption in vitro. Clin. Orthop. 1997, 342, 205–17.

[164]   Black J. Biological Performance of Materials: Fundamental of Biocompatibility. Marcel Dekker, New York, 1992.

[165]   Cunningham BW, Orbegoso CM, Dimitriev AE, Hallab NJ, Sefter JV, Asdourian P, McAfee PC. The effect of spinal particulate wear debris: an in vivo rabbit model and applied clinical study of retrieved instrumentation cases. Spine J. 2003, 3, 19–32.

[166]   Rhalmi S, Assad M, Leroux M, Charette S, Rivard CH. Spinal Evaluation of Porous Nitinol Particles: A Short-Term Study in Rabbits. In: Proceedings of the 49th Annual Meeting of Orthopedic Research Society. New Orleans, LA, USA, 2003; https://www.ors.org/Transactions/49/1088.pdf.

[167]   Rhalmi S, Charette S, Assad M, Coillard C, Rivard CH. The spinal cord dura mater reaction to nitinol and titanium alloy particles: a 1-year study in rabbits. Eur. Spine J. 2007, 16, 1063–72.

[168]   Oshida Y. Bioscience and Bioengineering of Titanium Materials. Elsevier, Oxford, UK, 1st edition, 2007.

[169]   Tan L, Crone WC, Sridharan K. Fretting wear study of surface modified Ni–Ti shape memory alloy. J Mater. Sci. Mater. Med. 2002, 12, 501–8.

[170]   Qian LM, Sun QP, Zhou ZR. Fretting wear behavior of superelastic nickel titanium shape memory alloy. Tribol. Lett. 2005, 18, 463–75.

[171]   Yang H, Qian L, Zhou Z, Ju X, Dong H. Effect of surface treatment by ceramic conversion on the fretting behavior of NiTi shape memory alloy. Tribol. Lett. 2007, 25, 215–24.

[172]   Dong H, Ju X, Yang H, Qian L, Zhou Z. Effect of ceramic conversion treatments on the surface damage and nickel ion release of NiTi alloys under fretting corrosion conditions. J. Mater. Sci. Mater. Med. 2008, 19, 937–46.

[173]   Siddiqui DA, Sivan S, Weaver JD, Di Prima M. Effect of wire fretting on the corrosion resistance of common medical alloys. J. Biomed. Mater. Res. B. Appl. Biomater. 2017, 105, 2487–94.

[174]   Landolt D, Mischler S, Stemp M. Electrochemical methods in tribocorrosion: a critical appraisal. Electrochim. Acta. 2001, 46, 3913–29.

[175]   Kosec T, Močnik P, Legat A. The tribocorrosion behaviour of NiTi alloy. Appl. Surf. Sci. 2013, 288, 727–35.

[176]   Močnik P, Kosec T, Kovač J, Bizjak M. The effect of pH, fluoride and tribocorrosion on the surface properties of dental archwires. Mater. Sci. Eng. C Mater. Biol. Appl. 2017, 78, 682–9.

Chapter 15
Dental/medical applications

Before getting detailed discussion on dental and medical applications of NiTi alloys in the subsequent two chapters, it would be meaningful and valuable to look through a variety of reviews or overview articles. Shape-memory alloys (SMAs), and in particular NiTi alloys, are characterized by two unique behaviors, thermally or mechanically activated phenomenon indicating the shape-memory effect (SME) and superelastic (SE) effect. These behaviors, due to the peculiar crystallographic structure of the alloys, assure the large recoverable strain of the original shape even after large deformations and the maintenance of a constant applied force in correspondence of significant displacements. Equiatomic NiTi, also known as nitinol, has a great potential for use as a biomaterial as compared to other conventional materials due to its shape-memory and superelastic properties [1–6]. Among various proposed ideas of applications of superelastic nitinol (which is recognized as a nearly equiatomic NiTi alloy), several new medical applications will be used to exemplify these points, including the quickly growing and technologically demanding stent applications. Stents are particularly interesting in that they involve new and complex manufacturing techniques, present a demanding and interesting fatigue environment, and most interestingly, take advantage of the thermoelastic hysteresis of nitinol [6–10]. NiTi SMAs showing superelastic behavior have great potential in dental and orthopedic applications where constant correcting loads may be required. In most of the clinical applications, the device may have been heat treated and during its life in service it will be cyclically deformed [11–19]. As dental applications, typical examples are endodontic instruments and orthodontic appliances; both of these applications utilize superelastic property of NiTi alloys.

Because of their unique mechanical properties and biocompatibility, SMAs have found many applications in minimally invasive surgery (MIS). Many novel surgical devices have been developed based on SMAs, which have become essential tools for MIS. Many minimally invasive techniques have been developed and gained widespread applications, they are now well established in surgery and routinely practiced, including cholecystectomy, Nissen fundoplication for gastro-esophageal reflux disease, appendicectomy, adrenalectomy, splenectomy, and many other advanced procedures. MIS offers great benefits to patients over conventional open surgery, the major benefits include reduced surgical trauma, reduced wound complications, shorter hospital stay, and accelerated recovery [20]. MIS is performed within a close space through small access entries (5 or 10 mm), surgeon's hands are outside the operative field, and surgical manipulations are performed remotely under visual control of monitor screen. In contrast to the larger movement of human hand-arm coordination in open surgery, only 4 degrees of freedom are available to surgeons because of the long and slender instruments used in MIS [21]. Many engineering technologies have

https://doi.org/10.1515/9783110666113-015

been brought to solve the problems associated with MIS, the use of SMAs in surgery is a wonderful example of all those successful stories. The unique materials properties of SMAs, that is, (SME)and (SE), have provided perfect answers to the strict design requirements in MIS, novel surgical devices have been developed based on SMAs, and they have become essential tools for many minimally invasive procedures. Cardiovascular surgery stent probably is the best example of SMAs in medicine and also the most successful product in terms of commercial and life-saving achievements. Nitinol stents have also been used in other parts of human body including stents for esophagus [22], gastrointestinal tract [23], ureter [24], tracheal airway [25], vascular anastomosis device [26, 27], radiofrequency ablation catheter [28], and prosthetic heart valves [29]. Laparoscopic and endoscopic surgery uses rigid or flexible endoscope to gain access to operative fields to diagnose and treat diseases [30, 31]. The early applications of nitinol in orthopedic surgery were staples and clamps to treat adolescent scoliosis [32], bone fractures [33], and also as ideal bone substitute [34, 35]. Equipment used for minimally invasive procedures depends on mechanical superelastic behavior, which is a significant departure from the original thermal shape-memory industrial uses of nitinol [36–38].

Biomedical applications of nitinol are related to transformation temperatures of nitinol that are close to body temperature (37 °C) [39]. Due to thermoelastic martensitic phase transformation and reverse transformation to parent austenite upon heating (SME) or upon unloading (SE), nitinol has a large number of biomedical applications [37]. Another important property of nitinol is its low elastic modulus close to natural bone material and compressive strength higher than natural bone material which makes it an ideal material for biomedical implant applications [9, 40, 42]. Biomedical applications of NiTi alloys should have a long list of devices and instruments; endodontic files and reamers; orthodontic brackets, palatal arch, and archwires; orthopedic head, spine, bone, muscles, hands/fingers, and legs; vascular aorta, arteries, vena cava filter, ventricular septal defect, vessels, and valves; for surgical instruments such as catheters/snares, scopes (ureteroscopy, endoscopy, and laparoscopy) and suture; and miscellaneous items including cardiology (heart), hepatology (liver, gallbladder, biliary tree, and pancreas), otorhinolaryngology (ear, nose, and throat), gastroenterology (gullet, stomach, and intestine), urology (kidneys, adrenal glands, ureters, urinary bladder, urethra, and the male reproductive organs), plastic, reconstructive and aesthetic surgery, and ophthalmology (eye).

Nitinol-based devices have been used in humans since mid-1970s [43]. Nitinol is now being practically used for various applications, for example, automotive applications, aerospace applications, eyeglasses, window frames, pipe couplings, antennae for cellular phones, actuators, sensors, seismic resistance, damping capacity, shock absorbers, automatic gas line shut-off valves, and SMA spring in water mixers [43–48].

In general, metallic materials, when compared to polymers and ceramics, possess (1) high strength, (2) high elasticity, (3) high fracture toughness, (4) a combination of high elasticity and stiffness, and (5) high electrical conductivity. Due to these properties, the use of metals has a long history with support by integrated discipline of materials science and engineering [49–51]. When considering the above properties, metals are generally superior to ceramics and polymers for dental/medical devices. Actually over 80% of implant devices are made of metals because of their strength, toughness, and durability. Among versatile dental/medical metallic devices and instruments, certain properties are required for certain types while there are still several common requirements over almost all devices such as good corrosion resistance and specific strength. Metals are essential for orthopedic implants, bone fixators, artificial joints, and external fixators since they substitute for the functions of hard tissues in orthopedics. Stents and stent grafts are placed in blood vessels for dilatation. Therefore, elasticity or plasticity for expansion and rigidity for maintaining dilatation are required in the devices. In dentistry, metals are used for restorations, orthodontic wire, and dental implants [1, 39]. Requirements for metals in medical devices are summarized in Table 15.1 [50, 51]. To meet these requirements, new alloy system should be designed and developed for appropriate surface technology and modifications should be applied [17, 52, 53].

Table 15.1: Requirements for metallic materials as biomedical devices [50, 51].

Required properties	Examples of medical devices	Biofunction or bioperformance
Elongation to fracture	Spinal fixation; maxillofacial plate	Improvement of durability
Elastic modulus (suitable)	Bone fixation; spinal fixation	Prevention of bone absorption by stress shielding; biomechanical compatibility
SE, SME	Multipurpose	Improvement of mechanical compatibility
Wear resistance	Artificial joint	Prevention of generation of wear debris; improvement of durability
Biodegradability	Stent; artificial bone; bone fixation	Elimination of materials after healing; no need of retrieval
Bone formation/bonding	Stem and cup of artificial hip joint; dental implant	Fixation of devices in bone
Prevention of bone formation	Bone screw; bone nail	Prevention of assimilation
Adhesion of soft tissue	Dental implant; trans skin device; external fixation; pacemaker housing	Fixation in soft tissue; prevention of inflectional disease
Inhibition of platelet adhesion	Devices contacting blood	Prevention of thrombus

Table 15.1 (continued)

Required properties	Examples of medical devices	Biofunction or bioperformance
Inhibition of biofilm formation	All implant devices; treatment tools and apparatus	Prevention of infectious disease
Low magnetic susceptibility	All implant devices; treatment tools and apparatus	No artifact in MRI
Passivation	All dental/medical devices	Excellent corrosion resistance
Formability (at surface layer)	Dental and orthopedic implants	Morphological compatibility
Porous formation	Dental and orthopedic implants	Biomechanical compatibility

References

[1]   Castleman LS, Motzkin SM, Alicandri FP, Bonawit VL. Biocompatibility of nitinol alloy as an implant material. J. Biomed. Mater. Res. 1976, 10, 695–731.
[2]   Andreasen GF, Morrow RE. Laboratory and clinical analyses of nitinol wire. Am. J. Orthod. 1978, 73. 142–51.
[3]   Wayman CM. Some applications of shape-memory alloys. JOM 1980, 32, 129–37.
[4]   El Feninat F, Laroche G, Fiset M, Mantovani D. Shape memory materials for biomedical applications. Adv. Eng. Mater. 2002, 4, 91–104.
[5]   Haga Y, Mizushima M, Matsunaga T, Esashi M. Medical and welfare applications of shape memory alloy microcoil actuators. Smart Mater. Struct. 2005, 14, S266–72.
[6]   Duerig T, Pelton A, Stöckel D. An overview of nitinol medical applications. Mater. Sci. Eng. A 1999, 273/275, 149–60.
[7]   Tosun G, Ünsaldi E, Özler L, Orhan N, Durmuş AS, Eröksüz H. Biocompatibility of NiTi alloy implants in vivo. Int. J. Biomed. Biol. Eng. 2013, 7, 271–4.
[8]   Hernández R, Polizu S, Turenne S, Yahia L'H. Characteristics of porous nickel-titanium alloys for medical applications. Bio-Med. Mater. Eng. 2002, 12, 37–45.
[9]   Biesjekierski A, Wang J, Gepreel MA-H, Wen C. A new look at biomedical Ti-based shape memory alloys. Acta Biomater. 2012, 8, 1661–9.
[10] Wang L, Liu RL, Majumdar T, Mantri SA, Ravi VA, Banerjee R, Birbills N. A closer look at the *in vitro* electrochemical characterisation of titanium alloys for biomedical applications using in-situ methods. Acta Biomater. 2017, 54, 469–78.
[11] Gil FX, Manero JM, Planell JA. Relevant aspects in the clinical applications of NiTi shape memory alloys. J. Mater. Sci. Mater. Med. 1996, 7, 403–6.
[12] Jackson MJ, Kopac J, Balazic M, Bombac D, Brojan M, Kosel F. Titanium and titanium alloy applications in medicine. Surf. Eng. Surg. Tools Med. Devices 2016, 533–76. [13] Sibum H, Titanium and titanium alloys – from raw material to semi-finished products. Adv. Eng. Mater. 2003, 55, 393–8.
[14] Wang K. The use of titanium for medical applications in the USA. Mater. Sci. Eng. A 1996, 213, 134–7.

[15] Rack HJ, Qazi JI. Titanium alloys for biomedical applications. Mater. Sci. Eng. C 2006, 26, 1269–77.
[16] Wayman CM. Some applications of shape-memory alloys. JOM 1980, 32, 129–37.
[17] Dunić V, Slavković R, Pieczyska EA. Properties and Behavior of Shape Memory Alloys in the Scope of Biomedical and Engineering Applications. In: Zivic F, et al. ed., Biomaterials in Clinical Practice. Springer, Cham 2017, 303–31.
[18] Dai K, Ning CQ. Shape Memory Alloys and Their Medical Application. In: Poitout DG, ed., Biomechanics and Biomaterials in Orthopedics. Springer, London, 2002, 179–184.
[19] Zhang L-C, Chen L-Y. A review on biomedical titanium alloys: recent progress and prospect. Adv. Eng. Mater. 2019, 21; https://doi.org/10.1002/adem.201801215.
[20] Song C. History and current situation of shape memory alloys devices for minimally invasive surgery. Open Med. Devices J. 2010, 2, 24–31.
[21] Cuschieri A. Minimal Access Surgery. In: Eremin O, ed., The Scientific and Clinical Basis of Surgical Practice. Oxford: Oxford University Press, 2001, 200–10.
[22] Cwikiel W, Stridbeck H, Tranberg K, Malignant K. Esophageal strictures: treatment with a self-expanding Nitinol stent. Radiology 1993, 187, 661–5.
[23] Morgan R, Adam A. Use of metallic stents and balloons in the esophagus and gastrointestinal tract. J. Vasc. Interv. Radiol. 2001, 12, 283–97.
[24] Kulkarni RP, Bellamy EA. A new thermo-expandable shape memory nickel-titanium alloy stent for the management of ureteric strictures. BJU Int, 1999, 83, 755–9.
[25] Vinograd I, Klin B, Brosh T. A new intratracheal stent made from nitinol an alloy with shape memory effect. J. Thorac. Cardiovasc. Surg. 1994, 107, 1255–61.
[26] Tozzi P, Corno AF, von Segesser L. Sutureless coronary anastomosis: revival of old concepts. Eur. J. Cardiothorac. Surg. 2002, 22, 565–70.
[27] Chan K, Godman M, Walsh K. Transcatheter closure of atrial septal defect and ineratrial communications with a new self expanding nitinol double disk device (Amplazer Septal Occluder): multicentre UK experience. Heart 1999, 82, 300–6.
[28] Himpens JM. Laparoscopic inguinal hernioplasty repair with a conventional vs. a new self-expandable mesh. Surg. Endosc. 1993, 7, 315–8.
[29] Rickers C, Hamm A, Stern H. Percutaneous closure of atrial septal defect with a new self centring device ("Angle Wings"). Heart 1998, 80, 517–21.
[30] Cuschieri A. Variable curvature shape-memory spatula for laparoscopic surgery. Surg. Endosc. 1991, 5, 179–81.
[31] Kirschniak A, Traub F, Kueper MA, Stuker D. Endoscopic treatment of gastric perforation caused by acute necrotizing pancretitis using over-the-scope clips: a case report. Endoscopy 2007, 39, 1100–2.
[32] Otsuka K, Wayman C. Shape Memory Materials. Cambridge University Press, Cambridge, 1998.
[33] Dai K, Chu Y. Studies and applications of NiTi shape memory alloys in the medical field in China. Biomed. Mater. Eng. 1996, 6, 233–40.
[34] Rhalmi S, Odin M, Assad M. Hard, soft tissue and in vitro cell response to porous nickel-titanium: a biocompatibility evaluation. Biomed. Mater. Eng. 1999, 9, 151–62.
[35] Kang SB, Yoon KS, Kim JS. In vivo result of porous TiNi shape memory alloy: bone response and growth. Mater. Trans. 2002, 43, SI1045–8.
[36] Petrini L, Migliavacca F. Biomedical applications of shape memory alloys. J. Metall. 2011; http://dx.doi.org/10.1155/2011/501483.
[37] Pelton A, Stöckel D, Duerig T. Medical use of nitinol. Mater. Sci. Forum 2000, 327/328, 63–70.
[38] Hunter JG, Sackier JM. Minimally Invasive High Tech Surgery; into the 21st Century. In: Hunter JG, et al. ed., Minimally Invasive Surgery. McGrawHill, New York, 1993, 3–6.

[39]   Duerig T, Pelton A, Stöckel D. An overview of nitinol medical applications. Mater. Sci. Eng. A
        1999, 273/275, 149–60.
[40]   Niinomi M. Recent research and development in titanium alloys for biomedical applications
        and healthcare goods. Sci. Technol. Adv. Mater. 2003, 4, 445–54.
[41]   Brailovski V, Prokoshkin S, Gauthier M, Inaekyan K, Dubinskiy S, Petrzhik M, Filonov M. Bulk
        and porous metastable beta Ti–Nb–Zr(Ta) alloys for biomedical applications. Mater. Sci. Eng.
        C 2011, 31, 643–57.
[42]   Machado LG, Savi MA. Medical applications of shape memory alloys Braz. J. Med. Biol. Res.
        2003, 36, 683–91.
[43]   Wadood A. Brief overview on nitinol as biomaterial. Adv. Mater. Sci. Eng. 2016;
        http://dx.doi.org/10.1155/2016/4173138.
[44]   Shabalovskaya SA, Tian H, Anderegg JW, Schryvers DU, Carroll WU, Humbeeck JV. The
        influence of surface oxides on the distribution and release of nickel from Nitinol wires.
        Biomaterials 2009, 30, 468–77.
[45]   Venugopalan R, Trépanier C. Assessing the corrosion behaviour of Nitinol for minimally-
        invasive device design. Minimal. Invasiv. Ther. All. Technol. 2000, 9, 67–73.
[46]   Shabalovskaya S, Anderegg J, Van Humbeeck J. Critical overview of Nitinol surfaces and their
        modifications for medical applications. Acta Biomater. 2008, 4, 447–67.
[47]   Morgan NB. Medical shape memory alloy applications-the market and its products. Mater.
        Sci. Eng. A 2004, 378, 16–23.
[48]   Yahia L'H, Ryhänen J. Bioperformance of shape memory alloys. Shape Mem. Implants
        2000, 3–23.
[49]   Oshida Y, Sachdeva R, Miyazaki S, Fukuyo S. Biological and chemical evaluation of TiNi alloys.
        Mater. Sci. Forum 1990, 56/58, 705–10.
[50]   Hanawa T. Research and development of metals for medical devices based on clinical needs.
        Sci. Technol. Adv. Mater. 2012, 13; https://www.tandfonline.com/doi/full/10.1088/1468-
        6996/13/6/064102.
[51]   Oshida Y. Bioscience and Bioengineering of Titanium Materials. Elsevier, Oxford, UK, 2007.
[52]   Elahinia MH, Hashemi M, Tabesh M, Bhaduri SB. Manufacturing and processing of NiTi
        implants: a review. Prog. Mater. Sci. 2012, 57, 911–46.
[53]   Oshida Y. Surface Engineering and Technology for Biomedical Implants. Momentum Press,
        New York, NY, USA, 2014.

Chapter 16
Dental applications

16.1 In general

More than 3,000 alloys are available in the market for dental applications, which are placed into long-term direct or indirect contact with the epithelium, the connective tissue, or the bone. Given this long-term intimate contact with the vital tissue, it is of paramount importance that dental metallic materials should exhibit safe bioperformance and biofunction [1–4]. By definition and general agreement, smart materials [including nickel–titanium and shape-memory alloy (NiTi SMA)] have properties that may be altered in a controlled manner by stimuli, such as stress, temperature, moisture, pH, and electric or magnetic fields. There is a strong trend in material science to develop and apply these intelligent materials in dentistry. These materials would potentially allow new and groundbreaking dental therapies with a significantly enhanced clinical outcome of treatments [5]. Near-equiatomic $Ni_{48-51}Ti$ (at%) alloys (also known as nitinol) possess superior strength and high corrosion resistance. These alloys exhibit three properties that are not commonly observed in other metallic materials, such as shape-memory effect (SME), superelasticity (SE), and high damping properties [6]. For SME, a large deformation can be completely recovered by slight heating to return to its original shape. This phenomenon has been well exploited for many biomedical devices at the surgical site that are activated by body heat or external heat sources. In dentistry, a highly successful application example is dental orthodontic wire that uses SE characterized by a constant stress upon loading and unloading [7–10]. The introduction of NiTi as a material for endodontic instruments about 15 years ago opened many new perspectives. Many dentists and scientists observe benefits in using NiTi files. Initial problems such as frequent fractures and the uncertainty of the best way to use them have been solved. Other challenges such as enhancing the cutting ability or optimizing the speed, torque, and fatigue are currently being addressed. Some clinicians are skeptical because they see this approach as too mechanical. Nevertheless, the combination of anatomical, biological, and pathophysiological knowledge with the use of NiTi instruments is a large step forward in optimizing the quality of root canal treatment worldwide [11, 12].

In this chapter, two major dental applications (endodontic instruments and orthodontic appliances) of smart NiTi alloys are discussed.

https://doi.org/10.1515/9783110666113-016

16.2 Endodontic applications

16.2.1 Development history of endodontic instruments

NiTi rotary instruments have become essential tools in endodontics due to their superior flexibility and higher cutting efficiency compared to those of conventional SS instruments [13]. In addition, due to its SE, NiTi instruments maintain the original canal shape with fewer irregularities, such as zip, ledge, or transportation during canal preparation [14]. Despite these advantages, NiTi instruments are not free from sudden file separation during the treatment, which may lead to a poor prognosis [15] as discussed in detail further. While studying the history of NiTi endodontic instruments, materials, R&D, heat treatment, alloying effects, fracture analysis, counterplan thereof, and so on are learnt. Hence, the history of NiTi endodontic instruments also implies the history of NiTi material development and application history in medicine. Root canal instrumentation may be performed using hand-held or engine-driven (rotary) instruments. Since their first appearance, the instrument design has changed considerably due to advanced manufacturing processes, metallurgical improvement, and thermomechanical treatments. One should always bear in mind that all file systems have benefits and disadvantages. Instrument properties and their clinical efficacy are controlled by various factors including the type of alloy, degree of taper, cross-sectional design, passive or active cutting manners, and operator's experience and skills [16, 17]. Hence, it is worthy to review the development history of NiTi endodontic instruments.

16.2.1.1 Incubation period

According to the historical recounted article [18], the serendipitous nature of the NiTi alloy was discovered in 1959 by William J. Buhler of the U.S. Naval Ordnance Laboratory, its subsequent development by Buhler and Frederick E. Wang. Civijan et al. [19] studied the mechanical behavior ("shape memory") associated with martensitic solid-state transformation in nearly equiatomic NiTi alloy (55-nitinol) and suggested potential endodontic applications of 55-nitinol (55 wt% Ni, 1.5 wt% Co, balance Ti) and 60-nitinol (60 wt% Ni). In 1988, Walia et al. [16], fabricated root canal files in size #15 and triangular cross sections (fluted structure of a K-type file) directly from 0.020-inch diameter (orthodontic) archwire blanks of nitinol. It was reported that the nitinol files were found to have two to three times more elastic flexibility in bending and torsion, as well as superior resistance to torsional fracture, compared with the size #15 SS files manufactured by the same process. In addition, the fracture surfaces were observed for clockwise (CW) and counterclockwise (CCW) torsions with the scanning electron microscope (SEM) and thus exhibited a largely flat morphology for files of both alloy types and torsional testing modes. It was possible to permanently precurve the nitinol files in the manner often used by clinicians with SS files.

These results suggest that the nitinol files may be promising for the instrumentation of curved canals, and evaluations of mechanical properties and in vitro cutting efficiency are in progress for size #35 instruments. Figure 16.1 shows intracanal files developed by Walia et al. [16].

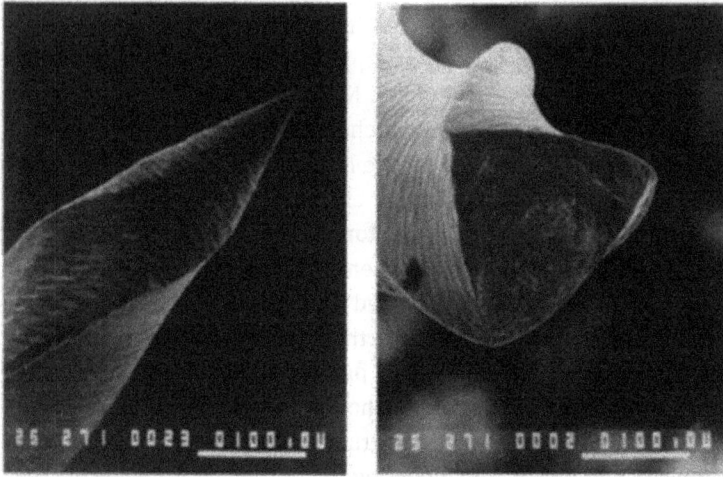

Figure 16.1: Intracanal files. [16].

16.2.1.2 The first generation (early 1990s)

In general, the first-generation NiTi files have passive cutting radial lands and fixed tapers of 4% and 6% over the length of their active blades [20]. The files are used for apical preparation and do not cut over most of the canal length because of the existence of a smooth small diameter shaft that also enhances the flexibility of the instrument. All the systems required a considerable number of files to achieve preparation objectives. By the mid-to-late 1990s, the Greta Taper (GT) files became available that provided a fixed taper on a single file of 6, 8, 10, and 12% [21]. The single most important design feature of the first-generation NiTi rotary file was passive radial lands, which encouraged a file to stay centered in canal curvatures during work. The ProFiles rotary NiTi instruments, introduced by Ben Johnson in 1994, are available in 0.02, 0.04, and 0.06 taper and in the International Standards Organization (ISO) diameters from 15 to 45 (0.02 taper) and from 15 to 80 (0.04 and 0.06 taper), as well as in 21-, 25-, and 30-mm lengths. All Pro-Files present a working surface of 16 mm and a round noncutting tip. Figure 16.2 shows typical three types of ProFile. Figure 16.3 shows the cross section of ProFile with U-shaped groove and radial land [22]. Figure 16.4 illustrates a ProFile with 20° of helix angle (angle between the cutting edge and long axis of file) and constant pitch.

(a)   (b)   (c)

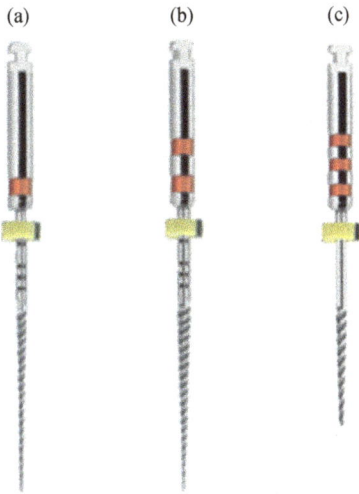

Figure 16.2: (a) ProFile 0.04 tapered series,
(b) ProFile 0.06 tapered series, (c) Orifice shaper.

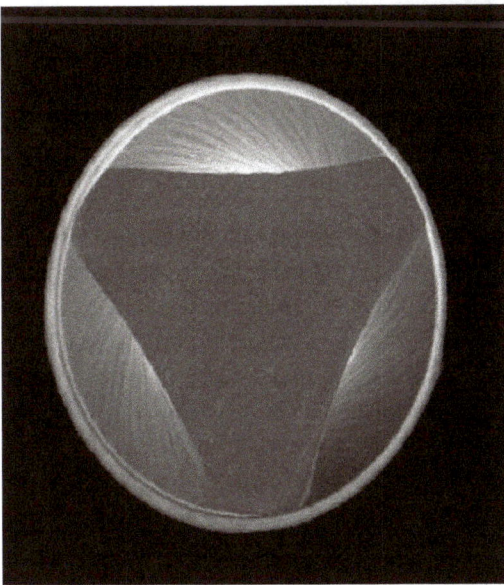

Figure 16.3: Cross section of ProFile with U-shaped groove and radial land. [22].

It is worthy to review the common nomenclatures related to the endodontic files [23, 24]. The concept of the radial lands is a critical design feature. It is a surface that projects axially from the central axis, between flutes, as far as the cutting edge. The best way to explain this is the blade support. Blade support is defined as the amount of material supporting the cutting blades of the instrument. This part of the file is called the radial land and the design feature is critical for the instrument. Lesser the

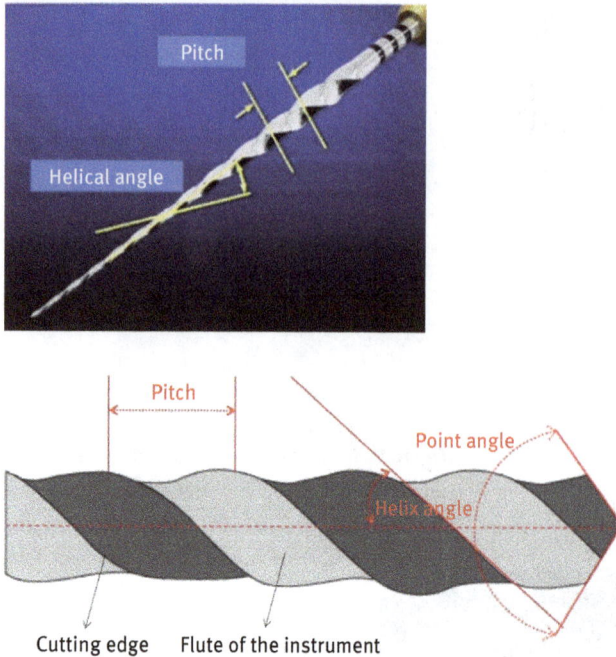

Figure 16.4: ProFile with 20° of helix angle and constant pitch.

blade support (the amount of metal behind the cutting edge), lesser is the resistance of the instrument to torsional or rotary stresses [25]. By increasing the contact surface area with the root canal wall, lateral stress increases, resulting in the reduction of flexibility of the file. It is the combination of a noncutting tip and radial land that keeps a file centered in the canal. Most rotary files derive their strength from the mass of material in the core. Peripheral strength can also be added to a file by extending the width of the radial land. Helical angle is the angle that the cutting edge makes with the long axis of the file. As a rotary file works in a canal, the dentinal debris needs to be removed quickly and effectively to achieve a satisfactory efficacy. Files with a constant helical flute angle allow debris to accumulate, particularly in the coronal part of the file. In addition, files that maintain the same helical angle along the entire working length (WL) are more susceptible to the effect of "screwing in" forces. By varying the flute angles, debris are removed in a more efficient manner and the file is less likely to screw into the canal [26]. *Pitch* is the number of spirals or threads per unit length. Historically, screws have had a constant pitch. The result of a constant pitch and constant helical angles is called a "pulling down" or "sucking down" into the canal. This is particularly significant in rotary instrumentation when using files with a constant taper [27, 28]. Taper is another feature of the file design and it is particularly important concerning "system concepts." There are two ways to shape a canal; (1) instrumentation of a root canal by using files of the same taper but with varying apical tip diameters and (2) instrumentation of root canal by varying or graduating tapers [29,30]. Rake angles are also important and affect the cutting efficiency of the instrument. However, there remains a confusion over what constitutes a rake angle and what is the cutting angle. The rake angle is the angle formed by the cutting edge and a cross section taken perpendicular to the long axis of the instrument. The cutting angle, on the other hand, is the angle formed by the cutting edge and a radius when the file is sectioned perpendicular to the cutting edge. Positive rake angles cut more efficiently than neutral rake angles, which scrap the inside of the canal. Most conventional endodontic files utilize a negative or "substantially neutral" rake angle. An overly positive rake angle results in digging and gouging of the dentin. This can lead to separation. ProFile has a negative rake angle. The clear differences among positive, neutral, and negative rake angles are illustrated in Figure 16.5 [31], and Figure 16.6 shows the cutting actions when applied to the endodontic files [32].

(a)   (b)   (c)

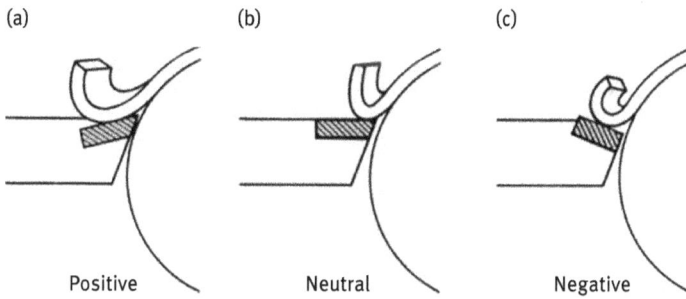

Positive   Neutral   Negative

Figure 16.5: Differences among positive, neutral, and negative rake angles.

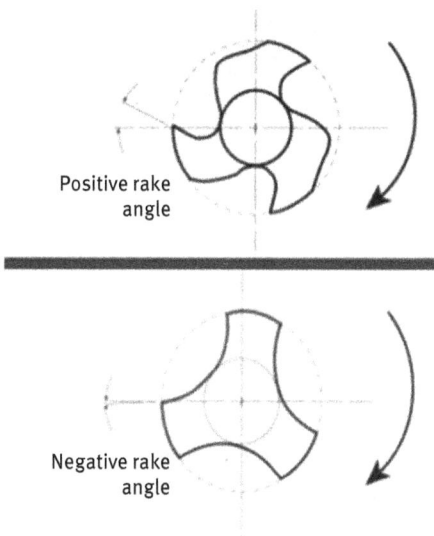

Positive rake angle

Negative rake angle

Figure 16.6: Positive and negative rake angle movements [32].

The LightSpeed (SybronEndo) instrument system is the only endodontic instrument designed to create root canal preparations that are biologically optimal. Unlike tapered endodontic instruments, which may cut indiscriminately, the LightSpeed endodontic instrument cuts only where you need to cut and thoroughly cleans the apical third without unnecessarily removing root structure coronally or in the mid-root, which can weaken the tooth [33, 34]. Figure 16.7 shows (a) the Light Speed type files with a thin and long noncutting shaft and (b) a front view of the cutting portion at the tip. Marending et al. [35] compared the apical fit in two dimensions of the first K-file versus the first LightSpeed LSX instrument binding at WL after an initial crown-down preparation. Twenty maxillary molars with fully developed roots and four separate root canals were selected. Canals were preflared with ProFile 0.04 taper instruments to three-quarters of the estimated WL. WL was electronically determined using a size 06 K-file. Progressively larger K-files were inserted passively to the WL. The first binding K-file was termed as the initial apical file (IAF). It was reported that: (i) IAF sizes ranged from 8 to 30 and were lowest in second mesiobuccal (MB2) and highest in palatal canals; (ii) the apical large canal diameter was assessed more accurately by the LSX instruments; however, the smallest available LSX instrument (i.e., size 20) did not reach WL in 39 of 80 canals; therefore, concluding that (iii) instruments with a flat widened tip were found to determine apical cross-sectional diameter better than round, tapered instruments.

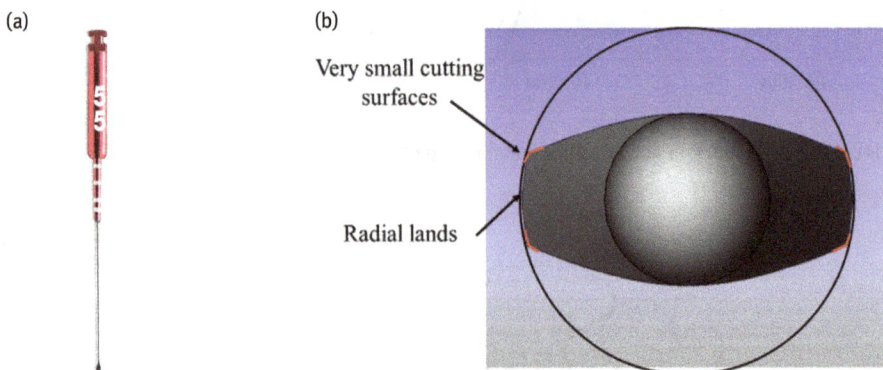

Figure 16.7: (a) The LightSpeed type files with a thin and long noncutting shaft and (b) a front view of the cutting portion at the tip [33,34].

In 1996, McSpadden [36] developed the Quantec (SybronEndo) system (composed of ten different types of files) designed to set the new standard in efficient and effective root canal preparation with continuous taper from orifice to apex. At SybronEndo, all the critical elements have been combined to achieve levels of predictability and performance unmatched in the dental industry. The system is characterized by having neutral radial lands with different width to enhance the rigidity of files. Figure 16.8 shows Quantec files.

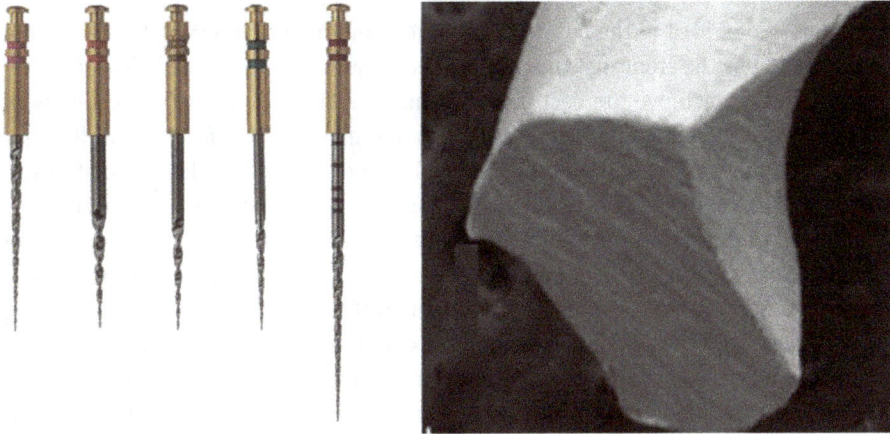

Figure 16.8: Quantec files [36].

In 1994, four kinds of GT files (with 0.06, 0.08, 0.10, and 0.12 tapers) were developed [21]: (1) characterized with a maximum diameter of 1.0 mm, (2) by increasing the taper, the length of cutting flute gets shorter, and (3) used mainly for crown down technique. Figure 16.9 shows the GT file.

Figure 16.9: GT file [21].

The development of NiTi rotary instruments is undoubtedly a quantum leap for the field of endodontics. Nevertheless, those who have gained some experience in the use of such instruments confirm that each file system has its own special advantages and disadvantages and particular rules for its usage need to be followed. With most NiTi systems, it is easy to reach the WL and prepare the apex to a small size such as ISO size 20. However, when the apex is prepared to larger sizes, the limits of a particular system quickly become apparent. Based on the said background, Walsch

[37] proposed an idea of the hybrid concept to combine instruments of different file systems and use different instrumentation techniques in managing individual clinical situations to achieve the best biomechanical cleaning and shaping results with least procedural errors. The hybrid concept combines the best features of different systems for safe, quick, and predictable results. The crown down idea is to step apically by using a series of files while decreasing the instrument size or taper. The next smaller file performs its cutting action deeper into the canal, leaving the engaging surface of each instrument minimal and, therefore, decreasing the torque load of each instrument. Repeating the use of such series of files also results in either gaining deeper access into the canal or enlarging the canal further by each sequence. Crown down minimizes coronal interference, eases instrument penetration, increases apical tactile awareness, reduces canal curvature, minimizes change in WL during apical instrumentation, allows irrigation penetration to preparation depth, removes bacteria before approaching the apical canal third (therefore preventing iatrogen apical contamination), reduces the contact area of each instrument (therefore reducing torque and increasing cutting efficiency and safety), and reduces the instrument tip contact and the incidence of procedural errors. As to preparation techniques using the coronal to apical approach, there are typically two techniques, known as the *step down technique* and *crown down pressureless technique*. The step down technique is to be recommended especially in the case of narrow and/or severely curved root canals. The first step, once the preparation length has been determined, is to (1) widen the coronal part of the root canal and then (2) to prepare the apical canal segment (b). The straight canal section is widened using Hedstroem files and smoothened with a Gates-Glidden's drill. Then the curved section is instrumented by means of a rotating/scraping working method using flexible instruments with a noncutting tip [38].

16.2.1.3 The second generation (late 1990s)

During this period, manufacturers began to focus on other methods to increase the resistance to file separation. Some manufacturers electropolished (EP) their files to remove surface irregularities caused due to the traditional grinding process. However, it has been clinically observed and scientifically reported that electropolishing dulls the sharp cutting edges. Excessive inward pressure, especially when utilizing fixed tapered files, invites taper lock, the screw effect, and excessive torque on the rotary file during work [39]. The critical distinctions of this generation of instruments are that they have actively cutting edges without radial lands and fewer instruments are required to fully prepare a canal. Second-generation systems were designed with positive rake angles (as shown in Figure 16.5 & Figure 16.6) having an active cutting edge without radial land for greater active cutting efficiency with small shear strain (Figure 16.10). It is also characterized by a smaller helix angle compared to that of the first generation, leading to a small screw-in effect. The angle between the cutting blade and the longitudinal axis of the instrument is lower than in first-generation files, which greatly reduces the tendency for a screwing effect. This generation of NiTi files includes the ProTaper rotary files that have multiple tapers of increasing and decreasing size on a single file [40]. It has been suggested that the surface condition of the NiTi instrument contributes to fatigue resistance because most fatigue failures nucleate from the surface, especially in the presence of high stress amplitude or surface defects. Figure 16.10 shows 0.04 and 0.06 taper ProFile rotary NiTi files [20].

(a)   (b)

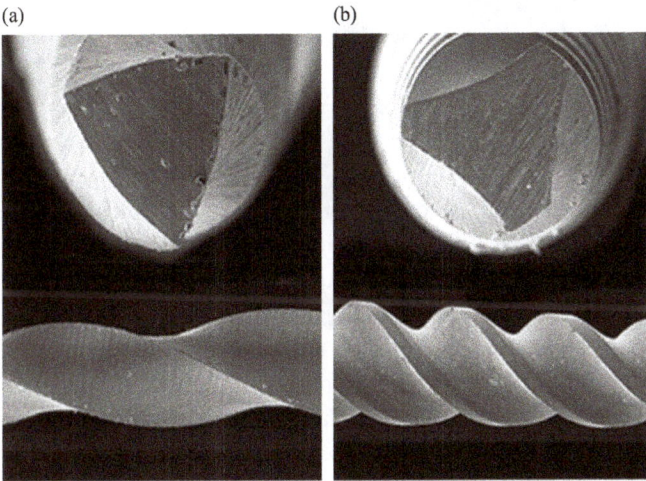

Figure 16.10: ProFile NiTi rotary file with: (a) 0.04 taper and (b) 0.06 taper [20].

EndoSequence (ES) is characterized by having an alternating contact point design (Figure 16.11), so that centering of the file can be accomplished without the radial land and unwanted screw-in phenomenon toward the root apex can also be prevented. Moreover, shaping the active cutting edge can improve the cutting efficiency and electropolishing [39] corrects the surface microdefects to enhance the file sharpness, resulting in preventing file fracture. The taper at the cutting portion is constant (Figure 16.12). It has been claimed that: (i) this file system for cleaning and shaping instruments should preserve the structural integrity of the tooth; (ii) the obturation materials should act as a conduit, not an inhibitor, to a successful long-term restoration; and (iii) the system should be as efficient as possible but the long-term prognosis of the tooth trumps simplicity [41].

Figure 16.11: EndoSequence with alternating contact point design.

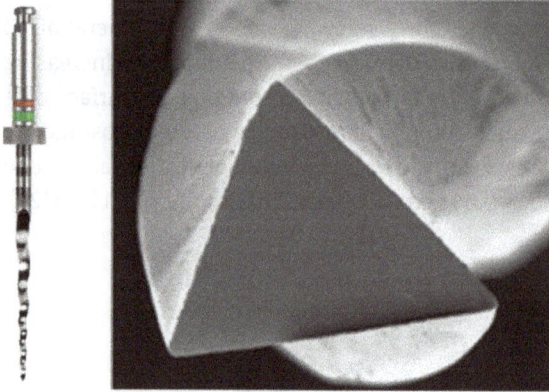

Figure 16.12:. EndoSequence file.

The ProTaper system is the alternative to all other file systems in which each file has a fixed taper. Originally, in 2001, Ruddle developed the ProTaper composed of three shaping files and five finishing files. Each ProTaper file has a variable changing taper over the length of its cutting blades. Specifically, the ProTaper shaping files have small-sized tips that act as guides to follow the path of the canal previously secured with hand files. Progressively tapered shaping files work away from their apical extents and, importantly, selectively cut dentin toward their larger, stronger, and more active blades [42]. By adding a function of progressively tapered design to single file, flexibility and cutting efficiency can be greatly improved and the recapitulation frequency can be reduced. To optimize safety and efficiency, shaping files are used like a brush to later-ally and selectively cut dentin on the outstroke. A brushing action creates lateral space that facilitates moving the larger and more active blades of the shaping file safely and progressively deeper into the canal. It is remarkable to note that, in this method of use, the files are essentially loose during the majority of their work within a canal. In the Pro-

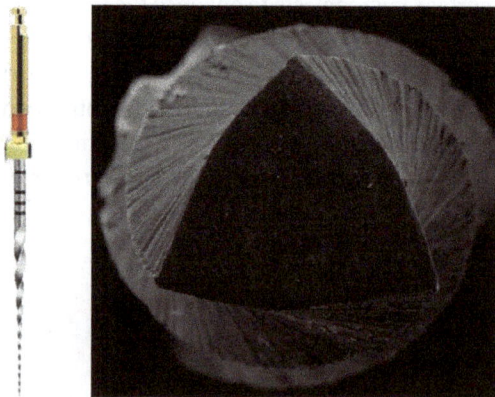

Figure 16.13: Typical ProTaper file [22].

Taper concept, more than the instrument design, the motion with which the file is used, creates the shape. Furthermore, as the cutting operation can be limited in a portion of the root canal, a risk for the torsional fatigue can be reduced to prolong the tool life [22].

K3 file system (which is after the Quantec and developed by McSpadden) possesses unique design of a triple-fluted and asymmetric cross-sectional feature. It was designed to cut quickly, efficiently, and safely with unparalleled debris removal. The system is further characterized by: (1) positive rake angle that provides the active cutting action of the K3 endo files, (2) wide radial land of the K3 rotary file that provides blade support while adding peripheral strength to resist torsional and rotary stresses, (3) the third radial land of K3 files that stabilizes and keeps the instrument centered in the canal and minimizes over engagement, (4) radial land relief that reduces friction on the canal wall, and (5) the system designed to maintain the centering of the file and to enhance the torsional fatigue strength [43].

As seen in Figure 16.14, to strengthen the file tip portion, variable pitch and core diameter are added.

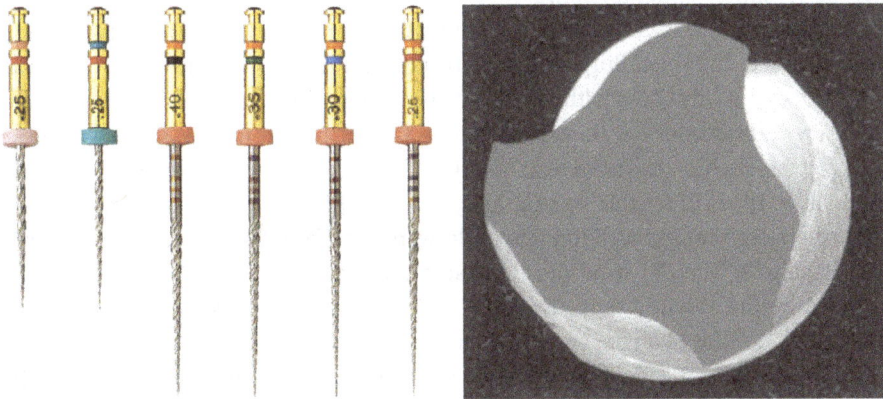

Figure 16.14: K3 endodontic file system.

16.2.1.4 The third generation (2000–2010)

In the third generation, there was tremendous improvement in NiTi metallurgy. Heat treatment is one of the most fundamental approaches toward adjusting the transition temperatures of NiTi alloys and affecting the fatigue resistance of NiTi endodontic files. In 2007, manufacturers began to focus on utilizing heat treatments to reduce cyclic fatigue and improve safety when rotary NiTi instruments work in more curved canals [44]. The desired phase transition point between martensite and austenite can be identified to produce a more clinically optimal metal than NiTi itself. It is produced by applying a series of heat treatments to NiTi wire blanks. M-wire was developed in 2005 [45]. According to the manufacturer, a thermal process allows twisting during a phase transformation into the so-called R-phase of NiTi, which is believed to exhibit about four times higher the cyclic fatigue strength [46], followed by the development of the R-phase instrument [47]. NiTi alloy has three different, temperature-dependent, microstructure phases: austenite, martensite, and the R-phase [15, 48, 49]. Austenitic NiTi is strong, hard, and exhibits SE, whereas martensitic and R-phase NiTi are soft, ductile, and can be easily subjected to plastic deformation. The mechanical characteristics of NiTi are influenced by the compositions of the three phases [50]. The conventional NiTi alloy is primarily in the austenite phase at room temperature. Thermomechanical treatments could maintain the alloy in the martensite phase, R-phase, or mixed form by altering the transformation temperature and consequently changing the characteristics of the alloy [15, 51, 52]. R-phase in NiTi alloys appears, depending on the temperature, between austenite and martensite phases as an intermediate phase and exhibits higher flexibility, stronger cyclic fatigue strength, and easier to follow the root canal pass compared with conventional NiTi file [49]; moreover, Santos et al. [48] mentioned that NiTi endodontic instruments containing only R-phase in their microstructure would show higher flexibility without compromising their performance under torsion.

The aim of this study was to investigate the effect of thermomechanical treatment on mechanical and metallurgical properties of NiTi rotary instruments. Eight kinds of NiTi rotary instruments with sizes of ISO #25 were selected: ProFile, K3, and One Shape for the conventional alloy; ProTaper NEXT (PTN), Reciproc (RPC), and WaveOne (WO) for the M-wire alloy; HyFlex CM (HFCM) for the controlled memory (CM) wire; and twisted file (TF) for the R-phase alloy. Torsional fracture and cyclic fatigue fracture tests were performed. Products underwent a differential scanning calorimetry (DSC) analysis. The CM wire and R-phase groups had the lowest elastic modulus, followed by the M-wire group. The maximum torque of the M-wire instrument was comparable to that of a conventional instrument, whereas those of the CM wire and R-phase instruments were lower. The angular displacement at failure (ADF) for the CM wire and R-phase instruments was higher than that of conventional instruments, whereas ADF of the M-wire instruments was lower. The cyclic fatigue resistance (CFR) of the thermomechanically treated NiTi instruments was higher. DSC plots revealed that NiTi instruments made with the conventional alloy were primarily composed of austenite at room temperature; stable martensite and R-phase were found in thermomechanically treated instruments.

The CM wire was introduced in 2010, which are manufactured using a special thermomechanical process that controls the memory of the material, making the files extremely flexible but without the shape memory of other NiTi files. Figure 16.15 demonstrates the high flexibility of the CM file [53].

The G-wire was developed in 2015, which is manufactured through several processes; (1) by producing a mixed phase of austenitic phase, R-phase, and martensitic phase by M-wire treatment, (2) by precipitating Ti_3Ni_4 in the austenitic phase, and

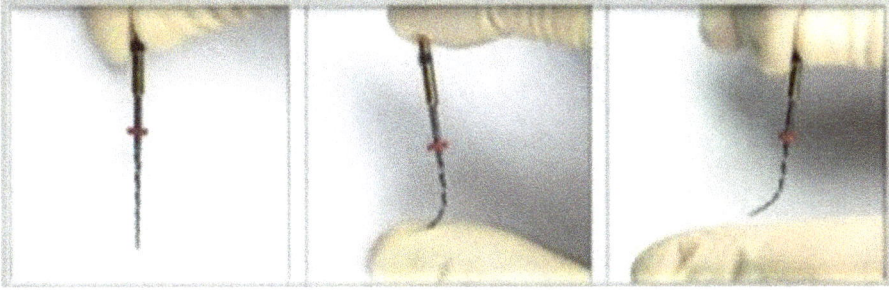

Figure 16.15: Flexible CM wire [53].

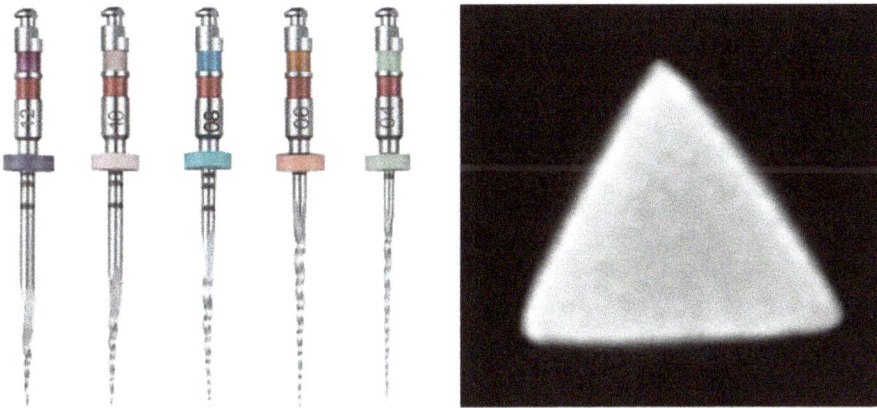

Figure 16.16: Twisted file [54].

(3) by controlling excessive growth of martensitic phase to promote the formation of R-phase (see Figure 16.16 [54]).

TF is manufactured by heat-treating austenitic phase to transform to R-phase, which is later deformed to produce files, resulting in improved SE and cyclic fatigue strength [54].

K3XF keeps original morphological feature of K3 file and manufactured through the R-phase technology, claiming that it is superior over K3 and TF, in terms of cyclic fatigue strength and flexibility [55–57]. Figure 16.17 shows K3XF file [55].

K3H is produced by thermomechanically treating K3 to obtain the similar characteristics of CM wire and exhibits higher cyclic fatigue strength than K3XF [57].

HyFlex (HF) is produced using CM wire to possess martensitic property at room temperature, resulting in no spring back and excellent root canal trackability. Precurve forming can be done to this type of file and the file can be applied to clinical cases such as Ledge syndrome. HF rotary files are NiTi instruments used for canal enlargement to prepare the root canal for irrigation and obturation in continuous

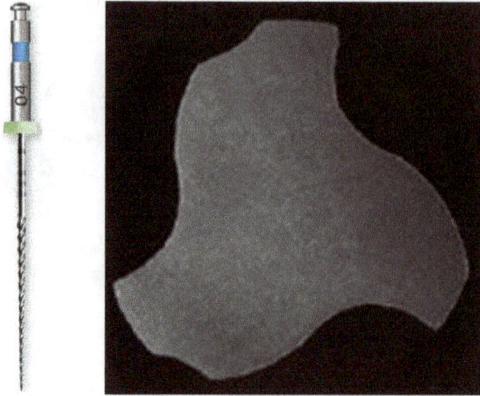

Figure 16.17: K3XF file [55].

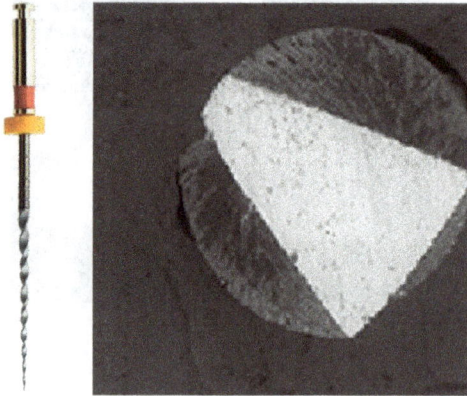

Figure 16.18: HyFlex file [58].

rotation or depending on the clinical situation also in reciprocal motion. The new generation of NiTi heat-treated files has been manufactured utilizing a unique process that controls the memory of the material, making the files extremely fracture resistant and flexible but without the shape memory of other NiTi files. The files are suitable for multiple use and can be restored during autoclaving. Figure 16.18 shows HF file [58].

ProFile Vortex (PV) rotary files are manufactured from an advanced NiTi alloy called the M-wire, and are designed with convex triangle (CT) shape without radial land and expanded pitch to reduce the likelihood of file separation from cyclic fatigue to improve flexibility [55, 59–61]. Figure 16.19 shows PV [55].

Vortex Blue (VB) is a generic product to PV. The distinctive color of VB rotary files is a visible titanium oxide (TiO_2) layer resulting from the processing of NiTi wire for optimum performance. Minimum of 65% improvement over CFR over M-wire and 99% over standard NiTi wire are observed. In rotary file design, normally increased

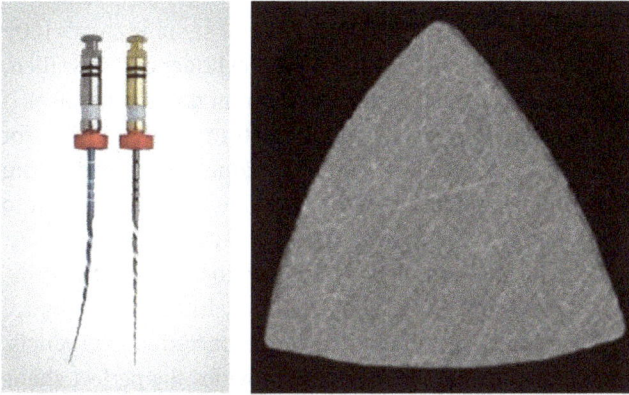

Figure 16.19: ProFile Vortex [55].

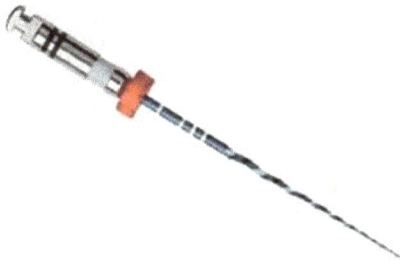

Figure 16.20: Vortex Blue file [62].

CFR usually comes with a trade-off in torque strength. VB rotary files are different. In addition to significantly greater resistance to cyclic fatigue, they offer at least a 42% higher peak torque strength increase over M-wire [65]. It should be noted that the color "blue" does not imply the color of TiO_2 thin layer, but is called as "interference color," which is simply a function of the thin-film thickness. Hence, if the heat treatment (oxidation) condition is strictly controlled to have a constant oxide film thickness, the color "blue" cannot be obtained regularly. Figure 16.20 shows VB file [62].

16.2.1.5 The fourth generation (2010)
Another advancement in canal preparation procedures utilizes reciprocation, which may be defined as any repetitive up-and-down or back-and-forth motion. All reciprocating systems in the market utilize smaller, yet equal, angles of CW/CCW rotation (degrees of rotation are absolutely equal). Compared to full rotation, a reciprocating file that utilizes an equal bidirectional movement requires more inward pressure to progress, does not cut as efficiently as a same-size rotary file, and is more limited in augering debris out of the canal. From these earlier experiences, innovation in recip-

rocation technology led to a fourth generation of instruments for shaping canals [40]. The self-adjusting file (SAF) is a hollow device, designed as a cylinder of thin-walled, delicate NiTi lattice with a lightly abrasive surface different from the traditional NiTi rotary files. The SAF system uses a hollow reciprocating instrument that allows for simultaneous irrigation throughout the mechanical preparation. When inserted into the root canal, the manufacturer claims that the SAF is capable of adapting itself to the canal shape three-dimensionally [63]. The new WO NiTi file system is a single-use, single-file system to shape the root canal completely from start to finish [64]. Shaping the root canal to a continuously tapering funnel shape not only fulfils the biological requirements for adequate irrigation to get rid of all bacteria, bacterial by-products, and pulp tissue from the root canal system [65], but also provides the perfect shape for 3-D obturation with gutta-percha [66].

WO is characterized by: (1) two-third of the shank side having CT cross-sectional feature and (2) one-third tip portion possessing concave triangle (TR) cross-section [67]. WO still keeps the basic design of the third-generation file concept used for root canal finishing under the reciprocating motor. Figure 16.21 shows WO file [55].

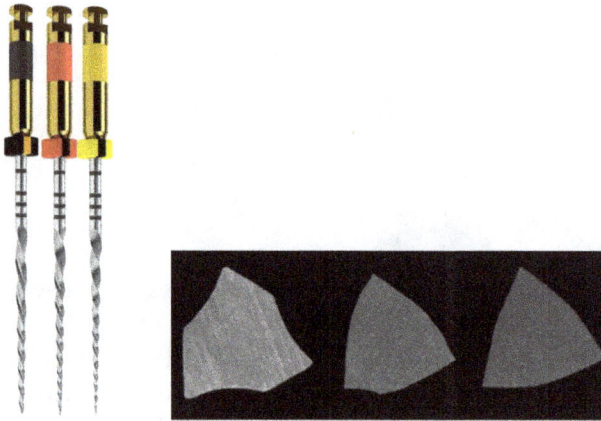

Figure 16.21: WaveOne file [55].

RPC is based on M-wire having an S-shape cross section. It is characterized with excellent flexibility, effective cutting efficiency, and possible augering debris out of the canal [68]. Figure 16.22 shows RPC file [68].

The SAF is a hollow device, designed as a cylinder of thin-walled, delicate NiTi lattice with a lightly abrasive surface different from the traditional NiTi rotary files. The SAF system uses a hollow reciprocating instrument that allows for simultaneous irrigation throughout the mechanical preparation. When inserted into the root canal, the manufacturer claims that the SAF is capable of adapting itself to the canal shape three-dimensionally under liquid supply [63, 69]. Figure 16.23 shows SAF [69].

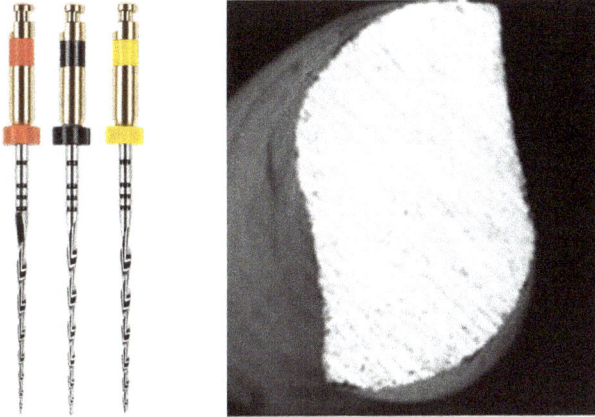

Figure 16.22: Reciproc file [68].

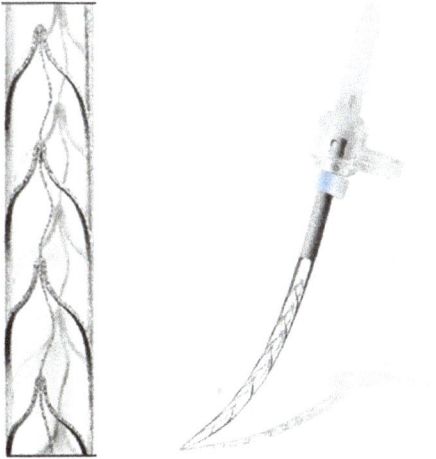

Figure 16.23: Self-adjusting file [69].

WaveOne Gold (WOG), a new generation of reciprocating files offering simplicity, safety, and single use in shaping canals. Through two-step heat treatments, M-wire crystalline structure is converted to structure of the G-wire to improve metallic properties and flexibility [70]. Advanced metallurgy WOG instruments are manufactured utilizing a new thermal process, producing a SE NiTi file. The gold process is a postmanufacturing procedure in which the ground NiTi files are heat-treated and slowly cooled. From a technical perspective, the heat treatment modifies the transformation temperatures (austenitic start and austenitic finish [A_F]) and this has a positive effect on the instrument properties. Although this process gives the file its distinctive gold finish (in author's [YO] opinion, the oxidation should have been at relatively high temperature), more importantly, it considerably improves its strength and flexibility far in excess of its predecessor. It has been reported that: (i) the CFR of WOG Primary is 50% greater than that of WO

Primary (which itself was twice as great as most standard rotary file systems), and (ii) the flexibility of WOG Primary is 80% greater than that of WO Primary. The cross section is revised to a shape of parallelogram as seen in Figure 16.24, the flexibility on short sides increases, screw-in to the root canal wall, and the load on cutting edge are reduced.

The RPC blue instruments have an S-shaped cross section, as seen in Figure 16.25. The instruments are used in conjunction with a motor at 10 cycles of reciprocation per second. The motor is programmed with the angles of reciprocation and speed for the three instruments. The values of the forward and reverse rotations are different. When the instrument rotates in the cutting direction (forward rotation), it advances in the canal and engage dentine to cut it. When it rotates in the opposite direction, that is, the reverse rotation (smaller than the forward rotation), the instrument immediately disen-

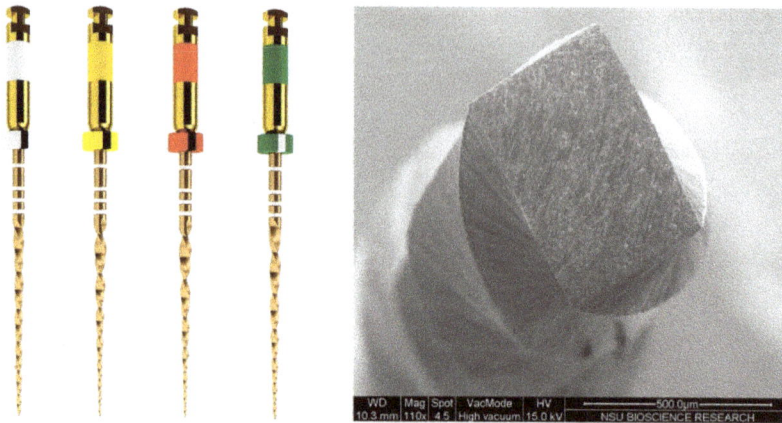

Figure 16.24: WaveOne Gold file [70].

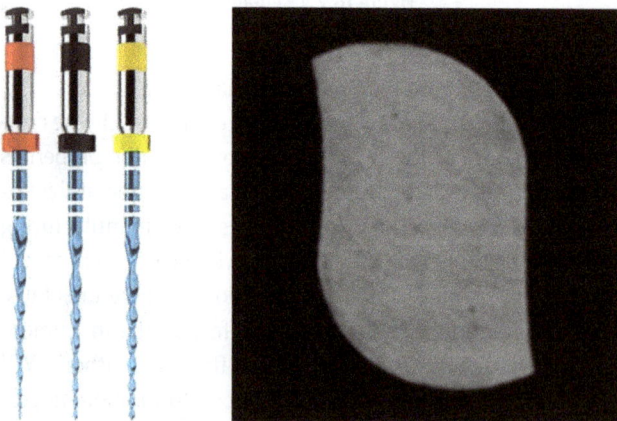

Figure 16.25: Reciproc blue file [72].

gages. The end result, related to the forward and reverse rotations, is an advancement of the instrument in the canal. The angles set on the reciprocating motor are specific to the RPC blue instruments. It was reported that: (i) the use of instruments in reciprocation with unequal forward and reverse rotations and with a limited pecking motion has been shown to be very safe; and (ii) the introduction of the RPC blue instruments with enhanced physical properties makes the procedure even safer with respect to instrument fracture and maintenance of canal curvature (internal evaluation) [71, 72].

16.2.1.6 The fifth generation (since 2013)

Over the decades, a wide array of NiTi instruments has been introduced for shaping root canals. Each generation of instruments has had something new to offer and has been intended to improve upon previous generations. Many variables and physical properties influence the clinical performance of NiTi rotaries. Ultimately, clinical experience, handling properties, safety, and case outcomes should decide the fate of a particular instrument design [73]. The fifth generation of shaping files has been designed such that the center of mass and/or the center of rotation are offset, so that a mechanical wave is produced. In rotation, files that have an offset design produce a mechanical wave of motion that travels along the active length of the file. Similar to the progressively percentage-tapered design of any given ProTaper file, this offset design serves to further minimize the engagement between the file and dentin. In addition, an offset design enhances augering debris out of a canal and improves flexibility along the active portion of a PTN file [40].

Revo-S is characterized by the asymmetrical cross section, which provides less stress on the instrument. The canal axis has three cutting edges located on three different radiuses: R1, R2, and R3 (Figure 16.26 [74]). Figure 16.27 shows SEM depicting detailed three cutting edges [75]. R1 and R3 have asymmetrical cross sections [80], leading to high flexibility and penetration capability by a "snake-like" movement. As a result, torsional fatigue strength is improved. The smaller section also offers a

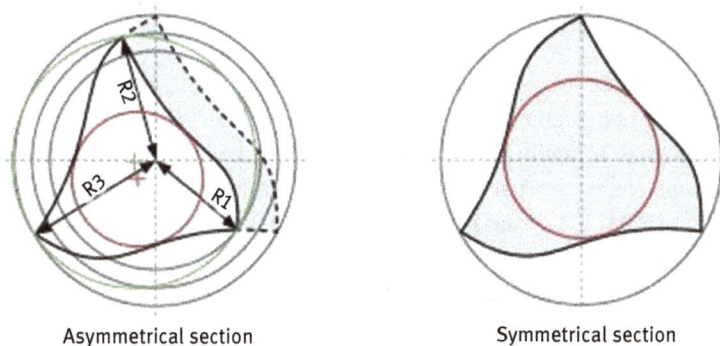

Asymmetrical section   Symmetrical section

Figure 16.26: Asymmetrical arrangement of three cutting edges [74].

Figure 16.27: SEM depicting detailed three asymmetric cutting edges [75].

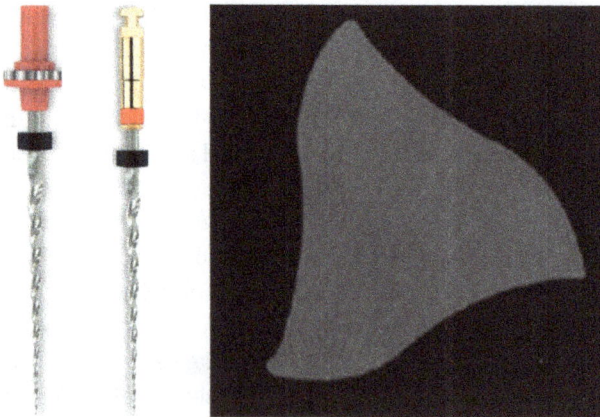

Figure 16.28: Revo-S file [55].

better ability to negotiate curves. The asymmetrical cross section increases the available volume for upward debris elimination [75]. Figure 16.28 shows Revo-S file [55].

One Shape belongs to a family of the single-file NiTi instruments, having three distinct cross-sectional zones; the first zone is designed as variable three-cutting-edge zone, the second zone is a transferring zone between the first and the third zone, and the third zone is again designed as two-cutting-edge zone [76]. It is stated that: (i) One Shape is the one and only NiTi instrument in continuous rotation for quality root canal preparations, (ii) it allows curved canal negotiation with an instrumental and easy dynamic, (iii) its nonworking (safety) tip ensures an effective apical progression avoiding obstructions, which are often preceded by instrument separation (IS), (iv) minimal fatigue along the length of the file virtually eliminates the risk of separation, and (v) it is recommended to use the One Shape® instrument for the treatment of only one tooth. Gernhardt et al. [82] overviewed the One Shape file preparation system. It was mentioned that: (i) in contrast to some other single-file systems, the One Shape file is used in continuous rotation, as known from many former multiple file rotary NiTi systems for root canal preparation, (ii) the unique design of the One Shape instrument incorporates a variety of different cross sections along the active length of the file, which offers an optimal and improved cutting action in the three zones of the root canal, (iii) each instrument has been electropolished to enhance cutting efficiency, (iv) flexibility and unique downward movement of One Shape ensures a highly effective apical progression, (v) it is delivered in a sterile blister, as done by many other manufacturers and should be used for one tooth only and then discarded, and (vi) it is designed for a maximum of three or four canals in one tooth and should not be sterilized, because the cutting efficiency decreases severely; concluding that (One Shape is optimally designed to prepare the majority of root canal configurations and is one of the few single-file instruments used in continuous CW rotation for a quick and probably safe root canal preparation. Figure 16.29 shows the One Shape file [76].

PTN is manufactured through the M-wire technology. It has a progressive taper and is constructed with five (X1–X5) files. The offset centroid of the file reduced screw-in action into the root canal wall. The off-centered rectangular cross section gives the file a snake-like "swaggering" movement as it moves through the root canal. The rotation of the off-centered cross section generates enlarged space for debris hauling. Optimization of canal tracking is also achieved due to the "swaggering effect." It is also stated that: (i) the unique swaggering movement and the greater flexibility files make it possible to shape more severely curved narrow canals than what was possible before with most NiTi systems; (ii) the risk of file breakage is significantly decreased;

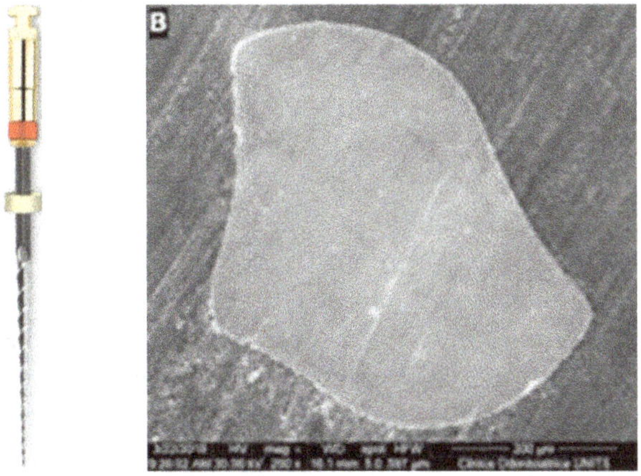

Figure 16.29: One Shape file [76].

at the same time, respect of the original root canal anatomy is greatly improved; and (iii) the high cutting efficiency reduces the shaping time [77–79]. Figure 16.30 shows Pro Taper NEXT [79].

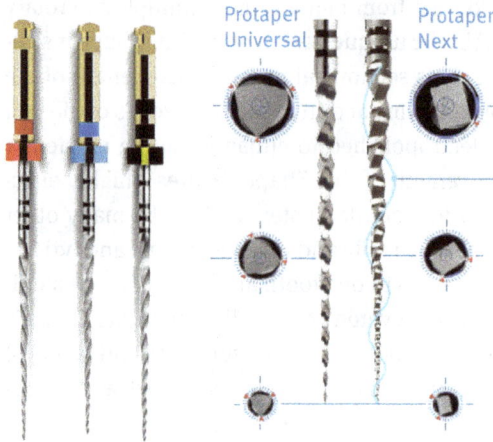

Figure 16.30: Pro Taper NEXT [79].

Summarizing this section, the introduction of automated instrumentation in endodontics represented a major advance in the progress of this specialty, with improvements in the quality and predictability of root canal preparation and a significant reduction in procedural errors. As discussed previously, in recent years, endodontic instruments have undergone a series of changes brought about by modifications in design, surface treatments, and thermal treatments. In addition, new movements have also been incorporated to offer greater safety and efficiency, optimizing the properties of the NiTi alloy, especially through eccentric rotary motion [80]. Although there are recognized advantages (such as SE) of the NiTi alloy, instrument fracture is still a clinical concern, as discussed further. Possible strategies to increase the efficiency and safety of NiTi rotary files include improvements in the manufacturing process or the use of new alloys that provide superior mechanical properties [81, 82]. The mechanical properties and behavior of the NiTi alloy vary according to its chemical composition and thermal/mechanical treatment during manufacturing [47, 83]. A timeline of these treatments, which indicates the development history of NiTi endodontic wires is illustrated in Figure 16.31 [80].

Single cross-sectional NiTi rotary instruments during continuous rotations are subjected to constant and variable stresses depending on the canal anatomy. Ahamed et al. [84] investigated two new experimental, theoretic single-file designs with combinations of triple U (TU), TR, and CT cross-sections and compared their bending stresses in simulated root canals with a single cross-sectional instrument

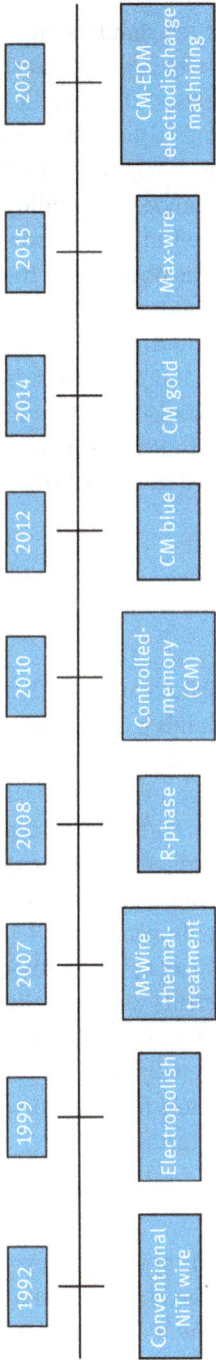

Figure 16.31: Development history of NiTi endodontic instruments [80].

using finite element analysis. A three-dimensional (3D) model of the simulated root canal with 45° curvature and NiTi files with five cross-sectional designs were created using Pro/ENGINEER Wildfire 4.0 software for finite element analysis. The NiTi files of three groups had single cross-sectional shapes of CT, TR, and TU designs, and two experimental groups had a CT, TR, and TU (CTU) design and a TU, TR, and CT (UTC) design. The file was rotated in simulated root canals to analyze the bending stress, and the von Mises stress value for every file was recorded in MPa. It was reported that: (i) the maximum bending stress of the rotary file was observed in the apical third of the CT design, whereas comparatively less stress was recorded in the CTU design, (ii) the TU and TR designs showed a similar stress pattern at the curvature, whereas the UTC design showed greater stress in the apical and middle thirds of the file in curved canals, and (iii) all the file designs showed a statistically significant difference. Based on these results, it was indicated that the CTU-designed instruments showed the least bending stress on a 45° angulated simulated root canal when compared with all the other tested designs.

16.2.2 Bioperformance and efficacy

The development history as discussed in the previous section, it should be noted that the growing use of NiTi rotary instruments in dental practice demands a good understanding of their concept of alloy and design in relation to improved properties and inherent limitations [85]. SE of NiTi allows more centered canal preparations with less transportation and a decreased incidence of canal aberrations [86]. Furthermore, the tapered files are used, resulting in the achievement of adequate irrigation and close adaptation of the filling material during endodontic treatment. Unique shaft and tip designs permit the use of a rotary handpiece allowing different tactile awareness [87]. On the other hand, special attention in maximizing the cutting efficiency and cutting control throughout instrumentation is given. NiTi rotary instruments are generally used in a crown-down approach and a continuous reaming motion. Consequently, rounder root canal preparations with less straightening and a smaller amount of apical extrusion are achievable [85, 87]. Three major important factors for evaluating efficacy and performance of endodontic instruments are reviewed further.

16.2.2.1 Shaping
Camps et al. [88] compared the machining efficiency of four brands of NiTi files and two brands of SS-K-type files under a linear motion simulating the clinical motion used to remove a file from the canal. It was reported that: (i) the SS instruments with a triangular cross-section were more efficient than the SS instruments with a square cross-section, (ii) there was a significant discrepancy between the machining ability of the NiTi K files, and (iii) the Maillefer instruments were the most efficient.

Thompson et al. [89] determined the shaping ability of NT Engine and McXim NiTi rotary instruments in simulated root canals. In all, 40 canals consisting of four different shapes in terms of angle and position of curvature were prepared by a combination of NT Engine and McXim instruments using the technique recommended by the manufacturer. It was found that: (i) the canals were smooth in the apical half of the canal in 33 specimens and in the coronal half of 39 specimens, (ii) all canals had good taper characteristics and 35 had good flow characteristics, (iii) NT Engine and McXim instruments prepared canals rapidly, with few deformations, no canal blockages, and with minimal change in the WL, and (iv) the 3D form of the canals demonstrated good flow and taper characteristics. The shaping characteristics of NiTi K-files and NiTi S-files manipulated by hand were compared [90]. A total of 60 extracted human roots were embedded in resin blocks. The embedded roots were divided into three groups: (1) roots with straight, (2) apically curved, and (3) continuously curved canals. Each of the three groups was randomly divided into two subgroups; one subgroup in each group was prepared with NiTi K-files and the other with NiTi S-files. It was reported that: (i) although canal preparation using NiTi K-files was quicker, there were no statistically significant differences between file types, (ii) the NiTi S-file removed significantly more material at the most coronal level, (iii) only in the proximal view of apically curved canals prepared with NiTi S-files was significantly more dentine removed from the outer aspect of the curvature, and (4) at the middle level (wide danger zone), the NiTi S-files removed more dentine from the inner aspect of the curvature in those roots with apically curved canals. In conclusion, preparation with NiTi K-files produced more appropriate shapes in roots with apically curved canals than NiTi S-files. Burroughs et al. [91] determined the shaping ability of three NiTi endodontic file systems by measuring canal transportation. Seventy-two S-shaped canals in resin blocks were randomly allocated into three groups ($n = 24$): the SAF group, the Typhoon (TYP) group, and the Vortex group (PV rotary files with M-wire NiTi). It was found that: (i) after adjusting for the level and canal wall side, the mean transportation was significantly higher for the TYP and Vortex groups compared with the SAF group, and (ii) the mean transportation was significantly higher for the TYP group versus the Vortex group; indicating that SAFs showed less canal transportation compared to the PV and TYP files in simulated S-shaped root canals. Giuliani et al. [92] compared the shaping effects of WO and full-sequence ProTaper NiTi files used in reciprocating and conventional movements in a simulated canal. Seventy-five S-shaped canals in resin blocks were randomly allocated to three groups ($n = 25$): WO (group 1), full sequence of ProTaper Universal (PTU) files in conventional movements (group 2), and full sequence of PTU files in reciprocating movements (group 3). It was obtained that group 3 maintained better coronal and apical canal curvature and less straightening of the simulated canals with respect to groups 1 and 2; concluding that when preparing S-shaped canals, full-sequence PTU NiTi files used in a reciprocating motion exhibited better shaping effects than full-sequence PTU NiTi files used in a conventional motion and WO.

Ozyurek et al. [93] compared the shaping ability of RPC, HyFlex EDM (HFEDM), and WOG NiTi files made of different NiTi alloys in S-shaped simulated canals. Sixty S-shaped canals in resin blocks were prepared to an apical size of 0.25 mm using RPC R25, WOG Primary, and HFEDM OneFile (n = 20 canal/group) systems. It was reported that: (i) NiTi file fracture was not observed during shaping of the simulated canals and (ii) there was no statistically significant difference between the WOG and HFEDM groups' apical, medial, and coronal regions; however, it was determined that the RPC group removed a statistically significantly higher amount of resin from all the canal regions when compared with the WOG and HFEDM groups. It was then concluded that all of the tested NiTi files caused various levels of resin removal; however, WOG and HFEDM NiTi files were found to cause a lower level of resin removal than RPC NiTi files. Duque et al. [94] evaluated the influence of the NiTi wire in Conventional NiTi (PTU) and CM NiTi (ProTaper Gold [PTG]) instrument systems on the quality of root canal preparation. It was indicated that the Conventional NiTi (PTU) and CM NiTi (PTG) instruments displayed comparable capabilities for shaping the straight mesial root canals of mandibular molars, although the PTG was better than the PTU at maintaining the centralization of the shape in the cervical portion. Using Twisted File Adaptive (TFA) files (SybronEndo, Orange, CA, USA) and Mtwo (Sweden & Martina, Padova, Italy) activated by continuous rotation or adaptive motion, the shaping ability of curved root canals was compared [95]. Thirty-two mandibular molars with two separate mesial canals and severe angles of curvature were selected. Each canal was randomly assigned to one of the four experimental groups (n = 16): TFA and Mtwo files used in continuous rotation (groups 1 and 3) or in adaptive motion (groups 2 and 4). It was found that: (i) volume and surface area increased less with TFA files in continuous rotation than in other groups, (ii) TFA files had significantly less transportation and higher centering ability than Mtwo both in continuous and adaptive motions, (iii) centering ratio, but not canal transportation, was improved by adaptive motion compared with continuous rotation for both instruments; however, no differences were found in canal transportation and centering ability in the apical third for both instruments and motions. In conclusion, no difference between the devices and kinematics was found in the apical third; TFA performed significantly better in the middle and coronal parts of the root canal. Espir et al. [19–96] evaluated the shaping ability and cleaning after oval root canal preparation using one or more instruments in reciprocating or rotary motion. Oval-shaped mandibular incisors were selected, based on the radiographic diameter (2 ≤ diameter ratio ≤ 4), and assigned according to root canal preparation: single-file (RPC R40); two reciprocating files (Unicone size 20 and 40, 0.06 taper), or Mtwo rotary files until a size 40, 0.06 taper instrument. Root canal preparations were performed using an open root canal model. It was obtained that: (i) the initial volume were similar among the groups, (ii) Unicone preparation was associated with higher debris, increase in root canal volume, and uninstrumented surface in entire root canal and in the middle third, (iii) Mtwo was associated with lower uninstrumented surface in the entire root canal and in the cervical third, and (iv) the apical third were similar for the three preparations. It was then concluded that Unicone system using two instruments in reciprocating motion resulted in higher increase in the volume; however, less remaining debris was observed when RPC single-file and Mtwo rotary systems were used. Huang et al. [97] evaluated the shaping ability of three heat-treated rotary NiTi systems including PTN, HFCM, and HFEDM during root canal preparation in simulated root canals. It was reported that: (i) HFEDM caused significantly greater volume increases than HFCM and PTN in the entire root canal and in the apical and middle thirds, (ii) HFCM removed the least amount of resin in the coronal third compared with HFEDM and PTN, and (iii) overall, HFCM caused significantly less transportation in the apical 2 mm and was better centered than PTN in the apical 3 mm; concluding that all systems prepared curved canals without significant shaping errors and instrument fracture, and PTN and HFCM cut less resin than HFEDM. HFCM stayed centered apically and cut the least material coronally. Camargo et al. [98] evaluated the influence of three engine-driven NiTi file systems manufactured from different NiTi alloys for the preparation of MB2 canals in extracted maxillary first molars. It was found that: (i) there was no significant difference among the groups regarding canal transportation and centering ability; however, the PDR size 25, 0.06 taper group had significantly lower canal volume and volume of dentine removal compared with a MO size 25, 0.06 taper and REC size 25, 0.08 taper; (ii) a root perforation was detected in MO size 25, .06 taper and REC size 25, 0.08 taper groups, respectively; and (iii) regarding the working time, the PDR size 25, 0.06 taper required a significantly longer time to achieve the WL than MO size 25, 0.06 taper and REC size 25, 0.08 taper. Based on these results, it was concluded that all NiTi systems had similar canal transportation, centering ability, and increase in apical volume after preparation of MB2 canals; however, the PDR size 25, 0.06 taper had less volume of dentine removal, absence of root canal perforation, and required a longer time to accomplish the root canal preparation.

16.2.2.2 Centering

During instrumentation of the root canal, it is important to develop a continuously tapered form and to maintain the original shape and position of the apical foramen. However, the presence of curvatures may cause difficulty in root canal instrumentation. The ability to keep the instruments centered is essential to provide a correct enlargement without excessive weakening of the root structure [99]. Several studies have shown that NiTi instruments remain significantly more centered and demonstrated less canal transportation compared to SS files. If there were any deviations from the original canal curvature, as a result, unwanted incidences might occur such as excessive and inappropriate dentin removal, straightening of the canal, and creation of a ledge in the dentinal wall, or overpreparation that weakens the tooth, resulting in fracture of the root [99]. Various rotary NiTi systems minimize accidents and facilitate the shaping process. Zarei et al. [100] compared the canal centering after instrumentation using the ProTaper system with Endo IT, electric torque-control motor, and NSK air-driven handpiece. Twenty-six mesial mandibular root canals with 10–35° curvature were involved. It was reported that: (i) the comparison of the rate of transportation toward internal or external walls between the two groups was not statistically significant, and (ii) comparison of the rate of transportation of sections within one group was not significant; concluding that the use of rotary NiTi file with either electric torque control motor or air-driven handpiece had no effect on canal centering. Ponti et al. [101] compared the ability of two NiTi rotary file systems (ProFile NiTi 0.06 taper Series 29 and ProFile GT) to maintain the original path of the canal by using a new split-mold design (the Endodontic Cube). It was reported that: (i) both systems remained centered within the canal with minimal deviation from the original canal path, and (ii) the Endodontic Cube was an effective tool for studying and comparing instrumentation techniques. Pinheiro et al. [102] evaluated apical transportation and centering ability during root canal preparation in mesial root canals of mandibular molars associated with PTG, ProDesign S (PDS), HCM, HFEDM and ProDesign Logic (PDL). It was reported that: (i) there were no significant differences in apical transportation amongst the rotary systems; (ii) all the systems created apical transportation; values ranging from 0.031 mm (PDL) to 0.072 mm (PTG), and enlargements between 39% (HCM) and 91.1% (PDS) were observed; and (iii) in relative to cervical transportation, significant differences were observed among the systems ($p < 0.05$). Mean transportation values between 0.07 mm (HCM) and 0.172 mm (PTG) were found, with enlargements between 35.4% (HCM) and 51.5% (PDS); concluding that all the thermally treated systems resulted in similar apical transportation, and in the cervical region, the HFCM and PDL systems were associated with more centered preparations.

16.2.2.3 Cutting

Cutting efficiency of 24 different types of endodontic hand instruments, which are primarily designed for a rotary (reaming) working action, was investigated under standardized conditions [103]. It was mentioned that: (i) nitinol K-files showed the least cutting efficiency, (ii) SS reamers and especially K-files showed better cutting efficiency than nitinol K-files, and (iii) flexible SS instruments displayed the best results; indicating that with regard to cutting efficiency, flexible SS instruments were clearly superior to SS reamers and K-files, and especially to nitinol K-files. Haïkel et al. [104] assessed the cutting efficiency of NiTi files in the presence and absence of sodium hypochlorite (NaOCl) treatment and compared them to a conventional SS K-type file (SS-K). NiTi files from four manufacturers were randomly selected and exposed to NaOCl for 12 or 48 h, or not at all. It was reported that: (i) in the absence of NaOCl, Brasseler and Maillefer NiTi files were most efficient, followed by JS Dental and McSpadden; (ii) these differences were significant, except for those between the latter two brands; (iii) moreover, NaOCl treatment did not alter the cutting efficiency of any brand of instruments significantly; and (iv) when compared with conventional SS files, all NiTi files tested were less efficient. Peters et al. [105] evaluated the cutting behavior of NiTi coronal flaring instruments. BioRaCe BR0 (BR), HFCM 1 (HY), ProFile OS No. 2, and ProTaper Sx (PT) instruments were used in simulated coronal flaring using a lateral action against bovine dentine blocks, at 250 and 500 rpm. Cutting efficiency was assessed by three methods: first, areas of notches produced by instruments were directly measured under a stereomicroscope; secondly, dentine specimens were then analyzed by surface profilometry to determine the maximum cutting depth; and finally by microcomputed tomography (MCT) to assess the volume of removed dentine. It was reported that: (i) for all three methods, HY and ProFile were the most and the least cutting-efficient instruments, respectively, (ii) significant differences were detected between 250 and 500 rpm for HY and PT (area); for BR, HY, and PT (depth); and for BR and HY (volume); and (iii) there were strong positive correlations between the results obtained with those three different methods with r-values ranging from 0.81 to 0.92. Based on these findings, it was concluded that measuring the amount of material removed in a specific time under stereomicroscopy is a simple and rational way to assess the cutting behavior of NiTi rotary instruments in lateral action. HF, manufactured with thermomechanically treated NiTi wire, was the most efficient instrument, and increased rotational speed was associated with increased cutting efficiency.

16.2.2.4 Glide path

3D cleaning, shaping, and obturation of root canal system can be considered as foundations for a predictable endodontic success. One guiding strategy that has emerged as a critical part of endodontic success is the creation and maintenance of a glide path [106]. The endodontic glide path is a smooth radicular tunnel from the canal orifice to the apical constriction. Its minimal size should be a, super loose No. 10 endodontic file. The glide path must be discovered if already present in the root canal anatomy or prepared if it is not present. The glide path can be short or long, narrow or wide, and essentially straight or curved. Without the endodontic glide path, the rationale of endodontics cannot be achieved; implying that cleaning and shaping become unpredictable as there is no guide for endodontic mechanics. Thus, a smooth glide path is the secret to safe and successful rotary shaping. The rationale states that any endodontically diseased tooth can be predictably saved if the root canal system can be nonsurgically or surgically sealed, the tooth is periodontally sound or can be made so, and if the tooth is restorable. A nonsurgical seal requires the creation of a radicular path that can be cleaned of viable and nonviable bacteria, vital and nonvital pulp tissue, biofilm, and smear layer and then shaped to a continuously tapering funnel that can be predictably and easily obturated [107, 108].

Pasqualini et al. [109] compared the ability of manual and mechanical glide path to maintain the original root canal anatomy, using the X-ray computed microtomography scanning for high-resolution 3D imaging of small objects. Eight extracted upper first permanent molars were scanned and buccal root canals of each specimen were randomly assigned to PathFile (PF) or SS-K to perform glide path at the full WL. It was found that: (i) specimens in the K group had a mean curvature of 35.4°± 11.5°; those in the PF group had a curvature of 38°± 9.9°, and (ii) the instrument factor (PF and K) were extremely significant for both the ratio of diameter ratios (RDR) and roughness average (Ra) parameters, regardless of the point of analysis; concluding that MCT scanning confirmed that NiTi rotary PF instruments preserve the original canal anatomy and cause less canal aberrations. D'Amario et al. [110] compared the maintenance of canal anatomy, the occurrence of apical transportation, and the working time observed using mechanized instrumentation with the new G-File rotary system (GF) with those observed using instrumentation with the PF system and manual instrumentation with K-type files (micro-mega) to create a glide path in curved root canals. The mesial canals of 45 mandibular molars (with curvature angles between 25° and 35°) were selected. It was reported that: (i) no statistically significant differences in the angle of canal curvature and apical transportation were found between the groups; (ii) however, concerning the working time, specimens from the group who underwent canal preparation using #12-17 GF rotary instruments achieved significantly lower mean values when compared with the other two groups, whereas the group who underwent canal preparation using the #10-15-20 SS-K manual files had the highest values. It was hence concluded that the GF rotary instruments, the PF system, and the manual instruments did not have any influence on the occurrence of apical transportation nor did they produce a change in the angle of canal curvature, and the GF instruments seemed to be the most rapid system in creating a safe glide path. Ajuz et al. [111] compared the incidence of deviation along S-shaped (double-curved) canals after glide path preparation with two NiTi rotary pathfinding instruments (rotary NiTi PF instruments: up to size 19, and rotary NiTi Scout RaCe instruments: up to size 20) and hand SS-K: up to size 20. It was found that: (i) intragroup analysis showed that all instruments promoted some deviation in virtually all levels; (ii) overall, regardless of the group, deviations were observed in the mesial wall at the canal terminus and at levels 4, 5, 6, and 7 mm and in the distal wall at levels 1, 2, and 3 mm; (iii) these levels corresponded to the inner walls of each curvature; (iv) both rotary NiTi instruments performed significantly better than hand K-files at all levels, except for PFs at the 0 mm level; and (v) ScoutRaCe instruments showed significantly better results than PFs at levels 0, 2, 3, 5, and 6 mm. Based on these results, it was concluded that rotary NiTi instruments are suitable for adequate glide path preparation because they promoted less deviation from the original canal anatomy when compared with hand-operated instruments. Moreover, of the two rotary pathfinding instruments, Scout RaCe showed an overall significantly better performance.

16.2.2.5 Handpiece versus engine-drive

Success of root canal treatment depends on several factors; among which, maintaining the original canal path during mechanical preparation is extremely important [112]. Such mechanical preparation is accompanied with movement, which is operated manually or engine-driven. Some handpieces that operate with hand files have reciprocating movements and were designed to simplify the process of root canal preparation. Labbaf et al. [113] mentioned that the level of pain and inflammation decreases during and after the use of reciprocating handpieces due to their optimum speed and high level of harmony of movements. Moreover, they decrease the risk of file anchorage or locking in the canal, which are commonly seen in complete rotational movements and results in the fracture of NiTi rotary files [118]. As proper root canal shaping and maintaining the original canal path play an important role in the success of root canal treatment, there are several studies to compare these two distinct mechanisms in terms of cutting / shaping efficacy.

Glossen et al. [114] compared root canals prepared by NiTi hand, NiTi engine-driven, and SS hand endodontic instruments. Sixty mesial canals of extracted human mandibular molars were randomly divided into five groups. In group A, canals were instrumented using a quarter turn / pull technique with K-Flexofiles. In group B, canals were prepared with NiTi hand files (Mity files) using the same technique as in group A. Group C was prepared with NT Sensor engine-driven files. Group D canals were prepared with NiTi Canal Master "U" hand instruments. Group E was prepared with engine-driven NiTi LightSpeed instruments. It was mentioned that: (i) engine-driven NiTi instruments and hand instrumentation with the Canal Master "U" caused significantly less canal transportation, remained more centered in the canal, removed less dentin ($p < 0.05$), and produced rounder canal preparations than K-Flex and Mity files; and (ii) engine instrumentation with LightSpeed and NT Sensor file was significantly faster than hand instrumentation. Guelzow et al. [13] compared ex-vivo various parameters of root canal preparation using a manual technique and six different rotary NiTi instruments (FlexMaster, System GT, HERO 642, K3, ProTaper, and RaCe). A total of 147 extracted mandibular molars were divided into seven groups ($n = 21$) with equal mean mesiobuccal root canal curvatures (up to 70°) and embedded in a muffle system. It was reported that: (i) no significant differences were detected between the rotary NiTi instruments for alteration of WL; (ii) all NiTi systems maintained the original curvature well, with minor mean degrees of straightening ranging from 0.45° (System GT) to 1.17° (ProTaper); (iii) ProTaper had the lowest numbers of irregular postoperative root canal diameters, the results were comparable between the other systems; (iv) instrument fractures occurred with ProTaper in three root canals, while preparation with System GT, HERO 642, K3, and the manual technique resulted in one fracture each; (v) NiTi instruments prepared canals more rapidly than the manual technique; and (vi) the shortest time for instrumentation was achieved with System GT (11.7 sec). Hence, it was concluded that all NiTi systems maintained the canal curvature, were associated with few instrument fractures, and were more rapid than a standardized manual technique. Moreover, ProTaper instruments created more regular canal diameters. Liu et al. [14] compared the cleaning efficacy and shaping ability of engine-driven ProTaper and GT files, and manual preparation using K-Flexofile instruments in curved root canals of extracted human teeth. Forty-five canals of maxillary and mandibular molars with curvatures between 25° and 40° were divided into three groups. The groups were balanced with regard to the angle and the radius of canal curvature. Canals in each group were prepared to an apical size of 25 with either the rotary ProTaper or GT system, or manually with K-Flexofile using the modified double-flared technique. Irrigation was done with 2 mL 2.5% NaOCl after each instrument and, as the final rinse, 10 mL 2.5% NaOCl, then 10 mL 17% ethylenediaminetetraacetic acid (EDTA), and finally 5 mL distilled water. It was reported that: (i) two GT files but none of the K-Flexofile and ProTaper instruments separated; (ii) for debris removal, the ProTaper group achieved a better result than GT but not the K-Flexofile group at all three regions (apical, middle, and coronal); (iii) K-Flexofiles produced significantly less smear layer than ProTaper and GT files only in the middle third of the canal; and (iv) both NiTi rotary instruments maintained the original canal shape better than the K-Flexofiles and required significantly less time to complete the preparation. Kakar et al. [115] assessed and compared the shaping potential of manual NiTi K-files and ProTaper Rotary instruments in narrow canals using CT. It was reported that: (i) there was a statistically significant difference between all the parameters judged for the study, (ii) instrumentation with ProTaper Rotary system took significantly less time than instrumentation with manual NiTi K-file, (iii) change in the canal volume following instrumentation with ProTaper Rotary was significantly greater than that produced by manual NiTi K-file, (iv) change in the cross-sectional area at 2 and 3 mm from the apex was significantly greater with manual NiTi K-file compared to ProTaper Rotary, and (v) change in cross sections at 4.5, 6, and 7.5 mm from the apex with ProTaper Rotary was significantly greater than that produced by manual NiTi K-file. Based on these findings, it was concluded that there is a drastic difference in the shaping ability of manual and rotary NiTi instruments used with step back and crown down technique respectively with the rotary instrumentation being faster and producing greater changes in the canal anatomy.

Subramaniam et al. [116] assessed the microflora of root canals in primary molars following the use of rotary NiTi files and conventional hand NiTi and SS files. This randomized clinical trial consisted of a total of 60 first and second primary molars requiring root canal treatment, who were selected from children aged 5–9 years and the teeth were randomly assigned to three groups of 20 teeth each; Group A: Rotary NiTi files, Group B: Hand NiTi files, and Group C: Hand SS files. It was obtained that in all three groups, there was a significant reduction in both aerobic and anaerobic mean microbial count following root canal instrumentation, concluding that rotary NiTi files were as efficient as conventional hand instruments in significantly reducing the root canal microflora. Radhika et al. [117] compared canal transportation, centering ability, and removed dentin thickness using hand NiTi K-flex files and rotary systems in primary molars. It was reported that: (i) rotary instruments caused significantly less canal transportation and showed more centering ability than hand NiTi instruments at all levels and in all directions, except at the cervical level in the mesiodistal direction, where the hand group showed significantly superior results, and (ii) the hand NiTi K-flex files removed significantly more dentin than rotary instrumentation at all levels and in all directions, except at the apical level in the buccodistal direction, where no significant difference between the two groups was observed, concluding that rotary files showed less canal transportation, more conservation of tooth structure, and superior centering ability compared to hand NiTi K-flex files. Apical transportation using RaCe NiTi rotary system and precurved SS hand files in a reciprocating handpiece was compared [112]. Mesiobuccal canals of 40 extracted human mandibular first and second molars with 20–45° curvatures and 3–7 mm curve radius were chosen for this study. After the determination of the WL, the teeth were divided into two groups (n = 20). Root canals were prepared with RaCe in group 1 and NSK handpiece and precurved SS hand files in group 2 up to #30 with 2% taper in both the groups. It was reported that: (i) no significant difference was noted between the two groups on buccolingual or mesiodistal views in degree of straightening and apical transportation on buccolingual view; (ii) however, on mesiodistal view, NSK reciprocating handpiece caused greater apical transportation at 0, 0.5, and 1 mm levels. It was then concluded that the RaCe system and precurved SS files in reciprocating handpiece were highly similar in terms of the degree of straightening and apical transportation, and thus, the engine-driven NSK reciprocating handpiece can be used as an efficient adjunct for root canal preparation [118].

16.2.2.6 Nickel titanium versus stainless steel

As to the material selection for fabricating endodontic instruments, there are mainly NiTi alloy and SS. There are several studies to compare their bioperformance. Esposito et al. [119] compared the maintenance of the original canal path of curved root canals during instrumentation with NiTi (Mac) hand files, NiTi engine-driven files, and SS K-Flexofiles. It was reported that: (i) NiTi hand and engine-driven instruments maintained the original canal path in all cases, (ii) the incidence of deviation from the original canal path during instrumentation with SS files increased with the file size, (iii) the difference between NiTi groups and SS became statistically significant with instruments larger than size 30, and (iv) NiTi files were more effective in maintaining the original canal path of curved root canals when the apical preparation was enlarged beyond size 30. Coleman et al. [120, 121] compared step-back preparations in curved canals using NiTi K-files and SS-K. Forty canals in mesial roots of mandibular molars were embedded in casting resin and cross-sectioned at three levels: 1–2 mm from the apical foramen, middle of the curve, and coronal. Direct digital computer images were recorded before and after instrumentation. It was obtained that: (i) NiTi files caused significantly less transportation and remained more centered at the apical level, (ii) area removed by NiTi and SS files was not significantly different, (iii) time of instrumentation was not significantly different for NiTi and SS instruments, and (iv) cross-sectional shape of the instrumented canal was not significantly different. Schafer et al. compared the cutting efficiency and the effects of instrumentation on curved canal shape of both SS and NiTi nonstandardized ProFile Series 29 hand instruments under standardized conditions [122] and compared the shaping ability in simulated curved canals of K3 rotary NiTi instruments with SS K-Flexofiles manipulated by hand [123]. It was concluded that: (i) flexible SS instruments with noncutting tips were superior to the nonstandardized ProFile Series 29 instruments with regard to cutting efficiency and instrumentation of curved canals [122], and (ii) K3 instruments prepared curved canals rapidly and with minimal transportation toward the outer aspect of the curve and fractures occurred significantly more often with K3 [123]. Lam et al. [124] examined the amount of apical and mid-curve transportation produced by a range of NiTi, titanium alloy, and SS files. It was noted that: (i) there were substantial differences in the amount and the pattern of apical and mid-curve transportation produced, (ii) the amount of transportation increased with each subsequent size of file, (iii) under the same conditions, NiTi files produced significantly less transportation than SS files, (iv) the least apical transportation was obtained with the NiTi Mity Turbo and the most by the SS-K and SS Hedstrom file, and (v) the least mid-curve transportation was produced by the NiTi Mity Turbo and the most by the SS Hedstrom file. Davis et al. [125] investigated pre- and postinstrumentation WL measurements in curved root canals under comparison with combinations of (a) SS hand files and Gates Glidden drills (SS) versus NiTi rotary files; and (b) early coronal flaring (flaring completed before WL determination) versus late coronal flaring (flaring completed after WL determination). It was noted that: (i) WL decreased for all canals as a result of canal preparation, (ii) the mean decrease in the WL was significantly greater for the SS group (−0.48 mm ±0.32) than for the NiTi group (- 0.22 mm ± 0.26), and (iii) less change in WL occurred in both groups when initial WL was determined after coronal flaring (SS: −0.12 mm ± 0.13, NiTi: −0.14 mm ±0.25). The comparison on effects of preparation with conventional SS Flexofiles and Gates Glidden burs versus NiTi GT rotary files in the shaping of mesial root canals of extracted mandibular molars was conducted [126]. A total of 54 canals from 27 mesial roots of mandibular molar teeth were prepared using one of the two methods by novice dental students. One canal in each root was prepared by a crown-down approach. using SS Flexofiles and Gates Glidden burs. The other canal was prepared using NiTi GT rotary files in a crown-down approach as recommended by the manufacturer. It was found that: (i) at the coronal and mid-root coronal one-third sections, the rotary GT files produced a significantly smaller postoperative canal area; (ii) in the mid-root sections there was significantly less transportation of the root canal toward the furcation, and less thinning of the root structure with GT files compared to the SS files; (iii) overall, there was greater conservation of structure coronally and more adequate shape in the mid-root level; (iv) the GT rotary technique was significantly faster than the SS hand-held file technique; and (v) two GT instruments fractured during the study. Based on these findings, it was concluded that novice dental students were able to prepare curved root canals with NiTi GT rotary files with less transportation and greater conservation of tooth structure, compared to canals prepared with hand instruments, and the rotary technique was significantly faster. Tasdemir et al. [127] compared ex-vivo root canal preparation with conventional SS-K and Hero 642 rotary NiTi instruments. Mesiobuccal canals of 20 maxillary first molars (with angles of curvature between 25° and 35°) were used. Amount of transportation and centering ability was assessed. Student's t-test was used for statistical analysis. It was obtained that: (i) less transportation occurred with Hero 642 rotary instruments than SS-K at the mid-root and coronal levels, and (ii) Hero 642 rotary instruments had better centering ability than K-files at all three levels. In conclusion, Hero 642 rotary instruments transported canals less, especially at the middle and coronal thirds of the root canals than SS-K, and have better centering ability. Nagatatna et al. [128] compared the NiTi rotary and K-files hand instrumentation on root canal preparation of primary and permanent molars for their efficiency in preparation time, instrument failure, and shaping the canals. About 20 primary mandibular second molar (I) and 20 permanent mandibular first molar (II) were selected. Each was further divided into 10 for K-files (a) and 10 for NiTi (b) groups, respectively. It was observed that: (i) preparation time Ib Ia and IIab<IIa was highly significant; (ii) in instrument failure, Ia (40%), IIa (30%) showed more deformation but not fracture and Ib (10%), IIb (20%) showed fracture, but not deformation; and (iii) ProFiles showed good canal taper and smoothness compared to the K-files. In conclusion, ProFile 0.04 taper 29 series, prepared canal rapidly than conventional K-file with good taper, smoothness though the flow was not satisfactory, and instrument failure with K-files was less. In primary teeth preparation time, instrument failure with ProFile was less compared to the permanent.

Matwychuk et al. [129] made a comparison of apical transportation, WL changes, and instrumentation time by using NiTi rotary file systems (crown-down method) or SS hand files (balanced-force technique) in mesiobuccal canals of extracted mandibular molars. The curvature of each canal was determined and teeth placed into three equivalent groups. Group 1 was instrumented with Sequence (Brasseler USA, Savannah, GA, USA) rotary files, group 2 with Liberator (Miltex Inc., York, PA, USA) rotary files, and group 3 with Flex-R (Union Broach, New York, NY, USA) files. It was reported that: (i) sequence rotary files, Liberator rotary files, and Flex-R hand files had similar effects on apical canal transportation and changes in WL, with no significant differences detected among the three groups, and (ii) hand instrumentation times were longer than with either NiTi rotary group, whereas the rotary NiTi groups had a higher incidence of fracture. Vaudt et al. [130] investigated the instrumentation time, working safety, and the shaping ability of two rotary NiTi systems (Alpha System and PTU) in comparison to SS hand instruments. A total of 45 mesial root canals of extracted human mandibular molars were selected. Based on the degree of curvature, the matched teeth were allocated randomly into three groups of 15 teeth each.

In group 1, root canals were prepared to size 30 using a standardized manual preparation technique; in groups 2 and 3, rotary NiTi instruments were used following the
manufacturers' instructions. It was mentioned that: (i) active instrumentation time
of the Alpha System was significantly reduced compared with PTU and hand instrumentation, and (ii) no instrument fractures occurred in any of the groups; concluding
that despite the demonstrated differences between the systems, an apical straightening effect could not be prevented; areas of uninstrumented root canal wall were left
in all regions using the various systems. Cheung et al. [131] compared the periapical
healing of molar root canal treatment using two instrumentation techniques. A total
of 225 maxillary and mandibular first and second permanent molars endodontically
treated by undergraduate or postgraduate students were randomly selected from a
computerized hospital database of which 110 molars had been prepared using a hybrid
rotary technique with NiTi instruments (group NR) and 115 with hand SS files (group
HF). Patients were recalled and the teeth were examined both clinically and radiographically for signs of periapical inflammation. It was reported that: (i) around 19%
and 39% of teeth in the NR and HF group, respectively, were judged to have some form
of procedural errors; (ii) a higher rate of periapical healing was noted for NR (77%)
than the HF group (60%); and (iii) factors contributing favorably to the treatment
outcome included the use of rotary technique, maxillary molar, experienced operator, and absence of preoperative radiolucent lesion. Based on these results, it was
concluded that there was a higher incidence of procedural errors and a lower success
rate for primary root canal treatment of teeth prepared with SS files compared with
the use of NiTi instruments in a continuous reaming action. Nordmeyer et al. [132]
compared various parameters of root canal preparation using FlexMaster rotary NiTi
and Endo-Eze anatomic endodontic technology (AET) SS instruments. It was noted
that: (i) the mean degree of straightening was significantly less for FlexMaster than
for AET, (ii) postoperative cross sections showed no significant differences between
the systems, (iii) neither of the systems completely eliminated debris and smear layer,
and (iv) no procedural incidents occurred with the instruments. Mean working time
was significantly shorter for FlexMaster than for AET; indicating that AET cannot
be recommended for the preparation of curved root canals due to unacceptable
straightening. Sandhu et al. [133] investigated the efficiency, behavior, and properties
of SE NiTi versus multi-stranded SS wires in Begg and preadjusted edgewise appliance (PEA) under moderate-to-severe crowding conditions. Ninety-six participants
(48 boys, 48 girls), aged 12–18 years old, with moderate-to-severe initial crowding
were distributed into four groups: SE NiTi PEA (n = 24), SE NiTi Begg (n = 24), multistranded (coaxial) SS PEA (n = 25), and multistranded (coaxial) SS Begg (n = 23).
In this study, 0.16-inch SE (austenitic active) NiTi and 0.175-inch multistranded (six
stranded, coaxial) SS wires were used in a 0.022-inch slot (Roth prescription) PEA and
Begg appliance with a follow-up of 6 weeks. It was reported that: (i) analysis of variance revealed no significant difference in reduction of crowding between SE NiTi PEA
and multistranded (coaxial) SS PEA groups, but reduction in crowding was signifi-

cantly greater in the SE NiTi Begg group compared with the multistranded (coaxial) SS Begg group; and (ii) linear regression demonstrated significant positive correlation between the amount of initial crowding and reduction in crowding in all groups except the multistranded (coaxial) SS Begg group, wherein a negative correlation did exist. It was hence concluded that SE NiTi performed significantly better than multistranded (coaxial) SS wire in the Begg appliance; however, in PEA, there was no significant difference.

16.2.2.7 Among Nickel titanium files

There are still comparison studies on cutting and shaping efficacy among various NiTi rotary files. Chuste-Guillot [134] compared the bacterial reduction of in-vitro infected root canals after instrumentation by three NiTi rotary files with different taper and diameter versus manual SS files. Sixty-four single-rooted human teeth were infected with a suspension of *Streptococcus sanguis* measured by optical density. Teeth were divided randomly into four groups of 16 and prepared with Flexofiles, GT rotary files, HERO 642, and ProFile. Bacterial samplings were performed before (S1), during (S2-S3), and after (S4) instrumentation. It was found that: (i) all techniques significantly reduced the number of bacterial cells in the root canals; (ii) there was no significant difference between NiTi and manual instrumentation at S2, S3, or S4; and (iii) concerning bacterial reduction, the results suggest that a manual SS file preparation is as efficient as a NiTi rotary instrumentation. This indicates, regardless of the root canal preparation technique, its taper, and diameter, the root dentin remained infected and was not bacteria-free at the end of the experiment. Burroughs et al. [135] determined the shaping ability of three NiTi endodontic file systems by measuring canal transportation. Seventy-two S-shaped canals in resin blocks were randomly allocated into three groups ($n = 24$): (1) the SAF group, (2) the TYP group (TYP rotary files with CM wire), and (3) the Vortex group (PV rotary files with M-wire NiTi). It was found that: (i) after adjusting for the level and canal wall side, the mean transportation was significantly higher for the TYP and Vortex groups compared with the SAF group, and (ii) the mean transportation was significantly higher for the TYP group versus the Vortex group; concluding that SAFs showed less canal transportation than PV and TYP files in simulated S-shaped root canals. Efforts to improve the performance of rotary NiTi instruments by enhancing the properties of NiTi alloy, or their manufacturing processes rather than changes in instrument geometries have been reported. Ba-Hattab et al. [136] compared in vitro, the shaping ability of three different rotary NiTi instruments produced by different manufacturing methods. Thirty simulated root canals with a curvature of 35° in resin blocks were prepared with three different rotary NiTi systems: AK-AlphaKite, GTX-GT, and TFs. It was reported that: (i) less canal transportation was produced by TF apically although the difference among the groups was not statistically significant; (ii) GTX removed the greatest amount of resin from the middle and coronal parts of the canal and the difference among the groups was statistically significant; (iii) the shortest preparation time was registered with TF (444 sec) and the longest with GTX (714 sec), the difference among the groups was statistically significant; (iv) during the preparation of the canals no instruments fractured; and (v) 11 instruments of TF and one of AK were deformed. Based on these results, it was concluded that all rotary NiTi instruments maintained the WL and prepared a well-shaped root canal. The least canal transportation was produced by AK. GTX displayed the greatest cutting efficiency, and TF prepared the canals faster than the other two systems.

Huang et al. [97] evaluated the shaping ability of three heat-treated rotary NiTi systems including PTN, HFCM, and HFEDM during root canal preparation in simulated root canals. A total of 45 simulated root canals were divided into three groups (n = 15) and prepared with PTN, HFCM, or HFEDM files up to size 25. MCT was used to scan the specimens before and after instrumentation. Volume and diameter changes, transportations, and centering ratios at 11 levels of the simulated root canals were measured and compared. It was found that: (i) HFEDM caused significantly greater volume increases than HFCM and PTN in the entire root canal and in the apical and middle thirds, (ii) HFCM removed the least amount of resin in the coronal third compared with HFEDM and PTN, and (iii) overall, HFCM caused significantly less transportation in the apical 2 mm and was better centered than PTN in the apical 3 mm. It was, therefore, concluded that all systems prepared curved canals without significant shaping errors and instrument fracture; in addition, PTN and HFCM cut less resin than HFEDM. HFCM stayed centered apically and cut the least material coronally. Drukteinis et al. [137] evaluated and compared the canal shaping ability of BioRace (BR), PTN, and Genius engine-driven NiTi file (GN) systems in extracted mandibular first molars using MCT. Sixty mesial root canals of mandibular first molars were randomly divided into three equal groups, according to the instrument system used for root canal preparation (n = 20): BR, PTN, or GN. Root canals were prepared to the full WL using a crown-down technique up to size 35, 0.04 taper instruments for BR and GN groups and size 30, 0.07 taper instruments for the PTN group. MCT was used to scan the specimens before and after canal instrumentation. It was reported that: (i) there were no significant differences between the three groups in the terms of dentine removed after preparation and determination of the root canal volume, or percentage of uninstrumented canal surface, and (ii) no significant differences were found between the systems for canal transportation in any canal third. This indicated that the shaping ability of the BR, PTN, and GN NiTi file systems was equally effective, and all instrumentation systems prepared curved root canal systems with no evidence of undesirable changes in 3D parameters or significant shaping errors. Poly et al. [138] compared canal transportation and centering ratio produced after instrumentation with a single heat-treated reciprocating system, WOG and a single heat-treated rotary instrument, XP-endo Shaper, using MCT imaging, and evaluated the ability of double-digital radiography (DDR) to detect canal transportation. It was obtained that (i) the MCT method showed that the shaping ability of XP-endo Shaper regarding the centering ability and canal transportation was significantly better than WOG only at the 7-mm level, and (ii) the DDR technique detected no difference in canal transportation between groups at any level; however, a significant difference between evaluation methods was detected at the 5-mm level in the WOG group; concluding that the MCT technique revealed a significantly better centering ability and less canal transportation with XP-endo Shaper compared to WOG, and the DDR technique was not capable of detecting the significant difference between the tested groups.

16.2.2.8 Affecting factors on efficacy

There are several factors that influence the cutting and shaping ability and efficacy. Jamleh et al. [139] studied NiTi instrument performance under different surrounding temperatures. Twenty-four SE NiTi instruments with a conical shape comprising a 0.30-mm diameter tip and 0.06 taper were equally divided into three groups according to the temperature employed. Using a specially designed cyclic fatigue testing apparatus, each instrument was deflected to give a curvature 10 mm in radius and a 30° angle. This position was kept as the instrument was immersed in a continuous flow of water under a temperature of 10, 37, or 50°C for 20 sec to calculate the deflecting load (DL). In the same position, the instrument was then allowed to rotate at 300 rpm to fracture, and the working time was converted to the number of cycles to fracture (NCF). It was reported that: (i) the mean DL (in N) and NCF (in cycles) of the groups at 10, 37, and 50°C were 10.16 ± 1.36 and 135.50 ± 31.48, 13.50 ± 0.92 and 89.20 ± 16.44, and 14.70 ± 1.21 and 65.50 ± 15.90, respectively; (ii) the group at 10°C had significantly the lowest DL that favorably resulted in the highest NCF; (iii) the surrounding temperature influences the CFR and DL of the SE NiTi instruments; (iv) lower temperatures are found to favorably decrease the DL and extend the lifetime of the SE NiTi instrument; and (v) NiTi instrument failure studies should be carried out under simulated body temperature.

Ha et al. [140] evaluated the effect of torsional preloads on the cyclic fatigue life of NiTi instruments with different history of heat treatments by manufacturers. WO (Primary) made of M-wire, K3XF of R-phase, and ProTaper of conventional NiTi alloy was used. Each file was preloaded at four conditions (0, 25, 50, and 75% of their mean ultimate torsional strength) before fatigue testing. The torsional preloads 10-, 30-, or 50-times were applied by securing 5 mm of the file tip, rotating it until the preset torque was attained before returning to the origin. At that point, the number of cycles to failure (N_F) was evaluated by rotational bending in a simulated canal. It was mentioned that: (i) by SEM observation, most WO after 75% preloading, regardless of repetitions, showed some longitudinal cracks parallel to the long axis of the file, which were rare for K3XF; (ii) the regression analysis revealed that the brand of instrument was the most critical factor; (iii) at up to 75% preloading, ProTaper and K3XF did not show any significant decline in NCF; (iv) for 30 repetition groups of WO, the 50 and 25% torsion preloaded groups showed a significantly higher NCF than the 0 and 75% groups; and (v) the alloy type of NiTi instrument have a significant effect on the phenomenon that a certain amount of torsional preload may improve the CFR of NiTi rotary instruments. Pereira et al. [141] investigated if NiTi instruments with similar designs manufactured by different thermal treatments would exhibit significantly different in vitro behavior. Thirty-six instruments each of PTU (F1; Dentsply Maillefer, Ballaigues, Switzerland), PV, VB, and TYP Infinite Flex NiTi (all size 25/0.06) were evaluated. It was reported that: (i) flexibility was significantly higher for TYP compared with the other three groups; (ii) with respect to the maximum torque at failure, PV group showed the highest resistance to twisting (torsional strength) among the analyzed instruments followed by VB, TYP, and PTU; (iii) the TYP group exhibited greater angular deflection at failure compared with the other groups; (iv) the mean dynamic torque scores during simulated canal preparation were highest for TYP (3.01 ± 0.71 Ncm) and lowest for PV (1.62 ± 0.79 Ncm); however, no significant differences were observed comparing groups PTU with TYP and VB and VB with PV, and (v) the highest mean forces were recorded with PTU (7.02 ± 2.36 N) and the lowest with TYP (1.22 ± 0.40 N). It was hence concluded that TYP instruments were significantly more flexible than the other instruments tested, and the PV group had the highest torsional strength and TYP, despite being the most flexible, showed similar torsional moments to the other instruments, whereas its angular deflection was the highest among the groups.

Qunsi et al. [142] compared photographic and MCT measurements and assessed if the repeated use of NiTi instruments affected the shape of canal preparation. Ten new sets of PTU instruments were used in 60 resin blocks simulating curved root canals. Groups 1 to 6 ($n = 10$) represented the first to sixth use of the instrument, respectively. It was obtained that: (i) two-way repeated-measures analysis of variance revealed significant differences between groups, (ii) regarding measurement type, there were no significant differences between buccolingual and mesiodistal measurements, but there were significant differences between MCT and buccolingual measurements and MCT and mesiodistal measurements, and (iii) significant differences were also noted between uses. It was, therefore, indicated that MCT scanning is more discriminative of the changes in canal space associated with repeated instrument use compared to photographic measurements, and canal preparations are significantly smaller after the third use of the same instrument.

16.2.2.9 Irrigation

Instrumentation of the root canal system must always be supported by the use of antimicrobial irrigating solutions. Despite technological advances in the ability to shape the root canals, there should be some root canal surfaces still remaining as uninstrumented. Hence, cleaning of the canal in terms of soft tissue removal and elimination of bacteria depends heavily on the adjunctive action of chemically active irrigating solutions due to the anatomical complexity of the pulp space. Irrigation is also necessary to suspend and rinse away debris created during instrumentation, to act as a lubricant for instruments and to remove the smear layer that forms on instrumented dentine surfaces. Normally, there are two types of irrigants; NaOCl and EDTA solution [143–146]. Haïkel et al. investigated the effect of NaOCl solution on mechanical properties of NiTi endodontic files [147] and on the cutting efficiency [104]. It was reported that: (i) JS Dental and McSpadden NiTi files were the most resistant to torsional fracture, but all NiTi files were inferior when compared with SS files from a previous study; however, NiTi files were superior in flexibility, and Maillefer and Brasseler instruments were the best of the instruments tested. NiTi files also had negligible permanent deformation angles. (ii) For all properties tested, NaOCl had no statistically significant effect [147]. (iii) In the absence of NaOCl, Brasseler (318 µg/J) and Maillefer (280 µg/J) NiTi files were most efficient, followed by JS Dental (71.4 µg/J) and McSpadden (40 µg/J). (iv) These differences were significant, except for those between the latter two brands. (v) NaOCl treatment did not alter the cutting efficiency of any brand of instruments significantly. (vi) When compared with conventional SS files, all tested NiTi files were less efficient [104]. Bramante et al. [148] examined the effect of the use of EDTA as a root canal irrigant in curved root canals instrumented with NiTi instruments. Twenty extracted maxillary molars were selected. MB roots were used. It was concluded that NiTi instruments used with EDTA were less effective in maintaining the original path of curved canals. Schafer [149] investigated the alterations in cutting efficiency when conventional and titanium nitride (TiN)-coated NiTi K-files were exposed to repeated sterilization using an autoclave. A total of 96 NiTi K-files (size 35) were randomly divided into two groups (A and B) of 48 instruments each. While the instruments of group B were exposed to physical vapor deposition (PVD) creating a coating of a TiN layer, the files of group A were not coated. The instruments of groups A and B were randomly divided into four subgroups of 12 instruments each. (1) A.1/B.1: Instruments were exposed to five cycles of sterilization; (2) A.2/B.2: Instruments were exposed to 10 cycles of sterilization; and (3) A.3/B.3: Instruments were immersed in NaOCl for 30 min, rinsed in water, and exposed to five cycles of sterilization. A.C/B.C: Instruments were not sterilized (controls). It was reported that: (i) the TiN-coated instruments of groups B.1, B.2, and B.3 showed no significant difference in comparison with the penetration depths of the controls, and (ii) the uncoated files of groups A.1, A.2, and A.3 showed significantly lower maximum penetration depths when compared to the control files. Grandini et al. [150] investigated the efficacy of four different irrigation techniques after canal preparation with ProFile NiTi rotary instruments. A modified technique for the use of Glyde File Prep is proposed. Forty anterior teeth were divided into four groups, instrumented, and irrigated as follows: physiological solution (group A), 2.5% NaOCl (group B), 2.5% NaOCl and Glyde File Prep (group C), and 2.5% NaOCl and Glyde File Prep applied at the end of the preparation with sterile paper points (group D). After SEM evaluation at three different levels, debris, smear layer, and dentinal tubules were scored. It was mentioned that: (i) groups A and B had significantly more smear layer and less open tubules on the canal walls compared with the groups C and D samples, and (ii) differences in the mean amount of debris between group A samples and other irrigation regimes were statistically significant. Slutzky-Goldberg et al. [151] measured root-dentin microhardness after instrumentation with two types of files and using irrigation with 2.5% NaOCl. Thirty roots were instrumented with irrigation: 10 roots had the pulp extirpated only; 10 roots were instrumented with SS files, and 10 roots were instrumented with rotary NiTi files. It was mentioned that: (i) significant differences were found between the microhardness at 500 μm and 1,000 μm in all groups, and (ii) instrumentation with NiTi rotary files affected dentin microhardness significantly to a lesser extent.

O'Hoy et al. [152] evaluated the effect of repeated cleaning procedures on fracture properties and corrosion of NiTi files. New NiTi instruments were subjected to two, five, and 10 cleaning cycles with the use of either diluted bleach (1% NaOCl) or Milton's solution (1% NaOCl plus 19% NaCl) as disinfectant. Files were tested for torsional failure and flexural fatigue. It was reported that: (i) up to 10 cleaning cycles did not significantly reduce the torque at fracture or number of revolutions to flexural fatigue although decreasing values were noted with increasing number of cleaning cycles using Milton's solution, (ii) no corrosion was detected on the surface of these files, (iii) files immersed in 1% NaOCl overnight displayed a variety of corrosion patterns, (iv) the extent of corrosion was variable among different brands of files and among files in each brand, and (v) overall, Milton's solution was much more corrosive than diluted bleach. It was, therefore, concluded that files can be cleaned up to 10 times without affecting fracture susceptibility or corrosion, but should not be immersed in NaOCl overnight, and Milton's solution is much more corrosive than bleach with the same NaOCl concentration. Passarinho-Neto et al. [153] evaluated the cleaning capacity of rotary NiTi instrumentation using Profile GT files, coupled with irrigation energized by ultrasound. Thirty-six human mandibular incisors were instrumented in vitro using the crown-down technique with the Profile GT system to 1 mm from the anatomical apex to a size 30 0.04 taper file. The instrumented teeth were then divided randomly into four groups in which various final irrigations were used: Group I: 100 ml of 1% NaOCl with a Luer-Lok syringe (control group); Groups II, III, and IV: final irrigation with 100 ml of 1% NaOCl energized by ultrasound for 1, 3, and 5 min, respectively. It was reported that: (i) group I (Luer-Lok syringe) showed the highest percentage of debris (35.81 ± 4.49) and was statistically different from the other groups energized by ultrasound: Group II (27.28 ± 4.49), Group III (24.39 ± 5.72), and Group IV (18.46 ± 5.25), and (ii) rotary instrumentation using NiTi files associated with final irrigation of 1% NaOCl energized by ultrasound leads to better debris removal from the apical third of mesiodistally flattened root canal. Paiva et al. [154] clinically evaluated the antibacterial effects of the root canal treatment procedures using molecular microbiology analyses. Samples were taken from 14 necrotic root canals of teeth with apical periodontitis before (S1) and after instrumentation with NaOCl irrigation (S2), a final rinse with chlorhexidine (CHX) (S3), and then 1-week interappointment medication with calcium hydroxide/CHX paste (S4). It was reported that: (i) all S1 samples were positive for bacteria in all tests; (ii) treatment procedures promoted a decrease in the microbial diversity and significantly reduced the incidence of positive results and the bacterial counts; (iii) in general, each subsequent treatment step improved disinfection; and (iv) no specific taxon or community pattern was associated with post-treatment samples. It was, therefore, concluded that supplementary steps consisting of a final rinse with CHX followed by calcium hydroxide interappointment medication promoted further decrease in the bacterial bioburden to levels significantly less than those achieved by the chemo-mechanical procedures alone, and because the long-term outcome of root canal treatment is dependent on maximal bacterial reduction, these results are of clinical relevance. Cai et al. [155] investigated the effect of irrigation on the surface roughness and fatigue resistance of HF and M3 CM wire NiTi instruments. Two new files of each brand were dynamically immersed in either 5.25% NaOCl or 17% EDTA solution for 10 min. Thirty-six files of each brand were randomly assigned to three groups ($n = 12$). Group 1 (the control group) was composed of new instruments. Groups 2 and 3 were dynamically immersed in 5.25% NaOCl and 17% EDTA solutions for 10 min, respectively. It was reported that: (i) for M3 files, the Ra, and the root mean square (RMS) values significantly increased after the immersion; (ii) for the HF file, the Ra and RMS values significantly increased only in EDTA, but not NaOCl; (iii) the resistance to cyclic fatigue of both HF and M3 files did not significantly decrease by immersing in 5.25% NaOCl and 17% EDTA solutions. Based on these findings, it was concluded that except the HF files immersed in NaOCl, the surface roughness of other files exposed to irrigants increased; however, a change in the surface tomography of CM wire instruments caused by contact with irrigants for 10 min did not trigger a decrease in CFR.

16.2.3 Fracture and separation

In general, it is well documented that NiTi rotary instruments have two basic mechanisms of fracture; (1) *cyclic fatigue fracture* or *flexural failure* may occur when a file rotates in a curved root canal and is caused by repeated tensile and compressive stresses, and (2) *torsional fracture* is related to a continuous rotation of the engine when the instrument binds in the root canal. Curved canals present an increased risk for transportation and, especially rotary NiTi instruments might be separated. Within a curvature, an instrument part at the inside of the curvature is compressed, whereas tension occurs on the outside of the curvature. Every half-turn rotation places the file portion formerly on the inside of the curvature on the outside and vice versa, with this cycle repeating itself continuously, resulting in the cyclic fatigue and causing rotary NiTi IS.

IS during canal preparation can be broadly divided into two types: torsional fracture and cyclic fatigue fracture [15, 156]. Torsional fracture occurs when the instrument tip is tightly bound in the canal and the handpiece continues to rotate over the maximum strain it can withstand, whereas cyclic fatigue fracture occurs when the instrument receives repeated stress during cyclic rotation in a curved canal [15]. In both fracture incidences, the instrument's fracture could take place without any remarkable indications and fracture occurs suddenly. To date, in particular, fracture due to cyclic fatigue, a nondestructive damage assessment has been studied. Fractographic observation might be also helpful to provide detailed fracture mechanisms for improving the mechanical properties and fracture toughness of the material. It is more essentially important to retrieve separated NiTi instruments, to which there are several studies conducted. Research has included various attempts by manufacturers to overcome this drawback. These efforts include: (i) modifying the design of the instrument, such as the cross-sectional form, taper, helical angle, and pitch length; (ii) changing the manufacturing process to use twisting instead of milling; and (iii) enhancing the surface of the instrument through special processes such as electropolishing [50, 157]. Recently, manufacturers have also focused their efforts not only on the modification of NiTi instruments as listed previously, but also on the improvement of the characteristics of the NiTi alloy [157, 158].

16.2.3.1 Fracture mechanism

Fracture of endodontic instruments are caused by various factors; including metallurgical structural stability, stability of ductility, manner of stress distribution, or overloading. Gil et al. [159] investigated the premature catastrophic fracture produced for different periods during clinical endodontic treatment of two brands of NiTi endodontic rotary instruments; three samples as received, six samples used with patients for 2 and 7 h and five samples fractured were studied for each brand of endodontic NiTi rotary instruments. Critical stresses until fracture ($\sigma^{\beta \to SIM}$, $\sigma^{SIM \to \beta}$) were obtained using an electromechanical testing machine. It was found that (i) calorimetric studies have shown an increase of the M_S and A_S transformation temperatures with time of use as well as a decrease of their stress transformations; (ii) reverse transformation enthalpies decreased along the time, because martensitic plates were anchored, which prevented their transformation to austenite; thus losing its SE effect; and (iii) the stabilization of the martensitic plates induced the collapse of the structure and so the main cause for the fracture. Kim et al. [160] evaluated the stress distribution of three NiTi instruments of various cross-sectional configurations under bending or torsional condition using a finite-element analysis model. Three NiTi files (ProFile, ProTaper, and PTU) were scanned using MCT to produce a 3D digital model. It was found that: (i) ProFile showed the greatest flexibility, followed by PTU and ProTaper; (ii) the highest stress was observed at the surface near the cutting edge and the base of (opposing) flutes during cantilever bending; (iii) concentration of stresses was observed at the bottom of the flutes in ProFile and PTU instruments in torsion; and (iv) the stress was more evenly distributed over the surface of ProTaper initially, which then concentrated at the middle of the convex sides when the amount of angular deflection was increased. It was concluded that incorporating a U-shaped groove in the middle of each side of the convex-triangular design, lowers the flexural rigidity of the origin ProTaper design; bending leads to the highest surface stress at or near the cutting edge of the instrument; and stress concentration occurs at the bottom of the flute when the instrument is subjected to torsion. Barbosa et al. [161] evaluated the presence and propagation of defects and their effects on surfaces of NiTi instruments. The flute surface areas of instruments from two commercial instrumentation systems, namely RPC R25 ($n = 5$) and WO Primary ($n = 5$), were assessed and compared before and after performing two instrumentation cycles in simulated root canals in clear resin blocks. A quantitative analysis was conducted before and after the first and second instrumentation cycles, using the Sa (average roughness over the measurement field), Sq (RMS roughness), and Sz (average height over the measurement field) amplitude parameters. All the data were submitted for statistical analysis at a 5% level of significance. It was reported that: (i) there was a significant increase in wear in both groups, especially between baseline and the second instrumentation cycle, with significantly higher wear values being observed on WO instruments (Sz median values of 33.68 and 2.89 µm, respectively, for WO and RP groups); (ii) a significant increase in the surface roughness was observed in both groups from the first to the second instrumentation cycle, mostly in WO specimens; and (iii) qualitative analysis revealed a greater number of defects on the flute topography of all the instruments after use. Accordingly, it was concluded that more defects were identified in WO Primary instruments compared to RPC R25, irrespective of the evaluation stage, and the study provided an accurate, repeatable, and reproducible assessment of NiTi instruments at different time points. The failure mechanism of ProTaper NiTi rotary instruments fractured under clinical conditions was studied [162]. A total of 46 ProTaper instruments that failed (fractured and/or plastically deformed) during the clinical use were collected from various dental clinics, and a new set of ProTaper instruments served as control. After inspection under stereomicroscopy, the instruments were classified into three categories: (1) plastically deformed but not fractured, (2) fractured with plastic deformation (ductile fracture), and (3) fractured without plastic deformation (brittle fracture). It was reported that (i) stereomicroscopic inspection showed that 17.4% of the discarded instruments were only plastically deformed, 8.7% were fractured with plastic deformation, and 73.9% were brittle-fractured; (ii) the micro-X-ray computed tomography revealed instruments without any surface or bulk defects along with a few files with crack development below the fracture surface; (iii) no defects were identified in the unused instruments; and (iv) fractographical examination of fractured surfaces demonstrated the presence of dimples and cones, a typical pattern of dimple rupture developed because of ductile failure. Based on these findings, it was suggested that a single overloading event causing ductile fracture of ProTaper instruments is the most common fracture mechanism encountered under the clinical conditions.

16.2.3.2 Surface characteristics

Fracture occurred at the site of highly and localized concentration of externally applied stress or internally residual tensile stress. Ferreira et al. [163] assessed nanoscale alterations in the surface topography of NiTi endodontic instruments using a high-resolution optical method and to verify the accuracy of the technique. Noncontact 3D optical profilometry was used to evaluate the defects on a size 25, 0.08 taper reciprocating instrument (WO), which was subjected to a cyclic fatigue test in a simulated root canal in a clear resin block. For the investigation, an original procedure was established for the analysis of similar areas located 3 mm from the tip of the instrument before and after canal preparation to enable the repeatability and reproducibility of the measurements with precision. It was reported that the 3D high-resolution image analysis showed clear alterations in the surface topography of the examined cutting blade and flute of the instrument, before and after use, with the presence of surface irregularities such as deformations, debris, grooves, cracks, steps, and microcavities. It was, therefore, concluded that the optical profilometry provided accurate qualitative nanoscale evaluation of similar surfaces before and after the fatigue test, and the stability and repeatability of the technique enables a more comprehensive understanding of the effects of wear on the surface of endodontic instruments.

16.2.3.3 Fractography

Despite improvements in mechanical properties of the endodontic instruments, NiTi endodontic instruments are still subject to be broken during the clinical use [131]. During the clinical use of rotary NiTi instruments, the files rotate at speeds much faster than 2 rpm with autoreverse mode, by which files undergo repetitive locking and unlocking at a prefixed level of torque. Thus, repeated locking and release of the rotary instruments by the torque-controlled motor are common in the clinic [164]. Particularly, as rotary instruments are subject to higher torsional stresses in narrow canals than in the wider canals, the risk of experiencing such repetitive torsional loads is relatively high [165]. The different modes of stress loading may result in different modes of mechanical failure and produce the different fractographic appearances [156, 166]. Therefore, the torsional resistance and the fracture modes of various NiTi instruments were compared using the dynamic repetitive torsional loading (RTL) method, incorporating the autoreverse motion and their fractured specimens were compared with the specimens from single continuous rotation test method. Fractured surfaces preserve the history of fracture process, for example, on fatigue-fractured surface, one can identify the location of crack initiation site and speed of crack propagation due to striation patterns.

Cheung et al. [167] investigated the mode of failure of a brand of NiTi instruments separated during clinical use, by detailed examination of the fracture surface. A total of 122 ProTaper S1 instruments were discarded from an endodontic clinic over a period of 17 months; 28 had been fractured. These fractured instruments were ultrasonically cleaned, autoclaved, and examined under a SEM. From the lateral view, the fracture was classified into "torsional" or "flexural." Twenty-seven separated instruments were available for analysis. It was noted that: (i) under low-power magnification, only two fell in the category of "torsional" failure when examined laterally; the others appeared to be "flexural"; (ii) close examination of the fracture surface revealed the presence of fatigue striations in 18 specimens; (iii) nine instruments (including the two putative "torsional" failures) fell into the shear fracture group, in which fatigue striations were absent or characteristics of shear failure of the material were found; and (iv) the mean length of fractured segments resulting from fatigue failure (4.3 ± 1.9 mm) was significantly greater than that for shear failure (2.5 ± 0.8 mm). It was then concluded that the examination of the fracture surface at high magnification is essential to reveal features that may indicate the possible origin of cracks and the mode of material failure; macroscopic or lateral examination of separated instruments would fail to reveal the true mechanism of failure; and fatigue seems to be an important reason for the separation of rotary instruments during clinical use. Alapati et al. [168] examined numerous discarded ProFile GT, ProFile, and ProTaper NiTi rotary instruments obtained from two graduate endodontic clinics by SEM. It was noted that (i) these instruments had an unknown history of clinical use and had fractured or experienced considerable permanent torsional deformation without complete separation; (ii) the failure processes generally exhibited substantial ductile character, evidenced by a dimpled rupture fracture surface; (iii) crack propagation at grain boundaries and cleavage surfaces indicative of transgranular fracture were observed for some specimens; (iv) it appeared that oxide particles from the manufacturing process served as the nucleating sites for the microvoids, leading to dimpled rupture; and (v) a previously unreported fracture mode was also observed, in which crack propagation, approximately parallel to the local flute orientation, connected pitted regions on the surface. Wei et al. [169] investigated the mode of fracture of ProTaper rotary instruments after clinical use and to compare stereomicroscopy with SEM to determine which is the better method for establishing the mode of material failure. Overall, 100 fractured ProTaper instruments were examined under stereomicroscope for the presence of plastic deformation along the cutting edge near the fracture site. It was reported that: (i) stereomicroscopy revealed 88 flexural cases and 12 torsional cases; (ii) fractography verified 91 flexural cases with fatigue striations and three torsional cases with circular abrasion marks; (iii) six instruments showed characteristics of both flexural and torsional failures; and (iv) cracks, microcracks, and pittings were common findings on longitudinal micrographs; indicating that inspecting the fractured surface at high-power magnification by SEM is a better method to reveal the mode of NiTi rotary IS. Shen et al. [170] investigated the mode of failure of three brands of NiTi instruments (ProTaper, ProTaper for Hand Use, and K3) that separated during clinical use. A total of 79 fractured instruments were collected from three endodontic clinics over 16 months. It was mentioned that: (i) most of the rotary instruments (78% of K3 and 66% of ProTaper) failed because of fatigue fracture, whereas 91% of NiTi hand instruments failed as a result of shear; (ii) the fracture mode of shaping files in rotary ProTaper was different between two different clinics; (iii) all surfaces with fatigue fracture ($n = 47$) revealed the presence of either one or two crack origins; (iv) the vast majority (86%) of K3 fatigue failure had two crack origins that could be found not only at the cutting edge, but also at various places along the flute compared with only 28% of ProTaper showing multiple crack origins; and (v) the latter showed one crack origin in 81% of the fatigued shaping files, but only 37% for finishing files, indicating that the failure mode of NiTi instruments is related to the preparation technique. Abu-Tahun et al. [171] compared the microscopic features of the fractured endodontic NiTi rotary instruments by two different torsional loadings: RTL and single torsional loading (STL). PTN, HFEDM, and V-Taper 2 were compared in this study. It was reported that (i) in the STL method, the torsional load was applied after fixing the 3 mm tip of the file, by continuous CW rotation (2 rpm) until fracture; (ii) in the RTL method, a preset rotational loading (0.5 N·cm) was applied and the CW loading to the preset torque and CCW unloading to original position were repeated at 50 rpm until the file fractured; and (iii) specimens from the RTL method showed ruptured aspects on cross sections, with multiple areas of initiated cracks, whereas the STL method showed the typical features of torsional failure, such as circular abrasion marks and fatigue dimples. This suggests that a new RTL method is much more clinically relevant and may result in a different fracture feature from STL method, and clinically accumulated torsional stresses may produce different topographic features on the instruments.

16.2.3.4 Factors influencing fracture phenomena

Fracture of endodontic instruments takes place under various influencing factors including operational conditions, NiTi material-related structures, synergistic effects of both torsional and fatigue actions, or anatomical factors. Troian et al. [172] evaluated, by SEM, the deformation and fracture of NiTi RaCe and K3 size 25, 0.04 taper instruments. Ten sets of instruments from RaCe and K3 NiTi rotary systems were used to prepare 100 simulated canals in epoxy resin (ER) blocks with 20 or 40° curvatures beginning 8 or 12 mm from the orifice. Three observers scored images of the instruments after each use for distortion of the spirals (no distortion, distortion of one spiral, or distortion of more than one spiral), wear (no wear, small, moderate, or severe wear), and fracture (yes or no). It was reported that: (i) no fractures occurred with K3 instruments, whereas six RaCe instruments were fractured; (ii) a statistically significant difference occurred between RaCe and K3 instruments in terms of distortion of spirals and surface wear; (iii) distortion of spirals and wear increased with progressive use of RaCe instruments, whereas K3 instruments remained relatively undamaged after their fifth use; and (iv) the simulated canals with smaller radii of curvature were positively associated with fracture of RaCe instruments; concluding that a significant difference was found between RaCe and K3 in terms of deformation and fracture of size 25, 0.04 taper instruments; K3 instruments had more favorable results. Setzer et al. [173] investigated the combined influence of cyclic fatigue and torsional stress on rotary NiTi instruments. The aim of this study was to determine possible differences in the fracture point of rotary NiTi instruments depending on the application of cyclic fatigue only or in combination with torsional stress. It was concluded that: (i) fatigue accompanied with torsional stress compared with fatigue-only resulted in statistically significantly different mean fragment lengths, and (ii) all fractures remained within the area of the curvature, but with the addition of a torsional load, the location of the fracture moved in the direction of the additionally applied torsional stress, suggesting that stress was distributed from the area in which the torsional load was applied toward the area undergoing cyclic fatigue.

Kuhn et al. [174] investigated the process history on fracture life of NiTi endodontics files. The results are based on microstructural investigations of NiTi engine-driven rotary instruments based on X-ray diffraction (XRD), SEM, and microhardness tests. It was mentioned that: (i) endodontic files are very work-hardened, and there is a high density of defects in the alloy that can disturb the phase transformation; (ii) the microhardness Vickers confirmed these observations (dislocations and precipitates); (iii) the X-rays show that the experimental spectrum lines are extended, typical of a distorted lattice, by showing the line broadening; and (iv) the surface state of the endodontic files (SEM) is an important factor in failure and fracture initiation. Kim et al. [175] compared the CFR and torsional resistance of rotary instruments with and without surface treatment. G6 A2 (group A2) with and G6 A2 without surface treatment after machining (group AN) were compared in this study. ProTaper F2 (group F2), which has similar dimension and shape, was also used for comparison. It was reported that (i) group A2 showed higher CFR than the groups AN and F2; (ii) although group A2 demonstrated lower ultimate torsional strength than the others, there were no significant differences in toughness among the groups; (iii) while obvious machining grooves were seen in groups AN and F2, group A2 showed smooth surface resulting from the surface treatment; and (iv) the specimens of fracture fragments showed typical features of cyclic failure such as microcracks, overloaded fast fracture zone, and torsional fracture such as unwinding helix, circular abrasion marks, and dimples; indicating that the surface-treated instruments may improve CFR while maintaining the torsional resistances and mechanical properties.

Kosti et al. [176] investigated the effect of root canal curvature on the failure incidence and fracture mechanism of ProFile rotary NiTi endodontic instruments. Three hundred mesial root canals of mandibular molars were instrumented using the ProFile system in a crown-down technique up to size 25 0.06 taper. Root canals were classified according to the angle and radius of curvature to: straight (group A: 0 + 10°, radius 0 mm), moderately curved (group B: 30 ± 10°, radius 2 ± 1 mm), and severely curved (group C: 60 ± 10°, radius 2 ± 1 mm). After each use, instruments were cleaned ultrasonically and autoclaved. Instruments that prepared 20 root canals, fractured or plastically deformed without fracture, were retrieved. It was reported that: (i) regardless of the size of instrument, fracture and the overall failure were significantly more frequent in group C, and (ii) SEM examination of the fracture surfaces revealed mainly the characteristic pattern of ductile failure, whereas examination under the metallographic microscope revealed no sign of cracks; concluding that the abruptness of root canal curvature negatively affected the failure rate of ProFile rotary NiTi instruments, and the fractographic results confirmed that the failure of NiTi files was caused by a single overload during chemo-mechanical preparation. Yared et al. [177] evaluated the influence of rotational speed, torque, and operator experience with a specific NiTi rotary instrumentation technique on the incidence of locking, deformation, and separation of instruments. ProFile NiTi rotary instruments (PRI) sizes 40-15 with a 6% taper were used in a crown-down technique. In one group of canals (n = 300), speeds of 150, 250, and 350 rpm (subgroups 1, 2, and 3) were used. Each one of the subgroups included 100 canals. In a second group (n = 300), torque was set at 20, 30, and 55 Ncm (subgroups 4, 5, and 6). In the third group (n = 300), three operators with varying experience (subgroups 7, 8, and 9) were also compared. Each subgroup included the use of 10 sets of PRI and 100 canals of extracted human molars. Each set of PRI was used in up to 10 canals and then sterilized before each case. NaOCl 2.5% was used as an irrigant. The number of locked, deformed, and separated instruments for the different groups, and within each part of the study was analyzed statistically for significance with chi-squared tests. It was reported that: (i) in group 1, only one instrument was deformed in the 150-rpm group and no instruments were separated or locked. In the 250-rpm group, IS did not occur; however, a high incidence of locking, deformation, and separation was noted in the 350-rpm group. (ii) In general, instrument sizes 30-15 were locked, deformed, and separated. Chi-squared statistics showed a significant difference between the 150 and 350 rpm groups but no difference between the 150 and 250 rpm groups with regard to IS. (iii) Overall, there was a trend toward a higher incidence of instrument deformation and separation in smaller instruments. (iv) Locking and separation occurred during the final passage of the instruments, in the last (tenth) canal in each subgroup. (v) In the second group, neither separation nor deformation and locking occurred during the use of the ProFile instruments, at 150 rpm, and at the different torque values. (vi) In the third group, chi-squared analysis demonstrated that significantly more instruments separated with the least experienced operator. (vii) Instrument locking, deformation, and separation did not occur with the most experienced operator. Based on these findings, it was concluded that preclinical training in the use of the PRI technique with crown-down at 150 rpm were crucial in avoiding IS and reducing the incidence of instrument locking and deformation.

16.2.3.5 Fracture rate

Daugherty et al. [178] compared the fracture rate between rotary endodontic instruments driven at 150 rpm and 350 rpm. Two groups of 30 mature molars each (S and F) were instrumented with ProFile 0.04 taper Series 29 rotary instruments. Group S molars were instrumented at 150 rpm and group F at 350 rpm. The number of fractures, deformed files, and instrumentation time were recorded for each tooth. It was mentioned that: (i) no instrument fractures occurred in either group; (ii) in group S, the mean deformation rate and instrumentation time were 1.1 deformed files and 8.0 min per molar; and (iii) in group F, they were 0.57 deformed files and 4.6 min per molar. Both differences were significant; indicating that ProFile 0.04 taper Series 29 rotary instruments should be used at 350 rpm to nearly double efficiency and half the deformation rate, compared with 150 rpm. As no instrument fractures occurred while instrumenting 60 mature molars, both speeds were considered safe. Parashos et al. [179] examined used, discarded rotary NiTi instruments obtained from 14 endodontists in four countries, and identified factors that may influence the defects produced during clinical use. A total of 7,159 instruments were examined for the presence of defects. It was mentioned that: (i) unwinding occurred in 12% of instruments and fractures in 5% (1.5% torsional, 3.5% flexural); (ii) the defect rates varied significantly among endodontists; (iii) instrument design factors also influenced the defect rate, but to a lesser extent; (iv) the mean number of uses of instruments with and without defects was 3.3 ± 1.8 (range: 1–10) and 4.5 ± 2.0 (range: 1–16), respectively; and (v) the most important influence on the defect rates was the operator, which may be related to clinical skill or a conscious decision to use instruments a specified number of times or until defects were evident. Patino et al. [180] evaluated the fracture rate of NiTi rotary instruments when following a manual glide path and using SS hand files before performing instrumentation by means of rotary files and, to compare the results in this study with those obtained in two previous analyses, in which the glide path technique was not used. A total of 208 canals obtained from a pool of freshly extracted human mandibular and maxillary molars was divided into three groups corresponding to K3, ProFile, and ProTaper. The coronal two-thirds of each tooth were used. In all three groups, the apical portion of the samples was prepared with size 10–20 SS K-type hand files. The apical stops were prepared using K3, ProFile, and ProTaper rotary instruments. It was found that: (i) logistic regression model analysis indicated that breakage was significantly associated with the angle of curvature of the canal and with the number of clinical uses, and (ii) the breakage rate obtained in this study is significantly lower than in our previous studies, in which the angle of curvature was also greater than 30° and rotational speed a constant 350 rpm, but in which the canals were not first prepared with hand files. Based on the results, the use of SS hand files to prepare the apical one-third of curved canals was recommended before introducing rotary files. Wolcott et al. [181] determined whether the number of uses affects the separation incidence of ProTaper rotary instruments. 4,652 consecutively treated root canals were performed in an endodontic group practice over a 17-month period. Both the separation incidence and the number of uses were tracked for each file. It was mentioned that the overall rate of instrument fracture in this study was 2.4% with no significant differences over the first four uses; indicating that the ProTaper rotary files may be safely reused at least four times, and the size of the rotary file, among other factors, determines how many times a particular file should be used.

Shen et al. studied the factors influencing the defect rates [182] and the common fracture pattern [194]. Three different types of NiTi systems (ProTaper, ProTaper for Hand Use and K3) that were discarded by three endodontic clinics were studied. The instruments were evaluated for defects and factors leading to instrument deformation or fracture. A total of 1,682 instruments were collected over 16 months and were examined. The location of the defect, if any, was recorded. It was reported that: (i) the overall prevalence of unwinding defects was 3% and fracture was 5%; the rates differed significantly between clinics; (ii) for one brand (ProTaper) used at two different clinics, a defect rate (fracture and distortion combined) of 7% (clinic A) versus 13% (clinic B) for shaping files, and about 4% versus 10% for finishing files was observed; (iii) fragments of broken shaping file were significantly longer in clinic A than in clinic B; and (iv) the lowest defect rate was found for K3 instruments: unwinding of 1%, and fracture of 3%; concluding that the defect rates of NiTi instruments were influenced by factors such as the operator, preparation technique, and the instrument design [182]. The same types of endodontic instruments that were discarded after single use by two endodontic clinics were studied. A total of 1,071 ProFile 0.04, 432 ProFile series 29 0.04, and 1,895 ProTaper rotary instruments were collected over 12 months and analyzed. These discarded files were ultrasonically cleaned and autoclaved. It was noted that (i) there were no fractures or deformations in the ProFile Series 29, (ii) the overall prevalence of deformation was 2.9% in ProTaper and 0.75% in ProFile, (iii) the incidence of IS was 0.26% in ProTaper, whereas no fractures occurred in ProFile instruments, (iv) the majority of instrument defects occurred in size 25 (6/8) for ProFile and in Sx for ProTaper (22/60), (v) the separated ProTaper instruments failed mostly because of shear stress, and (vi) some surface deposits and microcracks were found in single-use NiTi instruments; concluding that the risk of NiTi rotary instrument fracture in the canal is low when a new instrument is used by experienced endodontists, and the most common cause of failure, albeit rare, was shear failure [183]. Reciprocating instruments were developed to improve and simplify the preparation of the root canal system by allowing greater centralization of the canal and requiring a shorter learning curve. Despite the risk of IS, using a reciprocating instrument in more than one case is a relatively common clinical practice. Accordingly, Bueno et al. [184] evaluated the fracture resistance of RPC (R25) and WO (Primary) instruments according to the number of uses during the preparation of root canals in up to three posterior teeth. A prospective clinical study was conducted by three experienced specialists who performed the treatment of 358 posterior teeth (1,130 canals) over a period of 12 months using 120 reciprocating instruments, 60 of which were RPC R25 and 60 were WO Primary. The motion used during instrumentation followed the recommendations of the respective manufacturers. After each use, the instruments were observed under a dental operating microscope at 8× magnification. In the case of fracture or deformation, the instrument was discarded. It was reported that: (i) none of the instruments showed any signs of deformation, but three instruments fractured (0.26% of the number of canals and 0.84% of the number of teeth), and (ii) all fractures occurred in mandibular molars (one WO Primary file during the third use and two RPC R25 files, one during the first use and the other during the third use), concluding that there was a low incidence of fracture when reciprocating files were used in up to three cases of endodontic treatment in the posterior teeth.

16.2.3.6 Resistance and preventions

There are several modifications for improving fracture resistance and proposed ideas to prevent fractures of endodontic instruments. Anderson et al. [185] investigated the effect of electropolishing on cyclic flexural fatigue and torsional strength of rotary NiTi endodontic instruments. EP and non-EP EndoWave (EW), ProFile, and RaCe instruments from the same manufacturing batches were investigated. The number of rotations to fracture and torque at fracture were determined and compared among the instruments tested. It was reported that (i) overall, EP instruments performed significantly better than non-EP instruments in cyclic fatigue testing and, to a lesser extent, in static torsional loading; (ii) when viewing EP instruments with the SEM, milling grooves, cracks, pits, and areas of metal rollover were observed although they were more evident in the non-EP instruments; (iii) EP may have beneficial effects in prolonging the fatigue life of rotary NiTi endodontic instruments; and (iv) the benefits of electropolishing are likely to be caused by a reduction in the surface irregularities that serve as points for stress concentration and crack initiation. Chi et al. [186] assessed whether a novel surface treatment could increase the fatigue fracture resistance of dental NiTi rotary instruments. A 200- or 500-nm thick Ti-zirconium-boron (Ti-Zr-B) thin film metallic glass was deposited on PTU F2 files using PVD process. In cyclic fatigue tests, the files were performed in a simulated root canal (radius = 5 mm, angulation = 60°) under a rotating speed of 300 rpm. It was reported that (i) the amorphous structure of the Ti-Zr-B coating was confirmed by transmission electron microscopy and X-ray diffractometry; (ii) the surface of treated files presented smooth morphologies without grinding irregularity; (iii) for the 200- and 500-nm surface treatment groups, the coated files exhibited higher resistance of cyclic fatigue than untreated files; and (iv) in fractographic analysis, treated files showed significantly larger crack-initiation zone; however, no significant differences in the areas of fatigue propagation and catastrophic fracture were found compared to the untreated files. It was then concluded that the surface treatment of Ti-Zr-B thin film metallic glass on dental NiTi rotary files can effectively improve the fatigue fracture resistance by offering a smooth-coated surface with amorphous microstructure.

Shen et al. [187] examined the phase transformation behavior and microstructure of NiTi instruments from a novel CM NiTi wire to improve the fracture resistance of NiTi files. Instruments of ES, ProFile, PV (Vortex), TFs, TYP, and Typhoon™ CM (TYP CM), all size 25/0.04, were examined by DSC and XRD. Microstructures of etched instruments were observed by optical microscopy and SEM with X-ray energy-dispersive spectrometric (EDS) analyses. It was reported that (i) the DSC analyses showed that each segment of the TYP CM and Vortex instruments had an austenite transformation completion or A_F temperature exceeding 37°C, whereas the NiTi instruments made from conventional SE NiTi wire (ES, ProFile, and TYP) and TF had A_F temperatures substantially below mouth temperature; (ii) the higher A_F temperature of TYP CM instruments was consistent with a mixture of austenite and martensite structure, which was observed at room temperature with XRD; (iii) all NiTi instruments had room temperature martensite microstructures consisting of colonies of lenticular features with substantial twinning; and (iv) EDS analysis indicated that the precipitates in all NiTi instruments were Ti-enriched, with an approximate composition of Ti_2Ni. Based on these results, it was concluded that the TYP CM and Vortex instruments with heat treatment contribute to increased austenite transformation temperature, and the CM instrument has significant changes in the phase transformation behavior, compared with conventional SE NiTi instruments. With the increased use of NiTi rotary instruments for root canal preparation in endodontics, instrument fracture has become more prevalent. Instrument fracture is a serious iatrogenic incidence that can complicate and compromise the endodontic treatment. Di Fiore et al. [188] proposed 12 useful measures to prevent the fracture of endodontic instruments. In summary, it was mentioned that there are several measures that practitioners can take to prevent NiTi rotary instrument fracture during root canal preparation: (1) avoid subjecting NiTi rotary instruments to excessive stress; (2) use instruments that are less prone to fracture; (3) follow an instrument use protocol; (4) assess root canal curvatures radiographically and instrument them carefully; (5) ensure that the endodontic access preparation is adequate; (6) open orifices before negotiating canals; (7) enlarge root canals with fine hand instruments; (8) set rotational speed and torque at low levels; (9) use the crown-down technique; (10) irrigate and lubricate root canals during preparation; (11) manipulate rotary instruments with a pecking or pumping motion; and (12) if inexperienced, engage in preclinical training in the use of rotary instruments.

16.2.4 Fatigue and its related phenomena

16.2.4.1 Basic mechanisms
In general, if the amplitude of the total strain is significant plasticity, the fatigue lifetime is likely to be short [i.e., low cycle fatigue (LCF); strain life approach]. If the stresses are low enough that the strains are elastic, the lifetime is likely to be long [high cycle fatigue (HCF); stress-life approach]. At short component lifetimes, plastic strain dominates, and ductility controls the performance. At long component lifetimes, the elastic strain is more dominant and strength controls the performance. In most materials, improvements in the strength unfortunately lead to reductions in ductility and vice versa. This argument is true only if the surface is smooth so that the materials do not contain inherent defects, which are normally responsible for crack initiation site due to localized stress concentration. However, most materials contain inherent defects to some extent; therefore, these approaches can lead to overestima-

tion of useful life [189, 190]. The entire fatigue life comprises crack initiation period (N_I) and crack propagation period (N_P): $N_T = N_I + N_P$. In LCF, it is normal that N_I is approximately equal to N_P, whereas in HCF regime, N_I is much longer than N_P. During these periods, a wide variety of factors affect the behavior of a material under conditions of fatigue loading. The most obvious parameters are those that deal with the sign, magnitude, and frequency of loading, the geometry and material strength level, as well as ambient temperature. Often ill-considered are those processing and metallurgical factors that determine homogeneity of materials, the sign and distribution of residual stresses, and the surface finish.

Cheung et al. [191–193] compared the LCF behavior of some commercial NiTi instruments subjected to rotational bending, a deformation mode similar to an engine-file rotating in a curved root canal, using a strain-life analysis, in water. This study reported that (i) a linear strain-life relationship, on logarithmic scales, was obtained for the LCF region with an apparent fatigue-ductility exponent ranging from −0.40 to −0.56, and (ii) the LCF life of NiTi instruments declines with an inverse power function dependence on surface strain amplitude leading to final rupture (akin to the Griffith's criterion for brittle materials), but is not affected by the cross-sectional shape of the instrument. It was mentioned that the effect of the radius of curvature as an independent variable should be considered when evaluating studies of root canal instrumentation [194–197].

Studying fatigue of NiTi possesses unique feature as NiTi material experiences phase transformation (partially, strain-induced martensitic transformation). Figueiredo et al. [198] analyzed LCF under strain control of NiTi wires in bending-rotation tests. These were conducted on stable austenite, SE, and stable martensite wires, with strain amplitudes from 0.6% to 12%. It was reported that: (i) for strain amplitudes up to 4%, εa–Nf curves of SE wires are close to those values reported in the literature and to that of the stable austenite wire; (ii) for higher strain amplitudes, the fatigue life of SE wires increases with strain and approaches the fatigue life of stable martensite wire; and (iii) fatigue crack characteristics were studied by SEM, suggesting that the abnormal shape of the SE wire curve is associated with the changes in fatigue properties that occurs when the SE material transforms to martensite. Hieawy et al. [199], using PTU and PTG instruments, studied the rotational bending at a curvature of 40° and a radius of 6 mm. It was mentioned that (i) the DSC analyses showed that each segment of the PTG instruments had a higher A_F temperature (A_F: 50.1°C ± 1.7°C) compared to the PTU instruments (21.2°C ± 1.9°C); (ii) PTG instruments had a two-stage transformation behavior; and (iii) PTG exhibited different phase transformation behavior than PTU, which may be attributed to the special heat treatment history of PTG instruments. Arias et al. [200] assessed differences in cyclic fatigue life of contemporary heat-treated NiTi rotary instruments (HFEDM and TRU-Shape) at room (22°C ± 0.5°C) and body (37°C ± 0.5°C) temperatures and to document corresponding phase transformations. It was reported that (i) while TRU instruments lasted significantly longer at room temperature than at body temperature, temperature did not affect the HF behavior; (ii) HF instruments significantly outlasted TRU instruments at both temperatures; (iii) at body temperature, TRU was predominantly austenitic, whereas HF was martensitic or in R-phase; and (iv) TRU was in a mixed austenitic and martensitic phase at 22°C, whereas HF was in the same state as at 37°C, concluding that HFEDM had a longer fatigue life than TRU-Shape, which showed a marked decrease in the fatigue life at body temperature; neither the life span nor the state of the microstructure in the DSC differed for HFEDM between room or body temperature.

Most studies on fatigue behavior of endodontic instruments report that reciprocating motion improves the fatigue resistance of endodontic instruments, compared to continuous rotation, independent of other variables such as the speed of rotation, the angle or radius of curvature of simulated canals, geometry and taper, or the surface characteristics of the NiTi instruments [173, 194, 195, 201–206]. Fife et al. [197] mentioned that (i) the prolonged reuse of NiTi rotary instruments strongly affects instruments' fatigue, and (ii) the cyclic fatigue of each instrument at different levels of the shaft by altering the radius of curvature should be considered. Karamooz-Ravari et al. [207] demonstrated that NiTi SMA shows asymmetric material response in tension and compression, which can significantly affect the lifetime of the files fabricated from, and the material asymmetry can significantly affect the maximum von Mises equivalent stress as well as the force–displacement response of the tip of this file.

Haikel et al. [208, 209] studied the dynamic and cyclic fatigue fractures of various types of files, and reported that: (i) The K file, H file, Unifile, and the Helifile showed a distinct fracture starting point with crack striations and ductile fractures; (ii) the K-Flex and Flexofile showed only plastic deformations and axial fissures; and (iii) in all cases, fracture was found to be of a ductile nature, thus implicating cyclic fatigue as a major cause of failure and necessitating further analyses and setting of standards in this area. It was mentioned that: (i) rotary NiTi endodontic instruments manufactured by twisting the present greater resistance to cyclic fatigue compared with instruments manufactured by grinding, and (ii) the fracture mode observed in all instruments was of the ductile type [210]. Waver et al. [211] reported that: (i) the strain rate dependence might affect the fatigue properties of different alloys at different alternating strain values; and (ii) given the difference in loading rates between benchtop fatigue tests and in vivo deformations, the potential for strain rate dependence should be considered when designing durability tests for medical devices and in extrapolating results of those tests to in vivo performance.

16.2.4.2 Factors influencing fatigue behaviors

The major factors influencing the fatigue life of NiTi are temperature effect and corrosive environment. Taking these factors into consideration of the entire fatigue life, we will see how detrimental these factors are. Referring to Figure 16.32, (a) group explains during the crack initiation period, whereas (b) group illustrates crack propagation period. Normally, grain boundary is more chemically active, so that any

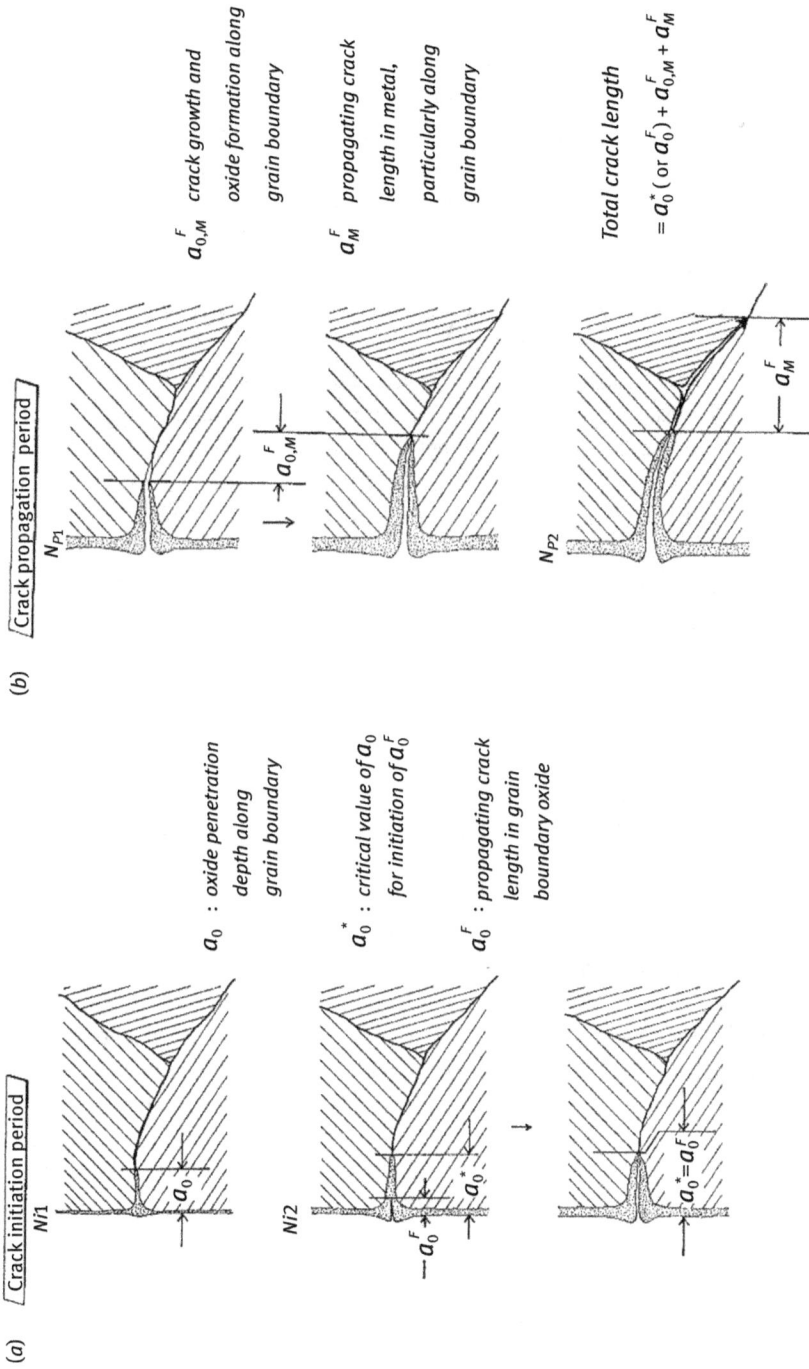

Figure 16.32: Environmental factors during the (a) crack initiation period and (b) crack propagation period.

corrosive agent or simply oxygen attacks first at the grain boundary, following an immediate formation of a corrosion product including oxide (or hydroxide). Because these products are ceramic and brittle in nature, they crack along the grain boundary, leading to further corrosion or oxidation advancement. The sequence is repeated at pluralities of grain boundaries at relatively surface area until some of these sites grow deep and long enough to recognize as a fatigue crack initiation (a). Thus, the created crack further propagates toward inside the material body during the crack propagation stage (b) until the fatigue failure. The rates for both crack initiation and propagation are totally dependent on the corrosiveness in corrosive environment or the temperature in oxidation environment.

Fatigue tests were conducted and compared at ambient temperature (22°C ± 0.5°C) and intracanal temperature (37°C ± 1°C) [200, 212–216]. The common findings and conclusions are: (i) the temperature had an effect on the fatigue behavior, (ii) immersion in water at body temperature was associated with a marked decrease in the fatigue life of all rotary instruments tested, and (iii) the body temperature treatment caused a marked reduction of the CFR for all reciprocating instruments tested. Grande et al. [217] conducted fatigue tests at 20°C (±2°C) for room temperature and –20°C (±2°C) for the cooled environment, and concluded that a low environmental temperature determines a drastic increase in the flexural fatigue resistance of NiTi endodontic instruments. There are several fatigue studies conducted at a wide range of temperatures: at temperatures 0, 10, 22, 37, and 60°C [218]; at temperatures 0, 20, 35, and 39°C [219]; at temperatures - 3, 22, 37, and 60°C [220], and at temperatures 25, 37, and 60°C in distilled water, 2.5% NaOCl, or 5.25% NaOCl [221]. The common conclusions are: (i) the fatigue resistance of tested files was proportionally lower as the environmental temperature increases, and (ii) an irrigating solution (NaOCl) at different concentrations and temperatures influenced the CFR of NiTi instruments.

In studies conducted in NaOCl and EDTA solutions, other environment effects on the fatigue behavior were reported. Common conclusions of reports using NaOCl with 5% concentration [222–224] and 5.25% concentration [225, 226] are: (i) NaOCl irrigant does not significantly affect the fatigue behavior of NiTi files, but (ii) Peters et al. [225] mentioned that NiTi rotary had reduced resistance to cyclic fatigue after contact with heated NaOCl (60°C) and may then be considered single-use instruments. About the effect of EDTA solutions (17% concentration), there are controversial conclusions; it was concluded that NaOCl and EDTA solutions did not have any effect on the CFR of WO and WOG files [227], whereas the type of NiTi metal alloy (CM files vs conventional SE NiTi files) influences the CFR [228].

16.2.4.3 New approach and analysis
Iacono et al. [229] evaluated the impact of a modified motion on the fatigue life of four brands of NiTi reciprocating instruments (WO, WOG, RPC, and RPC Blue) with two different motions: (1) X-Smart Plus (Denstply Maillefer) endodontic motor and (2)

a 4:1 contra-angle with an experimental motion (EVO) with different rotation angles and based on a sinusoidal acceleration. It was reported that: (i) RPC Blue resulted to be the most resistant instruments either with EVO or X-Smart; (ii) WOG lasted significantly longer than WO with EVO (probability of 91%), while no significant differences were found with X-Smart; and (iii) considering the numbers of fatigue cycles, RPC, WOG, and RPC Blue lasted significantly longer with EVO (probabilities of 66, 80, and 89%, respectively). WOG showed the highest beta parameter; concluding that the experimental motion was found to have a positive impact on the fatigue lifetime of reciprocating instruments. Herbst et al. [230] evaluated the CFR study networks, hypothesizing that indications and instrument properties (alloy, manufacturer) drive comparisons. A systematic review was conducted via Medline and Embase (1986–2017). In vitro studies analyzing the CFR of ≥2 engine-driven instruments in an artificial curved root canal system were included. Eighty-five studies on 56 different instruments (nine alloys, 17 manufacturers) were included. It was reported that (i) for instruments, a low-density (0.19), highly clustered (0.71) network with three separate subnetworks (for glide path, shaping, and retreatment instruments) emerged; (ii) certain instruments (PTU, Mtwo) served as hub nodes and possible gold standards; (iii) conventional NiTi was the most frequently used alloy; (iv) few manufacturers dominated the network; and (v) the diversity of tested instruments, alloys, and manufacturers increased in recent years. Based on these findings, it was concluded that comparisons of CFR were usually made along indications; some instruments and alloys (conventional NiTi) dominated the networks; overall risk of bias by comparator choice seems moderate; however, the evidence on certain, less often tested instruments may not be robust; and factors underlying the network geometry (e.g., sponsorship) should be explored.

16.2.5 Torsion

16.2.5.1 Torsional behaviors

The type of torsional (fatigue) failures occur in two modes, which are longitudinal or transverse along planes of maximum shear helical at 45° to the axis of the shaft along planes of maximum tension; namely, by the twisting of a file along the longitudinal axis at one end, while the other end is fixed. This can happen in straight or curved canals if the tip binds [231]. When the elastic limit of the metal is exceeded, the rotary instrument undergoes plastic deformation. Sattapan et al. [232] analyzed the type and frequency of defects in NiTi rotary endodontic files after routine clinical use (total of 378 files, discarded after normal use from a specialist endodontic practice over 6 months). It was mentioned that: (i) almost 50% of the files showed some visible defect; 21% were fractured and 28% showed other defects without fracture; and (ii) fractured files were divided into two groups according to the characteristics of the defects observed; torsional fracture occurred in 55.7% of all fractured files, whereas flexural fatigue occurred in 44.3%; indicating that torsional failure, which may be caused by using too much apical force during instrumentation, occurred more frequently than flexural fatigue. Torsional failure appears to be geometry-sensitive [231, 233–235].

Best et al. [164] evaluated the torsional cyclic fatigue characteristics and specifically the endurance limit of a NiTi rotary instrument. Size 30 0.06 taper ProFile instruments were evaluated. It was reported that: (i) instruments cycled at larger deflection angle consistently demonstrated fewer cycles to fracture than those cycled at smaller deflection angle, (ii) the differences among the mean log number of cycles of the different deflection angle were statistically significant, and (iii) cycles of 10^6 were completed without instrument fracture at 2.5°; concluding that a torsional fatigue ProFile was generated for a specific NiTi rotary instrument. Ninan et al. [236] investigated the torsion and bending properties of shape-memory files (CM Wire, HFCM, and Phoenix Flex) and compared them with conventional (ProFile ISO and K3) and M-wire (GT Series X and PV) NiTi files. Sizes 20, 30, and 40 (n = 12/size/taper) of 0.02 taper CM Wire, Phoenix Flex, K3, and ProFile ISO and 0.04 taper HFCM, ProFile ISO, GT Series X, and Vortex were tested in torsion and bending per ISO 3630-1 guidelines by using a torsiometer. It was found that: (i) significant interactions were present among factors of size and file; (ii) variability in maximum torque values was noted among the shape-memory files brands, sometimes exhibiting the greatest or least torque depending on brand, size, and taper; and (iii) in general, the shape-memory files showed a high angle of rotation before fracture but were not statistically different from some of the other files. However, the shape-memory files were more flexible, as evidenced by significantly lower bending moments. Based on these results, it was concluded that the shape-memory files show greater flexibility compared with several other NiTi rotary file brands.

16.2.5.2 Comparisons in torsional behaviors

There are several comparative studies on various NiTi files. Park et al. [237] compared the repetitive torsional resistance of five NiTi instruments (TF and RaCe, both with an equilateral triangular cross section, as well as the ProTaper, Helix, and FlexMaster, which had a convex triangular cross section). It was found that: (i) the TF had the lowest and FlexMaster the highest torsional resistance among the groups; (ii) SEM examination revealed a typical pattern of torsional fracture for TF, RaCe, and Pro-Taper that was characterized by circular abrasion marks and skewed dimples near the center of rotation; and (iii) in addition to these marks, Helix and FlexMaster presented a rough, torn-off appearance, indicating that files of same cross-sectional design may exhibit different resistance to fracture probably as a result of the manufacturing process. Yum et al. [238] compared the torsional strength, distortion angle, and toughness of five NiTi rotary files (TF and RaCe with equilateral TR, ProTaper with CT, ProFile with U-shape, and Mtwo with S-shape). It was reported that: (i) TF and RaCe had significantly lower yield strength (YS) than other systems; (ii) TF had a significantly lower ultimate strength than other files, whereas Mtwo showed the greatest; (iii) ProFile showed the highest distortion angle at break, followed by TF; (iv) ProFile also showed the highest toughness value, whereas TF and RaCe both showed a lower toughness value than the others; and (v) fractographic examination revealed typical pattern of torsional fracture for all brands, characterized by circular abrasion marks and skewed dimples near the center of rotation. Based on these results, it was concluded that the five tested NiTi rotary files showed a similar mechanical behavior under torsional load, with a period of plastic deformation before actual torsional breakage, but with unequal strength and toughness value. Lopes et al. [239] compared the torsional resistance of two brands of rotary NiTi instruments (Mtwo retreatment instruments #15 and #25 and PTU retreatment instruments D2 and D3) in CW rotation. The two parameters evaluated were maximum torque and angular deflection at failure. It was mentioned that: (i) the angular deflection at fracture decreased in the following order: Mtwo retreatment file #15 > Mtwo retreatment file #25 > PTU retreatment file D2 > PTU retreatment file D3; (ii) as for the maximum torque values, the results revealed the following descending order: PTU file D2 > Mtwo retreatment file #25 > PTU file D3 > Mtwo retreatment file #15; and (iii) SEM analysis revealed that plastic deformation occurred along the helical shaft of the fractured instruments. Fractured surfaces were of the ductile type; concluding that the instruments tested showed different torsional behavior depending on the parameter evaluated, and if one considers that high angular deflection values may serve as a safety factor, the Mtwo retreatment instruments showed significantly better results. Elnaghy et al. [240] compared the torsion and bending properties of OneShape (OS) and WO single-file systems. It was reported that: (i) WO had a significantly higher torsional resistance than OS, (ii) the average bending resistance as measured by the maximum force to bend instruments revealed that the WO had a significantly lower resistance to bend than OS, and (iii) SEM analysis of the fractured cross-sectional surfaces revealed typical features of torsional failure including skewed dimples near the center of the fracture surface and circular abrasion streaks. It was then concluded that the WO single-file system showed higher torsional resistance and flexibility than the OS single-file system, and different cross-sectional geometry and the alloys from which the instrument is manufactured could have significant influence on the torsional resistance and flexibility of the instruments. Arbab-CHirani et al. [241] compared numerically the bending and torsional mechanical behavior of five endodontic rotary NiTi instruments with equivalent size and various designs for tapers, pitch, and cutting blades [Hero (20/0.06), HeroShaper (20/0.06), ProFile (20/0.06), Mtwo (20/0.06), and ProTaper F1]. It was obtained that: (i) Protaper F1 presented the greatest level of bending stress and torque, and (ii) Hero and HeroShaper were more rigid than ProFile and Mtwo, indicating that the five endodontic instruments do not have the same bending and torsional mechanical behavior, and each clinician must be aware of these behavior differences so as to use the adequate file according to the clinical situation and to the manufacturer's recommendations. Alcalde et al. [242] compared the torsional properties of pathfinding NiTi rotary instruments [ProGlider

(M-wire), HF GPF (conventional NiTi wire and CM wire], Logic [conventional NiTi wire and CM wire) and Mtwo (conventional NiTi wire)]. It was reported that: (i) the Logic size 25, 0.01 taper had significantly higher torsional strength values, (ii) the ProGlider was significantly different when compared with HF GPF size 15, 0.01 taper and size 15, 0.02 taper, (iii) the Logic CM size 25, 0.01 taper had significantly higher torsional strength than HF GPF size 15, 0.01 taper and size 15, 0.02 taper, (iv) no difference was found among Mtwo size 10, 0.04 taper and HF GPF groups (size 15, 0.01 taper; size 15, 0.02 taper; size 20, 0.02 taper), (v) in relation to the angle of rotation, Logic CM size 25, 0.01 taper and HF GPF size 15, 0.01 taper had the highest angle values, (vi) the Pro-Glider had the lowest angle values in comparison with all the groups followed by Mtwo size 10, 0.04 taper, and (vii) the Logic size 25, 0.01 taper had significantly higher angle of rotation values than ProGlider and Mtwo size 10, 0.04 taper. Based on these findings, it was concluded that the Logic size 25, 0.01 taper instrument made of conventional NiTi alloy had the highest torsional strength of all instruments tested, and the ProGlider instrument manufactured from M-wire alloy had the lowest angle of rotation to fracture compared with the other instruments.

Torsional properties of the unused rotary NiTi files were compared with the used ones. Yared et al. [243, 244] compares torque (gcm) and angle of rotation (degrees) at fracture of new and used NiTi K3 0.04 rotary instruments. It was reported that: (i) the torque at fracture of the new instruments increased significantly with the diameter; (ii) the used instruments (sizes 15, 20, 30, 35, and 40) have lower torque at fracture compared to the new ones; (iii) the means of angle of rotation at fracture between the different sizes of new instruments were significantly different ($p < 0.0001$) except for sizes 15–20; (iv) the used instruments (sizes 20–40) had lower angle of rotation at fracture compared to the new ones; and (v) a linear relationship was found between the size of the file and the torque at fracture for the new instruments and the used ones; suggesting that, in general, the torque and the angle of rotation at fracture were significantly affected by the repeated use of 0.06 K3 instruments in resin blocks. Vieira et al. [245] assessed the influence of multiple clinical uses on the torsional behavior of PTU rotary NiTi instruments. Root canal treatments were performed on patients using the PTU rotary system to prepare canals. Ten sets of instruments were used by an experienced endodontist, each set being used in five molar teeth. After clinical use, S1, S2, F1, and F2 instruments were analyzed for damage by optical and SEM. It was reported that: (i) the use of the PTU rotary instruments by an experienced endodontist allowed for the cleaning and shaping of the root canal system of five molar teeth without fracture; (ii) the maximum torque for instruments S2, F1, and F2, and the angular deflection at fracture for instruments S2 and F1 were significantly lower following clinical use; (iii) the largest decrease in maximum torque was 18.6% for S2 instruments; and (iv) the same maximum percent decrease was found for angular deflection at fracture for F1 instruments; concluding that the torsional resistance and angular deflection of used instruments compared to that of new instruments, were reduced following clinical use.

There are some studies comparing torsional behaviors between NiTi files and SS files. Rowan et al. [246] compared the torsional properties of SS endodontic files (Flex-O-File) and recently developed NiTi endodontic files (quality dental products) in CW and CCW directions independently. It was shown that: (i) SS files had a significantly greater rotation to failure in the CW direction, whereas the NiTi files had a significantly greater rotation to failure in the CCW direction, and (ii) despite these differences in rotation to fracture, there was essentially no difference between the SS and NiTi instruments in the torque that it took to cause failure in both the CW and CCW directions; indicating that although the number of CW and CCW rotations to failure differed for the two instruments, the actual force that it took to cause that failure was the same. Wolcott et al. [247] evaluated and compared the torsional properties of SS K-type 0.02 taper and NiTi U-type 0.02 and 0.04 taper instruments. It was indicated that the SS instruments showed no significant difference between maximum torque and torque at failure, whereas both the NiTi instruments showed a significant differential between maximum torque and torque at failure. Kazemi et al. [248] conducted mechanical tests on SS and NiTi instruments of identical design to compare clinically relevant qualities of instruments manufactured from different alloys. Identical experimental endodontic files of H-type with double helix were fabricated from SS and NiTi alloys. It was obtained that (i) the torsional moment of the SS files was significantly higher than for the NiTi files, whereas the mean angular deflection for the NiTi files was significantly higher; (ii) the mean bending moment for the NiTi files was significantly lower than the mean value for the SS files; and (iii) the SEM demonstrated that the fracture surfaces of the SS files were brittle, whereas the NiTi files had a ductile fracture; concluding that when the design of endodontic instruments of different alloys is identical, the NiTi files are more flexible in bending; however, NiTi files require less force to deform to fracture.

16.2.5.3 Influencing factors on torsional properties

Bahia et al. [249] evaluated the influence of cyclic torsional loading on the flexural fatigue resistance and torsional properties of rotary NiTi instruments. Twelve sets of new K3 instruments, sizes 20, 25, and 30 with an 0.04 taper, and sizes 20 and 25 with an 0.06 taper, were torsion tested until rupture, to establish their mean values of maximum torque and angular deflection. It was reported that: (i) the cyclic torsional loading caused no significant differences in maximum torque or in maximum angular deflection of the instruments analyzed, but comparative statistical analysis between measured N_F values of new and previously cycled K3 instruments showed significant differences for all tested instrument; and (ii) longitudinal cracks, i.e., cracks apparently parallel to the long axis of the instruments cycled in torsion was observed, concluding that the cyclic torsional loading experiments in new K3 rotary endodontic instruments showed that the torsional fatigue decreased the resistance of these instruments to flexural fatigue although it did not affect their torsional resistance. Bahia et al. [233] also assesses the influence of cyclic deformation up to one-half and three-quarters of the fatigue life of NiTi rotary endodontic instruments on their behavior in torsion. Thirty sets of 0.04 and 0.06 ProFile instruments in sizes #20, #25, and #30 were randomly divided into three groups: group one contained 10 sets of new files, torsion tested to failure, whereas the other two groups, with 10 sets each, were first submitted to fatigue tests of up to one-half and three-fourths of their fatigue life, respectively, in geometrical conditions similar to those of the clinical practice and then torsion tested for rupture. It was shown that statistical analysis of the values obtained for the torque to failure and maximum angular deflection showed that the cyclic deformation of instruments of up to one-half and three-fourths of their fatigue life gives rise to a significant reduction in these parameters, indicating that the simulated clinical use of ProFile instruments for shaping curved canals adversely affects their torsional properties. Acosta et al. [250] evaluated the influence of cyclic deformation on the torsional resistance of CM NiTi files in comparison with SE instruments with similar geometric and dimensional characteristics. New 30/0.06 HF, TYP (CT), RaCe, and PTU F2 instruments (F2) were assessed. It was reported that: (i) new CM files had a significantly higher N_F when compared with SE instruments; HF exhibited the highest value; (ii) the mean torque value for F2 was the highest; (iii) CM files precycled to three-fourth N_F had a significantly lower torque than the new files, whereas the SE instruments displayed no significant differences; concluding that the cyclic flexural loading significantly reduced the torsional resistance of the CM instruments.

Ha et al. [140] evaluated the effect of torsional preloads on the cyclic fatigue life of NiTi instruments with different history of heat treatments (WO [Primary] made of M-wire, K3XF (#30/0.06) of R-phase, and ProTaper (F2) of conventional NiTi alloy). It was noted: (i) SEM showed that most WO after 75% preloading, regardless of repetitions, showed some longitudinal cracks parallel to the long axis of the file, which were rare for K3XF; (ii) regression analysis revealed that the brand of instrument was the most critical factor; (iii) at up to 75% preloading, ProTaper and K3XF did not show any significant decline in the N_F; and (iv) for 30 repetition groups of WO, the 50% and 25% torsion preloaded groups showed a significantly higher NCF than the 0% and 75% groups; concluding that the alloy type of NiTi instrument have a significant effect on the phenomenon that a certain amount of torsional preload may improve the CFR of NiTi rotary instruments. Oh et al. [251] evaluated the effect of torsional preloading on the torsional resistance of NiTi endodontic instruments. WO Primary and PTU F2 (Dentsply Maillefer) files were used. It was reported that: (i) all preloaded groups showed significantly higher ultimate strength than the unpreloaded groups; (ii) there was no significant difference among all groups for distortion angle and toughness; (iii) although WO had no significant difference between the repetition groups for ultimate strength, fracture angle, and toughness, ProTaper had a higher distortion angle and toughness in the 50 repetition group compared with the other repetition groups; and (iv) SEM examinations of the fractured surface showed typical features of torsional fracture. Based on these results, it was concluded that torsional preloading within the ultimate values could enhance the torsional strength of NiTi instruments, and the total energy until fracture was maintained constantly, regardless of the alloy type. The process of twisting has been used for decades to fabricate SS instruments, but it was previously thought to be an impractical method for NiTi instrument. Newly developed twisted NiTi files theoretically should overcome the problems associated with a grinding process, which previously limited the instrument strength. Gambarini et al. [252] evaluated the effects of a final deoxidation process on torsional resistance of TF instruments. Testing was performed in accordance with the ISO 3630-1 standard by comparing 20 TF instruments versus 20 TF prototype instruments produced without the final deoxidation process. It was found that TF instruments withstood 90% more torque (max torque) than TF prototype instruments manufactured without the deoxidation process (88.7 vs 53.3 g/cm); indicating that as design and dimensions of tested instruments were the same, the deoxidation process could be the only explanation of the improvement in the torsional resistance.

16.2.6 Separation

As Walia et al. [253] pointed out in 1988, the NiTi files were found to have 2–3 times more elastic flexibility in bending and torsion, as well as superior resistance to torsional fracture, compared with SS files; suggesting that the NiTi files may be promising for the instrumentation of curved canals. Currently, NiTi rotary instruments are part of the daily armamentarium of the endodontist and general practitioner with a great variety of instruments presenting different cross-sections, tapers, and new variations of the original alloy [199, 253–255]. Due to the synergistic effects of advanced manufacturing processes, appropriate heat treatments and skill-up of operators, as seen in Figure 16.33, the incidence of separation for rotary instruments ranged from 0 to 23% and the trend appears to be ever-decreasing. Obviously, the engine-driven instrumentation of root canal system aims to decrease the preparation

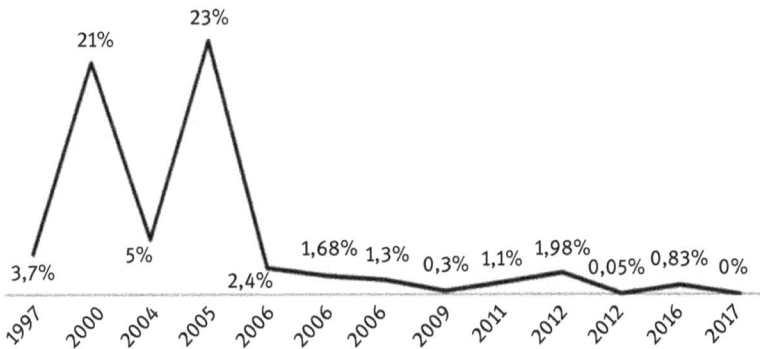

Figure 16.33: General trend of separation of NiTi rotary instruments [254].

time and simplification of root canal instrumentation. Nonetheless, in the early era of engine-driven instruments, there was a high risk of IS when compared to hand instruments, due to torsion, cyclic fatigue, or a combination of both forces, as discussed previously.

A single-file reciprocating system for NiTi instruments has been suggested, claiming to diminish the necessary steps for root canal preparation [254]. Besides, the single use of these instruments would decrease the risk of cross-contamination and IS [256, 257]. The critical areas of stress move progressively to new locations during the periodical change of the angle, thus effectively distributing the areas of stress to different points of the instrument decreasing the damage and increasing the life span of the instrument [254]. Clinical studies have demonstrated that these instruments present a low rate of separation when a single-use approach is applied. The separation incidence for reciprocating instruments ranged from 0% to 1.71% over reviewing from 2014 through 2017 [254]. Reciprocating instruments were mostly single used; one study reported their use up to three times. The separation rate in reciprocating instruments was similar in single use or in multiple uses 0.2%.

Tygesen et al. [258] determined whether there was a difference in the frequency of file distortion and separation between NiTi ProFile 0.04 ISO taper rotary files and NiTi Pow-R 0.04 ISO taper rotary files. For this study, 60 mesial root canals in 30 human mandibular first molars were chosen with the curvature angle of the MB and mesiolingual canals of the same tooth within 5°. It was concluded that there was no statistically significant difference in the incidence of instrument distortion or separation between the two file systems that were studied. Rosen et al. [259] compared the diagnostic ability to radiographically detect separated SS versus NiTi instruments located at the apical third of filled root canals with either AH 26 (Dentsply DeTrey GmbH, Konstanz, Germany) or Roth sealer (Roth International Ltd, Chicago, IL, USA). It was reported that: (i) there were no significant differences in the diagnostic ability between digital and conventional radiography or the different root canal sealers, and (ii) the sensitivity to detect fractured SS was significantly higher than NiTi; concluding that it may be difficult to radiographically detect a retained separated instrument, and it is easier to radiographically detect fractured SS than NiTi instruments retained at the apical third of the root canal. Kaul et al. [260] comparatively evaluated the physical surface changes and incidence of separation in rotary NiTi instruments using SEM. A total number of 210 freshly extracted human maxillary and mandibular first molars were selected and distributed between three groups. Three different systems of rotary NiTi instruments (ProFile, RaCe, and TF) were used to prepare the canals using crown-down technique. It was obtained that (i) RaCe showed the maximum wear of the surface followed by TF; (ii) ProFile showed the minimum wear except for its tip; and (iii) there was no correlation between electropolishing and file fracture. Insignificant difference was observed in the mean number of canals shaped by ProFile and TF before their separation. It is, therefore, concluded that clinically, TF performance was superior, followed by ProFile then RaCe, and RaCe fracture rate was the greatest after preparing the least number of canals.

16.2.7 Removal of separated instruments

When clinicians face with the situation of an unwanted incidence of the IS, a better approach for removal of separated instruments should be decided. There are several factors associated with the fractured instrument removal. Before discussing this topic in detail, it would be worthy to refer what has been documented about the risk management of fractured endodontic instruments [261], stating that (1) regardless of how sophisticated they may be, the use of an operating microscope and the appropriate armamentarium are required along with knowledge, training, and experience. (2) Any clinical procedure for the management of fractured instruments includes the risk of creating additional error(s) that may eventually jeopardize the prognosis of the tooth; therefore, continuous evaluation of the progress is needed to consider alternative options, if required. (3) Therapeutic options for the management of fractured instruments include no intervention, nonsurgical management, surgical management, and tooth extraction. (4) As a general rule of thumb, surgical management follows nonsurgical management of the error and is performed either immediately after the completion of the nonsurgical management or at a later time based on future evaluation; however, there are some cases where weight of risks versus benefits changes this rule and surgical procedures are performed without any nonsurgical attempt. (5) The timing and type of surgical procedure are determined by several interdependent factors that require evaluation and careful consideration on an individual case-by-case basis.

Hulsmann et al. [262] evaluated the influence of several factors on the success rate of removal procedures of fractured endodontic instruments postoperatively. In 105 teeth with 113 fragments, removal attempts were undertaken using a wide range of techniques and instruments. All cases were analyzed with special regard to the following factors: type of tooth and root canal, site of fragment in relation to root canal curvature, length of fragment, and type of fractured instrument. Success of treatment was defined as removal or complete bypassing of the fragment. It was reported that: (i) of 82 instruments in molars (maxillary: 32, mandibular: 50), 56 were removed or bypassed (max.: 26, mand.: 30), (ii) of 16 fragments in premolars (max.: 12, mand.: 4), eight could be removed or bypassed (max.: 6, mand.: 2), (iii) of 14 fragments in canines and incisors (max.: 7, mand.: 7), 13 could be removed completely (max.: 6, mand.: 7); (iv) when the fragment was localized before the curvature, two of 18 cases failed, when localized inside the curvature; 13 of 31 cases failed when localized beyond the curvature; and 15 of 33 cases failed, pointing out that the anatomical factors favorable for removal were: straight canals, incisors and canines, localization before the curvature, length of fragment >5 mm, localization in the coronal or mesial third of the root canal, and reamer or lentulo spirales. In molars, removal procedures were most successful in the palatal canals of maxillary molars. Shen et al. [263] evaluated the influence of various factors on the success or failure of attempts to remove fragments of separated NiTi instruments from root canals. Instrument removal attempts were undertaken on 72 teeth with a separated NiTi instrument using a variety of techniques and armamentarium. Factors including the type of tooth, degree of root canal curvature, location of fragment in relation to the root canal curvature, and radiographic length of fragment were analyzed. A success of treatment was defined as removal or complete bypassing of the fragments. It was mentioned that: (i) the overall success rate was 53%; (ii) the success rate for ProFile fragments was 41% and for the NiTi K-file was 60%; (iii) the success rate in maxillary teeth was higher than that in mandibular teeth; (iv) of 52 instruments in molars, 28 were successfully removed or bypassed; (v) of the 12 fragments in premolars, only two were removed; (v) all eight cases in the anterior teeth were retrieved completely; (vi) when the fragment was localized before the curvature, complete removal was achieved; (vii) when the fragments were located at and beyond the curvature, the success rates were 60% and 31%, respectively, (viii) in canals with a slight, moderate, and severe curvature, the success rates were 100%, 83%, and 43%, respectively; and (ix) in general, the longer the fragment, the greater the chance for successful removal or bypass. Based on these results, it was indicated that the favorable factors for the removal of separated NiTi fragments are straight root canals, anterior teeth, localization before the curvature, fragments longer than 5 mm, and hand NiTi K-file. Obviously, these factors directly influence the probability and success rate of removing fractured instruments. Suter et al. [264] evaluated the probability of removing fractured instruments and compared these findings with the results of a similar study. Within an 18-month period all referred endodontic cases involving fractured instruments within root canals were analyzed. The protocol for the removal of fractured instruments was to create straight-line access to the coronal portion of the fractured instrument and to create a ditched groove around the coronal aspect of the instrument using ultrasonic (US) files and/or to bypass it with K-Files. Subsequently, the fractured instrument was vibrated ultrasonically and flushed out of the root canal or an attempt was made to remove the instrument with the Tube-and-Hedström file method or similar techniques. The location of the fractured instrument and the time required for removal were recorded. Successful removal was defined as complete removal from the root canal without creating a clinically detectable perforation. It was reported that: (i) In total, 97 consecutive cases of instrument fracture were included in the time period. In all, 84 instruments (87%) were removed successfully. (ii) There was a significant correlation between the time needed to remove fractured instruments and a decrease in the success rate. (iii) Curved canals had significantly more fractured instruments than straight canals. (iv) Rotary instruments fractured significantly more often in curved canals compared with other instruments. (iv) Half of all instrument fractures occurred in mesial roots of lower molars and most often when using rotating instruments. (v) There was no statistically significant difference in the success rate with respect to the location of the fractured instrument (tooth/root type), the type of fractured instrument, or the different methods of instrument removal. Therefore, it was concluded that the curved canals are a higher risk for instrument fracture than straight canals; and in curved canals, rotary instruments (including lentulo spirals) fractured more often than other instruments. Totally, 87% of the fractured instruments were removed successfully. A decrease in the success rate was evident with increasing treatment time, and the use of an operating microscope was a prerequisite for the techniques used to remove the fractured instruments. Nevares et al. [265] evaluated the success rates of standardized techniques for removing or bypassing fractured instruments from root canals and in determining whether visualization of the fractured instrument with the aid of an operating microscope has any impact on the success rates. During the prospective study, attempts were made to visualize 112 fractured instruments under a dental operating microscope after creating a straight-line access to the fragment. By using US tips alone or associated with bypassing, the Ruddle technique was attempted to remove the 68 visible instruments. Bypassing was performed for the 44 nonvisible fragments. It was found that: (i) the overall success rate (removal and bypassing) was 70.5%; (ii) in the visible fragment group, the success rate was 85.3% ($n = 58$) and in the nonvisible fragment group, it was 47.7%; and (iii) success rates were significantly higher when the fragment was visible. Based on these results, it was concluded that the standardized techniques used in this study for removing or bypassing fractured instruments were effective, and approximately two times greater success rate was obtained when the fragment was visible inside the root canal compared with when it was nonvisible.

There are several ways to retrieve or treat separated instruments. Ormiga et al. [266] introduced a new concept of retrieval of fractured instruments from root canals based on an electrochemical process. Current register tests were used to evaluate the dissolution process of 25.04 NiTi K3 rotary files. A constant anodic potential was applied to the NiTi files, whereas the potentiostat registered the anodic current. After the tests, all files were observed by using an optical microscope. It was found that: (i) the current attained initial values of approximately 55 mA declined during the entire test, (ii) a good reproducibility of results was observed, and (iii) the optical microscopy analysis evidenced a progressive consumption of the files with increasing polarization time; concluding that the concept of fractured file retrieval by an electrochemical process is feasible, and the concept resulted in a consistent basis for the development of a method to remove fractured instruments from root canals. Takahashi [267] investigated the effects of two solutions on the corrosion of broken instruments and on root dentin. The two solutions are APF (composed of 2% sodium fluoride in phosphoric acid) and NCN (composed of 10% sodium hypochlorite + 19% sodium chloride). The morphological changes were also observed, and quantitative analysis was conducted using an X-ray microanalyzer. It was reported that: (i) in the broken NiTi file immersed in NCN, corrosion progressed from the broken edge resulting in ragged borders; (ii) in APF, pitting corrosion progressed gradually and uniformly all over the surface; (iii) in the dentin immersed in NCN or APF, there was a decrease in hardness; and (iv) the softened depth of the dentin immersed in NCN for 3 h was 100 µm, whereas that of dentin immersed in APF for 24 h was 200 µm. Based on these results, it was concluded that both NCN and APF are capable of corroding broken files while softening only the superficial layer of the root canal wall. Hence, the method may be clinically useful for facilitating the removal of broken root canal instruments. Rahimi et al. [268] reported on a conservative approach for removal of a fractured file in the severely curved apical portion of the distobuccal canal of a mandibular molar. It was indicated that with the assistance of SS hand files and a chloroform-dipped gutta-percha cone, a fractured rotary NiTi instrument was successfully removed. Also, the use of this technique may assist in the removal of loose instrument fragments that are not easily accessible to other removal techniques.

The US technique and the instrument removal system (iRS) in removing fractured NiTi rotary from root canals were evaluated [269]. A total of 30 extracted human molars with closed apices were collected. Profile 0.06 taper rotary endodontic instruments 25-mm long and ISO size 25 were driven into the mesial canal to be fractured in the curve. Angles and radii of curvature were measured for all the canals with the fractured fragments, and then canals were classified into slight, moderate, and severe subgroups according to their curvature angles. Within each subgroup, teeth were randomly distributed into US or iRS according to measured angles and radii; a total of 15 canals were assigned to both US and iRS techniques. It was reported that: (i) the overall success rate was 70% ($n = 21$); (ii) less curved canals and longer radii of curvature showed more success; (iii) the median time used for retrieval was 40 min using the US technique, whereas it was 55 min for iRS; and (iv) the removal of fractured rotary NiTi endodontic instruments was more successful with less curved canals and longer radius of curvature > 4.4 mm. Ward et al. [270] evaluated the use of the US technique to remove fractured rotary NiTi instruments from narrow, curved canals in both simulated (resin blocks) and mesiolingual canals of extracted mandibular first molars. It was mentioned that: (i) a technique using US tips, combined with the creation of a "staging platform" and the use of the dental operating microscope, was consistently successful and safe at removing fractured rotary NiTi instruments from curved root canals, when some part of the fractured instrument segment was located in the straight portion of the canal; (ii) when the fractured instrument segment was located entirely around the curve, the success rate was significantly decreased and major canal damage often occurred. A technique using modified Gates Glidden burs and USs has recently been advocated to remove fractured instruments from root canals. Varying extents of tooth structure are removed during this procedure, potentially leading to complications. Souter et al. [271] evaluated the in vitro and in vivo complications associated with the fractured file removal. Fractured instrument fragments were removed from three different levels (coronal, middle, or apical third) of mesiolingual canals of extracted human mandibular molars. The success rate, frequency of perforations, and root strength were recorded for each group. Perforations and unsuccessful file removal occurred only with fragments lodged in the apical third. It was reported that: (i) fracture resistance declined significantly with more apically located file fragments; (ii) a review of 60 clinical cases showed similar rates of successful file removal and rate of perforations; and (iii) removal of a fractured file fragment from the apical third of curved canals should not be routinely attempted. Arslan et al. [272] determined the effect of taper (0.08, 0.06, and 0.04) of separated K3XF instruments on duration taken for the secondary fracture formation during US activation. Ten 25/0.08 K3XF (SybronEndo), 10 25/0.06 K3XF, and 10 25/0.04 K3XF instruments were used for the study. The apical 5 mm of the instruments was cut to simulate the fragments in root canals. Fragments of the instruments were sandwiched between two straight dentin blocks. The US tip was used to cause a secondary fracture of the fragment. The time needed for the secondary fracture was recorded for each instrument. It was reported that: (i) the secondary fractures occurred in all instruments; (ii) in the 0.08 taper group, secondary fractures took longer than in the case of the 0.06 and the 0.04 taper groups; and (iii) there were no significant differences between the 0.06 and the 0.04 taper groups in terms of the time required for the occurrence of a secondary fracture; indicating that in the 0.08 taper group, secondary fracture took longer time than in the case of the 0.06 and the 0.04 taper groups due to its larger cross-sectional area involved.

It is important for endodontic instruments to have a low fracture rate. If a fracture does occur, it would be desirable to have the ability to bypass the broken segment and complete the root canal treatment [273]. A total of 162 root canals in 52 maxillary and mandibular first molars were cleaned and shaped with LightSpeed instruments by three endodontists in their private practices. The canals were instrumented using the technique recommended by the manufacturer. All canals were instrumented to at least a size 45 at the WL. It was reported that: (i) six instruments separated during treatment, (ii) all six had been used more times than recommended by the manufacturer; and (iii) five of the six were easily bypassed and the treatment was completed [293]. Usually, bypassing the instrument is the first step in instrument removal procedures. If this step is achieved, the outcome is not different from removed instruments, thus the separated instrument management can be considered successful [274]. Madarati et al. [274] reviewed separated instruments regarding treatment options, influencing factors, and complications and made some suggestions for decision-making process. For conducting article reviewing, the following keywords were used: instruments, files, obstructions, fractured, separated, broken, removal, retrieval, management, bypassing, and complications with or without root canal and endodontic. It was reported that: (i) there is a lack of high-level evidence on the management of separated instruments; (ii) conventional conservative management includes the removal of or bypassing the fragment or filling the root canal system to the coronal level of the fragment; (iii) a surgical intervention remains an alternative approach;

these approaches are influenced by a number of factors and may be associated with complications; and (iv) based on the current clinical evidence, a decision-making process for the management is suggested. Based on the review results, the formulated decisions on the management should consider the following items: (i) the constraints of the root canal accommodating the fragment, (ii) the stage of root canal preparation at which the instrument separated; (iii) the expertise of the clinician; (iv) the armamentaria available; (v) the potential complications of the treatment approach adopted; and (vi) the strategic importance of the tooth involved and the presence or absence of periapical pathosis. Penta et al. [275] demonstrated that the electrochemical impedance spectroscopy (EIS) could be a method for evaluating and predicting of ProTaper rotary file system clinical lifespan. The method used for quantification resides in the EIS theory and has in its main focus the characteristics of the surface TiO_2 layer [276, 277]. To assess the method, 14 ProTaper sets used on different patients in a dental clinic have been submitted for testing using EIS. The information obtained with regard to the surface oxide layer has offered an indication of use and proves that the said layer evolves with each clinical application. The feasibility of EIS method was recognized in being adapted for NiTi file investigation and correlation with surface and clinical aspects.

There are some studies to predictive approaches toward the tool separations. Young et al. [278] investigated the failure of $Ni_{55.8}Ti$ (wt%) wire, as a function of experimental parameters that include the clinically relevant regime. The effects of radius of curvature, angle of curvature, wire diameter, strain amplitude, cyclic frequency, volume under strain, and specific heat of the surrounding environmental fluid are considered systematically. It was indicated that the lifetime or cycles to failure N_F of a rotating NiTi wire can be predicted via a modified Coffin-Manson relation, which is a strong function of both strain amplitude and volume under strain and a weaker function of frequency and fluid-specific heat; suggesting that the resulting quantitative relation can be used to predict useful device lifetime under clinically relevant conditions and thereby reduce incidences of in vivo failure. Chakka et al. [279] conducted visual and microscopic evaluation of defects caused by torsional fatigue in hand and rotary NiTi instruments. Ninety-six NiTi greater taper instruments that were routinely used for root canal treatment only in anterior teeth were selected for the study, using ProTaper for hand use, ProTaper Rotary files, and EW rotary files. It was reported that: (i) the number of files showing defects were more under stereomicroscope than visual examination; (ii) however, the difference in the evaluation methods was not statistically significant; (iii) the different types of defects observed were bent instrument, straightening/stretching of twist contour, and partial reverse twisting; and (iv) EW files showed maximum number of defects followed by ProTaper for hand use and least in ProTaper Rotary. Based on these results, it was concluded that the visible defects due to torsional fatigue do occur in NiTi instruments after clinical use, and both visual and microscopic examinations were efficient in detecting defects caused due to torsional fatigue. This indicates that all files should be observed for any visible defects before and after every instrumentation cycle to minimize the risk of IS and failure of endodontic therapy. Stojanac et al. [280] examined the lifespan or N_F of tapered rotary NiTi endodontic instruments. Simulated root canals with different curvatures were used to determine a relation between canal curvature and instrument lifespan. Using a novel mathematical model for the deformation of SE NiTi alloy, it was shown that maximum stress need not necessarily occur at the outer layer. Based on this observation, the Coffin-Manson relation was modified with parameters determined from this experiment. It was shown that: (i) the N_F was influenced by the angle and radius of canal curvature and the size of instrument at the beginning of canal curvature, and (ii) the resulting quantitative mathematical relation could be used to predict the lifespan of rotary NiTi endodontic instruments under clinical conditions and thereby reduce the incidence of instrument failure in vivo. Cheung et al. [281] evaluated the bending fatigue lifetime of NiTi alloy and SS endodontic files using finite element analysis. The strain-life approach was adopted and two theoretical geometry ProFiles, the TR and the square cross-sections, were considered. Both LCF and HCF lifetime were evaluated. It was reported that: (i) the bending fatigue behavior was affected by the material property and the cross-sectional configuration of the instrument; (ii) both the cross-section factor and material property had a substantial impact on the fatigue lifetime; and (iii) the NiTi material and TR geometry ProFiles were associated with better fatigue resistance than that of SS and square cross-sections; indicating that the finite element models (FEMs) were established for endodontic files to prejudge their fatigue lifetime, a tool that would be useful for dentist to prevent premature fatigue fracture of endodontic files.

By either torsional action or LCF, damage process is accompanied with internal structural changes of NiTi materials. Such structural changes might include dislocation density, rigidity, electrical resistivity, and so on, as shown in Figure 9.1. Most of these changes can be detected by nondestructive measurements; for example, Weiss and Oshida et al. [282, 283] successfully conducted X-ray diffractogram analyses to find an excellent linear relationship between surface dislocation density increment and accumulated fatigue damage. There should be more techniques available to assess the cumulative damage at surface zone of NiTi alloys, so that any endodontic instruments can be avoided from subsequent clinical services. Li et al. [284] investigated the application of nondestructive testing in cyclic fatigue evaluation of endodontic PRIs. As-received ProFile instruments were made to rotate freely in sloped metal blocks by a contra-angle handpiece mounted on a testing machine. Rotation was interrupted periodically, and the instrument was removed and engaged onto a device to monitor its stiffness by using two strain gauges in four different directions. This monitoring method has the potential to be developed into a convenient, nondestructive turnkey system that allows in-situ assessment of the integrity of NiTi instruments in the clinic. Upon fracture, which was indicated by a change in instrument stiffness, the fractured surface would be examined under a SEM. Microscopic evaluation indicated a small area of fatigue fracture with a large area of final ductile fracture, whereby the latter was the major cyclic failure mode. Based on the results of the study, it was concluded that a potential nondestructive integrity assessment method for NiTi rotary instruments was developed. Hsieh et al. [285] used natural frequency for monitoring structural changes of a NiTi instrument during and after the instrumentation process. In the study, laboratory modal testing experiments were performed on cyclic fatigue-loaded NiTi rotary instruments with a natural frequency detecting device. In addition, 3D FEMs were established for assessing the structural changes that take place in repeatedly loaded NiTi instruments. It was reported that: (i) repeated rotational loading resulted in a significant decrease in natural frequency (with a decreasing ratio of 5.6%) when the tested instruments reached 77–85% of their total life limit; and (ii) in finite element analysis, a strong correlation between natural frequency and change in elastic modulus of the NiTi instrument was found; indicating that the natural frequency may represent an effective parameter for evaluating the micro-structural status of NiTi rotary instruments subjected to fatigue loadings.

16.2.8 Surface modification and mechanothermal modification

There are two major groups of materials' improvements: treatment on superficial layers (surface modification) and thermal or thermomechanical treatment affecting the entire body of the materials.

16.2.8.1 Surface modification

Oshida et al. [286] described the surface of NiTi alloy. The surface of NiTi instrument mainly consists of oxygen, carbon, and rutile-type TiO_2, and smaller amounts of nickel oxides (NiO and Ni_2O_3) and metallic nickel Ni; the thickness of the oxide layer varies between 2 and 20 nm. Depending on the preparation method, the surface chemistry and the amount of nickel may vary in a wide range and nickel may dissolve more easily than titanium because its oxide is not so stable; hence, surface layers of NiTi archwires have irregular features characterized by long island-like structures, which are indicative of selective dissolution of nickel [286]. Attempts to enhance the surface of NiTi instruments, minimize or eliminate their inherent defects, increase the surface hardness and flexibility, as well as improve the resistance to cyclic fatigue and cutting efficiency have been made. The cutting efficiency of endodontic instruments have resulted in a variety of surface modifications including the implantation of boron ions on the surface, thermal nitridation process, or PVD of TiN particles [287]. Besides these surface improvements in mostly mechanical performances, as Oshida et al. [286] pointed out, the surface layer of NiTi alloy is: (i) chemically or electrochemically manipulated so that harmful (if so) nickel element is selectively dissolved out, resulting in a Ti-rich layer of NiTi's surface; or (ii) as stated further, nickel ion is implanted to form TiN. On the other hand, the hard TiN coatings frequently have disadvantages due to the interface between the bulk and its coating. Ion implantation can solve this problem, as a continuous interface between the surface and the bulk is created [288].

During the plasma immersion ion implantation [289, 290], (nitrogen) ion implantation is a line-of-sight process in which ions are extracted from plasma, accelerated, and bombarded into a target material and has been extensively used to modify NiTi surface layers. Wolle et al. [291] investigated the effect of N ion and Ar ion implantation on morphologic alterations and fatigue resistance in ProTaper S1 NiTi rotary instruments. It was reported that: (i) no significant morphologic alterations were observed in the instruments after the preparation of five canals; (ii) crack density was similar in all groups; (iii) in the subsequent cyclic fatigue test, instruments implanted with nitrogen performed worse than those implanted with argon and the control group; and (iv) fracture faces show differences in the fracture modes. Based on these findings, it was concluded that Ar ion implantation improved the performance of S1 files moderately, whereas nitrogen ion-implanted files performed worse in the fatigue test, and a reduction in the file performance seems to be caused by nitrogen diffusion in the grain boundaries, instead of the desired improvement caused by TiN formation. Gavini et al. [292] assessed CFR in rotary NiTi instruments submitted to nitrogen ion implantation. by using a custom-made cyclic fatigue testing apparatus. It was found that ion-implanted instruments reached significantly higher cycle numbers before fracture (mean, 510 cycles) when compared to annealed (mean, 428 cycles) and nonimplanted files (mean, 381 cycles); indicating that the nitrogen ion implantation improves CFR in rotary NiTi instruments. The effects of nitrogen ion implantation on the flexibility of rotary NiTi instruments were assessed [293]. It was reported that: (i) the mean load required to bend instruments at a 30° angle was 376.26 g for implanted instruments and 383.78 g for nonimplanted instruments; and (ii) the difference was not statistically significant; suggesting that nitrogen ion implantation has no appreciable effect on the flexibility of NiTi instruments.

Boron implantation into nitinol alloy has a potential for developing improved nitinol root canal instruments with excellent cutting properties, without affecting their SE bulk-mechanical properties [294]. It was mentioned that: (i) the surface hardness of NiTi alloy was improved by ion-beam surface modification; (ii) with an implantation dose of 4.8×10^{17} boron/cm^2, a high concentration of boron (30 atm%) is incorporated into NiTi alloy by 110 keV boron ions at room temperature (25°C); (iii) boron-implanted and unimplanted NiTi alloys show surface hardness of 7.6 ± 0.2 and 3.2 ± 0.2 GPa, respectively, at the nanoindentation depth of 0.05 μm; and (iv) the ion beam-modified NiTi alloy exceeds the surface hardness of SS.

16.2.8.2 Thermal treatment

Remarkable benefits for treating materials through the cryogenic process have been recognized [295–297]. It was reported that the cold treatment of metals during manufacturing is advocated so as to improve the surface hardness and thermal stability of the metal. The optimum temperature range for cold treatment is between –60 and –80°C depending on the material and the quenching parameters involved [295]. There are some studies on the effects of cryogenic treatment on NiTi alloys. Vinoth-kumar et al. [298] evaluated the effects of cryogenic treatment on NiTi endodontic instruments. Ten rotary NiTi instruments (ProFile 30, 0.06) were subjected to deep dry cryogenic treatment at liquid nitrogen temperature (–185 °C, group A) and compared with that of nontreated instruments (group B). Cutting efficiency in rotary motion was assessed from weight loss of tooth samples after instrumentation. Wear resistance was assessed from surface defects on cutting heads pre- and postinstrumentation using SEM. It was reported that: (i) the weight loss was significantly higher in group A, (ii) the presence of surface defects after use was higher in group B, and (iii) the deep-dry cryogenic treatment increases the cutting efficiency significantly but not the wear resistance. George et al. [299] evaluated the effect of deep dry cryotherapy on the CFR of rotary NiTi instruments. Twenty K3, RaCe, and Hero Shaper NiTi instruments, size 25, 0.06 taper, were tested. It was obtained that a significant increase in the resistance to cyclic fatigue of deep-dry cryotreated NiTi files over untreated files was recognized and thus was concluded that the deep cryotherapy has improved the cyclic fatigue of NiTi rotary endodontic files. Kim et al. [300] investigated the effects of cryogenic treatment on NiTi endodontic instruments. Microhardness was measured on 30 NiTi K-files by a Vicker's indenter. Half of the specimens to be used for each analysis were subjected to cryogenic treatment in liquid nitrogen (-196 °C). Cutting efficiency was assessed by recording operator choice using 80 NiTi (ProFile® 20, 0.06), half of which had been cryogenically treated and had been distributed amongst 14 clinicians. It was reported that: (i) cryogenically treated specimens had a significantly higher microhardness than the controls, (ii) observers showed a preference for cryogenically treated instruments (61%), but this was not significant, and (iii) both treated and control specimens were composed of $Ni_{56}Ti$ (wt%) with a majority in the austenite phase; concluding that cryogenic treatment resulted in increased microhardness, but this increase was not detected clinically, and there was no measurable change in the elemental or crystalline phase composition.

Pereira et al. [141] investigated whether NiTi instruments with similar designs manufactured by different thermal treatments would exhibit significantly different in vitro behavior. Thirty-six instruments each of PTU (F1), PV, VB, and TYP Infinite Flex NiTi (all size 25/0.06) were evaluated. Bending resistance, torsion at failure, and dynamic torsional tests were performed with the instruments ($n = 12$). It was reported that: (i) flexibility was significantly higher for TYP compared with the other three groups; (ii) with respect to the maximum torque at failure, PV group showed the highest resistance to twisting (torsional strength) among the analyzed instruments followed by VB, TYP, and PTU; (iii) the TYP group exhibited greater angular deflection at failure compared to the other groups; (iv) the mean dynamic torque scores during simulated canal preparation were highest for TYP (3.01 ± 0.71 Ncm) and lowest for PV (1.62 ± 0.79 Ncm); however, no significant differences were observed comparing groups PTU with TYP plus VB and VB with PV; and (v) the highest mean forces were recorded with PTU (7.02 ± 2.36 Ncm) and the lowest with TYP (1.22 ± 0.40 Ncm). It was then concluded that the TYP instruments were significantly more flexible than the other instruments tested, and the PV group had the highest torsional strength and TYP, although despite being the most flexible, showed similar torsional movements to the other instruments, whereas its angular deflection was the highest among the groups. Gu et al. [301] assessed the performance of various heat-treated NiTi rotary instruments in S-shaped resin canals. Forty S-shaped resin canals were instrumented (10/group) with either TFs (R-phase), WO (M-wire), HFCM, or V Taper 2H (CM wire) with the same apical size and taper (25/0.08). Each S-shaped resin canal was scanned both before and after instrumentation with MCT. Changes in the canal volume and transportation were evaluated at regular intervals (0.5 mm). Differences between instruments at the apical curve, coronal curve, and straight portion of the canals were analyzed statistically. It was found that: (i) all tested instruments caused more transportation at the coronal rather than apical curvatures, with the exception of TFs for which apical transportation was the highest for any instrument or location; (ii) the transportation was mostly influenced by the alloy type rather than their cross-sectional characteristics; (iii) the volumetric increase after instrumentation was similar for all tested instruments at the apical curve, whereas HFCM created the most conservative preparations at the coronal curve; and (iv) at the straight portion, volumetric changes were largest for TFs and smallest for V Taper 2H. Based on these results, it was concluded that among heat-treated NiTi instruments, the CM wire-based instruments created the most favorable preparations in S-shaped resin canals.

There are three topics (mechanical properties, fatigue strength, and bending behaviors) being affected by thermal treatments. Miyara et al. [302] investigated the influence of heat treatment on the mechanical properties of NiTi file materials. Specimens were heated for 30 min at 300, 400, 450, 500, or 600 °C. It was reported that: (i) M_S and A_F for groups 400 and 450 were higher than those for others (ii) the load/deflection ratios of groups 400, 450, and 500 were lower than that of group 600; (iii) the bending load values at 2.0 mm deflection of groups 400, 450, and 500 were lower than those of group 300 and the control group; (iv) the NCFs of groups 400, 450, and 500 exceeded that of group 600; and (v) changes in the flexibility with heat treatment could improve the cyclic fatigue properties of NiTi instruments. Zinelis et al. [303] determined the effect of various thermal treatments on the fatigue resistance of a NiTi engine-driven endodontic file. Fifteen groups of five files each of ISO 30 and taper 0.04 were tested in this study. The cutting tip (5 mm from the end) of files from 14 groups were heat treated for 30 min in temperatures 250, 300, 350, 375, 400, 410, 420, 425, 430, 440, 450, 475, 500, and 550 °C, respectively, while group 1 was used as reference. The files were placed in a device that allowed the instruments to be tested for rotating bending fatigue inside an artificial root canal. The number of rotations to breakage was recorded for each file. It was found that the 430 and 440 °C groups showed the highest values, with fatigue resistance decreasing for thermal treatment at lower and higher temperatures; due to the result of metallurgical changes during annealing, indicating that the appropriate thermal treatment may significantly increase the fatigue resistance of the tested NiTi file. Braga et al. [304] assessed the influence of M-wire and CM technologies on the fatigue resistance of rotary NiTi files by comparing files made using these two technologies with conventional NiTi files. Files with a similar cross-sectional design and diameter were chosen for the study: new 30/0.06 files of the EW, HF, PV, and TYP systems together with PTU F2 instruments. It was reported that: (i) X-ray EDS analysis showed that, on average, all the instruments exhibited the same chemical composition, namely, 51% nickel and 49% titanium; (ii) the PV, TYP, and HF files exhibited increased transformation temperatures; (iii) the PTU F2, PV, and TYP files had similar D3 values, which were less than those of the EW and HF files; and (iv) the average N_F values were 150% higher for the TYP files compared with the PV files and 390% higher for the HF files compared with the EW files. It was, therefore, concluded that the M-wire and CM technologies increase the fatigue resistance of rotary NiTi files. Plotino et al. [305] evaluated the difference in CFR between VB and PV NiTi rotary instruments. Two groups of NiTi endodontic instruments, PV and VB, consisting of identical instruments in tip size and taper (15/0.04, 20/0.06, 25/0.04, 25/0.06, 30/0.06, 35/0.06, and 40/0.04) were tested. It was mentioned that: (i) when comparing the same size of the two different instruments, a statistically significant difference was noted between all sizes of VB and PV instruments except for tip size 15 and 0.04 taper, and (ii) no statistically significant difference was noted among all groups tested in terms of fragment length; concluding that VB showed a significant increase in the CFR when compared to the same sizes of PV.

There are still several researches on the effects of thermal treatments on bending behaviors of NiTi alloys. Miyai et al. [306] investigated the relationship between the functional properties and the phase transformation of NiTi endodontic instruments. Five types of rotary NiTi endodontic instruments (EW, HERO 642, K3, ProFile 0.06, and ProTaper) were selected to investigate the torsional and bending properties and phase transformation behavior. It was mentioned that: (i) the maximum torsional torque values of HERO, K3, and ProTaper were significantly higher than those of EW, ProFile, and K-file; (ii) the K-files had the lowest torque value; (iii) angular deflection at fracture was significantly higher for K-files than that for any NiTi instrument; (iv) the bending load values of HERO and K3 were significantly higher than those of EW, ProFile, ProTaper, and K-file; (v) the K-files had the lowest load value although residual deflection remained; and (vi) the transformation temperatures of HERO and K3 were significantly lower than those of EW, ProFile, and ProTaper. Based on these results, it was concluded that the functional properties of NiTi endodontic instruments, especially their flexible bending load level, were closely related to the transformation behavior of the alloys. Hayashi et al. [307] studied the bending properties of hybrid rotary NiTi endodontic instruments in relation to their transformation behavior. Four types of NiTi rotary endodontic instruments with different cross-sectional shapes (triangular-based and rectangular-based) and different heat treatment conditions (SE type and hybrid type with SME) were selected to investigate the bending properties and phase transformation behavior. It was found that: (i) the bending load values of the hybrid type that had undergone additional heat treatment at the tip were significantly lower than those of the SE type with no additional heat treatment, (ii) the bending load values of rectangular-based cross-sectional shaped instruments were significantly lower than those of triangular-based cross-sectional shaped instruments; (iii) phase transformation temperatures (M_S and A_F points) of the hybrid type were significantly higher than the SE type, and (iv) the M_F and A_S points of the tip part were significantly higher than those of the whole part of the hybrid instrument. Based on these findings, it was concluded that the additional heat treatment of hybrid NiTi instruments may be effective in increasing the flexibility of NiTi rotary instruments. Yahata et al. [308] investigated the effect of heat treatment on the bending properties of NiTi endodontic instruments in relation to their transformation behavior. NiTi SE alloy wire (1.00 mm ϕ) was processed into a conical shape with 0.30 mm diameter tip and 0.06 taper. The heat treatment temperature was set at 440 or 500 °C for 10 or 30 min. It was obtained that: (i) the transformation temperature was higher for each heat treatment condition compared with the control; (ii) two clear thermal peaks were observed for the heat treatment at 440 °C; (iii) the specimen heated at 440 °C for 30 min exhibited the highest temperatures for M_S and A_F, with subsequently lower temperatures observed for specimens heated at 440 °C for 10 min, 500 °C for 30 min, 500 °C for 10 min, and control (nontreated) specimens; (iv) the sample heated at 440 °C for 30 min had the lowest bending load values, both in the elastic range (0.5 mm deflection) and in the SE range (2.0 mm deflection); and (v) the influence of heat treatment time was less than that of heat treatment temperature. It was concluded that the change in the transformation behavior by heat treatment may be effective in increasing the flexibility of NiTi endodontic instruments. Ebihara et al. [309] evaluated the bending properties and shaping abilities of NiTi endodontic instruments processed by heat treatment. K3 files were heated for 30 min at 400 °C (group 400), 450 °C (group 450), and 500 °C (group 500). Files that were not heat treated served as controls. A cantilever-bending test was used to evaluate the changes in specimen flexibility caused by heat treatment. It was reported that in the cantilever-bending test, load values of the control group and group 500 were higher than those of groups 400 and 450 at the elastic range. At the SE range, the bending load of the control group was the highest among all groups. With regard to the shaping ability, in the control group, root canals at the apex were transported more to the outer side of the curvature compared with those of all heat-treated groups. Root canals of group 400 at 3 mm from the apex were transported less compared with those of other groups. No significant difference was found in the working time among the groups. In group 450, there was no plastic deformation or fracture of the file. Based on these results, it was concluded that heat treatment of files might improve their flexibility, making them more effective for preparation of curved canals.

As an alternative surface modification for NiTi alloys, thermal nitridation (to form TiN) has been investigated and developed. The nitriding method known as powder immersion reaction-assisted coating (PIRAC) produces TiN on NiTi [310]. The sequential processes are: (1) NiTi samples with a phase transformation temperature (at A_F of 15 °C) are annealed at 900 °C for 1.5 h, (2) 1,000 °C for an hour in sealed containers, and (3) nitrogen (N) atoms diffuse into the samples and atmospheric oxygen is stopped by a steel foil consisting of a notable amount of Cr. The modified surface consists of a thin outer layer of TiN and a thicker Ti_2Ni layer underneath [310, 311]. Shenhar et al. [312] studied the microhardness-microstructure and N concentration relationship for TiN-based PIRAC coatings grown at 850–1,100 °C. It was reported that a significant reduction in the friction coefficient and fretted areas was measured for the coated samples. Huang et al. [313] investigated the corrosion resistance and cell adhesion behavior of TiN-coated and ion-nitrided titanium substrates for dental applications. The TiN-coated specimen surface layer contained a TiN/Ti structure, whereas the ion-nitrided specimen contained a Ti_2N/TiN/Ti structure. It was mentioned that: (i) the polarization curves in artificial saliva (AS) showed that the corrosion rate and passive current for the specimens ranked as: untreated titanium > ion-nitrided titanium > TiN-coated titanium; (ii) the polarization resistance (R_P) obtained from the EIS ranked as: TiN-coated titanium > ion-nitrided titanium > untreated titanium; and (iii) after 24-h osteoblast-like U-2 OS cell incubation on the specimens, the attached cell number occurred in the order: TiN-coated titanium > ion-nitrided titanium > untreated titanium. The TiN-coating and ion-nitriding treatments can improve the corrosion resistance and cell adhesion behavior of titanium. Li et al. [314] investigated the effects of thermal nitriding surface modification of NiTi instruments on the wear resistance and cutting efficiency. In total, 25 ProFile 0.04 taper instruments were divided into five groups: two control groups and another three experimental groups that were subjected to thermal nitriding surface modification by treatment under three different conditions of 200, 250, and 300 °C. It was mentioned that: (i) nitrogen ions were successfully introduced onto the surface of NiTi instruments with gold appearance (indicating a formation of TiN), (ii) samples in the experimental groups showed increased wear resistance and cutting efficiency compared to the controls, and (iii) thermal nitriding surface treatments of NiTi rotary instruments produced higher wear resistance at a reasonable cost.

16.2.8.3 Thermomechanical treatment

Currently, the following types of endodontic instruments are on market (according to original NiTi types); conventional NiTi alloy (Mtwo, OS, ProFile, and PTU); EP NiTi alloy (RaCe, BioRaCe, iRace, F360, and F6 Skytaper); R-phase alloy (TF, TFA, K3XF); M-Wire alloy (PV, ProFile GT Series X, PTN, RPC, and WO); CM wire alloy (HFCM, TYP Infinite, Flex NiTi files, V-Taper 2H, and HFEDM); Gold heat-treated alloy (PTG); Blue heat-treated alloy (WOG, PV Blue, RPC Blue); and Max wire (XP-endo finisher, XP-endo shaper). Metallurgically, they can be classified into three phases: (1) austenitic phase (conventional NiTi alloy, EP NiTi alloy, M-wire, and R-phase), (2) martensitic phase (CM wire, Gold heat-treated alloy, and Blue heat-treated alloy), and (3) mixed phases (Max wire). Thermomechanically treated NiTi alloys have been reported to be more flexible with improved CFR and greater angle of deflection at failure when compared to conventional NiTi. These enhanced properties may be attributed to a modified phase composition containing varying amounts of R-phase and martensite. Endodontic instruments made of austenitic alloys possess SE properties because of stress-induced martensite transformation and consequently tend to spring back to their original form after deformation. In contrast, the martensitic instruments can easily be deformed due to the reorientation of the martensite variants and show a SME when heated. The use of martensitic alloy results in more flexible instruments, with an increased CFR compared to austenitic alloy [315]. Pereira et al. [316] compared physical and mechanical properties of one conventional and one thermomechanically treated NiTi wire used to manufacture rotary endodontic instruments. Two NiTi wires were characterized; one of them, C-wire, was processed in the conventional manner, and the other, termed M-wire received an additional heat treatment according to the manufacturer. It was reported that: (i) the two wires showed approximately the same chemical composition, close to the 1:1 atomic ratio, and the β-phase was the predominant phase present; (ii) B19′ martensite and the R-phase were found in M-wire, in agreement with the higher transformation temperatures found in this wire compared with C-wire, whose transformation temperatures were below the room temperature; (iii) average Vickers microhardness values were similar for M-wire and C-wire; and (iv) the stress at the transformation plateau in the tensile load–unload curves was lower and more uniform in the M-wire, which also showed the smallest stress hysteresis and apparent elastic modulus; concluding (v) the M-wire had physical and mechanical properties that can render more flexible and fatigue-resistant endodontic instruments compared to those made with conventionally processed NiTi wires. Zupanc et al. [315] concluded their extensive review by stating that: (i) the thermomechanical treatment of NiTi alloy allows a change in the phase composition leading to the appearance of martensite or R-phase under clinical conditions; whereas M-wire and R-phase instruments maintain an austenitic state, and CM wire and the Gold and Blue heat-treated instruments are composed of substantial amounts of martensite; (ii) the austenitic instruments possess SE properties and reveal high torque values at fracture, so that these files are appropriate to shape straight or slightly curved root canals; (iii) due to an increased amount of the martensite phase, martensitic instruments are more flexible with an enhanced resistance to cyclic fatigue and reveal a greater angle of rotation but lower torque at fracture; (iv) martensitic instruments should be preferred in cases of severely curved root canals or those with a double curvature; and (v) martensitic instruments are flexible enough to be bendable in advance, which can be useful when trying to bypass ledges. The performance and mechanical properties of NiTi instruments are influenced by factors such as cross-section, flute design, raw material, and manufacturing processes. Many improvements have been proposed by manufacturers during the past decade to provide clinicians with safer and more efficient instruments. In 2013, Shen et al. [317] mentioned that: (i) the mechanical performance of NiTi alloys is sensitive to their microstructure and associated thermo-mechanical treatment history; (ii) heat treatment or thermal processing is one of the most fundamental approaches toward adjusting the transition temperature in NiTi alloy, which affects the fatigue resistance of NiTi endodontic files; (iii) the NiTi instruments made from CM wire, M-wire, or R-phase wire represent the next generation of NiTi alloys with improved flexibility and fatigue resistance; (iv) the advantages of NiTi files for canal cleaning and shaping are decreased canal transportation and ledging, a reduced risk of file fracture, as well as faster and more efficient instrumentation; and (v) the clinician must understand the nature of different NiTi raw materials and their impact on the instrument performance because many new instruments are introduced on a regular basis. De-Deus et al. [318] evaluated the influence of Blue thermal treatment on the bending resistance and cyclic fatigue of conventional M-wire RPC files. Flexibility of standard RPC R25 files and the corresponding Blue prototypes was determined by 45° bending tests according to the ISO 3630-1 specification. Instruments were also subjected to CFR, measuring the time to fracture in an artificial SS canal with a 60° angle and a 5-mm radius of curvature. It was reported that: (i) RPC Blue instruments presented a significantly longer cyclic fatigue life and significantly lower bending resistance than the original RPC instrument; and (ii) regarding the roughness pattern, there was no significant difference between RPC Blue and the original RPC instruments, whereas RPC Blue revealed significantly lower microhardness than the original RPC instrument. It was then concluded that RPC Blue NiTi showed improved all-around performance when compared with the conventional M-wire SE NiTi, demonstrating improved flexibility and fatigue resistance, and reduced microhardness while maintaining similar characteristics of the surface.

16.2.8.4 Electropolishing

As Lopes et al. [319] evaluated the influence of surface grooves (peaks and valleys), resulting from machining during the manufacturing process of polished and unpolished NiTi BR4C endodontic files, on the fatigue life of the instruments, it was concluded that the depth of the surface grooves on the working part affected the NCF of the instruments tested; the smaller the groove depth, the greater is the NCF. The surface smoothness is an important factor to control the fatigue strength and life of NiTi instruments. After the machining process, instruments receive this surface treatment, which increases cutting efficiency while reducing defects resulting from the manufacturing process, thereby increasing fatigue resistance [61, 185]. Electropolishing is a standard surface treatment process employed as a final finish during manufacturing of NiTi instruments. An electric potential and current are applied, which result in ionic dissolution of the surface. In this process, the surface chemistry and morphology are altered, while surface imperfections are removed as dissolved metal ions.

During this process, anodic site of titanium is anodically oxidized to form TiO_2, which is a passive film; thus, it can protect the underlying material from further corrosion or oxidation. In this process, the amount of nickel on the surface decreases [185, 286, 320, 321]. The electrolyte for this process is most often concentrated acid solutions with a high viscosity, such as mixtures of sulfuric and phosphoric acid. Anderson et al. [185] mentioned that electropolishing may have beneficial effects in prolonging the fatigue life of NiTi endodontic instruments, due to the reduction in surface irregularities that serves as points for stress concentration and crack initiation. Bui et al. [322] investigated the effect of electropolishing PRIs on torque resistance, fatigue resistance, and cutting efficiency. It was concluded that: (i) electropolishing significantly reduced the resistance to cyclic fatigue but did not affect the torsional resistance; however, electropolishing reduced the angle at failure and amount of unwinding; and (ii) electropolishing did not significantly affect the cutting efficiency. However, the torque at failure and amount of unwinding were decreased. Cheung et al. [323] compared the LCF behavior of EP and non-EP NiTi instruments of the same design in hypochlorite. Forty-five EP and 62 non-EP NiTi engine files were subjected to rotational bending at various curvatures in 1.2% hypochlorite solution. It was reported that: (i) a linear relationship was found between LCF life and surface-strain amplitude for both groups, with no discernible difference between the two; (ii) no EP instrument showed more than one crack origin, significantly fewer than for the non-EP instruments; (iii) the square root of crack extension and strain amplitude were inversely related; and (iv) although surface smoothness is enhanced by electropolishing, this did not protect the instrument from LCF failure. Praisarnti et al. [324] examined the fatigue behavior, especially at the LCF region, of an experimentally EP FlexMaster and a commercial EP NiTi instrument (RaCe) in a corrosive environment. A total of 90 NiTi rotary instruments were subjected to rotational bending at various degrees of curvatures while immersed in 1.2% NaOCl solution until broken. The maximum surface strain amplitude, calculated from the curvature of the instrument and the diameter of the cross-section at break, was plotted against the LCF life. It was reported that: (i) the fatigue life of both instruments generally declined with increasing surface strain amplitude; there was a significant difference between the two instruments, and (ii) comparing the surface-treated FlexMaster with its commercially available non-EP counterpart, an improved resistance to fatigue breakage as a result of electropolishing was noted; concluding that the LCF life of a NiTi instrument rotating with a curvature in a corrosive environment is enhanced by electropolishing, and the design, both cross-sectional and longitudinal, appears to have an effect on the fatigue behavior of NiTi rotary instruments. Lopes et al. [325] evaluated the influence of electropolishing surface treatment on the NCF of BioRace rotary NiTi endodontic instruments. It was found that: (i) polished instruments displayed a significantly higher NCF when compared to nonpolished instruments; (ii) actually, NCF of a polished BR5C instrument was 124% higher than that of a nonpolished instrument; (iii) evaluation of the separated fragments after cyclic fatigue testing showed the presence of microcracks near the fracture surface; and (iv) polished instruments exhibited fine cracks that assumed an irregular path (zigzag crack pattern), whereas nonpolished instruments showed cracks running along the machining grooves; concluding that electropolishing surface treatment of BioRace endodontic instruments significantly increased the CFR. Condorelli et al. [326] assessed the failure mechanism of rotary NiTi instruments by chemical, structural, and morphological analyses to provide a rational explanation of the effects of surface and bulk treatments on their resistance to fatigue fracture. Thermal treatment (350–500 °C) was performed on EP and non-EP NiTi endodontic instruments. It was obtained that: (i) before thermal treatment, significant differences in the fatigue resistance between EP and non-EP instruments were attributed to differences in the surface morphology of the instruments; and (ii) the thermal treatment did not affect the surface morphology, however, resulted in significant changes in the instrument bulk with the appearance of an R-phase and an improved fatigue resistance; indeed after treatment at 500 °C, N_F increased up to 829 and 474 for EP and non-EP instruments, respectively; indicating that both thermal treatment and electropolishing improved the resistance of NiTi rotary instruments against fatigue fracture.

16.2.9 Sterilization

Reuse of endodontic instruments is common. During endodontic instrumentation, vital tissue, dentin shavings, necrotic tissue, bacteria, blood, blood byproducts, and other potential irritants are encountered with accumulation of the debris on the flutes of the instruments [327]. Transfer of these debris from patient to patient and to dental staff is highly undesirable, as these debris can act as antigens and infecting agents capable of transmitting diseases [328]. If the adequate infection control procedures were not performed, there was a high probability of transmitting pathogenic microorganism through endodontic instruments [329]. Sterilization is a process to render an object free from viable microorganisms, including bacterial spores and viruses. It is very important for endodontic instruments to be cleaned, disinfected, and sterilized effectively [330, 331]. Various methods for cleaning and sterilizing endodontic instruments have been proposed; including use of brushes and sponges, US cleaning, glutaraldehyde solution, NaOCl, glass-bead sterilization, dry heat, or steam sterilization [328, 332].

16.2.9.1 Cleaning efficiency and evaluation
Infection control procedures are essential for modern dental practice and are continually evolving to meet the dental profession's high standards. The procedures include assessing the efficacy of two cleaning procedures to reduce bacterial numbers on endodontic files and evaluating the effect of biological debris on the subsequent

sterilization of files [333]. SS and NiTi files were examined upon removal from the manufacturer's packaging, after instrumentation in root canals of human teeth inoculated with a broth containing two anaerobic species and one facultative anaerobic species of bacteria, and after instrumentation and cleaning with either an US bath or a thermal disinfector. For each file, the bacterial numbers were quantified using routine microbiological techniques in an anaerobic chamber. It was reported that: (i) no bacteria were detected from files direct from their packets, and (ii) no bacteria were detected from files that were subjected to steam sterilization, irrespective of the type of prior cleaning procedure; indicating that steam sterilization eliminated all bacteria from the endodontic files, irrespective of the presence of biological debris, and the majority of bacteria were eliminated from endodontic files after either US cleaning or using a thermal disinfector. Filho et al. [334] evaluated the efficacy of ultrasound in cleaning the surface of SS and NiTi endodontic instruments. Twenty NiTi instruments (10 Quantec files and 10 Nitiflex) and 20 SS-K (10 Maillefer-Dentsply and 10 Moyco Union Broach) were removed from their original packages and evaluated using a SEM. It was found that: (i) before cleaning, a greater amount of metallic debris was observed on the NiTi Quantec instruments, when compared to those made of SS, and (ii) statistical analysis showed that the use of ultrasound was effective for cleaning the instruments, regardless of the irrigating solution or the instruments type; concluding that the use of ultrasound proved to be an efficient method for the removal of metallic particles from the surface of SS and NiTi endodontic instruments.

There are several studies on cleaning efficacy by the autoclave method. Kumar et al. [335] determined the effectiveness of various disinfectants and sterilization techniques for disinfection and resterilization of dental burs and endodontic files. The materials used for the study were dental burs and endodontic files. Disinfectants used were: Quitanet plus, glutaraldehyde, glass-bead sterilizer, and autoclave. It was shown that: (i) the endodontic files and burs sterilized by autoclaving and glutaraldehyde showed complete sterilization, and (ii) burs and files immersed in glutaraldehyde (2.4%) for 12 h showed complete sterilization, whereas Quitanet plus solution and glass-bead sterilizer showed incomplete sterilization; indicating that autoclaving and glutaraldehyde (2.4%) showed complete sterilization, and other methods cannot be relied upon for sterilization. Spagnuolo [336] evaluated the effects of repeated autoclave sterilization cycles on surface topography of conventional NiTi and TiN-coated rotary instruments. A total of 60 NiTi rotary instruments, 30 ProTaper and 30 TiN-coated AlphaKite were analyzed. Instruments were evaluated in the as-received condition and after one, five, and 10 sterilization cycles. It was reported that: (i) SEM observations revealed the presence of pitting and deep milling marks in all instruments; (ii) EDS analysis confirmed that both types of instruments were composed mainly of NiTi, whereas AlphaKite had additional nitride; (iii) after multiple autoclave sterilization cycles, SEM examinations revealed an increase in surface alterations and EDS values indicated changes in chemical surface composition in all instruments; and (iv) Ra and RMS values of ProTaper significantly increased after five and 10 cycles with respect to the as-received instruments, whereas AlphaKite showed significant differences compared with the controls after 10 cycles. It was then concluded that multiple autoclave sterilization cycles modified the surface topography and chemical composition of conventional and TiN-coated NiTi rotary instruments. Razavian et al. [337] evaluated the effect of autoclave cycles on surface characteristics of S-file by SEM. Seventeen brand new S-files were used. The surface characteristics of the files were examined in four steps (without autoclave, one autoclave cycle, five autoclave cycles, and 10 autoclave cycles) by SEM. It was shown that new files had debris and pitting on their surfaces, and when the autoclave cycles were increased, the mean of surface roughness also increased at both magnifications; indicating that sterilization by autoclave increased the surface roughness of the files and this was directly related to the number of autoclave cycles.

16.2.9.2 Comparative studies among various sterilization methods

Luper et al. [338] investigated the effects of different sterilization methods on the fatigue life of finger pluggers. A total of 90 finger pluggers for each of four sizes (A, B, C, and D) were subdivided into subgroups of 10. Each subgroup was subjected to one, eight, or 15 cycles of steam autoclave, dry heat, or bead sterilization; 10 control pluggers for each size were not sterilized. After sterilization, experimental and control finger pluggers were subjected to cyclic bending until fracture. It was reported that: (i) only the "A" finger pluggers autoclaved for eight cycles had a significantly lower N_F compared with that of the controls, (ii) nine subgroups had significantly greater number of cycles before failure than did the control, and (iii) because all but one sterilized group had fatigue lifetimes statistically equal to or greater than nonsterilized controls, clinicians generally can use any of the three sterilization methods without fear of plugger failure. Linsuwanont et al. [339] evaluated an effective cleaning procedure for rotary NiTi endodontic instruments. New rotary instruments (ProFile size 25/0.04) were contaminated by preparing canals of extracted teeth. Three factors were evaluated to develop an effective cleaning sequence: dry or moist storage before cleaning; mechanical removal (brushing); and chemical dissolution in 1% NaOCl with ultrasonication. It was found that: (i) all new instruments showed metallic spurs and fine particulate debris on the surfaces; (ii) after contamination, brushing alone removed most particulate debris, but did not remove organic film; (iii) NaOCl effectively removed organic film; (iv) the sequential cleaning procedures (moist storage, brushing followed by immersion in 1% NaOCl, and US cleaning) completely removed the organic debris; (v) dry storage before cleaning or autoclaving with the present debris reduced cleaning effectiveness; and (vi) in three private practices, the cleaning protocol substantially reduced biological contamination, but complete cleaning was not always achieved (87% clean). Based on these findings, it was concluded that complete removal of organic debris from instruments is feasible using a combination of mechanical removal and chemical dissolution; however, it requires meticulous attention to details. Valois et al. [340] evaluated the surface of rotary NiTi files after multiple autoclave cycles. Two different types of rotary NiTi (Greater Taper and ProFile) were tested. It was mentioned that: (i) all topographic parameters were higher for both Greater Taper and ProFile after 10 cycles compared to the control, and (ii) ProFile also showed higher topographic parameters after five cycles compared to the control, indicating that multiple autoclave cycles increase the depth of surface irregularities located on rotary NiTi files. The efficacy of four methods of sterilizing endodontic instruments (autoclaving, carbon dioxide laser sterilization, chemical sterilization with glutaraldehyde, and glass-bead sterilization) were evaluated [341]. It was shown that: (i) the files sterilized by autoclave and lasers were completely sterile, and (ii) those sterilized using glass-bead were 90% sterile and those with glutaraldehyde were 80% sterile. Raju et al. [342] compared four different methods of sterilizing endodontic files in dental practice. Total 100 K-files, 21 mm long and of size 25 were divided into tow two groups, (1) 20 files were taken as control group and (2) the remaining 80 files were further divided into four groups of 20 files each, which further were tested for the efficacy of sterilization with different methods such as autoclave, glass-bead, glutaraldehyde, and CO_2 laser. It was obtained that (i) the files sterilized with autoclave and lasers were completely sterile and (ii) those sterilized with glass-bead were 90% sterile and those with glutaraldehyde were 80% sterile; indicating that autoclave or laser could be used as a method of sterilization in clinical practice and in advanced clinics and laser can be used also as a chair side method of sterilization.

The four recommended methods of sterilizing endodontic files in dental practice are: (1) autoclaving, (2) carbon dioxide laser sterilization, (3) chemical sterilization (with glutaraldehyde), and (4) glass-bead sterilization. It was concluded that autoclaving and exposing to laser give complete sterilization, whereas glass-bead sterilizer can be used as an alternative when these two methods are not available. Although autoclave is an effective method for sterilizing endodontic files, the time taken to sterilize is more. Shaha et al. [343] proposed a new method in which the files are initially soaked in gauze square for maximum 10 min, followed by US cleaning, and the files are placed in the sterilization device especially designed for endodontic files so that they are exposed to ultraviolet light for specific recommended period. The efficiency of the new method was compared with that of autoclave using culture method. It was concluded that: (i) its efficacy was equal to that of autoclave; (ii) considering the time consumed by both the methods, cleaning followed by autoclave requires 20–30 min, whereas, the new method requires 5–15 min; and (iii) the novel method can be done chair-side in between two consecutive appointments, indicating that the new device can help the clinician to work more efficiently in aseptic conditions. The sterilization efficacy of the rotary endodontic files using two methods of sterilization (autoclave and chemical sterilization using two solutions [CHX and glutaraldehyde]) was evaluated [344]. Sixty new presterilized rotary NiTi files contaminated with commercially available *Bacillus stearothermophillus* were used. The files were sterilized by the two methods and checked for sterility by incubating the files in test tubes containing thio-glycollate medium. It was obtained that: (i) the files sterilized by autoclave method were 100% sterile, and (ii) those sterilized with CHX showed 87% sterility and with glutaraldehyde showed only 60% sterility; concluding that the autoclave could be used as best reliable method of sterilization. Yenni et al. [345] compared the effectiveness of four different methods of sterilizing contaminated endodontic files (Group A: autoclave, Group B: glass-bead sterilization, Group C: glutaraldehyde, and Group D: Quitanet Plus [aldehyde-free solution]). A total of 48 SS-K were divided into four groups based on the sterilization method followed. In all the tested groups, half of the files were contaminated with *Escherichia coli* and remaining with *Enterococcus faecalis*. Further on, presterilization colony counts were recorded, followed by sterilization through respective methods. Later, the sterilized files were rinsed with distilled water and 100 l of the diluted concentration was transferred and cultured into the respective agar plates to determine the total microbial reduction. It was found that autoclave showed complete effectiveness in reducing the microbial count followed by Quitanet Plus, glass-bead sterilizer, and glutaraldehyde; indicating that autoclave can be considered as the best sterilization technique to prevent cross-infection in endodontic therapy.

16.2.9.3 Sterilization effects on cutting efficiency

Morrison et al. [346] evaluated the effects of steam sterilization and usage on sharpness on #25 endodontic files. Files were used to instrument one, five, and 10 molars. The control groups determined the effect of steam sterilization alone on cutting efficiency of unused files. A cutting efficiency test was conducted on an apparatus that compares sharpness of files when used in linear motion. It was reported that: (i) significant differences were found between experimental files used to instrument one molar and those used for five or 10 molars; (ii) the difference in the cutting efficiency between the second and third experimental groups was not significant, indicating that most of the decrease in sharpness occurred with use between one and five molars; and (iii) no significant difference was found between the control groups, indicating no decrease in cutting efficiency by sterilization alone. Schafer et al. [149] investigated alterations in the cutting efficiency when conventional and TiN-coated NiTi K-files were exposed to repeated sterilization using an autoclave. A total of 96 NiTi K-files were randomly divided into two groups (A and B) of 48 instruments each. The instruments of group B were exposed to PVD creating a coating of a TiN layer, whereas the files of group A were not coated. The instruments of groups A and B were randomly divided into four subgroups of 12 instruments each; (1) A.1/B.1: Instruments were exposed to five cycles of sterilization. (2) A.2/B.2: Instruments were exposed to 10 cycles of sterilization. (3) A.3/B.3: Instruments were immersed in NaOCl for 30 min, rinsed in water, and exposed to five cycles of sterilization. (4) A.C/B.C: Instruments were not sterilized (controls). The cutting efficiency of all files was determined by means of a computer-driven testing device. It was reported that: (i) the TiN-coated instruments of groups B.1, B.2, and B.3 showed no significant difference compared to the penetration depths of the controls; and (ii) the uncoated files of groups A.1, A.2, and A.3 displayed significantly lower maximum penetration depths when compared to the control files; concluding that repeated sterilization under autoclave or exposure to NaOCl prior to sterilization did not alter the cutting efficiency of PVD-coated NiTi K-files.

16.2.9.4 Effects on mechanical properties

The ability of heat treatment as a result of autoclave sterilization to extend the life of NiTi rotary endodontic instruments by reducing the effect of cyclic fatigue was evaluated [347], and it was concluded that the heat treatment as a result of autoclave sterilization does not extend the useful life of NiTi instruments. Viana et al. [348] evaluated the effect of repeated sterilization cycles on the mechanical behavior and fatigue resistance of rotary endodontic NiTi instruments in a dry oven or autoclave. It was reported that: (i) sterilization procedures resulted in no significant changes in Vickers microhardness, nor in the parameters describing the mechanical behavior of the wires; however, the N_F was statistically higher for all instruments after dry heat or autoclave sterilization cycles; and (ii) in the instruments previously fatigued to one-half of their fatigue life, autoclave sterilization gave rise to an increase of 39% in the remaining N_F. It was then concluded that the changes in the mechanical properties of NiTi endodontic instruments after five cycles of commonly used sterilization procedures were insignificant, and the sterilization procedures are safe as they produced a significant increase in the fatigue resistance of the instruments. Bergeron et al. [349] assessed multiple autoclave cycle effects on cyclic fatigue of GT Series X files and TFs. A jig using a 5-mm radius curve with 90° of maximum file flexure was used to induce cyclic fatigue failure. Files ($n = 10$) representing each experimental group (GT Series X 20/0.04 and 20/0.06; TFs 25/0.04 and 25/0.06) were first tested to establish the baseline mean cycles to failure (MCF). Experimental groups ($n = 20$) were then cycled to 25% of the established baseline MCF and then autoclaved. Additional autoclaving was accomplished at 50% and 75% of MCF, followed by continual testing until failure. Control groups ($n = 20$) underwent the same procedures, except autoclaving was not accomplished. It was reported that: (i) the GT Series X (20/0.04 and 20/0.06) files showed no significant difference in MCF for experimental versus control files, and (ii) the TFs (25/0.04) showed no significant difference in MCF between experimental and control groups; however, the TFs (25/0.06) experimental group showed a significantly lower MCF compared to the controls. Based on these results, it was concluded that autoclave sterilization significantly decreased CFR of one of the four file groups tested, and repeated autoclaving significantly reduced the MCF of 25/0.06 TFs; however, 25/0.04 TFs and both GT Series X files tested were not significantly affected by the same conditions. Plotino et al. [350] evaluated the effect of autoclave sterilization on CFR of rotary endodontic instruments made of traditional and new NiTi alloys. Four NiTi rotary endodontic instruments (K3, Mtwo, Vortex, and K3 XF prototypes) were tested. The sterilized instruments were subjected to 10 cycles of autoclave sterilization. Twelve files from each different subgroup were tested for CFR. Means and standard deviations of N_F and fragment length of the fractured tip were calculated for each group. It was obtained that: (i) comparing the results between unsterilized and sterilized instruments for each type of file, differences were statistically significant only between sterilized and unsterilized K3XF files (762 vs 651 NCF); (ii) the other instruments did not show significant differences in the mean NCF as a result of sterilization cycles (K3, 424 vs 439 NCF; Mtwo, 409 vs 419 NCF; Vortex, 454 vs 480 NCF); and (iii) comparing the results among the different groups, K3 XF (either sterilized or not) showed a mean NCF significantly higher than all other files. It was, therefore, concluded that the repeated cycles of autoclave sterilization do not seem to influence the mechanical properties of NiTi endodontic instruments, except for the K3 XF prototypes of rotary instruments that demonstrated a significant increase in CFR. Zhao et al. [351] compared the CFR of HFCM, TFs, K3XF, Race, and K3, and evaluated the effect of autoclave sterilization on the CFR of these instruments both before and after the files were cycled. It was found that: (i) HFCM, TFs, and K3XF had significantly higher CFR than Race and K3 in the unsterilized group 1; (ii) autoclave sterilization significantly increased the MCF of HFCM and K3XF ($p < 0.05$) both before and after the files were cycled; and (iii) SEM examination revealed a typical pattern of cyclic fatigue fracture in all instruments. Based on these results, it was indicated that HFCM, TFs, and K3XF instruments, composed of new thermal-treated alloy, were more resistant to fatigue failure than Race and K3, and autoclaving extended the cyclic fatigue life of HFCM and K3XF.

The effects of sterilization on torsional behaviors are also investigated. Iverson et al. [352] studied the effects of different sterilization methods on the torsional strength of two types of endodontic files. Kerr K-Flexofiles and Burns Unifiles were subjected to 10 cycles of autoclaving, bead, dry heat, and cold chemical sterilization. A torquemeter was used to determine the degrees of revolution and torque to failure for the files after repeated sterilization. It was obtained that: (i) repeated sterilization was found to have no effect on the torque resistance and degrees to failure for the Burns Unifile and no effect on degrees to failure for the K-Flexofile; (ii) dry heat and cold sterilization, however, slightly increased the torque resistance of the K-Flexofiles compared to the controls and other sterilization methods; and (iii) no differences were found among the files in the degrees of rotation to failure but the K-Flexofiles had a slightly greater torque resistance. Silvagio et al. [353] determined whether heat sterilization adversely effects the torsional properties of rotary NiTi files, making them more prone to fracture under torsional stress. A total of 900 sizes 2–10 Profile Series 29 0.04 taper files were divided into groups of 10 files each and sterilized 0, 1, 5, or 10 times in the steam autoclave, Statim autoclave, or dry heat sterilizer. They were subjected to torsional testing in a Torquemeter Memocouple. It was reported that: (i) 10 significant changes occurred for torsional strength and 10 for rotational flexibility; (ii) eight of 10 changes in the torsional strength were increased; and (iii) 52 of 54 (96.3%) comparisons for torsional strength and 47 of 54 (87%) for rotational flexibility showed a significant increase or no change; concluding that although heat sterilization of rotary NiTi files up to 10 times, it does not increase the likelihood of instrument fracture. Canalda-Sahli et al. [354] assessed the effect of dry heat or autoclave sterilization on the resistance to fracture in torque and angular deflection and the resistance to bending of K-type files made of NiTi (Nitiflex, Naviflex), titanium (Microtitane), or SS (Flexofile, Flex-R). Ten K-type files of each sort, from size 25 to 40, were tested. It was reported that: (i) sterilization with dry heat and autoclave slightly decreased the flexibility of files made of SS and NiTi for most of the sizes; (ii) the files made of titanium showed an increased flexibility after sterilization with autoclave and dry heat; (iii) resistance to fracture varied among the five groups of files tested as follows: it decreased in some sizes of SS instruments, decreased in all sizes of titanium files assessed by the torsional moment, and either increased or decreased in some sizes of NiTi files; and (iv) all files tested, however, satisfied relevant standards for angular deflection after being subjected to sterilization with an autoclave or dry heat. Hilt et al. [355] tested the hypothesis that multiple sterilizations of endodontic SS and NiTi files lead to a continuous decrease in the resistance of files to separation by torsion. A total of 100 SS and 100 NiTi #30 K-type files were divided into 20 groups of 10 and sterilized in increments of 10 cycles, using a full cycle and a fast cycle autoclave. These files were tested by twisting each of them in a CW direction until fracture (torque g-cm). Results indicated that neither the number of sterilization cycles nor the type of autoclave sterilization used affects the torsional properties, hardness, and microstructure of SS and NiTi files. Casper et al. [356] compared the effects of multiple autoclaving cycles on the torsional load resistance of three new rotary endodontic files, including M-wire (PV), TF, and CM wire. PV, TF, and CM files ($n = 100$; size 25/0.04) were divided into five groups ($n = 20$). Files were steam autoclaved for one, two, three, and seven sterilization cycles. It was reported that: (i) autoclave cycles had no significant overall effect on the file performance for any of the instrument systems tested; (ii) PV and CM displayed significantly greater resistance to torsional load than TF, but were not different from each other; and (iii) angular deflection values for TF and CM were significantly higher than for PV, with TF demonstrating greater rotational distortion than CM. Based on these findings, it was concluded that repeated steam autoclaving did not affect torsional resistance for unused files of the systems evaluated, and CM wire files might have a combined advantage of greater torsional strength and high deformation before failure.

16.2.10 Technical-sensitive endodontic treatment

Endodontics as a field of dentistry has made giant leaps in the past two to three decades [357]. Pioneering technological advancements include magnification, innovative material science, designs, and techniques for instrumentation, as well as obturation of the root canal systems. In contrast to this ascent in endodontic material, technique, and equipment innovations, a number of treatment outcome data reveal no statistically significant improvement in the overall endodontic success in the corresponding period [357–360].

Tickle et al. [361] described the quality and record the outcomes of root canal therapy on mandibular, first permanent molar teeth provided by general dental practitioners working according to the National Health Service (NHS) contracts. All patients aged 20–60 years attending the practices, had received an NHS-funded root filling in a mandibular first permanent molar between January 1998 and December 2003. The radiographic quality of root fillings in the teeth was assessed by an endodontic specialist and categorized into optimal, suboptimal, and teeth which had no radiograph or an unreadable radiograph. It was reported that: (i) 174 teeth were included in the study, of which 16 failed; (ii) the crude failure rates per 100 years with a root-filled tooth were very low and differed little for optimally (2.6), suboptimally (2.5) root-filled teeth, and for those with no or an unreadable radiograph (2.9), with approximately one in 37 root-filled mandibular first molar teeth failing each year; (iii) the majority of root fillings fail within the first 2 years (N = 10, 62.5%); and (iv) around 67 teeth (38.5%) were restored with a crown, none of which failed during the follow-up period compared to those with a plastic restoration. It was then concluded that the very low failure rates have significant implications for the design of research studies investigating the outcomes of endodontic therapy, and similar failure rates for teeth that had optimal and suboptimal root fillings suggest that endodontic treatment is not as technique sensitive as previously thought.

Even the endodontic treatment was considered as not technique sensitive [361], there are several clinical evidences indicating that the endodontic treatment is a technique-sensitive treatment, as Parashos et al. [179] indicated that the most important influence on defect rates was the operator, which may be related to clinical skill or a conscious decision to use instruments a specified number of times or until defects were evident.

16.2.10.1 Undergraduate students
Pettiette et al. [255] let inexperienced dental students use NiTi 0.02 hand files and SS-K to compare the 1-year success rate of endodontic treatment of the same teeth. Twelve-month follow-up radiographs were compared with the immediate follow-up radiographs. Both sets of radiographs were taken with the same customized stent. Sixty-seven percent of the patients returned with the 12-month follow-up radiographs

(19 NiTi vs 21 SS-K). It was reported that: (i) immediate postoperative periapical status was found to be similar, (ii) teeth instrumented with the NiTi files demonstrated a higher mean change in densitometric ratio compared with SS-K files, and (iii) further tests of success (values: > 0) and failure (value: < 0) with the Fisher-exact test showed more success (decreasing radiographic density) with NiTi files and more failures (increasing radiographic density) with SS-K files ($p < 0.03$); indicating that maintaining the original canal shape after instrumentation leads to a better prognosis of endodontic treatment. Sonntag et al. [362] investigated the root canal shaping with manual and rotary NiTi files performed by dental undergraduate students. Thirty undergraduate dental students prepared 150 simulated curved root canals in resin blocks with manual NiTi files with a stepback technique and 450 simulated curved canals with rotary NiTi files by crown-down technique. Incidence of fracture, preparation length, canal shape, and preparation time were investigated. Questionnaires were then issued to the students for them to note their experience of the two preparation methods. It was reported that: (i) Zips and elbows occurred significantly less frequently with rotary than with manual preparation. (ii) The correct preparation length was achieved significantly more often with rotary files than with manual files. (iii) Instrument fractures were recorded in only 1.3% of cases with both rotary and manual preparations. The mean time required for manual preparation was significantly longer than that required for rotary preparation. (iv) Prior experience with a hand preparation technique was not reflected in an improved quality of the subsequent rotary preparation. (v) Approximately 83% of the students claimed to have a greater sense of security in rotary than in manual preparation. (vi) Overall, 50% felt that manual and engine-driven preparation should be given equal status in undergraduate dental education. Based on these results, it was concluded that inexperienced operators achieved better canal preparations with rotary instruments than with manual files, and no difference in the fracture rate was recorded between the two systems.

Peru et al. [363] evaluated root canals instrumented by dental students using the modified double-flared technique, NiTi rotary system GT files, and NiTi rotary ProTaper files by MCT. A total of 36 root canals from 18 mesial roots of mandibular molar teeth were prepared; 12 canals were prepared with the modified double-flared technique, using K-flexofiles and Gates-Glidden burs; 12 canals were prepared using system GT; and 12 using ProTaper rotary files. It was reported that: (i) at the coronal and mid-root sections, System GT and ProTaper files produced significantly less enlarged canal cross-sectional area, volume, and perimeter than the modified double-flared technique; (ii) in the mid-root sections, there was significantly less thinning of the root structure toward the furcation with system GT and ProTaper; (iii) the rotary techniques were both three times faster than the modified double-flared technique; (iv) qualitative evaluation of the preparations showed that both ProTaper and system GT were able to prepare root canals with little or no procedural error compared to the modified double-flared technique; and (v) inexperienced dental students were able to prepare curved root canals with rotary files with greater preservation of tooth structure, low risk of procedural errors, and much quicker than with hand instruments. Shen et al. [364] analyzed the incidence and mode of ProFile IS during a predefined schedule of clinical use by the undergraduate students in a dental school for over 4 years. A total of 3,706 ProFile instruments discarded from the same undergraduate students' program between 2003 and 2007 were analyzed. The lateral and fracture surfaces of 12 separated instruments were examined by SEM, and the location of the fractures was recorded. It was found that: (i) the overall proportion of instrument defects was 1.3%; deformation without fracture occurred in 1%, and separation in 0.3%; (ii) the majority of instrument defects occurred in size 20 (34/48); and (iii) the ProFile instruments (10/12) failed mostly because of shear stress, whereas only two failed because of fatigue fracture; indicating that NiTi rotary instrument system was successfully introduced into an undergraduate endodontic program, and small-size files should be considered as single-use, disposable instruments because of the higher possibility of torsional deformation. Knowles et al. [365] assessed 3,543 canals treated in a 24-month interval by undergraduate students. The separation rate was 1.3% of LightSpeed instruments. Hanni et al. [366], in 87 cases among 40 undergraduate students, showed no file separation; the number of uses in this study is not specified. It is important to emphasize that those undergraduate students had an intense preclinical training, having opportunity to separate instruments without clinical consequences. A recent study has shown no IS for NiTi rotary and reciprocating instruments [367]. The cases were performed by third and fourth year undergraduate students. The study followed the American Association of Endodontists selection case guidelines, that is, the 715 cases treated were primary treatment of teeth, presenting no complex anatomies.

Martins et al. [368] evaluated the perceptions of Brazilian undergraduate dental students about the endodontic treatments performed using NiTi rotary instruments and hand SS. Data were collected using a questionnaire administered to the undergraduate dental students enrolled in endodontic disciplines. The students were divided into three groups: (1) Group 1, students who had treated straight canals with SS hand instruments; (2) Group 2, students who had treated curved canals with SS hand instruments; and (3) Group 3, students who had treated both straight and curved canals with NiTi rotary instruments. The number of endodontic treatments performed, types of treated teeth, students' learning, time spent, encountered difficulties, quality of endodontic treatment, and characteristics of the employed technique were analyzed. There was a 91.3% rate of return for the questionnaires. It was reported that: (i) mandibular molars were the most frequently treated teeth, followed by maxillary incisors; (ii) the Kruskal–Wallis test showed no differences in learning or in the characteristics of the technique used among the three groups; (iii) Group 3 students performed a greater number of endodontic treatments in a smaller time than Groups 1 and 2 students; (iv) difficulties were reported primarily by students in Groups 2 and 3 compared to Group 1; (v) the quality of endodontic treatments differed only between Group 1 and Group 2; and (vi) the use of NiTi rotary instruments should be included in undergraduate dental curriculum, contributing to the increase of patients assisted and consequently to improve the clinical experience of the students. Jungnickel et al. [369] evaluated the factors associated with the treatment quality of ex-vivo root canal treatments performed by undergraduate dental students using different endodontic treatment systems. Four students performed root canal treatment on 80 extracted human teeth using four endodontic treatment systems in designated treatment order following a Latin square design. Lateral seal and length of root canal fillings were radiographically assessed; for lateral seal, a graded visual scale was used. Treatment time was measured separately for access preparation, biomechanical root canal preparation, obturation, and for the total procedure. It was found that: (i) the use of machine-driven NiTi systems resulted in overall better quality scores for lateral seal than the use of the manual SS system; (ii) among systems with machine-driven files, scores did not significantly differ; (iii) the use of machine-driven instruments resulted in shorter treatment time than manual instrumentation; (iv) machine-driven systems with few files achieved shorter treatment times; and (v) with increasing number of treatments, root canal-filling quality increased and the treatment time decreased; a learning curve was plotted. Based on these findings, it was concluded that the use of endodontic treatment systems with machine-driven files led to higher quality lateral seal compared to the manual system; the three contemporary machine-driven systems delivered comparable results regarding the quality of root canal fillings, they were safe to use and provided a more efficient workflow than the manual technique; and (viii) increasing experience had a positive impact on the quality of root canal fillings while treatment time decreased.

Georgelin-Gurgel et al. [370] evaluated the prevalence of iatrogenic events during preclinical teaching of endodontics, comparing manual SS versus NiTi rotary techniques for shaping natural root canals. Two groups of 13 inexperienced dental students were randomly made up and asked to shape 104 canals in natural teeth. Group R used NiTi rotary files for shaping, while Group M used a sequence of five manual SS files. Occurrence of file breakage, loss of WL, and iatrogenic instrumentation on apical foramina were evaluated. It was found that: (i) the overall occurrence of adverse events during shaping did not differ between the groups, being 58% in Group R and 51% in Group M; (ii) inter-group distribution of type of event differed significantly; (iii) file breakage (7.7%) and loss of WL of >2 mm (6.7%) occurred only in Group R; and (iv) iatrogenic shaping on apical foramina showed the same frequency in each group. It was then concluded that manual instrumentation is safer than rotary instrumentation in the hands of inexperienced students, and acquiring skill in the use of NiTi rotary instrumentation requires specific preclinical training to avert file breakage. These findings argue for the rethinking of theoretical and practical coursework in endodontics teaching, especially in dentistry schools where students are required to treat patients during their training. Arbab-Chirani et al. [371] evaluated the impact of rotary NiTi instruments on undergraduate teaching and clinical use in French dental schools and to evaluate the impressions of dental students when learning and using these techniques. A questionnaire was mailed to all 16 French undergraduate dental schools. Data were gathered on a range of issues concerning teaching and use of NiTi endodontic techniques. It was reported that, based on 100% response rate, (i) the need for teaching NiTi techniques to dental students was agreed by all schools; (ii) lectures and laboratory courses for rotary NiTi techniques were organized in all schools; (iii) in 13 of the 16 teaching hospitals, students could use rotary NiTi techniques for canal preparation under the supervision of teaching staff; (iv) similarities were observed in the majority of responses, for example, type of rotary system taught and used clinically; and (v) some differences were also observed, for example, the association of hand files to rotary NiTi instruments and at what stage in the undergraduate curriculum the rotary instruments were introduced. Based on the responses, it was indicated that there was a national consensus over the need for undergraduate teaching of rotary NiTi systems in France, (vii) these techniques had made a substantial impact in endodontic teaching and were widely taught and used in French dental schools. Di Fiore et al. [44] prospectively determined the incidence of NiTi rotary instrument fracture in an endodontic clinical practice setting. Eleven second year endodontic residents, using four NiTi rotary instrument systems (ProFile, ProTaper, GTRotary, and K3Endo) according to the recommendations of the manufacturers, instrumented 3,181 canals in 1,403 teeth of 1,235 patients, in a dental school postgraduate endodontic clinic, in 1 year. The incidence of instrument fracture was determined based on the number of instruments used. When fracture occurred, data were collected concerning the type, size, taper, and prior use of the fractured instruments, as well as the length and location of the fragment within the root canal and the curvature of the canal. It was reported that: (i) the overall incidence of instrument fracture was 0.39%; (ii) the incidence of fracture for ProFile, ProTaper, GTRotary, and K3Endo files was 0.28, 0.41, 0.39, and 0.52%, respectively; (iii) there was no statistically significant difference between instrument systems; (iv) the percentage of teeth in which instruments fractured was 1.9% (0.28% for anterior teeth, 1.56% for premolars, and 2.74% for molars); (v) a total of 26 instruments fractured, of which 23 had tapers of 0.06 or greater; and (vi) most of the fragments were located in the apical third of the root canal, and both the median and mode among the fragment lengths were 2 mm; concluding that the low incidence of NiTi rotary instrument fracture supports the continued use of these instruments in the root canal treatment.

16.2.10.2 Experienced students and practitioners

Moving on from inexperienced dental students to experienced (to some extent) senior dental students and practitioners. Pettiette et al. [372] compared the effect of the type of instrument used by dental senior students on the extent of straightening and on the incidence of other endodontic procedural errors. NiTi 0.02 taper hand files were compared with traditional SS 0.02 taper K-files. Sixty molar teeth comprised of maxillary and mandibular first and second molars were treated by senior dental students. Instrumentation was with either NiTi hand files or SS-K files. Preoperative and post-operative radiographs of each tooth were taken using an XCP precision instrument with a customized bite block to ensure accurate reproduction of radiographic angulation. The radiographs were scanned and the images stored as TIFF files. By superimposing tracings from the preoperative over the postoperative radiographs, the degree of deviation of the apical third of the root canal filling from the original canal was measured. The presence of other errors, such as strip perforation and instrument breakage, was established by examining the radiographs. It was reported that: (i) in curved canals instrumented by SS-K files, the average deviation of the apical third of the canals was 14.44° (± 10.33°); (ii) the deviation was significantly reduced when NiTi hand files were used to an average of 4.39° (± 4.53°); and (iii) the incidence of other procedural errors was also significantly reduced by the use of NiTi hand files. Kfir et al. [373] compared procedural errors that occur in patients during root canal preparation by senior dental students using a new "eight-step method" versus the traditional "serial step-back technique." Senior dental students treated 221 root canals of maxillary and mandibular teeth. Instrumentation included coronal flaring with Gates-Glidden reamers and standardized SS-K files in all teeth. A new eight-step method was used to prepare 67 canals using standardized SS hand instruments (eight-step SS) and 69 canals using the rotary NiTi instruments (eight-step NiTi). The traditional serial step-back technique (step-back) was used for 85 root canals. In the apical third, reaming or filing motions were used up to sizes 25 and only reaming motion in sizes larger than 25 was used with the new 8-step method. A filing motion was used in the step-back for all sizes. Root canals of all groups were obturated with gutta-percha points and AH26 using a lateral condensation technique. Pre- and post-operative radiographs were taken of each tooth. Procedural errors were recorded and statistically analyzed using a binomic test for comparison of proportion. It was found that: (i) procedural errors detected consisted of two canals with transportation (3%) with the eight-step SS, and three canals (4%) with transportation with eight-step NiTi, (ii) there were no canal obstructions or ISs, (iii) with the step-back, 20 canals were transported (24%), seven canals had obstructions (8%), and in one canal instrument was separated (1%); concluding that the new eight-step method resulted in fewer procedural errors than the traditional serial step-back technique when senior students prepared root canals in patients either by hand with standardized K-files or by rotary NiTi instrumentation.

Iqbal et al. [374] investigated the incidence of hand and rotary IS in the endodontics graduate program at the University of Pennsylvania between 2000 and 2004. It was mentioned that: (i) in 4,865 endodontic resident cases, the incidence of hand and rotary IS was 0.25% and 1.68%, respectively; (ii) the odds for rotary IS were seven times more than for hand IS; (iii) the probability of separating a file in apical third was 33, and six times more likely when compared to coronal and middle thirds of the canals; (iv) the highest percentage of IS occurred in mandibular (55.5%) and maxillary (33.3%) molars; (v) furthermore, the odds of separating a file in molars were 2.9 times greater than premolars; (vi) among the ProFile series 29 rotary instruments, the 0.06 taper #5 and #6 files separated the most; and (vii) there was no significant difference in IS between the use of torque controlled versus nontorque-controlled handpieces, nor between first and second year residency. Mesgouez et al. [375] determined the influence of operator experience on the time needed for canal preparation when using a rotary NiTi system. A total of 100 simulated curved canals in resin blocks were used. Four operators prepared a total of 25 canals each. The operators included practitioners with prior experience of the preparation technique and practitioners with no experience. The WL for each instrument was precisely predetermined. All canals were instrumented with rotary NiTi PV Taper Series 29 engine-driven instruments using a high-torque handpiece. The time taken to prepare each canal was recorded. Significant differences between the operators were analyzed using the Student's t-test and the Kruskal–Wallis and Dunn nonparametric tests. It was reported that: (i) comparison of canal preparation times demonstrated a statistically significant difference between the four operators; (ii) in the inexperienced group, a significant linear regression between canal number and preparation time occurred; suggesting that the time required for canal preparation was inversely related to the operator experience. Al-Omari et al. [376] determined the influence of operator experience on the shaping ability of ProFile and K3 NiTi rotary instruments in simulated root canals. A total of 160 simulated canals consisting of four different shapes in terms of angle and position of curvature were prepared by experienced and inexperienced operators. One experienced operator prepared 80 canals and two inexperienced operators prepared 80 canals with a crown-down technique using either ProFile or K3 0.06 taper instruments. Images of the canals were taken, using a camera attached to a computer with image analysis software, before surgery and after preparation to sizes 20, 25, and 30 to WL. Postoperative images were combined with the preoperative image to highlight the amount and position of material removed during preparation, as well as the shape of the prepared canal. It was found that: (i) overall, there was a highly statistically significant difference between the instruments for the time taken to prepare the canals, with K3 instruments taking a mean of 4.9 min and ProFile of 6 min; (ii) of the six instruments fractured (three in each operator category); four were ProFile instruments; (iii) our instruments deformed, all with the inexperienced operators; three were K3; (iv) no perforations or zips were observed; however, one danger zone (created by the experienced operator using K3 instruments) and one ledge (created by an inexperienced operator using K3 instruments) were created; (v) 12 outer widenings were created with a highly significant difference between the operator and instrument used; and (vi) there was a highly significant difference by instrument, and experience regarding absolute transportation at the beginning of the curve and a statistically significant difference for the instrument used regarding absolute transportation half way to the orifice. Based on these results, it was concluded that the experienced operator prepared canals more quick and safe than the inexperienced operators when using K3 instruments; both used ProFile instruments quickly and safely; moreover, inexperienced operators should be advised to train using less aggressive instruments and when confident could progress to other instrument designs. Graduate students were assessed by Shen et al. [377], when the WO reciprocating instruments were used. That study showed that single-use of WO instruments separated in one of 85 cases (1.17%) and one of 90 cases (1.11%), in one of the specialist clinics assessed. Three endodontic specialists' clinics had no separation for the WO system.

Marending et al. [378] evaluated the impact of patient-related, tooth-related, and treatment-related factors on the therapy outcome in a series of consecutive patients. A total od 84 patients were included. Of these, 66 (79%) were available for recall after > 30 months (mean = 46 months). Root canal treatments were performed using a standard protocol. At recall, teeth were scored by means of the periapical index (PAI), which was the dependent variable (dichotomized to sound/unsound). Explanatory variables were patient age, integrity of the nonspecific immune system, smoking status, dichotomized PAI score before treatment, initial treatment versus retreatment, prior exposition of the root canal to saliva, SS hand versus NiTi rotary instrumentation, and quality of root filling. Data were analyzed using univariate tests and backward stepwise logistic regression analysis. It was found that: (i) after five steps with elimination of the least significant independent variable, status of the immune system, initial PAI (Pp = 0.04), and root-filling quality (p = 0.01) were found to be the indispensable predictors for the treatment outcome; (ii) using these three explanatory variables, the logistic regression model had a predictive value of 87%, compared to 91% with all eight variables; and (iii) the success rate at recall (PAI < 2 without symptoms) was 88% (95% CI: 78, 94); concluding that the integrity of a patient's nonspecific immune system, which had been neglected in earlier investigations, is a significant predictor for endodontic treatment outcome and should receive more attention in the future studies. Some other studies assessed the separation rate of instruments after clinical use by endodontic specialists. Satappan et al. [379] evaluated the Quantec NiTi rotary instruments collected during a 6-month interval. The separation rate registered was 21%. However, the authors could neither assure that the full sequence of instruments was used, nor the number of uses for each instrument was provided. An important aspect of this study was that the instruments were discarded when the cutting efficiency was clinically observed or when signs of distortion were noted. Other studies that evaluated the separation rate of rotary NiTi instruments presented better outcomes such as 1.98% for the Mtwo system after a glide path creation [380], 3.7% for Light Speed Instruments [381], 2.4% for ProTaper instruments [382], and 0.83% for TF Adaptive [383]. Three studies presented similar results with reciprocating instruments used by Endodontic specialists [384–386]. The rate of IS was low for both WO and RPC instruments evaluated in these recent studies. Berutti et al. [387] evaluated the influence of manual preflaring and torque on the failure rate of rotary NiTi Pro-Taper instruments Shaping 1 (S1), Shaping 2 (S2), Finishing 1 (F1), and Finishing 2

(F2). These factors were evaluated using an in vitro method by calculating the mean number of endo-training blocks shaped before file breakage under different conditions. Group A (S1 on simulators with no preflaring) shaped 10 blocks before failure, group B (S1 on manually preflared simulators) shaped 59 blocks ($p < 0.01$ vs group A), group C (S2 with low torque) shaped 28 blocks, group D (S2 with high torque) shaped 48 blocks ($p < 0.01$ vs group C), group E (F1 with low torque) shaped eight blocks, group F (F1 with high torque) shaped 23 blocks ($p < 0.01$ vs group E), group G (F2 with low torque) shaped four blocks, and group H (F2 with high torque) shaped 11 blocks ($p < 0.01$ vs group G). It was concluded that: (i) manual preflaring creates a glide path for the instrument tip and is a major determinant in reducing the failure rate of these rotary NiTi files and (ii) all instruments worked better at high torque. Peciuliene [388] analyzed collected information about the various aspects, technical and biological, of endodontic treatment as performed by Lithuanian general dental practitioners and compared their choices with the established endodontic treatment standards of undergraduate education. Questionnaires were sent to all 2,850 Lithuanian dental practitioners. The structured questionnaire included inquiries about gender, duration of professional activity, working environment, and details about instruments and materials. It was reported that: (i) from a total of 1,532 (53.8%) questionnaires returned, only responses from general dental practitioners (1,431) were included; (ii) of the respondents, 66% never used a rubber dam; (iii) most practitioners relied on conventional SS instruments and the NiTi hand files were often and routinely used by 32.2% of the respondents; (iv) NaOCl was the most popular choice as a root canal irrigant and calcium hydroxide paste was used as an inter-appointment medicament; and (v) cold-lateral condensation root-filling method was used by 72.8% of the respondents, whereas 15.6% used a paste for the root filling. Based on these results, it was indicated that the recently graduated dental practitioners were following the recommended standard of endodontic treatment better than those with a longer time from the graduation, and it is important to improve the quality of existing courses of continuous education in endodontology to ensure the necessary competency in clinical practice. Moreover, low use of a rubber dam and low adoption of new technologies in Lithuania is not acceptable and needs to be changed.

16.3 Orthodontic applications

Orthodontic mechanotherapy is vital for improving and maintaining good oral and dental health, as well as creating an attractive smile that contributes to the development of self-esteem. The mechanical foundation of orthodontic therapy is based on the principle that stored elastic energy can be converted into mechanical work by tooth movement, and the ideal control of tooth movement requires the application of a system of distinctive forces properly supported by accessories such as orthodontic wires [389, 390].

16.3.1 Past, present, and future of orthodontic mechanotherapy

Orthodontics is the branch of dentistry concerned with the prevention and correction of malocclusions. Appliances for aligning teeth go as far back as the Egyptians. In the early years (1990s), the goals of orthodontic treatment were to attain ideal occlusion without the extraction of teeth. As time went on, and with the introduction of cephalometrics, a number of orthodontists emphasized the importance of the relationship between the teeth and bones as well as soft tissues. Today, orthodontists evaluate a number of patient factors to individualize the treatment options. Advances in the field, including skeletal anchorage, digital radiography, improvements in bracket systems, and aligner therapy, have allowed orthodontists to provide patients with more options and better treatment than ever before. Skeletal anchorage provides the opportunity to make dental changes not previously possible. Mini-implants and mini-plates are often used today to move teeth more predictably and even to help modify the growth patterns. With all the current advances in technology, the field of orthodontics has grown in leaps and bounds in the past 20 years. By digital X-rays and 3D dental imaging, a more detailed examination at the structure of the mouth with better quality can be obtained. Because they are digital, orthodontists can gain immediate access to the images. Advances in cone-beam computed tomography (CBCT) allows orthodontists to obtain more information about tooth position and the craniofacial complex. CBCT provides a full set of X-rays in one 360° scan, creates 3D images of the teeth, and is used to plan surgical treatments in adults with skeletal problems to achieve optimum surgical outcomes, occlusal relationships, and facial esthetics. Another plus is that there is less radiation with this type of X-rays. Creating the braces has also gotten a lot of easier. The digital impressions allow orthodontists to create the teeth impressions digitally instead of manually. They simply scan the patient's mouth and then they can view the impressions on the computer just a few minutes later. In addition, digital impressions are more accurate reducing the number of fit issues with the braces. Moreover, CAD/CAM and robotic wire bending technology used in combination with digital scanner allows the orthodontist to position the brackets and wires more precisely, especially with lingual braces that are adhered to the backs of the teeth [391–393].

16.3.2 Orthodontic archwires

16.3.2.1 In general
The aim of orthodontic treatment is to move the teeth to a targeted position effectively and efficiently by the application of forces to them. An ideal force is the one that produces rapid tooth movement without damage to the teeth or periodontal tissues. Different biological and other factors, such as the type of movement and tooth size, are the important factors to be considered during application of the force, but it is

difficult to precisely determine the value of the ideal force [394]. Orthodontic/ortho-
pedic forces normally range from 1.5–5.0 N [395, 396]. Application of lower forces produces the optimal results and application of excessive force exceeding vascular blood
pressure reduces cellular activity in periodontal tissues and slows down or stops tooth
movement at least for a period of time [397]. The quality and efficacy of mechanotherapy can be improved by the application of lower forces and achieves a wider range
of movements between appointments. During orthodontic treatment, orthodontic
wires are used as fixed appliances to apply forces to the teeth. They release the energy
stored upon its placement by applying forces and torque to the teeth through the
appliances placed on them [398]. Therefore, an orthodontist should have adequate
knowledge of the biomechanical behavior and clinical applications of orthodontic
wires to design the treatment plan. To achieve these complicated mechanical movements and to select appropriate wire material(s), tension, bending, and torsional tests
are used to measure the properties of orthodontic wires and all these tests are completely different stress states investigating different characteristics related with wire
performance [399–401]. Kotha et al. [393] listed the ideal requirements of orthodontic
wires (Figure 16.34 [402, 403] as:

1. Should be biological nontoxic; biocompatibility.
2. Should be resistant to corrosion and tarnish; biostability.
3. Mechanical:

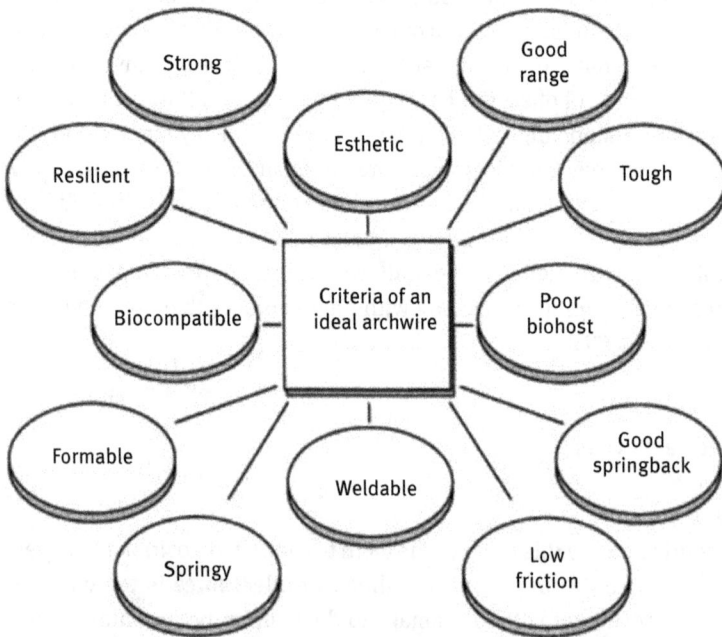

Figure 16.34: Requirements of an ideal orthodontic wire [402].

- Modulus of elasticity should be high, so that the wire can apply force for tooth movement.
  - Formability should be high so as to bend the wire into desired configuration without fracture.
- Spring back should be high, which results in an increase in its range of action. It is the measure of how far a wire can be deflected without causing permanent deformation.
- Stiffness should be lower. It provides the ability to apply lower forces constantly for a lower time.
- Resilience should be high. It increases the working range.
- It should be soldered or welded.
- Ductility should be sufficient enough to allow fabrication of appliance.
- Should provide least friction at bracket/wire interface. Otherwise, it leads to undue strain, which limits the tooth movement.

16.3.2.2 Cross-sectional shapes and material types

Orthodontic wires are classified according to the cross-sectional features; rounded, rounded rectangular, rectangular or square, co-axial, twisted, and woven, as seen in Figure 16.35 [402].

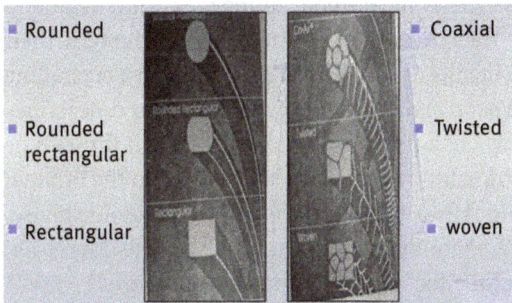

Figure 16.35: Classification of wires according to the cross- section [402].

Using a unique passivation phenomenon of NiTi alloy, wires are easily colored with variety of colors by controlling thickness of thin oxide film (color appearance is a function of thickness of oxide through the interference coloring action; Figure 16.36 [404]).

There are several types of materials chosen and fabricated orthodontic wires listed as:

(1) Austenitic SS: Fe-Cr (18–20%)-Ni(8–10%); it exhibits superior strength, higher modulus of elasticity, good resistance to corrosion, moderate costs, good biocompatibility, and excellent formability.

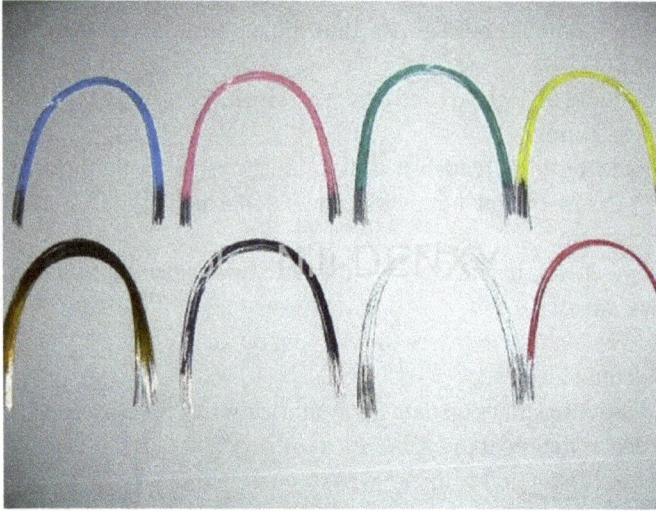

Figure 16.36: Colored orthodontic wires [430].

(2) Cobalt-chromium: Co(40%)-Cr(20%)-Ag(16%)-Ni(15%); it has greater resistance to fatigue and distortion, the mechanical properties of CoCr wires are very similar to those of SS.

(3) NiTi: Ni(55%)-Ti(45%) alloys; these have good springback and flexibility that allow large elastic deflections. Heat treatment of these alloys changes their crystallographic arrangement, producing the SME, which is resulted from a crystalline phase change known as the *thermoelastic martensitic transformation*, and describes the effect of restoring the original shape of a deformed wire by heating it through its transitional temperature range. This transformation from the distorted to the original shape involves a transformation of nitinol from the martensitic to the austenitic phase. However, these alloys present poor formability, which is a disadvantage. Moreover, the bending of these wires has a harmful effect on their springback property.

(4) Copper-NiTi alloys; the addition of copper to the alloy enhances the thermal-reactive properties of the wire and makes it highly resistant to permanent deformation.

(5) β-Titanium (β-Ti) alloys: β-Ti alloy is commercially available as TMA titanium–molybdenum alloy (TMA); β-Ti wires possess an excellent balance of properties suitable for many orthodontic applications such as good corrosion resistance, low potential for hypersensitivity, low stiffness, high springback, excellent formability, and good weldability, even compared with SS and Co-Cr-Ni orthodontic wires. Nowadays, β-Ti are used in many applications such as intrusion arches, uprighting molar spring, and cantilevers for intrusion or extrusion of teeth.

Table 16.1 compares major mechanical properties among SS wire, Co-Cr-Ni wire, TMA wire, and NiTi wire materials [405].

Table 16.1: Comparison of mechanical properties among four different wire materials [405].

Wire alloy	Modulus of elasticity (GPa)	Yield strength (MPa)	Springback
Stainless steel	160–180	1,100–1,500	0.0060–0.0094 AR 0.0065–0.0099 HT
Co-Cr-Ni alloy	160–190	830–1,000	0.0045–0.0065 AR 0.0054–0.0074 HT
TMA (β-Ti)	62–69	690–970	0.0094–0.011
NiTi alloy	34	210–410	0.0058–0.016

Co-Cr-Ni alloy: Cobalt-Chromium-Nickel alloy; NiTi: Nickel and titanium; TMA: titanium–molybdenum alloy.
Note: AR indicates as-received condition and HT refers to heat-treated condition.

Mechanical properties, tribological behaviors, intraoral corrosion issues, heat treatment, and other properties of NiTi wires are discussed further.

16.3.2.3 Mechanical properties in general

Three types of SE NiTi alloy orthodontic wires, Sentalloy-Blue (SB), Yellow (SY), and Red (SR), were studied regarding the change of properties, especially the reversible change of the load level, using three-point bending test and DSC [406]. It was reported that: (i) the heat treatment at 440 °C caused the largest reduction in mechanical properties of SB and SY, and the 400 °C treatment did that of SR; (ii) the load level of SR was less changeable than those of SB and SY by the difference of heat treatment temperature; (iii) mechanical properties of SY were almost reversible by the alternate heat treatment at 440 °C and 510 °C; (iv) 440 °C - heat treatment increased the transformation temperature, and 510 °C - heat treatment decreased them; (v) the variation of the mechanical properties completely corresponded to the change in the transformation temperature; (vi) changes in mechanical properties are observed by heat treatment at 360, 400, and 480 °C, but the reversibilities were not so clear; (vii) the change of the properties of SE wires were influenced by the kind of wire and the heat treatment condition; and (viii) the reversible change in the mechanical properties seems to have many advantages over SE NiTi alloy orthodontic wires. Wichelhaus et al. [407] defined the mechanical characteristics of several currently available closed-coil retraction springs and to compare these products. A universal test frame was used to acquire force-deflection diagrams of 24 NiTi closed-coil springs at body temperature. It was found that: (i) mechanical testing showed significant differences between the various spring types, but constant intrabatch behavior; (ii) four groups were formed according to the mechanical properties of the springs: strong SE without bias stress, weak SE without bias stress, strong SE with bias stress, and weak SE with bias stress; concluding that in sliding mechanics, the strongly SE closed-coil springs with preactivation are recommended; in addition, the oral environment seems to have only a minor influence on their mechanical properties. Sivarai [408] evaluated the SE property of eight groups of austenite-active NiTi wires. Eight groups of archwires bought from eight different manufacturers were studied. These wires were tested through mechanical tensile testing and electrical resistivity methods. It was reported that: (i) unloading curves were carefully assessed for SE behavior on deactivation; (ii) rankings of the wires tested were based primarily on the slope of the unloading curve; (iii) Ortho organizers wires ranked first and superior, followed by American Orthodontics and Ormco A wires; and (iv) Morelli and GAClowland NiTi wires were ranked last. Gravina et al. [409] compared eight types of SE and heat-activated NiTi archwires, by six trade companies (GAC, TP, Ormco, Masel, Morelli, and Unitek) to those with addition of copper (Cu-NiTi 27 °C and 35 °C, Ormco) [409]; the qualitative chemical compositions, and the surface morphology of fracture regions of eight types of NiTi conventional wires, SE, and heat-activated (GAC, TP, Ormco, Masel, Morelli, and Unitek), to the wires with addition of copper (Cu-NiTi 27 ºC and 35ºC, Ormco) after traction test [410]. It was concluded that: (i) the Cu-NiTi 35 °C archwires presented deactivation loadings biologically less favorable in relation to the other heat-activated NiTi archwires, associated with lower percentage of deformation, on the constant baselines of deactivation, showing less adequate mechanical behavior under traction, in relation to the other archwires [409]; and (ii) Cu-NiTi 35 ºC wires did not present better morphological characteristics compared to the other wires with regard to surfaces and fracture region [410]. Alobeid et al. [411] determined the mechanical properties of different esthetic and conventional orthodontic wires in three- and four-point bending tests, and in a biomechanical test using three-bracket systems. Tested wires were round wires with a diameter of 0.46 mm (0.018″) uncoated NiTi wires, surface modified NiTi wires; FLI® Orthonol Wire®, and glass fiber-reinforced plastic wires. The biomechanical bending test was performed using the following bracket types: metal brackets (Discovery®), ceramic brackets (Fascination®), and plastic brackets (Elegance®) at a temperature of 37 °C. It was reported that: (i) fiber-reinforced wires displayed lowest forces in three-point bending with values of 0.4 N at a displacement of 1 mm and 0.7 N at a 2-mm displacement; (ii) in four-point bending, the forces were 0.9 N and 1.4 N, respectively, at the same displacements; (iii) almost all the translucent wires showed fracture upon bending at displacements greater than 3 mm, independent of the bending test and bracket type; (iv) the different investigated NiTi wires, surface modified or conventional, only showed minor variation, for example, 2.2 N for rematitan® Lite White and 2.0 N for rematitan®, 2.1 N for FLI®-coated Orthonol®, and 1.7 N for Orthonol® in four-point bending; and (v) the rhodinized wire generated forces between these values (2.1 N). It was, therefore, indicated that the translucent wires had the lowest forces in all three bending tests; however, displacements > 3 mm resulted in increased risk of fracture, and forces of investigated NiTi wires were very high and, in part, greater than the clinically recommended values.

There are several studies on mechanical properties under the influence of temperature(s). Mullins et al. [412] examined the mechanical behavior of thermoresponsive nitinol archwires in flexure at 5 and 37 °C. Four same-sized (but different force level) rectangular archwires were examined using a three-point bend test. It was found that: (i) SE behavior was exhibited by all wires tested at 37 °C but not at 5 °C; (ii) permanent deformation was greater at 5 °C than 37 °C; (iii) the initial slope of the load-deflection data averaged 1,230 g/mm at 37 °C, which was significantly different from the average at 5 °C (500 g/mm); (iv) loads at the apparent yield point and the loads at 1, 2, and 3 mm of deflection were greater at 37 °C than at 5 °C; (v) while the slope and length of the SE region were not judged to be clinically and significantly different, the average load of the SE region was significantly different: F300 (340 g) > F200 (250 g) > F100 (180 g) and Bioforce (180 g); and (vi) when loaded at 5 °C and then unloaded at 37 °C, the mechanical hysteresis of the wires tested at 37 °C and the permanent deformation of the wires tested at 5 °C were reduced for all wires. Based on these findings, it was indicated that nitinol wires are available with a variety of mechanical properties; the different mechanical properties of thermoresponsive wires at 5 and 37 °C result in clinically useful shape-memory behavior; and (ix) using the SE and shape-memory features of thermoresponsive wires has clinical advantages. Meling et al. [413] studied the effect of these temperature changes associated with ingestion of cold or hot food on the torsional stiffness of NiTi alloys. Eight rectangular SE wires were activated to 20 °C, in longitudinal torsion at body temperature and subjected to cold (10 °C) or hot (80 °C) water with the strain held constant. It was mentioned that: (i) the torsional stiffness of some wires was strongly affected; (ii) the effect of hot water disappeared quickly, but the wires remained at a level of reduced torsional stiffness (up to 85% less than the baseline) after short applications of cold water; (iii) the most thermodynamic archwires showed incremental reductions in torsional stiffness when cold water was repeatedly applied; (iv) furthermore, the torsional stiffness remained low (up to 50% less than the baseline) and showed no tendency to increase even after 2 h of postexposure restitution; and (v) it is conceivable that some wires may provide inadequate forces for tooth movement after ingestion of cold liquids. The inducement of mechanical stress within NiTi wires can influence the transitional temperature range of the alloy, and thereby influence the expression of the SE properties. An analogous variation of the transitional temperature range may be expected during orthodontic therapy, when the archwires are engaged into the brackets. Santoro et al. [414] investigated this possibility. Samples of currently used orthodontic NiTi wires (Cu-NiTi SE at 27, 35, and 40 °C) were subjected to temperature cycles ranging between 4 and 60 °C. Electrical resistivity was used to monitor the phase transformations. It was reported that: (i) the results confirmed the presence of displacements of the transitional temperature ranges toward higher temperatures when stress was induced; and (ii) an alloy with a stress-related transitional temperature range corresponding to the fluctuations of the oral temperature expresses SE properties more consistently than others; indicating that Cu-NiTi

27 °C and nitinol heat-activated wires may be considered suitable alloys for the alignment stage. Iijima et al. [415] investigated the mechanical properties of SE NiTi orthodontic wires under controlled stress and temperature. Three different SE NiTi wires were examined using DSC, three-point bending test, and micro-XRD. The three-point bending test was conducted at constant temperature (23, 37, and 60 °C) and stepwise temperature changes (37 to 60 °C and to 37 °C; 37 to 2°C and to 37 °C). Five specimens of each wire were tested. Micro-XRD spectra were measured at the tension side of the wire when the temperature changed from 37 to 60 °C or 2 °C. It was found that: (i) the load during the stepwise temperature changes (37–2 °C and to 37 °C) was consistent with that measured at a corresponding constant temperature; (ii) the micro-XRD spectrum clearly showed that the austenite phase was transformed to martensite phase when the temperature is decreased from 37 to 2 °C; and (iii) in a stepwise temperature change (37 to 60 °C and to 37 °C), the load became higher than the original load at each corresponding constant temperature; however, there was no detectable change in the micro-XRD spectrum when the temperature was increased from 37 to 60 °C; indicating that the SE NiTi wires exhibited complicated and unexpected mechanical properties under stepwise temperature change, and the possibility of qualitative analysis using micro-XRD to understand mechanical properties of these NiTi wires was recognized. Lombardo et al. [416] investigated and compared the characteristics of commonly used types of traditional and heat-activated initial archwires at different temperatures by plotting their load/deflection graphs and quantifying three parameters describing the discharge plateau phase. Forty-eight archwires of cross-sectional diameters ranging from 0.010 to 0.016 inches were obtained from seven different manufacturers. A modified three-point wire-bending test was conducted on three analogous samples of each type of archwire at 55 °C and 5 °C, simulating an inserted archwire that is subjected to cold or hot drinks during a meal. For each resulting load/deflection curve, the plateau section was isolated and the mean value of each parameter for each type of wire was obtained [417]. It was reported that: (i) permanent strain was exhibited by all wires tested at 55 °C; (ii) statistically significant differences were found among almost all wires for the three considered parameters when tested at 55 °C and 5 °C; (iii) loads were greater at 55 °C than at 5 °C; (iv) differences were also found between traditional and heat-activated archwires, the latter of which generated longer plateaus at 55 °C, shorter plateaus at 5 °C, and lighter mean forces at both temperatures; and (v) the increase in average force seen with increasing diameter tended to be rather stable at both temperatures. Based on these results, it was concluded that all tested NiTi wires showed a significant change related to temperature in terms of behavior and force for both traditional and heat-activated wires, and stress under high temperatures can induce permanent strain, whereas the residual strain detected at low temperatures can be recovered as temperature increases [417].

Some intraoral environments may affect the mechanical properties of NiTi alloys. Walker et al. studied the effects of fluoride prophylactic agents on the mechanical properties of NiTi and Cu-NiTi orthodontic archwires. Preformed rectangular NiTi and

Cu-NiTi wires were immersed in either an acidulated fluoride agent, a neutral fluoride agent, or distilled water (control) for 1.5 h at 37 °C. After immersion, the loading and unloading elastic modulus and YS of the wires were measured with a three-point bend test in a water bath at 37 °C. It was found that: (i) unloading mechanical properties of NiTi orthodontic wires were significantly decreased after exposure to both fluoride agents; however, Cu-NiTi wire mechanical properties were not significantly affected by either fluoride agent; and (ii) corrosive changes in surface topography were observed for both wires, with Cu-NiTi appearing to be more severely affected; suggesting that using topical fluoride agents with NiTi wire could decrease the functional unloading mechanical properties of the wire and contribute to prolonged orthodontic treatment. Rerhrhaye et al. [418] investigated, in vitro, the impact of this acidic and fluoridated environment on the electrochemical behavior and the mechanical properties of orthodontic alloys in NiTi and in SS (controls) for the following parameters: Young's modulus (E), elastic limit (σe), and the maximum tensile load (σm). Six samples of each archwire alloy were used to assess these parameters. It was reported that: (i) for the NiTi archwires, immersion in the fluoridated and acidic medium showed a statistically significant reduction of the Young's modulus (E), the elastic limit (σe), and the maximum tensile load (σm); (ii) similarly, a higher level of released nickel proportionate to the increase in the fluoride concentration and acidity was observed in the immersion solutions; and (iii) SEM observations revealed the status of the surface of the different alloys and the presence of corrosive pitting. Lin et al. [419] investigated the influence of fluoride and an acidic environment on the mechanical properties of NiTi orthodontic wires in AS. Commercial, round 0.016-inch NiTi orthodontic wires were immersed in 0% or 0.05% natrium-fluoride-containing AS at a pH of 4 or 6 for 1 or 3 days, respectively. It was found that: (i) a pH of 4 increased microhardness and decreased the three-point bending strength significantly, whereas immersion time and fluoride concentration had no significant influence on the microhardness or on the three-point bending strength; and (ii) when examining the test group NiTi orthodontic wires after 3 days of immersion at a pH of 4, the SEM revealed a rough surface morphology, a damaged oxide layer, and signs of corrosion; concluding that the most influential factor for decreasing the unloading force and increasing the hardness seems to be the pH value, whereas immersion time and NaF addition do not have a major influence.

16.3.2.4 Load-deflection characteristics

Success of fixed orthodontic treatments depends on the type of wire used to exert force on teeth [420]. Light continuous force is considered physiologically appropriate to move the teeth although limited facts about optimal force are available [421]. The applied force must be beyond the biological threshold (0.5–0.7 N), but should not exceed the biological corridor (2–3 N) [422]. Due to these, the use of NiTi archwires became prevalent because of their ability to exert light continuous force and consequently improve the efficacy of treatment, particularly in aligning and leveling phase [420, 423]. Kapila et al. investigated the load-deflection characteristics of NiTi alloy wires under the clinical recycling [424], and clinical recycling and dry heat sterilization [425]. It was reported that: (i) recycling produced significant changes in both the loading and unloading characteristics of NiTi wires, but only with the loading forces associated with nitinol wires [424]; and (ii) dry heat sterilization alone, as well as clinical recycling, produced significant changes in the loading and unloading characteristics of nitinol and NiTi wires; however, the changes in the load-deflection characteristics of these wires after dry heat sterilization only were relatively small, and the clinical significance of these changes is open to question; (iii) in contrast, the force levels during loading and unloading were substantially increased for both types of wires after recycling; (iv) most of these changes in load-deflection characteristics occurred between T_0 (as received condition) and T_1 (after one cycle); and (v) clinical recycling appears to reduce the "pseudoplasticity" and "pseudoelasticity" of NiTi wires and increases the stiffness of both NiTi and nitinol wires [425]. Barwart [426] investigated the effect of temperature change on the force delivery of NiTi springs in their SE range. Japanese NiTi closed coil springs were heated and cooled between 20 °C and 50 °C, while held in constant extension. Commercially available SS closed coil springs were tested in the same way. It was reported that: (i) for all the springs examined, load values were found to increase with rising temperatures and to decrease with a drop in temperature; (ii) this relationship between temperature change and load was more pronounced in the case of NiTi than in the steel springs; (iii) the force measured at 37 °C was about twice as high as at 20 °C for one type of NiTi spring; and (iv) on cooling, the SE springs showed unusual behavior; immediately after the temperature started to drop, a rapid decrease in force occurred to levels below those found at rising temperatures; however, such a nonlinear decrease in load was not observed in the SS springs tested; indicating that only minimal changes in temperature can cause significant changes in the force delivery of SE NiTi springs. Wilkinson et al. [427] investigated the load-deflection characteristics of seven different 0.016-inches initial alignment archwires (Twistflex, NiTi, and five brands of heat-activated SE NiTi). Load-deflection tests were conducted on the wires with five different model designs, and data from selected points on the unloading phase of the generated graphs were statistically analyzed. Wire deflection was carried out at three temperatures (22.0, 35.5, and 44.0 °C) and to four deflection distances (1, 2, 3, and 4 mm). It was found that: (i) Twistflex and the five heat-activated SE NiTi wires produced a range of broadly comparable results, and NiTi gave the highest unloading values; and (ii) model rankings indicated that self-ligating brackets (SLBs) Twin-Lock produced lower friction than regular edgewise brackets; recommending using the rankings from the mechanical test simulations to predict possible clinical performance of archwires. Yanaru et al. [428] evaluated orthodontic forces of NiTi wires examined under the retrained condition on the dental arch model with the changes in temperature and deflection. The tested specimens were a commercially available SE

(W1) wire and two shape-memory wires with their nominal A_F points were 35 °C (W2) and 40 °C (W3), respectively. It was shown that: (i) they showed typical SE hysteresis loops under the restraint condition at 40 °C; (ii) the force levels were significantly larger than those generally obtained by simple three-bending test; (iii) the recovery forces in the plateau region at 1.0 mm deflection were much larger than desired in the clinical guidelines around oral temperatures; (iv) in the shape-memory wire W3, the recovery force rapidly decreased to zero by a small reduction of the deflection from its maximum; however, the wire again exerted the force with the remaining permanent deflection by temperature rising; and (v) it was small compared to the guidelines of desirable orthodontic force and seemed to be useful especially for the hypersensitive patients.

Kasuya et al. [429] examined the effect of ligation on the load-deflection characteristics of NiTi orthodontic wire. A modified three-point bending system was used for bending the NiTi round wire, which was inserted and ligated in the slots of three brackets, one of which was bonded to each of the three bender rods. Three different ligation methods, SS ligature (SSL), slot lid (SL), and elastomeric ligature (EL), were used, as well as a control with neither bracket nor ligation (NBL). It was reported that: (i) the load values of the ligation groups were two to three times greater than the NBL group at a deflection of 1,500 µm during unloading: 4.37 N for EL, 3.90 N for SSL, 3.02 N for SL, and 1.49 N for NBL; (ii) for the EL, a plateau region disappeared in the unloading curve; (iii) SL showed the smallest load; and (iv) the ligation of the bracket wire may make NiTi wire exhibit a significantly heavier load than that traditionally expected. NiTi wire exhibited the majority of its true SE with SL, whereas EL may act as a restraint on its SE. Bartzela et al. [430] determined the mechanical properties of commercially available thermodynamic wires and to classify these wires mathematically into different groups. A total of 48 NiTi wires (0.016 inch, 0.016- × 0.022 inch, 0.017 × 0.025 inch, and 0.018 × 0.025 inch) were tested. The orthodontic wires tested are classified as follows: (1) true SE wires, which presented a clinical plateau length of ≥ 0.5 mm; (2) borderline SE with a clinical plateau length of < 0.5 mm and > 0.05 mm; and (3) non-SE, with a clinical plateau length of ≤ 0.05 mm. It was shown that: (i) the range of products displays big variations in quantitative and qualitative behavior; (ii) a fraction of the tested wires showed weak SE, and others showed no SE; and (iii) some of the products showed permanent deformation after the three-point bending test; concluding that a significant fraction of the tested wires showed no or only weak SE; in addition, the practitioner should be informed for the load-deflection characteristics of the NiTi orthodontic wires to choose the proper products for the given treatment needs. The force-deflection behavior of selected initial alignment archwires by conducting three-point bending tests under controlled conditions was measured [431] using four wire designs: multistranded SS, conventional SS, SE NiTi, and thermoactivated NiTi archwires. The wires were ligated into SS brackets with steel ligatures. A testing machine recorded deactivations at 2.0 mm of deflection at 37 °C. Force-deflection measurements were recorded from only deactivation. It was obtained that:

(i) significant differences in the deactivation forces were observed among the tested wires; (ii) the multistranded SS wire had the lowest mean deactivation force (1.94 N), whereas the conventional SS group had the highest value (4.70 N); and (iii) the SE and thermoactivated (TA) NiTi groups were similar to the multistranded wire; indicating that both NiTi and multistranded steel archwires tested are potentially adequate for use during the leveling and aligning phase of orthodontics. Ahmadabadi [432] studied the load-deflection characteristics of SE NiTi commercial wires through three-point bending test. The SE behavior was also investigated by focusing on the bending time, temperature, and the number of cycles, which affect the energy dissipating capacity. It was indicated that: (i) the NiTi archwires are well suited for cyclic load-unload dental applications, and (ii) reduction in the SE property for used archwires after long-time static bending was recognized. Rose et al. [433] conducted the in vitro study to investigate the loads (forces), moments, and moment-to-force ratios (M:F) generated during the activation and deactivation of T closing loops made of rectangular NiTi and TMA wires incorporating either 0, 15, or 30° of preactivation. It was mentioned that: (i) nonpreactivated (0°) closing loops failed to produce an optimum M:F ratio for translational tooth movement; (ii) with increasing preactivation, the M:F ratio increased over the deactivation range for both alloys; (iii) the NiTi T-loops produced an M:F ratio of greater than 10:1 over a larger deactivation range (while still delivering a force of 50–150 g) than for the equivalent TMA T-loop; (iv) the difference in M:F between the 0° and 30° TMA loops was statistically significant, but not between the equivalent NiTi loops; and (v) there was no statistical difference between the NiTi wire alloys at any preactivation angulation. Based on these results, it was concluded that optimum M:F ratios for orthodontic translation can be achieved by using preactivated NiTi and TMA T-loops, with NiTi loops maintaining the optimum M:F ratio over a greater range of deactivation.

Lombardo et al. [434] investigated and compared the characteristics of commonly used types of traditional and heat-activated initial archwire by plotting their load/deflection graphs and quantifying three suitable parameters describing the discharge plateau phase. A total of 48 archwires (22 NiTi and 26 heat-activated NiTi) of cross-sectional diameter ranging from 0.010 to 0.016 inches were obtained from seven different manufacturers at a constant temperature (37.0 °C). It was obtained that: (i) statistically significant differences were found among almost all wires for the three parameters considered; (ii) statistically significant differences were also found between traditional and heat-activated archwires, the latter of which generated longer plateaus and lighter average forces; and (iii) the increase in average force seen with increasing diameter tended to be rather stable although some differences were noted between traditional and heat-activated wires. It was then concluded that although great variation was seen in the plateau behavior, heat-activated versions appear to generate lighter forces over greater deflection plateaus, and on average, the increase in the plateau force was roughly 50% when the diameter was increased by 0.002 inch (from 0.012 to 0.014 and from 0.014 to 0.016 inch) and about 150%

when the diameter was increased by 0.004 inch (from 0.012 to 0.016), with differences between traditional and heat-activated wires noted in this case. Gatto et al. [435] investigated the mechanical properties of SE and thermal NiTi archwires for correct selection of orthodontic wires. Seven different NiTi wires of two different sizes (0.014 and 0.016 inches), commonly used during the alignment phase, were tested. It was mentioned that: (i) the wire size had a significant effect on the forces produced with an increase in the archwire dimension, the released strength increased for both thermal and SE wires; (ii) SE wires showed, at a deflection of 2 mm, narrow and steep hysteresis curves in comparison with the corresponding thermal wires, which presented a wide interval between loading and unloading forces; (iii) during unloading at 4 mm of deflection, all wires showed curves with a wider plateau when compared with 2 mm deflection; and (iv) such a difference for the SE wires was caused by the martensite stress induced at higher deformation levels. Based on these findings, it was suggested that a comprehensive understanding of mechanical characteristics of orthodontic wires is essential and selection should be undertaken in accordance with the behavior of the different wires. It is also necessary to take into account the biomechanics used. In low-friction mechanics, thermal NiTi wires are to be preferred to SE wires, during the alignment phase due to their lower working forces, and in conventional straightwire mechanics, a low-force archwire would be unable to overcome the resistance to sliding (RS). Nucdra et al. [436] evaluated how different bracket-slot design characteristics affect the forces released by SE NiTi alignment (0.014-inches) wires at different amounts of wire deflection. The selected NiTi archwire was tested in association with three bracket systems: (1) conventional twin brackets with a 0.018-inch slot, (2) a SLB with a 0.018-inch slot, and (3) a SLB with a 0.022-inch slot. Wire specimens were deflected at 2 mm and 4 mm. It was reported that: (i) the use of a 0.018-inch slot bracket system, in comparison with the use of a 0.022-inch system, increases the force exerted by the SE NiTi wires at a 2-mm deflection; (ii) the use of a SLB system increases the force released by NiTi wires in comparison with the conventional ligated bracket system; and (iii) NiTi wires deflected to a different maximum deflection (2 mm and 4 mm) release different forces at the same unloading data point (1.5 mm). It was, therefore, concluded that the bracket design, type of experimental test, and amount of wire deflection significantly affected the amount of forces released by SE NiTi wires, and the phenomenon offers clinicians the possibility to manipulate the wire's load during alignment.

Higa et al. [437] evaluated deflection forces of rectangular orthodontic wires in conventional (MorelliTM), active (In-Ovation RTM), and passive (Damon 3MXTM) SLBs. Two brands of SS and NiTi wires (MorelliTM and GACTM), in addition to OrmcoTM Cu-NiTi wires were used. Results showed that: (i) there were lower forces in conventional brackets (CBs), followed by active and passive SLBs; and (ii) within the brands, only for NiTi wires, the MorelliTM brand presented higher forces than GACTM wires; concluding that the bracket systems provide different degrees of deflection force, with SLBs showing the highest forces. Arreghini et al. [438] determined and compared the relative stiffness of a large selection of commonly used square and rectangular steel, super-tempered steel, NiTi, and TMA orthodontic archwires of various cross-sections to provide the clinician with a useful, easy-to-consult guide to archwire sequence selection. It was found that: (i) a considerable difference in resistance to deflection was revealed between all the tested archwires; (ii) the resistance to deflection of archwires of the same cross-section was found to increase with increasing stiffness of their construction material; (iii) specifically, steel archwires can be as much as eight times stiffer than NiTi archwires of the same shape and cross-section, and super-tempered steel archwires are invariably stiffer than traditional steel versions; and (iv) marked differences in resistance to deflection were also found between NiTi archwires made of the same material but with different shape characteristics. It was then concluded that in archwires of the same cross-section, steel is always stiffer than TMA and NiTi and super-tempered steel is always stiffer than conventional steels, and in archwires of the same material, the stiffness increases with the cross-section, in particular with its height. Razali et al. [439] investigated the evolution of the forces released by a rectangular NiTi archwire toward possible intraoral temperature and deflection changes. A 3D FEM was developed to measure the force-deflection behavior of SE archwire. Finite element analysis was used to distinguish the martensite fraction and phase state of the archwire microstructure in relation to the magnitude of wire deflection. It was mentioned that: (i) the predicted tensile and bending results from the numerical model showed a good agreement with the experimental results; (ii) as contact developed between the wire and bracket, binding influenced the force-deflection curve by changing the martensitic transformation plateau into a slope; (iii) the archwire recovered from greater magnitude of deflection released lower force than one recovered from smaller deflection; and (iv) in contrast, it was observed that the plateau slope increased from 0.66 N/mm to 1.1 N/mm when the temperature was increased from 26 °C to 46 °C.

16.3.2.5 Bending behaviors

Cantilever bending properties were evaluated for several clinically popular sizes of three SE and three non-SE brands of NiTi orthodontic wires in the as-received condition, and for 0.016-inch diameter wires after heat treatment at 500 and at 600 °C, for 10 min and for 2 h [440]. It was reported that: (i) in general, the bending properties were similar for the three brands of SE wires and for the three brands of non-SE wires; (ii) for the three brands of SE wires, heat treatment at 500 °C for 10 min had minimal effect on the bending plots, whereas heat treatment at 500 °C for 2 h caused decreases in the average SE bending moment during deactivation; heat treatment at 600 °C resulted in the loss of SE; (iii) the bending properties for the three brands of non-SE wires were only slightly affected by these heat treatments; and (iv) the differences in the bending properties and heat treatment responses are attributed to the relative proportions of the austenitic and martensitic forms of NiTi alloy in the microstructures of the wire alloys. Yoneyama et al. [441] investigated the effect of heat treatment temperature on bending properties and transformation temperatures of a NiTi alloy wire, 1.0 mm in diameter, so that SE could be used in orthodontic appliances needing shape-memory processes. The heat treatment process was at 440 °C for 30 min and between 400 °C and 540 °C for 30 min. A three-point bending test and DSC were performed. It was found that: (i) the transformation temperatures of the wires were lowered with increasing the heat treatment temperature; (ii) the reverse transformation finishing temperature was below the body temperature with the treatment > 480 °C; (iii) residual deflection of the NiTi wire after bending was small with the secondary heat treatment > 460 °C; (iv) the load in the unloading process was less changeable and increased with the treatment temperature between 460 and 540 °C; and (v) secondary heat treatment in this range was suitable for using SE in expansion arch appliances. Lim et al. [442] examined the transverse stiffness of two aesthetic orthodontic archwires (0.018 inch Teflon-coated SS and 0.017 inch Optiflex) in a simulated clinical setting and to assess the influence of deflection direction on the bending stiffness. The aesthetic archwires were randomly divided into three equal groups: group 1, lingual deflection; group 2, labial deflection; and group 3, occlusal deflection. Each group consisted of six archwires of the same type. The control group consisting of 18 0.014 inch SS archwires were also subjected to the same grouping. A total of 54 archwires were tested in the study. It was reported that: (i) the mean stiffnesses of the archwires in the lingual, labial, and occlusal deflection groups were found to be 2.9, 0.8, and 2.5 mN/mm, respectively for 0.017 inch Optiflex; 13.2, 10.5, and 24.5 mN/mm, respectively for 0·018 inch Teflon-coated SS; and 26.6, 16.4, and 32.3 mN/mm, respectively for the control; (ii) springback was found to be poor for Optiflex and the archwire remained bent upon deactivation; (iii) the influence of arch curvature on the bending stiffness was significantly different for Optiflex, Teflon-coated SS, and the control group; (iv) stiffness for the Teflon-coated archwire was found to be higher and more in line with the stiffness for the control; (v) both the Teflon-coated and SS archwires displayed good springback property; however, Optiflex was found to have low stiffness and resilience; and (vi) due to the poor springback, the clinical efficacy of Optiflex is probably limited.

Yamamoto et al. [443] investigated the bending property of Cu-NiTi alloy castings (Ti-Ni$_{50.8}$ and Ti-Ni$_{40.8}$-Cu$_{10.0}$ [mol %]) in a three-point bending test for orthodontic application in relation to the phase transformation. Heat treatment was performed at 440, 480, or 520 °C for 30 min. It was mentioned that: (i) the bending load changed by the cross-sectional size and shape mainly because of the difference in the moment of inertia of area, but the load-deflection relation did not differ proportionally in the unloading process; (ii) the difference between the load values in the loading and the unloading processes was relatively small for Cu-NiTi alloy; (iii) with respect to the residual deflection, there was no significant difference between NiTi and Cu-NiTi alloys with the same treatment condition; (iv) the load values in the loading and the unloading processes decreased by each heat treatment for NiTi alloy; however, the decrease in the load values for Cu-NiTi alloy was not distinct; and (v) it was proved that Cu-NiTi alloy castings produce effective orthodontic force as well as stable low residual deflection, which is likely to be caused by the high and sharp thermal peaks during phase transformation. Fiscer-Brandies et al. [444] characterized five selected commercial NiTi archwires in terms of their transformation behavior, chemical composition, surface topography, and mechanical properties (at temperatures of 22, 37, and 60 °C). The rectangular orthodontic archwires investigated were Neo Sentalloy F80 (GAC, Central Islip, NY, USA), 35 °C Thermo-Active Copper NiTi, Rematitan "Lite," Titanol SE S, and Titanal in size 0.016" × 0.022". The transformation temperatures were measured by means of DSC in the range of −80 °C to +80 °C. The SEM analyses revealed abradant residues in virtually all archwires, whereas DSC revealed complex transformation properties. In addition to the martensitic and austenitic transformations, an R-phase transformation was also detected. The bending tests showed pronounced loading and unloading plateaus. The martensitic archwires (Neo Sentalloy F80, 35 °C Thermo-Active Copper NiTi) were found to have a lower strength than the martensitic-austenitic (Rematitan "Lite") and the austenitic archwires (Titanol SE S, Titanal). With increasing temperature (in the range from 22 to 60 °C), a linear rise in the plateau forces was recorded. Based on these results, it was concluded that when assessing the quality of archwires, account should be taken of the surface quality, as it is this that determines corrosion resistance, biocompatibility, and friction characteristics. The mechanical properties depend on the initial state; moderate plateau forces and plateau moments can only be achieved with martensitic archwires. In contrast to the conventional steel alloys, the strength characteristics are heavily dependent on temperature and need to be known if NiTi archwires are to be used to optimal effect; in addition, the SE plateau is used only partially, if at all, when minimum leveling is required.

Brauchi et al. [445] evaluated different bending methods in relation to the subsequent mechanical characteristics of the alloy. The mechanical behaviors of three archwires (Cu-NiTi 35 °C, Neo Sentalloy F 80, and Titanol Low Force) were investigated after heat treatment in a dental furnace at 550–650 °C. In addition, the change in A_F temperature was registered by means of DSC. It was obtained that: (i) heat treatment in the dental furnace as well as with the Memory-Maker led to widely varying force levels for each product; (ii) cold forming resulted in similar or slightly reduced force levels when compared to the original state of the wires; and (iii) A_F temperatures were in general inversely proportional to force levels. It was the concluded that the archwire shape can be modified by using either chair-side technique (Memory-Maker, cold forming) because the SE behavior of the archwires is not strongly affected; however, it is important to know the specific changes in force levels induced for each individual archwire with heat treatment; (v) cold forming resulted in more predictable forces for all products tested; therefore, cold forming is recommended as a chair-side technique for the shaping of NiTi archwires. Ballard et al. [446] evaluated the bending properties of fiber-reinforced polymeric composite (FRC) archwires compared with similarly sized NiTi archwires. It was reported that: (i) the 0.018-inch NiTi archwire demonstrated the highest force values at different deflection distances followed by translucent archwire II, 0.016-inch NiTi, translucent archwire I, and finally 0.014-inch NiTi; (ii) the 0.016-inch NiTi exhibited the highest modulus value, followed by 0.018-inch NiTi, 0.014-inch NiTi, translucent archwire II, and finally translucent archwire I; and (iii) during deactivation, the elastic recovery of 0.014-inch NiTi and 0.016-inch NiTi was significantly greater than translucent archwire II; concluding that the bending properties of BioMer's FRC archwires were found to be comparable to NiTi.

16.3.2.6 Biotribology

When straight-wire mechanics are used in orthodontics, the RS generated at the wire-bracket interface greatly influences the character of the force transmitted to teeth [447]. This resistance is believed to reduce the efficiency of orthodontic appliances and hence results in slower tooth movement [448]. Friction in orthodontics occurs at multiple contact points along the archwire. Variables affecting the friction between components of fixed appliance include the design of the appliance (of archwire and bracket), ligation, biological factors (of masticatory forces, saliva, variation of intraoral temperature, etc.), and/or oral hygiene agents (toothpaste, mouthwash, etc.) [447–449]. Friction is defined as the force that opposes a movement when an object moves tangentially against another [450]. All materials have two coefficients of friction; static friction coefficient and kinetic friction coefficient [448]. Sliding between bracket and wire in the oral cavity occurs at a low velocity in a sequence of short steps rather than as a continuous motion [447]. In such conditions, the distinction between static and kinetic frictional resistance is arbitrary, because these two forms of friction are dynamically related [451]. In orthodontics, a tooth undergoing a sliding movement along an archwire goes through many tipping and uprighting cycles, moving in small increments. Therefore, orthodontic space closure depends more on the static friction than on kinetic friction [452]. All surfaces are more or less irregular, and the physical explanation of friction is in terms of the true area of contact, which is determined by asperities, and the force with which the surfaces are forced together [453].

The level of kinetic frictional forces generated during in vitro translation at the bracket-wire interface were measured for two sintered SS brackets as a function of two slot sizes, four wire alloys, and five to eight wire sizes [449]. The two types of sintered SS brackets were tested in both 0.018-inch and 0.022-inch slots. Wires of four different alloy types, SS, Co-Cr, NiTi, and β-Ti were tested. There were five wire sizes for the 0.018-inch slot and eight wire sizes for the 0.022-inch slot. The wires were ligated into the brackets with ELs. It was found that: (i) for most wire sizes, lower frictional forces were generated with the SS of Co-Cr wires than with the β-Ti or NiTi wires; (ii) increase in the wire size generally resulted in increased bracket-wire friction; (iii) there were no significant differences between manufacturer for the sintered SS brackets; and (iv) the levels of frictional force in 0.018-inch brackets ranged from a low of 46 gm with 0.016-inch Co-Cr wire to a high of 157 gm with 0.016 × 0.025-inch β-Ti wire. Cacciafesta et al. [454] measured and compared the level of frictional resistance generated between conventional ceramic brackets; ceramic brackets with SS slot; conventional SS brackets, and three different orthodontic wire alloys: SS, NiTi, and β-Ti. It was reported that: (i) metal-insert ceramic brackets generated significantly lower frictional forces than the conventional ceramic brackets, but higher values than SS brackets; (ii) β-Ti archwires had higher frictional resistances compared to SS and NiTi archwires; (iii) no significant differences were found between SS and NiTi archwires; (iv) all the brackets showed higher static and kinetic frictional forces as the wire size increased; and (v) metal-insert ceramic brackets are not only visually pleasing, but also a valuable alternative to conventional SS brackets in patients with esthetic demands. Kapur et al. [455] compared in vitro the static and kinetic frictional resistances of ceramic brackets with metal-lined slots (clarity [CL]), SS brackets (miniature twin [MT]), and two ceramic brackets with different slot designs (contour [CO]; transcend [TR]). It was reported that: (i) there were no significant static or kinetic frictional differences when the smaller 0.018 × 0.025 inch wires were drawn through the brackets; (ii) there were no statistically significant static or kinetic frictional differences between the CL-CO, CL-MT, and CO-MT bracket pairs when 0.021 × 0.025 inch wires (SS, NiTi, β-Ti) were used; (iii) there were no significant kinetic frictional resistance differences between the CL-TR and MT-TR when the SS wires were used; (iv) in general, the static and kinetic resistances of 0.021 × 0.025 inch wires of NiTi wire < SS wire < β-Ti wire; (v) regardless of the wire type, some of the lowest kinetic resistances were found with the narrow CO brackets with the rounded slot bases; and (vi) the highest static and kinetic frictional resistances were found with the wide TR bracket, and with SS and β-Ti wires. Based on these findings, it was concluded that: (vii) the high static and kinetic frictional resistances of ceramic brackets can be reduced either by lining the slots with SS or by reducing the bracket width and rounding the slot base. Kao et al. [456] investigated and compared the levels of frictional resistance between metal brackets and orthodontic wires after immersion in an acidified phosphate fluoride (APF) agent. Three types of mandibular incisor SS metal brackets with β-Ti alloy wire (TMA), heat-activated NiTi, and two sizes of SS wires were immersed in 0.2% APF and pH 6.75 AS solutions for 24 h. The study included 480 bracket-wire specimens. It was obtained that: (i) in the APF-immersed group, the static frictional force was greater than the kinetic frictional force; (ii) the frictional forces of the orthodontic wires had statistically significant differences in this progressive order: TMA, NiTi, and SS wire; (iii) similar frictional force results were obtained in the pH-6.75 saliva group; and (iv) the frictional force values of the APF group were higher than those of the pH-6.75 saliva group; concluding that the frictional forces of orthodontic brackets and wires are influenced by contact with fluoride-containing solutions.

Stress-induced martensite formation and stress hysteresis, which changes the elasticity and stiffness of the NiTi wire, influences the sliding mechanics of archwire-guided tooth movement. Liaw et al. [457] investigated the frictional behavior of an improved SE NiTi wire with low-stress hysteresis, at a constant temperature of 35 °C. It was shown that: (i) compared with conventional Sentalloy (SE NiTi) wires, the L & H Titan (improved SE NiTi) wire had a narrower stress hysteresis including a lower loading plateau and a higher unloading plateau; (ii) in addition, L & H Titan wires were less stiff than the Sentalloy wires during most unloading stages; and (iii) the values of friction measured at deflections of 0, 2, and 3 mm were significantly increased in both types of wire; however, they showed a significant decrease in the friction from 3 to 4 mm of deflection. L & H Titan wires had less friction than Sentalloy wires at all bending deflections. It was hence concluded that the stress-induced martensite formation significantly reduced the stiffness and thus could be beneficial to decrease the binding friction of SE NiTi wires during sliding with large bending deflections, and austenitic-active alloy wires with low-stress hysteresis and lower stiffness and friction offer significant potential for further investigation. Hsu et al. [458] evaluated the static and kinetic frictional forces produced between different combination of orthodontic archwires and brackets. Three types of archwires were examined: (1) SS, (2) conventional NiTi alloy, and (3) improved SE NiTi alloy. Two types of brackets were tested: (1) SS and (2) plastic. It was shown that: (i) the static frictional force was significantly higher than the kinetic frictional force in all archwire-bracket combinations; (ii) the frictional force was lower for the SS bracket than for the plastic bracket with SS wire and the improved SE NiTi alloy wire; (iii) the frictional force was lower for the improved SE NiTi alloy wire than for NiTi wire with the SS bracket, but higher for NiTi wire with the plastic bracket; and (iv) the frictional force was lowest for SS wire for both types of bracket; demonstrating that the frictional forces of brackets are influenced by different combinations of bracket and archwire, and that the improved SE NiTi alloy wire does not exhibit low friction. Tecco et al. [459] compared friction of conventional and ceramic brackets (0.022-inch slot) using a model that tests the sliding of the archwire through 10 aligned brackets. Polycrystalline alumina (PCA) brackets, PCA brackets with a SS slot (PCA-M), and monocrystalline sapphire brackets (MCS) were tested under elastic ligatures using various archwires in dry and wet (saliva) states. Conventional SS brackets were used as controls. It was reported that: (i) in both dry and wet states, PCA and MCS brackets expressed a statistically significant higher friction value with respect to SS and PCA-M brackets when combined with the rectangular archwires, (ii) PCA brackets showed significantly higher friction than MCS brackets when coupled with 0.014 × 0.025-inch NiTi archwire; (iii) SEM analysis showed differences in the surfaces among SS, MCS, PCA-M, and PCA brackets; and (iv) in the wet state, the mean friction values were generally higher than in the dry state. Based on these results, it was concluded that PCA brackets showed significantly higher friction than MCS brackets only when combined with 0.014 × 0.025-inch NiTi archwires, and a 10 aligned-brackets study model showed similar results when compared to a single-bracket system, except for friction level with 0.014 × 0.025-inch NiTi archwires. Heo et al. [460] compared frictional properties according to the amounts of vertical displacement (VD) and horizontal displacement

(HD) of teeth and bracket types during the initial leveling and alignment stage. Combinations of SLBs (two active types: In-Ovation-R and In-Ovation-C; four passive types: Damon-3Mx, Damon-Q, SmartClip-SL3, and Clarity-SL) and 0.014-inch NiTi archwires (austenitic type, A-NiTi, and copper type, Cu-NiTi) were tested in a stereolithographically made typodont system that could simulate malocclusion status and periodontal ligament space. It was reported that: (i) in the gingival displacement of upper canines, Clarity-SL produced significantly lower frictional force, whereas Damon-3Mx, In-Ovation-R, and SmartClip-SL3 produced higher frictional force among SLBs; (ii) in the lingual displacement of upper lateral incisors, Damon-Q and Damon-3Mx produced significantly lower frictional force, whereas Clarity-SL produced the highest frictional force among SLBs; and (iii) Clarity-SL combined with A-NiTi and Cu-NiTi, Damon-3Mx combined with A-NiTi, and In-Ovation-C combined with Cu-NiTi showed differences in the frictional properties between VD and HD; indicating that as the frictional properties of SLBs would be different between VD and HD of teeth, it is necessary to develop SLBs with low friction in both VD and HD of teeth.

Alfonso et al. [461] analyzed the influence of the nature of the orthodontic archwires on the friction coefficient and wear rate against materials used commonly as brackets (Ti–6Al–4V and 316L SS). The materials selected as orthodontic archwires were ASI304 SS, NiTi, Ti, β-Ti, and Cu-NiTi. Wear tests were conducted in AS at 37 °C. It was shown that: (i) a linear relationship between the hardness of the materials and the friction coefficients was obtained; (ii) the material that showed lower wear rate was the ASI304 SS; and (iii) to prevent wear, the wire and the brackets have high hardness values and in the same order of magnitude. Meier et al. [462] investigated whether electrochemical surface treatment of NiTi and β-Ti reduces friction inside the bracket–archwire complex. It was obtained that: (i) force losses due to friction were reduced by 10% points (from 36 to 26%) in the NiTi and by 12% points (from 59 to 47%) in the β-Ti wire specimens; (ii) most of the other material properties exhibited no significant changes after surface treatment; and (iii) while the three-point bending tests revealed mildly reduced force levels in the β-Ti specimens due to diameter losses of roughly 2%, these force levels remained almost unchanged in the NiTi specimens; concluding that compared to untreated NiTi and β-Ti archwire specimens, the surface-treated specimens demonstrated reductions in friction loss by 10 and 12%, respectively. Carrion-Vilches et al. [463] determined the static and kinetic friction forces of the contact-bracket archwire with different dental material compositions to select those materials with lower RS. The static and kinetic friction forces under dry and lubricating conditions using an AS solution at 36.5 ºC were determined. The bracket-archwire pairs was as follows: SS-SS; SS-glass fiber composite; SS-Nitinol 60; sapphire-SS; sapphire-glass fiber composite; and sapphire-Nitinol 60. It was mentioned that the best performance is obtained for Nitinol 60 archwire sliding against a SS bracket, both under dry and lubricated conditions.

Self-ligating braces are placed much like conventional braces, with the exception of ligature. No elastic or rubber bands or metal ties are present as they are not needed. Instead, special clips or brackets are used to help the archwire move teeth into place. The clips allow greater freedom of tooth movement; thereby they might reduce the discomfort that is sometimes associated with traditional ligature. A simple adjustment is made to the brackets or clips, which saves time and causes less discomfort during the treatment session(s) [464, 465]. Depending on the case, self-ligating braces are also easier on your teeth and could reduce the amount of pressure and friction for those who wear them. There are two types of SLBs: passive brackets and active brackets. Passive brackets are applied and use a smaller-size archwire; the smaller archwire creates less friction and allows teeth to move more freely; while active brackets use a thicker archwire and delivers more pressure onto teeth, thereby moving them more efficiently to where they ideally should be [466]. Cacciafesta et al. [454] measured and compared the level of frictional resistance generated between SS SLBs (Damon SL II), polycarbonate SLBs (Oyster), and conventional SS brackets (Victory Series), and three different orthodontic wire alloys: SS, NiTi, and β-Ti. It was found that: (i) SS SLBs generated significantly lower static and kinetic frictional forces than both conventional SS and polycarbonate SLBs, which showed no significant differences between them; (ii) the β-Ti archwires had higher frictional resistances than SS and NiTi archwires; (iii) no significant differences were found between SS and NiTi archwires; and (iv) all brackets showed higher static and kinetic frictional forces as the wire size increased. Tecco et al. [467] evaluated the frictional resistance generated by conventional SS brackets, self-ligating Damon SL II brackets, Time Plus brackets, and low-friction ligatures (Slide) coupled with various SS, NiTi, and β-Ti (TMA) archwires. It was reported that: (i) coupled with 0.016 inch NiTi, Victory brackets generated the most friction and Damon SL II the least; with 0.016 × 0.022 inch NiTi, the SLBs (Time and Damon SL II) generated significantly lower friction than Victory Series and Slide ligatures; with 0.019 × 0.025 inch SS or 0.019 × 0.025 inch NiTi, Slide ligatures generated significantly lower friction than all other groups; and (ii) no difference was observed among the four groups when used with a 0.017 × 0.025-inch TMA archwire; suggesting that the use of an in vitro testing model that includes 10 brackets provides information about the frictional force of the various bracket-archwire combinations. Baccetti et al. [468] analyzed the forces released by passive SS SLBs and by a nonconventional EL-bracket system on CBs (slide ligatures on CBs [SLCB]) when compared with conventional ELs on conventional brackets (CLCB) during the alignment of apically or buccally malpositioned teeth in the maxillary arch. An experimental model consisting of five brackets was used to assess the forces released by the three different bracket-ligature systems with 0.012-inch SE NiTi wires in the presence of different amounts of apical or buccal canine misalignment of the canine (ranging from 1.5 to 6 mm). The forces released by each wire/bracket/ligature combination with the three different amounts of apical or buccal canine misalignment were tested 20 times. Comparisons between the different types of wire/bracket/ligature systems were carried out by means of analysis of variance on ranks with Dunnett's post hoc test ($p < 0.05$). No difference in the amount of force released in the presence of a misalignment of 1.5 mm was recorded among the three systems. At 3 mm of apical misalignment, a significantly greater amount of orthodontic force was released by SLB or SLCB when compared with CLCB, whereas no significant differences were found among the three systems at 3 mm of buccal canine displacement. When correction of a large amount of misalignment (6 mm) was attempted, a noticeable amount of force for alignment was still generated by the passive SLB and SLCB systems while no force was released in the presence of CLCB.

There are some studies on influencing the factors on tribological (particularly, friction) actions of NiTi alloys. Kusy et al. [469] investigated whether the surface roughness of opposing materials influence the coefficients of friction and ultimately the movement of teeth, archwires were slid between contact flats to simulate orthodontic archwire-bracket appliances. From laser specular reflectance measurements, the RMS surface roughness of these archwires varied from 0.04 μm for SS to 0.23 μm for NiTi. Using the same technique, the roughness of the contact flats varied from 0.03 μm for the 1 μm lapped SS, to 0.26 μm for the as-received alumina. After each of the archwire-contact flat couples was placed in a friction tester, 15 normal forces were systematically applied at 34 °C. It was found that: (i) from plots of the static and kinetic frictional forces versus the normal forces, dry coefficients of friction were obtained that were greater than those reported in the dental literature; (ii) all SS couples had lower kinetic coefficients (0.120–0.148) than the SS-polycrystalline alumina couple (0.187); (iii) when pressed against the various flats, the β-Ti archwire (RMS = 0.14 μm) had the highest coefficients of friction (0.445–0.658) although the NiTi archwire was the roughest (RMS = 0.23 μm), and energy dispersive X-ray analysis (EDX) verified that mass transfer of the β-Ti archwire occurred by adhesion onto the SS flats or by abrasion from the sharply faceted polycrystalline alumina flats. Wichelhaus et al. [470] investigated the friction and surface roughness of different commercially available SE NiTi wires before and after clinical use. Forty SE wires (Titanol Low Force, Titanol Low Force River Finish Gold, Neo Sentalloy, and Neo Sentalloy Ionguard) of diameter 0.016 × 0.022 inches were tested. It was reported that: (i) initially, the surface-treated wires demonstrated significantly less friction than the nontreated wires; (ii) the surface roughness showed no significant difference between the treated and the nontreated surfaces of the wires; and (iii) all 40 wires, however, showed a significant increase in the friction and surface roughness during clinical use; suggesting that surface-treated NiTi orthodontic archwires should only be used once. Bandeira et al. [471] evaluated the friction force in sliding systems composed of coated NiTi archwires, coated NiTi subjected to thermal cycling, and coated NiTi subjected to acid solution immersion, and compared them to NiTi and polymeric wires. Samples of NiTi ($n = 05$), coated NiTi ($n = 15$), and OPTIS ($n = 05$) 0.016 inches in diameter and 50 mm long, in conjunction with Metafasix ligatures and saliva in InVu brackets, were submitted to friction testing. Among the 15 coated NiTi samples, five were submitted to thermal cycling for 3,000 cycles; the other five samples were immersed in acid solution for

30 days. It was found that: (i) the mean maximum friction force for NiTi, coated NiTi, and OPTIS was 105.20 ± (2.63), 99.65 ± (0.64), and 59.76 ± (4.93), respectively; (ii) there was no significant difference in NiTi, coated NiTi, and acid-immersed–coated NiTi; (iii) among the thermal-cycled or acid-immersed– coated NiTi wires, there was lower friction force in those undergoing thermal cycling; (iv) the coated NiTi and the OPTIS presented homogeneous surfaces, whereas NiTi wires presented a heterogeneous surface; and (v) fractures were observed in the coated NiTi wires that underwent thermal cycling. Hosseinzadeh et al. [472] assessed the surface characterization and frictional resistance between SS brackets and two types of orthodontic wires made of SS and NiTi alloys after immersion in a CHX-containing prophylactic agent. SS orthodontic brackets with either SS or heat-activated NiTi wires were immersed in a 0.2% CHX and an AS environment for 1.5 h. The frictional force was measured on a universal testing machine with a crosshead speed of 10 mm/min over a 5-mm archwire. It was obtained that: (i) there was no significant difference in the frictional resistance values between SS and NiTi wires immersed in either CHX or AS; (ii) the frictional resistance values for the SS and NiTi wires immersed in 0.2% CHX solution were not significantly different from that in AS; and (iii) no significant difference in the average surface roughness for both wires before (as-received) and after immersion in either CHX or AS was observed. Based on these results, it was concluded that 1½-h immersion in 0.2% CHX mouthwash did not have significant influence on the archwires surface roughness or the frictional resistance between SS orthodontic brackets and archwires made of SS and NiTi; and (v) CHX-containing mouthwashes may be prescribed as nondestructive prophylactic agents on materials evaluated in this study for orthodontic patients. Dridi et al. [473] investigated the friction force of SS and NiTi rectangular archwires against SS brackets. Two types of brackets were used namely: SLB and CB. Human saliva, olive oil, aloe vera oil, sesame oil, and sunflower oil were used as biolubricants. The friction force was examined as a function of the ligation method and oil temperature. It was reported that: (i) under oil lubrication, the friction behavior in the archwire/bracket assembly were the best; (ii) the SLB ligation was better than the conventional ligation system; and (iii) the enhancement of the frictional behavior with natural oils was linked to their main components: fatty acids. The friction between the brackets and orthodontic wire during sliding mechanics inflicts difficulties such as decreasing the applied force and tooth movement and the loss of anchorage. Geramy et al. [474] assesses the effect of 0.05% sodium fluoride (NaF) mouthwash on the friction between orthodontic brackets and wire. Four types of orthodontic wires including rectangular standard SS, β-Ti, NiTi, and Cu-NiTi were selected. In each group, half of the samples were immersed in 0.05% NaF mouthwash and the others were immersed in AS for 10 h. It was found that: (i) the friction rate was significantly higher after immersion in 0.05% NaF mouthwash in comparison with AS; and (ii) Cu-NiTi wire showed the highest friction value followed by β-Ti, NiTi, and SS wires; concluding that 0.05% NaF mouthwash increased the frictional characteristics of all the evaluated orthodontic wires.

In sliding mechanics, frictional force is an important counter-balancing element to orthodontic tooth movement, which must be controlled to allow application of light continuous forces. Whitley et al. [475] evaluated the influence of interbracket distance (IBD) on the RS. Commercially pure titanium brackets (CP-Ti) were tested against rectangular SS, NiTi, and β-Ti archwires in the dry and wet (human saliva) states. With a custom testing apparatus that simulated a three-bracket system, the RS was measured at a normal force of 300 cN and at second-order angles ranging from -9° to +9°. Twenty-three pairs of IBDs (written as IBD1/IBD2) were varied to simulate clinically relevant biomechanical scenarios with IBD ranging from 16 to 7 mm. It was reported that: (i) in the dry state, the kinetic frictional coefficients were equal to 0.12, 0.23, and 0.24 for the SS, NiTi, and β-Ti archwires against the CP-Ti brackets, respectively; (ii) the presence of saliva slightly increased the kinetic frictional coefficient; (iii) the RS was inversely proportional to the total IBD (IBD(T) = IBD1 + IBD2) regardless of archwire alloy or bracket slot; and (iv) elastic binding (BI = RS - frictional force in the passive region) did not depend on the order of the IBDs in the IBD1/IBD2 pair. Based on these findings, it was concluded that for a specific archwire-bracket couple, the BI of an IBD1/IBD2 pair is equal to any other pair with an equal IBD(T), and the kinetic frictional coefficient was linearly related to IBD(T) and total archwire beam length. Carrion-Vilches et al. [476] determined the static and kinetic friction forces of the contact bracket-archwire with different dental material compositions to select those materials with lower RS using an AS solution at 36.5 ºC. The bracket-archwire pairs studied included: SS-SS; SS-glass fiber composite; SS-Nitinol 60; sapphire-SS; sapphire-glass fiber composite; and sapphire-Nitinol 60. It was mentioned that the best performance is obtained for Nitinol 60 archwire sliding against a SS bracket, both under dry and lubricated conditions. Choi et al. [477] investigated the effects of sliding on the ultrastructure of three representative esthetic SE 0.014-inch NiTi archwires. A combination of four different types of 0.014-inch metallic wires and two different types of 0.022 inch × 0.028 inch CBs were evaluated by in vitro sliding tests using a tensile-strength tester with a miniature load cell and syringe pump. The NiTi wires included an uncoated NiTi archwire (CO group), ER-coated NiTi archwire (ER group), Teflon-coated NiTi archwire (TF group), and Ag/biopolymer-coated NiTi archwire (AG group). The brackets included contained SS and ceramic (CE) brackets. It was reported that: (i) both ER and TF wire groups exhibited less surface roughness than CO wire groups; (ii) the AG group showed the highest surface roughness compared to the others because of its silver particles; (iii) the in vitro sliding tests led to a significant increase in the surface roughness of all 0.014-inch NiTi wires regardless of the bracket type; (iv) the wire groups combined with SS brackets were rougher than those of CE brackets regardless of the coating materials because of exfoliation of the coating materials; and (v) the TF-SS group showed the highest increase (five-fold) in surface roughness compared to the others, whereas the ER groups showed the lowest increase (1.4-fold) in surface roughness compared to the others. It was then suggested that the sliding-driven surface roughness of SE NiTi archwires is directly affected by coating materials, and although the efficiency of orthodontic treatment was affected by various factors, ER-coated archwires were best for both esthetics and tooth movement when only considering surface roughness. Savoldi et al. [478] investigated whether velocity influences the RS at values approximating the orthodontic movement. A SS SLB with a NiTi clip was fixed onto a custom-made model. Different shaped orthodontic SS wires of four sizes and two types (round, 0.020″ and 0.022″; rectangular, 0.016″ × 0.022″ and 0.017″ × 0.025″) were tested. It was obtained that: (i) RS was higher for rectangular wires, and for those with larger diameters; and (ii) lower sliding velocity (SV) was associated with lower RS, with wire type and size having an interaction effect; concluding that at very low SV and low normal forces, SV influences the RS of SS archwires in orthodontic brackets, and the proportionality is logarithmic; and (iv) although respecting these parameters in vitro is challenging, quantitative evaluations of RS should be carried out at clinically relevant velocities, if aiming at translational application in the clinical scenario. Phukaoluan et al. [479] compared the frictional forces between a SS bracket and five different wire alloys under dry and wet (AS) conditions. TiNi, Cu-TiNi, Co-TiNi, commercial wires A, and commercial wires B with equal dimensions of 0.016 × 0.022″ were tested. It was mentioned that: (i) the static and dynamic frictions in the wet condition tended to decrease more slowly than those in the dry condition; hence, the friction of Cu-TiNi and commercial wires B would increase; (ii) these results were associated with scarred surfaces, i.e., the increase in friction would result in a larger bracket microfracture; and (iii) copper addition resulted in an increase in friction under both wet and dry conditions; however, the friction in the wet condition was less than that in dry condition due to the lubricating effect of AS.

16.3.2.7 Chemical and electrochemical corrosion

Within orthodontics, nickel is one of the most commonly used metals in wires, as it is included in NiTi, SSs, and other alloys. However, nickel is the most common metal to cause contact dermatitis and to induce more cases of allergic reactions. Reports have suggested that a concentration of approximately 30 mg/L of nickel may be sufficient to prompt a cytotoxic response [480]. Moreover, some complexes of nickel have been considered carcinogenic, allergenic, and mutagenic [481]. NiTi alloys can have >50% of nickel content and, consequently, release sufficient nickel ions to cause allergic reactions [482]. In addition, there have been reports of nickel nonsensitive patients who have become nickel-sensitive after using NiTi wires [483, 484]. Because of its ionic, thermal, microbiological, and enzymatic properties, the oral environment is favorable to the biodegradation of metal wires and their alloys, with consequent release of metal ions in the oral cavity [485]. The major process of degradation of metals is corrosion. Oral conditions such as the temperature, the quantity and acidity of saliva, the presence of certain enzymes, and the physical and chemical properties of solid and liquid food may influence corrosion processes. The intraoral environment is hostile due to both corrosive and mechanical actions. It is continuously full of saliva, an aerated aqueous solution of chloride with varying amounts of Na, K, Ca, PO_4, CO_2, sulfur compounds, and mucin. The pH value is normally in the range of 6.5–7.5, but under plaque deposits it may be as low as 2.0. Temperatures can vary as ± 36.5 °C, and a variety of food and drink concentrations (with pH values ranging from 2.0 to 14.0) stay inside the mouth for short periods of time. Loads can go up to 1,000 Newtons (normally 200 N as a masticatory force), sometimes in an impact manner. Trapped food debris may decompose, and sulfur compounds discolor. Under these chemical and mechanochemical intraoral environments, materials in service in the mouth are still expected to last for relatively long period of time [486–489]. Along with the release of elements from metals or alloys, corrosion of orthodontic wires can lead to roughening of the surface and weakening of the appliances and can severely affect the ultimate strength of the material, leading to mechanical failure or even fracture of the orthodontic materials. Corrosion actions take place through a simple chemical reaction or an electrochemical reaction (which the latter the phenomenon is more localized). There is still a galvanic corrosion, which occurs when the dissimilar metals are electrically connected (not necessarily directly contacting rather through the electrolyte) and normally corrosion current density is relatively high, so that the resultant corrosion rate would be higher than the normal corrosion process. Moreover, there are corrosion-assisted failures such as stress-corrosion cracking, corrosion fatigue, etc. In this section, all these issues relevant to intraoral chemical and electrochemical activities of NiTi orthodontic appliances are discussed.

Mouth rinsing with fluoride-containing products is an effective method for prevention of caries because of the complicated morphologies of orthodontic appliances. Regular use of products containing fluoride during the course of orthodontic treatment is essential as fluoride ion can promote the formation of calcium fluoride globules that stimulate remineralization [490, 491]. Clinically available fluoride-containing products have a variety of fluoride concentrations (250–10,000 mg/L) and pH values (3.5–7) [492, 493]. Prophylactic fluoride gels with a low pH were found to be more effective in the increase of calcium fluoride (CaF_2) formation [490].

The influence of fluoride and chloride ions on the corrosion behavior of nearly equiatomic NiTi orthodontic wires was studied using conventional electrochemical measurement methods, including corrosion potential, potentiodynamic, and cyclic potentiodynamic polarization measurements [494]. It was mentioned that: (i) NiTi alloy is primarily susceptible to localized corrosion when exposed to a solution containing chloride, while it is susceptible to general corrosion when subjected to a solution containing fluoride; and (ii) the synergistic interaction of fluoride and chloride on corrosion of NiTi alloy is associated with their respective molar concentrations. Kwon et al. [495] investigated the effect of fluoride released from dental restoratives on orthodontic NiTi wires. Five different restoratives (four fluoride-containing and one nonfluoride-containing) and four different NiTi wires were examined in this study. The pH of AS was adjusted to 2.5 and 6. It was found that: (i) after immersion for 10 days, the initial microhardness of the wires decreased by 0.3–5.6% depending on the test solution; (ii) Dyract AP (DA) and F2000 (F2; compomers) released significantly more fluoride than the other resin products (composite resins) regardless of the test solution; (iii) in pH 2.5 solution, both DA and F2 released 40–45 ppm/day fluoride for 5–6 days; and (iv) as for the wires in contact, they did not show any visible modification in surface morphology; suggesting that despite the released fluoride, wires in contact with the fluoride-containing restoratives were not damaged regardless of the pH value of test solution. Abalos et al. [496] studied several topographical features and their influence upon fluoride corrosion. Four topographies (smooth, dimple, scratch, and crack) according to the main surface defect were characterized ($n = 40$). Static corrosion tests were performed in AS with fluorated prophylactic gel (12,500 ppm) for 28 days. It was observed that: (i) there was an increase in the surface defects and/or roughness of the cracked and scratched surfaces; (ii) these defects produced an important increase in corrosion behavior; (iii) the best surfaces for the orthodontic archwires were the smooth and dimpled surfaces, respectively; and (iv) the increase in defects was independent of roughness. Manufacturing processes that produce surface cracks should be avoided in orthodontic applications. Srivastava et al. [497] assessed the change in load deflection characteristics of NiTi, Cu- NiTi, SS, and β-Ti wires on immersing in fluoride mouth rinses of two types (Phosflur and neutral NaF) mouth rinse using a modified bending test and comparing it to control. It was obtained that there was no statistically significant reduction in load deflection characteristics of NiTi, Cu-NiTi, β-Ti, and SS wires on immersing in Phosflur mouth rinse and neutral NaF mouth rinses compared to control at 2.5 and 1 mm of deflection in unloading phase. Huang et al. [498] coated diamond-like carbon (DLC) films onto NiTi orthodontic archwires to protect against fluoride-induced corrosion and to improve orthodontic friction. It was indicated that: (i) superior nature of the DLC coating was noticed, with less surface roughness variation for DLC-coated archwires after immersion in a high fluoride ion environment; and (ii) friction tests also showed that applying a DLC coating significantly decreased the fretting wear and the coefficient of friction, both in ambient air and AS; concluding that the DLC coatings are recommended to reduce fluoride-induced corrosion and improve orthodontic friction. Pulikkottil et al. [499] evaluated the corrosion resistance of four different orthodontic archwires and determined the effect of 0.5% NaF (simulating high fluoride-containing toothpaste of about 2,250 ppm) on corrosion resistance of these archwires. Four different archwires (SS, NiTi, titanium TMA, and ion-implanted [L-TMA]) were considered for this study. It was reported: (i) the potentiostatic study reveals that the corrosion resistance of low-friction TMA (L-TMA) > TMA > NiTi > SS; (ii) AFM analysis showed the surface Ra of TMA > NiTi > L-TMA > SS; indicating that the chemical composition of the wire is the primary influential factor to have high corrosion resistance and surface Ra is only secondary; and (iii) the corrosion resistance of all wires had reduced significantly in 0.5% acidic fluoride-containing AS due to formation of fluoride complex compound. Based on these findings, it was concluded that the presence of 0.5% NaF in AS was detrimental to the corrosion resistance of the orthodontic archwires, and complete removal of residual high-fluorinated toothpastes from the crevice between archwire and bracket during tooth brushing is mandatory. Mocnik et al. [500] analyzed the influence of environmental effects on the corrosion and tribocorrosion properties of NiTi and SS dental alloys. The effects of pH and fluorides on the electrochemical properties were studied using the cyclic potentiodynamic technique. It was found that lowering the pH preferentially affects the corrosion susceptibility of NiTi alloys, whereas SS dental archwires are prone to local types of corrosion; (ii) the NiTi alloy is not affected by smaller increases of fluoride ions up to 0.024 M, whereas at 0.076 M (simulating the use of toothpaste) the properties are affected; (iii) a leaching test during wear-assisted corrosion showed that the concentrations of nickel ions released into the saliva exceeded the limit value of 0.5 μg/cm²/week; and (iv) the oxide films on the NiTi and SS alloys after the tribocorrosion experiment were thicker than those exposed to saliva only. Gupta et al. [501] assessed and evaluated the effects of fluoride prophylactic agents on mechanical properties of NiTi wires during orthodontic treatment using SEM. The commercially available round preformed NiTi orthodontic archwire (3 M company) and three different mouthwash solutions, i.e., Phos-Flur gel (1.1% sodium APF, 0.5% w/v fluoride, pH = 5.1; Colgate Oral Pharmaceuticals) and Prevident 5000 (1.1% NaF neutral agent, 0.5% w/v fluoride, pH = 7; Colgate Oral Pharmaceuticals) were used. It was reported that: (i) there was not much difference in the values of both modulus of elasticity and YS obtained after loading of stress on the wires in all the three experimental conditions; (ii) a significant difference in both modulus of elasticity and YS was observed during unloading of stress; and (iii) when the surface characteristics were observed for all the specimens using SEM images, it was observed that NiTi wires treated with Phos-Flur showed large surface defects, which appeared as round, pitted areas depicting corrosion, numerous white inclusions, and overall damaged surface structure of the wire compared to the control. Based on these results, it was concluded that fluoridated mouthwashes are essential to maintain good oral hygiene and decrease instance of caries in patients undergoing orthodontic treatment, and the prophylactic usage of topical fluoride agents on NiTi wire seems to diminish the mechanical properties of the orthodontic wire that could significantly affect future treatment outcomes. Recently, Mlinaric et al. [502] explored the influence of the interaction of oral antiseptics and various coatings on the corrosion behavior of NiTi orthodontic alloys. NiTi archwires with uncoated, rhodium-coated, and nitride-coated surface were exposed to AS pH 4.8 at the temperature of 37 °C, and to saliva with the addition of three commercial oral antiseptics (Curaspet, Gengigel, and Listerine). It was reported that: (i) EIS indicates the biggest corrosion resistance in Gengigel; (ii) rhodium-coated NiTi demonstrates higher corrosion rate in saliva and in all antiseptics than uncoated and nitride-coated NiTi; (iii) the highest tendency toward localized corrosion is seen in rhodium-coated NiTi in all media, and least in nitride-coated NiTi; and (iv) the release of titanium ions mainly supports findings of general corrosion rate; indicating that both antiseptics and coating modify corrosion.

There are several corrosion studies using different media such as Cl⁻ ion, including solution and AS. Iijima et al. [15–503] examined the corrosion behaviors of a commercial NiTi alloy orthodontic wire and a polished plate with same composition in 0.9% NaCl and 1% lactic acid (LA) solutions using an electrochemical technique, an analysis of released ions, and a surface analysis by X-ray photoelectron spectroscopy (XPS). It was reported that: (i) the XPS analysis demonstrated the presence of a thick oxide film mainly composed of TiO_2 with trace amounts of nickel hydroxide, which had formed on the wire surface during the heat treatment and subsequent pickling processes; (ii) the oxide layer contributed to the higher resistance of the as-received wire to both general and localized corrosion in 0.9% NaCl solution compared to that of the polished plate and the polished wire; and (iii) the thick oxide layer, however, was not stable and did not protect the orthodontic wire from corrosion in 0.1% LA solution. Yonekura et al. [504] investigated the corrosion characteristics of orthodontic alloy wires (both in as-received and grinded conditions) in 0.9% NaCl solution by atomic absorption spectrophotometry and potentiodynamic polarization measurements. It was reported that: (i) the amount of each metal ion released from most alloys was larger for the grinded wires than for the as-received wires; (ii) the β-Ti alloy wire (Ti-Mo-Zr) does not contain allergenic metals such as Ni, Co, and Cr, and the finding that resistance to both general and localized corrosion is the highest among the six wires investigated suggests that this wire is the most biocompatible orthodontic wire; and (iii) as a small amount of Ni, Cr, or Co ions were released from NiTi, CoCr, and SS wires, special attention should be paid during their clinical use for patients with allergic tendencies. Kao et al. [505] assesses the surface characteristics and to compare the corrosion potential of metal brackets and wires in environments containing different media. Four brands of metal brackets and two types of orthodontic wires (SS and NiTi) were investigated. The 37 °C test media were acidulated NaF and pH 4 and pH 6 AS solutions. It was recorded that: (i) the potentiodynamic curve showed that most brands of metal brackets were easily corroded in the NaF and pH 4 environments, whereas the NiTi and SS wires were easily corroded in the pH 4 AS; and (ii) SEM observations showed that defects or pitting corrosion occurred on the surfaces of the brackets and wires in all tested media.

16.3.2.8 Corrosion in artificial saliva

Huang et al. investigated the variation during in vitro corrosion resistance of commercial NiTi dental orthodontic wires from different manufacturers using the fast electrochemical technique wires, in acidic AS at 37 °C [506], as well as studied the surface characterizations and corrosion resistance of as-received commercial NiTi dental orthodontic archwires from different manufacturers using a cyclic potentiodynamic test in AS with various acidities [507]. It was reported that: (i) NiTi wires from different manufacturers had a statistically significant difference in R_P; (ii) the surface roughness of the commercial NiTi wires with similar surface chemical structure does not correspond with the difference in corrosion resistance; (iii) the surface structure of the passive film on the tested NiTi wires were identical, containing mainly TiO_2, with small amounts of NiO; and (iv) the difference in the corrosion resistance among these NiTi dental orthodontic archwires did not correspond with the surface roughness and preexisting defects. Wang et al. [508] investigated the electrochemical behavior of $Ti_{50}Ni_{47.2}Co_{2.8}$ alloy in deaerated AS solutions at 37°C with binary NiTi alloy as reference and to characterize the composition and structure of the passive film after polarization tests. It was found that: (i) potentiodynamic and potentiostatic tests results showed that the corrosion behavior of Co-NiTi was similar to that of NiTi alloy,; (ii) with the increase in pH value of the electrolytes, both corrosion potential and pitting corrosion potential decreased; (iii) XPS results revealed that the outmost passive film consisted mainly of TiO_2, which were identical with that of NiTi alloy; (iv) the nickel ion release amount of Co-NiTi was very close to that of NiTi alloy; and (v) neither titanium nor Co ion was detected due to the detection limitation. Liu et al. [509] studied the influence of bending stress on the nickel release of commercial NiTi orthodontic wires in vitro AS. It was reported that: (i) the stressed NiTi wires exhibited substantial increases in the nickel release compared with the unstressed specimens during all experimental periods; (ii) the highest dissolution rate during the 0–1 day incubation period was observed for all stressed specimens; however, a slight increase of nickel released as a function of time was observed in the three groups of stressed specimens after 3 days of immersion; and (iii) for the stressed specimens, it was hypothesized that the bending stress would induce buckling or cracking of the protective oxide film of the NiTi wires. It was then concluded that the bending stress influences the nickel release of NiTi wires, and the factor of loading condition with respect to corrosion behavior and passive film should be considered in view of the widespread use of NiTi wires for dental devices. The effect of fluoride-added AS solution on NiTi and β-Ti alloys, used for orthodontic wires, was investigated [510]. The orthodontic wires were immersed into fluoride-added AS solution of pH 4.8 at 37 °C. It was concluded that: (i) fluoride has an important effect on corrosion behavior of experimental wires that can be described as a type of wear; and (ii) EDS analysis denotes the existence of a thin oxide film layer in addition to the titanium-based oxide one on the surface of the metals. Trolić et al. [511] examined how probiotic supplements affect the corrosion stability of $Ni_{50.4}Ti_{49.6}$ orthodontic archwires. It was reported that: (i) rhodium-coated alloy in saliva has significantly higher general corrosion in saliva than nitrified alloy and uncoated alloy, with large effect size; (ii) in the presence of probiotics, the result was even more pronounced; (iii) probiotic supplement increases general and localized corrosion of rhodium-coated archwire and slightly decreases general corrosion and increases localized corrosion in uncoated archwire, whereas in the case of nitrified archwire, the probability of corrosion is very low; (iv) the differences in surface roughness between NiTi wires before corrosion are not significant; (v) exposure to saliva decreases Ra in rhodium-coated wire; and (vi) media do not significantly influence surface microgeometry in nitrified and uncoated wires.

Based on these results, it was concluded that probiotic supplement affects corrosion depending on the type of coating of the NiTi archwire, and it increases general corrosion of rhodium-coated wire and causes localized corrosion of uncoated and rhodium-coated archwire. Probiotic supplement does not have greater influence on the surface roughness compared to that of saliva.

16.3.2.9 Galvanic corrosion

There are two types of galvanic cell: the first type are the micro-cells (or local cells) and the second type are the macro-cells. In case of micro-cells, within a single piece of metal or alloy, there exist different regions of varying composition, and hence different electrode potentials. This is true for different phases in heterogeneous alloys [512]. The positive way of applying the phenomena is by etching the polished surface of metal to reveal the microstructures for metallographic observations. A severely deformed portion within one piece of a material can serve as an anodic site [513]. For example, a portion of a fully annealed nail was bent and the whole nail was immersed into a NaCl aqueous solution. After several hours, it was easily noticed that the localized plastic-deformed portion was rusted. For macro-cells, which take place at the contact point of two dissimilar metals or alloys with different electrode potentials, the so-called galvanic corrosion is probably the best known of all the corrosion types [512]. Titanium is the metal choice for implant material due to excellent tissue compatibility; however, superstructures of such titanium implants are usually made of different alloys. This polymetallism leads to detectable (in macro-level) galvanic corrosion. The intensity of the galvanism phenomenon is due to a number of factors, such as the electrode potential, extent of the polarization, the surface area ratio between anodic site and cathodic site, the distance between the electrodes, the surface state of the electrodes, conductivity of the electrolyte, and the passage of a galvanic current on the electrolytic diffusion, stirring, aeration or deaeration, temperature, pH and compositions of the electrolytic milieu, coupling manner, and an accompanying crevice corrosion. When two solid bodies are in contact, it is fair to speculate that there could be marginal gaps between them, because they are not subject to any bonding method like diffusion bonding or fusion welding. Therefore, when the galvanic corrosion behavior is studied in a dissimilar material couple, the crevice corrosion is another important localized corrosion measurement accompanied with galvanic corrosion [277, 513–516]. Among these numerous influencing factors, the surface area ratio between anodic surface and cathodic surface appears to be the most important factor to be considered. If the ratio of surface cathode area (A_C) and the anode area is large; the increased corrosion caused by coupling can be considerable. Conductivity of the electrolyte and geometry of the system enter the problem because only that part of the A_C is effective for which resistance between anode and cathode is not a controlling factor [516].

Schiff et al. [517] determined the influence of fluoride in certain mouthwashes on the risk of corrosion through galvanic coupling of orthodontic wires and brackets.

Two titanium alloy wires, NiTi and Cu-NiTi, and three commonly used brackets, Ti, FeCrNi (SS), and CoCr, were tested in a reference solution of Fusayama-Meyer AS and in two commercially available fluoride (250 ppm) mouthwashes, Elmex and Meridol. It was found that: (i) Meridol mouthwash, which contains stannous fluoride, was the solution in which the NiTi wires coupled with the different brackets, showed the highest corrosion risk, whereas in Elmex mouthwash, which contains sodium fluoride, the Cu-NiTi wires presented the highest corrosion risk; and (ii) such corrosion has two consequences: deterioration in mechanical performance of the wire-bracket system, which would negatively affect the final aesthetic result, and the risk of local allergic reactions caused by released nickel ions; suggesting that mouthwashes should be prescribed according to the orthodontic materials used. Iijima et al. [518] quantitatively assessed the galvanic corrosion behavior of orthodontic archwire alloys coupled to the orthodontic bracket alloys in 0.9% NaCl solution and to study the effect of surface area ratios. Two common bracket alloys, SSs and titanium, and four common wire alloys, NiTi, β-Ti, SS, and CoCrNi alloy, were used. Three different area ratios, 1:1, 1:2.35, and 1:3.64, were used; two of them assumed that the multibracket appliances consists of 14 brackets and 0.016 inch of round archwire or 0.016 × 0.022 inch of rectangular archwire. It was reported that: (i) when the NiTi alloy was coupled with titanium (1:1, 1:2.35, and 1:3.64 of the surface area ratio) or β-Ti alloy was coupled with titanium (1:2.35 and 1:3.64 of the surface area ratio),it initially was the anode and corroded; (ii) however, the polarity reversed in 1 h, resulting in corrosion of the NiTi or β-Ti; and (iii) the NiTi alloy coupled with SS or titanium exhibited a relatively large galvanic current density even after 72 h, suggesting that the coupling SS (Fe-18Cr-8Ni)-NiTi and Ti-NiTi may remarkably accelerate the corrosion of NiTi alloy, which serves as the anode, and the different anode and A_C ratios used in this study had little effect on galvanic corrosion behavior. Siargos et al. [519] compared the galvanic coupling of conventional and metal injection molded (MIM) brackets with commonly used orthodontic archwires. Six of each type of bracket were suspended in LA along with a sample of orthodontic wire (three NiTi and three copper-NiTi) for 28 days at 37 °C. It was found that: (i) the MIM brackets exhibited potential differences similar to those seen for the CBs; (ii) the greatest potential difference was found for MIM brackets with NiTi wires (512 mV), whereas MIM brackets with copper-NiTi wires had the smallest difference (115 mV); and (iii) the MIM bracket exhibited extensive internal porosity, whereas the CB was more solid internally. It was then concluded that the composition and manufacturing processes involved in fabricating MIM brackets impart corrosive properties similar to those seen in the bracket-wing area of CBs and may provide a measurable benefit when taking into account the increased corrosion between the bracket and brazing alloy of CBs. Bakhtari et al. [520] compared galvanic currents generated by different combinations of commonly used brackets and archwires. As-received SS, NiTi, and β-Ti wires were coupled to SS and titanium brackets in an AS medium. It was reported that: (i) two-way analysis of variance showed a significant difference in charge and galvanic currents when factored for the type of bracket, but no significant difference between them when factored by the type of wire;

and (ii) specifically, a brazed SS bracket was significantly greater in charge transferred and 10-h galvanic current than MIM SS and titanium brackets, which were not different from each other; concluding that the method of bracket manufacturing might be of equal or more relevance to galvanic corrosion susceptibility than bracket composition. Polychronis et al. [521] assessed galvanic behavior of lingual orthodontic brackets coupled with representative types of orthodontic wires. Three types of lingual brackets: Incognito (INC), In-Ovation L (IOV), and STb (STB) were combined with a SS and a NiTi orthodontic archwire. All bracket-wire combinations were immersed in acidic 0.1 M NaCl, 0.1 M LA, and neutral NaF 0.3wt% electrolyte, and the potential differences were continuously recorded for 48 h. It was obtained that: (i) the SEM/EDX analysis revealed that INC is a single-unit bracket made of a high gold alloy, whereas IOV and STB are two-piece appliances in which the base and wing are made of SS; (ii) the SS wire demonstrated austenite and martensite iron phase, whereas NiTi wire illustrated an intense austenite crystallographic structure with limited martensite; and (iii) all bracket-wire combinations showed potential differences below the threshold of galvanic corrosion (200 mV), except for INC and STB coupled with NiTi wire in NaF media. It was, therefore, concluded that all brackets tested demonstrated galvanic compatibility with SS wire, but fluoride treatment should be used cautiously with NiTi wires coupled with gold and SS brackets.

16.3.2.10 Metallic ion release

During the corrosion process, metallic ion(s) can be released from the substrate surface area, causing the dissolution into corrosive media or surrounding tissue (if the case is for implant). Hence, ion release is directly related to the stability of surface oxide (passive film) formed on the titanium substrate, and capability for repassivation when such a stable passive film is disrupted. On the polarization process, there should be two occasions when ions can be released (in other words; instability of passivation); active dissolution before the onset of passivation occurs at a relatively low potential and transpassivity (or breakdown of stable passive film) at relatively high potential. For example, higher than 1.5 V (vs SCE), depending on the electrolyte and type of titanium materials used. Such transpassivation is normally accompanied with the formation of pits [486]. The biological significance of the metal ion release is believed to be strongly related to cytocompatibility and hemocompatibility. It is also generally believed that corrosion resistance is responsible for the biocompatibility. The metallic ion release from dental alloys has been transformed into one of the principal problems in the health of dental patients who have metallic materials fitted in their mouths [522]. It is well known that metals are toxic in sufficient concentration and that they can cause inflammations [523], allergies, genetic mutations, or cancer [524]. The metallic elements that are released in oral implants have been detected in the tongue [524], the saliva [525], and in the gums adjacent to these alloys [526, 527]. On the other hand, the surface from which nickel element is selectively leached out should result in nickel-enriched surface of NiTi, leading to minimizing any potential risk of health-related concerns due to nickel element [286].

Huang et al. [528] investigated the corrosion resistance, in terms of ion release, of different NiTi orthodontic wires in AS with various acidities. Four types of as-received commercial NiTi orthodontic wires were immersed in AS (37 °C) at pH 2.5–6.25 for different periods (1–28 days). It was obtained that: (i) the manufacturer, pH value, and the immersion period, respectively, had a significantly statistical influence on the release amount of nickel and titanium ions; (ii) the amount of nickel ions released in all test solutions was well below the critical value necessary to induce allergy and below daily dietary intake level; (iii) the amount of titanium ions released in pH > 3.75 solution was mostly not detectable, representing that the TiO_2 film on NiTi wires exhibited a good protection against corrosion; (iv) preexisted surface defects on NiTi wires might be the preferred locations for corrosion; and (v) the NiTi wire with the highest release amount of metal ions had the maximal increase in surface roughness after immersion test, whereas a rougher surface did not correspond to a higher metal ion release. Karnam et al. [529] determined the amount of nickel, chromium, copper, cobalt, and iron ions released from simulated orthodontic appliance made of new archwires and brackets. A total of 60 sets of new archwire, band material, brackets, and ligature wires were prepared simulating fixed orthodontic appliance. These sets were divided into four groups of 15 samples each: (1) Group 1, SS rectangular archwires; (2) Group 2, rectangular NiTi archwires; (3) Group 3, rectangular copper NiTi archwires; and (4) Group 4: rectangular elgiloy archwires. It was reported that: (i) high levels of nickel ions were released from all four groups, compared to all other ions, followed by the release of iron ion levels; and (ii) there is no significant difference in the levels of all metal ions released in the different groups; indicating that the use of newer brackets and newer archwires confirms the negligible release of metal ions from the orthodontic appliance. Azizi et al. [530] evaluated the amount of titanium and nickel ions released from two wires with different shapes and a similar surface area. Forty round NiTi archwires with the diameter of 0.020 inches and 40 rectangular NiTi archwires with the diameter of 0.016 × 0.016 inches were immersed in 37 °C artificial saliva (AS) during a 21-day period. The surface area of both wires was 0.44 $inch^2$. Wires were separately dipped into polypropylene tubes containing 50 ml of buffer solution and were incubated and maintained at 37 °C. Inductively coupled plasma atomic emission spectrometry (ICP-AES) was used to measure the amount of ions released after exposure lengths of 1 h, 24 h, 1 week, and 3 weeks. It was indicated that: (i) the amount of nickel and titanium concentrations was significantly higher in the rectangular wire group; (ii) the most significant release of all metals was measured after the first hour of immersion; in the rectangular wire group, 243 ± 4.2 ng/ml of nickel was released after 1 h, whereas 221.4 ± 1.7 ng/ml of nickel was released in the round wire group; and (iii) similarly, 243.3 ± 2.8 ng/ml of titanium was released in the rectangular wire group and a significantly lower amount of 211.9 ± 2.3 ng/ml of titanium was released in the round wire group; concluding that the release of metal ions was influenced by the shape of the wire and increase of time. Katic et al. [531] determined the effect of pH, fluoride (F^-), and hydrofluoric acid concentration on dynamic of nickel (Ni^{2+}) and titanium (Ti^{4+}) ions release. NiTi wires with untreated surface (NiTi), rhodium (RhNiTi), and nitride (NNiTi) coating were immersed once a week for 5 min in remineralizing agents, followed by immersion in AS. It was reported that: (i) the release of Ni^{2+} from NiTi and NNiTi wires correlated highly linear and positively with hydrofluoric acid; for RhNiTi, the correlation was lower and negative; (ii) the prediction of Ti^{4+} release was significant for NiTi and NNiTi wires; and (iii) association with F^- was lower; for pH it was not significant. Hydrofluoric acid predicts the release of ions from the NiTi wires better than the pH and F^- of the prophylactic agents. Tahmasbi et al. [532] investigated the galvanic corrosion of brackets manufactured by four different companies coupled with SS or NiTi wires in an AS solution. A total of 24 mandibular central incisor Roth brackets of four different manufacturers (American Orthodontics, Dentaurum, Shinye, and ORJ) were used and these brackets were immersed in AS along with SS or NiTi orthodontic wires for 28 days. It was found that: (i) among ions evaluated, the release of nickel ions from Shinye brackets was significantly higher than that of other brackets; (ii) the mean potential difference was significantly lower in specimens containing a couple of Shinye brackets and SS wire compared with other specimens; (iii) no significant difference was observed in the mean corrosion rate of various groups; and (iv) microscopic evaluation showed corrosion in two samples only: Shinye bracket coupled with SS wire and American orthodontics bracket coupled with NiTi wire. Based on these results, it was concluded that the Shinye brackets coupled with SS wire showed more susceptibility to galvanic corrosion, and there were no significant differences among specimens in terms of the corrosion rate or released ions, except the release of nickel ions, which was higher in Shinye brackets.

16.3.2.11 Nickel ion release

Eliades et al. [533] assessed the nickel content of as-received and retrieved SS and NiTi archwires alloys. New and used brand-matched, composition-matched, and cross-section-matched archwires were subjected to SEM and energy-dispersive electron probe microanalysis. Elemental analysis was performed on three randomly selected areas, and the nickel content with an expression of ratios of Ni/Ti (in NiTi wires) or Ni/Fe (in SS). It was found that: (i) no changes were detected with respect to nickel content ratios between as-received and retrieved NiTi or SS wires, suggesting an absence of nickel release; and (ii) wear and delamination phenomena on the wire surface and the formation of galvanic couple between the SS wires and bracket brazing materials intraorally may modify the corrosion susceptibility of the wire alloys in clinical conditions. Arndt et al. [534] analyzed seven NiTi levelling arches, one titanium molybdenum, a cobalt chromium, and three sSS wires with respect to their corrosion behavior under realistic conditions in AS and static immersion tests in AS or LA, as well as immersion tests with mechanical, thermal, and combined mechanical and thermal stresses were performed. It was mentioned that the obtained results yield information not only about the relative corrosion tendency of the wires under in vitro conditions, but also give a quantitative estimation about the nickel ion release of the orthodontic wires during in vivo treatment. It was also reported that: (i) generally, the maximum release of nickel ions was two orders of magnitude below the daily dietary intake level; (ii) mechanical and thermal loading increases nickel release in the immersion tests by a factor of 10–30; and (iii) two NiTi wires (Dentaurum Tensic and Forestadent Titanol Low Force), which were examined showed lower rupture potentials and a higher tendency toward corrosion in the immersion tests than the others due to their surface composition; however, these differences are levelled off by long-term mechanical and thermal loading. The patterns of nickel release from NiTi vary depending on the type of material (NiTi alloys with low or no processing vs commercial wires or sheets). A thick TiO_2 layer generated on the wire surface during processing is often considered as a reliable barrier against nickel release. Shabalovskaya et al. [535] studied nitinol wires with surface oxides resulting from production to identify the sources of nickel release and its distribution in the surface sublayers. It was reported that: (i) wire samples in the as-received state showed low breakdown potentials (200 mV); the improved corrosion resistance of these wires after treatment was not affected by strain; and (ii) nitinol wires with the thickest surface oxide TiO_2 showed the highest nickel release, attributing to the presence of particles of essentially pure nickel whose number and size increased while approaching the interface between the surface and the bulk. Petoumenou et al. [536] examined whether NiTi archwires cause an increase of nickel concentration in the saliva of 18 orthodontic patients to estimate the possible risk of these archwires in patients who have nickel hypersensitivity. Saliva samples were collected before orthodontic treatment, after placement of the bands and brackets, 2 weeks later and before placing the NiTi archwires, immediately after placing the NiTi archwires, 4 weeks after placing the wires, and 8 weeks after placing the wires. It was reported that: (i) by using mass spectrometry, no statistically significant differences were found in the nickel concentrations in the samples taken without appliances, in those obtained 2 weeks after placement of the bands and brackets, and 4 and 8 weeks after placement of the archwires; and (ii) samples taken immediately after placement of the bands and brackets and the NiTi archwires showed slight but significant increases in the nickel concentration of 78 and 56 mg/l, respectively, compared to the pretreatment value of 34 mg/l. Based on these findings, it was concluded that nickel leaching occurred after placement of the bands and brackets and after placement of the NiTi archwires, associated with an increase of the nickel ion concentration in the patient's saliva, and this effect decreased within 10 weeks. Suárez et al. [537] studied surface topographic changes and nickel release in lingual orthodontic archwires in vitro. SS, NiTi, and Cu-NiTi lingual orthodontic archwires were studied using atomic absorption spectrometry for nickel release after immersion in a saline solution. It was mentioned that: (i) statistically significant changes in roughness were seen in all archwires, except

NiTi; (ii) surface changes were most severe in the Cu-NiTi alloy; and (iii) SS archwires released the highest amount of nickel; concluding that only roughness changes in Cu-NiTi archwires seemed to be clinically significant; and (v) the amount of nickel released for all archwires tested is below the levels known to cause cell damage. Espinar et al. [538] studied NiTi orthodontic archwires that have been treated using a new oxidation treatment for obtaining nickel-free surfaces. The TiO_2 on the surface significantly improves corrosion resistance and decreases nickel ion release, while barely affecting the transformation temperatures. The stable passive oxide film formation and a certain oxidation treatment can avoid the allergic reactions or toxicity in the surrounding tissues produced by the chemical degradation of the NiTi. On the other hand, the lack of low friction coefficient for the NiTi SE archwires makes it difficult for the optimal use of these materials in orthodontic applications. It was mentioned that the decrease of this friction coefficient has been achieved by means of oxidation treatment. Ousehal et al. [539] evaluated the levels of nickel released into saliva by fixed orthodontic appliances on 16 patients (eight boys and eight girls). After 8 weeks placement: (i) a significant increase in the nickel levels was seen just after NiTi archwire insertion; however, the difference was nonsignificant; (ii) certain studies concur with ours, showing appreciable changes in concentration, but with no significant difference; and (iii) others although have shown a statistically significant difference; indicate that orthodontic appliances release nickel ions mainly at the start of orthodontic treatment. Jamilian et al. [540] measured the levels of nickel and chromium ions released from 0.018" SS and NiTi wires after immersion in three solutions (Oral B®, Orthokin®, and AS) for 1, 6, 24 h, and 7 days. It was obtained that the amount of chromium and nickel significantly increased in all solutions during all time intervals; concluding that Cr and nickel ions were released more in NiTi wire in all solutions compared with SS wire, and the lowest increase rate was also seen in AS. Mlir hashemi et al. [541] assessed the release of nickel and chromium ions from NiTi and SS orthodontic wires following the use of four common mouthwashes available on the market in Oral B, Oral B 3D White Luxe, Listerine, and Listerine Advance White for 1, 6, 24, and 168 h. It was found that: (i) nickel ions were released from both wires at all time points; the highest amount was in Listerine and the lowest in Oral B mouthwashes; (ii) the remaining two solutions were in-between this range; and (iii) the process of release of chromium from the SS wire was the same as that of nickel; however, the release trend in NiTi wires was not uniform; indicating that Listerine caused the highest release of ions. Listerine Advance White, Oral B 3D White Luxe, and distilled water were the same in terms of ion release, and Oral B showed the lowest amount of ion release.

16.3.2.12 Nickel allergy issue

Nickel is a component of NiTi and SS alloys, which are widely used in orthodontic appliances. Level of nickel in saliva and serum increases significantly after the insertion of fixed orthodontic appliances. A threshold concentration of approximately 30 ppm of nickel may be sufficient enough to elicit a cytotoxic response. Experimental and clinical studies indicate that oral exposure to nickel containing alloys may reduce the chance of nickel sensitization by later exposure to the metal, i.e., induce a certain tolerance. Alternatives like TwistFlex SS, fiber-reinforced composite archwires, TMA, pure titanium, and gold-plated wires may also be used without risk. Although no cytotoxic effect caused by NiTi alloy has been reported [542], information concerning the biological side effects of nickel is available in literature [543–546]. Nickel is capable of eliciting toxic and allergic responses [543]. It can produce more allergic reactions than all other metal elements [544]. In the in vivo study, NiTi alloys show cytotoxic reactions [545]. Cases also show the conversion of Ni-nonsensitive subjects into Ni-sensitive subjects following the use of NiTi wires [546].

Titanium brackets are used in orthodontic patients with an allergy to nickel and other specific substances. Harzer et al. [547] investigated the corrosive properties of fluoride-containing toothpastes with different pH values. The in vivo study tested how the surfaces of titanium brackets react to the corrosive influence of acidic fluoride-containing toothpaste during orthodontic treatment. Molar bands were placed on 18 orthodontic patients. In these patients, titanium brackets were bonded on the left quadrants and SS brackets on the right quadrants of the upper and lower arches. Fifteen patients used Gel Kam containing soluble tin fluoride (pH - 3.2), whereas three used fluoride-free toothpaste. The brackets were removed for the evaluation by light microscopy and SEM 5.5–7 months and 7.5– 17 months after bonding. Macroscopic evaluation showed the matte gray color of titanium brackets dominating over the silver gleam of the steel brackets. The plaque accumulation on titanium brackets is high because of the very rough surface. Pitting and crevices were observed in only three of the 165 brackets tested. The in vivo investigation confirms the results of the in vitro studies, but the changes are so minor that titanium brackets can safely be used for up to 18 months [547]. Genelhu et al. [548] conducted the retrospective analysis to investigate the roles of age, sex, previous allergic history, and time of exposure to fixed orthodontic appliances in the etiopathogeny of nickel-induced allergic contact stomatitis NiACS). Forty-four orthodontic patients (range, 10–44 years) were divided into two groups, depending on their NiACS clinical manifestations. It was found that young patients, especially females with a history of allergic reactions, had a greater predisposition to NiACS clinical manifestations; time of exposure to orthodontic appliances was not a significant factor; concluding that a previous allergic reaction should be considered a predictive factor of NiACS clinical manifestations and should be noted in the patient's medical history. Pazzini et al. [549] determined the prevalence of nickel allergy in a sample of orthodontic patients and longitudinally compare the clinical periodontal status of these individuals with that of a group of nonallergic patients. The initial sample consisted of 96 patients selected randomly from a databank of patients who sought orthodontic care at a teaching institution. Following the selection and beginning of treatment, periodontal status was assessed over a 12-month period [one evaluation every 3 months-T(1), T(2,) T(3,) T(4)] using the Loe index. The evaluations were performed blindly by a single, calibrated examiner and were followed by prophylaxis and orientation regarding oral hygiene. The prevalence of nickel allergy was determined by the patch test 9 months after the beginning of treatment and occurred in 16 individuals (17.2%). Two groups were then established: the allergic group (n - 16) and the age-paired nonallergic control group (n - 16). It was found that: (i) significant differences were present between groups at the T(3) and T(4) evaluations for the Loe index, with allergic individuals showing higher mean values than nonallergic individuals (hyperplasia, change in color, and bleeding), and no significant differences were found in the intragroup evaluations between the four evaluations; suggesting that (ii) a cumulative effect of nickel throughout the orthodontic treatment was associated with clinically significant periodontal abnormalities in allergic individuals over time. Narmada et al. [550] investigated the effect of different artificial salivary pH on the ions released and the surface morphology of SS brackets-NiTi and archwire combinations. The immersion test was performed at artificial salivary pH levels of 4.2, 6.5, and 7.6 at 37 °C for 28 days. It was reported that: (i) the chemical composition of all orthodontic appliances contained nickel element; (ii) in addition, XRD was depicted phases not only NiTi but also nickel, titanium, silicon, and zinc oleate; (iii) the immersion test showed that the highest release of nickel ions was observed at a pH of 4.2, with no significant difference at various levels of pH; and (iv) at a pH of 4.2, the surfaces of orthodontic appliances become unhomogenous and rough compared to those at other pH concentrations. Based on these results, it was concluded that the reduction of pH in the AS increases the amount of released nickel ions, as well as causing changes to the surface morphology of brackets and archwires.

16.3.2.13 Cytotoxicity

As stated before, the biological significance of the metal ion release is believed to be strongly related to the cytocompatibility and hemocompatibility. Eliades et al. [551] qualitatively and quantitatively characterized the substances released from orthodontic brackets and NiTi wires, and comparatively assessed the cytotoxicity of the ions released from these orthodontic alloys. Two full sets of SS brackets of 20 brackets and two groups of NiTi archwires of 10 wires each were immersed in 0.9% saline solution for a month. The cytotoxic or cytostatic activity of the media was investigated with the MTT and the DNA synthesis assays. It was reported that: (i) the results of the cytotoxicity assay indicated no ionic release for the NiTi alloy aging solution, whereas measurable nickel and traces of chromium were found in the SS bracket-aging medium; and (ii) concentrations of the nickel chloride solution > 2 mM were found to reduce by more than 50% the viability and DNA synthesis of fibroblasts; however, neither orthodontic materials-derived media had any effect on the survival and DNA synthesis of either cells. David et al. [552], using murine cortical cell cultures, examined the in vitro neurotoxicity of commonly used orthodontic metallic archwire alloys. The materials examined included NITi,

Cu-NiTi, Ti-Mo, Elgiloy, and SS archwire alloys. It was reported that: (i) NiTi, Cu-NiTi, and Ti-Mo alloys were not neurotoxic, whereas SS and Elgiloy were significantly toxic; and (ii) washing the archwires for 7 days in a saline solution did not alter the toxicity; however, the free radical scavenger, trolox, blocked the toxicity of both SS and Elgiloy, indicating that the death was free radical mediated. Kao et al. [485] investigated the toxicity of fluoride corrosion extracts of SS and NiTi wires on a human osteosarcoma cell line (U2OS). The SS and NiTi wires were corroded by an electrochemical method with the application of three kinds of electrolytes: 0.2% pH 3.5 APF (NaF) in AS, and pH 4 and pH 6.75 AS solutions. The extracts were analyzed for nickel, chromium, and titanium ions by the atomic absorption method. The extracts were diluted with medium-to-different concentrations (1, 0.1, and 0.01 l/ml). The cell survival rate was determined by the ability of test cells to cleave the tetrazolium salt to form a formazan dye. It was obtained that: (i) the release of ionic nickel was different in different extract groups; (ii) the SS and NiTi wires in the 0.2% pH 3.5 NaF AS group caused a dose-dependent decrease in the survival rate; and (iii) survival rates of cells in the groups exposed to extracts of SS and NiTi wires in pH 4 and pH 6.75 AS solutions showed no statistical differences; concluding that orthodontic wires in acidulated fluoride saliva solution can cause U2OS cell toxicity. Sodor et al. [553] evaluated the biocompatibility of some orthodontic biomaterials such as SS archwires, brackets, and NiTi coil springs, using human fibroblasts cultures on which the orthodontic materials were applied. It was reported that: (i) the materials used in the study caused cell alterations of variable intensity; (ii) the metallic brackets represented an important cell stress factor causing shape changes; (iii) for the metallic brackets, a preferential tropism for different areas of the bracket was also obvious; (iv) a preferential tropism for the areas between the NiTi coil spring spirals was observed; and (v) for the SS archwires, we observed for 6 days, a decay of cell density and a higher amount of cells near the archwire areas on which bends were performed. Based on these results, it was concluded that all biomaterials analyzed in this study caused cellular changes of varying intensity without necessarily showing a cytotoxic character. SE and TA NiTi archwires are used in everyday orthodontic practice, based on their acceptable biocompatibility and well-defined shape-memory properties. However, the differences in their surface microstructure and cytotoxicity have not been clearly defined, and the standard cytotoxicity tests are too robust to detect small differences in the cytotoxicity of these alloys, all of which can lead to unexpected adverse reactions in some patients. Čolić et al. [554] examined the hypothesis that the differences in manufacture and microstructure of commercially available SE and TAed NiTi archwires may influence their biocompatibility. The archwires were studied in as-received conditions and after conditioning for 24 h or 35 days in a cell culture medium under static conditions. It was found that: (i) all of the tested archwires, including their conditioned medium, were noncytotoxic for L929 cells, but Rematitan SE (both as received and conditioned) induced the apoptosis of rat thymocytes in a direct contact; (ii) in contrast, TruFlex SE and Equire TA increased the proliferation of thymocytes; and (iii) the cytotoxic effect of Rematitan SE correlated with the higher release of nickel ions in conditioned medium, higher concentration of surface nickel, and an increased oxygen layer thickness after the conditioning. It was, therefore, concluded that the apoptosis assay on rat thymocytes, in contrast to the less sensitive standard assay on L929 cells, revealed that Rematitan SE was less cytocompatible compared to other archwires, and the effect was most probably associated with a higher exposition of the cells to nickel on the surface of the archwire, due to the formation of unstable oxide layer.

16.3.2.14 Fracture

Dreschder et al. [555] reported the long-term fracture resistance of orthodontic NiTi wires. The tested material comprised nine NiTi wires (dimensions 0.016", round and 0.016" × 0.022", rectangular) as well as a SS and a β-Ti wire that were included as reference. It was reported that: (i) compared to the SS wire, the NiTi wires exhibited two- to five-fold higher yield forces in bending; (ii) at a specified deflection angle, the generated bending forces of the NiTi wires reached one-half to one-fourth of the values of SS; (iii) the fracture resistance under long-term loading was determined using the Wöhler method; (iv) after 10^5 loadings, 0.016" NiTi wires were subject to break failure, if forces exceed values greater than 1.2–3.1 N; (v) SS and β-Ti wires could be loaded with forces of up to 4.4 and 3.7 N, respectively; (vi) the 0.016" × 0.022" rectangular wires allowed forces of approximately twice this magnitude; (vii) elastic fatigue of the SE specimens "Memorywire," "Rematitan Lite," and "Sentalloy medium" showed up as hardening of the wire by up to 70%; and (viii) material degradation lead to a severe deformation of the hysteresis loop and to plastic deformation. Work-hardened martensitic NiTi wires did not show these effects to this extent. Yokoyama et al. [556] mentioned that in view of the galvanic current in the mouth, the fracture of the NiTi alloys might be attributed to the degradation of the mechanical properties due to hydrogen absorption. Zinelis et al. [557] characterized intraorally fractured NiTi archwires, determined the type of fracture, assessed changes in the alloy's hardness and structure, and proposed a mechanism of failure. Eleven NiTi SE 200 and 19 Cu-NiTi intraorally fractured archwires were collected. The location of fracture (anterior or posterior), wire type, cross-section, and the period of service before fracture were recorded. The retrieved wires and brand-, type-, and size-matched specimens of unused wires were subjected to SEM to assess the fracture type and morphological variation of fracture site of retrieved specimens. It was reported that: (i) the fracture site distribution showed a preferential location at the midspan between the premolar and the molar, suggesting that masticatory forces and complex loading during engagement of the wire to the bracket slot and potential intraoral aging might account for fracture incidence; and (ii) all retrieved wires had the distinct features of brittle fracture without plastic deformation or crack propagation, whereas no increase in hardness was observed for the retrieved specimens. Based on these observations, it was concluded that most fractures sites were in the posterior region of the arch, probably because of the high-magnitude masticatory forces; (iv) brittle fracture without plastic deformation was observed in most NiTi wires regardless of archwire composition; (v) there was no increase in the hardness of the intraorally exposed specimens regardless of the wire type; and (vi) this contradicts previous in-vitro studies and rules out hydrogen embrittlement as the cause of fracture. Per-inetti et al. [558] evaluated the surface corrosion and fracture resistance of two com-mercially available NiTi-based archwires, as induced by a combination of fluoride, pH, and thermocycling. A total of 110 rectangular-section NiTi-based archwires were used, 55 of each of the following: thermally activated Thermaloy® and SE NeoSental-loy® 100 g. It was mentioned that: (i) significant effects in terms of surface corrosion, but not fracture resistance, were seen mainly for the Thermaloy® group at the lowest pH, with no effects of thermocycling, irrespective of the group or pH condition; and (ii) different NiTi-based archwires can have different corrosion resistance although the effects of surface corrosion and fracture resistance appear not to be significant in clinical situations, especially considering that thermocycling had no effect on these parameters. Guzman et al. [559] determined the incidence and location of fracture in round NiTi and round SS orthodontic archwires, both commonly used in orthodon-tics, and determined whether there is any correlation between archwire fracture and gender, diameter of the archwire, arch type (maxillary/mandibular), or the bracket used. A total of 1,000 orthodontic patients (1,434 archwires) were evaluated during regular treatment visits to assess archwire fracture and location. The patient's gender, age, type of archwire (round NiTi and round SS), diameter of the archwire, arch type, location of fracture (anterior or posterior), and period of service before fracture were recorded. (i) Twenty-five archwire failures were reported (1.7%) of the total sample size; (ii) all fractured archwires were NiTi, and 76% of the fractures were located in the posterior region; and (iii) no statistical significance was found between archwire fracture and gender, arch type (maxillary/mandibular), archwire diameter, or bracket type. It was then concluded that the frequency of archwire fracture during regular orthodontic visits is very low, and the most common archwire fracture site is the pos-terior region. NiTi wires are the most commonly fractured archwire, and no statisti-cally significant correlation exists between archwire fracture and gender, arch type, bracket type, or diameter of archwire.

16.3.2.15 Fatigue
Prymak et al. [560] analyzed in vitro the fatigue resistance of NiTi and Cu-NiTi ortho-dontic wires when subjected to forces and fluids, which are present intraorally. The wires were subjected to dynamic mechanical analysis while they were immersed into different fluids with mechanical loading parameters similar to those that are sub-jected in the mouth. The characteristic temperatures of transitions and a rough surface structure on the perimeter of the wires were determined by DSC and SEM, respectively, before and after the dynamic mechanical analysis experiments. SS wires were used for comparison. It was mentioned that: (i) in general, NiTi wires fractured earlier than

the SS specimens; (ii) survival times were lower for the NiTi wires when immersed in fluids (water, citric acid, NaCl solution, AS, and fluoridated AS) than in air; (iii) SEM surface analysis showed that the NiTi and Cu-NiTi wires had a rougher surface than the SS wires; and (iv) the fracture occurred within a short number of loading cycles. Until fracture occurred, the mechanical properties remained mostly constant. Bourauel et al. [561] investigated the fracture resistance of as-received and retrieved NiTi archwires. NiTi archwires (German Orthodontics, CA, USA) of various cross-sections were retrieved from orthodontic patients and brand-, type-, and size-matched wires were included as controls. Specimens prepared from the selected wires were subjected to bending deformation at a loading frequency of 2 Hz and cyclic loading was applied either until fatigue failure or with a maximum number of loading cycles of 2×10^6. For each wire cross-section, a minimum of five specimens were loaded; the cycle number of the fatigue fracture was recorded and the mean of the cycles was calculated from the five specimens to provide the corresponding point in the Wöhler diagram. It was reported that: (i) retrieved wires fractured at a significantly lower number of cycles compared to their as-received matches; (ii) the size of the wire played a role in determining fracture with larger cross-sections showing reduced fatigue failure properties and 0.30- and 0.035-mm specimens showing no fracture at the selected strain and number of cycles; (iii) the SEM investigations revealed evidence of smoother fractured surfaces in large-sized, rectangular cross-sectioned retrieved wires; and (iv) Kaplan-Meier survival analysis[*] with log-rank test demonstrated the effect of aging and size in predicting survival of NiTi wires. It was emphasized as clinical significance that the extended life expectancy of NiTi archwires proposed in current treatment trends is associated with higher probability of fatigue fracture of wires, large diameter and square or rectangular cross-sections possess an increased chance of failure relative to smaller diameter wires, and a necessity may emerge to monitor patients to identify wire failures despite the commonly used practice of increased time intervals between appointments in patients treated with NiTi wires.

[*]The Kaplan–Meier estimator [A], also known as the product limit estimator, is a nonparametric statistic used to estimate the survival function from lifetime data. In medical research, it is often used to measure the fraction of patients living for a certain amount of time after treatment. [A] Kaplan EL, Meier P. Nonparametric estimation from incomplete observations. J. Amer. Statist. Assoc. 1958, 53, 457–481.

16.3.2.16 Torsional action
Filleul et al. [562] examined the stress–strain behavior of two NiTi and two Cu–NiTi orthodontic wires in induced torsion under controlled conditions of moment and temperature. It was shown that: (i) there is a diversity of behavior of these wires, (ii) the loading and unloading curves and plateau regions were found to be closely related to temperature with stiffness varying dramatically over mouth temperature range

under identical stress, and (iii) diversity of reaction to stress is linked to the crystalline structure of the alloys. Bolender et al. [563] reproduced and compared the intraoral torsional behavior of 10 commonly used preformed upper NiTi 0.017 × 0.025 archwires in 0.018-slot brackets at 20, 35, and 55 °C. Ten upper preformed NiTi archwires were compared to a multibraided SS wire. An original testing bench was used to reproduce palatal root torque applied onto an upper central incisor with a maximum value of 1,540 g × mm. It was reported that: (i) loading and unloading at 20 °C revealed three categories of wires: a group of four NiTi wires of relative stiffness bereft of any SE, a group of six NiTi wires displaying some horizontal plateau, and finally the SS wire of lesser stiffness; (ii) testing at the average oral temperature of 35 °C produced the same three categories of wires, with only two of 10 NiTi wires displaying a SE effect (Cu-NiTi 35 °C and 40 °C); (iii) none of the NiTi wires was SE at 55 °C; and (iv) moments increased with temperature as the martensite was replaced by the more rigid austenite. Based on three findings, it was concluded that most NiTi wires did not exhibit the SE effect in torsion traditionally described in bending; (vi) the combination of straight-wire prescriptions and rectangular SE NiTi archwires did not provide optimal constant moments necessary to gain third-order control of tooth movement early in treatment; and (vii) a braided SS rectangular archwire displayed better torsional behavior at 35 °C than most NiTi archwires of the same dimensions. Partowi et al. [564] compared various NiTi and SS wires in a pure torsion experiment. To experimentally examine the torque characteristic of orthodontic wires, the following materials were used; NiTi wires manufactured by Dentaurum and ODS measuring 0.40 × 0.40 mm, 0.40 × 0.56 mm, 0.43 × 0.64 mm, 0.46 × 0.64 mm, 0.48 × 0.64 mm, 0.51 × 0.51 mm, and steel wires made by Dentaurum, 3 M Unitek measuring 0.40 × 0.40 mm, 0.40 × 0.56 mm, 0.43 × 0.56 mm, 0.43 × 0.64 mm, 0.46 × 0.64 mm, 0.48 × 0.64 mm, and 0.51 × 0.51 mm. Torque was measured at an ambient temperature of 37 °C for the NiTi wires. The distance between the bearing points was 10 mm for all measurements. The torque moments of the steel wires were determined at a torsion angle of 20° and the torque moments of the NiTi wires were investigated on the plateau. We compared the results of the means and the standard deviations of all wires. Starting and end points of the loading and unloading plateau were determined for all NiTi wires. It was found that: (i) the torque moment and torsion angle diagrams of the NiTi wires by different manufacturers displayed the curves typical of SE wires; (ii) the torque moments of the wires with a smaller cross-section yielded values < 5 Nmm and thus appear obviously too small (e.g., Dentaurum Tensic, 0.40 × 0.40 mm: 1.3 Nmm); and (iii) the mean values of the starting point of loading plateau yielded values of about 20°. Hence it was concluded that as the torsional play in the bracket slot of the wires with a smaller cross-section is typically about 10°, it was questionable whether these wires reach the SE plateau, and production-associated variations in the material properties of various NiTi wires were observed; indicating that it is difficult for the practitioner to draw a correlation between a wire's cross-section and the actual torque moment delivered. The force moment providing rotation of the tooth around the x-axis (buccal-lingual) is referred to as torque expression in orthodontic literature. Many factors affect the torque expression, including the wire material characteristics. Archambault et al. [565] compared the torque expression between wire types. With a worm-gear–driven torque apparatus, wire was torqued while a bracket mounted on a six-axis load cell was engaged. Three 0.019 × 0.0195 inches wires (SS, TMA, and Cu-NiTi] and three 0.022 inches slot bracket combinations (Damon 3 MX, In-Ovation-R, and SPEED) were compared. It was obtained that: (i) at low twist angles (<12°), the differences in torque expression between wires were not statistically significant; (ii) at twist angles over 24°, SS wire yielded 1.5–2 times the torque expression of TMA and 2.5–3 times that of NiTi; and (iii) at high angles of torsion (>40°) with a stiff wire material, loss of linear torque expression sometimes occurred; concluding that SS has the largest torque expression, followed by TMA and NiTi. The torque moment generated by third-order bends is important for tooth movement. Hirai et al. [566] measured the torque moment that can be delivered by various archwire and bracket combinations at the targeted tooth. SS upper brackets with 0.018 and 0.022-inches slots, two sizes of NiTi alloy wires, and three sizes of SS wires for each bracket were used. The wire was ligated with elastics or wire. It was mentioned that: (i) the torque moment increased as the degree of torque and wire size increased; (ii) there was no significant difference in the torque moment between the SS and NiTi wires at lower or higher than 40° torque; and (iii) the torque moment with wire ligation was significantly larger than that with elastic ligation with 0.016 × 0.022 and 0.017 × 0.025 inch NiTi wires in the 0.018-inch slot brackets and the 0.017 × 0.025 and 0.019 × 0.025 inch SS and NiTi wires in the 0.022 inch slot brackets; however, there was no significant difference in the torque moment between either ligation method when using the full slot size wires.

16.3.2.17 Joining
Iijima et al. [567] investigated laser welding 0.016 × 0.022 inch β-Ti, NiTi, and CoCrNi orthodontic wires by measuring the joint tensile strength, the laser penetration depth, determining metallurgical phases using micro-XRD, and examining microstructures with a SEM. It was found that: (i) the mean tensile strength for NiTi groups was significantly lower than for most other groups of laser-welded specimens; (ii) although mean tensile strength for β-Ti and CoCrNi was significantly lower than for control specimens joined by silver soldering, it was sufficient for clinical use; (iii) the β-Ti orthodontic wire showed deeper penetration depth from laser welding than the NiTi and CoCrNi orthodontic wires; (iv) micro-XRD patterns of laser-welded β-Ti and NiTi obtained 2 mm from the boundary were similar to as-received specimens, indicating that original microstructures were maintained; (v) when output voltages of 190 V and higher were used, most peaks from joint areas disappeared or were much weaker, perhaps because of a directional solidification effect, evidenced by SEM observation of fine striations in welded β-Ti; and (vi) laser welding β-Ti and CoCrNi wires may be acceptable clinically, as joints had sufficient strength and metallurgical phases in the

original wires were not greatly altered. Tam et al. [568] examined the microstructure and mechanical properties of resistance microwelded (RMW) Ni-rich Nitinol wires at different applied currents. It was shown that: (i) a solid-state bonding mechanism consisted of six main stages, including cold collapse, dynamic recrystallization, interfacial melting, squeeze out, excessive flash, and surface melting; (ii) the joint strength and fracture mechanism were linked closely to the metallurgical properties of the welds; (iii) through DSC testing, it was found that the weld metal underwent phase transformation at lower temperatures compared with the base material; (iv) the pseudoelastic property of Nitinol was found to have a large effect on the contact resistance during the onset of welding current; (v) compression tests with varying temperatures confirmed distinct differences in displacement (i.e., cold collapse) under the welding load of 5 kg-f caused by changes in the thermodynamic stability of the austenite phase; and (vi) both the dynamic resistance and displacement measurements were found to be significantly different during the RMW Nitinol crossed wires compared with welding 316 low-vacuum melted SS crossed wires. Matsunaga et al. [569] investigated the possibility of electrical and laser welding to connect titanium-based alloy (β-Ti and NiTi) wires and SS or Co-Cr alloy wires for fabrication of combination archwires. Four kinds of straight orthodontic rectangular wires (0.017 × 0.025 inch) were used: SS, Co-Cr, β-Ti, and NiTi. Homogeneous and heterogeneous end-to-end joints (15 mm long each) were made by electrical and laser welding. It was reported that: (i) the SS/SS and Co-Cr/Co-Cr specimens showed significantly higher values of the maximum load at fracture and elongation than those of the NiTi/NiTi and β-Ti/β-Ti specimens for electrical welding and those of the S-S/S-S and Co-Cr/Co-Cr specimens welded by laser; (ii) on the other hand, the laser-welded NiTi/NiTi and β-Ti/β-Ti specimens exhibited higher values of the maximum load at fracture and elongation compared to those of the corresponding specimens welded by electrical method; (iii) in the heterogeneously welded combinations, the electrically welded NiTi/SS, β-Ti/SS, and β-Ti/Co-Cr specimens showed significantly higher maximum load at fracture and elongation than those of the corresponding specimens welded by laser; (iv) electrical welding exhibited the higher values of maximum load at fracture and elongation for heterogeneously welded combinations compared to laser welding. Mesquita et al. [570] identified the appropriate power level for electric welding of three commercial brands of NiTi wires. A total of 90 pairs of 0.018-inch and 0.017 × 0.025-inch NiTi wires were divided into three groups according to their manufacturers – GI (Orthometric), GII (3 M OralCare), and GIII (GAC) – and welded by electrical resistance. Each group was divided into subgroups of five pairs of wires, in which welding was done with different power levels. In GI and GII, power levels of 2.5, 3, 3.5, 4, 4.5, and 5 were used, whereas in GIII 2.5, 3, 3.5, and 4 were used (each unit of power of the welding machine representing 500 W). The pairs of welded wires underwent a tensile strength test on a universal testing machine until rupture and the maximum forces were recorded. It was reported that the 2.5 power exhibited the lowest resistance to rupture in all groups (43.75 N for GI, 28.41 N for GII, and 47.57 N for GIII), the 4.0 power provided the highest resistance in GI and GII (97.90 N and 99.61 N, respectively), and in GIII (79.28 N), the highest resistance was achieved with a 3.5 power welding; concluding that the most appropriate power for welding varied for each brand, being 4.0 for Orthometric and 3 M, and 3.5 for GAC NiTi wires.

16.3.2.18 Heat treatment

A thorough knowledge of the mechanical behavior of an orthodontic archwire is required to select one of suitable size and material that will provide optimal and predictable treatment results [571, 572]. Many different materials for archwires such as NiTi, β-Ti, and SS are used during an orthodontic mechanotherapy. NiTi archwires are the most used at the beginning of the treatment when more of the dental movement is needed. For this reason, their outstanding clinical characteristics as flexibility and springback make them ideal for alignment and levelling the teeth. Such required properties can be obtained by thermal manipulation of NITi materials. Kawashima et al. [573] conducted simple three-point bending tests on NiTi wires with three different A_F points (1, 13, and 34 °C) to clarify the relationship between A_F temperature and load changes under constant deformation. Each wire was deformed at 37 °C and thermal changes were imposed by temperatures of 2 °C or 60 °C. The following were recorded: (i) the load changes with thermal changes from 37 °C to 2 °C or 60 °C showed the same tendency on the wires with different A_F points; in the loading stage, the load became lower than the initial level at 37 °C and in the unloading stage, the load became higher than the initial load. (ii) The largest load change in the unloading stage was measured with the 13 °C A_F point wire. (iii) Care must be taken when handling NiTi wire with an A_F point of less than 1 °C to prevent it from reaching the limit of critical stress of slip deformation when the temperature in the mouth rises to > 40 °C.

16.3.2.19 Coating

The coating thickness of four brands of as-received esthetic-coated rectangular archwires was evaluated, and their surface characteristics and coating stability after 21 days of oral exposure compared to those of conventional SS and NiTi ones were also studied [574]. The labial surface of the selected archwires was observed with a stereoscope and in a SEM, and surface roughness was assessed with an atomic force microscope. It was found that: (i) all groups showed an average coating thickness of <0.002 inches; (ii) after oral exposure, archwires from two groups lost all coating on the labial surface; and (iii) on average, 28.71 and 72.90% of the coating was lost in each of the other two groups, and the surface roughness of the remaining coating was higher than postclinical control wires. It was, therefore, concluded that coated archwires had a low esthetic value as they presented a nondurable coating, and the remaining coating showed a severe deterioration and a greater surface roughness than postclinical control counterparts (conventional SS and NiTi wires).

Boccaccini et al. [575] fabricated polyetheretherketone (PEEK) and PEEK/Bioglass® coatings on shape-memory NiTi alloy wires using electrophoretic deposition (EPD). It was reported that: (i) best results were achieved with suspensions of PEEK powders in ethanol (1–6 wt%), using a deposition time of 5 min and applied voltage of 20 V; (ii) EPD using these parameters led to high-quality PEEK coatings with a homogeneous microstructure along the wire length and a uniform thickness of up to 15 μm without development of cracks or the presence of large voids; (iii) suspensions of PEEK powders in ethanol with addition of Bioglass® particles (0.5–2 wt%; size < 5 μm) were used to produce PEEK/Bioglass® coatings; (iv) sintering was performed as a post-EPD process to densify the coatings and to improve the adhesion of the coatings to the substrate; (v) the sintering temperature was 340 °C, sintering time 20 min and heating rate 300 °C/h and (vi) the bioactive behavior of PEEK/Bioglass® composite coatings was investigated by immersion in acellular simulated body fluid (SBF) for up to 2 weeks and hydroxyapatite crystals formed on the surface of the coated wires after 1 week in SBF, confirming the bioactive character of the coatings; demonstrating that EPD is a very convenient method to obtain homogeneous and uniform bioactive PEEK and PEEK/Bioglass® coatings on Nitinol® wires for biomedical applications. Bravo et al. [576] treated 20 $Ni_{55.5}Ni_{44.8}$ (wt%) orthodontic archwires by dipping treatment to coat the NiTi surface by a polyamide polymer. It has been selected as Polyamide 11 due to its remarkable longlasting performance. The transformation temperatures as well as the transformation stresses of the NiTi alloy were determined to know whether the coating process can alter its properties. It was mentioned that: (i) the adhesive wear tests have demonstrated that the wear rates as well as the dynamic friction coefficients μ of polymer-coated wires are much lower than metallic wires; (ii) the corrosion studies have shown that the use of this polymer, as coating, seals the NiTi surface to prevent corrosion and the release of nickel ions; and (iii) the average decrease of nickel ions release due to this coating is around 85%.

Liu et al. [577] developed TiN/Ti coating on orthodontic NiTi wires to study the stress corrosion of specimens, simulating the intraoral environment in as realistic manner as possible. TiN/Ti coatings were formed on orthodontic NiTi wires by PVD. It was reported that: (i) from the potentiodynamic polarization and SEM results, the untreated NiTi wires showed localized corrosion compared with the uniform corrosion observed in the TiN/Ti-coated specimen under both unstressed and stressed conditions; (ii) the bending stress influenced the corrosion current density and breakdown potential of untreated specimens at both pH 2 and pH 5.3; (iii) although the bending stress influenced the corrosion current of the TiN/Ti-coated specimens, stable and passive corrosion behavior of the stressed specimen was observed even at 2.0 V (Ag/AgCl); and (iv) it should be noted that the surface properties of the NiTi alloy could determine clinical performance. For orthodontic application, the mechanical damage destroys the protective oxide film of NiTi; however, the self-repairing capacity of the passive film of NiTi alloys is inferior to titanium in chloride-containing solutions; indicating that the TiN coating was found to be able to provide protection against mechanical damage, whereas the titanium interlayer improved the corrosion properties in an aggressive environment. Sugisawa et al. [578] evaluated the corrosion behaviors and mechanical properties of TiN-coated SS and NiTi orthodontic wires prepared by ion plating. It was mentioned that: (i) TiN coating by ion plating improves the corrosion resistance of orthodontic wires; (ii) the corrosion pitting of the TiN-coated wire surface become small; (iii) the tensile strength and stiffness of SS wire were increased after TiN coating; (iv) in contrast, its elastic force, which is a property for NiTi wire, was decreased. In addition, TiN coating provided small friction forces; and (v) the low level of friction may increase tooth movement efficiently; suggesting that TiN-coated SS wire could be useful for orthodontics treatment.

Several different materials are recognized as coating species onto NiTi surfaces. Huang et al. [579] coated DLC films onto NiTi orthodontic archwires. The influence of a fluoride-containing environment on the surface topography and the friction force between the brackets and archwires were investigated. It was recorded that: (i) the superior nature of the DLC coating, with less surface roughness variation for DLC-coated archwires after immersion in a high fluoride ion environment was recognized, and (ii) friction tests also showed that applying a DLC coating significantly decreased the fretting wear and the coefficient of friction, both in ambient air and AS; concluding that DLC coatings are recommended to reduce fluoride-induced corrosion and to improve orthodontic friction. Anuradha et al. [580] developed surface coatings on NiTi archwires capable of protection against nickel release and to investigate the stability, mechanical performance, and prevention of nickel release of titanium sputter-coated NiTi archwires. Coated and uncoated specimens immersed in AS were subjected to critical evaluation of parameters such as surface analysis, mechanical testing, element release, friction coefficient, and adhesion of the coating. It was reported that: (i) titanium element coatings exhibited high reliability on exposure even for a prolonged period of 30 days in AS; (ii) the coatings were found to be relatively stable on linear scratch test with reduced frictional coefficient compared to uncoated samples; (iii) titanium sputtering adhered well with the NiTi substrates at the molecular level, this was further confirmed by inductive coupled plasma emission spectroscopy analysis, which showed no dissolution of nickel in the AS; and (iv) titanium sputter coatings seem to be promising for nickel sensitive patients; indicating that it was confirmed the superior nature of the coating was evident as reduced surface roughness, friction coefficient, good adhesion, minimal hardness, and elastic modulus variations in AS over a given time period. The effects of nickel allergy due to NiTi alloys were reduced due to the TiN coating obtained by nitrogen gas diffusion at different temperatures. Gil et al. [581] treated the samples to immerse in AS at 37 °C for different periods demonstrating that the TiN coating prevents biodegradation. In addition, it was noted that: (i) lack of low friction coefficient for the NiTi SE archwires makes the optimal use of these materials in orthodontic applications difficult, and (ii) the reduction in this friction coefficient was achieved by nitrogen diffusion heat treatments. Katic et al. [582] analyzed the effect of various fluoride formulations in commercially available agents on working properties of various NiTi orthodontic wires. Uncoated (NiTi), rhodium-coated (Rh-NiTi), and nitrified (N-NiTi) wires were immersed to dH$_2$O, MiPaste, Elmex, and Mirafluor for 1 h. Unloading slope characteristics (average force, bending action of the force, and average plateau length) and the percentage of useable constant force during unloading were observed. Surface Ra was measured. SEM and EDS were used for observation of the surface. It was reported that: (i) NiTi had decreased loading and unloading elastic modulus and YS after immersion to MIPaste and Mirafluor; (ii) the unloading YS decreased in the Rh-NiTi by the MIPaste; (iii) the loading and unloading YS of the N-NiTi increased in Elmex and increased average plateau force; (iv) Rh-NiTi showed higher-average plateau length and the percentage of useful constant force during unloading in Mirafluor and the average plateau force lowered after immersion to MIPaste; (v) the unloading slope characteristics for NiTi were affected by all three prophylactic agents, mostly by Mirafluor, and produced significantly lower forces during both loading and unloading, similarly to the NNiTi wires; (vi) the Rh-NiTi had the lowest forces during both loading and unloading in MIPaste; and (vii) all results were significant; difference in Ra was observed for RhNiTi after immersion to the MIPaste; concluding that the NiTi and N-NiTi wires lose less working force when combined with Elmex, and the Rh-NiTi improve their working properties with Mirafluor and deteriorate when combined with MiPaste.

16.3.3 Orthodontic brackets

The bracket is an orthodontic attachment secured to a tooth for the purpose of attaching an archwire. Normally, the bracket is cemented directly onto the tooth, which eliminates the need for a band. Brackets with an attached archwire are parts that make the appliance or brace as seen in Figure 16.37 [583].

Various materials have been used as a form of dental braces throughout history, including gold, platinum, silver, steel, rubber, vulcanite, wood, ivory, zinc, copper, and brass [584, 585]. Nowadays, SS, titanium materials, plastics, polycarbonates, or ceramics are chosen as a bracket material. Among these, metal braces and ceramic braces are frequently used. Metal braces have two basic components; the metal bracket that is applied to the teeth and the bendable metal wire that is threaded through the brackets to apply pressure to the teeth and ultimately move them. This is the type of braces that most teenagers get, because they are the least expensive version and are often the fastest way of moving teeth into their ideal locations. On the other hand, ceramic braces are very popular alternative to metal braces. As the brackets are made of a tooth-color ceramic material that has the same texture as teeth, it would be very difficult to see the ceramic brackets on the teeth. However, like the real teeth, the ceramic brackets could easily stain if the patients were not diligent.

Sfondrini et al. [586] compared the nickel released from three kinds of orthodontic brackets: new conventional SS, recycled SS, and nickel-free brackets. The in vitro study

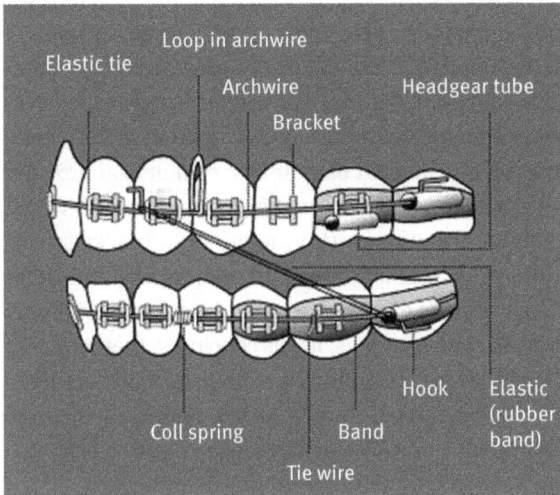

Figure 16.37: General arrangement for orthodontic mechanotherapy [583].

was carried out by using a classic batch procedure. Samples were immersed in AS at various acidities (pH 4.2, 6.5, and 7.6) over an extended time interval (0.25, 1, 24, 48, and 120 h). The amount of nickel released was determined by using an atomic absorption spectrophotometer and an ICP-AES. Statistical analysis included a linear regression model for repeated measures, with calculation of Huber-White robust standard errors to account for intra-bracket correlation of data. For post hoc comparisons, the Bonferroni correction was applied. It was found that: (i) the recycled brackets released maximum nickel (74.02 ± 170.29 µg/g); the new SS brackets released 7.14 ± 20.83 µg/g; (ii) the nickel-free brackets released the least nickel (0.03 ± 0.06 µg/gm); and (iii) all the differences among the groups were statistically significant; concluding that reconditioned brackets released the most nickel, and the highest nickel release was recorded in the two experiments performed at pH 4.2; it was lower at pH 6.5 and 7.6, while conversely, no relevant differences were observed overall between the maxillary and mandibular arches. Although all manufacturers of orthodontic brackets label these products for single use, there are commercial providers offering bracket reconditioning (or "recycling"). Reimann et al. [587] investigated the effects of different recycling techniques on material-related parameters in orthodontic brackets, aiming to derive indications for clinical use and conclusions about the biocompatibility, longevity, and application of recycled brackets. New metal brackets (equilibrium®) were compared to brackets recycled by different techniques, including direct flaming with a Bunsen burner, chemical reconditioning in an acid bath, a commercial unit, and outsourcing to a company. Material-related examinations included the following: (1) corrosion behavior by static immersion testing and use of a mass spectrometer to determine nickel-ion concentrations in the corrosive medium, (2) surface features in scanning electron micrographs before and after corrosion testing, (3) Vickers hardness using a hardness testing machine, (4) shear bond strength as defined in DIN 13990-1, (5) dimensional stability of the bracket slots by light microscopy, and (6) frictional loss as assessed by an orthodontic measurement and simulation system. Each examination was performed on 10 brackets. Student's *t*-test was used for statistical analysis. It was reported that: (i) compared to the new brackets, those recycled in an acid bath or by a commercial provider revealed significant dimensional changes; (ii) corrosion on the recycled brackets varied according to the administered recycling techniques; (iii) the group of brackets recycled by one company revealed hardness values that differed from those of all the other groups; and (iv) no significant differences were observed in nickel-ion release, frictional loss, and shear bond strength. Based on these findings, it was concluded that recycling was found to significantly reduce the corrosion resistance and dimensional stability of orthodontic brackets, and as the savings generated by recycling do not justify the risks involved, the practice of labeling orthodontic brackets for single use remains a responsible precaution that safeguards patients and orthodontists against definite risks.

As the function of the bracket clearly differs from that of the archwire (which requires certain level of elasticity and stores energy), material selection for bracket is limited and NiTi is seldom used, rather SS is frequently chosen. However, such rend should be changed if futuristic approach of nanotechnology in orthodontics is taken into considerations. Govindankutty et al. [588] described that nanotechnology is manipulating matter at nanometer level. This concept can be applied to the field of medicine and dentistry with the terms *nanomedicine* and *nanodentistry* being used respectively. Nanotechnology holds promise in many areas such as advanced diagnostics, targeted drug delivery, and biosensors. It has several applications in dentistry as well, from diagnosis of pathological conditions to local anesthesia, orthodontic tooth movement, and periodontics. Lapatki et al. [589, 590] had foreseen that quantitative knowledge of the 3D force-moment systems applied for orthodontic tooth movement is of utmost importance for the predictability of the course of tooth movement as well as the reduction of traumatic side effects. Nanomechanical integrated sensors can be fabricated and be incorporated into the base of orthodontic brackets to provide real-time feedback about the applied orthodontic forces. This real-time feedback allows the orthodontist to adjust the applied force to be within a biological range to efficiently move teeth with minimal side effects [588]. It has been reported that a smart bracket for multidimensional force and moment control is introduced on a large-scale prototype bracket that used a microsystem chip encapsulated into small low-profile contemporary bracket systems with reduced dimensions to allow clinical testing of this technology [589, 590]. Lekhadia [591] mentioned about the concept of nanotechnology and its applications in the field of orthodontics. Orthodontics is a specialty field of dentistry that deals primarily with malpositioned teeth and the jaws. It was mentioned that futuristic approach to nanotechnology in orthodontics include the use of the nanoelectromechanical systems to modify the bioelectric potential, orthodontic mini-screws with greater stability as well as ease of retraction, delivering low-intensity pulsed ultrasound to the target site of the teeth being moved, and use nano-mechanical sensors for 3D force and moment measurement built-in bracket system.

16.3.4 Mini-orthodontic implants

In orthodontics, normally there are two types of mini-scale devices to promote and support tooth movement safely and effectively; (1) temporary anchorage device (TAD), which has a pin-shape device made of titanium materials like NiTi alloy and (2) mini-orthodontic implants made of Ti-based alloys including Ti-6Al-4V and NiTi.

16.3.4.1 Temporary anchorage pins

TADs are small titanium pins used in certain orthodontic cases to help achieve quicker tooth movement with more efficiency and comfort. TADs may be used in addition to braces or as an alternative to headgear. Figure 16.38 [592] shows typical NiTi pin as TADs and Figure 16.39 illustrates anchorage function during the orthodontic treatment [25].

Orthodontists are accustomed to using teeth and auxiliary appliances, both intra-oral and extraoral, to control anchorage. There are several criteria for these tempo-

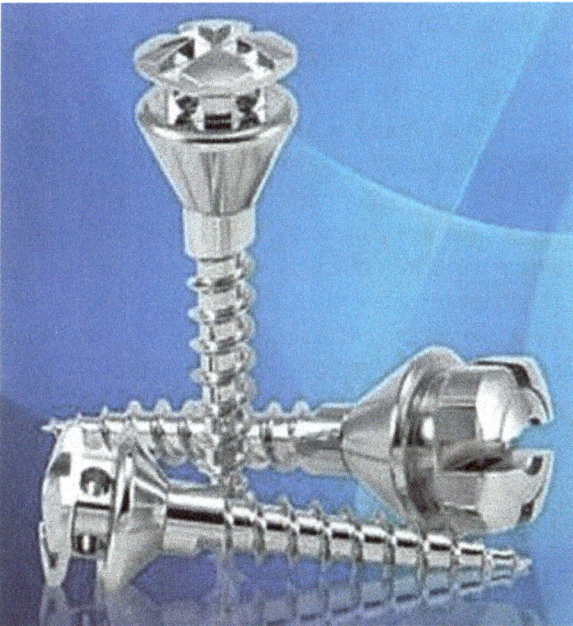

Figure 16.38: Typical nickel and titanium pins as a temporary anchorage device [592].

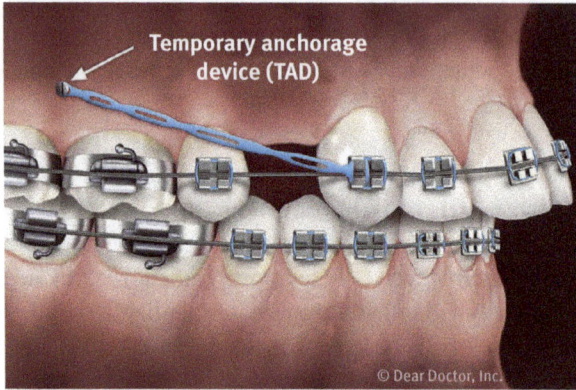

Figure 16.39: Anchorage function of temporary anchorage NiTi pin connected to one of brackets through a rubber band [593].

rary devices. The material must be nontoxic and biocompatible, possesses excellent mechanical properties, and provides resistance to stress, strain, and corrosion. Commonly used materials can be divided into three categories: biotolerant (SS, Cr–Co alloy), bioinert (titanium, carbon), and bioactive (hydroxyapatite, ceramic oxidized aluminum). Due to the characteristics of titanium (no allergic and immunologic reactions and no neoplasm formation), it is considered an ideal material and is widely used. The osseointegration mechanism should be the same as ordinal dental implants, which will be discussed later. Major difference from the ordinal dental implants is the size and shape of the orthodontic implants [594]. Implant fixtures must achieve primary stability and withstand mechanical forces. The maximum load is proportional to the total bone–implant contact (BIC) surface. Factors that determine the contact area are length, diameter, shape, and surface design (rough vs smooth surface, thread configuration). The ideal fixture size for orthodontic anchorage remains to be determined. Various sizes of implants, from mini-implants (6 mm long, 1.2 mm in diameter) to standard dental implants (6–15 mm long, 3–5 mm in diameter), have proved to effectively improve anchorage. Therefore, the dimension of implants should be congruent with the bone available at the surgical site and the treatment plan. Implant shape should determine the BIC area available for stress transfer and initial stability. The design must limit surgical trauma and allow good primary stability [594]. The most commonly used shape is cylindrical or cylindrical-conical, with a smooth or threaded surface. Studies have shown that the degree of surface roughness is related to the degree of osseointegration.

16.3.4.2 Mini-orthodontic implants

Mini-orthodontic implant is shown in Figure 16.40, where the head is the platform that connects to the orthodontic appliances or elastic traction. The neck is the part that traverses the mucosa. The body is the endosseous section with threads around a core and a tapered tip and threaded body within the cortical and cancellous bone.

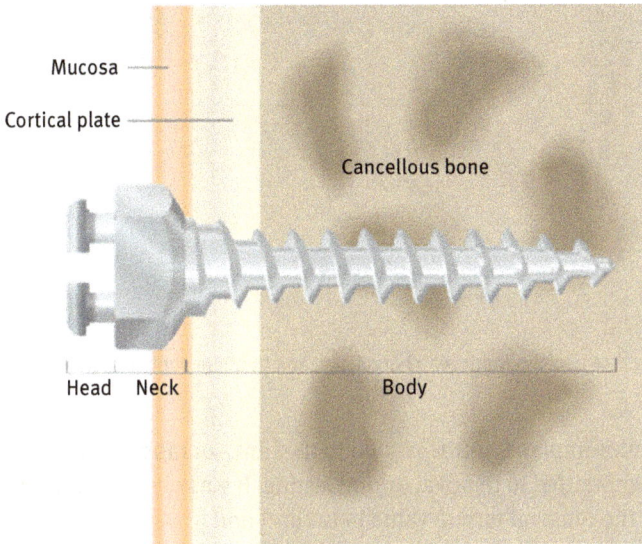

Figure 16.40: Mini-implants with three major constituent parts (the head, neck and body) [595].

Currently, many self-drilling screws are commonly used and these have a tapered body shape with sharp tips and threads, and are inserted in a corkscrew-like manner. Full-depth predrilling is avoided although shallow perforation of the cortex is still advantageous where the cortex is thick or dense, for example, the posterior mandible and palate [595]. Figure 16.41 shows typical NiTi mini-self-drilling screws.

Tseng et al. [597] explored the use of mini-implants for skeletal anchorage and assessed their stability and the causes of failure. Forty-five mini-implants were used in orthodontic treatment. The diameter of the implants was 2 mm, and their lengths were 8, 10, 12, and 14 mm. The drill procedure was directly through the cortical bone without any incision or flap operation. Two weeks later, a force of 100–200 g was applied by an elastometric chain or NiTi coil spring. Risk factors for the failure of mini-implants were examined statistically using the Chi-square or Fisher exact test as applicable. It was reported that: (i) the average placement time of a mini-implant was about 10–15 min; (ii) four mini-implants loosened after orthodontic force loading; (iii) the overall success rate was 91.1 %; and (iv) the location of the implant was the significant factor related to failure; indicating that the mini-implants are easy to insert for skeletal anchorage and could be successful in the control of tooth movement. Removable osseointegrated titanium mini-implants were successfully used as anchorage devices in orthodontics. The early load is necessary to simplify the mini-implant methodology, but can lead to failure during osseointegration. The Ti-6Al-4V alloy was used instead of CP-Ti due to its superior strength. However, the corrosion resistance is low, allowing for metal ion release. Morais et al. [598] analyzed the immediately loaded mini-implant fixation and gauged the vanadium ion release during the healing

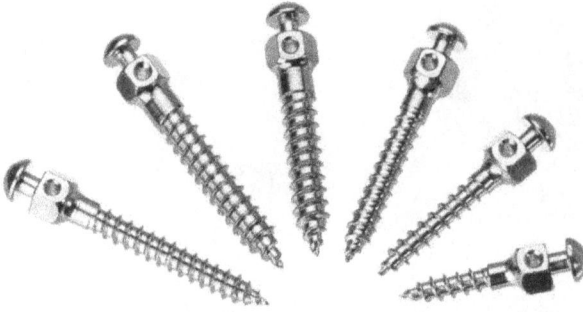

Figure 16.41: Mini self-drilling thread orthodontic matching nickel and titanium implants screw [596].

process. Titanium alloy mini-implants were inserted in the tibiae of rabbits. After 1, 4, and 12 weeks, they were submitted to removal torque testing. It was observed that: (i) there was no increase in the removal torque value between 1 and 4 weeks of healing, regardless of the load; (ii) nevertheless, after 12 weeks, a significant improvement was observed in both groups, with the highest removal torque value for the unloaded group; (iii) in comparison with the control values, the content of vanadium increased slightly after 1 week, significantly increased after 4 weeks, and decreased slightly after 12 weeks, without reaching the 1 week values; (iv) a stress analysis was carried out, which enables both the prediction of the torque at which CP-Ti and Ti-6Al-4V deform plastically and the shear strength of the interface; and (v) the analysis reveals that the removal torques for CP-Ti dangerously approach the yield stress; indicating that titanium alloy mini-implants can be loaded immediately with no compromise in their stability, and the detected concentration of vanadium did not reach toxic levels in the animal model. One notable complication of mini-implants that are used to provide anchorage in orthodontic treatment is loosening. Uemura et al. [599] evaluated the relationship between mini-implant mobility during the healing phase and the prognosis for implant stability. Twenty male Wistar rats (aged 20 weeks) were used. Drills with diameters of 0.8, 0.9, 1.0, and 1.1 mm were used to make pilot holes in the rat tibiae. The inserted mini-implants (diameter 1.4 mm; spearhead 1.2 mm; halfway between maximum and minimum 1.3 mm; length 4.0 mm) were subjected to an experimental traction of force for 3 weeks. BIC was observed histologically. Another 20 male rats (aged 20 weeks) underwent an identical procedure, and the stability of the mini-implants was measured using the Periotest before and after traction. It was reported that: (i) the BIC ratios of the 0.9 and 1.0 mm groups were significantly greater than those of the other groups; (ii) the Periotest values measured 3 weeks after implant insertion were significantly lower than those measured at insertion, except in the 1.1 mm group; (iii) to obtain mini-implant stability, the hole diameter should be between 69% and 77% of the diameter of the mini-implant; and (iv) a significant decrease in the mobility of the mini-implants 3 weeks postinsertion implies a good prognosis for the subsequent mini-implant stability. Orthodontic space opening is a common treatment for congenitally missing teeth. Dental implants are often used to replace the missing tooth to establish ideal esthetics without restoring the adjacent teeth. Abduljabbar et al. [600] investigated the outcome of implants placed in orthodontically created bone. A total of 50 dental implants were used for rehabilitation of lost tooth/teeth in 30 patients after space creation using orthodontic treatment. Patients with congenital and traumatic missing tooth/teeth, who lost the space for accommodation of normal size crown, and patients with spacing were enrolled in the study. NiTi open coil spring was used for space creation. The patients were followed up clinically and radiographically. It was recorded that: (i) of 50 implants, the success rate was 78% (39 implants); (ii) 11 implants (22%) failed to get osseointegration; (iii) mean age of the patients was 26.33 (female patients constituted 63.33%, and male patients 36.67%); and (iv) most of the failures were in the mandible; concluding that the orthodontic treatment can be used as a type of osteodistraction method for provision of an acceptable amount of bone for dental implants, and the quality of the gained bone can be improved by offering sufficient time for healing.

16.4 Dental implants

There are three common types of dental implants; endosteal type, subperiosteal type, and zygomatic type [601, 602]. Endosteal is the safest and the most common type of dental implant. They are suited for most patients but, they require a good, healthy jawbone for the post to fuse to (if not, bone grafting or other supplemental pretreatment is required). Endosteal implants are placeholder posts that are shaped like screws and put into the jaw that the false teeth are fitted onto. Once the procedure is done, it takes a little time to heal. It requires the time to fuse together and create a stronghold – osseointegration. Once it is healed, the denture teeth can be placed onto the post to fit in with the surrounding teeth – prosthesis. Figure 16.42 shows three types of endosteal implants – cylinder type, screw type, and blade type.

The second type of dental implant is subperiosteal implants as seen in Figure 16.43 [601]. This type of dental implant can be considered as the main alternative to endosteal implants. Instead of being fixed into the jawbone, subperiosteal implants rest on top of the bone but still is under the gum. A metal frame is placed under the gum with a post attached to it. The gum then heals around the frame to hold it in place. The denture teeth as a superstructure are secured to the poles that come from the gum. This procedure is only used if the patient does not have enough jawbone for an implant to be placed.

The third and least common type of dental implants is zygomatic implants as seen in Figure 16.44 [601]. It is the most complicated procedure and should only be done if the patient does not have enough jawbone for the endosteal implant. The implant is placed in the patient's cheekbone rather than the jawbone.

In dentistry, there are several applications of NiTi alloys including blade-type implant, fixator, and prosthesis.

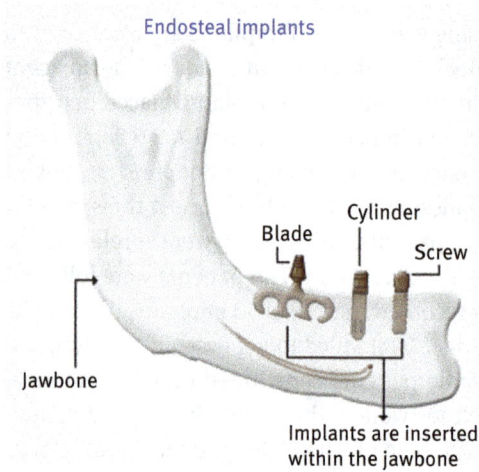

Endosteal implants

Blade   Cylinder   Screw

Jawbone

Implants are inserted
within the jawbone

Figure 16.42: Three typical endosteal implants [601].

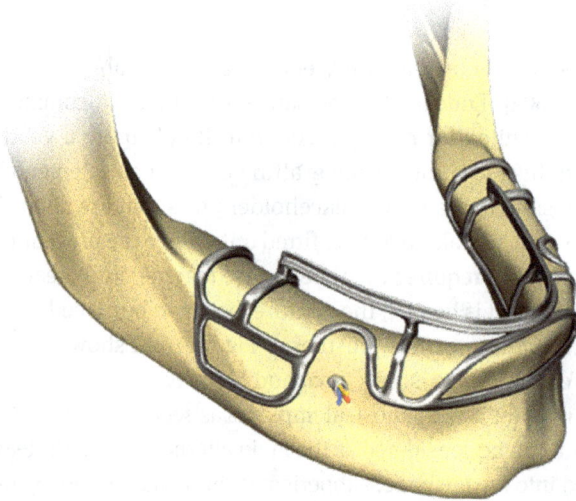

Figure 16.43: Subperiosteal implant as an alternative to endosteal type [601].

16.4.1 Blade type dental implant

According to literatures [602–606], the advantages of blade-type implants are: (1) possibility to insert blades in the narrowest alveolar crests, (2) adaptability to the majority of anatomical conformations, (3) avoiding bone regeneration surgery, (4) mechanical correction of parallelism during implant surgery, (5) easy adaptation to

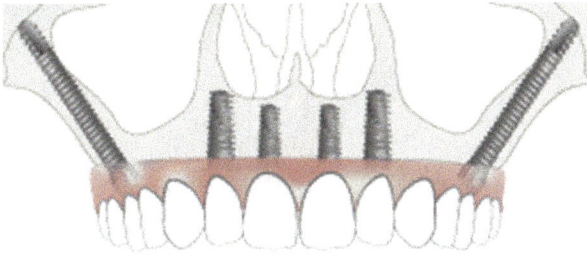

Figure 16.44: Typical zygomatic dental implants [601].

the deep anatomical structures by modifying the implant, (6) presence of numerous contacts with deep cortical layer, (7) possibility of inserting a part of the implant below the intact cortex (compared to the endosseous distal extension technique), (8) adequate management of attached gingiva during implant surgery, and (9) simple surgical technique performed with standard instruments. On the other hand, disadvantages of blade implants are listed as follows: (1) invasion of adjacent bone sites with mesiodistally positioned blades and (2) poor adaptability to postextraction alveolar sites. Carlo et al. [606] conducted retrospect researches on a total of 522 blades that were inserted in 20 years (1989–2009): 309 in females and 213 in males. The median age was 59 ± 11 years (min–max: 24–80 years). Implants were inserted in deep and atrophic narrow crests. It was reported that: (i) the success rate was 93.4% globally, 98.9% at 5 years, 89% at 8 years, and 86.2% at 10 years; and (ii) these data show very good results at 5 years, but slightly more failures at 8 and 10 years. In conclusion, the blade implant is a valid therapeutic device useful for treating cases such as narrow bone crest and scarce spongy bone in the lower distal sector; demonstrating a long-term survival; nonetheless, to prevent failure, practitioners should be aware that blade implants are not indicated in wide alveolar crests or in areas where bone density is insufficient and that the implant cannot be positioned in the deep cortical layer. Gil et al. [607] mentioned that: (i) an artificial tooth root can, after application in a deformed shape, return to its required shape after reaching body temperature and, so, is tightened in the mandible. As seen in Figure 16.45 [608], the single wing-type shape-memory implant (SMI); part of the apex opens buccolingually after insertion into the jawbone. The advantages of the SMIs are that they have a good initial fixation in the jaw bone, are easily installed in a simple operation, and have a good stress distribution to the sur-rounding bone compared with the ordinary non-opening blade-type implant [609–611].

The NiTi SMI is an endosseous blade-type implant that is used as an abutment for reconstruction of the dentition in the partially edentulous individual. The special design of the SMI better enables it to withstand masticatory loading the currently used implants [612]. Shape-memory blade type implant that was developed by Fukuyo et al. [609, 612] possesses uniqueness, and particularly, it was claimed that the initial fixation strength is remarkably high. Jawbone was cut into a rectangular shape that is

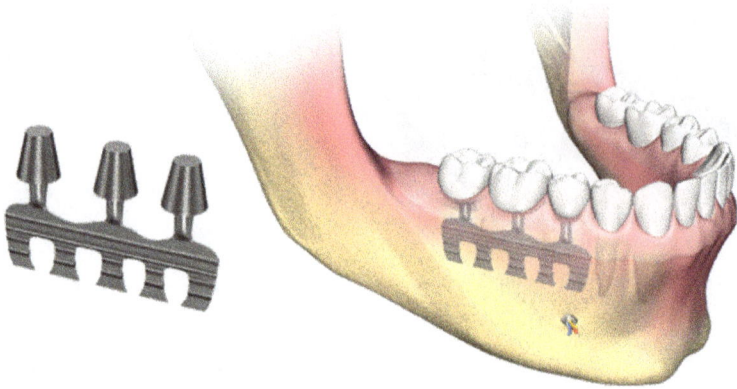

Figure 16.45: Blade-type dental implant [608].

just similar size for the blade in fold-position of every individual blades. The blades were previously trained in such a way that each blade can bend into positive directions, so that when the implant is exposed to body temperature (or slightly heated externally) to revisit the temperature at which the open-blade shape was memorized, these memorized shapes will be recovered and each blade tries to open (toward the preeducated direction) inside the very narrow precut groove. As a result, pushing force of each blades can create a tremendous fixing force as illustrated in Figure 16.46.

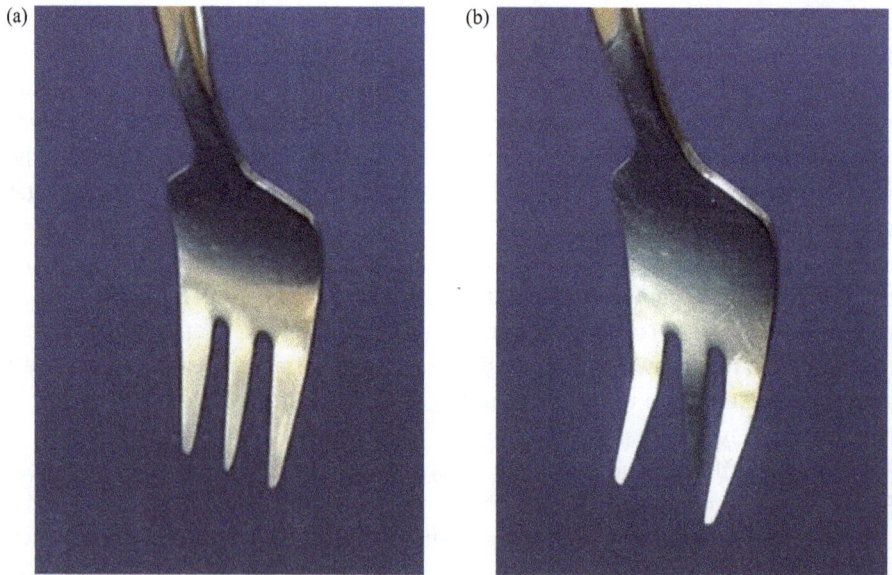

Figure 16.46: Positions of closed-blade implant (a), and open-blade creating fixation force upon heating due to recovering memorized shapes (b).

Instead of using whole solid implant structure, graded functional design concept was proposed [486] as seen in Figure 16.47. Inside core of the implant body should be solid to bear the external force, whereas the surface case layer should be biologically and biomechanically compatible to the receiving hard tissue. The case structure can be formed as foam structure, as controlled porous structure as mentioned by Nouri [613].

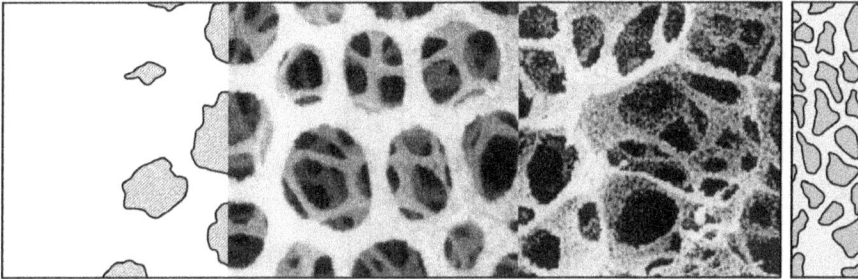

Ti implant						bone
body	sub-surface zone	surface layer		bony growth zone		
strong	←	Biomechanical strength		←		weak
		Modulus of Elasticity [GPa]				
250~200 ← 150~100	←	100~50	←	50~20	←	20~10
weak →		Biological and Biochemical reactions		→		strong

Figure 16.47: Schematic and conceptual titanium implant with graded functions [486].

In recent years, significant progress has been made toward the development of titanium foams, or porous titanium scaffolds, for orthopedic and dental applications [613]. Furthermore, an attempt has also been made to categorize titanium dental implants according to their type and design, including implants with fully porous structures, implants with porous surfaces and a solid core, and implants with a thin porous surface. As seen in Figure 16.48, surface case of NiTi implant can be fabricated [613].

Due to their unique properties, NiTi alloys have also received increasing attention as potential materials for porous orthopedic implants [614–617]. Porous

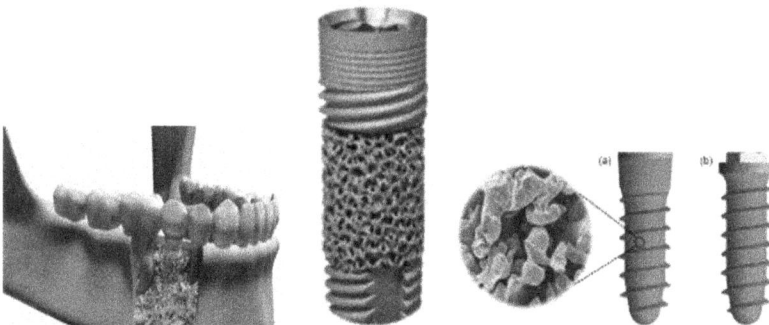

Figure 16.48: Porous NiTi implant surfaces [613].

NiTi alloys usually have a low elastic modulus (61–69 GPa), which can be further reduced by SE effect while maintaining high strength [614, 618]. Porous NiTi with a 30–80% porosity exhibits elastic modulus values as low as that of cortical bone (12–17 GPa) or even cancellous bone (<3 GPa) [619]. Nitinol with the SE property can be fabricated to have 70% porosity [620]. It was reported that the porous NiTi materials bond very well to newly formed bone tissues and are suitable for clinical applications under load-bearing conditions [621]. Accordingly, foam structure or porosity-controlled NiTi structure appears to be the appropriate structure as an implant case structure.

16.4.2 Fixators

Another significant orthodontic application is represented by NiTi fixators for mounting bridgework [622]. Fixators are, in general, based on a small SMA element that is notched on both sides. As the temperature increases, the notched area expands, causing a permanent hold of bridgework. These fixators can also be used to prevent a loose tooth from falling out (Figure 16.49). The basic concept for this fixators' design is similar to the one just described under SMI.

Fixing mechanism for these fixators should be the same as previous blade-type implants. The implants are manufactured to have open blades when in the austenite phase (during the transformation temperature). At room temperature, when the material is in the martensite phase, the implants are deformed to obtain flat blades, which are installed with a simple incision, thus facilitating the insertion. After installation, the implant is heated (by body temperature) up to more than the austenite transformation temperature to allow recovery of the customized shape and anchoring into bone.

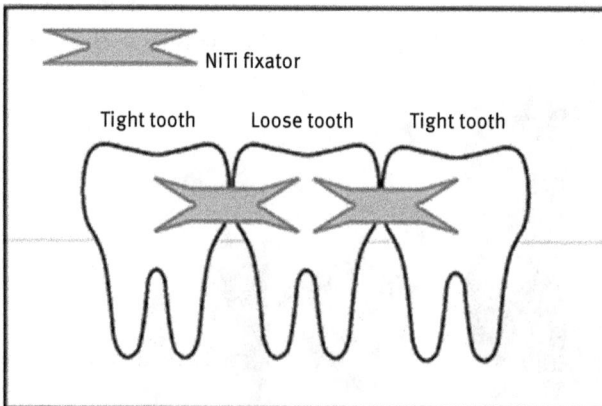

Figure 16.49: Shape-memory dental fixators [622].

16.4.3 Prosthesis

NiTi alloy used in dental prosthetic fixators shows SE behavior and exhibits a great potential in dental and orthopedic applications where constant correcting loads are required. To use such materials in dental prosthetic fixators, where the device is cyclically deformed, Sabrià et al. [623] investigated the effect of the cyclic straining upon the transformation stresses and temperatures of the material, and to study the load cycling of a SE NiTi SMA to be applied in the making of dental prosthetic fixators. Favot et al. [624] studied full dental prosthetic reconstruction on four implants by analyzing the influence of material parameters on the mechanical behavior of the restored mandible compared to the natural mandible. FEM of an edentulous mandible with prosthetic rehabilitation was established. Four materials were investigated for the framework of the prosthesis (zirconia, titanium, gold, and NiTi), as well as three cortical bone thicknesses. Various muscles were used to simulate the main stages of mastication. Three distinct phases of mastication were modelled: maximum intercuspation, incisal clench, and unilateral molar clench. It was reported that: (i) the zirconia (ZrO_2 ceramic) framework demonstrated the highest stresses and NiTi the weakest; (ii) the highest stresses in the framework were obtained during maximum intercuspation; (iii) the highest stresses at the bone-implant interface were recorded on the working axial implant during unilateral molar clench and on tilted implants during maximum intercuspation; (iv) the influence of the material stiffness of the framework on the stresses at the bone-implant interface was insignificant for axial implants and slightly more significant for tilted implants, and (v) mandibular flexion decreased with an increase of the cortical bone thickness and the stiffness of the prosthetic framework material. Based on these findings, it was concluded that among all materials, NiTi allowed a better preservation of the mandibular flexure, during all the mastication stages, and compared to stiffer materials, NiTi also permitted physiological and mechanical conditions at the bone-implant interface, in almost all mastication stages.

References

[1]   Moore BK, Oshida Y. Materials Science and Technology in Dentistry. In: Wise DL, ed., Encyclopedic Handbook of Biomaterials and Bioengineering, Part B. Marcel Dekker, Co., 1995, 1325–430.

[2]   Geurtsen W. Biocompatibility of dental casting alloys. Crit. Rev. Oral Biol. M. 2002, 13; https://doi.org/10.1177/154411130201300108.

[3]   Yoneyama T, Doi H, Kobayashi E, Hamanaka H. Super-elastic property of Ti-Ni alloy for use in dentistry. Front Med. Biol. Eng. 2000, 10, 97–103.

[4]   Wataha J, Schmalz G. Dental Alloys. In: Schmalz G, et al. ed., Biocompatibility of Dental Materials. Springer, 2008, 221–54.

[5]   Badami V, Ahuja B. Biosmart materials: breaking new ground in dentistry. Sci. World J. 2014; http://dx.doi.org/10.1155/2014/986912.

[6] Duerig TW, Melton KN, Stockel D, Wayman CM. Engineering Aspects of Shape Memory Alloys. Butterworth-Heinemann, London, 1990.

[7] Miyazaki S. Medical and Dental Applications of Shape Memory Alloys. In: Otsuka K, et al. ed., Shape Memory Materials. Cambridge University Press, 1998, 267–81.

[8] Farzin-nia F, Yoneyama T. Orthodontic Devices Using Ti-Ni Shape Memory Alloys. In: Yoneyama T, et al. ed., Shape Memory Alloys for Biomedical Applications. Woodhead Publishing, Cambridge, UK, 2009, 257–96.

[9] Nouri A. Titanium Foam Scaffolds for Dental Applications. In: Wen C, ed., Metallic Foam Bone: Processing, Modification and Characterization and Properties. Woodhead Pub., Elsevier 2017, 131–60; https://doi.org/10.1016/B978-0-08-101289-5.00005-6.

[10] Torrisi L. The Ni–Ti superelastic alloy application to the dentistry field. Biomed. Mater. Eng. 1999, 1, 39–47.

[11] Baumann MA. Nickel-titanium: options and challenges. Dent. Clin. North Am. 2004, 48, 55–67.

[12] Thompson SA. An overview of nickel-titanium alloys used in dentistry. Int. Endod. J. 2000, 33, 297–310.

[13] Guelzow A, Stamm O, Martus P, Kielbassa AM. Comparative study of six rotary nickel-titanium systems and hand instrumentation for root canal preparation. Intl. Endod. J. 2005, 38, 743–52.

[14] Liu S-B, Fan B, Cheung GS, Peng B, Fan MW, Gutmann JL, Song YL, Fu Q, Bian Z. Cleaning effectiveness and shaping ability of rotary ProTaper compared with rotary GT and manual K-Flexofile. Am. J. Dent. 2006, 19, 353–8.

[15] McGuigan MB, Louca L, Duncan HF. Endodontic instrument fracture: causes and prevention. Br. Dent. J. 2013, 214, 341–8.

[16] Walia HM, Brantley WA, Gerstein H. An initial investigation of the bending and torsional properties of Nitinol root canal files. J. Endod. 1988, 14, 346–51.

[17] Goel A, Rastogi R, Rajkumar B, Manisha T, Boruah L, Gupta V. An overview of modern endodontic NiTi systems. Int. J. Sci. Res. 2015, 4, 595–7.

[18] Kauffman GB. The story of nitinol: the serendipitous discovery of the memory metal and its applications. Chem. Educ. 1997, 2, 1–21.

[19] Civjan S, Huget EF, DeSimon LB. Potential applications of certain nickel-titanium (nitinol) alloys. J. Dent. Res. 1975, 54, 89–96.

[20] Bryant ST, Dummer PM, Pitoni C, Bourba M, Moghal S. Shaping ability of .04 and .06 taper ProFile rotary nickel-titanium instruments in simulated root canals. Int. Endod. J. 1999, 32, 155–64.

[21] Kramkowski TR, Bahcall J. An in vitro comparison of torsional stress and cyclic fatigue resistance of ProFile GT and ProFile GT series X rotary nickel-titanium files. J. Endod. 2009, 35, 404–7.

[22] Kim JG, Kum KY, Kim ES. Comparative study on morphology of cross-section and cyclic fatigue test with different rotary NiTi files and handling methods. J. Korean Acad. Conserv. Dent. 2006, 31, 96–102.

[23] Sanghvi Z, Mistry K. Design features of rotary instruments in endodontics. J. Ahmedabad Dent. Coll. Hosp. 2011, 2, 6–11.

[24] Eleazer PD, et al. ed. American Association of Endodontists. Glossary of endodontic terms. 2019; https://www.aae.org/specialty/download/glossary-of-endodontic-terms/.

[25] Schäfer E, Tepel J. Relationship between design features of endodontic instruments and their properties. Part III. Resistance to bending and fracture. J. Endod. 2001, 27, 299–303.

[26] Diemer F, Calas P. Effect of pitch length on the behavior of rotary triple helix root canal instruments. J. Endod. 2004, 30, 716–8.

[27] Park H. A comparison of Greater Taper files, Profiles and stainless steel files to shape curved root canals. Oral Surg. Oral Med. Oral Pathol. Oral Radiol. Endod. 2001, 91, 715–8.

[28] Young GR, Parashos P, Messer HH. The principles of techniques for cleaning root canals. Aus. Dent. J. Suppl. 2007, 52, 52–60.

[29] Hulsmann M, Schade M, Schafer's F. A comparative study of root canal preparation using Profile 0.04 and Light Speed NiTi instruments. Int. Endod. J. 2001, 34, 538–46.

[30] Bergmans L, Van CJ, Beullens M, Wevers M, Van MB, Lambrechts P. Progressive versus constant tapered shaft design using NiTi rotary instruments. Int. Endod. J. 2003, 36, 288–95.

[31] https://www.quora.com/What-is-the-significance-of-providing-rake-angle-for-a-cutting-tool.

[32] https://www.slideshare.net/deembandar/biomechanical-instrumentation.

[33] Wildey WL, Senia ES, Montgomery S. Another look at root canal instrumentation. Oral Surg. Oral Med. Oral Pathol. 1992,74, 499–507.

[34] Senia ES. What's new in Ni-Ti rotary instrumentation: part 1. Dent. Today 2007, 26, 80–3.

[35] Marending M, Schicht OO, Paque F. Initial apical fit of K-files versus LightSpeed LSX instruments assessed by micro-computed tomography. Intl. Endod. J. 2011; doi: 10.1111/j.1365-2591.2011.01967.x.

[36] McSpadden JT. Mastering Endodontic Instrumentation. Cloudland Institute, 2007, 71; https://nanoendo.com/MasteringEndo.pdf.

[37] Walsch H. The hybrid concept of nickel-titanium rotary instrumentation. Dent. Clin. North Am. 2004, 48, 183–202.

[38] http://www.med-college.de/en/wiki/artikel.php?id=302.

[39] Boessler C, Paque F, Peters OA. The effect of electropolishing on torque and force during simulated root canal preparation with ProTaper shaping files. J. Endod. 2009, 35, 102–6.

[40] Ruddle CJ, Machtou P, West JD. The shaping movement 5th generation technology. Adv. Endod. 2013, 7 pages; https://www.endoruddle.com/tc2pdfs/144/ProTaperNEXT_Apr2013.pdf.

[41] https://brasselerusadental.com/products/endosequence-endodontic-file-system/.

[42] https://www.qedendo.co.uk/acatalog/ProTaper_System.html.

[43] Lopes HP, Elias CN, Siqueira JF. Instrumentos Endodonticos. In: Lopes HP, Siqueira JF, ed., Endodontia: Biologia e Tecnica. 2a ed., Guanabara Koogan, Rio de Janeiro, 2004.

[44] Di Fiore PM, Genov KA, Komaroff E, Li Y, Lin L. Nickel–titanium rotary instrument fracture: a clinical practice assessment. Int. Endod. J. 2006, 39, 700–8.
[45] Frick C, Ortega A, Tyber J, Maksound A, Maier H, Liu Y, Gall K. Thermal processing of polycrystalline NiTi shape memory alloys. Mater. Sci. Eng. A 2005, 405, 34–49.

[46] Johnson E, Lloyd A, Kuttler S, Namerow K. Comparison between a novel nickel-titanium alloy and 508 nitinol on the cyclic fatigue life of ProFile 25/.04 rotary instruments. J. Endod. 2008, 34, 1406–9.

[47] Gutmann JL, Gao Y. Alteration in the inherent metallic and surface properties of nickel–titanium root canal instruments to enhance performance, durability and safety: a focused review. Int. Endod. J. 2012, 45, 113–28.

[48] Santos Lde A, Resende PD, Bahia MG, Buono VT. Effects of R-phase on mechanical responses of a nickel-titanium endodontic instrument: structural characterization and finite element analysis. Sci. World J. 2016, 2016, 7617493; doi: 10.1155/2016/7617493.

[49] Shim K-S, Oh S, Kum KY, Kim Y-C, Jee K-K, Chang SW. Mechanical and metallurgical properties of various nickel-titanium rotary instruments. Biomed. Res. Int. 2017; doi: 10.1155/2017/4528601.

[50] Peters OA, de Azevedo Bahia MG, Pereira ESJ. Contemporary root canal preparation: innovations in biomechanics. Dent. Clin. North Am. 2017, 61, 37–58.

[51] Hou XM, Yahata Y, Hayashi Y, Ebihara A, Hanawa T, Suda H. Phase transformation behaviour and bending property of twisted nickel-titanium endodontic instruments. Int. Endod. J. 2011, 44, 253–8.

[52] Braga LC, Magalhães RRS, Nakagawa RKL, Puente CG, Buono VTL, Bahia MGA. Physical and mechanical properties of twisted or ground nickel-titanium instruments. Int. Endod. J. 2013, 46, 458–65.

[53] https://www.slideshare.net/shadanAltayar/niti-in-endodontics-75307318.

[54] Gu Y, Kum KY, Perinpanayagam H, Kim C, Kum DJ, Lim SM, Chang SW, Baek SH, Zhu Q, Yoo YJ. Various heat-treated nickel-titanium rotary instruments evaluated in S-shaped simulated resin canals. J. Dent. Sci. 2017, 12, 14–20.

[55] Lopes HP, Gambarra-Soares T, Elias CN, Siqueira JF Jr, Inojosa IF, Lopes WS, Vieira VT. Comparison of the mechanical properties of rotary instruments made of conventional nickel-titanium wire, M-wire, or nickel-titanium alloy in R-phase. J. Endod. 2013, 39, 516–20.

[56] Perez-Higueras JJ, Arias A, de la Macorra JC. Cyclic fatigue resistance of K3, K3XF, and twisted file nickel-titanium files under continuous rotation or reciprocating motion. J. Endod. 2013, 39, 1585–8.

[57] Choi J, Oh S, Kim YC, Jee KK, Kum K, Chang S. Fracture resistance of K3 nickel-titanium files made from different thermal treatments. Bioinorg. Chem. Appl. 2016, 2016; doi: 10.1155/2016/6374721.

[58] Iacono F, Pirani C, Generali L, Bolelli G, Sassatelli P, Lusvarghi L, Gandolfi MG, Giorgini L, Prati C. Structural analysis of HyFlex EDM instruments. Int. Endod. J. 2017, 50, 303–13.

[59] Shen Y, Coil JM, Zhou H-M, Tam E, Zheng Y-F, Haapasalo M. ProFile vortex instruments after clinical use: a metallurgical properties study. J. Endod. 2012, 38, 1613–7.

[60] AlShwaimi E. Comparing ProFile vortex to ProTaper next for the efficacy of removal of root filling material: an ex vivo micro-computed tomography study. Saudi Dent. J. 2018, 30, 63–9.

[61] Gao Y, Gutmann JL, Wilkinson K, Maxwell R, DanAmmon D. Evaluation of the impact of raw materials on the fatigue and mechanical properties of ProFile vortex rotary instruments. J. Endod. 2012, 38, 398–401.

[62] https://www.dentsplysirona.com/en-us/categories/endodontics/vortex-blue.html.

[63] Metzger Z, Teperovich E, Zary R, Cohen R, Hof R. The self-adjusting file (SAF). Part 1: respecting the root canal anatomy – a new concept of endodontic files and its implementation. J. Endod. 2010, 36, 679–90.

[64] https://endoexperience.com/documents/WaveOne.pdf.

[65] Sjogren U, Figdor D, Persson S, Sundqvist G. Influence of infection at the time of root filling on the outcome of endodontic treatment of teeth with apical periodontitis. Int. Endod. J. 1997, 30, 297–306.

[66] West JD. Endodontic Predictability – "Restore or Remove: How Do I Choose?" In: Cohen M, ed., Interdisciplinary Treatment Planning: Principles, Design, Implementation. Quintessence Publishing Co., 2008, 123–64.

[67] Lopes WSP, Lopes HP, Elias CN, Vieira MVB, Batista MMD, Cunha RS. Resistance to bending and buckling of WaveOne and Reciproc instruments. ENDO (Lond Engl) 2014, 8, 153–6.

–[68] Cassim I. A novel use of the Reciproc R25 endodontic file for root canal obturation. SADJ 2014, 69, 460–2.

[69] Metzger Z. The self-adjusting file (SAF) system: an evidence-based update. J. Conserv. Dent. 2014, 17, 401–19.

[70] Webber J. Shaping canals with confidence: WaveOne gold single-file reciprocating system. Roots 2015, 1, 34–40.

[71] Zehnder M. Root canal irrigants. J. Endod. 2006, 32, 389–98.

[72] Yared G. Reciproc blue: the new generation of reciprocation. Giornale Italiano di Endodonzia 2017, 31, 96–101.

[73] Haapasalo M, Shen Y. Evolution of nickel–titanium instruments: from past to future. Endod. Top. 2013, 29, 3–17.

[74] https://vishesh.dentalkart.com/micro-mega-revo-s-start-me-kit.html.

[75] Basrani B, Roth K, Sas G, Kishen A, Peters OA. Torsional profiles of new and used Revo-S rotary instruments: an in vitro study. J. Endod. 2011, 37, 989–92.

[76] Serafin M, De Biasi M, Franco V, Angerame D. In vitro comparison of cyclic fatigue resistance of two rotary single-file endodontic systems: OneCurve versus OneShape. Odontology 2019, 107, 196–201.

[77] Gernhardt CR. One Shape – a single file NiTi system for root canal instrumentation used in continuous rotation. ENDO – Endod. Pract. Today 2013, 7, 211–6.

[78] http://www.dentsplymaillefer.com/product-category/glide-path-shaping/protaper-next.

[79] Silva EJ, Vieira VC, Tameirao, Belladonna FG, Neves Ade A, Souza EM, DE-Deus G. Quantitative transportation assessment in curved canals prepared with an off-centered rectangular design system. Braz. Oral Res. 2016, 30; http://dx.doi.org/10.1590/1807-3107BOR-2016.vol30.0043.

[80] Gavini G, Santos MD, Caldeira CL, Machado MEL, Freire LG, Iglecias EF, Peters OA, Candeiro GTM. Nickel-titanium instruments in endodontics: a concise review of the state of the art. Braz. Oral Res. 2018, 18; doi: 10.1590/1807-3107bor-2018.vol32.0067.

[81] Gambarini G, Plotino G, Grande NM, Al-Sudani D, De Luca M, Testarelli L. Mechanical properties of nickel-titanium rotary instruments produced with a new manufacturing technique. Int. Endod. J. 2011, 44, 337–41.

[82] Shen Y, Qian W, Abtin H, Gao Y, Haapasalo M. Fatigue testing of controlled memory wire nickel-titanium rotary instruments. J. Endod. 2011, 37, 997–1001.

[83] Ye J, Gao Y. Metallurgical characterization of M-Wire nickel-titanium shape memory alloy used for endodontic rotary instruments during low-cycle fatigue. J. Endod. 2012, 38, 105–7.

[84] Ahamed SBB, Vanajassun PP, Rajkumar K, Mahalaxmi S. Comparative evaluation of stress distribution in experimentally designed nickel-titanium rotary files with varying cross sections: a finite element analysis. J. Endod. 2018, 44, 654–8.

[85] Bergmans L, Van Cleynenbreugel J, Wevers M, Lambrechts P. Mechanical root canal preparation with NiTi rotary instruments: rationale, performance and safety. Status report. Am. J. Dent. 2001, 14, 324–33.

[86] Pedrinha VF, Brandão JMDS, Pessoa OF, Rodrigues PA. Influence of file motion on shaping, apical debris extrusion and dentinal defects: a critical review. Open Dent. J. 2018, 28, 189–201.

[87] Metzger Z, Solomonov M, Kfir A. The role of mechanical instrumentation in the cleaning of root canals. Endod. Top. 2013, 29, 87–109.

[88] Camps JJ, Pertot WJ. Machining efficiency of nickel-titanium K-type files in a linear motion. Int. Endod. J. 1995, 28, 279–84.

[89] Thompson SA, Dummer PMH. Shaping ability of NT Engine and McXim rotary nickel-titanium instruments in simulated root canals. Part 1. Int. Endod. J. 1997, 30, 262–9.

[90] Dobó-Nagy C, Serbán T, Szabó J, Nagy G, Madléna M. A comparison of the shaping characteristics of two nickel-titanium endodontic hand instruments. Int. Endod. J. 2002, 35, 283–8.

[91] Burroughs JR, Bergeron BE, Roberts MD, Hagan JL, Himel VT. Saping ability of three nickel-titanium endodontic file systems in simulated S-shaped root canals. J. Endod. 2012, 38, 1618–21.

[92] Giuliani V, Di Nasso L, Pace R, Pagavino G. Shaping ability of waveone primary reciprocating files and ProTaper system used in continuous and reciprocating motion. J. Endod. 2014, 40, 1468–71.

[93] Özyürek T, Yılmaz K, Uslu G. Shaping ability of reciproc, WaveOne GOLD, and HyFlex EDM single-file systems in simulated S-shaped canals. J. Endod. 2017, 43, 805–9.

[94] Duque JA, Vivan RR, Cavenago BC, Amoroso-Silva PA, Bernardes RA, Vasconcelos BC, Duarte MA. Influence of NiTi alloy on the root canal shaping capabilities of the ProTaper Universal and ProTaper gold rotary instrument systems. J. Appl. Oral Sci. 2017, 25, 27–33.

[95]   Pedullà E, Plotino G, Grande NM, Avarotti G, Gambarini G, Rapisarda E, Mannocci F. Shaping ability of two nickel-titanium instruments activated by continuous rotation or adaptive motion: a micro-computed tomography study. Clin. Oral Invest. 2016, 20, 2227–33.

[96]   Espir CG, Nascimento-Mendes CA, Guerreiro-Tanomaru JM, Cavenago BC, Duarte MAH, Tanomaru-Filho M. Shaping ability of rotary or reciprocating systems for oval root canal preparation: a micro-computed tomography study. Clin. Oral Invest. 2018, 22, 3189–94.

[97]   Huang Z, Quan J, Liu J, Zhang W, Zhang X, Hu X. A microcomputed tomography evaluation of the shaping ability of three thermally-treated nickel-titanium rotary file systems in curved canals. J. Int. Med. Res. 2019, 47, 325–34.

[98]   Camargo EJ, Duarte MAH, Marques VAS, Só MVR, Duque JA, Alcalde MP, Vivan RR. The ability of three nickel-titanium mechanized systems to negotiate and shape MB2 canals in extracted maxillary first molars: a micro-computed tomographic study. Int. Endod. J. 2019, 52, 847–56.

[99]   Kandaswamy D, Venkateshbabu N, Porkodi I, Pradeep G. Canal-centering ability: an endodontic challenge. J. Conserv. Dent. 2009, 12, 3–9.

[100]   Zarei M, Javidi M, Erfanian M, Lomee M, Afkhami F. Comparison of air-driven vs electric torque control motors on canal centering ability by ProTaper NiTi rotary instruments. J. Contemp. Dent. Pract. 2013, 14, 71–5.

[101]   Ponti TM, McDonald NJ, Kuttler S, Strassler HE, Dumsha TC. Canal-centering ability of two rotary file systems. J. Endod. 2002, 28, 283–6.

[102]   Pinheiro SR, Alcalde MP, Vivacqua-Gomes N, Bramante CM, Vivan RR, Duarte MAH, Vasconcelos BC. Evaluation of apical transportation and centering ability of five thermally treated NiTi rotary systems. Int. Endod. J. 2018, 51, 705–13.

[103]   Tepel J, Schäfer E, Hoppe W. Properties of endodontic hand instruments used in rotary motion. Part 1. Cutting efficiency. J. Endod. 1995, 21, 418–21.

[104]   Haïkel Y, Serfaty R, Wilson P, Speisser JM, Allemann C. Cutting efficiency of nickel-titanium endodontic instruments and the effect of sodium hypochlorite treatment. J. Endod. 1998, 24, 736–9.

[105]   Peters OA, Morgental RD, Schulze KA, Paqué F, Kopper PM, Vier-Pelisser FV. Determining cutting efficiency of nickel-titanium coronal flaring instruments used in lateral action. Int. Endod. J. 2014, 47, 505–13.

[106]   Passi S, Kaler N, Passi N. What is a glide path? Saint's Intl. Dent. J. 2016, 2, 32–7.

[107]   West JD. The endodontic glidepath: secret to rotary safety. Dent. Today 2010, 29, 86–93.

[108]   Tomson PL, Simon SR. Contemporary cleaning and shaping of the root canal system. Prim. Dent. J. 2016, 5, 46–53.

[109]   Pasqualini D, Bianchi CC, Paolino DS, Mancini L, Cemenasco A, Cantatore G, Castellucci A, Berutti E. Computed micro-tomographic evaluation of glide path with nickel-titanium rotary PathFile in maxillary first molars curved canals. J. Endod. 2012, 38, 389–93.

[110]   D'Amario M, Baldi M, Petricca R, De Angelis F, El Abed R, D' Arcangelo C. Evaluation of a new nickel-titanium system to create the glide path in root canal preparation of curved canals. J. Endod. 2013, 39, 1581–4.

[111]   Ajuz NC, Armada L, Gonçalves LS, Debelian G, Siqueira JF Jr. Glide path preparation in S-shaped canals with rotary pathfinding nickel-titanium instruments. J. Endod. 2013, 39, 534–7.

[112]   Mohammadian F, Sadeghi A, Dibaji F, Sadegh M, Ghoncheh Z, Kharrazifard MJ. Comparison of apical transportation with the use of rotary system and reciprocating handpiece with precurved hand files: an *in vitro* study. Iran Endod. J. 2017, 12, 462–7.

[113]   Labbaf H, Shakeri L, Orduie R, Bastami F. Apical extrusion of debris after canal preparation with hand-files used manually or installed on reciprocating air-driven handpiece in straight and curved canals. Iran Endod. J. 2015, 10, 165–8.

[114] Glossen CR, Haller RH, Dove SB, del Rio CE. A comparison of root canal preparations using Ni-Ti hand, Ni-Ti engine-driven, and K-Flex endodontic instruments. J. Endod. 1995, 21, 146–51.

[115] Kakar S, Dhingra A, Sharma H. Shaping potential of manual NiTi K-File and rotary ProTaper and analyzing the final outcome of shaped canals using CT. J. Contemp. Dent. Pract. 2013, 14, 451–5.

[116] Subramaniam P, Tabrez TA, Babu KL. Microbiological assessment of root canals following use of rotary and manual instruments in primary molars. J. Clin. Pediatr. Dent. 2013, 38, 123–7.

[117] Radhika E, Reddy ER, Rani ST, Kumar LV, Manjula M, Mohan TA. Cone beam computed tomography evaluation of hand nickel-titanium K-files and rotary system in primary teeth. Pediatr. Dent. 2017, 39, 319–23.

[118] Shaikh SM, Goswami M. Evaluation of the effect of different root canal preparation techniques in primary teeth using CBCT. J. Clin. Pediatr. Dent. 2018, 42, 250–5.

[119] Esposito PT, Cunningham CJ. A comparison of canal preparation with nickel-titanium and stainless steel instruments. J. Endod. 1995, 21, 173–6.

[120] Coleman CL, Svec TA, Rieger MR, Suchina JA, Wang MM, Glickman GN. Analysis of nickel-titanium versus stainless steel instrumentation by means of direct digital imaging. J. Endod. 1996, 22, 603–7.

[121] Coleman CL, Svec TA. Analysis of Ni-Ti versus stainless steel instrumentation in resin simulated canals. J. Endod. 1997, 23, 232–5.

[122] Schäfer E, Lau R. Comparison of cutting efficiency and instrumentation of curved canals with nickel-titanium and stainless-steel instruments. J. Endod. 1999, 25, 427–30.

[123] Schäfer E, Florek H. Efficiency of rotary nickel-titanium K3 instruments compared with stainless steel hand K-Flexofile. Part 1. Shaping ability in simulated curved canals. Int. Endod. J. 2003, 36, 199–207.

[124] Lam TV, Lewis DJ, Atkins DR, Macfarlane RH, Clarkson RM, Whitehead MG, Brockhurst PJ, Moule AJ. Changes in root canal morphology in simulated curved canals over-instrumented with a variety of stainless steel and nickel titanium files. Aust. Dent. J. 1999, 44, 12–9.

[125] Davis RD, Marshall JG, Baumgartner JC. Effect of early coronal flaring on working length change in curved canals using rotary nickel-titanium versus stainless steel instruments. J. Endod. 2002, 28, 438–42.

[126] Gluskin AH, Brown DC, Buchanan LS. A reconstructed computerized tomographic comparison of Ni-Ti rotary GT files versus traditional instruments in canals shaped by novice operators. Int. Endod. J. 2001, 34, 476–84.

[127] Taşdemir T, Aydemir H, Inan U, Unal O. Canal preparation with Hero 642 rotary Ni-Ti instruments compared with stainless steel hand K-file assessed using computed tomography. Int. Endod. J. 2005, 38, 402–8.

[128] Nagaratna PJ, Shashikiran ND, Subbareddy VV. In vitro comparison of NiTi rotary instruments and stainless steel hand instruments in root canal preparations of primary and permanent molar. J. Indian Soc. Pedod. Prev. Dent. 2006, 24, 186–91.

[129] Matwychuk MJ, Bowles WR, McClanahan SB, Hodges JS, Pesun IJ. Shaping abilities of two different engine-driven rotary nickel titanium systems or stainless steel balanced-force technique in mandibular molars. J. Endod. 2007, 33, 868–71.

[130] Vaudt J, Bitter K, Neumann K, Kielbassa AM. Ex vivo study on root canal instrumentation of two rotary nickel-titanium systems in comparison to stainless steel hand instruments. Int. Endod. J. 2009, 42, 22–33.

[131] Cheung GS, Liu CS. A retrospective study of endodontic treatment outcome between nickel-titanium rotary and stainless steel hand filing techniques. J. Endod. 2009, 35, 938–43.

[132] Nordmeyer S, Schnell V, Hülsmann M. Comparison of root canal preparation using Flex Master Ni-Ti and Endo-Eze AET stainless steel instruments. Oral Surg. Oral Med. Oral Pathol. Oral Radiol. Endod. 2011, 111, 251–9.

[133] Sandhu SS, Shetty VS, Mogra S, Varghese J, Sandhu J, Sandhu JS. Efficiency, behavior, and clinical properties of superelastic NiTi versus multistranded stainless steel wires. Angle Orthod. 2012, 82, 915–21.

[134] Chuste-Guillot MP, Badet C, Peli JF, Perez F. Effect of three nickel-titanium rotary file techniques on infected root dentin reduction. Oral Surg. Oral Med. Oral Pathol. Oral Radiol. Endod. 2006, 102, 254–8.

[135] Burroughs JR, Bergeron BE, Roberts MD, Hagan JL, Himel VT. Shaping ability of three nickel-titanium endodontic file systems in simulated S-shaped root canals. J. Endod. 2012, 38, 1618–21.

[136] Ba-Hattab R, Pröhl AK, Lang H, Pahncke D. Comparison of the shaping ability of GT® series X, twisted files and AlphaKite rotary nickel-titanium systems in simulated canals. BMC Oral Health 2013, 13, 72; doi: 10.1186/1472-6831-13-72.

[137] Drukteinis S, Peciuliene V, Dummer PMH, Hupp J. Shaping ability of BioRace, ProTaper NEXT and genius nickel-titanium instruments in curved canals of mandibular molars: a MicroCT study. Int. Endod. J. 2019, 52, 86–93.

[138] Poly A, AlMalki F, Marques F, Karabucak B. Canal transportation and centering ratio after preparation in severely curved canals: analysis by micro-computed tomography and double-digital radiography. Clin. Oral Invest. 2019, May, 1–8; doi: 10.1007/s00784-019-02870-8.

[139] Jamleh A, Yahata Y, Ebihara A, Atmeh AR, Bakhsh T, Suda H. Performance of NiTi endodontic instrument under different temperatures. Odontology 2016, 104, 324–8.

[140] Ha JH, Kim SK, Cheung GS, Jeong SH, Bae YC, Kim HC. Effect of alloy type on the life-time of torsion-preloaded nickel-titanium endodontic instruments. Scanning 2015, 37, 172–8.

[141] Pereira ÉS, Viana AC, Buono VT, Peters OA, Bahia MG. Behavior of nickel-titanium instruments manufactured with different thermal treatments. J. Endod. 2015, 41, 67–71.

[142] Ounsi HF, Franciosi G, Paragliola R, Huzaimi KA, Salameb Z, Tay FR, Ferrari M, Grandini S. Comparison of two way techniques for assessing the shaping efficacy of repeatedly used nickel-titanium rotary instruments. J. Endod. 2011, 37, 847–50.

[143] Busslinger A, Sener B, Barbakow F. Effects of sodium hypochlorite on nickel-titanium Lightspeed® instruments. Int. Endod. J. 1998, 31, 290–4.

[144] Haapasalo M, Shen Y, Wang Z, Gao Y. Irrigation in endodontics. Br. Dent. J. 2014, 216, 299–303.

[145] Park E, Shen Y, Khakpour M, Haapasalo M. Apical pressure and extent of irrigant flow beyond the needle tip during positive-pressure irrigation in an *in vitro* root canal model. J. Endod. 2013, 39, 511–5.

[146] Howard RK, Kirkpatrick TC, Rutledge RE, Yaccino JM. Comparison of debris removal with three different irrigation techniques. J. Endod. 2011, 37, 1301–5.

[147] Haïkel Y, Serfaty R, Wilson P, Speisser JM, Allemann C. Mechanical properties of nickel-titanium endodontic instruments and the effect of sodium hypochlorite treatment. J. Endod. 1998, 24, 731–5.

[148] Bramante CM, Betti LV. Comparative analysis of curved root canal preparation using nickel-titanium instruments with or without EDTA. J. Endod. 2000, 26, 278–80.

[149] Schäfer E. Effect of sterilization on the cutting efficiency of PVD-coated nickel-titanium endodontic instruments. Int. Endod. J. 2002, 35, 867–72.

[150] Grandini S, Balleri P, Ferrari M. Evaluation of glyde file prep in combination with sodium hypochlorite as a root canal irrigant. J. Endod. 2002, 28, 300–3.

[151] Slutzky-Goldberg I, Liberman R, Heling I. The effect of instrumentation with two different file types, each with 2.5% NaOCl irrigation on the microhardness of root dentin. J. Endod. 2002, 28, 311–2.

[152] O'Hoy PYZ, Messer HH, Palamara JEA. The effect of cleaning procedures on fracture properties and corrosion of NiTi files. Int. Endod. J. 2003, 36, 724–32.

[153] Passarinho-Neto JG, Marchesan MA, Ferreira RB, Silva RG, Silva-Sousa YT, Sousa-Neto MD. In vitro evaluation of endodontic debris removal as obtained by rotary instrumentation coupled with ultrasonic irrigation. Aust. Endod. J. 2006, 32, 123–8.

[154] Paiva SS, Siqueira JF Jr, Rôças IN, Carmo FL, Leite DC, Ferreira DC, Rachid CT, Rosado AS. Clinical antimicrobial efficacy of NiTi rotary instrumentation with NaOCl irrigation, final rinse with chlorhexidine and interappointment medication: a molecular study. Int. Endod. J. 2013, 46, 225–33.

[155] Cai JJ, Tang XN, Ge JY. Effect of irrigation on surface roughness and fatigue resistance of controlled memory wire nickel-titanium instruments. Int. Endod. J. 2017, 50, 718–24.

[156] Cheung GSP. Instrument fracture: mechanisms, removal of fragments, and clinical outcomes. Instrument fracture: mechanisms, removal of fragments, and clinical outcomes. Endod. Top. 2009, 16, 1–6.

[157] Kaul R, Farooq R, Kaul V, Khateeb SU, Purra AR, Mahajan M. Comparative evaluation of physical surface changes and incidence of separation in rotary nickel-titanium instruments: an in vitro SEM study. Iran. Endod. J. 2014, 9, 204–9.

[158] Choi J, Oh S, Kim Y-C, Jee K-K, Kum K, Chang S. Fracture resistance of K3 nickel-titanium files made from different thermal treatments. Bioinorg. Chem. Appl. 2016; http://dx.doi.org/10.1155/2016/6374721.

[159] Gil J, Rupérez E, Velasco E, Aparicio C, Manero JM. Mechanism of fracture of NiTi superelastic endodontic rotary instruments. J. Mater. Sci. Mater. Med. 2018, 29; doi: 10.1007/s10856-018-6140-7.

[160] Kim TO, Cheung GS, Lee JM, Kim BM, Hur B, Kim HC. Stress distribution of three NiTi rotary files under bending and torsional conditions using a mathematic analysis. Int. Endod. J. 2009, 42, 14–21.

[161] Barbosa I, Ferreira F, Scelza P, Neff J, Russano D, Montagnana M, Zaccaro Scelza M. Defect propagation in NiTi rotary instruments: a noncontact optical profilometry analysis. Int. Endod. J. 2018, 51, 1271–8.

[162] Spanaki-Voreadi AP, Kerezoudis NP, Zinelis S. Failure mechanism of ProTaper Ni-Ti rotary instruments during clinical use: fractographic analysis. Int. Endod. J. 2006, 39, 171–8.

[163] Ferreira F, Barbosa I, Scelza P, Russano D, Neff J, Montagnana M, Zaccaro Scelza M. A new method for the assessment of the surface topography of NiTi rotary instruments. Int. Endod. J. 2017, 50, 902–9.

[164] Best S, Watson P, Pilliar R, Kulkarni GGK, Yared G. Torsional fatigue and endurance limit of a size 30 .06 ProFile rotary instrument. Int. Endod. J. 2004, 37, 370–3.

[165] Peters OA, Peters CI, Schönenberger K, Barbakow F. ProTaper rotary root canal preparation: assessment of torque and force in relation to canal anatomy. Int. Endod. J. 2003, 36, 93–9.

[166] Chang SW, Shim KS, Kim YC, Jee KK, Zhu Q, Kum KY. Cyclic fatigue resistance, torsional resistance, and metallurgical characteristics of V taper 2 and V taper 2H rotary NiTi files. Scanning 2016, 38, 564–70.

[167] Cheung GSP, Peng B, Bian Z, Shen Y, Darvell BW. Defects in ProTaper S1 instruments after clinical use: fractographic examination. Int. Endod. J.. 2005, 38, 802–9.

[168] Alapati SB, Brantley WA, Svec TA, Powers JM, Nusstein JM, Daehn GS. SEM observations of nickel-titanium rotary endodontic instruments that fractured during clinical Use. J. Endod. 2005, 31, 40–3.

[169] Wei X, Ling J, Jiang J, Huang X, Liu L. Modes of failure of ProTaper nickel-titanium rotary instruments after clinical use. J. Endod. 2007, 33, 276–9.

[170] Shen Y, Cheung GS, Peng B, Haapasalo M. Defects in nickel-titanium instruments after clinical use. Part 2: fractographic analysis of fractured surface in a cohort study. J. Endod. 2009, 35, 133–6.

[171] Abu-Tahun IH, Kwak SW, Ha J-H, Kim H-C. Microscopic features of fractured fragment of nickel-titanium endodontic instruments by two different modes of torsional loading. Scanning 2018; https://doi.org/10.1155/2018/9467059.

[172] Troian CH, Só MV, Figueiredo JA, Oliveira EP. Deformation and fracture of RaCe and K3 endodontic instruments according to the number of uses. Int. Endod. J. 2006, 39, 616–25.

[173] Setzer FC, Böhme CP. Influence of combined cyclic fatigue and torsional stress on the fracture point of nickel-titanium rotary instruments. J. Endod. 2013, 39, 133–7.

[174] Kuhn G, Tavernier B, Jordan L. Influence of structure on nickel-titanium endodontic instruments failure. J. Endod. 2001, 27, 516–20.

[175] Kim BH, Ha JH, Lee WC, Kwak SW, Kim HC. Effect from surface treatment of nickel-titanium rotary files on the fracture resistance. Scanning 2015, 37, 82–7.

[176] Kosti E, Zinelis S, Molyvdas I, Lambrianidis T. Effect of root canal curvature on the failure incidence of ProFile rotary Ni-Ti endodontic instruments. Int. Endod. J. 2011, 44, 917–25.

[177] Yared GM, Bou Daugher FE, Machtou P. Influence of rotational speed, torque and operator's proficiency on ProFile failures. Int. Endod. J. 2001; 34, 47–53.

[178] Daugherty DW, Gound TG, Comer TL. Comparison of fracture rate, deformation rate, and efficiency between rotary endodontic instruments driven at 150 rpm and 350 rpm. J. Endod. 2001, 27, 93–5.

[179] Parashos P, Gordon I, Messer HH. Factors influencing defects of rotary nickel-titanium endodontic instruments after clinical use. J. Endod. 2004, 30, 722–5.

[180] Patiño PV, Biedma BM, Liébana CR, Cantatore G, Bahillo JG. The influence of a manual glide path on the separation rate of NiTi rotary instruments. J. Endod. 2005, 31, 114–6.

[181] Wolcott S, Wolcott J, Ishley D, Kennedy W, Johnson S, Minnich S, Meyers J. Separation incidence of protaper rotary instruments: a large cohort clinical evaluation. J. Endod. 2006, 32, 1139–41.

[182] Shen Y, Haapasalo M, Cheung GS, Peng B. Defects in nickel-titanium instruments after clinical use. Part 1: relationship between observed imperfections and factors leading to such defects in a cohort study. J. Endod. 2009, 35, 129–32.

[183] Shen Y, Coil JM, Mclean AGR, Hemerling DL, Haapasalo M. Defects in nickel-titanium instruments after clinical use. Part 5: single use from endodontic specialty practices. J. Endod. 2009, 35, 1363–7.

[184] Bueno CSP, de Oliveira DP, Pelegrine RA, Fontana CE, Rocha DGP, Bueno CEDS. Fracture incidence of WaveOne and Reciproc files during root canal preparation of up to 3 posterior teeth: a prospective clinical study. J. Endod. 2017, 43, 705–8.

[185] Anderson ME, Price JW, Parashos P. Fracture resistance of electropolished rotary nickel-titanium endodontic instruments. J. Endod. 2007, 33, 1212–6.

[186] Chi CW, Deng YL, Lee JW, Lin CP. Fracture resistance of dental nickel-titanium rotary instruments with novel surface treatment: thin film metallic glass coating. J. Formos. Med. Assoc. 2017, 116, 373–9.

[187] Shen Y, Zhou HM, Zheng YF, Campbell L, Peng B, Haapasalo M. Metallurgical characterization of controlled memory wire nickel-titanium rotary instruments. J. Endod. 2011, 37, 1566–71.

[188] Di Fiore PM. A dozen way to prevent nickel-titanium rotary instrument fracture. J. Am. Dent. Assoc. 2007, 138, 196–201.

[189] Weiss V, Oshida Y. Fatigue damage characterization by x-ray diffraction line analysis and failure probability. EPRI AP-4477 1986, EPRI, 16-1/16-18.

[190] Suresh S. Fatigue of materials. Cambridge University Press, New York, NY, USA, 1991.

[191] Cheung GS, Darvell BW. Fatigue testing of a NiTi rotary instrument. Part 1: strain-life relationship. Int. Endod. J. 2007, 40, 612–8.

[192] Cheung GS, Darvell BW. Fatigue testing of a NiTi rotary instrument. Part 2: fractographic analysis. Int. Endod. J. 2007, 40, 619–25.

[193] Cheung GS, Darvell BW. Low-cycle fatigue of NiTi rotary instruments of various cross-sectional shapes. Int. Endod. J. 2007, 40, 626–32.

[194] Pruett JP, Clement DJ, Carnes DL Jr. Cyclic fatigue testing of nickel-titanium endodontic instruments. J. Endod. 1997, 23, 77–85.

[195] Plotino G, Grande NM, Cordaro M, Testarelli L, Gambarini G. Influence of the shape of artificial canals on the fatigue resistance of NiTi rotary instruments. Int. Endod. J. 2010, 43, 69–75.

[196] Gao Y, Shotton V, Wilkinson K, Phillips G, Johnson WB. Effects of raw material and rotational speed on the cyclic fatigue of ProFile vortex rotary instruments. J. Endod. 2010, 36, 1205–9.

[197] Fife D, Gambarini G, Britto Lr Lr. Cyclic fatigue testing of ProTaper NiTi rotary instruments after clinical use. Oral Surg. Oral Med. Oral Pathol. Oral Radiol. Endod. 2004, 97, 251–6.

[198] Figueiredo AM, Modenesi P, Buono V. Low-cycle fatigue life of superelastic NiTi wires. Int. J. Fatigue 2009, 31, 751–8.

[199] Hieawy A, Haapasalo M, Zhou H, Wang ZJ, Shen Y. Phase transformation behavior and resistance to bending and cyclic fatigue of ProTaper gold and ProTaper universal instruments. J. Endod. 2015, 41, 1134–8.

[200] Arias A, Macorra JC, Govindjee S, Peters OA. Correlation between temperature-dependent fatigue resistance and differential scanning calorimetry analysis for 2 contemporary rotary instruments. J. Endod. 2018, 44, 630–4.

[201] Ferreira F, Adeodato C, Barbosa I, Aboud L, Scelza P, Zaccaro Scelza M. Movement kinematics and cyclic fatigue of NiTi rotary instruments: a systematic review. Int. Endod. J. 2017, 50, 143–52.

[202] Jamleh A, Kobayashi C, Yahata Y, Ebihara A, Suda H. Deflecting load of nickel titanium rotary instruments during cyclic fatigue. Dent. Mater. J. 2012, 31, 389–93.

[203] Versluis A, Kim HC, Lee W, Kim BM, Lee CJ. Flexural stiffness and stresses in nickel-titanium rotary files for various pitch and cross-sectional geometries. J. Endod. 2012, 38, 1399–403.

[204] Grande NM, Plotino G, Pecci R, Bedini R, Malagnino VA, Somma F. Cyclic fatigue resistance and three-dimensional analysis of instruments from two nickel-titanium rotary systems. Int. Endod. J. 2006, 39, 755–63.

[205] Lee MH, Versluis A, Kim BM, Lee CJ, Hur B, Kim HC. Correlation between experimental cyclic fatigue resistance and numerical stress analysis for nickel-titanium rotary files. J. Endod. 2011, 37, 1152–7.

[206] Inan U, Aydin C, Tunca YM. Cyclic fatigue of ProTaper rotary nickel-titanium instruments in artificial canals with 2 different radii of curvature. Oral Surg. Oral Med. Oral Pathol. Oral Radiol. Endod. 2007, 104, 837–40.

[207] Karamooz-Ravari MR, Dehghani R. The effects of shape memory alloys' tension-compression asymmetry on NiTi endodontic files' fatigue life. Proc. Inst. Mech. Eng. H 2018, 232, 437–45.

[208] Haikel Y, Gasser P, Allemann C. Dynamic fracture of hybrid endodontic hand instruments compared with traditional files. J. Endod. 1991, 17, 217–20.

[209] Haïkel Y, Serfaty R, Bateman G, Senger B, Allemann C. Dynamic and cyclic fatigue of engine-driven rotary nickel-titanium endodontic instruments. J. Endod. 1999, 25, 434–40.

[210] Rodrigues RC, Lopes HP, Elias CN, Amaral G, Vieira VT, De Martin AS. Influence of different manufacturing methods on the cyclic fatigue of rotary nickel-titanium endodontic instruments. J. Endod. 2011, 37, 1553–7.

[211]   Waver JD, Gutierrez EJ. Comparing rotary bend wire fatigue test methods at different test speeds. J. Mater. Eng. Perform. 2015, 24, 4966–74.

[212]   Arias A, Hejlawy S, Murphy S, de la Macorra JC, Govindjee S, Peters OA. Variable impact by ambient temperature on fatigue resistance of heat-treated nickel titanium instruments. Clin. Oral Invest. 2019, 23, 1101–9.

[213]   de Vasconcelos RA, Murphy S, Carvalho CA, Govindjee RG, Govindjee S, Peters OA. Evidence for reduced fatigue resistance of contemporary rotary instruments exposed to body temperature. J. Endod. 2016, 42, 782–7.

[214]   Plotino G, Grande NM, Mercadé Bellido M, Testarelli L, Gambarini G. Influence of temperature on cyclic fatigue resistance of ProTaper gold and ProTaper universal rotary files. J. Endod.. 2017, 43, 200–2.

[215]   Klymus ME, Alcalde MP, Vivan RR, Só MVR, de Vasconselos BC, Duarte MAH. Effect of temperature on the cyclic fatigue resistance of thermally treated reciprocating instruments. Clin. Oral Invest. 2018; doi: 10.1007/s00784-018-2718-1.

[216]   Klymus ME, Alcalde MP, Vivan RR, Só MVR, de Vasconselos BC, Duarte MAH. Effect of temperature on the cyclic fatigue resistance of thermally treated reciprocating instruments. Clin. Oral Invest. 2019, 23, 3047–52.

[217]   Grande NM, Plotino G, Silla E, Pedullà E, DeDeus G, Gambarini G, Somma F. Environmental temperature drastically affects flexural fatigue resistance of nickel-titanium rotary files. J. Endod. 2017, 43, 1157–60.

[218]   Shen Y, Huang X, Wang Z, Wei X, Haapasalo M. Low environmental temperature influences the fatigue resistance of nickel-titanium files. J. Endod. 2018, 44, 626–9.

[219]   Plotino G, Grande NM, Testarelli L, Gambarini G, Castagnola R, Rossetti A, Özyürek T, Cordaro M, Fortunato L. Cyclic fatigue of reciproc and reciproc blue nickel-titanium reciprocating files at different environmental temperatures. J. Endodo. 2018, 44, 1549–52.

[220]   Dosanjh A, Paurazas S, Askar M. The effect of temperature on cyclic fatigue of nickel-titanium rotary endodontic instruments. J. Endod. 2017, 43, 823–6.

[221]   Alfawaz H, Alqedairi A, Alsharekh H, Almuzaini E, Alzahrani S, Jamleh A. Effects of sodium hypochlorite concentration and temperature on the cyclic fatigue resistance of heat-treated nickel-titanium rotary instruments. J Endod. 2018, 44, 1563–6.

[222]   Pedullà E, Grande NM, Plotino G, Pappalardo A, Rapisarda E. Cyclic fatigue resistance of three different nickel-titanium instruments after immersion in sodium hypochlorite. J. Endod. 2011, 37, 1139–42.

[223]   Pedullà E, Grande NM, Plotino G, Palermo F, Gambarini G, Rapisarda E. Cyclic fatigue resistance of two reciprocating nickel-titanium instruments after immersion in sodium hypochlorite. Int. Endod. J. 2013, 46, 155–9.

[224]   Pedullà E, Benites A, La Rosa GM, Plotino G, Grande NM, Rapisarda E, Generali L. Cyclic fatigue resistance of heat-treated nickel-titanium instruments after immersion in sodium hypochlorite and/or sterilization. J. Endod. 2018, 44, 648–53.

[225]   Peters OA, Roehlike JO, Baumann MA. Effect of immersion in sodium hypochlorite on torque and fatigue resistance of nickel-titanium instruments. J. Endod. 2007, 33, 589–93.

[226]   Huang X, Shen Y, Wei X, Haapasalo M. Fatigue resistance of nickel-titanium instruments exposed to high-concentration hypochlorite. J. Endod. 2017, 43, 1847–51.

[227]   Gülşah Uslu G, Özyürek T, Yılmaz K, Plotino G. Effect of dynamic immersion in sodium hypochlorite and EDTA solutions on cyclic fatigue resistance of WaveOne and WaveOne gold reciprocating nickel-titanium files. J. Endod. 2018, 44, 834–7.

[228]   Shen Y, Qian W, Abtin H, Gao Y, Haapasalo M. Effect of environment on fatigue failure of controlled memory wire nickel-titanium rotary instruments. J. Endod. 2012, 38, 376–80.

[229] Iacono F, Pirani C, Arias A, de la Macorra JC, Generali L, Gandolfi MG, Prati C. Impact of a modified motion on the fatigue life of NiTi reciprocating instruments: a Weibull analysis. Clin. Oral Invest. 2019, 23, 3095–102.

[230] Herbst SR, Krois J, Schwendicke F. Comparator choice in studies testing endodontic instrument fatigue resistance: a network analysis. J. Endod. 2019, 45, 784–90.

[231] Peters OA, Barbakow F. Dynamic torque and apical forces of ProFile .04 rotary instruments during preparation of curved canals. Int. Endod. J. 2002, 35, 379–89.

[232] Sattapan B, Nervo GJ, Palamara JE, Messer HH. Defects in rotary nickel titanium files after clinical use. J. Endod. 2000, 26, 161–5.

[233] Bahia M, Melo MC, Buono VT. Influence of simulated clinical use on the torsional behavior of nickel-titanium rotary endodontic instruments. Oral Surg. Oral. Med. Oral Pathol. Oral Radiol. Endod. 2006, 101, 675–80.

[234] Berutti E, Chiandussi G, Gaviglio I, Ibba A. Comparative analysis of torsional and bending stresses in two mathematical models of nickel-titanium rotary instruments: ProTaper versus ProFile. J. Endod. 2003, 29, 15–9.

[235] Xu X, Eng M, Zheng Y, Eng D. Comparative study of torsional and bending properties for six models of nickel-titanium root canal instruments with different cross-sections. J. Endod. 2006, 32, 372–5.

[236] Ninan E, Berzins DW. Torsion and bending properties of shape memory and superelastic nickel-titanium rotary instruments. J. Endod. 2013, 39, 101–4.

[237] Park SY, Cheung GS, Yum J, Hur B, Park JK, Kim HC. Dynamic torsional resistance of nickel-titanium rotary instruments. J. Endod. 2010, 36, 1200–4.

[238] Yum J, Cheung GS, Park JK, Hur B, Kim HC. Torsional strength and toughness of nickel-titanium rotary files. J. Endod. 2011; 37: 382–6.

[239] Lopes HP, Elias CN, Vedovello GA, Bueno CE, Mangelli M, Siqueira JF Jr. Torsional resistance of retreatment instruments. J. Endod. 2011, 37, 1442–5.

[240] Elnaghy AM, Elsaka SE. Torsion and bending properties of OneShape and WaveOne instruments. J. Endod. 2015, 41, 544–7.

[241] Arbab-Chirani R, Chevalier V, Arbab-Chirani S, Calloch S. Comparative analysis of torsional and bending behavior through finite-element models of 5 Ni-Ti endodontic instruments. Oral Surg. Oral Med. Oral Pathol. Oral Radiol. Endod. 2011, 111, 115–21.

[242] Alcalde MP, Duarte MAH, Bramante CM, Tanomaru-Filho M, Vasconcelos BC, Só MVR, Vivan RR. Torsional fatigue resistance of pathfinding instruments manufactured from several nickel-titanium alloys. Int. Endod. J. 2018, 51, 697–704.

[243] Yared G, Kulkarni GK, Ghossayn F. Torsional properties of new and used rotary K3 NiTi files. Aust. Endod. J. 2003, 29, 75–8.

[244] Yared G, Kulkarni GK, Ghossayn F. An in vitro study of the torsional properties of new and used K3 instruments. Int. Endod. J. 2003, 36, 764–9.

[245] Vieira EP, Nakagawa RK, Buono VT, Bahia MG. Torsional behaviour of rotary NiTi ProTaper universal instruments after multiple clinical use. Int. Endod. J. 2009, 42, 947–53.

[246] Rowan MB, Nicholls JI, Steiner J. Torsional properties of stainless steel and nickel-titanium endodontic files. J. Endod. 1996, 22, 341–5.

[247] Wolcott J, Himel VT. Torsional properties of nickel-titanium versus stainless steel endodontic files. J. Endod. 1997, 23, 217–20.

[248] Kazemi RB, Stenman E, Spångberg LS. A comparison of stainless steel and nickel-titanium H-type instruments of identical design: torsional and bending tests. Oral Surg. Oral Med. Oral Pathol. Oral Radiol. Endod. 2000, 90, 500–6.

[249] Bahia MG, Melo MC, Buono VT. Influence of cyclic torsional loading on the fatigue resistance of K3 instruments. Int. Endod. J. 2008, 41, 883–91.

[250] Acosta EC, Resende PD, Peixoto IF, Pereira ÉS, Buono VT, Bahia MG. Influence of cyclic flexural deformation on the torsional resistance of controlled memory and conventional nickel-titanium instruments. J. Endod. 2017, 43, 613–8.

[251] Oh SH, Ha JH, Kwak SW, Ahn SW, Lee W, Kim HC. The effects of torsional preloading on the torsional resistance of nickel-titanium instruments. J. Endod. 2017, 43, 157–62.

[252] Gambarini G, Testarelli L, Galli M, Tucci E, De Luca M. The effect of a new finishing process on the torsional resistance of twisted nickel-titanium rotary instruments. Minerva Stomatol. 2010, 59, 401–6.

[253] Walia HM, Brantley WA, Gerstein H. An initial investigation of the bending and torsional properties of Nitinol root canal files. J. Endod. 1988, 14, 346–51.

[254] Coelho MS, de Azevêdo Rios M, da Silveira Bueno CE. Separation of nickel-titanium rotary and reciprocating instruments: a mini-review of clinical studies. Open Dent. J. 2018, 12, 864–72.

[255] Pettiette MT, Delano EO, Trope M. Evaluation of success rate of endodontic treatment performed by students with stainless-steel K-files and nickel-titanium hand files. J. Endod. 2001, 27, 124–7.

[256] Yared G. Canal preparation using only one Ni-Ti rotary instrument: preliminary observations. Int. Endod. J. 2008, 41, 339–44.

[257] De-Deus G, Moreira EJ, Lopes HP, Elias CN. Extended cyclic fatigue life of F2 ProTaper instruments used in reciprocating movement. Int. Endod. J. 2010, 43, 1063–8.

[258] Tygesen YA, Steiman HR, Ciavarro C. Comparison of distortion and separation utilizing profile and Pow-R nickel-titanium rotary files. J. Endod. 2001, 27, 762–4.

[259] Rosen E, Azizi H, Friedlander C, Taschieri S, Tsesis I. Radiographic identification of separated instruments retained in the apical third of root canal-filled teeth. J. Endod. 2014, 40, 1549–52.

[260] Kaul R, Farooq R, Kaul V, Khateeb SU, Purra AR, Mahajan R. Comparative evaluation of physical surface changes and incidence of separation in rotary nickel-titanium instruments: an in vitro SEM study. Iran Endod. J. 2014, 9, 204–9.

[261] Lambrianidis T. Therapeutic options for the management of fractured instruments. Management of fractured endodontic instruments 2018, 75–195; doi: 10.1007/978-3-319-60651-4.

[262] Hülsmann M, Schinkel I. Influence of several factors on the success or failure of removal of fractured instruments from the root canal. Endod. Dent. Traumatol. 1999, 15, 252–8.

[263] Shen Y, Peng B, Cheung GS. Factors associated with the removal of fractured NiTi instruments from root canal systems. Oral Surg. Oral Med. Oral Pathol. Oral Radiol. Endod. 2004, 98, 605–10.

[264] Suter B, Lussi A, Sequeira A. Probability of removing fractured instruments from root canals. Int. Endod. J. 2005, 38, 112–23.

[265] Nevares G, Cunha RS, Zuolo ML, Bueno CE. Success rates for removing or bypassing fractured instruments: a prospective clinical study. J. Endod. 2012, 38, 442–4.

[266] Ormiga F, da Cunha Ponciano Gomes JA, de Araújo MC. Dissolution of nickel-titanium endodontic files via an electrochemical process: a new concept for future retrieval of fractured files in root canals. J. Endod. 2010, 36, 717–20.

[267] Takahashi T. Basic research on chemical removal method of fractured NiTi file in root canal. J. Jap. Soc. Endod. 2010, 3, 8–22.

[268] Rahimi M, Parashos P. A novel technique for the removal of fractured instruments in the apical third of curved root canals. Int. Endod. J. 2009, 42, 264–70.

[269] Alomairy KH. Evaluating two techniques on removal of fractured rotary nickel-titanium endodontic instruments from root canals: an in vitro study. J. Endod. 2009, 35, 559–62.

[270] Ward JR, Parashos P, Messer HH. Evaluation of an ultrasonic technique to remove fractured rotary nickel-titanium endodontic instruments from root canals: an experimental study. J. Endod. 2003, 29, 756–63.

[271] Souter NJ, Messer HH. Complications associated with fractured file removal using an ultrasonic technique. J. Endod. 2005, 31, 450–2.

[272] Arslan H, Yıldız ED, Taş G, Akbıyık N, Topçuoğlu HS. Duration of ultrasonic activation causing secondary fractures during the removal of the separated instruments with different tapers. Clin. Oral Invest. 2019, May, 5 pages; doi: 10.1007/s00784-019-02936-7.

[273] Ramirez-Salomon M, Soler-Bientz R, de la Garza-González R, Palacios-Garza CM. Incidence of Lightspeed separation and the potential for bypassing. J. Endod. 1997, 23, 586–7.

[274] Madarati AA, Hunter MJ, Dummer PM. Management of intracanal separated instruments. J. Endod. 2013, 39, 569–81.

[275] Penta V, Pirvu C, Demetrescu I. Electrochemical impedance spectroscopy investigation on the clinical lifetime of ProTaper rotary file system. Biomed. Res. Int. 2014, 2014, 754189; doi: 10.1155/2014/754189.

[276] Oshida Y, Sellers CB, Mirza K, Farzin-Nia F. Corrosion of dental metallic materials by dental treatment agents. Mater. Sci. Eng. C 2005, 25, 343–8.

[277] Al-Ali S, Oshida Y, Andres CJ, Barco MT, Brown DT, Hovijitra S, Ito M, Nagasawa S, Yoshida T. Effects of coupling methods on galvanic corrosion behavior of commercially pure titanium with dental precious alloys. Bio-Med. Mater. Eng. 2005, 15, 307–16.

[278] Young JM, Van Vliet KJ. Predicting in vivo failure of pseudoelastic NiTi devices under low cycle, high amplitude fatigue. J. Biomed. Mater. Res. B Appl. Biomater. 2005, 72, 17–26.

[279] Chakka NV, Ratnakar P, Das S, Bagchi A, Sudhir S, Anumula L. Do NiTi instruments show defects before separation? Defects caused by torsional fatigue in hand and rotary nickel-titanium (NiTi) instruments which lead to failure during clinical use. J. Contemp. Dent. Pract. 2012, 13, 867–72.

[280] Stojanac I, Drobac M, Petrovic L, Atanackovic T. Predicting in vivo failure of rotary nickel-titanium endodontic instruments under cyclic fatigue. Dent. Mater. J. 2012, 31, 650–5.

[281] Cheung GS, Zhang EW, Zheng YF. A numerical method for predicting the bending fatigue life of NiTi and stainless steel root canal instruments. Int. Endod. J. 2011, 44, 357–61.

[282] Weiss V, Oshida Y, Wu A. Towards practical non-destructive fatigue damage indicators, fatigue and fracture of engineering. Mater. Struct. 1979, 1, 333–41.

[283] Weiss V, Oshida Y, Wu A. A note on fatigue damage assessment by x-ray diffraction techniques for a 304 stainless steel specimen. J. Nondestruct. Eval. 1980, 1, 207–13.

[284] Li UM, Shin CS, Lan WH, Lin CP. Application of nondestructive testing in cyclic fatigue evaluation of endodontic Ni-Ti rotary instruments. Dent. Mater. J. 2006, 25, 247–52.

[285] Hsieh S-C, Lee S-Y, Ciou C-Y, Huang H-M. Non-destructive natural frequency tests of cyclic fatigue-loaded nickel–titanium rotary instruments. Med. Biol. Eng. Comput. 2010, 48, 555–60.

[286] Oshida Y, Sachdeva RC, Miyazaki S. Microanalytical characterization and surface modification of TiNi orthodontic archwires. Biomed. Mater. Eng. 1992, 2, 51–69.

[287] Mohammadi Z, Soltani MK, Shalavi S, Saeed Asgary S. A review of the various surface treatments of NiTi instruments. Iran Endod. J. 2014, 9, 235–40.

[288] Shevchenko N, Pham M-T, Maitz M. Studies of surface modified NiTi alloy. Appl. Surf. Sci. 2004, 235, 126–31.

[289] Conrad J, Dodd R, Worzala F, Qiu X. Plasma source ion implantation: a new, cost-effective, non-line-of-sight technique for ion implantation of materials. Surf. Coat. Technol. 1988, 36, 927–37.

[290] Tendys J, Donnelly I, Kenny M, Pollock J. Plasma immersion ion implantation using plasmas generated by radio frequency techniques. Appl. Phys. Lett. 1988, 53, 2143–5.

[291]   Wolle CF, Vasconcellos MA, Hinrichs R, Becker AN, Barletta FB. The effect of argon and nitrogen ion implantation on nickel-titanium rotary instruments. J. Endod. 2009, 35, 1558–62.

[292]   Gavini G, Pessoa OF, Barletta FB, Vasconcellos MA, Caldeira CL. Cyclic fatigue resistance of rotary nickel-titanium instruments submitted to nitrogen ion implantation. J. Endod. 2010, 36, 1183–6.

[293]   dos Santos M, Gavini G, Siqueira EL, da Costa C. Effect of nitrogen ion implantation on the flexibility of rotary nickel-titanium instruments. J. Endod. 2012, 38, 673–5.

[294]   Lee DH, Park B, Saxena A, Serene TP. Enhanced surface hardness by boron implantation in Nitinol alloy. J. Endod. 1996, 22, 543–6.

[295]   Molinari A, Pellizzari M, Gialanella S, Straffelini G, Stiasny K. Effect of deep cryogenic treatment on the mechanical properties of tool steels. J. Mater. Process. Technol. 2001, 118, 350–5.

[296]   Mohan Lal D, Renganarayanan S, Kalanidhi A. Cryogenic treatment to augment wear resistance of tool and die steels. Cryogenics 2001, 41, 149–55.

[297]   Huang J, Zhu Y, Liao X, Beyerlein I, Bourke M, Mitchell T. Microstructure of cryogenic treated M2 tool steel. Mater. Sci. Eng. A 2003, 339, 241–4.

[298]   Vinothkumar TS, Miglani R, Lakshminarayananan L. Influence of deep dry cryogenic treatment on cutting efficiency and wear resistance of nickel-titanium rotary endodontic instruments. J. Endod. 2007, 33, 1355–8.

[299]   George GK, Sanjeev K, Sekar M. An in vitro evaluation of the effect of deep dry cryotreatment on the cutting efficiency of three rotary nickel titanium instruments. J. Conserv. Dent. 2011, 14, 169–72.

[300]   Kim JW, Griggs JA, Regan JD, Ellis RA, Cai Z. Effect of cryogenic treatment on nickel-titanium endodontic instruments. Int. Endod. J. 2005, 38, 364–71.

[301]   Gu Y, Kum KY, Perinpanayagam H, Kim C, Kum DJ, Lim SM, Chang SW, Baek SH, Zhu Q, Yoo YJ. Various heat-treated nickel-titanium rotary instruments evaluated in S-shaped simulated resin canals. J. Dent. Sci. 2017, 12, 14–20.

[302]   Miyara K, Yahata Y, Hayashi Y, Tsutsumi Y, Ebihara A, Hanawa T, Suda H. The influence of heat treatment on the mechanical properties of Ni-Ti file materials. Dent. Mater. J. 2014, 33, 27–31.

[303]   Zinelis S, Darabara M, Takase T, Ogane K, Papadimitriou GD. The effect of thermal treatment on the resistance of nickel-titanium rotary files in cyclic fatigue. Oral Surg. Oral. Med. Oral Pathol. Oral Radiol. Endod. 2007, 103, 843–7.

[304]   Braga LC, Faria Silva AC, Buono VT, de Azevedo Bahia MG. Impact of heat treatments on the fatigue resistance of different rotary nickel-titanium instruments. J. Endod. 2014, 40, 1494–7.

[305]   Plotino G, Grande NM, Cotti E, Testarelli L, Gambarini G. Blue treatment enhances cyclic fatigue resistance of vortex nickel-titanium rotary files. J. Endod. 2014, 40, 1451–3.

[306]   Miyai K, Hayashi Y, Ebihara A, Doi H, Suda H, Yoneyama T. Influence of phase transformation on the torsional and bending properties of nickel-titanium rotary endodontic instruments. Int. Endod. J. 2006, 39, 119–26.

[307]   Hayashi Y, Yoneyama T, Yahata Y, Miyai K, Doi H, Hanawa T, Ebihara A, Suda H. Phase transformation behavior and bending properties of hybrid nickel-titanium rotary endodontic instruments. Int. Endod. J. 2007, 40, 247–53.

[308]   Yahata Y, Yoneyama T, Hayashi Y, Ebihara A, Doi H, Hanawa T, Suda H. Effect of heat treatment on transformation temperatures and bending properties of nickel-titanium endodontic instruments. Int. Endod. J. 2009, 42, 621–6.

[309]   Ebihara A, Yahata Y, Miyara K, Nakano K, Hayashi Y, Suda H. Heat treatment of nickel-titanium rotary endodontic instruments: effects on bending properties and shaping abilities. Int. Endod. J. 2011, 44, 843–9.

[310]   Starosvetsky D, Gotman I. Corrosion behavior of titanium nitride coated Ni-Ti shape memory surgical alloy. Biomaterials 2001, 22, 1853–9.

[311]   Liu X, Chu PK, Ding C. Surface modification of titanium, titanium alloys, and related materials for biomedical applications. Mater. Sci. Eng., R: Rep. 2004, 47, 49–121.

[312]   Shenhar A, Gotman I, Radin S, Ducheyne P. Microstructure and fretting behavior of hard TiN-based coatings on surgical titanium alloys. Ceram. Int. 2000, 26, 709–13.

[313]   Huang H-H, Hsu C-H, Pan S-J, He J-L, Chen C-C, Lee T-L. Corrosion and cell adhesion behavior of TiN-coated and ion-nitrided titanium for dental applications. Appl. Surf. Sci. 2005, 244, 252–6.

[314]   Li U-M, Chiang Y-C, Chang W-H, Lu C-M, Chen Y-C, Lai T-M, Lin C-P. Study of the effects of thermal nitriding surface modification of nickel titanium rotary instruments on the wear resistance and cutting efficiency. J. Dent. Sci. 2006, 1, 53–8.

[315]   Zupanc J, Vahdat-Pajouh N, Schäfer E. New thermomechanically treated NiTi alloys – a review. Int. Endod. J. 2018, 51, 1088–103.

[316]   Pereira ESJ, Peixoto IFC, Viana ACD, Oliveira II, Gonzalez BM, Buono VTL, Bahia MGA. Physical and mechanical properties of a thermomechanically treated NiTi wire used in the manufacture of rotary endodontic instruments. Int. Endod. J. 2011, 45, 469–74.

[317]   Shen Y, Zhou H-M, Zheng Y-F, Peng B, Haapasalo M. Current challenges and concepts of the thermomechanical treatment of nickel-titanium instruments. J. Endod. 2013, 39, 163–72.

[318]   De-Deus G, Silva EJ, Vieira VT, Belladonna FG, Elias CN, Plotino G, Grande NM. Blue thermomechanical treatment optimizes fatigue resistance and flexibility of the reciproc files. J. Endod. 2017, 43, 462–6.

[319]   Lopes HP, Elias CN, Vieira MV, Vieira VT, de Souza LC, Dos Santos AL. Influence of surface roughness on the fatigue life of nickel-titanium rotary endodontic instruments. J. Endod. 2016, 42, 965–8.

[320]   Haider W, Munroe N, Tek V, Pulletikurthi C, Gill PK, Pandya S. Surface modifications of nitinol. J. Long Term Eff. Med. Implants 2009, 19, 113–22.

[321]   Trepanier C, Tabrizian M, Yahia LH, Bilodeau L, Piron DL. Effect of modification of oxide layer on NiTi stent corrosion resistance. J. Biomed. Mater. Res. 1998, 43, 433–40.

[322]   Bui TB, Mitchell JC, Baumgartner JC. Effect of electropolishing ProFile nickel-titanium rotary instruments on cyclic fatigue resistance, torsional resistance, and cutting efficiency. J. Endod. 2008, 34, 190–3.

[323]   Cheung GS, Shen Y, Darvell BW. Does electropolishing improve the low-cycle fatigue behavior of a nickel-titanium rotary instrument in hypochlorite? J. Endod. 2007, 33, 1217–21.

[324]   Praisarnti C, Chang JW, Cheung GS. Electropolishing enhances the resistance of nickel-titanium rotary files to corrosion-fatigue failure in hypochlorite. J. Endod. 2010, 36, 1354–7

[325]   Lopes HP, Elias CN, Vieira VT, Moreira EJ, Marques RV, de Oliveira JC, Debelian G, Siqueira JF Jr. Effects of electropolishing surface treatment on the cyclic fatigue resistance of BioRace nickel-titanium rotary instruments. J. Endod. 2010, 36, 1653–7.

[326]   Condorelli GG, Bonaccorso A, Smecca E, Schafer E, Cantatore G, Tripi TR. Improvement of the fatigue resistance of NiTi endodontic files by surface and bulk modifications. Int. Endod. J. 2010, 43, 866–73.

[327]   Mustafa M. Knowledge, attitude and practice of general dentists towards sterilization of endodontic files: a cross – sectional study. Indian J. Sci. Tech. 2016, 9, 11–6.

[328]   Enabulele JE, Omo JO. Sterilization in endodontics: knowledge, attitude, and practice of dental assistants in training in Nigeria – a cross-sectional study. Saudi Endod. J. 2018, 8, 106–10.

[329]   Aslam A, Panuganti V, Nanjundarethy JK, Halappa M, Krishna VH. Knowledge and attitude of endodontic post graduate students towards sterilization of endodontic files: a cross – sectional study. Saudi Endod. J. 2016, 4, 18–22.

[330]   Limbhore M, Saraf A, Medha A, Jain D, Mattigatti S, Mahaparale R. Endodontic hand
        instrument sterilization procedures followed by dental practitioners. Unique J. Med. Dent.
        Sci. 2014, 2, 106–11.
[331]   Ferreira MM, Michelotto AL, Alexandre AR, Morgantio R, Carnillo EV. Endodontic files:
        sterilize or discard. Dent. Press Endod. 2012, 2, 46–51.
[332]   Hurtt CA, Rossman LE. The sterilization of endodontic hand files. J. Endod. 1996, 22,
        321–2.
[333]   Van Eldik DA, Zilm PS, Rogers AH, Marin PD. Microbiological evaluation of endodontic files
        after cleaning and steam sterilization procedures. Aust. Dent. J. 2004, 49, 122–7.
[334]   Filho MT, Leonardo MR, Bonifácio KC, Dametto FR, Silva AB. The use of ultrasound for
        cleaning the surface of stainless steel and nickel-titanium endodontic instruments. Int.
        Endod. J. 2001, 34, 581–5.
[335]   Kumar KV, Kumar KSK, Supreetha S, Raghu KN, Veerabhadrappa AC, Deepthi S. Pathological
        evaluation for sterilization of routinely used prosthodontic and endodontic instruments.
        Therm. Anal. Polym. 2015, 5, 232–6.
[336]   Spagnuolo G, Ametrano G, D'Antò V, Rengo C, Simeone M, Riccitiello F, Amato M. Effect
        of autoclaving on the surfaces of TiN-coated and conventional nickel-titanium rotary
        instruments. Int. Endod. J. 2012, 45, 1148–55.
[337]   Razavian H, Iranmanesh P, Mojtahedi H, Nazeri R. Effect of autoclave cycles on surface
        fharacteristics of S-file evaluated by scanning electron microscopy. Iran Endod. J. 2016, 11,
        29–32.
[338]   Luper WD, Eichmiller FC, Doblecki W, Campbell D, Li SH. Effect of three sterilization
        techniques on finger pluggers. J. Endod. 1991, 17, 361–4.
[339]   Linsuwanont P, Parashos P, Messer HH. Cleaning of rotary nickel-titanium endodontic
        instruments. Int. Endod. J. 2004, 37, 19–28.
[340]   Valois CR, Silva LP, Azevedo RB. Multiple autoclave cycles affect the surface of rotary nickel-
        titanium files: an atomic force microscopy study. J. Endod. 2008, 34, 859–62.
[341]   Venkatasubramanian R, Jayanthi, Das UM, Bhatnagar S. Comparison of the effectiveness of
        sterilizing endodontic files by 4 different methods: an *in vitro* study. J. Indian Soc. Pedod.
        Prev. Dent. 2010, 28, 2–5.
[342]   Raju TBVG, Garapati S, Agrawal R, Reddy S, Razdan A, Kumar SK. Sterilizing endodontic files
        by four different sterilization methods to prevent cross-infection – an in-vitro study. J. Int.
        Oral Health 2013, 5, 108–12.
[343]   Shaha SN, Deepa S, Pratima S, Yogesh R, Ameya P, Sonalb D. To compare the new method of
        sterilization of endodontic instruments with autoclave: an in vitro study. J. for Dent. Photon
        2015, 111, 239–43.
[344]   Kommmineni NK, Dappili SR, Prathyusha P, Vanaja P, Reddy KK, Vasanthi D. Comparative
        evaluation of sterilization efficacy using two methods of sterilization for rotary endodontic
        files: an in vitro study. J. NTR Univ. Health Sci. 2016, 5, 142–6.
[345]   Yenni M, Bandi S, Avula SS, Margana PG, Kakarla P, Amrutavalli A. Comparative evaluation of
        four different sterilization methods on contaminated endodontic files. CHRISMED J. Health
        Res. 2017, 4, 194–7.
[346]   Morrison SW, Newton CW, Brown CE Jr. The effects of steam sterilization and usage on cutting
        efficiency of endodontic instruments. J. Endod. 1989, 15, 427–31.
[347]   Mize SB, Clement DJ, Pruett JP, Carnes DL Jr. Effect of sterilization on cyclic fatigue of rotary
        nickel-titanium endodontic instruments. J. Endod. 1998, 24, 843–7.
[348]   Viana AC, Gonzalez BM, Buono VT, Bahia MG. Influence of sterilization on mechanical
        properties and fatigue resistance of nickel-titanium rotary endodontic instruments. Int.
        Endod. J. 2006, 39, 709–15.

[349]   Bergeron BE, Mayerchak MJ, Roberts MJ, Jeansonne BG. Multiple autoclave cycle effects on cyclic fatigue of nickel-titanium rotary files produced by new manufacturing methods. J. Endod. 2011, 37, 72–4.

[350]   Plotino G, Costanzo A, Grande NM, Petrovic R, Testarelli L, Gambarini G. Experimental evaluation on the influence of autoclave sterilization on the cyclic fatigue of new nickel-titanium rotary instruments. J. Endod. 2012, 38, 222–5.

[351]   Zhao D, Shen Y, Peng B, Haapasalo M. Effect of autoclave sterilization on the cyclic fatigue resistance of thermally treated nickel-titanium instruments. Int. Endod. J. 2016, 49, 990–5.

[352]   Iverson GW, von Fraunhofer JA, Herrmann JW. The effects of various sterilization methods on the torsional strength of endodontic files. J. Endod. 1985, 11, 266–8.

[353]   Silvaggio J, Hicks ML. Effect of heat sterilization on the torsional properties of rotary nickel-titanium endodontic files. J. Endod. 1997, 23, 731–4.

[354]   Canalda-Sahli C, Brau-Aguadé E, Sentís-Vilalta J. The effect of sterilization on bending and torsional properties of K-files manufactured with different metallic alloys. Int. Endod. J. 1998, 31, 48–52.

[355]   Hilt BR, Cunningham CJ, Shen C, Richards N. Torsional properties of stainless-steel and nickel-titanium files after multiple autoclave sterilizations. J. Endod. 2000, 26, 76–80.

[356]   Casper RB, Roberts HW, Roberts MD, Himel VT, Bergeron BE. Comparison of autoclaving effects on torsional deformation and fracture resistance of three innovative endodontic file systems. J. Endod. 2011, 37, 1572–5.

[357]   Kottoor J, Albuquerque D. Questioning the spot light on Hi-tech endodontics. Restor. Dent. Endod. 2016, 41, 80–2.

[358]   Ricucci D, Russo J, Rutberg M, Burleson JA, Spångberg LS. A prospective cohort study of endodontic treatments of 1,369 root canals: results after 5 years. Oral Surg. Oral Med. Oral Pathol. Oral Radiol. Endod. 2011, 112, 825–42.

[359]   Pirani C, Chersoni S, Montebugnoli L, Prati C. Long-term outcome of non-surgical root canal treatment: a retrospective analysis. Odontology 2015, 103, 185–93.

[360]   Kwak Y, Choi J, Kim K, Shin S-J, Kim S, Kim E. The 5-year survival rate of nonsurgical endodontic treatment: a population-based cohort study in Korea. J. Endod. 2019, 45, 1192–9.

[361]   Tickle M, Milsom K, Qualtrough A, Blinkhorn F, Aggarwal VR. The failure rate of NHS funded molar endodontic treatment delivered in general dental practice. Brit. Dent. J. 2008, 204; doi: 10.1038/bdj.2008.133.

[362]   Sonntag D, Delschen S, Stachniss V. Root-canal shaping with manual and rotary Ni-Ti files performed by students. Int. Endod. J. 2003, 36, 715–23.

[363]   Peru M, Peru C, Mannocci F, Sherriff M, Buchanan LS, Pitt Ford TR. Hand and nickel-titanium root canal instrumentation performed by dental students: a micro-computed tomographic study. Eur. J. Dent. Educ. 2006, 10, 52–9.

[364]   Shen Y, Coil JM, Haapasalo M. Defects in nickel-titanium instruments after clinical use. Part 3: a 4-year retrospective study from an undergraduate clinic. J. Endod. 2009, 35, 193–6.

[365]   Knowles KI, Hammond NB, Biggs SG, Ibarrola JL. Incidence of instrument separation using Lightspeed rotary instruments. J. Endod. 2006, 32, 14–6.

[366]   Hänni S, Schönenberger K, Peters OA, Barbakow F. Teaching an engine-driven preparation technique to undergraduates: initial observations. Int. Endod. J. 2003, 36, 476–82.

[367]   Coelho MS, Card SJ, Tawil PZ. Safety assessment of two hybrid instrumentation techniques in a dental student endodontic clinic: a retrospective study. J. Dent. Educ. 2017, 81, 333–9.

[368]   Martins RC, Seijo MO, Ferreira EF, Paiva SM, Ribeiro Sobrinho AP. Dental students' perceptions about the endodontic treatments performed using NiTi rotary instruments and hand stainless steel files. Braz. Dent. J. 2012, 23, 729–36.

[369] Jungnickel L, Kruse C, Vaeth M, Kirkevang LL. Quality aspects of ex vivo root canal treatments done by undergraduate dental students using four different endodontic treatment systems. Acta Odontol. Scand. 2018, 76, 169–74.

[370] Georgelin-Gurgel M, Devillard R, Lauret ME, Diemer F, Calas P, Hennequin M. Root canal shaping using rotary nickel-titanium files in preclinical teaching. Odontostomatol. Trop. 2008, 31, 5–11.

[371] Arbab-Chirani R, Vulcain JM. Undergraduate teaching and clinical use of rotary nickel-titanium endodontic instruments: a survey of French dental schools. Int. Endod. J. 2004, 37, 320–4.

[372] Pettiette MT, Metzger Z, Phillips C, Trope M. Endodontic complications of root canal therapy performed by dental students with stainless-steel K-files and nickel-titanium hand files. J. Endod. 1999, 25, 230–4.

[373] Kfir A, Rosenberg E, Zuckerman O, Tamse A, Fuss Z. Comparison of procedural errors resulting during root canal preparations completed by senior dental students in patients using an '8-step method' versus 'serial step-back technique'. Oral Surg. Oral Med. Oral Pathol. Oral Radiol. Endod. 2004, 97, 745–8.

[374] Iqbal MK, Kohli MR, Kim JS. A retrospective clinical study of incidence of root canal instrument separation in an endodontics graduate program: a PennEndo database study. J. Endod. 2006, 32, 1048–52.

[375] Mesgouez C, Rilliard F, Matossian L, Nassiri K, Mandel E. Influence of operator experience on canal preparation time when using the rotary Ni-Ti ProFile system in simulated curved canals. Int. Endod. J. 2003, 36, 161–5.

[376] Al-Omari MA, Aurich T, Wirtti SW, Jordan I. Shaping canals with ProFiles and K3 instruments: does operator experience matter? Oral Surg. Oral Med. Oral Pathol. Oral Radiol. Endod. 2010, 110, e50–5.

[377] Shen Y, Coil JM, Mo AJ, Wang Z, Hieawy A, Yang Y, Haapasalo M. WaveOne rotary instruments after clinical use. J. Endod. 2016, 42, 186–9.

[378] Marending M, Peters OA, Zehnder M. Factors affecting the outcome of orthograde root canal therapy in a general dentistry hospital practice. Oral Surg. Oral Med. Oral Pathol. Oral Radiol. Endod. 2005, 99, 119–24.

[379] Sattapan B, Nervo GJ, Palamara JE, Messer HH. Defects in rotary nickel-titanium files after clinical use. J. Endod. 2000, 26, 161–5.

[380] Ehrhardt IC, Zuolo ML, Cunha RS, De Martin AS, Kherlakian D, Carvalho MC, Bueno CE. Assessment of the separation incidence of two files used with preflaring: prospective clinical study. J. Endod. 2012, 38, 1078–81.

[381] Ramirez-Salomon M, Soler-Bientz R, de la Garza-González R, Palacios-Garza CM. Incidence of Lightspeed separation and the potential for bypassing. J. Endod. 1997, 23, 586–7.

[382] Wu J, Lei G, Yan M, Yu Y, Yu J, Zhang G. Instrument separation analysis of multi-used ProTaper Universal rotary system during root canal therapy. J. Endod. 2011, 37, 758–63.

[383] Gambarini G, Piasecki L, Di Nardo D, Miccoli G, Di Giorgio G, Carneiro E, Al-Sudani D, Testarelli L. Incidence of deformation and fracture of twisted file adaptive instruments after repeated clinical use. J. Oral Maxillofac. Res. 2016, 7; doi: 10.5037/jomr.2016.7405.

[384] Cunha RS, Junaid A, Ensinas P, Nudera W, Bueno CE. Assessment of the separation incidence of reciprocating WaveOne files: a prospective clinical study. J. Endod. 2014, 40, 922–4.

[385] Shen Y, Coil JM, Mo AJ, Wang Z, Hieawy A, Yang Y, Haapasalo M. WaveOne rotary instruments after clinical use. J. Endod. 2016, 42, 186–9.

[386] Plotino G, Grande NM, Porciani PF. Deformation and fracture incidence of Reciproc instruments: a clinical evaluation. Int. Endod. J. 2015, 48, 199–205.

[387] Berutti E, Negro AR, Lendini M, Pasqualini D. Influence of manual preflaring and torque on the failure rate of ProTaper rotary instruments. J. Endod. 2004, 30, 228–30.

[388]   Peciuliene V, Rimkuviene J, Aleksejuniene J, Haapasalo M, Drukteinis S, Maneliene R. Technical aspects of endodontic treatment procedures among Lithuanian general dental practitioners. Stomatologija 2010, 12, 42–50.

[389]   Quintão CCA, Brunharo IHVP. Orthodontic wires: knowledge ensures clinical optimization Dental Press. J. Orthod. 2009, 14, 144–57.

[390]   Castro SM, Ponces MJ, Lopes JD, Vasconcelos M, Pollmann MCF. Orthodontic wires and its corrosion – the specific case of stainless steel and beta-titanium. J. Dent. Sci. 2015, 10, 1–7.

[391]   Dolce C, Alfonso M. Orthodontics: a review. Hist. Orthod.; https://www.dentalcare.com/ en-us/professional-education/ce-courses/ce202/references.

[392]   Kusy RP. Orthodontic biomaterials: from the past to the present. Angle Orthod. 2002, 72, 501–12.

[393]   Gemmi C. A brief history of orthodontics, orthodontic treatment 2018; https://www. orthodonticslimited,com/orthodontics/orthodontics-history/.

[394]   Kotha RS, Alla RK, Mohammed S, Ravi RK. An overview of orthodontic wires. Trends Biomater. Artif. Organs 2014, 28, 32–6.

[395]   Shetty V, Caridad JM, Caputo AA, Chaconas SJ. Biomechanical rationale for surgical-orthodontic expansion of the adult maxilla. J Oral Maxillofac. Surg. 1994, 52, 742–9.

[396]   Holberg C, Holberg N, Rudzki-Janson I. Sutural strain in orthopaedic headgear therapy: a finite element analysis. Am. J. Orthod. Dentofacial Orthop. 2008, 134, 53–9.

[397]   Leach HA, Ireland AJ, Whaites EJ. Radiographic diagnosis of root resorption in relation to orthodontics. Br. Dent. J. 2001, 190, 16–22.

[398]   Rudolph DJ, Willes PMG, Sameshima GT. A finite element model of apical force distribution from orthodontic tooth movement. Angle Orthod. 2001, 71, 127–31.

[399]   Kusy RP, Greenberg AR. Effects of composition and cross section on the elastic properties of orthodontic arch wires. Angle Orthod. 1981, 51, 325–41.

[400]   Asgharnia MK, Brantley WA. Comparison of bending and tension tests for orthodontic wires. Am. J. Orthod. Dentofacial Orthop. 1986, 89, 228–35.

[401]   Kapila S, Sachdeva R. Mechanical properties and clinical applications of orthodontic wires. Am. J. Orthod. Dentofacial Orthop. 1989, 96, 100–9.

[402]   Indian Dental Academy. Wires in orthodontics; https://www.slideshare.net/indiandenta-lacademy/wires-in-orthodontics.

[403]   Khamatkar A. Ideal properties of orthodontic wires and their clinical implications – a review. J. Dent. Med. Serv. 2015, 14, 47–50.

[404]   https://www.diytrade.com/china/pd/9699598/coated_Niti_arch_wires_Color_dental_wire. html.

[405]   https://www.slideshare.net/skaziz13/wires-in-othodontics.

[406]   Hamanaka H, Yoneyama T, Doi H, Okamoto Y, Mogi M, Miura F. Mechanical properties and phase transformation of super-elastic Ni-Ti alloy wires. Part 2: changes of properties through heat treatment. Shika Zairyo Kikai 1989, 8, 216–23.

[407]   Wichelhaus A, Brauchli L, Ball J, Mertmann M. Mechanical behavior and clinical application of nickel-titanium closed-coil springs under different stress levels and mechanical loading cycles. Am. J. Orthod. Dentofacial Orthop. 2010, 137, 671–8.

[408]   Sivarai A. Comparison of superelasticity of nickel titanium orthodontic arch wires using mechanical tensile testing and correlating with electrical resistivity. J. Int. Oral Health 2013, 5, 1–12.

[409]   Gravina MA, Brunharo IH, Canavarro C, Elias CN, Quintão CC. Mechanical properties of NiTi and CuNiTi shape-memory wires used in orthodontic treatment. Part 1: stress-strain tests. Dental Press J. Orthod. 2013, 18, 35–42.

[410]   Gravina MA, Canavarro C, Elias CN, das Graças Afonso Miranda Chaves M, Brunharo IH, Quintão CC. Mechanical properties of NiTi and CuNiTi wires used in orthodontic

treatment. Part 2: microscopic surface appraisal and metallurgical characteristics. Dental Press J. Orthod. 2014, 19, 69–76.

[411] Alobeid A, Dirk C, Reimann S, El-Bialy T, Jäger A, Bourauel C. Mechanical properties of different esthetic and conventional orthodontic wires in bending tests: an in vitro study. J. Orofac. Orthop. 2017, 78, 241–52.

[412] Mullins WS, Bagby MD, Norman TL. Mechanical behavior of thermo-responsive orthodontic archwires. Dent. Mater. 1996, 12, 308–14.

[413] Meling TR, Odegaard J. The effect of short-term temperature changes on the mechanical properties of rectangular nickel titanium archwires tested in torsion. Angle Orthod. 1998, 68, 369–76.

[414] Santoro M, Beshers DN. Nickel-titanium alloys: stress-related temperature transitional range. Am. J. Orthod. Dentofacial Orthop. 2000, 118, 685–92.

[415] Iijima M, Ohno H, Kawashima I, Endo K, Mizoguchi I. Mechanical behavior at different temperatures and stresses for superelastic nickel-titanium orthodontic wires having different transformation temperatures. Dent. Mater. 2002, 8, 88–93.

[416] Lombardo L, Toni G, Stefanoni F, Mollica F, Guarneri MP, Siciliani G. The effect of temperature on the mechanical behavior of nickel-titanium orthodontic initial archwires. Angle Orthod. 2013, 83, 298–305.

[417] Walker MP, White RJ, Kula KS. Effect of fluoride prophylactic agents on the mechanical properties of nickel-titanium-based orthodontic wires. Am. J. Orthod. Dentofacial Orthop. 2005, 127, 662–9.

[418] Rerhrhaye W, Bahije L, El Mabrouk K, Zaoui F, Marzouk N. Degradation of the mechanical properties of orthodontic NiTi alloys in the oral environment: an in vitro study. Int. Orthod. 2014, 12; doi:10.1016/j.ortho.2014.06.006.

[419] Lin J, Han S, Zhu J, Wang X, Chen Y, Vollrath O, Wang H, Mehl C. Influence of fluoride-containing acidic artificial saliva on the mechanical properties of nickel-titanium orthodontics wires. Indian J. Dent. Res. 2012, 23, 591–5.

[420] Aghili H, Yassaei S, Ahmadabadi MN, Joshan N. Load deflection characteristics of nickel titanium initial archwires. J. Dent., Tehran Univ. Med. Sci. 2015, 12, 695–704.

[421] Gurgel JA, Kerr S, Powers JM, LeCrone V. Force-deflection properties of superelastic nickel-titanium archwires. Am. J. Orthod. Dentofacial Orthop. 2001, 120, 378–82.

[422] Proffit WR, Fields HW. Contemporary Orthodontics. St. Louis, MI, Mosby, 2013, 319–21.

[423] Schumacher HA, Bourauel C, Drescher D. The effect of the ligature on the friction between bracket and arch. Fortschr Kieferorthop 1990, 51, 106–16.

[424] Kapila S, Reichhold GW, Anderson S, Watanabe LG. Effects of clinical recycling on mechanical properties of nickel-titanium alloy wires. Am. J. Orthod. Dentofacial Orthop. 1991, 100, 428–35.

[425] Kapila S, Haugen JW, Watanabe LG. Load-deflection characteristics of nickel-titanium alloy wires after clinical recycling and dry heat sterilization. Am. J. Orthod. Dentofacial Orthop. 1992, 102, 120–6.

[426] Barwart O. The effect of temperature change on, the load value of Japanese NiTi coil springs in the superelastic range. Am. J. Orthod. Dentofacial Orthop. 1996, 110, 553–8.

[427] Wilkinson PD, Dysart PS, Hood JA, Herbison GP. Load-deflection characteristics of superelastic nickel-titanium orthodontic wires. Am. J. Orthod. Dentofacial Orthop. 2002, 121, 483–95.

[428] Yanaru K, Yamaguchi K, Kakigawa H, Kozono Y. Temperature- and deflection-dependences of orthodontic force with Ni-Ti wires. Dent. Mater. J. 2003, 22, 146–59.

[429] Kasuya S, Nagasaka S, Hanyuda A, Ishimura S, Hirashita A. The effect of ligation on the load deflection characteristics of nickel titanium orthodontic wire. Eur. J. Orthod. 2007, 29, 578–82.

[430] Bartzela TN, Senn C, Wichelhaus A. Load-deflection characteristics of superelastic nickel-titanium wires. Angle Orthod. 2007, 77, 991–8.

[431] Quintão CC, Cal-Neto JP, Menezes LM, Elias CN. Force-deflection properties of initial orthodontic archwires. World J. Orthod. 2009, 10, 29–32.

[432] Ahmadabadi MN, Shahhoseini T, Habibi-Parsa M, Haj-Fathalian M, Hoseinzadeh-NikT, Ghadirian H. Static and cyclic load-deflection characteristics of NiTi orthodontic archwires using modified bending tests. J. Mater. Eng. Perform. 2009, 18, 793–6.

[433] Rose D, Quick A, Swain M, Herbison P. Moment-to-force characteristics of preactivated nickel-titanium and titanium-molybdenum alloy symmetrical T-loops. Am. J. Orthod. Dentofacial Orthop. 2009, 135, 757–63.

[434] Lombardo L, Marafioti M, Stefanoni F, Mollica F, Siciliani G. Load deflection characteristics and force level of nickel titanium initial archwires. Angle Orthod. 2012, 82, 507–21.

[435] Gatto E, Matarese G, Di Bella G, Nucera R, Borsellino C, Cordasco G. Load-deflection characteristics of superelastic and thermal nickel-titanium wires. Eur. J. Orthod. 2013, 35, 115–23.

[436] Nucera R, Gatto E, Borsellino C, Aceto P, Fabiano F, Matarese G, Perillo L, Cordasco G. Influence of bracket-slot design on the forces released by superelastic nickel-titanium alignment wires in different deflection configurations. Angle Orthod. 2014, 84, 541–7.

[437] Higa RH, Semenara NT, Henriques JF, Janson G, Sathler R, Fernandes TM. Evaluation of force released by deflection of orthodontic wires in conventional and self-ligating brackets. Dental Press J. Orthod. 2016, 21, 91–7.

[438] Arreghini A, Lombardo L, Mollica F, Siciliani G. Load deflection characteristics of square and rectangular archwires. Int. Orthod. 2016, 14, 1–14.

[439] Razali MF, Mahmud AS, Mokhtar N. Force delivery of NiTi orthodontic arch wire at different magnitude of deflections and temperatures: a finite element study. J. Mech. Behav. Biomed. Mater. 2018, 77, 234–41.

[440] Khier SE, Brantley WA, Fournelle RA. Bending properties of superelastic and nonsuperelastic nickel-titanium orthodontic wires. Am. J. Orthod. Dentofacial Orthop. 1991, 99, 310–8.

[441] Yoneyama T, Doi H, Hamanaka H, Yamamoto M, Kuroda T. Bending properties and transformation temperatures of heat treated Ni-Ti alloy wire for orthodontic appliances. J. Biomed. Mater. Res. 1993, 27, 399–402.

[442] Lim KF, Lew KKK, Toh SL. Bending stiffness of two aesthetic orthodontic archwires: an *in vitro* comparative study. Clin. Mater. 1994, 16, 63–71.

[443] Yamamoto M, Kuroda T, Yoneyama T, Doi H. Bending property and phase transformation of Ti-Ni-Cu alloy dental castings for orthodontic application. J. Mater. Sci. Mater. Med. 2002, 13, 855–9.

[444] Fischer-Brandies H, Es-Souni M, Kock N, Raetzke K, Bock O. Transformation behavior, chemical composition, surface topography and bending properties of five selected 0.016" x 0.022" NiTi archwires. J. Orofac. Orthop. 2003, 64, 88–99.

[445] Brauchli LM, Keller H, Senn C, Wichelhaus A. Influence of bending mode on the mechanical properties of nickel-titanium archwires and correlation to differential scanning calorimetry measurements. Am. J. Orthod. Dentofacial Orthop. 2011, 139, e449–54.

[446] Ballard RW, Sarkar NK, Irby MC, Armbruster PC, Berzins DW. Three-point bending test comparison of fiber-reinforced composite archwires to nickel-titanium archwires. Orthodontics (Chic.) 2012, 13, 46–51.

[447] Frank CA, Nikolai RJ. A comparative study of frictional resistances between orthodontic bracket and arch wire. Am. J. Orthod. 1980, 78, 593–609.

[448] Prashant PS, Nandan H, Gopalakrishnan M. Friction in orthodontics. J. Pharm. Bioallied Sci. 2015, 7, S334–8.

[449] Vaughan JL, Duncanson MG Jr, Nanda RS, Currier GF. Relative kinetic frictional forces between sintered stainless steel brackets and orthodontic wires. Am. J. Orthod. Dentofacial Orthop. 1995, 107, 20–7.

[450] Rabinowicz E. Friction and Wear of Materials. John Wiley and Sons, New York, 1965.

[451] Rossouw PE, Kamelchuk LS, Kusy RP. A fundamental review of variables associated with low velocity frictional dynamics. Semin. Orthod. 2003, 9, 223–35.

[452] Omana HM, Moore RN, Bagby MD. Frictional properties of metal and ceramic brackets. J. Clin. Orthod. 1992, 26, 425–32.

[453] Jastrebski ZB. The Nature and Properties of Engineering Materials. Wiley, New York, 3rd edition, 1987.

[454] Cacciafesta V, Sfondrini MF, Scribante A, Klersy C, Auricchio F. Evaluation of friction of conventional and metal-insert ceramic brackets in various bracket-archwire combinations. Am. J. Orthod. Dentofacial Orthop. 2003, 124, 403–9.

[455] Kapur Wadhwa R, Kwon HK, Close JM. Frictional resistances of different bracket-wire combinations. Aust. Orthod. J. 2004, 20, 25–30.

[456] Kao CT, Ding SJ, Wang CK, He H, Chou MY, Huang TH. Comparison of frictional resistance after immersion of metal brackets and orthodontic wires in a fluoride-containing prophylactic agent. Am. J. Orthod. Dentofacial Orthop. 2006, 130, e1–9.

[457] Liaw YC, Su YY, Lai YL, Lee SY. Stiffness and frictional resistance of a superelastic nickel-titanium orthodontic wire with low-stress hysteresis. Am. J. Orthod. Dentofacial Orthop. 2007, 131, e12–8.

[458] Hsu J-T, Wu L-C, Chang Y-Y, Weng T-N, Huang H-L, Yu C-H. Frictional forces of conventional and improved superelastic NiTi-alloy orthodontic archwires in stainless steel and plastic brackets. IFMBE Proc. 2009, 312–5.

[459] Tecco S, Teté S, Festa M, Festa F. An in vitro investigation on friction generated by ceramic brackets. World J. Orthod. 2010, 11, e133–44.

[460] Heo W, Baek SH. Friction properties according to vertical and horizontal tooth displacement and bracket type during initial leveling and alignment. Angle Orthod. 2011, 81, 653–61.

[461] Alfonso MV, Espinar E, Llamas JM, Rupérez E, Manero JM, Barrera JM, Solano E, Gil FJ. Friction coefficients and wear rates of different orthodontic archwires in artificial saliva. J. Mater. Sci.: Mater. Med. 2013, 24, 1327–32.

[462] Meier MJ, Bourauel C, Roehlike J, Reimann S, Keilig L, Braumann B. Friction behavior and other material properties of nickel–titanium and titanium–molybdenum archwires following electrochemical surface refinement. J. Orofac. Orthop. 2014, 75, 308–19.

[463] Carrion-Vilches FJ, Bermudez M-D, Fructuoso P. Static and kinetic friction force and surface roughness of different archwire bracket sliding contacts. Dent. Mater. J. 2015, 34, 648–53.

[464] Pizzoni L, Ravnholt G, Melsen B. Frictional forces related to self-ligating brackets. Eur. J. Orthodont. 1998, 20, 283–91.

[465] Ehsani S, Mandich M-A, El-Bialy TH, Flores-Mir C. Frictional resistance in self-ligating orthodontic brackets and conventionally ligated brackets. Angle Orthod. 2009, 79, 592–601.

[466] Chen SSH, Greenlee GM, Kim J-E, Smith CL, Huang GJ. Systematic review of self-ligating brackets. Am. J. Orthod. Dentofacial Orthop. 2010, 137, 726–6.

[467] Tecco S, Di Iorio D, Cordasco G, Verrocchi I, Festa F. An in vitro investigation of the influence of self-ligating brackets, low friction ligatures, and archwire on frictional resistance. Eur. J. Orthod. 2007, 29, 390–7.

[468] Baccetti T, Franchi L, Camporesi M, Defraia E. Orthodontic forces released by low-friction versus conventional systems during alignment of apically or buccally malposed teeth. Eur. J. Orthod. 2011, 33, 50–4.

[469]   Kusy RP, Whitley JQ. Effects of surface roughness on the coefficients of friction in model orthodontic systems. J. Biomech. 1990, 23, 913–25.

[470]   Wichelhaus A, Geserick M, Hibst R, Sander FG. The effect of surface treatment and clinical use on friction in NiTi orthodontic wires. Dent. Mater. 2005, 21, 938–45.

[471]   Bandeira AM, dos Santos MP, Pulitini G, Elias CN, da Costa MF. Influence of thermal or chemical degradation on the frictional force of an experimental coated NiTi wire. Angle Orthod. 2011, 81, 484–9.

[472]   Hosseinzadeh Nik T, Hooshmand T, Farazdaghi H, Mehrabi A, Razavi ES. Effect of chlorhexidine-containing prophylactic agent on the surface characterization and frictional resistance between orthodontic brackets and archwires: an in vitro study. Prog. Orthod. 2013, 14, 1–8.

[473]   Dridi A, Bensalah W, Mezlini S, Tobji S, Zidi M. Influence of bio-lubricants on the orthodontic friction. J. Mech. Behav. Biomed. Mater. 2016, 60, 1–7.

[474]   Geramy A, Hooshmand T, Etezadi T. Effect of sodium fluoride mouthwash on the frictional resistance of orthodontic wires. J. Dent. (Tehran) 2017, 14, 254–8.

[475]   Whitley JQ, Kusy RP. Influence of interbracket distances on the resistance to sliding of orthodontic appliances. Am. J. Orthod. Dentofacial Orthop. 2007, 132, 360–72.

[476]   Carrion-Vilches FJ, Bermudez M-D, Fructuoso P. Static and kinetic friction force and surface roughness of different archwire bracket sliding contacts. Dent. Mater. J. 2015, 34, 648–53.

[477]   Choi S, Park DJ, Kim KA, Park KH, Park HK, Park YG. In vitro sliding-driven morphological changes in representative esthetic NiTi archwire surfaces. Microsc. Res. Tech. 2015, 78, 926–34.

[478]   Savoldi F, Visconti L, Dalessandri D, Bonetti S, Tsoi JKH, Matinlinna JP, Paganelli C. In vitro evaluation of the influence of velocity on sliding resistance of stainless steel arch wires in a self-ligating orthodontic bracket. Orthod. Craniofac. Res. 2017, 20, 119–25.

[479]   Phukaoluan A, Khantachawana A, Kaewtatip P, Dechkunakorn S, Anuwongnukroh N, Santiwong P, Kajornchaiyakul J. Comparison of friction forces between stainless orthodontic steel brackets and TiNi wires in wet and dry conditions. Int. Orthod. 2017, 15, 13–24.

[480]   Rahilly G, Price N. Current products and practice: nickel allergy and orthodontics. J. Orthod. 2003, 30, 171–4.

[481]   Petoumeno E, Kislyuk M, Hoederath H, Keilig L, Bourauel C, Jäger A. Corrosion susceptibility and nickel release of nickel titanium wires during clinical application. J. Orofac. Orthop. 2008, 69, 411–23.

[482]   Agarwal P, Upadhyay U, Tandon R, Kumar S. Nickel allergy and orthodontics. Asian J. Oral Health All. Sci. 2011, 1, 61–3.

[483]   Huang H-H, Chiu Y-H, Lee T-H, Wu S-C, Yang H-W, Su K-H, Hsu C-C. Ion release from NiTi orthodontic wires in artificial saliva with various acidities. Biomaterials 2003, 24, 3585–92.

[484]   Sfondrini MF, Cacciafesta V, Maffia E, Scribante A, Alberti G, Biesuz R, Klersy C. Nickel release from new conventional stainless steel, recycled, and nickel-free orthodontic brackets: an in vitro study. Am. J. Orthod. Dentofacial Orthop. 2010, 137, 809–15.

[485]   Kao C-T, Ding S-J, He H, Chou MY, Huang T-H. Cytotoxicity of orthodontic wire corroded in fluoride solution in vitro. Angle Orthod. 2007, 77, 349–54.

[486]   Oshida Y. Bioscience and Bioengineering of Titanium Materials. Elsevier, Oxford, UK, 2007.

[487]   Brockhurst PJ. Dental materials: new territories for materials science. Met. Forum. Austr. Inst. Met. 1980, 3, 200–10.

[488]   Martinez JR, Barker S. Ion transport and water movement. Arch. Oral Biol. 1987, 32, 843–7.

[489]   Maijer R, Smith DC. Biodegradation of the orthodontic bracket system. Am. J. Orthod. Dentofac. Orthop. 1986, 90, 195–8.

[490] Lussi A, Hellwig E, Klimek J. Fluorides – mode of action and recommendations for use. Schweiz Monatsschr Zahnmed 2012, 122, 1030–6.

[491] Schiff N, Grosgogeat B, Lissac M, Dalard F. Influence of fluoridated mouthwashes on corrosion resistance of orthodontics wires. Biomaterials 2004, 25, 4535–42.

[492] Muguruma T, Iijima M, Brantley WA, Yuasa T, Kyung, H-M, Mizoguchi I. Effects of sodium fluoride mouth rinses on the torsional properties of miniscrew implants. Am. J. Orthod. Dentofacial Orthop. 2011, 139, 588–93.

[493] Schiff N, Grosgogeat B, Lissac M, Dalard F. Influence of fluoride content and pH on the corrosion resistance of titanium and its alloys. Biomaterials 2002, 23, 1995–2002.

[494] Li X, Wang J, Han EH, Ke W. Influence of fluoride and chloride on corrosion behavior of NiTi orthodontic wires. Acta Biomater. 2007, 3, 807–15.

[495] Kwon YH, Jang C-M, Jang J-H, Park J-H, Kim T-H, Kim H-I. Effect of fluoride released from fluoride-containing dental restoratives on NiTi orthodontic wires. Dent. Mater. J. 2008, 27, 133–8.

[496] Abalos C, Paúl A, Mendoza A, Solano E, Gil FJ. Influence of topographical features on the fluoride corrosion of Ni-Ti orthodontic archwires. J. Mater. Sci.: Mater. Med. 2011, 22, 2813–21.

[497] Srivastava K, Chandra PK, Kamat N. Effect of fluoride mouth rinses on various orthodontic archwire alloys tested by modified bending test: an in vitro study. Indian J. Dent. Res. 2012, 23, 433–4.

[498] Huang SY, Huang JJ, Kang T, Diao DF, Duan YZ. Coating NiTi archwires with diamond-like carbon films: reducing fluoride-induced corrosion and improving frictional properties. J. Mater. Sci.: Mater. Med. 2013, 24, 2287–92.

[499] Pulikkottil VJ, Chidambaram S, Bejoy PU, Femin PK, Paul P, Rishad M. Corrosion resistance of stainless steel, nickel-titanium, titanium molybdenum alloy, and ion-implanted titanium molybdenum alloy archwires in acidic fluoride-containing artificial saliva: an *in vitro* study. J. Pharm. Bioallied Sci. 2016, 8, S96–9.

[500] Močnik P, Kosec T, Kovač J, Bizjak M. The effect of pH, fluoride and tribocorrosion on the surface properties of dental archwires. Mater. Sci. Eng. C Mater. Biol. Appl. 2017, 78, 682–9.

[501] Gupta AK, Shukla G, Sharma P, Gupta AK, Kumar A, Gupta D. Evaluation of the effects of fluoride prophylactic agents on mechanical properties of nickel titanium wires using scanning electron microscope. J. Contemp. Dent. Pract. 2018, 19, 283–6.

[502] Mlinaric MR, Kanizaj L, Zuljevic D, Katic V, Spalj S, Curkovic HO. Effect of oral antiseptics on the corrosion stability of nickel-titanium orthodontic alloys. Mater. Corros. 2018, 69, 510–8.

[503] Iijima M, Endo K, Ohno H, Yonekura Y, Mizoguchi I. Corrosion behavior and surface structure of orthodontic Ni-Ti alloy wires. Dent. Mater. J. 2001, 20, 103–13.

[504] Yonekura Y, Endo K, Iijima M, Ohno H, Mizoguchi I. In vitro corrosion characteristics of commercially available orthodontic wires. Dent. Mater. J. 2004, 23, 197–202.

[505] Kao CT, Huang TH. Variations in surface characteristics and corrosion behaviour of metal brackets and wires in different electrolyte solutions. Eur. J. Orthod. 2010, 32, 555–60.

[506] Huang HH. Variation in corrosion resistance of nickel-titanium wires from different manufacturers. Angle Orthod. 2005, 75, 661–5.

[507] Huang HH. Surface characterizations and corrosion resistance of nickel-titanium orthodontic archwires in artificial saliva of various degrees of acidity. J. Biomed. Mater. Res. A. 2005, 74, 629–39.

[508] Wang QY, Zheng YF. The electrochemical behavior and surface analysis of Ti50Ni47.2Co2.8 alloy for orthodontic use. Dent. Mater. 2008, 24, 1207–11.

[509] Liu J-K, Lee T-M, Liu I-H. Effect of loading force on the dissolution behavior and surface properties of nickel-titanium orthodontic archwires in artificial saliva. Am. J. Orthod. Dentofacial Orthop. 2011, 140, 166–76.

[510]   Ünal Hİ. Effect of fluoride added artificial saliva solution on orthodontic wires. Prot. Met. Phys. Chem. S. 2012, 48, 367–70.

[511]   Trolić IM, Turco G, Contardo L, Serdarević NL, Otmačić H, Ćurković, Špalj S. Corrosion of nickel-titanium orthodontic archwires in saliva and oral probiotic supplements. Acta Stomatol. Croat. 2017, 51, 316–25.

[512]   Wranglén G. An Introduction to Corrosion and Protection of Metals. Chapman and Hill, New York, 1985, 85–7.

[513]   Van Black LH. Elements of Materials Science and Engineering.Addison-Wesley Pub., Reading, MA, 1989, 509–11.

[514]   Fontana MG, Greene ND. Corrosion Engineering. McGraw-Hill Book Co., New York, 1967, 20–5, 330–8.

[515]   Tomashov ND. Theory of Corrosion and Protection of Metals. MacMillan Co., New York, 1966, 212–8.

[516]   Uhlig HH, Revie RW. Corrosion and Corrosion Control. John Wiley & Sons, New York, 1985, 101–5.

[517]   Schiff N, Boinet M, Morgon L, Lissac M, Dalard F, Grosgogeat B. Galvanic corrosion between orthodontic wires and brackets in fluoride mouthwashes. Eur. J. Orthod. 2006, 28, 298–304.

[518]   Iijima M, Endo K, Yuasa T, Ohno H, Hayashi K, Kakizaki M, Mizoguchi I. Galvanic corrosion behavior of orthodontic archwire alloys coupled to bracket alloys. Angle Orthod. 2006, 76, 705–11.

[519]   Siargos B, Bradley TG, Darabara M, Papadimitriou G, Zinelis S. Galvanic corrosion of metal injection molded (MIM) and conventional brackets with nickel -titanium and copper -nickel-titanium archwires. Angle Orthod. 2007, 77, 355–60.

[520]   Bakhtari A, Bradley TG, Lobb WK, Berzins DW. Galvanic corrosion between various combnations of orthodontic brackets and archwires. Am. J. Orthod. Dentofacial Orthop. 2011, 140, 25–31.

[521]   Polychronis G, Al Jabbari YS, Eliades T, Zinelis S. Galvanic coupling of steel and gold alloy lingual brackets with orthodontic wires: is corrosion a concern? Angle Orthod. 2018, 88, 450–7.

[522]   Cortada M, Giner L, Costa S, Gil FJ, Rodríguez D, Planell JA. Metallic ion release in artificial saliva to titanium oral implants coupled with different superstructures. J. Biomed. Mater. Eng. 1997, 7, 213–20.

[523]   Goyer RA. Toxic Effects of Metals. In: Klaassen CD, Amdur MO, Doull J, ed., Cassarett and Doull's Toxicology. Mcmillan, New York, 1986, 582–635.

[524]   Beach JD. Nickel carbonyl inhibition of RNA synthesis by a chromatin-RNA polymerase complex from hepatitic nuclei. Cancer Res. 1970, 30, 48–50.

[525]   Stenburg T. Release of cobalt from cobalt-chromium alloy constructions in the oral cavity of man. Scand. J. Dent. Res. 1982, 90, 472–9.

[526]   Rechmann P. LAMMS and ICP-MS detection of dental metallic compounds in non-discolored human gingival. Dent. Res. 1992, 71, 599–603.

[527]   Hao SQ, Lemons JE. Histology of dog dental tissues with Cu-based crowns. Dent. Res. 1989, 68, 322–7.

[528]   Huang HH, Chiu YH, Lee TH, Wu SC, Yang HW, Su KH, Hsu CC. Ion release from NiTi orthodontic wires in artificial saliva with various acidities. Biomaterials 2003, 24, 3585–92.

[529]   Karnam SK, Reddy AN, Manjith CM. Comparison of metal ion release from different bracet archwire combinations: an in vitro study. J. Contemp. Dent. Pract. 2012, 13, 376–81.

[530]   Azizi A, Jamilian A, Nucci F, Kamali Z, Hosseinikhoo N, Perillo L. Release of metal ions from round and rectangular NiTi wires. Prog. Orthod. 2016, 17, 10. doi: 10.1186/s40510-016-0123-3.

[531]   Katic V, Curkovic L, Ujevic Bosnjak M, Peros K, Mandic D, Spalj S. Effect of pH, fluoride and hydrofluoric acid concentration on ion release from NiTi wires with various coatings. Dent. Mater. J. 2017, 36, 149–56.

[532]   Tahmasbi S, Sheikh T, Hemmati YB. Ion release and galvanic corrosion of different orthodontic brackets and wires in artificial saliva. J. Contemp. Dent. Pract. 2017, 18, 222–7.

[533]   Eliades T, Zinelis S, Papadopoulos MA, Eliades G, Athanasiou AE. Nickel content of as-received and retrieved NiTi and stainless steel archwires: assessing the nickel release hypothesis. Angle Orthod. 2004, 74, 151–4.

[534]   Arndt M, Brück A, Scully T, Jäger A, Bourauel C. Nickel ion release from orthodontic NiTi wires under simulation of realistic *in-situ* conditions. J. Mater. Sci. 2005, 40, 3659–67.

[535]   Shabalovskaya SA, Tian H, Anderegg JW, Schryvers DU, Carroll WU, Van Humbeeck J. The influence of surface oxides on the distribution and release of nickel from nitinol wires. Biomaterials 2009, 30, 468–77.

[536]   Petoumenou E, Arndt M, Keilig L, Reimann S, Hoederath H, Eliades T, Jäger A, Bourauel C. Nickel concentration in the saliva of patients with nickel-titanium orthodontic appliances. Am. J. Orthod. Dentofacial Orthop. 2009, 135, 59–65.

[537]   Suárez C, Vilar T, Gil FJ, Sevilla P. In vitro evaluation of surface topographic changes and nickel release of lingual orthodontic archwires. J. Mater. Sci.: Mater. Med. 2010, 21, 675–83.

[538]   Espinar E, Llamas JM, Michiardi A, Ginebra MP, Gil FJ. Reduction of Ni release and improvement of the friction behaviour of NiTi orthodontic archwires by oxidation treatments. J. Mater. Sci. Mater. Med. 2011, 22, 1119–25.

[539]   Ousehal L, Lazrak L. Change in nickel levels in the saliva of patients with fixed orthodontic appliances. Int. Orthod. 2012, 10, 190–7.

[540]   Jamilian A, Moghaddas O, Toopchi S, Perillo L. Comparison of nickel and chromium ions released from stainless steel and NiTi wires after immersion in Oral B®, Orthokin® and artificial saliva. J. Contemp. Dent. Pract. 2014, 15, 403–6.

[541]   Mlir hashemi A, Jahangiri S, Kharrazifard M. Release of nickel and chromium ions from orthodontic wires following the use of teeth whitening mouthwashes. Prog. Orthod. 2018, 19; doi: 10.1186/s40510-018-0203-7.

[542]   Ryhanen J, Niemi E, Serlo W, Niemela E, Sandvik P, Pernu H, Salo T. Biocompatibility of nickel titanium shape memory metal and its corrosion behavior in human cell cultures. J. Biomed. Mater. Res. 1997, 35, 451–7.

[543]   McKay GC, Macnair R, MacDonald C, Grant MH. Interactions of orthopaedic metals with an immortalized rat osteoblast cell line. Biomater. 1996, 17, 1339–44.

[544]   Kerosuo H, Kullaa A, Kerosuo E, Kanerva L, Hensten-Ptterson A. Nickel allergy in adolescents in relation to orthodontic treatment and piercing of ears. Am. J. Orthod. Dentofacial Orthop. 1996, 109, 148–54.

[545]   Berger-Gorbet M, Broxup B, Rivard C, Yahia L'H. Biocompatibility testing of NiTi screw using immuno histochemistry on sections containing metallic implants. J. Biomed. Mater. Res. 1996, 32, 243–8.

[546]   Bass JK, Fine H, Cisnero GJ. Nickel hypersensitivity in the prosthodontics patient. Am. J. Orthod. Dentofacial Orthop. 1993, 103, 280–5.

[547]   Harzer W, Schröter A, Gedrange T, Muschter F. Sensitivity of titanium brackets to the corrosive influence of fluoride-containing toothpaste and tea. Angle Orthod. 2001, 71, 318–23.

[548]   Genelhu MC, Marigo M, Alves-Oliveira LF, Malaquias LC, Gomez RS. Characterization of nickel-induced allergic contact stomatitis associated with fixed orthodontic appliances. Am. J. Orthod. Dentofacial Orthop. 2005, 128, 378–81.

[549] Pazzini CA, Júnior GO, Marques LS, Pereira CV, Pereira LJ. Prevalence of nickel allergy and longitudinal evaluation of periodontal abnormalities in orthodontic allergic patients. Angle Orthod. 2009, 79, 922–7.

[550] Narmada IB, Sudarno NT, Sjafei A, Setiyorini Y. The influence of artificial salivary pH on nickel ion release and the surface morphology of stainless steel bracket-nickel-titanium archwire combinations. Dent. J. Majalah Kedokteran GiGi 2017, 50, 80–5.

[551] Eliades T, Pratsinis H, Kletsas D, Eliades G, Makou M. Characterization and cytotoxicity of ions released from stainless steel and nickel-titanium orthodontic alloys. Am. J. Orthod. Dentofacial Orthop. 2004, 125, 24–9.

[552] David A, Lobner D. In vitro cytotoxicity of orthodontic archwires in cortical cell cultures. Eur. J. Orthod. 2004, 26, 421–6.

[553] Sodor A, Ogodescu AS, Petreuş T, Şişu AM, Zetu IN. Assessment of orthodontic biomaterials' cytotoxicity: an in vitro study on cell culture. Rom. J. Morphol. Embryol. 2015, 56, 1119–25.

[554] Čolić M, Tomić S, Rudolf R, Marković E, Šćepan I. Differences in cytocompatibility, dynamics of the oxide layers' formation, and nickel release between superelastic and thermo-activated nickel-titanium archwires. J. Mater. Sci. Mater. Med. 2016, 27, 128; doi: 10.1007/s10856-016-5742-1.

[555] Drescher D, Bourauel C, Sonneborn W, Schmuth GP. The long-term fracture resistance of orthodontic nickel-titanium wires. Schweiz Monatsschr Zahnmed 1994, 104, 578–84.

[556] Yokoyama K, Hamada K, Moriyama K, Asaoka K. Degradation and fracture of Ni-Ti superelastic wire in an oral cavity. Biomaterials 2001, 22, 2257–62.

[557] Zinelis S, Eliades T, Pandis N, Eliades G, Bourauel C. Why do nickel-titanium archwires fracture intraorally? Fractographic analysis and failure mechanism of in-vivo fractured wires. Am. J. Orthod. Dentofacial Orthop. 2007, 132, 84–9.

[558] Perinetti G, Contardo L, Ceschi M, Antoniolli F, Franchi L, Baccetti T, Di Lenarda R. Surface corrosion and fracture resistance of two nickel-titanium-based archwires induced by fluoride, pH, and thermocycling. An in vitro comparative study. Eur. J. Orthod. 2012, 34, 1–9.

[559] Guzman U, Jerrold L, Abdelkarim A. An in vivo study on the incidence and location of fracture in round orthodontic archwires. J. Orthod. 2013, 40, 307–12.

[560] Prymak O, Klocke A, Kahl-Nieke B, Epple M. Fatigue of orthodontic nickel-titanium (NiTi) wires in different fluids under constant mechanical stress. Mater. Sci. Eng. A 2004, 378, 110–4.

[561] Bourauel C, Scharold W, Jäger A, Eliades T. Fatigue failure of as-received and retrieved NiTi orthodontic archwires. Dent. Mater. 2008, 24, 1095–101.

[562] Filleul MP, Jordan L. Torsional properties of Ni-Ti and copper Ni-Ti wires: the effect of temperature on physical properties. Eur. J. Orthod. 1997, 19, 637–46.

[563] Bolender Y, Vernière A, Rapin C, Filleul MP. Torsional superelasticity of NiTi archwires. Angle Orthod. 2010, 80, 1100–9.

[564] Partowi S, Keilig L, Reimann S, Jäger A, Bourauel C. Experimental analysis of torque characteristics of orthodontic wires. J. Orofac. Orthop. 2010, 71, 362–72.

[565] Archambault A, Major TW, Carey JP, Heo G, Badawi H, Major PW. A comparison of torque expression between stainless steel, titanium molybdenum alloy, and copper nickel titanium wires in metallic self-ligating brackets. Angle Orthod. 2010, 80, 884–9.

[566] Hirai M, Nakajima A, Kawai N, Tanaka E, Igarashi Y, Sakaguchi M, Sameshima GT, Shimizu N. Measurements of the torque moment in various archwire-bracket-ligation combinations. Eur. J. Orthod. 2012, 34, 374–80.

[567] Iijima M, Brantley WA, Yuasa T, Muguruma T, Kawashima I, Mizoguchi I. Joining characteristics of orthodontic wires with laser welding. J. Biomed. Mater. Res. Part B: Appl. Biomater. 2008, 84B, 147–53.

[568]   Tam B, Pequegnat A, Khan MI, Zhou Y. Resistance Microwelding of Ti-55.8 wt pct Ni nitinol wires and the effects of pseudoelasticity. Metall. Mater. Trans. A 2012, 43, 2969–78.

[569]   Matsunaga J, Watanabe I, Nakao N, Watanabe E, Elshahawy W, Yoshida N. Joining characteristics of titanium-based orthodontic wires connected by laser and electrical welding methods. J. Mater. Sci.: Mater. Med. 2015, 26; doi: 10.1007/s10856-015-5391-9.

[570]   Mesquita TR, Martins LP, Martins RP. Welding strength of NiTi wires. Dental Press J. Orthod. 2018, 23, 58–62.

[571]   Juvvadi SR, Kailasam V, Padmanabhan S, Chitharanjan AB. Physical, mechanical, and flexural properties of 3 orthodontic wires: an in-vitro study. Am. J. Orthod. Dentofacial Orthop. 2010, 138, 623–30.

[572]   Krishnan V, Kumar KJ. Mechanical properties and surface characteristics of three archwire alloys. Angle Orthod. 2004, 74, 825–31.

[573]   Kawashima I, Ohno H, Sachdeva R. Relationship between Af temperature and load changes in Ni-Ti orthodontic wire under different thermomechanical conditions. Dent. Mater. J. 1999, 18, 403–12.

[574]   da Silva DL, Mattos CT, Simão RA, de Oliveira Ruellas AC. Coating stability and surface characteristics of esthetic orthodontic coated archwires. Angle Orthod. 2013, 83, 994–1001.

[575]   Boccaccini AR, Peters C, Roether JA, Eifler D, Misra SK, Minay EJ. Electrophoretic deposition of polyetheretherketone (PEEK) and PEEK/Bioglass® coatings on NiTi shape memory alloy wires. J. Mater. Sci. 2006, 41, 8152–9.

[576]   Bravo LA, de Cabañes AG, Manero JM, Rúperez E, Gil FJ. NiTi superelastic orthodontic archwires with polyamide coating. J. Mater. Sci.: Mater. Med. 2014, 25, 555–60.

[577]   Liu J-K, Liu I-H, Liu C, Chang C-J, Kung K-C, Liu Y-T, Lee T-M, Jou J-L. Effect of titanium nitride/titanium coatings on the stress corrosion of nickel–titanium orthodontic archwires in artificial saliva. Appl. Surf. Sci. 2014, 317, 974–81.

[578]   Sugisawa H, Kitaura H, Ueda K, Kimura K, Ishida M, Ochi Y, Kishikawa A, Ogawa S, Takano-Yamamoto T. Corrosion resistance and mechanical properties of titanium nitride plating on orthodontic wires. Dent. Mater. J. 2018, 37, 286–92.

[579]   Huang SY, Huang JJ, Kang T, Diao DF, Duan YZ. Coating NiTi archwires with diamond-like carbon films: reducing fluoride-induced corrosion and improving frictional properties. J. Mater. Sci.: Mater. Med. 2013, 24, 2287–92.

[580]   Anuradha P, Varma NK, Balakrishnan A. Reliability performance of titanium sputter coated Ni-Ti arch wires: mechanical performance and nickel release evaluation. Biomed. Mater. Eng. 2015, 26, 67–77.

[581]   Gil FJ, Solano E, Mendoza A, Pena J. Inhibition of Ni release from NiTi and NiTiCu orthodontic archwires by nitrogen diffusion treatment. J. Appl. Biomater. Biomech. 2004, 2, 151–5.

[582]   Katić V, Mandić V, Ježek D, Baršić G, Špalj S. Influence of various fluoride agents on working properties and surface characteristics of uncoated, rhodium coated and nitrified nickel-titanium orthodontic wires. Acta Odontol. Scand. 2015, 73, 241–9.

[583]   https://www.amazingsmilesorthodontist.com/braces-diagram/.

[584]   Evans D. The history of braces & orthodontics; https://www.davidevansdds.com/history-of-braces/.

[585]   Smith Y. History of dental braces, 2019; https://www.news-edical.net/health/History-of-Dental-Braces.aspx.

[586]   Sfondrini MF, Cacciafesta V, Maffia E, Scribante A, Alberti G, Biesuz R, Klersy C. Nickel release from new conventional stainless steel, recycled, and nickel-free orthodontic brackets: an in vitro study. Am. J. Orthod. Dentofacial Orthop. 2010, 137, 809–15.

[587]   Reimann S, Rewari A, Keilig L, Widu F, Jäger A, Bourauel C. Material testing of reconditioned orthodontic brackets. J. Orofac. Orthop. 2012, 73, 454–66.

[588]   Govindankutty D. Applications of nanotechnology in orthodontics and its future implications: a review. Int. J. Appl. Dent. Sci. 2015, 1, 166–71.
[589]   Lapatki BG, Paul O. Smart backets for 3D-force-moment measurement in orthodontic research and therapydevelopmental status and prospects. J. Orofac. Orthop. 2007, 68, 377–96.
[590]   Lapaiki HG, Bartholomeyczik J, Ruther P, Jonas IE, Paul O. Smart bracket for multi-dimensional force and moment measurement. J. Dent. Res. 2007, 86, 73–8.
[591]   Lekhadia DR. Nanotechnology in Orthodontics – Futuristic Approach. In: Chaughule RS, ed., Dental Applications of Nanotechnology. Springer, 2018, 155–75.
[592]   https://www.youtube.com/watch?v=gCcws1EDo3c.
[593]   http://www.montereyorthodontics.com/treatment/temporary-anchorage-devices-tads.
[594]   Singh K, Kumar D, Jaiswal RK, Bansal A. Temporary anchorage devices – mini-implants. Natl. J. Maxillofac. Surg. 2010, 1, 30–4.
[595]   Self-drilling mini-implants. Br. Dent. J. 2008, 24; doi: 10.1038/bdj.2007.1210.
[596]   https://www.walmart.ca/en/ip/Micro-Size-Self-Drilling-Thread-Orthodontic-Matching-Implants-Screw-NiTi-Mini-Screws/1216HVUFDDPO.
[597]   Tseng YC, Hsieh CH, Chen CH, Shen YS, Huang IY, Chen CM. The application of mini-implants for orthodontic anchorage. Int. J. Oral. Maxillofac. Surg. 2006, 35, 704–7.
[598]   Morais LS, Serra GG, Muller CA, Andrade LR, Palermo EF, Elias CN, Meyers M. Titanium alloy mini-implants for orthodontic anchorage: immediate loading and metal ion release. Acta Biomater. 2007, 3, 331–9.
[599]   Uemura M, Motoyoshi M, Yano S, Sakaguchi M, Igarashi Y, Shimizu N. Orthodontic mini-implant stability and the ratio of pilot hole implant diameter. Eur. J. Orthod. 2012, 34, 52–6.
[600]   Abduljabbar OF, Shihab O, Omer OA. Dental implants in an orthodontically created spaces using NiTi open coil spring in anterior and premolar regions. Zanco J. Med. Sci., 2017, 21; https://doi.org/10.15218/zjms.2017.026.
[601]   Types of implants; http://www.dentaledu.tv/types-of-implants/.
[602]   Alison A. Types of dental implants; https://www.dentalimplantcostguide.com/types-of-dental-implants/.
[603]   Pasqualini U, Pasqualini M. The Blade Implant. In: Pasqualini U, et al. ed., Treatise of Implant Dentistry: The Italian Tribute to Modern Implantology. Ariesdue srl Carimate, Como, 2009.
[604]   Grafelmann HL. The latest developments in blade implant clinical applications. Dent. Implantol Update 1993, 4, 22–5.
[605]   Misch C. Blade vent implant: still viable. Dent. Today 1989, 834–42.
[606]   Carlo LD, Pasqualini ME, Carinci F, Corradini M, Vannini F, Linkow LI. A brief history and guidelines of blade implant technique: a retrospective study on 522 implants. Ann. Oral Maxillofacial Surg. 2013; doi: 10.13172/2052-7837-1-1-390.
[607]   Gil FJ, Planell JA. Shape memory alloys for medical applications. Proc. Instn. Mech. Engrs. Part H 1998, 212, 473–88.
[608]   http://www.1888implant.com/mobile/dental_implants.html.
[609]   Fukuyo S, Suzuki Y, Suzuki K, Sairenji E. Shape memory implants. In: FDuering TW, et al. ed., Shape Memory in Engineering Aspects of Shape Memory Alloys. Butterwotth-Heinemann, London, 1990, 470–2.
[610]   Gil FJ, Fernández E, Manero JM, Planell JA, Sabrià J, Cortada M, Giner L. A study of load cycling in a NiTi alloy with pseudoelastic behaviour used indental rosthetic fixators. Biomed. Mater. Eng. 1996, 6, 153–8.
[611]   Sabrià J, Cortada M, Giner L, Fernández E, Gil FJ, Manero JM, Planell JA. A study of load cyclingin a NiTi alloy with pseudoelastic behaviour used in dentalprosthetic fixators. Biomed. Mater. Eng, 1995, 5, 161–7.

[612]   Fukuyo S, Sachdeva R, Oshida Y, Sairenji E. The Perio Root Implant. In: Fukuyo S, et al. ed., Perio Root Implant and Medical Application of Shape Memory Alloy. Japan Medical Culture Center, 1992, 65–72.

[613]   Nouri A. Titanium Foam Scaffolds for Dental Applications. In: Wen C, ed., Metallic Foam Bone: Processing, Modification and Characterization and Properties. Woodhead Pub., Elsevier, 2017, 131–60.

[614]   Bansiddhi A, Dunand DC. Shape-memory NiTi foams produced by replication of NaCl space-holders. Acta Biomater. 2008, 4, 1996–2007.

[615]   Bansiddhi A, Dunand DC. Shape-memory NiTi foams produced by solid-state replication with NaF. Intermetallics 2007, 15, 1612–22.

[616]   Köhl M, Habijan T, Bram M, Buchkremer HP, Stöver D, Köller M. Powder metallurgical near-net-shape fabrication of porous NiTi shape memory alloys for use as long-term implants by the combination of the metal injection molding process with the space-holder technique. Adv. Eng. Mater. 2009, 11, 959–68.

[617]   Li DS, Zhang YP, Eggeler G, Zhang XP. High porosity and high-strength porous NiTi shape memory alloys with controllable pore characteristics. J. Alloys Compd. 2009, 470, 1–5.

[618]   Dunand DC, Mari D, Bourke MAM, Roberts JA. NiTi and NiTi-TiC composites: part IV. Neutron diffraction study of twinning and shape-memory recovery. Metall. Mater. Trans. A 1996, 27, 2820–36.

[619]   Gibson LJ, Ashby MF. Cellular Solids: Structure and Properties. CambridgeUniversity Press, UK, 2nd edition, 1997.

[620]   Gotman I, Ben-David D, Unger RE, Böse T, Gutmanas EY, Kirkpatrick CJ. Mesen-chymal stem cell proliferation and differentiation on load-bearing trabecular Nitinol scaffolds. Acta Biomater. 2013, 9, 8440–8.

[621]   Liu X, Wu S, Yeung KW, Chan YL, Hu T, Xu Z, et al. Relationship between osseoin-tegration and superelastic biomechanics in porous NiTi scaffolds. Biomaterials 2011, 32, 330–8.

[622]   Auricchio F, Boatti E, Conti M. SMA Biomedical Applications. In: Lecce L., et al. ed., Shape Memory Alloy Engineering. Elsevier, 2015, 307–41.

[623]   Sabrià J, Cortada M, Giner L, Gil FJ, Fernández E, Manero JM, Planell JA. A study of load cycling in a NiTi shape memory alloy with pseudoelastic behaviour used in dental prosthetic fixators. Biomed. Mater. Eng. 1996, 6, 153–7.

[624]   Favot LM, Berry-Kromer V, Haboussi M, Thiebaud F, Ben Zineb T. Numerical study of the influence of material parameters on the mechanical behaviour of a rehabilitated edentulous mandible. J. Dent. 2014, 42, 287–97.

Chapter 17
Medical applications

17.1 In general

Shape-memory alloys (SMA) constitute a group of metallic materials with the ability to recover a previously defined length or a shape when subjected to an appropriate thermomechanical load [1, 2]. When there is a limitation of shape recovery, these alloys promote high restitution forces. Because of these properties, there is a great technological interest in using SMA for different applications [1, 3]. NiTi (nickel–titanium) alloy is just a type of metallic alloy exhibiting SME (shape-memory effect), and others include Ag–Cd (44/49 at%) with transformation temperature ranging from –190 to –50 °C, Au–Cd (47/50 at%) from 30 to 100 °C, Cu–Al (15 wt%)–Ni (5 wt%) from –140 to 100 °C, Cu–Sn (15 at%) from –120 to 30 °C, Cu–Zn (40 wt%) from –180 to –10 °C, C–Zn–Si (or Sn, Al) from –180 to 200 °C, In–Ti (20 at%) from 60 to 100 °C, Ni–Al (37 at%) from –180 to 100 °C, Ni–Ti (50 at%) from –50 to 100 °C, Fe–Pt (25 at%) to about –130 °C, Mn–Cu (5/35 at%) from –250 to 180 °C, and Fe–Mn (30 wt%)–Si (6 wt%) from –200 to 150 °C [3]. Although a relatively wide variety of alloys present SME, only those that can recover from a large amount of strain or generate an expressive restitution force are of commercial interest. Particularly, the most important among these alloys are NiTi-based alloys, and are most frequently used in commercial applications because they combine good mechanical properties with shape memory.

SMA present typical thermomechanical behaviors like superelasticity (or pseudoelasticity) and SMEs (one-way and two-way). Pseudoelasticity occurs whenever an SMA is at a temperature above A_F (the temperature above which only the austenitic phase is stable for a stress-free specimen). The second thermomechanical behavior that can be observed in SMA is the SME, exhibiting at a low temperature (less than M_F, the temperature below which only the martensitic phase is stable). Due to these two functional properties, biomedical applications of SMA have been extremely successful; besides their dental applications in endodontic instruments and orthodontic appliances as discussed in the previous chapter, increasing both the possibility and the performance of minimally invasive surgeries [4–6], in addition to these biofunctions, their excellent biocompatibility and their biomedical applications as orthopedic implants [7], cardiovascular devices [4], and surgical instruments [5] have been extensively developed. In medical applications of NiTi alloys, there are mainly three typical groups of applications: a variety of cardiovascular devices, orthopedic applications, and various types of surgical instruments. All are discussed in this chapter.

https://doi.org/10.1515/9783110666113-017

17.2 Cardiovascular applications

17.2.1 Cardiovascular catheter

In the cardiac catheterization, a very small, flexible, hollow tube (called a catheter) is inserted into a blood vessel in the groin, arm, or neck. Then the catheter is threaded through the blood vessel into the aorta and into the heart (see Figure 17.1) [8, 9]. Once the catheter is in place, several tests will be performed. The catheter will be guided into the coronary arteries (which are vessels carrying blood to the heart muscle) and the contrast dye will be injected to check blood flow, called as the coronary angiography [10] (see Figure 17.2). These are some of the other procedures that may be done during or after the cardiac catheter insertion.

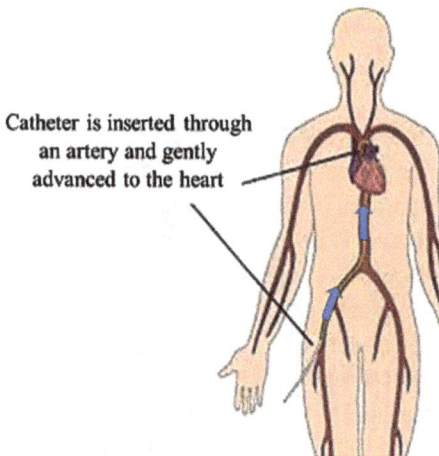

Catheter is inserted through
an artery and gently
advanced to the heart

Figure 17.1: Catheter insertion from groin [8].

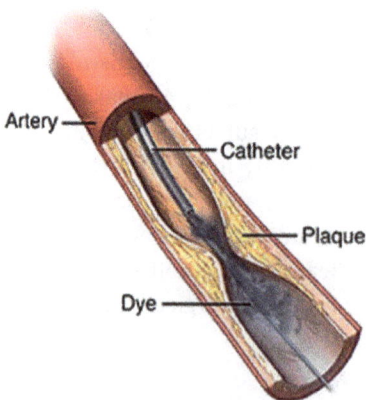

Artery

Catheter

Plaque

Dye

Figure 17.2: Insertion of cardiovascular catheter [10].

Sometimes, a cardiovascular guidewire is inserted, which is used to navigate vessels to target lesions; hence, larger, less maneuverable treatment devices can be pushed over the wire to rapidly reach the lesion (see Figure 17.3) [11].

Figure 17.3: Cardiovascular guidewire [11].

During the angioplasty procedure, a tiny balloon is inflated at the tip of the inserted catheter, so that generated pressure presses any plaque buildup against the artery wall and improves blood flow through the artery (see Figure 17.4) [8, 12].

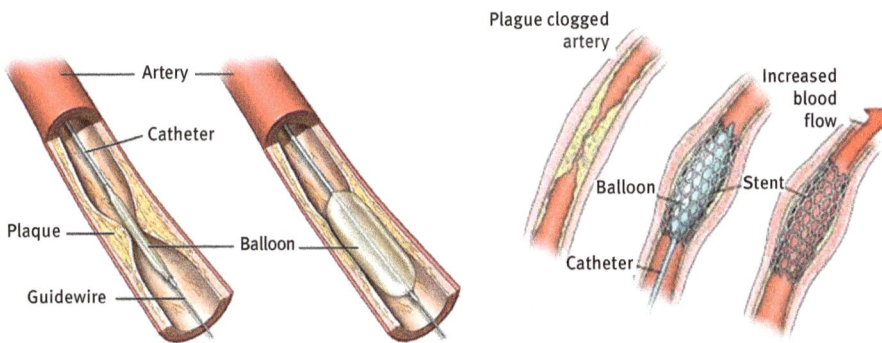

Figure 17.4: Demonstration of biofunctions of stent and balloon [8, 12].

17.2.2 Inferior vena cava filter

Thromboembolic disease continues to be a cause of morbidity and mortality [13]. The inferior vena cava (IVC) is the largest vein in the body. It carries blood from the lower body to the heart and lungs. A blood clot in a vein below the heart blocks the normal blood flow and may cause swelling, redness, and pain in the area. Placing a filter in the IVC is an important way to prevent significant pulmonary embolism (PE) arising from a deep vein thrombosis (DVT). This procedure is currently performed under radiologic guidance via femoral vein or jugular vein access. Figure 17.5 shows typical vena cava filters [14, 15]. The Simon vena cava filter traps these clots that in time are dissolved by the bloodstream. The insertion of the filter inside the human body is done by exploiting the SME. From its original shape in the martensitic state, the filter is deformed and placed on a catheter tip. Saline solution flowing through the catheter is used to keep a low temperature, while the filter is placed inside the body. When the catheter releases the filter, the flow of the saline solution is stopped. As a result, the bloodstream promotes the heating of the filter that returns to its former shape [5].

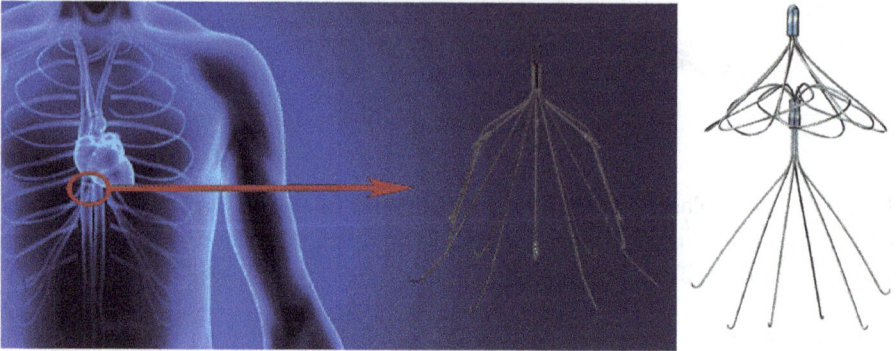

Figure 17.5: Typical inferior vena cava filter [14, 15].

Engmann et al. [16] and Poletti et al. [17] evaluated the clinical efficacy, mechanical stability, and safety of the Simon nitinol IVC filter (SNF). It was concluded that (i) antecubital venous insertion of the SNF is a safe and effective method for the prevention of PE in patients who cannot be managed with traditional anticoagulation, and offers the option of inserting a peripherally inserted central catheter with no added complications [16], and (ii) regardless of long-term anticoagulation, the rate of caval thrombosis is acceptably low and except for occasional access-site thrombosis, no other filter-related morbidity was observed [17]. Asch et al. [18] evaluated preliminary clinical experience in humans with the recovery nitinol filter (RNF) for the IVC, especially the efficacy of the device and safety of its retrieval. Thirty-two patients were followed up to assess for filter efficacy and for ability to remove the filter. Sixteen men and 16 women aged 18–83 years (mean, 53 years) underwent treatment with the RNF. Indications for placement were recent PE ($n = 16$), recent deep venous thrombosis ($n = 20$), and/or prophylaxis ($n = 2$). It was reported that (i) four patients had contraindications to anticoagulant therapy, and four had complications from anticoagulant therapy; (ii) the filter was successfully placed in 32 patients; (iii) in 24 (100%) of 24 patients, the filter was successfully retrieved with a jugular approach; (iv) the mean implantation period was 53 days (range, 5–134 days); and (v) there were no episodes of PE or insertion-site thrombosis. It was then concluded that (vi) the efficacy of the RNF was confirmed and (ii) the feasibility and safety of retrieval up to 134 days after implantation was demonstrated. Bruckheimer et al. [19] evaluated the clot-trapping ability, stability, and migration of a new low-profile, retrievable IVC filter in an in vitro model. The SafeFlo IVC filter consists of two superelastic nitinol wires that form a double-ring platform and spiral filter. The filter is collapsed into a 5-6-F catheter and delivered into the IVC model. The in vitro model closely simulates the physical parameters of flow in the human IVC. Human blood clots of 2 and 4 mm diameters and 3 cm lengths were injected into the flow system in sets of five clots. Filter delivery and retrieval were performed in every series. Filtration was evaluated in IVC models of 20 and 24 mm lumen diameter in vertical and horizontal positions. Stability and migration of the filter were evaluated by direct vision of maintenance of position and shape before and after clot trapping. It was reported that (i) filter delivery and retrieval were straightforward and repeatable in a total of 20 procedures, (ii) the filters maintained shape and position throughout the study, (iii) a total of 248 clots were injected and 225 (90.7%) were trapped, and (iv) the individual tests in horizontal and vertical positions with either clot size demonstrated trapping rates of 85.7–97.1%. Based on these results, it was concluded that (v) the SafeFlo IVC filter is a stable and effective filter in an in vitro model, and (vi) the filter design is amenable to simple delivery and retrieval.

17.2.3 Atrial septal occluder

Atrial septal defect (ASD) is among the most common congenital heart defects. *Atrial septal occlusion devices* are implantable cardiac *devices* used in patients with certain types of *ASDs*. They are used in cases of *ASDs* with right *atrial* or ventricle enlargement to prevent paradoxical embolism, left-to-right shunting, and platypnea-orthodeoxia syndrome. The atrial septal occlusion device is employed to seal the atrial hole. The atrial hole is located between the two upper heart chambers upon the surface that splits the upper part of the heart into the right and left atria. The anomaly occurring when this hole is open can reduce life expectancy. The traditional surgery that fixes this anomaly is extremely invasive and dangerous. The thorax of the patient is opened and the atrial hole is sewn. Because of the intrinsic risks of this surgery, several problems might occur. The atrial septal occlusion device is an alternative to this surgery. Normally, the device is composed of SMA wires and a waterproof film of polyurethane [5] (see Figure 17.6 for occluder and insertion manner thereof). As is the case for the Simon filter, the surgery to place this device exploits the SME, being much less invasive than the traditional one. First, one half of the device is inserted through a catheter by the vena cava up to the heart, in its closed form. Then, it is placed on the atrial hole and opened, recovering its original shape. Next, the second half of the device is placed by the same route as the first one, and then both halves are connected. This procedure seals the hole, avoiding blood flow from one atrium to the other. It is expected that the device will stay in the heart for an indefinite period of time since the heart tissue regenerates.

Roy et al. [22] mentioned that the patent foramen ovale (PFO) is a normal fetal communication between the right and left atria that persists postpartum and is

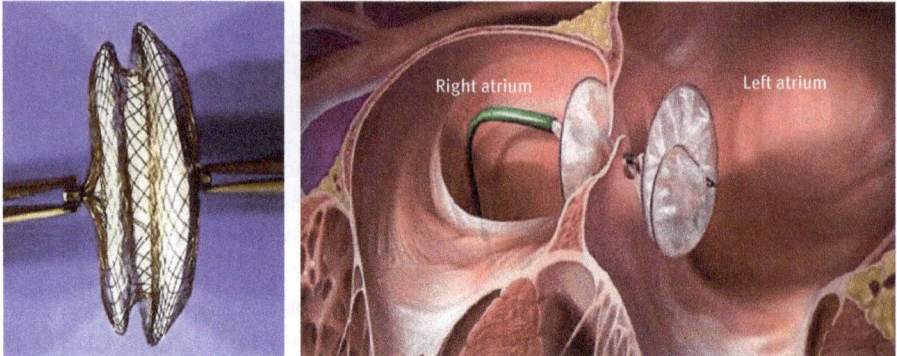

Figure 17.6: Atrial septal occlude [20] and insertion operation [21].

common in 20–40% of the population. Occasionally, PFOs can lead to paradoxical embolism that can manifest as a stroke or systemic arterial embolism [23, 24]. ASD are far less prevalent than PFO; however, it remains one of the most common congenital cardiac defects and the majority of these are ostium secundum defects. These occur either from excessive resorption of the septum primum or from deficient growth of the septum secundum. Patients with ASDs often present with exercise intolerance or palpitations, but occasionally present with overt right ventricular failure, especially in older patients and paradoxical embolism [25]. Percutaneous closure of both PFOs and secundum ASDs is increasingly becoming the treatment of choice where closure is indicated, and while there is good evidence for ASD closure, the closure of symptomatic and asymptomatic PFOs remains highly controversial. A multitude of devices are available for percutaneous closure of both defects, but there are limited data on the comparative efficacy of these devices. The chapter reviews the indications and percutaneous device options for PFO and ASD closure. Turner et al. [26] prospectively evaluated the risk of hemodynamic compromise and obtain medium-term survival data on patients implanted with the AMPLATZER Septal Occluder for percutaneous closure of secundum ASDs. Subjects were enrolled prospectively at 50 US sites and followed for 2 years. Between 2008 and 2012, ASD closure with the AMPLATZER Septal Occluder was attempted in 1,000 patients (aged 0.3–83.6 years, mean 21±22 years). It was reported that (i) procedural closure occurred in 97.9%, with 1 month and 2 year closure of 98.5% and 97.9%, respectively; (ii) hemodynamic compromise occurred in six subjects (0.65%), because of dysrhythmia in two, device embolization in one, and cardiac erosion in three; and (iii) the rate of cardiac erosion was 0.3% (average 83, range 12–171 days from implant). It was therefore concluded that (iv) closure of ASD with the AMPLATZER Septal Occluder is safe and effective, (vi) the rate of hemodynamic compromise and cardiac erosion is rare, and (vii) the risk factors for cardiac erosion after device closure are not yet clear.

17.3 Various bioapplications of stents

17.3.1 In general

A stent is a tiny tube that is inserted into a blocked passageway to keep it open. The stent restores the flow of blood or other fluids, depending on where it is placed. Self-expanding stents (named after the dentist C.T. Stent) are another important cardiovascular application that is used to maintain the inner diameter of a blood vessel. Actually, these devices are used in several situations in order to support any tubular passage such as the esophagus and bile duct, and blood vessels such as the coronary, iliac, carotid, aorta, and femoral arteries. In this type of application, a cylindrical scaffold with shape memory is placed, for example, inside a blood vessel through a catheter. Initially, this scaffold is precompressed in its martensitic state. As the scaffold is heated, due to the body temperature, it tends to recover its original shape, expanding itself. This device can be used not only in the angioplasty procedure, in order to prevent another obstruction of a vessel, but also in the treatment of aneurysms for the support of a weakened vessel. Stents are usually needed when plaque blocks a blood vessel. Plaque is made of cholesterol and other substances that attach to the walls of a vessel.

Nitinol (NiTi-SMA) exhibits a combination of properties (SME and superelasticity), which makes these alloys particularly suitable for self-expanding stents. Some of these properties cannot be found in engineering materials used for stents presently. Nitinol stents are manufactured to a size slightly larger than the target vessel size and delivered constrained in a delivery system. After deployment, they position themselves against the vessel wall with a low, chronic outward force. They resist outside forces with a significantly higher radial resistive force [27]. Superelastic properties of NiTi-SMAs permit self-expanding stents to be crimped without plastic deformation, but their nonlinear properties can contribute toward stent buckling. McGrath et al. [28] investigated the axial buckling of a prototype tracheobronchial nitinol stent design during crimping, with the objective of eliminating buckling from the design. To capture the stent buckling mechanism, a computational model of a radial force test is simulated, where small geometric defects are introduced to remove symmetry and to allow buckling. With the buckling mechanism ascertained, a sensitivity study is carried out to examine the effect that the transitional plateau region of the nitinol loading curve has on stent stability. It was mentioned that the transitional plateau region can have a significant effect on the stability of a stent during crimping, and by reducing the amount of transitional material within the stent hinges during loading the stability of a nitinol stent can be increased.

Besides, axial bucking issue is associated with NiTi stent design; Allie et al. [29] concerned about fractures. Percutaneous interventional treatments for superficial femoral artery (SFA) disease have long suffered from excessively high restenosis rates regardless of treatment with PTA alone or with stenting. Earlier SFA stent studies reported 1-year patency rates between 29% and 68% using the Wallstent and >50% with use of the Palmaz stent. More recent SFA stent reports using self-expanding nitinol designs have shown improved 1- to 2-year primary and secondary patency rates, but secondary reintervention rates for occlusion or in-stent restenosis (ISR) remain high (20–30% at 1 year). Stent fractures have been reported with balloon-expandable stainless steel stents and Wallstents, and nitinol stent fractures have been reported in a variety of clinical settings, including the esophagus, biliary tract, congenital heart disease, dialysis grafts, coronary arteries, iliac arteries, popliteal arteries, and subclavian arteries and veins [30–34]. Wattam et al. [30] mentioned that in a series of 66 patients who had palliation of malignant obstructive jaundice by percutaneous placement of Memotherm expanding metal stents, four cases of stent fracture were reported. Geschwind et al. [31] reported the case of a liver transplant patient who developed a biliary stricture 3 years postoperatively and was treated with an endostent. During endoscopic removal, the stent fractured, and a portion of it lodged itself within the intrahepatic portion of a portal vein branch. Internal drainage with transhepatically or endoscopically placed endoprostheses has been used for many years as a temporary or definitive treatment for biliary tract obstruction. As a late complication, stent migration may occur. Diller et al. [32] reviewed their records to identify patients who were operated on for a migrated endoprosthesis that was causing complications. In all, five such patients were identified. It was found that (i) one patient had a large bowel perforation; (ii) bowel penetration led to an interenteric fistula in one patient and to a biliocolic fistula formation in another; (iii) small bowel distension was found in two patients; (iv) surgical treatment consisted of local excision in three patients, segmental resection in one patient, and a bypass operation in the patient with biliocolic fistula; and (v) postoperatively, four patients recovered without problems, but one patient died during a complicated postoperative course, concluding that if a stent becomes stuck in the gastrointestinal tract and is not accessible for endoscopic removal, early operative revision is mandatory to prevent further complications. Knirsch et al. [33] mentioned that (i) fracture of a 12 mm Palmaz stent after implantation in the left pulmonary artery for palliation of postoperative stenosis in a 9-year-old child is described, and (ii) successful management by implantation of a second Palmaz stent revealed immediate stabilization and no signs of significant restenosis during 1-year follow-up. Zhou et al. [34] assessed the clinical features of biliary stent fractures and evaluate associated factors and reported that (i) stent fractures following Malignant Biliary Obstruction (MBO) treatment constitute a relatively rare long-term complication, and (ii) though there were no factors found to be significantly associated with self-expandable metal stents (SEMSs) fracture, a trend could be observed toward more fractures in multistent, transpapillary, and balloon dilation groups. Meir et al. [35] and Phipp et al. [36] reported a >50% stent fracture rate with subclavian vein stents placed for Paget–Schroder syndrome or exertional axillosubclavian vein thrombosis and recommend stenting cautiously and only after first rib resection. Multiple stent fractures are reported with aortic endografts for Abdominal

Aortic Aneurysm (AAA) repair, and they can be associated with clinical sequelae [37]. Dowling et al. [38] reported a complete transverse linear nitinol stent fracture in the popliteal artery at 6 months after stent deployment presenting with acute occlusion requiring surgery for limb salvage. Kroger et al. [39] have reported three cases of popliteal stent fracture; all were symptomatic and required reintervention.

Biliary stents are commonly used in the palliative treatment of obstructive jaundice from pancreatic neoplasms. Complications associated with biliary stents include bleeding, infection, obstruction, and migration, which can occur any time after placement. As we have seen earlier on stent's fractures, they may present early or late after placement and can be identified using radiographic imaging. Lohr et al. [40] described the case of an elderly man who presented with jaundice after his biliary stent fractured. A fragment migrated to his bowel and caused obstruction, requiring surgical intervention. It was concluded that (i) biliary stents are commonly used as palliative treatment for obstructive jaundice due to pancreatic neoplasms; (ii) a longer survival period increases the risk that the patient will experience complications, including stent fracture; (iii) prophylactic measures are recommended for such patients to prevent complications that could require surgical intervention; and (iv) close follow-up using radiographs, and if there are complications with the stent, cholangiography, and CT scanning, may be helpful for evaluating these patients.

17.3.2 For coronary artery disease

It is well known that across all populations (based on geographic location, race, ethnicity, age, and sex), coronary artery disease (CAD) is the single most common cause of death. The commonly performed revascularization procedures for the treatment of symptomatic CAD are percutaneous transluminal coronary angioplasty (PTCA) by itself or followed by the deployment of either a bare-metal stent or a drug-eluting stent (DES) [41]. In the latter type, a drug that is either embedded in polymeric or nonpolymeric coating(s) on the stent surface or directly attached to the stent surface elutes into the blood stream at a controlled rate over a period of time, typically 14–30 days. Over the years, there has been a steady decline in the use of PTCA and a concomitant sharp increase in the use of stents, with DESs being the predominant choice in the last 3 years. Coronary angioplasty (or called percutaneous coronary intervention) is a procedure used to open clogged heart arteries. Angioplasty uses a tiny balloon catheter that is inserted in a blocked blood vessel to help widen it and improve blood flow to your heart. Angioplasty is often combined with the placement of a small wire mesh tube called a stent. The stent helps prop the artery open, decreasing its chance of narrowing again. Most stents are coated with medication to help keep the artery open (DESs). Angioplasty can improve symptoms of blocked arteries, such as chest pain and shortness of breath. Angioplasty is also often used during a heart attack to quickly open a blocked artery and reduce the amount of damage to your heart. Figure 17.7 shows a typical nitinol stent and detailed view of coronary angioplasty [42, 43].

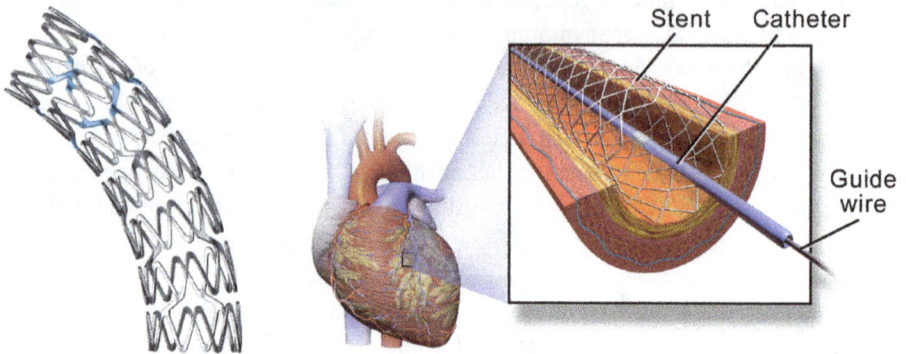

Figure 17.7: Typical stent [42] and coronary angioplasty [43].

There are reports on the amplatzer occlusion devices. Thanopoulos et al. [44] examined the efficacy and safety of the Amplatzer, a new self-centering septal occluder that consists of two round disks made of nitinol wire mesh and linked together by a short connecting waist. Sixteen patients with secundum ASD met established two-dimensional and three-dimensional (3D) echocardiographic and cardiac catheterization criteria for transcatheter closure. The Amplatzer's size was chosen to be equal to or 1 mm less than the stretched diameter. The device was advanced transvenously into a 7F long guiding sheath and deployed under fluoroscopic and ultrasound guidance. Once its position was optimal, it was released. It was reported that (i) the mean ASD diameter by transesophageal echocardiography was 14.1 ± 2.3 mm and was significantly smaller than the stretched diameter of the ASD (16.8 ± 2.4 mm), (ii) the mean device diameter was 16.6 ± 2.3 mm, and (iii) no complications were observed. After deployment of the prosthesis, there was no residual shunt in 13 (81.3%) of 16 patients, and (iv) in three patients there was trivial residual shunt immediately after the procedure that had disappeared in two of them at the 3-month follow-up. It was concluded that the Amplatzer is an efficient prosthesis that can be safely applied in children with secundum ASD; however, a study including a large number of patients and a longer follow-up period are required before this technique can be widely used. Walsh et al. [45] analyzed their experiences using the Amplatzer device in 150 patients with interatrial communications. Of these, 104 had a defect within the oval fossa, 33 a patent oval foramen, and 13 had undergone fenestration of a Fontan procedure. Of those with defects within the oval fossa, a device was implanted in 100 patients, and 2 of these patients subsequently required surgical intervention, 1 because of migration and the other because of malformation of the device. Of the remaining 98 patients, complete occlusion has been achieved in 90% at 1 year. Any residual leaks are either trivial or small. In those with a patent oval foramen, the septal occluder was used to close 20, while the device designed specifically for this purpose was used in 13. On follow-up contrast echocardiography, only two patients have a small residual right-to-left shunt. Complete occlusion was achieved for all the Fontan fenestrations, although one patient later underwent surgery for baffle dehiscence. Other significant complications occurred in two patients who developed DVT, and three patients who suffered transient supraventicular arrhythmias. It was summarized that although the Amplatzer device has been in clinical use for only 3 years, its unique design and ease of use have resulted in its widespread adoption by many centers, indicating that the results to date were encouraging, but it must be remembered that there is, as yet, no long-term follow-up data available for this lifelong implant.

Chlanda et al. [46] introduced a term "metallosis," which is defined as a potential nickel ion release from NiTi-SMAs that are believed to be biocompatible and possess the self-passivating capability. Besides, it is also well known that cardiac implants require addressing the issue of blood clotting on the surface. Treatment in glow-discharge low-temperature plasma makes it possible to produce titanium layers with structure and properties that are controlled via process parameters. In addition, antithrombogenic properties can be improved by depositing a carbon coating via the radio frequency chemical vapor deposition process. Based on these background, Chlandra et al. [46] investigated the structure, surface topography, adhesive properties, wettability, and surface free energy, and evaluated metallosis after producing TiO_2 and a-C:N:H + TiO_2 composite layers on NiTi alloy. It was reported that (i) the produced surface layers are capable of significantly reducing metallosis; (ii) in contrast to NiTi in its initial state, layers of nanocrystalline TiO_2 (rutile) with a homogeneous structure demonstrate greater adhesion strength and more developed surface in the microscale, which facilitates the formation of an a-C:N:H coating, indicating that the formation of a coating of a-C:N:H amorphous carbon on NiTi alloy that has previously been oxidized in low-temperature plasma may prove to be a favorable solution in terms of using NiTi alloy to produce cardiac implants.

Using the similar concept, there is a wide variety of stents used for different purposes, from expandable coronary, vascular, and biliary stents, to simple plastic stents used to allow the flow of urine between the kidney and the bladder. The term "stent" is also used as a verb to describe the placement of such a device, particularly when a disease such as atherosclerosis has pathologically narrowed a structure such as an artery.

17.3.3 Stents for artery

Tyagi et al. [47] evaluated the use of self- and balloon-expandable stents in patients with suboptimal response to balloon angioplasty (BA). Twenty-one hypertensive patients (aged 18–61 years; mean, 28.6 ± 11.2 years) with native CoA (coarctation of the aorta) and in whom results of BA were suboptimal [namely, residual peak systolic gradient (PSG) >20 mm Hg] underwent stent implantation. Balloon-expandable Palmaz stents were implanted in five patients (group A) and self-expandable nitinol aortic stents in the remaining 16 patients (group B). It was reported that (i) in group A, PSG decreased from 62.8 ± 10.6 (53–80) mm Hg to 28.1 ± 6.3 (22–39) mm Hg after BA; (ii) systolic gradient further decreased to 8.3 ± 3.9 (2–16) mm Hg after implantation of the balloon-expandable Palmaz stent, and (iii) in group B, PSG decreased from 70.2 ± 24.6 (40–110) mm Hg to 28.4 ± 9.8 (22–42) mm Hg after BA and further reduced to 9.0 ± 5.5 (4–16) mm Hg. One of these patients had a nitinol self-expandable stent implanted after a Palmaz stent embolized immediately after deployment. Nitinol stents were easier to deploy and conformed better to aortic anatomy compared with balloon-expandable stents. It was further mentioned that in group A, the diameter of the coarcted segment increased from 3.8 ± 0.8 mm to 13.3 ± 0.8 mm after stent implantation and in group B it increased from 4.5 ± 1.1 mm to 14.1 ± 2.1 mm, and none of the patients showed aneurysm formation. Hence, it was concluded that (i) stent implantation is safe and effective in improving suboptimal results after BA for CoA, and (ii) self-expandable stents were easier to implant, adapted better to the wall of the aorta, and most of the patients had similar efficacy in reducing coarctation as balloon-expandable stents. Hausegger et al. [48] determined the effectiveness of a prototype nitinol stent in the iliac arteries. Fourteen patients with arteriosclerotic lesions of the iliac arteries (nine stenoses, five occlusions) were treated percutaneously with a prototype nitinol vascular stent (Cragg stent) after unsuccessful percutaneous transluminal angioplasty. It was reported that in 13 patients available for follow-up, the mean ankle–brachial index (ABI) increased from 0.4 ± 0.32 (standard deviation) before the procedure to 0.81 ± 0.27 after the procedure. After 6, 12, and 24 months, the mean ABI was 0.97 ± 0.2, 0.87 ± 0.15, and 0.89 ± 0.1, respectively. In two patients, radial stiffness of the stent was too low to completely eliminate a high-grade, calcified stenosis; however, long-term patency of the vessel was preserved. In all other patients, stent placement achieved good vascular reconstruction, and no stent occlusion or restenosis was observed, concluding that the results with the Cragg stent were similar to those with other commercially available stents. Nematzadeh et al. [49], using the finite element method, evaluated the effects of material properties on the mechanical performance of the new geometry designed for the Z-shaped open-cell femoral artery self-expanding stent, made of nitinol wire, by application of crushing force. The behavior of the stents, having two sets of properties, was compared. It was mentioned that the stents with higher A_F temperature show better clinical behavior due to lower chronic outward force, higher radial resistive strength, and more suitable superelastic behavior. Model calculations show that a large change of A_F temperature could exert a substantial effect on the practical performance of the stent.

Despite the improvements afforded by intracoronary stenting, restenosis remains a significant problem. The optimal physical properties of a stent have not been defined. Therefore, Carter et al. [50] compared the vascular response to a thermoelastic self-expanding nitinol stent with a balloon-expandable tubular slotted stainless steel stent in normal porcine coronary arteries. Twenty-two stents (11 nitinol and 11 tubular slotted) were implanted in 11 miniature swine. The nitinol stents were deployed using the intrinsic thermal properties of the metal, without adjunctive balloon dilation. The tubular slotted stents were implanted using a noncompliant balloon with a mean inflation pressure of 12 atm. Intravascular ultrasound and histology were used to evaluate the vascular response to the stents. Based on the obtained results, it was concluded that (i) a thermoelastic nitinol stent exerts a more favorable effect on vascular remodeling, with less neointimal formation, than a balloon-expandable design; (ii) progressive intrinsic stent expansion after implant does not appear to stimulate neointimal formation and, therefore, may provide a mechanical solution to prevent ISR. Acute and subacute stents thrombosis along with thrombus-mediating neointimal proliferation within the stent struts remain major concerns in coronary stenting. Up to date, there is an obvious lack of data on the thrombogenicity of stent materials in physiological conditions. Accordingly, Thierry et al. [51] compared the thrombogenicity of TiNi versus stainless steel stents in an ex vivo shunt porcine model and the results showed that TiNi stents represent only a few of thrombus mainly located at the strut intersections while stainless steel stents clearly exhibit more thrombus. It was reported that (i) scanning electron observations showed different thrombus morphologies for nitinol and stainless steel, and (ii) along with the unique mechanical properties of nitinol, its promising hemocompatibility demonstrated in the study may promote their increasing use for both peripheral and coronary revascularization procedures.

17.3.4 Stents for colorectal and bowel obstruction, and gastrointestinal strictures

Bonin et al. [52] mentioned that the gastrointestinal SEMS placement is an effective endoscopic technique as a nonsurgical approach in the treatment of malignant colonic obstruction. Up to 20% of patients with colorectal cancer (CRC) experience an acute symptomatic obstruction, and (https://www.hindawi.com/journals/tswj/2014/651765/ – B3) considering the high incidence of CRC, SEMS placement has to be done with a valid endoscopic treatment not only in patients with incurable malignant obstruction but also as a bridge to surgery [53, 54]. Di Mitri et al. [55] described that SEMS placement can be a rescue therapy in those with acute severe colonic obstruction having significant improvement of overall quality of life (QOL) and QOL related to gastrointestinal symptoms. It was further mentioned that 20–30% of patients experience early or late SEMS-related complications, such as migration, obstruction, or perforation [56–58]. Chun et al. [15–59] mentioned that advances in stent design have led to a substantial increase in the use of stents for a variety of malignant and benign strictures in the gastrointestinal tract and biliary system. The majority of contemporary stents are SEMSs that are composed of either nitinol or stainless steel, and these stents are able to exert an adequate expansile force and, at the same time, are highly flexible and biocompatible. And it was also described that the advent of new nickel–titanium (nitinol)-conformable stents with an SMA has improved the treatment of malignant biliary and gastrointestinal strictures.

SEMS are a nonsurgical option for treatment of malignant colorectal obstruction and also as a bridge to surgery approach. The new nitinol-conformable stent has improved clinical outcomes in these kinds of patients. Nagula et al. [60] reported a pilot experience on the use of nitinol SEMS as a bridge to surgery treatment in patients with colorectal obstruction. Between April and August 2012, data on colonic nitinol-conformable SEMS placement in a cohort of consecutive symptomatic patients with malignant colorectal obstruction were collected, who were then treated as a bridge to surgery. Technical success, clinical success, and adverse events were recorded. Ten patients (7 male: 70%), with a mean age of 69.2 ± 10.1, were evaluated. The mean length of the stenosis was 3.6 ± 0.6 cm. Five patients (50%) were treated on an emergency basis. The median time from stent placement to surgery was 16 days (interquartile range 7–21). It was reported that technical and clinical success was achieved in all patients with a significant early improvement of symptoms, and no adverse events due to the SEMS placement were observed. It was concluded that the pilot study confirmed the important role of nitinol-conformable SEMS as a bridge to surgery option in the treatment of symptomatic malignant colorectal obstruction. CRC is the second most prevalent cancer in the world with incidence of one million new cases per year and mortality of about 529,000 deaths [61]. In advanced stages, the tumor tends to grow inside the colon lumen, causing stenosis that may block the passage of stool. Obstruction has been reported in 7–29% of patients with CRC [54]. Patients with malignant large-bowel obstruction tend to have advanced disease and be poor surgical candidates. The traditional method of managing complete or subtotal cancer colonic obstruction is surgical, but in the emergency setting, surgery carries a high mortality (15–20%) and high morbidity (45–50%) with increased prevalence of intensive care stay, infections, and complications related to stomas [62]. Moreover, even when the tumor is surgically resectable, after the tumor excision, the surgeon should make a temporary colostomy, because a dirty colon cannot be shunted. Therefore, the initial surgery (resection of the primary tumor and colostomy) must be followed by a second intervention to perform the intestinal anastomosis sometime after. To avoid these disadvantages, endoscopic placement of SEMS to relieve colonic obstruction has been introduced [63]. Puértolas et al. [64] studied the behavior and functionality of a bell-shaped colonic self-expandable NiTi stent by the finite element simulation. Catheter introduction, releasing at position, and the effect of peristaltic wave were simulated. To check the reliability of the simulation, a clinical experimentation with porcine specimens was carried out. It was reported that (i) the stent presented a good deployment and flexibility, and (ii) stent behavior was excellent, expanding from the very narrow lumen corresponding to the maximum peristaltic pressure to the complete recovery of operative lumen when the pressure disappears. Tack et al. [65] evaluated the use of a new type of self-expandable nitinol stent in the palliation of rectosig-moidal carcinoma. In 10 patients with advanced obstructing rectosigmoidal carci-noma, initial Nd:YAG laser treatment was performed if necessary to allow passage of a gastroscope. Subsequently, a self-expanding nitinol stent with flanged ends was inserted under combined fluoroscopic and endoscopic control. Endoscopic and clin-ical follow-up was carried out at regular intervals. It was reported that (i) after 2 ± 0.4 sessions of initial laser therapy, minimal lumen diameter was 9 ± 1 mm; (ii) stent insertion was successful in nine patients, increasing minimal lumen diameter to 14 ± 1.2 mm; (iii) in one patient, stent deployment was complicated by a sigmoid perfo-ration, requiring surgery; (iv) after insertion, colorectal stents remained adequately positioned and free of obstruction for 103 ± 31 days; and (v) patient survival after stent placement was 204 ± 43 days. Stent migration occurred in three patients, after 38 ± 10 days, and obstruction of the stent because of tumor ingrowth was observed after 268 days in only one patient. It was hence concluded that insertion of self-expandable nitinol stents in patients with rectosigmoidal carcinoma is technically feasible, and metallic stents are effective in the palliation of malignant rectosigmoid obstruction. They provide an alternative to repeated palliative laser therapy or palli-ative surgery.

There are studies on the clips for colonic anastomosis. Nudelman et al. [66] reported their successful experience with a compression anastomosis clip (CAC) in an animal model. The study sample included 20 patients scheduled for colonic resections, of whom 10 underwent anastomosis with the CAC and 10 with staplers. It was reported that neither group had anastomotic or other complications, except for one patient in the CAC group in whom a subphrenic infected hematoma developed after left hemicolectomy with splenectomy, concluding that the CAC is safe and simple to use, coming close to the "no touch surgery concept" and is of low cost compared with the staples used today. Song et al. [67] investigated the design and performance of a shape-memory colonic anastomosis clip (CAC). The thermomechanical proper-ties of the SMA material were studied and the data were used to derive a nonlinear material model. This enabled the development of computer-aided design models and finite element analysis of the clip and tissue compression. It was reported that the maximum strain of the anastomosis clip was within the recoverable range, and it exerted parallel compression of the colonic walls with a uniform pressure distribu-tion. The design of the anastomosis clip was optimized for safe, simple, and effective use in colon surgery.

17.3.5 Stents for ureteropelvic junction

Ureteropelvic junction (UPJ) obstruction is a functional or anatomic obstruction to urine flow from the renal pelvis to the ureter that, if left untreated, results in symptoms or renal damage. Most of the patients with UPJ obstruction are diagnosed in the perinatal period by widespread use of antenatal ultrasound; however, diagnosis is sometimes delayed until adulthood, when lumbar pain, infection, stones, and hematuria might occur [68]. UPJ obstruction was traditionally repaired surgically by open pyeloplasty [69]. Shalhav et al. [70] evaluated experiences with endopyelotomy for UPJ obstruction by stratifying the results of an antegrade versus a retrograde approach for primary, secondary, calculi-related, high insertion, and impaired renal function-related obstruction, individually. It was concluded that antegrade endopyelotomy is the preferred approach in patients with primary UPJ obstruction and concomitant renal calculi (13.4% of cases), and may also be preferable in patients with high insertion obstruction (6.7%). For all other primary and secondary UPJ obstruction, antegrade and retrograde endopyelotomy is an effective therapy, yet retrograde endopyelotomy results in less operating room time, shorter hospital stay, fewer complications, and significantly less expense to achieve the desired outcome. Although the success rate of the open surgical approach is reported to be 72–98% [71], up to one-quarter of patients require at least one additional intervention. For rare patients, where multiple surgical attempts have failed and symptoms of UPJ obstruction persist, a viable alternative is double-J catheter placement with replacement at regular time intervals. In the 1990s, permanent implantable metallic mesh stents were introduced to the treatment of ureteric stenosis [72], which promised to offer a permanent solution; however, their use in patients with expected long-term survival was largely abandoned due to very high rate of stone incrustations and difficulty of their removal when complications emerged. Three types of urethral stents can be used in the treatment of recurrent strictures: Urolume/Wallstent is a self-expanding mesh, which is incorporated into the urethral epithelium; the ASI titanium stent is a short rigid mesh of titanium wire which is also incorporated into the urethra; nitinol is a flexible spring in one or two parts connected by a steel wire. Yachia [73] inserted 65 Niticol UroCoil system stents in 56 patients and the stent was removed after one year in 41 patients. It was reported that the use of these stents has considerably decreased the number of repeated dilatations and urethrotomies. Mori et al. [74] placed a new urethral stent made of SMA to relieve prostatic obstruction in 17 patients in whom other approaches were contraindicated. All patients were unable to tolerate intervention with sedation and positioning. Placement of the SMA stent mounted on a 16F Foley catheter is similar to insertion of a urethral catheter except for the heat-sensitive expansion. It was found that each device was easily implanted with the patient on a flat examination table, and there was no migration or incrustation of the SMA stent during the indwelling period, indicating that clinical results demonstrate that the SMA urethral stent might be the only choice for management of prostatic obstruction in debilitated patients. Smrkolj et al. [69] reported a rare case of a patient with a large stone encrusted on a nitinol mesh stent in the UPJ. The stent was inserted in the year 2000 after failure of two pyeloplasty procedures performed due to symptomatic UPJ stenosis. By combining minimally invasive urinary stone therapies – extracorporeal shock wave lithotripsy, semirigid ureterorenoscopy with laser lithotripsy, and percutaneous nephrolithotomy – it was possible to completely remove the encrusted stone and nitinol mesh stent that was implanted for 15 years, rendering the patient symptom and obstruction free. Burgos et al. [75] evaluated the safety and efficacy of nitinol stents and the Detour extra-anatomical ureteral bypass graft in treatment of ureteral stenosis after kidney transplantation. Eighteen kidney transplant recipients with complex stenosis caused by failure of primary treatment or with high surgical risk or a poorly functioning graft (serum creatinine concentration >2.5 mg/dL) were treated using antegrade percutaneous implantation of nitinol stents (n = 16) or extra-anatomical ureteral bypass grafts (n = 3); one patient was treated with both techniques. It was found that mean (range) follow-up of ureteral stents was 51.2 (3–118) months, and patency rate at last follow-up, resumption of dialysis therapy, or death was 75% (12 of 16 patients). In four patients (25%), stent occlusion developed, which was treated using a double-J catheter in two patients, stent removal and pyeloureterostomy using the native ureter in one patient, and implantation of an extra-anatomical bypass graft in one patient. The mean follow-up in patients with extra-anatomical ureteral bypass grafts was 32 (8–64) months. One patient developed a urinary tract infection, and another had encrustation with obstruction. Based on these results, it was concluded that the use of nitinol ureteral stents and extra-anatomical ureteral bypass grafts is a safe and effective alternative to surgery for treatment of postkidney transplantation ureteral stenosis in patients with chronic graft dysfunction, those at high surgical risk, and those in whom previous surgical treatment has failed.

17.3.6 Stents for biliary obstruction

SEMSs are claimed to prolong biliary stent patency, although no formal comparative trial between plastic and expandable stents has been done. In a prospective randomized trial, Davids et al. [76] assigned 105 patients with irresectable distal bile duct malignancy to receive either a metal stent (49) or a straight polyethylene stent (56). It was reported that median patency of the first stent was significantly prolonged in patients with a metal stent compared with those with a polyethylene stent (273 vs 126 days). The major cause of stent dysfunction was tumor ingrowth in the metal stent group and sludge deposition in the polyethylene stent group. Treatment after any occlusion included placement of a polyethylene stent. In the metal stent group none of 14 second stents occluded, whereas 11 of 23 (48%) second stents clogged in the polyethylene stent group. The overall median survival was 149 days and analysis showed that the initial placement of a metal stent results in a 28% decrease of endoscopic procedures. SEMSs have a longer patency than polyethylene stents and offer adequate palliation in patients with irresectable malignant distal bile duct obstruction.

17.3.7 Intratracheal stents

Temporary or permanent tracheal splinting in pediatric patients may be indicated in tracheomalacia or bronchomalacia, repair of congenital tracheal stenosis, and after tracheal resection. Vinograd et al. [77] indicated the results of the development of a new intraluminal airway stent made from nitinol alloy. The stent, connected to a small electric power supply, was introduced into 20 young rabbits with the use of a 2.5 cm rigid bronchoscope. After implantation in the martensitic state, the stent was warmed to 40 °C, and the austenitic state by an electric current of 1.5–3 A for 1–2 s. After a period of 8–10 weeks, the stent was removed (in its martensitic state) through the same sized bronchoscope after being cooled with 3–4 mL of 80% alcohol solution at 6 °C. It was reported that no signs of airway obstruction developed in any of the animals after implantation or extraction of the stent. The biomechanical properties of the trachea, as shown by strain measurements with the use of incremental forces, showed significant differences between the stented and unstented segments. The titanium alloy intratracheal stent adequately fulfilled the requirements of a temporary intraluminal airway splint, and because of its unique feature of SME the stent could be inserted, fixed, and removed easily, even in very small airways. A self-expanding nitinol stent was used in two patients with inoperable tracheal stenosis due to invasive malignant tumor of the trachea [78]: one was a 70-year-old man with recurrent tumor from adenocarcinoma of the left lung, and the other was a 63-year-old man with recurrent tumor in mediastinal lymph nodes from esophageal cancer. It was mentioned that the self-expanding nitinol stent was very useful and effective in inoperable tracheal stenosis due to intraluminal tumor invasion.

17.3.8 Stents for esophageal strictures

The reaction of the normal esophageal wall to inserted self-expanding nitinol stents was studied in pigs [79]. An inflammatory reaction with increasing fibrotic activity and degeneration of the muscular layers in the esophageal wall was demonstrated. Five patients with severe dysphagia secondary to benign esophageal strictures also underwent insertion of self-expanding nitinol stents. It was reported that (i) all of the stents expanded completely with subsequent regression of dysphagia; (ii) one treated esophagus was resected and showed deep implantation of the stent meshwork in the esophageal wall; and (iii) significant stenoses secondary to tissue hyperplasia, located at the edges of the stent, occurred in two patients, indicating that (iv) self-expanding nitinol stents may be used for palliation of dysphagia in patients with benign esophageal strictures, and (v) because of the observed reactions in the esophageal wall, such treatment should be restricted to selected patients until more experience has been gained. Pocek et al. [80] evaluated technical and clinical results of self-expanding esophageal stent implanted in patients with malignant esophageal strictures and clinically significant dysphagia. From June 1992 to September 1994, 27 patients with inoperable tumors of the esophagus or gastric cardiac were treated by placement of 37 self-expanding nitinol stents. Water-soluble contrast and endoscopy studies were performed after the procedure and during the follow-up period. It was reported that (i) successful stenting of the stricture was achieved in 27 patients; (ii) the mean dysphagia grade dropped from 2.3 to 1 (SD ± 0.54) immediately after the procedure; (iii) after the insertion of the stent, 16 patients died in a period of time ranging from 0 to 13 months (mean 5.6 months), whereas at the end of the study 11 patients were alive 4–15 months after the procedure (mean 8.3 months); and (iv) no major complications were observed, which indicate that the esophageal stent placement was technically easy and clinically effective. Watanabe et al. [81] developed an artificial esophagus simulating peristaltic movement with the use of a NiTi-SMA actuator. Serial pairs of NiTi-SMAs were placed around a Gore-Tex vascular graft in a helical position such that they obliterated the lumen of the vessel when they contracted. In an animal experiment using a goat, the cervical esophagus was resected over a length of approximately 20 cm. The artificial esophagus was anastomosed with the remaining cervical esophagus. When a direct current of 500 mA at 5 V was applied to the NiTi-SMAs, the first pair of the NiTi-SMA contracted. The following pairs of the NiTi-SMAs contracted consecutively. The entire contraction of the artificial esophagus was similar to the esophageal peristaltic movement observed by x-ray examination in humans. Hence, there is the possibility that the artificial esophagus could function as an artificial esophagus having peristaltic movement.

17.3.9 Stents for otosclerosis

The limitations of manual prosthesis crimping in hearing restoration surgery for otosclerosis are thought to have a key role in the occurrence of incomplete postoperative elimination of conductive hearing loss and postoperative recurrences of conductive hearing loss. To eliminate manual crimping, that is, self-crimping, Rajan et al. [82] introduced SMA nitinol stapes piston in nine otosclerosis patients. The results were compared with those in a database of surgeries performed with conventional titanium pistons. The effects of the self-crimping nitinol prosthesis on the postoperative elimination of conductive hearing loss and its postoperative variations were investigated. It was reported that (i) the variations of postoperative residual conductive hearing loss were significantly smaller and the extent of conductive hearing loss elimination is greater in the nitinol group; (ii) the mean postoperative residual conductive hearing loss was smaller in the nitinol group; and (iii) the postoperative stability of conductive hearing loss elimination was similar in both patient groups, suggesting that the self-crimping SMA nitinol staples piston overcomes the drawbacks of manual crimping in hearing restoration surgery for otosclerosis.

Closing this section, it should be noted that a stent should be differentiated from a shunt. A shunt is a tube that connects two previously unconnected parts of the body to allow fluid to flow between them. Stents and shunts are made of similar materials but they perform two different tasks [83].

17.4 Orthopedic applications

17.4.1 Spinal vertebra spacer

NiTi-SMAs possess a variety of orthopedic applications. The spinal vertebra spacer is one of them to surgically correct the scoliosis problem [84]. Sanders et al. [84] treated in vivo six goats with experimental scoliosis instrumented with 6 mm nitinol rods. The rods were transformed, and the scoliosis corrected in the awakened goats by 450 kHz radio frequency induction heating. The curves averaged 41° before instrumentation, 33° after instrumentation, and 11° after rod transformation. It was reported that the animals tolerated the heating without discomfort, neurologic injury, or evidence of thermal injury to the tissues or the spinal cord. Then the in vitro test with nitinol rods was conducted under both constant deflection and constant loading conditions and plotted temperature versus either force or displacement. The 6 mm rod generated forces of 200 N, and the 9 mm rod generated up to 500 N. It was mentioned that the use of SMAs allows continuous neurologic monitoring during awake correction, true rotational correction by rod torsion, and the potential option of periodic correction to take advantage of spinal viscoelasticity and the potential of true rotational correction by rod torsion, indicating that the use of a shape-memory spacer permits the application of a constant load regardless of the position of the patient, who preserves some degree of motion. Wever et al. [85] investigated the biocompatibility and functionality of a new scoliosis correction device based on the properties of the shape-memory metal NiTi alloy. With this device, the shape recovery forces of a shape-memory metal rod are used to achieve a gradual 3D scoliosis correction. In the experimental study, the action of the new device was inverted: the device was used to induce a scoliotic curve instead of the corrected one. Surgical procedures were performed in six pigs. An originally curved squared rod, in the cold condition, was straightened and fixed to the spine with pedicle screws. Peroperatively, the memory effect of the rod was activated by heating the rod to 50 °C by a low-voltage, high-frequency current. After 3 and 6 months the animals were sacrificed. The first radiographs, obtained immediately after surgery, showed in all animals an induced curve of about 40° Cobb angle – the original curve of the rod. This curve remained constant during the follow-up. The postoperative serum nickel measurements were around the detection limit and were not significantly higher compared to the preoperative nickel concentration. It was reported that (i) macroscopic inspection after 3 and 6 months showed that the device was almost overgrown with newly formed bone; (ii) biocorrosion and fretting processes were not observed; (iii) histologic examination of the sections of the surrounding tissues and sections of the lung, liver, spleen, and kidney showed no evidence of a foreign body response; and (iv) in view of the initiation of the scoliotic deformation, it is expected that the shape-memory metal-based scoliosis correction device also has the capacity to correct a scoliotic curve. Therefore, the new device will show good biocompatibility in clinical application. Figure 17.8 shows three groups of spine structures (C-group, T-group, and L-group) [86], and Figure 17.9 depicts cervical plating system [87] and inserted spinal vertebra spacers [88].

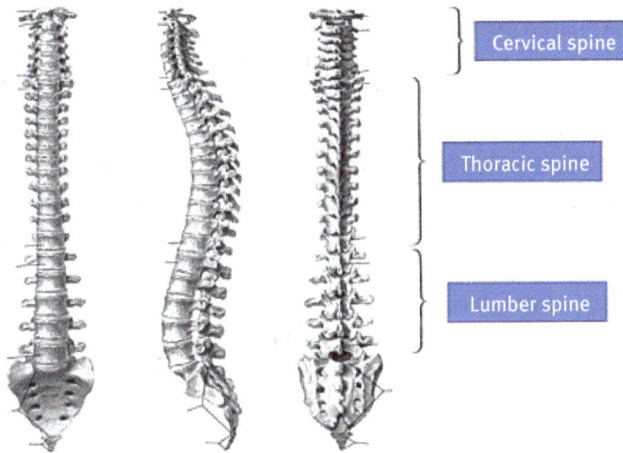

Cervical spine

Thoracic spine

Lumber spine

Figure 17.8: Side, front, and back views of C-series, T-series, and L-series of spine structures [85].

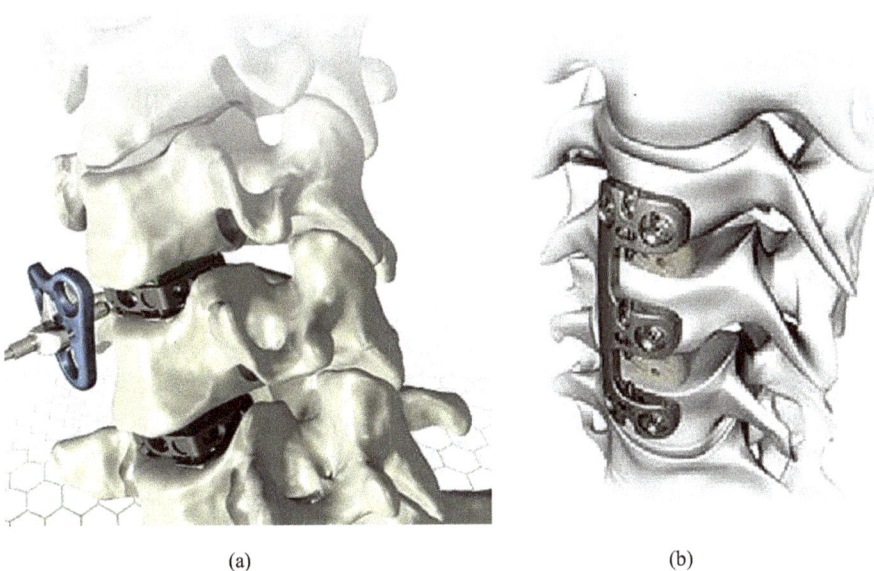

(a)   (b)

Figure 17.9: (a) Cervical plating system [87] and (b) inserted spinal vertebra spacers [88].

17.4.2 Orthopedic bone plate

Apart from biological factors, the bone healing process strongly depends on the mechanical properties of the implant applied. The stiffness of the implant influences the micromotion, one of the most important stimuli of the healing process within the fracture gap. Pfeifer et al. [89] mentioned that both the "ideal" stiffness of an implant and the alteration of the mechanical properties over the course of healing are still a topic of research. In order to adapt the implantation to the actual healing situation, NiTi SME alloy (SMA)-based implants that exhibit a variable middle piece were developed. Due to contactless induction heating, the transition temperature of the SMA is reached and triggers the one-way SME, leading to a slight modification of the second moment of inertia and to an adaption of the bending stiffness of the implant, respectively. Currently, the implant design is being adapted in order to meet the requirements for human applications. Custom-made NiTi-SMA with a transition temperature (A_F) of around 45–50 °C could continue to reduce the impact of the heating process and expand this concept to further medical applications, for example, compression, anchoring, or expansion of implants or devices [91]. Diaphyseal fractures of the radius and ulna present specific problems not encountered in the treatment of fractures of the shafts of other long bones. The adaptive modular implants based on smart materials represent a superior solution in the osteosynthesis of the fractured bones over the conventional implants known so far [90]. The small sizes of the modules enable the surgeon to make small incisions, using surgical techniques minimally invasive, having the following advantages: reduction of soft tissues destruction; eliminating intra-operator infections; reduction of blood losses; the reduction of infection risk; and the reduction of the healing time. All these concerns can be satisfied by nitinol devices. Figure 17.10 shows a typical NiTi orthopedic bone plate [90] and demonstrates application of plates and screws on fractured cadaver mandibular jaw [1].

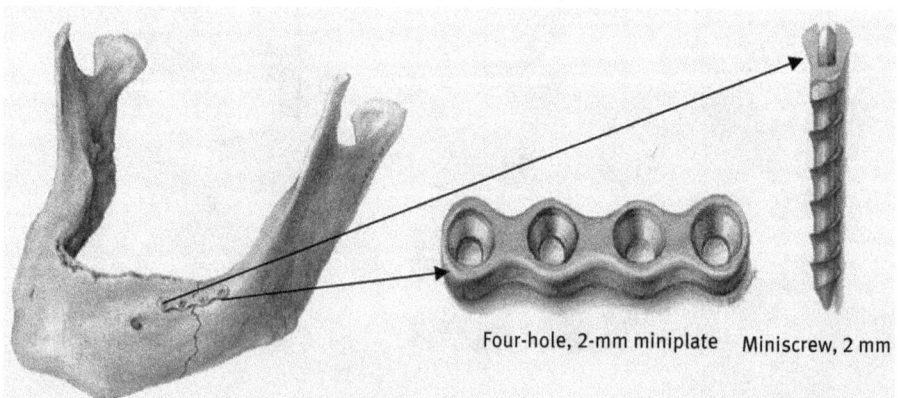

Four-hole, 2-mm miniplate   Miniscrew, 2 mm

Figure 17.10: Typical nitinol orthopedic bone plate [90] and application to broken cadaver mandibular jaw [1].

17.4.3 Compression staples

The field of orthopedics is a constantly evolving discipline. Despite the historical success of plates, pins, and screws in fracture reduction and stabilization, there is a continuing search for more efficient and improved methods of fracture fixation. Another application in the orthopedic area is related to the healing process of broken and fractured bones. Several types of shape-memory orthopedic staples are used to accelerate the healing process of bone fractures, exploiting the SME. The shape-memory staple, in its opened shape, is placed at the site where one desires to rebuild the fractured bone. Through heating, this staple tends to close, compressing the separated part of bones. It should be pointed out that an external device performs this heating, and not the temperature of the body. The force generated by this process accelerates healing, reducing the time of recovery. Nitinol bone staples are generally inserted through a process of being chilled, opened, and inserted into pre-drilled holes. Upon insertion, they recover their preprogrammed shape either through springing back after removal of a constraint (using superelasticity) or thermal triggering (using thermal shape memory). In general, as appointed out by Russell et al. [91], nitinol bone staples fall into one of three categories: (1) those that are superelastic at room temperature, (2) those that recover their shape upon heating to the body temperature, or (3) those that recover their shape upon heating above the body temperature with the application of an external heat source. These three different design approaches – room temperature superelastic, body temperature activated, and heat activated – have different performance characteristics. A version of the heat-activated staple that uses a controlled heat source appears to have the best combination of clinical forces and procedural control. Figure 17.11 shows a typical nitinol compression staples, which are applied to the carpal bone joint portion [92].

There are numerous reports on successful applications of nitinol compression staples. Dai et al. [93] mentioned that although the management of intra-articular fractures remains difficult, shape-memory compression staples fulfil nearly all the special treatment requirements of intra-articular fractures. It was reported that (i) early bone union and 93.5% satisfactory function were achieved in a series of 121 cases, and (ii) stable fixation, "early" movement, and continuous compressive force produced by the staple are the main factors contributing to good results. Laster et al. [94] reported that a simple method using a staple was successfully used to treat a 74-year-old woman with a fractured frontozygomatic suture.

In order to control the localized corrosion of NiTi alloy (in other words, Ni element dissolution from the substrate), Villermaux et al. [95] employed plasma-polymerized tetrafluoroethylene (PPTFE) coating to improve the corrosion resistance of NiTi SMA plates and corresponding NiTi stables. It was reported that (i) the scratch test indicates a good surface adhesion of the film but that it lacks cohesiveness, (ii) electrochemical potentiodynamic tests in physiological Hank's solution show that PPTFE coating improved the pitting corrosion resistance, (iii) the passivation range

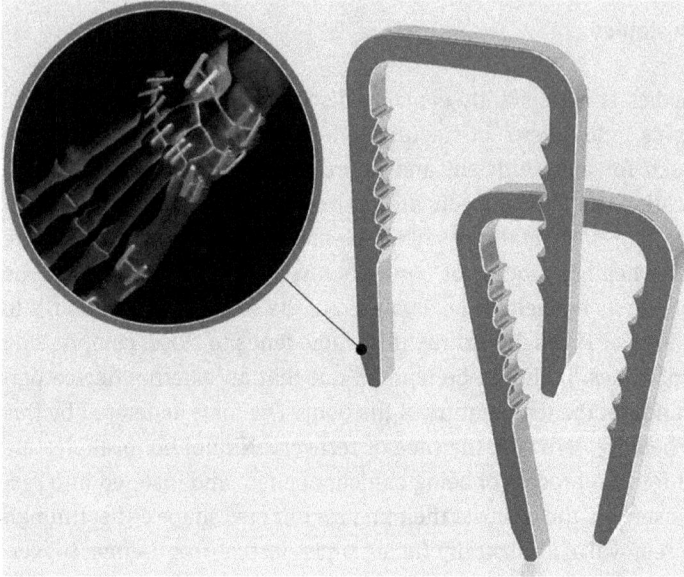

Figure 17.11: Nitinol compression staples, applied to carpal bone joint portion [92].

is increased from 35% to 96% compared to the untreated sample and the pit diameter is decreased from 100 to 10 μm, (iv) the uniformity of the deposited film is a very important parameter, and (v) when the film is damaged, the corrosion seems to increase in comparison to the untreated samples, suggesting that, if the staple is carefully manipulated, the coating follows the large deformations induced by the memory effect of the alloy without cracking, and then protects efficiently the staple from pitting. Mei et al. [96] evaluated the biomechanical effect and clinical application of the shape-memory expansion clamp. The study involves three phases: a clinical study of 30 patients, a biomechanical study to assess the expansion force of a recovering shape-memory expansion clamp, and a biomechanical study using cadaveric specimens to assess the pullout strength of the shape-memory expansion clamp as a function of the shape of the clamp. The major complication of anterior cervical decompression and fusion was graft dislodgement and pseudarthrosis. Improvement of the shape of bone graft and fusion technique could not eliminate the problem completely. The authors designed the shape-memory expansion clamp by using the characteristic of NiTi SMA and first applied it in an anterior cervical operation. The expansion force to recover the shape-memory expansion clamp was measured in a biomechanical study. Eight fresh human cadaver cervical spine specimens were used to assess the pullout strength of the shape-memory expansion clamp as a function of the shape of the clamp. The shape-memory expansion clamp had been used to fix rotated circular grafts in 30 cases with cervical spine injuries or cervical spondylosis after anterior decompression. It was obtained that (i) the expansion force of the shape-memory expansion clamp in recovery was 4.65–27.96 N, and the pullout strength was 8.82–20.58 N; (ii) bone fusion was achieved in all 30 cases; and (iii) dislocation of bone graft, loosening of clamp, or kyphotic deformity had not occurred. Based on these results, it was concluded that the expansion force and the pullout strength of the partially recovered shape-memory expansion clamp were determined biomechanically, and the clinical study demonstrated that the shape-memory expansion clamp is safe and effective. Hoon et al. [97] evaluated shape-memory staples and compared them to a currently used implant for internal fracture fixation. Multiplane bending stability and interfragmentary compression were assessed across a simulated osteotomy using single- and double-staple fixation and compared to a bridging plate. Transverse osteotomies were made in polyurethane blocks (20 × 20 × 120 mm), and repairs were performed with one ($n = 6$) or two ($n = 6$) 20 mm nitinol staples, or an eight-hole 2.7 mm quarter-tubular plate ($n = 6$). A pressure film was placed between fragments to determine the contact area and compressive forces before and after loading. Loading consisted of multiplanar four-point bending with an actuator displacement of 3 mm. Gapping between segments was recorded to determine loads corresponding to a 2 mm gap and residual postload gap. It was found that staple fixations showed statistically significant higher mean compressive loads and contact areas across the osteotomy compared to plate fixations. Double-staple constructs were superior to single-staple constructs for both parameters, and constructs were significantly stiffer and endured significantly larger loads before 2 mm gap formation compared to other constructs in the dorsoventral plane; however, both staple constructs were significantly less stiff and tolerated considerably lower loads before 2 mm gap formation when compared to plate constructs in the ventrodorsal and right-to-left lateral loading planes. Loading of staple constructs showed significantly reduced permanent gap formation in all planes except ventrodorsally when compared to plate constructs. It was therefore concluded that although staple fixations were not as stable as plate fixations in particular loading planes, double-staple constructs demonstrated the most consistent bending stiffness in all planes, and placing two perpendicular staples is suggested instead of single staples whenever possible, with at least one staple applied on the compression side of the anticipated loading to improve construct stability. Overall, nitinol compression implants are fast and simple to insert and have a high radiographic union rate for midfoot and hindfoot arthrodesis, and applications of nitinol technology in orthopedic surgery are rapidly expanding with the improved and broadened portfolio of implants available [98].

17.4.4 Intramedullary nail

Intramedullary (IM) nailing is surgery to repair a broken bone and keep it stable. The most common bones fixed by this procedure are the thigh, shin, hip, and upper arm. A permanent nail or rod is placed into the center of the bone. It will help a patient be able to put weight on the bone. An IM rod, also known as an IM nail or inter-locking nail or Küntscher nail (without proximal or distal fixation), is a metal rod forced into the medullary cavity of a bone. IM nails have long been used to treat fractures of long bones of the body. IM nails resulted in earlier return to activity for the soldiers, sometimes even within a span of a few weeks, since they share the load with the bone, rather than entirely supporting the bone.

Figure 17.12 shows a proximal femur nail with locking and stabilization screws for treatment of femur fractures of left thigh [99]. Zelle et al. [100] suggested placement of interlocking screws as an important part of the IM nailing procedure. While percutaneous screw placement is typically safe, surgeons need to be aware of the surrounding soft tissue structures at risk. For most tibial shaft fractures two proximal and two distal interlocking screws provide sufficient stability. Proximal and distal third tibial fractures may benefit from placement of additional interlocking screws in different planes in order to increase the stability of the construct (as seen in Figure 13 [100], demonstrating uneventful healing).

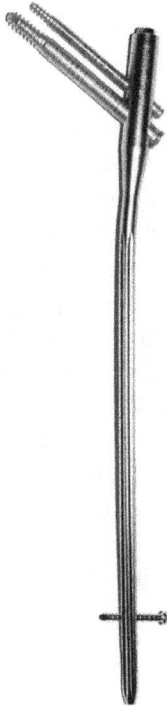

Figure 17.12: Intramedullary nail with stabilizing screws [99].

Kujala et al. [101] examined whether if bone modeling can be controlled with a functional IM NiTi nail. Preshaped IM NiTi nails (length 26 mm, thickness 1.0–1.4 mm) with a curvature radius of 25–37 mm were implanted in the cooled martensite form

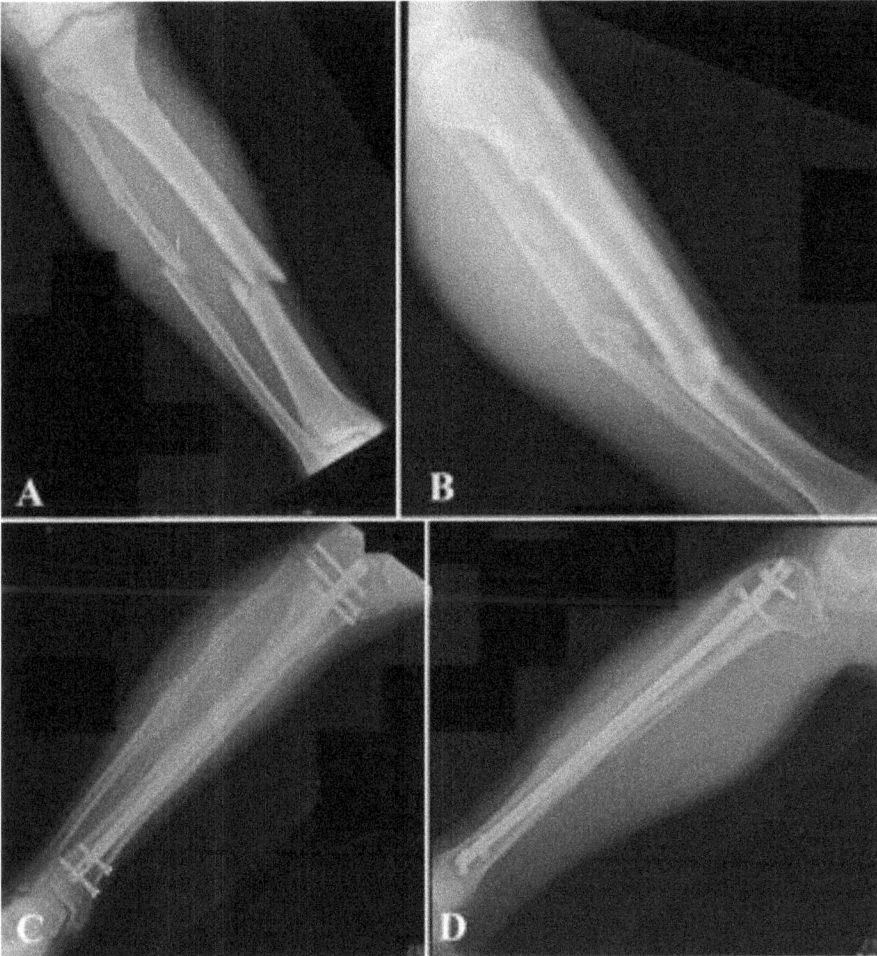

Figure 17.13: Segmental tibia fracture (A,B) and treatment with intramedullary nailing with two distal and three proximal interlocking screws (C,D) [100].

in the medullary cavity of the right femur in eight rats, where they restored their austenite form, causing a bending force. After 12 weeks, the operated femurs were compared with their non-operated contralateral counterpairs. It was reported that (i) anteroposterior radiographs demonstrated significant bowing, as indicated by the angle between the distal articular surface and the long axis of the femur; (ii) significant retardation of longitudinal growth and thickening of operated femurs were also seen; (iii) quantitative densitometry showed a significant increase in the average cross-sectional cortical area and cortical thickness, which were most obvious in the mid-diaphyseal area; (iv) cortical bone mineral density increased in the proximal part of the bone and decreased in the distal part; and (v) polarized light microscopy of the histological samples revealed that the new bone induced by the functional IM nail was mainly woven bone. Hence, it was concluded that (vi) bone modeling can be controlled with a functional IM nail made of NiTi SMA.

Statically locked, reamed IM nailing remains the standard treatment for displaced tibial shaft fractures. Establishing an appropriate starting point is a crucial part of the surgical procedure. Zelle et al. [100] mentioned that, recently, suprapatellar nailing in the semiextended position has been suggested as a safe and effective surgical technique and numerous reduction techiques are available to achieve an anatomic fracture alignment and the treating surgeon should be familiar with these maneuvers. There remains a compelling biological rationale for both reamed and unreamed IM nailing for the treatment of tibial shaft fractures. Previous small trials have left the evidence for either approach inconclusive. Bhandari et al. [102] compared reamed and unreamed IM nailing with regard to the rates of reoperations and complications in patients with tibial shaft fractures. A possible benefit for reamed IM nailing in patients with closed fractures was demonstrated. No difference was found between approaches in patients with open fractures. Delaying reoperation for nonunion for at least 6 months may substantially decrease the need for reoperation. IM nail fixation remains the treatment of choice for unstable and displaced tibial shaft fractures in the adult. The goals of surgical treatment are to achieve osseous union and to restore length, alignment, and rotation of the fractured tibia. IM nailing carries the advantage of minimal surgical dissection with appropriate preservation of blood supply to the fracture. Moreover, the surgical implant offers appropriate biomechanical fracture stabilization and acts as a load sharing device allowing for early postoperative mobilization. Recent advances in nail design and reduction techniques have expanded the indications for IM nail fixation to include proximal and distal third tibial fractures. As of today, IM nail fixation represents a well-described and commonly performed surgical procedure for both the community orthopedic surgeon as well as the subspecialized orthopedic trauma surgeon. Despite its popularity, IM nail fixation of displaced tibial shaft fractures remains challenging and is associated with multiple potential pitfalls. The surgical technique continues to develop and numerous recent investigations have contributed significant advances in this area. The goal of this chapter is to describe the current concepts of IM nail fixation of tibial shaft fractures and to summarize recent developments in this field [100]. Singh [103] mentioned that most of the modern nails come with a locking mechanism. Locking is a process of fixing the nail to the bone using premade holes in the nail on both the proximal and distal ends of the nail for a stable fixation. Nails that can be locked are called interlocking nails. Modern nailing is a technique whereby the nail is inserted into the bone from one end while not disturbing the fracture site at all under an X-ray image intensifier. It was, furthermore, indicated that femur and tibia are bones where nails are most commonly and successfully used. The ideal indications for nailing are: (1) transverse and short oblique fractures of the tibial and femoral shafts; (2) comminuted fractures of tibia and femur, provided cross-locking facilities are available; (3) pathological fractures; (4) delayed or nonunion of femur or tibia; and (4) selected open fractures. Besides these indications, it was noted that in children, the nail may damage growth plates and should be avoided. Arthrodesis has been shown to be a successful operative procedure for the treatment of intractable pain, instability, and deformity of the distal interphalangeal joint of the fingers, and interphalangeal joint of the thumb. Multiple fixation techniques have been used in the past, including Kirschner wires, tension bands, headless screws, or lag screws. Clinical results are generally acceptable, but complication rates have been reported to vary between 10% and 20%. Complications include nonunion, pain, malunion, infection, nerve injury, and protruding hardware. Based on this background, Seitz et al. [104] described an alternative technique for arthrodesis of the distal interphalangeal joint of the fingers or interphalangeal joint of the thumb using nitinol implants for IM fixation with a minimal complication rate.

17.5 Scaffold structure

17.5.1 In general

In tissue engineering, a highly porous artificial extracellular matrix or scaffold is required to accommodate mammalian cells and guide their growth and tissue regeneration in three dimensions. Scaffolds are materials that have been engineered to cause desirable cellular interactions to contribute to the formation of new functional tissues for medical purposes. Cells are often "seeded" into these structures capable of supporting 3D tissue formation. Scaffolds mimic the extracellular matrix of the native tissue, recapitulating the in vivo milieu and allowing cells to influence their own microenvironments. They usually serve at least one of the following purposes: allow cell attachment and migration, deliver and retain cells and biochemical factors, enable diffusion of vital cell nutrients and expressed products, and exert certain mechanical and biological influences to modify the behavior of the cell phase. The existing 3D scaffolds for tissue engineering were proved to be less than ideal for actual applications, not only because they lack mechanical strength, but they also do not guarantee interconnected channels [105]. To achieve the goal of tissue reconstruction, scaffolds must meet some specific requirements. High porosity and adequate pore size are necessary to facilitate cell seeding and diffusion throughout the whole structure of both cells and nutrients. Biodegradability is often an essential factor since scaffolds should preferably be absorbed by the surrounding tissues without the necessity of surgical removal. The rate at which degradation occurs has to coincide as much as possible with the rate of tissue formation: this means that while cells are fabricating their own natural matrix structure around themselves, the scaffold is able to provide structural integrity within the body and eventually it will break down leaving the newly formed tissue which will take over the mechanical load. Injectability is also important for clinical uses. Recent research on organ printing is showing how crucial a good control of the 3D environment is to ensure reproducibility of experiments and offer better results

[106]. The tissue engineering is multidisciplinary that combines biology, biochemistry, clinical medicine, and materials science whose application in cellular systems such as organ transplantation serves as a delivery vehicle for cells and drug. The scaffold fabrication techniques can be classified into two main categories: conventional and modern techniques. These tissue engineering fabrication techniques are applied in the scaffold building which later on are used in tissue and organ structure [107].

17.5.2 Scaffold structure

There seems to be four major structures designed and fabricated: mesh, fibrous, foam, and porous (with and without controlled porosity, either closed pore or contentious pore).

17.5.2.1 Mesh scaffold structure
The epidemiology of valvular heart disease has significantly changed in the past few decades with aging as one of the main contributing factors. The available options for replacement of diseased valves are currently limited to mechanical and bioprosthetic valves, while the tissue-engineered ones that are under study are currently far from clinical approval. The main problem with the tissue-engineered heart valves is their progressive deterioration that leads to regurgitation and/or leaflet thickening a few months after implantation. The use of bioresorbable scaffolds is speculated to be one factor affecting these valves' failure. Accordingly, Alavi et al. [108] developed a non-degradable superelastic nitinol mesh scaffold concept that can be used for heart valve tissue engineering applications, based on the hypothesis that the use of a nondegradable superelastic nitinol mesh may increase the durability of tissue-engineered heart valves, avoid their shrinkage, and accordingly prevent regurgitation.

17.5.2.2 Fibrous scaffold structure
Shi et al. [109] mentioned that scaffold along with seed cells and microenvironment is an essential component of tissue engineering, and plays an important role in engineered tissue regeneration. Fibrous scaffold is an important part of scaffold materials due to its ability to mimic the structure of native extracellular matrix. With the advancement of materials science and related techniques, fibers can be modified either chemically or physically to gain special physicochemical and biological properties that can provide an artificial niche environment for cell adhesion, proliferation, and differentiation and ultimately lead to tissue regeneration.

17.5.2.3 Foam scaffold structure
Gotman et al. [110] evaluated the bone regeneration capability of novel load-bearing nitinol scaffolds. High-strength trabecular nitinol scaffolds were prepared

by PIRAC (powder immersion reaction-assisted coating) annealing of the highly porous Ni foam in Ti powder at 900 °C, followed by PIRAC nitriding to mitigate the release of potentially toxic Ni ions. New Zealand white rabbits received bone defect in right radius and were divided into four groups randomly. In the control group, nothing was placed in the defect. In other groups, nitinol scaffolds were implanted in the defect: (i) as produced, (ii) loaded with bone marrow aspirate (BMA), and (iii) biomimetically CaP coated. The animals were sacrificed after 12 weeks. The forelimbs with scaffolds were resected, fixed, sectioned, and examined in scanning electron microscopy (SEM). It was reported that bone ingrowth into the scaffold was observed in all implant groups, mostly next to the ulna. New bone formation was strongly enhanced by BMA loading and biomimetic CaP coating, the bone penetrating as much as 1–1.5 mm into the scaffold, indicating that the newly developed high-strength trabecular nitinol scaffolds can be successfully used for bone regeneration in critical size defects. Figure 17.14 shows two types of mesh scaffold structures [110]. Wu et al. [111] used titanium hydride powders to enhance the foaming process in the formation of orthopedic nitinol scaffolds during capsule-free hot isostatic pressing. In order to study the formation mechanism, the thermal behavior of titanium hydride and hydrogen release during the heating process are systematically investigated in air and argon and under vacuum by X-ray diffraction, thermal analysis, including thermogravimetric analysis and differential scanning calorimetry, energy-dispersive X-ray spectroscopy, and transmission electron microscopy (TEM). It was found that hydrogen is continuously released from titanium hydride as the temperature is gradually increased from 300 to 700 °C. Hydrogen is released in two transitions: $TiH_{1.924} \rightarrow TiH_{1.5}/TiH_{1.7}$ between 300 and 400 °C and $TiH_{1.5}/TiH_{1.7} \rightarrow \alpha\text{-}Ti$ between 400 and 600 °C. In the lower temperature range between 300 and 550 °C the rate of hydrogen release is slow, but the decomposition rate increases sharply above 550 °C. The X-ray diffraction patterns obtained in air and under vacuum indicate that the surface oxide layer can deter hydrogen release. The holding processes at 425, 480, 500, 550, and 600 °C are found to significantly improve the porous structure in the nitinol scaffolds due to the stepwise release of hydrogen. Nitinol scaffolds foamed by stepwise release of hydrogen are conducive to the attachment and proliferation of osteoblasts and the resulting pore size also favor ingrowth of cells.

17.5.3 Porous scaffold structures

The most widely used scaffold materials for bone tissue engineering applications are calcium phosphate ceramics (e.g., hydroxyapatite and tricalcium phosphate, as known as the bone cement) due to their osteoinductive and osteoconductive properties [112, 113]. However, these materials generally lack the tensile strength required for initial load bearing and primary stability and, as bulk material, do not match the mechanical properties of the surrounding bone (known as mismatching for biomechanical compatibility [114, 115]), limiting their application to nonload-bearing

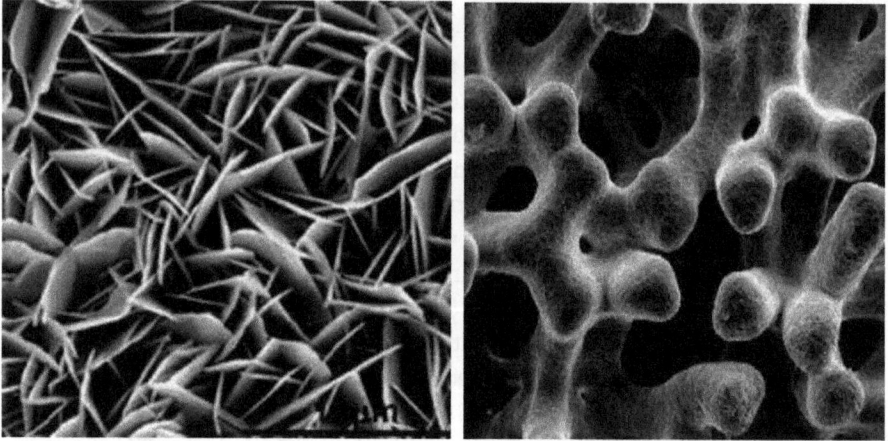

Figure 17.14: Trabecular nitinol foam scaffold structures [110].

situations or requiring long periods of immobilization during bone healing. As an alternative to ceramics, metals have been used for prostheses in orthopedics and orthodontics for decades due to their superior mechanical properties. In particular, titanium has traditionally been one of the most commonly used metallic implant materials, demonstrating biocompatibility and osseointegration [116]. However, the Young's modulus of Ti (100 GPa) significantly exceeds that of cortical bone (3–20 GPa), which can result in stress shielding, as discussed in previous chapters. As the Ti implant absorbs most of the applied mechanical load, the surrounding bone is shielded from the applied stress, ultimately leading to bone resorption [117]. As compared to pure metals, metallic alloys allow the tuning of the particular mechanical properties toward specific medical needs (e.g., Young's modulus). In particular, NiTi alloys have particularly low Young's moduli, which are comparable to that of bone, are pseudoelastic, have a high damping capacity, and exhibit shape-memory properties. Due to this unique combination of mechanical properties, NiTi possesses great promise as a next-generation scaffold material for bone repair [118]. The superelastic nature of bones requires matching biomechanical properties from the ideal artificial biomedical implants in order to provide smooth load transfer and foster the growth of new bone tissues. The interest in using porous SMA scaffolds as implant materials has been growing in recent years due to the combination of their unique mechanical and functional properties, that is, SME and superelasticity, low elastic modulus combined with new bone tissue ingrowth ability and vascularization. These attractive properties are of great benefit to the healing process for implant applications.

Wu [119] reported the large-scale direct growth of nanostructured bioactive titanates on 3D microporous Ti-based metal (NiTi and Ti) scaffolds via a facile low-temperature hydrothermal treatment. The nanostructured titanates show characteristics of 1D nanobelts/nanowires on a nanoskeleton layer. Besides resembling cancelous bone structure on the micro/macroscale, the 1D nanostructured titanate on the exposed surface is similar to the lowest level of hierarchical organization of collagen and hydroxyapatite. It was mentioned that the resulting surface displays superhydrophilicity and favors deposition of hydroxyapatite and accelerates cell attachment and proliferation. The remarkable simplicity of this process makes it widely accessible as an enabling technique for applications from engineering materials treatment including energy absorption materials and pollution-treatment materials to biotechnology. Similar results were reported on porous SMA scaffolds [120]. Liu et al. [121] determinee the biomechanical characteristics of porous NiTi implants and investigated bone ingrowth under actual load-bearing conditions *in vivo*. Porous NiTi, porous Ti, dense NiTi, and dense Ti were implanted into 5 mm diameter holes in the distal part of the femur/tibia of rabbits for 15 weeks. It was reported that the porous NiTi materials bond very well to newly formed bone tissues and the highest average strength of 357 N and best ductility are achieved from the porous NiTi materials. The bonding curve obtained from the NiTi scaffold shows similar superelasticity as natural bones with a deflection of 0.30–0.85 mm thus shielding new bone tissues from large load stress. This indicates that new bone tissues can penetrate deeply into the porous NiTi scaffold compared to the one made of porous Ti. Histological analysis reveals that new bone tissues adhere and grow well on the external surfaces as well as exposed areas on the inner pores of the NiTi scaffold. The *in vitro* study indicates that the surface chemical composition and topography of the porous structure leads to good cytocompatibility. In conjunction with the good cytocompatibility, the superelastic biomechanical properties of the porous NiTi scaffold bodes well for fast formation and ingrowth of new bones, and porous NiTi scaffolds are thus suitable for clinical applications under load-bearing conditions. With the ultimate goal of fabricating porous implants for spinal, orthopedic and dental applications, Hoffmann et al. [118] fabricated NiTi scaffolds were fabricated by means of selective laser melting (SLM). The response of human mesenchymal stromal cells to the NiTi substrates was compared to mesenchymal stromal cells cultured on clinically used titanium. Selective laser-melted titanium as well as surface-treated NiTi and Ti served as controls. It was reported that mesenchymal stromal cells had similar proliferation rates when cultured on selective laser-melted NiTi, clinically used titanium, or controls. Osteogenic differentiation was similar for mesenchymal stromal cells that are cultured on the selected materials, as indicated by similar gene expression levels of bone sialoprotein and osteocalcin. Mesenchymal stromal cells seeded and cultured on porous 3D-selective laser-melted NiTi scaffolds homogeneously colonized the scaffold, and following osteogenic induction, filled the scaffold's pore volume with extracellular matrix. Hence, it is concluded that the combination of bone-related mechanical properties of selective laser-melted NiTi with its cytocompatibility and support of osteogenic differentiation of mesenchymal stromal cells highlights its potential as a superior bone substitute as compared to clinically used titanium. Figure 17.15 shows CAD model of entire cylindrical 3D porous scaffold and SEM micrograph of the 3D nitinol scaffold fabricated by the SLM process [118].

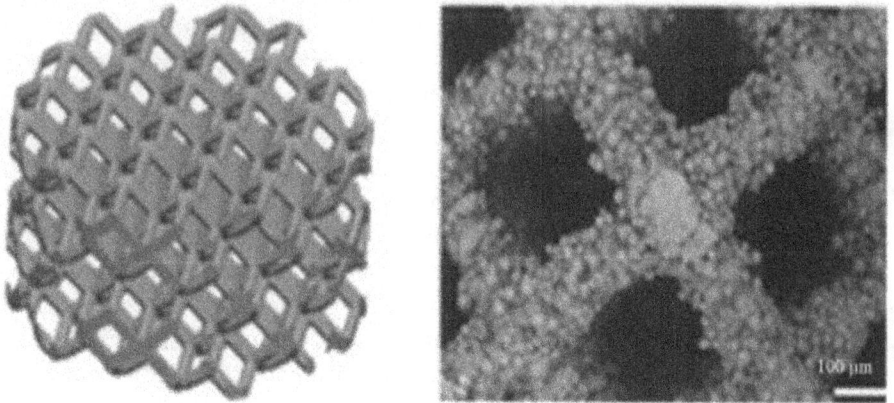

Figure 17.15: CAD model of porous scaffold and SEM image of SLMed 3D nitinol scaffold [118].

17.6 Bone implant

The field of biomaterials has become a vital area, as these materials can enhance the quality and longevity of human life, and the science and technology associated with this field has now led to multimillion dollar business [122]. Common metals for stable long-term implants (e.g., stainless steel, commercially pure titanium, Ti-6Al-4V, and Ti-6Al-7Nb) are much stiffer than spongy cancellous and even stiffer than cortical bone. When bone and implant are loaded, this stiffness mismatch results in stress shielding and, as a consequence, degradation of surrounding bony structure can lead to disassociation of the implant. From this background, Rahmanian et al. [123] mentioned that due to its lower stiffness and high reversible deformability that is associated with the superelastic behavior, NiTi is an attractive biomaterial for load-bearing implants; however, the stiffness of austenitic nitinol is closer to that of bone but still too high. Hence, the feasibility study was conducted on the additive manufacturing that can provide, in addition to the fabrication of patient-specific implants, the ability to solve the stiffness mismatch by adding engineered porosity to the implant [123] and it was indicated that different nitinol scaffolds can be produced by additive manufacturing. Filip et al. [124] characterized the surface and the bulk structure of TiNi implants using SEM, TEM, X-ray photoemission spectroscopy, and scanning Auger microprobe analysis. TiNi implants were compared with otherwise identically prepared nonimplanted specimens, and sputter-cleaned and reoxidized samples. It was reported that nonimplanted and implanted samples had essentially the same surface topography and microstructure. Ti, O, and C were the dominant elements detected on the surface. Trace amounts (~1 at%) of Ni and Ca, N, Si, B, and S were also detected. Ti was present as TiO_2 on the surface, while nickel was present in metallic form. A significant difference in Ni peak intensity was observed when retrieved or nonimplanted control samples (a very low nickel content) were compared with sputter-cleaned and repassivated samples (well-detected nickel). No major changes occurred in the TiNi samples bulk structure or in the surface oxide during the implantation periods investigated. In order to adapt the implant to the actual healing situation, NiTi SMA-based implants that exhibit a variable middle piece were developed [89]. It was mentioned that due to contactless induction heating, the transition temperature of the SMA is reached and triggers the one-way SME, leading to a slight modification of the second moment of inertia and to an adaption of the bending stiffness of the implant, respectively. Based on laser cutting and welding, NiTi-SMA–based implants were manufactured by using thin, commercially available NiTi (thickness: 0.5 and 1.0 mm; transition temperature A_F between 55 and 65°C) sheets. Depending on the implant design, four-point bending measurements showed feasible bending stiffness alterations around a range of 50%, with regard to the original stiffness, indicating that increasing the bending stiffness and decreasing the bending stiffness have been realized. It is therefore concluded that the implant design is being adapted in order to meet the requirements for human applications, and custom-made NiTi-SMA with a transition temperature around 45–50°C could continue to reduce the impact of the heating process and expand this concept to further medical applications, for example, compression, anchoring, or expansion of implants or devices. Shishkovsky et al. [125] conducted the comparative microstructural analyses and histomorphological studies of tissue reactions to porous titanium and nitinol implants synthesized by selective laser sintering (SLS) for a rat model for bone implants. It was obtained that the surface of porous pegs of titanium and nitinol made by SHS/SLS (a technology combined by selective heat sintering and SLS) exhibit a significantly favorable structure to the mechanical interlocking with bone and soft tissues. Histological analysis of decalcified paraffin sections after implant removal could only show that trabecular bone structures and marrow cavities were observed around the porous implants. In the connective tissue of the remaining implant beds, cells such as macrophages, fibroblasts, adipocytes, and lymphocytes are discernible, and the nitinol synthesized by combined SHS/SLS technique has a developed and ordered microstructure.

Synthetic meshes are now widely used for a repair of various tissue defects. Polypropylene meshes (PPMs) that are most frequently used for repair of abdominal wall defects possess some disadvantages; therefore, the development of new mesh materials for reconstructive surgery is urgently required. Zaworonkow et al. [126] compared a TiNi-based alloy superelastic mesh implant [TiNi-alloy mesh (TNM)] with a PPM in the Wistar rat model. A midline resection abdominal wall defect was treated with a mesh implant. Three groups of 20 animals were assessed: (1) a TNM group with the defect managed by a TiNi-based wire woven implant, (2) a PPM group with the defect repaired by Optomesh®, and (3) a no-mesh (NM) group treated without the use of any prosthetic material for systemic reaction to implant control. Evaluation of both mesh implantations was carried out on days 14, 28, 56, and 90 after surgery. It was found that (i) all the animals from NM group developed hernias; (ii) the PPM group showed satisfactory results, although some complications like implant dislocation (6/20) and hernia recurrence (5/20) were noted; (iii) in the TNM group neither shrinkage nor implant dislocation were observed; and (iv) compared with the PPM group, the TNM group was characterized by no inflammation reaction and better integration with tissues. Based on these findings, it was concluded that suppression of inflammation response together with a more physiological wall remodeling process than with PPM makes TNM an attractive concept for abdominal wall defect reconstruction. Since there was neither shrinkage nor dislocation of the TNM itself it may be suitable for intraperitoneal management. The essence of enhanced biocompatibility is due to stress–strain mechanical behavior of TNM and its surface oxycarbonitride (TiCNO) layer, which facilitate its incorporation. It is summarized that NiTi alloy mesh was found to be a very promising material for the repair of abdominal wall defects in clinical practice, and ideally NiTi alloy meshes would be applied in circumstances when there are indications for the management of large hernias by PPMs. Figure 17.16 shows difference of tissue coverage between PPM and TNM. From SEM observation, it was clearly shown that the PPM was not homogeneously covered by tissues, and flaws (especially around mesh knots) were noted, and defects were explored on all surfaces for the whole observation period; on the contrary, the TNM implant was evenly covered by the surrounding tissue and no flaws were found.

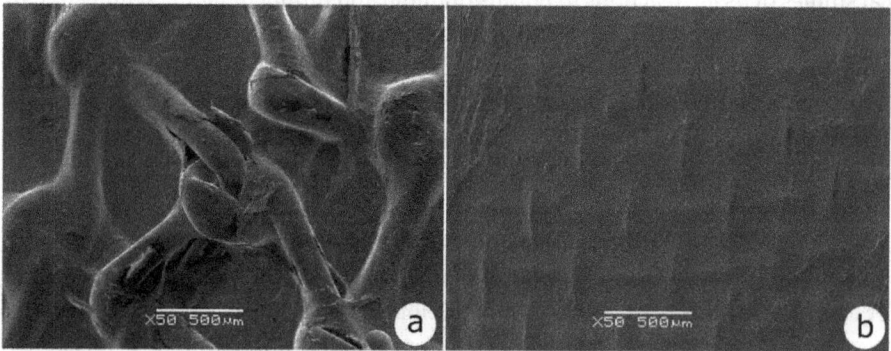

Figure 17.16: Difference of tissue coverage between (a) polypropylene meshes (PPM) and (b) TiNi-alloy meshes (TNM) [126].

Porous nitinol has recently attracted attention in clinical surgery because it is a very interesting alternative to the more brittle and less machinable conventional porous Ca-based ceramics; hence, this material becomes a promising new material for a bone graft substitute with good strength properties and an elastic modulus closer to that of bone than any other metallic material. Kujala et al. [127] evaluated the effect of porosity on the osteointegration of NiTi implants in rat bone. The porosities (average void volume) and the mean pore size (MPS) were 66.1% and 259 ± 30 μm (group 1, n = 14),

59.2% and 272 ± 17 μm (group 2, $n = 4$) and 46.6% and 505 ± 136 μm (group 3, $n = 15$), respectively. The implants were implanted in the distal femoral metaphysis of the rats for 30 weeks. It was reported that (i) the proportional bone-implant contact was best in group 1 (51%) without a significant difference compared to group 3 (39%), (ii) group 2 had lower contact values (29%) than group 1, and (iii) fibrotic tissue within the porous implant was found more often in group 1 than in group 3, in which 12 samples out of 15 showed no signs of fibrosis. This concludes that porosity of 66.1% (MPS 259 ± 30 μm) showed best bone contact (51%) of the porosities tested here; however, the porosity of 46.6% (MPS 505 ± 136 μm) with bone contact of 39% was not significantly inferior in this respect and showed lower incidence of fibrosis within the porous implant. Recognizing that there are some limitations with NiTi alloy coming from the chemical homogeneity of the as-processed porous NiTi alloys, which always contain undesired secondary Ti- and Ni-rich phases. These are known to weaken the NiTi products, to favor their cavitation corrosion, and to decrease their biocompatibility. Elemental nickel must also be avoided because it could give rise to several adverse tissue reactions. Therefore, the synthesis of porous single-phase NiTi alloys by using a basic single-step sintering procedure is an important step toward the processing of safe implant materials. Based on this background, Bertheville et al. [128] used the sintering process based on a vapor-phase calciothermic reduction operating during the NiTi compound formation. It was found that the as-processed porous NiTi microstructure is single phase and shows a uniformly open pore distribution with porosity of about 53% and pore diameters in the range 20–100 μm, and (ii) due to the process, fine CaO layers grow on the NiTi outer and inner surfaces, acting as possible promoting agents for the ingrowth of bone cells at the implantation site. Barrabés et al. [129] examined NiTi foams that have been treated using a new oxidation treatment for obtaining Ni-free surfaces that could allow the ingrowth of living tissue, thereby increasing the mechanical anchorage of implants. It was mentioned that a significant increase in the real surface area of these materials can decrease corrosion resistance and favor the release of Ni element. This chemical degradation can induce allergic reactions or toxicity in the surrounding tissues. It was demonstrated that the new surface oxidation treatment reduced amount of Ni release into the simulated body fluid since the oxidation treatment re-formed surface zone as Ni-free surfaces. These foams have pores in an appropriate range of sizes and interconnectivity, and thus their morphology is similar to that of bone. Their mechanical properties are biomechanically compatible with bone. The titanium oxide on the surface significantly improves corrosion resistance and decreases nickel ion release, while barely affecting transformation temperatures.

NiTi has been shown to be of great interest for bone implant applications. Introducing porosity to NiTi bone implants is an effective technique to tune their equivalent modulus of elasticity in order to acquire similar value to that of cortical bone. Moreover, such porous implants allow for better tissue ingrowth due to the interconnecting open pore structure. The effect of porosity percentage on the NiTi equivalent modulus of elasticity is well understood. Like most bone deficits, mandibular segmental defects may result from surgical reconstruction due to congenital deformity, tumor resection, other pathologies, senescence, trauma, or infection. The goals of mandibular recon-struction are to restore the mandible's function and normal appearance. Mandibu-lar segmental defect reconstruction is most often necessitated by tumor resection, trauma, infection, or osteoradionecrosis. Moghaddam et al. [130, 131] investigated mandibular segmental defect reconstruction. An important issue is the biomechani-cal compatibility and stress shielding issue [114, 115]. Although Ti-6Al-4V has a lower stiffness (110 GPa) than other common materials (e.g., stainless steel, tantalum), it is still much stiffer than the cancellous (1.5–4.5 GPa) and cortical portions of the man-dible (17.6–31.2 GPa). Surgical grade 5 titanium (Ti-6Al-4V) is commonly used to fab-ricate the fixture plate due to its low density, high strength, and high biocompatibil-ity. One of the potential problems with mandibular reconstruction is stress shielding caused by a stiffness mismatch between the titanium fixation plate and the remaining mandible bone and the bone grafts. A highly stiff fixture carries a large portion of the load (e.g., muscle loading and bite force); therefore the surrounding mandible would undergo reduced stress. As a result, the area receiving less strain would remodel and may undergo significant resorption. This process may continue until the implant fails. To avoid stress shielding it is ideal to use fixtures with stiffness being similar to that of the surrounding bone. Although Ti-6Al-4V has a lower stiffness (110 GPa) than other common materials (e.g., stainless steel, tantalum), it is still much stiffer than the cancellous (1.5–4.5 GPa) and cortical portions of the mandible (17.6–31.2 GPa). As a solution, Moghaddam et al. [130] employed a nitinol in order to reduce stiffness of the fixation hardware to the level of mandible. To this end, the finite element analysis was conducted to look at strain distribution in a human mandible in three different cases: (i) healthy mandible, (ii) resected mandible treated with a Ti-6Al-4V bone plate, and (iii) resected mandible treated with a nitinol bone plate. It was reported that the stress–strain trajectories of the mandibular reconstruction using nitinol fixation is closer to normal than if grade 5 surgical titanium fixation is used. Moghaddam et al. [131] reviewed the safety, design, and efficacy of metal implants in general and specif-ically for the repair of mandibular segmental defects with major concerns including implant incorporation, implant failure mechanisms (e.g., stress concentration, stress shielding), corrosion and toxicity, infection, and muscle reattachment. It was also mentioned that (i) the porous NiTi implants for the repair of skeletal defects through the example of mandibular segmental defects is highly feasible, (ii) resorbable mag-nesium and porous NiTi provide new options which may better match the material properties, and if they can be 3D printed, better match the shape of surrounding host tissue, reducing engrafted bone and metal implant failure and restore musculo-skeletal function for the long term. Amerinatanzi et al. [132] extensively studied the effect of porosity type on NiTi bone implant's performance, in terms of the geometri-cal structure and other mechanical properties, by simulating three porous structures made of shape-memory Ti-rich $NiTi_{50.09}$ (at%) alloy. The effect of porosity type on the NiTi implant's geometrical structure and mechanical properties was studied using numerical tests. The purpose is to compare three NiTi implants with different kinds of porosities, at a similar level of porosity (i.e., 69%). The assigned porosity types are Schwartz-type, Gyroid-type, and Diamond-type. It was reported that anisotropic behavior within the three different porous structures was noted. With the same level of porosity (i.e., 69%), equivalent modulus of elasticity was observed to be 48.9, 34.8, and 30.2 GPa for Schwartz-type, Gyroid-type, and Diamond-type, respectively. The Schwartz-type scaffold was seen to offer the highest stress at plateau start and the lowest residual strain after unloading, in comparison with the other two types of structure; indicating that if 3D designing and printing are appropriately operated, the ideal NiTi alloy implant with perfect porosity and geometry can be fabricated.

Figure 17.17 shows SEM micrographs of porous NiTi produced under different fabrication methods: (a) SHS (selective heating sintering) process (65 ± 10% porosity, 100–360 lm) [133] and (b) metal injection molding and sintering with NaCl spaceholder (prealloyed powders, 70% porosity, 355–500 lm) [138], where "lm" refers to the sediment particle-size percentile.

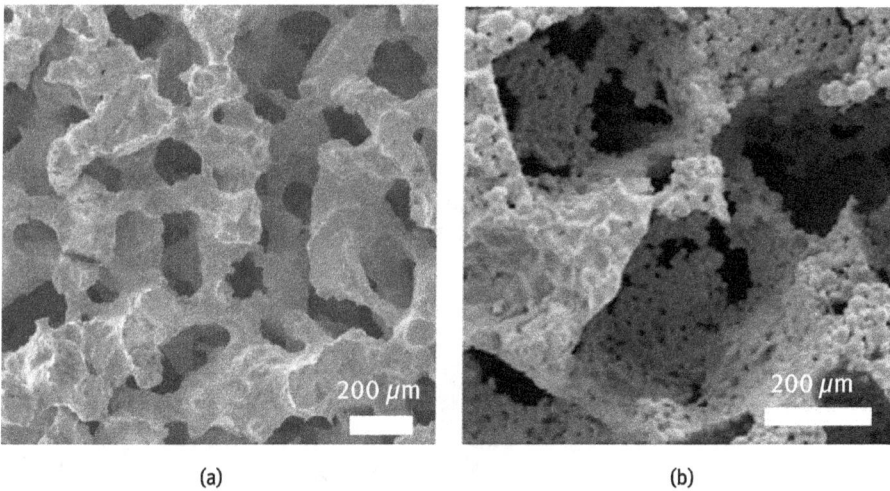

(a)   (b)

Figure 17.17: SEM micrographs of porous NiTi produced under different fabrication methods: (a) SHS (selective heating sintering) process [133] and (b) MIM (metal injection molding) and sintering with NaCl spaceholder [134].

Bansiddhi et al. [135] reviewed NiTi foams for bone replacement, focusing on three specific topics: (i) surface modifications designed to create bio-inert porous NiTi surfaces with low Ni release and corrosion, as well as bioactive surfaces to enhance and accelerate biological activity, (ii) in *vitro* and in *vivo* biocompatibility studies to confirm the long-term safety of porous NiTi implants, and (iii) biological evaluations for specific applications, such as in intervertebral fusion devices and bone tissue scaffolds; suggesting that with the simultaneous consideration of processing methods, mechanical properties, NiTi composition, microstructural features and biological performance, more rapid optimization of porous NiTi for biomedical applications is likely to occur in the near future. Laser spray cladding in N_2-rich atmosphere is a competitive technique for manufacture of porous NiTi SMA and TiN surface film. In order to gain expected porous coating and porosity size, TiH_2 powder was considered to be used as porosity initiating agent, and nitrogen was used for shielding and active nitriding gas [140]. Yang et al. manufactured a gradient porous coating of NiTi alloy by laser spraying and reported that there are pores with size of 100–200 μm near the surface, and X-ray diffraction analysis shows that the coating consists of intermetallic compounds NiTi and $NiTi_2$, and nitrides TiN and Ti_2N [136].

Gradation of porosity (from dense solid structure at core portion to high porosity at case layer) is an ideal implant structure. Figure 17.18 [114] illustrates a schematic and conceptual Ti material implant which possesses a gradual function of mechanical and biological behaviors, so that mechanical compatibility and biological compatibility can be realized. Since microtextured Ti surfaces and/or porous Ti surfaces promote fibroblast apposition and bone ingrowth, the extreme left side representing the solid Ti material implant body should have gradually increased internal porosities toward the right side which is in contact with the vital hard/soft tissue. Accordingly, mechanical strength of this implant system decreases gradually from left to right, whereas biological activity increases from left to right. Therefore, the biomechanical compatibility can be completely achieved. Porosity-controlled surface zones can be fabricated by an electrochemical technique [137], polymeric sponge replication method [138], powder metallurgy technique, superplastic diffusion bonding method [139], or foamed metal structure technique [140].

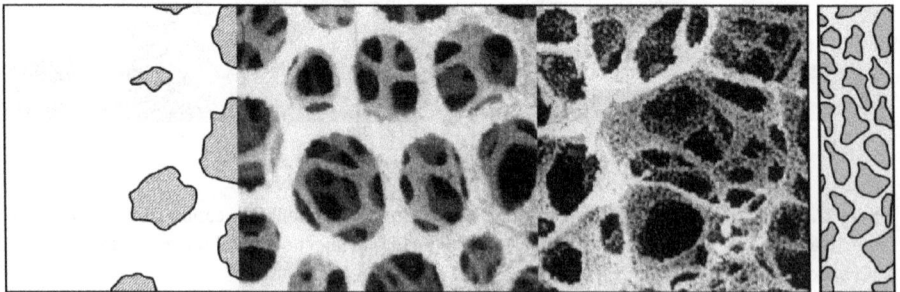

Ti implant				bone
body	sub-surface zone	surface layer	bony growth zone	
strong	←	Biomechanical strength	←	weak
Modulus of Elasticity [GPa]				
250~200 ← 150~100 ←		100~50 ←	50~20	← 20~10
weak	→	Biological and Biochemical reactions	→	strong

Figure 17.18: Schematic and conceptual porous implant with gradient mechanical and biological functions [114].

17.7 Other applications

Settings

Due to ever-increasing risk of aging population, there are a variety of unforeseen risks jeopardizing individual's QOL, which should be including a higher incidence of CAD, diabetes, systemic disease, injury-related frailty, etc. Hence there is a need for more reliable and safer treatments [141]. Limiting our discussion to orthopedic medicine, for traditional bone implant manufacturing processes, the metal material selection depends on both performance and cost. Additive manufacturing techniques, however, are no longer impacted by the choice of materials. They tailor materials according to the specific needs of the customized implant. Metals have been used for orthopedic implants for a long time due to their excellent mechanical properties [142]. With the rapid development of additive manufacturing technology, studying customized implants with complex microstructures for patients has become a trend of various bone defect repair. Although we have been discussing major orthopedic applications of NiTi alloys, there are still more applications left. Just for a reference, Figure 17.19 shows the different implants for numerous medical applications which are manufactured using mainly NiTi biomaterials as a part thereof or entire structure [143, 144].

With respect to the healing of fractured bones, one can also point out shape-memory plates for the recovery of bones. These plates are primarily used in situations where a cast cannot be applied to the injured area, i.e., maxillofacial areas, nose, jaw and eye socket. They are placed on the fracture and fixed with screws, maintaining the original alignment of the bone and allowing cellular regeneration under orthopedic and/or cosmetic surgery treatment. Because of the SME, when heated these plates tend to recover their former shape, exerting a constant force that tends to join parts separated by fractures, helping with the healing process.

17.7.1 Nitinol suture

It was reported [145] that (i) nitinol 0.007-inch suture wire is five times stronger than prolene and 38 times stronger than nylon, (ii) proven in neurosurgery and cardiac stent applications, it is highly biocompatible and can be delivered onehanded through needles as small as 30 gauge for intraocular procedures, (iii) unlike filament sutures, the wire clip can be bent and returned to its original shape, (iv) in animal tests, pupilloplasty was performed 16 times faster with the nitinol suture than with a modified Seipser slip knot, and the need for limbal-to-limbal passes was eliminated, making the procedure less technically challenging, (v) similarly, IOL (intraocular lens) iris fixation times were reduced 20-fold and while nitinol can be formed into any shape, for intraocular use a circular double coiled clip may be most useful, (vi) the suture can be manipulated with forceps after injection, and (vii) pre-set sutures can be sprung open to grasp haptics, glaucoma shunts, rings or other devices. Kujala

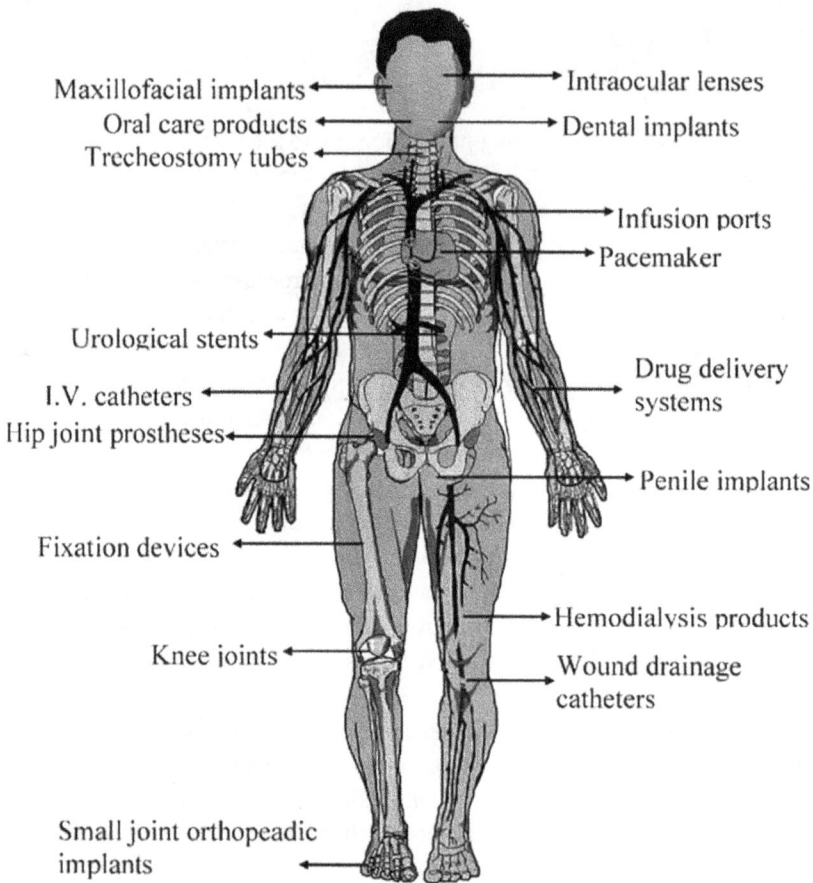

Maxillofacial implants
Oral care products
Trecheostomy tubes
Intraocular lenses
Dental implants
Infusion ports
Pacemaker
Urological stents
I.V. catheters
Hip joint prostheses
Drug delivery systems
Penile implants
Fixation devices
Hemodialysis products
Knee joints
Wound drainage catheters
Small joint orthopeadic implants

Figure 17.19: Different medical implants (both dental and orthopedic) [144].

et al. [146] mentioned that nitinol is a promising new tendon suture material with good strength, easy handling and good superelastic properties. NiTi sutures were implanted for biocompatibility testing into the right medial gastrocnemius tendon in 15 rabbits for 2, 6 and 12 weeks. Additional sutures were implanted in subcutaneous tissue for strength measurements in order to determine the effect of implantation on strength properties of NiTi suture material. Braided polyester sutures (Ethibond) of approximately the same diameter were used as control. It was reported that (i) encapsulating membrane formation around the sutures was minimal in the case of both materials, (ii) the breaking load of NiTi was significantly greater compared to braided polyester, and (iii) implantation did not affect the strength properties of either material.

Self-closing nitinol U-clips have primarily been used by vascular surgeons to facilitate vessel anastomosis. The advantage of this device is that it obviates the need for suturing when manipulation of the tissue or needle is prohibitive. Campbell et al. [147] described a novel use of the U-clip as an aid in endoscopic reconstruction of the skull base. Two patients underwent U-clip assisted skull base repair. The first patient underwent transnasal endoscopic repair of an ethmoidal meningoencephalocele requiring primary dural closure and the second underwent odontoidectomy for basilar invagination via an endoscopic transnasal approach requiring primary closure of the soft tissues. It was obtained that (i) the defects were successfully closed using the U-Clip anastomotic device, (ii) no cerebrospinal fluid leaks were experienced postoperatively, and (iii) intraoperative evaluation combined with postoperative imaging confirmed a satisfactory reconstruction. It was therefore concluded that (iv) the U-clip device may be used as an adjunct in select cases to facilitate endoscopic skull base reconstruction, (v) the use of this tool may prevent tissue migration and enhance dural closure, and (vi) it may be particularly useful in cases where narrow operative corridors make conventional suturing technically difficult. Figure 17.20 shows (a) self-closing nitinol U-clip device consisting of a needle attached to a malleable suture with the terminal nitinol coil and (b) clip loaded on a needle-driver [147].

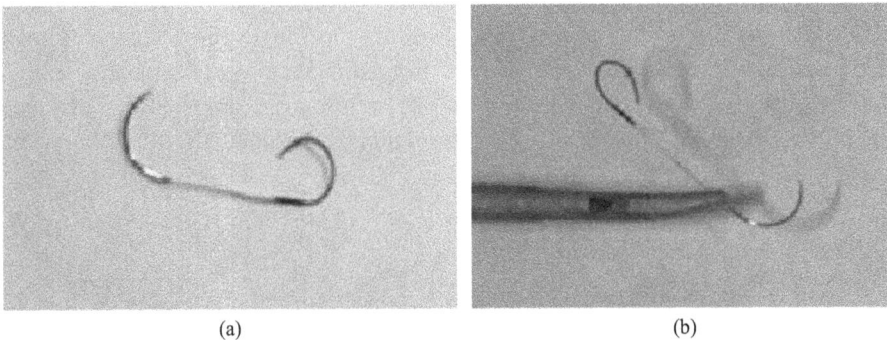

Figure 17.20: (a) self-closing nitinol U-clip and (b) clip with a needle-driver [147].

Lamprakis et al. [148] evaluated the biomechanical results of meniscal repair *in vitro* by a nitinol suture and compare them with Ethibond and nylon. The 6 testing groups consisted of nitinol, Ethibond and nylon sutures (No. 2-0 and No. 3-0). Sixty bovine menisci with a vertical longitudinal tear were repaired with 1 horizontal mattress suture and were fully immersed into a water bath, adjusted to a tensile testing machine. All specimens were subjected to tensile testing, and force/displacement curves were obtained. Load to 5-mm gap, load to failure, tensile strength, stiffness, and mode of failure for each suture group were recorded. Statistical analysis included analysis of variance with Bonferroni correction for the post hoc multiple comparisons. It was found that (i) nylon sutures achieved the lowest scores in all measurements, (ii) nitinol achieved better scores, but not significantly better scores, than Ethibond in load to 5-mm gap and stiffness, (iii) the No. 2-0 and No. 3-0 nitinol suture repair showed the highest mean tensile strength and load to failure, with significant differences, being 36% and 45% stronger, respectively, than Ethibond, and (iv) modes of failure included pulling through the inner segment of meniscus and rupture of the suture at the knot. Based on these results, it was concluded that (v) the superior load-to-failure and tensile strength characteristics of nitinol was recognized; however, in terms of stiffness and gap resistance force, the results were equivalent to those of Ethibond, and (vi) nitinol is an interesting and promising suture. As to the clinical relevance, it was also mentioned that (vii) nitinol can be elongated and become soft and flexible for proper suturing at low temperature, (viii) at body temperature, it can contract to its original length, providing stronger knots, and (ix) this may result in more efficient primary stability of meniscal repair, minimizing the chances of loosening during healing and allowing earlier rehabilitation. Nespoli et al. [149] examined the proposed idea of new suturing procedure based on self-accommodating suture points. Each suture point is made of a commercial NiTi wire hot-shaped in a single loop ring; a standard suture needle is then fixed at one end of the NiTi suture. According to this simple geometry, several NiTi suture stitches have been prepared and tested by tensile test to verify the closing force in comparison to that of commercial sutures. Further experimental tests have also been performed on anatomic samples from animals to verify the handiness of the NiTi suture. It was reported that (i) the NiTi suture expresses high stiffness and a good surface quality, and (ii) the absence of manual knotting allows for a simple, fast and safe procedure.

17.7.2 Muscle

Orthopedic treatment also exploits the properties of SMA in the physiotherapy of semi-standstill muscles. These wires reproduce the activity of hand muscles, promoting the original hand motion. The two-way SME is exploited in this situation. When the glove is heated, the length of the wires is shortened. On the other hand, when the glove is cooled, the wires return to their former shape, opening the hand. As a result, semi-standstill muscles are exercised. Webb et al. [150] employed micromanipulators and robotic actuators to mimic the smooth movement of human muscles. Figure 17.21 depicts a conceptional idea of using multiple shape-memory nitinol spring as artificial muscles [151].

An air muscle is a simple pneumatic device, like biological muscles, air muscles contract when activated. Robotists find it interesting that air muscles provide a reasonable working copy of biological muscles. When a human skeleton with air muscles is attached to the skeleton at primary biological muscle locations, biomechanics and low-level neural properties of biological muscles can be investigated. Air muscles have applications in robotics, biorobotics, biomechanics, artificial limb replacement and industry, due to several unique characteristics (1) air muscles are

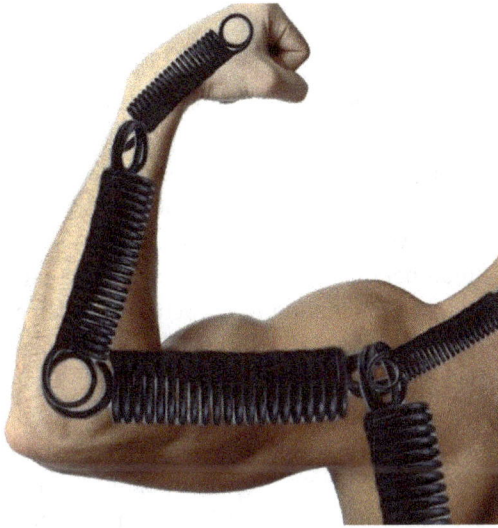

Figure 17.21: Shape-memory nitinol muscles [151].

ease of use and simple construction, (2) air muscles are soft, lightweight and compliant, (iii) it has a high power to weight ratio (400:1), can be twisted axially and used on unaligned mounting and provide contractive force around bends, and (iv) air muscles may also be used underwater [152]. Figure 17.22 demonstrates; (a) insert the rubber inside the braided sleeve, align one end of the sleeve with the bottom of the head on the 10–24 screw in the rubber tube, wrap a piece of 24-gauge wire three or four times around the end, capturing the sleeve, tubing and threaded portion of the 10–24 screw, and then twist the ends of the wire together. Use a pair of pliers to make this a straight as possible; (b) push down the sleeve until it is aligned with the rubber tube on the air coupling, wrap a piece of 24-gauge wire around this end, tighten wire with pliers then cut off any excess wire, pressurize (about 20 psi) the air muscle to insure the two fittings do not leak.

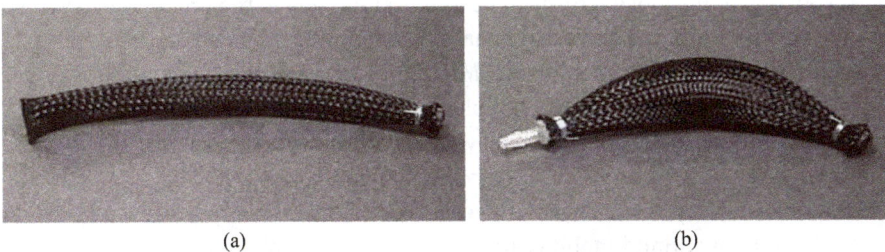

(a)   (b)

Figure 17.22: Air muscle; (a) insertion of rubber inside the nitinol air muscle tube, and (b) pressurize (about 20 psi) the rubber [152].

17.7.3 Glove

Disability to move hands perfectly is one of the most severe human physical disabilities, and it is mostly common among adults or those who have experienced serious accidents. It is desired to find a methodology to restore the motion of the hand. To date, various hand exoskeleton devices have been proposed. Most of them, however, contain linkage mechanisms which are relatively weighty. Hadi et al. [153] developed a wearable lightweight hand exoskeleton robot (or smart glove). The glove actuation mechanism is tendon driven and the glove is SMA tendon-based actuated which can be utilized for both rehabilitation exercises and assistance for people with hand disability. For every finger, two active degrees of freedom (DOFs) are supported in design. Consequently, four tendons are considered for activating each DOF in order to complete opening and closing phases of the fingers. Totally, twenty tendons are used for rehabilitation of a hand through the glove. Using kinematic relations between tendon length and finger movement, the required deflection of each tendon is extracted. Since a short length of SMA wires cannot provide an enough displacement of tendons in the glove, therefore, an extra mechanism was embedded to the developed glove to support the required length of SMA wires. The SMA actuators were selected and mounted on the system to support the tendons of the mechanism effectively. Moreover, the gripping force provided by the developed glove was also studied. To this end, an analysis was accomplished to extract the relationship between tendon actuation and gripping force of the glove. It was reported that (i) the obtained results offered a proper model for such a tendon driven glove, (ii) coupling the model of the SMA actuators to that of the tendon driven glove, a composite model of smart glove was extracted [153]. A biomechanical glove was proposed to assist hand movement through the controlled heating and cooling of SMA that laced throughout the glove [154]. The assistive glove will be powered by a battery that induces an electrical current through the wires of the glove, and the natural resistance within the wires provides a safe way to heat the wires. When the wires are heated, the SMA wires contract in length, and the wires running along the fingers in the glove flex with it. Thus, the glove assists a disabled hand in flexing the fingers by electrical resistive heating of the SMA wires. As the wires cool down, the glove returns to its idle position due to the property of SMA, so the person who wears the glove can release the originally intended object at the task. The controlled heating and cooling of SMA in the glove will assist in actuating finger movement so that daily tasks and hobbies can be performed at an ideal level [154]. Figure 17.23 shows (A) the open position when the nitinol wire was at low temperature and (B) the closed position when the wire at thigh temperature [117-1].

Physical therapy is an important resource for the recovery process of several medical conditions. Hand mobility impairment, for example, affects patients' QOL, making it a need to develop aid devices that improve the results of hand therapy, quickening the recovery process. Accordingly, Jiménez et al. [155] developed a wearable and portable rehabilitation glove, based on the use of muscle wires (or nitinol

<table>
<tr><td>(a)</td><td>(b)</td></tr>
</table>

Figure 17.23: Smart glove at (A) open position and (B) close position [1].

wire) and specially designed flex sensors. It was mentioned that the automated control of the device was performed based on pulse width modulation, its working cycle, and the feedback provided by the flex sensors; allowing the controlled movement of the different joints in each finger through the use of an interactive graphical user interface, and simplifying the phases of measuring the bending angles of each joint before and after each session was recognized.

17.7.4 Drug delivery system

One of the most commercially successful examples of utilization of nitinol is the production of expandable stents, which are applied in the palliation treatment of malignant obstruction of esophagus [156, 157], bile duct [158], gastric outlet [159] or trachea [160]. The occlusion can be caused by many kinds of cancers, such as gallbladder carcinoma, esophagus carcinoma, cholangiocarcinoma, gastric carcinoma, etc. However, tumor ingrowth and overgrowth into the nitinol stents often causes reocclusion, leading to durability reduction and function degradation subsequently [161]. Therefore, developing nitinol stents with antitumor abilities is highly desirable. Wang et al. [160] prepared an intelligent and biocompatible drug-loading platform, based on a gold nanorods-modified butyrate-inserted NiTi-layered double hydroxides film on the surface of nitinol alloy. The prepared films function as drug-loading sponges that which pump butyrate out under near-infrared (NIR) irradiation and resorb drugs in water when the NIR laser is shut off. It was reported that (i) the in vitro and in vivo studies reveal that the prepared films possess excellent biosafety and high efficiency in synergistic thermochemo tumor therapy, showing a promising application in the construction of localized stimuli-responsive drug-delivery systems, and (ii) the newly designed drug-loading film shows excellent biocompatibility and synergistic chemothermal therapy to tumors in vitro and in vivo, which we believe, will find a new avenue for potential applications in localized drug-eluting systems.

For purposes of reduction of the ISR and enhancing the drug load and the sustaining release for promoting the hemocompatibility of NiTi substrate, Lai et al. [162] co-deposited the heparin (Hep) combined with calcium phosphate (CaP) and gelatin (Gel), without any additive or solvent, on hydroxyapatite (HA) coated NiTi alloy. It was reported that (i) the consequences indicate that heparin accompanied respectively with CaP, and Gel through ionic bonds can be loaded on the NiTi alloy, (ii) the porous post-HA coating can dramatically enhance the heparin content from 148 for the single layer coating (CaP-Hep) to 325 μg/cm^2 for the tri-layer coating (HA/CaP-Hep/Gel-Hep), also resulting in the heparin release duration from 1 to >35 days, supposed to meet the requirement to prevent the proliferation of VSMCs (vascular smooth muscle cells), (iii) both the drug content and releasing time are remarkable, (iv) as the result of clotting tests *in vitro*, drug loaded composite coatings reveal good anticoagulant property which is proportional to the cumulative content of drug release in an hour; indicating no denaturalization of heparin found during the electrochemical process.

17.8 New FDA testing guidelines for nitinol materials

The following article was introduced by Donovan [163] in This Week Orthopedics.

The Food and Drug Administration (FDA) – Center for Devices and Radiological Health – announced April 18, 2019 that it is proposing new guidelines for pre-market testing of Nitinol, a titanium-nickel alloy widely used in medical implants, although nothing has been found to be wrong with devices made from this metal. Their intent is to require more and deeper pre-market tests of all the newer materials going into implants. It's part of FDA's plan for addressing how to regulate new devices made with new materials but which use predicates made from older materials.

Nitinol, an alloy of titanium and nickel, is used in orthopedics for "internal fixation by the use of fixatives, compression bone stables used in osteotomy and fracture fixation, rods for the correction of scoliosis, shape-memory expansion staples used in cervical surgery, staples in small bone surgery, and fixation systems for suturing tissue in minimal invasive surgery," according to the article "Biomechanical Properties of Nitinol Staples: Effects of Troughing, Effective Leg Length, and 2-Staple Constructs" in *The Journal of Hand Surgery*, October 18, 2018. At least one lumbar cage is made from Nitinol, the Actipore™ PLFx from a Canadian company, Biorthex Inc., which says it's made from a "biologically and biomechanically compatible porous Nitinol material" which "has an elasticity almost identical to cancellous bone that results in load sharing. The isotropic interconnected porous structure of the device promotes rapid tissue ingrowth in conjunction with bone cell survival."

It was also stated that "Devices made with Nitinol provide many important benefits to patients, but we need to be able to assess whether, among other things, there are any health risks when the material comes into contact with various parts of the body for extended periods of time. To ensure that the benefits patients receive from these devices outweigh any risks resulting from their use, the FDA needs to receive the right information as part of the premarket review process." However, they didn't say why they chose Nitinol to be the first metal subjected to new proposed testing guidelines. Their March 15 statement says FDA has seen that some patients have powerful allergic or inflammatory reactions to metal implants, but they didn't point to any actual horror stories. One possibility is that Nitinol has been widely used in cardiovascular stents, which would raise the stakes considerably if the metal in a device were a problem.

The April 18 proposed guidelines are detailed in a 17-page paper, "Technical Considerations for Non-Clinical Assessment of Medical Devices containing Nitinol – Draft Guidance for Industry and Food and Drug Administration Staff." The guidelines propose new requirements for mechanical stress testing, computational stress/strain analysis, corrosion testing (including pitting, nickel ion release and galvanic corrosion) and biocompatibility, as well as new labeling to warn surgeons that: "Persons with allergic reactions to these metals may suffer an allergic reaction to this implant. Prior to implantation, patients should be counseled on the materials contained in the device, as well as potential for allergy/hypersensitivity to these materials."

It was also mentioned that "We also have our own team of FDA scientists and engineers conducting research to better understand device materials in our Center for Devices and Radiological Health's (CDRH) Office of Science and Engineering Laboratories (OSEL)," in the March 15 announcement. "We are also working to fully implement the National Evaluation System for health Technology (NEST) that will link and synthesize data from different sources including clinical registries, electronic health records and medical billing claims; this will help improve the quality of real-world evidence that will empower the FDA to more quickly identify, communicate and act on new or increased medical device safety concerns," they said. By combining data from all these sourc es, NEST is intended to be a faster and more thorough reporting system for medical device problems than the current Manufacturer and User Facility Device Experience (MAUDE) database.

References

[1] Machado LG, Savi MA. Medical applications of shape memory alloys. Braz. J. Med. Biol. Res. 2003, 36, 683–91.
[2] Hodgson DE, Wu MH, Biermann RJ. Shape Memory Alloys. In: Metals Handbook. Vol. 2: Properties and Selection: Nonferrous Alloys and Special-Purpose Materials. ASM International, Ohio, 1990, 897–902.
[3] Tadaki T, Oysuka K, Shimizu K. Shape memory alloys. Ann. Rev. Mater. Sci. 1988, 18, 25–45.
[4] Mantovani D. Shape memory alloys: properties and biomedical applications. J. Min. Metal. Mater. Soc. 2000, 52, 36–44.
[5] Duerig TM, Pelton A, Stöckel D. An overview of nitinol medical applications. Mater. Sci. Eng. A 1999, 273/275, 149–60.

[6]   Pelton AR, Stöckel D, Duerig TW. Medical uses of nitinol. Mater. Sci. Forum 2000, 327/328, 63–70.

[7]   Chu Y, Dai K, Zhu M, Mi X. Medical application of NiTi shape memory alloy in China. Mater. Sci. Forum 2000, 327/328, 55–62.

[8]   Auricchio F, Boatti E, Conti M, SMA Cardiovascular Applications and Computer-Based Design. In: Lecce L, et al. ed., Shape Memory Alloy Engineering for Aerospace, Structural and Biomedical Applications. Elsevier, 2015, 343–67; http://dx.doi.org/10.1016/B978-0-08-099920-3.00012-7.

[9]   https://www.google.com/url?sa=i&source=images&cd=&ved=&url=https%3A%2F%2Fwww.thecardiacinstitute.com%2Fcardiac-diagnostic-services-north-palm-beach%2Fcoronary-angioplasty-palm-beach-gardens%2F&psig=AOvVaw0pQiasGZq8xE4FM3aMhMDP&ust=1575530914242100.

[10]   https://www.hopkinsmedicine.org/health/treatment-tests-and-therapies/cardiac-catheterization.

[11]   https://www.industryjournal.info/143331/cardiovascular-catheter-guidewires-market-evaluation-geographical-analysis-and-revenue-by-regions2019-2024/.

[12]   https://images.app.goo.gl/22Cif1GYzwaeNC2Q7.

[13]   Crowther MA. Inferior vena cava filters in the management of venous thromboembolism. Am. J. Med. 2007, 120, S13–7.

[14]   https://www.riaendovascular.com/services/inferior-vena-cava-filter/.

[15]   https://www.crbard.com/Peripheral-Vascular/en-US/Products/SIMON-NITINOL-Vena-Cava-Filter.

[16]   Engmann E, Asch MR. Clinical experience with the antecubital Simon nitinol IVC filter. J. Vasc. Intervent. 1998, 9, 774–8.

[17]   Poletti PA, Becker CD, Prina L, Ruijs P, Bounameaux H, Didier D, Schneider PA, Terrier F. Long-term results of the Simon nitinol inferior vena cava filter. Eur. Radiol. 1998, 8, 289–94.

[18]   Asch MR. Initial experience in humans with a new retrievable inferior vena cava filter. Radiology 2002, 225, 835–44.

[19]   Bruckheimer E, Judelman AG, Bruckheimer SD, Tavori I, Naor G, Katzen BT. In vitro evaluation of a retrievable low-profile nitinol vena cava filter. J. Vasc Intervent. Radiol. 2003, 14, 469–74.

[20]   https://herzzentrum.immanuel.de/en/services-offered/therapy-options/atrial-septal-defect-repair/.

[21]   www.gore.com/en_gb/index.html.

[22]   Roy D, Sharma R, Bunce N, Ward D, Brecker SJ. Selecting the optimal closure device in patients with atrial septal defects and patent foramen ovale. Intervantional Cardol. 2012, 4; https://www.openaccessjournals.com/articles/selecting-the-optimal-closure-device-in-patients-with-atrial-septal-defects-and-patent-foramen-ovale.html.

[23]   Sardesai SH, Marshall RJ, Mourant AJ. Paradoxical systemic embolization through a patent foramen ovale. Lancet 1989, 1, 1732–3.

[24]   Mas JL, Arquizan C, Lamy C, Zuber M, Cabanes L, Derumeaux G, Coste J. Recurrent cerebro-vascular events associated with patent foramen ovale, atrial septal aneurysm, or both. N. Engl. J. Med. 2001, 345, 1740–6.

[25]   Thomas JD, Tabakin BS, Ittleman FP. Atrial septal defect with right to left shunt despite normal pulmonary artery pressure. J. Am. Coll. Cardiol. 1987, 9, 221–4.

[26]   Turner DR, Owada CY, Sang CJ Jr, Khan M, Lim DS. Closure of secundum atrial septal defects with the AMPLATZER septal occluder: a prospective, multicenter, post-approval study. Circ. Cardiovasc. Interv. 2017, 10; doi: 10.1161/CIRCINTERVENTIONS.116.004212.

[27]   Stöckel D, Pelton A, Duerig T. Self-expanding Nitinol stents: material and design consid-erations. Eur. Radiol. 2004, 14, 292–301.

[28] McGrath DJ, O'Brien B, Bruzzi M, McHugh PE. Nitinol stent design – understanding axial buckling. J. Mechan. Behav. Biomed. Mater. 2014, 40, 252–63.

[29] Allie DE, Hebert CJ, Rcis RT, Walker CM. Nitinol stent fractures in the SFA. Endovasc. Today 2004, July/August, 22–34.

[30] Wattam PR Jr. Fracture of Memotherm metallic stents in the biliary tract. Cardiovasc. Intervent. Radiol. 2000, 23, 55–6.

[31] Geschwind J-F H, Dagli MS, Vogel-Claussen J, Arepally AA, Venbrux AC. Percutaneous removal of a fractured endostent remnant from the portal vein. Cardiovasc Intervent Radiol 2002, 25, 152–4.

[32] Diller R, Senninger N, Kautz G, Tübergen D. Stent migration necessitating surgical intervention. Surg. Endoscv. 2003, 17, 1803–7.

[33] Knirsch W, Haas NA, Lewin MA, Uhlemann F. Longitudinal stent fracture 11 months after implementation in the left pulmonary artery and successful management by a stent-in-stent maneuver. Cathet. Cardiovasc. Intervent. 2003, 58, 116–8.

[34] Zhou C, Wei B, Wang J, Huang Q, Li H, Gao K. Self-expanding metallic stent fracture in the treatment of malignant biliary obstruction. Gastroenterol Res. Pract. 2018, 2018; doi: 10.1155/2018/6527879.

[35] Meier GH, Pollak JS, Rosenblatt M, Dickey KW, Gusberg RJ. Initial experience with stent in exertional maxillary subclavian vein thrombosis. J. Vasc. Surg. 1996, 24, 974–83.

[36] Phipp LH, Scott DJ, Kessel D, Robertson I. Subclavian stents and stent-grafts: cause for concern? J. Endovasc. Surg. 1999, 6, 223–6.

[37] Carpenter JP, Anderson WN, Brewster DC, Kwolek C, Makaroun M, Martin J, McCann R, McKinsey J, Beebe HG. Multicenter pivotal trial results of the Lifepath System for endovascular aortic aneurysm repair. J. Vasc. Surg. 2004, 39, 34–43.

[38] Dowling R, Mitchell P, Cox GS, Thomson KR. Complication of a venous Wallstent. Austr. Radiol. 1999, 43, 246–8.

[39] Kroger K, Santosa F, Goyen M. Biomechanical incompatibility of popliteal stent placement. J. Endovasc. Ther. 2004, 11, 686–94.

[40] Lohr CE, Trang H, Bansal S. Using radiography to detect a biliary stent fracture and related bowel obstruction. HCP Live Network 2008; https://www.mdmag.com/journals/surgical-rounds/2008/2008-02/2008-02_03.

[41] Lewis G. Materials, fluid dynamics, and solid mechanics aspects of coronary artery stents: a state-of-the-art review. J. Biomed. Mater. Res. Part B 2008, 86, 569–90.

[42] https://www.esticastresearch.com/report/embolic-protection-device-market/.

[43] Blausen.com staff. Medical gallery of Blausen Medical 2014. WikiJournal of Medicine 2014, 1; doi: 10.15347/wjm/2014.010.

[44] Thanopoulos BD, Laskari CV, Tsaousis GS, Zarayelyan A, Vekiou A, Papadopoulos GS. Closure of atrial septal defects with the amplatzer occlusion device: preliminary results. J. Am. Coll. Cardiol. 1998, 31, 1110–6.

[45] Walsh KP, Maadi IM. The Amplatzer septal occlude. Cardiol. Young 2000, 10, 493–501.

[46] Chlanda A, Witkowska J, Morgiel J, Nowińska K, Choińska E, Swieszkowski W, Wierzchoń T. Multi-scale characterization and biological evaluation of composite surface layers produced under glow discharge conditions on NiTi shape memory alloy for potential cardiological application. Micron 2018, 114, 14–22.

[47] Tyagi S, Singh S, Mukhopadhyay S, Kaul UA. Self- and balloon-expandable stent implantation for severe native coarctation of aorta in adults. Am. Heart J. 2003, 146, 920–8.

[48] Hausegger KA, Cragg AH, Lammer J, Lafer M, Flückiger F, Klein GE, Sternthal MH, Pilger E. Iliac artery stent placement: clinical experience with a nitinol stent. Radiology 1994, 190, 199–202.

[49] Nematzadeh F, Sadrnezhaad SK. Effects of material properties on mechanical performance of Nitinol stent designed for femoral artery: finite element analysis. Scientia Iranica 2012, 19, 1564–71.

[50] Carter AJ, Scott D, Laird JR, Bailey L, Kovach JA, Hoopes TG, Pierce K, Heath K, Hess K, Farb A, Virmani R. Progressive vascular remodeling and reduced neointimal formation after placement of a thermoelastic self-expanding nitinol stent in an experimental model. Catheter. Cardio. Diag. 1998, 44, 193–201.

[51] Thierry B, Merhi Y, Bilodeau L, Trépanier C, Tabrizian M. Nitinol versus stainless steel stents: acute thrombogenicity study in an ex vivo porcine model. Biomaterials 2002, 23, 2997–3005.

[52] Bonin EA, Baron TH. Update on the indications and use of colonic stents. Curr. Gastroent. Reports 2010, 12, 374–82.

[53] Ansaloni L, Andersson RE, Bazzoli F, Catena F, Cennamo V, Di Saverio S, Fuccio L, Jeekel H, Leppäniemi A, Moore E, Pinna AD, Pisano M, Repici A, Sugarbaker PH, Tuech J-J. Guidelenines in the management of obstructing cancer of the left colon: consensus conference of the world society of emergency surgery (WSES) and peritoneum and surgery (PnS) society. World J. Emergency Surg. 2010, 5; doi: 10.1186/1749-7922-5-29.

[54] Deans GT, Krukowski ZH, Irwin ST. Malignant obstruction of the left colon. Brit. J. Surg. 1994, 81, 1270–6.

[55] Di Mitri R, Mocciaro F. The new nitinol conformable self-expandable metal stents for malignant colonic obstruction: a pilot experience as bridge to surgery treatment. Sci. World J. 2014; doi: 10.1155/2014/651765.

[56] Fernández-Esparrach G, Bordas JM, Giráldez MD, Ginès A, Pellisé M, Sendino O, Martínez-Pallí G, Castells A, Llach J. Severe complications limit long-term clinical success of self-expanding metal stents in patients with obstructive colorectal cancer. Am. J. Gastroenterol. 2010, 105, 1087–93.

[57] Small AJ, Coelho-Prabhu N, Baron TH. Endoscopic placement of self-expandable metal stents for malignant colonic obstruction: long-term outcomes and complication factors. Gastrointest. Endosc. 2010, 71, 560–72.

[58] Lee HJ, Hong SP, Cheon JH, Kim TI, Min BS, Kim NK, Kim WH. Long-term outcome of palliative therapy for malignant colorectal obstruction in patients with unresectable metastatic colorectal cancers: endoscopic stenting versus surgery. Gastrointest. Endosc. 2011, 73, 535–42.

[59] Chun HJ, Kim ES, Hyun JJ, Kwon YD, Keum B, Kim CD. Gastrointestinal and biliary stents. J. Gastroenterol. Hepatol. 2010, 25, 234–43.

[60] Nagula S, Ishill N, Nash C, Markowitz AJ, Schattner MA, Temple L, Weiser MR, Thaler HT, Zauber A, Gerdes H. Quality of life and symptom control after stent placement or surgical palliation of malignant colorectal obstruction. J. Am. Coll. Surg. 2010, 210, 45–53.

[61] Parkin DM, Bray F, Ferlay J, Pisani P. Global cancer statistics, 2002. CA: A Cancer J. Clin. 2005, 55, 74–108.

[62] Leitman IM, Sullivan JD, Brams D, DeCosse JJ. Multivariate analysis of morbidity and mortality from the initial surgical management of obstructing carcinoma of the colon. Surg. Gynecol. Obstet. 1992, 174, 513–8.

[63] Dohmoto M. New method: endoscopic implantation of rectal stent in palliative treatment of malignant stenosis. Endoscopia Digestiva 1991, 3, 1507–12.

[64] Puértolas S, Bajador E, Puértolas JA, López E, Ibarz E, Gracia L, Herrera A. Study of the behavior of a bell-shaped colonic self-expandable NiTi stent under peristaltic movements. BioMed Res. Int. 2013; http://dx.doi.org/10.1155/2013/370582.

[65] Tack J, Gevers AM, Rutgeerts P. Self-expandable metallic stents in the palliation of rectosigmoidal carcinoma: a follow-up study. Gastrointest. Endosc. 1998, 48, 267–71.

[66] Nudelman L, Fuko V, Greif F, Lelcuk S. Colonic anastomosis with the nickel-titanium temperature-dependent memory-shape device. Am. J. Surg. 2002, 183, 697–701.

[67] Song C, Frank T, Cuschieri A. Shape memory alloy clip for compression colonic anastomosis. J. Biomech. Eng. 2005, 127, 351–4.

[68] Park JM, Bloom DA. The pathophysiology of UPJ obstruction: current concepts. Urol. Clin. N. Am. 1998, 25, 161–9.

[69] Smrkolj T, Šalinović D. Endoscopic removal of a nitinol mesh stent from the ureteropelvic junction after 15 years. Case Rep. Urol. 2015; http://dx.doi.org/10.1155/2015/273614.

[70] Shalhav AL, Giusti G, Elbahnasy AM, Hoenig DM, McDougall EM, Smith DS, Maxwell KL, Clayman RV. Adult endopyelotomy: impact of etiology and antegrade versus retrograde approach on outcome. J. Urol. 1998, 160, 685–9.

[71] Streem SB. Ureteropelvic junction obstruction. Open operative intervention. Urol. Clin. N. Am. 1998, 25, 331–41.

[72] Lugmayr H, Pauer W. Self-expanding metallic stents in malignant ureteral stenosis. Deut. Med. Wochenschr. 1991, 116, 573–6.

[73] Yachia D. The use of urethral stents for the treatment of urethral strictures. Ann. d'Urologie 1993, 27, 245–50.

[74] Mori K, Okamoto S, Akimoto M. Placement of the urethral stent made of shape memory alloy in management of benign prostatic hypertrophy for debilitated patients. J. Urol. 1995, 154, 1065–8.

[75] Burgos FJ, Bueno G, Gonzalez R, Vazquez JJ, Diez-Nicolás V, Marcen R, Fernández A, Pascual J. Endourologic implants to treat complex ureteral stenosis after kidney transplantation. Transpl. P. 2009, 41, 2427–9.

[76] Davids PHP, Groen AK, Rauws EAJ, Tytgat GNJ, Huibregtse K. Randomised trial of self-expanding metal stents versus polyethylene stents for distal malignant biliary obstruction. Lancet 1992, 340, 1488–92.

[77] Vinograd I, Klin B, Brosh T, Weinberg M, Flomenblit Y, Nevo Z. A new intratracheal stent made from nitinol, an alloy with 'shape memory effect'. J. Thorac. Cardiovasc. Surg. 1994, 107, 1255–61.

[78] Yanagihara K, Mizuno H, Wada H, Hitomi S. Tracheal stenosis treated with self-expanding nitinol stent. Ann. Thorac. Surg. 1997, 63, 1786–90.

[79] Cwikiel W, Willen R, Stridbeck H, Lillo-Gil R, Von Holstein CS. Self-expanding stent in the treatment of benign esophageal strictures: experimental study in pigs and presentation of clinical cases. Radiology 1993, 187, 667–71.

[80] Pocek M, Maspes F, Masala S, Squillaci E, Assegnati G, Moraldi A, Simonetti G. Palliative treatment of neoplastic strictures by self-expanding nitinol Strecker stent. Eur. Radiol. 1996, 6, 230–5.

[81] Watanabe M, Sekine K, Hori Y, Shiraishi Y, Maeda T, Honma D, Miyata G, Saijo Y, Yambe T. Artificial esophagus with peristaltic movement. ASAIO J. 2005, 51, 158–61.

[82] Rajan GP, Eikelboom RH, Anandacoomaraswamy KS, Atlas MD. In vivo performance of the nitinol shape-memory stapes prosthesis during hearing restoration surgery in otosclerosis: a first report. J. Biomed. Mater. Res. Part B 2005, 72, 305–9.

[83] US FDA. Cerebral Spinal Fluid (CSF) shunt systems. 2018; https://www.fda.gov/medical-devices/implants-and-prosthetics/cerebral-spinal-fluid-csf-shunt-systems.

[84] Sanders JO, Sanders AE, More R, Ashman RB. A preliminary investigation of shape memory alloys in the surgical correction of scoliosis. Spine 1993, 18, 1640–6.

[85] Wever D, Elstrodt J, Veldhuizen A, Horn J. Scoliosis correction with shape-memory metal: results of an experimental study. Eur. Spine J. 2002, 11, 100–6.

[86] Tarniță D, Tarniță D, Bizdoaca, N, Mîndrilă I, Vasilescu M. Properties and medical applications of shape memory alloys. Rom. J. Morphol. Embryol. 2009, 50, 15–21.

[87]   https://www.odtmag.com/contents/view_breaking-news/2019-09-16/atlas-spine-receives-510k-clearance-for-cervical-plating-system/.

[88]   https://www.prweb.com/releases/2015/04/prweb12619750.htm.

[89]   Pfeifer R, Müller CW, Hurschler C, Kaierle S, Wesling V, Haferkamp H. Adaptable orthopedic shape memory implants. Procedia CIRP 2013, 5, 253–8.

[90]   Tarniţă D, Tarniţă DN, Hacman L, Copiluş C, Berceanu C. In vitro experiment of the modular orthopedic plate based on Nitinol, used for human radius bone fractures. Rom. J. Morphol. Embryol. 2010, 51, 315–20.

[91]   Russell S. Design considerations for Nitinol bone staples. J. Mater. Eng. Perform. 2008, 18, 831–5.

[92]   https://neosteo.com/osteotomie/super-elastic-memory-staples/.

[93]   Dai KR, Hou XK, Sun YH, Tang RG, Qiu SJ, Ni C. Treatment of intra-articular fractures with shape memory compression staples. Injury 1993, 24, 651–5.

[94]   Laster Z, MacBean AD, Ayliffe PR, Newlands LC. Fixation of a frontozygomatic fracture with a shape-memory staple. Brit. J. Oral Maxillofac. Surg. 2001, 39, 324–5.

[95]   Villermaux F, Tabrizian M, Yahia L, Czeremuszkin G, Piron DL. Corrosion resistance improvement of NiTi osteosynthesis staples by plasma polymerized tetrafluoroethylene coating. Biomed. Mater. Eng. 1996, 6, 241–54.

[96]   Mei F, Ren X, Wang W. The biomechanical effect and clinical application of a NiTi shape memory expansion clamp. Spine 1997, 22, 2083–8.

[97]   Hoon QCJ, Pelletier MH, Christou C, Johnson KA, Walsh WR. Biomechanical: evaluation of shape-memory alloy staples for internal fixation – an in vitro study. J. Exp. Orthop. 2016, 3; doi: 10.1186/s40634-016-0055-3.

[98]   Schipper ON, Ellington JK. Nitinol compression staples in foot and ankle surgery. Orthop. Clin. 2019, 50, 391–9.

[99]   https://en.wikipedia.org/wiki/Intramedullary_rod#/media/File:Proximal_femur_nail.jpg.

[100]   Zelle BA, Boni G. Safe surgical technique: intramedullary nail fixation of tibial shaft fractures. Patient Saf. Surg. 2015, 9; doi: 10.1186/s13037-015-0086-1.

[101]   Kujala S, Ryhänen J, Jämsä T, Danilov A, Saaranen J, Pramila A, Tuukkanen J. Bone modeling controlled by a nickel-titanium shape memory alloy intramedullary nail. Biomaterials 2002, 23, 2535-43.

[102]   Bhandari M, Guyatt G, Tornetta P III, Schemitsch EH, Swiontkowski M, Sanders D, Walter SD. Randomized trial of reamed and unreamed intramedullary nailing of tibial shaft fractures. J. Bone Joint Surg. Am. 2008, 90, 2567–78.

[103]   Singh AP. Intramedullary nailing of fractures; https://boneandspine.com/intramedullary-nailing/.

[104]   Seitz WH Jr, Marbella ME. Distal interphalangeal joint arthrodesis using nitinol intramedullary fixation implants: X-fuse implants for DIP arthrodesis. Tech. Hand Up Extrem. Surg. 2013, 17, 169–72.

[105]   Yang S, Leong K-F, Du Z, Chua C-K. The design of scaffolds for use in tissue engineering. Part I. Traditional factors. Tissue Eng. 2001, 7, 679–89.

[106]   Mertsching H, Schanz J, Steger V, Schandar M, Schenk M, Hansmann J, Dally I, Friedel G, Walles T. Generation and transplantation of an autologous vascularized bioartificial human tissue. Transplantation 2009, 88, 203–10.

[107]   Eltom A, Zhong G, Muhammad A. Scaffold techniques and designs in tissue engineering functions and purposes: a review. Adv. Mater. Sci. Eng. 2019; https://doi.org/10.1155/2019/3429527.

[108]   Alavi SH, Soriano Baliarda M, Bonessio N, Valdevit L, Kheradvar A. A tri-leaflet nitinol mesh scaffold for engineering heart valves. Ann. Biomed. Eng. 2017, 45, 413–26.

[109]   Shi Y, Wang Y, Zhang P, Liu W. Fibrous scaffolds for tissue engineering. Inflammation Regener. 2014, 34, 23–32.

[110]   Gotman I, Zaretzky A, Psakhie SG, Gutmanas EY. Effect of a novel load-bearing trabecular Nitinol scaffold on rabbit radius bone regeneration. AIP Conf. Proceed. 2015, 1683; https://doi.org/10.1063/1.4932933.

[111]   Wu S, Liu X, Yeung KWK, Hub T, Xu Z, Chung JCY, Chu PK. Hydrogen release from titanium hydride in foaming of orthopedic NiTi scaffolds . Acta Biomater. 2011, 7, 1387–97.

[112]   Hutmacher DW, Schantz JT, Lam CXF. State of the art and future directions of scaffold-based bone engineering from a biomaterials perspective. J. Tiss. Eng. Regen. Med. 2007, 1, 245–60.

[113]   Rezwan K, Chen QZ, Blaker JJ. Biodegradable and bioactive porous polymer/inorganic composite scaffolds for bone tissue engineering. Biomaterials 2006, 27, 3413–31.

[114]   Oshida Y. Bioscience and Bioengineering of Titanium Materials. Elsevier, Oxford, UK, 2007.

[115]   Oshida Y. Surface Engineering and Technology for Biomedical Implants. Momentum Press, New York, NY USA, 2014.

[116]   Nag S, Banerjee R. Fundamentals of Medical Implant Materials. In: Narayan R, ed., ASM Handbook, Volume 23: Materials for Medical Devices. ASM International, Materials Park, OH, 2012.

[117]   Huiskes R, Weinans H, van Rietbergen B. The relationship between stress shielding and bone resorption around total hip stems and the effects of flexible materials. Clin. Orthop. Relat. Res. 1992, 274, 124–34.

[118]   Hoffmann W, Bormann T, Rossi A, Müller B, Schumacher R, Martin I, de Wild M, Wendt D. Rapid prototyped porous nickel–titanium scaffolds as bone substitutes. J. Tissue Eng. 2014; https://doi.org/10.1177/2041731414540674.

[119]   Wu S, Liu X, Tao Hu T, Chu PK, Ho JPY, Chan YL, Yeung KWK, Chu CL, Hung TF, Huo KF, Chung CY, Lu WW, Cheung KMC, Luk KDK. A biomimetic hierarchical scaffold: natural growth of nanotitanates on three-dimensional microporous Ti-based metals. Nano Lett. 2008, 8, 3803–8.

[120]   Wen CE, Xiong JY, Li YC, Hodgson PD. Porous shape memory alloy scaffolds for biomedical applications: a review. Phys. Scr. 2010, 39; https://iopscience.iop.org/article/10.1088/0031-8949/2010/T139/014070/meta.

[121]   Liu XM, Wu SL, Yeung KWK, Chan YL, Hu T, Xu ZS, Liu XY, Chung JCY, Cheung KMC, Chu PK. Relationship between osseointegration and superelastic biomechanics in porous NiTi scaffolds. Biomater 2011, 32, 330–8.

[122]   Geetha M, Singh AK, Asokamani R, Gogia AK. Ti based biomaterials, the ultimate choice for orthopaedic implants – a review. Prog. Mater. Sci. 2009, 54, 397–425.

[123]   Rahmanian R, Moghaddam NS, Haberland C, Dean D, Miller M, Elahinia M. Load bearing and stiffness tailored NiTi implants produced by additive manufacturing: a simulation study. Proc. SPIE 9058, Behav. Mechan. Multifunct. Mater. Composites 2014; https://doi.org/10.1117/12.2048948.

[124]   Filip P, Lausmaa J, Musialek J, Manzanec K. Structure and surface of TiNi human implants. Biomaterials 2001, 22, 2131–8.

[125]   Shishkovsky IV, Kuznetsov MV, Morozov IG. Porous titanium and nitinol implants synthesized by SHS/SLS: microstructural and histomorphological analyses of tissue reactions. Int. J. Self-Propag. High-Temp. Synth. 2010, 19, 157–67.

[126]   Zaworonkow D, Chekan M, Kusnierz K, Lekstan A, Grajoszek A, Lekston Z, Lange D, Chekalkin T, Kang J-H, Gunther V, Lampe P. Evaluation of TiNi-based wire mesh implant for abdominal wall defect management. Biomed. Phys. Eng. Express 2018, 4; https://iopscience.iop.org/article/10.1088/2057-1976/aaa0b0.

[127]   Kujala S, Ryhänen J, Danilov A, Tuukkanen J. Effect of porosity on the osteointe-
         gration and bone in growth of a weight-bearing nickel-titanium bone graft
         substitute. Biomaterials 2003, 24, 4691–7.
[128]   Bertheville B. Porous single-phase NiTi processed under Ca reducing vapor for use as a
         bone graft substitute. Biomaterials 2006, 27, 1246–50.
[129]   Barrabés M, Michiardi A, Aparicio C, Sevilla P, Planell JA, Gil FJ. Oxidized nickel–titanium
         foams for bone reconstructions: chemical and mechanical characterization. J. Mater. Sci.
         Mater. Med. 2007, 18, 2123–9.
[130]   Moghaddam NS, Miller M, Elahinia M, Dean D. Enhancement of bone implants by
         substituting Nitinol for titanium (Ti-6Al-4V): a modeling comparison. Conference: ASME
         2014 Conference on Smart Materials, Adaptive Structures and Intelligent Systems, SMASIS
         2014; doi: 10.1115/SMASIS20147648.
[131]   Moghaddam NS, Andani MT, Amerinatanzi A, Haberland C, Huff S, Miller M, Elahinia M,
         Dean D. Metals for bone implants: safety, design, and efficacy. Biomanuf. Rev. 2016, 1;
         https://doi.org/10.1007/s4089.
[132]   Amerinatanzi A, Moghaddam NS, Ibrahim H, Elahinia M. The effect of porosity type on the
         mechanical performance of porous NiTi bone implants. Smart Mater. Adapt. Struct. Intell.
         Syst. 2016; https://doi.org/10.1115/SMASIS2016-9293.
[133]   Gu YG, Tay BY, Lim CS, Yong MS. Nanocrystallite apatite formation and its growth kinetics
         on chemically treated porous NiTi. Nanotechnology 2006, 17; https://iopscience.iop.org/
         article/10.1088/0957-4484/17/9/023/meta.
[134]   Assad M, Jarzem P, Leroux MA, Coillard C, Chernyshov AV, Charette S, Rivard C-H. Porous
         titanium-nickel for intervertebral fusion in a sheep model: part 1. Histomorphometric and
         radiological analysis. J. Biomed. Mater. Res. Part B, Appl. Biomater. 2003, 64, 107–20.
[135]   Bansiddhi A, Sargeant TD, Stupp SI, Dunand DC. Porous NiTi for bone implants: a review.
         Acta Biomater. 2008, 4, 773–82.
[136]   Yang YQ, Man HC. Laser spray cladding of porous NiTi coatings on NiTi substrates. Surf.
         Coat. Technol. 2007, 201, 6928–32.
[137]   Aziz-Kerrzo M, Conroy KG, Fenelon AM, Farrell AT, Breslin CB. Electro-chemical studies on the
         stability and corrosion resistance of titanium-based implant materials. Biomaterials 2001,
         22, 1531–9.
[138]   Larsson C, Esposito M, Liao H, Thomsen P. The Titanim-Bone Interface In Vivo. In: Brunette
         DM, Tengvall P, Textor M, Thomsen ed. Titanium in Medicine: Materials Science, Surface
         Science, Engineering, Biological Responsese, and Medical Applications. Springer,
         New York, 2001, 587–648.
[139]   Cook SD, Thongpreda N, Anderson RC, Haddad RJ. The effect of post-sintering heat
         treatments on the fatigue properties of porous coated Ti-6Al-4V alloy. J. Biomed. Mater.
         Res. 1988, 22, 287–302.
[140]   Kohn DH, Ducheyne P. A parametric study of the factors affecting the fatigue strength of
         porous coated Ti-6Al-4V implant alloy. J. Biomed. Mater. Res. 1990, 24, 1483–501.
[141]   Pulletikurthi C. Biocompatibility assessment of biosorbable polymer coated Nitinol alloys.
         Mater. Sci. 2014; https://digitalcommons.fiu.edu/etd/1552.
[142]   Bai L, Gong C, Chen X, Sun Y, Zhang J, Cai L, Zhu S, Xie SQ. Additive manufacturing of
         customized metallic orthopedic implants: materials, structures, and surface modifications.
         Metals 2019, 9, 1004; https://doi.org/10.3390/met9091004.
[143]   Dutta RC, Dutta AK, Basu B. Engineering implants for fractured bones-metals to tissue
         constructs. J. Mater. Eng. Appl. 2017, 1, 9–13.
[144]   Park J, Lakes RS. Biomaterials: An Introduction. Springer, New York, NY, USA, 2007, 564.
[145]   EuroTimes. Nitinol suture. 2013; https://www.eurotimes.org/nitinol-sutures/.

[146] Kujala S, Pajala A, Kallioinen M, Pramila A, Tuukkanen J, Ryhänen J. Biocompatibility and strength properties of nitinol shape memory alloy suture in rabbit tendon. Biomaterials 2004, 25, 353–8.

[147] Campbell PG, Sharma P, Yadla S, Luginbuhl AJ, Rosen M, Evans JJ. Endoscopic skull base suturing with a self-closing U-clip: a report of two cases and review of the literature. Medicine 2010; doi: 10.29046/jhnj.005.1.007.

[148] Lamprakis AA, Fortis AP, Kostopoulos V, Vlasis K. Biomechanical testing of a shape memory alloy suture in a meniscal suture model. Arthroscopy 2009, 25, 632–8.

[149] Nespoli A, Dallolio V, Villa E, Passaretti F. A new design of a Nitinol ring-like wire for suturing in deep surgical field. Mater. Sci. Eng. C 2015, 56, 30–6.

[150] Webb G, Wilson L, Lagoudas DC, Rediniotis O. Adaptive control of shape memory alloy actuators for underwater biomimetic applications. AIAA J. 2000, 38, 325–34.

[151] https://www.youtube.com/watch?v=-K57cbOhA5g.

[152] http://enc4u.blogspot.com/2010/03/robotic-air-muscle-actuator-biorobotics.html.

[153] Hadi A, Alipour K, Kazeminasab S, Elahinia M. ASR glove: a wearable glove for hand assistance and rehabilitation using shape memory alloys. J. Intell. Mater. Syst. Struct. 2018, 5; https://doi.org/10.1177/1045389X17742729.

[154] Herrera SV. Development of Shape Memory Alloy (SMA) actuated glove; http://www.eng.auburn.edu/sites/aubelab/research1/sma-glove.html.

[155] Jiménez C, Mora R, Pérez J, Quirós K. Nitiglove: Nitinol-Driven Robotic Glove Used to Assist Therapy for Hand Mobility Recovery. In: González-Vargas J, et al. ed., Wearable Robotics: Challenges and Trends. Biosystems & Biorobotics, vol. 16. Springer, Cham, 2017, 370–83.

[156] Liu J, Wang Z, Wu K, Li J, Chen W, Shen Y, Guo S. Paclitaxel or 5-fluorouracil/esophageal stent combinations as a novel approach for the treatment of esophageal cancer. Biomaterials 2015, 53, 592–9.

[157] Homs MYV, Steyerberg EW, Eijkenboom WMH, Tilanus HW, Stalpers LJA, Bartelsman J, van Lanschot JJB, Wijrdeman HK, Mulder CJJ, Reinders JG, Boot H, Aleman BMP, Kuipers E, Siersema PD, Dutch SSG. Single-dose brachytherapy versus metal stent placement for the palliation of dysphagia from oesophageal cancer: multicentre randomised trial. Lancet 2004, 364, 1494–504.

[158] Smith AC, Dowsett JF, Russell RCG, Hatfield ARW, Cotton PB. Randomised trial of endoscopic stenting versus surgical bypass in malignant low bileduct obstruction. Lancet 1994, 344, 1655–60.

[159] Tezuka S, Kobayashi S, Ueno M, Moriya S, Irie K, Goda Y, Ohkawa O, Aoyama T, Morinaga S, Morimoto M, Maeda S. Sa1528 endoscopic duodenal stenting followed by systemic chemotherapy: management and survival impact on advanced pancreatic cancer with gastric outlet obstruction. Gastroenterology 2016, 150, S333.

[160] Wang D, Ge N, Yang T, Peng F, Qiao Y, Li Q, Liu X. NIR-triggered crystal phase transformation of NiTi-layered double hydroxides films for localized chemothermal tumor therapy. Adv. Sci. 2018, 5; https://doi.org/10.1002/advs.201700782.

[161] Park S, Son H, Park M, Jung G, Koo J. Use of a polyurethane covered nitinol stent for malignant gastroduodenal obstruction. Gastroenterology 2000, 118, A524; doi: https://doi.org/10.1016/S0016-5085(00)84223-7.

[162] Lai Y-L, Cheng P-Y, Yang C-C, Yen S-K. Electrolytic deposition of hydroxyapatite/calcium phosphate-heparin/gelatin-heparin tri-layer composites on NiTi alloy to enhance drug loading and prolong releasing for biomedical applications. Thin Solid Films 2018, 649, 192–201.

[163] Donovan B. New FDA testing for Nitinol implants. This Week Orthopedics, May 3rd, 2019; https://ryortho.com/breaking/new-fda-testing-for-nitinol-implants/.

Chapter 18
Future perspective

It would be valuable and meaningful to review what was foreseen on "Future of NiTi alloys," which was addressed about 10 years ago and to compare what has been actually happening nowadays and realized. Although many dreams and expectations on nitinol materials had been described 10 years ago, particularly in areas of industrial applications such as actuators, robots, R-phase-related applications, and R&D of new materials in high-temperature shape-memory alloys (SMA), this chapter will limit to the future perspective of medical/dental fields.

18.1 What was foreseen, 10 years ago

Russell [1], in 2007, mentioned that (i) nitinol will continue to be an important material for the medical devices market as procedures become less invasive. The success of nitinol in the peripheral vascular space has spurred development in a number of other areas, including general surgery, structural heart disease and orthopedics and development of the superficial femoral artery and (ii) many new applications will require the advances of materials technology in emerging areas such as thin-film nitinol, porous Nitinol and new alloys with improved mechanical and physical properties (better radiopacity, higher strength, higher purity, and tighter transformational hysteresis with certain improved compositions which reduce the sensitivity of transformation temperatures to composition without a negative effect on mechanical properties or biocompatibility). Echoing to what Russel foresaw, Duerig et al. [2] emphasized similar important points. With other reports, summering what was predicted about 10 years ago is as follows [1–5]:

(1) Thin-film devices or devices with thin-film deposition will certainly have an impact in many areas. Even though this material is not biodegradable like Mg element, nitinol offers a rare combination of being robust in thin-film form and is quickly endothelialized [1].

(2) Nitinol with controlled porosity offers potential as an extraordinarily compliant material that promotes boney ingrowth. Using self-propagating high-temperature synthesis and capsule-free hot isostatic pressing technologies, pore sizes can be controlled in a range from 100 to 500 μm and porosity in a range from 30% to 60% [3]. These structures are sometimes recognized as foam structures. Acceptable cytocompatibility, genotoxicity, irritation, toxicity reaction, and sensitization are reported, as well as high bone ingrowth with good fixation. Furthermore, surface layer can be manipulated to be Ni-depleted layer by oxygen plasma immersion ion implantation (PIII) [4]. Hence, structures with controlled porosities or with foam shape will be found in versatile applications.

https://doi.org/10.1515/9783110666113-018

(3) New alloys, both more radiopaque and stiffer, can be developed. Alloys of greater purity may also be of value as devices with a smaller size. Probably more importantly, the explosion in usage radiates from a very narrow field of application: the treatment of peripheral vascular disease. There is every reason to believe that the same advantages will eventually be realized in other areas, such as orthopedics, cardiology, and endoscopy [2]. Bansiddhi et al. [3] also mentioned that although the study of porous NiTi has developed rapidly over the past few years, biological studies are still at an early research stage compared to other porous Ti and Ti-6Al-4V alloys and to dense NiTi. More comprehensive studies are needed on biological performance in order to develop the biomedical applications of porous NiTi. For this purpose, optimization and long-term in vivo studies on surface-modified systems are still needed [3].

(4) Because of their unique mechanical properties and biocompatibility, SMAs have found many applications in minimally invasive surgery (MIS) or noninvasive surgical technologies. Many novel surgical devices have been developed based on SMAs, which have become essential tools for MIS. Song et al. [4] reviewed the historical development of SMAs for MIS, including devices for cardiovascular surgery, laparoscopic and endoscopic surgery, orthopedic surgery, and other minimally invasive procedures. As we gain more understanding of the shape-memory technologies through collaboration between scientists and clinicians, we are capable of developing more advanced and more robust devices for the next generation of surgery [4].

(5) More studies of surface-coating systems will be performed to improve drug-eluting performance [2].

18.2 Current status and future perspective

As we have seen in previous chapters, all items listed 10 years ago have been remarkably advanced and still under further developments. Besides these ongoing projects, we can see new dimension and concept proposed and some of them have already initiated.

Dutta et al. [6] mentioned that one of the remarkable features of biological materials is their structural hierarchy. Extracellular matrix of bone shows even more complex structural features due to continuous bone remodeling dictated by the stress and strain to which it is exposed to. As we have seen already, the density, strength, and rigidity of implantable materials normally differ from those of receiving hard tissues such as cortical, cancellous, and cartilaginous bones. Additional factors making the situation more complicated should include the location and overall biomechanistic environments. In human body, the type, flexibility, and extent of movements possible in different directions vary from joint to joint (see Figure 18.1). Each joint possesses different three-dimensional mechanistic modes (tension, compression, and/or

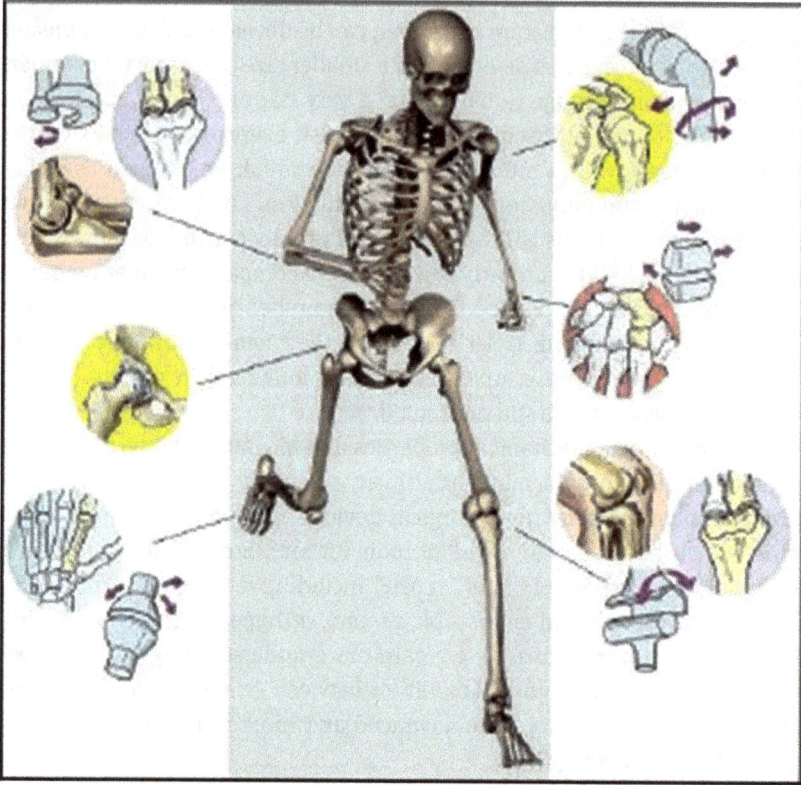

Figure 18.1: Schematic illustrations of various joints with varieties in movements of skeletal system in biomechanistic environments [6].

torsion) and translational and rotational movement as well. Since it should be more than affordable efforts to accommodate these infinite variations in biomechanistic environments, in order to avoid revision surgery and the associated trauma, unrelenting efforts are on to develop smart biomaterials for implants that can overcome the issues of infection, integration while managing the strength at desired level [7] or even dynamically in gradient manner [8].

Moghaddam et al. [9] mentioned that (i) the porous NiTi implants for the repair of skeletal defects through the example of mandibular segmental defects is highly feasible, (ii) resorbable magnesium and porous NiTi provide new options which may better match the material properties, and if they can be 3D printed, better match the shape of surrounding host tissue; reducing engrafted bone and metal implant failure and restore musculoskeletal function for the long term. It can be expected that development of new materials which includes composites and hybrid shape-memory materials, fabrication technologies and treatment process which makes stable, durable, and also develops robust computation models of SMA behavior

[10, 11]. Using advanced computer-aided design computer-aided manufacturing (CAD/CAM) and 3D-printing technologies, a chair-side fabrication of custom-made implants might be enabled.

Surface modification (as discussed in Chapter 12) plays a crucial role in various surface activities including biotribological actions (friction and wear), biocompatibility (in terms of corrosion resistance), surface energy alterations (for osseointegration), and so on. Hanawa [12] reviewed the evolution of surface modification techniques to improve hard-tissue compatibility at the research level, with reference to the stems of artificial hip joints and the fixtures of dental implants, as seen in Figure 18.2.

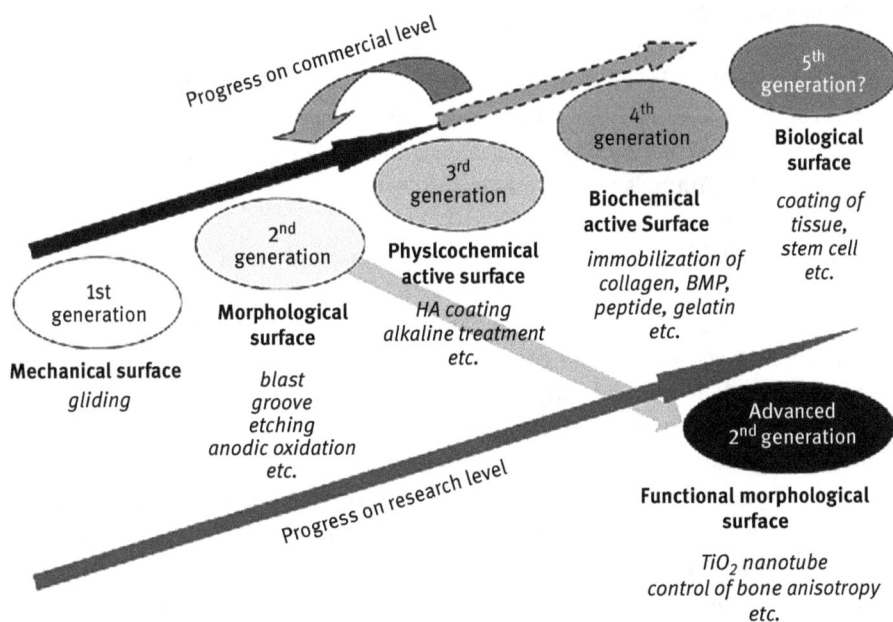

Figure 18.2: History of surface modification techniques to improve hard-tissue compatibility [12].

It was further explained by following generations; the first generation was characterized by machined surface (evidence showed that osteoblast cells grow along the grooves as the tool marks), the second generation was emphasized by increasing effective surface area by sand blasting, acid treatment or laser abrasion, or atmosphere-controlled oxidation, the third generation is typified by chemical treatment and HA coating (to serve dual purposes of chemical similarity to bone chemistry and stress buffering effect acting as if a periodontal membrane), the fourth generation is characterized by treating surface by immobilization of biofunctional molecules (collagen, bone morphogenetic protein, and peptide) and the fifth generation is the current stage of the surface development by coating stem cell and tissues. Meanwhile, during the oxidation treatment in the second generation, titania nanotubes were developed to function the micromorphological compatibility. Macromorphological compatibility was achieved by the blasting and acid treatment [13]. Hanawa [12] described that (i) most of commercialized goods are categorized into the second generation, a few belong to the third generation, and there is no prospect for the commercialization of the fourth generation, at present and (ii) the commercialization went faster for the second than the third generation possibly because materials employing mechanical anchoring are more practical than materials employing chemical bonding with bone. As Shao et al. [14] mentioned the design and creation innovative forms of SMAs, such as functionally graded SMAs, architecture SMAs, and SMA-based metallic composites will be further developed, based on the recent discovery of the principle of lattice strain matching between the SMA matrix and superelastic nanoinclusions embedded in the matrix. Based on this principle, different SMA-metal composites have been designed to achieve extraordinary shape-memory performances, such as complete pseudoelastic behavior at as low as 77 K and stress plateau as high as 1,600 MPa, and exceptional mechanical properties, such as tensile strength as high as 2,000 MPa and Young's modulus as low as 28 GPa [14], indicating versatile applications available.

Artificial intelligence (AI) is a general term that implies the use of a computer to model intelligent behavior with minimal human intervention [15]. AI is generally accepted as having started with the invention of robots. AI was not even mentioned 10 years ago even if the concept of AI had been already established. The term is applicable to a broad range of items in medicine such as robotics, medical diagnosis, medical statistics, and human biology – up to and including today's "omics." The physical branch is best represented by robots used to assist the elderly patient or the attending surgeon. Also embodied in this branch are targeted nanorobots, a unique new drug delivery system [15, 16]. With its origins in the mid- to late-1900s, today (in 2019), AI is used in a wide range of medical fields for varying purposes; some of the most current applications of machine learning (ML) in medicine according to the following four specific categories: (1) its use in assessing the risk of disease onset and in estimating treatment success, (2) its use in managing or alleviating complications, (3) its role in ongoing patient care, and (4) its use in ongoing pathology and treatment efficacy research [17]. AI in medicine can be versatilely employed for diagnostics, drug development, treatment personalization, and gene-editing ML has made great advances in pharmaceutical and biotechnological efficiency. Buch et al. [18] mentioned that informing clinical decision-making through insights from past data is the essence of evidence-based medicine. Traditionally, statistical methods have approached this task by characterizing patterns within data as mathematical equations, for example, linear regression suggests a line of best fit. Through the ML, AI can provide techniques that uncover complex associations which cannot easily be reduced to an equation. Activities of AI in medicine possess two phases; virtual and physical [19]. The virtual part ranges from applications such as electronic health record systems to neural network-based guidance in treatment decisions. The physical part deals with robots assisting in performing surgeries, intelligent prostheses for handicapped people, and elderly care. Analytically and logically machines may be able to translate human behavior, but certain human traits such as critical thinking, interpersonal and communication skills, emotional intelligence, and creativity cannot be honed by the machines. Amisha et al. [19] concluded that (i) it is important that primary care physicians get well versed with the future AI advances and the new unknown territory the world of medicine is heading toward, (ii) the goal should be to strike a delicate mutually beneficial balance between effective use of automation and AI and the human strengths and judgment of trained primary care physicians and (iii) it is essential because AI completely replacing humans in the field of medicine is a concern that might otherwise hamper the benefits which can be derived from it.

AI in dental care is no different from those currently practiced and foreseen in medicine. As mentioned before, activities of AI in medicine possess two phases; virtual and physical [19]. Phillips [20] described that (i) AI/ML technologies offer care providers an opportunity to accelerate radiology processes and more easily process the large quantities of data generated through digital images, (ii) the FDA has begun approving the marketing of AI algorithms designed to interpret medical images, and over the next few years they will make a huge impact in dentistry and hence (iii) with the help of larger datasets, ML algorithms will only become more sophisticated over time. Shuman et al. [21, 22] and Kwon et al. [23] listed up the most prevalent subfields of AI including (1) ML: machines can use neural networks, physics, and statistics to find insights in data without being explicitly programmed to do so, (2) deep learning: this is the learning of complex patterns in large amounts of data and think image and speech recognition, (3) cognitive computing: this aims to simulate human processes through image and speech interpretation and responding appropriately, (4) computer vision: this recognizes content in photos and videos, and (5) natural language processing: this is the ability of computers to analyze and generate human speech. Gupta [24] pointed out that with the newest capabilities enabled by deep learning techniques, AI will begin to impact dentistry on a clinical level as well.

AI's applicable values in everyday practice are as follows [24–28]:

(1) AI software can document all necessary data and present it to the dentist much faster and more efficiently than a human counterpart by collecting all necessary dental records, extra oral photographs, and radiographs necessary for diagnosing any dental condition. For machines to perform tasks such as reading radiographs, they must be trained on huge data sets to recognize meaningful patterns. Computers consider only the data being provided. Machines also do not get tired. Human being can work for four or five hours straight before getting fatigued; machines work 24/7 without coffee breaks. Another advantage is that machines do not get bored. With the ability to analyze vast numbers of diagnostic images such as x-rays, CT scans, and MRIs, systems like this can

point doctors and radiologists to the most probable areas of concern, increasing both the speed and probability of detection. In the very future, we foresee deep learning analysis tools for images, assisting in diagnosing and treatment planning of periodontal disease by enabling early detection of bone loss and changes in bone density.

(2) In the field of the orthodontics, the software can perform a number of analyses on radiographs and photographs that aid in diagnosis and treatment planning. For example, to determine if extractions are necessary before orthodontic treatment. AI is now available for orthodontic diagnosis, treatment planning and treatment monitoring. With precise 3D scans and virtual models, it is easy to 3D-print the aligners with customized treatment plan. As the vast data get computed, it creates an algorithm which in terms intelligently decides how a patient's tooth or teeth should be moved, with how much pressure, even identifying pressure points for that particular tooth or teeth. AI should be a promising tool for predicting the sizes of unerupted canines and premolars with greater accuracy in the mixed dentition period. The AI-aided aligners not only deliver precise treatment execution but also helps in monitoring the progress as well and claim to reduce treatment time as well as appointment schedules.

(3) To classify patients into aggressive periodontitis and chronic periodontitis group based on their immune response profile. In the very future, we foresee deep learning analysis tools for images, assisting in diagnosing, and treatment planning of periodontal disease by enabling early detection of bone loss and changes in bone density. Detection of peri-implantitis and early intervention is a likely benefit in implant dentistry.

(4) First-hand experience with development stage technologies (i.e., caries detection) has already demonstrated AI's potential value in everyday practice. It is confirmed that these tools can recognize things on images that even the most experienced dentist may otherwise miss. Furthermore, it is seen the results returned in near real time, fast enough to be incorporated into a busy practice workflow. Maybe early detection of oral cancer can be achieved.

(5) Precision for the design and chairside manufacturing of dental prostheses can be done, based on digital image acquisition following tooth cusps assessment. Another breakthrough in the field of restorative and prosthetic dentistry is the use of CAD/CAM technology for precision fit of prosthesis, but with innovation in generative adversarial networks, laboratories are using AI to automatically generate advanced dental restoration, designed to perfect fit and ideal function while exceeding aesthetic expectations. This not only will help dentistry but will have a huge potential and impact on orofacial or craniofacial prosthesis.

(6) Robotic surgeons are available nowadays, who perform semiautomated surgical tasks with increasing efficiency under the guidance of an expert surgeon.

(7) Investigating the properties of dental materials such as effective biofunctionality, chemical stability, wear resistance, and flexural strength, material design for new alloys can be achieved by AI-assisted approach.

References

[1]   Russell S. Metal with memory: the future of Nitinol. Med. Device, 2007; https://www.medicaldevice-network.com/features/feature1338/.
[2]   Duerig T, Pelton A. The future, 2010; https://nitinol.com/reference/the-future/.
[3]   Bansiddhi A, Sargeant TD, Stupp SI, Dunand DC. Porous NiTi for bone implants: a review. Acta Biomater. 2008, 4, 773–82.
[4]   Ho JPY, Wu SL, Poon RWY, Chung CY, Tjong SC, Chu PK, Yeung KWK, Lu WW, Cheung MC, Luk KDK. Oxygen plasma treatment to restrain nickel out-diffusion from porous nickel titanium orthopedic materials. Surf. Coat. Technol. 2007, 201, 4893–6.
[5]   Song C. History and current situation of shape memory alloys devices for minimally invasive surgery. Open Med. Dev. J. 2010, 2, 24–31.
[6]   Dutta RC, Dutta AK, Basu B. Engineering implants for fractured bones-metals to tissue constructs. J. Mater. Eng. Appl. 2017, 1, 9–13.
[7]   Sheikh Z, Najeeb S, Khurshid Z, et al. Biodegradable materials for bone repair and tissue engineering applications. Materials 2015, 8, 5744–94.
[8]   Sobral JM, Caridade SG, Sousa RA, et al. Three-dimensional plotted scaffolds with controlled pore size gradients: effect of scaffold geometry on mechanical performance and cell seeding efficiency. Acta Biomater. 2011, 7, 1009–18.
[9]   Moghaddam NS, Andani MT, Amerinatanzi A, Haberland C, Huff S, Miller M, Elahinia M, Dean D. Metals for bone implants: safety, design, and efficacy. Biomanuf. Rev. 2016, 1; https://doi.org/10.1007/s4089.
[10]   Stupkiewicz S, Petryk H. A robust model of pseudoelasticity in shape memory alloys. Int. J. Numer. Meth. Eng. 2013, 93, 747–69.
[11]   Arghavani J, Auricchio F, Naghdabadi R, Reali A. On the robustness and efficiency of integration algorithms for a 3D finite strain phenomenological SMA constitutive model. Int. J. Numer. Meth. Eng. 2011, 85, 107–34.
[12]   Hanawa T. Surface treatment and modification of metals to add biofunction. Dent. Mater. J. 2017, 36, 533–8.
[13]   Oshida Y. Bioscience and Bioengineering of Titanium Materials. Elsevier, Oxford, UK, 2007.
[14]   Shao Y, Guo F, Ren Y, Zhang J, Yang H, Jiang D, Hao S, Cui L. NiTi-enabled composite design for exceptional performances. Shape Memo Superelast. 2017, 3, 67–81.
[15]   Hamlet P, Tremblay J. Artificial intelligence in medicine. Metabolism 2017, 69S, 36–40.
[16]   Loh E. Medicine and the rise of the robots: a qualitative review of recent advances of artificial intelligence in health. BMJ Leader 2018, 2, 59–63.
[17]   Becker A. Artificial intelligence in medicine: what is it doing for us today? Health Policy Technol. 2019, 8, 198–205.
[18]   Buch VH, Ahmed I, Maruthappu M. Artificial intelligence in medicine: current trends and future possibilities. Brit. J. Gen. Pract. 2018, 68, 143–4.
[19]   Amisha, Malik P, Pathania M, Rathaur VK. Overview of artificial intelligence in medicine. J. Family Med. Prim. Care. 2019, 8, 2328–31.
[20]   Phillips A. AI-powered innovations that are transforming dentistry, 2018; https://becominghuman.ai/ai-innovations-transforming-dentistry-aef03479664d.

[21]   Shuman L. How artificial intelligence is shaping dentistry, 2019; https://www.
        dentaleconomics.com/macro-op-ed/article/16386252/how-artificial-intelligence-is-
        shaping-dentistry.
[22]   Shuman L, Parashar A, Angam SA. The AI revolution coming to dentistry. Dental Products
        Report. May 2018, 40.
[23]   Kwon HB, Park YS, Han JS. Augmented reality in dentistry: a current perspective. Acta
        Odontol. Scand. 2018, 76, 497–503.
[24]   Gupta S. The future of artificial intelligence in dentistry, 2018. https://dzone.com/articles/
        the-future-of-artificial-intelligence-in-dentistry.
[25]   Vashisht A, Choudhary E. Artificial intelligence; mutating dentistry. Int. J. Res. Anal. Rev. IJRAR
        2019, 6; https://www.researchgate.net/publication/331465771.
[26]   Khanna SS, Dhaimade PA. Artificial intelligence: transforming dentistry today. Indian J. Basic
        Appl. Med. Res. 2017, 6, 161–7.
[27]   Raith S, Vogel EP, Anees N, Keul C, Güth JF, Edelhoff D, Fischer H. Artificial neural networks
        as a powerful numerical tool to classify specific features of a tooth based on 3D scan data.
        Comput. Biol. Med. 2017, 80, 65–76.
[28]   Deshmukh SV. Artificial intelligence in dentistry. Intl. Clin. Dent. Res. Org. 2019; https://www.
        researchgate.net/publication/330013707_Artificial_intelligence_in_dentistry.

Epilogue

The topics that we did not cover in this book are related to the litigation on dental practices dealing with NiTi alloys. As we are aware of, there was a class action lawsuit for inflammatory reactions on the soft tissue when wearing orthodontic appliances and several lawsuit cases for unretrievable separated endodontic files. For both cases, it is believed that nickel element was claimed to respond. Besides these cases, there are global dental negligence cases in almost each dental sectors including oral surgery (e.g., extraction), endodontics (e.g., nerve and sinus perforations), dental implants (e.g., postoperative infection), periodontics (e.g., failure to diagnose or treat periodontal disease in a timely fashion), orthodontics (e.g., root resorption), prosthodontics (e.g., poor treatment plan), and adverse drug reactions (e.g., contraindicated drug administration). Although all these issues are so important for not only practitioners but also patients, it is our great regret not to discuss these in the main text of the book, which is originally planned not to include these complications in the scope of this book.

As society is ever-aging with the increasing population of the elderly, special care and attention should be paid for managing diseases and fully understanding the frailty, as well as treating these populations. A geriatric dentist needs to be fine-tuned to accommodate such social changes and challenges. Orthopedic and dental implants still have a window left to be further developed to facilitate the aged society. But we have a great hope that by analyzing numerous articles, since the research and development in these sectors has been so advanced and well developed, sooner or later public health services will become much easier and cheaper in the coming years. Besides, research and development of smart materials should be well advanced. As we see in the future perspective, AI-supported treatment plans, AI-assisted minimum invasive surgery, and many other AI-related medical software and hardware will be improved. There could be a variety of new materials, ideas, concepts, and technologies to be introduced and proposed in the coming years. The revised edition of this book (if needed) cannot wait for another 10 years, maybe 5 years are still long enough to wait. If a certain percentage of aged population becomes a patient, they are definitely older than their physicians who have not had experienced such pain, anxiety, fear, and symptom that the patients are having and feeling. AI-assisted treatment plan and AI-suggested diagnosis might be helpful. And it is our hope that sooner or later such activities will be seen routinely at our clinics.

Closing this book, we admit to have a very long list of individual's names to be mentioned with our memories and appreciation: from our teachers and mentors who inspired us, friends and colleagues who challenged us, and most importantly former graduate students (some of them are now our colleagues) who continue to both inspire and challenge us. To all of them, both of us owe our knowledge and capability to comprehend the cited articles presented in this book. These people,

https://doi.org/10.1515/9783110666113-019

who were and are valuable to us and to this book, should at least include the late T. Nakayama, T. Koizumi, late S. Iguchi, V. Weiss, late I.R. Kramer, late C.D. Monsson, H.W. Liu, late J.A. Schwartz, D.L. Yahia, A.R. Pelton, T.W. Duerig, C.M. Wayman, late B.K. Moore, Y. Miwa, N. Saotome, T. Nishihara, F. Farzin-Nia, R. Sachdeva, S. Miyazaki, T. Yoneyama, T. Hanawa, P.-C. Chen, G.K. Stookey, C.J. Andres, T.M. Barco, A. Hashem, A. Wu, M. Yapchulay, S. Isikbay, C. Kuphasuk, C-M. Lin, I. Garcia, W. Panyayong, Y-J. Lim, M. Reyes, W.C. Lim, J-C. Chang, S. Al-Ali, B. Anbari, D. Bartovic, F. Hernandez, I.W. Koh, Z. Khabbaz, R. Xirouchaki, E.B. Tuna, H. Yasukawa, P. Agarwal, N. Al-Nasr, C.B. Sellers, J. Dunigan-Miler, S. Al-Johancy, C.S. Wang, E. Matsis, R. Rani, Y-C. Wu, K. Mirza, I. Katsilieri, M.A. Dirlam, C.A. Beckner, E. Tada, and K. Takahira.

Index

https://doi.org/10.1515/9783110666113-020

www.ingramcontent.com/pod-product-compliance
Lightning Source LLC
Chambersburg PA
CBHW080332220326
41598CB00030B/4490